Tropical Marine Ecology

Tropical Marine Ecology

DANIEL M. ALONGI
Tropical Coastal and
Mangrove Consultants
Victoria, Australia

WILEY Blackwell

This edition first published 2022

Registered Offices
John Wiley & Sons, Inc., 111 River Street, Hoboken, NJ 07030, USA
John Wiley & Sons Ltd, The Atrium, Southern Gate, Chichester, West Sussex, PO19 8SQ, UK

Editorial Office
9600 Garsington Road, Oxford, OX4 2DQ, UK

For details of our global editorial offices, customer services, and more information about Wiley products visit us at www.wiley.com.

Wiley also publishes its books in a variety of electronic formats and by print-on-demand. Some content that appears in standard print versions of this book may not be available in other formats.

Library of Congress Cataloging-in-Publication Data

Names: Alongi, D. M. (Daniel M.) author.
Title: Tropical marine ecology / Daniel Michael Alongi.
Description: Hoboken, NJ : Wiley-Blackwell, 2022. | Includes
bibliographical references and index.
Identifiers: LCCN 2021031560 (print) | LCCN 2021031561 (ebook) | ISBN
9781119568865 (cloth) | ISBN 9781119568889 (adobe pdf) | ISBN
9781119568926 (epub)
Subjects: LCSH: Marine ecology.
Classification: LCC QH541.5.S3 A453 2022 (print) | LCC QH541.5.S3 (ebook)
| DDC 577.7–dc23
LC record available at https://lccn.loc.gov/2021031560
LC ebook record available at https://lccn.loc.gov/2021031561

Cover Design: Wiley
Cover Images: © Chris & Monique Fallows/OceanwideImages.com, Mangrove photo courtesy of Dan Alongi

Set in 10/12pt STIXTwoText by Straive, Pondicherry, India
Printed and bound by CPI Group (UK) Ltd, Croydon, CR0 4YY

C9781119568865_221121

Contents

PART 3 FUNCTION 237

Preface

No realm on earth elicits thoughts of paradise more than the tropics. Such ideas often spring to mind when living through a snowy and icy winter. Many people living in temperate and boreal regions fulfil such dreams by holidaying in iconic places such as the islands of the Caribbean, the Great Barrier Reef, the Mediterranean and truly exotic locales such as Bali. When as a young man I left the United States to first visit Australia to work on the Great Barrier Reef, I truly felt that I had arrived at a tropical paradise. And the Great Barrier Reef is paradisiacal, being one of the greatest natural wonders on earth. Thousands of people the world over come to immerse themselves in its clear azure waters to observe the beauty and grandeur of many of its coral reefs. Also, if you think tropical rainforests are beautiful, then like me you can also enjoy the remarkable geometry and asymmetry of tropical mangrove forests.

But like all preconceived notions, dreams, and thoughts, the tropics also has a dark side, often subtle, but lurking in the shadows. Tropical waters give rise to cyclones, hurricanes, and typhoons, and the summer months can be unbearably hot and humid; closer to the equator, it is sticky, hot, and humid year-round. This reality can best be understood in the first instance by reading any of the classic tales of early explorers of the tropics, such as James Cook or the part-time pirate but intrepid explorer William Dampier or the great German scientist Alexander von Humboldt.

But as I will show in this book, the tropical marine realm is special in myriad ways and for many reasons from seas of higher latitude, in housing iconic habitats such as coral reefs, snow white beaches, crystal clear waters, mangrove forests, extensive and rich seagrass meadows and expansive river deltas, such as the exemplar, the Amazon. The reader will learn that from a global perspective it is in fact the great tropical rivers that have the most significant role to play in the cycles of nutrient and materials that help to foster life in tropical seas. These great conduits of mud, freshwater, and nutrients are the pumps that fuel the primary producers sustaining complex and beautifully intricate food webs. It is an irony that if it wasn't for these least photogenic of habitats, no (or exceedingly little) tropical marine life would exist. Even coral reefs have a tenuous, if important, functional connection to tropical rivers and estuaries; reefs are not quite as self-sufficient as we once thought, and many rely on connectivity with life in adjacent coastal muddy waters. In a nutshell, coral reefs are not divorced from the waters that bathe them. Nowadays this connection is unfortunately becoming more of a curse than a blessing; destructive human activities on land such as land clearing and overuse of chemical fertilizers, pesticides, and herbicides are on the rise with concomitant increases in human population growth along tropical coastlines.

The purpose of this book is to document the structure and function of tropical marine populations, communities, and ecosystems in relation to environmental factors including climate patterns and climate change and patterns of oceanographic phenomena such as tides and currents and major oceanographic features, as well as chemical and geological drivers. The book focuses on estuarine, coastal, shelf, and open ocean ecosystems. No such book on the tropical marine realm exists for the advanced undergraduate and postgraduate student, researcher, or manager. Another reason for writing this book is to reorient and expand the knowledge base of marine ecology. Several excellent textbooks exist on marine biology and ecology, but they are inadequate in describing life in the tropics; iconic habitats such as coral reefs and mangroves are usually covered only briefly. Until recently, this perfunctory treatment was understandable considering that the study of marine ecology has focused on boreal and temperate seas near where the major oceanographic institutes and universities reside. Since the 1980s, however, there has been a drastic rise in the number of journal articles published on aspects in tropical marine ecology to the extent that a textbook focusing on the tropics is now warranted.

Such an authoritative work is timely given the increasing concern of the problems associated with rapid population growth in developing nations – nearly all of which reside in the tropics – and a growing awareness of the role of the tropical ocean as the heat engine for global climate and in regulating earth's biogeochemical cycles. Many students are still being taught basic principles of marine ecology based on research conducted primarily in high latitudes. This is unfortunate because the tropical ocean is in many ways different from colder seas both structurally and functionally. The tropical ocean contains the centre for marine biodiversity, is a major driver of earth's climate, and is where most freshwater and sediment from land are discharged into the sea, greatly impacting ocean chemistry, geology, and the structure and function of biota. Many major environmental characteristics and adaptive flora and fauna are endemic to or dominant in tropical seas.

A basic understanding of marine biology and ecology is assumed so the reader may be tempted to skip the first part of the book dealing with the climate, physics, geology, and chemistry of the tropical marine environment. I urge the reader not to do so as one cannot properly understand what drives tropical organisms without understanding the uniqueness of the physical milieu in which they live.

The second section focuses on the origins, diversity, biogeography, and the structure and distribution of tropical biota. The tropical marine realm started in the Tethys Sea where most phyla first evolved and radiated through time to produce the major latitudinal patterns we see today. Populations of organisms from the size of microbes to whales will be examined in terms of their population regulation, growth dynamics, fluctuations, and cycles over time, as well as life history traits and strategies, including aggregation and refugia, territoriality, and behaviour. Pelagic and benthic community structure and their drivers, such as adaptations to stress, competition, predation, symbiosis, and other trophic factors, will be dealt with to underscore the fact that ecosystems are not simply 'black boxes' but consist of a wide array of complex trophic groups and communities. The ecosystem chapter will deal with not only classification of types (sandy beaches, mangroves, coral reefs, continental shelf, open ocean) but also how they developed over time and how they connect to one another.

The third part explores the rates and patterns of primary and secondary productivity, their drivers, and the characteristics of food webs. All organisms play important roles in the cycling of carbon and macro- and micronutrients, and these biogeochemical cycles are considered from the intertidal zone out to the open ocean.

The fourth part examines how humans are altering tropical ecosystems via unsustainable fisheries and the decline and loss of habitat and fragmentation; pollution is altering an earth already in the throes of climate change. This book ends with a hopefully not-too-long list of dot points highlighting how tropical biota and their ecosystems are different to those of higher latitude and how their future is in our hands.
I would like to acknowledge the staff of Wiley for doing such a wonderful, professional job in stitching this book together. I thank colleagues Bob Aller, Josie Aller, Michelle Burford, Erik Kristensen, Janice Lough, Matsui Mazda, Dave McKinnon, John (Charlie) Veron, Gullaya Wattayakorn, and Bob Wasson for critically commenting

on various chapters. I am grateful to Morgan Pratchett and Ciemon Caballes for the photos of crown-of-thorns and coral bleaching. Finally, I thank my loving wife Fiona for her beautiful illustrations that have made this book much better than I had hoped and both my daughters for reminding me that there is indeed life after science. Of course, any errors are mine and I would be grateful for students, faculty, and other readers to bring any errors to my attention.

Daniel M. Alongi, PhD
Email: dmalongi@outlook.com

CHAPTER 1

Introduction

1.1 Definition of the Tropics

There is no standard definition of the tropics. It has been defined in so many ways, as a reflection of its complexity, that only an operational definition can suffice; there have been notable climatological and oceanographic exceptions to all definitions. No one definition meets with universal approval, and there have been many attempts to define it, first most simply, by the patterns of the trade winds of the "torrid zone" (Dampier 1699) to a rigid definition of the region between the Tropic of Capricorn and the Tropic of Cancer (Townsend 2012), that is, the most northerly and southerly position at which the sun may appear directly overhead at its zenith. In fact, the word 'tropical' comes from the Greek *tropikos*, meaning 'turn' referring to the fact that these latitudes mark where the sun appears to turn annually in its motion across the sky. Recent evidence indicates that the tropics have expanded due to climate change (Seidel et al. 2008).

Other definitions have recognised that the boundaries of the tropics *sensu lato* do not equate with rigid zones and have classified the tropics on the basis of terrestrial vegetation (the Köppen-Geiger system) or seasonal patterns in rainfall, where the zonation is identified as 'humid,' 'wet and dry,' and 'dry.' Such definitions are functional, but none fit our requirement for an ocean climate-based scheme.

The marine tropics is defined here as the area of ocean and coastline included within the annual isotherms of sea surface temperature (SST) of 25 °C (Figure 1.1). This area encompasses (i) most of the Indian Ocean including most of the east coast of Africa to Mozambique and the southern tip of Madagascar, (ii) the Red Sea and the Gulf of Aden, (iii) the Arabian Sea, (iv) the Bay of Bengal, (v) the waters of Southeast Asia, New Guinea, and northern Australia (the South China Sea, Java Sea, Coral Sea, Philippine Sea, Timor and Arafura Seas, and the Gulf of Carpentaria), (vi) most of the small island arcs of the northern and southern Pacific Ocean to the west coast of Mexico and down the Central American coast to Ecuador, (vii) most of the Caribbean Sea and the Gulf of Mexico and the coasts of Central and South America down to central Brazil, and (viii) a large portion of the West African coastline from Guinea-Bissau to Gabon (Gulf of Guinea). The marine tropics is thus not a uniform or fixed region. There is a considerable degree of plasticity to these boundaries considering differences between the extremes of winter and summer which foster biological plasticity. The West African coast from Gabon to the Congo and down to the north coast of Angola, for instance, has an essentially tropical benthic biota (Longhurst 1959). Such variations are caused in part by the asymmetrical form and unequal size of the

Tropical Marine Ecology, First Edition. Daniel M. Alongi.

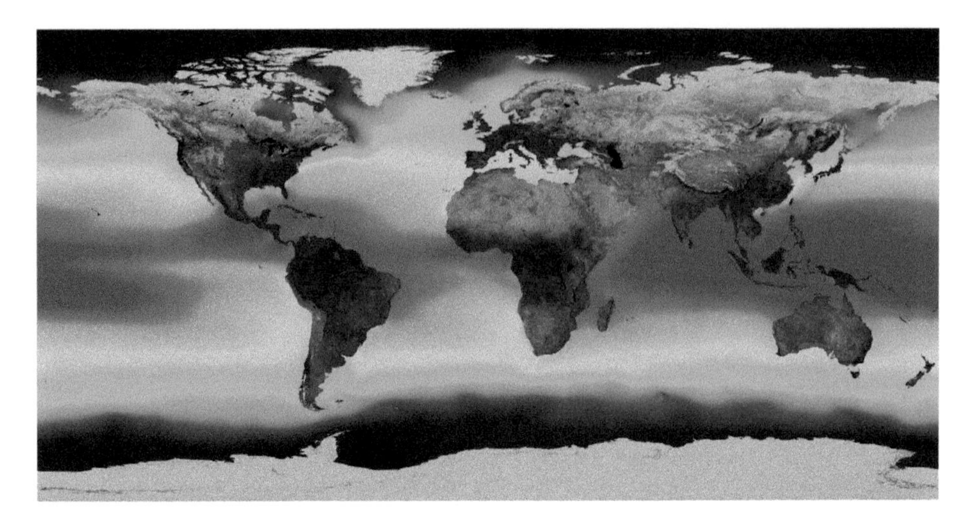

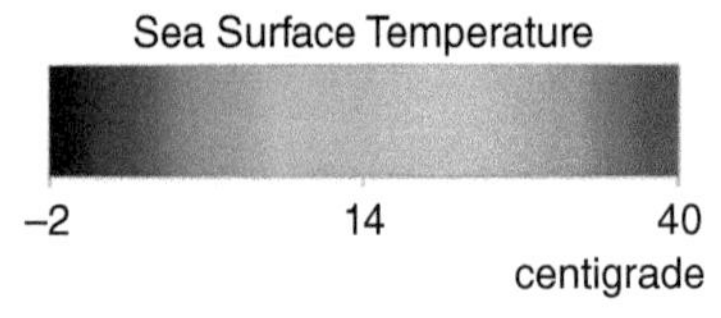

FIGURE 1.1 Annual mean sea surface temperatures in the global ocean, 2005–2017.
Source: Image retrieved via public access from the NASA Scientific Visualization Studio. https://sus.gsfc.nasa.gov/3652 (accessed 7 June 2020). © John Wiley & Sons.

ocean margins, which strongly influences sea surface temperatures and current and nutrient regimes (Webster 2020).

1.2 What Makes the Tropics Different?

What makes the marine tropics unique compared to seas of higher latitude? Tables 1.1 and 1.2 summarise many of the characteristics that this book will cover; clearly, there are many environmental attributes that are either unique to or are more common in the tropics. Several habitats attain peak luxuriance in the tropics, namely, mangrove forests, seagrass meadows, and coral reefs. Both 'wet' (or 'humid') and 'dry' tropical regions occur as do areas that undergo distinct 'wet' and 'dry' seasons. More research has tended to focus on what at first glance appears to be richer, wetter ecosystems, but areas and periods of aridity are more common than are reflected in the literature.

Spatial and temporal variations in rainfall and temperature are large in the tropics; daily thermal and precipitation changes increase away from the equator. The western boundaries of the tropical oceans are warmer, wetter, and more stable climatically than the eastern boundaries, caused by the asymmetrical form and unequal size of the ocean margins, which in turn strongly affect sea surface temperatures, currents, and nutrient regimes (Webster 2020) These geographic differences are of considerable ecological importance, influencing the distribution and abundance of shallow water habitats.

TABLE 1.1 **Major hydrological and climatological characteristics unique to or dominant in the tropical oceans. Summary from Chapters 2 and 3.**

Hydrology	Climatology
37% of world ocean area	High and stable solar radiation
69.1% of freshwater discharge to the world ocean	Absorbed solar radiation exceeds long-wave radiation so net radiation balance is positive
Lower mean tidal amplitudes	High and stable temperatures
Small Coriolis parameter in proximity to the equator	Lowest and highest rates of evaporation and precipitation
Large Rossby radius	Trade winds (easterlies and westerlies)
Weak rotational constraint on bottom boundary layer	Absent/uncommon frontal storms within 5° of equator
Large buoyancy flux	Interannual variation > seasonal variation
Wind-produced homogenous layer deepest in equatorial waters	Monsoons (dry–wet or arid): Asian, African, Indo-Australian, and South American systems
$D_{CRITICAL\ DEPTH} > D_{WATER\ DEPTH}$	Tropical ocean absorbs most incoming solar energy
Seasonal upwelling	Tropical ocean-atmospheric system is the heat engine of the global climate system
Permanently stratified thermoclines and haloclines; oxygen minimum layers	Hadley Circulation distributes equatorial winds in the low latitudes
Salinity and pH highly variable; acidic and hypersaline conditions common	Intertropical Convergence Zone, a belt of convective cloud about the equator. Zone of rising air and intense precipitation (accounts for 32% of global precipitation)
Estuarization of shelves by river plumes	Indo-Pacific Warm Pool, an oceanographic/climatological phenomenon in the western Pacific Ocean; heat engine of the planet
Strong tidal fronts	Formation of tropical cyclones (typhoons, hurricanes)
Lutoclines (a front between two layers of comparatively high and low suspended sediment concentration) and high-salinity plugs in estuaries and nearshore waters in dry season/arid regions	El Niño-Southern Oscillation, large-scale, global, coupled atmosphere–ocean system resulting in major surface climate anomalies throughout tropics
Tidal mixing, trapping, and complex small-scale circulation in mangrove tidal waters	Indian Ocean Dipole, coupled ocean–atmosphere differences in convection, winds, sea surface temperatures, and thermocline causing large-scale differences in rainfall patterns
Highly complex, small-scale circulation on coral reefs and in hypersaline lagoons	Madden-Julian Oscillation, a phenomenon that is a major source of intra-annual variability in the tropical atmosphere, affecting monsoonal and cyclonic patterns

(*continued*)

TABLE 1.1 *(Continued)*

Hydrology	Climatology
Indonesian Throughflow, unique feature passing warm and fresh Pacific waters into the Indian Ocean via the Indonesian Archipelago	Pacific Decadal Oscillation, dominant year-round pattern of North Pacific sea surface temperature variability. Complex aggregate of different atmospheric and oceanographic forcings spanning the extratropical and tropical Pacific

The tropics form a band around the equator that comprises nearly 40% of the world's open ocean area and over one-third of its continental shelves (Table 1.1). As aforementioned, mangrove forests, coral reefs, and seagrass meadows constitute the richest habitats, but the drier tropical regions have hypersaline lagoons, stromatolites, and carbonate-dominated shelf margins, the exemplar of the latter being the Great Barrier Reef shelf. Hydrological and climatological characteristics of marine tropical seas are *in toto* unique, reflecting proximity to the equator. Such physical characteristics include high and stable solar radiation and temperature, highest and lowest rates of rainfall, easterly trade winds, a large Rossby radius with low mean tidal amplitudes (one notable exception is the NW coast of Australia where mean tidal ranges can exceed 10 m), a small Coriolis effect resulting in a lack of cyclones, hurricanes, and typhoons close to the equator, permanently stratified shelf waters with strong tidal fronts, but with estuaries having high salinity plugs and lutoclines during the dry periods; interannual variation is greater than seasonal variability despite some regions having distinct 'wet' and 'dry' intervals.

At least 69% of all freshwater and 60% of all sediment discharged to the world's coastal ocean do so via tropical rivers (Milliman and Farnsworth 2011; Laruelle et al. 2013). This phenomenon occurs primarily in the wet tropics and plays an important role as a driver of geological characteristics, oceanographic processes, and the structure and function of pelagic and benthic food webs. For instance, coastal waters receiving river water have a large buoyancy flux, and there are several regions where upwelled waters mix in a complex manner with discharged river water and associated materials, e.g. the southeast (the Gulf of Papua) and SW (the Aru Sea) coasts of New Guinea (Aller et al. 2004, 2008b; Alongi et al. 2012) producing an 'estuarisation' of the shelf margin with oxygen minimum layers and strong tidal fronts.

Geologically, intensely weathered silt and clay particles form muddy facies that dominate many inner and middle shelf margins, especially close to river deltas (Table 1.2); such shelves are ordinarily wide and shallow, and inshore areas have massive mud banks ('chakara') that migrate seasonally and annually as well as varying over decadal time scales (Gratiot and Anthony 2016). In drier regions where river discharge is small and/or highly seasonal, erosion is also highly seasonal but there is a high level of resolution of the geological record in which sedimentary facies are either carbonate-dominated or mixed carbonate–terrigenous deposits, or both. Coral material and debris from other calcium carbonate-bearing benthic organisms are most abundant in these areas; the extreme of this phenomenon is cementing dunes ('sabhka') in hypersaline lagoons. Throughout the 'wet' and 'dry' tropics, there are thus extremes of sediment accumulation and of burial of carbon and other elements. In many tropical regions, high rates of sediment erosion due to no or poor land-use practices have resulted in rivers that are highly eroded with beds that are wide, shallow, and sand- and/or gravel-dominated.

TABLE 1.2 Major geological and chemical characteristics unique to or dominant in the tropical oceans. Summary from Chapter 4.

Geology	Chemistry
60% of world's sediment discharge from tropical rivers	Lowest organic carbon and nitrogen content in carbonate deposits
Mud and coral most abundant on inner shelves	Highest organic carbon and nitrogen content in mangrove muds, mud banks, and off river plumes
Intense physicochemical weathering of bedrock and soils	Low (μM) concentrations of dissolved inorganic nutrients
Many shelves wide, shallow (<120 m depth), and carbonate-dominated open shelves or protected ('rimmed') lagoons	NO_2^- + NO_3^- and SO_4^- present in interstitial waters
Mixed carbonate–terrigenous sedimentary facies on shelf margins	Low O_2 conditions (<5 mg/l) in estuaries, lagoons, and inshore waters
Migrating fluid mud banks ('chakara')	Benthic nutrient regeneration rates low
Cementing dunes ('sabhka') in hypersaline lagoons	Low interstitial water content, especially in carbonates
Highly seasonal erosion/deposition cycles	Dominance of iron and manganese reduction in sediment suboxic diagenesis
River beds highly eroded, wide, shallow, and gravel-dominated	Particle coastings enriched in Fe-, Mn-, and Al-oxides
Bioturbation mostly at sediment surface	Kaolinite and gibbsite common clay minerals
High resolution of geological record	Intense scavenging of dissolved oceanic components
Extremes in sediment accumulation and carbon burial rate	Photochemical processes important

Intense chemical and physical weathering of tropical soils results in their transfer to the marine environment, with the result being sediment particles rich in iron, manganese, and aluminium oxide coatings. The most common minerals in such highly weathered environments are kaolinite and gibbsite, and in the water column, there is intense scavenging of dissolved oceanic components as well as important photochemical processes. When fuelled by highly weathered but low concentrations of carbon and nitrogen (lignin-rich) debris, this combination favours microbial decomposition pathways in sediments that are dominated by metal reduction. The latter is also fostered by high-disturbance events especially near large tropical rivers where the seabed is shallow, and the benthos is dominated by near-surface bioturbation and by small, opportunistic benthic infauna noticeably lacking in large, deep-dwelling, equilibrium species of annelids and molluscs (Aller et al. 2008a), but with an abundant epifauna.

Distal to rivers, phytoplankton production and respiration can be just as high as in higher latitudes, but such production (mainly by small-sized picoplankton rather than by larger diatoms and chlorophytes) is often displaced offshore due to high turbidity within plumes (Smith and DeMaster 1996; McKinnon et al. 2007).

These rapid rates of productivity occur despite low (μM) concentrations of dissolved nutrients, comparatively low (≤5 mg/l) oxygen concentrations, and low rates of benthic nutrient regeneration. Pelagic food chains are arguably dominated by abundant macrozooplankton, mostly crustaceans such as penaeid shrimp, whose abundance and productivity yield a high percentage of crustaceans to finfish catch off tropical fishing grounds. Why crustaceans are so predominant in the low latitudes may lie in their genetics, competitive abilities with finfish or with life histories being simpatico with tropical oceanographic or climatological peculiarities, the latter of which we will explore in Chapter 2.

References

Aller, R.C., Hannides, A., Heilbrun, C. et al. (2004). Coupling of early diagenetic processes and sedimentary dynamics in tropical shelf environments, the Gulf of Papua deltaic complex. *Continental Shelf Research* 24: 2455–2486.

Aller, J.Y., Alongi, D.M., and Aller, R.C. (2008a). Biological indicators of sedimentary dynamics in the central Gulf of Papua: seasonal and decadal perspectives. *Journal of Geophysical Research: Earth Science* 113: F01S08. https://doi.org/10.1029/2007JF000823.

Aller, R.C., Blair, N.E., and Brunskill, G.J. (2008b). Early diagenetic cycling, incineration, and burial of sedimentary organic carbon in the central Gulf of Papua (Papua New Guinea). *Journal of Geophysical Research, Earth Science* 113: F10S09. https://doi.org/10.1029/2006JF000689.

Alongi, D.M., Wirasantosa, S., Wagey, T. et al. (2012). Early diagenetic processes in relation to river discharge and coastal upwelling in the Aru Sea, Indonesia. *Marine Chemistry* 140: 10–23.

Dampier, W. (1699). *Voyages and Descriptions, Volume II, Part 3, A Discourse of Trade winds, Breezes, Storms, Seasons of the Year, Tides and Currents of the Torrid Zone throughout the World; with an Account of Natal in Africa, its Product, Negro's. etc.* London: J. Knapton.

Gratiot, N. and Anthony, E.J. (2016). Role of flocculation and settling processes in development of the mangrove-colonized, Amazon-influenced mud-bank coast of South America. *Marine Geology* 373: 1–10.

Laruelle, G.G., Dürr, H.H., Laurerwald, R. et al. (2013). Global multi-scale segmentation of continental and coastal waters from the watersheds to the continental margins. *Hydrology and Earth System Sciences* 17: 2029–2051.

Longhurst, A. (1959). Benthos densities off tropical west Africa. *Journal Conseil International pour l' Exploration de la Mer* 25: 21–28.

McKinnon, A.D., Carleton, J.H., and Duggan, S. (2007). Pelagic production and respiration in the Gulf of Papua during May 2004. *Continental Shelf Research* 27: 1643–1655.

Milliman, J.D. and Farnsworth, K.L. (2011). *River Discharge to the Coastal Ocean: A Global Synthesis.* Cambridge, UK: Cambridge University Press.

Seidel, D.J., Fu, Q., Randel, W.J., and Reichler, T.J. (2008). Widening of the tropical belt in a changing climate. *Nature Geoscience* 1: 21–24.

Smith, W.O. Jr. and DeMaster, D.J. (1996). Phytoplankton biomass and productivity in the Amazon River plume: correlation with seasonal river discharge. *Continental Shelf Research* 16: 291–319.

Townsend, D.W. (2012). *Oceanography and Marine Biology: An Introduction to Marine Science.* Sunderland, USA: Sinauer.

Webster, P.J. (2020). *Dynamics of the Tropical Atmosphere and Oceans.* Hoboken, USA: Wiley-Blackwell.

PART 1

PHYSICAL ENVIRONMENT

CHAPTER 2

Weather and Climate

2.1 Tropical Heat Engine

Due to the inequitable distribution of solar insolation, the tropical ocean absorbs most of the incoming solar energy and is the heat engine of Earth's climate (Webster 2020). The oceans receive more than half of the energy (mostly in the upper 100 m) absorbed by the planet and balanced by evaporative cooling, making the ocean the primary source of water vapour and heat for the atmosphere. As the oceans have great capacity to store heat energy, seasonal cycles in surface temperatures tend to be small in the tropics compared to higher latitudes. The mixed layer of the upper ocean tends to be thinner in the tropics, where the ocean is being heated and thicker at higher latitudes where the ocean gives up its energy via a complex series of atmospheric and oceanographic processes (Webster 2020).

As the sun's apparent position migrates north and south through the year, the zone of maximum tropical heating also migrates to either side of the equator (Figure 2.1). In January, most solar insolation is received south of the equator when it is summer in the Southern Hemisphere (Figure 2.1 top); in April, maximum tropical heating shifts north of the equator during summer in the Northern Hemisphere (Figure 2.1 bottom). The low latitudes are the focus of most heat as they receive the sun's rays at the least angle compared with the high latitudes where the angle of the earth's axis to the sun lengthens summer days and shortens them in winter. Net radiation is greatest over the tropical ocean where the surface albedo is low, and the surface temperature is moderate compared with the land masses.

Absorbed solar radiation exceeds outgoing long-wave radiation in the tropics so that the net radiation balance is positive (Figure 2.2). This latitudinal gradient in annual mean net radiation is balanced by a poleward flux of energy in the atmosphere and ocean. This flux is about twice as great in the Southern Hemisphere due to the greater area of ocean and smaller land areas mass compared to the Northern Hemisphere.

Evaporation driven by solar insolation is the boiler driving the circulation of the atmosphere and earth's hydrological cycle. The net surface heat flux (i.e. flux of heat energy from the atmosphere into the ocean) is large and negative over the western boundary currents of the ocean basins; it is positive along the equator where the air is heating the water, and along the eastern margins of the oceans where upwelling brings cold, nutrient rich water to the surface (Figure 2.3).

Tropical Marine Ecology, First Edition. Daniel M. Alongi.

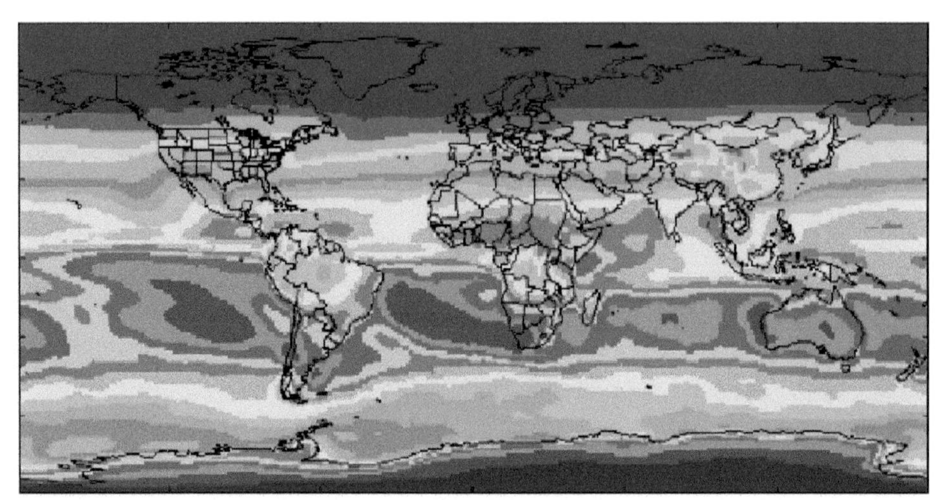

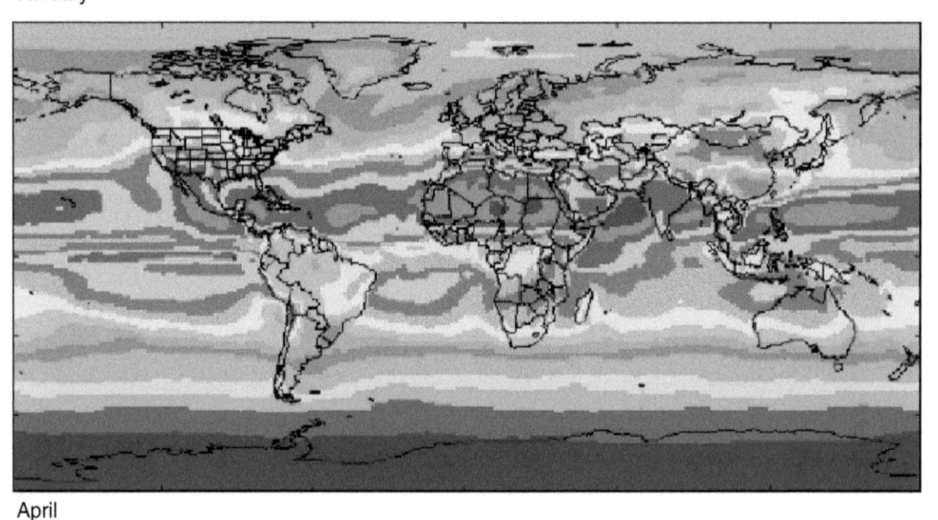

FIGURE 2.1 Mean daily solar insolation (kWh m^{-2} d^{-1}) in the global ocean (top) from January 1984 to 1993 (bottom) from April 1984 to 1993. *Source*: Image in the public domain courtesy of Roberta DiPasquale, Surface Meteorology and Solar Energy Project, NASA Langley Research Center and the ISCCP Project. http://eoimages.gsfc.nasa.gov/images/imagerecords/1000/1355/insolation.gif (accessed 12 February 2018). © NASA.

In upwelling systems, such as the summer upwelling in the South China Sea, there are prominent ocean–atmosphere interactions with the ocean driving the atmosphere, with wind with steady direction sustaining these interactions (Yu et al. 2020). Interannual variabilities in air–sea coupling are largely impacted by El Niño events changing the regional wind. The influence of a strong El Niño event on summer upwelling is characterised by warmer sea surface temperatures (SSTs) and declining winds, which are associated with weakened SST gradients and mesoscale air–sea coupling.

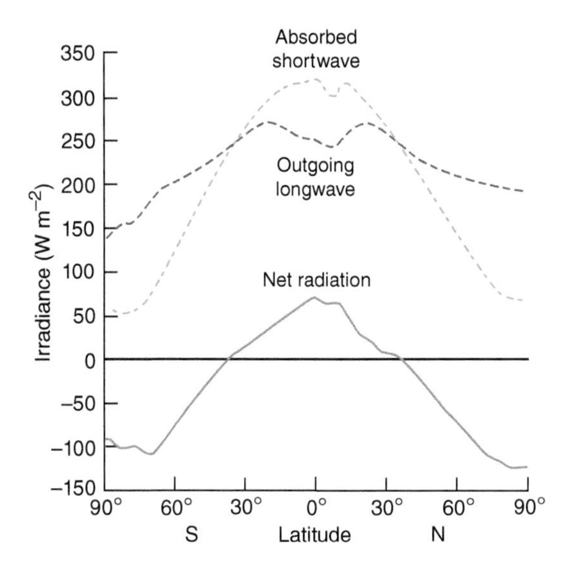

FIGURE 2.2 Annual mean absorbed solar radiation, outgoing longwave radiation and net radiation (Wm^{-2}) as a function of latitude, 2000–2013. *Source*: Hartmann (2016), figure 2.12, p. 44. © Elsevier.

2.2 Tropical Winds and the Intertropical Convergence Zone (ITCZ)

The key to understanding wind patterns in the low latitudes is a feature known as the Hadley Circulation (Nguyen et al. 2013), a thermally driven atmospheric circulation that features ascent of equatorial air to a height of about 15 km, with transportation aloft towards the poles, descent at the subtropics and a return flow near the surface. This circulation pattern explains the persistence and extent of the trade winds and the subtropical high-pressure belt that dominates the climate at low latitudes. There is an overall strengthening trend in the trade winds in the western equatorial Pacific and an overall weakening trend in the eastern equatorial Pacific (Li et al. 2019). This trend can be primarily attributed to a cold tongue mode, an out-of-phase relationship in SST anomaly variability between the Pacific cold tongue region in the east and elsewhere in the Pacific. The cold tongue region in the eastern equatorial Pacific (Liu et al. 2019) is characterized by a strong atmospheric subsidence that exerts powerful controls on global circulation patterns and the position of the intertropical convergence zone.

In the tropics, the air mass is barotropic (horizontally homogenous) so these large horizontal differences are unable to drive large-scale weather systems. Instead, winds and their associated weather are driven by vertical motion, which is a result of both heating by the sun and the northerly and southerly convergence of air. Over the Atlantic and most of the Pacific, these winds are westerlies and are often relatively light at upper levels but are highly variable seasonally at the lowest layer of the earth's atmosphere, the trophosphere. Over the Indian Ocean, Africa, and the western Pacific, winds are easterly close to the equator. These patterns are the trade winds that originate in the subtropical high-pressure systems centred near 30 °N and 25 °S. They persist to such an extent that early explorers such as William Dampier (Dampier 1699)

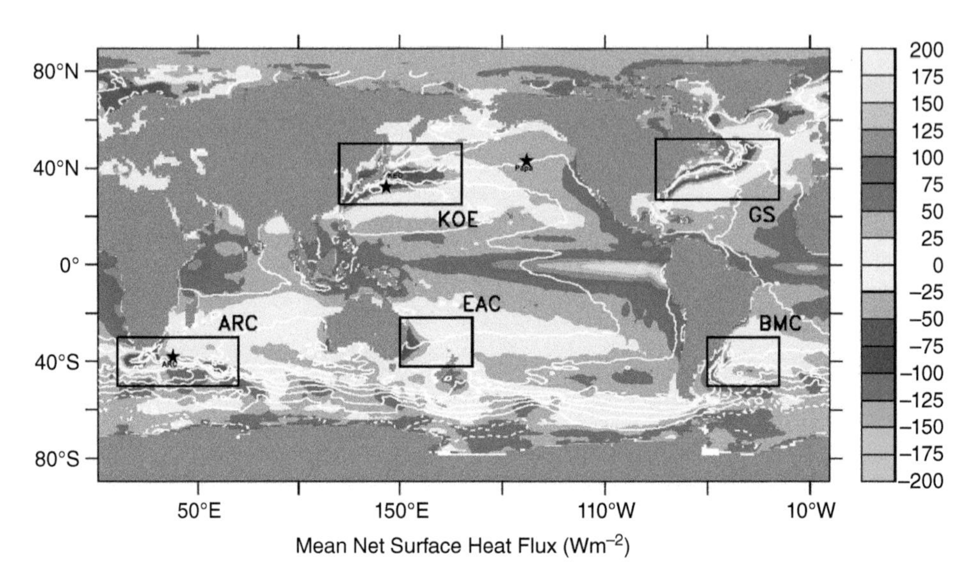

FIGURE 2.3 Annual mean net surface heat flux (W m^{-2}) into the global ocean. White contours show the mean dynamic sea surface height. Boxes are western boundary current regions. Ocean Climate Station moorings of the Pacific Marine Environmental Laboratory, NOAA, are indicated by stars. Abbreviations: KOE = Kuroshio Current; ARC = Agulhas Return Current; EAC = Eastern Australia Current; GS = Gulf Stream; BMC = Brazil-Malvinas Confluence. *Source*: Public access at https://www.pmel.noaa.gov/ocs/air-sea-fluxes (accessed 15 April 2019). © United States Department of Commerce.

were able to delimit the global trade wind patterns in the late-seventeenth century (Figure 2.4). In the Northern Hemisphere, northeasterlies converge near the equator with southeasterlies converging in the Southern Hemisphere. The trade winds provide enough forcing for deep convection to form a belt of convective cloud about the equator called the inter-tropical convergence zone or ITCZ.

The ITCZ is a narrow band of rising air and intense precipitation. The latter in the ITCZ is driven by moisture convergence associated with the northerly and southerly trade winds that collide at the equator. The ITCZ accounts for 32% of global precipitation (Kang et al. 2018) and moves north and south across the equator following the seasonal cycle of solar insolation and is intimately connected to seasonal monsoon circulations. On an annual average, the ITCZ lies a few degrees north of the equator. The location of the ITCZ has not changed significantly over the past three decades, but there has been a narrowing and strengthening of precipitation in the ITCZ over recent decades in both the Atlantic and Pacific Oceans (Byrne et al. 2018). Climate models project further narrowing and a weakening of the average ascent within the ITCZ as the climate continues to warm.

Wind speed and direction over the global tropical ocean are thus the result of a balance of forces that vary with distance from the equator. As the Coriolis force is weak at the equator this balance breaks down, although there is reasonable balance until about 6° latitude where some momentum usually carries the wind in the direction it is moving when in near-geostrophic balance, that is, wind in equilibrium between the pressure gradient and Coriolis forces thus blowing parallel to isobars or contours of height. Thus, there is a strong effect of latitude on the wind field patterns. Vertical wind motion takes place at a range of scales throughout the tropics. This vertical motion is facilitated by the Hadley Circulation, the main means by which the atmosphere tries to move energy from the equator to the poles (Nguyen et al. 2013)

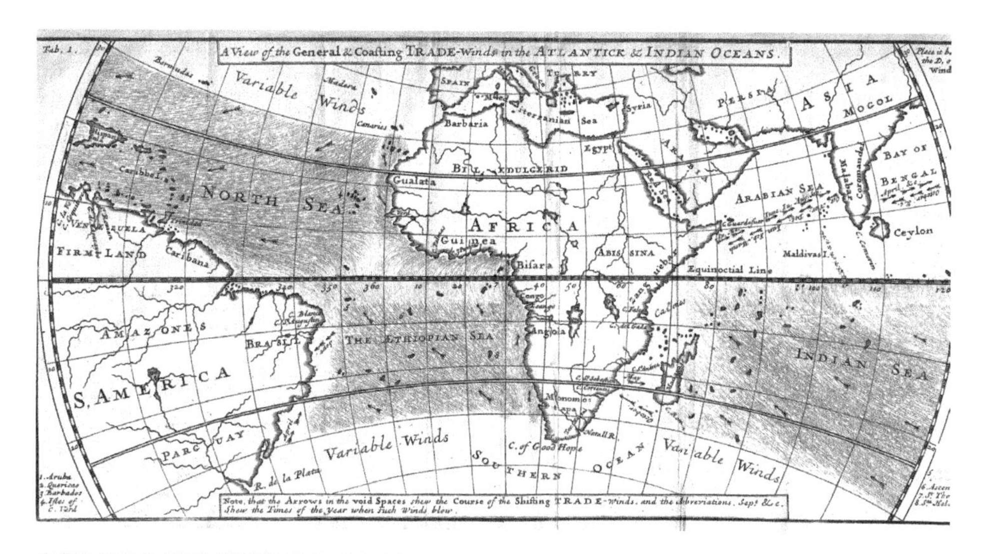

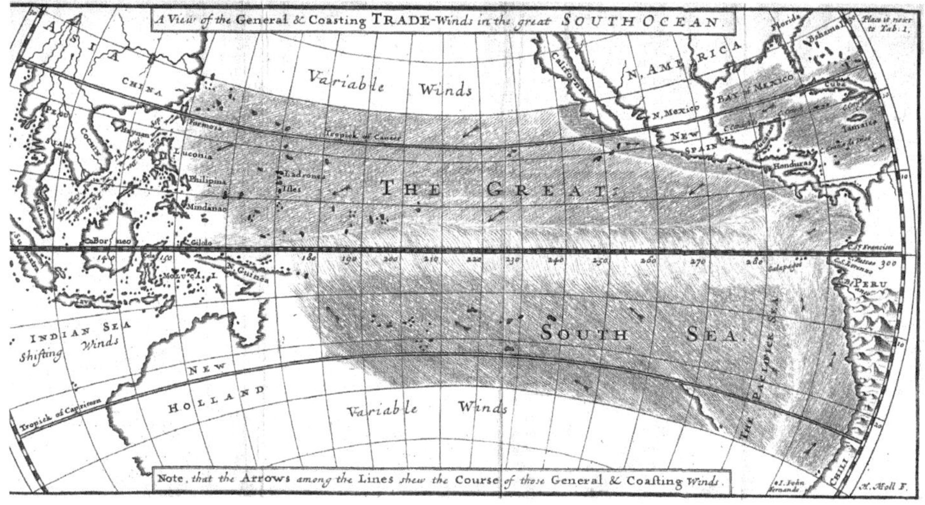

FIGURE 2.4 Late-seventeenth century chart of global trade winds. (top) Atlantic and Indian Oceans and (bottom) Pacific Ocean. *Source*: Modified from Dampier (1699), figures b2 and table 1, p 134 and 156. © John Wiley & Sons.

2.3 Tropical Rainfall and Temperature Patterns

Precipitation peaks about the equator and heavy rainfall are associated with the ITCZ where the trade winds converge. Moisture-laden air near the earth's surface flows towards the equator from both hemispheres and converges about the equator where it is released. Evaporation varies more smoothly than precipitation with a broad maximum in the tropics (Figure 2.5); precipitation exceeds evaporation in the belt from 15°–40° latitude. The runoff ('P-E' in Figure 2.5) implies transport of water vapour from the subtropics to the equatorial and high latitude zones. A return flow in oceans and rivers carries the water back towards the subtropics.

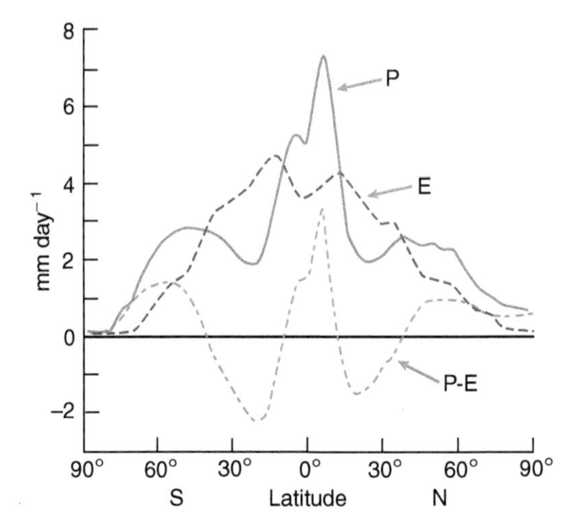

FIGURE 2.5 Latitudinal distribution of the annually averaged surface water balance, showing evaporation, E, precipitation, P, and P-E (runoff), 1979–2009. *Source*: Hartmann (2016), figure 5.2, p. 134. © Elsevier.

These patterns are reflected in the seasonal distribution of global precipitation (Figure 2.6), which shows a heavy band of precipitation about the equator (excluding the Arabian Sea region), but especially in the regions of northern South America and Southeast Asia where convective instability and moisture content of the air are high. These high rates of precipitation result in high rates of river discharge to the coastal ocean. These global averages mask great variability. For example, there are tropical coastal areas of high aridity, such as the Red Sea and Arabian Sea as well as the NW coast of Australia. In these areas, rates of evaporation exceed precipitation and temperatures often exceed the global average, having devastating effects on coral growth and reproduction and lower biodiversity, as fewer species can tolerate these conditions.

In the tropical Atlantic, atmospheric circulation anomalies interact with ocean circulation to produce anomalous SST and precipitation patterns (Figures 1.1 and 2.6). These anomalies originate off the African coast and expand westwards. Indeed, precipitation is low in these regions, and SSTs are correspondingly high year-round. A large area of the tropical southeast Atlantic is low in rainfall, south of the equator to about 30 °S, like the eastern boundary of South America. The El Niño-Southern Oscillation (ENSO) phenomenon is the primary source of short-term climate variability and is associated with distinct and different atmospheric and oceanic climate anomalies that affect rainfall, river flow, temperature, tropical cyclone activity, and shifts in the position of the major convergence zones (see Section 2.6).

SSTs (Figure 1.1) vary across tropical seas, but one of the most unique features is the existence and persistence of the Indo-Pacific Warm Pool (IPWP), a large area ($>30 \times 10^6$ km^2) of the western Pacific Ocean that is characterised by permanent SSTs >28 °C and is therefore called the 'heat engine' of the planet (De Deckker 2016), playing an important role in the seasonal monsoon and interannual variability such as ENSO. The IPWP is called the 'steam engine' of the globe because of high convective clouds which can reach altitudes up to 15 km, generating much

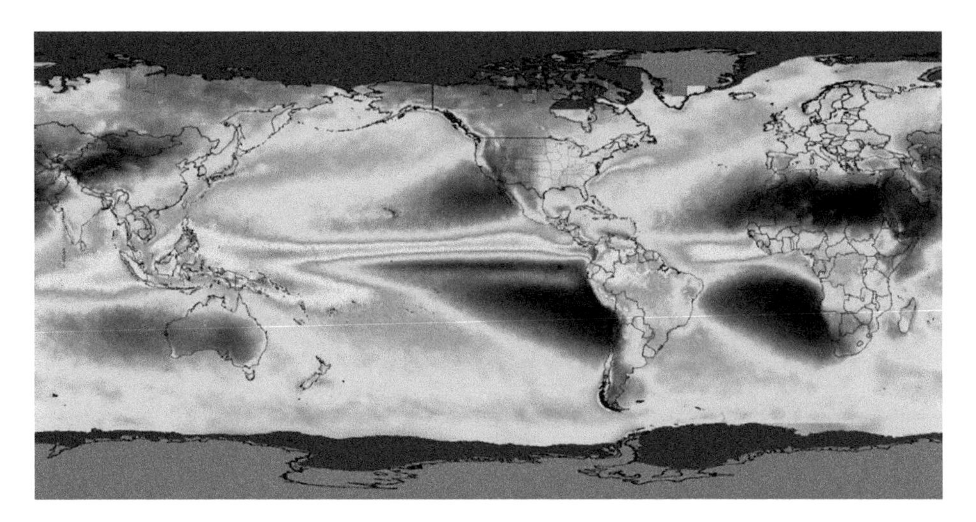

FIGURE 2.6 Annual mean global precipitation (mm per year) from the IMERG climatology dataset, NASA, June 2000-May 2019. Colours grading to yellow -red indicate greater precipitation. *Source*: Image in the public domain courtesy of the NASA Scientific Visualization Studio. https://sus.gsfc.nasa.gov/4760 (accessed 10 June 2021). © NASA.

latent heat in the process of convection. Broad seasonal change in surface salinities is caused by seasonal and contrasting monsoonal activity throughout the region. This area lies along the path of the Great Ocean Conveyor Belt and is a 'dilution' basin due to high incidence of tropical rain; away from the equator tropical cyclones contribute to a significant decline in ocean salinity. The IPWP has been expanding on average by 2.3×10^5 km^2 a^{-1} during 1900–2018 and at an accelerated average rate of 4×10^5 km a^{-1} during 1981–2018. Most of this expansion has occurred in the Indian Ocean rather than in the western Pacific Ocean and has been attributed to increasing concentrations of greenhouse gases and natural fluctuations associated with the Pacific Decadal Oscillation (Roxy et al. 2019)

2.4 Monsoons

Monsoons are a central component of global climate and are critical to the global transport of atmospheric energy and water vapour. More than 70% of the world's population lives in monsoonal regions and these systems have a profound effect on society and the global economy. Zhisheng et al. (2015) proposed the following definition of the global monsoon:

> *The global monsoon is the significant seasonal variation of three-dimensional planetary-scale atmospheric circulations forced by seasonal pressure system shifts driven jointly by the annual cycle of solar radiative forcing and land-sea interactions, and the associated surface climate is characterised by a seasonal reversal of prevailing wind direction and a seasonal alternation of dry and wet conditions.*

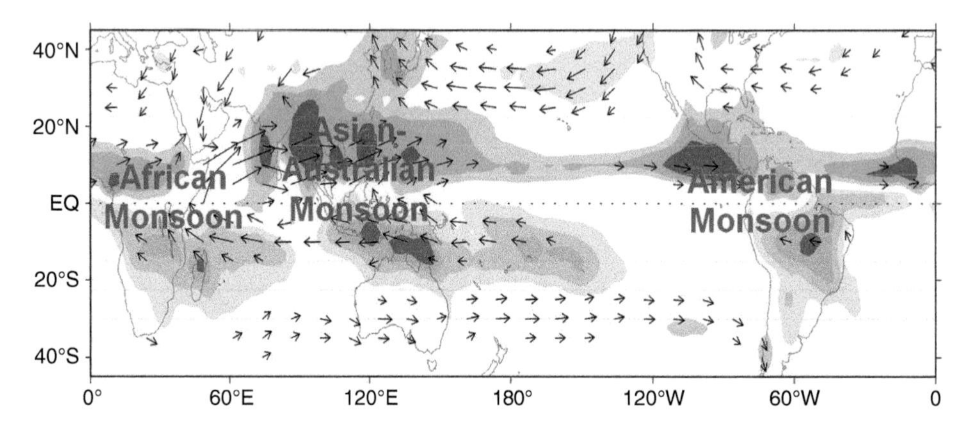

FIGURE 2.7 Global distribution of monsoon domains and their local components, including by differences of 850 hPa wind and precipitation between the June-July-August and December-January-February mean. *Source:* Zhou et al. (2016), figure 1, p. 3590. Licensed under CC BY 4.0. © Copernicus Publications.

The tropical monsoon primarily lies between the seasonal migration boundaries of the ITCZ (Figure 2.7). The seasonal migration of the ITCZ, caused by cross-equatorial pressure gradients, produces a strong tropical monsoon under the annual cycle of solar radiation. The global monsoon is distributed primarily across several major subtropical and tropical regions: tropical Asia, Indonesia-Australia, Africa, and South America (Figure 2.7). Although West Africa, northern Australia and parts of South America depend on much of their annual rainfall from seasonal monsoons, the archetypal monsoon system is the tropical Asian-Australian monsoon which consists of several main subsystems: the tropical Australian monsoon, the Maritime Continent monsoon, the South China Sea monsoon, the Indochina Peninsula, and western North Pacific monsoon and the Indian or South Asian monsoon.

In the Southern and Northern Hemispheres, subtropical monsoons (Figure 2.8) are caused by the seasonal shift of the subtropical high and land-sea distribution and are closely related to several features: large-scale topography, the Rossby radius of deformation, the jet stream, and the interaction between the jet stream and large-scale topography. (The Rossby radius of deformation is the length scale at which rotational effects become as important as buoyancy or gravity wave effects in the evolution of the flow about some disturbance.) The Southern Hemisphere subtropical monsoon includes the Southern Australian, South African, and subtropical South Pacific monsoons. The Northern Hemisphere subtropical monsoon consists of the East Asian, North American, North African, Tibetan Plateau, subtropical North Atlantic, and North Pacific monsoons.

2.4.1 The Asian Monsoon

The Asian monsoon system reaches from the western Arabian Sea through East Asia and North Australia (Figure 2.8). This system is composed of the Indian (or South Asian) and East Asian subsystems, roughly divided at about 105°E. Both subsystems are linked to varying degrees by regions of strong sensible heating (Indo-Asian landmass) and strong latent heat export (the western Pacific Warm Pool and the southern

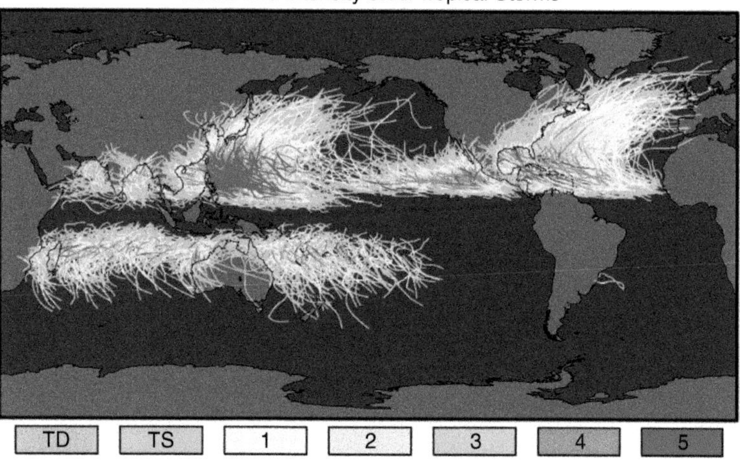

FIGURE 2.8 Global tracks and intensity of all tropical storms, 1856–2006. *Source*: Image retrieved via public access from NASA Earth Observatory. http://earthobservatory.nasa.gov/IOTD/view.php?id=7079. (accessed 25 June 2019). © NASA.

subtropical Indian Ocean). However, they also have significant differences determined by the contrasting sea-land distributions.

The Indian system is characterised by land in the north and ocean in the south, whereas the East Asian system has land in the north and south, a maritime continent in the west and open ocean to the east. Summer heating of the Asian continent, especially around the Tibetan Plateau, generates low-pressure cells and thus summer rains in South and East Asia. In the winter, a reversed high-pressure system is established, with dry, cool to cold winds blowing out of Asia. Waters of the northern Indian Ocean and South China Sea are subjected to seasonal changes in SSTs and salinity patterns and current strength and direction.

These dynamics lead to seasonal upwelling in the NW Arabian Sea. The monsoon is a result of northward seasonal migration of the east–west-oriented precipitation belt (Tropical Convergence Zone) from the Southern Hemisphere in winter to the Northern Hemisphere in summer. The largest northward movement of the rain belt takes place over the Indian monsoon region where it moves from a mean position of about 5°S in winter to about 20°N in summer. Equatorial easterlies in the upper atmosphere are weak and confined between 5°N and 10°S, while the subtropical westerlies intrude up to 10°N during winter. The subtropical westerlies recede to north of 30°N during summer and a strong easterly jet characterises the equatorial upper atmosphere. Seasonal rainfall is highly variable with prolonged spells of dry and wet conditions often lasting for two–three weeks. Periods of wet conditions are generally associated with an increase in cyclonic activity and decrease in surface pressure over the central Indian monsoon trough region (northern Bay of Bengal to West India) and strengthening of the low-level westerly jet over the Arabian Sea.

The East Asian monsoon features powerful breaks of cold air from the Eurasian continent during winter. During summer, the Tibetan Plateau acts as an elevated heat source and fosters an imposing subtropical monsoon front extending from the Bay of Bengal and the South China Sea all the way to the North Pacific, bringing abundant rainfall to East Asia, and transporting water and energy from the tropics to the mid-latitudes. In contrast to warm trends elsewhere on earth, an interdecadal surface cooling has occurred in some regions of East Asia in the past few decades. Such interdecadal changes have been attributed to tropical ocean forcing. For example, in the late 1970s, summer rainfall anomalies in China were consistent with an intensification and southward extension of the western Pacific subtropical high over the subtropical regions of East Asia. Changes in SSTs over the Indian Ocean and tropical western Pacific may have partly accounted for these interdecadal changes.

2.4.2 The Indo-Australian Monsoon

The Indo-Australian monsoon is a component of the large-scale Asian-Australian monsoon system. The monsoon is associated with the seasonal migration of the ITCZ, which is located 10–15° north of the equator in austral winter and migrates southward over the north of the Australian continent during austral summer, driven by the development of intense heat lows over NW Australia and NW Queensland. North Australia experiences dry low-level southeast trade winds in winter which undergo a reversal to dominant moist NW winds in summer. The onset of the monsoon usually occurs between late November and early January, but the monsoon season includes periods when southeast winds bring temporarily dry conditions.

The southern limit of the monsoon is defined by areas that receive >85% of the rainfall between November and April. Main rainfall events are associated with cyclones and other disturbances in the monsoon trough. Such conditions do not necessarily translate into high annual rainfall as there is a strong latitudinal gradient of rainfall with large regions along the northern and eastern coastlines having median rainfalls exceeding 1500 mm, but with most of the inland experiencing annual medians from 300 to 600 mm.

The Indo-Australian wet monsoon season is characterised by excessively wet conditions usually referred to as 'bursts' interspersed with relatively dry 'breaks'; both can sometimes extend for several weeks (Moise et al. 2020). Both Indonesia and Australia are linked in the summer monsoon. These wet and dry periods have a significant impact on recharge of freshwater ecosystems and river discharge on estuarine and marine ecosystems. A 'burst' in the Indo-Australian monsoon is preceded by the development of a well-defined extratropical wave packet in the Indian Ocean which propagates towards the Australian continent in the few days leading up to the onset of heavy rainfall in the tropics. These extratropical disturbances propagate equatorward over the continent and are accompanied by lower-tropospheric air mass boundaries which also propagate into low latitudes. Ahead of these boundaries, relatively warm moist air is advected from the surrounding seas. Monsoon bursts are more likely to occur when the active phase of the Madden-Julian Oscillation (Section 2.6) is in the vicinity of Australia (Duan et al. 2019).

2.4.3 The African Monsoons

The main two components of the African Monsoons are the West African Monsoon (WAM), which prevails during the Northern Hemisphere summer (June through September), and the East African Monsoon (EAM) with occurs during spring

(March–May) and autumn (October–December). The combined influence of the Indo-Pacific and the Atlantic Oceans drive the interannual and decadal monsoon variability over these regions.

The key feature of the WAM is low-level SW flow from the Atlantic Ocean and the Intertropical Convergence Zone (ITCZ) north of the equator. The WAM is characterised by the migration of zonally banded rainfall from the Guinea coast to the Sahel (a 3860-km arc-like landmass immediately south of the Sahara Desert stretching east–west across the breadth of the African continent) and back again, resulting in two rainy seasons per year in the south and one in the north. The dynamics of the WAM circulation are linked to the existence of the African easterly jet, a mid-tropospheric flow with peak amplitude of about 10–20 $m\,s^{-1}$ in August at latitude 15°N. This jet is a basic part of the momentum balance of the WAM circulation and has a meteorological impact on the patterns of rainfall in the Sahel. Unlike the Australian monsoon, over West Africa the monsoon trough moves much further inland, some 15° or more in latitude from the Guinea coast, and therefore the pressure gradients over much of the Sahel are directed from the humid south towards the dry north. There is a seasonal cycle of the WAM with two phases: the 'pre-onset' and the 'onset' (Janicot et al. 2011). In the 'pre-onset' phase, the average (10°W – 10°E) of the intertropical front (the confluence of monsoonal winds from the south and dry winds from the north) crosses 15°N on its northward migration, while the ITCZ is still in the south. The first rainy season along the Guinean coast occurs between mid-April and the end of June. In the 'onset' phase during the summer monsoon season, the ITCZ moves from 5°N to 10°N bringing rain over the Sahel.

The formation of the Atlantic cold tongue is the dominant seasonal SST signal in the eastern equatorial Atlantic and influences the onset of the WAM (Caniaux et al. 2011). From March to mid-June ('onset phase'), the cold tongue results from the intensification of the southeast trade winds associated with the anticyclone system located off the island of St. Helena. Steering of surface winds by the basin shape of the eastern equatorial Atlantic imparts optimal wind stress for generating the maximum upwelling south of the equator. From mid-June to August, wind speeds north of the equator increase due to the northward progression of the intensifying trade winds and due to significant surface heat flux gradients produced by the differential cooling between the Atlantic cold tongue and the tropical waters circulating in the Gulf of Guinea. Thus, there is a close link between the eastern equatorial Atlantic and the West African monsoon.

The East African monsoon system corresponds with the Greater Horn of Africa region, extending from Tanzania in the south to Yemen and Sudan in the north. During winter, when NE trade winds flow across the NW Indian Ocean and equatorial moisture moves over the Indian Ocean exhibiting strong westerly flows over the equatorial Indian Ocean, East African rainfall is limited to a few highland areas (Funk et al. 2016). During spring, the Indian monsoon circulation transitions, the trade winds over the NW Indian Ocean reverse, and East African moisture convergence supports the 'long' rains. In summer, the SW Somali Jet intensifies over East Africa. As subsidence forms along the westward flank of this jet, precipitation shuts down over eastern portions of East Africa. In autumn, the Jet subsides but easterly moisture supports rainfall in limited regions of the eastern Horn of Africa. Thus, Indian Ocean SSTs drive East African rainfall variability by altering the local Walker circulation (see Section 2.6). The influence of the Pacific Ocean appears to be minimal (Tierney et al. 2013). SST anomalies were the dominant forcing of the severe drought of the 1970s and 1980s (Biasutti 2019). However, ENSO showed a positive relationship with the East African 'short' rain during the

1949–2016 period. Only during a recent period (2000–2016) was a significant relationship found between ENSO and East African 'long' rain. The strengthened interannual relationship between ENSO and East African' long' rain is associated with an Indian Ocean Walker Cell in spring, which implies that their relationship could be affected by either multidecadal natural variability or anthropogenic forcing (Park et al. 2020).

2.4.4 The South American Monsoon

The South American Monsoon is part of the monsoon system of the Americas, encompassing most of the continent and featuring strong seasonal variability in a region lying between the Amazon and La Plata Basin (Silva Dias and Carvalho 2011). In the upper troposphere, the wet summer season is characterised by an anticyclonic circulation over Bolivia and a trough over the tropical and subtropical South Atlantic, near the coast of NE Brazil. There are several prominent low-level features, including (i) surface high-pressure systems and anticyclonic circulation over the subtropical Atlantic and Pacific Oceans; (ii) the Chaco thermal low centred over northern Argentina; (iii) the South Atlantic Convergence Zone (SACZ); and (iv) the South American Low-Level Jet (SALLJ) east of the Andes, a low-level NW flow east of the Andes that extends from the SW Amazon to southeast South America (Marengo et al. 2012).

Transient moisture flow from the Amazon is important for maintenance of the SACZ, and the location of the SACZ is influenced by the topography in central-east Brazil, while it has a strong influence on the position and intensity of the Bolivian High. Climate variability associated with the SACZ and the SALLJ dipole-like pattern has been observed on intra-seasonal, inter-seasonal, and decadal time scales. One phase of the dipole is characterised by an enhanced SACZ and suppressed convection to the south, whereas the other phase is characterised by a suppressed SACZ and increased convection in the subtropical plains. A strengthening of the SALLJ and associated transports of massive amounts of moisture from the Amazon Basin into the subtropics accompanies the latter phase.

The annual cycle of precipitation features distinct wet and dry seasons between the equator and 25°S. Rates of rainfall in the wet season are somewhat less than those in other monsoon systems but like those over the ocean. The annual cycle of rainfall in most pronounced in the southern Amazon, where some of the largest seasonal rainfall occurs, and extends to the southeast in the South Atlantic convergence zone, which is an extension of the wet season maximum at 10°S. Precipitation in this region is out of phase with that further south and is driven synoptically. The Madden-Julian Oscillation (Section 2.6) increases the average daily precipitation by >30% and doubles the frequency of extreme events over central-east South America, including the South Atlantic convergence zone (Grimm 2019).

ENSO plays an important role in the dynamics of the South American monsoon: during the warm phase of ENSO, precipitation decreases near the equator during the summer wet season and increases rainfall in SE South America. Large sub-seasonal changes also occur within an ENSO cycle, but long-term trends of precipitation are small in the Amazon Basin. However, downstream convergence of moisture advected from the Amazon Basin by a strong low-level jet flowing southward along the eastern flank of the Andes results in some of the most frequent and large mesoscale systems on earth over the northern part of La Plata Basin, the second largest drainage basin in South America after the Amazon Basin. Those powerful systems contribute a larger proportion of total rainfall than any other region on earth.

The onset of the monsoon progresses towards the equator from an area in southern Brazil. Rainfall is at near minimum along the equator in December–February, while the wettest season occurs in April and May. At about 5°S, it begins to rain earlier than at 10°S, but rainfall stays below average until after the heavy rains begin farther south. Interannual variability in rainfall is related to variations in either the onset or ending of the wet season. More rainfall occurs when Atlantic SST anomalies are positive south of the equator, resulting in a southward displacement of the ITCZ. This may be a mechanism by which Atlantic SST influences monsoon rainfall, to the extent that these anomalies associated with a displaced ITCZ occur early and extend westward into the Amazon. Long-term trends are not well understood due to lack of observation sites, but records of river discharge indicate alternating decadal trends. For example, the Amazon had low flows in the late 1960s, high flows in the mid-1970s, and low flows again in the early 1980s. Such trends in rates of river discharge have significant impacts on marine processes.

2.5 Tropical Weather Systems

Earth averages about 85 revolving storms per year, with an average of 12 in the Atlantic, 17 in the NE Pacific, 26 in the NW Pacific, five in the north Indian, nine in the SW Indian, seven in the southeast Indian, and nine in the SW Pacific Ocean (Figure 2.8). These systems are known as hurricanes in the NE Pacific, North Atlantic and Caribbean, typhoons in the western Pacific, and cyclones in the Indian Ocean and SW Pacific.

Tropical cyclones form within regions of pre-existing deep precipitating convection. Before cyclone formation, the convection is only loosely organised into mesoscale areas of enhanced cloudiness that are often called tropical disturbances. Only a small fraction of such disturbances become cyclones (Figure 2.8). Such storms initially form as a cloud mass to one side of the equator over the sea when temperatures are at or above 27 °C and occur far enough away from the equator for the Coriolis force to induce 'twist' on the storm; storms such as these do not develop on or within 5 °N or S of the equator.

Cyclogenesis is not well understood but considering that 80–90% of all tropical cyclones form within 20° of the equator (Webster 2020), the genesis of cyclones may be modulated by the family of equatorial and near-equatorial waves that propagate zonally in this band (Sharkov 2012). The waves could cause cyclones to form by organising deep convection and/or by altering the flow in preferred areas. Mixed Rossby-gravity waves, tropical depression-type, or easterly waves, equatorial Rossby waves (large-scale waves of low amplitude that move along the thermocline), and the Madden-Julian Oscillation may play a role in cyclogenesis. These waves appear to enhance the local circulations by increasing the forced upward vertical motion, increasing the low-level vorticity at the site of formation, and by modulating the vertical steer. Most convection occurs near the centre and this usually allows an 'eye' to develop due to convective subsidence; a 'wall' also develops around the centre due to dynamical forcing and development of a ring of cumulonimbus cloud as well as altostratus and nimbostratus clouds. Gale force winds develop due to falling pressure.

Even when conditions are favourable for cyclone formation, storms may not occur. For instance, hurricanes rarely spawn off the Brazilian coast despite the existence of favourable conditions because wind shear in the upper troposphere is too strong to permit storm development. Convective heating must be strong enough and upper-level winds must be weak enough to permit development of a warm core.

These evolving storm systems deepen over the NW Pacific most commonly between March and December, between November and April in the south Indian Ocean, in the SW Pacific between December and April, and in the NE Pacific, North Atlantic, and Caribbean between June and November. Fewer such storms occur over the northern Indian Ocean, rarely during the SW monsoon and during the inter-monsoon season, while no storms form in the SE Pacific where SSTs are too low. These storms all have one common characteristic: the need for a very warm water source. Such storms rapidly dissipate over land, so an immense supply of constant latent heat over the tropical ocean is required. Not surprisingly, rain bands are spiral in these revolving storms and are formed by cooler air and bring heavy rainfall and frequent thunderstorms. These rain bands are complex and can re-energise over land near swamps and large lakes, or if they reach another parcel of warm water. For instance, cyclones spawned in the Gulf of Carpentaria west of Cape York peninsula in northern Australia have been observed to weaken considerably crossing the cape but then regenerate once back over the warm water of the Coral Sea. Similar regeneration can occur for storms moving westward from the South China Sea into the Bay of Bengal, and storms originating in the Bay of Bengal have been known to cross the Indian subcontinent and reform in the Arabian Sea.

In the western North Pacific, typhoons are more frequent than in other ocean basins (seasonal average of 26), with peak activity from June to October. More intense typhoons are found in the Central Pacific compared to eastern Pacific El Niño years. The higher occurrence of intense typhoons in the Central Pacific is related to longer typhoon lifespan, maximum potential intensity, ocean heat content (OHC), vertical shear of the zonal wind, outgoing long-wave radiation and moist static energy (Zhang et al. 2015). A longer typhoon lifespan is caused by the westward shift of the subtropical high, which tends to steer typhoons to the west and NW, usually towards the Philippines, Pacific Islands, Taiwan, and southern China. Typhoons tend to obtain more energy from the warm SSTs along such tracks. The evolution of SST anomalies over the north Indian Ocean and tropical Pacific plays a crucial role in typhoon formation following strong El Niño events.

In the eastern Pacific, hurricanes occur mostly from May to November with a seasonal average of 16.6, making this region the second most active cyclone region. Oceanic control through meridional redistribution of subsurface heat is the main driver of tropical cyclone activity following eastern Pacific El Niño events (Boucharel et al. 2016a). Equatorial Kelvin waves control the sub-annual and intra-seasonal variability of thermocline depth in the eastern Pacific region (Boucharel et al. 2016b), affecting ocean subsurface temperature which in turn fuels hurricane intensification. (An equatorial Kelvin wave is a wave in the ocean that balances the Coriolis force.)

The Indian Ocean accounts for about 20% of global cyclonic activity. Cyclones in the northern Indian Ocean (seasonal average of 4.8) mainly occur in the western and central part of the Bay of Bengal. The strong vertical shear during the SW monsoon inhibits cyclones, leading to a bimodal seasonal distribution with peak activity during the pre- and post-monsoon (Lengaigne et al. 2019). Cyclones that form in the northern Indian Ocean are among the world's deadliest, inundating coastal north and NE India, Bangladesh, and Myanmar. In the southern Indian Ocean, cyclones (seasonal average of 9.3) occur over an elongated band centred on

15°S from November to April, with enhanced cyclone occurrence over the SW Indian Ocean around Mauritius, La Réunion and Madagascar. Considerably different upper ocean thermohaline structures in the southern and northern Indian Ocean may result in a different sensitivity of cyclones to ocean–atmosphere coupling (Lengaigne et al. 2019). The Bay of Bengal is characterised by strong stratification that may limit cooling promoting cyclone intensification.

In contrast, the SE Indian Ocean is one of the rare oceanic regions where warm SSTs coexist with a shallow thermocline favouring an enhanced cooling below the storm and a strengthened negative feedback at the early stages of storm intensification. The south Indian Ocean warms following a strong El Niño affecting Indo-Pacific climate in early summer. Warming of this region induces an anomalous meridional circulation with descending motion over the NE Indian Ocean. The SE Indian Ocean warming lags the SW Indian Ocean warming by one season. South of the equator, El Niño-forced Rossby waves are reflected as eastward-propagating Kelvin waves along the equator on the western boundary. The Kelvin waves subsequently depress the thermocline and develop the warming of the southeast tropical Indian Ocean (Chen 2019).

Like cyclones in the northern Indian Ocean, tropical cyclones in the South China Sea are prone to landfall resulting in huge losses in human life and infrastructure along the southeast coast of China. Cyclones in the South China Sea form in summer (May–August) and in winter (September–December) with significant interannual variations in the two seasons and relatively clear inter-decadal variability in summer (Wang et al. 2019). In winter, vertical circulation is obvious due to a strong south Indian Ocean Dipole (Section 2.6) that induces intensified ascending air and high SSTs over the northern South China Sea with enough moisture to favour tropical cyclone formation. The impact of the south IOD is weaker in the summer, but a convergent zone with upwards motion also can be found over the NE South China Sea. The south IOD and ENSO are two primary factors impacting on cyclone formation in winter and numbers of cyclones increase when La Niña and positive IOD events occur simultaneously. However, the IOD plays a dominant role in tropical cyclone formation compared with the influence of ENSO (Wang et al. 2019).

The Australian region is unique as about 50% of all tropical cyclones form within approximately 300 km of the north coast. The region is climatologically active for tropical cyclones, with a typical season average of 12.5 cyclones, with six forming in the eastern Indian Ocean, three forming in the Timor and Arafura Seas and the remaining three and one- half cyclones developing in the Coral Sea. About five cyclones cross the coastline each season doing considerable damage. These cyclones are embedded within the monsoonal trough, although there is considerable variability in areas of low pressure during the summer. Heavy rains and flash floods are common during summer to the extent that many rivers that are reduced to dust in the winter dry season can burst their banks in the wet season. The strength of the summer monsoon is strongly associated with variations in cyclonic activity. Tropical cyclones form closer to the north coast during El Niño years than La Niña years, owing to an equatorward shift of the monsoon trough, warmer SSTs, and weaker vertical wind shear. Tropical cyclones are more likely to make landfall in Western Australia and the Northern Territory during El Niño years. There is a significant correlation between the number of tropical cyclones and the Southern Oscillation Index (SOI), which is defined as the difference between the standardised sea-level pressure anomalies of Tahiti and Darwin in the Northern Territory. Increases in tropical cyclone frequency and in landfall impacts have been noted for strongly positive SOI values, indicating anomalously low sea-level pressures over northern Australia.

Hurricanes in the North Atlantic develop from African easterly waves which originate along the southern and northern flanks of the African easterly jet. Hurricanes develop from June to November with peak activity in August–September and seasonally average 12.1 systems per season (Burn and Palmer 2015). These systems occur within the Atlantic warm pool, a region of warm water (> 28.5 °C) encompassing the Caribbean, the Gulf of Mexico, and the western tropical North Atlantic and varying spatially on interannual and interdecadal timescales. Atlantic cyclonic activity is influenced primarily by ENSO and the Atlantic Multidecadal Oscillation (AMO), which combine to modulate SSTs and vertical wind shear. ENSO affects Atlantic climate interannually, and its amplitude modulates on interdecadal (50–90 years) timescales (Burn and Palmer 2015). ENSO influences climate in the tropical Atlantic by controlling variations in sea-level pressure across the equatorial Pacific, which modulate the strength of the Pacific Walker Circulation and upper-level westerly wind flow into the region where Atlantic hurricanes develop. Corresponding changes in vertical wind shear in the tropical Atlantic parallel changes in the strength of the Atlantic NE trade winds caused by fluctuations in the surface pressure gradient between the North Atlantic and equatorial Pacific. El Niño (La Niña) events are typically associated with stronger (weaker) vertical wind shear, which suppresses (enhances) rainfall and the formation of hurricanes in the tropical Atlantic. The thermodynamic environment that underpins hurricane development is controlled by the AMO which is an index of Atlantic SST anomalies varying periodically on interdecadal timescales. However, in the Caribbean Sea, much more cyclonic activity occurs with La Niña conditions than with El Niño events.

2.6 The El Niño- Southern Oscillation (ENSO), the Indian Ocean Dipole (IOD), the Madden-Julian Oscillation (MJO), and the Pacific Decadal Oscillation (PDO)

Seasonal variations in climate are modulated on intra-annual, interannual, and decadal time scales by climate phenomena such as ENSO, the IOD, the MJO, and the PDO. How they affect each other is less well-known than how they impact Earth's climate, but it is reasonably clear that most or all of Earth's climate systems are somehow interlinked, especially within subtropical and tropical latitudes. These phenomena are accompanied by changes in atmospheric and oceanic circulation, affecting global climate, marine and terrestrial ecosystems, and human activities.

The ENSO phenomenon is a large-scale, global, coupled atmosphere–ocean phenomenon that results in major surface climate anomalies throughout much of the tropics (Timmermann et al. 2018). The extreme phases of ENSO can bring either warmer (an El Niño event) or colder (a La Niña event) than normal SSTs to the central and eastern tropical Pacific that then affect global atmospheric circulation and weather patterns on timescales ranging from weeks to decades. Most ENSO events occur on a timescale of three-five years. The occurrence of droughts, floods, cyclones, and other weather phenomena across the planet have been associated with ENSO periods. The term 'Southern Oscillation' refers to the variability in the pressure difference across the Pacific Ocean (between Darwin and Tahiti), referred to previously as SOI which has been developed to measure the strength of El Niño.

The dynamics of ENSO are complex and variable from event to event (Timmermann et al. 2018), but a composite evolution of El Niño events captures the typical evolution of ocean and atmosphere conditions from the early spring initiation of El Niño, to its wintertime peak and transition to La Niña during the subsequent summer (Figure 2.9). Onset of an ENSO cycle begins with warm and deep SSTs in April with random forcing and downwelling Kelvin waves followed by growth by August and an eastward shift with nonlinear feedbacks (positive feedback along the equator in which a weakened equatorial SST gradient weakens trade winds which in turn further reduces the SST gradient). By December, the El Niño matures with warm SSTs in a narrow west–east band and a southward shift of wind anomalies. By April of the following year, the event decays as heat content in the deep ocean is discharged. A transition to colder SSTs in the eastern Pacific occurs by August followed by maturation of the La Niña (Figure 2.9) by December. Strong El Niño conditions typically last for one year, but La Niña events can persist for up to several years.

The El Niño event of 2015/2016 was initiated in boreal spring by a series of westerly wind events. This wind forcing triggered downwelling oceanic Kelvin waves, thus reducing the upwelling of cold subsurface waters in the eastern Pacific cold tongue, leading to surface warming in the central and eastern Pacific. The positive SST anomaly shifted atmospheric convection from the western Pacific Warm Pool to the central equatorial Pacific, causing a reduction in equatorial trade winds, which in turn intensified surface warming through positive feedback. Termination of the 2015/2016 event was associated with ocean dynamics and the slow discharge of equatorial heat into off-equatorial regions. The event started to decline in early 2016 and transitioned into a weak La Niña in mid-2016.

The most dramatic example of the impact of an El Niño event on oceanographic processes is the Peruvian upwelling system (Chapter 10, Section 10.9.2). During El Niño, the thermocline and nutricline deepen significantly during the passage of coastal-trapped waves within the Peruvian upwelling system. While the upwelling-favourable wind increases, the coastal upwelling is compensated by a shoreward geostrophic near-surface current. The depth of upwelling source waters remains unchanged during El Niño, but their nutrient content decreases dramatically, which along with a mixed layer depth increase, impacting phytoplankton growth. Offshore of the coastal zone, enhanced eddy-induced subduction during El Niño plays a potentially important role in nutrient loss.

Another dramatic example of the effect of an El Niño event are mass coral bleaching events (Section 13.2). A relatively modern phenomenon first reported in the 1980s, bleaching of corals has been unequivocally linked to SSTs above the upper thermal tolerance limits of corals and is widespread typically during El Niño events.

Each ENSO event is different and such variability may arise from climate phenomena outside the tropical Pacific, including the North and South Pacific meridional modes, extra-tropical atmospheric circulation patterns, and tropical Atlantic variability. The negative phase of the North Pacific Oscillation tends to favour the development of positive SST anomalies in the central Pacific by weakening the trade winds in the Northern Hemisphere, while the positive phase of the South Pacific Oscillation tends to weaken the trade winds in the Southern Hemisphere, thereby

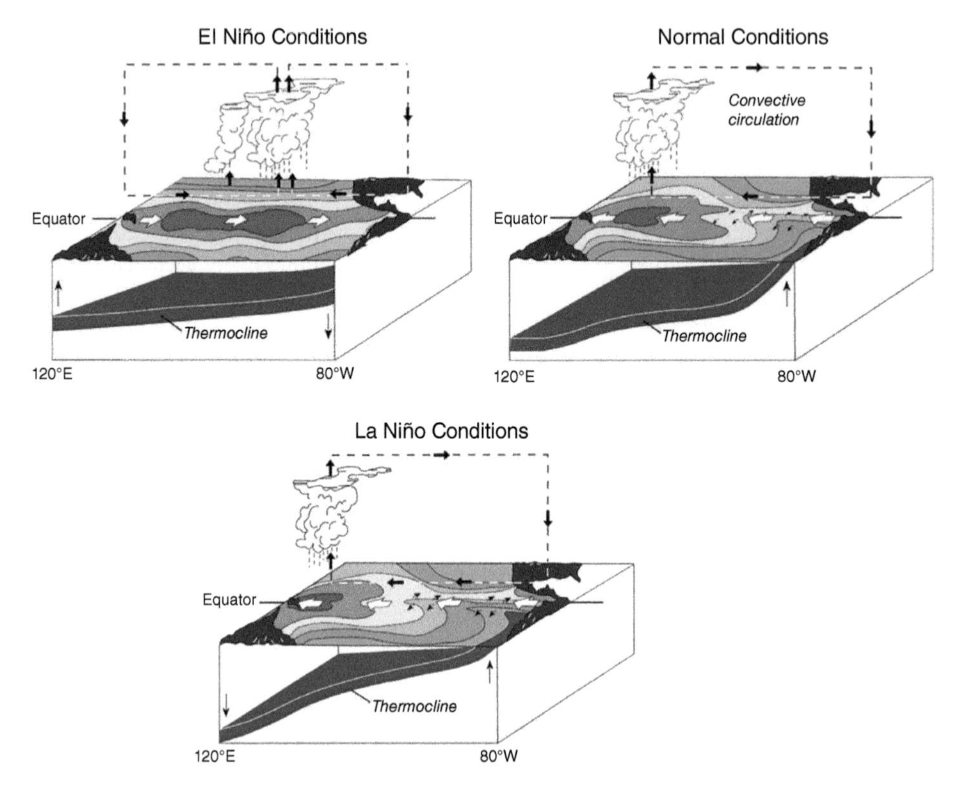

FIGURE 2.9 The three phases of the El Niño-Southern Oscillation (ENSO). During an El Niño event (upper left panel), trade winds weaken or may reverse, allowing the area of warmer than normal water to move into the central and eastern tropical Pacific Ocean with the thermocline shallowing eastwards. During normal (or neutral) conditions (upper right panel), trade winds blow east to west bringing warm moist air and warmer surface waters towards the western Pacific and keeping the central Pacific relatively cool with the thermocline much deeper in the west. During a La Niña event (lower panel), the Walker circulation intensifies with greater convection over the western Pacific and stronger trade winds. *Source*: Public domain image from https://www.pmel.noaa.gov/elnino (accessed November 13, 2020). © United States Department of Commerce.

favouring the development of positive SST anomalies in the eastern Pacific. Western wind anomalies in the western equatorial Pacific tend to favour El Niño in the central Pacific, whereas westerly wind anomalies in the central and eastern Pacific tend to favour El Niño in the eastern Pacific. There is some evidence of an increasing trend in ENSO amplitude during the past century (Christensen et al. 2013).

Although ENSO is global in scope and effect, it is linked to a series of other climatological phenomena, such as the Indian Ocean Dipole (Fan et al. 2017). The equatorial Indian Ocean is warmer in the east with a deeper thermocline and mixed layer and supports a more convective atmosphere than in the west. During September–October, the eastern equatorial Indian Ocean becomes usually cold with equatorial winds blowing from east to west and from the SW off the coast of Sumatra, facilitating coastal upwelling. At the same time, the western equatorial Indian Ocean becomes warm and enhances atmospheric convection. Sea-level decreases in the eastern equatorial Indian Ocean and rises in the central region. This coupled ocean–atmosphere phenomenon of interacting convection, winds, SSTs, and thermocline is known as the Indian Ocean Dipole (IOD). The state described above is referred to as a positive IOD and the reverse, a negative IOD is characterised by warmer STT anomalies, enhanced convection, higher sea level, a deeper thermocline in the east and cooler SSTs, lower sea level, a shallower thermocline, and suppressed convection in the west. A negative IOD event is an intensification of the normal state whereas positive IOD represents the opposite to the normal state.

The IOD results in two large-scale patterns in countries bordering the Indian Ocean: (i) anomalously high land temperature and rainfall in the western Indian Ocean and low land temperature and rainfall in the east and (ii) enhanced rainfall over the Asian monsoonal trough, extending from Pakistan up to southern China. During IOD events, biological productivity of the eastern Indian Ocean increases as does the frequency of coral deaths. The IOD also affects rainfall over Australia and eastern Africa. The IOD is characterised by (i) a Walker cell anomaly over the equator in the Indian Ocean, (ii) deep modulation of the monsoonal westerlies, and (iii) a Hadley cell anomaly over the Bay of Bengal (Fan et al. 2017). The IOD is important to global climate as about 50% of IOD events over the past century have co-occurred with ENSO events.

Both ENSO and the IOD interact with a phenomenon called the Madden-Julian Oscillation or MJO (Zhang 2005). The MJO is a major source of intra-annual variability in the tropical atmosphere and often results in breaks and bursts of monsoonal activity, helping to invigorate tropical cyclones (e.g. Cyclone Winston in 2016). The MJO also modulates cyclone development in the Caribbean. Even though it originates in the equatorial Indian and western Pacific Oceans, the MJO affects equatorial surface winds in the tropical Atlantic. The MJO interacts with the underlying ocean to influence weather and climate, especially over the Pacific Islands, monsoonal Asia and Australia, South America, and Africa. The interannual variability of SSTs associated with ENSO affects interannual variability in the MJO in the Pacific.

The MJO elicits ocean responses that have some bearing for life in tropical seas. The oceanic mixed layer, for instance, is a direct consequence of the surface cooling in convection centres of the MJO, and warming outside, fluctuations in SST propagate eastwards in tandem. SST differences can be about 0.5 °C. Strong surface wind of the MJO forces ocean currents and hence possible effects of horizontal advection. Strong winds force eastward equatorial currents of $\approx 1\,\mathrm{m\,s^{-1}}$ near the surface which may penetrate to 100 m depth, affecting the movements of pelagic organisms.

The pulse-like structure of the MJO forces pulses of downwelling Kelvin waves (Zhang 2005). They propagate from their origin in the western Pacific to the eastern Pacific where the MJO is weak or absent. Vertical displacement of the thermocline thus occurs, typically to a depth of 20–30 m. This can affect ENSO events. That is, in the central Pacific near the eastern edge of the Indo-Pacific Warm Pool, the eastward surface current of the Kelvin wave results in advection of warmer water eastward. In the eastern Pacific, the displacement of the thermocline associated with the downwelling Kevin waves weakens the cooling of equatorial upwelling, leading to warmer equatorial SSTs.

The MJO disturbs the upper ocean through surface fluxes of momentum, latent, and sensible heat, radiation and freshwater, with the latter three accounting for buoyancy flux. The net freshwater flux into the ocean (P-E) is mainly controlled by rainfall, as strong evaporation in convective centres of the MJO compensates only slightly for the frcshwater input. The net result is that perturbations in solar radiation flux (controlled by cloudiness) and latent heat flux (mostly controlled by surface winds) have similar amplitudes (25–30 $\mathrm{W\,m^{-2}}$). The intra-seasonal amplitude of the net heat flux (mostly composed of radiation and latent heat) depends on the relative phase of different components of the MJO.

The Pacific Decadal Oscillation (PDO) is the dominant year-round pattern of monthly North Pacific SST variability and is often described as a long-lived El Niño-like pattern in the tropical Pacific (Vishnu et al. 2018). The PDO is not a single phenomenon but is instead a complex aggregate of different atmospheric and oceanographic forcing spanning the extratropical and tropical Pacific (Newman et al. 2016). The PDO's amplitude is greatest from November–June, with weak maxima both in mid-winter and late spring and a pronounced late summer-early autumn minimum (Newman et al. 2016).

Positive (negative) phases of the PDO are associated with warming (cooling) of the tropical Pacific Ocean. The PDO modulates climate variability in various parts of the globe, such as drought frequency in the United States and summer monsoon rainfall in south China. Positive (negative) phases of the PDO are associated with the deficit (excess) Indian summer monsoon rainfall and enhance (suppress) the teleconnection between the rainfall in India and ENSO (Vishnu et al. 2018). The frequency of tropical cyclones over the western North Pacific also shows a decadal variability associated with the PDO. The number of tropical cyclones across the Pacific is less (high) in the warm (cold) phases of the PDO. There is also an out-of-phase variation in the number of monsoon depressions over the Bay of Bengal and the PDO. Vishnu et al. (2018) postulate that the variation in SSTs in the western equatorial Indian Ocean associated with the PDO could be one of the reasons for the changes in the moisture advection over the Bay of Bengal and hence the variation in the number of monsoon depressions on an interdecadal timescale.

The positive and negative phases of the PDO may have an impact on the expansion of the poorly oxygenated regions of the eastern Pacific Ocean (Duteil et al. 2018). During a 'typical' positive phase of the PDO, modelling indicates that the volume of the suboxic regions expanded by 7% over a 50 year period due to a slowdown of the large-scale circulation related to the decrease in the intensity of the trade winds. The model suggested that the prevailing positive phase conditions of the PDO since 1975 may explain a significant part of the current deoxygenation of the eastern Pacific Ocean.

2.7 Climate Change: Physical Aspects

Humankind has had and is still having a direct impact on earth's climate. Since the beginning of the Industrial Age there has been an increase in the concentrations of carbon dioxide (Figure 2.10 top), methane and nitrous oxide in the atmosphere, the direct result of the burning of fossil fuels, and to a lesser extent, deforestation. The increases in atmospheric greenhouse gases have had a direct impact on the global ocean, warming the earth's seas which in turn has resulted in a rise in sea level via thermal expansion. The uptake of CO_2 has resulted in a decrease in ocean pH (Figure 2.10 bottom) and CO_3^{2-} concentration, a process termed ocean acidification. Other impacts of human-induced climate change include increases in air and ocean temperature, an increase in OHC and changes in precipitation patterns.

Warming of the global ocean is the largest near the surface, the upper 75 m warmed by 0.11 °C per decade over the period 1971–2010 (IPCC 2014). It is likely that the ocean warmed from the 1870s to 1971. Regions of naturally high salinity

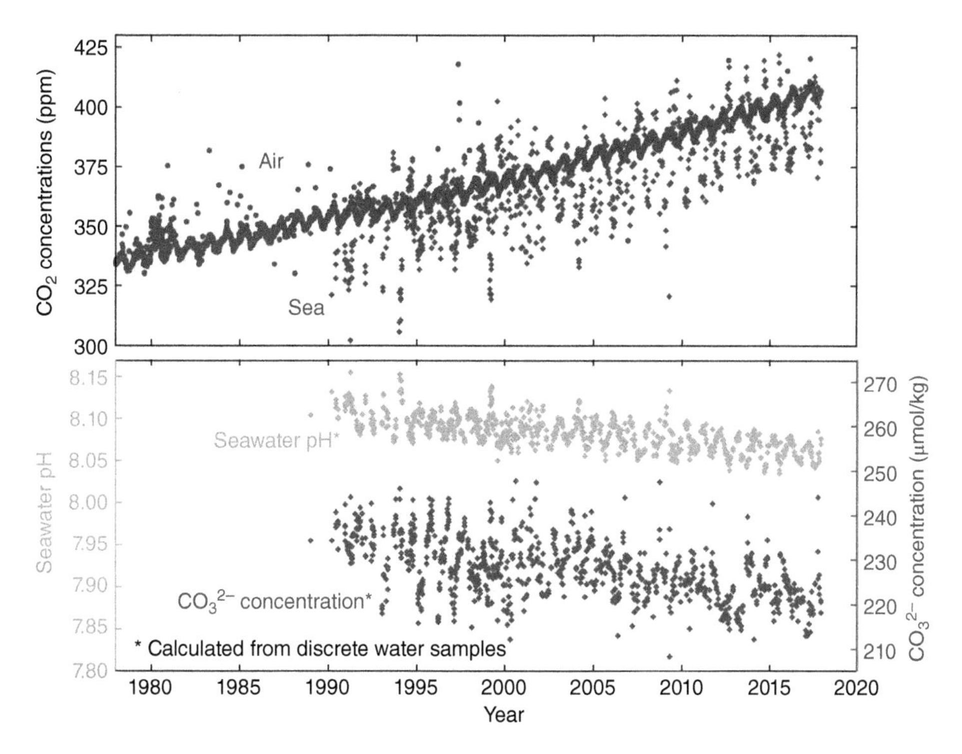

FIGURE 2.10 Trends in surface (< 50 m depth) ocean carbonate chemistry calculated from observations obtained during the Hawaii Ocean Time-Series Program in the North Pacific from 1988 to 2017. The upper graph shows the concomitant increase in CO_2 concentrations in both the atmosphere (red) and surface ocean (blue), presented as CO_2 concentration in air (ppm). The bottom graph shows a decline in ocean pH (light blue, primary y-axis) and carbonate ion (CO_3^{2-}) concentration (green, secondary y-axis on right). Source: Doney et al. (2020), figure 1, p. 87. Licensed under CC BY 4.0. © Annual Reviews.

have become more saline, while regions of low salinity have become fresher since the 1950s. Such regional trends in ocean salinity provide indirect evidence for changes in evaporation and precipitation over the ocean and for changes in the global water cycle (IPCC 2014). Since the beginning of the Industrial Era, oceanic uptake of CO_2 has resulted in a decline in surface ocean pH (Section 2.7.2). There is medium confidence (IPCC 2014) that in parallel to warming, oxygen concentrations have declined in coastal waters and in the open ocean thermocline since the 1960s. OMZs are progressively expanding in the tropical Pacific, Atlantic, and Indian Oceans due to reduced ventilation and oxygen solubility in warmer, more stratified oceans (Stramma et al. 2011).

Over the period 1901–2010, global mean sea level rose by 0.19 m. The rate of sea-level rise (SLR) since the mid-nineteenth century has been larger than the mean rate during the previous two millennia (IPCC 2014). The mean rate of SLR was 1.7 mm a^{-1} between 1901 and 2010 with an increase to 3.2 mm a^{-1} between 1993 and 2010. Over this period, global mean SLR has been consistent with the sum of the observed contributions from ocean thermal expansion, the Greenland ice sheet, the Antarctic ice sheet, and land water storage. Rates of mean SLR vary over different regions due to fluctuations in ocean circulation. It is extremely likely (IPCC 2014) that more than half of the observed increase in global average surface temperature and mean SLR from 1951 to 2010 was caused by the anthropogenic increase in greenhouse gas concentrations and other anthropogenic forcing.

The recent climatological forecasts by the Intergovernmental Panel on Climate Change (IPCC) for until the end of this century (Church et al. 2013; Collins et al. 2013; Bindoff et al. 2019; Oppenheimer et al. 2019) predict that globally: (i) SSTs will rise by 1–3 ° C; (ii) oceanic pH will decline by 0.07–0.31 units; and (iii) mean atmospheric CO_2 concentrations will increase to 441 ppm (from 391 ppm in 2011). Regional differences (Table 13.1) will occur for some parameters, such as (i) sea-level, which will continue to rise globally at an average rate between 1.8–3.6 mm a^{-1}, (ii) precipitation will increase in some regions of the wet tropics and decrease the dry tropics, and (iii) salinity will change in tandem with changes in precipitation.

Rising atmospheric CO_2 and climate change are associated with shifts in temperature, circulation, stratification, nutrient input, oxygen content, and ocean acidification, with potentially wide-ranging effects on the biology and ecology of marine organisms and their communities (Chapter 13).

2.7.1 Rising Atmospheric CO_2

Mean atmospheric CO_2 concentrations will increase to 441 ppm over the 2081–2100 period (Church et al. 2013; Collins et al. 2013). This projected increase is the net result of complex atmospheric, land, and ocean forces. While tropical atmospheric CO_2 concentrations are rising, they respond to climatic events, such as El Niño. The tropical Pacific Ocean, for instance, plays an important role in modulating changes in atmospheric CO_2 concentrations during El Niño events. ENSO is correlated with large interannual variability in global atmospheric CO_2 concentrations. During the 2015–2016 ENSO event, there was a negative CO_2 change during the development phase (spring–summer 2015) which was likely due to a reduction in local CO_2 outgassing from the tropical Pacific Ocean; a positive CO_2 change was measured during the mature phase (autumn 2015–2016), likely a reflection of an increase in atmospheric CO_2 concentrations due to a combination of reduced vegetation uptake across pan-tropical regions and enhanced biomass burning in Southeast Asia (Chatterjee et al. 2017). Thus, although atmospheric CO_2 levels are increasing, there are still notable variations and oscillations over time and space and probably explain some of the observed variability in pCO_2 concentrations in the world ocean.

2.7.2 Ocean Acidification

The global ocean has a large capacity to absorb atmospheric CO_2 because CO_2 dissolves and reacts with seawater to form bicarbonate (HCO_3^-) ions and protons (H^+). Between one-quarter and one-third of the CO_2 emitted into the atmosphere from the burning of fossil fuels, cement manufacturing and land-use changes have been absorbed by the ocean (Kleypas and Langdon 2006; Turley and Findlay 2016). Over thousands of years, the changes in pH have been buffered by bases such as carbonate ions (CO_3^{2-}). However, the rate at which CO_2 is currently being absorbed is too rapid to be buffered sufficiently to prevent substantial changes in ocean pH. Con sequently, the relative seawater concentrations of CO_2, HCO_3^-, CO_3^{2-} and pH have been altered. Since the Industrial Revolution, ocean pH has decreased globally by 0.1

unit (Figure 2.10 bottom), and it is predicted that ocean pH will decline by 0.4–0.5 unit by 2100 (Kleypas and Langdon 2006). It will take tens of thousands of years for these changes in ocean chemistry to be buffered through neutralization by carbonate sediments, and the level at which the ocean pH will eventually stabilize will be lower than it currently is. pH is sensitive to changes in salinity and total alkalinity and DIC concentrations, so organisms in coastal waters receiving river inputs are ordinarily exposed to greater variability in pH than organisms in the open ocean.

CO_3^{2-} concentration directly influences the saturation, and consequently, the rate of dissolution of calcium carbonate ($CaCO_3$) minerals in the ocean. Laboratory experiments and field observations indicate that ocean acidification is a threat to the survival of many marine organisms, especially organisms that use $CaCO_3$ to produce shells, tests, and skeletons, such as corals (Andersson and Glenhill 2013). A shift in pH alters the saturation state of $CaCO_3$ (called the 'aragonite saturation state') in seawater. The saturation state is expressed as:

$$\Omega_{\text{arg}} = \left[\text{Ca}^{2+}\right]\left[\text{CO}_3^{\ 2-}\right] / K_{sp}^{\ *}$$

where *Ksp** is the solubility product for $CaCO_3$ (as the mineral aragonite, arg) and $[Ca^{2+}]$ and $[CO_3^{2-}]$ are the calcium and carbonate concentrations, respectively. When $\Omega_{arg} > 1$, seawater is supersaturated with respect to mineral $CaCO_3$ and the larger this value the more suitable the environment will be for organisms that produce $CaCO_3$ shells and skeletons. When $\Omega_{arg} < 1$, seawater is undersaturated and is corrosive to $CaCO_3$. When pH decreases, Ω_{arg} decreases. The surface waters of the tropical oceans are currently supersaturated with respect to aragonite (mean $\Omega_{arg} = 4.0 \pm 0.2$). However, Ω_{arg} steadily declined from a calculated 4.6 ± 0.2 one hundred years ago and is expected to continue declining to 2.8 ± 0.2 by 2100 (Kleypas and Langdon 2006). The aragonite saturation state (Ω_{arg}) is sensitive to the partial pressure of CO_2 above the ocean as well as ocean temperature. In the latter case, the aragonite saturation of warm tropical oceans is higher than that of polar oceans because CO_2 is more soluble in cold water. As atmospheric CO_2 has increased, Ω_{arg} of the world's oceans has decreased.

The acidification of the world's oceans is a complex process that is only now being understood. In the western tropical Pacific Warm Pool, acidification is proceeding, with declines in pH (-0.0013 ± 0.0001 a^{-1}) and Ω_{arg} (-0.0083 ± 0.0007 a^{-1}) over the 1986–2016 period (Ishii et al. 2020). Oceanographic modelling indicates that the acidification of the Warm Pool occurs primarily through uptake of anthropogenic CO_2 in the extra tropics which is then transported to the tropics through the Equatorial Undercurrent from below (Ishii et al. 2020). The rate of Warm Pool acidification can be expected to be modulated by the contribution of a long interior residence time (years to decades) acting in concert with accelerating CO_2 increases in the atmosphere.

Acidification is an even more complex process in estuarine and coastal environments as carbonate chemistry is strongly regulated by changes in biological activity related to eutrophication and the delivery of nutrients by rivers and groundwater (Wallace et al. 2014). The increased loading of nutrients into estuaries and shallow inshore waters causes the accumulation of algal biomass and subsequent decomposition of this organic material, decreasing dissolved O_2 levels and contributing towards hypoxia. Hypoxia increases pCO_2 concentrations and upwelling processes can bring CO_2-enriched water in contact with coastal waters, amplifying the effects of acidification. Land-use changes such as deforestation and fossil fuel combustion also produce increased dissociation products of strong acids (HNO_3 and H_2SO_4) and bases (NH_3) in coastal waters, causing decreases in surface water alkalinity, pH, and DIC. River discharge further reduces alkalinity as river waters, especially in the tropics, are typically more acidic than receiving coastal waters. Further, the decomposition of organic matter, especially in bottom sediments and wetland soils, can reduce pH and alter carbonate chemistry, producing an increase in pCO_2 mostly due to microbial respiration. Nearly all estuarine and nearshore waters in the tropics naturally exhibit wide variations in salinity, pH, and carbonate chemical parameters, especially pCO_2 and $[CO_3^{2-}]$ (Alongi 2020). Carbonate chemistry in the coastal zone thus responds more strongly to eutrophication than ocean acidification (Borges and Gypens, 2010).

2.7.3 Rising Temperatures, Increased Storms, Extreme Weather Events, and Changes in Precipitation

The rate of ocean heating has increased within the 0–700 m layer by 3.22 ± 1.61 ZJ (ZJ = 10^{21} joules) from 1969 to 1993 and 6.28 ± 0.48 ZJ from 1993 to 2017, representing a twofold increase in heat uptake (Bindoff et al. 2019). This has resulted in temperature rises as noted earlier. Warming has also strengthened vertical stratification, inhibiting exchange between surface and deep ocean waters. Redistribution of heat accounts for 65% of heat storage at low latitudes. Tropical warming results from the interplay between increased stratification and equatorward heat transport by the subtropical gyres, which redistributes heat from the subtropics to the tropics (Dias et al. 2020).

Projections (Bindoff et al. 2019) are that by 2100 the ocean is very likely to warm by 2–4 times as much for low emissions (IPCC scenario RCP2.6) and 5–7 times as much for the IPCC's 'business-as-usual' scenario (RCP8.5) compared with the observed changes since 1970. The top 200 m of the upper ocean will continue to stratify to 2100 in the very likely range of 1–9% and 12–30% for scenarios RCP2.6 and RCP8.5, respectively.

Projections of future precipitation (Table 2.1) in the tropics indicate widely variable changes with regions. Rainfall is forecast to decline in northern South America, the Caribbean and western Central America, South Asia, and northern and eastern Australia. Increased precipitation is expected in all other tropical areas, except in the Arabian Sea where little change is expected. Much of these forecasted changes are linked to forecasted changes in the intensity and frequency of tropical cyclones (Knutson et al. 2020). There is at least medium-to-high confidence in an increase globally in tropical cyclone precipitation, with a median projected increase of 14% or close to the rate of tropical water valour increase with warming, at constant relative humidity (Knutson et al. 2020); cyclone intensity will increase with medium-to-high confidence. Knutson et al. (2020) indicate that expert opinion was more mixed and confidence levels lower for (i) a further poleward expansion of the latitude of maximum cyclone intensity in the western North Pacific; (ii) a decrease in global tropical cyclone frequency; and (iii) an increase in very intense global tropical cyclone frequency (category 4–5). These changes in tropical precipitation are linked to changes in surface land-ocean temperature and changes in near-surface relative humidity (Lambert et al. 2017). High precipitation under climate change is associated with the highest surface relative humidity and temperatures.

TABLE 2.1 IPCC projected changes for tropical regions in salinity, precipitation, and sea-level rise for 2081–2100 (relative to the 1986–2005 reference period).

Region	Salinity[a]	Precipitation[b]	Sea-level rise[c]
N. South America	0 to 5 ↑	−10 to 40% ↓	0.22 to 0.24 m
E. South America	0 to 5 ↑	0 to 10% ↑	0.18 to 0.20 m
Caribbean and W. Central America	0.5 to 1.0 ↑	−20 to 10% ↓	0.18 to 0.20 m
Central West Africa	0 to 1.5 ↓	10 to 20% ↑	0.20 to 0.24 m
Central East Africa	0 to 2.0 ↓	10 to 50% ↑	0.20 to 0.24 m
Red Sea/Arabian Peninsula	No change	−10 to 10% ↔	0.22 to 0.24 m
South Asia	0 to 5 ↑	−40 to 10% ↓	0.18 to 0.24 m
SE Asia	0 to 1.0 ↓	0 to 20% ↑	0.18 to 0.20 m
N. Australia	No change	0 to 10% ↓	0.18 to 0.20 m
E. Australia	No change	−10 to 0% ↓	0.18 to 0.20 m
Oceania	No change	0 to 10% ↑	0.18 to 0.22 m

[a] Range of projected sea surface salinity changes for 2081–2100 relative to the 1986–2005 reference period.
[b] Range of projected changes in December to February precipitation for 2081–2100 relative to 1986–2005. Data from Collins et al. (2013).
[c] Range of ensemble mean projections of the time-averaged dynamic and steric sea-level changes for the period 2081–2100 relative to 1986–2005. Data from Church et al. (2013) and Oppenheimer et al. (2019).
Source: Church et al. (2013), Collins et al. (2013), Bindoff et al. (2019) and Oppenheimer et al. (2019).

2.7.4 Changes in Ocean Circulation

Increasing ocean heat is resulting in changes in ocean circulation. Models of ocean heat content (OHC) indicate that the poleward OHC substantially reduce (increase) in the Northern (Southern) Hemisphere, with circulation changes varying among subtropical gyres and among western and eastern boundary currents (Dias et al. 2020). In the North Atlantic Current, ocean heat transport (OHT) weakens at all depths, whereas it strengthens at the surface and weakens at mid-depth in the subtropical gyre. The Gulf Stream has weakened but the Canary and North Equatorial Currents have increased. Changes in the North Atlantic subtropical gyre and associated OHT reduction suggest that heat moving poleward with the Gulf Stream/North Atlantic Current has reduced, and the extra heat, stored passively in the gyres, transported equatorward via eastern boundary currents, and equatorial currents. Similar changes have also been observed in the Pacific and Indian Ocean. The intensification of the equatorial currents should transport extra heat both westward and eastward via the complex system of equatorial currents and counter currents. OHT associated with the Brazil Current should intensify, contributing to the poleward OHT in the Southern Hemisphere. In the South Atlantic, a shift from equatorward OHT to poleward is predicted to result in an intensification of the Brazil Current.

Some studies have suggested that the surface warming trend in the open ocean is opposite to the trends nearshore and that the poleward expansion of the Hadley cells would increase upwelling at the poleward end and decrease it equatorward (Rykaczewski et al. 2015). Although stratification has intensified over most subtropical gyres and the global ocean, the model results of Dias et al. (2020) indicate an opposite trend in tropical regions, especially in the eastern Pacific and Atlantic basins.

Whether there is an emerging trend of global ocean circulation is not yet clear, but there is a significant increasing trend in the globally integrated, ocean kinetic energy since the early 1990s, indicating a substantial acceleration of global mean ocean circulation (Hu et al. 2020). This increase in kinetic energy is especially prominent in the global tropical ocean, reaching depths of several thousand meters, and induced by a planetary intensification of surface winds. However, regional trends are diverse; the Agulhas Current has not intensified, but shallow cells in the Pacific Ocean have accelerated in response to intensified trade winds since the early 2000s, contributing to a recent warming hiatus and leading to an increased leakage of heat and freshwater in the Indian Ocean via the Indonesian Throughflow (Hu et al. 2020). Other models project a slowdown of south Indian Ocean circulation, an important region as it modulates marine life and global climate through important oceanic connections between the Pacific, Atlantic and Southern Oceans (Stellema et al. 2019). A weakening of the Leeuwin Current and Undercurrent off the west coast of Australia is projected due to reduced onshore flow and downwelling; the reduced flow is related to changes in the alongshore pressure gradient which is consistent with the projected weakening of the Indonesian Throughflow. A strong weakening in the SW Indian Ocean of the NE and SE Madagascar Currents, Agulhas Current and flow through the Mozambique Channel is predicted, as this reduced western boundary flow is partly associated with a weaker Indonesian Throughflow (Stellema et al. 2019).

2.7.5 Sea-Level Rise (SLR)

Global mean sea-level is rising and accelerating. Data from tide gauges and altimetry observations indicate that global mean sea-level increased from 1.4 mm a^{-1} over the period 1901–1190 to 2.1 mm a^{-1} over the period 1970–2015 to 3.2 mm a^{-1} over the period 1993–2015 to 3.6 mm a^{-1} over the period 2006–2015 (Oppenheimer et al. 2019). SLR is projected to rise between 0.43 m (likely range: 0.29–0.59 m) and 0.84 m (likely range: 0.61–1.10 m) by 2100 relative to 1986–2005 (Oppenheimer et al. 2019). Sea-level is projected to continue to rise for centuries beyond 2100 due to continuing deep ocean heat uptake and mass loss of the Greenland and Antarctic ice sheets and will remain elevated for thousands of years. Under the 'business-as-usual' scenario (RCP8.5), the rate of SLR will be 15 mm a^{-1} (likely range: 10–20 mm a^{-1}) in 2100 and could exceed several cm a^{-1} in the twenty-second century.

SLR involves a significant anthropogenic component, mainly induced by global ocean thermal expansion and the melting of land ice. SLR patterns relative to land are also influenced by geological processes such as glacial isostatic adjustment. In response to a changing climate, SLR will not be spatially uniform but show complex patterns. As a result, some regions could experience local SLRs considerably greater and larger than the global average, whereas the local SLR elsewhere may be well below the global mean or even negative.

Multiple factors impact global and regional SLR: oceanic net mass change related to an increase/decrease of ice sheets and glaciers, groundwater mining and dam building, glacial isostatic adjustment (vertical movement of the earth's crust due to ice sheet mass changes), changes in ocean water temperature and salinity, changes of the earth's gravitational field related to the melting of ice sheets and ocean dynamics associated with variations in wind-driven or buoyancy-driven ocean circulation (Webster 2020). There is significant regional heterogeneity, as SLR and its long-term trends due to steric and dynamic components (steric sea-level is rising and falling of sea level as the temperature and salinity of the water column varies; dynamic sea-level is sea-level changing as water mass is redistributed within the ocean or is added or removed) deviate significantly from the global mean and between different ensemble members.

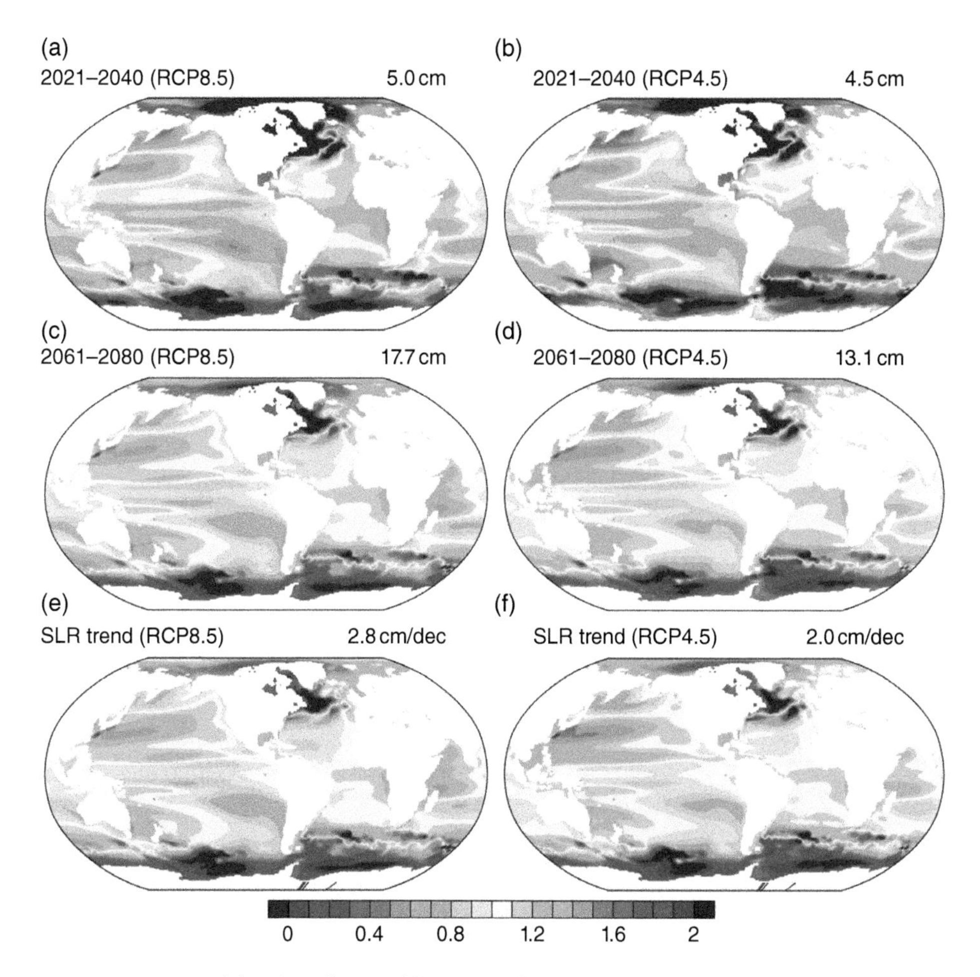

FIGURE 2.11 Model ratios of ensemble averaged 20 year mean sea-level rise and the decadal trend of sea-level rise and the global mean. Mean sea-level rise over 2021–2040 under a (a) 'business-as-usual' emission scenario (RCP8.5) and (b) a 'lower emission' scenario (RCP4.5), mean sea-level rise over 2061–2080 under a (c) 'business-as-usual' scenario and (d) under a 'lower emission' scenario relative to mean sea-level over 1986–2005. The decadal sea-level rise trend, the average 10 year trend over the period 2006–2080, under a (e) 'business-as-usual' scenario and (f) a 'lower emissions' scenario. *Source*: Hu and Bates (2018), figure 1, p.3. Licensed under CC BY 4.0. © Springer Nature Switzerland AG.

Model projections of regional and global SLR based on the RCP4.5 (lower emission scenario) and the RCP8.5 (the 'business-as-usual' scenario) assumptions (Figure 2.11) show that the pattern of the ensemble mean mid-century (Figure 2.11a and b) and late-century (Figure 2.11c and d) regional SLR and the pattern of long-term (Figure 2.11e and f) trends are similar between the two scenarios (Hu and Bates 2018). Higher than average SLR is forecast for the subtropical Pacific, South Atlantic, Arctic, part of the subpolar North Atlantic, equatorial Pacific, southeast part of the South Pacific and subpolar North Pacific in both scenarios. The similarity of both forecasts suggests that the underlying internal processes are similar for both scenarios and scaled by the strength of forcing by greenhouse gases (Hu and Bates 2018). There is less certainty in the SLR projections in the regions that might experience the largest SLR. The steric and dynamic variability of SLR in the twenty-first century increases in most regions for both scenarios compared with in the late twentieth century, with SLR variance in the long term showing an increase in the Indian Ocean, west of Australia, the west and east coastal regions of the Pacific, west coast of Europe and the Atlantic sector of the Southern Oceans; decreases in sea level are forecast to occur in many other regions for both scenarios.

References

Alongi, D.M. (2020). Vulnerability and resilience of tropical coastal ecosystems to ocean acidification. *Examines in Marine Biology and Oceanography* **3**: EIMBO.000562.2020. https://doi.org/10.31031/EIMBO.2020.03.000562.

Andersson, A.J. and Glenhill, D. (2013). Ocean acidification and coral reefs: effects of breakdown, dissolution, and net ecosystem calcification. *Annual Review of Marine Science* 5: 321–348.

Biasutti, M. (2019). Rainfall trends in the African Sahel: characteristics, processes, and causes. *WIREs Climate Change* 10: e591. https://doi.org/10.1002/wcc.591.

Bindoff, N.L., Cheung, W.W.L., Kairo, J.G. et al. (2019). Changing ocean, marine ecosystems, and dependent communities. In: *IPCC Special Report on the Ocean and Cryosphere in a Changing Climate* (eds. H.-O. Pörter, D.C. Roberts, V. Masson-Delmotte, et al.), 447–587. Geneva, Switzerland: Intergovernmental Panel on Climate Change.

Borges, A.V. and Gypens, N. (2010). Carbonate chemistry in the coastal zone responds more strongly to eutrophication than to ocean acidification. *Limnology and Oceanology* 55: 346–353.

Boucharel, J., Jin, F.-F., Lin, I.I. et al. (2016a). Different controls of tropical cyclone activity in the Eastern Pacific for two types of El Niño. *Geophysical Research Letters* 43: 1679–1686.

Boucharel, J., Jin, F.-F., England, M.H. et al. (2016b). Influence of oceanic intraseasonal Kelvin waves on Eastern Pacific hurricane activity. *Journal of Climate* 29: 7941–7955.

Burn, M.J. and Palmer, S.E. (2015). Atlantic hurricane activity during the last millennium. *Scientific Reports* 5: 12838. https://doi.org/10.1038/srep12838.

Byrne, M.P., Pendergrass, A.G., Rapp, A.D. et al. (2018). Response of the Intertropical Convergence Zone to climate change: location, width, and strength. *Current Climate Change Reports* 4: 355–370.

Caniaux, G., Giordani, H., Redelsperger, J.-L. et al. (2011). Coupling between the Atlantic cold tongue and the West African monsoon in boreal spring and summer. *Journal of Geophysical Research: Oceans* 116: c04003. https://doi.org/10.1029/2010JC006570.

Chatterjee, A., Gierach, M.M., Sutton, A.J. et al. (2017). Influence of El Niño on atmospheric CO2 over the tropical Pacific Ocean: findings from NASA's OC-2 mission. *Science* 358: eaam5776. https://doi.org/10.1126/science.aam5776.

Chen, Z. (2019). Evolution of south tropical Indian Ocean warming and the climatic impacts following strong El Niño events. *Journal of Climate* 32: 7329–7347.

Christensen, J.H., Krishna Kumar, K., Aldian, S.-I. et al. (2013). Climate phenomena and their relevance for future regional climate change. In: *Climate Change 2013, The Physical Basis, Contribution of Working Group I to the Fifth Assessment Report of the Intergovernmental Panel on Climate Change* (eds. T.F. Stocker, D. Qin, G.-K. Platter, et al.), 1217–1271. Cambridge, UK: Cambridge University Press.

Church, J.A., Clark, P.U., Cazenave, A. et al. (2013). Sea level change. In: *Climate Change 2013, The Physical Basis, Contribution of Working Group I to the Fifth Assessment Report of the Intergovernmental Panel on Climate Change* (eds. T.F. Stocker, D. Qin, G.-K. Platter, et al.), 1137–1226. Cambridge, UK: Cambridge University Press.

Collins, M., Knutti, R., Arblaster, J. et al. (2013). Long-term climate change, projections, commitments and irreversibility. In: *Climate Change 2013, The Physical Basis, Contribution of Working Group I to the Fifth Assessment Report of the Intergovernmental Panel on Climate Change* (eds. T.F. Stocker, D. Qin, G.-K. Platter, et al.), 1029–1136. Cambridge, UK: Cambridge University Press.

Dampier, W. (1699). *Voyages and Descriptions, Volume II, Part 3, A Discourse of Trade winds, Breezes, Storms, Seasons of the Year, Tides and Currents of the Torrid Zone throughout the World; with an Account of Natal in Africa, its Product, Negro's. etc.* London: J. Knapton.

De Deckker, P. (2016). The Indo-Pacific Warm Pool: critical to world oceanography and world climate. *Geoscience Letters* 3: 20. https://doi.org/10.1186/s40562-016-0054-3.

Dias, F.B., Fiedler, R., Marsland, S.J. et al. (2020). Ocean heat storage in response to changing ocean circulation processes. *Journal of Climate* 33: 9065–9082.

Doney, S.C., Shallin Busch, D., Cooley, S.R. et al. (2020). The impacts of ocean acidification on marine ecosystems and reliant human communities. *Annual Review of Environment and Resources* 45: 83–112.

Duan, Y., Liu, H., Yu, W. et al. (2019). The onset of the Indonesian-Australian summer monsoon triggered by the first branch eastward-propagating Madden-Julian Oscillation. *Journal of Climate* 32: 5453–5470.

Duteil, O., Oschlies, A., and Böning, C.W. (2018). Pacific Decadal Oscillation and recent oxygen decline in the eastern tropical Pacific Ocean. *Biogeosciences* 15: 7111–7126.

Fan, L., Liu, Q., Wang, C. et al. (2017). Indian Ocean Dipole modes associated with different types of ENSO development. *Journal of Climate* 30: 2233–2249.

Funk, C., Hoell, A., Shulkla, S. et al. (2016). The East African monsoon system: seasonal climatologies and recent variations. In: *The Monsoons and Climate Change* (eds. L.M.V. de Carvalho and C. Jones), 163–185. Basel, Switzerland: Springer International.

Grimm, A.M. (2019). Madden-Julian Oscillation impacts on South American summer monsoon season: precipitation anomalies, extreme events, teleconnections, and the role of the MJO cycle. *Climate Dynamics* 53: 907–932.

Hartmann, D.L. (2016). *Global Physical Climatology*, 2e. Amsterdam: Elsevier.

Hu, A. and Bates, S.C. (2018). Internal climate variability and projected regional steric and dynamic sea level rise. *Nature Communications* 9: 1068. https://doi.org/10.1038/s41467-018-03474-8.

Hu, S., Sprintall, J., Guan, C. et al. (2020). Deep-reaching acceleration of global mean ocean circulation over the past two decades. *Science Advances* 6: eaax7727. https://doi.org/10.1126/sciadv.aax7727.

IPCC (2014). *Climate Change 2014, Synthesis Report. Contribution of Working Groups I, II and III to the Fifth Assessment Report of the Intergovernmental Panel on Climate Change.* Geneva, Switzerland: IPCC.

Ishii, M., Rodgers, K.B., Inoue, H.Y. et al. (2020). Ocean acidification from below in the tropical Pacific. *Global Biogeochemical Cycles* 34: e2019GB006368. https://doi.org/10.1029/2019GB006368.

Janicot, S., Caniaux, G., Chauvin, F. et al. (2011). Intraseasonal variability of the West African monsoon. *Atmospheric Science Letters* 12: 58–66.

Kang, S.M., Shin, Y., and Xie, S.P. (2018). Extratropical forcing and tropical rainfall distribution: energetics framework and ocean Ekman advection. *npj Climate and Atmospheric Science* 1: 1–10.

Kleypas, J.A. and Langdon, C. (2006). Coral reefs and changing seawater carbonate chemistry. In: *Coral Reefs and Climate Change: Science and Management* (eds. J.T. Phinney, O. Hoegh-Guldberg, J. Kleypas, et al.), 73–110. Washington, DC: American Geophysical Union.

Knutson, T., Camargo, S.J., Chan, J.C.L. et al. (2020). Tropical cyclones and climate change assessment. Part 2, Projected response to anthropogenic warming. *Bulletin of the American Meteorological Society* 101: 303–322.

Lambert, F.H., Ferraro, A.J., and Chadwick, R. (2017). Land-ocean shifts in tropical precipitation linked to surface temperature and humidity change. *Journal of Climate* 30: 4527–4545.

Lengaigne, M., Neetu, S., Samson, G. et al. (2019). Influence of air-sea coupling on Indian Ocean tropical cyclones. *Climate Dynamics* 52: 577–598.

Li, X., Xu, J., Shi, Z. et al. (2019). Regulation of protist grazing on bacterioplankton by hydrological conditions in coastal waters. *Estuarine, Coastal and Shelf Science* 218: 1–8.

Liu, J., Tian, J., Liu, Z. et al. (2019). Eastern equatorial Pacific cold tongue evolution since the late Miocene linked to extratropical climate. *Science Advances* 5: eaau6060. https://doi.org/10.1126/sciadv.aau.aau6060.

Marengo, J.A., Liebmann, B., Grimm, A.M. et al. (2012). Recent developments on the South American monsoon system. *International Journal of Climatology* 32: 1–21.

Moise, A., Smith, I., Brown, J.R. et al. (2020). Observed and projected intra-seasonal variability of Australian monsoon rainfall. *International Journal of Climatology* 40: 2310–2327.

Newman, M., Alexander, M.A., Ault, T.R. et al. (2016). The Pacific Decadal Oscillation, revisited. *Journal of Climate* 29: 4399–4427.

Nguyen, H., Evans, A., Lucas, C. et al. (2013). The Hadley circulation in reanalyses: climatology, variability, and change. *Journal of Climate* 26: 3357–3376.

Oppenheimer, M., Glavovic, B.C., Hinkel, J. et al. (2019). Sea level rise and implications for low-lying islands, coasts and communities. In: *IPCC Special Report on the Ocean and Cryosphere in a Changing Climate* (eds. H.-O. Pörter, D.C. Roberts, V. Masson-D, et al.), 321–445. Geneva, Switzerland: Intergovernmental Panel on Climate Change.

Park, S., Kang, D., Yoo, C. et al. (2020). Recent ENSO influence on East African drought during rainy seasons through the synergistic use of satellite and reanalysis data. *ISPRS Journal of Photogrammetry and Remote Sensing* 162: 17–26.

Roxy, M.K., Dasgupta, P., McPhaden, M.J. et al. (2019). Twofold expansion of the Indo-Pacific warm pool warps the MJO life cycle. *Nature* 575: 647–651.

Rykaczewski, R.R., Dunne, J.P., Sydeman, W.J. et al. (2015). Poleward displacement of coastal upwelling-favorable winds in the ocean's eastern boundary currents through the 21st century. *Geophysical Research Letters* 42: 6424–6431.

Sharkov, E.A. (2012). *Global Tropical Cyclogenesis*, 2e. Berlin: Springer.

Silva Dias, M.A.F. and Carvalho, L.M.V. (2011). The South American monsoon system. In: *The Global Monsoon System: Research and Forecast*, 3e (eds. C.-P. Chang, H.-C. Kuo, N.-C. Lau, et al.), 137–157. Singapore: World Scientific.

Stellema, A., Sen Gupta, A., and Taschetto, A.S. (2019). Projected slowdown of South Indian Ocean circulation. *Scientific Reports* 9: 17705. https://doi.org/10.1038/s41598-019-54092-3.

Stramma, L., Prince, E.D., Schmidtko, S. et al. (2011). Expansion of oxygen minimum zones may reduce available habitat for tropical pelagic fish. *Nature Climate Change* 2: 33–37.

Tierney, J.E., Smerdon, J.E., Anchukaitis, K.J. et al. (2013). Multidecadal variability in East African hydroclimate controlled by the Indian Ocean. *Nature* 493: 389–392.

Timmermann, A., An, S.-I., Kug, J.-S. et al. (2018). El Niño-Southern Oscillation complexity. *Nature* 550: 535–545.

Turley, C. and Findlay, H.S. (2016). Ocean acidification. In: *Climate Change: Observed Impacts on Plant Earth*, 2e (ed. T.M. Letcher), 271–293. Amsterdam: Elsevier.

Vishnu, S., Francis, P.A., Shenoi, S.C. et al. (2018). On the relationship between the Pacific Decadal Oscillation and monsoon depressions over the Bay of Bengal. *Atmospheric Science Letters* 19: e825. https://doi.org/10.1002/asl.825.

Wallace, R.B., Baumann, H., Grear, J. et al. (2014). Coastal ocean acidification: the other eutrophication problem. *Estuarine, Coastal and Shelf Science* 148: 1–13.

Wang, T., Lu, X., and Yang, S. (2019). Impact of south Indian Ocean Dipole on tropical cyclone genesis over the South China Sea. *International Journal of Climatology* 39: 101–111.

Webster, P.J. (2020). *Dynamics of the Tropical Atmosphere and Oceans*. Hoboken, USA: Wiley-Blackwell.

Yu, Y., Wang, Y., Cao, L. et al. (2020). The ocean-atmosphere interaction over a summer upwelling system in the South China Sea. *Journal of Marine Systems* 208: 103360. https://doi.org/10.1016/j.jmarsys.2020.103360.

Zhang, C. (2005). Madden-Julian Oscillation. *Reviews of Geophysics* 43: RG2003. https://doi.org/10.1029/2004RG000158.

Zhang, W., Leung, Y., and Fraedrich, K. (2015). Different El Niño types and intense typhoons in the western North Pacific. *Climate Dynamics* 44: 2965–2977.

Zhisheng, A., Guoxiong, W., Jianping, L. et al. (2015). Global monsoon dynamics and climate change. *Annual Review of Earth and Planetary Sciences* 43: 2.1–2.49.

Zhou, T., Turner, A.G., Kinter, J.L. et al. (2016). GMMIP (v1.0) contribution to CMIP6: global monsoons model inter-comparison project. *Geoscientific Model Development* **9**: 3589–3604.

CHAPTER 3

Tropical Marine Hydrosphere

3.1 Introduction

The circulation patterns of the oceans are closely intertwined with the atmosphere. Most heat and precipitation occur in the tropics, and these factors are important drivers of tropical ocean circulation (Webster 2020). Indeed, sea surface temperatures (SSTs) are sufficiently high that deep atmospheric convection occurs over it. Small changes in SSTs result in movements of deep convection globally, underscoring the connection of the oceans to the atmosphere and vice versa. Circulation patterns in the tropics are complex, and much of this complexity occurs due to inherent thermal instability and mixing, which is a highly non-linear function of the mean circulation such that it may vary considerably with seasonal and non-seasonal circulation changes. There is evidence that global warming is weakening tropical ocean circulation (Vecchi and Soden 2007).

Tidal mixing and horizontal mixing are also non-linear and research into them is still in its infancy. Below, we will describe both large-scale oceanic and small-scale coastal and estuarine circulation patterns and how the small-scale processes fit into the scheme of oceanic processes which have major impacts on the ecology of the tropical ocean.

3.2 Large-Scale Circulation Patterns

Equatorial flows within the Pacific Ocean are complex, driven mostly by equatorial heat influx (Johnson et al. 2001). NE trade winds north of the equator and SE trade winds south of the equator drive the North and South Equatorial Currents (NEC and SEC) westward at the surface, pushing warm water into the western Pacific (Figure 3.1). To counteract these currents, the Equatorial Under Current (EUC) is driven eastward by an along equatorial pressure gradient which develops over the upper 250 m to roughly balance wind stress. It is this current system and its linkage to ENSO that results in the Peruvian upwelling system (Chapter 11). The EUC thus shoals and upwells, supplying the bulk of the surface water that diverges from the equatorial east Pacific. The SEC

Tropical Marine Ecology, First Edition. Daniel M. Alongi.

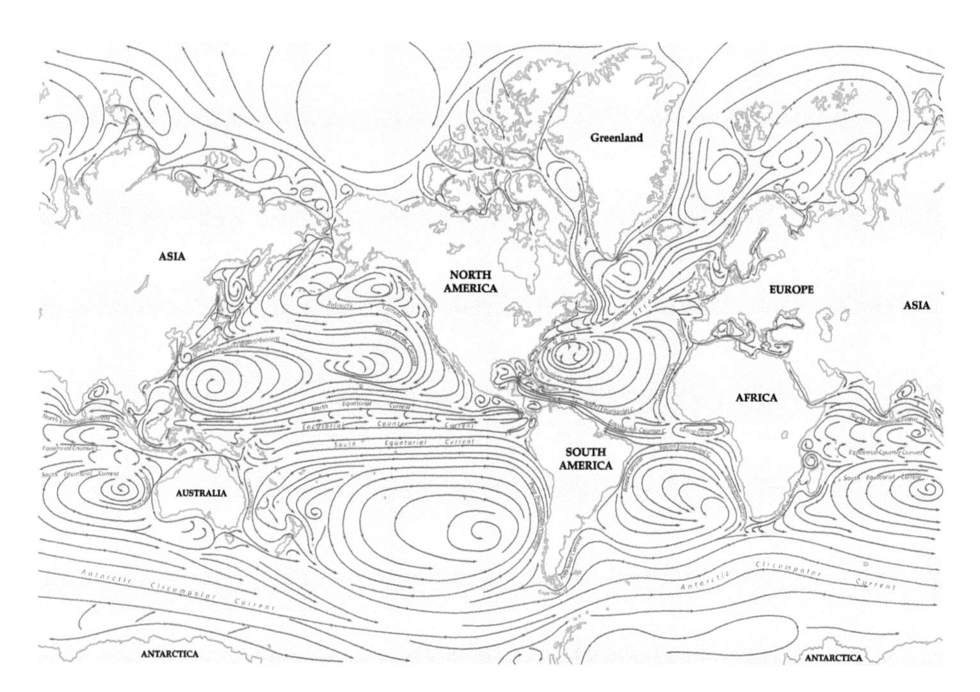

FIGURE 3.1 Main currents of the world ocean showing the main cold and warm flows. *Source*: Image obtained from Alamy Australia Pty. Ltd., Brisbane and constructed by Rainer Lesniewski/Alamy stock vector and reproduced under royalty-free license agreement (accessed 10 June 2021).

is a broad shallow (upper 200 m) current extending from the subtropical south Pacific to 2–5°N and its width is set by patterns of wind curl which also generate the North Equatorial Counter Current (NECC) found north of 2°N in the west and 5°N in the east Pacific. These flows combine with surface Ekman flows and the equatorial western boundary currents seasonally through the ENSO cycle to carry mean heat and freshwater inputs out of the equatorial Pacific. Two other flows feed into the EUC: The Mindanao Current and the New Guinea Coastal Current. These western boundary flows are in turn fed by subduction in subtropical latitudes and are characterised by high oxygen and high salinity. The NECC encounters the Philippines where it forms the Mindanao Current that partly flows into the Celebes Sea; most flows northwards and becomes the warm Kuroshio Current. In the west, the SEC flows mostly southward along the coast of Australia to become the Eastern Australian Current, but it also partially flows into the shallow Arafura Sea north of Australia.

Internal flows exist across both sides of the equator, that is, a surface poleward flow and mixing resulting in flow towards the equator below the thermocline. The equatorward flow is well defined in the south Pacific being between 17°S and 7°S, nearly all in the East and Central Pacific, and directly feeds the EUC (Johnson et al. 2001). The situation in the North Pacific is more complex and less well understood, but it is unlikely that the EUC is fed as strongly as in the south Pacific. Convergence towards the equator occurs below the surface mixed layer on the trough between the SEC and the NECC. An excess of heat energy that is needed to supply the EUC is presumably warmed in the westward-flowing SEC, subducted in the east or central Pacific, and then upwelled again to flow out in the surface mixed layer further west. This phenomenon is referred to as the ‘Tropical Cell’ (Godfrey et al. 2001).

The complex oceanographic features of the equatorial Pacific are the foundation for a rich yellowfin and skipjack tuna fishery, which accounts for roughly 40% of the world's annual tuna catch (Chapter 9). There is a paradoxical link between the tuna catch and a strong divergent equatorial upwelling in the central Pacific called the 'cold tongue', which is favourable to the development of high phytoplankton production (Lehodey 2001). This 'cold tongue' is contiguous with the Indo-Pacific Warm Pool (IPWP), which is characterised by lower rates of primary production. Consistent with the observed movements of tuna, there is a clear out-of-phase pattern linked to ENSO between the western Pacific region and the 'cold tongue'.

These equatorial Pacific currents are linked to the Indonesian Throughflow (ITF) (Figure 3.2) but precisely how is uncertain (Feng et al. 2018). The ITF is a unique feature at the crossroads between the Indian and Pacific Oceans, carrying warm and fresh Pacific waters through the Indonesian Archipelago into the Indian Ocean. It is the only series of channels in the tropics through which water passes from one ocean to another. The ITF contributes to circulation and thermal structure around northern and eastern Australia and the southern Indian Ocean (Tillinger 2011). Blockage of the ITF weakens the Indian Ocean South Equatorial Current and Agulhas Current and strengthens the Eastern Australian Current. It also maintains a stronger New Guinea Coastal Undercurrent, enhancing tropical-subtropical exchange in the south. Blockage or weakening of the ITF reduces thermocline fluctuations and increases SSTs in the central and equatorial Pacific and raises the mean thermocline of the Indian Ocean and decreases SSTs in the southern Indian Ocean.

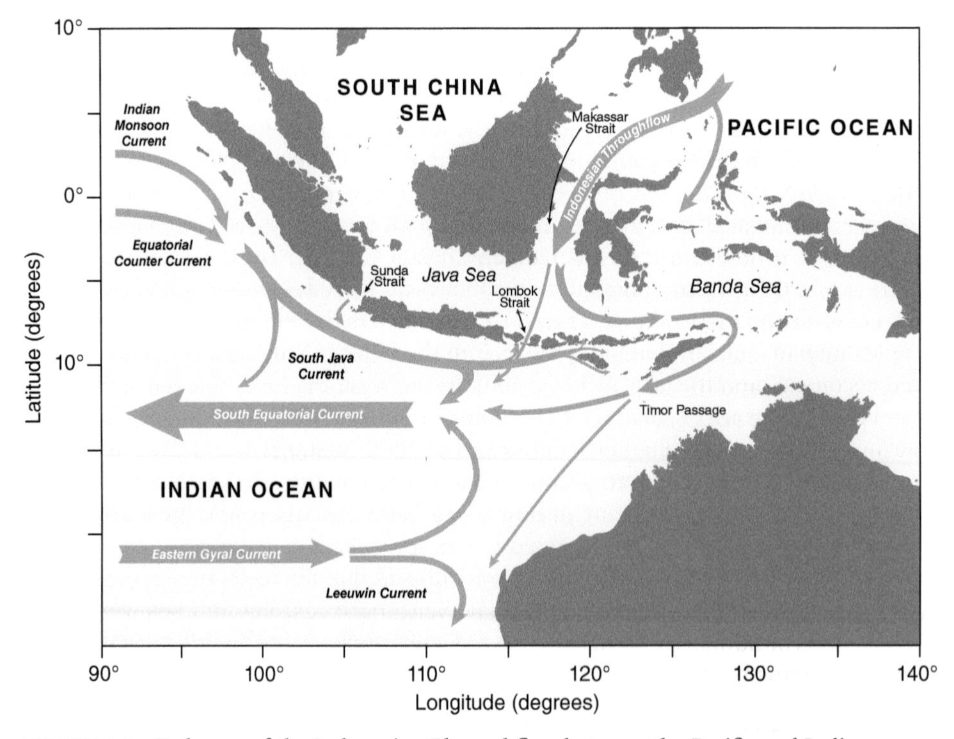

FIGURE 3.2 Pathways of the Indonesian Throughflow between the Pacific and Indian Oceans and linkage to other major ocean currents. *Source*: Feng et al. (2018), figure 1, p.3. Licensed under CC BY 4.0. © Springer Nature Switzerland AG.

The ITF varies both seasonally and annually (Tillinger 2011) as about 60–90% of sea-level variability, and 70% of thermocline variability can be understood in terms of free Kelvin waves and Rossby waves generated by remote zonal winds along the equator in the Indian and Pacific Oceans. Variations in zonal Pacific equatorial winds force a response along the Arafura Sea/Australia shelf break through Pacific equatorial Rossby waves exiting coastally trapped waves off the western end of New Guinea which propagate poleward along the Australian west coast as the Leeuwin Current. The regional circulation off west Australia is thought to be embedded in a subtropical 'super gyre' that connects the Indo-Pacific via south Australia ('the Tasman Gateway') and the passages of the Indonesian archipelago (Lambert et al. 2016). There is also an effect of the ITF felt by energy radiating westward across the Banda Sea and into the southern Indian Ocean. Wind energy across the equatorial Indian Ocean propagates along the south coasts of the islands of Sumatra, Java, and Nusa Tenggara to penetrate the Savu Sea, the western Banda Sea, and Makassar Strait. Thus, the ITF affects nearly the entire ocean field of the Indonesian Archipelago, as well as coastal New Guinea and Australia.

The New Guinea Coastal Current (NGCC) likely sets up a strong shear flow in austral summer when the surface flow of the NGCC is towards the southeast against the mean NW flow. Tidal mixing may also play a role in producing vertical eddies and coastal upwellings throughout the Indonesian Archipelago, implying that surface heat fluxes are carried through the mixed layer. It is possible that tidally enhanced eddies are widely distributed throughout the west Pacific especially near reef complexes. As noted in Chapter 2, the MJO, ENSO, as well as small-scale seasonal cycles play a strong role in large-scale water circulation in the equatorial Pacific.

Once the ITF passes through the many islands of the Indonesian archipelago, it circulates through the Indian Ocean back into the Pacific south of Australia (Lambert et al. 2016). Somewhat reminiscent of the equatorial Pacific, although the wind system is greatly different, the South Indian Ocean Counter Current (SICC) flows from west to east across the Indian Ocean against the wind-driven circulation.

Circulation in the Indian Ocean is driven not only by the SICC but also by ENSO, IOD, and the MJO (Chapter 2). Wind-driven upwelling occurs mainly in the seasonally reversing, western boundary currents rather than in the eastern equatorial region; a completely different set of mechanisms drives heat and freshwater absorption (Hood et al. 2017). In the north, the Indian Ocean has two large water bodies west and east of India: the Arabian Sea and the Bay of Bengal. In the Arabian Sea, there is upwelling of cold, nutrient-rich water during the SW monsoon (SWM) along Somalia, Oman, and the west coast of India. In the western Bay of Bengal, upwelling occurs during the NE monsoon (NEM), whereas south of Sri Lanka, there is coastal upwelling where upwelling blooms are swept into the Bay of Bengal by the SW Monsoon Current (McCreary et al. 2009). In the tropical south Indian Ocean, there is a weak surface plankton bloom during boreal summer when new phytoplankton production is enhanced by nutrient entrainment. In boreal winter, the mixed layer is thinner resulting in less plankton production as the thermocline is deeper and nutrient entrainment is weaker. ENSO/IOD events can cause plankton blooms south of the islands of Sumatra and Java, while upwelling further east is driven by entrainment and mixing of the ITF with other currents such as the Java Current.

In the Atlantic, the average circulation bears some resemblance to the equatorial Pacific (Figure 3.1). An Equatorial Under Current (EUC) in the equatorial thermocline is surrounded by westward currents, the Southern Equatorial Current (SEC). North of 5°N, a seasonal surface-trapped Northern Equatorial Counter Current (NECC) occurs.

At the equator, the EUC usually overlies a westward current bounded by eastward currents at 4°N and 3–4°S. The current structure is more variable than in the Pacific with some suggestion of eastward currents near 2–3°S (South Intermediate Counter Current) and 2–3°N (North Intermediate Counter Current). Both the EUC and SEC derive their physical properties from the Southern Hemisphere via the North Brazil Under Current (NBUC) with some seasonal input from the Northern Hemisphere. Currents carry low oxygen water to the western boundary, whereas the eastward currents of the Antarctic Intermediate Water (AAIW) often carry oxygen-rich water.

In the western tropical Atlantic Ocean, fresh surface waters from the Amazon may induce a strong halocline in the 3–30 m depth range, which in turn induces a pycnocline that acts as a barrier layer for mixing between the surface and subsurface waters. Following maximum Amazon discharge, the river plume and resultant barrier layer extends over a large part of the equatorial basin north of the equator in boreal summer and autumn (Varona et al. 2019). This anomaly due to the river discharge is powerful enough to contribute to a northward shift in the ITCZ during this period. The Amazon plume is great enough to drive spatial and temporal variations in oceanic primary productivity (Gouveia et al. 2019).

Even in the open sea in the tropics, high rates of evaporation and precipitation and upwelling can destroy the permanently stratified thermocline, unlike in temperate and polar oceans where water masses turnover by cooling in autumn and winter. North of the equator, the eastern Pacific and eastern Atlantic Oceans are eddy-dominated, with counter currents impinging upon inshore waters and estuaries fed by major rivers and wide shelf areas.

3.3 Coastal Circulation

Tides and wind-generated waves play in important role in the circulation of water close to shore, although the large-scale circulation patterns set the characteristic signatures of nearshore water masses (Masselink and Hughes 2014). Water circulation in embayments, bays, and other nearshore water bodies are greatly influenced by daily tidal cycles and wind waves and are also affected by long-shore currents that are in turn influenced at the macro- and meso-scale. Coastal circulation is ultimately driven by energy derived from solar heating or gravity, barometric pressure, and the density of oceanic waters that impinge on the coastal zone. Mixing results from tides and waves and buoyancy effects from river runoff, if any. Water mixing and circulation are greatly affected by geometry and bathymetry of the coastal zone.

Regional variability of precipitation and high solar insolation produces very sharp gradients in temperature, salinity, and other properties, such as dissolved nutrient concentrations, in tropical coastal waters. Sharp thermoclines and haloclines coincide with strong vertical discontinuity maintained throughout most of the year, except where equatorial upwellings force cooler water to the surface, or where waters from central oceanic gyres intrude into humid regions to become warmer and more dilute. Lower salinities are characteristic of surface waters of the wet tropics, and conversely, surface waters in arid tropical regions are hypersaline. Great variability in salinity and its ability to adjust rapidly to changes in wind-induced motion and temperature characterises tropical surface coastal waters (Webster 2020).

Three main types of coastal circulation are recognised: (i) gravitational (due to river runoff), (ii) tidal, and (iii) wind-driven. Tidal circulation is usually the most prevalent with interaction by coastal boundaries generating turbulence, advective mixing, and longitudinal mixing and trapping, the latter setting up coastal boundary layers or fronts. All three circulation patterns may, however, operate simultaneously in a water body. While bays and lagoons are tidal- or wave-dominated systems, some coastal systems such as small, restricted inlets and large, semi-enclosed bays, defy simple classification. Depending on several factors noted above, coastal embayments and bays and open coastal waters may or may not be stratified. Some waters are seasonally stratified.

Coastal waters are greatly affected by the larger oceanic currents. Off East Africa, seasonal circulation patterns are generated by the behaviour of the ITCZ, which creates two distinct seasons, the NE and SE monsoons. During the southeast monsoon, coastal waters are characterised by cool water, a deep thermocline, high water column mixing and wave energy, and fast currents and low salinity due to high precipitation. These characteristics are reversed during the NE monsoon. The 1998 ENSO event produced heavy rains with resulting large sand bars deposited off river mouths, with a persistent decrease in salinity and temperature in inshore waters indicating a coastal boundary layer. ENSO rains also produced semi-permanent flood channels serving as tidal inlets leading to tidal flooding of low-lying areas.

Inshore waters bathe, nurture, and are critical to the development and persistence of mangrove forests, coral reefs, and seagrass meadows. For example, in Gazi Bay, Kenya, a shallow, coastal bay, the main forcing function for water circulation are semi-diurnal tides which generate strong reversing currents in the deep, narrow channels in the mangrove zone, but not in the seagrass and coral reef zones (Kitheka 1997). The peak ebb and flood currents are symmetrical in the seagrass and coral zones with equal duration and magnitude, unlike in the mangrove zone where tidal asymmetry results in ebb currents being slightly stronger than flood tides. Current speeds are slower in the seagrass and coral reef areas of the bay, but the tidal asymmetry in the mangrove zone promotes export of mangrove detritus to the seagrass zone. There is spatial variation in salinity due to evaporation, freshwater, and ocean water inflow. Ocean water is driven out of the bay during ebb tide. The mixing of the different water masses is especially noticeable in the dry season when the freshwater outflow is negligible. The influx of oceanic water into the bay often leads to a slight lowering of water temperatures. The coral zone is dominated by cooler temperatures, higher oxygen concentrations, and higher salinity due to turbulent mixing promoted by wave breaking. Overall, due to the orientation of the bay with respect to dominant tidal water circulation patterns, the lack of sills, and an open entrance, the rate of exchange between inshore and offshore waters is high, about 60–90% of the volume per tidal cycle (Kitheka 1997).

While each bay and embayment are unique, findings as in Gazi Bay have been found in other wet tropical regions where it is common to find a clear gradient from mangrove-lined creeks to seagrass meadows to coral reefs with some mixing between estuarine and offshore waters (e.g. Sawi Bay in southern Thailand, Ayukai et al. 2000). Such physical gradients are even more distinct in the dry tropics where strong salinity gradients persist year-round, especially in hypersaline lagoons. Good examples of such lagoons lie along the east coasts of Mexico, Brazil, and Sri Lanka. The northern Yucatán coastal zone of Mexico is a complex environment characterised by extreme evaporation and high seasonal rainfall (Enriquez et al. 2013). It has water exchange with numerous coastal lagoons ranging from brackish to hypersaline and receives intense groundwater discharges. Two main water masses persist along the coast: (i) the Caribbean Subtropical Underwater mass (CSUW) which is upwelled from the Caribbean and hugs the bottom along the coast and (ii) the Yucatan Sea Water mass

(YSW) which is a mass of warm hypersaline water which originates in Yucatán due to high temperature and evaporation rates. Permeable karstic geology prevents the surface discharge of riverine water and instead water permeates directly to the aquifer and travels to the ocean via caves and fractures. Similarly, the Lagoa de Arauama is a hypersaline lagoon in coastal Brazil and persists due to semi-arid conditions, but with a small drainage basin and a choked entrance channel (Kjerfve et al. 1996). For at least 450 a, the lagoon has been hypersaline, although the average salinity has varied in response to the difference between evaporation and precipitation. There is a long-term trend of decreasing salinity due to constant pumping of freshwater from an adjacent watershed. A salt budget indicates that there is a delicate balance between the import of salt from the coastal ocean and eddy diffusive export of salt to the ocean.

The hypersaline conditions of coastal lagoons can also be affected by climate change and by reduced river discharge due to anthropogenic alterations in water flow (Kennish and Paerl 2010). In the Puttalam Lagoon, a large, shallow water body on the west coast of Sri Lanka, salinity averages 37 with maximum values exceeding 50 during drought periods. The salinity regime is seasonal with rapidly decreasing salinities during the rainy season and increasing salinities during drought. Salinities were lower than oceanic water in 1960–1961 due to high freshwater discharge prior to human-induced changes in river flow.

Open coastal waters can also be hypersaline in the dry season. Salinities of 37 have been recorded in the Great Barrier Reef lagoon due to evaporation exceeding precipitation. These hypersaline waters are not flushed out by salinity-driven baroclinic currents because lagoon waters are vertically well-mixed and are transported by a longshore residual current, thus forming a coastal boundary layer (Andutta et al. 2011) exhibiting both longshore and cross-shelf characteristics. The dynamics of the coastal boundary layer reaches steady-state in about 100 days which is the average length of the dry season and differs from other coastal boundary layers that often are one-dimensional with a dominant along-channel salinity gradient. Although distinctive in its two-dimensional nature, coastal waters of the tropics often have similar types of boundary layers with clear salinity gradients. These boundary conditions often disappear and break down completely during the wet season when heavy rains and storms help to mix inshore and offshore coastal waters.

At the opposite extreme, continental shelf waters in the tropics commonly undergo 'estuarisation', especially near rivers during the wet season (Longhurst and Pauly 1987). This phenomenon occurs on the inner and middle portions of continental shelves and consists of low-salinity waters usually exhibited as discrete plumes of discharged river water. The transport of river plumes onto continental shelves is a prime example of such 'estuarisation' and the exemplar is the Amazon plume, the low salinity (32–34) of which floats as far away as 2000 km from the mouths of the Amazon and Orinoco Rivers as a shallow plume of 20–30 m depth (Hu et al. 2004). 'Estuarisation' is also important in the Bay of Bengal, the Gulf of Panama, the South China Sea, and the Gulf of Guinea, in addition to discernible plumes proximal to other major tropical rivers. 'Estuarisation' on many tropical shelves is seasonal, depending on the onset of the monsoon, and the transition from unstratified to stratified conditions is sharp, delimited by a tidal front.

Even small- and medium-sized rivers can produce impressive shelf 'estuarisation'. For example, in the Great Barrier Reef lagoon during and immediately after the wet season, diluted seawater migrates out to the Great Barrier Reef proper (Schroeder et al. 2012). These plumes can persist for weeks especially after a severe rainy season or a cyclone.

Subsurface water masses below the thermocline in the eastern tropical coastal oceans frequently contain an oxygen minimum layer. Several explanations have been offered to account for this poorly understood phenomenon, including minimal circulation or mixing of water to replenish oxygen consumed or that detritus accumulates in stagnant areas because of increases in water density with depth, leading to the depletion of oxygen (Stramma et al. 2008). Irrespective of the cause, low oxygen concentrations have important consequences for demersal fish and the benthic fauna (Chapter 7). Small rivers, creeks, and estuaries are also characterised by waters of low oxygen content with presumably similar biological consequences. These oxygen-minimum zones are expanding as a direct result of global warming (Stramma et al. 2008).

Coastal upwelling is another major feature of the tropical oceans. Such events occur at all latitudes, but within the tropics and subtropics, physicochemical differences between upwelled and surface water masses are greatest. Upwelling is dominant along the subtropical-tropical boundary coasts of Peru-Chile (Peru Current), Morocco-Mauritania (Canary Current), Angola-Namibia (Benguela Current) and California-Mexico. Upwelling events also occur on the Malabar coast of India, off the Andaman Islands, Western Australia, the Gulf of Panama (the Costa Rica Dome), the Gulf of Nicoya and Tehuantepec, off NE Venezuela and Brazil south of Cabo Frio, from Ghana to Togo (Gulf of Guinea), on the Somali coast, and off southern Arabia (Longhurst and Pauly 1987). Not all upwelling occurs off eastern boundaries of continents, as some upwelling events are driven by events an entire ocean away. For example, in the Gulf of Guinea and along the Somali coast of Africa, a diversity of mechanisms drives coastal upwelling (Valsala 2009). Seasonality of upwelling in the tropics is well described, but the actual circulatory patterns are poorly understood. Seasonal changes in current patterns occur, driven mainly by the movement of the ITCZ across the equator every six months. Upwelling events and monsoons are thus ultimately linked to seasonal changes in the equatorial climate.

3.4 Estuarine Circulation

The shores of many tropical estuaries are inhabited by mangrove forests. Their presence results in unique circulation patterns, which lead to distinct chemical and biophysical characteristics that are quite different from those in temperate estuaries (Mazda et al. 2007). Water flows in mangrove forests are strongly influenced by the presence of the trees and their aboveground roots as well as by the geomorphology of the tidal creeks. These unique characteristics have important ecological consequences. Tidal and wave energy in any estuary constitutes an auxiliary energy subsidy as tides allow mangrove forests to store and pass on new fixed carbon and benefits animals adapted to make use of subsidised energy. Tides thus do the work of bringing nutrients, food, and sediments to mangroves and their food webs as well as exporting their waste products. This subsidy is an advantage in that organisms do not have to expend energy on these processes and can shunt more energy into growth and reproduction.

A few tropical estuaries are driven by macro-tides, although these are the exception rather than the rule as most tropical estuaries are micro- or meso-tidal. Along the Brazilian coast south of the Amazon, there are four different types of macro-tidal estuary: (i) 'typical' macro-tidal, (ii) estuaries with large fluvial discharge, (iii) shallow, frictionally dominated macro-tidal estuaries, and (iv) estuaries with structural control (Asp et al. 2013). In the 'typical' macro-tidal estuary, ebb tides are longer than flood tides, and this condition is like that found in macro-tidal estuaries in northern Australia (Wolanski et al. 2006). In the Daly estuary in northern Australia, freshwater becomes dominant up to the mouth and tides can be suppressed during the wet season (Wolanski et al. 2006). The Daly estuary is a 'leaky' sediment trap with its trapping efficiency varying with season and between years. In contrast, Darwin Harbour, a wider and much larger macro-tidal estuary NE of the Daly, is poorly flushed, especially in the dry season; much sediment remains trapped on intertidal flats and in mangroves with little loss to the sea (Wolanski et al. 2006). Within the harbour, complex bathymetry results in the generation of jets, eddies, and stagnation zones that can trap sediments inshore. There may be a feedback between tidal circulation and bathymetry as tidally averaged circulation appears to control the formation and movement of sand banks.

In micro-tidal estuaries, circulation is similarly complex and greatly influenced by the presence or absence of discharging rivers and the width of the connection between the estuary and the adjacent coastal ocean. Along the western Gulf of Mexico, tropical micro-tidal estuaries share many characteristics, including a narrow connection between the estuary and the adjacent continental shelf (Salas-Monreal et al. 2020). In the Jamapa River estuary, for example, surface horizontal displacements of the salinity and temperature fronts during the dry season occur, while during the wet season, the salinity and temperature gradients are observed in the vertical at about 1 m depth. A cyclonic recirculation at the mouth of the estuary occurs when the ratio between the mouth and the estuary width is below 0.4. This should hold true for all tropical micro-tidal estuaries in the western Gulf of Mexico (Salas-Monreal et al. 2020).

Not all tropical estuaries are driven solely by tides year round. In some northern Australian estuaries, a salinity maximum zone develops during in the dry season that is driven by high rates of evaporation (Wolanski 1986). This zone occurs near the mouth of each estuary where downwelling occurs, and a classical and an inverse estuarine circulation prevails upstream and downstream of the salinity maximum. This zone acts as a 'high salinity plug' inhibiting the mixing of estuarine and open ocean water to the extent that, in some cases, freshwater does not leave the estuary. Similar conditions have been found in estuaries along the SW coast of Ghana (Dzakpasu and Yankson 2015), in the Konkouré estuary of Guinea (Capo et al. 2009) and in the Gulf of Fonseca estuary on the Pacific coast of Central America (Valle-Levinson and Bosley 2003).

Some tropical estuaries are so complex as to defy simple classification. Good examples of such complexity can be found along the north coast of Brazil (Medeiros et al. 2001; Schettini et al. 2013). These estuaries have multiple riverine systems feeding into a larger lagoon which is ordinarily fronted by coral reefs or coral-fringed barrier islands. In the Itamaracá estuarine system, there are several estuarine waterways than feed into an 'inner sea' via a series of inlets, each considerably different from the other (Medeiros et al. 2001). Most of the freshwater enters the northern branch of the Santa Cruz Channel through the Catuama, Carrapicho, do Congo, Arataca, Botafogo, and Igarassu Rivers, the last three being the main source of freshwater. During the dry season, hypersaline conditions exist at both entrances in the "inner sea" due to evaporation, evapotranspiration by mangroves, and reduced exchange between the channels and reef shelf waters; a series of coral reefs fringe the outer edge of the estuary.

Tidal circulation within most mangrove waterways is characterised by a pronounced asymmetry between ebb and flood tides, with the ebb tide being shorter, but with stronger current velocity than the flood tide (Cavalcante et al. 2013). Current velocities in tidal creeks can often exceed 1 $m s^{-1}$ but only rarely approach 0.1 $m s^{-1}$ within the forest proper (Cavalcante et al. 2013). This asymmetry results in self-scouring of the tidal waterways to the extent that the bottom of most channels is composed of bedrock, gravel, and sand, with little or no accumulation of fine sediment. The ecological implication of this asymmetry is that there tends to be export of particulate nutrients, rather than net import.

The velocity of tidal circulation ultimately depends on the geometry of the waterway, that is, the ratio of the forest area to the waterway area as well as the slope of the forest. The ratio appears to be on the order to 2–10 (although few such measurements have been made) with a very small forest slope (Wolanski 2007). The tidal prism of a mangrove estuary thus increases greatly with an increase in the ratio between forest area to waterway area.

Numerical modelling has determined the importance of the interaction between tidal creek geometry and mangrove forests in causing asymmetry of tides. The dominance of the ebb tide is due to friction in the mangrove forest which is in turn controlled by the density of the forest (Mazda et al. 1995). Inside the forest, the water level and the current velocity are strongly controlled by drag force due to the vegetation (Norris et al. 2019). The denser the forest, the greater the drag resulting in slower current velocity and greater tidal asymmetry in the creek. This relationship, however, is not straightforward. The peak velocity in the waterway decreases at flood tide and increases at ebb tide for increasing levels of drag force. But when the drag force is excessive, the ebb flow is reduced allowing the waterway to silt. There is a natural feedback relationship among the vegetation, water, and sediments. This phenomenon is unique to mangrove estuaries.

An additional asymmetry of the currents in mangroves is the direction of the currents in relation to forest position (Li et al. 2012). At rising tide, the currents flow perpendicular into the forest, while at falling tide they are oriented at an angle, typically 30–60° to the bank. This lengthens the pathways of water at falling tide reducing the chance that materials such as mangrove detritus can escape the forest.

Another characteristic feature of mangrove hydrodynamics is the mixing and lateral trapping of water (Wolanski 2007; Mazda and Wolanski 2009). Lateral trapping of water within the forest is a dominant process controlling longitudinal mixing in mangrove waterways. The trapping phenomenon occurs when some of the water flowing in and out of the estuary is temporarily retained in the mangrove forest to be returned to the main water channel later. Trapping of water is enhanced in the dry season when there is little, if any, freshwater to cause buoyancy effects on water circulation. In the wet season, the buoyancy effect is important as freshwater is trapped in the forest at high tide, and as a floating lens or boundary layer hugging the riverbank at low tide. This effect means that the forests control the runoff of freshwater, especially at the end of a flood tide. The evaporation of water and the build-up of salt generated by the physiological activities of the trees helps to generate gradients of salt and other materials, both laterally and longitudinally, especially during a long dry season.

A significant lateral gradient within mangrove creeks during the dry season can be attributed to high evapotranspiration (Le Minor et al. 2021). Weak stratification usually prevails at the headwaters of mangrove waterways in the wet season as the result of freshwater input from rivers. Such gradients have been observed in Gazi Bay in Kenya (Kitheka 1997) and in the Konkoure River delta in Guinea (Wolanski and Cassagne 2000). In arid-zone estuaries, the salinity structure is inverse due to the lack of freshwater input and the high evaporation rate, especially in relation to salt flats and mangrove bordering the estuary; salinities can be >50 (Kitheka 1997).

The behaviour of tidal water is also longitudinally complex. Longitudinal diffusion is proportional to the square of the water velocity, which means that mixing rates are very small at the headwaters of mangrove creeks where currents are also very small. Water speed decreases from the mouth to the headwaters along the length of a waterway. The longitudinal and cross-sectional gradients in current speed are partly the result of shear dispersion processes that are magnified by the presence of the forest. This diffusion process drives the intensity of mixing and trapping. These complex processes translate into long residence times for water near the head of the waterway, especially in the dry season.

All estuaries, including those inhabited by mangroves, exhibit secondary circulation patterns superimposed on the primary tidal circulation. This phenomenon is responsible for the often observed trapping of floating detritus in density-driven convergence fronts during a rising tide (Stieglitz and Ridd 2001). These fronts occur in well-mixed estuaries due to the interaction between the velocity of water across the estuary and the density gradient up the estuary. Due to friction, the velocity is slower near the riverbanks than in the centre of the estuary, thus causing on flood tides a greater density mid-channel than at the banks. A two-cell circulation pattern results from the sinking of water in the centre of the estuary. The existence of these cells has ecological consequences. A net upstream movement of floating debris occurs, on the order of several km per day; mangrove propagules are unlikely to enter the mangrove forest when these cells are present and will accumulate in large numbers in 'traps' upstream from the convergence and upstream from the mangrove fringe (Stieglitz and Ridd 2001). Trapping of propagules is not conducive to the natural strategy of maximizing dispersal of seeds.

Within the mangrove forest, trees, roots, animal burrows and mounds, timber, and other decaying vegetation exert a drag force on the movement of tidal waters (Le Minor et al. 2021). The drag force of the trees can be simplified to a balance between the slope of the surface water and the flow resistance due to the vegetation (Mullarney et al. 2017). Water flow in the forest depends on the volume of the trees relative to the total forest area. The momentum of tidal forces is greater than the shear stress induced by the presence of the trees, including friction with the soil surface. Even the presence of dense pneumatophore roots induces turbulent friction near the forest floor (Norris et al. 2019). The presence of mangrove seedlings results in alteration of tidal flow by modifying the vertical velocity and the magnitude of turbulent energy (Chang et al. 2020). The dynamics of tidal forces in mangrove forests changes in relation to different tree species, density of the vegetation, and state of the tides.

Currents in the forest itself are not negligible, and a secondary circulation pattern is usually present due to the vegetation density and the overflow of water into the forest at high tide (Mazda et al. 2007; Mullarney et al. 2017). This secondary circulation enhances the trapping effect of tides. The drag force has two main influences: (i) inundation of the forest is inhibited and this decrease in water volume results in smaller dispersion and (ii) the trapping of water in the forest is enhanced, favouring dispersion. Thus, the magnitude of tidal trapping depends on the drag force due to the vegetation, so the magnitude of dispersion depends ultimately on the vegetation density.

Animal structures also influence water circulation in mangrove forests (Le Minor et al. 2021). Benthic organisms, especially crabs, produce numerous burrows and tubes in the forest floor through which tidal water flows. Tidal water flows through a labyrinth of interconnected crab burrows in the same direction as the surface current (Ridd 1996; Stieglitz et al. 2013). The flow through the burrows is caused by a pressure difference between multiple burrow openings. The total quantity of water that flows through crab burrows can range from 1000 to 1 010 000 m^3 representing from 0.3 to 3% of the total volume of water moving through the forest. This percentage is not large but is enough for the burrows to play an important role in flushing salt away from mangrove roots. The transport of salt derived from the tree roots results in variations in the density of water flowing through the burrows, having an impact on flushing time. Burrows, tubes, and other biogenic structures impart some significant delay in water flow, assisting in the trapping of water in the forest (Stieglitz et al. 2013).

Not all water entering an estuary leaves via the surface. Some water leaves via subterranean pathways such that mangrove estuaries often have significant groundwater flow. This flow can have significant biogeochemical and biological effects, such as removing excess salt from mangrove roots and removing high concentrations of respired carbon from the forest to the adjacent coastal zone (Gleeson et al. 2013). Crab burrows and other biogenic structures can facilitate groundwater flow. The flow of groundwater in mangroves usually has three components (Mazda and Ikeda 2006): (i) a near-steady flow towards the open ocean due to the pressure gradient induced by the difference in height between water levels in the forest and the open sea, (ii) a reversing tidal flow with a damped amplitude and delayed phase towards the forest, and (iii) a residual flow towards the forest caused by the damped tidal flow. This residual flow reduces the outflow of water from the forest towards the sea.

Mangroves often receive a significant amount of wave action, even in an estuary. Mangroves attenuate wave energy via two primary mechanisms: (i) multiple interactions of waves with mangrove trunks and roots and (ii) bottom friction. The latter is not well understood, but a significant amount of attention has focused on the effect of the presence of tree trunks and roots. Forces induced by waves on tree stems and roots are inertial and drag-type forces, with drag force dominating for most mangroves (Hashim et al. 2013). The degree of wave attenuation increases with increasing tree diameter, although interactions between tree stems can influence the extent of drag. Waves within a mangrove forest are strongly dissipated by these interactions. Dissipation of wave energy is a function of total tree area which is in turn a function of both tree diameter and forest density. Water depth can also play a role in wave dissipation. For a very dense forest, wave energy is almost totally dissipated within 40–50 m from the mangrove-sea boundary, but in less dense forests, about 35% of the incident wave energy is still extant behind the forest (Hashim et al. 2013). In mangrove forests that are small in area due to urban disturbance, such as in Singapore, the percentage of wave height reduction is higher under storm events compared to normal conditions, with vegetation drag being the main mechanism of wave dissipation; mangrove density and width were positively correlated to the percentage of wave height reduction during a storm (Lee et al. 2021). Mangrove roots contributed to a larger percentage of wave height reduction than trunks and canopies, although there were no significant differences in the extent of wave height reduction between forest types, incident wave heights, and water levels. Thus, even comparatively small, disturbed mangrove forests can offer some protection from wave energy.

Mangroves grow best where wave energy is low and tidal flow and subsequent attenuation within the forest result in the deposition of fine particles from the incoming tidal water. The transport and deposition of suspended sediment is controlled by several interrelated processes: tidal pumping; baroclinic circulation; trapping of small particles in the turbidity maximum zone; flocculation; the mangrove tidal prism; physicochemical reactions that destroy flocs of cohesive sediment; and microbial production of mucus (McLachlin et al. 2020). The relative importance of these processes is site-specific. Along the river's edge, mud banks form not just as the result of baroclinic circulation but also as the result of tidal pumping and mixing, especially in the turbidity maximum zone. A turbidity maximum zone is formed within an estuary where the residual inward bottom flow meets the outflow river flow. This zone is usually at the most landward point reached by the saline water flow. The water is generally shallowest at this point because this is a convergence point where sediment accumulates, the bulk of which creeps along the bottom. Some suspended matter settles here as the net current velocities in the zone are low; the zone is not stationary but moves with tidal ebb and flood.

In relation to the turbidity maximum, flocculation of particles begins at salinities often <1; the largest flocs remain near the river bottom. The small flocs and unflocculated particles move further downstream with the currents where they aggregate with local particles (Gratiot and Anthony 2016). As floc size increases, they move towards the riverbed where they are entrained upstream by the baroclinic circulation. Due to tidal pumping, these flocs are carried further upstream at flood tide rather than downstream at ebb tide. The flocs are a loose matrix of clay and silt particles, typically a few micrometres in diameter, with their small size controlled somewhat by the strength of the tidal currents. Disaggregation starts when tidal velocities exceed $1\,m\,s^{-1}$. During spring tides, flocs are typically between 15 and 40 μm in diameter and are larger during neap tides. These flocs are colonised by bacteria, protists, and fungi and their extracellular mucus and threads, which help to cement the flocs and to maintain size when subjected to turbulence. Within the forest, flocs can remain in suspension owing to turbulence generated by flow around the trees. The settling of flocs occurs for a short period when the tides turn from rising to falling and the waters are quiescent. Settling is facilitated by the sticking of microbial mucus and by formation of invertebrate faecal pellets.

It is thus correct to state that mangroves actively help to settle fine particles and are not just passive importers. The size, shape, and distribution patterns of mangrove trees have a profound impact on sedimentation. Tree species with large above-ground root systems, such as *Rhizophora*, facilitate the deposition of particles to a much greater degree than tree species without extensive roots. The flocculation of particles results in faster settling velocities; most flocs settle within 30 minutes just before high slack tide. Until slack water, turbulent wakes created by tree trunks, roots, and pneumatophores maintain particles in suspension. Once in the forest, however, conditions are unfavourable for them to be resuspended as the high vegetation density inhibits water motion.

3.5 Coral Reef Hydrodynamics

The circulation of water in and around an individual coral reef is similarly complex and is central to understanding the high biodiversity and grandeur of these beautiful structures and their food webs. From individual coral colonies to entire coral reefs, hydrodynamics regulate their structure and function. Unlike mangroves, coral reefs thrive in highly physical conditions where water rapidly circulates and where residence times are short (Monismith 2007).

At the scale of an individual branching coral, denser branching tends to divert more water flow to the exterior, whereas coral geometries with less dense branching allow more flow through the interior (Reidenbach et al. 2006). At a finer scale, branches create wakes that lead to blocking of interior flows. In the presence of waves, the flow through a coral colony is different, with the velocities inside the colony like those outside the colony. As colonies are living structures, flow variations inside the colony may induce localized calcification and a specific growth form or may lead to branch orientation that optimises functions such as the capture of prey or the uptake of nutrients. With higher flow velocities, the coral skeleton invariably becomes uniformly thicker, reducing the spaces between branches. This phenomenon suggests the possible importance of internal translocation of nutrients, etc., in the tissue layer that connects the polyps of a colony. Coral skeletal growth is often positively related to mean current velocity and to velocity fluctuations (Lenihan et al. 2015). Corals thus benefit directly from increased currents, probably through enhanced autotrophy (i.e. photosynthesis of coral symbionts) and/or heterotrophy (feeding on particles and living organisms) and indirectly by reduced feeding on corals by corallivorous fish.

At the scale of 1–10 m, bottom drag is critical in water circulation. The bottom drag coefficient, C_D, for coral reefs is roughly 10 times greater than for a sandy unconsolidated seabed (Rosman and Hench 2011). Because of this roughness and because of the complex nature of corals, reefs have much higher transfer rates of water flow than predicted by simple engineering models. Waves are the dominant water feature of corals at this scale. Indeed, nutrient uptake is linked directly to the wave regime of the reef. However, as roughness increases, mass transfer is reduced, that is, water movement decreases. Lower parts of the rough coral surface are sheltered from wave action, that is, velocities of water movement are higher near the tops of coral colonies than in the troughs between colonies. As greater roughness leads to greater water motion, this has the practical result of making grazing by corals more efficient. Turbulence produced by the coral reef also affects the uptake and kinetics of the rate that particulates suspended in the water column can be grazed by the reef community. As a reef grows and becomes rougher, it can support denser benthic assemblages, such as sponges, which in turn leads to a richer reef community.

At the scale of 100–1000 m, waves dominate coral reefs. The basic pattern is that incident waves break over the offshore face of the reef front and pushes water over the reef flat and then into the lagoon (if there is one). There are two regimes of flow behaviour when water flows over the front and the reef flat: (i) reef-top control, when the reef flat flow is in a friction–pressure gradient balance when the mean water level is sufficiently above the reef flat and (ii) reef-rim control, when the lagoon end of the reef flat determines the depth of the overall flow when the mean water level is at or below the reef flat (Gourlay 1996a,b). In the latter situation, wave breaking occurs on the reef front and wave-driven flow is due to swash running up and onto the reef flat.

When the depth over the reef is shallow, flows will be weak due to the friction of the reef flat, whereas if water is deep over the reef, flows will be weak because of limited wave breaking (Hearn 2011). There will thus be a water depth at which the wave-driven flow is maximum. Tidal variations also induce variations in mean flows over a reef as tides will vary the water depth over the reef front and reef flat thereby affecting water flow. For instance, in the Puerto Morelos fringing reef lagoon off Mexico where tides are small, surface wave-induced flow is the predominant flux with some influence from the Yucatán Current (YC) and trade winds (Coronado et al. 2007). The wave flow regime is modulated by the presence of the YC offshore. Similarly, cool, clear water flow up onto the flat of the fringing coral reef off the south coast of Molokai, Hawaii, during flood tide (Storlazzi et al. 2004). At high tide, sediment resuspension increases due to greater wave energy being available to propagate onto the reef flat. Trade-wind driven surface currents and wave breaking at the reef front cause increasing water depth on the reef flat, enhancing the development of depth-limited waves and sediment resuspension. As the tide ebbs, water and suspended sediment flow off the reef flat and is advected offshore and to the west by trade winds and tidally driven currents. Thus, the muddy, shallow reef flat is linked to suspended sediments on the deeper reef front where coral growth is maximal.

Each coral reef is unique, but water flow on all reefs conforms roughly to the same general pattern of water flow impinging on the reef front and flowing over the reef flat and attenuating into a lagoon or back reef. Water circulation in reef lagoons is rapid compared to adjacent oceanic waters, but reef lagoons are quiescent and usually deep enough to accumulate unconsolidated carbonate mud and sand deposits.

Water circulation in and around coral reefs thus depends greatly on reef type and geomorphology. Reefs fringing a shoreline such as the fringing reefs of Pago Bay, Guam, have a simpler hydrology than an individual oceanic reef; water flow on the Pago Bay reefs is driven mainly by wave height in the reef flat and channel (Comfort et al. 2019). Wind and wave directions that are directly across-shore contribute to faster flow speeds throughout the bay. On the fore reef, wave height is the strongest predictor of current speed, but wind direction has a strong influence on current direction. Flow in the channel and on the reef flat is more tightly tied to environmental factors of wind and waves than flow on the fore reef. Cold pulses around the slopes of the island indicate large internal waves that result in periodically cool deeper areas of the fore reef.

3.6 Fluid Mechanics in Seagrass Meadows

Seagrasses, like their mangrove and coral reef counterparts, are ecosystem engineers capable by their very existence of reducing the velocity of currents and attenuating waves to the extent that sediment particles can deposit on surfaces and on the seabed in quiescent zones (Peterson et al. 2004). Other factors play important roles in helping to accumulate carbon in seagrass meadows, such as canopy complexity, turbidity, wave height, and water depth (Samper-Villarreal et al. 2016). But the essence of what drives the accumulation of sediment particles and associated carbon is fluid dynamics. The movement of water among, between and around seagrass blades is the key feature of carbon capture (Koch et al. 2006).

The main source of energy required to move water is the sun that causes winds that lead to waves and thermal gradients that in turn lead to expansion, mixing, and instabilities in water gradients and thus flow. Seawater, being an incompressible fluid, moves at a flow rate that is defined by the velocity of the fluid that passes through a cross-sectional area, *A*. Water flow leads to both hydrostatic and dynamic pressures which are a constant. What this means in practical terms is that the sum of the pressures helps to explain lift that occurs within, around and under seagrass canopies. Drag is another force that operates in the case of water motion and has two components: (i) viscous drag that exists due to the interaction of the seagrass surface with the water and (ii) dynamic or pressure drag that exists under high flows when flows separate from boundaries (Koch et al. 2006).

Water flow can be either smooth and regular (laminar flow) or rough and irregular (turbulent flow), depending on the velocity and temporal and spatial scale under investigation as defined by the Reynolds number:

$$R_e = lu / v \tag{3.1}$$

where l is the length scale under observation and v is the kinematic viscosity. R_e defines four flow regimes that may occur: (i) creeping flow where $R_e << 1$ which occurs at very slow flows and spatial scales such as those experienced by microbes, (ii) laminar flow ($1 < R_e < 10^3$) which is smooth and regular, (iii) transitional flow ($R_e \approx 10^3$) which involves the production of eddies and disturbances in the flow, and (iv) fully turbulent flow ($R_e >> 10^3$). These flows are scale-dependent; flow is almost always turbulent across entire seagrass meadows, but laminar at the scale of individual seagrass leaves (Koch et al. 2006).

Flow conditions become more complex when water approaches a boundary, such as the seagrass canopy or seafloor. Water cannot penetrate such boundaries, but slips by it, a condition which leads to the development of a velocity gradient perpendicular to the boundary as the velocity at the boundary will be zero relative to the stream velocity. As water flows downstream, the velocity gradient will get larger and a slower moving layer of water will develop next to the boundary. Vertically, there is a sublayer in which the forces are largely viscous. Consequently, mass transfer in this layer is slow, dominated by diffusion, in what is called a diffusive boundary layer. Such boundary layers can become embedded within one another such that it is possible to define boundary conditions around blade epiphytes, flowers, leaves, and the canopy.

At the molecular level, a boundary layer develops on the sediment surface as well as on each leaf, shoot, or flower as water flows through a meadow. The faster the water movement, the thinner the diffusive boundary layer and thus the transfer of molecules (e.g. CO_2) is faster from the boundary layer to the water column. When currents are weak, the flux of molecules may be diffusion-limited, but after a critical velocity is reached, the transfer is no longer limited by diffusion but by the rate of assimilation capacity (i.e., biological or biochemical activity). The mass transfer of molecules also depends on other factors, such as the thickness of the periphyton layer on the seagrass leaves, reactions within the periphyton layer, and the concentration of molecules in the water adjacent to the leaf-periphyton assemblage.

At the scale of shoots (mm to cm), a feedback mechanism operates as individual shoots are affected by other shoots and its position within the canopy (that is, edge versus centre of the entire meadow). As water velocity increases, shoots bend minimising drag, but the forces exerted on individual shoots are more complex when waves are involved, as a shoot is exposed to unsteady flows in different directions. This is confirmed by the fact that in wave-swept environments, seagrass leaves become longer as wave exposure increases (de Boer 2007). Flow around shoots results in bending but also pressure gradients on the leeward side of the leaf such that a vertical ascending flow is generated downstream of the shoot. This water then disperses horizontally at the point where the leaves bend over with the flow; interstitial water is also flushed out at the base of the shoot due to the pressure gradients generated on the sediment surface.

At the whole-canopy level, reduced flows occur within the canopy due to the deflection of the current over the canopy and loss of momentum within the canopy. Water speed as a result can be two to >10 times slower than outside the meadow. It is this process that allows water and sediment particles to be trapped during low tide; even short seagrass canopies can reduce water velocity. Vertically, however, water flow intensifies at the height of the sheath or stem as these parts are much less effective at reducing water velocity compared with the leaf component. Canopy flow is nevertheless complex because it is a function of the drag or resistance of the leaves on the water.

Seagrass meadows are where sediments and carbon deposit and accumulate largely due to the reduction in velocity and intensity of turbulence, that is, a reduction in flow strength that leads to a reduction in resuspension within the canopy (de Boer 2007; Gullström et al. 2018). Accumulation of sediment may be seasonal, especially during summer when seagrasses are at their maximum density and in winter then resuspension may be greater than accumulation when seagrasses are minimal, although roots and rhizomes may alone be enough to stabilize the accumulated deposits (Gullström et al. 2018). Epiphytes on seagrass leaves may foster the accumulation of sediment particles by increasing the roughness of the canopy and increasing the thickness of the boundary layer on the leaf surface. However, in highly wave-exposed locations, seagrasses may not accumulate fine sediments due to resuspension. Indeed, in some cases, sediment may be coarser beneath seagrass patches due to turbulence generated by the leaves themselves.

3.7 Tides

A variety of tidal types and ranges exist across the tropical ocean as a result not just of the pull of the moon and the sun, but because of the presence of continents and shorelines, and the relative size scales of these coastal features (Townsend 2012). Various tidal types and ranges exist because some areas are more tied to the daily cycle of the sun's gravitational attraction than others. In these areas, diurnal tides exist; in areas more tied to the moon's gravitation, semi-diurnal tides exist. And still other areas exhibit a combination of both influences, producing mixed tides. Macro-tides such as in northern Australia result when the resonant frequency of a body of water closely matches the lunar tidal frequency and is modified by geomorphology; lesser tides are produced with a less close match between lunar frequency and basin shape.

Semi-diurnal tides predominate along the coast of northern South America, the Pacific coast of Central America, most of tropical Africa and along stretches of northern Australia (Figure 3.3). Mixed tides occur everywhere else (Figure 3.3), except some areas of Southeast Asia (western Borneo, eastern Sumatra, Gulf of Thailand), and the Gulf of Mexico (Figure 3.3) where diurnal tides occur.

With respect to tidal ranges, macro-tides (> 4 m tidal range) predominate along a portion of the northern Australia coast, the southern Great Barrier Reef shelf, the Pacific coast of southern Central America, and a portion of East Africa (Figure 3.4). Meso-tidal (2–4 m tidal range) regions include the northern Arabian Sea, parts of East Africa and the eastern Bay of Bengal, and some portions of Southeast Asia and NE Brazil; micro-tides (< 2 m tidal range) dominate everywhere else in the tropics.

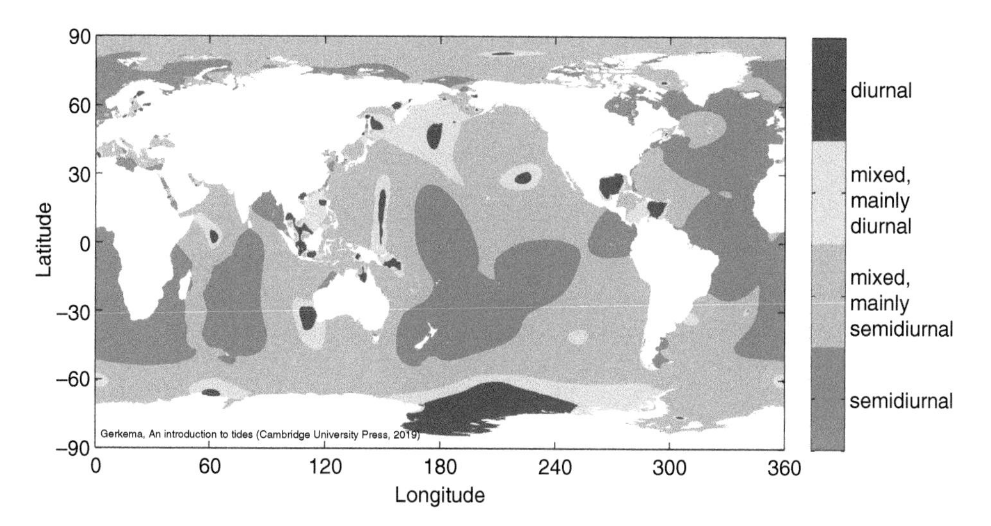

FIGURE 3.3 Global distribution of semi-diurnal, diurnal, and mixed tides. *Source*: Gerkema (2019), figure 1, p.14. © Cambridge University Press.

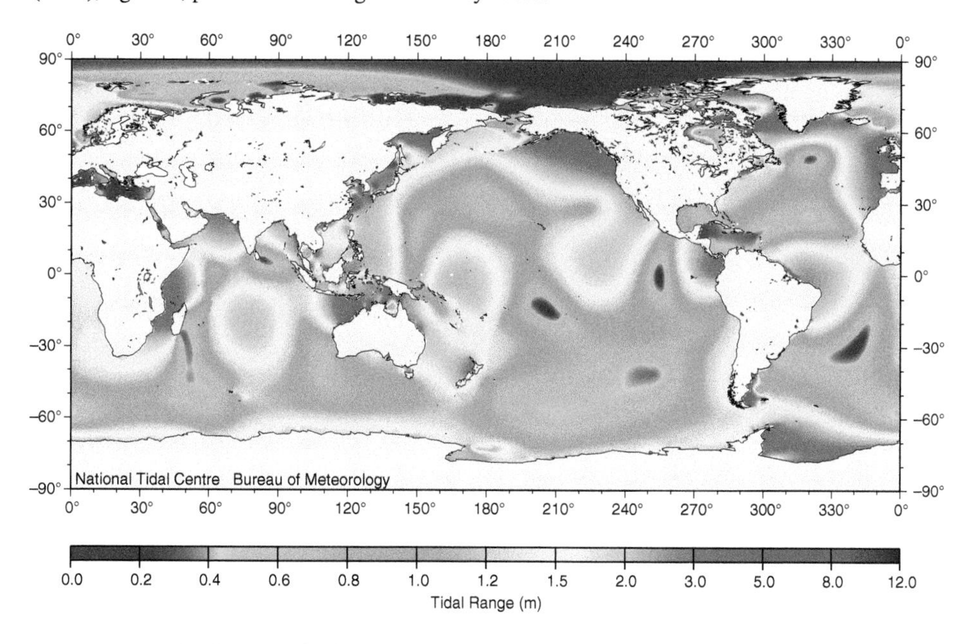

FIGURE 3.4 Distribution of tidal ranges across the world ocean. *Source*: Image provided by and used with permission of James Chittleborough, Australian Bureau of Meteorology. © Australian Bureau of Meteorology.

References

Andutta, F.P., Ridd, P.V., and Wolanski, E. (2011). Dynamics of hypersaline coastal waters in the Great Barrier Reef. *Estuarine, Coastal and Shelf Science* 94: 299–305.

Asp, N.E., de Freitas, P.T., Gomes, V.J.C. et al. (2013). Hydrodynamic overview and seasonal variation of estuaries at the eastern sector of the Amazonian coast. *Journal of Coastal Research* 65: 1092–1097.

Ayukai, T., Wolanski, E., Wattayakorn, G. et al. (2000). Organic carbon and nutrient dynamics in mangrove creeks and adjacent coastal waters of Sawi Bay, southern Thailand. *Phuket Marine Biological Center Special Publication* 22: 51–62.

de Boer, W.F. (2007). Seagrass-sediment interactions, positive feedbacks and critical thresholds for occurrence: a review. *Hydrobiologia* 591: 5–24.

Capo, S., Brenon, I., Sottolichio, A. et al. (2009). Tidal sediment transport versus freshwater flood events in the Konkoure Estuary, Republic of Guinea. *Journal of African Earth Sciences* 55: 52–57.

Cavalcante, G.H., Feary, D.A., and Kjerfve, B. (2013). Effects of tidal range variability and local morphology on hydrodynamic behaviour and salinity structure in the Caeté River estuary, north Brazil. *International Journal of Oceanography* 2013: 315328. https://doi.org/10.1155/2013/315328.

Chang, Y., Chen, Y., and Wang, Y.P. (2020). Field measurements of tidal flows affected by mangrove seedlings in a restored mangrove swamp, southern China. *Estuarine, Coastal and Shelf Science* 235: 106561. https://doi.org/10.1016/j.ecss.2019.106561.

Comfort, C.M., Walker, G.O., McManus, M.A. et al. (2019). Physical dynamics of the reef flat, channel, and fore reef areas of a fringing reef embayment: an oceanographic study of Pago Bay, Guam. *Regional Studies in Marine Science* 31: 100740. https://doi.org/10.1016/j.rsma.2019.100740.

Coronado, C., Candela, J., Iglesias-Prieto, R. et al. (2007). On the circulation in the Puerto Morelos fringing reef lagoon. *Coral Reefs* 26: 149–163.

Dzakpasu, M.F.A. and Yankson, K. (2015). Hydrographic characteristics of two estuaries on the south western coast of Ghana. *New York Science Journal* 8: 60–69.

Enriquez, C., Mariño-Tapia, I., Jeronimo, G. et al. (2013). Thermohaline processes in a tropical coastal zone. *Continental Shelf Research* 69: 101–109.

Feng, M., Zhang, N., Liu, Q. et al. (2018). The Indonesian throughflow, its variability and centennial change. *Geoscience Letters* 5: 3. https://doi.org/10.1186/s40562-018-0102-2.

Gerkema, T. (2019). *An Introduction to Tides*. Cambridge University Press: Cambridge, UK.

Gleeson, J., Santos, I.R., Maher, D.T. et al. (2013). Groundwater–surface water exchange in a mangrove tidal creek: evidence from natural geochemical tracers and implications for nutrient budgets. *Marine Chemistry* 156: 27–37.

Godfrey, J.S., Johnson, G.C., McPhaden, M.J. et al. (2001). The tropical ocean circulation. In: *Ocean Circulation and Climate: Observing and Modelling the Global Ocean* (eds. G. Siedler, J. Church and J. Gould), 215–246. New York: Academic Press.

Gourlay, M.R. (1996a). Wave set-up on coral reefs. 1. Set-up and wave-generated flow on an idealized two-dimensional reef. *Journal of Coastal Engineering* 27: 161–193.

Gourlay, M.R. (1996b). Wave set-up on coral reefs. 2. Wave set-up on reefs with various profiles. *Journal of Coastal Engineering* 28: 17–55.

Gouveia, N.A., Gherardi, D.F.M., Wagner, F.H. et al. (2019). The salinity structure of the Amazon River plume drives spatiotemporal variation of oceanic primary productivity. *Journal of Geophysical Research: Biogeosciences* 124: 147–165.

Gratiot, N. and Anthony, E.J. (2016). Role of flocculation and settling processes in development of the mangrove-colonized, Amazon-influenced mud-bank coast of South America. *Marine Geology* 373: 1–10.

Gullström, M., Lyimo, L.D., Dahl, M. et al. (2018). Blue carbon storage in tropical seagrass meadows relates to carbonate stock dynamics, plant–sediment processes, and landscape context: insights from the western Indian Ocean. *Ecosystems* 21: 551–566.

Hashim, A.M., Pheng, S.M.C., and Takajudin, H. (2013). Effectiveness of mangrove forests in surface wave attenuation: a review. *Research Journal of Applied Sciences, Engineering and Technology* 5: 4483–4488.

Hearn, C.J. (2011). Perspectives in coral reef hydrodynamics. *Coral Reefs* 30: 1. https://doi.org/10.1007/s00338-011-0752-4.

Hill, D.F. (2016). Spatial and temporal variability in tidal range: evidence, causes and effects. *Current Climate Change Reports* 2: 232–241.

Hood, R.R., Beckley, L.E., and Wiggert, J.D. (2017). Biogeochemical and ecological impacts of boundary currents in the Indian Ocean. *Progress in Oceanography* 156: 290–325.

Hu, C., Montgomery, E.T., Schmitt, R.W. et al. (2004). The dispersal of the Amazon and Orinoco River water in the tropical Atlantic and Caribbean Sea: observation from space and S-PALACE floats. *Deep-Sea Research II* 51: 1151–1171.

Johnson, G.C., McPhaden, M.J., and Firing, E. (2001). Equatorial Pacific Ocean horizontal velocity, divergence, and upwelling. *Journal of Physical Oceanography* 31: 839–849.

Kennish, M.J. and Paerl, H.W. (eds.) (2010). *Coastal Lagoons. Critical Habitats of Environmental Change*. Boca Raton, FL: CRC Press.

Kitheka, J.U. (1997). Coastal tidally-driven circulation and the role of water exchange in the linkage between tropical coastal ecosystems. *Estuarine, Coastal and Shelf Science* 45: 177–187.

Kjerfve, B., Schettini, C.A.F., Knoppers, B. et al. (1996). Hydrology and salt balance in a large, hypersaline coastal lagoon, Lagoa de Araruma, Brazil. *Estuarine, Coastal and Shelf Science* 42: 701–725.

Koch, E.W., Ackerman, J.D., Verduin, J. et al. (2006). Fluid dynamics in seagrass ecology – from molecules to ecosystems. In: *Seagrasses: Biology, Ecology and Conservation* (ed. A.W.D. Larkum), 193–225. Dordrecht, The Netherlands: Springer.

Lambert, E., Le Bars, D., and de Ruijter, W.P.M. (2016). The connection of the Indonesian throughflow, south Indian Ocean Countercurrent and the Leeuwin current. *Ocean Sciences* 12: 771–780.

Le Minor, M., Zimmer, M., Helfer, V. et al. (2021). Flow and sediment dynamics around structures in mangrove ecosystems – a modelling perspective. In: *Dynamic Sedimentary Environments of Mangrove Coasts* (eds. F. Sidik and D.A. Friess), 83–120. Amsterdam: Elsevier.

Lee, W.K., Tay, S.H.X., Ooi, S.K. et al. (2021). Potential short wave attenuation function of disturbed mangroves. *Estuarine, Coastal and Shelf Science* 248: 108767. https://doi.org/10.1016/j.ecss.2020.106747.

Lehodey, P. (2001). The pelagic ecosystem of the tropical Pacific Ocean: dynamic spatial modelling and biological consequences of ENSO. *Progress in Oceanography* 49: 439–468.

Lenihan, H.S., Hench, J.L., Holbrook, S.J. et al. (2015). Hydrodynamics influence coral performance through simultaneous direct and indirect effects. *Ecology* 96: 1540–1549.

Li, Z., Saito, Y., Tamura, T. et al. (2012). Mid-Holocene mangrove succession and its response to sea-level change in the upper Mekong River delta, Cambodia. *Quaternary Research* 78: 386–399.

Longhurst, A. and Pauly, D. (1987). *Ecology of Tropical Oceans*. San Diego, USA: Academic Press.

Masselink, G. and Hughes, M.G. (2014). *An Introduction to Coastal Processes and Geomorphology*. London: Routledge.

Mazda, Y. and Ikeda, Y. (2006). Behaviour of the groundwater in a riverine-type mangrove forest. *Wetlands Ecology and Management* 14: 477–488.

Mazda, Y. and Wolanski, E. (2009). Hydrodynamics and modelling of water flow in mangrove areas. In: *Coastal Wetlands: An Integrated Ecosystem Approach* (eds. G.M.E. Perillo, E. Wolanski, D.R. Cahoon, et al.), 231–262. Amsterdam: Elsevier.

Mazda, Y., Kanazawa, N., and Wolanski, E. (1995). Tidal asymmetry in mangrove creeks. *Hydrobiologia* 295: 51–58.

Mazda, Y., Wolanski, E., and Ridd, P.V. (2007). *The Role of Physical Processes in Mangrove Environments. Manual for the Preservation and Utilization of Mangrove Ecosystems*. Tokyo: Terrapub.

McCreary, J.P., Murtugudde, R., Vialard, J. et al. (2009). Biophysical processes in the Indian Ocean. In: *Indian Ocean Biogeochemical Processes and Ecological Variability* (eds. J.D. Wiggert, R.R. Hood, S.W.A. Naqvi, et al.), 9–32. Washington, D.C.: American Geophysical Union.

McLachlin, R.L., Ogston, A.S., Asp, N.E. et al. (2020). Impacts of tidal-channel connectivity on transport asymmetry and sediment exchange with mangrove forests. *Estuarine, Coastal and Shelf Science* 233: 106524. https://doi.org/10.1016/j.ecss.106524.

Medeiros, C., Kjerfve, B., Araujo, M. et al. (2001). The Itamaracá estuarine ecosystem, Brazil. In: *Coastal Marine Ecosystems in Latin America* (eds. U. Seeliger and B. Kjerfve), 71–81. Berlin: Springer.

Monismith, S.G. (2007). Hydrodynamics of coral reefs. *Annual Review of Fluid Mechanics* 39: 37–55.

Mullarney, J.C., Henderson, S.M., Reyns, J.A. et al. (2017). Spatially varying drag within a wave-exposed mangrove forest and on the adjacent tidal flat. *Continental Shelf Research* 147: 102–113.

Norris, B.K., Mullarney, J.C., Bryan, K.R. et al. (2019). Turbulence within natural mangrove pneumatophore canopies. *Journal of Geophysical Research: Oceans* 124: 2263–2288.

Peterson, C.H., Luettich, R.A., Micheli, F. et al. (2004). Attenuation of water flow inside seagrass canopies of differing structure. *Marine Ecology Progress Series* 268: 81–92.

Reidenbach, M.A., Koseff, J.R., Monismith, S.G. et al. (2006). Effects of waves, unidirectional currents, and morphology on mass transfer in branched reef corals. *Limnology and Oceanography* 51: 1134–1141.

Ridd, P.V. (1996). Flow through animal burrows in mangrove creeks. *Estuarine, Coastal Shelf Science* 43: 617–625.

Rosman, J.H. and Hench, J.L. (2011). A framework for understanding drag parameterizations for coral reefs. *Journal of Geophysical Research: Oceans* 116 https://doi.org/10.1029/2010JC006892.

Salas-Monreal, D., Riveron-Enzastiga, M.L., de Salas-Perez, J. et al. (2020). Bathymetric flow rectification in a tropical micro-tidal estuary. *Estuarine, Coastal and Shelf Science* 235: 106562. https://doi.org/10.1016/j.ecss.2019.106562.

Samper-Villarreal, J., Lovelock, C.E., Saunders, M.I. et al. (2016). Organic carbon on seagrass sediments is influenced by seagrass canopy complexity, turbidity, wave height, and water depth. *Limnology and Oceanography* 61: 938–952.

Schettini, C.A.F., Pereira, M.D., Siegle, E. et al. (2013). Residual fluxes of suspended sediment in a tidally-dominated tropical estuary. *Continental Shelf Research* 70: 27–35.

Schroeder, T., Devlin, M.J., Brando, V.E. et al. (2012). Interannual variability of wet season freshwater plume extent into the Great Barrier Reef lagoon based on satellite coastal ocean colour observations. *Marine Pollution Bulletin* 65: 210–223.

Stieglitz, T. and Ridd, P.V. (2001). Trapping of mangrove propagules due to density-driven secondary circulation in the Normanby River estuary, NE Australia. *Marine Ecology Progress Series* 211: 131–142.

Stieglitz, T.C., Clark, J.F., and Hancock, G.J. (2013). The mangrove pump: the tidal flushing of animal burrows in a tropical mangrove forest determined from radionuclide budgets. *Geochimica et Cosmochimica Acta* 102: 12–22.

Storlazzi, C.D., Ogston, A.S., Bothner, M.H. et al. (2004). Wave- and tidally-driven flow and sediment flux across a fringing coral reef, southern Molokai, Hawaii. *Continental Shelf Research* 24: 1397–1419.

Stramma, L., Johnson, G.C., Sprintall, J. et al. (2008). Expanding oxygen-minimum zones in the tropical oceans. *Science* 320: 655–657.

Tillinger, D. (2011). Physical oceanography of the present day Indonesian Throughflow. In: *The SE Asian Gateway: History and Tectonics* (eds. R. Hall, R. Cottam and M.E.J. Wilson), 267–281. London: Geological Society.

Townsend, D.W. (2012). *Oceanography and Marine Biology: An Introduction to Marine Science*. Sunderland, USA: Sinauer.

Valle-Levinson, A. and Bosley, K.T. (2003). Reversing circulation patterns in a tropical estuary. *Journal of Geophysical Research: Oceans* 108: 3331. https://doi.org/10.1029/2003JC001786.

Valsala, V. (2009). Different spreading of Somali and Arabian coastal upwelled waters in the northern Indian Ocean: A case study. *Journal of Oceanography* 65: 803–816.

Varona, H.L., Veleda, D., Silva, M. et al. (2019). Amazon River plume influence on western tropical Atlantic dynamic variability. *Dynamics of Atmospheres and Oceans* 85: 1–15.

Vecchi, G.A. and Soden, B.J. (2007). Global warming and the weakening of the tropical circulation. *Journal of Climate* 20: 4316–4340.

Webster, P.J. (2020). *Dynamics of the Tropical Atmosphere and Oceans*. Hoboken, USA: Wiley-Blackwell.

Wolanski, E. (1986). An evaporation-driven salinity maximum zone in Australian tropical estuaries. *Estuarine, Coastal and Shelf Science* 22: 415–424.
Wolanski, E. (2007). *Estuarine Ecohydrology*. Amsterdam: Elsevier.
Wolanski, E. and Cassagne, B. (2000). Salinity intrusion and rice farming in the mangrove-fringed Konkoure River delta, Guinea. *Wetlands Ecology and Management* 8: 29–36.
Wolanski, E., Williams, D., and Hanert, E. (2006). The sediment trapping efficiency of the macro-tidal Daly estuary, tropical Australia. *Estuarine, Coastal and Shelf Science* 69: 291–298.

CHAPTER 4

Tropical Marine Geosphere

4.1 Major Sedimentary Patterns

Because most riverine discharge to the oceans occurs in the tropics and that glacial sedimentation occurs at the poles, there are clear latitudinal patterns in sediment types and composition on the world's continental shelves (Figure 4.1). The major factors responsible for these global patterns are weathering, the presence or absence of large rivers and coral reefs, glaciation, tectonics, authigenic mineral and detrital formation, and ice-rafting (Flemming 2011). Recent changes in sedimentary habitats are due to extensive degradation from coastal development, reduced sediment delivery from rivers due to damming, increased coastal erosion, and sea-level rise. Mud and coral/carbonate deposits of biogenic origin are most abundant in the tropics, whereas sand is globally dominant, decreasing with higher latitudes to be proportionally displaced by glacial rock and gravel (Figure 4.1).

The latitudinal pattern of inner shelf sediment types is somewhat deceptive because a large proportion of mud in the tropics occurs in proximity to the major rivers, particularly the deltaic systems of the Amazon, Orinoco, Mekong, Ganges, and Brahmaputra Rivers. Muds along coastlines near major river plumes are stirred up sufficiently due to cyclones that disrupt wave trains and form mud banks. This phenomenon occurs predominantly in the tropics and is well documented along the coast of SW India (Muraleedharan et al. 2018), northern South America (Proisy et al. 2021), and northern Australia (Caitcheon et al. 2012). On the Kerala coast of SW India, certain inshore areas produce zones of calm water by dampening wave action. Large quantities of riverine-derived matter in colloidal suspension lead to the dissipation of wave energy. These mud banks, locally known as 'chekara', extend over areas of at least 25 km^2 and are characterised by silty-clay sediments, oxygen-deficient bottom waters, and possibly by generation of gases. These 'chekara' are generally 1–2 m thick.

Formation and migration of the Kerala coast mud banks are associated with high-period waves and their refraction pattern along the sea bottom (Muraleedharan et al. 2018). The size of each bank is determined by the location of converging intertidal currents and offshore flow. The mud bank is supplied continuously with mud from both directions and compensates for losses due to settlement and from export by

Tropical Marine Ecology, First Edition. Daniel M. Alongi.

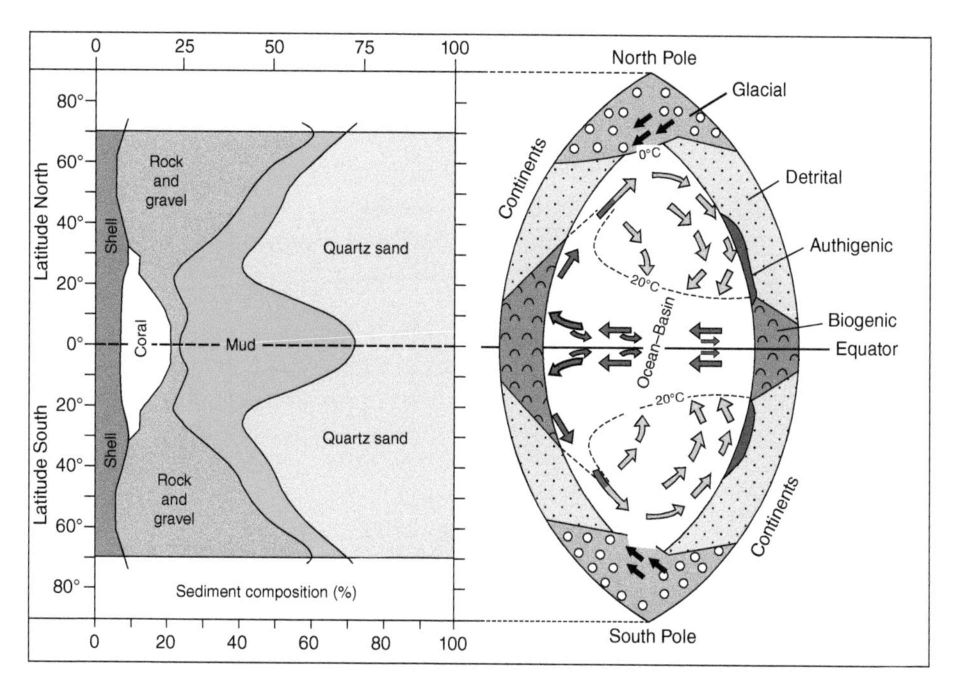

FIGURE 4.1 Latitudinal distribution of sediment types and composition on the world's inner continental shelves. *Source*: Flemming (2011), figure 9, p. 15. © Elsevier.

currents moving offshore. Particulate nutrient concentrations in mud bank deposits are generally high and positively correlated with decreasing grain size.

The coast of the Guianas (French Guiana, Surinam, British Guiana) of South America is also characterised by migrating mud banks. The coast is bordered by the sources of mud in the region, in the east by the mouth of the Amazon, and in the west by the Orinoco River (Jolivet et al. 2019). Mud banks attached to the shore are gigantic (about 200 km^2) and are composed of fluid mud, forming a temporary storage for silt and clay, and colonised and recolonised by mangroves after alternating episodes of accretion and erosion in relation to river discharge (Proisy et al. 2021). As off the SW coast of India, the subtidal nearshore seabed is shallow, gently sloped, and fronted by mangrove forests at the water's edge, leading to similar wave and tidal current patterns which foster accretion and migration of fluid mud. These linear mud shoals change rapidly in space and time and are transported westward by wave-induced currents on the inner shelf and by the Guiana Current on the outer shelf (Jolivet et al. 2019).

Land-use change is an important factor in changing mud coastlines. For example, along the Mahin mud section of the Nigerian coast, Gulf of Guinea, 58% of the coastline experienced serious erosion, with a rapid rate of coastline retreat triggering a land loss of 10.6 km^2 to the Atlantic over the last three decades (Dada et al. 2019). Although the changing wave climate had a strong influence on the observed patterns of erosion and accretion along the Mahin mud coast, both marine and anthropogenic processes act in unison to influence coastal retrogradation along the coast; mangrove deforestation has contributed greatly to the increased rate of retreat.

Tropical coastlines characterised by migrating mud banks are the exception rather than the rule. Excluding these areas and other regions such as off tropical river deltas, the Peruvian upwelling system, and the Bight of Biafra where highly reducing, sulphidic blue mud persists, the largest area of tropical shelves is sand dominated. Several shelves are dominated by carbonates and in many instances, bordered landward by extensive tidal flats, mangroves, and seagrasses or fringed seawards by coral reefs (Martini 2014).

Modern carbonate shelves in the subtropics and tropics fall into two categories: (i) protected shelf lagoons (the Bahamas, Florida, Belize, Cuba, and the Great Barrier Reef) and (ii) open shelves (Yucatan, western Florida, the Persian Gulf, and northern Australia). Shelf lagoons are characterised by the presence of fringing barrier reefs, islands, and shoals, and commonly have across-shelf gradients of mixed terrestrial-carbonate deposits. On the Belize shelf, the Grand Bahamas Bank, and the Great Barrier Reef, the lagoons consist of gradients from inshore terrigenous quartz sand and mud, grading to mixed terrigenous/carbonate deposits and then to carbonate sand and mud out to the edge of the shelf (Vieira et al. 2019). Carbonate shelves are not limited to the low latitudes, but protected shelves are latitudinally restricted because only in the low latitudes has the production of carbonate at the shelf margin sufficient to keep pace with Holocene sea-level rise. On many tropical shelves, soft sandy-mud sediments derived from continental drainage dominate inshore areas with varying mixtures of quartz sand and carbonate sand deposits dominating the middle and outer shelf areas, the extent of which vary with shelf width (Vieira et al. 2019).

Many coastal lagoons which formed behind barrier islands are either hypersaline in arid regions due to excessive evaporation (e.g. the Persian Gulf) or form gigantic interconnecting waterways in the humid tropics, as along the Gulf of Guinea off Africa. Sediment composition varies greatly among lagoons depending on their openness to the sea and the presence or absence of rivers and coastal vegetation. Lagoons in highly arid regions are often trapped with aeolian dunes that have been cemented by the precipitation of calcium carbonate, within which biogenic material is rapidly produced. Microbial stromatolites develop particularly in arid Indo-Pacific areas (e.g. Shark Bay, Western Australia) where other biota is excluded by accretion of precipitated inorganic carbonates. The formation of cyanobacterial mats constitutes the final stages in the formation of coastal gypsum lakes.

On sandy beaches, ecosystem structure and function are highly dependent on the beach type. Beaches are defined by the interactions among wave energy, tides, and the nature of the available sand (McLachlan and Defeo 2018). The beach slope is the simplest index of beach state and is the product of all three variables; beach face slopes flatten as wave energy or tidal range increases or particle size decreases, assuming other factors are kept constant. The flattest beach thus occurs in macro-tidal regions of high wave energy and fine sand while the steepest beaches occur in micro-tidal regions with low wave energy and coarse sand. A range of beach morphological types can be distinguished between these extremes.

In a micro-tidal regime, beaches are wave dominated, and three beach types can be recognised: reflective, intermediate, and dissipative (Figure 4.2). A reflective beach is characterised by a steep face and absence of a surf zone with gentle wave and coarse sand. Dissipative beaches are characterised by a flat beach face and wide surf zone; waves break far out from the beach face and dissipate their energy while traversing the surf zone before expiring as swash on the beach face. Dissipative beaches are thus the product of large waves moving over fine sand. Between these two extremes are intermediate beaches distinguished by the presence of surf zones

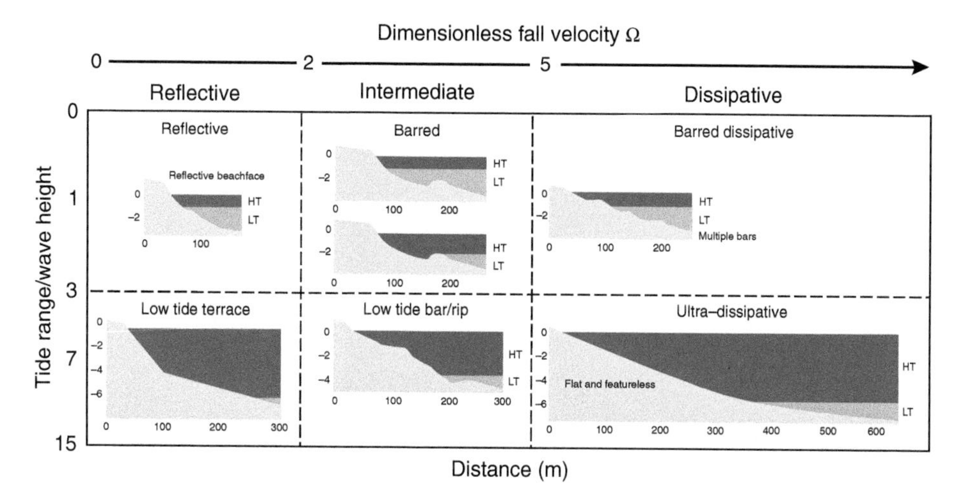

FIGURE 4.2 Model of sandy beaches from reflective to dissipative types with differences with tidal and/or wave heights. Distances are in metres. *Source*: McLachlan and Defeo (2018), figure 2.16, p. 26. © Elsevier.

that are smaller than for dissipative beaches and are generally 20–100 m wide. The surf zone of an intermediate beach has well-developed sand banks and channels with rip currents.

A beach type can be altered by storms, moving towards dissipative conditions over such circumstances and towards reflective conditions during calm weather (McLachlan and Defeo 2018). Tides also play a role in these transformations as spring tides during storms can foster dissipative conditions and neap tides can permit the development of a reflective beach. Simply, sand erodes or accretes on the beach face as wave height and tide range rises or drops.

A useful index to describe the state of a micro-tidal beach is called Dean's parameter (McLachlan and Defeo 2018):

$$\Omega = \text{wave energy} / \text{sand fall velocity}$$

wave energy is given by breaker height (cm) divided by wave period (seconds) and sand fall velocity is the sinking rate (cm per second) of the mean sand particle size on the beach. Values for $\Omega < 2$ indicate reflective beaches and values > 5 indicate dissipative beaches.

In macro-tidal regions, the beach type is more complex as tides play a role that is like waves in that increasing tide range tends to make beaches even more dissipative because increasing tide range allows the surf zone to work back and forth over a wider area. In fact, fully reflective beaches will not occur when the tide range exceeds 1–1.5 m. Reflective beaches only occur on beaches with larger tides at the top of the shore between the neap and spring high-water swash lines. Under large tidal regimes, beaches are generally tide dominated whereas in intermediate beaches they are mixed and either waves or tides can dominate.

A useful index of the relative importance of waves and tides is the relative tide range (RTR) which is derived by the mean spring tide range divided by the breaker height. Thus, a two-dimensional model (Figure 4.2) is produced of beach states of Ω versus RTR which span the entire range of tidal and wave conditions.

4.2 Distribution of Major Habitat Types

Wide variations in tropical rainfall, hydrography, geomorphology, and tectonics lead to the formation of many sedimentary habitats peculiar to the tropics. Expansive sandy beaches, mud banks, green and blue anoxic mud regions, mixed terrigenous-carbonate bedforms, hypersaline lagoons, stromatolites and, more generally, extensive intertidal sand- and mud flats, mangroves, coral reefs, and seagrass meadows are characteristic of shallow, tropical seas. These habitats are created and altered by processes peculiar to themselves and linked to climate and oceanographic factors and the rate of terrigenous sedimentation.

Extensive sandy beaches and flats, mud flats, mangrove forests, coral reefs, and seagrass meadows are among the most iconic of estuarine and marine habitats and are distributed widely throughout subtropical and tropical latitudes. Intertidal sand and mud flats develop in conditions more quiescent than sandy beaches, fostering deposition of fine-grain sediment (Eisma 1997). The global distribution of sandy shorelines (Figure 4.3) shows that 31% of the ice-free world shoreline is sandy, with Africa having the highest presence (66%) of sandy beaches (Luijendijk et al. 2018). The global distribution shows a distinct relation with latitude and hence to climate, while there is no relation with longitude. The relative occurrence of sandy shorelines increases in the subtropics and from 20 to 40° latitude with maxima near 30 °S and 25 °N. They are relatively less common (<20%) in the humid tropics where muddy substrates are most abundant because of high river discharge and precipitation. The global distribution of sandy shorelines agrees with the earlier determination of latitudinal variation of sediments on the inner continental shelf (Figure 4.1).

Tidal flats have a range of complex sedimentary structures, such as cross bedding, lenticular bedding, and mud/silt couplets that reflect depositional history. Mud flats can be sheltered or moderately exposed and are commonly found in tropical estuaries, tidal inlets, and river deltas. Tidal flats occur in macro-tidal settings where local areas of deposition occur where sedimentologic processes are active and stratigraphic sequences are developing. There are several types of tidal flats in macro-tidal regions: low tidal sand flats, mud/sand slopes encompassing the low to mid intertidal zones, mangrove-fringed mud flats, and high intertidal and supratidal salt flats. Hypersaline tidal flats have recently been found to be important storage sites for salt, sediments, carbon, and nutrient elements (Brown et al. 2021). Unlike tidal flats in micro- and meso-tidal settings, physical processes dominate biological processes such as bioturbation in salt flats.

While tidal flats occur throughout the marine geosphere, roughly 60% (about 75000 km^2) of sand, rock, and mud flats occur within the low latitudes. Of a global total area of 127921 km^2 (Murray et al. 2019), 49.2% of the world's tidal flats are in Indonesia, China, Australia, the United States, Canada, India, Brazil, and Myanmar; about 70% are found on three continents: Asia (44% of total), North America (15.5%), and South America (11%). Tidal flats are declining in global area; 16% of tidal flats were lost between 1984 and 2016 (Murray et al. 2019). In addition to direct losses from coastal development, increased subsidence and compaction of intertidal sediments, reductions in sediment supply, altered sediment deposition and erosion rates, vegetation loss, coastal eutrophication, and sea-level rise are also likely drivers of intertidal flat loss. Tidal flats have responded by local migration, but not quickly enough to offset ongoing losses. For example, the highly dynamic tidal flats of the Meghna River estuary in Bangladesh have migrated extensively since 1984, but now

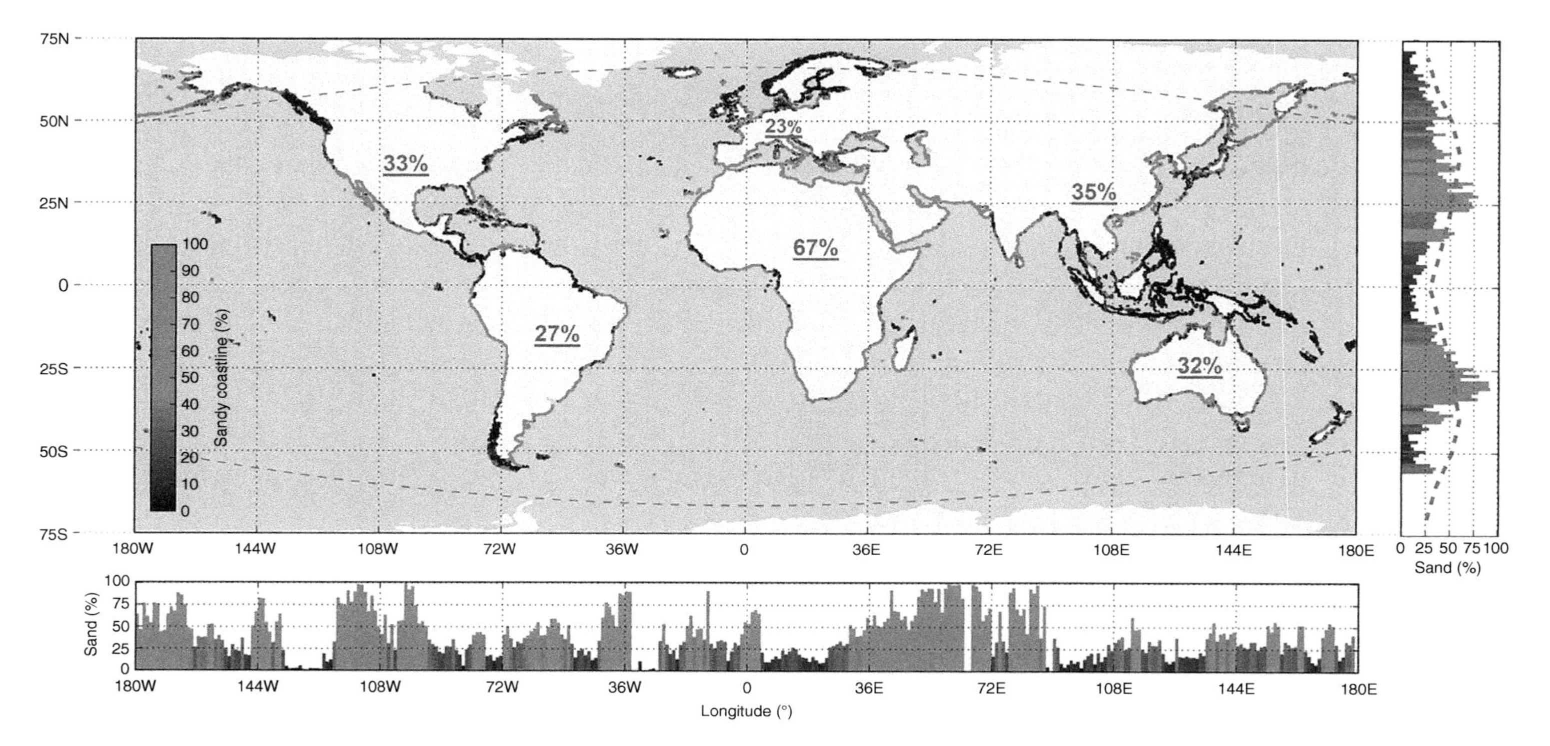

FIGURE 4.3 Global distribution of sandy shorelines. The coloured dots along the world's shores represent the local percentage of sandy shorelines (light brown is sand, dark brown is non-sand). The subplot to the right presents the relative occurrence of sandy shores per degree latitude where the dashed line shows the latitudinal distribution. The lower subplot presents the occurrence of sandy shores per degree longitude. The curved grey lines in the main plot represent the boundaries of the ice-free shorelines. The underlined percentages indicate the percentages of sandy shores averaged per continent. *Source*: Luijendijk et al. (2018), figure 1, p. 4. Licensed under CC BY 4.0. © Springer Nature Switzerland AG.

occur within only 17% of their initial extent despite expanding in area by 21% due to the rate of sediment delivery exceeding the rates of subsidence and sea-level rise; seaward migration of tidal flats has been slow, influenced by altered sediment deposition patterns due to coastal development and local expansion of mangroves (Murray et al. 2019).

The global distribution of mangroves (Figure 4.4) indicates a tropical dominance with major latitudinal limits relating best to major ocean currents and the 20 °C seawater isotherm in winter (Bunting et al. 2018) with most mangroves occurring in Southeast Asia and the Americas, including the Caribbean. Both mangroves and seagrasses grow best in quiescent environments where hydrology is favourable for their development (Chapter 2). Estimates of global mangrove area range from 83 495 km^2 (Hamilton and Casey 2016) to 135 870 km^2 (Worthington et al. 2020). A global typology of mangroves has found that as of 2016, 40.5% of mangrove ecosystems were deltaic, 27.5% were estuarine, and 21.0% were located on open coasts, with lagoonal mangroves occupying only 11% of global mangrove area (Worthington et al. 2020); mangroves in carbonate settings represent just 9.6% of global coverage.

In contrast, the known area of tropical seagrass meadows is poorly constrained by large areas remaining unmapped and inconsistent methodology being used (McKenzie et al. 2020). The global area of seagrasses most likely is 160 387 km^2 with a moderate to high confidence (Figure 4.5), but possibly 266 562 km^2 with lower confidence (McKenzie et al. 2020). Seagrass meadows in the tropical Atlantic (44 222 km^2 with moderate–high confidence) and in the tropical Indo-Pacific (87 791 km^2 with moderate–high confidence) make up 82% of the global total and tropical seagrass meadows make up 85% of the global total if the low confidence estimates are included (McKenzie et al. 2020).

Like seagrasses and mangroves, the global distribution of coral reefs (Figure 4.6) reflects the influence of long-term environmental conditions that are most suitable for establishment and growth. Species richness of corals is greatest in the Coral Triangle in Southeast Asia (Figure 4.6, dark red area). Corals are found in three principal areas: the Caribbean, the Red Sea, including the Indian Ocean islands such as the Seychelles, and in the Indo-West Pacific. Corals are found in a broad band throughout the tropics, although there are areas out of this band where warm currents permit the existence of corals, such as on the west and east coasts of Australia and as far north as the southernmost islands of Japan. Coral reefs cover about 600 000 km^2 of the global ocean, comprising only about 0.17% of total ocean area and roughly 15% of

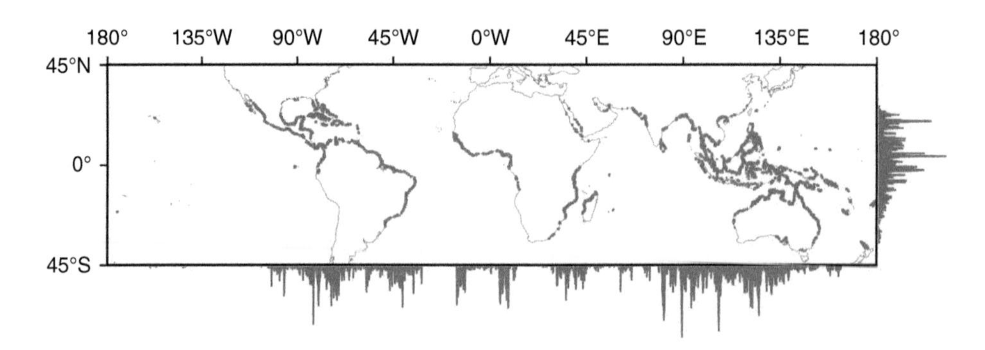

FIGURE 4.4 Global distribution of mangrove forests. The green bars outside the box indicate relative distribution of mangrove with latitude (right) and longitude (bottom). *Source:* Bunting et al. (2018), figure 4, p. 10. Licensed under CC BY 4.0. © MDPI.

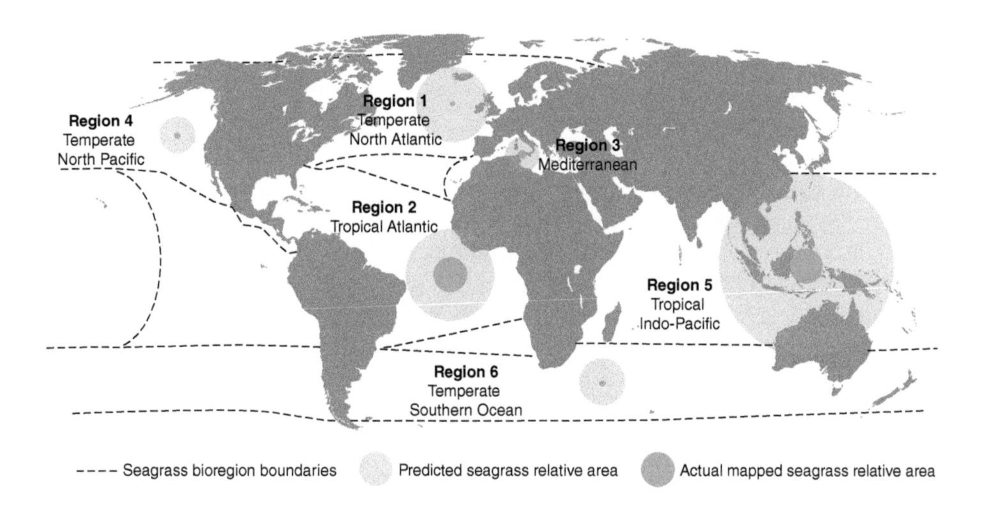

FIGURE 4.5 Global seagrass area relative to the maximum potential seagrass area within each of the global seagrass bioregions which are represented by scaled circles. Seagrass area in each bioregion: 1. Temperate North Atlantic = 3229 km^2; 2. Tropical Atlantic = 44 222 km^2; 3. Mediterranean = 14 167 km^2; 4. Temperate North = 1866 km^2; 5. Tropical Indo-Pacific = 87 791 km^2; 6. Temperate Southern = 9112 km^2. *Source*: McKenzie et al. (2020), figure 4, p. 7. Licensed under CC BY 4.0. © IOP Publishing.

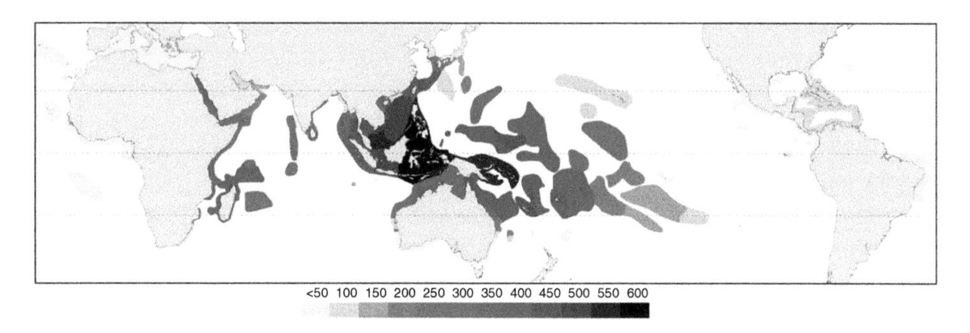

FIGURE 4.6 Global distribution of coral reefs. The coloured areas indicate species richness of hermatypic coral reefs in each region. *Source*: www.coralsoftheworld.org (accessed 3 January 2021). © Japanese Coral Reef Society.

the world's sea floor <30 m (Spalding et al. 2001). This small area belies their importance to the structure and function of the tropical coastal ocean as they are important areas of fisheries, marine biodiversity, and carbonate production.

Six major physical factors limit the development of coral reefs: temperature, light, depth, salinity, sedimentation, and emersion (Montaggioni and Braitwaite 2009). Hermatypic corals are found in waters bounded by the 20 °C isotherm with the lower temperature set at 18 °C for reef formation. Optimum reef development occurs at mean temperatures of 23–25 °C, although some corals can tolerate higher temperatures. The absence of reefs from the west coasts of Central and South America and the west coast of Africa is due to the upwelling of cold water. Reefs are also depth and thus light-limited and do not form in waters > 50–100 m. Most grow in 25 m or less and are restricted to continental margins or islands due to light limitation. Light compensation depth is the water depth where light intensity is 1–2% of incident light at the ocean surface. Corals do not grow below 50–60 m in the Pacific but grow as deep as 100 m in the Caribbean due to greater light penetration.

Another factor limiting the development of coral reefs is salinity as corals have a narrow tolerance for salt. Freshwater runoff can occur in proximity to reefs which they can tolerate for short periods of time, but generally corals thrive in areas where there is little if any decreases in salinity and increases in sedimentation that clogs feeding structures and reduces available light by turbidity or mixing (Montaggioni and Braitwaite 2009). Corals can also tolerate short periods of exposure to air, but generally their growth is limited to the tide mark of mean low water.

There are four types of coral reefs: fringing reefs, barrier reefs, patch reefs, and atolls (Sheppard et al. 2018). While they all differ in their geomorphology, they are all part of a series of forms that develop in the same basic manner. Corals will grow where conditions are suitable, especially in clear shallow waters and they can grow along tropical rocky coasts to about 45 m depth. Corals grow upward until limited by emersion into air and begin to spread outward. Fringing and barrier reefs are found along continental coasts and off islands while atolls are mostly found in the Indo-Pacific area. Atolls are oceanic and circular in shape with a series of sandy cays enclosing a deep lagoon. They form when a submarine volcano develops a fringing reef and as it sinks over time the coral will grow upward. The top of the volcano then subsides to eventually form a deep lagoon in the centre of a group of coral reefs.

Barrier reefs can be located further offshore with a broad, wide lagoon compared to fringing reefs. Patch reefs are generally oval along the axis of the prevailing winds and may have a sandy cay on the leeward side. In some areas where there is enough shelter, patch reefs can develop into islands where they become low wooded islands and may even have mangroves and seagrasses in a patchy lagoon.

Reefs display a variety of zonation patterns depending on the water depth, wave action, and exposure (Sheppard et al. 2018) but the 'classical' zonation pattern is of a reef front or slope culminating in a wave break zone, followed by a reef crest then a reef flat which leads to a back reef or lagoon (Figure 4.7). The reef front or slope extends from the low tide mark to deep water and it is here that coral growth is most rapid; the slope is dominated by large corals such as *Acropora* and *Monastrea* within the upper 15–25 m. Wave action and light intensity are reduced below this depth; light is reduced to only about 25% of the surface so only small branching corals predominate. At about 25–40 m depth, corals become patchy as light becomes scarce and there is some accumulation of sediment. Gorgonian corals can dominate at this depth range. The wave-break zone and reef crest bear the full brunt of the waves and there is often a pattern of groove and spurs which forms because of the constant wave action. The reef crest zone is exposed at low tide and varies in width from a few m to tens of m and is dominated by very hardy coral species that can withstand strong wave action. The reef flat can be tens of m in length and is one of the largest areas of the reef by area. It receives less wave action than the more forward zones but is still exposed at low tide and consists of a wide mixture of corals and turf algae and can often have quiescent pockets of sandy patches where a variety of invertebrates and fish exist. The reef flat deepens into the back lagoon where unconsolidated sediment prevails and where there can be 'bommies' or hummocks of massive coral skeletons on which grow a variety of organisms, including young corals. The back reef can be exposed at low tide and often has a dominant biota of calcified green algae, such as *Halimeda*, along with various species of seagrasses and hummocks of corals, such as *Porites*. It extends outward from the shore to the lagoon and reef flat and may be any

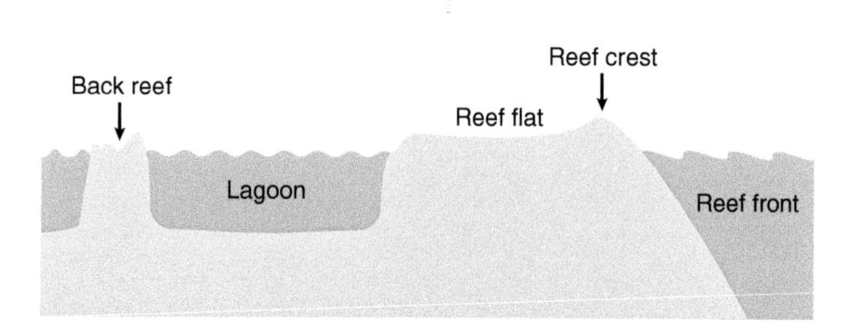

FIGURE 4.7 Schematic of an idealised coral reef showing various reef zones from the reef front to the back reef. Zones are not to scale.

area in size from a few tens to hundreds of m in length. The back reef is shallow and sheltered from wave action. Here, water circulation is less rapid, and sediment tends to accumulate, contributing to poor coral growth; benthic invertebrates are common.

The geological development of coral reefs is controlled by temperature, nutrient availability, hydrology, and changes in sea-level and ocean chemistry. Most research has focused on sea-level changes in relation to ancient reef development and evolution (Montaggioni and Braitwaite 2009). Changes in sea-level are related to the availability of habitats suitable for coral reef development and such changes, when large enough, have triggered mass extinctions (Chapter 5).

Biotic controls play a role in reef development (Montaggioni and Braitwaite 2009). The evolutionary history of coral reefs shows an increase in biological disturbance such that there was an increase during the Cretaceous and Cenozoic in predators specialised for corals, including bioeroders and herbivores. These specialised organisms influenced the community structure of coral reef ecosystems. Such organisms limit the distribution and abundance of sessile organisms, such as corals, which require a stable substrate and quiescent sedimentological conditions.

Coastal lagoons can be most simply defined as natural enclosed or semi-enclosed water bodies parallel to the shoreline. Lagoons are sometimes confused with other coastal ecosystems, such as estuaries and coral reef lagoons. Thus, coastal lagoons can be most precisely defined as 'shallow aquatic ecosystems that develop at the interface between coastal terrestrial and marine ecosystems and can be permanently open or intermittently closed off from the adjacent sea by depositional barriers' (Esteves et al. 2008). The waters of coastal lagoons can span the range of salinities from fresh to hypersaline depending on the balance of hydrological drivers, including local precipitation, river inflow, evaporation, groundwater discharge, and seawater intrusion through or directly via the depositional barrier.

The geophysical characteristics that contribute to the formation and maintenance of a coastal lagoon are important and help in identifying different types of lagoons (Eisma 1997). The first characteristic is whether the coastal lagoon has a connection to the sea. Some lagoons are lentic non-tidal, that is, without permanent connection to the sea or lentic micro-tidal, permanently connected to the sea. The second characteristic of a coastal lagoon is its origins. Most lagoons have originated from the flooding of lowland coastal areas due to the global rise in sea-level during the Late Quaternary marine transgression (Esteves et al. 2008). Lagoons originating in this way generally have large surface areas and are located parallel to the coastline which increases the probability for marine intrusions through or over the depositional barrier. Other lagoons have originated by the build-up of sediments at the mouths of rivers due to the working of waves and tides to form a barrier. Such lagoons have a branched configuration and a high perimeter to area ratio and are formed by the flooding of river valleys. Due to their geomorphology, high levels of dissolved and particulate materials from land enter such coastal lagoons.

Perhaps no other coastal environments are as complex as coastal lagoons. The heterogeneity of geomorphologies observed among coastal lagoons has created a vast array of physicochemical and ecological gradients and microhabitats crucial in supporting fisheries and humans. Coastal lagoon complexes exist in many dry tropical regions, originating as wave-cut terraces when sea-levels were lower during the Pleistocene glaciations (Eisma 1997). In the Arabian Gulf, for instance, marine terraces or 'sabkhas' surround these high salinity lagoons. Aeolian dunes migrate across the terraces under the influence of NW or 'shamal' winds. Other high salinity lagoonal pools are equally ancient, formed by similar sea-level changes isolating areas behind raised coral reefs receiving a subterranean supply of seawater seeping through coral stone.

Not all coastal lagoons are hypersaline. Large stretches of the Pacific coast of Mexico consist of lagoons frequently lying between rivers and connected by 'esteros', narrow and winding sea channels which permit ocean water to enter as a typical salt wedge and having all the characteristics of stratified estuaries. Salinities vary in relation to the dry and wet seasons. Lagoons in the wet tropics are frequently oligohaline for long periods of time. The lagoons along the north coast of the Gulf of Guinea (Ivory Coast) are situated in an equatorial climate where the annual rainfall is about 2000 mm. In Ebrié Lagoon, the largest of three main gulf systems, temperature varies little, but salinity varies with season and in different parts of the lagoon, ranging from euryhaline to oligohaline (Albaret and Laé 2003). The lagoon, like most lagoons worldwide, is frequently deoxygenated by pollution and by lack of circulation in the deeper areas. Coastal lagoons experience forcing from river inputs, wind stress, tides, the balance of precipitation to evaporation, different salinity regimes, and many human-induced changes, all of which make each lagoon unique. This is probably why no universal classification scheme for coastal lagoons has ever been developed.

Abiotic factors are central to understanding the myriad properties of coastal lagoons. Flushing of a lagoon maintains water quality and physicochemical conditions and provides a mechanism for the import and export of nutrients, plankton, and fish. The overall characteristics of a lagoon are determined by salt and heat fluxes controlling warming and cooling. Geomorphological factors that play important roles in coastal lagoons include inlet and outlet configuration, lagoon size and orientation with respect to wind direction, bottom topography, and water depth. The size of the inlet/outlet controls the exchange of water and associated dissolved and suspended material and biota. The effects of sand bar openings can have a significant effect on physicochemical variables but can also have effects on the biota. For instance, the spatial variation in pH, dissolved oxygen, and nutrients in a hypertrophic coastal lagoon in Brazil (the Grussai lagoon) was linked to anoxic and nutrient-rich groundwater discharge, the development of aquatic macrophytes, the biological activities of the phytoplankton community, and marine inputs (Suzuki et al. 1998). Whenever the sand bar closes, and the lagoon is cut off from the sea, the lagoon water becomes supersaturated with dissolved oxygen, exhibiting high pH and chlorophyll *a*, and low levels of dissolved nutrients. When the passage re-opens, there is an enrichment of dissolved inorganic nutrients and a decrease in pH and in dissolved oxygen. Within a few days, marine conditions return suggesting that biological mechanisms in the lagoon are highly efficient. Groundwater can play an equally important role in forcing physiochemical conditions in some lagoons. For example, there are two different types of groundwater in the Celestun Lagoon, Mexico: one derived from springs within the lagoon and a second characterised by moderate salinities compared to the low-salinity groundwater, mixed lagoon water, and seawater (Young et al. 2008). Groundwater discharge occurs through small and large springs scattered throughout the lagoon and the relative proportions of low versus moderate salinity groundwater vary over the tidal cycle. Substantial groundwater discharges can occur during both the dry and rainy seasons and can have a huge impact on nutrient concentrations and salinity in the lagoon.

The main boundaries of the coastal ocean (Figure 4.8) encompass the upper limit at the tidal freshwater zone (1) down to the river, estuary, and adjacent inner shelf waters, (2) with the seaward limit at the coastal boundary layer, (3) which is often delineated by a tidal front. These areas comprise the coastal zone where the seaward limit is dynamic, oscillating over time and space, especially in the wet season when it is displaced further seawards and the actual boundary layer breaks down. Boundary layers are formed when turbid coastal waters are mixed and trapped along the coast during calm conditions. These boundary layers break down not only during periods of high river discharge but also during periods of sustained strong winds. The coastal zone varies greatly in length and breadth depending on the strength and characteristics of local coastal circulation, river discharge, shelf width, climate, and latitude. On a semi-arid or arid continental shelf, the coastal zone may not be located close to shore as such shelves are often macro-tidal, with mixing of inshore and offshore waters extending to mid-shelf. In the wet tropics, the coastal zone often extends beyond the shelf edge, especially in proximity to large rivers, such as the Amazon.

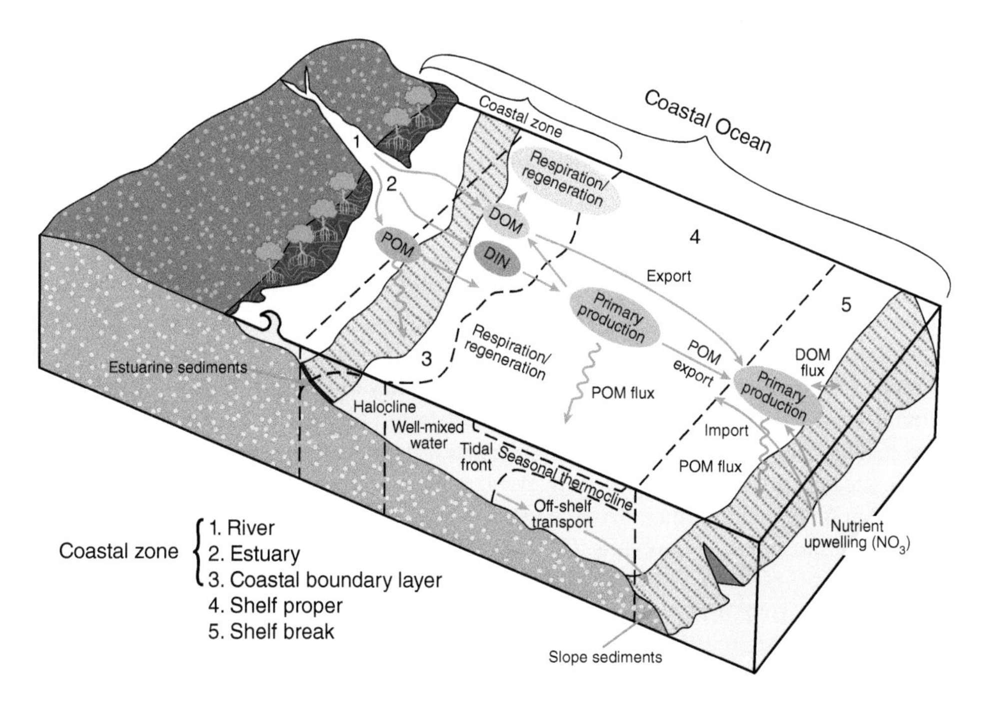

FIGURE 4.8 Idealised scheme defining the coastal ocean and the coastal zone with some key biogeochemical fluxes linking land and sea and pelagic and benthic processes. The latter are not to scale. *Source*: Alongi (1998), figure 6.15, p. 184. © Taylor & Francis Group LLC.

Coastal circulation is driven by energy derived from solar heating or gravity, barometric pressure, and the density of oceanic waters (Section 3.3). Mixing results from tides, wind-driven waves and buoyancy effects from river runoff, and mixing and circulation are thus greatly affected by geomorphology and bathymetry of the coastal zone. There are three main types of estuarine and coastal circulation: gravitational (due to river runoff), tidal (tidal pumping), and wind-driven (Walsh 1988). Tidal circulation is usually the most important, with interaction by coastal boundaries generating turbulence, advective mixing, and longitudinal mixing and trapping, with the latter setting up coastal boundary layers. Coastal systems may be classified as tide-dominated, wave-dominated, or river-dominated or a mixture of each, depending upon coastal geomorphology and local hydrography.

The boundaries of the coastal ocean are somewhat arbitrary, driven by the energetics of a very dynamic sea. The coastal zone can extend to the shelf edge under extreme circumstances, but for the most part extends to the inner shelf. Oceanic and estuarine waters intermingle on the shelf proper and tongues of oceanic water regularly or irregularly intrude onto the outer shelf but can sometimes intrude as far as the middle of the continental shelf (Walsh 1988).

4.3 Nutrients

Very sharp gradients in temperature, salinity, dissolved oxygen, and nutrients exist in tropical waters, partially reflecting high local and regional variability in precipitation and high solar insolation. Sharp thermoclines and haloclines coincide with strong vertical discontinuity maintained throughout most of the year, except where equatorial and coastal upwelling force cooler and more nutrient-rich water to the surface, or where waters from central oceanic gyres intrude into humid regions to become warmer and more dilute. Vertical stratification often breaks down in shallow coastal waters, especially during the wet season, and during the dry season when trade winds are sustained. Great variability in salinity and its ability to adjust rapidly to changes in wind-induced motion and temperature characterises tropical surface waters that are always warm (25–28 °C) and often less saline (33–34).

The global distribution of sedimentary organic carbon and nitrogen is not related to latitude but dependent on water depth, grain size, terrestrial runoff, and hydrography (Alongi 1990; Burdige 2006). The highest concentrations of organic matter in sediments, as in the water column, are in regions of coastal upwelling and in proximity to rivers, and more generally contributes to patterns of pelagic primary productivity. While there are no latitudinal trends, highest sediment carbon and nitrogen have been measured in mangroves and seagrass meadows and the lowest in carbonate, mainly reef, deposits. In nearshore subtidal sediments, the highest values have been measured off the east and west coasts of India where mud banks occur and where organic pollution prevails. Carbon concentrations > 5% and nitrogen levels > 1% by sediment DW are not uncommon in tropical inshore muds and in vegetated deposits. Total phosphorous concentrations are also frequently high in areas of domestic waste, such as in many of the polluted estuaries of Southeast Asia.

Seasonal variations in particulate organic matter, particularly in estuaries, are greatly influenced by monsoonal rains. In Indian estuaries (Gireeshkumar et al. 2013) total organic matter and organic carbon (C) and nitrogen (N) levels decline during the monsoon season due to scouring of sediments during flood discharge. Low levels of organic matter and nutrients are common in carbonate deposits and are generally lower than in quartz sand and mud of equivalent grain size. The ratios of C:N and N:P (phosphorus) vary greatly in tropical sediments, as they do in other latitudes; these variations reflect the relative importance of terrestrial compared to the marine origin of the deposited organic matter (Gireeshkumar et al. 2013).

Concentrations of dissolved inorganic nutrients are normally lower in tropical sediments than in temperate sediments of equivalent grain size. In tropical sediments, concentrations of all constituents are in the micromolar range, whereas interstitial nutrients are usually in the millimolar range in temperate sediments. Lower nutrient concentrations in sediments as well as in the water column reflect the fact that microbial decomposition and thus turnover of the nutrient pools are faster in the tropics due to warmer temperatures and highly productive microbial assemblages (Alongi 1990; Pratihary et al. 2009). It may also be partly due to phytoplankton communities that are of generally smaller size than the net-sized phytoplankton of temperate waters, with generally less deposition of phytoplankton-derived detritus to the seabed (Alongi 1990; Pratihary et al. 2009). This is reflected also in the fact that nutrient regeneration in the seabed is low compared with regeneration from temperate deposits. As in terrestrial ecosystems in the tropics, it is likely that nutrients in tropical marine ecosystems are tied up in living plant and microbial biomass.

4.4 Tropical River Loads, Plumes, and Shelf Margins

Tropical rivers of various sizes occupy a significant fraction of the world's coastlines, but tidal rivers located in the wet tropics have the most impact on the geology and hydrology of the global ocean. 69.1% of the world's river water ($26\,084 \times 10^3$ km a^{-1}; Laruelle et al. 2013), laden with nearly 60% of the world sediment discharge (10×10^9 tonnes a^{-1}), enters the tropical coastal ocean annually, mostly from the largest rivers (Table 4.1). The Amazon is the world's largest river, but most tropical river water and sediment enters the global ocean from the Indo-Pacific archipelago where high relief and rainfall produce high freshwater and sediment yields. Other major river/ocean boundary regions are in north-eastern South America and west-central Africa. The Amazon alone accounts for a disproportionate amount of the global flux, but the smaller mountainous rivers in Southeast Asia, ignored until recently, account for a greater proportion (43%) of the present sediment discharge estimates.

Tropical estuarine (349.4×10^3 km^2) and watershed ($58\,707 \times 10^3$ km^2) areas constitute 34.5% and 52.0% of the world's totals, respectively (Laruelle et al. 2013). Tropical continental shelf area ($11\,094 \times 10^3$ km^2) and volume ($720\,576$ km^3) constitute 36.6 and 18.7% of the world's totals. The small percentage of shelf volume is due to the fact the tropical shelves are on average narrower and shallower than shelves of higher latitude (Laruelle et al. 2013).

The dissolved loads of wet tropical rivers constitute about 65% of the world's total (Huang et al. 2012). The proportion of water and sediment discharged from tropical rivers are a likely underestimate as many small- and medium-sized tropical rivers remain ungauged (Latrubesse et al. 2005).

The relative importance of small mountainous rivers to the coastal ocean is exemplified on the islands of New Guinea and Timor. On the island of New Guinea,

TABLE 4.1 **Estimates of water ($km^3\ a^{-1}$) and suspended sediment (10^9 tonnes a^{-1}) discharge from gauged tropical rivers. Rivers are ranked by water discharge.**

River	Country	Water discharge	Sediment yield
Amazon	Brazil	6300	1200
Zaire	Zaire	1300	43
Orinoco	Venezuela	1100	210
Brahmaputra	Bangladesh	630	540
Mekong	Vietnam	550	110
Ganges	Bangladesh	490	520
Ayeyarwady	Burma	430	360
Tocantins	Brazil	370	75
Pearl	China	300	69
Zambesi	Mozambique	220	20
Salween	Burma	210	180
Fly	New Guinea	170	120
Niger	Nigeria	160	40
Magdalena	Colombia	140	140
Hungho	Vietnam	123	130
Song Hong	Vietnam	120	50
Rajang	Borneo	110	30
Purari	New Guinea	105	105
Sao Francisco	Brazil	97	1
Godavari	India	92	47
San Juan	Colombia	82	16
Atrato	Colombia	76	11
Sepik	New Guinea	72	45
Essequibo	Guyana	70	4.5
Usumacinta	Mexico	67	6.2
Sanaga	Cameroon	65	6
Marowijne	French Guinea	57	1.4
Corantijn	Guyana	47	1.1
Mahanadi	India	47	16
Tigris-Euphrates	Iraq	46	1

TABLE 4.1 (*Continued*)

River	Country	Water discharge	Sediment yield
Doce	Brazil	45	10
Papaloapan	Mexico	44	6.9
Kikori	New Guinea	43	33
Sai Gon	Vietnam	42	3
Patia	Colombia	40	21
Volta	Ghana	40	1.6
Parnaiba	Brazil	40	3
Brahmani	India	36	7
Coco	Honduras	36	6.5
Rufiji	Tanzania	35	17
Tsiribihina	Madagascar	31	12
Araguari	Brazil	30	0.5
Chao Phraya	Thailand	30	3
Nile	Egypt	30	0.2
Grande de Matagalpa	Nicaragua	29	5.3
Oyapoc	French Guiana	28	0.5
Paraiba	Brazil	28	4
Mira	Colombia	27	9.7
Escondido	Nicaragua	26	4.7
Grijalva	Mexico	23	1.3
Jequitinhonha	Brazil	23	5
Narmada	India	23	15
Ca	Vietnam	22	4
Senegal	Senegal	22	3
Cauweri	India	21	0.4
Pampanga	Philippines	21	1.4
Prinza Polka	Nicaragua	21	3.6
Mangoky	Madagascar	20	10
Panuco	Mexico	19	6.6
Ikopa	Madagascar	19	15
San Juan	Nicaragua	18	4.9

(*continued*)

TABLE 4.1 (*Continued*)

River	Country	Water discharge	Sediment yield
Pahang	Malaysia	18	3
Kelantan	Malaysia	18	2.5
Ma	Vietnam	17	3
Mearim	Brazil	17	0.7
Bengawan Solo	Indonesia	15	19
Coppename	Suriname	15	0.4
Suriname	Suriname	14	0.3
Thu-Bon	Vietnam	14	2
Gurupi	Brazil	14	10
Lempa	El Salvador	14	7
Mae Klong	Thailand	13	8.1
Balsas	Mexico	13	11
Cavaly	Liberia	13	5.3
Perak	Malaysia	12	0.9
Sinu	Colombia	12	4.2
Approuague	French Guiana	12	0.2
Mitchell	Australia	12	0.4
Porong	Indonesia	12	6.2
Krishna	India	12	1
Mana	French Guiana	12	0.1
Berbice	Guyana	11	0.2
Damodar	India	10	28
Sassandra	Ivory Coast	10	2.9
Grande de Terraba	Costa Rica	10	1.9
Cross	Nigeria	10	7.5
Subarnarekha	India	10	3
Indus	Pakistan	5	10
Limpopo	Mozambique	5	33

Source: Alongi (1990), Milliman and Farnsworth (2011) and Liu et al. (2020). © John Wiley & Sons.

the ten largest rivers contribute only 35% to the island's total river yield of 1.7×10^9 t of sediment to the adjacent coastal zone as discharge from roughly 240 smaller rivers make up the balance (Milliman 1995). There are no large rivers on the much smaller island of Timor and the island discharges much smaller amounts of water ($170\,km^3\,a^{-1}$) and sediment ($133 \times 10^6\,t\,a^{-1}$), but area-specific rates, including carbon and dissolved and particulate nutrients, are much higher than in New Guinea due to very high rates of deforestation and land degradation (Alongi et al. 2013). Borneo is a special case where despite moderate relief the sediment yield is high (even under rainforest), probably because of continuous uplift.

Regardless of size, a great variety of physical processes, ultimately driven by climate, make these tropical coastal margins unique compared to coastal settings of higher latitude. The climate of the equatorial region is characterised by high rates of rainfall, solar insolation, and temperature. By virtue of these characteristics and global position, Coriolis forces are small, and winds are dominated by easterly trade winds (Chapter 2). These physical forces, coupled with the enormous loads of freshwater and sediment draining from the land, produce extensive buoyant plumes, in some instances, extending beyond the shelf edge. Rapid rates of sediment accumulation and high rates of nutrient flux and primary productivity are but a few of the unique characteristics of these river-dominated coastal systems.

Tropical rivers worldwide drain a variety of geologic/geomorphologic settings: (i) orogenic mountain belts, (ii) sedimentary and basaltic plateau/platforms, (iii) cratonic areas, (iv) lowland plains in sedimentary basins, and (v) mixed terrains (Latrubesse et al. 2005). These types of rivers show high but variable peak discharges during the rainy season and a period of low flow during the dry season; some rivers show two flood peaks during the year, a main one and a secondary flood peak.

Tropical rivers exhibit a variety of channel forms and, consequently, a variety of different delta and mouth morphologies (Latrubesse et al. 2005). In most cases, rivers morph from one form to another over time so they are difficult to classify. Two main settings are rivers that discharge onto a tectonically active margin and those that discharge onto a passive margin (Leithold et al. 2016). Active margins are narrow and passive margins are wide.

Exactly how and where the discharged sediment deposits onto the adjacent shelf are not well understood, although it appears that sediments being discharged onto a wide, passive margin deposit in shallow waters often proximal to the river mouths, while a significant, if highly variable, proportion of sediments discharging onto narrow, active margins are transported to the adjacent shelf slope and deep sea (Wright and Nittrouer 1995; Leithold et al. 2016). The fate of sediment seaward of river mouths involves five stages: (i) supply via plumes, (ii) initial deposition, (iii) resuspension and transport by marine hydrographic processes, (iv) sediment that comes back on shore from far away and/or via tidal pumping, and (v) long-term net accumulation (Figure 4.9). These processes vary greatly with river regime and coastal physics. The Amazon plume extends along the NW portion of the coast and far seaward. And although tide range is large and mixing processes are relatively intense, the enormous volume of outflow results in the effluent filling the entire water column beyond the mouth, before ascending above the seawater. It continues to expand as a thick (5–10 m) buoyant plume which reaches more than 300 km offshore and about 1000 km to the NW entrained by the North Brazil Current (NBC). However, the plume is highly variable over time and space due to wind forcing and tidal variations in bottom drag and vertical mixing. There is also seasonality in

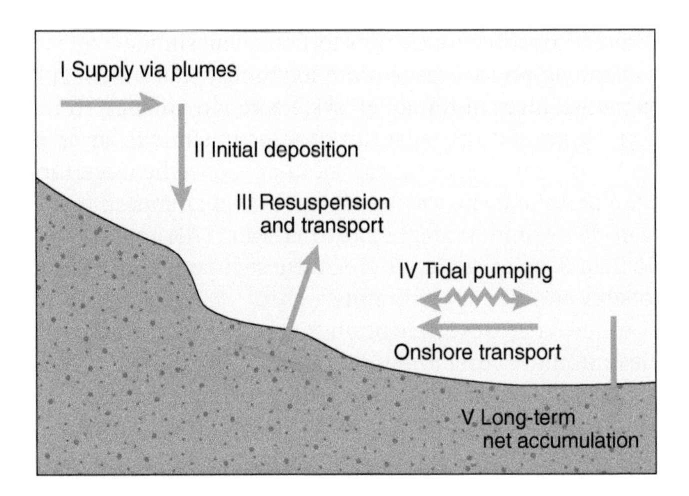

FIGURE 4.9 Major stages in sediment dispersal of river sediments in the coastal ocean. *Source*: Wright and Nittrouer (1995), figure 3, p. 503. © Springer Nature Switzerland AG.

discharge, with maximum discharge during May–June; peak sediment discharge precedes peak water discharge by about a month or more, so the volume-specific rates of discharge vary considerably. Strong tidal currents and waves generated by the easterly trade winds dominate the Amazon shelf resulting in variable spatial and temporal distribution of sediments on the shelf. Intense reworking of sediment on the inner shelf allows only temporary deposition. Once resuspended, the NBC carries sediment far to the NW. Despite high alongshore current flux, erosion occurs along the shore although erosion–deposition episodes depend on the strength of the NBC. At least 50% of sediment accumulation takes place on the mid-shelf (depth 30–50 m) seaward and NW of the mouth. Remaining sediment is probably stored in the tidal reaches of the lower river.

The Purari River system on New Guinea is different, being much smaller (Table 4.1) and having a mountainous watershed (Wright and Nittrouer 1995). The Purari delta is heavily vegetated by mangroves and crossed by an intricate network of interconnected channels which trap most of the river sediment load. Saltwater intrusion is prevented by large shallow and mobile sand banks within and outside the river mouth. Fine sediments are carried onto the inner Gulf of Papua shelf as muddy, low-salinity plumes that are broken up by the coastal oceanographic regime which is dominated by onshore-directed southeast trade winds for most of the year. Thus, much of the sediment remains trapped relatively close inshore as a turbid band and is advected alongshore. Plumes enter tidal channels on flood tides supporting the extensive mangroves within the delta. Some sediment is transported directly offshore especially during summer when the trade winds are weak, and rainfall is at its peak.

In the Ayeyarwady River system, little modern sediment accumulates on the shelf immediately off the delta. Instead, a major mud wedge with a distal depocentre of up to 60 m thickness has been deposited seaward of the adjacent Gulf of Martaban, extending to about 130 m into the Martaban Depression (Liu et al. 2020). However, no river-derived sediment has been found in the adjacent deep Martaban Canyon. There is a mud drape wrapping around the narrow western Myanmar shelf in the eastern Bay of Bengal. Unlike other large river systems in Asia, such as the Yangtze and Mekong, there is a bidirectional transport and depositional pattern controlled by local currents and tides, and seasonally varying monsoonal winds and waves.

In the Ganges–Brahmaputra River system, approximately one-third of total sediment discharge is sequestered within the flood plain and delta plain (Rahman et al. 2018). The remaining load appears to be apportioned between the accumulating subaqueous delta and the deep-sea Bengal fan via a nearshore canyon. The roughly equal partitioning of sediment among the flood plain, shelf, and deep sea reflects the respective influence of an inland subsiding tectonic basin, a wide shelf, and a deeply incised canyon system (Rahman et al. 2018).

Plumes of other large tropical river systems may be dispersed laterally due to local coastal currents and hydrography. For instance, the typical seasonal orientation of the Zaire River plume is northward for most of the year, except during February–March when the plume has a large westward extension onto the narrow shelf (Denamiel et al. 2013). The northward extension of the plume is explained by a buoyancy-driven upstream coastal flow and the combined influences of the ambient ocean currents and the wind. During February–March, the surface ocean circulation drives the westward expansion of the plume and the presence of the deep Congo canyon increases the intrusion of seawater into the river mouth.

Off the Mekong delta, a similar lateral plume occurs throughout most of the year, with a net deposition SW of the river mouth down the Ca Mau peninsula (Szczuciński et al. 2013). In summer, a large amount of fluvial sediment is deposited near the Mekong River mouth, but in the following winter, strong mixing and coastal currents lead to resuspension and south-westward dispersal of previously deposited sediment. Strong wave mixing and downwelling-favourable coastal current associated with the more energetic NE monsoon in the winter are the main factors controlling post-depositional dispersal to the SW.

For tropical river plumes, coastal hydrography plays an important role in governing how the discharged sediment is dispersed onto the adjacent shelf (Wright and Nittrouer 1995; Hetland and Hsu 2014). Oceanic processes that resuspend and transport sediment act in concert with maximum plume outflow, causing sediments to be dispersed farther from the delta mouth. Sediments can be dispersed over relatively wide areas that extend considerable distances from the delta; some rivers do not have a subaerial delta or do not protrude much beyond the regional coastline. Alongshore dispersal of sediments takes place primarily on the inner shelf but near the delta mouth the depth is too shallow and energetic for fine-grain sediments to be deposited, except temporarily (Hetland and Hsu 2014). Much of the material transported alongshore becomes sequestered within tidal currents, and off the Amazon, muds that are transported alongshore ultimately accumulate, forming expensive, accumulating mud banks hundreds of km to the NW of the mouth. Despite a high level of instability, these mud banks are colonised by mangroves. However, the Amazon is characterised by accumulations of sediments as subaqueous delta deposits at relatively large distances on the mid-shelf. A large fraction of Amazon sediments reaches the mid-shelf due to the energetic currents and waves that sustain sediments in suspension until they reach relatively deep water (Wright and Friedrichs 2006). Much of the observed diversity of sediment dispersal and accumulation is attributable to variations in coastal energy regimes and to the temporal sequencing of river discharge relative to oceanographic transport processes. Sediment transport in the southeast section of the Amazon coastal zone is greatly affected by tidal asymmetry, seasonal variation of the wind and wave regime, and river discharge (Gomes et al. 2020). Climate and geological configuration have resulted in numerous estuaries and large-mangrove-lined coastal plains that partially divided the estuarine basins, which are connected by tidal channels within the Amazon delta. Convergence of transported sediment occurs within a channel connecting the estuaries, resulting in mud retention and further delivery to the mangrove-covered plains, with a net flux of suspended sediments between estuaries. The connectivity between estuaries via channels is a key process to redistribute muddy sediments along this coastal sector, which helps to explain the evolution and maintenance of the relatively homogenous and widespread progradation of mangroves along the coast.

Mechanisms that dominate the short-term spreading and mixing of riverine sediment may differ from the mechanisms that determine the longer-term dispersal of sediment. Sediment records from the South China Sea show that strong monsoons are associated with intensified reworking of pre-existing floodplain sediment over millennial timescales (Clift 2020). Strong monsoons result in deposition of more altered material that is also delivered at higher rates than during drier periods. Millennial-scale changes in monsoon strength result in changes in the weathering regime but not fast enough to account for the changes seen in the sediments preserved in Asian deltas; instead, monsoon-modulated recycling dominates. Over longer time periods ($>10^6$ year) strengthening of the monsoon is linked to faster bedrock erosion and increased sediment flux to the ocean.

An additional complication is the fact that most tropical river systems are heavily affected by humans; few, if any, tropical rivers are pristine. Human disturbances such as the construction of dams and deforestation can greatly impact water and sediment discharge. Increased greenhouse gas emissions are projected to impact twenty-first century precipitation distribution, altering riverine water and sediment discharge. Modelling indicates that increasing global warming will lead to more extreme changes and greater rates of increasing or decreasing changes in fluvial discharge (Moragoda and Cohen 2020). At the end of the twenty-first century under all IPCC climate change scenarios, mean global river discharge will increase by 2–11% relative to the 1950–2005 period, while global suspended sediment flux will increase by 11–16% under pristine conditions. Combining the effects of climate change with natural and anthropogenic impacts, tropical rivers have and will continue to be affected greatly in future. For example, natural and human-induced factors have greatly altered the discharge of the Patía River of Colombia (Restrepo and Kettner 2012). In 1972, the river flow was diverted to an adjacent river resulting in several environmental changes, such as coastal retreat along the abandoned delta, the formation of barrier islands with exposed peat soils in the surf zone, abandonment of former active distributions in the southern delta plain with associated closing of inlets and formation of ebb tidal inlets, breaching events on the barrier islands, and accretion on the northern delta plain.

Tropical deltas are sensitive to human encroachment due to regional water management, global sea-level rise, and climatic extremes (Shearman et al. 2013; Darby et al. 2020). Vertical change within a delta is a function of the change in delta surface elevation relative to sea-level (Darby et al. 2020), derived as the sum of the rates of natural compaction and anthropogenic subsidence, eustatic sea-level change, rate of crustal deformation due to local geodynamics, and the rate of surface aggradation (Figure 4.10). A significant fraction of the world's deltas is subsiding and at risk of imminent drowning, having significant impacts on changes in mangrove area (Darby et al. 2020). In the Ayeyarwady delta, the loss of mangrove habitats and its conversion to cultivation has led to increased salinity intrusion, coastline retreat, and increased flood risk (Kroon et al. 2015). Trends in other rivers of the Asia-Pacific

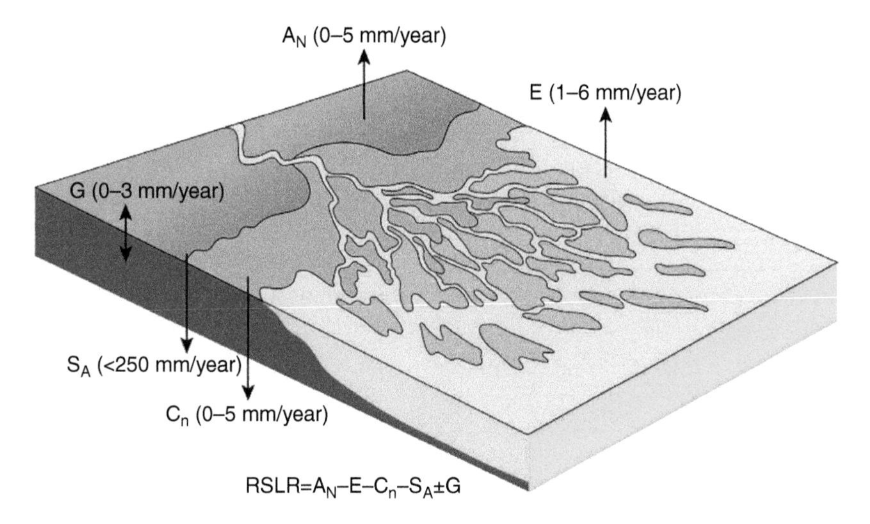

FIGURE 4.10 Idealised scheme of the factors and processes contributing to vertical changes within river deltas in the face of relative sea-level rise (RSLR). Most of the world's deltas currently have low rates of natural sediment supply and high rates of eustatic sea-level rise ($E = 1–6\,mm\,a^{-1}$) and often higher rates of human-induced subsidence ($S_A < 250\,mm\,a^{-1}$), meaning that many deltas are subsiding or in danger of drowning as sediment accretion ($A_N = 0–5\,mm\,a^{-1}$) is the only factor that can offset relative sea-level rise. Natural compaction ($C_N = 0–5\,mm\,a^{-1}$) and crustal deformation ($G = 0–3\,mm\,a^{-1}$) are also important factors. *Source*: Darby et al. (2020), figure 5.1, p. 105. Licensed under CC BY 4.0. © Springer Nature Switzerland AG.

region indicate deforestation and subsequent destabilisation of coastline (Shearman et al. 2013). Overall, Shearman et al. (2013) observed a net contraction of 76 km^2, but trends varied among different river systems. Further, some systems such as the Ganges–Brahmaputra are naturally subsiding, resulting in a high-risk situation in relation to sea-level rise. Thus, most tropical river systems are at moderate to high risk of anthropogenic change. With increasing rates of sea-level rise and more intense cyclones, tropical river systems will increasingly undergo environmental and ecological change into the foreseeable future. A model to estimate such future impacts on the Mekong delta (Bussi et al. 2021) indicates that climate change will play a secondary role compared to dams; planned dams will reduce suspended sediment fluxes to the delta by up to 50% over the next two decades.

The geomorphology of continental margins influences the biosphere by helping to mediate genetic connectivity of populations during sea-level change (Dolby et al. 2020). Combining genetic data, geographical information systems-based estuarine habitat modelling, and paleobiologic and recent effects of sea-level change on evolution, Dolby et al. (2020) tested the relation between overall shelf area and species richness using data of 1721 fish species. They found an 82% global reduction of estuarine habitat abundance at low-stand relative to high-stand periods and found that large habitats change in size much more than small habitats. Narrow continental margins have significantly less habitat at high-stand and low-stand than wide margins and narrow margins significantly associate with active settings, effectively linking tectonic setting to habitat abundance. Narrow margins host greater species richness. Dolby et al. (2020) offer three possible explanations for this finding. First, physical isolation imposed by narrow margins may facilitate the formation of new species over time. Second, the size stability of small habitats, which disproportionately occur on narrow margins, may increase and retain species extirpated in the more variable habitats on wide margins. Third, the smaller habitats on narrow margins may facilitate greater species richness through greater habitat heterogeneity. The concept of narrow shelf margins as a 'diversification pump' is in opposition to previous paleontological information that generally argued areal restrictions were unimportant and/or regressions led to extinctions. However, these results support the idea that the complex and peculiar relation between habitat and sea-level change depends on the inherent geomorphic properties of the coastline. These results remain to be tested but do illustrate the crucial role of the geosphere in marine ecological processes.

References

Albaret, J.J. and Laé, R. (2003). Impact of fishing on fish assemblages in tropical lagoons: the example of the Ebrié lagoon, west Africa. *Aquatic Living Resources* 16: 1–9.

Alongi, D.M. (1990). The ecology of tropical soft-bottom benthic ecosystems. *Oceanography and Marine Biology: An Annual Review* 28: 381–496.

Alongi, D.M. (1998). *Coastal Ecosystem Processes*. Boca Raton, FL: CRC Press.

Alongi, D.M., da Silva, M., Wasson, R.J. et al. (2013). Sediment discharge and export of fluvial carbon and nutrients into the Arafura and Timor Seas: a regional synthesis. *Marine Geology* 343: 146–158.

Brown, D.R., Marotta, H., Peixoto, R.B. et al. (2021). Hypersaline tidal flats as important "Blue Carbon" systems: a case study from three ecosystems. *Biogeosciences* **18**: 2527–2538.

Bunting, P., Rosenqvist, A., Lucas, R.M. et al. (2018). The Global Mangrove Watch – a new 2010 global baseline of mangrove extent. *Remote Sensing* 10: 1669. https://doi.org/10.3390/rs10101669.

Burdige, D.J. (2006). *Geochemistry of Marine Sediments*. Princeton, NJ: Princeton University Press.

Bussi, G., Darby, S.E., Whitehead, P.G. et al. (2021). Impact of dams and climate change on suspended sediment flux to the Mekong delta. *Science of the Total Environment* 755: 142468. https://doi.org/10.1016/j.scitotenv.2020.142468.

Caitcheon, G.G., Olley, J.M., Pantus, F. et al. (2012). The dominant erosion processes supplying fine sediment to three major rivers in tropical Australia, the Daly (NT), Mitchell (Qld) and Flinders (Qld) rivers. *Geomorphology* 151-152: 188–195.

Clift, P.D. (2020). Asian monsoon dynamics and sediment transport in SE Asia. *Journal of Asian Earth Sciences* 195: 104352. https://doi.org/10.1016/j.jseaes.2020. 104352.

Dada, O.A., Agbaje, A.O., Adesina, R.B. et al. (2019). Effect of coastal land use change on coastline dynamics along the Nigerian Transgressive Mahin mud coast. *Ocean & Coastal Management* 168: 251–264.

Darby, S.E., Addo, K.A., Hazra, S. et al. (2020). Fluvial sediment supply and relative sea-level rise. In: *Deltas in the Anthropocene* (eds. R.J. Nicholls, W. Neil Adger, C.W. Hutton, et al.), 103–126. Cham, Switzerland: Palgrave Macmillan.

Denamiel, C., Budgell, W.P., and Toumi, R. (2013). The Congo River plume: impact of the forcing on the far-field and near-field dynamics. *Journal of Geophysical Research, Oceans* 118: 964–989.

Dolby, G.A., Bedolla, A.M., Bennett, S.E.K. et al. (2020). Global physical controls on estuarine habitat distribution during sea level change: consequences for genetic diversification through time. *Global and Planetary Change* 187: 103128. https://doi.org/10.1016/j.gloplacha.2020.103128.

Eisma, D. (1997). *Intertidal Deposits: River Mouths, Tidal Flats and Coastal Lagoons*. Boca Raton, FL: CRC Press.

Esteves, F.A., Caliman, A., Santangelo, J.M. et al. (2008). Neotropical coastal lagoons: an appraisal of their biodiversity, functioning, threats and conservation management. *Brazilian Journal of Biology* 68: 967–981.

Flemming, B.W. (2011). Geology, morphology, and sedimentology of estuaries and coasts. In: *Treatise on Estuarine and Coastal Science*, vol. 3 (eds. E. Wolanski and D.S. McLusky), 7–38. Waltham, USA: Academic Press.

Gireeshkumar, T.R., Deepulal, P.M., and Chandramohanakumar, N. (2013). Distribution and sources of sedimentary organic matter in a tropical estuary, south west coast of India (Cochin estuary): a baseline study. *Marine Pollution Bulletin* 66: 239–245.

Gomes, V.J.C., Asp, N.E., Siegle, E. et al. (2020). Connection between macrotidal estuaries along the southeastern Amazon coast and its role in coastal progradation. *Estuarine, Coastal and Shelf Science* 240: 106794. https://doi.org/10.1016/j.ecess.2020.106794.

Hamilton, S.E. and Casey, D. (2016). Creation of a high spatio-temporal resolution global database of continuous mangrove forest cover for the 21st century (CGMFC-21). *Global Ecology and Biogeography* 25: 729–738.

Hetland, R.D. and Hsu, T.J. (2014). Freshwater and sediment dispersal in large river plumes. In: *Biogeochemical Dynamics at Major River-Coastal Interfaces: Linkages with Global Change* (eds. T.S. Bianchi, M.A. Allison and W.-J. Cai), 55–85. Cambridge, UK: Cambridge University Press.

Jolivet, M., Gardel, A., and Anthony, E.J. (2019). Multi-decadal changes on the mud dominated coast of western French Guiana: implications for mesoscale shoreline mobility, river-mouth deflection, and sediment sorting. *Journal of Coastal Research* 88: 185–194.

Kroon, M.E.N., Rutten, M.M., Stive, M.J.F. et al. (2015). Scoping study on coastal squeeze in the Ayeyarwady Delta. *E-Proceedings of the 36th IAHR World Congress,* The Hague, The Netherlands (28 June–3 July 2015). Red Hook, NY: Curran Associates Inc.

Laruelle, G.G., Dürr, H.H., Laurerwald, R. et al. (2013). Global multi-scale segmentation of continental and coastal waters from the watersheds to the continental margins. *Hydrology and Earth System Sciences* 17: 2029–2051.

Latrubesse, E.M., Stevaux, J.C., and Sinha, R. (2005). Tropical rivers. *Geomorphology* 70: 187–206.

Leithold, E.L., Blair, N.E., and Wegmann, K.W. (2016). Source-to-sink sedimentary systems and global carbon burial: a river runs through it. *Earth-Science Reviews* 153: 30–42.

Liu, J.P., Kuehl, S.A., Pierce, A.C. et al. (2020). Fate of Ayeyarwady and Thanlwin Rivers sediments in the Andaman Sea and Bay of Bengal. *Marine Geology* 423: 106137. https://doi.org/10.1016/j.margeo.2020.106137.

Luijendijk, A., Hagenaars, G., Ranasinghe, R. et al. (2018). The state of the world's beaches. *Scientific Reports* 8: 6641. https://doi.org/10.1038/s41598-018-24630-6.

Martini, I.P. (2014). General considerations and highlights of low-lying coastal zones: passive continental margins from the poles to the tropics. *Sedimentary Coastal Zones from High to Low Latitudes: Similarities and Differences.* London: Geological Society Special Publications 388, pp. 1–32.

McKenzie, L.J., Nordlund, L.M., Jones, B.L. et al. (2020). The global distribution of seagrass meadows. *Environmental Research Letters* 15: 074041. https://doi.org/10.1088/1748-9326/ab7d06.

McLachlan, A. and Defeo, O. (2018). *The Ecology of Sandy Shores*, 3e. London: Academic Press.

Milliman, J.D. (1995). Sediment discharge to the ocean from small mountainous rivers: the New Guinea example. *Geo-Marine Letters* 15: 127–133.

Milliman, J.D. and Farnsworth, K.L. (2011). *River Discharge to the Coastal Ocean: A Global Synthesis.* Cambridge, UK: Cambridge University Press.

Montaggioni, L.F. and Braitwaite, C.J.R. (2009). *Quaternary Coral Reef Systems, Volume 5: History, Development and Controlling Factors.* Amsterdam: Elsevier.

Moragoda, N. and Cohen, S. (2020). Climate-induced trends in global riverine water discharge and suspended sediment dynamics in the 21st century. *Global and Planetary Change* 191: 103199. https://doi.org/10.1016/j.gloplanch.2020.103199.

Muraleedharan, K.R., Dinesh Kumar, P.K., Prasanna Kumar, S. et al. (2018). Formation mechanism of mud bank along the SW coast of India. *Estuaries and Coasts* 41: 1021–1035.

Murray, N.J., Phinn, S.R., DeWitt, M. et al. (2019). The global distribution and trajectory of tidal flats. *Nature* 565: 222–225.

Pratihary, A.K., Naqvi, S.W.A., Naik, H. et al. (2009). Benthic flux in a tropical estuary and its role in the ecosystem. *Estuarine, Coastal and Shelf Science* 85: 387–398.

Proisy, C., Walcker, R., Blanchard, E. et al. (2021). Mangroves: a natural early-warning system of erosion on open muddy coasts in French Guiana. In: *Dynamic Sedimentary Environments of Mangrove Coasts* (eds. F. Sidik and D.A. Friess), 47–66. Amsterdam: Elsevier.

Rahman, M., Dustegir, M., Karim, R. et al. (2018). Recent sediment flux to the Ganges-Brahmaputra-Meghna delta system. *Science of the Total Environment* 643: 1054–1064.

Restrepo, J.D. and Kettner, A. (2012). Human induced discharge diversion in a tropical delta and its environmental implications, the Patía River, Colombia. *Journal of Hydrology* 424-425: 124–142.

Shearman, P., Bryan, J., and Walsh, J.P. (2013). Trends in deltaic change over three decades in the Asia-Pacific region. *Journal of Coastal Research* 29: 1169–1183.

Sheppard, C., Davy, S., and Pilling, G. (2018). *The Biology of Coral Reefs*, 2e. Oxford, UK: Oxford University Press.

Spalding, M.D., Ravilious, C., and Green, E. (2001). *World Atlas of Coral Reefs*. Berkeley, USA: University of California.

Suzuki, M.S., Ovalle, A.R.C., and Pereira, E.A. (1998). Effects of sand bar openings on some limnological variables in a hypertrophic coastal lagoon of Brazil. *Hydrobiologia* 368: 111–122.

Szczuciński, W., Jagodziński, R., Hanebuth, T.J.J. et al. (2013). Modern sedimentation and sediment dispersal pattern on the continental shelf off the Mekong River delta, South China Sea. *Global and Planetary Change* 110: 195–213.

Vieira, F.V., Bastos, A.C., Quaresma, V.S. et al. (2019). Along-shelf changes in mixed carbonate-siliciclastic sedimentation patterns. *Continental Shelf Research* 187: 103964. https://doi.org/10.1016/j.csr.2019.103963.

Walsh, J.J. (1988). *On the Nature of Continental Shelves*. San Diego, USA: Academic Press.

Worthington, T.A., zu Ermgassen, P.S.E., Friess, D.A. et al. (2020). A global biophysical typology of mangroves and its relevance for ecosystem structure and deforestation. *Scientific Reports* 10: 14652. https://doi.org/10.1038/s41598-020-71194-5.

Wright, L.D. and Friedrichs, C.T. (2006). Gravity-driven sediment transport on continental shelves: a status report. *Continental Shelf Research* 26: 2092–2107.

Wright, L.D. and Nittrouer, C.A. (1995). Dispersal of river sediments in coastal seas: six contrasting cases. *Estuaries* 18: 494–508.

Young, M.B., Gonneea, M.E., Fong, D.A. et al. (2008). Characterizing sources of groundwater to a tropical coastal lagoon in a karstic area using radium isotopes and water chemistry. *Marine Chemistry* 109: 377–394.

PART 2

STRUCTURE

CHAPTER 5

Biogeography and Origins

5.1 Tropical Biogeography

A latitudinal diversity gradient in the marine biosphere has long been recognised, with an increase in species richness with decreasing latitude from the poles to the tropics (Ekman 1953; Briggs 1974). The circumtropical belt of high diversity is not uniform within this gradient, as some fauna, such as in mangrove-lined estuaries, show very variable diversity. However, this gradient holds true for many vertebrates and invertebrates in inshore and shelf ecosystems and has been attributed to high water temperatures and maximum solar irradiation in proximity to the equator (Jablonski et al. 2017; Crame 2020). Within the tropics, there is also much longitudinal variation within faunal groups.

Although recognized much earlier, it was Ekman (1953) who popularised the idea of there being a distinct 'warm-water' fauna. This fauna was split into two provinces: 'The Indo-West Pacific' (IWP) and 'The Atlanto-East-Pacific' with the former encompassing the islands of the Central Pacific, the Indo-Malayan region, Hawaii, subtropical and tropical Australia, the Indian Ocean, and subtropical Japan and the latter encompassing 'Subtropical and Tropical America' and 'Subtropical and Tropical West Africa.' This idea was further refined in 1974 by Briggs who categorised the tropical ocean into four regions: 'The IWP,' 'The Eastern Pacific,' 'The Western Atlantic,' and 'The Eastern Atlantic.' It has long been understood that the richest and most diverse fauna is found in the shallow (<200 m) waters of the tropics. The tropics was defined by Briggs (1974) by the 20 °C isotherm for the coldest month of the year, and longitudinally, it was recognized that barriers exist that are effective in separating one region from another with a high degree of endemism.

Today, the separation of tropical faunas is more complex due to advances in our knowledge of species distributions, fossil evidence and evidence that has been provided by major advances in genetics (Jablonski et al. 2017). Briggs and Bowen (2013) and Veron et al. (2015) have, respectively, summarized the tropical faunal provinces based on the distribution patterns of fish and corals.

An improved knowledge base has advanced our understanding of the distribution and evolution of the Atlantic fauna (Briggs and Bowen 2013), which is now identified as consisting of four origins of Atlantic genera: (i) relict (Tethys Sea) origins

Tropical Marine Ecology, First Edition. Daniel M. Alongi.

prior to the collision of Africa and Europe about 12–20 million years ago (Ma); (ii) origins in the New World (West Atlantic-East Pacific) prior to the closure of the Isthmus of Panama about 3.1 Ma; (iii) radiations within the Atlantic; and (iv) invasions from the Indo-Pacific via southern Africa (Figure 5.1).

The relationship among the provinces highlights the geographic origin of species and the effect of both soft and hard biogeographic barriers. For example, one soft barrier is the freshwater discharge of the Amazon River which separates the boundaries between the Caribbean (CA) and Brazilian (BR) provinces. This boundary was identified by the fact that 348 reef fish species are shared between the CA and BR provinces which represent about 42% of the species diversity of the CA and 74% of the BR. The fauna is much more diverse in the CA (Luiz et al. 2012). Another soft barrier is the open-water expanse of the mid-Atlantic; the CA shares 105 reef fish species with the Tropical Eastern Atlantic (TEA). These transatlantic species account for about 27% of the shallow TEA fish fauna. The BR shares a similar fauna with the TEA. About 112 fish species are shared between them and may be considered as transatlantic.

The open sea to the east and west of the mid-Atlantic ridge provinces of Ascension and St. Helena is the next most permeable barrier within the Atlantic. Both islands share 64 and 71 fish species with the eastern and western Atlantic, respectively. Most

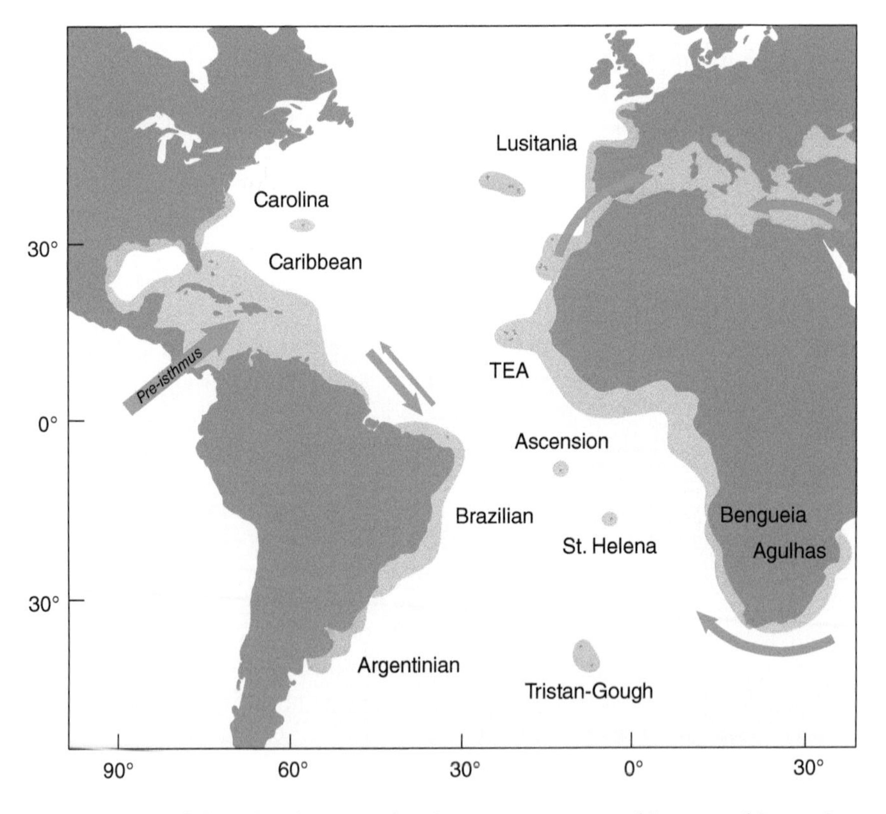

FIGURE 5.1 Map of the Atlantic Ocean showing warm-temperate biogeographic provinces (orange), tropical biogeographic provinces (lime green), and the biogeographic pathways that contribute to biodiversity in these provinces (blue arrows). Parallel arrow sizes indicate relative size of migratory flows. Acronym: TEA: Tropical Eastern Atlantic. *Source*: Briggs and Bowen (2013), figure 1, p. 1024. © John Wiley & Sons.

of these species are transatlantic in character, having relatively large latitudinal ranges; they are known to be associated with floating debris in the open ocean, which is most likely the main conduit of connectivity.

Another soft barrier is the relatively cool Benguela Current that separates the Benguela (Atlantic Ocean) from the Agulhas (Indian Ocean) provinces. At least 47 fish species have colonised the Atlantic from the Indian Ocean with at least 38 of these species also found in the TEA where they account for about 10% of the total number of species (Floeter et al. 2008). How they move from the Indian to the Atlantic is a matter of some speculation, but two perspectives have arisen (Rocha et al. 2005a; Reese et al. 2010). As indicated by phylogeographic and paleontological studies on the distributions of some molluscs and fish, there may be a colonisation route through the Agulhas Province west to Brazil. The second hypothesis is that several individual events that may be attributed to the cessation of upwelling at the end of the glacial cycles, or to warm cyclonic eddies that periodically cross the Benguela Current, fostered multiple pulses of colonisation into the Atlantic. Ocean circulation between the Atlantic and Indian Oceans is thought to be minor compared with the connectivity between the Indian and Pacific Oceans, but the minor circulation appears to be sufficient over time to result in some transfer of fauna.

West to east dispersal has been resolved based on phylogeny as the transatlantic distribution of 112 fish species shows a clear connection across the Atlantic (Bowen et al. 2006) or possibly recent colonisation between Atlantic provinces (Rocha et al. 2005b). Between the CA and BR, species flow is mainly from north to south across the soft Amazon barrier, although there has been some dispersal in the opposite direction (Rocha et al. 2008). The CA has 150 fish genera with 24 endemic and the BR has 117 genera with only 3 endemic, which suggests that most of the BR fish fauna originated in the CA.

The warm-temperate provinces of the Atlantic have been called the 'impoverished outposts of adjacent tropical areas' (Floeter et al. 2008). The Lusitania Province that borders the TEA shows that most fish species originated from the adjacent tropical province. Similarly, most of the fauna of the warm-temperate Carolina Province originates from the Caribbean. The Gulf Stream plays an important role as a mode of transport as shown by the islands of Bermuda which are clearly populated from the Caribbean. The CA has been functioning as the centre of origin for the warm provinces of the Atlantic as indicated by its great diversity and the dispersal of fauna to north, east, and south. The greatest diversity of coastal invertebrate species within the CA occurs from Cuba through the Antilles to Colombia and Venezuela (Miloslavich et al. 2010).

The tropical Atlantic continues to gain species from other sources. Evolutionary separation is stimulated by the three soft barriers of the Amazon, mid-Atlantic, and the Benguela. A few species have colonised north from the BR to add to the species richness of the CA (Rocha et al. 2008). Parapatric speciation (two subpopulations of a species evolve reproductive isolation from one another while continuing to exchange genes) may predominate due to the softness of these barriers while allopatric speciation (speciation that happens when two populations of the same species become isolated from each other due to geographic changes) may predominate across wide stretches of the open ocean between the West, Central, and East Atlantic (Briggs and Bowen 2013).

The Atlantic contains numerous genera that were once part of a 'general New World (Western Atlantic and Eastern Pacific) fauna' prior to the rise of the isthmus of Panama about 3.1 Ma; at least three tropical Atlantic fish species are thought to be relicts from the Tethys Sea (Briggs and Bowen 2013). Trans-Pacific migrations may have been the key to the fact that about 20 genera had apparently reached the New World. Caribbean reef fish are more closely related to those of the eastern Pacific than to those of the eastern Atlantic as many New World genera do not occur in the eastern Atlantic. The Caribbean Province may be the centre of 'evolutionary innovation' (Briggs and Bowen 2013) given that it has 24 endemic genera and 272 endemic species.

The greatest richness of invertebrate and vertebrate species lies in the IWP. This has been acknowledged since at least the time of Ekman (1953) and the reasons for this richness continue to be a source of rich debate (Bellwood et al. 2012; Veron et al. 2015). The IWP spans an immense area, and this is reflected in the unique distribution patterns of fish (Allen 2008; Mundy et al. 2010). Allen (2008) calculated an average range for reef fish of 9 357 070 km^2, and Mundy et al. (2010) found that of the fish fauna of the US Phoenix and Line Islands nearly 70% ranged from the Indian Ocean to their study area at the eastern edge of the SW Pacific. Obviously, a high level of connectivity is maintained across vast expanses of ocean (Eble et al. 2011) including the barriers between the eastern and western Indian Ocean and across the far eastern Pacific. In contrast, there is considerable flow of fauna between the western Pacific and the eastern Indian Ocean that are connected via the Indonesian Throughflow (ITF). Considerable research has focused on this connectivity (Williams et al. 2002; Gaither et al. 2010). The bridge between both oceans was not nearly so open as it is now; during the last glaciation about 18 000 years ago, sea levels were considerably lower (about 130 m) than at present, although there was still a narrow connection. Nevertheless, during the Pliocene–Pleistocene glaciations, the ITF was reduced, lessening the chance for ecological connectivity. Gaither et al. (2010) have provided some evidence of genetic distinctions within 15 of 18 species of fish, crustaceans, and echinoderms between both oceans, although these genetic breaks are no larger than found elsewhere in the sea (e.g., Horne et al. 2008). On the other hand, Veron (1995) considered the coral fauna of the eastern Indian Ocean to be nearly identical to that of the western Pacific underscoring the fact that there have been, and continues to be, connections between the faunas of the Indian and Pacific Oceans.

There are also connections between the western Pacific with the Hawaiian archipelago and the central Pacific with the eastern Pacific. However, the eastern Indian and western Pacific may eventually be distinguished. Briggs and Bowen (2013) point out two points that must be considered in this regard: (i) the unique possibility of overlap by distinct fauna by the fact that the Indo-Pacific barrier is different from other barriers in that it switches on and off in 100 000 year oscillations (Rocha et al. 2005b) and (ii) the Pacific fauna is expanding westwards due to the presence of Indian and Pacific taxa hybridising in the Indian Ocean (Hobbs et al. 2009). The barrier between the Indian and Pacific Oceans thus appears to be diffuse in the current day and ephemeral over evolutionary time, being spread across 25° longitude from the Sunda shelf to the Cocos/Keeling Island group. Briggs and Bowen (2013) concluded that the distinction between the western Pacific and the eastern Indian Oceans has been blurred due to repeated invasions of species, although the eastern Indian Ocean at the height of the last glaciation 18 000 years ago may have been a distinct biogeographic province.

Mangrove species diversity is higher in the IWP than in the East Pacific and Atlantic, with a global total of 80 species currently. Mangroves first appeared on the shores of the Tethys Sea, having diverged from terrestrial forbearers during the mid-Cretaceous (Duke 2017). Mangrove dispersal of the 32 genera in 18 families began during the early- to mid-Eocene (Duke 2017). Their evolution has been closely related to sea-level changes throughout geological time (Srivastava and Prasad 2019). Two biogeographic subregions for mangroves are recognised: the Indo-Malaysia from the Indian Ocean to Southeast Asia and Australasia from New Guinea and Australia to the islands of the West Pacific, fostering the notion that mangroves originated in Southeast Asia and expanded across the Pacific to the west coast of the Americas and westward to East Africa and then to the east and west coasts of the Atlantic. A further secondary hotspot for mangroves occurs in the Caribbean Central American area (Duke 2017). The floras of these two relatively rich subregions constitute a centre of diversity at the convergence of the Indo-Pacific encompassing the seas around Sumatra and the southern half of peninsular Malaysia to the easternmost point of New Guinea.

Seagrasses are also composed of relatively few species: currently 58 species in 12 genera. They appear to have evolved more than once and have an evolutionary history that is still the subject of debate. Seagrasses are divided into 5 families and 12 genera: Hydrocharitaceae (*Halophila, Enhalus, Thalassia*), Ruppiaceae (*Ruppia*), Zosteraceae (*Zostera, Phyllospadix*), Posidoniaceae (*Posidonia*), and Cymodoceaceae (*Amphibolis, Cymodocea, Halodule, Syringodium, Thalassodendron*). Globally, seagrass distribution is divided into six regional floras: temperate North Atlantic, tropical Atlantic, Mediterranean, temperate North Pacific, tropical Indo-Pacific, and temperate Southern Ocean. Species richness is positively correlated with decreasing latitude with the greatest richness occurring in Southeast Asia (Hogarth 2015).

5.2 The Coral Triangle

The greatest concentration of marine biodiversity lies in the Coral Triangle (Veron et al. 2009). The area encompasses the nations of Malaysia, Indonesia, Papua New Guinea, the Philippines, and the Solomon Islands (Figure 5.2) and is composed of 21 ecoregions encompassing the eastern Philippines, Palawan/North Borneo, Sulawesi Sea/Makassar Strait, NE Sulawesi, Halmahera, Papua, Banda Sea, Lesser Sunda, Bismark Sea, Solomon Sea, and the Solomon Archipelago. These ecoregions are delineated into four groups that have their own unique (Figure 5.2, violet, brown, orange, and purple) characteristics regarding species richness and connectivity. The Coral Triangle has 605 zooxanthellate corals including 15 regional endemics, amounting to 76% of the world's total coral species (Veron et al. 2009). Within the triangle, the area of highest biodiversity is the Bird's Head Peninsula of West Papua (the Papua ecoregion) which hosts 574 species with individual reefs supporting up to 270 species ha^{-1}. Reef fish have a remarkably similar pattern.

The plankton of the open ocean contains relatively few species in diverse groups, owing to the scarcity of credible barriers. Pelagic organisms have widespread distributions through dispersal; allopatric speciation seems to be unimportant. They seem to live quite well at the edges of their range. However, very rapid species turnover has been established from the fossil record and is most likely the result of sympatric speciation (Norris 2000). Thus, it is unsurprising that there is no unambiguous centre of biodiversity, although planktonic Foraminifera show some evidence of a large circumtropical belt of diversity (Tokeshi 1999).

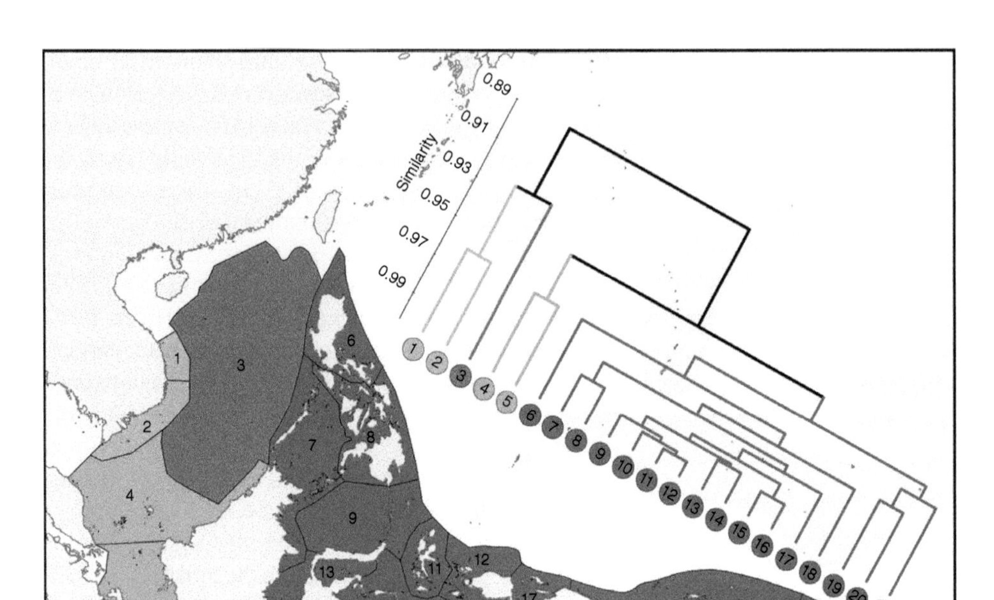

FIGURE 5.2 Ecoregions of zooxanthellate corals of the Coral Triangle with dendrogram indicating similarity between ecoregions. Numbers of species in each ecoregion can be found in Veron et al. (2009). *Source*: www.coralsoftheworld.org (accessed 4 January 2021). © Japanese Coral Reef Society.

Although the taxonomy is weak, there are at least 35 centres of pelagic fauna in the sea, without distinct boundaries, and the Coral Triangle is only one of these centres (Spalding et al. 2012). There is a latitudinal increase in pelagic species richness from the polar regions to the tropics, but most studies have concentrated on open ocean rather than coastal waters (Angel 1997). Many tropical species have ocean-wide or circum-equatorial ranges (Norris 2000). Pelagic biogeographic zones reflect specific hydrographic features, such as particular current regimes, and large-scale oceanic circulation patterns (Longhurst 2007).

Corals are not the only benthic organisms to attain peak species richness in the Coral Triangle. Both mangroves and seagrasses have their highest biodiversity in the IWP with significant overlap with the Coral Triangle. That is, mangroves, seagrasses, and corals all show similar distributions as delineated by the 20 °C winter isotherm.

The biogeography of seagrasses is complex with nine floras, including an Indo-Pacific and a West Pacific flora whose ranges overlap with the Coral Triangle where the highest concentration of seagrass species is found (Larkum et al. 2018). There are five areas of high seagrass species diversity in the IWP: insular Southeast Asia, Japan and Korea, SW Australia, East Africa, and Southeast India (Green et al. 2003). Green et al. (2003) contend that the Philippines, New Guinea, and Indonesia constitute the centre of biodiversity of seagrasses. The biogeography of seagrasses suggests that they evolved in the Tethys Sea during the late Cretaceous (Larkum et al. 2018).

With respect to benthic infauna and epifauna, many groups show high biodiversity in or overlapping with the Coral Triangle. Molluscs, for instance, have high concentrations of species in the Indo-Malayan region or a larger part of the central IWP (Briggs 1999). Members of the genus *Strombus* attain highest diversity in the Philippines and in eastern Indonesia. For gastropods, the highest species numbers have been found in New Guinea and to a lesser degree in Australia (Wells 1990). Only a few species in most gastropod families show widespread Indo-Pacific distributions.

Bivalve molluscs show three large regions of over 500 species in the East Indian-West Pacific, the southern Caribbean, and the tropical eastern Pacific (Clarke and Crame 1997). In the East Indian-West Pacific, there are two smaller biodiversity hotspots of over 1000 species in which the Philippines and Indonesia are the richest. Species richness patterns of most benthic invertebrates, such as the bivalve molluscs, are reef-associated in that their biodiversity patterns mirror those of their coral reef hosts. Mollusc families show highest biodiversity in or near the Coral Triangle: the Cerithiidae (*Cerithium, Clypeomorus*), the Conidae, the Cypraeidae, the Haliotidae, the Littorinidae (*Littoraria*), the Muricidae (*Murex-Haustellum*), the Olividae (*Oliva*), and the Strombidae. Diversity patterns may differ between coral reef-associated molluscs and those associated with mangroves and seagrass meadows; on average, there is greater species richness on coral reefs than in the other two habitats.

Coral-associated barnacles, decapod crustaceans, sipunculids, fish, and many symbiotic taxa all show biodiversity patterns that mirror corals (Baeza et al. 2013). The decapods are among the most diverse crustacean taxa associated with the coral reefs of the IWP but are outnumbered by members of the Alpheidae (de Grave 2001). Most range overlap occurs in both the Indo-Philippine and in the Indo-Malayan regions where highest biodiversity also occurs. Like many other invertebrates, the decapods show a high level of restricted distribution with their coral reef hosts. Thus, their ranges and diversity patterns depend completely on coral reefs.

The richest marine fish fauna is found in eastern Indonesia, New Guinea, and the Philippines and is directly attributable to high habitat diversity. Allen (2002) considers the eastern Indonesia–southern Philippines corridor as the region of highest biodiversity. Indonesia is an especially rich region with a high concentration of endemic and rare fish species (Roberts et al. 2002). Based on the 20°C winter isotherm, the northernmost boundary for species-rich faunas is the region between the Indo-Malayan and the Sino-Japanese subtropical zone.

The highest concentration of large Foraminifera species is found in the Indo-Malayan area where a hotspot has been identified from southern Japan to the Sahul shelf, which includes the Philippines and most of Indonesia (Langer and Hottinger 2000). Foraminifera at the generic level have a biodiversity centre from Borneo to the northern coast of New Guinea. Hoeksema (2007) indicated that the centre of biodiversity for larger benthic invertebrates not only includes the Philippines, eastern Indonesia, and southern Japan but also northern New Guinea.

There have been five major extinction episodes (Veron 1995) involving coral reefs: (i) at the end of the Ordovician (444 Ma), (ii) the late Devonian (375 Ma), (iii) the end of the Permian (251 Ma), (iv) the end of the Triassic (200 Ma), and (v) the end of the Cretaceous (65 Ma). By the end of the Cretaceous, about 70% of corals had become extinct; 10 families of coral survived from this period and 185 genera evolved recently over the last 8 Ma. When the Tethys Sea closed 16 Ma, relict species survived. Following these mass extinctions, a decline in corals and many other faunas occurred. Some important reef-building corals suffered during the late Devonian with extinctions of stromatoporoids and tabulate corals. The extinction at the end of the Permian brought huge changes to coral history as well as further evolutionary development. The stony corals of the Palaeozoic became extinct, but several genera of calcareous sponges survived to build reefs in the Permian and in the Triassic.

Modern reefs show great diversity since their last extinction at the end of the Cretaceous with representatives of all phyla and classes found. Corals belong to the class Anthozoa within the phylum Cnidaria and there are more than 6000 species of anthozoans. There are about 1000 species of hermatypic corals worldwide with the centre of biodiversity being in the IWP which houses over 70 genera and about 500 species, 400 of which are found in the Philippines alone. In contrast, there are about 20 genera and over 60 species in the Caribbean. Modern corals are the product of 6000 years of growth during recent sea-level rises as sea level has risen by about 135 m over the past 10 000 years. The oceans are presently experiencing an interglacial period, which is one of the warmest periods for the past 850 000 years.

As noted earlier, the species records are most complete for the stony corals and it is with this group that the Coral Triangle is best delineated (Veron et al. 2009, 2015). The global patterns of species richness for zooxanthellate corals show peak biodiversity within the ecoregions of the Coral Triangle. Sixteen regions of the world have greater than 500 species, and, in total, the Coral Triangle has 605 zooxanthellate coral species of which 66% are common to all ecoregions (Figure 5.3). This diversity amounts to 76% of the world's total species. More than 80% of all Coral Triangle species are found in at least 12 of the 16 Coral Triangle ecoregions (Figure 5.3). Ninety-five percent of Coral Triangle species are found in one or more adjacent ecoregions, notably other parts of Southeast Asia, Japan, Micronesia, the Great Barrier Reef, Vanuatu, New Caledonia, and Fiji.

Mushroom corals (Scleractinia, Fungiidae), based predominantly on a taxonomic revision, have a concentration of species in the Coral Triangle, especially in the area comprising Indonesia, the Philippines, and New Guinea (Hoeksema 2007). The shape of this centre overlaps with the position of the continental margin as during low sea-level stands this margin represented part of the continental coastline where coral reefs occurred. The centre of origin in the Coral Triangle was completely established only about 10 Ma, but its predecessor had existed in the Tethys Sea between Africa and Eurasia since the early Cretaceous (Briggs 2006).

The Coral Triangle, like most of the marine tropics, is threatened by human influence. Pollution, overfishing, habitat destruction, ocean acidification, and climate change are having a severe impact on this heavily populated region (Hoegh-Guldberg et al. 2009; Peñaflor et al. 2009; Mcleod et al. 2010; McManus et al. 2020).

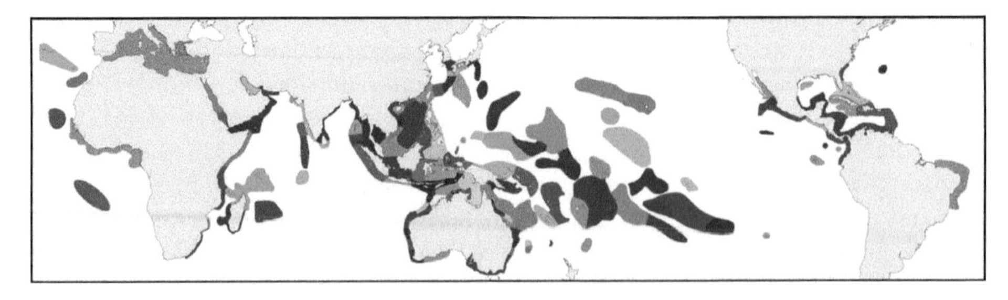

FIGURE 5.3 Ecoregions of zooxanthellate corals delineated based on known faunal and/or environmental uniformity. Numbers of species in each ecoregion can be found in Veron et al. (2009). *Source*: www.coralsoftheworld.org (accessed 5 January 2021). © Japanese Coral Reef Society.

Over the past two decades, SSTs have risen by 0.4 °C (Peñaflor et al. 2009), and much of the area is highly susceptible to rising sea levels (Mcleod et al. 2010), although impacts may differ between subregions (McManus et al. 2020).

It is not possible to understand the development of mangroves, seagrass, coral reefs, and the Coral Triangle without understanding the development of the Tethys Sea and its subsequent history (Table 5.1). By about 132 Ma, Eastern and Western Mediterranean sub-provinces could be distinguished from what was an Indo-Mediterranean Province. A separate Caribbean Province was formed about 124 Ma, and the endemism that distinguished it reflects the increasing distance between the Old and New World as the Atlantic Ocean became wider (Briggs 2006). During the early Palaeogene (65–45 Ma), many extant families and genera, including the earliest coral reef fish assemblages, evolved in the Indo-Mediterranean Province. As the climate grew colder, the tropical biota trapped in the Mediterranean was gradually eliminated. This happened after the early Miocene collision between Africa and Eurasia eliminated the Tethys Sea and formed the Mediterranean.

The Tethyan fauna that had become isolated to form the Caribbean Province became divided into the West Central American and Antillean Provinces; the late Cretaceous subdivision of the Caribbean Province may have been a response to the formation of a Central American archipelago, and an early Central American isthmus may have formed at that time. The modern, high-diversity fauna of the Southern Caribbean was largely derived from the Caribbean Province of the Tethys Sea, although as Briggs (2006) points out, some of it came from the Western Pacific across the East Pacific Barrier before the formation of the isthmus. The East Indies fauna was rich and inherited from the Indo-Mediterranean Province via the Indian Ocean.

The origin of the great majority of our present species probably took place in the Pliocene and Pleistocene with a large proportion of species richness originating within the two tropical centres. What this means in practical terms is that more than 75% of the Indo-Pacific reef fish and about 450 species of hermatypic corals were also present in the Coral Triangle.

5.3 Origins Explained

At least six main hypotheses have been put forward to explain the centre of biodiversity. The 'Centre of Origin' hypothesis considers the centre of biodiversity to be the centre of origin, with speciation occurring inside the centre with successive outward dispersal to adjacent areas (Briggs 1999). Successive periods of glaciations and low sea-level stands caused the emergence of barriers and the isolation of populations in deep sea basins in between the island groups of Indonesia and the Philippines, although this idea does not explain the high biodiversity along northern New Guinea. As the hypothesis indicates, however, it is the species that are successful that have given rise to phyletic lines leading to new genera and families. It is relatively easy for species of probable origins within the Coral Triangle to penetrate other tropical regions. Indeed, many species have achieved circumtropical distributions. There is some support based on molecular phylogenetic and biogeographic data to support the centre of origin hypothesis. Molecular data for reef dwarf gobies (Gobiidae: *Eviota*) indicate that two species complexes contain multiple genetically distinct, geographically restricted, colour morphs indicative of recently diverged species originating in the Coral Triangle (Tornabene et al. 2015). These data also suggest

TABLE 5.1 Key tectonic events of the Cenozoic and their effects on the oceans and paleocurrents.

Event	Effects
Isolation of Antarctica	
Early Eocene (50 Ma)-full deep-water separation of South Tasman Rise	First indications of global cooling at 50–40 Ma and significant 2°C temperature drops in both the late–middle Eocene and middle–late Eocene boundary. Further isolation of Antarctic marine fauna.
Eocene–Oligocene boundary (37 Ma)	Major cooling of both surface and bottom waters by 5°C. Onset of widespread Antarctic glaciation.
Opening of Drake Passage (36–23 Ma)	Almost complete isolation of Antarctic marine fauna.
Mid-Miocene (15 Ma)-full establishment of Antarctic Circumpolar Current	Latitudinal temperature gradient like that of today. Development of Polar Frontal Zone.
Closure of Tethyan Seaway	
End of Cretaceous period (75–65 Ma)-vast circumpolar-equatorial tropical ocean	Major westerly flowing equatorial current system. Some faunal differentiation but no clear high-diversity loci.
Paleogene (65–23 Ma)-continuity of the tropical Tethyan Ocean	Largely homogenous tropical fauna. Major pulse in coral reef development at the end of Oligocene (23 Ma); marked similarities between western Tethys (Mediterranean) and Caribbean/Gulf of Mexico.
Early Miocene (20 Ma)-closure of Tethyan Seaway by northward movement of Africa/Arabia landmass	Westerly flowing tropical current drastically curtailed. Mediterranean Sea excluded from reef belt. Caribbean and eastern Pacific regions become progressively isolated marking beginning of the distinction between IWP and Atlantic Caribbean–East Pacific (ACEP) foci. Further development through the Neogene (<20 Ma) sees relative impoverishment of ACEP and enrichment of IWP.
Collision of Australia/New Guinea with Southeast Asia	
Beginning of Cenozoic era (65 Ma)-Australia/New Guinea separated from SE Asia by deep-water gateway	Single tropical Tethyan ocean. No differentiation between Indian and Pacific Oceans.
Paleogene (65–23 Ma)-progressive closure of Indo-Pacific gateway; northward subduction of Indian–Australian lithosphere beneath the Sunda–Java–Sulawesi arcs	
Mid-Oligocene (30 Ma)-gap narrowed but still a clear deep-water passage formed by oceanic crust	

TABLE 5.1 *(Continued)*

Event	Effects
Latest Oligocene (25 Ma)-New Guinea collides with leading edge of eastern Philippines–Halmahara–New Guinea arc system	
Early Miocene (20 Ma)-deep water passage between Indian and Pacific Oceans closes	Major reorganisation of tropical current systems; new shallow-water habitats appear in Indonesian region.
Early-late Miocene (20–10 Ma)-continued northward movement of Australia/New Guinea. Rotation of several plate boundaries and formation of tectonic provinces that are recognizable today	Widespread growth of coral reefs in IWP region; huge rise in numbers of reef and reef-associated taxa; many modern genera and species evolve.
Later Neogene (10 Ma–present)-continued northern movement of Australia/New Guinea; in Early Pliocene (4 Ma)-a critical point when close contact is made with the island of Halmahara	Warm Pacific waters deflected eastward at the Halmahera eddy to form Northern Equatorial Counter Current. Warm waters in ITF are replaced by relatively cold ones from the North Pacific. These changes affect heat balance between East and West Pacific and help promote onset of Northern Hemisphere glaciation.
Uplift of Central American Isthmus (CAI)	
Mid Miocene (15–13 Ma)-sedimentary evidence of earliest phases of shallowing	Still a deep-water connection at CAI.
Latest Miocene–early Pliocene (6–4 Ma)-continued shallowing to <100 m depth	Major effect on oceanic circulation. Gulf Stream begins to deflect warm shallow waters northward to eventually initiate the major "conveyor belt" of deep ocean circulation.
Mid-Pliocene (3.6 Ma)-further closure of CAI	North Atlantic thermohaline circulation system intensifies; Arctic Ocean isolated from warm Atlantic leading eventually to onset of Northern Hemisphere glaciation at 2.5 Ma.
Late Pliocene (3 Ma)-complete closure of CAI	Final separation of shallow water Atlantic–Caribbean and Eastern Pacific Provinces. Reverse flow of water through the Bering Strait; this influences Pliocene–Pleistocene patterns of thermohaline circulation in Arctic and North Atlantic Oceans.

Source: Briggs (1974, 2006) and Lomolino et al. (2016).

that regional isolation due to sea-level fluctuations may explain some speciation, but other species show no evidence of physical isolation, thus both allopatric and sympatric speciation may have taken place within the region.

There are two basic versions of the centre of origin hypothesis. The first is what Bellwood et al. (2012) have called the 'spreading dye model' in which species originating within the Coral Triangle expanded their geographic range by chance. The second version states that new species with superior competitive abilities displaced older, inferior competitors towards the periphery of the region's boundaries. Thus, in theory, newer species will be at the centre of the boundary and that there would be a "peripheral halo" of predominantly older species that will also have greater geographic range.

The second hypothesis is the 'Centre of Overlap' (or Vicariance) model. This idea states that the centre of biodiversity consists of overlapping distribution ranges that extend into either the Pacific or the Indian Ocean resulting from either larval dispersal or ancient plate tectonics. These biogeographic boundaries are well recognised, but any subsequent expansion or movement of ranges would lead to an overlap of geographic ranges and a localised increase in species richness in the overlap zone (Santini and Winterbottom 2002). The inherent complexity of the physical environment within the Coral Triangle makes it highly likely that this area exhibited many such divisions that were susceptible to frequent changes through time. It may thus be considered a 'dynamic mosaic' (Bellwood et al. 2012) of constantly changing distributions driven by continual climatic, geologic, and oceanographic processes. This hypothesis has found some support from reef fish (Woodland 1986), corals (Wallace 2002), crustaceans (Fransen 2007), and gastropods (Reid et al. 2006).

The third hypothesis is the 'Centre of Accumulation' model that is the opposite of the centre of origin hypothesis. It proposes that species arose in peripheral locations, around, or at some distance from the margin of the centre and that they subsequently moved into the centre. New species can be from anywhere outside the centre. It does not require overlap with related taxa. There are two distinct versions of this hypothesis. First, the centre of accumulation by individuals places the emphasis on isolation on peripheral oceanic islands in underpinning the speciation process. Second, the centre of accumulation by faunas reflects the accumulation of entire faunas on moving land masses. In this version, species richness may be enhanced by the merging of entire faunas via the merging of island arcs on the north coast of New Guinea with the numerous land fragments from Asia, Gondwana, and Australia.

The fourth hypothesis is the 'Centre of Survival' idea that is a composite hypothesis emphasising persistence and survival in an area rather than on origin of the species in question. The key aspect of this hypothesis is regional variation in the relative rates of extinction, and it makes no assumptions about the rates or location of origin of species. Speciation may thus have occurred anywhere. The centre of biodiversity is just an area of survival with species extinctions outside the centre's boundaries. The maintenance of habitat diversity and the availability of sufficient abundance for each species is an important condition for this hypothesis.

The fifth hypothesis is the 'Centre of Mid-Domain Overlaps' idea that is a variation on the second hypothesis. In this version, a maximum in species richness in the middle of a geographical area has formed by overlying randomised distributions of the locations of individual geographic ranges within species groups. The maximum is predicted to occur where the probability of maximum of species range overlaps is highest, that is, if the ranges are randomly placed within the Indo-Pacific, the resultant pattern of species richness forms a peak in the middle. The hotspot is thus the result of random placement of species' geographic ranges (Bellwood et al. 2012).

The sixth hypothesis is 'Reticulate Evolution' that is arguably the main mechanism of evolutionary change in most marine taxa (Veron et al. 2009). It recognises "(a) that currents are both genetic barriers (as in vicariance) and paths of genetic connectivity, (b) that species fuse as well as divide over time and space, (c) that species are generally not isolated units and (d) that evolution is driven by the physical environment (especially ocean currents) rather than biological mechanisms (e.g., competition). Furthermore, reticulate evolution does not deny the existence of Darwinian evolution which could become uppermost were genetic mixing to weaken sufficiently to create isolated gene pools. This would allow evolution to occur through biological selection. Conditions which promote reticulate evolution are at a maximum in the Coral Triangle (sometimes referred to as an 'evolutionary cauldron') because of habitat diversity and the ever-changing complexity of ocean surface currents."

Another hypothesis based on fossil and molecular evidence is the idea of 'hopping hotspots' (Renema et al. 2008). There is good evidence for the fact that during the past 50 Ma, there have been at least three marine biodiversity hotspots that have moved across half the globe, with their timing and locations coinciding with major movements of the earth's plates. Based on generic diversity of large benthic foraminifera, three successive movements of biodiversity hotspots have been identified: (a) in the late Middle Eocene (42–39 Ma), (b) in the early Miocene (23–16 Ma), and (c) in the present day (Renema et al. 2008). These are known as the West Tethys, Arabian, and Indo-Australian Archipelago biodiversity hotspots. During the Eocene, diversity peaked in SW Europe, NW Africa, and along the eastern shore of the Arabian Peninsula, Pakistan, and West India. The fossil record of mangroves and reef corals suggests maximal global diversity in the West Tethyan hotspot. By the late Eocene, the highest diversity was recorded in the Arabian hotspot that has an overlapping taxonomic composition with the earlier West Tethys and the later Indo-Australian Archipelago hotspots. The Miocene is the most diverse period in Southeast Asia for both large benthic foraminifera and mangroves. Regional uplift during the Arabia–Eurasia collision resulted in the demise of the Arabian hotspot during the middle to late Miocene. The present hotspot appeared with the disappearance of the Arabian hotspot and has been extant at least since the early Miocene (20 Ma) as shown by fossil records of foraminifera, corals, mangroves, and gastropods (Renema et al. 2008).

It is unlikely that there is a single explanation to account for the Coral Triangle hotspot, although for corals and most major taxa equatorial temperatures and habitat diversity are highly explanatory. The relative importance of any one hypothesis will change with increases in data and methodology. Some faunal groups, such as the stomatopods and the bryopsidale algae, do not conform to any of the hypotheses. What is clear is that there is a biodiversity hotspot in the Coral Triangle in which species richness patterns differ little from random expectations. Also, there are strong correlations between total species richness and a limited range of key environmental factors; the consensus among taxa is that there is a shared evolutionary and geological history with most taxa in the Coral Triangle being primarily of Miocene origin with ancestors from the Cenozoic hotspots of West Tethys and Arabia (Bellwood et al. 2012).

With respect to the origins of tropical marine biodiversity, it is clear that (i) physical isolation (allopatric speciation) is not the sole mechanism for speciation, (ii) oceanic archipelagos that were thought to be peripheral habitats for speciation are in fact regions that can export biodiversity, and (iii) opportunities are fewer for allopatric speciation in the oceans leaving greater opportunity for speciation along ecological boundaries (Di Martino et al. 2018). Bowen et al. (2013) have emphasised that areas such as the Coral Triangle and the Caribbean produce and export species but can also accumulate biodiversity produced in marginal habitats. This benefit has been dubbed the "biodiversity feedback" (Bowen et al. 2013).

5.4 Marine Ecoregions and Provinces

Marine realms (Figure 5.4a) and ecoregions (Figure 5.4b) have been identified across the marine biome representing what is an ecological characterisation based on water temperature, depth, and substrate rather than historical distributions based on long-term climatic patterns (Figure 5.4). Some history, however, is preserved in some benthic taxa with limited dispersal ability and some show a latitudinal pattern in species diversity and composition. These latitudinal patterns still appear to represent modern differences rather than historical ones, thus the boundaries between regions have shifted latitudinally with changing ocean temperature during the Pleistocene, and the composition of local assemblages has changed as species' ranges have expanded or contracted (Spalding et al. 2007).

In the Tropical Atlantic Realm, there are six ecoregions: Tropical NW Atlantic, North Brazil Shelf, Tropical SW Atlantic, St. Helena and Ascension Islands, the West African Transition, and the Gulf of Guinea. The Western Indo-Pacific Realm consists of seven ecoregions: Red Sea and Gulf of Aden, Somali/Arabian, Western Indian Ocean, West and South India Shelf, Central Indian Ocean Islands, Bay of Bengal, and the Andaman Sea. The Central Indo-Pacific Realm has several tropical ecoregions: South China Sea, The Sunda Shelf, Java Transitional, Tropical NW Pacific, Western Coral Triangle, Eastern Coral Triangle, Sahul Shelf, NE Australian Shelf, NW Australian Shelf, and the Tropical SW Pacific. The Eastern Indo-Pacific Realm consists of six island ecoregions: Hawaii, Marshall, Gilbert and Ellis Islands, Central Polynesia, Cook Islands, and Southeast Polynesia, Marquesas and Easter Island. The Tropical East Pacific Realm consists of both the Tropical East Pacific and Galapagos ecoregions.

There is a level of provincialism that does reflect the influence of tectonic and oceanographic history on the distribution and origin of lineages. Tropical oceans have been a barrier to the distribution of cold-water organisms, but there has also obviously been some level of distinction among tropical provinces, at least for some taxa. For instance, there has been a historical subdivision between some Pacific and Indian Ocean species of sea horses (*Hippocampus*; Lourie et al. 2005) and the mantis shrimp (*Haptosquilla pulchella*; Barber et al. 2000). Thus, there are true biogeographic divisions across the tropical marine biome that may not necessarily reflect history but certainly reflect current or recent environmental cues and are separated by zones of rapidly changing species composition.

5.5 The Latitudinal Diversity Gradient

Tropical ecosystems hold more than three-quarters of all species on earth. One of the most pervasive features of life on earth is the increase in species diversity from the poles to the tropics. Prime examples of this gradient in the marine biosphere are the diversity gradients of marine prosobranch gastropods in the eastern Pacific and western Atlantic (Figure 5.5) and zooxanthellate (Veron et al. 2015). The gastropod example (Roy et al. 1998) is based on an analysis of the geographic ranges of 3916 species of gastropods living on the shelves of the western Atlantic and eastern Pacific from the tropics to the Arctic Ocean; western Atlantic and eastern Pacific diversities are similar despite many important historical and physical differences between the oceans. This

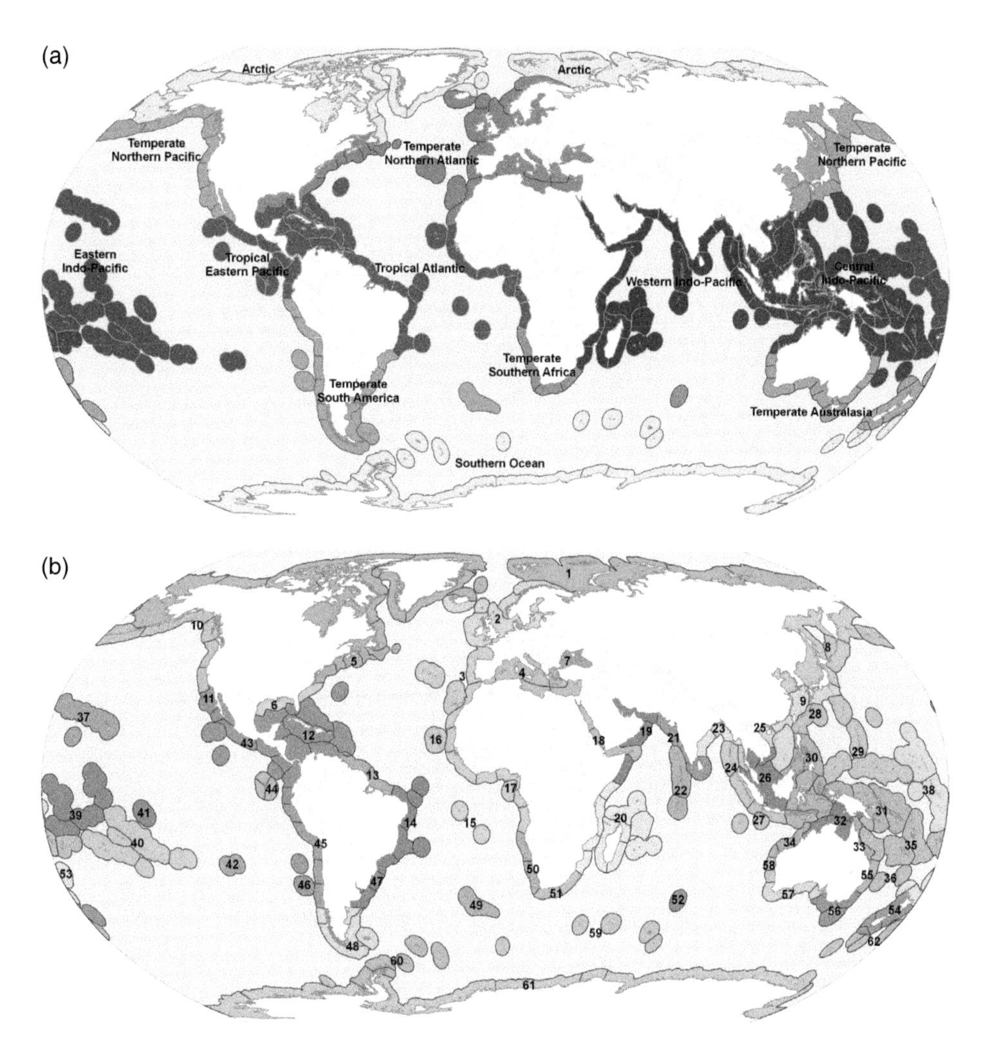

FIGURE 5.4 Marine realms (a) with ecoregions (b) across the global ocean. 1 = Arctic, 2 = Northern European Seas, 3 = Lusitanian, 4 = Mediterranean Sea, 5 = Cold-Temperate Northwest Atlantic, 6 = Warm-Temperate Northwest Atlantic, 7 = Black Sea, 8 = Cold-Temperate Northwest Pacific, 9 = Warm-Temperate Northwest Pacific, 10 = Cold-Temperate Northeast Pacific, 11 = Warm-Temperate Northeast Pacific, 12 = Tropical Northwestern Atlantic, 13 = North Brazil Shelf, 14 = Tropical Southwestern Atlantic, 15 = St. Helena and Ascension Islands, 16 = West African Transition, 17 = Gulf of Guinea, 18 = Red Sea and Gulf of Aden, 19 = Somali/Arabian, 20 = Western Indian Ocean, 21 = West and South Indian Shelf, 22 = Central Indian Ocean Islands, 23 = Bay of Bengal, 24 = Andaman, 25 = South China Sea, 26 = Sunda Shelf, 27 = Java Transitional, 28 = South Kuroshio, 29 = Tropical Northwestern Pacific, 30 = Western Coral Triangle, 31 = Eastern Coral Triangle, 32 = Sahul Shelf, 33 = Northeast Australian Shelf, 34 = Northwest Australian Shelf, 35 = Tropical Southwestern Pacific, 36 = Lord Howe and Norfolk Islands, 37 = Hawaii, 38 = Marshall, Gilbert, and Ellis Islands, 39 = Central Polynesia, 40 = Southeast Polynesia, 41 = Marquesas, 42 = Easter Island, 43 = Tropical East Pacific, 44 = Galapagos, 45 = Warm-Temperate Southeastern Pacific, 46 = Juan Fernández and Desventuradas, 47 = Warm-Temperate Southwestern Atlantic, 48 = Magellanic, 49 = Tristan Gough, 50 = Benguela, 51 = Agulhas, 52 = Amsterdam–St. Paul, 53 = Northern New Zealand, 54 = Southern New Zealand, 55 = East Central Australian Shelf, 56 = Southeast Australian Shelf, 57 = Southwest Australian Shelf, 58 = West Central Australian Shelf, 59 = Subantarctic Islands, 60 = Scotia Sea, 61 = Continental High Antarctic, 62 = Subantarctic New Zealand. *Source*: Spalding et al. (2007), figure 2, p. 577. © Oxford University Press.

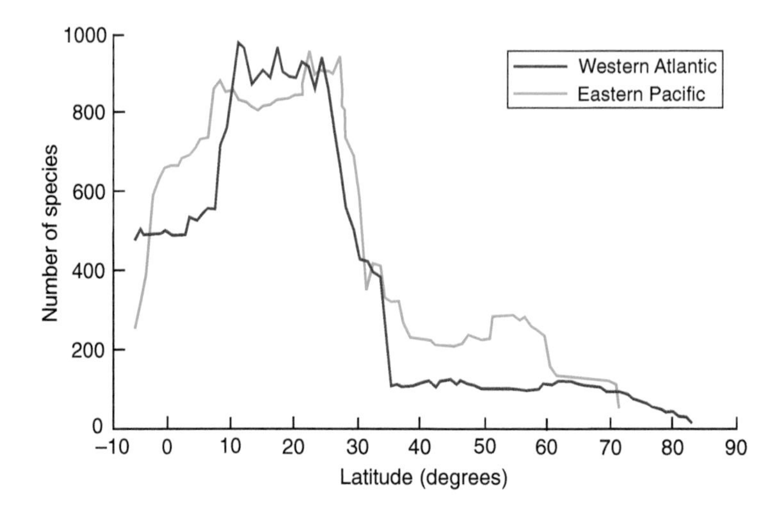

FIGURE 5.5 Latitudinal diversity gradient of eastern and western Pacific marine prosobranch gastropods per degree of latitude. *Source*: Roy et al. (1998), figure 1, p. 3700. © National Academy of Sciences, USA.

diversity pattern cannot be explained simply by latitudinal differences in species range-length, species-area effects, or recent geologic histories. As pointed out by Roy et al. (1998), one parameter that does correlate significantly with diversity in both oceans is solar energy input, that is, the diversity pattern may be linked to sea surface temperatures; warmer temperatures result in a faster pace of evolution, including speciation.

Using an extensive database of published gradients, Hillebrand (2004) found that the overall strength and slope of the gradient for marine organisms were significantly negative and of similar magnitude compared with gradients for terrestrial organisms. The marine gradients were stronger than those for freshwater organisms, although there were clear differences in gradients between biogeographical regions. At the local scale, gradients were weakest for autotrophs and strongest for carnivores. The gradient parameters also differed between oceans and between different habitats, with steeper declines related to the open ocean pelagic rather than coastal benthic fauna. Weaker gradients were found for the Pacific than in the Atlantic, although the Pacific is older than the Atlantic. The Indian Ocean demonstrated an interesting pattern: the gradient was very steep, but the strength was weak. This latter pattern conforms to the notion that biodiversity hotspots dominate the spatial pattern of diversity in the Indian Ocean from which diversity radially decreases. Radial rather than latitudinal gradients and high overall diversity correspond well to the steep slope and weak strength of the latitudinal gradient.

Several marine organisms were characterised by weak gradients due to low body mass (diatoms, protozoa), organism type (macrophytes), and life form (infauna). Nevertheless, marine gradients were as strong as or even stronger than terrestrial gradients. Body mass is an important factor influencing the strength and slope of the latitudinal gradient, and this may be related to dispersal chance, energetic constraints, and/or population size. The Analysis of Reef Life survey data for 4127 marine species at 2406 coral and rocky sites worldwide confirms that total richness of vertebrate species peaks in the low latitudes between 15°N and 15°S, but richness of large mobile invertebrates is highest at high latitudes (Edgar et al. 2017). Species richness correlated with temperature for fish and nutrients for macroinvertebrates. The pattern for mobile macroinvertebrates was attributed to constraints imposed by temperature-mediated fish predation and herbivory across the tropics.

Not all organisms of different kingdoms or phyla show clear latitudinal trends. For example, pelagic marine microbial assemblages did not show a latitudinal gradient in one study (Moss et al. 2020), while another study found a clear increase in diversity towards the tropics for planktonic archaea, bacteria, eukaryotes, and major virus clades (Ibarbalz et al. 2019). These conflicting results underscore the fact that there are limits to our understanding of the latitudinal gradient and microbial speciation.

The causal mechanisms of the latitudinal diversity gradient have been the subject of intense debate (Roy et al. 1998; Brown 2014; Veron et al. 2015; Pontarp et al. 2018). Hypotheses proposed to explain latitudinal biodiversity gradients include temperature, primary productivity, area, natural disturbance regime, climatic stability, fragmentation and connectivity, and human disturbance (Edgar et al. 2017). Recent discussions have centred around two main phenomena: phylogenic niche conservatism and ecological productivity. The former refers to the fact that tropical species appear to occur in narrower niches than species of higher latitude. The latter refers to the fact that productivity (and subsequent metabolism) of a species tends to be greater at higher temperatures. These two factors undoubtedly play important roles, but accumulating evidence suggests that the single most important factor is kinetics, that is, the temperature dependence of ecological and evolutionary rates as relatively higher temperatures in the tropics generate and maintain high diversity. Nonetheless, over 32 different hypotheses have been advanced to explain the latitudinal gradient (Brown 2014; Pontarp et al. 2018). The search for a primary cause has led to some consideration that there may not be any one cause but that many causes co-vary depending on location and more importantly, with taxonomic group.

Compelling evidence exists that most lineages originated in the tropics via reticulate evolution (Jablonski et al. 2013; Veron et al. 2015). What is fairly clear is that (i) rates of origination of new species are highest in the tropics, (ii) higher rates of speciation than extinction generate high diversity of species and clades within the tropics, (iii) most species and clades of tropical origin remain confined to the tropics due to environmental constraints, (iv) a small number of tropical species overcome such constraints to expand their ranges into higher latitudes, and (v) high rates of extinction result in low standing stocks of species and clades at higher latitudes (Brown 2014).

A central causal theory proposed by Brown (2014) offers that there is a framework showing how historical events and environmental conditions affect the dynamics of fundamental ecological and evolutionary processes to generate and maintain variation in standing stocks of biodiversity. The fossil record, phylogenetic reconstructions, analyses of variation in genomes and niche traits, and metabolic theory all play a role in the search for a central explanatory hypothesis. Brown (2014) states that the latitudinal gradient is so ancient and pervasive because "The relationship of the earth to the sun and the variation in solar energy input creates a gradient of environmental temperature. Temperature affects the rate of metabolism and all biological activity including the rates of ecological interactions and coevolution. Diversity begets diversity because the Red Queen runs faster when she is hot." What this means most simply is that species diversity is higher in the tropics than towards the poles because temperatures are warmer in the lower latitudes and that productivity and speciation are correspondingly greater. This is the simplest explanation and does not preclude other causal mechanisms from operating to explain the observed patterns of latitudinal diversity.

References

Allen, G.R. (2002). Indo-Pacific coral reef fish as indicators of conservation hotspots. *Proceedings Ninth International Coral Reef Symposium*, Bali, Indonesia (23–27 October 2000), vol. 2, pp. 921–926.

Allen, G.R. (2008). Conservation hotspots of biodiversity and endemism for Indo-Pacific coral reef fish. *Aquatic Conservation, Marine and Freshwater Ecosystems* 18: 541–556.

Angel, M.V. (1997). Pelagic biodiversity. In: *Marine Biodiversity, Patterns and Processes* (eds. R.F.G. Ormond, J.D. Gage and M.V. Angel), 35–68. Cambridge, UK: Cambridge University Press.

Baeza, J.A., Ritson-Williams, R., and Fuentes, M.S. (2013). Sexual and mating system in a caridean shrimp symbiotic with the winged pearl oyster in the Coral Triangle. *Journal of Zoology* 289: 172–181.

Barber, P.H., Palumbi, S.R., Eddmann, M.V. et al. (2000). Biogeography – a marine Wallace's line? *Nature* 406: 692–693.

Bellwood, D.R., Renema, W., and Rosen, B.R. (2012). Biodiversity hotspots, evolution and coral reef biogeography: a review. In: *Biotic Evolution and Environmental Change in Southeast Asia* (eds. D. Gower, K.G. Johnson, J.E. Richardson, et al.), 216–245. Cambridge, UK: Cambridge University Press.

Bowen, B.W., Bass, A.L., Muss, A.J. et al. (2006). Phylogeography of two Atlantic squirrelfish (family, Holocentridae): exploring pelagic larval distribution and population connectivity. *Marine Biology* 149: 899–913.

Bowen, B.W., Rocha, L.A., Toonen, R.J. et al. (2013). The origins of tropical marine biodiversity. *Trends in Ecology and Evolution* 28: 359–366.

Briggs, J.C. (1974). *Marine Zoogeography*. New York: McGraw-Hill.

Briggs, J.C. (1999). Coincident biogeographic patterns: Indo-West Pacific Ocean. *Evolution* 53: 326–335.

Briggs, J.C. (2006). Proximate sources of marine biodiversity. *Journal of Biogeography* 33: 1–10.

Briggs, J.C. and Bowen, B.W. (2013). Marine shelf habitat, biogeography, and evolution. *Journal of Biogeography* 40: 1023–1035.

Brown, J.H. (2014). Why are there so many species in the tropics? *Journal of Biogeography* 41: 8–28.

Clarke, A. and Crame, J.A. (1997). Diversity, latitude and time: patterns in the shallow sea. In: *Marine Biodiversity, Patterns and Processes* (eds. R.F.G. Ormond, J.D. Gage and M.V. Angel), 122–147. Cambridge, UK: Cambridge University Press.

Crame, J.A. (2020). Early Cenozoic evolution of the latitudinal diversity gradient. *Earth-Science Reviews* 202: 103090 https://doi.org/10.1016/j.earscirev.2020.103090.

Di Martino, E., Jackson, J.B., Taylor, P.D. et al. (2018). Differences in extinction rates drove modern biogeographic patterns of tropical marine biodiversity. *Science Advances* 4: eaaq1508. https://doi.org/10.1126/sciadv.aaq1508.

Duke, N.C. (2017). Mangrove floristics and biogeography revisited: further deductions from biodiversity hot spots, ancestral discontinuities, and common evolutionary processes. In: *Mangrove Ecosystems: A Global Biogeographic Perspective* (eds. V.H. Rivera-Monroy, S.Y. Lee, E. Kristensen and R.R. Twilley), 17–53. Cham, Switzerland: Springer International.

Eble, J.A., Rocha, L.A., Craig, M.T. et al. (2011). Not all larvae stay close to home: long-distance dispersal in Indo-Pacific reef fish, with a focus on the brown surgeonfish (*Acanthurus nigrofuscus*). *Journal of Marine Biology* 2011: 1–12.

Edgar, G.J., Alexander, T.J., Lefcheck, J.S. et al. (2017). Abundance and local-scale processes contribute to multi-phyla gradients in global marine diversity. *Science Advances* 3: e1700419. https://doi.org/10.1126/sciadv.1700419.

Ekman, S. (1953). *Zoogeography of the Sea*. London: Sidgwick and Jackson.

Floeter, S.R., Rocha, L.A., Robertson, D.R. et al. (2008). Atlantic reef fish biogeography and evolution. *Journal of Biogeography* 35: 22–47.

Fransen, C.H.J.M. (2007). The influence of land barriers on the evolution of pontoniine shrimps (Crustacean, Decapoda) living in association with molluscs and solitary ascidians.

In: *Biogeography: Time and Space, Distributions, Barriers and Islands* (ed. W. Remena), 103–116. Dordrecht, The Netherlands: Springer.
Gaither, M.R., Toonen, R.J., Robertson, R.R. et al. (2010). Genetic evaluation of marine biogeographic barriers: perspectives from two widespread Indo-Pacific snappers (*Lutjanus kasmina* and *Lutjanus fulvus*). *Journal of Biogeography* 37: 133–147.
de Grave, S. (2001). Biogeography of Indo-Pacific Pontoniinae shrimps (Crustacea, Decapoda): a PAE analysis. *Journal of Biogeography* 28: 1239–1253.
Green, E.P., Short, F.T., and Frederick, T. (2003). *World Atlas of Seagrasses*. Oakland: University of California.
Hillebrand, H. (2004). Strength, slope and variability of marine latitudinal gradients. *Marine Ecology Progress Series* 273: 251–267.
Hobbs, J.P.A., Frisch, A.J., Allen, G.R. et al. (2009). Marine hybrid hotspot at Indo-Pacific biogeographic border. *Biology Letters* 5: 258–261.
Hoegh-Guldberg, O., Hoegh-Guldberg, H., Veron, J.E.N. et al. (2009). *The Coral Triangle and Climate Change: Ecosystems, People and Societies*. Sydney: WWF Australia.
Hoeksema, B.W. (2007). Delineation of the Indo-Malayan centre of maximum marine biodiversity, The Coral Triangle. In: *Biogeography: Time, and Place, Distributions, Barriers, and Islands* (ed. W. Renema), 117–178. Berlin: Springer.
Hogarth, P.J. (2015). *The Biology of Mangroves and Seagrasses*. Oxford, UK: Oxford University Press.
Horne, J.B., van Herwerden, L., Choat, H.J. et al. (2008). High population connectivity across the Indo-Pacific: congruent lack of phylogeographic structure in three reef fish congeners. *Molecular Phylogenetics and Evolution* 49: 629–638.
Ibarbalz, F.M., Henry, N., Brandão, M.C. et al. (2019). Global trends in marine plankton diversity across kingdoms of life. *Cell* 179: 1084–1097.
Jablonski, D., Belanger, C.L., Berke, S.K. et al. (2013). Out of the tropics, but how? Fossils, bridge species, and thermal ranges in the dynamics of the marine latitudinal diversity gradient. *Proceedings of the National Academy of Sciences* 110: 10487–10494.
Jablonski, D., Huang, S., Roy, K. et al. (2017). Shaping the latitudinal diversity gradient: new perspectives from a synthesis of paleobiology and biogeography. *The American Naturalist* 189: 1–12.
Langer, M.R. and Hottinger, L. (2000). Biogeography of selected 'larger' foraminifera. *Micropaleontology* 46: 105–126.
Larkum, A.W.D., Waycott, M., and Conran, J.G. (2018). Evolution and biogeography of seagrasses. In: *Seagrasses of Australia* (eds. A.W.D. Larkum, G.A. Kendrick and P.J. Ralph), 3–29. Cham, Switzerland: Springer International.
Lomolino, M.V., Riddle, B.R., and Whittaker, R.J. (2016). *Biogeography*, 5e. Sunderland, USA: Sinauer.
Longhurst, A.L. (2007). *Ecological Geography of the Sea*, 2e. London: Academic Press.
Lourie, S.A., Green, D.M., and Vincent, A.C.J. (2005). Dispersal, habitat preferences and comparative phylogeography of southeast Asian seahorses (*Syngnathidae, Hippocampus*). *Molecular Ecology* 14: 1073–1094.
Luiz, O.J., Madin, J.S., Robertson, D.R. et al. (2012). Ecological traits influencing range expansion across large oceanic dispersal barriers: insights from tropical Atlantic reef fish. *Proceedings of the Royal Society B: Biological Sciences* 279: 1033–1040.
Mcleod, E., Hinkel, J., Vafeidis, A.T. et al. (2010). Sea-level rise vulnerability in the countries of the Coral Triangle. *Sustainability Science* 5: 207–222.
McManus, L.C., Vasconcelos, V.V., Levin, S.A. et al. (2020). Extreme temperature events will drive coral decline in the Coral Triangle. *Global Change Biology* 26: 2120–2133.
Miloslavich, P., Diaz, J.M., Klein, E. et al. (2010). Marine biodiversity in the Caribbean: regional estimates and distributional patterns. *PLoS One* 5: e11916.
Moss, J.A., Henriksson, N.L., Pakulski, J.D. et al. (2020). Oceanic microplankton do not adhere to the latitudinal diversity gradient. *Microbial Ecology* 79: 511–515.
Mundy, B.C., Wass, R., Demartini, E. et al. (2010). Inshore fish of Howland Island, Baker Island, Jarvis Island, Palmyra Atoll, and Kingman Reef. *Atoll Research Bulletin* 585: 1–133.
Norris, R.D. (2000). Pelagic species diversity, biogeography, and evolution. *Paleobiology* 26: 236–258.

Peñaflor, E.L., Skirving, W.J., Strong, A.E. et al. (2009). Sea-surface temperatures and thermal stress in the Coral Triangle over the past two decades. *Coral Reefs* 28: 841–850.

Pontarp, M., Bunnefeld, L., Cabral, J.S. et al. (2018). The latitudinal diversity gradient: novel understanding through mechanistic eco-evolutionary models. *Trends in Ecology and Evolution* 34: 211–223.

Reese, J.S., Bowen, B.W., Smith, D.G. et al. (2010). Molecular phylogenetics of moray eels (Muraenidae) demonstrates multiple origins of shell-crushing jaw (*Gymnomuraena, Echidna*) and multiple colonizations of the Atlantic Ocean. *Molecular Phylogenetics and Evolution* 57: 829–835.

Reid, D.G., Lal, K., Mackenzie-Dodds, J. et al. (2006). Comparative phylogeography and species boundaries in *Echinolittorina* snails in the central Indo-West Pacific. *Journal of Biogeography* 33: 990–1006.

Renema, W., Bellwood, D.R., Braga, J.C. et al. (2008). Hopping hotspots: global shifts in marine biodiversity. *Science* 321: 654–657.

Roberts, C.M., McClean, C.J., Veron, J.E.N. et al. (2002). Marine conservation hotspots and conservation priorities for tropical reefs. *Science* 295: 1280–1284.

Rocha, L.A., Robertson, D.R., Rocha, C.R. et al. (2005a). Recent invasion of the tropical Atlantic by an Indo-Pacific coral reef fish. *Molecular Biology* 14: 3921–3928.

Rocha, L.A., Robertson, D.R., Roman, J. et al. (2005b). Ecological speciation in tropical reef fish. *Proceedings of the Royal Society B: Biological Sciences* 272: 573–579.

Rocha, L.A., Rocha, C.R., Robertson, D.R. et al. (2008). Comparative phylogeography of Atlantic reef fish indicates both origin and accumulation of diversity in the Caribbean. *BMC Evolutionary Biology* 8: 157. https://doi.org/10.1186/1471-2148-8-157.

Roy, K., Jablonski, D., Valentine, J.W. et al. (1998). Marine latitudinal diversity gradients: tests of causal hypotheses. *Proceedings of the National Academy of Sciences* 95: 3699–3702.

Santini, F. and Winterbottom, R. (2002). Historical biogeography of Indo-western Pacific coral reef biota: is the Indonesian region a centre of origin? *Journal of Biogeography* 29: 189–205.

Spalding, M.D., Fox, H.E., Allen, G.R. et al. (2007). Marine ecoregions of the world: a bioregionalization of coastal and shelf areas. *Bioscience* 57: 573–583.

Spalding, M.D., Agostini, V.N., Rice, J. et al. (2012). Pelagic provinces of the world: a biogeographic classification of the world's surface pelagic waters. *Ocean & Coastal Management* 60: 19–30.

Srivastava, J. and Prasad, V. (2019). Evolution and paleobiogeography of mangroves. *Marine Ecology* 40: e12571. doi:10.111/maec.12571.

Tokeshi, M. (1999). *Species Coexistence: Ecological and Evolutionary Perspectives*. Oxford, UK: Blackwell.

Tornabene, L., Valdez, S., Eerdmann, M. et al. (2015). Support for a 'center of origin' in the Coral Triangle: cryptic diversity, recent speciation, and local endemism in a diverse lineage of reef fish (Gobiidae, *Eviota*). *Molecular Phlyogenetics and Evolution* 82: 200–210.

Veron, J.E.N. (1995). *Corals in Space and Time*. Ithaca, USA: Cornell.

Veron, J.E.N., Devantier, L.M., Turak, E. et al. (2009). Delineating the Coral Triangle. *Galaxea* 11: 91–100.

Veron, J., Stafford-Smith, M., DeVantier, L. et al. (2015). Overview of distribution patterns of zooxanthellate Scleractinia. *Frontiers in Marine Science* 1: 81. https://doi.org/10.3389/fmars.2014.00081.

Wallace, C.C. (2002). Journey to the heart of the centre-origins of high marine faunal diversity in the central Indo-Pacific from the perspective of an acropologist. *Proceedings Ninth International Coral Reef Symposium*, Bali, Indonesia (23–27 October 2000), vol. 1, pp. 33–39.

Wells, F.E. (1990). Comparative zoogeography of marine molluscs from northern Australia, New Guinea, and Indonesia. *The Veliger* 33: 140–144.

Williams, S.T., Jara, J., Gomez, E. et al. (2002). The marine Indo-West Pacific break: contrasting the resolving power of mitochondrial and nuclear genes. *Integrative and Comparative Biology* 42: 941–952.

Woodland, D.J. (1986). Wallace's Line and the distribution on marine inshore fish. In: Indo-Pacific Fish Biology. In: *Proceedings of the Second International Conference on Indo-Pacific Fish* (eds. T. Uyeno, R. Arai, T. Taniuchi, et al.), 453–460. Tokyo: Ichthyological Society of Japan.

CHAPTER 6

Populations and Communities

6.1 Introduction

The existence of a latitudinal gradient in species diversity raises an interesting question: Is there a latitudinal gradient in the importance of biotic interactions? Dobzhansky (1950) proposed that the benign, predictable climate in the tropics led to a greater importance of biotic interactions than in the more variable climate of the temperate regions. Schemske (2009) revisited Dobzhansky's views and proposed a mechanism whereby biotic interactions may increase rates of adaptation and speciation in the tropics. The prevalence of strong biotic interactions in the tropics promotes coevolution, and that faster adaptation and speciation may be the result of the optimum phenotype constantly changing due to strong interacting species. However, this idea has its detractors as some evidence shows that in terrestrial ecosystems, species interactions and specialisation are not greater in the tropics (Moles and Ollerton 2016).

Several early hypotheses suggested that stronger biotic interactions in the tropics lead to greater species richness (Paine 1966; Connell 1971). One hypothesis suggested that the greater stability of production in the tropics sustains a disproportionate number of predator species that enhances diversity by ameliorating competitive exclusion among prey (Paine 1966). Another hypothesis proposed that competitive exclusion is prevented by high density dependent predation (Connell 1971). The evidence is not exhaustive, but there is considerable evidence pointing to a higher degree of interaction among and between organisms throughout the tropics (Schemske et al. 2009). Some of the best-supported examples are higher herbivory and insect predation, the greater predominance of tropical mutualisms, such as cleaning symbioses, and intense plant–herbivore interactions. In the marine realm, a recent examination of fish feeding pressure on the benthos across 61° of latitude in the western Atlantic revealed that biotic interactions, especially plant–herbivore interactions, are more intense in the tropics (Longo et al. 2019).

This chapter examines biotic interactions such as density dependent and density independent population growth and intraspecific competition, interspecific competition, predator–prey interactions, mutualism, host–parasite interactions, plant–herbivore interactions, and multi-trophic interactions, including the evidence for trophic and facilitation cascades and intra-guild predation.

Tropical Marine Ecology, First Edition. Daniel M. Alongi.

6.2 Density Independence, Density Dependence, and Intraspecific Competition

Populations tend to grow exponentially, and all populations with discrete or continuous generations as well as populations with age structures obey the exponential law. All populations show exponential growth if the environment does not affect the population in a systematic manner. Populations however do not grow forever. Eventually, individuals begin to run out of food, water, space, or other resources and/or become increasingly subject to disease or predation. That is, self-limitation comes into effect in all populations. This phenomenon is known as density dependence.

A density-dependent population is one in which there is an effect of present and/or past population sizes on the population growth rate. There is a net input either via birth and/or immigration (or in the case of open populations, settlement, or recruitment) and a net loss via death or emigration (Figure 6.1). Each demographic rate at each population size is the sum of both a density independent component (defined by the y-intercept) and a density dependent component (defined by the slope). Where the input and loss lines intersect is when the population is in equilibrium. This equilibrium point varies in nature and may never be attained. Density dependence nevertheless is the essence of population regulation.

As in all regions of the marine biome, density dependent behaviour has been observed in tropical populations across a wide variety of size spectra. The best-documented examples of density dependence in tropical marine organisms involves coral reef fish (Hixon and Webster 2002). Most demersal species have a pelagic larval stage followed by nektonic juvenile and adult stages that are associated with a benthic area on a coral reef. Demersal fish may form meta-populations which are 'populations of populations' linked by dispersal (Fobert et al. 2019). In contrast, a local population is defined as a group of juvenile and adult conspecifics meeting two

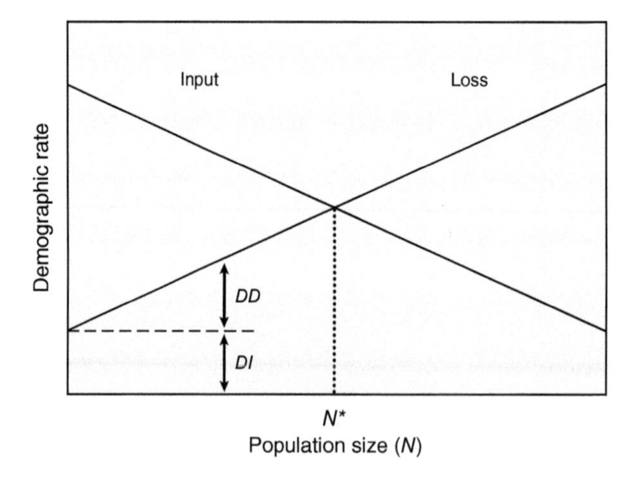

FIGURE 6.1 Demographic density dependence showing an input rate (birth, immigration, settlement, recruitment) and a loss rate (death, emigration) that are both density dependent. N* is where the population is in equilibrium.

criteria: (i) mature members of the group spawn only with other group members and (ii) between-group movement is demographically negligible due to sufficient spatial isolation (Hixon and Webster 2002). The local population can be as small as a single coral head or patch reef as some reef fish have extremely limited ranges. A meta-population, in contrast, is a group of local populations linked only by larval dispersal.

There are four possible sources of demographic density dependence in a demersal meta-population: (i) fecundity, (ii) pre-settlement mortality of eggs and larvae, (iii) mortality during settlement, and (iv) post-settlement mortality of juveniles and adults. In most marine fish, 'birth' occurs via spawning with external fertilisation. Egg or zygote production is a possible source of density dependence because of two relationships: (i) growth is often density dependent and (ii) fecundity is a function of body size (Hixon and Webster 2002). These relationships are well known for coral reef fish. Demographic 'birth' occurs by the process of larval settlement which is the transition from a pelagic larval stage to a demersal juvenile stage. The success of this 'birth' depends on larval supply and the suitability of the benthic habitat. Mortality can be high during this process. Settlement is measured indirectly as recruitment because it is so difficult to observe directly. For a local population, recruitment can be calculated by dividing the density of new recruits by the density of residents during a specific time interval. Hixon and Webster (2002) noted that, in these cases, recruitment appeared to be density dependent, as each recruit represented a decreasing proportion of the total population as local population size increased. For a meta-population, it is better to examine the total recruitment rate which is the density of new recruits appearing during a set time interval.

Post-settlement mortality of reef fish is much better understood and documented compared with the mortality of eggs and larvae (Hixon and Webster 2002; Goatley and Bellwood 2016). The mortality rate is calculated on a per-capita basis as a proportion of the population dying during a set time interval. Hixon and Webster (2002) examined several reef studies to determine whether recruitment is density dependent or not. They found four different patterns in recruitment: (i) density dependence, (ii) density independence, (iii) inverse density dependence (when numbers increase with a larger local population size), and (iv) a simple relationship between total recruitment and local population size. Thus, no clear consistent patterns were observed in recruitment. Why this is so is not clear and may be species-specific.

Compared to recruitment, mortality is better documented. A meta-analysis of published size-specific mortality rates with increasing body size for coral reef fish revealed an exponential decline in mortality rate with increasing body size (Goatley and Bellwood 2016). There are three distinct phases within this broad relationship. The first phase is characterised by recruitment of native fish which suffer extremely high mortality rates; in this phase, fish must learn quickly to survive predation. After a few days surviving fish enter the second phase in which small increases in body size result in pronounced increases in lifespan. About 50% of reef fish remain in this phase throughout their lives. Fish reaching a size threshold of about 43 mm total length enter the third phase, where mortality rates are relatively low and the pressure to grow is significantly reduced.

In experimental studies that explicitly tested whether post-settlement mortality was density dependent or not, Hixon and Webster (2002) found that 7 of 13 species always showed density dependent behaviour, whereas three species showed density dependence at times and density independence at other times; two species showed inverse density independence. The 13th species also showed inverse density dependence apparently due to anti-predatory behaviour. A meta-analysis of the available data indicates that there is a high correlation between local population density and instantaneous daily per-capita mortality rate (Hixon and Webster 2002). This correlation suggests that post-settlement mortality is density dependent and widespread among coral reef fish.

What causes density dependence in coral reef fish populations? The mechanisms causing density dependence are poorly understood but are possibly related to recruitment and mortality (Hixon et al. 2012). The total recruitment rate can be density dependent, density independent or inversely density dependent. Density dependence in the total recruitment rate indicates that increasing densities of resident fish inhibit subsequent establishment of new recruits presumably due to intraspecific competition or cannibalism or both. Density independence suggests no effect of resident fish on recruitment, whereas inverse density dependence indicates facilitation whereby settling fish are attracted to conspecifics or enjoy enhanced survival in larger groups, or both (Hixon et al. 2012). Facilitation probably operates at low population densities whereas inhibition occurs at high densities, apparently as a function of increasing intraspecific competition.

Intraspecific competition has been directly measured among reef fish and can be intense. For example, the per-capita effect of intraspecific competition within two species of reef fish in the Bahamas was at least twice as great as the effect of interspecific competition (Forrester et al. 2006). Growth rate was better predicted by density measures that incorporated body size rather than numerical density, suggesting interference competition.

Predation is the major cause of mortality in coral reef fish, especially shortly after settlement. Predation has been directly implicated as the source of density dependent mortality. Habitat complexity and the synergistic effects of various groups of predators and interference competitors have also been shown to play important roles in reef fish dynamics. Predation has been implicated or demonstrated to play the ultimate or proximate mechanism causing post-settlement density dependent mortality. Evidence is limited, but the importance of an aggregative response of transient predators congregating at sites where new recruits are abundant and ignoring sites where new recruits are of low density may play a structuring role in fish dynamics on a coral reef (Beukers-Stewart et al. 2011). Of course, predation will play an important role in structuring reef fish populations and may have some bearing on whether density dependence is involved. For example, small juveniles of the planktivorous damselfish, *Chromis cyanea*, are known to suffer heavy mortality that is spatially density dependent only in the presence of two suites of predators: transient piscivores attacking from above and reef-resident piscivores attacking from below. In the absence of predators, the early mortality of the damselfish is virtually density independent (Hixon and Carr 1997).

Recruitment is not a consistent source of demographic density dependence, but post-settlement mortality is often density dependent, especially shortly after settlement, and is caused mainly by predation (Hixon and Webster 2002). Seventeen of 20 species examined by Hixon and Webster (2002) experienced density dependent per capita mortality over their lifetime; density independent mortality was reported for older juvenile fish suggesting that regulation via mortality is an early post-settlement phenomenon. The fact that post-settlement mortality is common has two consequences: (i) an important source of regulation for the entire meta-population may be the local populations suggesting that the mechanisms causing local density dependence should be conserved and (ii) the hypothesis that post-settlement mortality will be density dependent has been falsified in most cases, but evidence exists documenting that some local reef fish populations are recruitment limited (Sandin and Pacala 2005).

Apparent contradictions in the literature regarding density dependent and inversely density independent mortality may be a question of scale. White et al. (2010) noted that the observed pattern depends on predator and prey behaviour, the spatial configuration of the reef habitat, and the spatial and temporal scales of observation. For instance, predators tend to produce density dependent mortality at their characteristic scale of foraging, but prey mortality is inversely density dependent at smaller spatial scales due to 'attack-abatement effects.' For non-social, non-aggregating species, and species that aggregate to take advantage of spatially clumped refuges, inversely density dependent mortality is possible, but likely superseded by direct density dependent refuge competition. For species that aggregate socially, mortality should be inversely density dependent at the scale of individual aggregations, but directly density dependent at larger scales.

There may be other causal mechanisms of density dependence. For instance, Barneche et al. (2016) have argued that energetic constraints play an important role in reef fish dynamics. They tested a model that yields predictions linking population density to the physiological constraints of body size and temperature on individual metabolism, and on the ecological constraints of trophic structure and species richness on energy partitioning among species. From a model analysis of over 5600 fish populations, they found population density declines markedly with increases in community species richness and that, after accounting for richness, energetic constraints are manifest most strongly for the most abundant species, which generally are of small body size and occupy the lower trophic groups. Their findings suggest that factors associated with community species richness are the major drivers of variation in population density, at least at the global scale. At the local scale, such species can be susceptible to extinction given that populations of species-rich tropical ecosystems exhibit much lower maximum densities.

Populations of other reef organisms show evidence of density dependent regulation. The early life stages of the Caribbean coral, *Siderastrea radians,* showed profound effects of local densities on early survivorship (Vermeij and Sandin 2008). Settlement of planulae correlated negatively with the abundance of turf algae, but positively with crustose coralline algae. Adult density showed independent effects on coral settlement and early post-settlement survivorship. Settlement rates increased across low levels of adult cover and became saturated at a maximum cover of about 10%. Early post-settlement survivorship decreased with adult cover showing density dependence in coral settlers. The early benthic phase of *S. radians* is defined by a severe demographic bottleneck with appreciable density dependent and density independent effects on survivorship. A similar level of density dependence has been observed for two hermatypic populations on the outer reef of Moorea, French Polynesia (Bramanti and Edmunds 2016). Following a catastrophic disturbance in which coral cover was reduced by 93%, high rates of coral recruitment facilitated population recovery. Coral recruitment was associated with high overall coral cover, but the relationship differed between two genera. Acroporids recruited at low densities and the density of recruits was positively associated with cover of *Acropora*, whereas pocilloporids recruited at high densities and densities of their recruits were negatively associated with cover of *Pocillopora*. These results suggest that associations between adult cover and density of juveniles and recruits can mediate rapid coral recovery following a large disturbance. They also suggest that different genera respond differently to disturbance. For example, spawning corals such as *Acropora* tend to recover from disturbance much faster than brooding corals, such as *Pocillopora* (Doropoulos et al. 2015).

Density dependent and density independent behaviour has been observed in non-reef organisms such as barnacles, sea urchins, the queen conch, *Strombus gigas*, the green turtle, *Chelonia mydas,* and the southern bluefin tuna, *Thunnis maccoyii* (Sutherland 1990; Tiwari et al. 2006; Miller et al. 2007). A well-documented example is the tropical intertidal barnacle, *Chthamalus fissus*, which inhabits tropical rocky shores along the central Pacific coast of Costa Rica (Sutherland 1990). Sutherland (1990) studied the relative importance of pre- and post-settlement processes in determining patterns of distribution and abundance. Mortality after settlement was density independent for at least 90-days; abundance appeared limited by sparse recruitment. Intraspecific competition was weak, except for occasional peaks of dense recruitment. Demography of *C. fissus* was regulated by severe physical forces (storms, strong waves, etc.) affecting settlement and subsequent recruitment.

6.3 Populations with Age Structure

Most invertebrate and vertebrate populations have a distinct age structure with clear age classes, but many populations have complex age distributions. Age distribution can govern a population, at least in the short term. Knowledge of the age-specific survivorship and fertility patterns of a population allows us to understand what age categories are most important to the persistence of the population.

Arguably, the best evidence for tropical marine populations with a distinct age structure comes from fish, crustacean, mollusc, and coral populations on sandy beaches, reefs, and mangroves. On sandy beaches, short-lived and rapidly growing species tend to be the norm, persisting in dense populations, such as the wedge clam, *Donax lubricus*. This species is commonly found on tropical sandy beaches and has a distinct age structure (Tenjing 2019).

Size structure of coral populations is generally highly skewed with a preponderance of the smallest size class. Such population structure may be altered in degraded reef environments. Bak and Meesters (1999) found that coral colony size distributions are affected in degraded reefs, implying that changes in mortality patterns or recruitment result in relatively fewer small and more large colonies. They hypothesise that deterioration of local conditions caused by climate change may similarly lead to coral populations becoming increasingly skewed towards larger colonies.

Other forms of disturbance are known to have an observable impact on the structure of coral colonies. The population structure of the clonal gorgonian, *Plexayra kuna,* responded in different ways to different levels of disturbance (Coffroth and Lasker 1998). Genotypic diversity decreased at intermediate disturbance and increased again at the highest levels of disturbance. It was hypothesised that disturbance-insensitive clones that do not rely on vegetative propagation generally do not disperse and form aggregated clones, but among gorgonians, disturbance had a greater effect on individual survival than on propagation. Genotypic diversity then declined at the highest levels of disturbance.

Examination of corals in harvested habitats versus those living in a marine reserve revealed a significant difference in population age structure. The Mediterranean red coral, *Corallium rubrum,* is a long-lived octocoral that is harvested commercially (Tsounis et al. 2006). 89% of red corals in the harvested area were <10 years old and 96% of all colonies had not yet grown more than second-order branches. The age structure of harvested populations was highly skewed towards younger and smaller colonies compared to those in the marine reserve (Tsounis et al. 2006). A similar response to disturbance has been observed in massive *Porites lutea* corals at Luhuitou fringing reef in the South China Sea (Zhou et al. 2014). After several anthropogenic disturbances, live coral cover declined by 80% and surviving populations were skewed towards a high abundance of small colonies. Age structure analysis indicated that most corals were <50 years old, with 55% of the coral species on the reef flat having been recruited following the establishment of a marine reserve.

Numerous studies have examined the age and size structure of reef fish populations (Newman et al. 1996; Meekan et al. 2001; Schwamborn and Ferreira 2002; Marriott et al. 2007; Waldrop et al. 2016). Significant differential growth between the sexes in the tropical snappers, *Lutjanus adetii and L. quinquelineatus,* was observed, with males growing larger than females on the Great Barrier Reef (Newman et al. 1996). The shape of the growth curves was steep for the first few years and then became asymptotic. The annual instantaneous rate of natural mortality indicated an annual survivorship of 79 and 86% for *L. adetii* and *L. quinquelineatus*, respectively. Despite their small size, the protracted longevity and low mortality imply that both species are vulnerable to overfishing. Three other species of red snappers, *Lutjanus erythropterus, L. malabaricus* and *L. sebae*, were relatively long-lived and grew slowly after reaching reproductive maturity. Age estimates up to 32 years for *L. erythropterus,* 20 years for *L. malabaricus*, and 22 years for *L. sebae*, together with their low mortality rates, underscore the possibility of overfishing.

Spatial variation may play a role in population structure for coral reef fish. The age structure of five species of *Stegastes* damselfish on coral reefs at the centre (Panama), at the northern edge (Baja California) and at the southern edge (Galapagos) in the tropical eastern Pacific Ocean were analysed and populations of widespread species in the Galapagos and Baja grew to larger adult sizes, had long life spans and lower rates of mortality, once asymptotic mean sizes were attained than populations of the same species in Panama (Meekan et al. 2001). Long life spans and low mortality rates were shared by endemic species in the Galapagos and Baja. Furthermore, strong year-classes in the age structures of both widespread and endemic species in the Galapagos corresponded to the timing of ENSO events.

Long-lived and slow-growing reef fish have been observed in the tropical Atlantic. The dusky damselfish, *Stegastes fuscus,* for instance can live for 15 years (Schwamborn and Ferreira 2002). The Brazilian endemic greenback parrotfish, *Scarus trispinosus*, the largest herbivorous reef fish in the South Atlantic, has populations in the Abrolhos Bank region with 1–22 annual growth rings in otoliths, indicative of many individuals being long-lived and slow-growing (Freitas et al. 2019).

A large, long-lived reef fish, the red bass, *Lutjanus bohar,* is harvested throughout the Indo-Pacific region and has unique population characteristics compared with other reef fish (Marriott et al. 2007). The maximum estimated age of 55+years is the oldest recorded for any tropical snapper, and sampling at different depths revealed that many of the oldest red bass reside at the greatest depths. This species unsurprisingly shows relatively slow growth with no significant differences between males and females. Female bass matured at a much larger size and older age than males and were reproductively active from August to April. Compared to most other tropical reef fish, this species has a relatively *K*-selected life history strategy which makes it susceptible to over-exploitation.

In mangroves, reproduction in most crab species is extremely diversified, with breeding activity following a continuous or seasonal pattern or a very variable reproductive strategy. Breeding activity in most mangrove crabs is generally associated with variations in extrinsic and intrinsic factors, including sex ratio and larval development time. Extrinsic factors such as temperature, salinity, and rainfall are important variables that trigger reproduction. The influence of these factors on breeding activity can vary in different species inhabiting different ecological localities. Mangrove crabs thus have distinct age classes depending on the species and local environment. The fiddler crab, *Uca annulipes,* has a unimodal size frequency distribution in East African mangroves, with half of the crabs being males and the other half female of which half were ovigerous (Litulo 2005). Breeding took place year-round with two spawning peaks in summer (December and January). Recruits were present year-round with peak reproductive capacity during summer. This fiddler crab follows a rapid breeding cycle accompanied by rapid larval development and settlement in Mozambique (Litulo 2005) and has a seasonal growth pattern in Iranian mangroves with maximum growth during autumn and early summer, with growth ceasing during winter due to low temperatures (Mokhtari et al. 2008). The population biology of the crab, *Ucides cordatus,* in Brazilian mangroves indicated maximum longevities of about 18 years for males and about 16 years for females, with both sexes showing similar growth rates (de Macedo Costa et al. 2014). In another Brazilian mangrove forest, the reproductive period of *Uca maracoani* was based on a low rate of ovigerous females found throughout the year, occurring from June to August, with continuous but more intense recruitment in spring (Azevedo et al. 2017).

Grapsid crabs contribute the largest number of species among the six Brachyura families that are associated with mangroves, but their population ecology is poorly understood. Lee and Kwok (2002) studied the population dynamics of the sesarmid crabs, *Parasesarma affinis* and *Parisesarma bidens,* in *Avicennia marina* and *Kandelia candel* forests and observed that both species were more abundant in the *K. candel* forest with higher gonosomatic and hepatosomatic indices than those in the *A. marina* forest. There were, however, no differences in male-biased sex ratio, growth and reproductive periodicity, and size-specific reproductive output for these two species between forest types. The monthly size–frequency histograms for both species at both forest types were polymodal, with up to seven cohorts represented with two cohorts from every year. The populations comprised between 1 and 3 age groups. The modal shift of the cohorts in the size–frequency histograms suggests that the crabs usually attained an increase of 1.2 mm in carapace width per month and took usually about one year to mature. The sesarmid crabs, *Aratus pisonii, Armases rubripes, Armases angustipes*, and *Sesarma rectum*, are the four species of sesarmids that inhabit Brazilian mangroves (Ribeiro and Bezerra 2014). These species reproduce continuously throughout the year, with peaks coincident with rainy periods or high temperatures. *Sesarma rectum* showed higher fecundity than the other sesarmids (Ribeiro and Bezerra 2014). The narrowback mud crab, *Panopeus americanus*, reproduces continuously in remnant Brazilian mangroves at Araçá, Sáo Paulo (Vergamini and Mantelatto 2008). The size–frequency distribution was bimodal, reflecting the occurrence of more than one recruitment pulse and the dominance of adults.

Seasonal reproduction in mangrove crabs is as common as continuous reproduction. *Ucides cordatus cordatus* showed seasonal reproduction and followed a strict lunar rhythm in Brazilian mangroves (Diele 2000; Sant'anna et al. 2014). Each year, four to five events of mate searching take place within four days following new moons between December and April with a peak occurring in either January or February. Shortly thereafter, females extrude eggs and incubate them until they spawn within four days around the following new moon. Zoaea larval release is also precisely timed. In flooded mangrove forest, females spawn around slack spring tides and peak spawning always occur one day before the new moon. *Ucides cordatus cordatus* is a long-lived brachyuran species with a maximum life span of <10 years (Diele 2000). The mangrove horseshoe crab, *Carcinoscorpius rotundicauda*, reproduces seasonally on the equator in Singapore (Cartwright-Taylor et al. 2009). The size cohorts indicated recruitment to the smallest size classes from November to March and recruitment to the larger size classes from March to July. Juveniles were not found in June indicating that there may be a 'rest period' of low to no breeding from May to July. The breeding activity of the sesarmid crab, *Parasesarma plicatum*, in mangroves of southern India was linked to seasonality of the monsoon. The frequency of ovigerous females correlated positively with salinity and temperature and negatively with rainfall (Jasmine and Shyla Suganthi 2012).

Some crabs, however, do show variable reproductive strategies, even close to the equator. Populations of four species of fiddler crabs showed different reproductive trends in the Caeté estuary in north Brazil (Ribeiro and Bezerra 2014). Three species (*Uca cumulanta, U. maracoani, U. rapax*) reproduced mainly during the dry season, but *Uca vocator* had a clear spawning peak towards the end of the wet season. For the latter three species, breeding was prolonged and sometimes year-round. Co-generic species can thus show different life history strategies even within the same forest. *Goniopsis cruentata*, a common mangrove crab throughout the western Atlantic, was found to breed both seasonally and continuously with peaks in the Brazilian dry period which were not associated with monthly variations in salinity, rainfall, or temperature (de Lira et al. 2013). However, breeding intensified in periods of higher salinity and temperatures and lower rainfall. *Sesarma quadratum* bred continuously in Indian mangroves, but spawning and larval release was linked to lunar rhythms (Syama and Anilkumar 2011). Another sesarmid, *Sesarma rectum*, reproduced continuously, but the highest frequency of ovigerous females was recorded in spring and summer. The major pulse of recruitment occurred during autumn and winter, which was related to greater reproductive activity during the warmer months of the year (da Silva Castiglioni et al. 2011).

6.4 Meta-populations

The concept of a meta-population as a 'population of populations that go extinct and recolonize' has become a major paradigm in ecology, especially conservation biology (Hanski and Simberloff 1997). The concept applies to situations where habitat is not homogenous, but form as discrete patches, so populations tend to be separated from one another and thus have their own population dynamics. Such patches are not completely isolated, as individuals can cross barriers to connect populations. In fact, if small enough, some patches may die off or a new patch may be colonised by dispersers from occupied patches. Grimm et al. (2003) have questioned the applicability of the meta-population concept to marine ecology. While there is good evidence of meta-population structure in terrestrial systems, the only evidence for discrete patch dynamics in marine populations comes from coral reefs. It is thus not surprising that most papers on marine meta-population dynamics have dealt with coral reef fish populations. It is from this database that we will discuss the concept of marine meta-populations.

The spatial structure of meta-populations of coral reef fish is inextricably linked to the spatial structure of reef habitat. Patch size is one of the spatial characteristics of meta-populations, and the maximum possible extent of any meta-population is set by a species' range (Lett et al. 2015). The presence or absence of dispersal barriers, coupled with the availability and continuity of suitable habitat, sets the limits for all species' ranges with a given biogeographical province. A species' range is set by its competitive abilities, evolutionary origin and subsequent dispersal ability, and the range of tolerance to environmental conditions (Lett et al. 2015). Thus, there is tremendous variability in species' range. Unfortunately, the boundaries for any reef fish meta-population are not yet known. Genetic data may eventually be able to discern such boundaries. It is possible, however, that reef fish meta-populations might function across the full extent of species' ranges, although whether this is true or not depends upon dispersal. In terms of patch size within a given coral reef, it is assumed that there are suitable hatches of habitat distributed within a wider field of unsuitable habitat space. Each occupied patch is inhabited by a local population of organisms that randomly interbreed within the patch. Many species' patches will be driven largely by microhabitat structure or by individual or social behaviour or a combination of both.

The local breeding group is clearly the most definable grouping of reef fish, where there is mating with one another, and involving different individuals at different spatial scales. A good example is provided by Kritzer and Sale (2006): a single large coral formation may have several colonies of branching *Acropora* corals, each housing several goby species of the genus *Gobiodon*. Each of these groups will constitute a local breeding group and each will be restricted to its own coral. Atop the coral formation (or 'bommy') is a school of planktivorous *Chromis* damselfish which are unlikely to stray from the bommy and constitute a local breeding group. The bommy and several others may also be the territory of a dominant male *Scarus* parrotfish which with his harem constitute another local breeding group. Across the entire coral reef is a large *Plectropomus* grouper that will gather at a strategic location with other groupers to spawn at certain times of the year. Thus, a small patch of coral can contain several localised breeding groups of different species.

The meta-population concept has fostered the notion of identifying patterns and connections among reefs (Horne et al. 2013). Habitat area is not a 'perfect predictor' of population size due partly to the influence of habitat structure on local density. Although the relationship between habitat area and population size can be quite variable among reefs, the usefulness of habitat area across reefs as a predictor of relative population size seems greatest for local populations. For example, a fourfold difference in population size of the stripey bass, *Lutjanus carponotatus*, can exist between different islands within an island group as found on the Great Barrier Reef (Kritzer 2002), but the habitat area of these populations differs about sevenfold. The disparity between the relative difference in habitat size and the relative difference in population size might be sufficiently small to allow the former to serve as a useful proxy for the latter, given the uncertainty in the density estimates on which these estimates of population size are based, coupled with the potential for temporal variability in population size and the 'imperfect relationship' between reproductive output and population size (Kritzer and Sale 2006).

Spatial patterns in demographic rates are also important for meta-populations (Sutherland et al. 2014). Spatial patterns determine the inequities in local reproductive output and population dynamics that can create 'source-sink' dynamics and 'rescue effects.' Variation in different demographic traits can have unequal consequences for reproductive output and meta-population dynamics. For example, a sea urchin can have varying mortality, age at maturity, and growth rate that can have important consequences for the intrinsic rate of population growth, but asymptotic body size does not. From a meta-population context, this suggests that spatial patterns in the former traits and not the latter might lead to "inequities in the production of offspring, and therefore could be the roots of source-sink status and other inter-population disparities" (Kritzer and Sale 2006). Spatial variation in a demographic trait is not necessarily indicative of significant differences among populations in terms of reproductive output, but "also that the importance of spatial variation in a given demographic trait can be very context-specific."

Density dependence can also be important in a meta-population context. If a local population replenishes with offspring without relying on another population, that population still appears to be reliant upon recruits from other local populations, although it would fare well without such external replenishment. Density dependence would still be operative for early post-settlement life stages.

Meta-population dynamics is highly dependent on processes governing its applicability, such as extinction-recolonisation events, anthropogenic extinctions, and synchrony of population fluctuations (Fowler 1991; Bay et al. 2008). At One Tree Reef in the Great Barrier Reef, Fowler (1991) found that two species of butterflyfish were abundant but did not show any signs of mature females. Thus, these populations must have been established and maintained by colonisation from elsewhere. These two butterflyfish populations nevertheless persist for years without a 'rescue effect' from other local populations, indicating connectivity with some external source of recruits. It is not clear how and where this external connectivity operates, but it is clear there is a source–sink relationship between the fish of this reef with other reefs within the Great Barrier Reef. As Kritzer and Sale (2006) point out it may be argued that a non-reproducing population is "effectively irrelevant" from a simple population dynamics perspective, but the meta-population dynamics that establish even non-reproducing populations can affect "multi-species interactions, including fisheries, in the outlying areas, because fish will eat and be eaten at their destination reef."

The concept of meta-populations requires some degree of asynchrony in fluctuations among local populations, primarily because of partial closure, and therefore partial independence, of those populations (Kritzer and Sale 2006). No such data yet exists, but data does exist for asynchrony of recruitment of reef fish. Despite the limitations of recruitment data, recruitment is undoubtedly important in determining population size and represents the best way to assess the degree of asynchrony in population fluctuations of reef fish. Asynchrony in meta-populations is the result of the independent dynamics within local populations, but "the existence of asynchronous recruitment fluctuations alone is insufficient to argue for meta-population structure among coral reef fish." Recruitment to a local population may be mainly of nonlocal origin, but a variety of environmental conditions near a reef may be the ultimate determinant of settlement success. Thus, whether asynchrony arises from meta-population dynamics or not is unclear, but the recruitment data do suggest the necessity of asynchrony in characteristic population fluctuations.

Whether or not the concept of meta-populations applies to sessile coral invertebrates, such as corals is problematical as a classic meta-population includes a large variety of habitat patches that are either occupied by a population or not.

Unoccupied patches become occupied by colonisation from occupied patches which can become unoccupied through local extinctions. Coral populations differ from this model in several ways including the fact that local extinctions of corals are rare at the scale of reef, whereas colonisation describes a multi-stage process of production of larvae, dispersal, arrival, settlement, and establishment. The original model implies equal chance of each unoccupied patch to be open to colonisation (Mumby and Dytham 2006).

One of the interesting questions regarding what is a meta-population is: to what extent are sub-populations different? Are they populations different genetically or are they the same? A good example of a coral meta-population is the green turtle (*Chelonia mydas*) of the southern Great Barrier Reef. Chaloupka et al. (2004) studied the growth dynamics of four separate groups of turtles which are from the same genetic stock and found that mean maturity age varied from 25 to 50 years of age, depending on the foraging ground where there were significant differences in juvenile recruitment between the four groups. There was significant temporal variability in growth rates among the groups, possibly due to temporal differences in the dynamics of local food stocks. However, despite these variations, the juveniles of all four groups showed a growth spurt at 15–20 years of age with females tending to grow faster than males after the growth spurt. Thus, there are spatially and temporally distinct differences among the four groups of this same genetic stock. Utilizing genetic markers to determine the degree of difference among sub-populations of the panda clownfish, *Amphiprion polymnus*, along the coast of Papua New Guinea, Saenz-Agudelo et al. (2011) found that >82% of the juveniles were immigrants, while the remainder were progeny of parents genotyped from the local meta-population. Only 6% of the immigrants were likely to be genetically distinct, suggesting connectivity among sub-populations from the same genetic pool.

6.5 Interspecific Competition

There is a long tradition in marine ecology that has emphasised the importance of interspecific competition in population regulation and in shaping population and community relationships among species. More recently, it has been recognised that where competitive forces modulate species abundances, 'ecological drift' can be a major driver in structuring marine communities (Ford and Roberts 2020). Ecological drift, which refers to stochastic changes in the relative abundances of different species within a community over time due to the inherent random processes of birth, death, and reproduction, causes species abundances to fluctuate randomly, lowering diversity within communities and increasing differences among otherwise equivalent communities. Ecological drift has been little studied in the marine tropics.

Arguably, the best examples of interspecific competition in the marine tropics have been found on coral reefs where such interactions can be most readily observed. Hermatypic reef-building corals are known to be fierce competitors for space as it is necessary for them to settle and metamorphose into primary coral polyps (Connell et al. 2004; Chadwick and Morrow 2011). The space they occupy must be exposed to enough irradiance and to currents carrying food and carrying away waste products because of their dependence on zooxanthellae (endosymbiotic algae) for energy and their need to consume zooplankton for essential nutrients. Suitable space is thus often a limiting resource for the settlement, metamorphosis, and growth of reef corals.

Competition between corals and other sessile benthos has become more severe due to the impacts of global warming on coral reefs, with widespread declines in coral cover due in part in mass bleaching, disease, cyclones, and the loss of grazers, exposing large areas of reef and contributing to colonisation by other benthos, such as microalgae and sponges (Hughes et al. 2007).

The evidence to date points to strong competition among and between corals and other benthic organisms with complex networks of competitive interactions, rather than a simple linear competitive hierarchy between species (Chadwick and Morrow 2011). This allows for the coexistence of numerous species with space occupied by each coral or other benthic organism to change slightly over years or decades as individuals and colonies are damaged by neighbours while overgrowing others. This competitive networking creates a competitive mosaic of competing organisms in which space is maintained by the greatest number of species. This balance is disturbed by cyclones and other climatic events, such as global warming, disease, or predator outbreaks, that free up space for colonisation by new organisms (Chadwick and Morrow 2011).

Not all competitive interactions between and among corals or other sessile benthos are necessarily complex. A field study in Japan (Rinkevich and Sakai 2001) revealed a linear hierarchy of dominance among five species of *Porites*. Similarly, a study of colonies of the thin leaf lettuce coral, *Agaricia tenuifolia* on Belize reefs concluded that they may not necessarily benefit from competition (Chornesky 1991). Other studies, however, do point to more complex competitive hierarchies among corals. A long-term study on Heron Island, Great Barrier Reef, revealed a complex interplay among corals and with their environment to sustain high level of species richness (Connell et al. 2004). They found that reductions in species were linked to damage by storms and cyclones and elimination by competition. Differences in the size of the species pool of recruits influenced recovery after such reductions. There were variable responses to effects of recruitment with the species pool of coral larval recruits low in some places, but high in others. Recovery times varied greatly, between 3 and 25 years and this variable recovery time was attributed to the differential effects of cyclone damage, as some reefs sustained much more damage than others. High species richness was maintained in shallow water because superior competitors could not overlap their competitors without exposing themselves at low tide. At other times and in other places, species richness and diversity did decline over time. One cause for this decline was a concomitant decline in the physical environment. Species richness and diversity fell because the wave action during cyclones either killed corals in whole or part, or changed the drainage patterns over the reef crest, leaving corals to become dry at low tide (Connell et al. 2004). Species richness and diversity declined at deeper sites due either to heavy wave action or interspecific competition. What all this leads to is a hierarchy of competitive interactions that is highly dependent and structured by position on the coral reef as well as by the impact of disturbance (Figure 6.2).

Competitive networks are thus the norm for most coral reefs. A stony coral network was found on reefs in Taiwan (Dai 1990) as well as on reefs in the Red Sea (Abelson and Loya 1999). This does not mean that the outcome of such interactions is pre-ordained. The aggressive rank of some corals is similar between the Pacific Ocean and the Red Sea, but the outcome varies widely for some species among regions. For instance, a field survey of the brain coral, *Platygyra daedalea,* in the Red Sea revealed that about half of all interactions involved damage to other corals (Lapid et al. 2004). The clonal corallimorpharian, *Rhodactis rhodostoma,* was dominant over

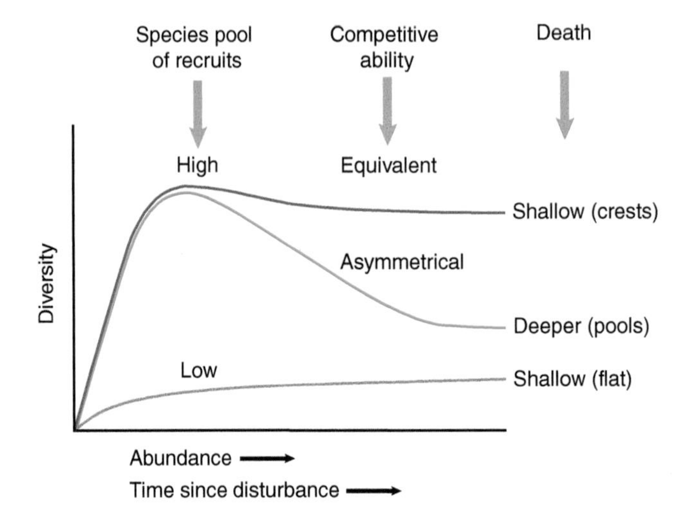

FIGURE 6.2 Model of the relationship of species diversity to variation in abundance in different coral reef zones following a disturbance. *Source:* Connell et al. (2004) figure 11, p.208. © John Wiley & Sons.

most stony and soft corals, but some massive faviid corals caused unilateral damage to the corallimorpharian polyps (Langmead and Chadwick-Furman 1999). In Jamaica, competitive contacts with an encrusting octocoral damaged stony corals, but not sea anemones or algae (Sebens and Miles 1988). Growth and survival of juvenile *Pocillopora verrucosa* and *Acropora retusa*, the two dominant branching coral species in the lagoons of Moorea, French Polynesia, were impacted by predation and competition with vermetid gastropods (Lenihan et al. 2011). Within two years, 20–60% of juvenile corals suffered partial predation by corallivorous fish and injured corals experienced reduced growth and survival. Competition with the gastropods reduced growth of both branching species, but also provided an associational defence against corallivory. Overall, biotic interactions, especially corallivory, had a greater negative effect on *Acropora* than *Pocillopora*.

Competition between reef-building corals and benthic algae is widespread with the outcome varying greatly among coral reefs (O'Brien and Scheibling 2018). Widespread replacement of corals by algae often reflects coral mortality due to disturbances such as nutrient pollution or coral bleaching, rather than competitive overgrowth, but may lead to competitive inhibition of coral recruitment with severe consequences for reef recovery. Algal inhibition of corals may involve one or more of the following: overgrowth, shading, abrasion, chemical pre-emption/recruitment barrier, and epithelial sloughing (Diaz-Pulido et al. 2010; Barott and Rohwer 2012). Coral inhibition of algae may likewise involve overgrowth, shading, abrasion, and chemical as well as stinging defence, space pre-emption, and mucus secretion (Chadwick and Morrow 2011; Barott et al. 2012; Swierts and Vermeij 2016). Some of these mechanisms constitute direct interference competition (e.g. overgrowth, abrasion), while others are indirect and exploitative (e.g. shading).

Phase shifts in all major reef regions between stony corals and non-algal alternate dominants (e.g. sea anemones, soft corals, ascidians, sponges, sea urchins) have been documented by Nordström et al. (2009) from case studies of long-term observations on the outcomes of competition among sessile reef organisms. Positive feedback may occur between the high abundance of alternate dominants and the continued mortality of reef-building corals (Figure 6.3), thus reinforcing an alternative state. Macroalgal cover on coral reefs can vary considerably as in the study of Connell et al. (2004) in which it varied from 15 to 85% and in the study of other Australian reefs in which it fluctuated from 41 to 56% cover in just over 18 months (Tanner 1995).

Macroalgal dominance can persist on coral reefs due to several processes (Figure 6.4) such as the level of eutrophication and herbivory. Cyanobacteria and other microbes also play an important role in maintaining algal dominance (Barott and Rohwer 2012) as turf algae and microalgae promote microbial overgrowth of coral. Complex flow patterns also transport organic matter and pathogens from algae to downstream corals. Cyanobacterial epiphytes may protect some red microalgae from herbivory, allowing the algae to persist for several years following an ENSO disturbance that killed previously dominant corals (Fong et al. 2006). The Littler et al. (2006) model (Figure 6.4) illustrates not just the impact of eutrophication and herbivory, but also the fact that each factor can mediate the others. For instance, high rates of herbivory can delay the impact of elevated nutrients, and low levels of nutrients can offset the impact of reduced herbivory. Coral recovery from phase shifts can be inhibited further by large-scale stochastic disturbances, such as cyclones, global warming, disease, and predator outbreaks (Littler et al. 2009).

The studies by Jompa and McCook (2002a, b) help shed some light on the competition between corals and macroalgae in the face of disturbance. They experimentally tested the effects of herbivory and nutrients on the competitive interactions between the hard coral, *Porites cylindrica*, and the brown alga, *Lobophora variegata,* on the Great Barrier Reef. The results showed that coral tissue mortality was strongly enhanced by the presence of the alga and this effect was significantly greater when herbivores were excluded; coral growth was unaffected. The addition of nutrients did not have a significant effect on the coral but had a small effect on algal growth and consequent coral tissue mortality when herbivores were excluded. One of their experiments utilised exclusion cages to test for effects of herbivores, and the removal

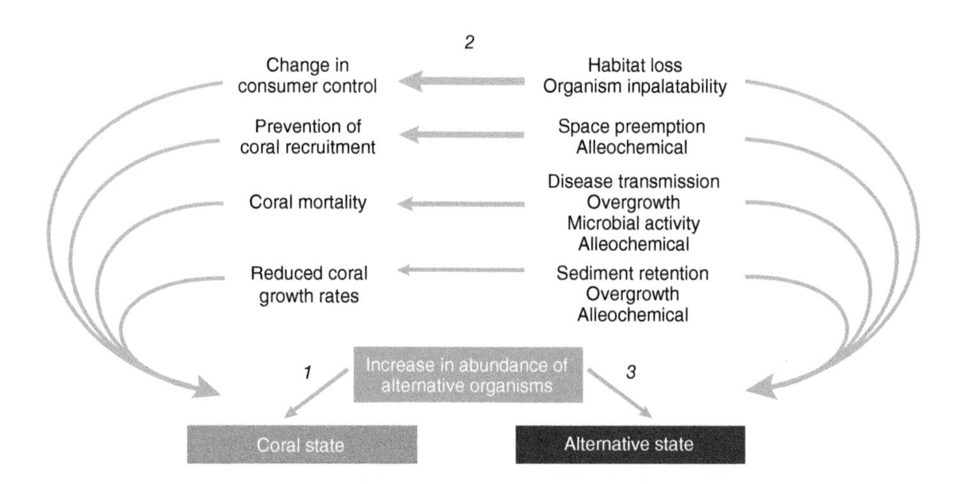

FIGURE 6.3 Potential feedback mechanisms supporting competitive phase shifts on coral reefs to dominance by nonreef builders: (1) opening of space due to coral mortality leads to an increase in the abundance of alternative benthic organisms; then (2) various mechanisms of positive feedback; and (3) directly or indirectly reinforce the alternative states. *Source:* Nordström et al. (2009), figure 3, p. 302. © Inter-Research Science Publisher.

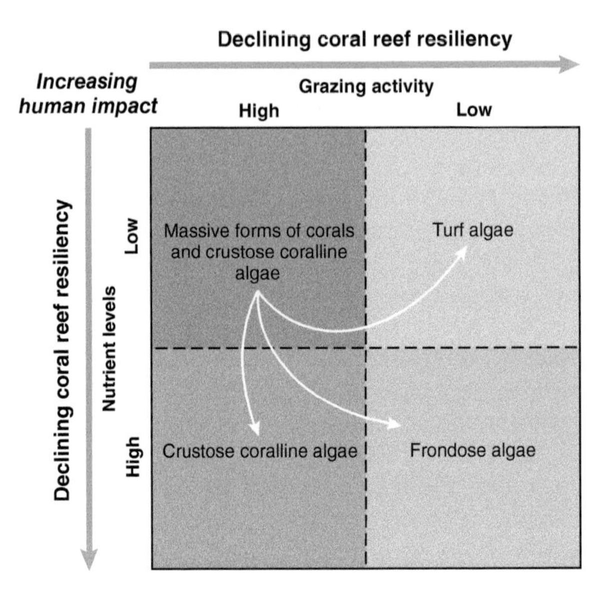

FIGURE 6.4 Relative dominance model in which the abundance of four major functional groups of space occupiers varies with level of human impact. *Source:* Littler et al. (2006), figure 1, p. 566. © Elsevier.

of algae and coral tissue at their boundary, to test for inhibition of the competitors. Comparisons of coral branches with coral tissue unmanipulated or damaged showed that the coral inhibited the overgrowth of the brown alga, but that the algae were superior competitors. Reduced herbivory resulted in faster algal growth and consequent overgrowth and mortality of coral tissue demonstrating the importance of herbivory to the outcome of the competition between coral and algae.

Bruno et al. (2009) disagree that coral-algal phase shifts are becoming more common. They compiled data from multi-year field surveys between 1996 and 2006 and concluded that most reefs worldwide lie between the extremes of algal- and coral-dominated cover and are not representative of phase shifts between two stable states. They noted that most cover on coral reefs in both the Caribbean and Indo-Pacific consists of organisms other than hard corals and macroalgae. Sponges, soft corals, and other sessile benthos are in fact becoming more common as replacement organisms (Nordström et al. 2009).

Chemical combat may also be a means of competition between corals (Dizon and Yap 2005), including soft corals that can damage stony corals. Langmead and Chadwick-Furman (1999) transplanted individuals of the soft coral, *Rhodactis rhodostoma* adjacent to stony corals at Eilat, Red Sea, and after 1.5 years observed that the soft corals damaged and overgrew branching *Acropora* corals but were killed by massive brain corals. Competition between soft and stony corals may be affected in the future by ocean acidification. In a series of experiments subjecting reef-building corals from the Gulf of Aqaba, Red Sea, to competitive interactions under present-day ocean pH (8.1) and future pH 7.6, coral growth was significantly less at lower pH under intraspecific competition (Horwitz et al. 2017). Shifts in the competitive hierarchy and a decrease in overall coral cover under low pH were observed, suggesting modified competitive performance under ocean acidification.

Corals exhibit different mechanisms of competition. One mechanism is pre-empting the settlement of competitors through rapid reproduction, or in the case of soft corals, rapid asexual reproduction of large clonal aggregations, and sexual production of dispersive propagules (Chadwick-Furman and Spielgel 2000). Another mechanism in corals is interference competition, and this form of competition occurs among corals and in interactions with sponges and algae. The main physical mechanisms for interference competition are skeletal overgrowth, redirection of growth, locomotion away from competitors, and physical interception of food particles. Corals employ a complex variety of intraspecific competitive mechanisms that include such physical responses. Caribbean reefs, for instance, are becoming increasingly dominated by non-hermatypic organisms like sponges, gorgonians, and zoanthids (Ladd et al. 2019) with the common hermatypic coral species, *Porites porites* and *Siderastrea siderea*, outcompeted by the aggressive encrusting gorgonian, *Erythropodium caribaeorum*.

A wide variety of cnidarians employ chemical mechanisms in interference competition. Aqueous toxic compounds occur in many coral species (Chadwick and Morrow 2011). The presence of such broad-spectrum compounds suggests significant use in mediating competitive interactions among corals, including the larvae of other competitive corals. Many stony corals thus produce toxic halos around themselves to reduce encroachment by coral competitors during larval settlement and thereafter. Another mechanism is the deployment of mucus to damage competitors. This has been observed in mobile mushroom corals. However, mushroom corals release nematocysts into their surface mucus, and these nematocysts may be the major mechanism causing damage to neighbouring stony corals. Like corals, sponges can also produce noxious chemicals that can give them a competitive advantage (Thakur and Singh 2016). These chemicals can be released in sponge mucus or directly into the water column. The bio-eroding sponge, *Siphonodictyon* spp., is known to bore deeply into living coral heads to secrete toxic mucus that acts as a carrier for the secondary metabolite siphondictidine, which inhibits coral growth. Thus, chemically active invertebrates such as sponges may complete with corals by undermining the skeletal integrity of corals and reducing their photosynthetic efficiency and respiration (Chadwick and Morrow 2011).

Biological mechanisms employed during competition mostly involve behavioural and morphological modifications. Elongated polyps or tentacles (sweeper polyps, sweeper tentacles, and bulbous marginal tentacles) are deployed by various species of cnidarians to defend space. It is possible that such elongated structures serve other functions, such as protection against predation, and such wide competitor-free zones visible around coral colonies likely are the net result of such aggressive behaviour.

Other sessile reef invertebrates have mechanisms to compete. Sponges can excavate and encrust carbonate substratum and can also spread laterally at rapid rates, overtopping hard corals (Ramsby et al. 2017). Some sponges send out filaments that perforate coral skeletons and excavate below the surface layer, undermining coral skeletal support and induce polyp retraction and death. Invasion of corals by sponges may take place under environmental conditions favourable for the growth of sponges, such as low irradiance or high levels of sedimentation, or following episodes of coral bleaching, disease, or attacks by crown-of-thorns starfish (COTS). This competitive advantage can lead to reduced coral cover and eventual replacement by these alternative dominants (Figure 6.4).

Ascidians are usually a minor component of coral communities, but they compete for space by their rapid growth rates, early sexual maturity, high fecundity, and lack of predators. They can spread rapidly on coral reefs subjected to high nutrient loads (Tebbett et al. 2019). Thus, sessile reef invertebrates such a sponges, ascidians, gastropods, and bryozoans may be able to complete successfully with stony corals by using both physical and chemical mechanisms.

Various factors may alter competition among sessile reef organisms, including both the relative age and time since contact between interacting organisms, which strongly influence competitive outcomes. Water motion may also alter competition by deflecting aggressive organs, thus reducing the aggressive reach of corals, allowing them to grow closer together. Water depth is another factor as macroalgae are more important competitors with corals in shallow well-lit waters, whereas ascidians compete frequently with corals in deeper waters. Aggregation of inferior coral competitors can also reduce the negative impacts of competition with superior competing corals (Chadwick and Morrow 2011).

Large-scale disturbances can ameliorate competition. Disturbances can remove certain types of organisms to shift the relative dominance of competing organisms. Rates of colonisation and spread of macroalgae on some reefs depend primarily on the extent of prior damage to coral colonies from disturbances. Human disturbances such as nutrient loading, sedimentation, and warmer temperatures can also alter competitive outcomes (Figure 6.4).

Interspecific competition among corals and with other organisms often results in diminished growth and reproduction in corals. Coral growth, fecundity, and survival are all reduced during competition among corals and with macroalgae (Idjani and Karlson 2007; Vieira 2020). Short-term growth rates of solitary fungoid corals were impaired, and their rates of mucus production and mobility enhanced, when they were surrounded by other fungiid species on Yakai Reef on Okinawa (Elahi 2008). Soft coral polyps are significantly larger and contain more ovaries at the centres of aggregations than along the edges where they contact other corals (Chadwick-Furman and Spielgel 2000). Interspecific competition with corals appears to reduce both body size and sexual reproductive rates in these soft coral species.

The growth of both corals and macroalgae is similarly affected when competing, particularly along the zone of contact (Jompa and McCook 2002b). Macroalgae may reduce coral fitness by inducing energy reallocation from reproduction to tissue repair along zones of contact and in the centre of coral patches. Competition with macroalgae likely diminishes the energy reserves that are critical to normal immune function and resilience of healthy corals, potentially contributing to chronic bleaching events and disease infection (Ritson-Williams et al. 2009). Macroalgae can impact coral replenishment through reduction of larval survival and coral fecundity, pre-emption of space for larval settlement, abrasion and overgrowth of new recruits, dislodgment of recruits settled on crustose algae, and changes to habitat conditions (Birrell et al. 2008). Some brown algae cause recruitment inhibition or avoidance behaviour of coral larvae, while others reduce the survival of new recruits of corals (Kuffner et al. 2006). These interactions can be highly species-specific.

Competition has been examined in reef fish (Forrester 2015), with an emerging generalisation that most deaths of reef fish are due to predation. However, competition for structural refuges within the reef matrix is often the "underlying mediator" of predation risk (Forrester 2015). Food shortages, in contrast to shortage of refuge space, result in less fish growth. Competition is influenced by body size, order of arrival on the reef, predator and prey behaviour, the spatial configuration of the reef habitat, habitat quality, and the spatial and temporal level of observation.

High species richness of reef fish populations is maintained by a variety of mechanisms. Munday (2004) tested the mechanism for coexistence between two closely related species of coral-dwelling goby, *Gobiodon histrio* and *G. erythrospilus*, at Lizard Island, Great Barrier Reef. These two species exhibited similar patterns of habitat use and nearly identical ability to compete for vacant coral space. The first species to occupy a vacant coral excluded the other species of similar body size (Munday 2004). These two species appeared to co-exist by means of a competitive lottery for vacant space. Additional experiments with six closely related goby species (Munday et al. 2001) found complex interactions among the species, especially with the competitive dominant, *Gobiodon histrio*. Two species competed for space with *G. histrio* while three other species did not, but for different reasons. Different mechanisms account for different competitive or non-competitive outcomes. There is thus no single relationship between overlap in resource use and the occurrence of interspecific competition.

Not all populations compete, and there are many adaptations for populations to avoid competition. For example, five species of flatfish in a tropical Brazilian bay partition resources to avoid competitive overlap (Guedes and Araújo 2008). Two species (*Achirus lineatus* and *Trinectes paulistanus*) have a specialised feeding strategy of feeding on polychaete worms mostly in the inner bay. The other three species (*Citharichthys spilopterus, Etropus crossotus,* and *Symphurus tessellatus*) fed on a wider range of benthic prey items. Trophic resource partitioning and spatial differences were enough to allow for the coexistence of all five species in the bay. The juveniles of the reef fish, *Lethrinus harak, L. obsoletus, Lutjanus gibbus*, and *Parupeneus indicus*, in a Taiwanese coral lagoon were found to utilize different habitats to avoid competition (Lee and Lin 2015). Large juveniles of *L. obsoletus* and most juvenile *L. gibbus* preferred coral reefs, while *P. indicus* was a dietary generalist in bare sand, coral, and seagrasses.

Resource partitioning occurs among tropical seagrass fish communities (Nagelkerken et al. 2006). The fish fauna of a typical tropical seagrass meadow consists of species feeding at three trophic levels: (i) herbivores, (ii) omnivores, zoobenthivores, and zooplanktivores, and (iii) piscivores. The herbivores partition food by specialising on seagrass epiphytes, seagrass leaves, or macroalgae from the seagrass bed with some species likely feeding also in adjacent mangrove waters. Fish of the second trophic level show temporal segregation in feeding habits, while the third trophic group consists of few piscivorous species. There can be seasonality in resource partitioning. For example, 13 fish species showed ontogenetic and/or seasonal changes in food-use patterns in seagrasses in Trang, southern Thailand (Horinouchi et al. 2012). Small fish generally preyed on small plankton, such as copepod larvae or harpacticoid copepods, but switched to different food items such as shrimps, crabs, detritus, and filamentous algae as they grew larger. This seagrass fish community comprised eight feeding guilds: large benthic/epiphytic crustaceans; detritus; planktonic animals; small benthic/epiphytic crustaceans; molluscs; invertebrate eggs; polychaetes; and fish feeders.

Resource partitioning of space occurs in tropical seagrass meadows. In seagrass beds surrounding Dongsha Island, South China Sea, tidal cycles were observed to play an important role in niche partitioning in fish communities (Lee et al. 2014). The density of herbivores, large-sized carnivores, and piscivores increased, and detritivore density decreased, during flood tides. Deeper water seagrasses during flood tides supported more space for herbivores and carnivores to forage than in intertidal meadows, whereas shallower seagrasses formed important temporary refuges for fish during ebb tides.

Spatial partitioning also occurs among seagrass plants (Ooi et al. 2014). In multispecies meadows in Pulau Tinggi, Malaysia, the distribution of smaller species, such as *Halophila* spp. and *Halodule uninervis,* was influenced by hydrodynamics and water depth, whereas the distribution of large, localised species (*Syringodium isoetifolium* and *Cymodocea serrulata*) were influenced by sediment burial, seagrass colonisation and growth, and physical disturbance. Thus, there was enough partitioning of resources over time and space to permit coexistence.

Competition occurs in seagrass ecosystems but has rarely been examined in tropical meadows. Seagrasses must compete with other macroalgae for limiting resources. In seagrasses of Southeast Asia, Duarte et al. (2000) tested competitive interactions by removing shoots of seagrass species in order of decreasing and increasing resource requirements for plant growth. When shoots of the dominant species, *Thalassia hemprichii,* were removed, the shoot density of *Enhalus acoroides* decreased and that of *Syringodium isoetifolium* increased when the shoots of all species with greater resource requirements were removed. The size of *Halophila ovalis* shoots declined when both *T. hemprichii* and *E. acoroides* shoots were removed. The density of *T. hemprichii* shoots increased only when the shoots of all other species were removed, suggesting that species interactions among seagrasses can be asymmetric. Investigating interspecific interactions between the seagrass, *Thalassia testudinum,* and the calcareous alga, *Halimeda incrassata,* in a mature seagrass meadow in south Florida, Davis and Fourqurean (2001) manipulated densities of both species and were able to determine that the presence of seagrass decreased the size of algal thalli and the growth rate, but the presence of *H. incrassata* had no effect on the seagrass. In algal removal experiments, the seagrass leaf tissue N:P ratio was significantly lower, whereas there was no impact for irradiance, suggesting that competition for N was the mechanism for the competition.

Amongst mangrove trees, competition can be severe, resulting in complex spatial patterns (Berger and Hildenbrandt 2000, 2003). In laboratory experiments between the mangroves, *Ceriops tagal* and *Ceriops australis, C. tagal* was the superior competitor at lower salinities and *C. australis* was superior at higher salinities, in agreement with field observations that both species are often segregated in their intertidal distribution (Smith 1988a). Laboratory experiments between the mangroves, *Sonneratia apetala* and *Aegiceras corniculatum*, for available light found that under full light, *S. apetala* grew faster than *A. corniculatum* (Chen et al. 2013). Under various shade treatments, growth was reduced in *S. apetala* but not in *A. corniculatum,* suggesting that under shaded conditions *A. corniculatum* can outcompete the other species, but not under full sunlight.

Competition between mangroves can be mediated by pollinators (Landry 2013). Observing pollinator activities between the mangroves, *Avicennia germinans* and *Laguncularia racemosa*, Landry (2013) found that insect visitation was significantly greater to *A. germinans* when both species were co-flowering. Visitation from bees, wasps, flies, and butterflies to *L. racemosa* increased significantly when the other species stopped flowering. When co-flowering, *A. germinans* outcompeted *L. racemosa* for pollinators. However, *L. racemosa* hermaphrodites self-pollinated autogamously when not visited by insects limiting male reproductive success.

Crabs are dominant organisms in mangrove forests, so it is unsurprising that a wide array of competitive interactions have been observed among species (Fratini et al. 2000, 2011; Murai et al. 2002; Wellens et al. 2015; Cannicci et al. 2018). Interference competition was observed between grapsid crabs and gastropods in a Kenyan mangrove forest (Fratini et al. 2000) with considerable interference by the crab, *Neosarmatium smithi,* on the gastropod, *Terebralia palustris,* when foraging on decaying mangrove leaves. Once the crabs found a leaf, they quickly dragged it back into their burrow, possibly to reduce intraspecific competition. Crabs will drag away a leaf even if it is inhabited by snails; crabs will strongly pull the leaf away or push the snails off the leaf. The success of the crab depends on both crab size and the density of the snails on the leaf. Competition for food was high in another East African forest as the grapsid crab, *Neosarmatium meinerti,* lost about half of its leaves to the crab, *Cardisoma carnifex. N. meinerti* may be regarded as a facultative kleptoparasitic species (Fratini et al. 2011). Four sesarmid species, *Neosarmatium smithi, N. asiaticum, N. malabaricum*, and *Muradium tetragonum*, inhabiting a Sri Lankan mangrove forest collected significantly more leaves of *Bruiguiera* spp. and *Rhizophora apiculata* than those of *Excoecaria agallocha* (Cannicci et al. 2018) There was no temporal segregation in feeding activity among the four species, resulting in high interference competition for leaves. *N. smithii* was always successful in winning interspecific contests.

Competition is fierce among crabs for leaves, propagules, and other litter, but competition does not always occur. Wellens et al. (2015) studied potential competition between the crabs, *Goniopsis cruentata* and *Ucides cordatus,* in mangroves in subtropical Paranaguá Bay, Brazil. They hypothesised that the herbivory rates of *G. cruentata* would be lower in mangroves where it coexists with the other species than in mangroves where *U. cordatus* is absent. Contrary to expectation, while herbivory rates for both species were high, they did not differ significantly between mangroves with and without the potential competitor, *U. cordatus*. Both species thus seem to have evolved means to avoid competition.

Scramble competition among males occurs in the fiddler crab, *Uca paradussumieri,* as found in the Kelang River estuary, Malaysia (Murai et al. 2002). When previously courted female crabs were approached by other males, the initial courter attempted to forcefully disrupt the courtship, perhaps serving to allow males the exclusive use of information on female reproductive condition. Before each semilunar cycle mating period, males were more likely to court females with whom they would later mate than other nearby females with whom they would not mate, suggesting scramble competition among males for females and that males were able to locate receptive females. The ability of males to assess female reproductive status may thus be a competitive advantage because it increases male mating success.

Estuarine fish seem to partition resources in mangrove estuaries to minimise competition. In a tropical mangrove estuary in Senegal, West Africa, 10 fish species were classified into five trophic groups: detritivores, zooplanktivores, benthivores, piscivores, and macrocarnivores, suggesting effective partitioning of food resources (Faye et al. 2012). Nine carnivorous fish species in the Paraíba do Norte River estuary, Brazil, consumed similar food items, but did so in varying amounts and proportions, such that niche overlap among species was low (da Silva Garcia and Vendel 2016). Partitioning of food by dominant species of estuarine juvenile fish in the Rio Mamanguape estuary, Brazil, into six trophic groups (zooplanktivores, benthivores, omnivores, detritivores, macrocarnivores, insectivores) probably facilitates coexistence among these fish, especially when they are abundant (Pessanha et al. 2015).

The interspecific competition of aquatic macrophytes may be density dependent and its outcome has importance for the zonation of monospecific and mixed stands in tropical estuaries. However, interspecific competition of two macrophytes was found not to be density dependent in the Itanhae′m River estuary, Brazil (Correia Nunes and Monteiro Camargo 2020). The salt marsh grass, *Spartina alterniflora*, had greater competitive ability than the string lily, *Crinum americanum,* with *S. alterinflora* limiting the aboveground biomass of *C. americanum* in lower estuary sediments regardless of their planting densities in laboratory experiments. In the lower estuary,

S. alterinflora might competitively exclude *C. americanum* regardless of their colonisation densities, while in the middle estuary, the arrival order may be important in organising mixed stands.

6.6 Mutualism

Mutualism is an interaction in which both species benefit. The apex of mutualism in the marine tropics is undoubtedly the symbiotic relationship between the photosynthetic dinoflagellate symbiont (especially of the genus *Symbiodinium*) and its reef-building coral host (Figure 6.5). The relationship is pivotal and obligate for the coral as evidenced by the impact of coral bleaching events in which the coral ejects its photosynthetic alga when seawater temperatures are excessively warm; corals may take years to, and often never, recover from such a catastrophic event. Recent evidence points to a third player in this symbiotic relationship: several lineages of uncharacterised apicomplexans (Kwong et al. 2019), poorly known microbes that are known in terrestrial environments to be parasites infecting various animals. Kwong et al. (2019) identified an apicomplexan lineage which they informally named "corallicolids" that was found in high prevalence across all groups of corals. Corallicolids are the second most abundant coral-associated microeukaryotes after the dinoflagellates and are therefore core members of the coral microbiome.

The precise nature of the relationship between corals and these symbiotic microbes is still being debated, but it is currently accepted that coral calcification and growth is dependent upon the metabolic processes of photosynthesis by the endosymbionts. Corals acquire carbon fixed by algal photosynthesis (glycerol and lipids)

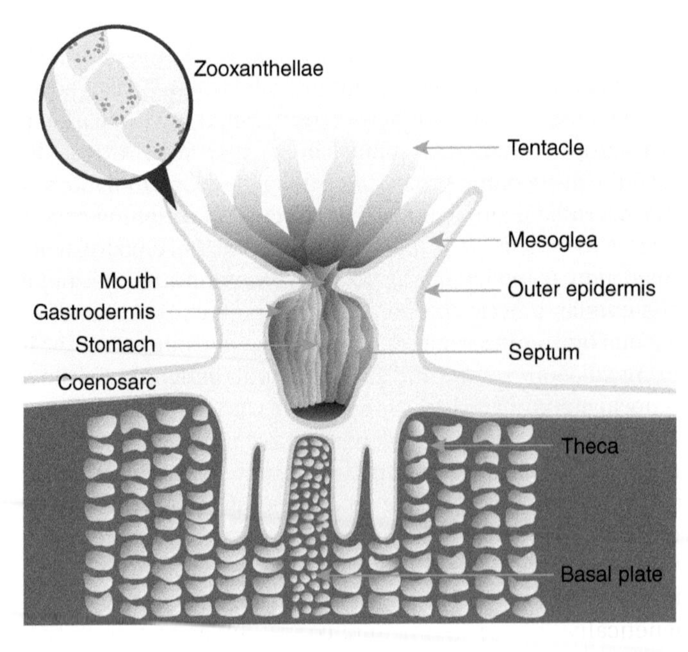

FIGURE 6.5 Rendering of the physical structure of a living coral polyp, zooxanthellae, and the massive skeleton beneath.

for metabolic needs, and the symbionts take up inorganic nutrients and CO_2 released as metabolic waste products by the coral (Sheppard et al. 2018). Densities of symbionts are normally within a narrow range of 1.0–5.0 × 10^6 cells cm^{-2} of coral surface. There is a wide range of rates of gross primary production and respiration for corals (Section 8.5), such that some corals are net autotrophic while others are net heterotrophic. Calcification rates increase coincident with increases in photosynthesis in relation to available light. The obligate relationship restricts corals to shallow well-lit waters where they modify their shape to maximize sunlight. Even in this endeavour the symbionts help in manufacturing 'sunscreen' to protect corals from the damaging effects of UV light.

Vigorous biotic turnover in the Late Triassic culminated in the evolution of coral reefs (Chapter 5), and skeletal stable isotopes and organic analyses support the hypothesis for the coevolution of the coral-zooxanthellate symbiosis (Stanley and Swart 1995; Muscatine et al. 2005). Given the mass extinctions of reefs (and other biota) at the end of the Triassic, their eight to ten-million-year eclipse and subsequent reappearance in the fossil record, it has been argued that algae and not the corals are the true "masters" of the reef (Stanley 2006).

Reef-building corals associate with a diverse array of eukaryotic and non-eukaryotic microbes, in addition to their relationship with *Symbiodinium* (Knowlton and Rohwer 2003). These dinoflagellates are now recognised as several, large genetically diverse groups (at least nine clades) while corals host diverse assemblages of bacteria, some of which seem to have species-specific associations. The diversity of these bacterial associates may have important ecological and evolutionary implications and such associations most likely co-evolved similarly to the zooxanthellae–coral relationship.

Molecular and genetic studies have shed new light on the nature of the symbioses between coral and their microbial symbionts (Goulet and Coffroth 2003; Putnam et al. 2012; Silverstein et al. 2012; Byler et al. 2013; Fabina et al. 2013; Baums et al. 2014; Kwong et al. 2019) and its level of flexibility and stability. A few studies have pointed to the stability of the symbiosis (Goulet and Coffroth 2003; Putnam et al. 2012). Ancient DNA recovered from century-old, dry specimens of octocoral show that the symbiotic association has been static over the past century, suggesting that at least for these gorgonians, adaptive shifts to new symbiont types is uncommon. The relationship between the octocoral, *Plexaura kuna,* and its dinoflagellate symbionts remained unchanged over a decade, despite transplanting the coral to new environments (Goulet and Coffroth 2003). The relationship, however, with an octocoral can be more complex. There was little differentiation in dinoflagellate composition, but significant differentiation was observed among *Symbiodinium* hosted by sympatric octocoral, *Gorgonia ventalina,* colonies (Andras et al. 2011). This result suggests that *Symbiodinium* may have an epidemic population structure, whereby corals recruit locally dominant strains that can change over time.

Such flexibility has been observed for several coral species. Corals of the genera *Acropora* and *Pocillopora* exhibited high intra- and interspecies flexibility in their dinoflagellate assemblages (Putnam et al. 2012). These coral genera are among the most environmentally sensitive. In contrast, massive *Porites* colonies are specialists, exhibiting low flexibility and harbouring taxonomically narrow dinoflagellate groups. Surveying *Symbiodinium* clades A to D from 39 species of geographically and phylogenetically diverse hermatypic corals, Silverstein et al. (2012) found at least two *Symbiodinium* clades (C and D) in at least one sample each of all the coral species; all four clades were discovered in over half of 26 coral species thought to be restricted to hosting a single clade. On average, 68% of all corals hosted two or more clades suggesting that the ability to associate with multiple clades is common in hard corals and that specificity is rarely absolute. Using a similar approach, Byler et al. (2013) identified the dinoflagellate symbionts hosted by the Red Sea coral, *Stylophora pistillata,* and found that juvenile coral colonies utilised transmission via both parent to offspring (vertical transmission) and acquired anew each generation from the environment (horizontal transmission). Coral colonies may acquire physiologically advantageous novel symbionts that are then perpetuated via vertical transmission by releasing progeny with maternally derived symbionts capable of subsequent horizontal transmission. This mode of symbiont inheritance can provide a mechanism for coral adaptation.

Physiological variation may be provided by genotype-to-genotype interactions, influencing the adaptive potential of symbiotic reef corals to selection during times of environmental change. The Caribbean elkhorn coral, *Acropora palmata,* associates mainly with one symbiont species (*Symbiodinium fitti*); sexual recombination was a source of genetic variation (Baums et al. 2014). When these data were examined at the level of the entire distribution of the coral, gene flow was an order of magnitude greater in the coral colonies than in the symbiont, suggesting that evolutionary processes between coral and symbiont are independent of one another. This raises some questions regarding the extent of co-evolution during times of environmental stress, such as with climate change. Fabina et al. (2013) simulated such "local extinctions" and the results suggest four factors greatly increasing coral-*Symbiodinium* community stability in response to climate change: (i) potential symbiotic unions are maximised by the survival of generalist hosts and symbionts; (ii) redundant or complementary symbiotic functions are provided for by elevated symbiont diversity; (iii) local recolonisation may be created by compatible symbiotic assemblages; and (iv) the persistence of certain traits associate with the diversity and redundancy function of the symbiont. These were only simulations, but they do point to the potential for *Symbiodinium* to facilitate coral persistence through environmental change.

Corals benefit by other mutualistic relationships. For instance, trapeziid crabs help to maintain coral health by clearing sediments off corals (Stewart et al. 2006). In a field experiment, mortality rates of two species of branching corals were significantly reduced by the presence of the crabs. All corals planted with crabs survived, whereas 45–80% of corals died within a month without them. Growth was slower and both tissue bleaching and sediment loads were higher for corals that survived without crabs. Laboratory experiments showed that crabs removed the more damaging, finer sediment particles from the corals (Stewart et al. 2006).

Mutualist crustaceans also play an important role in helping corals to survive. "Guard" crustaceans defend their coral host from predatory COTS (Sheppard et al. 2018). Predation was reduced by 15–45% in volume of coral consumed, by the presence of the crustaceans. The existence of a cooperative synergy in defensive behaviour of "guard" crustaceans was suggested by the fact that in the presence of both crustaceans, the volume of coral tissue lost declined by 73% which was significantly more than the expected 38% reduction from independent defensive efforts.

Mutualism benefits sponges on coral reefs, although their associations can be complex (Wulff 1997, 2008a). Three species of common Caribbean reef sponges (*Iotrochota birotulata*, *Amphimedon rubens*, and *Aplysina fulva*) had intimate associations that were mutually beneficial, although they were sufficiently different to be classified in different sponge orders (Wulff 1997). They also differed, despite their close association, in their susceptibility to such hazards as predation by angelfish, trunkfish, and starfish, breakage by storms, smothering by sediment, fragment mortality, and pathogens. Despite these differences, the sponges benefited from their mutualistic relationships and appeared to be able to decrease their loss rates by adhering to species that differ from them in terms of biochemistry, tissue density, and skeletal structure. A fourth sponge species (*Desmapsamma anchorata*) exploited these mutualisms by growing faster, fragmenting more easily, and suffering higher mortality rates (Wulff 2008a). The cost was high for the three asexually fragmenting sponges but low for this fourth "weedy" species, which benefited from association with the other three species by relying on its superior tensile strength and ability to reduce damage by physical disturbance. Life-history differences among sponges have a large role of play in determining the strength of their mutualism.

Mutualism is presumably common in other tropical marine habitats as it is in coral reefs, but evidence is lacking for many communities, such as estuarine, coastal and shelf benthos, and pelagic food webs. Mutualistic relationships occur in seagrass meadows and mangroves. Sponges, for instance, show mutualistic relationships in tropical seagrass meadows (Wulff 2012). Sponges typical of coral reefs are generally inhibited from living in seagrass meadows because they are vulnerable to predation by the large starfish, *Oreaster reticulatus*. However, the large coral reef sponge, *Lissodendoryx colombiensis,* expanded its range into meadows of the seagrass, *Thalassia testudinium* in Belize (Wulff 2008b). This sponge was a poor competitor among seagrasses because areas of many individual sponges were colonised by sponge species native to the seagrass meadow. This adherence resulted in the ability of *L. colombiensis* to overgrow and escape predation by the starfish. *L. colombiensis* has a naturally rapid growth rate and sizes of individuals fluctuate widely. These fluctuations were not evident at the population level as losses due to the starfish were balanced by a combination of rapid regeneration and growth, efficient recruitment, and protection of portions of individuals by the overgrowth of seagrass sponge species. The seagrass sponges benefit by acquiring a stable platform for growth within the meadow.

Facultative mutualism occurs in seagrass meadows in the Gulf of Mexico and in the Caribbean. The suspension-feeding mussel, *Modiolus americanus*, is usually associated with the seagrass, *T. testudinium* (Peterson and Heck 2001). Density manipulations of mussels resulted in the seagrass significantly increasing leaf widths and lengths and showed higher productivity than in the absence of mussels. In a predation experiment, predation on mussels was significantly lower in the presence of the seagrass. Thus, there were clear mutualistic interrelationships between the seagrass and the mussel.

There are other such mutualistic relationships between molluscs and seagrasses, and these can have a strong impact on growth and production of both groups. A facultative mutualism exists between the tropical seagrass, *Zostera noltii,* and the small lucinid bivalve, *Loripes lucinalis,* in intertidal meadows at Banc d'Arguin, Mauritania (de Fouw et al. 2016, 2018). In a healthy seagrass meadow, the bivalve facilitates the consumption of sulphides that are toxic to the plant and the plant provides food to the bivalve (Figure 6.6, left panel). When drought is triggered in the region, this facultative relationship breaks down and amplifies the degradation of seagrass (Figure 6.6 right panel). Field measurements comparing degraded patches of seagrass with patches that remained healthy demonstrated that bivalves declined dramatically in degraded patches with high sediment sulphide concentrations, confirming the breakdown of this facultative relationship. Large-scale environmental disturbances can thus result in the sudden breakdown of facultative mutualistic feedbacks in tropical marine ecosystems.

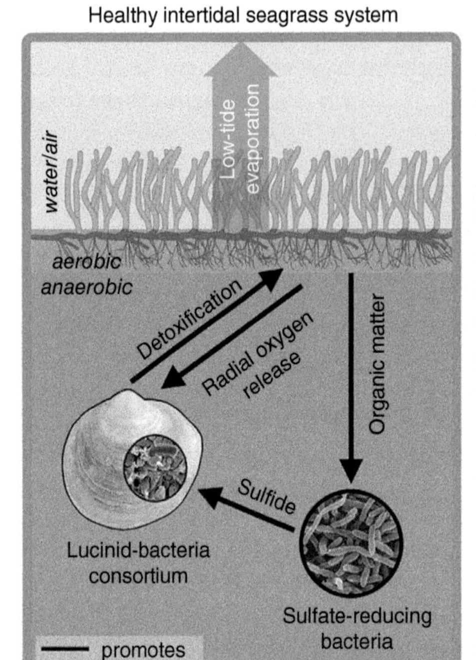

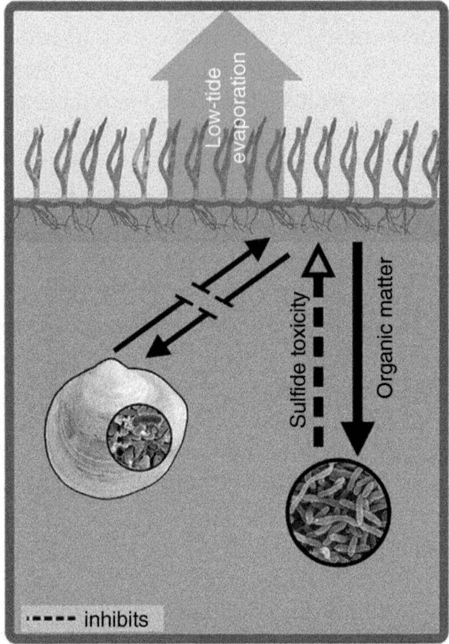

FIGURE 6.6 Mutualism between healthy seagrass and the lucinid bivalve, *Loripes lucinalis* and their functional relationship (left panel) and breakdown of the facultative relationship and amplification of seagrass die off when drought is triggered (right panel). *Source:* de Fouw et al. (2016), figure 1, p. 1051. © Elsevier.

Mutualism occurs amongst mangrove biota, but such interactions are poorly understood (Saenger 2002). The relationships between mangrove rhizomes and microbes are mutualistic and are crucial to the growth and survival of both plant and microbial assemblages. On and in mangrove roots and rhizomes lie a rich and diverse array of bacteria, archaea, protists, and fungi that weave a complex set of relationships with the trees, thriving off mangrove exudates, but also in turn decomposing complex organic molecules and fixing nitrogen (Inoue et al. 2020). Manglicolous fungi constitute a mutualistic consortium, with abundant species numbers and functional types within mangrove roots and rhizomes (Sarma and Raghukumar 2013). There is also mutualism among microbial groups, even among bacterial functional types, such as N_2-fixers and phosphate solubilisers. Even epiphytic algae on the surface of roots such as the pneumatophores of the mangrove, *A. marina,* likely have a strong mutualistic relationship (Naidoo et al. 2008).

Ants, molluscs, and sponges have important mutualistic relationships with mangroves. Ants belonging to the subfamilies Formicinae, Myrmicinae, Dolichoderinae, and Pseudomyrmecinae are common inhabitants of leaves of the mangroves, *A. marina* and *R. mucronata,* in eastern India (Gawas and Yogamoorthi 2015). These ants are mutualists as the plant provides habitat and the ant preys on herbivorous insects. Such mutualism is common among different ant subfamilies and mangroves (Saenger 2002). These relationships are like those found in tropical rainforests. Teredinid bivalves are mutualists and 'ecosystem engineers' in mangrove forests

(Hendy et al. 2014), tunnelling through substantial amounts of dead wood on the forest floor. These tunnels provide a refuge and food not only for the bivalves but also a refuge for many other organisms. In Indonesian mangrove forests, temperatures within the tunnels were significantly cooler than ambient air, and animal abundances were greater at cooler temperatures (Hendy et al. 2014). Animals co-existing with the bivalves included a diverse array of centipedes, spiders, crickets, fish, octopods, and polychaete worms. Many juveniles of these animals were found, suggesting that breeding was occurring within the tunnels. This mechanism is a means to increase intertidal biodiversity, facilitate wood decomposition, and to provide food for the bivalves.

As in seagrasses, facultative mutualism with sponges extends to mangroves (Ellison et al. 1996; Hunting et al. 2010). Sponge and ascidian epibionts act as barriers to isopod attack of roots of the mangrove, *Rhizophora mangle,* in Belize. The relationship between the root-fouling sponges, *Tedania ignis* and *Haliclonia implexiformis,* and *R. mangle*, is mutualistic (Ellison et al. 1996). Live sponges transplanted onto bare prop roots increased root growth two to fourfold relative to controls. The roots fouled by these and other sponges produced adventitious fine rootlets that ramified throughout the sponge tissue, enabling these fine rootlets to facilitate nutrient uptake. In turn, sponges growing on mangrove roots grew many times faster than those on PVC control tubes. Stable isotope analyses suggested that the mangrove roots obtained dissolved inorganic nitrogen and the sponges obtained carbon from the roots. The relationship between the sponge (*Tedania ignis*) and *R. mangle* may involve a chemical cue. In a field experiment, tannin concentrations were positively correlated with larval recruitment of the sponge. Roots significantly enhanced tannin and polyphenolic content in response to sponge fouling. The tannins may have elicited a positive behavioural response in sponge larvae and/or affected the structure of the fouling microbial biofilm on the mangrove bark (Hunting et al. 2010).

6.7 Commensalism

Commensalism is an interaction in which one species benefits and the other remains unaffected. It is often difficult in practice to discern if a given trophic relationship is a form of mutualism or commensalism. Commensalism has been infrequently studied in the marine tropics, and most of the literature focuses on the systematics of the commensal rather than on a clear delineation of the relationship. Much of the literature has focused on commensals in seagrass and other benthic habitats.

In one of the few mangrove examples of commensalism, Wasserman and Mostert (2014) found that the gobiid fish, *Redigobius dewaali*, lives in burrows of the mud crab, *Scylla serrata,* probably acting as an 'early warning signal' for the crab. Presumably, this fish was too small to be a prey item for the mud crab and thus the burrow offered a refuge from possible predation by piscivorous fish common to mangrove waters. The other advantage to living within a crab burrow is that the fish has a shorter distance to swim to adjacent intertidal feeding grounds.

Whether a relationship is commensalistic or not can be ascertained from an analysis of the gut contents of the supposed host. Several pearlfish found inside the holothurian, *Bohadschia argus,* and the starfish, *Culcita novaeguinea,* in Moorea Lagoon, French Polynesia were examined to determine if the associations were parasitic or commensalistic (Parmentier and Das 2004). Some associations were parasitic, some were commensalistic. Two species of pearlfish did not feed on host holothurian tissue while other pearlfish did so. Gobiid fish act as commensals with a variety of invertebrates and plants, such as soft corals, gorgonians, seagrasses, and algae. The gobiid fish, *Australethops wardi,* is a commensal in the burrow of the thalassinidean shrimp, *Neaxius acanthus,* in seagrass beds of the Spermonde Archipelago, Indonesia (Liu et al. 2008). The fish feeds almost exclusively on seagrass. Given the fish's diet, it is unlikely that the shrimp benefits from the relationship. The shrimp's burrow housed many species, all probable commensals: shrimps, the bivalve, *Barrimysia cumingii,* gammarid amphipods, the goby, *A. wardi*, a palaemonid shrimp and tube-building polychaetes (Kneer et al. 2008). The diet of *N. acanthus,* its commensal *A. wardi*, and another thalassinid shrimp, *Coralianassa coutierei,* consisted mostly of seagrass leaves, sedimentary organic matter, and seagrass epiphytes. Thalassinid shrimp burrows are thus common sites for commensals.

Alpheid and pontoniine shrimps are common commensals in tropical seagrass beds (Anker and Dworschak 2007; Kou et al. 2015). Alpheid shrimps are commensal to mud shrimps in various benthic habitats in the southern Caribbean Sea (Venezuela and Colombia) and in Philippine seagrasses where they live in the burrows of mud shrimps, stomatopods, other alpheids, and echiurans (Anker and Dworschak 2007). Both alpheid and pontoniine shrimps represent highly diverse groups in terms of their morphology, lifestyle, and habitat. Using molecular phylogenetic analyses, Kou et al. (2015) were able to distinguish between two groups of pontoniine shrimps: a "primitive group" and a "derived group" which separated those shrimps that were commensal with sponges, cnidarians, echinoderms, molluscs, polychaetes, ascidians, and other crustacea.

One of the difficulties in determining commensalistic relationships is that they can easily metamorphose into parasitism. In the Bahamas, the sponge, *Halichondria melanadocia,* is a commensal on the seagrass, *Thalassia testudinum*, benefiting from the presence of the seagrass up to medium shoot densities (Archer et al. 2015). The "net neutral effect" of the sponge on the seagrass is likely a balance between the negative effects of shading by the sponge with the positive effects of N and P supplied to the seagrass by the sponge. Environmental changes such as eutrophication have the potential to shift the relationship from commensalism to parasitism. Archer et al. (2015) simulated this effect and showed that if this relationship becomes parasitic, there is a reduction in the above-ground production of the seagrass. Commensalism may be as common as other trophic interactions in the marine tropics and is potentially a rich field for further research.

6.8 Parasitism

Parasitism may be more common than most other trophic interactions in the marine tropics (de Meeús and Renaud 2002). Marine parasites are associated a wide range of organisms from microbes to whales. Parasites are ecologically important, shaping host community structure through their effects on trophic interactions, food webs, competition, behaviour, population size, fitness, and biodiversity. It is often a challenge to decipher the ecological interactions of parasites as many live their lives secretly in intimate contact with their host. Parasites can function as both predators and prey and can be highly diverse and abundant. A study of trematode parasites on the Great Barrier Reef predicted that there may be as many as 1100–1800 species

(Cribb et al. 2014). Parasites even occur in mangroves, including on the trees themselves (Saenger 2002).

Parasitism is particularly well studied in coral reef fish. Justine (2010) listed over 580 host–parasite combinations within fish of New Caledonian reefs and stated that this list probably represents only about 2% of the possible number of metazoan parasites on reef fish. There were about 10 parasite species per fish (Justine 2010). Parasites include Platyhelminthes, copepods, nematodes, cestodes, and isopods. The most common hosts are serranid, lethrinid, and lutjanid fish. Many fish species from other tropical coastal and marine habitats have parasites. The mangrove red snapper, *Lutjanus argentimaculatus*, for instance, can become infected with the acanthocephalan, *Tenuiproboscis* sp. This fish is highly valued along the SW coast of India where Sanil et al. (2011) found 100% infestation by the acanthocephalan in the fish from those collected along the coast. Although it lodges in the fish's intestines, there appears to be no effect on fish health. The lack of health effects also occurs for many coral reef fish. The bridled monocle bream, *Scolopsis billineata*, is parasitised by a large isopod, *Anilocra nemipterid*, which attaches above the eye. This ectoparasite is associated with increased energy costs and decreased endurance but has no impact on the overall metabolic performance of the fish (Binning et al. 2014).

The role of ecological and phylogenetic processes is basic to our understanding of how parasite communities are structured. Muñoz et al. (2007) analysed the structure of communities of ecto-and endoparasites of 14 wrasse species from Lizard Island, Great Barrier Reef. They examined 264 individual fish from which 37 000 individual parasites and 98 parasite categories were recorded. The most prevalent and abundant parasites were gnathiid and cestode larvae. After controlling for host phylogeny, mean richness, abundance and biovolume of ectoparasites per fish species were positively correlated with host body size, whereas no such relationship was found for any of the endoparasites. This pattern may be explained by the fact that increased space (host body size) may increase the colonisation and recruitment of ectoparasites.

Various host–parasite interactions can combine with predator–prey and competitive interactions to cause density dependence in population growth. Forrester and Finley (2006) tested the hypothesis that parasitism and a shortage of refuges jointly influence the strength of density-dependent mortality in the bridled goby, *Coryphopterus glaucofraenum*. They used mark-recapture analysis to estimate host mortality of gobies infected with a copepod gill parasite, *Pharodes tortugensis*, and found that 63 marked gobies were infected while 188 were uninfected. Using the spatial scale at which gobies cluster naturally as an "ecologically relevant neighbourhood," they found that goby survival declined with increasing density, and this decline was sharper for gobies with access to few refuges than for gobies with more refuges. The negative effects of high density and a shortage of refuges were more severe for gobies that were parasitised than for gobies free of parasites. The parasite thus affects the fitness of the goby.

Parasites affect the early life history strategies of tropical fish. In the coral reef damselfish, *Pomacentrus moluccensis*, parasite prevalence increases with age, with the most marked increase occurring immediately after settlement of larval fish, suggesting that settling on the reef exposes young fish to parasites (Grutter et al. 2010). When larval and recently settled stages of the damselfish, *Neopomacentrus azysron* (which has pelagic larvae) and of the young damselfish, *Acanthochromis polyacanthus* (which does not have a pelagic larval stage) were exposed to micropredatory isopods, the proportion of infections did not vary significantly among the different host types (Grutter et al. 2008). However, mortality was greater in larval *N. azysron*, but mortality did not differ significantly between larval *N. azysron* and juvenile *A. polyacanthus* suggesting that a fish with a pelagic larval stage is no more susceptible to infection than a fish without a pelagic larval stage, but that all fish settling on the reef are exposed to potentially deadly parasites.

Relationships between parasites and corals can be just as complex as those with fish. Encystment by the trematode *Plagioporus* sp. in the hermatypic coral, *Porites compressa,* leads to an altered appearance of the coral polyp (Aeby 1991). The polyp's change in appearance and behaviour is thought to increase their vulnerability to predation. Parasitised coral showed a 50% reduction in growth compared to normal unparasitised *P. compressa* (Aeby 1991). In cage experiments, no significant differences were found in growth of corals compared with uncaged corals, although uncaged corals showed a marked reduction in the number of parasitic cysts, with the infected polyps being replaced by healthy polyps. The reduction in the number of cysts suggests that the rate of transmission of the parasite is enhanced by exposure of infected polyps to fish predation. Thus, the parasite does affect the ecological relationships of its host within the coral reef system.

Parasites can also affect mutualistic relationships. Cheney and Côté (2005) studied the exact nature of the interactions between the Caribbean cleaning goby, *Elacatinus evelynae,* and the longfin damselfish, *Stegastes diencaeus,* in relation to the availability of ectoparasites. They observed that cleaning interactions at sites with more ectoparasites were characterised by greater reductions in ectoparasite loads on damselfish clients and lower rates of removal of scales and mucus by cleaning gobies, whereas the opposite was observed at sites where ectoparasite numbers were lower, indicating that for the damselfish, cleaning was mutualistic in some locations but neutral or parasitic in others. The outcome of cleaning symbioses was thus quite variable in outcome and conflicting conclusions may be explained by parasite abundance.

Parasite infections can play an important role in population dynamics and carbon flow within food webs (Dunne et al. 2013) and appear to do so in at least one tropical marine ecosystem (Salomon et al. 2009). Infection of marine dinoflagellates (especially *Ceratium falcatiforme*) by the parasitic dinoflagellate, *Amoebophyrya* spp., showed that prevalence averaged only 2% in the tropical south Atlantic, but that the infection escalated to 7% over a six-day period concomitant to a 94% decrease in host cell numbers (Salomon et al. 2009). Infection was estimated to have killed 11% of the host cell population. About 7% of the carbon in the decaying *C. falcatiforme* population was transferred to parasitic dinospores which in turn became available to tintinnid ciliates. This represents a significant trophic link as marine food webs in the tropical South Atlantic are microbially dominated.

6.9 Predation

Vermeij (1977, 1978) was one of the first to suggest that predation is more intense in the tropics. He suggested that a noticeable increase in the sturdiness of snail shells in the Early Cretaceous was possibly in response to an increase in the evolution of powerful shell-destroying predators. Predation is intense in tropical marine environments, with countless predator–prey interactions. Arguably, the most spectacular examples of predation in the tropics are predation on living coral during outbreaks of the COTS and predation on seeds by sesarmid crabs in mangrove forests. Both have severe impacts on the structure and function of these habitats. There are other spectacular examples of predation, such as whale shark feeding on zooplankton, but the impacts of such predation on pelagic community structure and function are poorly understood (Rohner et al. 2015).

The crown-of-thorns starfish, *Acanthaster* spp. (Figure 6.7), is a complex of several sibling species and has been reported throughout the tropical Indo-Pacific from the Red Sea to the Pacific coast of Central America (Pratchett et al. 2014) but has never been reported in the Atlantic or the Caribbean. The enormous reproductive potential of the starfish is one of its most salient features and is responsible for its notorious population fluctuations. *Acanthaster* spp. can release millions of eggs and is a gonochoristic species in which male and female individuals must be close together and spawn simultaneously to reproduce as fertilisation rates decline with distance between the sexes. Males exude milky clouds of sperm while females exude translucent spherical eggs. It is unclear if the starfish spawn once or several times per year. Breeding and spawning tend to occur at temperatures above 27 °C in summer and usually in the late afternoon, but there is no apparent link between the timing of spawning and lunar cycles (Pratchett et al. 2014).

Typical of most asteroids, the life cycle of *Acanthaster* spp. begins with a pelagic-feeding larval stage and three brachiolaria stages during which brachiolar arms appear and become well developed. During the larval stages, phytoplankton is the main food, although they may exploit other foods. Predation on larval starfish may be high. After about 11 days, the juveniles settle by flexing the anterior body dorsally to orient the brachiolar arms against the substratum to test its suitability for settlement. Juveniles can be quite particular about where they settle, although they prefer habitats that are topographically complex. Some work suggests that biofilms covering the substratum influence settlement rather than the rugosity of the benthos (Pratchett et al. 2014).

FIGURE 6.7 The crown of thorns starfish, *Acanthaster* spp., preying on live corals. Image taken and used with permission of Ciemon Caballes, ARC Centre of Excellence for Coral Reef Studies, James Cook University, Australia. The photograph was taken in November 2014 at an unnamed reef NE of Helsdon Reef, northern Great Barrier Reef.

Following settlement, metamorphosis occurs in which a five-armed juvenile emerges. The juvenile now has two pairs of tube feet, a terminal tentacle, and a red optic cushion on each arm. The juvenile starfish adds arms at two-week intervals, and the body turns pink after three weeks. The starfish do not feed on coral during this phase but become cryptic, possibly as a mechanism to avoid predation. After around 20 months when they are about 15 cm in diameter, the starfish become more active during daylight. High rates of post-settlement mortality and the likely movement of larger juvenile starfish as they switch to eating corals result in patterns that are quite different to the initial abundance and distribution at settlement.

The growth and longevity of adult *Acanthaster* spp. are very plastic in that they are strongly dependent on local environmental cues, such as temperature, wave exposure, and food availability. In outbreak populations, the starfish can growth well beyond 35 cm and can live for more than eight years. *Acanthaster* spp. can resolve into distinct cohorts with a range of sizes and ages, even in outbreak populations. The growth pattern of the starfish is sigmoid with slow growth both when starfish first settle and when they start to feed on calcareous algae and when they attain sexual maturity at approximately 2+ years of age. Wide ranges in the size of individuals reflect marked differences in ages rather than variation in growth.

The capacity of COTS to devastate living coral cover is what separates them from many other coral predators, as most corallivores are limited by feeding on the relatively thin cover of living coral tissue or on the larvae (Rotjan and Lewis 2008). *Acanthaster* spp. feed by everting their stomachs through their oral opening and spreading it over the surface of living corals which can be digested within three to five hours by enzymes secreted through the gastric tissue. Many factors influence feeding preferences, including the nutritional content and growth form of the coral prey, coral defences, coral defences by commensal infauna, coral distribution, local environmental conditions, and prior state of individual starfish. The stony corals, *Acropora* and *Montipora,* are the preferred prey items. These dietary preferences can result in drastic changes in the structure of coral communities during outbreaks.

COTS can produce outbreaks in which overall abundance can increase over six orders of magnitude within one to two years. Outbreaks of *Acanthaster* spp. are thought to occur in two ways. First, some outbreaks consist of starfish that are of only a single cohort or year-class suggesting a single massive influx of new recruits. Second, outbreak populations include starfish from at least five or six different cohorts with a gradual increase over several years, indicating a progressive accumulation of individuals over several successive recruitments. This latter phenomenon may represent a mechanism by which primary outbreaks are initiated, which then give rise to massive numbers of offspring which lead in turn to secondary outbreaks on nearby or downstream reefs.

At least 246 outbreaks have been reported since 1990 throughout the Indo-Pacific region which is three times the number recorded for outbreaks prior to 1990. This increased frequency may reflect better monitoring and increased reporting rather reflecting a true increase in outbreaks. Outbreaks follow a pattern of progression in accord with the prevailing currents that can carry starfish larvae. On the Great Barrier Reef, the recurring pattern is a southward progression of outbreaks from 17 to 20 °S at a rate of 1° latitude every three years. The outbreaks also appear to spread northwards, but the progression is slower and less consistent; southerly and northerly progression has been attributed to movement of adult starfish between reefs, but the size frequency on different reefs strongly implies larval dispersal.

The rapid and dramatic increases of COTS can cause extensive and dramatic depletion of hard corals, although this is not always the case. Large aggregations of the starfish have been recorded throughout the Indian and Pacific Oceans, but destruction has been most evident in the southern and western Pacific, specifically in southern Japan, Micronesia, the Great Barrier Reef, and French Polynesia (Pratchett et al. 2014). This pattern mimics that of its preferred prey, *Acropora* spp., which tend to dominate reefs in the southern and western Pacific.

At the local scale, the greatest depletion of coral cover tends to occur on the leeward side of coral reefs. This is due to reduced water flow on the backside of reefs preventing dislodgement of starfish by strong currents or to high abundance of *Acropora* in these areas. Other environmental variables must be at play because the starfish are rarely found in semi-enclosed lagoons which often support their favoured prey items. *Acanthaster* spp. are rarely found on reefs deeper than 30 m even though such reefs are extensive and contain *Acropora*.

Outbreaks of the starfish can lead to indirect effects on coral reefs and have been linked to increased abundance of soft corals, algae, sea urchins, herbivorous fish, and motile invertebrates. Extensive coral loss can result in the decline in abundance of many coral organisms. Aside from herbivores, most reef fish decline in abundance in the aftermath of an outbreak, especially those that depend directly on corals for food or habitat. Starfish outbreaks also lead to severe declines in habitat heterogeneity as dead, exposed skeletons of hard corals are susceptible to erosion and form a monotonous habitat.

Several hypotheses have been offered to explain *Acanthaster* outbreaks, which generally fall into two groups that place importance either on factors affecting recruitment rates or on changes in the behaviour or survivorship of post-settlement individuals. No single hypothesis has universal support (Pratchett et al. 2014) as it is likely that several factors operate synergistically or antagonistically to foster outbreaks. It is harder to explain low densities rather than outbreaks given the fact that this starfish is predisposed to major fluctuations in population abundance. Several interrelated factors may keep population densities in check: (i) low and fragmentary coral cover, (ii) hydrodynamic conditions, and (iii) relatively high populations of predators. Outbreaks can partly be due to natural and anthropogenic causes, but increasing effects of outbreaks are not necessarily evidence of unnatural causes. Instead, it has been proposed that anthropogenic disturbances have predisposed coral reefs to be more susceptible to outbreaks, eroding their resilience to disturbance. Pratchett et al. (2014) have stated that "even more likely is that the pervasive effects of humans on coastal ecosystems have fundamentally altered the structure and function of both COTS populations and reef ecosystems, forever altering any semblance of a natural system."

Much of the current discussion on outbreaks is focused on two hypotheses: nutrient enrichment and predatory release. The first hypothesis holds that outbreaks of COTS arise due to enhancement of larval survivorship through nutrient enrichment. A pulse of nutrients leading to a phytoplankton bloom leads to an increase in food for starfish larvae and increased survivorship. Further, it has been suggested that outbreaks are related to river flood events. However, there are extended delays between such events and reported outbreaks and this has been attributed to limitations in the detection of outbreaks, although this linkage is tenuous. Spatial patterns of outbreaks do not relate to the location of flood events as outbreaks have been not been found on coral reefs close to riverine floods, but on more distal reefs, making it difficult to discern a link between events. The southward progression of waves of outbreaks on the Great Barrier Reef do not match with the pattern of northwards progression of flood waters. Also, outbreaks on atolls and low islands cannot be readily linked to terrestrial runoff. Despite these inconsistencies, terrestrial runoff may lead indirectly to a greater propensity of outbreaks as the nutrient enrichment hypothesis provides one of the more plausible means by which human influences cause outbreaks. Modelling analysis of COTS on the Great Barrier Reef (GBR) indicates that temperature and flood plume exposure are the best predictors of COTS prevalence and larval connectivity potential and minimum SST are the best predictors of COTS outbreaks. Major hotspots of COTS activity are on the mid shelf of the central GBR and the southern Swains Reefs (Matthews et al. 2020).

The predator removal hypothesis assumes that COTS are normally regulated by high levels of post-settlement or adult predation. This idea has gained credence since the publication of studies (Dulvy et al. 2004; Sweatman 2008) that reported decreased incidence or severity of outbreaks in areas subject to fisheries regulation. Sweatman (2008) compared the rates of occurrence of starfish outbreaks on individual reefs of the Great Barrier Reef that were open to fishing and on reefs where fishing was prohibited: 75% of reefs that were open to fishing suffered outbreaks compared to 20% for reefs that had been closed to fishing. This result implies that harvesting of coral reef fish may increase the likelihood that outbreaks of COTS will occur by removing predation pressure on the starfish. The heavily fished emperor (family Lethrinidae) preys on COTS as do triggerfish and pufferfish, and normal densities of these fish may regulate COTS numbers.

Outbreaks of COTS typically end with population collapse, due to starvation or movement to find food further afield. An alternative explanation is that the starfish becomes more susceptible to disease. COTS are especially prone to disease and exhibit symptoms of pathogenesis.

Climate change is an additional cofactor that must be considered in COTS outbreaks. A laboratory study found that increasing the incubation temperature 2 °C resulted in faster development of late-development starfish larvae (Uthicke et al. 2015). While food was the main driver, temperature shortened development time by 30% and increased the probability of survival to settlement by 240%. COTS may also be affected by the impacts of ocean acidification (Kumya et al. 2017). Experiments were conducted to partition the direct effects of ocean acidification on juvenile starfish and indirect effects through changes in their crustose coralline algal (CCA) food. *Acanthaster planci* juveniles were grown in three pH levels (7.9, 7.8, 7.6) and fed CCA grown at identical pH. Consumption of CCA increased when the juveniles were grown at low pH and when they were fed CCA grown at low pH. COTS fed CCA grown at pH 7.6 grew fastest, but the pH/pCO_2 that they were reared in had no direct effect on growth. Increased growth may be driven by enhanced palatability, increased nutritive state, and reduced defences of their CCA food. These results suggest that ocean acidification may increase the success of juvenile starfish and facilitate development.

The other spectacular example of predation in the marine tropics is predation by sesarmid crabs on mangrove propagules. Seed predation is a major factor controlling the structure and function of all forested ecosystems (Perry et al. 2008), and mangroves are no exception. Consumption of mangrove propagules (mangroves are viviparous so the unit of dispersal is a propagule, not a true seed) greatly affects natural regeneration and influences the distribution of species across the intertidal zone (Smith 1992; Krauss et al. 2008). In a series of field experiments in which mangrove propagules were tethered in the forest and then the amount of consumption was determined over time (Smith 1987), an inverse relationship was found between the dominance of the species in the canopy (*A. marina, R. stylosa, Bruguiera gymnorrhiza,* and *B. exaristata*) and the amount of predation on its propagules (Figure 6.8). This relationship was not found for *Ceriops australis*. Subsequent caging experiments to study the establishment and growth of *A. marina* in the mid-intertidal zone (a region where the species is usually not found) found that when protected from crabs, *A. marina* propagules survived and grew, but virtually 100% of the uncaged propagules were consumed by crabs, illustrating the importance of seed predation by sesarmid crabs on species composition and structure of mangrove forests. Smith (1992) presented additional data indicating that predation on propagules may effectively preclude the establishment of *A. germinans* and *Laguncularia racemosa* in forests dominated by *Rhizophora mangle* and *Pelliciera rhizophorae* along the Pacific and Caribbean coasts of Panama. However, the amount of predation on *R. mangle* and *P. rhizophorae* propagules in forests dominated by *A. marina* and *L. racemosa* was not high, suggesting that other factors, such as competition, play a role in structuring mangrove communities.

Predation on propagules may have an impact on mangrove forest succession. For example, no *A. marina* saplings were observed in a forest in which the canopy size class was dominated by this same species (Smith 1988b). The sapling size class was dominated by *B. gymnorrhiza, B. exaristata*, and *C. australis*. Smith's (Smith 1988b) predation study showed that >95% of the *A. marina* propagules were consumed in this forest, but less than 25% of the propagules of the other species were eaten, raising the interesting question as to how this forest came to be dominated by *A. marina*.

The relationship between sesarmid crabs and mangroves is a close one, and subsequent studies have confirmed the importance of crabs in regulating mangrove species composition and community structure (McGuiness 1996; Clarke and Kerrigan 2002; Bosire et al. 2005; Lee 2008). Lee (2008) found a simple but significant relationship between the number of crab species and the number of mangrove species

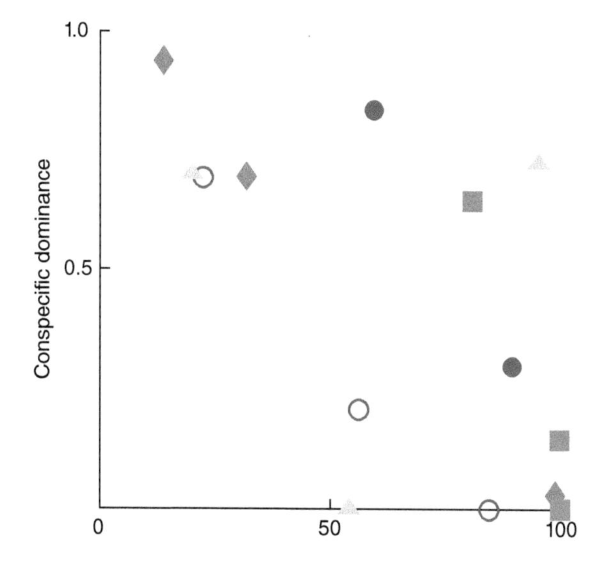

FIGURE 6.8 Relationship between conspecific dominance and cumulative amount of predation (%) on five mangrove species from north Queensland, Australia. Square = *Avicenna marina*, Solid circle = *Bruguiera exaristata*, Open circle = *B. gymnorrhiza*, Triangle = *Ceriops tagal*, Diamond = *Rhizophora stylosa*. *Source:* Smith (1987), figure 3, p. 271. © John Wiley & Sons.

(Figure 6.9), underscoring the close-knit relationship between crabs and mangroves. Like Smith (Smith 1988b, 1992), McGuiness (1996) found heavy predation on *C. tagal* propagules with 83% being damaged or consumed within three-months; poor dispersal and establishment were the main reasons why this species was limited in its distribution. Clarke and Kerrigan (2002) further tested under what conditions mangrove propagules are negatively affected by crabs by conducting experiments testing whether planting, intertidal position, and/or canopy gaps influence propagule predation. They found that predation ranged from 22 to 100% with the species order of predation being *Aegiceras corniculatum* > *A. marina* > *Bruguiera parviflora* > *Aegialitis annulata* > *B. exaristata* > *C. australis* > *Ceriops decandra* = *B. gymnorrhiza* > *R. stylosa*. The major treatment effects were whether the propagules were prone or implanted; canopy type had the largest magnitude of effects across all species. Propagules in the prone position were preyed upon more readily than those that were implanted or in canopy gaps. No species had significant effects that supported the hypothesis that dominance-predation was operative. Predation by crabs showed no significant effects of intertidal position although predation changed often with position across the intertidal zones. In contrast to Smith's (1987,1992) results, it was concluded that crab predation has no major influence on mangrove zonation.

Crab predation can affect the reforestation of mangroves. Seedling recruitment in a nine-year old *R. mucronata* plantation was examined by Bosire et al. (2005) to test the effects of pruning, position, and species found to be growing spontaneously in the plantation. Predation intensity was greater in the pruned part of the plantation and on prone rather than planted propagules, while *R. mucronata* was preyed upon least compared with *C. tagal* and *B. gymnorrhiza*, suggesting that predation favours the establishment of *R. mucronata*.

One of the hypotheses to explain the community structure and species composition of mangrove forests is the 'tidal sorting' hypothesis, which states that smaller propagules will be carried farther inland than larger propagules, stranding and establishing greater numbers in the high-intertidal zone, with crab predation having no or little impact on mangrove species zonation. In an explicit test of this hypothesis, Sousa et al. (2007) measured dispersal patterns of propagules of *A. germinans*, *R. mangle*, and *L. racemosa* in a series of experiments conducted along the Caribbean coast of Panama. Contrary to the hypothesis, all three species established best in the low-intertidal zone and poorly in the high-intertidal zone. But in accordance with the

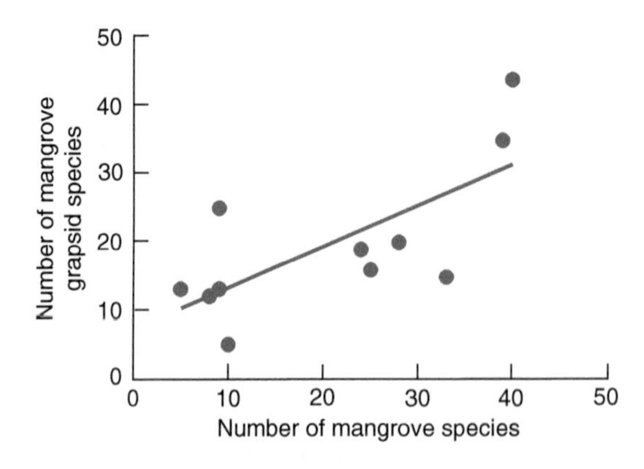

FIGURE 6.9 Relationship between the species richness of grapsid crabs and mangroves. Regression has an r^2 value of 0.508, $p < 0.02$. *Source:* Lee (2008), figure 1, p. 20. © Elsevier.

hypothesis, the large propagules of *Rhizophora* grew better than the other two species in a turbulent back-reef habitat. However, they rejected the tidal sorting hypothesis.

Although sesarmid crabs are crucial in shaping mangrove forest community structure and functioning, other invertebrates can play an important structuring role. Ocypodid crabs play the same role as sesarmid crabs in terms of predation and organic matter processing in Neotropical mangroves (Cannicci et al. 2008). Ocypodid crabs process large amounts of organic matter in both Neotropical and Indo-Pacific mangroves and function as ecosystem engineers.

The feeding ecology of mangrove crabs is more complex than was initially thought. van Nedervelde et al. (2015) tested whether (i) crab density influences propagule predation rate; (ii) crab size influences food competition and predation rate; (iii) a crab preys at different rates according to propagule and canopy species cover; (iv) vegetation density is correlated with crab density; and (v) food preferences of crabs are determined by size, shape, and nutritional value. They found that in mangroves of Gazi Bay, Kenya, crab density correlated with the density of propagules, but crab competitive abilities were unrelated to their size and crabs preferred smaller food items with a lower C:N ratio. *A. marina* propagules were consumed more rapidly than those of *C. tagal,* except within forests of *C. tagal*; crab density negatively correlated with the density of *A. marina* trees and pneumatophores. There is thus a mutual relationship in which vegetation density influences crab density and crab density affects propagule availability and hence vegetation recruitment rate.

Leaf-eating and wood boring insects and herbivorous insects can play a strong role in structuring some mangrove forests (Cannicci et al. 2008). Early work on the role of insects was problematical and conflicting (Saenger 2002), but later work stressed the importance of insects on sublethal effects of predation. For instance, Sousa et al. (2003) investigated the influence of propagule size and pre-dispersal insect damage on the establishment and growth of *A. germinans*, *L. racemosa*, and *R. mangle* on the Caribbean coast of Panama. All three species suffered moderate to high levels of attack by larval insects and exhibited considerable intraspecific variation in mature propagule size. Insect predation was independent of propagule size. The effect of sublethal tissue damage or loss on the subsequent growth of established seedlings varied among the three mangrove species. For an *Avicennia* propagule of a given size, the more tissue lost, the slower the growth of the seedling. The response for *Laguncularia* was all- or-none, whereas the growth of *Rhizophora* was like the effect on *Avicennia*: if an infestation did not completely girdle the propagule the seedling survived but grew at a reduced rate. Establishment and early growth of mangroves thus depend on significant differences in seedling performance due to natural levels of variation in propagule size and pre-dispersal damage by insects.

Some pelagic communities in mangrove estuaries exhibit predator–prey interactions. In the warm temperate mangrove estuaries of eastern South Africa, late-stage larvae of the estuarine round herring, *Gilchristella aestuaria*, were most dense in relation to mangrove presence, turbidity, copepod prey density, and competition pressures by predatory mysids (Bornman et al. 2019). The round herring larvae negatively selected the calanoid copepod, *Paracartia longipatella*, despite co-occurring at high densities, but larvae consumed more of the larger calanoid copepod, *Pseudodiaptomus hessei*, within two mangrove estuaries than in two non-mangrove estuaries, despite other prey species occurring at similar densities. This selective feeding behaviour may be the consequence of the relatively low prey densities as well as the lower turbidity in the mangrove estuaries.

Seed predation occurs in seagrass meadows, but information on variations in predation pressure and the extent to which it alters seagrass community structure and composition is scarce. In a series of tethering experiments, Orth et al. (2007) examined seed predation on the Western Australian seagrass, *Posidonia australis,* and found that a diverse group of crustaceans were responsible for the predation. They found that the seeds of a sub-dominant seagrass, *Halophila ovalis,* were also preyed upon by gammaridean amphipods. The predation was intense enough to greatly influence seagrass community composition and structure.

Predation occurs in many forms in seagrass beds and most predation episodes can be intense (Heck and Orth 2006). The presence of seagrass can affect the intensity of predator–prey interactions as seagrass shoots significantly reduce the effectiveness of fish predation on a variety of epifauna relative to their success on adjacent bare substrates. One issue that remains unstudied is whether predators above a seagrass meadow might be influenced by the height of seagrass leaves and the degree to which they overlap each other which could have significant impacts on the ability of down-looking predators to detect prey. Another issue that remains controversial is whether there is a threshold density at which predation becomes more efficient; some evidence points to a non-linear, monotonic decrease in the relationship between seagrass density and predation rate (Heck and Orth 2006). Predation rates have generally not been found to differ significantly between light and dark periods, although different predators forage at different times of the day and there are seasonal and annual differences as predator composition changes.

Most studies support the notion that predation intensity is negatively correlated with seagrass density. However, as suggested by Canion and Heck (2009), the problem with most studies that have examined the issue is that the same number of predators and prey across a gradient of seagrass density treatments has been used. Canion and Heck (2009) subsequently conducted a series of experiments in which they increased the number of predators and prey as seagrass density increased. They found that the only significant difference in predation rate was between the unvegetated treatment and the seagrass treatments, suggesting that increasing density of seagrass does not necessarily provide increased protection for prey.

A meta-analysis found that growth of macrofaunal species was significantly greater in seagrass meadows than in bare substrates (Heck et al. 2003). Models of small fish growth in relation to seagrass density suggest two possible modes of predation. The first model is a bell-shaped curve in which growth is maximal at intermediate seagrass density, whereas the second model suggests that the response is a non-linear decline in growth with increasing seagrass density.

While there is good evidence to support the idea that structural complexity increases the chance of prey survival, there is much less information known about the effects of seagrass complexity on large predators. Heck and Orth (2006) postulated that larger, more mobile predators may respond more to landscape features such as patchiness or fragmentation, size of bed and proximity, and connectivity to other habitats, such as coral reefs or mangroves, than to seagrass complexity. The evidence from several studies (Irlandi and Crawford 1997; Hovel and Lipcius 2002) supports this idea.

The issue of scale is an important one and has drastic outcomes for predator–prey relations (Figure 6.10). At the cm scale, the important structural issues are shoot density, leaf surface area, leaf morphology, and biomass (Heck and Orth 2006). At the m scale, patch size and shape and distance are the important spatial parameters, while at the km scale, the numbers of patches, connectivity with other habitats,

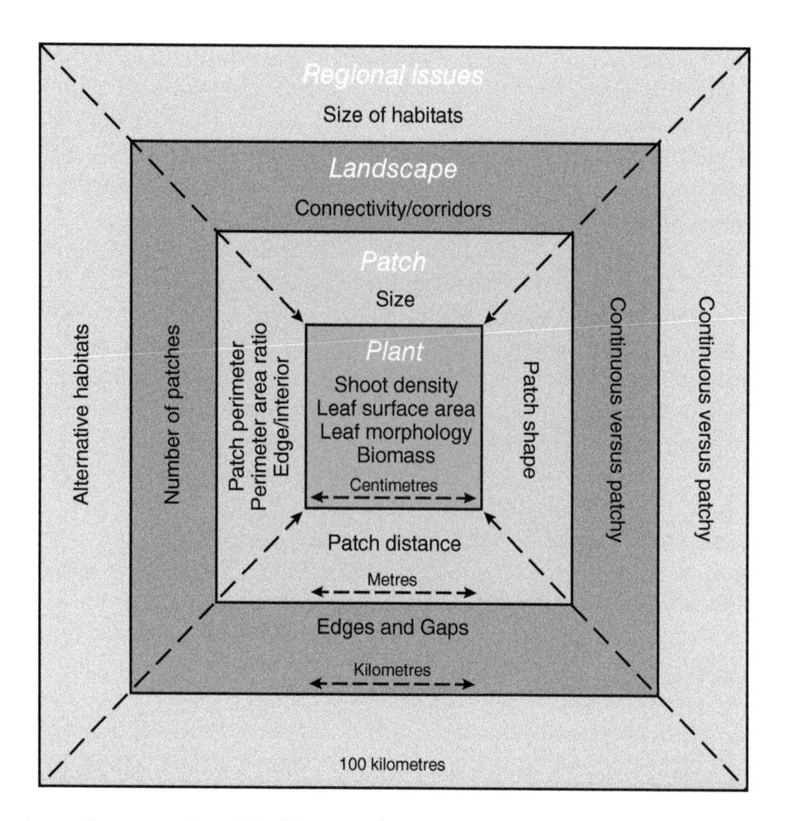

FIGURE 6.10 Conceptual model of how predator–prey interactions in seagrass beds may vary at different spatial scales. *Source:* Heck and Orth (2006), figure 5, p. 544. © Springer Nature Switzerland AG.

edges, and gaps in patches and whether seagrass beds are continuous or patchy are the main spatial features. At the scale of hundreds of km, the important spatial features are whether there are alternative habitats and the sizes of these habitats relative to seagrass habitat.

Predation can be intense amongst coral reef fish and invertebrate populations. Mid-trophic level piscivores (meso-predators) target new recruits on coral reefs and such predation is so intense that it can be a bottleneck for successful recruitment of some fish species (Hixon 2015). Because of intense predation, reef fish display several morphologies and behaviours associated with both evasion and capture. While foraging, different predators can interact with each other, cooperatively or negatively, and different prey species can similarly interact in ways that can increase or decrease their being caught. Direct density dependent mortality in reef fish, a net result of predation, can regulate local populations. This can involve interactions among different species of predators (synergistic predation) and forcing prey species to compete for space on the reef. Thus, piscivores can affect the structure and composition of reef fish communities by affecting the relative abundance of prey species. The relative recruitment rates of prey species can also be affected by the differential colonisation of coral reefs. Multiple piscivores can consume the same prey species or different species affecting abundance at lower trophic levels.

Predators commonly select the most abundant prey, but such is not always the case. Almany and Webster (2004) found that generalist predators had a greater impact on less abundant fish species, thereby reducing species richness. Disproportionate effects were evident during a narrow window between settlement and one to two days later, underscoring how predator-prey interactions following settlement can influence prey community structure. Thus, predators can reduce prey diversity. Early post-settlement mortality can indeed be a bottleneck for prey species. Using meta-analysis techniques, Almany and Webster (2006) found that for 24 taxonomically diverse species, an estimated 56% of juveniles were consumed within one to two days of settlement.

A series of elegant experiments indicate that predation is indeed stronger throughout the marine tropics (Freestone et al. 2011, 2013, 2020). Predator-exclusion experiments conducted on sessile invertebrates in the western Atlantic Ocean and the Caribbean showed that after three months, there was no effect on species richness in the temperate region, but in the tropics, communities were 2–10x richer in species in the absence of predators than when predators were present (Freestone et al. 2011). In other predator-exclusion experiments focusing on non-native tunicates in a heavily invaded habitat, the effect of predation on species richness was more than 3x greater in tropical Panama than in temperate Connecticut (Freestone et al. 2013). In Panama, there were large reductions in abundance and often exclusion of non-native tunicates from experimental communities, whereas predation reduced the abundance of only one non-native tunicate but not the abundances or exclusion of any other non-native tunicates in Connecticut. Similar experiments with both tropical and temperate seagrass communities found that predation decreased sessile invertebrate abundance, richness, and diversity in tropical but not temperate seagrass habitats (Freestone et al. 2020). Further observations of predators showed higher but variable consumption rates on invertebrates at tropical relative to temperate latitudes. These experiments show that predation can both limit local species abundance and shape patterns of regional coexistence in the tropics. Species richness was always reduced in the tropics whether the experiments were conducted for one day or for several days. These experiments thus demonstrate the magnitude to which variation in predation pressure can contribute to the maintenance of tropical species diversity (Freestone et al. 2011, 2013, 2020).

Predator–prey interactions are intense on tropical rocky intertidal shores. In one of the earliest studies of predation in the tropical rocky intertidal, Bertness (1981) examined predation by fish and crabs on the hermit crab populations in the Bay of Panama. He found that predation was most intense in the low-intertidal zone and least in the high-intertidal where physical stress (high temperatures, desiccation) overwhelms biota. The hermit crab community was spread throughout the intertidal shore and subject to other pressures such as competition; the gastropod, *Clibanarius albidigitus,* was found throughout the mid- and high-intertidal partly due to competition with the hermit crab, *Calcinus obscurus*, from the low-intertidal zone. Predation was avoided by these crabs as they have a behavioural escape response that is well-developed, although *C. albidigitus* can withstand thermal stress more than *C. obscurus*. Therefore, the intertidal distribution of hermit crabs is the net result of the interplay of predation, competition, and physical stress and the skills of individual species at minimizing exposure to all three.

The effects of predation, competition, and physical stress can be both direct and indirect. Garrity and Levings (1981) examined interactions between the predatory gastropod, *Purpura pansa,* and its gastropod prey, *Nerita scabricosta,* along the Pacific coast of Panama. They found that both snails have evolved avoidance behaviour in response to heavy predation pressure by fish and harsh physical conditions to the extent that both species restrict movement and limit foraging time which results in a cyclic overlap of their distributions. *Purpura pansa* was unable to control the distribution of its prey as it elicits an avoidance response in *N. scabricosta*. In a heterogeneous habitat, *N. scabriscosta's* distribution was restricted and its local abundance further reduced by avoidance responses to stable aggregations of *P. pansa* (Garrity and Levings 1981). These aggregations were hypothesised to result in the creation of local patches of substrate that are free from the effects of *N. scabriscosta*. These aggregations are centred on microhabitat that is safe from fish predation. This tropical rocky shore is therefore subject to a mosaic of interactions and effects induced by predation by fish and by *P. pansa*, leading to a four trophic-level community. Further experiments showed that foraging patterns between the herbivorous snails, *Nerita scabricosta* and *N. funiculata,* on the Pacific coast of Panama were constrained by fish predation and by the physical effects of overheating and desiccation (Levings and Garrity 1983). *N. funiculata* affected local, discrete areas of substrate by causing small-scale patchiness in algal cover in the mid-intertidal zone, whereas *N. scabricosta* was found in the high-intertidal zone while inactive at high tide. As the tide fell, this snail moved down from the high- to the low-intertidal zone to graze and retreated when the tide began to rise again. These snails affected barnacle settlement, the size structure and density of the co-occurring species of *Littorina*, and the abundance of crustose algae across the intertidal seascape.

Predation affects the recruitment and life history of the barnacle, *Chthamalus fissus,* on the Pacific coast of Costa Rica (Sutherland and Ortega 1986). Periods of heavy recruitment of the barnacle were observed in areas infrequently visited by the predatory gastropod, *Acanthina brevidentata*. The barnacle grew around and imprisoned populations of the limpet, *Siphonaria gigas*. Limpets lost weight when surrounded by barnacles until they were removed by the gastropod. Limpets had no effect on the barnacle. Heavy barnacle recruitment was, however, unpredictable in time and space.

Arguably the most comprehensive experiments on predation in the tropical rocky intertidal were conducted by Menge, Lubchenco, and colleagues on Taboguilla Island on the Pacific coast of Panama (Menge and Lubchenco 1981; Lubchenco et al. 1984; Menge et al. 1985, 1986a, b). Menge and Lubchenco (1981) conducted consumer exclusion experiments on the island and found that (i) predation and herbivory were severe year-round, (ii) many types and species of predators and herbivores function to create cumulative consumer pressure, (iii) most prey was restricted to holes and crevices to escape predation pressure, and (iv) abundances of the prey were kept low by small consumers. Holes and crevices were important three-dimensional spaces, which served as a refuge from predation, while escaping from predators assumed secondary importance. This is contrary to results on temperate rocky shores where escapes from consumers in body size, time or two-dimensional space are of prime importance.

On Taboguilla shores, animals and erect macroalgae were rare, making these shores appear to be barren. In the high-intertidal zone, 92–98% of surfaces were bare, except for the small barnacles, *Chthamalus fissus* and *Euraphia imperatrix*. In the mid-and low-intertidal zones, sessile animals were scarce with encrusting algae dominating most (26–93%) of the available space. Diverse and abundant consumers (i.e. limpets, gastropods, crabs, chitons, and fish) existed in all zones. Both species richness and diversity increased with decreasing tidal level reflecting the fact that these shores were physically dominated. Despite seasonality in the physical environment, there was little, if any, seasonality in the biota compared with those occupying temperate shores. Compared with temperate rocky shores, this Panama shore is dominated by desiccation and heat stress, and intense, consistent grazing and predation as well as possible inhibition of settlement by crustose algae.

In a series of further experiments, Menge et al. (1985, 1986a, b) examined the influences on substrate heterogeneity and consumers on patterns of species diversity. Species richness and diversity was greater on irregular than on smooth surfaces, and most sessile benthos occurred in holes and crevices. Predation was intense on open spaces, but variable in time and space. In predator exclusion experiments, algal crusts were replaced by foliose algae, hydrozoans, and sessile invertebrates, mainly bivalves, which became more abundant and diverse. Predators maintained low diversity by keeping prey scarce and causing local extinctions on small spatial scales, while on larger spatial scales, they maintained high diversity through their interaction with substrate heterogeneity. Possibly low dispersal rates of sessile invertebrate at large spatial scales were affected by higher consumers. Fish and crabs reduced the abundance of predaceous snails, herbivorous molluscs, foliose algae, and sessile invertebrates, while predaceous gastropods reduced the abundance of herbivorous molluscs and sessile invertebrates, and herbivorous molluscs reduced the abundance of foliose algae and young sessile invertebrates. These experiments underscore the importance of physical harshness of the environment, and predation and competition as biological forces structuring the tropical rocky intertidal zone.

Not all predatory events are more intense in the tropics. A number of marine bird and mammal species and individuals are greater at the poles than at the equator (Grady et al. 2019). Analysing a comprehensive dataset of nearly 1000 species of shark, fish, reptiles, mammals, and birds, Grady et al. (2019) concluded that predation on ectothermic (cold-blooded) prey is easier in colder waters, which generates a larger resource base for large endothermic (warm-blooded) predators in polar regions. From the equator to the poles, consumption of available food by marine mammals increases by a factor of about 80 after controlling for productivity. Thus, richness, phylogenetic diversity, and abundance of marine predators diverge systematically with thermoregulatory strategy and water temperature, reflecting metabolic differences between endotherms and ectotherms that drive trophic and competitive interactions.

6.10 Plant–Herbivore Interactions

Plant–herbivore interactions are intense in nearly all tropical marine habitats. Herbivores are common consumers mostly in shallow-water habitats where enough light sustains rates of algal production. One of the most dramatic examples of the effects of grazing by tropical herbivores was the demise of the long-spined sea urchin, *Diadema antillarum,* throughout the Caribbean, Florida, and Bermuda in 1983–1984. Mass die-off of the sea urchin, most likely to a host-specific pathogen, resulted in a phase shift on some Caribbean reefs from coral-dominated to alga-dominated communities (Lessios 2016).

Sea urchins can also drastically affect the structure and function of seagrass meadows (Vonk et al. 2008a). Overgrazing by sea urchins occurs globally (Eklöf et al. 2008) with most episodes affecting a small area ($<0.5\,km^2$) and recovery often occurring within a few years; overgrazing can have a range of large, long-term indirect effects, such as sediment destabilisation and loss of fauna. What causes overgrazing is not known, but several ideas have been raised (Eklöf et al. 2008), including nutrient enrichment, reduced predation on sea urchins due to overfishing, changes in water temperature, and populations fluctuations (Figure 6.11). It is likely that multiple disturbances affect the population dynamics of sea urchins, leading to a release from predation pressure and increased sea urchin recruitment.

Mega-herbivores similarly play a dramatic role in affecting seagrass species composition and community structure and function. Subtropical and tropical seagrasses provide food for several marine vertebrates, such as the green turtle, *Chelonia mydas*, the West Indian manatee, *Trichechus manatus*, the West African manatee, *Trichechus senegalensis*, and the dugong, *Dugong dugong* (Aragones and Marsh 2000). Dugongs and manatees graze destructively by uprooting seagrasses when the rhizomes are accessible; when not accessible, they graze by cropping seagrass leaves, which also happens to be the feeding mode for the green turtle. The degree of impact and rate of recovery of seagrass depends upon two factors: (i) the intensity and time spent grazing and (ii) the species composition of the seagrass meadows. Under the right conditions, they can graze a significant fraction (50–80%) of seagrass biomass. Dugongs in Australia consume biomass equivalent to nearly 30% of total seagrass production (Preen 1992). Not only do dugongs remove a substantial amount of seagrass but they can also alter the subsequent nutritional quality of regrowing seagrasses. In tropical Australian meadows, seagrasses that regrew months after cropping exhibited increased whole-plant N concentrations.

Whether a small or large herbivore, grazing in tropical seagrass meadows can be intense (Heck and Valentine 2006; Valentine and Duffy 2006; Unsworth et al. 2007a, b; Vonk et al. 2008a, b, c; Lee et al. 2015) with from 3 to 100% of seagrass net primary production entering the food web via the grazing pathway. Grazing varies greatly over space and time. This conclusion is contrary to an earlier paradigm that seagrasses are dominated by the detrital pathway. Heck and Valentine (2006) suggested that the species-rich seagrass assemblages in the Indo-Pacific may be a product of the past grazing activities of large vertebrate herbivores. Given the importance of grazing by mega-herbivores, it is possible that large grazers have played in important role in determining the life history characteristics of modern-day seagrasses. Grazers continue to be strong interactors in many present-day seagrass meadows, as sea urchins and herbivorous fish in the tropics ingest large amounts of seagrass aboveground and can force the local extinction of seagrasses. Herbivorous fish can graze down a meadow as readily as a dugong.

Herbivorous fish play an important role on coral reefs as demonstrated by differences between reefs where fishing is permitted and prohibited. Herbivory was greater where fishing was prohibited on Kenyan reefs after the 1998 ENSO event (McClanahan 2008). Despite high levels of herbivory, the effect of the ENSO event was larger than within fishery closures, with reefs undergoing a temporary transition from coral-dominance to algal-dominance. Where fishing was permitted, there was little change but there was greater cover of algae and sponges after the ENSO event. In the fishery closures, there was greater herbivory to the extent that the abundance and persistence of algae was reduced, allowing for corals to dominate.

Herbivory on reefs can affect algal succession, although different guilds of herbivorous fish have different effects on algal growth and succession. Territorial damselfish can have a dramatic impact on succession on the Great Barrier Reef, stopping the development of the algal community to where filamentous algae were dominant and fleshy macroalgae were excluded (Ceccarelli et al. 2011). Roving herbivores did not suppress fleshy algae but slowed succession and promoted a diverse algal community. In contrast to reefs dominated by roving herbivores, damselfish

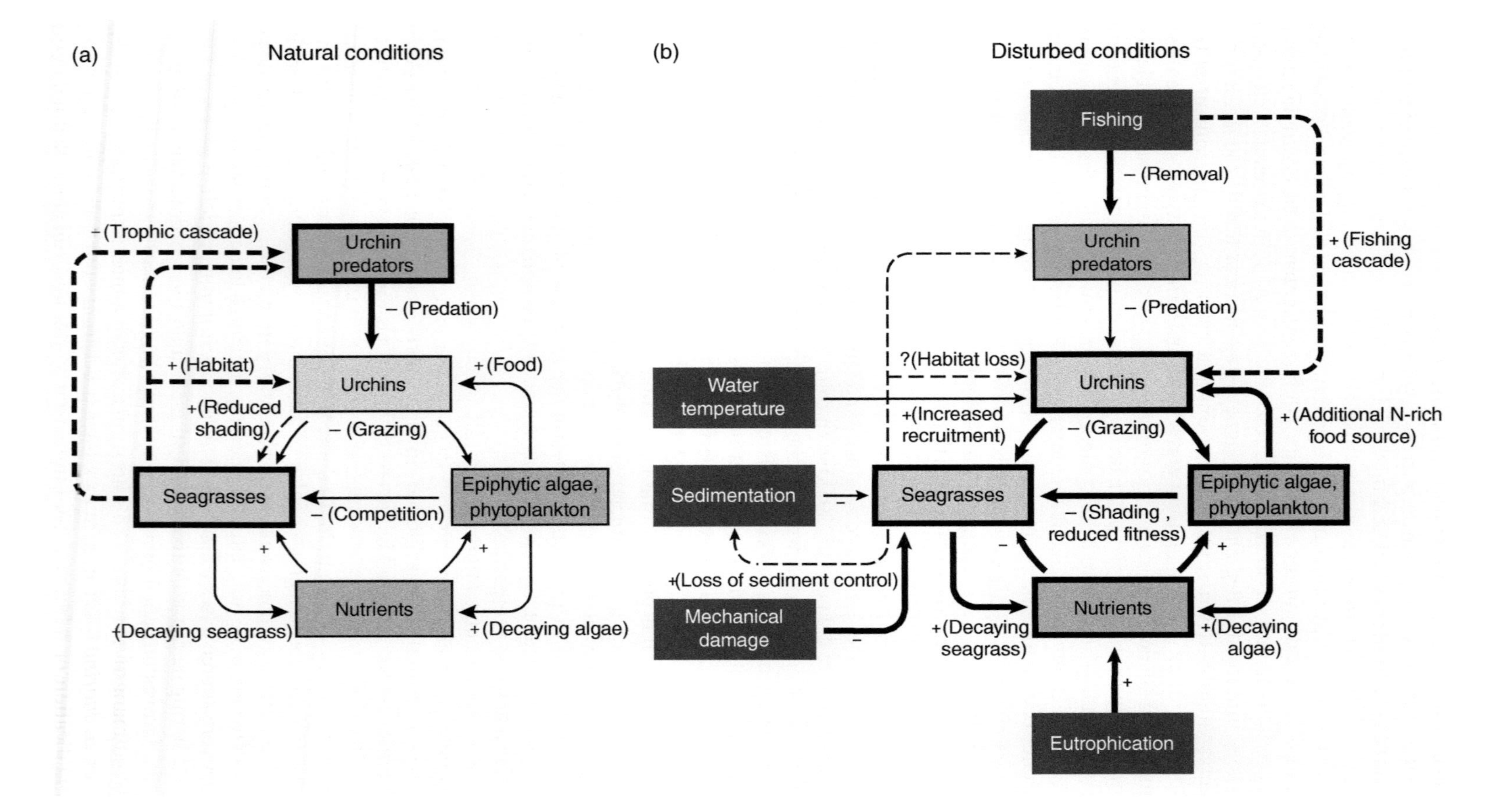

FIGURE 6.11 Conceptual model of a seagrass food web under (a) normal and (b) disturbed conditions. Solid arrows indicate direct links, dotted arrows indicate potential feedback mechanisms, and the sign (+ or −) indicates the type of effect on the next component. The thickness of boxes and arrows indicate relative importance for the food web. *Source:* Eklöf et al. (2008), figure 4, p. 576. © Elsevier.

territories do not appear to function solely as roving herbivore exclusion areas or promote algal diversity due to intermediate grazing pressure.

What drives the spatial distribution of herbivory on coral reefs is poorly understood. Vergés et al. (2011) studied this problem on Ningaloo Reef, Western Australia and found that spatial patterns in macroalgal consumption were best explained by differences in structural complexity among habitats. On the reef crest and reef flat, herbivore biomass and rates of herbivory were greater than in structurally simple habitats such as in reef lagoons. Within 75 m of structurally complex habitat, macroalgal consumption was negligible and algae were most abundant in the reef lagoon. Herbivory patterns were not influenced by the distribution, productivity, or nutritional quality of macroalgae. This contrasts with terrestrial ecosystems, where there is a direct link between herbivory and the structure and nutritional quality of plants.

Although not as dramatic as seagrass grazers, mangrove grazers can exert a powerful effect on the structure and function of mangrove forests. The two primary grazers are insects and sesarmid crabs. Once thought to have a negligible effect, insect grazers can graze significant quantities of mangrove foliage, although the amount varies greatly from one forest to another. Several factors regulate the degree of severity of herbivory in Belizean forests, including canopy cover, tree age, leaf age, tidal height, and nutrient enrichment (Farnsworth and Ellison 1991). Herbivores of various kinds and phyla (insects, crabs, etc.) damaged 4–45% of *Rhizophora* leaf area and 8–36% of *Avicennia* leaf area, with significant differences dependent on tidal height between seedlings and trees. Under a canopy of adult conspecifics, *Rhizophora* seedlings suffered twice as much damage as seedlings growing in areas without an adult canopy. Nutrient enrichment had no impact on herbivory, but in another study of the effect of nutrient enrichment on herbivore responses, the opposite conclusion was reached (Feller and Chamberlain 2007). In Belizean mangroves nearby, the periderm-mining insect, *Marmara* sp., responded positively to nutrient enrichment, although other herbivores, such as moths did not. In contrast, a leaf-mining, *Marmara* sp., was controlled by parasites and predators that killed nearly all its larvae. Nutrient enrichment thus alters patterns of herbivory of some, but not all, herbivores.

Mass defoliation of the mangrove, *A. germinans,* to insect larvae has been observed. On Brazil's Amazon coast, defoliation was observed every two years for about one month when the foliage of *A. germinans* was consumed completely by the insect, *Hyblaea puera* (Fernandes et al. 2009); at other times, the insect can consume an average of 13% of total foliage.

Wood-boring insects can play a crucial role in mangrove herbivory. On offshore mangrove islands in Belize, woodborers killed > 50% of the *Rhizophora mangle* canopy by girdling, pruning, and hollowing (Feller 2002). Leaf-feeding insects, in contrast, removed less than 6% of the canopy. The patterns of herbivory were heterogeneous with no relationship with tidal height. The branches that were pruned by stem girdlers created numerous holes and gaps in the mangrove canopy. In *R. mangle* trees in which 50% of their branches were experimentally girdled, shoot growth and flowering increased. However, because leaves were lost as green fall, the quantity and quality of leaf litter were altered when a leaf-bearing branch was girdled or hollowed, suggesting that wood-boring insects play an important role in altering nutrient cycling processes.

In mangroves enriched by nutrient pollution, insect herbivory can be intense. In polluted mangroves of the Indus delta, insects consumed as much as 22% of the foliage of *A. marina* with herbivory showing a vertical trend in the canopy, and an intensification of herbivory from the upper to the lower canopy (Saifullah and Ali 2004). In similarly polluted mangroves in southern India, this author observed *A.*

marina trees that were heavily defoliated by insects, with at least 50% of the foliage consumed. The role of leaf chemistry may play a role in the degree of herbivory as some mangrove species are not affected by insects. Tong et al. (2006) found that leaf chemistry plays such a role in contrasting *Kandelia obovata* forests in Hong Kong where leaves in a forest containing higher amounts of soluble tannins, ash, and crude fibre were fed upon less than in a forest with *K. obovata* leaves containing less of these constituents. Herbivory may vary by differences between mangrove species in leaf chemistry, although this remains to be more fully explored.

Large quantities of leaf litter are shredded in mangrove forests by molluscs and especially sesarmid crabs in the Indo-Pacific. This activity profoundly changes the chemical nature of the litter (Cannicci et al. 2008; Lee 2008). Mangrove leaf litter is a rich source of carbon but a poor source of N and P, and it is thought that these herbivores obtain these additional nutrients from the consumption of other invertebrates and carrion. While this may be true for crabs, it is unlikely to be true for molluscs, such as *Terebralia palustris,* which is unlikely to feed on such supplemental foods, except perhaps for bacteria and benthic microalgae. The consumption of mangrove litter is a paradox as these animals seem to have developed the ability to profitably consume material of little nutritional value. However, stable isotopes and other tracers indicate that crabs apparently do not depend heavily on leaf litter for their nutritional needs (Lee 2005, 2008; Hall et al. 2006). Several hypotheses have been put forward to explain this apparent paradox (Thongtham and Kristensen 2005), such as supplemental consumption of other foods rich in N and other essential nutrients.

Some crab species living in both Neotropical and Indo-Pacific mangroves have evolved a tree-climbing habit in addition to consuming leaf litter (Cannicci et al. 2008). One of the most studied tree climbers is the sesarmid crab, *Aratus pisonii,* a common inhabitant in mangroves of both the Pacific and Atlantic tropical and subtropical coasts of the Americas. This species is known to inhabit the mangrove canopy as an adult and to rely almost exclusively on fresh leaves (Erickson et al. 2003). This crab removes the top layers of the leaves by scaping the leaf and by this means can remove up to 30% of an individual leaf. This tree climber has a heavy impact on forests of *Rhizophora mangle* in Florida with damage of up to 30–40% of total foliage (Erickson et al. 2003). Reliable data on tree-climbing crab species in the Indo-Pacific are scant even though diversity is undoubtedly high, with most data extant only for *Parasesarma leptosoma*, a tree-climber along the East African coast (Fratini et al. 2005) where their average density can be as high as 200–300 crabs per tree and damaging up to 50–60% of the foliage of their preferred food source, *R. mucronata.*

In the Indo-Pacific region, sesarmids are clearly the main agents of high leaf litter turnover rates in mangroves (Robertson and Daniel 1989; Poovachiranon and Tantichodok 1991; Olafsson et al. 2002). While their influence can vary, sesarmid crabs can consume from 9 to 100% of leaf litter on a mangrove forest floor; herbivory is low in areas where crab numbers are low. Ocypodid crabs of the genus *Ucides* are similarly responsible for high rates of litter turnover. Twilley et al. (1997) found that the ocypodid crab, *Ucides occidentalis,* greatly affected the litter dynamics of mangrove forests in Ecuador with similarly high rates of leaf consumption as the sesarmid crabs in the Indo-Pacific. Indo-Pacific Ocypodidae are among the main consumers of microbial communities, such as benthic microalgae and bacteria, as are crabs of the genera *Uca* and *Dotilla* (Bouillon et al. 2002).

The role of leaf consumption and leaf-burying by crabs appears to be a mechanism to retain litter within the forest, and this may have energetic consequences.

For instance, Robertson (1986) estimated that crabs could remove about 28% of the annual leaf fall in a mixed *Rhizophora* forest in northern Australia. Such retention may serve to help recycle organic matter within the forests even if the leaves were not readily consumed by the crabs; bacteria, fungi, and other microbes presumably help to decompose this litter buried by crabs in the forest floor. Nevertheless, little is known about the fate of this material or of the leaf litter they consume.

While other foods must be necessary for the crabs to maintain a balanced diet, little is known about the nutritional complexity of this material. Leaf litter is assimilated at a low (<50%) rate and about 60% of the dry mass consumed is egested as faecal matter. The digestion process of crabs suggests that the chemical composition of this material changes considerably. Lee (1997) showed that the faeces of *Perisesarma messa* was significantly rich in N and less rich in tannins than unprocessed leaf litter. In follow-up experiments, aged faeces proved to be a better source of nutrition for the amphipod, *Parhyallela* sp., than mangrove leaf litter. Similarly, the faecal material of the crab, *Parasesarma erythrodactyla,* was found to be colonized by a rich flora of bacteria with a rise in N content concomitant with the growth of the bacteria. Feeding by the ocypodid, *Ucides cordatus,* similarly produced finely fragmented faecal material enriched in N, C and bacterial biomass compared with mangrove sediment (Nordhaus and Wolff 2007).

6.11 Trophic Cascades

A trophic cascade is when changes in the relative abundances of multiple species in a community are the result of changes in abundance of one organism. In the simplest example (Figure 6.12), a trophic cascade occurs when a predator eats herbivores,

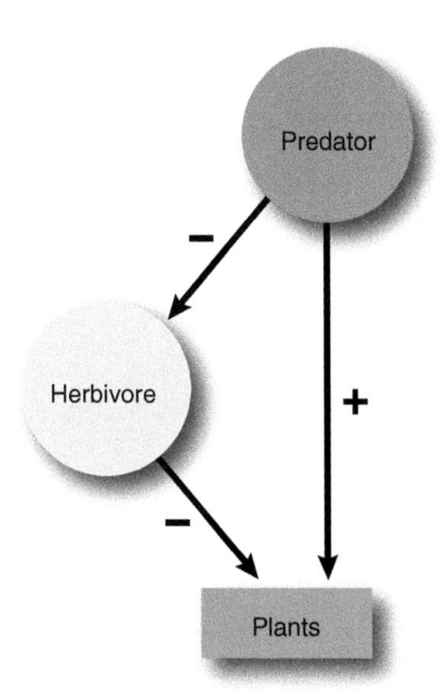

FIGURE 6.12 Simplified diagram of a trophic cascade. A trophic cascade occurs when a predator eats herbivores, releasing the plants from intense grazing resulting in an increase in plant abundance.

releasing the plants in question from intense grazing resulting in an increase in plant abundance. Trophic cascades thus ensue from both direct predation and risk effects of predators. Indeed, it is possible that a food web can be 'fished down' such that trophic cascades are relaxed. A good example exists from the Caribbean where a rapid shift in fish community structure in Belize was recorded in which a drastic decline in grouper and snapper abundance resulted in a switch towards smaller, 'less desirable' herbivorous parrotfish. Over a six- to seven-year period (2002–2008/2009), Mumby et al. (2012) recorded marked declines of large-bodied grouper and carnivorous snappers with a concomitant increase of herbivorous parrotfish. As a result, the biomass of predators increased dramatically by 880% as compared to 2002. This response was attributed to a release from predation and constraints to foraging behaviour imposed by groupers. There was also a noticeable decline in adult damselfish (Mumby et al. 2012). Where fishing was prohibited, there was no change in parrotfish density, but parrotfish declined dramatically due to fishing, the net result

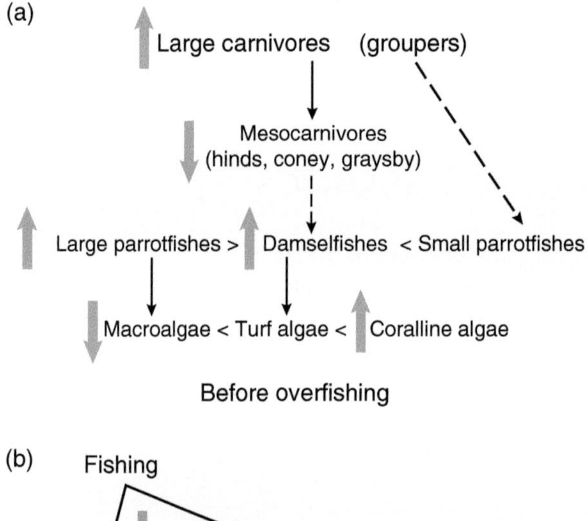

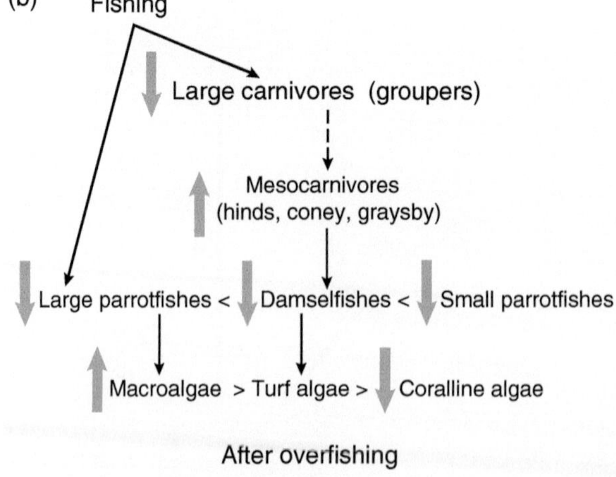

FIGURE 6.13 Effects of fishing impacts on reef food web structure and function. Blue arrows: abundance trends, Black arrows: strong ecological interactors, Dotted arrows: weak or non-existent ecological interactors. *Source:* Mumby et al. (2012), figure 9, p.21. © Inter-Research Science Publisher.

being an increase in the alga *Halimeda* spreading to microhabitats that have been previously grazed intensively by parrotfish (Figure 6.13). This contrasts with the situation prior to overfishing in which the groupers and parrotfish helped to maintain coral-dominated reefs.

Trophic cascades exist on coral reefs and are important in structuring populations. For instance, sea urchins tend to become more abundant due to intensive fishing and feature strongly in most documented trophic cascades. The best documented cascade is one in which algae have been released from heavy grazing pressure by predators feeding on herbivorous fish. In Kenya, the triggerfish, *Balistapus undulatus,* is the top predator on sea urchins on coral reefs to the extent that it controls sea urchin numbers (McClanahan 1995). Due to overfishing this fish and other urchin predators become depleted to the extent that only one sea urchin species remains to be the dominant grazer. This results in the coral reef shifting to filamentous algae becoming more abundant because they can withstand sea urchin grazing; hard corals decline in percentage cover. The cascade effect is most likely mediated by the triggerfish.

Another prime example of the effects of fishing on trophic cascades has occurred in Caribbean reefs where the mass mortality of the sea urchin *D. antillarum* in 1983 resulted in a shift in grazer dominance to parrotfish. As noted earlier, Mumby et al. (2006) compared the effects of the negative impacts of increased predation on parrotfish with the positive impacts of a decline in fishing mortality on parrotfish inside marine reserves. They found that because large parrotfish escape the risk of predation from the Nassau grouper, the predation effect reduced grazing only by a negligible amount. However, the increase in density of large parrotfish resulted in a net doubling in grazing which resulted in turn in a fourfold reduction in macroalgal cover. In a subsequent study, Mumby et al. (2007) found that while reduced fishing pressure and weak predator–prey relationships within marine reserves can create trophic cascades that increase the abundance of grazing fish and reduce the cover of macroalgae, the trophic cascades within the reserves enhance the process of coral recruitment. Increased fish grazing was negatively correlated with macroalgal cover and resulted in a twofold increase in the density of coral recruits.

On coral reefs, the removal of predators can have a strong impact on primary producers, including crustose coralline algae (O'Leary and McClanahan 2010). For Kenyan reefs, O'Leary and McClanahan (2010) hypothesised that herbivorous fish benefit coralline algae, abundant sea urchins heavily graze the algae, and that fishing indirectly reduces algal cover by removing sea urchin predators. Using closures and fished reefs as well as a caging experiment, they found that coralline algal cover declined with increasing fish and sea urchin abundances, but the negative impact of sea urchins was much stronger than that of fish. Sea urchins reduced algal growth rates to zero, preventing the build-up of crustose algae. An ENSO event during the study resulted in a short-term positive effect on the algae but was less strong than the effect of sea urchin grazing. They concluded that on Kenyan reefs, predator removal leads to a trophic cascade that is expected to reduce net calcification of reefs, thus reducing reef stability.

Trophic cascades have been recorded in seagrass meadows (Burkholder et al. 2013). Using long-term exclusion cages to examine the effects of dugongs and sea turtles on seagrasses in Florida, Burkholder et al. (2013) conducted experiments in habitats with low and high risk of predation by the tiger shark, *Galeocerdo cuvier,* to determine if the presence of this roving predator structures herbivore impacts on seagrasses. In low-risk habitats, the exclusion of dugongs and sea turtles resulted in an increased leaf length for the seagrasses, *Cymodocea angustata* and *Halodule uninervis,* while shoot densities of *C. angustata* nearly tripled; *H. uninervis* nearly disappeared from inside the exclusion cages. In high-risk meadows, tiger sharks altered the grazing behaviour of the large herbivores indicating mediation by risk-sensitive foraging and a behaviour-mediated trophic cascade.

Evidence for prey release and more complex three-level trophic cascades is mixed in tropical nearshore ecosystems, with little information for mangroves (Sandin et al. 2010; Atwood and Hammill 2018). Herbivores exert strong control over algae, but the effect of predators is inconsistent. The evidence for trophic cascades is limited, possibly due to flexibility in trophic relationships within nearshore food webs and that the effect of human-induced disturbance can play some role in ameliorating trophic cascades (Sandin et al. 2010). They summarized the evidence showing trophic cascades in the tropics on both coral reefs and in seagrass beds (Figure 6.14). In the Kenyan situation (Figure 6.14a), an increase in the abundance of the red-lined triggerfish, *Balistapus undulatus,* resulted in the decline of its sea urchin prey with a subsequent increase in reef-building corals and coralline algae. In Fiji (Figure 6.14b), overfishing of predatory fish resulted in an increase in COTS numbers, cascading to a decline in reef-building corals. In seagrass meadows of the Gulf of Mexico (Figure 6.14c), large piscivorous fish regulated the abundance of smaller predators, such as pinfish, without which grazers such as caridean shrimp reached high densities and efficiently grazed epiphytic algae. Without large piscivores, pinfish densities increased, causing overgrowth of epiphytes on seagrass blades. On Caribbean coral reefs (Figure 6.14d), the presence of a pathogen produced mass mortality of the sea urchin, *D. antillarum*, leading to an increase in macroalgae and shifting the trophic state of coral reefs.

Despite the limited evidence for trophic cascades, especially in seagrass and mangrove ecosystems (Atwood and Hammill 2018), there is abundant evidence that there have been numerous alterations to natural food webs by human activity. Such alterations have in turn altered the composition and nature of food webs to the extent that they may be said to have undergone 'phase shifts' or 'alternate stable states.' Our understanding of the role of trophic interactions in tropical marine ecosystems is based for the most part on altered populations and communities as exceedingly few systems have remained untouched by humans.

6.12 Facilitation Cascades

The importance of positive interactions in fostering the structure of marine communities has emerged as an important concept in community ecology. It is now recognized that spatially dominant foundation species, such as algae, mangroves, corals and seagrasses, support biodiversity by defining the physical architecture of the ecosystem. The physical structure of some ecosystem engineers, which may or may not be space-dominant, shape biodiversity by creating and maintaining habitats. These two groups are collectively referred to as "habitat-forming species" (Gribben et al. 2019). A subset of interactions between habitat-formers (referred to as facilitation cascades) is hierarchical in which a main habitat-former promotes a secondary habitat-former which in turn supports an inhabitant community. Facilitation cascades arise through three major pathways: (i) the basal habitat-former provides resources such as attachment substrate to the secondary habitat-former, (ii) the basal habitat-former facilitates the secondary habitat-former by reducing environmental stress, and (iii) the basal habitat-former reduces consumer or competition pressure on the secondary

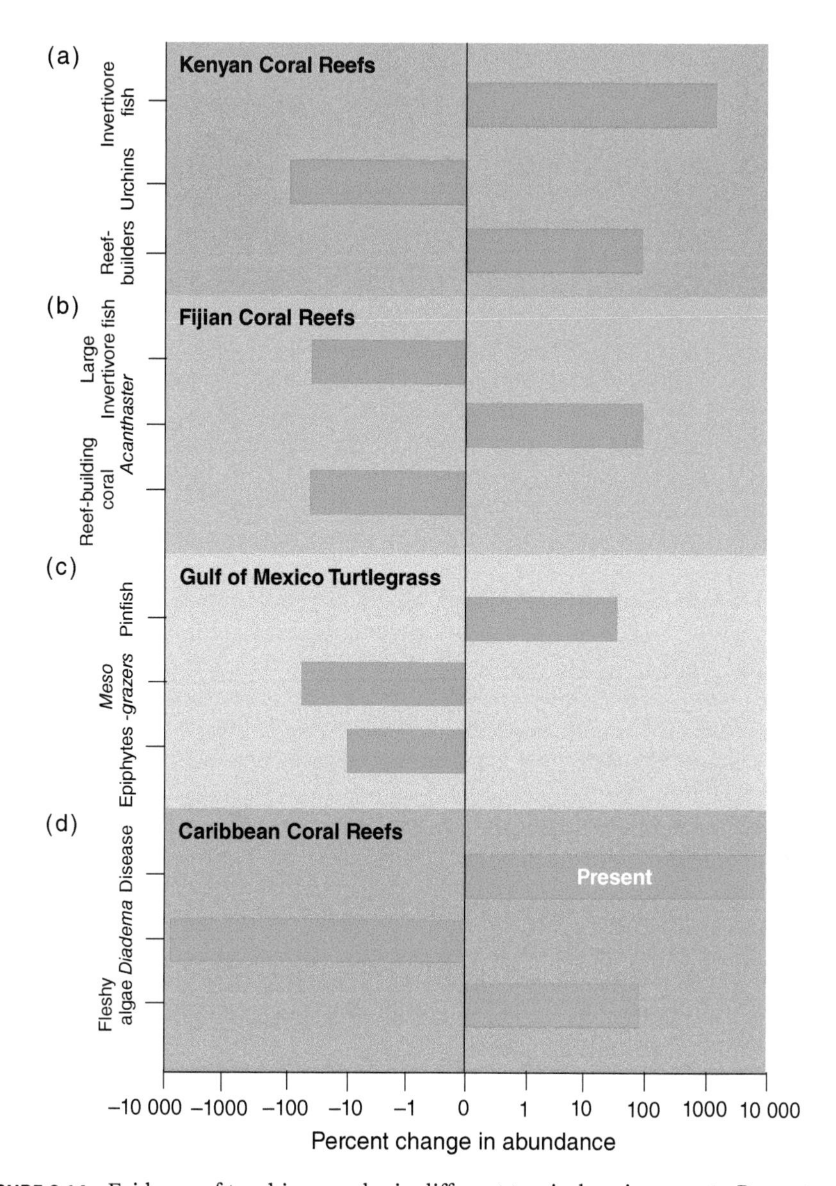

FIGURE 6.14 Evidence of trophic cascades in different tropical environments. Percentage change in abundance for each of three trophic levels for (a) protected Kenyan coral reefs: the invertivorous fish, *Balistapus undulatus,* sea urchins, and reef-builders; (b) fished Fijian coral reefs: predatory fish, *Acanthaster planci,* and reef-builders; (c) experimental manipulations of Gulf of Mexico turtlegrass community: pinfish, mesograzers, and epibionts; and (d) Caribbean coral reefs post-1983: *Diadema antillarum* and fleshy algae. *Source:* Sandin et al. (2010), figure 5.1, p.77. © Island Press.

habitat-former (Gribben et al. 2019). The recruitment and persistence of secondary habitat-formers requires the presence of the primary habitat-former.

Facilitation cascades are just as important as other interactions among biota and are best documented in marine ecosystems, especially in intertidal habitats (Thomsen et al. 2018). For example, epiphytes growing on trees and freshwater plants or drifting algae entrapped by mangrove pneumatophores enhance biodiversity relative to habitat provided by only a single or no habitat-former (Angelini et al. 2011; Gribben et al. 2017). There are a host of other marine examples (Gribben et al. 2019) such as seagrasses (basal habitat-former) promoting the recruitment of bivalves (secondary foundation species) possibly by reducing predation pressure or other biotic stress, which in turn provide a surface for attached epibionts. Coral reefs buffer seagrasses and mangroves from offshore wave energy, allowing them to establish under calm conditions, whereas mangroves and seagrasses trap terrestrial sediment and nutrients that would otherwise smother reefs. Another common example is the positive effect of multiple, structurally distinct, reef-building corals on fish and invertebrate diversity.

Most studies describing facilitation cascades have been conducted in subtropical (49%) and temperate (35%) regions, with relatively few studies in tropical (10%) and polar (6%) regions (Gribben et al. 2019), with most studies conducted in coastal benthic habitats. The most common basal habitat-formers to date (in decreasing order) are seagrasses, macroalgae, mangroves, reef-forming bivalves, and salt marsh plants. Algae were the most common secondary habitat-formers, followed by bivalves as both accounted for >90% of all studies. Most studies have focused on macroinvertebrates as secondary habitat-formers that are physically attached to the basal habitat-formers.

The number of studies that demonstrate facilitation cascades is growing, but the mechanisms by which hierarchical positive interactions among species are established and maintained are not well understood. What is known it that associations between basal and secondary habitat-formers and between secondary habitat-formers and inhabitant communities may involve either active or passive processes. Facilitation may involve resource provisioning, amelioration of biotic and abiotic stressors, and variation between intertidal and subtidal habitats. Facilitation commonly involves the provision of limiting resources by habitat-formers. Such resources commonly include food and a stable or suitable settlement substrate. Substrate is a resource strongly reinforced in intertidal habitats, such as rocky shores, and in mangroves and seagrasses where roots and shoots trap and retain floating algae, whereas the search for food is universal. In some cases, other resources are provided, such as by pool-excavating crabs that modify the habitat in West African seagrasses and facilitate colonisation by other invertebrates (van der Zee et al. 2016).

Amelioration of environmental stress is one important role that basal habitat-formers are involved in. In intertidal rocky shores, reduction of heat and desiccation by habitat-forming species, such as macroalgal canopies, barnacles, and bivalves, promotes the colonisation of more stress-sensitive macroalgae and invertebrates (Bulleri 2009). Amelioration of biotic stressors, such as predation and competition, can be facilitated by the provision of the physical structure provided by the habitat-formers. For example, drift algae entrapped in seagrass (Adams et al. 2004) and algal cover on mangrove prop roots (Jaxion-Harm and Speight 2012) facilitated post-settlement fish and crab survival by reducing foraging by predators. Mechanisms underpinning facilitation cascades vary between different environments. Positive interactions generated through the amelioration of physical stress are more common in intertidal habitats, whereas those arising from a reduction in predation or grazing are more common in subtidal environments (Bulleri 2009). For example, facilitation of invertebrates by the oyster, *Saccostrea glomerata,* in mangroves along the east Australian coast was helped by the fact that the oysters lived under the mangrove canopy where temperatures were cooler and of greater humidity than outside the canopy (McAfee et al. 2016). An example of reduced predation in a subtidal habitat is when colonies of the fire coral, *Millepora platyphylla*, provide a refuge to other coral taxa against COTS, as discovered in coral reefs of French Polynesia (Kayal and Kayal 2017).

Facilitation cascades can vary spatially and temporally, operating in different seasons and at local and landscape scales. The traits and density of a basal habitat-former may influence facilitation cascades by determining whether the environment is suitable for the secondary habitat-former to colonise and form habitat and modifying traits of the secondary habitat-former. There is some evidence from temperate Australian mangroves that multiple traits of both basal and secondary habitat-formers show independent and additive effects on biodiversity. Both the density and height of pneumatophores of the basal habitat-former, the mangrove, *A. marina*, had effects on the trapping and retention of the secondary habitat-former, the fucoid alga, *Hormosira banksii* (Bishop et al. 2013). Tall pneumatophores at high density initially served as a physical barrier to algal colonisation but over the long term, they enhanced retention of colonised algae. The increase in algal biomass in turn facilitated the colonisation of epifauna. The densities of two secondary habitat-formers, *H. banksii* and the oyster, *Saccostrea glomerata*, each dependent on pneumatophores, had independent and additive effects on inhabitant biodiversity. Both the height of pneumatophores and the thallus length of *H. banksii* independently influenced the biomass of the algae retained by the *A. marina* forest (Bishop et al. 2013).

Patch size and landscape scale effects are known to occur, although there are few data from tropical marine environments. Habitat-formers that initially establish in another ecosystem may form a facilitation cascade if one species is dislodged and then trapped, for example, among mangrove pneumatophores, facilitating an invertebrate community (Bishop et al. 2009). Modification of environmental conditions by facilitation cascades can extend into surrounding areas beyond the patches where habitat-formers co-occur. Coral reefs, for example, are built through a network of facilitation cascades in which various corals, algae, and sponges facilitate one another and build the reef structure, buffering shorelines against waves (Ferrario et al. 2014). This allows for the development of diverse coastal communities hundreds of metres inshore of the reef crest.

References

Abelson, A. and Loya, Y. (1999). Interspecific aggression among stony corals in Eilat, Red Sea: a hierarchy of aggression ability and related parameters. *Bulletin of Marine Science* 65: 851–860.

Adams, A.J., Locascio, J.V., and Robbins, B.D. (2004). Microhabitat use by a post-settlement stage estuarine fish: evidence from relative abundance and predation among habitats. *Journal of Experimental Marine Biology and Ecology* 299: 17–33.

Aeby, G.S. (1991). Behavioral and ecological relationships of a parasite and its hosts within a coral reef system. *Pacific Science* 45: 263–269.

Almany, G.R. and Webster, M.S. (2004). Odd species out as predators reduce diversity of coral reef fish. *Ecology* 85: 2933–2937.

Almany, G.R. and Webster, M.S. (2006). The predation gauntlet: early post-settlement mortality in reef fish. *Coral Reefs* 25: 19–22.

Andras, J.P., Kirk, N.L., and Harvell, D. (2011). Range-wide population genetic structure of *Symbiodinium* associated with the Caribbean Sea fan coral, *Gorgonia ventalina*. *Molecular Ecology* 20: 2525–2542.

Angelini, C., Altieri, A.H., Silliman, B.R. et al. (2011). Interactions among foundation species and their consequences for community organization, biodiversity, and conservation. *Bioscience* 61: 782–789.

Anker, A. and Dworschak, P.C. (2007). *Jengalpheops rufus* gen. nov., sp. nov., a new commensal alpheid shrimp from the Philippines (Crustacea, Decapoda). *Zoological Studies* 46: 290–302.

Aragones, L. and Marsh, H. (2000). Impact of dugong grazing and turtle cropping on tropical seagrass communities. *Pacific Conservation Biology* 5: 277–288.

Archer, S.K., Stoner, E.W., and Layman, C.A. (2015). A complex interaction between a sponge (*Halichondria melanadocia*) and a seagrass (*Thalassia testudinum*) in a subtropical coastal ecosystem. *Journal of Experimental Marine Biology and Ecology* 465: 33–40.

Atwood, T.B. and Hammill, E. (2018). The importance of marine predators in the provisioning of ecosystem services by coastal plant communities. *Frontiers in Plant Science* 9: 1289. https://doi.org/10.3389/fpls.2018.01289.

Azevedo, D.S., Silva, J.V.C.L., and Castiglioni, D.S. (2017). Population biology of *Uca maracoani* in a tropical mangrove. *Thalassas: An International Journal of Marine Sciences* 33: 1–13.

Bak, R.P.M. and Meesters, E.H. (1999). Population structure as a response of coral communities to global change. *American Zoologist* 39: 56–65.

Barneche, D.R., Kulbicki, M., Floeter, S.R. et al. (2016). Energetic and ecological constraints on population density of reef fish. *Proceedings of the Royal Society B: Biological Sciences* 283: 2015218.

Barott, K.L. and Rohwer, F.L. (2012). Unseen players shape benthic competition on coral reefs. *Trends in Microbiology 20*: 621–628.

Barott, K.L., Williams, G.J., Vermeij, M.J. et al. (2012). Natural history of coral – algae competition across a gradient of human activity in the Line Islands. *Marine Ecology Progress Series* 460: 1–12.

Baums, I.B., Devlin-Durante, M.K., and Lajeunesse, T.C. (2014). New insights into the dynamics between reef corals and their associated dinoflagellate endosymbionts from population genetic studies. *Molecular Ecology* 23: 4203–4215.

Bay, L.K., Caley, M.J., and Crozier, R.H. (2008). Meta-population structure in a coral reef fish demonstrated by genetic data on patterns of migration, extinction and re-colonisation. *BMC Evolutionary Biology* 8: 248. https://doi.org/10.1186/1471-2148-8-248.

Berger, U. and Hildenbrandt, H. (2000). A new approach to spatially explicit modelling of forest dynamics, spacing, aging and neighbourhood competition of mangrove trees. *Ecological Modelling* 132: 287–302.

Berger, U. and Hildenbrandt, H. (2003). The strength of competition among individual trees and the biomass-density trajectories of the cohort. *Plant Ecology* 167: 89–96.

Bertness, M.D. (1981). Predation, physical stress, and the organization of a tropical rocky intertidal hermit crab community. *Ecology* 62: 411–425.

Beukers-Stewart, B.D., Beukers-Stewart, J.S., and Jones, G.P. (2011). Behavioural and developmental responses of predatory coral reef fish to variation in the abundance of prey. *Coral Reefs 30*: 855–864.

Binning, S.A., Barnes, J.I., Davies, J.N. et al. (2014). Ectoparasites modify escape behaviour, but not performance, in a coral reef fish. *Animal Behaviour* 93: 1–7.

Birrell, C.L., McCook, L.J., Willis, B.L. et al. (2008). Effects of benthic algae on the replenishment of corals and the implications for the resilience of coral reefs. *Oceanography and Marine Biology: An Annual Review* 46: 25–63.

Bishop, M.J., Morgan, T., Coleman, M.A. et al. (2009). Facilitation of molluscan assemblages in mangroves by the fucalean alga *Hormosira banksii*. *Marine Ecology Progress Series* 392: 111–122.

Bishop, M.J., Fraser, J., and Gribben, P.E. (2013). Morphological traits and density of foundation species modulate a facilitation cascade in Australian mangroves. *Ecology* 94: 1927–1936.

Bornman, E., Nadine, A.S., and Tris, H.W. (2019). Predator-prey interactions associated with larval *Gilchristella aestuaria* (Family: Clupeidae) in mangrove and non-mangrove estuaries. *Estuarine, Coastal and Shelf Science* 228: 106391. https://doi.org/10.1016/j.ecss.2019.106391.

Bosire, J.O., Kairo, J.G., Kazungu, J. et al. (2005). Predation on propagules regulates regeneration in a high-density reforested mangrove plantation. *Marine Ecology Progress Series* 299: 149–155.

Bouillon, S., Koedam, N., Raman, A.V. et al. (2002). Primary producers sustaining macro-invertebrate communities in intertidal mangrove forests. *Oecologia* 130: 441–448.

Bramanti, L. and Edmunds, P.J. (2016). Density-associated recruitment mediates coral population dynamics on a coral reef. *Coral Reefs* 35: 543–553.

Bruno, J.F., Sweatman, H., Precht, W.F. et al. (2009). Assessing evidence of phase shifts from coral to macroalgal dominance on coral reefs. *Ecology* 90: 1478–1484.

Bulleri, F. (2009). Facilitation research in marine systems: state of the art, emerging patterns and insights for future developments. *Journal of Ecology* 97: 1121–1130.

Burkholder, D.A., Heithaus, M.R., Fourqurean, J.W. et al. (2013). Patterns of top-down control in a seagrass ecosystem: could a roving apex predator induce a behaviour-mediated trophic cascade? *Journal of Animal Ecology* 82: 1192–1202.

Byler, K.A., Carmi-Veal, M., Fine, M. et al. (2013). Multiple symbiont acquisition strategies as an adaptive mechanism in the coral *Stylophora pistillata*. *PLoS One* 8: e59596. https://doi.org/10.1371/jounral.pone.0059596.

Canion, C.R. and Heck, K.L. Jr. (2009). Effect of habitat complexity on predation success: re-evaluating the current paradigm in seagrass beds. *Marine Ecology Progress Series* 393: 37–46.

Cannicci, S., Burrows, D., Fratini, S. et al. (2008). Faunal impact on vegetation structure and ecosystem function in mangrove forests: a review. *Aquatic Botany* 89: 186–200.

Cannicci, S., Fusi, M., Cimó, F. et al. (2018). Interference competition as a key determinant for spatial distribution of mangrove crabs. *BMC Ecology* 18: 8. https://doi.org/10.1186/s12898-018-0164-1.

Cartwright-Taylor, L., Lee, J., and Hsu, C.C. (2009). Population structure and breeding pattern of the mangrove horseshoe carb *Carcinoscorpius rotundicauda* in Singapore. *Aquatic Biology* 8: 61–69.

Ceccarelli, D.M., Jones, G.P., and McCook, L.J. (2011). Interactions between herbivorous fish guilds and their influence on algal succession on a coastal coral reef. *Journal of Experimental Marine Biology and Ecology* 399: 60–67.

Chadwick, N.E. and Morrow, K.M. (2011). Competition among sessile organisms on coral reefs. In: *Coral Reefs: An Ecosystem in Transition* (eds. Z. Dubinsky and N. Stambler), 347–371. Dordrecht, The Netherlands: Springer.

Chadwick-Furman, N.E. and Spielgel, M. (2000). Sexual reproduction in the tropical corallimorpharian *Rhodactis rhodostoma*. *Invertebrate Biology* 119: 361–369.

Chaloupka, M., Limpus, C., and Miller, J. (2004). Green turtle somatic growth dynamics in a spatially disjunct Great Barrier Reef metapopulation. *Coral Reefs* 23: 325–335.

Chen, L., Peng, S., Li, J. et al. (2013). Competitive control of an exotic mangrove species: restoration of native mangrove forests by altering light availability. *Restoration Ecology* 21: 215–223.

Cheney, K.L. and Côté, I.M. (2005). Mutualism or parasitism? The variable outcome of cleaning symbioses. *Biology Letters* 1: 162–165.

Chornesky, E.A. (1991). The ties that bind: inter-clonal cooperation may help a fragile coral dominate shallow high-energy reefs. *Marine Biology* 109: 41–51.

Clarke, P.J. and Kerrigan, R.A. (2002). The effects of seed predators on the recruitment of mangroves. *Journal of Ecology* 90: 728–736.

Coffroth, M.A. and Lasker, H.R. (1998). Population structure of a clonal gorgonian coral: the interplay between clonal reproduction and disturbance. *Evolution* 52: 379–393.

Connell, J.H. (1971). On the role of natural enemies in preventing competitive exclusion in some marine animals and in rain forest trees. In: *Dynamics of Populations* (eds. P.J. den Boer and G.R. Gradwell), 298–312. Wageningen, The Netherlands: Center for Agricultural Public Documents.

Connell, J.H., Hughes, T.P., Wallace, C.C. et al. (2004). A long-term study of competition and diversity of corals. *Ecological Monographs* 74: 179–210.

Correia Nunes, L.S. and Monteiro Camargo, A.F. (2020). The interspecific competition of tropical estuarine macrophytes is not density-dependent. *Aquatic Botany* 164: 103233. https://doi.org/10.1016/j.aquabot.2020.103233.

Cribb, T.H., Bott, N.J., Bray, R.A. et al. (2014). Trematodes of the Great Barrier Reef, Australia: emerging patterns of diversity and richness in coral reef fish. *International Journal of Parasitology* 44: 929–939.

Dai, C.F. (1990). Interspecific competition in Taiwanese corals with specific reference to interactions between alcyonaceans and hermatypics. *Marine Ecology Progress Series* 60: 291–297.

Davis, B.C. and Fourqurean, J.W. (2001). Competition between the tropical alga, *Halimeda incrassata*, and the seagrass, *Thalassia testudinum*. *Aquatic Botany* 71: 217–232.

Diaz-Pulido, G., Harii, S., McCook, L.J. et al. (2010). The impact of benthic algae on the settlement of a reef-building coral. *Coral Reefs* 29: 203–208.

Diele, K. (2000). Life history and population structure of the exploited mangrove crab, *Ucides cordatus cordatus* (Linnaeus, 1763) (Decapoda, brachyura) in the Caeté estuary, north Brazil. PhD dissertation. University of Bremen.

Dizon, R.M. and Yap, H.T. (2005). Coral responses in single-and mixed-species plots to nutrient disturbance. *Marine Ecology Progress Series* 296: 165–172.

Dobzhansky, T. (1950). Evolution in the tropics. *American Scientist* 38: 209–221.

Doropoulos, C., Ward, S., Roff, G. et al. (2015). Linking demographic processes of juvenile corals to benthic recovery trajectories in two common reef habitats. *PLoS One* 10: e0128535. https://doi.org/10.1371/journal.pone.0128535.

Duarte, C.M., Terrados, J., Agawin, N.S.R. et al. (2000). An experimental test of the occurrence of competitive interactions among SE Asian seagrasses. *Marine Ecology Progress Series* 197: 231–240.

Dulvy, N.K., Freckleton, R.P., and Polunin, N.V. (2004). Coral reef cascades and the indirect effects of predator removal by exploitation. *Ecology Letters* 7: 410–416.

Dunne, J.A., Lafferty, K.D., Dobson, A.P. et al. (2013). Parasites affect food web structure primarily through increased diversity and complexity. *PLoS Biology* 11: e1001579. https://doi.org/10.1371/journal.pbio.1001579.

Eklöf, J.S., de la Torre-Castro, M., Gullström, M. et al. (2008). Sea urchin overgrazing of seagrasses: a review of current knowledge on causes, consequences, and management. *Estuarine, Coastal and Shelf Science* 79: 569–580.

Elahi, R. (2008). Effects of aggregation and species identity on the growth and behaviour of mushroom corals. *Coral Reefs* 27: 881–885.

Ellison, A.M., Farnsworth, E.J., and Twilley, R.R. (1996). Facultative mutualism between red mangroves and root-fouling sponges in Belizean mangal. *Ecology* 77: 2431–2444.

Erickson, A.A., Saltiss, M., Bell, S.S. et al. (2003). Herbivore feeding preferences as measured by leaf damage and stomatal ingestion: a mangrove crab example. *Journal of Experimental Marine Biology and Ecology* 289: 123–138.

Fabina, N.S., Putnam, H.M., Franklin, E.C. et al. (2013). Symbiotic specificity, association patterns, and function determine community responses to global changes: defining critical research areas for coral-*Symbiodinium* symbioses. *Global Change Biology* 19: 3306–3316.

Farnsworth, E.J. and Ellison, A.M. (1991). Patterns of herbivory in Belizean mangrove swamps. *Biotropica* 23: 555–567.

Faye, D., Le Loc'h, F., Thiaw, O.T. et al. (2012). Mechanisms of food partitioning and ecomorphological correlates in ten fish species from a tropical estuarine marine protected area (Bamboung, Senegal, West Africa). *African Journal of Agricultural Research* 7: 443–455.

Feller, I.C. (2002). The role of herbivory by wood-boring insects in mangrove ecosystems in Belize. *Oikos* 97: 167–176.

Feller, I.C. and Chamberlain, A. (2007). Herbivore responses to nutrient enrichment and landscape heterogeneity in a mangrove ecosystem. *Oecologia* 153: 607–616.

Fernandes, M.E.B., Nascimento, A.A.M., and Carvalho, M.L. (2009). Effects of herbivory by *Hyblaea puera* (Hyblaeidae, Lepidoptera) on litter production in the mangrove on the coast of Brazilian Amazonia. *Journal of Tropical Ecology* 25: 337–339.

Ferrario, F., Beck, M.W., Storlazzi, C.D. et al. (2014). The effectiveness of coral reefs for coastal hazard risk reduction and adaptation. *Nature Communications* 5: 1–9.

Fobert, E.K., Treml, E.A., and Swearer, S.E. (2019). Dispersal and population connectivity are phenotype dependent in a marine metapopulation. *Proceedings of the Royal Society B: Biological Sciences* 286: 20191104. https://doi.org/10.1098/rspb.2019.1104.
Fong, P., Smith, T.B., and Wartian, M.J. (2006). Epiphytic cyanobacteria maintain shifts to macroalgal dominance on coral reefs following ENSO disturbance. *Ecology* 87: 1162–1168.
Ford, B.M. and Roberts, J.D. (2020). Functional traits reveal the presence and nature of multiple processes in the assembly of marine fish communities. *Oecologia* 192: 143–154.
Forrester, G.E. (2015). Competition in reef fish. In: *Ecology of Fish on Coral Reefs* (ed. C. Mora), 34–40. Cambridge, UK: Cambridge University Press.
Forrester, G.E. and Finley, R.J. (2006). Parasitism and a shortage of refuges jointly mediate the strength of density dependence in a reef fish. *Ecology* 87: 1110–1115.
Forrester, G.E., Evans, B., Steele, M.A. et al. (2006). Assessing the magnitude of intra- and interspecific competition in two coral reef fish. *Oecologia* 148: 632–640.
de Fouw, J., Govers, L.L., van de Koppel, J. et al. (2016). Drought, mutualism breakdown, and landscape-scale degradation of seagrass beds. *Current Biology* 26: 1–6.
de Fouw, J., van der Heide, T., van Belzen, J. et al. (2018). A facultative mutualistic feedback enhances the stability of tropical intertidal seagrass beds. *Scientific Reports* 8: 12988. https://doi.org/10.1038/s41598-018-31060-x.
Fowler, A.J. (1991). Reproductive biology of bisexual and all-female populations of chaetodontid fish from the southern Great Barrier Reef. *Environmental Biology of Fish* 31: 261–274.
Fratini, S., Cannicci, S., and Vannini, M. (2000). Competition and interaction between *Neosarmatium smithi* (Crustacea; Grapsidae) and *Terebralia palustris* (Mollusca, Gastropoda) in a Kenyan mangrove. *Marine Biology* 137: 309–316.
Fratini, S., Vannini, M., Cannicci, S. et al. (2005). Tree-climbing mangrove crabs: a case of convergent evolution. *Evolutionary Ecology Research* 7: 219–233.
Fratini, S., Sacchi, A., and Vannini, M. (2011). Competition for mangrove leaf litter between two east African crabs, *Neosarmatium meinerti* (Sesarmidae) and *Cardisoma carnifex* (Gecarcinidae): a case of kleptoparasitism? *Journal of Ethology* 29: 481–485.
Freestone, A.L., Osman, R.W., Ruiz, G.M. et al. (2011). Stronger predation in the tropics shapes species richness patterns in marine communities. *Ecology* 92: 983–993.
Freestone, A.L., Ruiz, G.M., and Torchin, M.E. (2013). Stronger biotic resistance in tropics relative to temperate zone: effects of predation on marine invasion dynamics. *Ecology* 94: 1370–1377.
Freestone, A.L., Carroll, E.W., Papacostos, K.J. et al. (2020). Predation shapes invertebrate diversity in tropical but not temperate seagrass communities. *Journal of Animal Ecology* 89: 323–333.
Freitas, M.O., Previero, M., Leite, J.R. et al. (2019). Age, growth, reproduction and management of SW Atlantic's largest and endangered herbivorous reef fish, *Scarus trispinosus* Valenciennes, 1840. *PeerJ* 7: e7459. https://doi.org/10.7717/peerj.7459.
Garrity, S.D. and Levings, S.C. (1981). A predator-prey interaction between two physically and biologically constrained tropical rocky shore gastropods: direct, indirect and community effects. *Ecological Monographs* 51: 267–286.
Gawas, H. and Yogamoorthi, A. (2015). Studies on ant-plant interaction in tropical mangroves, *Rhizophora mucronata* and *Avicennia marina* from Pondicherry region, south India. *Asia-Pacific Journal of Energy and the Environment* 2: 53–58.
Goatley, C.H.R. and Bellwood, D.R. (2016). Body size and mortality rates in coral reef fish: a three-phase relationship. *Proceedings of the Royal Society B: Biological Sciences* 283: 20161858. https://doi.org/10.1098/rspb.2016.1858.
Goulet, T.L. and Coffroth, M.A. (2003). Stability of an octocoral-algal symbiosis over time and space. *Marine Ecology Progress Series* 250: 117–124.
Grady, J.M., Maitner, B.S., Winter, A.S. et al. (2019). Metabolic asymmetry and the global diversity of marine predators. *Science* 363: eaat4220. https://doi.org/10.1126/science.aat4220.
Gribben, P.E., Kimbro, D.L., Vergés, A. et al. (2017). Positive and negative interactions control a facilitation cascade. *Ecosphere* 8: e02065. https://doi.org/10.1002/ecs2.2065.

Gribben, P.E., Angelini, C., Altieri, A.H. et al. (2019). Facilitation cascades in marine ecosystems: a synthesis and future directions. *Oceanography and Marine Biology: An Annual Review* 57: 127–168.

Grimm, V., Reise, K., and Strasser, M. (2003). Marine metapopulations: a useful concept? *Helgoland Marine Research* 56: 222–228.

Grutter, A.S., Pickering, J.L., McCallum, H. et al. (2008). Impact of micropredatory gnathiid isopods on young coral reef fish. *Coral Reefs* 27: 655–661.

Grutter, A.S., Cribb, T.H., McCallum, H. et al. (2010). Effects of parasites on larval and juvenile stages of the coral reef fish *Pomacentrus moluccensis*. *Coral Reefs* 29: 31–40.

Guedes, A.P.P. and Araújo, F.G. (2008). Trophic resource partitioning among five flatfish species (Actinopterygii, Pleuronectiformes) in a tropical bay in south-eastern Brazil. *Journal of Fish Biology* 72: 1035–1054.

Hall, D., Lee, S.Y., and Meziane, T. (2006). Fatty acids as trophic tracers in an experimental estuarine food chain: tracer transfer. *Journal of Experimental Marine Biology and Ecology* 336: 42–53.

Hanski, I. and Simberloff, D. (1997). The metapopulation approach, its history, conceptual domain, and application to conservation. In: *Metapopulation Biology: Ecology, Genetics and Evolution* (eds. I. Hanski and M.E. Gilpin), 5–26. New York: Academic Press.

Heck, K.L. Jr., Hays, G., and Orth, R.J. (2003). Critical evaluation of the nursery role hypothesis for seagrass meadows. *Marine Ecology Progress Series* 253: 123–136.

Heck, K.L. Jr. and Orth, R.J. (2006). Predation in seagrass beds. In: *Seagrasses: Biology, Ecology and Conservation* (ed. A. Larkum), 537–550. Dordrecht, The Netherlands: Springer.

Heck, K.L. Jr. and Valentine, J.F. (2006). Plant-herbivore interactions in seagrass meadows. *Journal of Experimental Marine Biology and Ecology* 330: 420–436.

Hendy, I.W., Michie, L., and Taylor, B.W. (2014). Habitat creation and biodiversity maintenance in mangrove forests: teredinid bivalves as ecosystem engineers. *PeerJ* 2: e591. https://doi.org/10.7717/peerj.591.

Hixon, M.A. (2015). Predation, piscivory and the ecology of coral reef fish. In: *Ecology of Fish on Coral Reefs* (ed. C. Mora), 41–52. Cambridge, UK: Cambridge University Press.

Hixon, M.A. and Carr, M.H. (1997). Synergistic predation, density dependence, and population regulation in marine fish. *Science* 277: 946–949.

Hixon, M.A. and Webster, M.S. (2002). Density dependence in marine fish: coral reef populations as model systems. In: *Coral Reef Fish: Dynamics and Diversity in a Complex Ecosystem* (ed. P.F. Sale), 303–330. New York: Academic Press.

Hixon, M.A., Anderson, T.W., Buch, K.L. et al. (2012). Density dependence and population regulation in marine fish: a large-scale, long-term field manipulation. *Ecological Monographs* 82: 467–489.

Horinouchi, M., Tongnunui, P., Furumitsu, K. et al. (2012). Food habits of small fish in seagrass habitats in Trang, southern Thailand. *Fisheries Science* 78: 577–587.

Horne, J.B., van Herwerden, L., and Abellana, S. (2013). Observations of migrant exchange and mixing in a coral reef fish metapopulation link scales of marine population connectivity. *Journal of Heredity* 104: 532–546.

Horwitz, R., Hoogenboom, M., and Fine, M. (2017). Spatial competition dynamics between reef corals under ocean acidification. *Scientific Reports* 7: 40288. https://doi.org/10.1038/srep40288.

Hovel, K.A. and Lipcius, R.N. (2002). Effects of seagrass habitat fragmentation on juvenile blue crab survival and abundance. *Journal of Experimental Marine Biology and Ecology* 271: 75–98.

Hughes, T.P., Rodrigues, M.J., Bellwood, D.R. et al. (2007). Phase shifts, herbivory, and the resilience of coral reefs to climate change. *Current Biology* 17: 360–365.

Hunting, E.R., van der Geest, H.G., Krieg, A.J. et al. (2010). Mangrove-sponge associations: a possible role for tannins. *Aquatic Ecology* 44: 679–684.

Idjani, J.A. and Karlson, R.H. (2007). Spatial arrangement of competitors influences coexistence of reef-building corals. *Ecology* 88: 2449–2454.

Inoue, T., Shimono, A., Akaji, Y. et al. (2020). Mangrove–diazotroph relationships at the root, tree and forest scales: diazotrophic communities create high soil nitrogenase activities in *Rhizophora stylosa* rhizospheres. *Annals of Botany* 125: 131–144.

Irlandi, E.A. and Crawford, M.K. (1997). Habitat linkages: the effects of intertidal saltmarshes and adjacent subtidal habitats on abundance, movement, and growth of an estuarine fish. *Oecologia* 110: 222–230.

Jasmine, E. and Shyla Suganthi, A. (2012). Influence of external factors on the breeding activity of a brachyuran crab, *Parasesarma plicatum* Latreille. *Journal of Basic and Applied Biology* 6: 21–24.

Jaxion-Harm, J. and Speight, M.R. (2012). Algal cover in mangroves affects distribution and predation rates by carnivorous fish. *Journal of Experimental Marine Biology and Ecology* 414-415: 19–27.

Jompa, J. and McCook, L.J. (2002a). The effects of nutrients and herbivory on competition between a hard coral (*Porites cylindrica*) and a brown alga (*Lobophora variegata*). *Limnology and Oceanography* 47: 527–534.

Jompa, J. and McCook, L.J. (2002b). Effects of competition and herbivory on interactions between a hard coral and a brown alga. *Journal of Experimental Marine Biology and Ecology* 271: 25–39.

Justine, J.L. (2010). Parasites of coral reef fish, how much do we know? With a bibliography of fish parasites in New Caledonia. *Belgian Journal of Zoology* 140: 155–190.

Kayal, M. and Kayal, E. (2017). Colonies of the fire coral *Millepora platyphylla* constitute hermatypic survival oases during *Acanthaster* outbreaks in French Polynesia. *Marine Biodiversity* 47: 255–258.

Kneer, D., Asmus, H., and Vonk, J.A. (2008). Seagrass as the main food source of *Neaxius acanthus* (Thalassinidae, Strahlaxiidae), its burrow associates, and of *Corallianassa coutierei* (Thalassinidea, Callianassidae). *Estuarine, Coastal and Shelf Science* 79: 620–630.

Knowlton, N. and Rohwer, F. (2003). Multispecies microbial mutualisms on coral reefs: the host as a habitat. *The American Naturalist* 162: S51–S62.

Kou, Q., Li, X.Z., Chan, T.Y. et al. (2015). Divergent evolutionary pathways and host shifts among the commensal pontoniine shrimps: a preliminary analysis on selected Indo-Pacific species. *Organisms, Diversity and Evolution* 15: 369–377.

Krauss, K.W., Lovelock, C.E., McKee, K.L. et al. (2008). Environmental drivers in mangrove establishment and early development: a review. *Aquatic Botany* 89: 105–127.

Kritzer, J.P. (2002). Variation in the population biology of striped bass *Lutjanus carponotatus* within and between island groups on the Great Barrier Reef. *Marine Ecology Progress Series* 243: 191–207.

Kritzer, J.P. and Sale, P.F. (2006). The metapopulation ecology of coral reef fish. In: *Marine Metapopulations* (eds. J.P. Kretzer and P.F. Sale), 31–67. Amsterdam: Elsevier.

Kuffner, I.B., Walters, L.J., Becerro, M.A. et al. (2006). Inhibition of coral recruitment by macroalgae and cyanobacteria. *Marine Ecology Progress Series* 323: 107–117.

Kumya, P.Z., Byrne, M., Mos, B. et al. (2017). Indirect effects of ocean acidification drive feeding and growth of juvenile crown-of-thorns starfish, *Acanthaster planci*. *Proceedings of the Royal Society B: Biological Sciences* 284: 20170778. https://doi.org/10.1098/rspb.2017.0778.

Kwong, W.K., del Campo, J., Mathur, V. et al. (2019). A widespread coral-infecting apicomplexan with chlorophyll biosynthesis genes. *Nature* 568: 103–107.

Ladd, M.C., Shantz, A.A., and Burkepile, D.E. (2019). Newly dominant benthic invertebrates reshape competitive networks on contemporary Caribbean reefs. *Coral Reefs* 38: 1317–1328.

Landry, C.L. (2013). Pollinator-mediated competition between two co-flowering neotropical mangrove species, *Avicennia germinans* (Avicenniaceae) and *Laguncularia racemosa* (Combretaceae). *Annals of Botany* 111: 207–214.

Langmead, O. and Chadwick-Furman, N.E. (1999). Marginal tentacles of the corallimorpharian *Rhodactis rhodostoma*. I. Role in competition for space. *Marine Biology* 134: 479–489.

Lapid, E.D., Wielgus, J., and Chadwick-Furman, N.E. (2004). Sweeper tentacles of the brain coral *Platygyra daedalea*, induced development and effects on competitors. *Marine Ecology Progress Series* 282: 161–171.

Lee, S.Y. (1997). Potential trophic importance of the faecal material of the mangrove sesarmine crab *Sesarma messa*. *Marine Ecology Progress Series* 159: 275–284.

Lee, S.Y. (2005). Exchange of organic matter and nutrients between mangroves and estuaries: myths, methodological issues and missing links. *International Journal of Ecology and Environmental Science* 31: 163–175.

Lee, S.Y. (2008). Mangrove macrobenthos: assemblages, services, and linkages. *Journal of Sea Research* 59: 16–29.

Lee, S.Y. and Kwok, P.W. (2002). The importance of mangrove species association to the population biology of the sesarmine crabs *Parasesarma affinis* and *Perisesarma bidens*. *Wetlands Ecology and Management* 10: 215–226.

Lee, C.-H. and Lin, H.-J. (2015). Ontogenetic habitat utilization patterns of juvenile reef fish in low-predation habitats. *Marine Biology* 162: 1799–1811.

Lee, C.-L., Huang, Y.-H., Chung, C.-Y. et al. (2014). Tidal variation in fish assemblages and trophic structures in tropical Indo-Pacific seagrass beds. *Zoological Studies* 53: 56. https://doi.org/10.1186/s40555-014-0056-9.

Lee, C.-L., Huang, Y.-H., Chung, C.-Y. et al. (2015). Herbivory in multi-species, tropical seagrass beds. *Marine Ecology Progress Series* 525: 65–80.

Lenihan, H.S., Holbrook, S.J., Schmitt, R.J. et al. (2011). Influence of corallivory, competition, and habitat structure on coral community shifts. *Ecology* 92: 1959–1971.

Lessios, H.A. (2016). The great *Diadema antillarum* die-off: 30 years later. *Annual Review of Marine Science* 8: 267–283.

Lett, C., Nguyen-Huu, T., and Cuif, M. (2015). Linking local retention, self-recruitment, and persistence in marine metapopulations. *Ecology* 96: 2236–2244.

Levings, S.C. and Garrity, S.D. (1983). Diel and tidal movement of two co-occurring neritid snails: differences in grazing patterns on a tropical rocky shore. *Journal of Experimental Marine Biology and Ecology* 67: 261–278.

de Lira, J.J.P.R., dos Santos Calado, T.C., and de Araújo, M.S.L.C. (2013). Breeding period in the mangrove crab *Goniopsis cruentata* (Decapoda: Grapsidae) in NE Brazil. *Revista de Biología Tropical* 61: 29–38.

Littler, M.M., Littler, D.S., and Brooks, B.L. (2006). Harmful algae on tropical coral reefs: bottom-up eutrophication and top-down herbivory. *Harmful Algae* 5: 565–585.

Littler, M.M., Littler, D.S., and Brooks, B.L. (2009). Herbivory, nutrients, stochastic events, and relative dominances of benthic indicator groups on coral reefs: a review and recommendations. *Smithsonian Contributions in Marine Science* 38: 401–414.

Litulo, C. (2005). Population biology of the fiddler crab *Uca annulipes* (Brachyura, Ocypodidae) in a tropical east African mangrove (Mozambique). *Estuarine, Coastal and Shelf Science* 62: 283–290.

Liu, H.T.H., Kneer, D., Asmus, H. et al. (2008). The feeding habits of *Austrolethops wardi*, a gobiid fish inhabiting burrows of the thalassinidean shrimp *Neaxius acanthus*. *Estuarine, Coastal and Shelf Science* 79: 764–767.

Longo, G.O., Hay, M.E., Ferreira, C.E.L. et al. (2019). Trophic interactions across 61 degrees of latitude in the Western Atlantic. *Global Ecology and Biogeography* 28: 107–117.

Lubchenco, J., Menge, B.A., Garrity, S.D. et al. (1984). Structure, persistence, and role of consumers in a tropical rocky intertidal community (Taboguilla Island, Bay of Panama). *Journal of Experimental Marine Biology and Ecology* 78: 23–73.

de Macedo Costa, T.M., Pitombo, F.B., and Soares-Gomes, A. (2014). The population biology of the exploited crab *Ucides cordatus* (Linnaeus, 1763) in a southeastern Atlantic Coast mangrove area, Brazil. *Invertebrate Reproduction & Development* 58: 259–268.

Marriott, R.J., Mapstone, B.D., and Begg, G.A. (2007). Age-specific demographic parameters, and their implications for management of the red bass, *Lutjanus bohar* (Forsskal, 1775), a large, long-lived reef fish. *Fisheries Research* 83: 204–215.

Matthews, S.A., Mellin, C., and Pratchett, M.S. (2020). Larval connectivity and water quality explain spatial distribution of crown-of-thorns starfish outbreaks across the Great Barrier Reef. *Advances in Marine Biology* 87: 223–258.

McAfee, D., Cole, V.J., and Bishop, M.J. (2016). Latitudinal gradients in ecosystem engineering by oysters vary across habitats. *Ecology* 97: 929–939.

McClanahan, T.R. (1995). Fish predators and scavengers of the sea urchin *Echinometra mathaei* in Kenyan coral-reef marine parks. *Environmental Biology of Fish* 43: 187–193.

McClanahan, T.R. (2008). Response of the coral reef benthos and herbivory to fishery closure management and the 1998 ENSO disturbance. *Oecologia* 155: 169–177.

McGuiness, K.A. (1996). Dispersal, establishment and survival of *Ceriops tagal* propagules in a north Australian mangrove forest. *Oecologia* 109: 80–87.

Meekan, M.G., Ackerman, J.L., and Wellington, G.M. (2001). Demography and age structure of coral reef damselfish in the tropical eastern Pacific Ocean. *Marine Ecology Progress Series* 212: 223–232.

de Meeús, T. and Renaud, F. (2002). Parasites within the new phylogeny of eukaryotes. *Trends in Parasitology* 18: 247–251.

Menge, B.A. and Lubchenco, J. (1981). Community organization in temperate and tropical rocky intertidal habitats: prey refuges in relation to consumer pressure gradients. *Ecological Monographs* 51: 429–450.

Menge, B.A., Lubchenco, J., and Ashkenas, L.R. (1985). Diversity, heterogeneity and consumer pressure in a tropical rocky intertidal community. *Oecologia* 65: 394–405.

Menge, B.A., Lubchenco, J., Gaines, S.D. et al. (1986a). A test of the Menge-Sutherland model of community organization in a tropical rocky intertidal food web. *Oecologia* 71: 75–89.

Menge, B.A., Lubchenco, J., Ashkanas, L.R. et al. (1986b). Experimental separation of effects of consumers on sessile prey in the low zone of a rocky shore in the Bay of Panama: direct and indirect consequences of food web complexity. *Journal of Experimental Marine Biology and Ecology* 100: 225–269.

Miller, R.J., Adams, A.J., Ebersole, J.P. et al. (2007). Evidence of positive density-dependent effects in recovering *Diadema antillarum* populations. *Journal of Experimental Marine Biology and Ecology* 349: 215–222.

Mokhtari, M., Savari, A., Rezai, H. et al. (2008). Population ecology of fiddler crab, *Uca lactea annulipes* (Decapoda, Ocypodidae) in Sirik mangrove estuary, Iran. *Estuarine, Coastal and Shelf Science* 76: 273–281.

Moles, A.T. and Ollerton, J. (2016). Is the notion that species interactions are stronger and more specialized in the tropics a zombie idea? *Biotropica* 48: 141–145.

Mumby, P.J. and Dytham, C. (2006). Metapopulation dynamics of hard corals. In: *Marine Metapopulations* (eds. J.P. Kritzer and P.F. Sale), 157–203. Amsterdam: Elsevier.

Mumby, P.J., Dahlgren, C.P., Harborne, A.R. et al. (2006). Fishing, trophic cascades, and the process of grazing on coral reefs. *Science* 311: 98–100.

Mumby, P.J., Harborne, A.R., Williams, J. et al. (2007). Trophic cascade facilitates coral recruitment in a marine reserve. *Proceedings of the National Academy of Sciences* 104: 8362–8367.

Mumby, P.J., Steneck, R.S., Edwards, A.J. et al. (2012). Fishing down a Caribbean food web relaxes trophic cascades. *Marine Ecology Progress Series* 445: 13–24.

Munday, P.L. (2004). Competitive coexistence of coral-dwelling fish: the lottery hypothesis revisited. *Ecology* 85: 623–628.

Munday, P.L., Jones, G.P., and Caley, M.J. (2001). Interspecific competition and coexistence in a guild of coral-dwelling fish. *Ecology* 82: 2177–2189.

Muňoz, G., Grutter, A.S., and Cribb, T.H. (2007). Structure of the parasite communities of a coral reef fish assemblage (Labridae): testing ecological and phylogenetic host factors. *Journal of Parasitology* 93: 17–30.

Murai, M., Koga, T., and Yong, H.S. (2002). The assessment of female reproductive state during courtship and scramble competition in the fiddler crab, *Uca paradussumieri*. *Behavioral Ecology and Sociobiology* 52: 137–142.

Muscatine, L., Goiran, C., Land, L. et al. (2005). Stable isotopes (δ^{13}C and δ^{15}N) of organic matrix from coral skeleton. *Proceedings of the National Academy of Sciences* 102: 1525–1530.

Nagelkerken, I., van der Velde, G., Verberk, W.C.E.P. et al. (2006). Segregation along multiple resources axes in a tropical seagrass fish community. *Marine Ecology Progress Series* 308: 79–89.

Naidoo, Y., Steinke, T.D., Mann, F.D. et al. (2008). Epiphytic organisms on the pneumatophores of the mangrove *Avicennia marina*: occurrence and possible function. *African Journal of Plant Science* 1: 12–15.
van Nedervelde, F., Cannicci, S., Koedam, N. et al. (2015). What regulates crab predation on mangrove propagules? *Acta Oecologia* 63: 63–70.
Newman, S.J., Williams, M.B., and Russ, G.R. (1996). Age validation, growth and mortality rates of the tropical snappers (Pisces, Lutjanidae) *Lutjanus adetii* (Castelnau, 1873) and *L. quiquelineatus* (Bloch, 1790) from the central Great Barrier Reef, Australia. *Marine and Freshwater Research* 47: 575–584.
Nordhaus, I. and Wolff, M. (2007). Feeding ecology of the mangrove crab *Ucides cordatus* (Ocypodidae): food choice, food quality and assimilation efficiency. *Marine Biology* 151: 1665–1681.
Nordström, A.V., Nyström, M., Lokrantz, J. et al. (2009). Alternative states on coral reefs: beyond coral-macroalgal phase shifts. *Marine Ecology Progress Series* 376: 295–306.
O'Brien, J.M. and Scheibling, R.E. (2018). Turf wars: competition between foundation and turf-forming species on temperate and tropical reefs and its role in regime shifts. *Marine Ecology Progress Series* 590: 1–17.
O'Leary, J.K. and McClanahan, T.R. (2010). Trophic cascades result in large-scale coralline algae loss through differential grazer effects. *Ecology* 91: 3584–3597.
Olafsson, E., Buchmayer, S., and Skov, M.W. (2002). The east African decapod crab *Neosarmatium meinerti* (de Man) sweeps mangrove floors clean of leaf litter. *Ambio* 31: 569–573.
Ooi, J.L.S., Van Niel, K.P., Kendrick, G.A. et al. (2014). Spatial structure of seagrass suggests that size-dependent plant traits have a strong influence on the distribution and maintenance of tropical multispecies meadows. *PLoS One* 9: e86782. https://doi.org/10.1371/journal.pone.0086782.
Orth, R.J., Kendrick, G.A., and Marion, S.R. (2007). *Posidonia australis* seed predation in seagrass habitats of Two Peoples Bay, Western Australia. *Aquatic Botany* 86: 83–85.
Paine, R.T. (1966). Food web complexity and species diversity. *American Naturalist* 100: 65–75.
Parmentier, E. and Das, K. (2004). Commensal vs. parasitic relationship between Carapini fish and their hosts: some further insight through $\delta^{13}C$ and $\delta^{15}N$ measurements. *Journal of Experimental Marine Biology and Ecology* 310: 47–58.
Perry, D.A., Oren, R., and Hart, S.C. (2008). *Forest Ecology*, 2e. Baltimore, USA: Johns Hopkins.
Pessanha, A.L.M., Araújo, F.G., Oliveira, R.E.M.C.C. et al. (2015). Ecomorphology and resource use by dominant species of tropical estuarine juvenile fish. *Neotropical Ichthyology* https://doi.org/10.1590/1982-0224-2010080.
Peterson, B.J. and Heck, K.L. (2001). Positive interactions between suspension-feeding bivalves and seagrass- a facultative mutualism. *Marine Ecology Progress Series* 213: 143–155.
Poovachiranon, S. and Tantichodok, P. (1991). The role of sesarmid crabs in the mineralisation of leaf litter of *Rhizophora apiculata* in a mangrove, southern Thailand. *Phuket Marine Biological Centre Research Bulletin* 56: 63–74.
Pratchett, M.S., Caballes, C.F., Rivera-Posada, J.A. et al. (2014). Limits to understanding and managing outbreaks of crown-of-thorns starfish (*Acanthaster* spp.). *Oceanography and Marine Biology: An Annual Review* 52: 133–200.
Preen, A. (1992). Interactions between dugongs and seagrasses in a subtropical environment. PhD dissertation. James Cook University.
Putnam, H.M., Stat, M., Pochon, X. et al. (2012). Endosymbiotic flexibility associates with environmental sensitivity in hermatypic corals. *Proceedings of the Royal Society B: Biological Sciences* 279: 4352–4361.
Ramsby, B.D., Hoogenboom, M.O., Whalan, S. et al. (2017). A decadal analysis of bioeroding sponge cover on the inshore Great Barrier Reef. *Scientific Reports* 7: 1–10.
Ribeiro, F.B. and Bezerra, L.E.A. (2014). Population ecology of mangrove crabs in Brazil: sesarmid and fiddler crabs. In: *Crabs: Global Diversity, Behavior and Environmental Threats* (ed. C. Ardovini), 19–56. New York: Nova.
Rinkevich, B. and Sakai, K. (2001). Interspecific interactions among species of the coral genus *Porites* from Okinawa, Japan. *Zoology* 104: 1–7.

Ritson-Williams, R., Arnold, S.N., Fogarty, N.D. et al. (2009). New perspectives on ecological mechanisms affecting coral recruitment on reefs. *Smithsonian Contribution in Marine Science* 38: 437–458.

Robertson, A.I. (1986). Leaf-burying crabs: their influence on energy flow and export from mixed mangrove forests (*Rhizophora* spp.) in NE Australia. *Journal of Experimental Marine Biology and Ecology* 102: 237–248.

Robertson, A.I. and Daniel, P.A. (1989). The influence of crabs on litter processing in high intertidal mangrove forests in tropical Australia. *Oecologia* 78: 191–198.

Rohner, C.A., Armstrong, A.J., Pierce, S.J. et al. (2015). Whale sharks target dense prey patches of sergestid shrimp off Tanzania. *Journal of Plankton Research* 37.

Rotjan, R.D. and Lewis, S.M. (2008). Impact of coral predators on tropical reefs. *Marine Ecology Progress Series* 367: 73–91.

Saenger, P. (2002). *Mangrove Ecology, Silviculture and Conservation*. Dordrecht, The Netherlands: Kluwer.

Saenz-Agudelo, P., Jones, G.P., Thorrold, S.R. et al. (2011). Connectivity dominates larval replenishment in a coastal reef fish metapopulation. *Proceedings of the Royal Society B: Biological Sciences* 278: 2954–2961.

Saifullah, A.S.M. and Ali, M.S. (2004). Insect herbivory in polluted mangroves of the Indus delta. *Pakistan Journal of Botany* 36: 127–131.

Salomon, P.S., Granéli, E., Neves, M.H.C.B. et al. (2009). Infection by *Amoebophrya* spp. parasitoids of dinoflagellates in a tropical marine coastal area. *Aquatic Microbial Ecology* 55: 143–153.

Sandin, S.A. and Pacala, S.W. (2005). Demographic theory of coral reef fish populations with stochastic recruitment: comparing sources of population regulation. *The American Naturalist* 165: 107–119.

Sandin, S.A., Walsh, S.M., and Jackson, J.B.C. (2010). Prey release, trophic cascades, and phase shifts in tropical nearshore ecosystems. In: *Trophic Cascades: Predators, Prey and the Changing Dynamics of Nature* (eds. J. Terborgh and J.A. Estes), 71–90. New York: Island Press.

Sanil, N.K., Asokan, P.K., John, L. et al. (2011). Pathological manifestations of the acanthocephalan parasite, *Tenuiproboscis* sp. in the mangrove red snapper (*Lutjanus argentimaculatus*) (Forsskål, 1775), a candidate species for aquaculture from southern India. *Aquaculture* 310: 259–266.

Sant'anna, B.S., Borges, R.P., Hattori, G.Y. et al. (2014). Reproduction and management of the mangrove crab *Ucides cordatus* (Crustacea, Brachyura, Ucididae) at Iguape, São Paulo, Brazil. *Anais da Academia Brasileira de Ciências* 86: 1411–1421.

Sarma, V.V. and Raghukumar, S. (2013). Manglicolous fungi from Chorao mangroves, Goa, west coast of India: observations on fungi species consortia. *Kavaka* 41: 18–22.

Schemske, D.W. (2009). Biotic interactions and speciation in the tropics. In: *Speciation and Patterns of Diversity* (eds. R.K. Butlin and D. Schluter), 219–239. Cambridge, UK: Cambridge University Press.

Schemske, D.W., Mittelbach, G.G., Cornell, H.V. et al. (2009). Is there a latitudinal gradient in the importance of biotic interactions? *Annual Review in Ecology and Systematics* 40: 245–269.

Schwamborn, S.H.L. and Ferreira, B.P. (2002). Age structure and growth of the dusky damselfish, *Stegastes fuscus*, from Tamandaré reefs, Pernambuco, Brazil. *Environmental Biology of Fish* 63: 79–88.

Sebens, K.P. and Miles, J.S. (1988). Sweeper tentacles in a gorgonian octocoral: morphological modifications for interference competition. *Biological Bulletin* 179: 378–387.

Sheppard, C., Davy, S., and Pilling, G. (2018). *The Biology of Coral Reefs*, 2e. Oxford, UK: Oxford University Press.

da Silva Castiglioni, D., de Oliveira, P.J.A., da Silva, J.S. et al. (2011). Population dynamics of *Sesarma rectum* (Crustacea: Brachyura: Grapsidae) in the Ariquindá River mangrove, north-east of Brazil. *Journal of the Marine Biological Association of the United Kingdom* 91: 1395–1401.

da Silva Garcia, A.F. and Vendel, A.L. (2016). Dietary overlap and food resource partitioning among fish species of a tropical estuary in NE Brazil. *Gaia Scientia* 10: 86–97.

Silverstein, R.N., Correa, A.M.S., and Baker, A.C. (2012). Specificity is rarely absolute in coral-algal symbiosis: implications for coral response to climate change. *Proceedings of the Royal Society B: Biological Sciences* 279: 2609–2618.

Smith, T.J. III (1987). Seed predation in relation to tree dominance and distribution in mangrove forests. *Ecology* 68: 266–273.

Smith, T.J. III (1988a). Differential distribution between sub-species of the mangrove *Ceriops tagal*: competitive interactions along a salinity gradient. *Aquatic Botany* 32: 79–89.

Smith, T.J. III (1988b). Structure and succession in tropical, tidal forests: the influence of seed predators. *Proceedings of the Ecological Society of Australia* 15: 203–211.

Smith, T.J. III (1992). Forest structure. In: *Tropical Mangrove Ecosystems* (eds. A.I. Robertson and D.M. Alongi), 101–136. Washington, DC: American Geophysical Union.

Sousa, W.P., Kennedy, P.G., and Mitchell, B.J. (2003). Propagule size and predispersal damage by insects affect establishment and early growth of mangrove seedlings. *Oecologia* 135: 564–575.

Sousa, W.P., Kennedy, P.G., Mitchell, B.J. et al. (2007). Supply-side ecology in mangroves: do propagule dispersal and seedling establishment explain forest structure? *Ecological Monographs* 77: 53–76.

Stanley, G.D. (2006). Photosymbiosis and the evolution of modern coral reefs. *Science* 312: 857–858.

Stanley, G.D. and Swart, P.K. (1995). Evolution of the coral-zooxanthellae symbiosis during the Triassic: a geochemical approach. *Paleobiology* 21: 179–199.

Stewart, H.L., Holbrook, S.J., Schmitt, R.J. et al. (2006). Symbiotic crabs maintain coral health by clearing sediments. *Coral Reefs* 25: 609–615.

Sutherland, J.P. (1990). Recruitment regulates demographic variation in a tropical intertidal barnacle. *Ecology* 71: 955–972.

Sutherland, J.P. and Ortega, S. (1986). Competition conditional on recruitment and temporary escape from predators on a tropical rocky shore. *Journal of Experimental Marine Biology and Ecology* 95: 155–166.

Sutherland, C.S., Elston, D.A., and Lambin, X. (2014). A demographic, spatially explicit patch occupancy model of metapopulation dynamics and persistence. *Ecology* 95: 3149–3160.

Sweatman, H.P.A. (2008). No-take reserves protect coral reefs from predatory starfish. *Current Biology* 18: R598–R599.

Swierts, T. and Vermeij, M.J.A. (2016). Competitive interactions between coral and turf algae depend on coral colony form. *PeerJ* 4: e1984. https://doi.org/10.17717/peerj.1984.

Syama, V.P. and Anilkumar, G. (2011). Lunar rhythm-dependent spawning and larval release in the continuously breeding mangrove crab, *Sesarma quadratum* (Decapoda: Brachyura). *2011 2nd International Conference on Environmental Science and Technology IPCBEE*, IACSIT Press, Singapore 6: V1-188-V190.

Tanner, J.E. (1995). Competition between hermatypic corals and macroalgae: an experimental investigation of coral growth, survival and reproduction. *Journal of Experimental Marine Biology and Ecology* 190: 151–168.

Tebbett, S.B., Streit, R.P., and Bellwood, D.R. (2019). Expansion of a colonial ascidian following consecutive mass coral bleaching at Lizard Island, Australia. *Marine Environmental Research* 144: 125–129.

Tenjing, Y. (2019). Population dynamics of edible wedge clam, *Donax lubricus* Hanley 1845 (Bivalvia: Donacidae): a first study in Asia. *Regional Studies in Marine Science* 32: 100819. doi :10.1016.j.rsma.2019.100819.

Thakur, N.L. and Singh, A. (2016). Chemical ecology of marine sponges. In: *Marine Sponges: Chemicobiological and Biomedical Applications* (eds. R. Pallela and H. Ehrlich), 37–52. Springer: New Delhi.

Thomsen, M.S., Altieri, A.H., Angelini, C. et al. (2018). Secondary foundation species enhance biodiversity. *Nature Ecology & Evolution* 2: 634–639.

Thongtham, N. and Kristensen, E. (2005). Carbon and nitrogen balance of leaf-eating sesarmid crabs (*Neoepisesarma versicolor*) offered different food sources. *Estuarine, Coastal and Shelf Science* 65: 213–222.

Tiwari, M., Bjorndal, K.A., Bolten, A.B. et al. (2006). Evaluation of density-dependent processes and green turtle *Chelonia mydas* hatchling production at Tortuguero, Costa Rica. *Marine Ecology Progress Series* 326: 283–293.

Tong, Y.F., Lee, S.Y., and Morton, B. (2006). The herbivore assemblage, herbivory and leaf chemistry of the mangrove *Kandelia obovata* in two contrasting forests in Hong Kong. *Wetlands Ecology and Management* 14: 39–52.

Tsounis, G., Rossi, S., Gill, J.M. et al. (2006). Population structure of an exploited benthic cnidarian: the case study of red coral (*Corallium rubrum* L.). *Marine Biology* 149: 1059–1070.

Twilley, R.R., Pozo, M., Garcia, V.H. et al. (1997). Litter dynamics in riverine mangrove forests in the Guayas River estuary, Ecuador. *Oecologia* 111: 109–122.

Unsworth, R.K.F., Wylie, E., Smith, D.J. et al. (2007a). Diel trophic structuring of seagrass bed fish assemblages in the Wakatobi Marine Natural Park, Indonesia. *Estuarine, Coastal and Shelf Science* 72: 81–88.

Unsworth, R.K.F., Taylor, J.D., Powell, A. et al. (2007b). The contribution of scarid herbivory to seagrass ecosystem dynamics in the Indo-Pacific. *Estuarine, Coastal and Shelf Science* 74: 53–62.

Uthicke, S., Logan, M., Liddy, M. et al. (2015). Climate change as an unexpected co-factor promoting coral eating sea star (*Acanthaster planci*) outbreaks. *Scientific Reports* 5: 8402.

Valentine, J.F. and Duffy, J.E. (2006). The central role of grazing in seagrass ecology. In: *Seagrasses: Biology, Ecology and Conservation* (ed. A.W.D. Larkum), 463–501. Dordrecht, The Netherlands: Springer.

Vergamini, F.G. and Mantelatto, F.L. (2008). Continuous reproduction and recruitment in the narrowback mud crab *Panopeus americanus* (Brachyura, Panopeidae) in a remnant human-impacted mangrove area. *Invertebrate Reproduction & Development* 51: 1–10.

Vergés, A., Vanderklift, M.A., Doropoulos, C. et al. (2011). Spatial patterns in herbivory on a coral reef are influenced by structural complexity but not by algal traits. *PLoS One* 6: e17115. https://doi.org/10.1371/journal.pone.0017115.

Vermeij, G.J. (1977). The Mesozoic marine revolution: evidence from snails, predators and grazers. *Paleobiology* 3: 245–258.

Vermeij, G.J. (1978). *Biogeography and Adaptation: Patterns of Marine Life*. Cambridge, USA: Harvard.

Vermeij, M.J.A. and Sandin, S.A. (2008). Density-dependent settlement and mortality structure the earliest life phases of a coral population. *Ecology* 89: 1994–2004.

Vieira, C. (2020). Lobophora–coral interactions and phase shifts: summary of current knowledge and future directions. *Aquatic Ecology* 54: 1–20.

Vonk, J.A., Pijnappels, M.H.J., and Stapel, J. (2008a). In situ quantification of *Tripneustes gratilla* grazing and its effects on three co-occurring tropical seagrass species. *Marine Ecology Progress Series* 360: 107–114.

Vonk, J.A., Christianen, M.J.A., and Stapel, J. (2008b). Redefining the trophic importance of seagrasses for fauna in tropical Indo-Pacific meadows. *Estuarine, Coastal and Shelf Science* 79: 653–660.

Vonk, J.A., Middelburg, J.J., Stapel, J. et al. (2008c). Dissolved organic nitrogen uptake by seagrasses. *Limnology and Oceanography* 53: 542–548.

Waldrop, E., Hobbs, J.P.A., Randall, J.E. et al. (2016). Phylogeography, population structure and evolution of coral-eating butterflyfish (Family Chaetodontidae, genus Chaetodon, subgenus Corallochaetodon). *Journal of Biogeography* 43: 1116–1129.

Wasserman, R.J. and Mostert, B.P. (2014). Intertidal crab burrows as a low-tide refuge habitat for a specific gobiid: preliminary evidence for commensalism. *Marine and Freshwater Research* 65: 333–336.

Wellens, S., Sandrini-Neto, L., Gonzalez-Wangüemert, M. et al. (2015). Do the crabs *Goniopsis cruentata* and *Ucides cordatus* compete for mangrove propagules? A field-based experimental approach. *Hydrobiologia* 757: 117–128.

White, J.W., Samhouri, J.F., Stier, A.C. et al. (2010). Synthesizing mechanisms of density dependence in reef fish: behaviour, habitat configuration, and observational scales. *Ecology* 91: 1949–1961.

Wulff, J.L. (1997). Mutualisms among species of coral reef sponges. *Ecology* 78: 146–159.
Wulff, J.L. (2008a). Life-history differences among coral reef sponges promote mutualism or exploitation of mutualism by influencing partner fidelity feedback. *The American Naturalist* 171: 597–609.
Wulff, J.L. (2008b). Collaboration among sponge species increases sponge diversity and abundance in a seagrass meadow. *Marine Ecology* 29: 193–204.
Wulff, J.L. (2012). Ecological interactions and the distribution, abundance, and diversity of sponges. *Advances in Marine Biology* 61: 273–344.
van der Zee, E.M., Angelini, C., Govers, L.L. et al. (2016). How habitat-modifying organisms structure the food web of two coastal ecosystems. *Proceedings of the Royal Society B: Biological Sciences* 283: 20152326. https://doi.org/10.1098/rspb.2015.2326.
Zhou, M.X., Yu, K.F., Zhang, Q.M. et al. (2014). Age structure of massive *Porites lutea* corals at Luhuitou fringing reef (northern South China Sea) indicates recovery following severe anthropogenic disturbance. *Coral Reefs* 33: 39–44.

CHAPTER 7

Ecosystems

7.1 Introduction

An ecosystem is an integrated system composed of interacting biotic and abiotic components. Ecosystems have boundaries and structure which implies that we can distinguish between the ecosystem and its surrounding environment, beyond the boundaries of the system itself. Thus, all systems that encompass biotic and abiotic components may be considered as an ecosystem.

There are at least five basic properties of ecosystems: (i) they are open, (ii) they have directionality, (iii) they have connectivity, (iv) they have emergent hierarchies, and (v) they have complex dynamics. Ecosystems are open in that external variables determine the state of the ecosystem. Ecosystems have directionality in that that they grow and develop over time and space. That the biotic and abiotic components of an ecosystem are connected in a network shows that ecosystems have connectivity. Ecosystems are self-organising and self-regulated and cycle energy, matter, and information and are organised hierarchically, reflecting the fact that they have emergent hierarchies. Ecosystems are adaptive systems that grow and develop and can cope with disturbances, have high buffer capacities, particularly under natural conditions, and usually recover rapidly after disturbance.

Ecosystems have structure, and in this chapter, we examine that structure, most simply depicted by a list of species present, as well as the relationship of biota with the local environment. There are, of course, other aspects of ecosystem structure including food web structure and hierarchies and measures of other properties such as species diversity in relation to the environment. We will start first with those ecosystems that are subjected to the power of tides.

7.2 Rocky Shores

Tropical rocky shores, unlike their temperate counterparts, appear superficially to be barren due to a lack of foliose macroalgae and a low abundance of sessile invertebrates. The dominant space-occupying species on many tropical rocky shores are low-lying forms, such as encrusting algae and turf algae (Mayakun et al. 2020). The importance of encrusting algae as dominants or being abundant on tropical rocky shores has been observed on shores in Central America, Africa, and in the tropical western Pacific.

Tropical Marine Ecology, First Edition. Daniel M. Alongi.

Despite their abundance on many tropical shores, encrusting algae are usually classified as a functional group or under generic names rather than by species (Mayakun et al. 2020). For example, eight common species of encrusting algae were recorded on rocky shores of varying exposure in Hong Kong (Kaehler and Williams 1996) with greatest abundance recorded on shores of intermediate exposure where four distinct zonation bands were discerned: (i) a cyanobacterial '*Kyrtuthix* zone' in the upper intertidal; (ii) a 'bare zone' below this; (iii) a 'mixed zone' in the low and mid intertidal; and (iv) a 'coralline zone' in the lower fringe. On shores of greater and lower exposure to wave action, abundances declined, but it was on these two exposed shores that bivalves and barnacles were competitively dominant. The abundance of encrusting algae was greatest towards the lowest sections of the shoreline where physical stress was less and where herbivore density was greatest. Most species increased in abundance during the cool season while the cover and vertical extent of encrusting algae decreased during summer when temperatures were maximal. Thus, at least on the shores of Hong Kong, encrusting algae have a high species richness and exhibit spatial and temporal variability, regulated in part by seasonal variations in temperature and herbivory.

Comparing structures between tropical and temperate rocky intertidal shores, Menge and Lubchenco (1981) found that in comparison to temperate shores, the rocky intertidal of Panama was characterised by extremely low abundances of non-crustose algae and sessile animals, indistinct vertical zonation patterns, and the occurrence of most invertebrates and algae, except barnacles, in holes and crevices. In the Panama ecosystem, in contrast to temperate rocky ecosystems, escapes from consumers in time, body size, and two-dimensional space are secondary in importance for most prey. Fast-moving consumers such as herbivorous and predaceous fish and herbivorous crabs are rare or absent on temperate rocky shores.

The relative importance of recruitment and other causes of variation in community structure between temperate and tropical shores were examined by Menge (1991), who observed that rates of increase of prey abundance in predator exclusion experiments in New England were at least an order of magnitude greater than in Panama. And while recruitment densities of sessile invertebrates and algae were highly variable at both locations, they were lower by an order of magnitude and less synchronous in Panama than in New England. The proportionate contributions of predation, competition and recruitment differed between locations, with recruitment explaining 11% while predation and competition explained 50–78% of variation, respectively, in sessile invertebrate abundance in New England. In contrast, in Panama, recruitment explained 39–87% while predation and competition explained 8–10% of variation respectively in sessile invertebrate abundance.

Along similar lines, Chan (2006) examined whether there was a latitudinal gradient of barnacles on temperate to tropical rocky intertidal zones from the NW Pacific to the South China Sea. The high intertidal zone was occupied by different species, but the mid-intertidal zones were occupied by the genus *Tetraclita*. Barnacles in Japan were observed to have a shorter reproductive period than those in Hong Kong; barnacles in Hong Kong followed a seasonal trend of early gonad development and settlement followed by a regular summer die-off due to high temperatures. While some of the barnacle life history patterns along the gradient can be explained by the differences in latitude, some differences can be explained by different ocean currents between Hong Kong and the NW Pacific.

Although tropical, Hong Kong experiences a seasonal climate with cool winters and hot, wet summers in which typhoons are common and have a great influence on rocky intertidal ecosystems (Hutchinson and Williams 2003). In August 1997, Typhoon Victor affected Hong Kong coastal waters bringing high seas, heavy rain, and strong winds. The effects of the typhoon were patchy, with an increase in free space and a decrease in the abundance of the dominant molluscan grazer, *Monodonta labio,* occurring in one area but not in other semi-exposed areas. Long-term impacts of this disturbance were negligible as during this time of year (summer) there is ordinarily little algae cover and grazer densities are low due to the high temperatures and subsequent desiccation. Thus, pulse disturbances such as typhoons are of less consequence than the annual disturbance effects of summer die-off.

The most complete study of the structure of a tropical rocky intertidal shore was conducted by Lubchenco et al. (1984) on Taboguilla Island in the Bay of Panama. Studying the high, mid, low, and very low-intertidal zones, they found that temperature and rainfall varied seasonally, with most of the latter falling from May to mid-December. The shores of Taboguilla appear barren throughout the year compared with temperate rocky shores because of the lack of epibenthic animals and plants, including erect macroalgae. This is most clearly seen in the high-intertidal zone where > 90% of the space is bare basaltic rock. On temperate shores, there are very clearly defined horizontal bands corresponding to the intertidal zonation of the shore; although not as prominent, there was vertical zonation in this tropical rocky ecosystem. The most obvious sessile animals in each tidal zone were barnacles or oysters, including *Chthamalus fissus, Euraphia imperatrix,* or both, in the high-intertidal zone, *Tetraclita panamensis* in the mid-intertidal zone, and *Balanus inexpectatus* or *Chama echinata* in the low-intertidal zone. Only 115 macrofaunal species occurred in this ecosystem which is low relative to number of species recorded in temperate rocky shores.

Primary space was dominated by algal crusts in the mid, low, and very low-intertidal zones and by bare rock in the high-tidal zone. Sessile animals were scarce as most species were rare or patchily abundant. Although abundant in temperate areas, barnacles and large bivalves were scarce on Taboguilla Island and the cover of erect algae was low. Species diversity and species richness were greatest in the lower zones.

The paucity of macroalgae and the abundance of herbivorous fish and crabs and predaceous fish distinguish tropical and temperate rocky shores. Benthic-feeding fish seem to be minor food web components in temperate rocky intertidal communities. The scarcity of erect sessile invertebrates and foliose algae was at least partly due to predation and herbivory. Thus, consumers appear to obscure zonation patterns on Taboguilla by preventing the development of prominent horizontal bands of sessile animals or erect macroalgae. For example, in the high intertidal zone, a prominent barnacle zone so distinct on temperate rocky shores was indistinct at Taboguilla. Consumer removal experiments suggested that barnacle abundance was strongly affected by consumers. Feeding data indicate that whelks (*Thais melones, Acanthina brevidentata*) and several fish species were probably the major predators. Grazers may also interfere with barnacle settlement.

In the mid-intertidal zone, fleshy encrusting algae covered most of rock surfaces and barnacles whereas foliose macroalgae and mussels that are typically abundant on temperate shores, were scarce on Taboguilla, although mobile invertebrates were common. In the high-intertidal zone, mortality from heat and desiccation, larval scarcity, and recruitment variability all contributed to the maintenance of mostly bare rock. Consumers such as limpets, predaceous gastropods, crabs, chitons, and fish were abundant and diverse across the intertidal zones.

Disturbance and top-down and bottom-up drivers can influence early algal succession patterns on tropical rocky shores. Seasonality of algal patterns was examined on rocky shores at Koh Pling, Thailand, where caging experiments found an early succession pattern involving 17 species of which three were dominant: *Ulva paradoxa*, *Padina* in the *Vaughaniella* stage and *Polysiphonia sphaerocarpa*. *U. paradoxa* was the earliest coloniser and occupied cleared patches within the first month (Mayakun et al. 2020). It was then replaced by the late successional algae, *Padina* in the *Vaughaniella* stage and *P. sphaerocarpa*. Neither herbivory nor nutrient enrichment affected *U. paradoxa* during the dry season, but in the wet season, *U. paradoxa* was influenced by nutrient enrichment, with the abundance of most algae unaffected by disturbance.

A tropical intertidal shore in an upwelling region was examined to determine the effect, if any, of upwelling on intertidal biota (Sibaja-Cordero and Cortés 2008). Two sheltered and two exposed shores along the north Pacific coast of Costa Rica during the upwelling (April) and during the non-upwelling (October) seasons of Costa Rica were sampled to determine the vertical zonation of algae and invertebrates. Some seasonality was found in the percent cover of the algal assemblage and the presence of herbivores related to upwelling. The dominant *Ulva* and the encrusting *Gymnogongrus* decreased during an ENSO event, possibly due to reduced nutrients, high temperatures, and strong wave action. In contrast, sessile invertebrates showed no change between seasons, although they did show a patchy structure in sheltered sites, while in exposed locations, invertebrates formed horizontal bands. Predation and space monopolisation were cited as possible causes of lower cover of barnacles in sheltered sites and lower abundance of molluscs in exposed areas. The high cover of sessile invertebrates compared with the Bay of Panama may be explained by the regional differences in intensity of upwelling.

7.3 Sandy Beaches and Tidal Flats

Many tropical beaches and sand and mud flats located adjacent to mangrove forests, seagrass beds and reefs, receive significant amounts of organic matter from these habitats, and are important sites for nutrient regeneration and energy flow (Chapter 10). Microalgae are a central part of ecosystem structure on sandy beaches and tidal mud and sand flats. Diatom assemblages are species-rich on sediment surfaces and are selectively ingested quite commonly by meiofauna, macroinfauna, and mudskippers (Yang et al. 2003; Dinh et al. 2020). A complex ecosystem exists on sand and mud flats as less dynamic physical forces facilitate the development of abundant and often diverse benthic communities that are in turn fed upon by seabirds.

Shorebirds are a crucial structural component of tidal flat and beach ecosystems. Avian predators of intertidal benthos are common, and the fraction of macrofaunal prey consumed tends to be high. The harvestable fractions of bivalves in two West African tidal flats (Banc d'Arguin, Mauritania and Guinea-Bissau) were high with 23–84% of total biomass in six species consumed mostly by the knot, *Calidris canutus*

(Piersma et al. 1993). Shorebirds feeding on the extensive tidal flats of Banc d'Arguin were able to deplete benthic food stocks during their winter migration to the flats (Salem et al. 2014), driving seasonal changes in faunal abundances and richness of the bivalves, *Loripes lucinalis* and *Dosinia isocardia*.

Benthic species diversity varies widely among different tropical tidal flats for many reasons, including variable wave and tidal action as well as pollution. Species diversity of macrofauna was low and densities were highly variable, but ordinarily low (8–1583 ind. [individuals] m^{-2}) in mud flats of the Niger delta, Nigeria (Zabbey and Hart 2014) that suffered oil spills. Low species diversity and richness were also found in the dynamic, tropical tidal flats on the Guiana coast of South America (Jourde et al. 2017). These latter findings are not surprising as these tidal flats are very unstable and migrate along the coast, with inherently unstable, fluidised muds (Section 4.1).

Sandy beaches vary in type from reflective to dissipative (Figure 4.2) although most beaches reflect a continuum between both extremes which reflect morphological and oceanographic factors controlling the form of sandy beach as well as the biota. A comparison of latitudinal trends in faunal species richness of sandy beaches shows that the number of species (Figure 7.1) significantly decreases linearly from tropical to temperate regions in the southern hemisphere, whereas a linear decreasing pattern is barely significant for the northern hemisphere (Defeo and McLachlan 2013). For the same degree of latitude, species richness is significantly greater in the northern than the southern hemisphere. These results must be considered cautiously as northern beaches are under-represented compared with southern hemisphere beaches. Also, there was no "hemisphere × latitude" interaction effect, meaning that the highest species range in the northern hemisphere covered the entire latitudinal range.

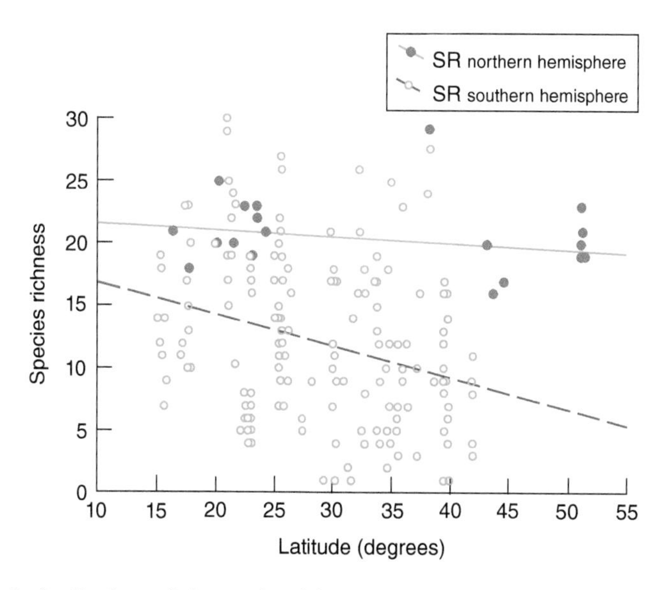

FIGURE 7.1 Latitudinal trends in species richness in sandy beach macrofauna discriminated by hemisphere and using latitude as the continuous covariate. *Source:* Defeo and McLachlan (2013), figure 2, p. 108. © Elsevier.

Species richness for macrofauna significantly deceases with beach slope as shown in a semi-log model of tropical and temperate regions (Figure 7.2a). A comparison of latitudinal trends in species richness using beach slope as a co-variate shows that for the same beach slope, the number of species is significantly greater in the tropics (Defeo and McLachlan 2013; McLachlan and Defeo 2018). For abundance, there is evidence (Figure 7.2b) that it is significantly higher in the tropics for slopes > 0.05 and that temperate sandy beaches with slopes ≤ 0.05 have higher abundance than their tropical counterparts. For biomass, almost identical trends between regions were seen (Figure 7.2c), denoting similar declining rates in biomass towards steeper slopes in tropical and temperate sites. For body size, estimated as the ratio between biomass and abundance for each beach, a significant increasing pattern with beach slope was detected only for temperate regions, even though for a same slope, body size trends

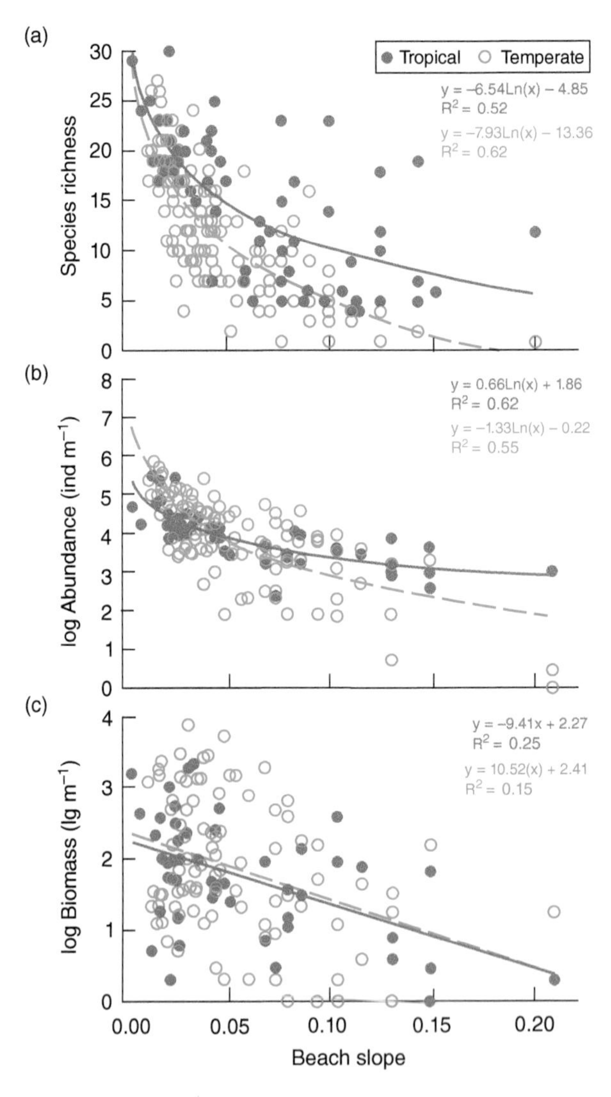

FIGURE 7.2 Patterns of (a) species richness, (b) abundance and (c) biomass in temperate versus tropical sandy beaches using beach slope as a covariate. All models fitted are highly significant. *Source:* Defeo and McLachlan (2013), figure 3, p. 109. © Elsevier.

tend to be higher in temperate beaches, but with no significant differences between tropical and temperate.

Data from hundreds of beaches (Defeo and McLachlan 2013; McLachlan and Defeo 2018) show that controlling for the effect of beach morphology, species richness is greatest in the tropics. Tropical regions have more reflective beaches compared with temperate and subtropical regions, whereas temperate beaches are mostly dissipative. Defeo and McLachlan (2013) argue that this probably reflects higher wave energy, but it is also possible that this might reflect bias in site selection rather than true latitudinal differences. The tropical zone of South America on both the Pacific and Atlantic coasts is dominated by a "narrow discontinuous stretch of sheltered and pocket sand beaches, alternating with rocky shores and mangrove swamps" (Defeo and McLachlan 2013). Coral reefs front some tropical sand beaches where they further dissipate wave energy.

Abundance patterns are more complex than those for species richness, with higher abundances on steep-slope tropical beaches than on gentle-slope temperate beaches (McLachlan and Defeo 2018). Biomass shows the same pattern suggesting decreasing carrying capacity towards reflective beaches. Abundance trends are less clear due to variable patterns for molluscs and polychaetes, and it is possible that this variability could be related to surf zone productivity and beach wrack as influenced by wave energy. Similar patterns of increasing species richness towards the tropics were found along eastern and southern Australian beaches (Hacking 2007), although abundance and biomass appeared to differ in magnitude between biogeographical regions as a function of surf zone processes and climate. The dominance of physical forces on sandy beach ecosystems does not preclude the importance of the role of biological forces. Identification of large-scale patterns of species abundance in relation to competitors or predators is often limited by the short time series of coverage of most studies as well as the use of different methodologies.

A series of studies of the mole crab, *Emerita brasiliensis,* and the cirolanid isopod, *Excirolana braziliensis,* in South America (Defeo and Cardoso 2002, 2004; Cardoso and Defeo 2003, 2004) revealed several changes from subtropical (ST) to warm temperate (WT) beaches (Figure 7.3): a decrease in abundance, growth, and mortality rates, increased predominance of females to males, a shift from continuous to seasonal reproduction, and increases in individual size of ovigerous females, fecundity, length at maturity, mass, and longevity. Some of these patterns were different than expected. For example, the greater total densities in subtropical reflective beaches contradict worldwide trends that show greater macrofaunal abundance occurring in dissipative beaches. Thus, such differences may be attributed to biological factors or to physical forces that remain unexamined. Departures from expected patterns can also be ascribed to the response to local variations in physical and biological (e.g. competition) characteristics rather than to biogeographical effects. Indeed, converse latitudinal trends suggest that sandy beach populations and communities are regulated by density dependent processes operating at the mesoscale in addition to acting with environmental factors.

Biological interactions have an important role in structuring sandy beach macrofauna. The tube-dwelling polychaete, *Diopatra cuprea*, is an ecosystem engineer as more than 30 taxa have been found associated exclusively with their tube structures (Santos and Aviz 2018) on a macrotidal beach of the Amazonia coast, suggesting intimate interrelationships between the polychaete and the other macrofauna. A significant increase in the abundance and richness of the macrofauna was observed in the rainy season when the density of worm tubes increased. The presence of

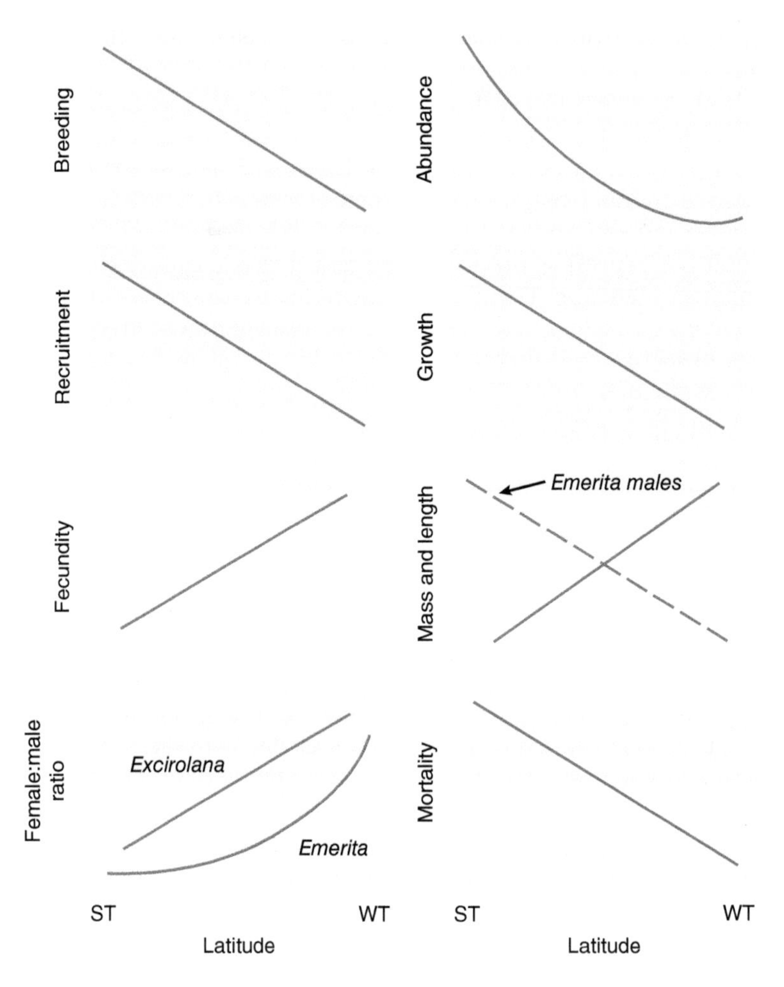

FIGURE 7.3 Latitudinal patterns in life history traits of the mole crabs, *Emerita brasiliensis* and *Emerita brasiliensis* on South American sandy beaches. ST = subtropical, WT = warm temperate. *Source:* Defeo and McLachlan (2005), figure 5, p.6. © Inter-Research Science Publisher.

even a single tube can have a significant influence on environmental conditions for other fauna.

Sandy beach populations can show a high degree of plasticity over latitudinal gradients. Such plasticity may lead to separate populations displaying differences in life history patterns and in behaviour. Phenotypic plasticity and genetic stasis have been observed in several sandy beach species, such as the whelk, *Bullia digitalis,* and the surf clam, *Donax serra* (Soares et al. 1999; Laudien et al. 2003).

A more sophisticated analysis of global diversity patterns in sandy beach macrofauna has been conducted in which oceanographic variables and historical processes were considered (Barboza and Defeo 2015). Temperature, salinity, and primary productivity were considered as possible predictors of community structure. Biogeographic units (realms, provinces, and ecoregions) were used to incorporate historical factors. The results indicate that ecoregions best represent trends in species richness globally, while temperature is the main predictor of species richness which increased from temperate to tropical beaches (Figure 7.4). Species richness increased with

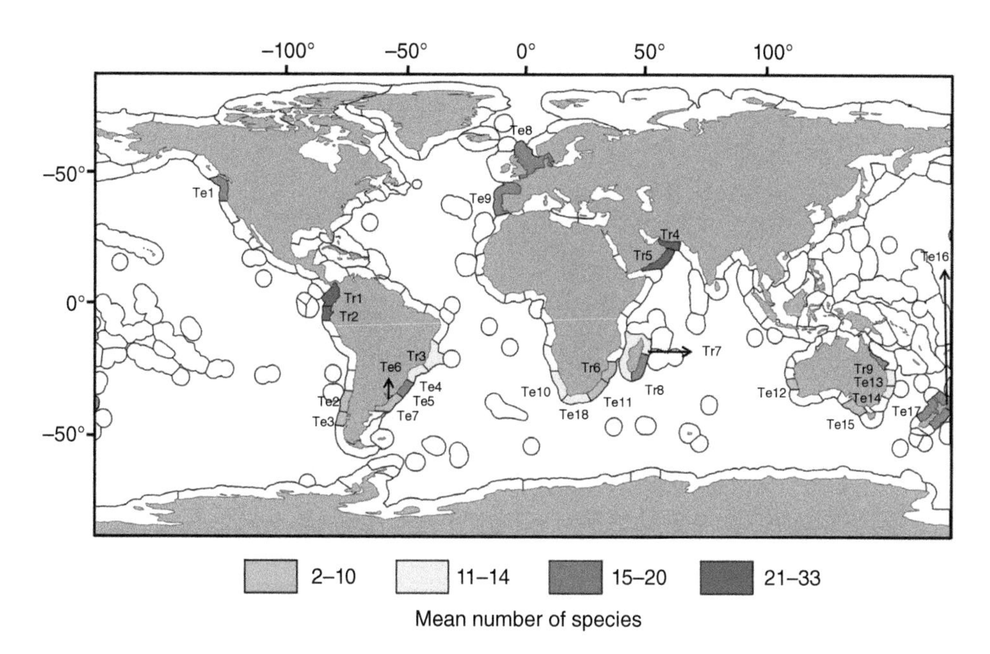

FIGURE 7.4 Species richness on tropical sandy beaches defined as ecoregions. Tr1 = Panama Bight, Tr2 = Guayaquil, Tr3 = Eastern Brazil, Tr4 = Gulf of Oman, Tr5 = Western Arabian Sea, Tr6 = Delagoa, Tr7 = Western and Northern Madagascar, Tr8 = Southeast Madagascar, Tr9 = Central and Southern Great Barrier Reef. *Source:* Barboza and Defeo (2015), figure 2, p. 4. Licensed under CC BY 4.0. © Springer Nature Switzerland AG.

tide range and towards wide beaches with gentle slopes and fine sand (Barboza and Defeo 2015). This latter result is consistent with the idea that habitat availability has an important role in structuring sandy beach macrofauna. The analysis of ecoregions for species richness (Figure 7.4) shows that NW South America (Panama Bight and Guayaquil) and Gulf of Oman and the western Arabian Sea have the most species, while the next highest ecoregions of richness are southeast Madagascar, the Rio Grande of Brazil, and the central and southern Great Barrier Reef. There was some temperate versus tropical overlap in ecoregions in terms of species richness as some temperate regions were species-rich (e.g. the North Sea), but clearly, the least species-rich regions were temperate and not tropical.

The community structure of tropical sandy beaches and tidal flats can be complex, with beaches and sand and mud flats geographically close to one another being significantly different (Dittmann 2002). Tropical beaches usually support richer faunas that temperate beaches. Macrofauna on tropical sandy beaches exhibit complex zonation patterns and a high species richness and relate to basic physical parameters, such as grain size, beach type, and slope. For instance, the ecological patterns of macrofauna in sandy beaches between the Caribbean and Pacific coasts of Costa Rica show similar vertical zonation with polychaetes occupying the low-intertidal, crustaceans and molluscs the mid-intertidal and isopods (Cirolanidae) dominating the upper littoral (Sibaja-Cordero et al. 2019). However, there were differences in faunal richness, abundance, and composition between both coasts that were explained by differences in tidal range, beach slope, sediment organic matter, and sediment composition. Describing both the macro- and meiofauna of sandy beaches, Filho et al. (2011) found that across shore, the meiofauna attained peak numbers in the mid-intertidal zone, whereas macrofauna abundance and richness increased from high- to low-intertidal zones. Richness and abundance of the benthic fauna were related to mean particle diameter and the percentage of fine sediments. Molluscs dominated an exposed, low-tide terrace-rip beach (i.e. an intermediate-type beach) in Ecuador (Aerts et al. 2004) with the fauna occupied by different suites of species depending on tidal zone. In the upper-intertidal zone (Figure 7.5a), the fauna was dominated by low densities of the isopod, *Excirolana braziliensis*. The middle beach assemblage (Figure 7.5b) was dominated by high densities of the gastropod, *Olivella semistriata,* and the amphipod, *Haustorius* sp. The lower beach fauna (Figure 7.5c) was composed mostly of *O. semistriata* and several other species and groups such as bivalve larvae.

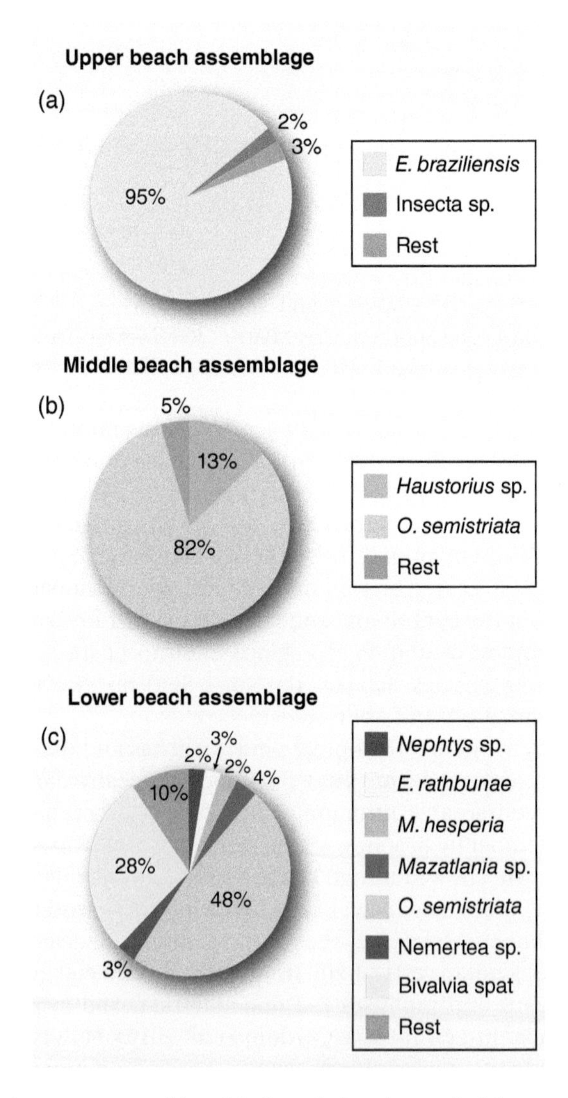

FIGURE 7.5 Relative percentage of benthic faunal abundances in (a) upper (b) middle and (c) lower beach zones of the Bay of Valdivia in Ecuador. *Source:* Aerts et al. (2004), figure 5, p. 22. Reproduced under the Creative Commons Attribution License (http://creativecommons.org/licenses/by/3.0/). © Belgian Journal of Zoology.

ENSO events may affect sandy beach fauna. Total macrobenthos densities on a sandy beach in Ecuador were 300% lower during the La Niña phase compared with months during the normal, non-ENSO period (Vanagt et al. 2006). Crustaceans and molluscs showed a marked increase in densities towards the normal period. However, polychaetes and echinoderms showed higher densities during La Niña. There are two possible explanations for these patterns: (i) low densities during La Niña could be the result of the effects of the strong El Niño prior to sampling suggesting that the fauna was still recovering, and (ii) La Niña is the actual cause of the low densities due to warm water (Vanagt et al. 2006).

Other exogenous factors can play a role in determining sandy beach zonation and abundance. For example, monsoonal rains may lead to a decline in faunal numbers, depending on the severity of the monsoon. This phenomenon has been documented in India where macrofauna and meiofauna have been found to be negatively affected by fluctuations in salinity caused by the onset of the monsoon (Sivadas et al. 2012). Similarly, river inputs may play an important role in structuring sandy beach fauna. For instance, macrofaunal densities and species richness at Higuerote, Venezuela, were found to be highest at sites furthest from actual river inputs (Herrera and Bone 2011). Near river mouths, the plume modified the functioning of the system by discharging large amounts of nutrients and sediments which appear to smother benthic fauna.

Like many stretches of temperate coastline, allochthonous macroalgae deposited on tropical shores greatly influence faunal distribution and structure. Such algal inputs can even affect subtidal fish communities off tropical sandy beaches. For example, fish biomass, density, and species richness increased from pre-drift to drift periods along the surf zones of two Brazilian beaches and species composition differed between periods (Andrades et al. 2014). Early fish stages were most obviously affected by the presence of algae, and the gut contents of the two most abundant species during the drift period, *T.* and *Trachinotus goodei,* demonstrated the importance of drift-related amphipods in their fish diets showing that this beach drift material is being used as feeding habitat. Drift algae may also provide both food and shelter for young fish, especially during autumn and winter.

Detailed studies on the ecology and behaviour of individual species on tropical sandy beaches and tidal flats remain scarce, making it difficult to generalise about their reproductive patterns and general behaviour especially in avoiding the harsh environment of a tropical beach or tidal flat. Early studies (references in Alongi 1990) conducted in India suggested continuous reproduction, but such studies were preliminary in that they often sampled only a few times per year or only at one site. In a detailed study, the shallow-water mysid shrimp, *Mesopodopsis orientalis,* living on a sand flat in Malaysia was found to exhibit continuous reproduction (Hanamura et al. 2009). Males accounted for 33–57% of the population with females predominating throughout the year. The number of eggs/embryos carried by females exhibited a positive correlation with female size. Reproduction of other sandy beach invertebrates often correlates with the onset of the monsoon, thus reproduction is often discontinuous in the tropics, especially with distance from the equator where reproductive conditions are optimal year-round. The population dynamics of the hermit crab, *Clibanarius symmetricus*, for instance, was linked to seasonal rainfall patterns on an exposed beach and sand flat of the Amazonia coast (Danin et al. 2020). Although salinity did not inhibit the simultaneous presence of age and sex categories, crab densities reached high values and increased with increasing salinity in the dry season. Reproduction is probably continuous given the constant presence of ovigerous females and juveniles and explains the high production: biomass ratio (P:B = 2.44).

Activity patterns, like reproductive patterns, vary greatly upon distance from the equator and whether the beach in question lies in the wet or dry tropics. In the wet tropics, most sandy beach communities suffer increased mortality or migrate during monsoons to escape sediment erosion and low salinities. In the dry tropics, population densities of most benthic organisms vary in response to seasonal changes in temperature, high salinity, and desiccation. Nearly all organisms in tropical intertidal environments will optimise their behaviour and activities to avoid the heat of the day, regardless of whether they are on a beach in the wet or dry tropics. The ghost crab, *Ocypode quadrata,* for instance, like most tropical intertidal organisms remains inactive from about 1100–1600 h, coinciding with peak daytime temperatures (Valero-Pacheco et al. 2007), identical to the pattern found for the Pacific mole crab, *Hippa pacifica* (Lastra et al. 2002). Most intertidal fauna are therefore active during the night or in the early daylight hours and burrow during the heat of the day.

Climate-induced range shifts in mud flat fauna from the tropics to higher latitude may be occurring. Caswell et al. (2020) found that up to eight species of nominally subtropical-tropical molluscs, polychaetes, crustaceans, and foraminifera from intertidal mud flats on the SE coast of Australia have moved poleward at speeds comparable with that of other Australian marine fauna. Movement was particularly fast for polychaetes and molluscs (about 70–300 km decade^{-1}), perhaps assisted by the southward flow of the East Australian Current. Modelling indicated that the replacement of temperate by tropical fauna may likely result in changes in functioning, especially in nutrient and carbon cycling.

7.4 Coastal Lagoons

Biological forces have a crucial role to play in the structure and function of coastal lagoon faunal communities. In the Açu Lagoon in Brazil, for instance, primary production and decomposition processes are coupled, as evidenced by the positive correlation between pH and dissolved oxygen and the negative correlation between pH and dissolved CO_2 (Chagas and Suzuki 2005). Coastal lagoons often have high densities of microbes, depending upon salinity and other factors, such as the degree of pollution. Bacterioplankton community composition in the Chilika Lagoon, India, was found to be related to seasonal and spatial variations in salinity, pH, dissolved O_2, and phosphate (Mohapatra et al. 2020). In four coastal lagoons in the state of Rio de Janeiro, Brazil, no correlations were found between bacterioplankton activity and DOC concentrations, as P appeared to be the main limiting nutrient (Farjalla et al. 2002). Bacteria consumed from 2 to 8% of the bulk DOC in the lagoons suggesting that a large proportion of the bulk DOC may accumulate, although there was an increase in DOC consumption by bacteria and in bacterial growth efficiency after addition of nutrients. It was concluded that low P concentrations and low quality of bulk DOC regulate bacterial growth in the lagoons.

Large seasonal variations in abundance, biomass, and composition of bacterioplankton in relation to phytoplankton and in the dynamics among microbial groups in the microbial hub occur in tropical lagoons. In the hypertrophic lagoon of Ciénaga Grande de Santa Marta, Colombia, bacterioplankton densities and biomass did not correlate with changes in salinity, phytoplankton or seston concentrations, but bacterial numbers and biomass were among the highest ever reported for natural coastal waters (Gocke et al. 2004). Bacterial and phytoplankton protein accounted for 24–57%

of the total seston protein, respectively. At low phytoplankton abundance, bacterial carbon was almost equal to phytoplankton biomass, while at high phytoplankton abundance bacterial biomass was low. The lack of a correlation between bacteria and phytoplankton in these shallow lagoon waters was probably due to resuspension of benthic bacteria and microalgae caused by frequent wind-induced mixing and to allochthonous organic material that served as food for the bacteria.

Phytoplankton communities can also vary greatly in lagoonal waters in relation to water depth, links to the sea, pollution, and salinity as well as the amount of watershed input. For instance, the Imboassica Lagoon, Brazil, ordinarily receives multiple marine inputs, resulting in stark variations in salinity, nutrients, and phytoplankton biomass throughout the year (Melo et al. 2007). Phytoplankton biomass varied spatially, being lowest nearest to the channel to the sea and highest near a bank of macrophytes. Diatoms and cryptomonads dominated some areas while euglenoids, cryptomonads, and dinoflagellates dominated others. High fluctuations in salinity and reduced nutrient availability were considered as the main cause of the temporal and spatial variability of the phytoplankton community.

Some lagoons are fringed with macroalgae, seagrasses, and mangroves, and these plants can have important consequences for phytoplankton and nutrient dynamics. In a Mexican coastal lagoon receiving considerable amounts of mangrove litter, concentrations of PO_4^{3-}, NH_4^+, and phenolic compounds were highest in the dry season when maximum mangrove litter fall was recorded (Aké-Castillo and Vázquez 2008). Variations in these nutrients depended on the internal biogeochemical processes of the lagoon, with blooms of diatoms and dinoflagellates occurring seasonally and in the different mangrove zones. High phytoplankton cell densities in mangroves within the lagoon suggest that plant organic matter derived from macrophytes contributed to phytoplankton dynamics.

Zooplankton show similar responses to environmental changes in tropical lagoons. In Imboassica Lagoon, Brazil, prior to the sandbar opening, the zooplankton community was dominated by the rotifers, *Brachionus calyciflorus* and *B. havanaensis,* these species being typical of freshwater to oligohaline conditions. After the opening of the sandbar, salinity increased and the zooplankton community structure shifted to the marine rotifer, *Brachionus plicatilis*, marine copepods, and meroplanktonic larvae. Two years after this opening, the community structure of the zooplankton remained the same indicating a low resilience of the original zooplankton community to this disturbance (Santangelo et al. 2007). However, the entry of marine water via the opening of the sandbar resulted in the establishment not only of a salinity gradient but a decrease in dilution of the nutrient load.

Zooplankton show high variability in response to environmental conditions in coastal lagoons compared with offshore waters. In the partially enclosed Nichupté Lagoon of the Yucatán Peninsula, zooplankton densities were higher than offshore, with the bulk of the catch composed of decapod larvae and copepods (Alvarez-Cadena and Segura-Puertas 1997). The copepod, *Acartia tonsa,* dominated the copepod assemblage within the lagoon. The highest number of copepod species occurred offshore where conditions were more stable than inside the lagoon.

In shallow water lagoons, wind may cause significant wave action and mixing of water to the bottom, resuspending sediments, and disturbing the benthic community. Changes in salinity appear to be the other factor structuring the shallow water benthos. During the season of strong northerly winds ('nortes') in the Celestun Lagoon on the Yucatán Peninsula, (Hernández-Guevara et al. 2008), molluscs dominated the fauna and polychaetes almost disappeared. During the dry season, the polychaetes recovered, becoming as common as crustaceans and molluscs. During the rainy season, polychaetes became the dominant group and molluscs declined. During the rainy season, high freshwater input by rainfall submerged the entire lagoon, inducing a marked salinity gradient to which the benthos responded by varying in relative dominance. Molluscs took advantage of these conditions and increased in the more oligohaline zone and polychaetes moved to the seaward zone (Figure 7.6, blue box). During the 'nortes' season (Figure 7.6, yellow box), wind intensity increased inducing more mixing in the water column. The resulting resuspension resulted in a decline in polychaetes but benefited filter-feeding molluscs. Migrant birds in search of winter shelter, feeding partially in polychaetes, contributed further to water column mixing and resuspension. During the dry season, evaporation increased salinity and induced stratification in the water column. During this time, the abundance of opportunistic species increased (Figure 7.6, orange box) and therefore the input of marine water to the lagoon allowed for the slow recovery of the polychaete populations which again attained high abundance (Hernández-Guevara et al. 2008). Thus, weather and climate are the ultimate factors driving the Celestun lagoon ecosystem, while physical habitat characteristics (resuspension, wind) are the proximal forces controlling benthic community structure.

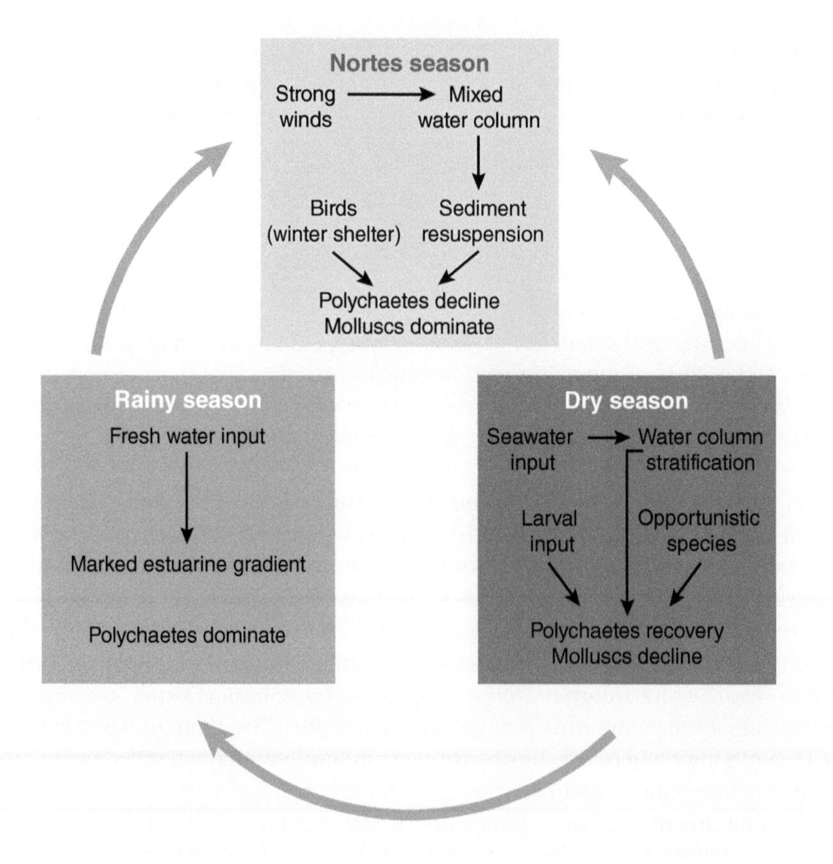

FIGURE 7.6 The benthic response to seasonal variations influencing the Celestun coastal lagoon (Gulf of Mexico) during the rainy, 'nortes' and dry seasons. *Source:* Hernández-Guevara et al. (2008), figure 8, p. 778. © CSIRO.

Seasonality in benthic community structure and distribution has been observed in African coastal lagoons. The macrobenthos in the large, hypersaline, but shallow Keta Lagoon, Ghana, was characterised by low densities and species diversity and numerically dominated by bivalves and capitellid polychaetes (Lamptey and Armah 2008). This benthic community was capable of withstanding physical disturbances such as when water levels become extremely low and when salinities become extremely high. Species richness and diversity were higher in the wet season than in the dry season with salinity, percent clay, pH, and turbidity being the major variables structuring the benthos. In the Ebrié Lagoon, Ivory Coast, the lagoon is flushed by seawater through tidal action and freshwater inputs via three rivers. The lagoon is very shallow and large and connected to the sea via a single canal. The lagoon is fringed with mangrove forest, and the climate is characterised by two rainy seasons and two dry seasons. The taxonomic composition of the benthic community was characterised by gastropods, bivalves, amphipods, isopods, and polychaetes and was typical of those in other African lagoons, except that the biota of this lagoon appeared to be richer in diversity and abundance (Kouadio et al. 2011). Spatially, the central part of the lagoon was characterised by more diverse assemblages due to the opening with the sea. The eastern and western parts of the lagoon had lower species richness and diversity, perhaps due to organic pollution. Temporally, the benthos increased in species richness in the long dry season when salinity favours the intrusion of marine species. Low values during the long rainy season were attributed to high concentrations of suspended sediment due to river runoff entering the lagoon. The major environmental variables that affected benthic community structure were dissolved oxygen concentrations and the percentage of mud and coarse sand.

Tropical lagoons are characterised by rich fisheries that are commercially exploited and, in most cases, overfished. For example, the density and spatial distribution of fish communities in the Cienaga Grande de Santa Marta Lagoon, Colombia, were found to be heavily affected by commercial fishing (Rueda 2001; Rueda and Defeo 2003). Three commercially important fish species showed different spatial correlations that matched different patch distributions among seasons. The strongest predictors of species occurrence were salinity, depth, and substrate type; temperature was not a significant factor. For all fish species, spatial distribution of fish richness differed between rainy and dry seasons. Associations were based on salinity regime: a marine, marine-estuarine, and a freshwater species group.

Fish assemblages in most tropical lagoons are also heavily impacted by pollution as well as natural variations in environmental conditions. The fish assemblages in Ebrié Lagoon, for instance, can clearly be distinguished between the eastern and western sections. To the east, the assemblage forms two different entities: one assemblage with clear freshwater affinities and the other with great seasonal variability under the influence of pollution and salinity changes in the dry and wet seasons. To the west, the fish community is characterised by strong spatial uniformity and low seasonal variability (Ecoutin et al. 2003). However, some sections are more heavily fished than others and this may affect fish community structure. Most heavily fished sections have an increased catch yield, a lowering of catch diversity, fish biomass, average catch length and trophic level of catches (Albaret and Laé 2003).

The opening of sand banks can have just as much an effect on fish communities as it can on planktonic and benthic communities. The change in salinity linked to the opening of the sandbank of Imboassica Lagoon was the main factor influencing the fish fauna (Sánchez-Botero et al. 2009). These changes were indicated by the recolonisation of primary freshwater species, oligohaline salinity, and small variations in the composition and abundance of the fish community. The resilience and resistance of the fish fauna were linked to the magnitude and frequency of the disturbance due to opening.

Salinity fluctuations play a central role in structuring fish communities in tropical coastal lagoons. For example, in coastal lagoons on the Yucatán Peninsula, Mexico, fish assemblages related best to differences in salinity (Arceo-Carranza and Vega-Cendejas 2009). Spatially, differences in species composition were observed between areas of different salinity. However, both salinity and temperature were the best predictors of the abundance of some pelagic species, especially in relation to the location of some freshwater seeps. In the Terminos Lagoon, fish species richness decreased with salinity, particularly for the river runoff zones with high variation in salinity throughout the year. This result may be due partly to the penetration of freshwater fish into estuarine areas following freshwater discharges and partially to the dominance of estuarine fauna more able to tolerate variable salinity levels (Sosa-López et al. 2007). The longer-term impacts of human influence have also played an important role in structuring the fish fauna of this lagoon. There were major shifts in the trophic structure of the system when sampled 18 years apart: biomass of the omnivorous, estuarine species in the middle of the food web were replaced by carnivorous and herbivorous-detritivorous species, with a long-term increase in salinity. A combination of interacting drivers may have played a role in the trophic shift: increased marine conditions as well as artificial reefs constructed in adjacent zones may have enhanced biomass of marine predators and detritivorous species; an attenuation of estuarine influences may have led to a decrease in estuarine generalists, and the establishment of a marine protected area may have led to an increase in predators and a subsequent decline in prey species (Sosa-López et al. 2005). Such changes are common in tropical coastal lagoons as nearly all have been severely affected by human encroachment.

7.5 Mangrove Forests

Mangroves are woody trees and scrubs that grow above mean sea level to form intertidal forests along subtropical and tropical coastlines. Mangroves occur in a variety of coastal settings from fully marine to estuarine and develop and persist over timescales in which morphological evolution of coastlines occurs (Fromard et al. 1998, 2004; Walcker et al. 2015). The standing crop of mangroves is, on average, greater than that of any other aquatic ecosystem, with a decline in biomass with increasing latitude. Equatorial mangrove forests can be immense, rivalling the biomass of tropical rainforests (Alongi 2009). They are pioneers that colonise newly formed mud flats, but they can shift their intertidal position in the face of environmental change. Forest establishment involves positive feedback in that saplings and adult trees trap silt and clay particles brought in by tides and rivers, helping to consolidate the deposits on which they grow. This feedback continues for the life of the forest until, eventually, the forest floor lies above the reach of tides. Terrestrial plants in theory eventually outcompete and replace the mangroves, but mangrove forests are rarely in geological and ecological equilibrium.

Long-term patterns indicate that mangrove forests are a mosaic of small to large patches of different successional stage. Mangroves are often subject to frequent disturbance so intermediate disturbances would be unlikely to culminate in a classic climax community (Lugo 1980). Alongi (2008) hypothesised that "for disturbed mangrove forests, stand composition and structure are the result of a complex interplay of physiological tolerances and competitive interactions leading to a mosaic of interrupted or arrested succession sequences in response to physical/chemical gradients and changes in geomorphology." Climatic disturbances such as tropical cyclones play an important role in structuring mangrove forests in many cyclone-prone regions (Krauss and Osland 2020).

Mangroves possess morphological, physiological, and ecological characteristics that make them structurally and functionally unique: aerial roots, viviparous embryos, tidal dispersal of propagules, frequent absence of an understorey, wood with narrow, densely distributed vessels, highly efficient nutrient retention mechanisms, and the ability to cope with salt and to maintain water and carbon balance. Mangroves exhibit different strategies to deal with salt, such as salt avoidance and regulation, tolerance, and resistance strategies, such as exclusion, extrusion, storage, succulence, compartmentalisation, and osmoregulation (Lovelock and Ball 2002; Saenger 2002). These various salt coping mechanisms vary greatly among species, and there are wide species-specific variations in growth responses. Water relations can be constrained by the presence of salt as a positive water balance and photosynthesis can only be maintained if the potentials in the plant are lower than in the soil. At high salinities, it becomes difficult for the plant to maintain a proper water balance due to the problem of trying to take up essential inorganic ions to maintain osmotic balance while trying to avoid the adverse impacts of high ionic levels in the cytoplasm (Lovelock and Ball 2002).

Mangroves display a range of features to minimise water loss due to the high metabolic cost of trying to maintain water balance, including low transpiration rates and thick-walled leaves usually with a multilayered epidermis covered by a thick, waxy cuticle to minimise evaporation. On the under surface of leaves are dense hairs or scales to reduce water loss from salt glands and stomata. Mangrove plants follow a very conservative water-use strategy; when salinity is high, transpiration rates are low, and when salinity is lower, transpiration rates can be high, especially during the wet season. Conserving water is a trade-off between the need for the stomata to open to maintain intercellular CO_2 concentrations and the simultaneous loss of water vapour. Carbon gain is balanced by some water loss. CO_2 uptake is restricted by the need to conserve water, resulting in low intercellular CO_2 concentrations, low assimilation rates, and high water-use efficiencies (the ratio of carbon assimilated to water used). High water-use efficiency is achieved by adaptations such as specialised leaf and stomatal anatomy, high levels of photo-oxidative protection, small vessels and dense wood, and greater investment in roots than in aboveground tree parts (Feller et al. 2010). Many species can adjust the angle of their leaves to avoid high temperatures and maximise heat loss to maximise water-use efficiency and to minimise physiological stress. However, a favourable leaf angle comes at the expense of light harvesting and assimilative capacity.

Morphological and physiological adaptations to maximise root aeration are features to deal with lack of oxygen and the presence of potentially toxic metabolites in waterlogged saline soils. These adaptations include relatively high root/shoot ratios, pneumatophores, stilt root, knee roots, buttress roots, and aerial roots that originate from the trunk or lower branches but usually do not reach the soil. Some species possess one or more of these root types, but some mangrove species have no such specialised roots.

Mangrove roots are composed mostly of aerenchymatous tissue, honeycombed with open gas spaces that run down the longitudinal axis. The more waterlogged the tree, the more gas spaces exist. Further evidence of root aeration is the presence of lenticels which permit gas transport in synchrony with the tide. During tidal inundation, oxygen concentrations decline inside the roots with a concomitant reduction in gas pressure (Saenger 2002). When exposed at low tide, the low gas pressure induces the flow of air back into the roots leading to a return of oxygen concentration. Soil anoxia leads to several other physiological changes, such as reducing water stress which may lead to reduced growth rates (due to the accumulation of ethylene or imbalance of gibberellin in the plant), depressed stomatal conductance, photosynthesis and oxygen transport via the roots, and increased leaf sodium levels.

Mangrove trees have architecture that allows for efficient light interception and stability for growth on soft, waterlogged soils (Saenger 2002). For *Rhizophora* forests there are unique patterns of tree expansion based on prop root development, drop roots from branches, and multiple stems under certain conditions, all of which result in a dense canopy with little or no understorey. Anatomical adaptations to life in waterlogged conditions are many and include complex leaf anatomy with conspicuous xerophytic and halophytic features, specialised epidermal cells with glandular hairs that function in salt secretion, the leaves of some species contain tannin-containing cells in the hypodermis, and calcium oxalate druses (Alongi 2009).

Most mangroves have flowers and are hermaphroditic with species-specific variations in flower and pollen structure (Saenger 2002). Canopy dwellers (birds, bats, bees, other insects) are responsible of most of the pollination. As with terrestrial plants, there are various types of germination, including a unique type common to mangroves in which the cotyledons remain in the fruit and the root tip and hypocotyl are exposed. True vivipary to crypto-vivipary occurs in nearly one-third of species. Vivipary is advantageous as it allows for rapid establishment of seedlings (Krauss et al. 2008). Most mangroves appear to be self-fertilising for this reason. Buoyant propagules are another characteristic of mangroves, dispersing usually at the end of the wet season. Passive dispersal of buoyant propagules (fruits and seeds) of mangroves results in large dispersal potential, as wind and currents transport propagules.

Like other forests, undisturbed mangroves follow a natural series of phases over time, from an initial pioneering stage through to rapid early growth and development, to later maturity, senescence, and death. This natural progression is supported by long-term data from French Guiana where Fromard et al. (1998) measured the changes in structure, biomass, and stand dynamics of several mangrove species. These data show a natural development of mangroves with a negative correlation between estimated forest age and stem density. Like most other forests, tree densities declined with forest age and stands became less dense due to self-thinning, as evidenced by an increase in tree girth. Less dense but larger mature trees led to an increase in total above-ground biomass with increasing age. Changes in species composition occurred, especially in light gaps, and with increasing distance upland.

Factors influencing the structure of mangrove forests vary in relation to global, regional, and local scales over different time scales (Smith 1992; Friess 2017). At the global scale, mangroves are ultimately limited by temperature, but at the regional scale, the area and biomass of mangrove forests vary in relation to rainfall, tides, waves, and rivers. At the local scale, various schemes have been developed to classify mangroves (Smith 1992; Friess 2017). Most forests represent a continuum of geomorphological types based on their location within broader settings classified as river-dominated, tide-dominated, wave-dominated, composite wave- and river-dominated, drowned bedrock, and carbonate (Woodroffe et al. 2016).

Mangroves, typically distributed from mean sea level to highest spring tide, exhibit zonation as a conspicuous feature in which there is a sequential change of tree species parallel to shore (Friess 2017). Many factors have been suggested to account for the apparent zonation of tree species across the intertidal seascape, including gradients in salinity, soil type and chemistry, nutrient content, physiological tolerances, seed predation, and competition. The evidence to date indicates that there is no one driver regulating species zonation, prohibiting generalisations about the mechanisms governing zonation. It is more likely that a few of these factors in combination come into play over different spatial and temporal scales to control mangrove distribution.

Several factors operate in tandem to regulate plant growth, including temperature, salinity, nutrients, solar radiation, O_2, and water. Natural changes in mangrove forests occur on the scale of minutes to hours for microbial and physiological processes, months to years for tree growth and replacement, and decades to centuries for regional forest changes (Friess et al. 2019). Mangrove forests share many of the same basic physical and ecological attributes with terrestrial forests, but other mangrove attributes appear to be unique, challenging concepts such as the old-growth or late successional forest (Lugo 1997). The apparent paradox that mangroves appear to be in steady state despite exhibiting characteristics of establishment, thinning, and transitional stage forests can be explained by the periodic nature of disturbances (Lugo 1997). A variety of ecosystem states can develop due to mangrove growth and development being altered by changes in sea level, lightning strikes, cyclones, and other disturbances, resulting in a forest exhibiting a mosaic of successional characteristics (Krauss and Osland 2020).

Historical sediment records show that mangrove forests change greatly over geological time at a given location in relation to changes in regional climate, local geomorphology, and sea level. On the western coast of the Gulf of Mexico, for instance, the establishment of dense mangrove stands took place around 3700 cal BP when regional sea levels stabilised resulting in an increase in organic matter and carbon stored in the sediments (Cordero-Oviedo et al. 2019). However, the mangrove ecological succession that started at about 6000 cal BP was interrupted by a regional drought that extended from about 5400 to 3700 cal BP; from 3700 cal BP to the present, the gulf coastal zone has been characterised by relatively stable substratum and sea level that together have facilitated the establishment of mangrove forests.

Mangroves do not always establish in isolation as they have often evolved and developed with other ecosystems, such as seagrasses and coral reefs. At Iriomote Island, SW Japan, exposed fossil microatolls and core samples from a mangrove forest and a coral reef at the Yutsun River mouth have shown that sea level reached its present position before 5100 cal BP and a relative sea-level high stand of 1.1–1.2 m above present sea level occurred from 5100 to 3600 cal BP, followed by a gradual fall in relative sea level (Yamano et al. 2019). At 6500–3900 cal BP, a nearshore reef dominated by massive *Porites* and arborescent *Acropora* initially developed, but development was terminated by relative fall in sea level and discharge from the Yutsun River. An offshore coral reef reached present-day sea level after 1000 cal BP, forming a wave break that enabled the development of mangrove forest in the calm backreef environment. These geological records illustrate the chronology and causal relationship between coral reef and mangrove development that often occurred due to changes in Holocene sea level and river discharge.

Mangroves have long been classified as 'land-builders,' and while this is an over-exaggeration, mangroves do colonise and accrete soils where quiescent conditions permit the entrapment of fine particles (Section 3.4). Mangroves can efficiently trap silt and clay particles, but other factors play a role in slowing tidal water movements to facilitate sedimentation. The area of forest is ordinarily large in relation to the area of waterway. This phenomenon serves to slow water movement as it spreads out over a large area and water volume. Tree trunks, roots, pneumatophores, and even animals colonising aboveground tree parts add to the friction created by the soil surface, forest floor slope, as well as the numerous burrows, cracks, tubes, and fissures that pockmark the forest floor (Section 3.4). Microbes living on surfaces also help to bind sediment with the mucus they produce.

Gap dynamics in mangrove forests is poorly understood but represents a cycle of natural mortality and replacement that must be considered in long-term changes of mangrove forest structure (Krauss et al. 2008). Traditional measurement of tree species abundance, size, and structure over time has been one approach in determining the natural changes in forest structure over time. Modelling methods have been used to delimit changes in forest structure, and these include methods of simulating competition, spacing, and ageing of trees (Berger and Hildenbrandt 2000, 2003; Crase et al. 2015). Mangrove forests are commonly regulated by intra- and inter-specific competition for light, space, and soil nutrients, factors that are also patchy in space and time (Lugo 1997).

The intertidal mangrove zone is highly dynamic and ever-changing, disturbed often enough by weather events such as storms and cyclones, disease, pests, and anthropogenic intrusions that the progression to terrestrial forest occurs infrequently along most coastlines (Bomer et al. 2020). Mangroves occupy a harsh environment, subjected daily to tidal and seasonal variations in temperature, salinity, and anoxic soils, but are robust and highly adaptable or tolerant to such changes. Mangrove forests can adapt to surface elevation changes and sedimentation dynamics, as exemplified in the Ganges–Brahmaputra delta (Bomer et al. 2020).

Mangrove ecosystem characteristics include comparatively simple food webs containing a mixture of marine, estuarine, and terrestrial organisms; acting as nursery grounds and breeding sites for birds, reptiles, and mammals; and acting as accumulation sites for sediment, some contaminants, carbon, and nutrients. Trees and bacteria constitute the bulk of forest biomass, but many other organisms originating from adjacent terrestrial and marine environments are found in mangroves. Birds, bats, monkeys, tigers, insects, fish amphibians, reptiles, and a rich fauna of estuarine and marine plankton and benthic invertebrates spend all or part of their life cycle in the forest and tidal waterways (Kathiresan and Bingham 2001). Populations and communities overlap as mangroves are ecotones having a high level of connectivity with adjacent environments. As noted earlier, the most conspicuous organisms in mangroves are sesarmid and grapsid crabs, being ecosystem engineers in many forests (Aschenbroich et al. 2016, 2017). Crabs, like nearly all other dominant fauna in mangroves, exhibit strong patterns of intertidal zonation in genera across the globe.

Mangrove structure and their food webs are shaped by a variety of organisms. For example, gastropods do so by consuming large volumes of mangrove litter, algae, and wood; wood-boring isopods help to facilitate fungal decomposition of wood. Above the substratum, mangrove roots are often overgrown by epibionts such as tunicates, sponges, algae, and bivalves while the forest canopy is rich in fauna. Both the canopy and epibiotic communities are diverse with close associations between tree and animals. The functional significance of some of these organisms, especially the vertebrates, is largely unknown.

Penaeid shrimps and fish have received the most attention from researchers due to their role as food items. Strong correlations exist between the abundance and biomass of prawns and the extent of the surrounding mangrove area. The primary prawn targets are species of the genera *Penaeus, Metapenaeus, Parapeneopsis*, and *Macrobrachium;* a large amount of research effort has focused on their life history strategies, distribution, abundance, degree of habitat dependence, and catch per unit effort (Manson et al. 2005). Fish life histories are similarly well known, with species richness of temporary and permanent residents being a function of salinity, tides, water depth and clarity, water currents, microhabitat diversity, and proximity to seagrass meadows and coral reefs (Faunce and Serafy 2006). The number of fish species in any given mangrove estuary can vary from as little as 10 to nearly 200 with a tendency for more species in larger estuaries. Density and biomass estimates are, of course, variable, ranging from 1 to 160 fish m^{-2} and 0.4 to 29 g m^{-2} and generally greater than in temperate estuaries (Blaber 2002). Five feeding guilds constitute the bulk of fish trophic groups: herbivorous, iliophagous, zooplanktivorous, piscivorous, and benthic invertebrate feeders. Many fish species shift their trophic preferences as they age.

Mangroves have often been touted as major nursery grounds for fish and prawns since the late 1960s; the links between mangrove and fish and prawns have been known by indigenous peoples much farther back in time. The connection between mangroves and coastal fisheries has received much attention, and three hypotheses have been offered to explain the link. First, the food hypothesis suggests that mangroves offer an abundant variety of food items. The second hypothesis is the refuge idea which suggests that mangroves offer refuge from predation. The third hypothesis is the shelter idea which suggests that mangroves provide shelter from a variety of disturbances (Manson et al. 2005). The true situation is likely that all three factors may be operating simultaneously as none of these hypotheses are mutually exclusive. Evidence supports the idea that coral reef fish use mangroves and seagrasses as juvenile habitat (Kimirei et al. 2013). There is little evidence to explain the relationship but the stable isotope study of Le et al. (2020) indicates that juvenile reef fish exploit mangroves and seagrasses for food. In a shallow-water Malaysian lagoon, reef fish were highly associated with seagrass beds as feeding areas and preyed in both mangrove and seagrass habitats in the central lagoon, whereas all juveniles showed preferential foraging within seagrasses near the lagoon mouth (Le et al. 2020).

Mangroves are not always prime fish habitats. A recent study comparing fish inhabiting mangrove habitats and adjacent mud flats in the Gulf of Paria, Trinidad (Marley et al. 2020), discovered that the composition of fish communities in mangrove waters and on the mud flat at high tide was distinct; mean species richness, total species richness, and juvenile species richness were significantly greater on the mud flat, indicating that mud flats may be as important as repositories of biodiversity and nursery function as adjacent mangroves.

Mangrove forests are particularly important habitats for land birds, shore birds, and waterfowl. Birds inhabiting mangroves may be permanent residents that forage and nest in the trees or they can be temporary visitors. Migratory birds may fly long distances to find food and nesting places in a mangrove canopy. In the Pacific mangroves of Colombia, 77 species were counted of which 43% were permanent residents, 22% were regular visitors, and 18% were temporary migrants (Naranjo 1997). Birds are highly dependent on the survival of mangrove habitat. Forest destruction and fragmentation can result in a drastic decline in bird numbers. In fragmented mangroves of Florida, Bancroft et al. (1995) found reduced populations of mangrove cuckoos. Similar findings have been observed in other reduced mangrove habitats (Kathiresan and Bingham 2001).

Mammals also make their homes in mangroves, and this includes dolphins, monkeys, otters, and bats. Like birds, mammal numbers have been reduced by habitat destruction and fragmentation. For instance, 32 mammal species once lived in the Sundarbans mangroves of Bangladesh. Of these, four species have gone extinct: the Javan rhinoceros, wild buffalo, swamp deer, and hog deer (Kathiresan and Bingham 2001). Mammals in mangroves are transients, but this does not make them any less important to the structure and function of mangrove food webs.

Humans are of course also inhabitants of mangroves, and it is safe to say that they are the most ubiquitous and the most destructive. Deforestation and land use change (Section 14.4) have a severe effect on mangrove food webs. For example, fish densities, species numbers, diversity, and numbers of trophic groups were significantly less in a highly disturbed Tanzanian mangrove creek than in an undisturbed creek (Mwandya 2019). Omnivorous fish were the most common feeding guild in the disturbed creek, whereas zoobenthivores/piscivores were the most diverse guild in the undisturbed creek. Mangrove deforestation combined with land-use change had a greater impact on the trophic structure of fish in mangrove creeks than deforestation only. Extreme weather events caused by human-induced climate change can restructure mangrove ecosystems (Harada et al. 2020). Mangrove dieback induced by extreme weather in the Gulf of Carpentaria, Australia, resulted in significantly fewer crabs that rely on mangrove litter for food in forests suffering dieback, but more crabs that rely on microphytobenthos for food. This was likely to be the result of greater microalgal growth and biomass on the forest floor due to more light availability within a canopy.

7.6 Seagrass Meadows

Seagrasses are not true grasses but are monocotyledonous plants with typical grass-like morphological traits. They grow from the intertidal zone down to depths where enough light penetrates for them to grow (≈50 m depth). Intertidal seagrasses can be short unlike subtidal seagrasses which can be composed of quite large plants physically supported by the water. A seagrass meadow, like a mangrove forest, physically and biologically dominates its habitat space and can actively facilitate accumulation of fine silt and clay particles. Seagrasses support a rich and diverse fauna and flora, and these organisms often show considerable connectivity with adjacent mangroves and coral reefs (Hogarth 2015).

Except for members of the genus *Phyllospadix* which grow on rocky shores, seagrasses grow in mud and sand and various mixtures of both. Like terrestrial monocotyledonous plants, seagrasses have rhizomes which ensure adequate anchorage in soft sediments and assist in the transport of nutrients and gases to the leaves. The rhizomes are horizontal, close to the surface, where at intervals they form a vertical rhizome which develops leaves from a basal meristem and adventitious roots and rootlets. This vertical unit is called a ramet, which is the basic unit of seagrass growth. Nutrients and growth hormones travel along the rhizomes and ramets to facilitate growth. Seagrasses can vary in size from small intertidal forms, such as

Zostera asiatica, to intermediate size, such as *Enhalus acoroides* which can produce leaves of up to 1 m to gigantic species, such as *Posidonia*, whose roots may exceed 5 m in length. The upper limit of distribution is set by the ability to tolerate wave action.

Seagrass meadows quite often comprise large areas of a single species because extension and branching of the rhizome act like a 'clone' to occupy space. While the rhizomes can be used in this way, about 50–60% of seagrass biomass is underground; this percentage may be higher in nutrient-poor sediments. Seagrasses may also increase the proportion of below-ground biomass for anchorage in wave-dominated habitats. The allocation of biomass reflects the availability of resources, such as light, nutrients, and other features of the environment as indicated by the response to wave action. Leaf size does tend to increase with diminishing light with increasing water depth (Hogarth 2015).

Seagrasses trap and consolidate sediment so they have a major effect on the substrate. Seagrasses are vulnerable to sediment movement, but in quiescent environments, seagrasses may become carbon- or nutrient-limited due to a diffusive boundary layer around the leaf. This effect is species-specific depending mostly on leaf structure (de Boer 2007). Increasing current speeds can improve the transfer of CO_2 and nutrients to the leaf, breaking down the diffusive boundary.

For optimal growth, seagrasses require about 11% of available light at the water–air boundary (Hogarth 2015). Thus, most seagrasses are found to depths < 50 m although they have been found photosynthesising to depths of about 90 m. Seagrass leaves lack stomata and are surrounded by a thin and porous cuticle through which gas exchange and nutrient uptake take place. Unlike terrestrial plants which utilise CO_2, seagrasses utilise HCO_3^- (bicarbonate) for their carbon requirements, and this ion can become limiting under quiescent and light-saturated conditions. A network of gas-filled lacunae like mangrove aerenchyma tissue penetrates seagrass organs, facilitating exchange of gases with the surrounding seawater. These lacunae are used to transport CO_2 produced by respiration to the leaves for photosynthesis and conduct oxygen to the roots for respiration. O_2 concentrations in the seagrass fronds can exceed that of atmospheric concentrations due to photosynthesis. Like mangroves, seagrasses translocate O_2 to their roots and rhizomes, and this O_2 can be used to oxidise toxic anoxic substances and to bathe the below-ground parts in oxygen to avoid anoxia. Gases in the leaves also make the leaves buoyant, helping them to be raised up to facilitate light interception.

Seagrasses prefer full seawater, although estuarine seagrasses survive in brackish water and lagoonal seagrasses tolerate hypersaline conditions. Seagrasses are therefore mostly euryhaline, dying in salinities < 5 and > 45 (Hemminga and Duarte 2000). Seagrasses have two mechanisms to cope with salt: exclusion and tissue tolerance. Seagrasses can actively exclude sodium ions from the leaf and import potassium ions. How the plants exclude salt is not well understood but may involve an ATPase enzyme in the plasma membrane of epidermal cells. Most intracellular enzymes of seagrasses are not salt-tolerant and are significantly inhibited by rather low salt concentrations. As in mangroves and other halophytes, sodium ions are stored within the cell vacuole and excluded from the cytoplasm. Organic solutes probably maintain osmotic balance between vacuole and cytoplasm (Touchette 2007).

Like mangroves, seagrasses differ in their tolerances to salt though to a lesser extent. Some species such as *Thalassia testudinum* can tolerate a broad range of salinities, whereas othcrs such as *Syringodium filiforme* have a narrow tolerance (Lirman and Cropper 2003). These species-specific differences may go a long way in explaining species distribution patterns as species with a broad salinity tolerance may have a competitive advantage over a species with narrow tolerances in habitats where salinities are ever-changing, whereas the opposite is true in environments where salinity and other environmental cues are constant or nearly so.

Salinity is only one factor that can account for the distribution of species. Nutrients are often a limiting factor for seagrass growth, with some evidence that P is limiting in carbonate-rich sediments and N is limiting in terrigenous deposits or by both elements in siliceous deposits. Nutrient availability to seagrass depends on several factors, such as the degree of anoxia and the state of microbial processes, such as N_2-fixation and denitrification. P availability also depends on the ability of seagrasses to exude organic acids making the element available for uptake. Seagrasses can take up nutrients via their leaves. This is especially important in the tropics where nutrient levels are ordinarily low; the majority of nutrient uptake in tropical seagrasses can occur via this pathway. Like mangroves, seagrasses have highly efficient nutrient uptake and retention mechanisms.

Resource limitation plays a key role in competition among species, whether for light, inorganic carbon, space, or nutrients. Indeed, the abundance of a limiting element, such as P, can alter the species composition of a given area. When nutrient supply was artificially manipulated in a Floridan meadow, a seagrass bed dominated by *T. testudinum,* was progressively replaced by *Halodule wrightii* after considerable enrichment by bird droppings from the placement of artificial roosts for seabirds (Taplin et al. 2005). This effect lasted for up to eight years after the end of nutrient supplementation. Under normal nutrient-limiting conditions, the superior competitor *T. testudinum* dominates to the exclusion of other species (Taplin et al. 2005). Seagrasses can also compete with algae.

An experimental comparison of the relative availability of N and P in terrigenous and carbonate sediments in Indonesia revealed that neither was limiting to seagrasses (Erftemeijer et al. 1994). Dissolved phosphate concentrations were high which was attributed to the limited adsorption capacity of the coarse-grained sediments that these seagrasses inhabited, suggesting that P limitation is a function of sediment grain size with P becoming limiting with decreasing sediment particle size and increased adsorption. Conversely, P adsorption to iron oxyhydroxides in carbonate sediments may result in iron (Fe) limitation. Fertilisation of a mixed Caribbean seagrass bed with soluble Fe stimulated growth of *Thalassia testudinum* and increased the chlorophyll a concentration in *T. testudinum* and *S. filiforme* tissues (Duarte et al. 1995). This finding was attributed to the low natural levels of Fe in carbonate sediments and in overlying waters coupled with the plant's requirement for Fe for tissue production. Fe, N, and P may thus act in synchrony to limit tropical seagrass growth.

Seagrasses vary in the portioning of their biomass. They can store a considerable fraction of their biomass as rhizomes, but they can also store (depending on species) much of their biomass aboveground. Total standing crops vary greatly among seagrasses and with age of meadows from 50 to 850 g DW m^{-2} with a grand mean of 315 g DW m^{-2}; biomass varies seasonally, and there are no clear patterns of biomass with latitude (Duarte 1989).

Growth via rhizomes enables seagrasses to expand their space in the seabed. The rate of expansion varies greatly with smaller species, such as *Halophila,* capable of growing up to rates of 5 m a^{-1}. Seagrasses also propagate vegetatively by dispersal of fragments of rhizome. Over longer distances, dispersal of viable fragments of leaf or rhizome can occur. For example, fragments of *H. wrightii* remain viable for up to a month; other species however survive only a few days (Hall et al. 2006).

Sexual reproduction occurs only intermittently as most species flower only rarely with only a small fraction of members of a meadow flowering at any one time. There is also great variation between years and within a species. For example, subtidal *Zostera* reproduces asexually leading to the formation of large clones which can be over 1000 years old, while in the intertidal zone, *Zostera* may flower relatively frequently in response to widely fluctuating temperature and salinity. Seagrass flowers are inconspicuous and in 9 of 13 seagrass genera are dioecious, with separate male and female plants. Without animal vectors, pollination is by current transport with pollen grains shed by the male anthers into the water and in some cases, the anther may be shed to enable pollen to rain down on female flowers. Pollen grains vary greatly in size among genera, but there is little information on how far the pollen is transported by the parent plant, although the distance is thought to be small. While generally there are no animal vectors, some small crustaceans feed on seagrass pollen resulting in some pollen sticking to their legs and being carried onwards. A major factor in pollination success is probably the hydrodynamic effect of a continuous seagrass meadow in reducing current speeds and water movement, thus increasing the efficiency of pollen trapping (Vermaat et al. 2004).

Seed dispersal varies greatly among seagrass species. Some seeds are buoyant, such as for *Enhalus,* while some are small and disperse only a short distance from the parent, such as *Halodule* and *Cymodocea*. Although dispersal of up to 100 km has been recorded, most dispersal occurs at considerably shorter distances. The buoyant seeds of *Enhalus acoroides* have been recorded to disperse over distances of more than 300 km. Vivipary is common in some seagrass species, such as *Thalassia* whose non-viviparous seedlings may travel much greater distances after germination than for the negatively buoyant seed before germination. Settling may take a variety of mechanisms. For example, the bell-shaped seeds of *Thalassia hemprichii* are cued to land upwards only on coral grains, enabling differential survival depending on environment. After settling, seeds may be buried due to bioturbation, and in fact some seeds rely on shallow burial for germinating. Germination quickly follows dispersal without an extended dormant phase. Some seeds persist for only a few days, but those of other species may survive for months. Overall, for seagrasses, sexual reproduction is a low investment.

Once started as either a propagule or as a viable fragment from existing tissue, seagrasses colonise a given area quickly as the propagule grows vegetatively and expands in all directions to spread out over the seabed. The extent of expansion depends on environmental conditions, and its composition will largely be determined by interspecific competition. The presence of intertidal species is probably determined in part by competition with algae. Exposure to ultraviolet light is another factor limiting distribution as UV light depresses photosynthesis and damages sensitive cellular organelles. Seagrasses compensate by producing flavonoids which are pigments that absorb UV light and by producing thickened leaves with a pigment-containing epidermis which may absorb up to 99% of the UV radiation (Hogarth 2015).

While the upper limit of intertidal species is unlikely to depend on the level of UV protection (desiccation can be the major problem), for subtidal species the upper limit can be set by tidal height; wave action and turbulence can uproot seagrasses. The lower limit is set by light penetration, but water turbidity is much more limiting than competition for light among neighbours. Soil oxygenation also affects species composition and the competitive interactions between species; this is especially true for tropical seagrasses.

Seagrass meadows generally comprise only a few species. Up to seven species may be found in a specific meadow in the Indo-Pacific, with *T. hemprichii* often the most abundant (Hogarth 2015). Competition amongst species probably limits species composition whereby weaker competitors are eliminated "culminating in domination by a single climax species. Homogeneity is restored" (Hogarth 2015).

Gaps in a seagrass meadow are colonised by horizontal extension of the rhizome, as normally happens when a species colonises a given area. A gap may be colonised by a different species, either by the germination of buried seeds or by the random settling of viable propagules. The gap is however more likely to be recolonised by regeneration of buried rhizomes. The composition of the gap will ultimately depend on the availability of species present and the dispersal ability of the species in question. Some species are better dispersers than others, while others are more effective competitors for resources. Like mangrove gaps, several species may initially appear until the effects of competition for space, light, and nutrients reduce the number of viable survivors.

Seagrass meadows experience a variety of small and large disturbances, and how they respond depends on the intensity, frequency, and duration of the disturbance. Storms and heavy grazing constitute the most common forms of natural disturbance, but there are also human-induced disturbances. Some meadows are rarely interrupted by disturbances and at the other extreme there are meadows that are continuously disrupted. Under these conditions, the seagrass community will consist largely of frequently disturbed patches that rapidly redevelop post-disturbance. Most seagrass meadows experience intermediate levels of disturbance that permit them to exist as patches of mosaics of competitively superior species.

Seagrasses can dampen wave and tidal action and can suppress water circulation by up to 90% (Short et al. 2018). This slowing of water flow facilitates the accumulation of sediment and acts as a positive feedback mechanism to help maintain and establish new vegetative growth. This mechanism also helps to decrease turbidity and increase light penetration, benefiting seagrasses.

Currents enhance rates of seagrass primary productivity by mixing and distributing nutrients and gases and removing wastes; movement of water is necessary to replenish nutrient and gas pools and to remove metabolites to sustain growth. Seagrasses respond phenotypically to long-term changes in water movement, while affecting hydrodynamics in several ways, but only rarely has this phenomenon been examined and considered as a key factor in the growth and sustainability of seagrasses. Enhancement effects of water flow are attained at a narrow window of current speed; increased turbulence above a given threshold results in suspended sediments reducing light availability and lowering of production and growth. The thickness of the boundary layer surrounding leaf blades varies over space and time. Seagrass beds can modify ambient currents and waves leading to important ecological effects on the distribution of organisms and their food supply within the understorey, as well as fluxes of nutrients and dispersal of gametes, spores, and larvae.

Seagrass meadows introduce habitat complexity to what would otherwise be barren seabed. Seagrass blades are home to a rich assortment of epiphytes and the seabed is inhabited by a rich epifauna and infauna, not to mention a rich variety of visitors and permanent residents, such as fish, crustaceans, dugongs, turtles, and manatees (Short et al. 2018). Subsurface roots and rhizomes oxygenate the soil, resulting in soil favourable for the continued development of clonal growth of seagrasses. The feedbacks of the seagrass community can be both positive and negative, as some species such as *Zostera* can decrease or increase the settlement of silt and clay particles leading to major changes to soil texture. Hogarth (2015) referred to these decreases and increases as "sandification" and "muddification," respectively. The ability of seagrasses to promote sedimentation and the interplay of factors such as wave action and disturbance can result in the formation of seagrass "bands" instead of a continuous seagrass community.

7.7 Coral Reefs

Coral reefs are masterpieces of nature, often described as oases in an oceanic desert. However, the water bathing coral reefs are not deserts, but in fact teem with microbial life. Concentrations of nutrients are low, but turnover is high as nutrients are processed and recycled rapidly through microbial networks which are swept into the framework and food webs of the coral reef (Gattuso et al. 1996). The supply of nutrients supporting rich coral life comes from a variety of sources: advection, upwelling, groundwater, rain, terrestrial runoff, guano, and immigrant organisms (Rougerie and Wauthy 1988). The fact that these sources are external underscores to the fact that coral reefs are dependent on adjacent waters for their existence. Coral reef ecosystems are highly efficient at recycling nutrients (O'Neil and Capone 2008) so that their primary productivity is at least 30–250 × greater than in the adjacent open ocean (Kinsey 1985; Hatcher 1988, 1990).

Vital for a coral reef are the connections between various sections of a reef. Seawater and associated nutrients impinging on the front of a coral reef move across the reef to the back lagoon, thus forming a close connection between various reef parts (Monismith 2007). This interlinkage helps to maintain a balance between production and consumption of energy. Reef geomorphology plays an important role in the transfer of waters and materials throughout a coral reef, determining the residence time of water and organisms and thus the flux of energy.

Both the hermatypic (hard) corals and the octocorals (soft corals) have the basic body form of all anthozoans, a polyp with a gastrovascular cavity and tentacles (Sheppard et al. 2018). Hermatypic corals have an exoskeleton of $CaCO_3$ that builds the spectacular reefs of the tropics. Their polyps are smaller than most anthozoans, and they are mostly found in colonies. They secrete a $CaCO_3$ skeleton of about 1–3 mm in diameter from the base of the polyp called the basal plate through which corals are connected. Coral polyps, like anemones, have stinging nematocysts. Some hermatypic corals are low in profile and encrust over the substrate and grow slowly at about 0.5–2 cm a^{-1} while others, such as the staghorn coral, are higher in profile, branching and growing more quickly at about 10 cm a^{-1} (Sheppard et al. 2018). Such corals can reproduce asexually via budding of new individuals, but also reproduce sexually by producing planktonic larvae that disperse and form new colonies.

Hundreds of coral species reproduce in mass spawning events that are synchronised to the lunar cycle. Hybridisation among mass spawning species in the Caribbean is common, whereas some rare Indo-Pacific *Acropora* are probably hybrids. During mass spawning, corals are usually hermaphrodites, spawning together for 5–8 nights after a full moon (Kaniewska et al. 2015). As spawning is synchronised, there is no relationship between the release of new coral recruits and the abundances of adults. Mass spawning also results in the release of large amounts of C and N derived from the spawn which can result in intense blooms of dinoflagellates (Glud et al. 2008).

The octocorals include the gorgonian and soft corals, and it is the gorgonian corals that are commonly found in the Indo-West Pacific. These corals bend with the currents and their branching rods collect plankton for food; some grow to over 2 m in length.

Corals have a symbiotic relationship with zooxanthellae and other microeukaryotes which are central to the growth and production of coral reefs (Section 6.6). Coral reefs can sustain high rates of gross primary productivity because of this relationship, but the contribution of various sources, including net inputs of allochthonous carbon, and variations in productivity among reef zones, is still uncertain. It is difficult to separate out the various contributions from such a high degree of spatial and organismal diversity as a rich variety of symbiotic zooxanthellae, phytoplankton, seagrasses, and micro- and macroalgae living in and on the hard substrata and in unconsolidated sediments contribute to production of fixed carbon on a coral reef (Section 8.5). Other organisms also produce inorganic carbon, including foraminifera, sponges, bryozoans, molluscs, and echinoderms, and fix inorganic carbon as calcium carbonate in the process of calcification.

Like the corals themselves, reef ecosystems consist of a high diversity of non-coral species, and they are truly as species rich as any other ecosystem on earth (Hatcher 1990; Sheppard et al. 2018). While the symbiotic relationships between dinoflagellates, apicomplexans, and corals are central to the existence of modern coral reefs, there are many other autotrophs that play a key role in the coral reef processes. There are four forms of coral reef algae: phytoplankton, micro-filamentous or turf algae (chlorophytes, rhodophytes, cyanobacteria), coralline algae (mostly rhodophytes but also some chlorophytes and phaeophytes), and macroalgae. The algal flora of coral reefs is typically species rich as they thrive in an ecosystem with many microhabitats and with a complex physical nature. Algae on coral reefs are not zoned as they are in some other habitats, such as rocky shores.

Coralline algae are tough with calcite crystals embedded in their cells. A hard, outer crust makes these cells resistant to wave action and herbivores. These algae play a key role in the formation and structure of coral reefs as they can colonise parts of reefs that are most exposed to waves and tides as well as storm damage. The growth of coralline algae can be slower than other algae as they calcify, an energy-consuming process. Coralline algae grow well where herbivory is intense, and they are often the most abundant algae on some sections of a coral reef.

Filamentous or turf algae in contrast grow quickly, and while they colonise exposed reef surfaces, they are a source of food for a wide variety of herbivores and are often grazed down to the point where their biomass is low. Coralline algae avoid being overgrown by these rapidly growing turf algae by continuously losing and renewing an outer layer called the epithallium. Colonisers are in this way sloughed off with this outer layer.

Nutrients play an important role in maintaining these algae although concentrations are exceedingly low on coral reefs but recycling and retention are high (Capone et al. 1992; O'Neil and Capone 2008; Sheppard et al. 2018). Much of the N required by a coral reef is fixed by the cyanobacteria with light largely determining the depth distribution of algae (O'Neil and Capone 2008). P can also be limiting, and it is often the rate of nutrient input from surrounding waters that determines the level of limitation for algae.

The major factor affecting algal abundance on coral reefs is grazing (Sheppard et al. 2018). Coralline algae are resistant to grazing, and hence their abundance is not determined by grazing intensity, but the interactions between coralline and other algae can play a major role in the distribution and abundance of all algae on a reef. Turf algae thrive where nutrient availability and herbivory are low; if nutrient concentrations increase, they are replaced by macroalgae. Despite their low abundance, but due to their high turnover, algae contribute high rates of primary productivity which derive from high rates of nutrient cycling and turnover. Coral algae clearly facilitate the transfer of fixed carbon to the next trophic level.

Roughly 25% of the world's fish species are found only on coral reefs (Sheppard et al. 2018). Most fish are specialised feeders. Some are herbivores, some feed on plankton, and others feed on other fish (piscivores) or are carnivores which feed on invertebrates. Feeding can be intense and can greatly impact the structure and function of the coral reef food web. Fish can also play a role in determining the zonation of a coral reef. By grazing on algae, for instance, they keep certain species in check and prevent them from dominating the corals. Parrotfish also graze on the corals themselves which opens new space for other corals, but they also contribute to the formation of sandy areas due to their production of calcareous remains which can be colonised by deposit feeders and burrowers. There is diurnal activity on coral reefs with many organisms, including fish, being active only during either the day or night.

7.8 Continental Shelves

The continental margin occupies only about 8 and 0.5% respectively of the surface area and volume of the global ocean, but the high productivity of coastal seas equates to nearly 30% of total oceanic net primary productivity and most of the global fish catch (Alongi 1998, 2005). Modelling of climate change impacts on marine capture fisheries shows large-scale redistribution of global catch potential with an average of 30–70% increase in high latitudes and a drop of up to 40% in the tropics (Section 13.7). Even with changes in climate, productivity is driven by nutrient inputs from rivers and groundwater, upwelling, exchanges at the continental shelf edge, and atmospheric inputs. The coastal ocean thus acts as a highly efficient trap for materials from land. The tight coupling at the land-sea boundary reveals how crucial the coastal ocean is to the structure and function of estuarine and marine food webs, including fisheries.

Energy flow through food webs on the shelf proper differs greatly among shelf seas, driven largely by differences in carbon fixation which are determined ultimately by the confluence of local- and ocean-scale patterns of water circulation, chemistry, and shelf geomorphology. At the shelf edge, cold, nutrient-rich water intrudes onto the shelf as upwelling fronts. Such exchanges are often rapid, promoting conditions favourable for high primary productivity and the subsequent rapid deposition of phytoplankton detritus onto the seabed. Higher fertility in upwelling zones accounts for the fact that approximately 90% of the world's fish catch is harvested on the continental shelves rather than in the open sea (Alongi 1998, 2005).

Ecological categorisation of the world's continental shelves (Alongi 1998, 2005) reveals that: (i) low latitude upwelling areas on the eastern boundaries of the ocean are usually the most productive shelves; (ii) the subtropical and tropical river-dominated shelves on the ocean's western boundaries are nearly as productive as shelves at temperate latitudes; and (iii) the shelves of many cold temperate seas are comparatively unproductive. Further, Alongi (1998) offered three generalisations of many shelf ecosystems: (i) outwelling of materials from land is usually restricted to the coastal zone by complex physical, chemical, and geological processes, and this material significantly influences inshore food webs and nutrient cycles, (ii) impingement of nutrient-rich, open ocean water enhances primary and secondary productivity on the outer shelf and shelf edge, including important fisheries especially along boundary currents, and (iii) comparatively little organic matter is exported from the continental margins. There are several important factors which determine to what extent these generalisations are validated or violated: the presence or absence of rivers, the presence or absence of upwelling, the location of ocean boundaries and shelf width. These factors are influenced by climate.

The first studies of the benthos of tropical continental shelves began off the western edge of Africa in the 1950s. However, it was not until the 1970s that there was a discernible increase in the study of tropical pelagic and benthic communities on continental shelves. Most studies were conducted off the east and west coasts of India and along the west coast of Africa where tropical demersal fisheries first developed (Longhurst and Pauly 1987). Meiofauna on tropical shelves have been sporadically investigated with most information obtained from the Indian Ocean. The available data suggest low-to-moderate densities compared with those from higher latitudes (Alongi 1990; Aller et al. 2008a, b). Macroinfauna densities and biomass are generally low compared with macroinfauna in higher latitudes, with densities often ranging from 50 to 600 ind. m^{-2} and biomass ranging from <1 to 90 g wet weight (WW) m^{-2} reflecting the small size of individuals (Ansari et al. 2012; Manokaran et al. 2014; Raja et al. 2014; Pabis et al. 2020; Velayudham et al. 2020; Vital et al. 2020). Greater densities and biomass are found under conditions of upwelling. For example, macrofauna densities averaged 5770 (range: 413–12 553) ind. m^{-2} in the upwelling area in the NE Arabian Sea (Parulekar and Wagh 1975); similar values were recorded off Baja California (Smith et al. 1974) and within the Benguela upwelling zone off north Namibia (Zettler et al. 2009). Measurements of microbenthic diversity and species richness are not available. Most available information has focused on megafauna, such as echinoderms, corals, molluscs, and fish.

Several environmental factors appear to mitigate against the development of high densities of macroinfauna including the fact that tropical shelf habitats are warm, wide, and shallow and susceptible to stress from climatic disturbances such as monsoons or by infringements of mass water movement and lack of seasonal water column turnover, both of which facilitate the development of OMZs (Naqvi et al. 2000). Food limitation may also be a contributing factor for low benthic densities and biomass, as well as physical disturbances caused by wave action, especially in such shallow waters (Aller et al. 2008a, b). Unlike high latitudes where phytoplankton detritus settles to the benthos, in the low latitudes most detritus is of low-to-moderate nutritional quality, composed of a wide assortment of material, but little phytoplankton, comparatively little of which settles on the seabed.

The plankton of tropical continental shelves has similarly been little studied with most studies examining coastal and upwelling areas. Tropical plankton are rich in species compared with those of higher latitudes but differ from temperate plankton in several important aspects. Tropical seas tend to be more oligotrophic, with micro-flagellates and cyanobacteria dominating tropical phytoplankton assemblages compared with temperate seas where diatoms tend to flourish (Heng et al. 2017). This difference has important consequences for grazing, as micro-flagellates are often thought to be too small for many metazoan grazers to efficiently exploit, and some types of cyanobacteria are known to be toxic or otherwise inimical to certain grazers. Thus, it is problematical to extrapolate information on zooplankton grazing from temperate to tropical habitats.

The biomass of tropical phytoplankton communities is often influenced by different climate phenomena. In Indonesian waters, chlorophyll *a* concentration declined during concurrent positive IOD and ENSO events (Siswanto et al. 2020), whereas they increased during concurrent negative IOD and La Niña events. In coastal waters, chlorophyll *a* variability was associated with varying rainfall/river discharge and in more open ocean waters, climate influenced phytoplankton biomass by varying the occurrence of upwelling and downwelling. The dominant climate mode determining chlorophyll *a* concentration shifted from the IOD in the eastern Indian Ocean to ENSO in the western Pacific Ocean. The impact of the IOD in the eastern Indian Ocean was more than one order of magnitude greater than the impact of ENSO, implying that phytoplankton and high-trophic level organisms such as fish may be more affected by the IOD than ENSO.

The role of plankton on tropical shelves is best seen from the perspective of their food web dynamics, especially their role in the microbial machinery of tropical waters (Figure 7.7). A large proportion of energy and material flow is funnelled through highly diverse, growing assemblages of archaea, bacteria, ciliates, flagellates, amoebae, and viruses, many of mixed trophic states, and subsequently transferred to higher consumers via a chain of small protistan grazers in what is essentially a 'microbial loop.' The concept of the 'microbial loop' has evolved to recognise the functional complexity of the intricate microbial machinery of food webs and functions more as a 'microbial hub' than loop in that microbes are tightly integrated with the classical food web (Figure 7.7). Production and consumption within the hub are largely passed to higher consumers via multiple trophic transfers or lost via remineralisation within the eutrophic zone. Microzooplankton grazing can be intense accounting for nearly all microbes and phytoplankton consumed within a water parcel. For instance, grazing by herbivorous zooplankton (<200 μm fraction) in coastal and slope regions of the South Brazil Bight accounted for most phytoplankton mortality (McManus et al. 2007). Grazing accounted for >80% of phytoplankton production in oligotrophic slope water. Estimates from microscopic counts found that most of the microzooplankton were copepod nauplii.

Abundances of tropical bacterioplankton range from 10^5 to 10^6 cells ml^{-1} which is within the range of cell densities found in other tropical and temperate coastal waters (Ducklow 2000). Bacterioplankton densities are usually within a comparatively narrow range, and this phenomenon has been attributed to grazing pressure as this narrow density range may represent a threshold below which capture by bacteriovores becomes functionally and energetically inefficient (Thelaus et al. 2008). Cell densities vary with changes in tidal flow, concentrations of suspended particles, dissolved organic matter (DOM), and dissolved inorganic nutrients as well as the onset of monsoons, temperature, predation, and phytoplankton productivity. For example, most bacteria are free in the water column in inshore waters, but densities of bacteria attached to particles vary in synchrony with the tide (Alongi et al. 2015). When tidal waters are at maximum rates of ebb flow inshore, attached bacterial densities also peak, especially in the wet season when suspended matter is at maximum concentration; at slack tide, numbers decline as particle–bacteria complexes settle. Substrate availability can be a key factor controlling bacterial numbers inshore where tides still exert considerable influence. This appears to be true for other eutrophic water bodies such as off India where high suspended solid loads and water-column anoxia co-occur with high densities of bacteria (Fernandes et al. 2008). Phytoplankton associated with seasonal oxygen depletion off the western shelf of India also greatly alter their community structure: microplankton dominate oxic waters

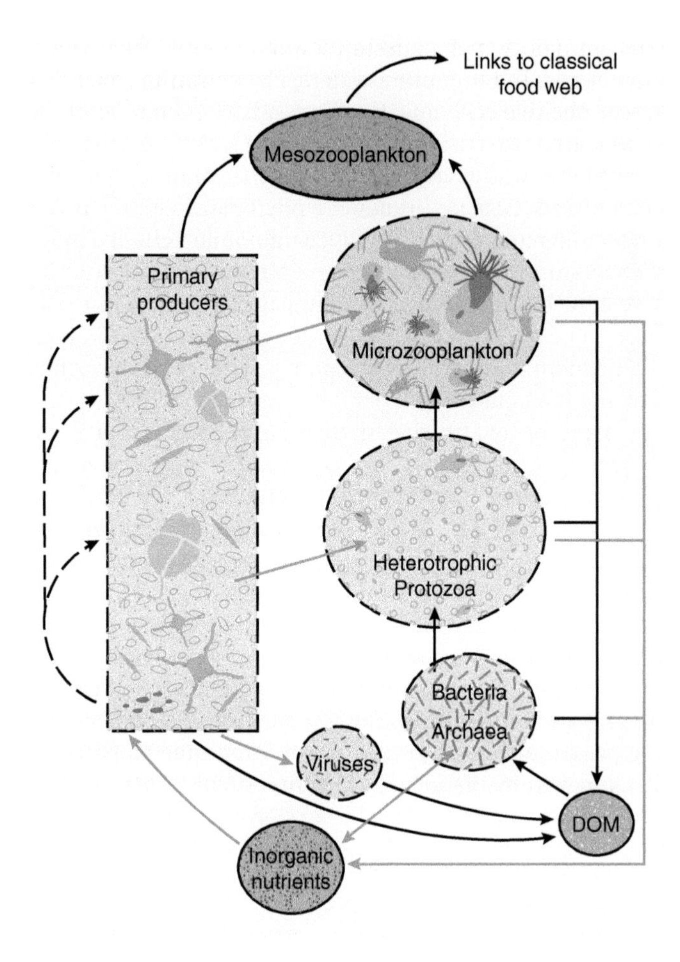

FIGURE 7.7 A simplified visualisation of the microbial consortia and their interrelationships in the microbial hub.

and pico-autotrophs and pennate diatoms dominate the autotrophic biomass in oxygen-depleted waters (Gauns et al. 2020).

Nevertheless, the intensity of trophic interactions and recycling within the microbial hub is poorly understood. There is some evidence that viruses can exceed the abundance of bacteria by an order of magnitude, but the functional significance of viruses in tropical waters is virtually unknown. A good example of the significance of virioplankton was described for the southern Great Barrier Reef shelf (Alongi et al. 2015). Virioplankton numbers and biomass varied significantly across shelf with water depth and with season. There were only two discernible patterns: in coral reef lagoons, virus numbers varied seasonally, and virus biomass was significantly lower at shallower coastal stations (2.2–5.7 mg C m^{-2}) than at other locations (20.4–113.1 mg C m^{-2}) across the shelf. Virioplankton numbers and biomass correlated with bacterioplankton numbers and production and bacterial growth and respiration correlated with net primary production, suggesting close virus–bacteria–phytoplankton interactions and a significant impact on bacterial dynamics by viral infection. The virus:bacteria ratio averaged 7, within the range measured in other tropical reef waters and at the low end of the range observed on other shelf waters, suggesting tight regulatory mechanisms between groups. Such close relationships can be a mechanism facilitating intense recycling within the microbial hub. When bacterial cells lyse, the released phages and resultant cellular debris are made up of easily digested protein and nucleic acids that can be readily reutilised by bacteria as food, representing a partly closed loop whereby bacterial biomass is consumed mostly by other bacteria. This bacterial–viral loop would have the net effect of oxidising organic matter and regenerating dissolved inorganic nutrients, thus linking these microbes to the phytoplankton. This has been shown recently in a South Pacific coral reef where lytic viruses removed a significant fraction of the bacterioplankton standing stock (Payet et al. 2014).

Significant mortality from viruses greatly increases bacterial growth rates (Breitbart et al. 2008). Although not measured in the Great Barrier Reef study, micro- and mesozooplankton likely play a key role in bacterial grazing, as ciliates, nano- and dinoflagellates, and crustacean nauplii are abundant enough throughout the southern Great Barrier Reef to be significant bacterial consumers. Close bacteria–plankton relationships were similarly observed in the Gulf of Aqaba, with the same drivers controlling microbial interrelationships (Bar-Zeev et al. 2009). Moreover, they found that bacteria were linked to phytoplankton by mutual coagulation and utilisation of transparent exopolymer particles and mucus. Such a process may be operating on the southern Great Barrier Reef shelf.

Ordinarily there is close coupling between primary producers and secondary consumers in the coastal zone, including on the continental shelf. This is because pelagic bacteria are largely fuelled by dissolved and particulate carbon derived from phytoplankton exudates and phytodetritus. As noted earlier, this material is degraded rapidly by the microbial consortia present, but mostly by the heterotrophic bacterial community. Up to 60% of carbon fixed by phytoplankton flows through bacteria, either lost as respiration or assimilated into new biomass (Calbet and Landry 2004). Picoplankton assemblages are numerically dominant in pelagic food webs, accounting for much of the pelagic biomass in tropical coastal seas (Crosbie and Furnas 2001).

Aside from viruses, what controls bacterial assemblages in the tropical coastal zone? As with all populations, coastal bacteria are regulated by availability of resources, environmental cues, and biological interactions. Organisms may be regulated within the food web structure as a trophic cascade. The idea of the trophic cascade (Section 6.11) indicates that regulation works in two ways. First, there are some 'bottom-up' controls in which a change in resources at the base of the food web will propagate up the food web with decreasing level of control. For coastal bacteria, this would mean that an increase in organic detritus would result in a concomitant increase in bacterial biomass and activity, but smaller increases in zooplankton or fish biomass. Second, some regulation works by 'top-down' control in which changes in a consumer population result in a decline in prey immediately below, but with an increase in the prey population immediately below that. For shelf phytoplankton, this would mean that an increase in the carnivorous copepod population would lead to a decrease in protozooplankton abundance, but an increase in bacterioplankton biomass caused by a release from immediate grazing control. Bacteria and other pelagic organisms in the food web are thus regulated by a combination of 'bottom-up' and 'top-down' controls (Azam and Malfatti 2007).

Pelagic bacteria are regulated by changes in environmental factors (e.g., temperature, light) and by interactions, including predation, within the microbial hub. Physical processes have until recently been ignored when considering trophic energetics within the microbial hub in tropical waters. Tidal fluctuations in plankton abundance and microbial activity have been noted in macro-tidal estuarine systems (McKinnon et al. 2015). Within this milieu operates both 'bottom-up' and 'top-down' controls on planktonic food webs.

Protozooplankton have a significant influence in transferring heterotrophic bacterial biomass to higher trophic levels. Protists are highly abundant (10^8–10^{10} flagellates m^{-3} and 10^3–10^8 ciliates m^{-3}) and feed on other protists, bacteria, and phytoplankton at rates partly determined by prey abundance and cell size. The proportion of bacterial and phytoplankton biomass that is consumed by microzooplankton varies greatly, but often exceeds 50% and in some instances, accounts for the biomass of all prey. Protists may remove a greater portion of phytoplankton biomass than larger zooplankton, exerting considerable grazing impact on phytoplankton activity. Unfortunately, there is little such data for subtropical and tropical waters.

The degree of predatory control by metazoan zooplankton on protozoan communities in tropical waters has received relatively little attention. Copepods, rotifers, cladocerans, fish larvae, chaetognaths, euphausiid and mysid shrimps, and other pelagic metazoans all ingest protists to varying degrees, as suggested by laboratory experiments and gut content analyses (Azam and Malfatti 2007). Reciprocal relationships observed in the field between protozoan and metazoan population densities suggest some predatory control, but no hard estimates exist for the proportion of protozooplankton abundance lost to predation in tropical coastal and shelf waters.

The significance of meso- and macro-zooplankton to food web structure in tropical shelf waters is not well understood but is presumably like that for temperate waters where their significance has been long established. Zooplankton feed on phytoplankton, protozooplankton, smaller metazoans, and larvae, but they may feed extensively on detritus particles and associated microbes. These plankton communities can maintain high densities throughout the year in many tropical shelf and coastal waters given their dietary habits as generalists and tolerance to changes in physicochemical conditions. Although the trophic roles of meso- and macro-zooplankton are generally well known, how their life cycles and distribution are affected by shelf hydrodynamics has received little attention in tropical waters. In proximity to estuaries, these communities are determined by tidal regime, the extent of freshwater discharge, and the intrusion of more saline waters. Zooplankton may be imported or exported from a given coastal water body depending upon tidal regime, mobility, and the location of the coastal boundary layer. Other factors such as oxygen and nutrient concentrations play a strong regulatory role. Zooplankton abundances are therefore a function of their responses to changes in water residence times and movements, climate, and their physiological limitations, although these conditions have not received much attention in the tropics (Ambriz-Arreola et al. 2018). Nevertheless, it is probable that biological conditions in tropical waters are not divorced from changes in environmental conditions but are in fact ultimately controlled by them.

The structure of benthic food webs is greatly influenced by changes in climate, sediment type, salinity, and other environmental conditions, including changes in available food supply (Gray and Elliot 2009). Organic matter in sediment is a rich consortium of materials derived from different sources and identifying and distinguishing these sources is complex and still in its infancy. The organic matter in sediments is a rich mixture of dissolved and particulate materials derived from autochthonous and allochthonous sources. Dead and decaying phytoplankton, epiphytes, benthic microalgae, macrophytes, and metazoans constitute the richest sources of proteins, carbohydrates, lipids, fatty acids, amino acids, and other organic compounds that sustain benthic communities. Detritus derived from land or advected from offshore frequently comprise a large portion of standing amounts of sediment organic matter. Living organisms make up only a small percentage of the total sedimentary organic matter pool.

Benthic deposit-feeders are fuelled by complex mixtures of organic material (Gray and Elliot 2009). Clear temporal changes in benthic detrital food webs may or may not occur in tropical coastal and shelf waters, but there are undoubtedly seasonal changes in the availability of seagrass detritus and in other foods that make up a continually changing diet for benthic deposit-feeders. Thus, benthic communities live in an ever-changing environment, responding to seasonal and likely spatial inputs of various forms of detritus. Infauna, including microbes, respond positively to sedimentation of high-quality detritus. Arguments have raged over the relative importance of microbes versus detritus as food sources for benthic detritivores, including the role of coprophagy and optimal foraging (Gray and Elliot 2009). A more realistic concept of life in sediments has emerged: diets vary greatly among detritivores as some species feed mainly on bacteria and microalgae while other species feed mostly on a variety of living and dead organic matter to obtain a balanced diet. However, on average, bacteria alone are insufficient to meet the nutritional requirements of most benthic detritivores. Benthic microalgae are a higher quality food for benthic organisms than bacteria, owing in part to their more complex lipid and fatty acid contents. No one factor regulates detrital food webs, but rather an interactive hierarchy of mechanisms: quality and quantity of food resources, physiological constraints and tolerances, and behavioural and life history strategies (Campanyà-Llovet et al. 2017). These mechanisms operate within the confines of environmental cues.

Benthic filter-feeders also consume of wide variety of food particles, including phytoplankton. Evidence suggests that organic aggregates precipitated from macrophyte-derived DOM can serve as food for a variety of filter-feeders, including inshore bivalves. Dietary plasticity in which free bacterial and phytoplankton cells, detrital particles and attached microbes, and organic aggregates are consumed is well known for benthic suspension-feeders (Campanyà-Llovet et al. 2017). Physical forces alone therefore do not determine the distribution of benthic suspension-feeders. Constraints at the local scale suggest why they are patchily distributed. The minimum and maximum densities of suspension-feeders are set by physical forces such as turbulence, vertical mixing, and horizontal advection, which minimises or maximises food intake, but the limits of median densities of suspension-feeders are set by intra- and interspecific interactions, predation, and reproductive strategies. Suspension-feeders can serve to stabilise comparatively open shelf ecosystems with short residence times by retaining and thus conserving nutrients within the system, minimising losses of organic matter. The biomass of benthic suspension-feeders is greatest in coastal and shelf habitats where water residence times are shortest implying that clearance rates of food are a function not of particle concentration, but of the rate of water flow. A good example of this phenomenon is the benthic 'garden' of suspension-feeders found at the shelf edge of the southern Great Barrier Reef where water residence times are short, current velocities are rapid, and suspension of detritus particles is high (Alongi et al. 2011).

Benthic food webs on tropical river-dominated shelves are linked to sedimentary structures as controlled by factors such as rates of freshwater and sediment discharge and bottom topography. Off the Amazon, lowest bacterial and faunal densities tend to occur within the physically reworked, inshore fluid muds, but with no clear bathymetric patterns. Abundance varies with changes in river discharge, with peak densities occurring during falling and low discharge periods and lowest densities occurring during rising discharge (Aller and Stupakoff 1996; Aller and Aller 2004; Aller et al. 2008a, b). Off the rivers of southeast Papua New Guinea, highest densities of infauna and bacteria occur within a large mudbank located at mid-shelf southeast of the Fly River (Alongi et al. 1992). Benthic faunal and microbial activity are greater in the Gulf of Papua than within the Fly River, where even intertidal muds are too unstable to support benthic organisms larger than microbes. The benthic infauna of both the Amazon and the Fly Rivers shares several characteristics: (i) small size (most animals are < 2 cm in length), (ii) numerical dominance of polychaetes over crustaceans and bivalves, (iii) presence of protobranch mollusc shell layers indicating recent death assemblages, and (iv) absence of suspension-feeding bivalves.

A model of responses of benthic infauna and changes in sediment fabric to discharge from large tidal rivers was proposed by Rhoads et al. (1985) based on their work off the Changjiang River, China. The responses of benthos off the Amazon and Fly Rivers follow the same general pattern. Close to the river delta, sediments undergo episodes of deposition and erosion to the extent that the seabed is too unstable to sustain benthic life other than bacteria (Figure 7.8, diagram 1). Farther from the delta, a deepening water-column and algal blooms result in a gradient of conditions increasingly favourable for some infauna populations (Figure 7.8, diagrams 2 and 3). Phytodetritus settles to the sea floor fuelling higher densities of surface deposit feeders and increased microbial decomposition results in high rates of dissolved nutrient release to the overlying water column. Bioturbation is maximal in this zone, obliterating physical sedimentary structures (Figure 7.8, diagram 4). Farther offshore, primary productivity and biomass are lower, and rates of sediment accumulation are minimal, resulting in low densities of benthos (Figure 7.8, diagram 5). The location of various sedimentary structures varies in each shelf settling, but

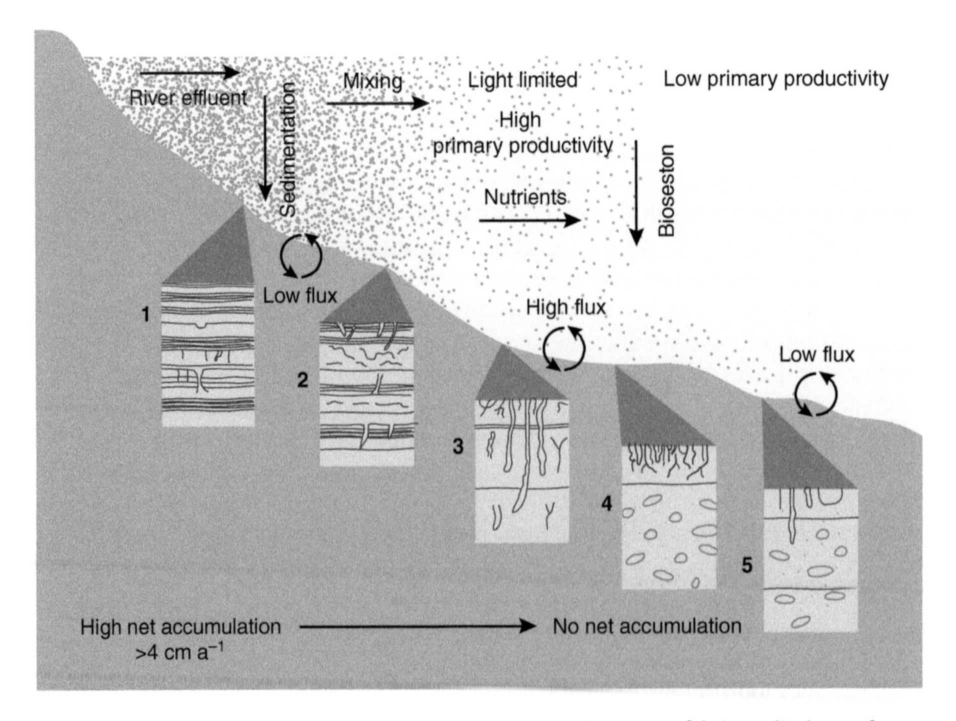

FIGURE 7.8 Idealised model of infaunal responses and sedimentary fabric to discharge from large tidal rivers, from processes 1 to 7. *Source:* Rhoads et al. (1985), figure 13, p. 205. © Elsevier.

the general scheme for the most part applies to the benthos off most tropical river/shelf boundaries.

Benthic communities on these shelves appear to be regulated by physical disturbance, food limitation, and predation. All three factors may be operating at any one time and place. Physical disturbance of the seabed plays the major role influencing the distribution, abundance, and structure of tropical shelf assemblages. X-radiographs of sediment slabs taken from the inshore muds of the Amazon and the interbedded mud and sand facies off the Fly River reveal the dominance of primary physical structures (Alongi et al. 1992; Aller and Aller 2004; Aller et al. 2008a, b). Erosional or burial contacts in subsurface sediments are common, with evidence of truncated biogenic structures (tubes, burrows) suggesting past episodes of benthic colonisation and elimination. Radiochemical data support the radiographic evidence and indicate deep scouring and re-depositional episodes in the inshore areas off these river deltas. Laminated deposits form the most common sedimentary profile, and laminated and interbedded deposits likely reflect alternating episodes of erosion and deposition caused by fluctuations in near-bottom current velocity; high shear stress results from strong tidal currents.

Analysis of seasonal and spatial patterns of invertebrate death assemblages buried in deep subsurface deposits off the Amazon (Aller 1995) provides further evidence of periodic physical disturbance and subsequent elimination of infauna. Aller (1995) found evidence of cycles of surface exposure and burial of mollusc shells out to a water depth of 40 m, indicating that physical disturbance is massive in scale. Evidence was also found for large-scale, onshore transport of shells and fragments of relict coral and bryozoans from the outer shelf, followed by deposition and downward mixing in the inshore deposits. Boring of shells was comparable to other shelves, indicating some predation despite the overpowering impact of physical disturbance.

Predation was initially discounted as a prime regulatory factor because of the absence or low abundance of demersal fish or other obvious large predators and epibenthos being composed of phyla unlikely to prey on infauna. However, there are shrimp fisheries off both the Amazon and Fly River deltas with trawling grounds coinciding with areas of highest benthic densities and highly bioturbated structures. This suggests a trophic connection between commercially important shrimp fisheries and benthic communities. Alongi and Robertson (1995) hypothesised that the main trophic pathway for benthic food webs off the Amazon and Papuan shelves is:

$$\text{Detritus} + \text{microbes} \rightarrow \text{shrimp}$$

The fact that penaeid species are detritus and micro-invertebrate feeders suggests that this is a reasonable conjecture. The larger infauna appears to be a minor trophic pathway to finfish predators as finfish landing off both shelves is minor. The scenario of a short food chain is plausible, but clearly more information is needed on the diets of commercially important species. It is unknown if such a scenario applies to other tropical shelves as there are large bathymetric differences among shelves and there are also large latitudinal differences in the composition of demersal fisheries within tropical latitudes (Sharp 1988; Fitzpatrick et al. 2012).

The dominance of pioneering infauna assemblages by bacteria and small surface deposit feeders may be an adaptation to a regime of episodic physical disturbance but also suggests food limitation. Infauna biomass relates positively to phytoplankton production with bacteria accounting for 75% of total benthic biomass and at least 80% of benthic carbon production (Alongi et al. 1992; Aller et al. 2008a, b). Compared with temperate benthos, proportionately more production and biomass are vested in bacteria than in metazoans, likely reflecting the fact that microbes, being smaller and having shorter generation times than metazoans, can respond more rapidly to, and are less affected by, disturbance (Aller and Aller 2004; Aller et al. 2008a, b).

The benthos of upwelling regions in contrast to those off tropical rivers are highly abundant due to high rates of primary productivity. Upwelling occurs in some tropical bays and gulfs which only intermittently receive intrusions of oxygen-rich waters, causing anoxia for long periods. A good example is the Golfo Dulce on the west coast of Costa Rica which is usually anoxic (León-Morales and Vargas 1998). The area has unusually low abundances, diversity, and biomass of macroinfauna despite high concentrations of organic matter. Along an inner sill within the gulf, the fauna is rich (650–9240 ind. m^{-2}), but biomass is still low. The fauna is composed of small opportunists which are most capable of repeated colonisation when intermittent intrusions of oxygenated water permit the flushing of stagnant waters.

In upwelling regions where conditions do not lead to the development of anoxic sediments, benthic standing stocks are high. Benthic biomass and densities were high off NW Africa and ranging 2.4–94.4 g WW m^{-2} and 1635–35 200 ind. m^{-2}, respectively, over the entire continental shelf (Buchanan 1958; Thiel 1982). Megafauna were most abundant along the shelf slope boundary as were plankton standing crops where shelf-break upwelling produces considerable amounts of organic matter. Meiofaunal and bacterial densities were also high in this region, generally higher than on non-upwelled, continental shelves. These relatively high densities are probably the direct result of high food availability from upwelling. Latitudinal differences along the NW African coast appear to be related to the intensity of upwelling (Buchanan 1958; Thiel 1982).

Seabed comparisons between upwelled and non-upwelled sites are rare. In the Gulf of Panama, upwelling occurs in the dry season when primary productivity nearly doubles compared with during the wet season. Recruitment of macroinfauna occurs mainly during the latter half of the upwelling resulting in high seasonal densities. After upwelling, total densities and species richness decrease rapidly. Most of the species are opportunistic, responding rapidly to seasonal changes in food availability (Alongi 1990). In the nearby Gulf of Chiriqui where there is no upwelling, the benthos fluctuates much less than in upwelling regions. Recruitment may occur during upwelling phases because of increased food availability; more food increases the probability of larval and juvenile survival. Species-rich, dense communities can be maintained in an unstable environment where competition and physical factors are not able to account for seasonal changes. Predation by epibenthic predators, mainly portunid crabs, appears to be the major factor regulating communities in the upwelling region off the west coast of South America, but the generally high densities of both infaunal prey and epibenthos must ultimately be regulated by food supplied from upwelling.

There is little data on the impact of upwelling events on planktonic food webs except for a few areas such as the Gulf of Panama (MacKenzie et al. 2019) and on the southern Great Barrier Reef shelf where high densities of bacteria, viruses, and phytoplankton presumably translate into high rates of trophic transfer within the microbial hub and within the classical food chains (Alongi et al. 2015). The pelagic ecosystem of the Gulf of Panama has a relatively simple, linear structure from primary producers to apex predators, such as yellowfin tuna and mahi, and has a predator: prey ratio of 376:1 which is low compared with other upwelling ecosystems (MacKenzie et al. 2019).

A phenomenon known as 'chakara' are mud banks that appear and disappear off the Kerala coast of SW India (Mathew and Gopinathan 2000). These mud banks are destroyed during the SW monsoon (July–September). The benthic communities do not fully recolonise the mud banks for at least 6–8 months. Abundances of macroinfauna are very variable depending upon the migratory stage of the mud banks and the season of the year. Low abundances of macroinfauna have been recorded on some banks (Harkantra and Parulekar 1987). Off Cochin, the fauna was composed mainly of polychaetes (>80% of total fauna) dominated by *Nephthys, Glycera, Nereis, Aphrodita, Owenia, Onuphis, Diopatra*, and *Pectinaria*. The shelf was sampled during the pre-monsoon season when densities would be expected to be low corresponding to the period of minimum primary production. However, in the mud banks south of Cochin, benthic biomass was high (400 g WW m^{-2}) during June–July but declined to <50 g WW m^{-2} by September–October after the monsoon season. Biomass of macroinfauna was low further north of Cochin, ranging from 30 to 100 g WW m^{-2}. Dense populations of tube-dwelling polychaetes (e.g. *Sabellaria cementarium*) developed when the mud banks first formed, but quickly vanished when the mud banks disappeared.

Meiofauna of the mud banks are abundant (>1000 ind.10 cm^{-2}) during pre-monsoon months but become rapidly depleted during the SW monsoon period when bottom muds are resuspended, and the banks are destroyed. Recolonisation of meiofauna was faster than for the macroinfauna with complete recovery attained in a few months. Crustaceans appeared to be the most affected by destruction of the mud banks, probably the result of sulphides released into the water column. On more stable mud banks closer to the coast, high population densities of meiofauna appear to be maintained throughout the year (Mathew and Gopinathan 2000).

Phytoplankton and zooplankton appear to be significantly influenced by the formation and destruction of these mud banks. Phytoplankton were highly productive just after the period of resuspension, with an average rate of production of 1 g C m^{-2} d^{-1} (Jyothibabu et al. 2018). During other monsoon periods, productivity was low to moderate. Plankton blooms showed a gradually increasing trend from June onwards reaching a maximum in August, primary due to the abundance of the dinoflagellate, *Noctiluca miliaris*. After August, there was a gradual decrease in the volume of plankton, reaching its lowest in December. After December, there was a rise in phytoplankton biomass through the succeeding months and reaching its peak at the period of the next mud bank formation. Diatoms dominated during the formation of the mud bank while dinoflagellates were most abundant during mud bank dissipation. The average mud bank, despite its limited primary production potential due to the shallow euphotic zone, supports higher standing crops during mud bank formation and is presumably favoured by abundant rainfall and enrichment of nutrients from the seabed. Zooplankton biomass was low during pre- and post-mud bank periods (Jyothibabu et al. 2018). The major groups in order of decreasing abundance were copepods, appendicularians, fish eggs and larvae, prawn larvae, and crab larvae. This composition was unusual in that fish eggs constitute a major fraction of the zooplankton which could be due to the large-scale migration of the spawning population of fish and prawns into the mud bank area. These mud banks are heavily fished, suggesting significant fisheries production (Mathew and Gopinathan 2000).

Another unique characteristic of tropical shelves is the fact that many continental margins are dominated by carbonate deposits and by various mixtures of terrigenous and carbonate sediments (Mutti and Hallock 2003). The early studies of Buchanan

(1958) and Longhurst (1959) off West Africa were the first to indicate faunal differences between shallow continental shelves and their quartz sand-dominated counterparts in higher latitudes. Buchanan (1958) recognised four distinct animal communities off the coast of Ghana: (i) a transition zone between the surf zone and shallow subtidal dominated by infauna, (ii) an inshore fine sand community (13% $CaCO_3$) dominated by the bivalve, *Cultellus tenuis,* and the tube-dwelling polychaete, *Diopatra neapolitana,* (other dominant forms are patchily distributed as '*Owenia* beds'), 'Accra Bay silt patches or *Macoma* beds' and '*Dentalium* zones,' (iii) a silty-sand community with varying amounts of carbonate (30–50% $CaCO_3$); two species of large Foraminifera dominated this transitional seabed and epifaunal organisms were abundant, particularly ophiuroids and holothurians, and (iv) an offshore coarse sand community (68–80% $CaCO_3$) composed mostly of carbonates with outcrops of solid limestone colonised by an abundant epifauna.

Longhurst (1959) recognised that the benthos off the West African shelf bore a close resemblance to that of the Mediterranean fauna. Off the Gambia and Sulima Rivers at the boundary between Sierra Leone and Liberia, the shelf contained various proportions of muds: black reducing muds in the estuarine areas, grey or blue muds in the offshore oxidised areas, and olive-brown muds in other inshore areas. At depths > 100 m offshore, carbonate sands predominated, composed of remains of planktonic Foraminifera. Hard stony substrata occurred across the shelf, generally at shallow depths; a zone of yellow coral, *Dendrophyllia* sp., occurred on rocky bottom at depths of 80–100 m and were described variously as 'madreporic sills,' 'fonds coralligenes' and 'massifs coralens.' The benthic communities across the shelf were partitioned on the basis of sediment type and dominant species: (i) shallow soft-deposit communities dominated by *Venus* on shelly sands and *Amphioplus* on muds and muddy sands on the Guinea and Senegal shelves, (ii) shallow hard substrata communities such as inshore reef, estuarine gravel, and coastal rock communities, (iii) deep soft-deposit communities, on muddy sands from 80 m to below the shelf break, dominated by ophiuroids, cnidarians, and holothurians, and (iv) deep hard substrata communities occurring on rock ground (yellow corals), very soft rock or 'marl' at the shelf edge off Guinea. These studies supported the original contention of Thorson (1957) that no increases in infaunal densities and species richness occur from the Arctic to the tropics, but an increase does occur in the richness and diversity of epifauna. In truth, there is not enough shelf data for benthos to conclude one way or the other if tropical shelf communities are richer or poorer than their counterparts in higher latitudes.

Carbonate-bearing shelves in the Caribbean have complex benthic communities like those off West Africa. Lewis (1965) described three community zones off Barbados in the West Indies based on water depth: (i) a sponge and coral community (50–150 m), (ii) a diverse and abundant community of molluscs, echinoderms, and coelenterates (100–300 m), and (iii) a mollusc-dominated community between 300 and 400 m. These communities are like the West African fauna, namely in the abundance of coelenterates and the sporadic occurrence of hard substrata composed mostly of limestone. Polychaetes dominated both shelves between depths of 40–250 m, and ophiuroids were common to depths of 250 m as were small bivalves such as *Venus* and *Astarte*.

One characteristic of Thorson's (1957) scheme which may still hold true is the increase in abundance and diversity of large epibenthic invertebrates from the poles to the tropics. Epifaunal assemblages are rich and diverse on the continental shelves off West Africa (Buchanan 1958; Longhurst 1959), Madagascar (Picard 1967), the

Gulf of Mexico (Hedgpeth 1955), Brazil (Aller and Aller 1986), and the Great Barrier Reef (Birtles and Arnold 1988). The fauna of many carbonate-dominated continental shelves is dominated by lancelets (e.g. *Branchiostoma*) and large Foraminifera, such as *Jullienella, Marginopora*, and *Alveolinella*. Several reasons have been put forward to explain this dominance, including attraction to hard-bottom areas of limestone and the lack of anoxic muddy conditions on the sandy, compacted bottoms.

Several continental shelves are characterised by gradients in terrigenous to carbonate sedimentation resulting in various mixed terrigenous-carbonate sediment deposits, such as off Belize, the Grand Banks off the Bahamas, and the Great Barrier Reef shelf. These variations occur across shelves that are lagoonal in character and rimmed by relict or living coral reefs. The shallow continental shelf of the Great Barrier Reef is characterised by such an across-shelf sedimentary transition and is commonly subjected to various physical disturbances such as cyclones, storms, and commercial trawling. Standing stocks of both benthos and zooplankton do not vary significantly across the shelf or at least in any consistent pattern, although macroinfaunal densities decrease significantly. Small, tube-building deposit and suspension-feeding polychaetes and amphipods dominate the infauna; bivalve molluscs are notably rare. However, like other tropical shelves, there is a rich and diverse epibenthos.

Tropical and subtropical continental shelves are generally shallow, driven by intermittent intrusions of nutrient-rich, upwelled water and/or by seasonal discharge of estuarine water and detritus (Alongi 1998). Benthic communities thrive in tropical upwelling regions off Panama and Arabia and in areas where there is massive, large-scale riverine discharge, such as off the Amazon. These shelves, however, are commonly subjected to anoxia with deleterious effects on the benthos if inputs of river effluent and upwelling episodes occur on a massive scale, too frequently or occur simultaneously with periods to stagnant water masses. These conditions probably account for the wider variations of benthic abundances, biomass, and diversity on tropical shelves than on continental shelves of higher latitude.

Another common feature of tropical and subtropical shelves is abundant demersal fisheries. As in higher latitudes, tropical fisheries potential is generally related to both water column production and benthic standing crop. Longhurst and Pauly (1987) summarised some general principles of demersal fish communities on tropical shelves: (i) an increase in diversity towards the equator, (ii) an east-west gradient in diversity with more taxa in the western Atlantic than the eastern Atlantic, and (iii) more families in the Indo-West Pacific region. Benthic communities and demersal fish assemblages appear to be influenced by similar environmental and biogeographic barriers. Similar fish families occur throughout the tropics chiefly in response to the same factors, such as the presence of reefs and hard rocky bottoms, sediment type, and the degree of brackish conditions in lagoons and rivers. For example, species richness and structure of demersal fish communities on the NE Brazilian shelf were linked to season, depth, latitude, longitude, and distance from the coast (Nóbrega et al. 2019). Fish biomass peaked during the rainy season between depths of 45–65 m, but maximum biomass shifted to 35–75 m in the dry season.

Pelagic fish are thought to be relatively more abundant and diverse than demersal assemblages in tropical seas (Longhurst and Pauly 1987), but many genera such as many clupeids and engraulids are benthopelagic, so the distinction between pelagic and demersal fish communities is somewhat blurred. Some groups appear to be particularly abundant in the tropics. About a dozen families of the Perciformes dominate tropical demersal fish assemblages and may be grouped by sediment type preferences. On rocky grounds, the Serranidae, Lutjanidae, and Lethrinidae (groupers and snappers) are common. On inshore muddy areas, Sciaenidae often dominate. Commonly known as croakers and drums, they frequently co-occur with threadfins and spadefish. Many attain large size and have specialised feeding habits. Breams, grunts, threadfin breams, ponyfish, and other families occur on sandy grounds, feeding mainly on the benthic epifauna, particularly decapod crustaceans and molluscs. Flatfish range throughout the tropics. Generally, there appears to be four types of communities: nearshore and estuarine soft-bottom areas dominated by sciaenids, sandy bottoms characterised by sparids, rocky bottoms dominated by lutjanids, and reef areas dominated by no one family. The wealth of benthopelagic scombroids and clupeoids is one peculiar characteristic of tropical demersal assemblages compared with temperate fish faunas. Tropical demersal fisheries possess characteristics different from fisheries of other latitudes (see Table 4 in Pauly [1979] and Table 9 in Sharp [1988]) and are at least as dependent on the benthic environment as temperate demersal fish, if not more so.

7.9 Open Ocean

The tropical open ocean comprises 37% of the area of the global ocean. The epipelagic zone is a thin layer of lighted and wind-mixed water lying atop the cooler mass of the deep ocean. The change in physical, chemical, and biological properties between the two depth zones is more frequently abrupt than gradual. When abrupt, the changes in ecological conditions that occur usually over a few tens of metres over the euphotic zone are greater than across most vertical fronts that intersect the sea surface. This phenomenon is most striking in the eastern parts of the tropical oceans (Longhurst 2007). Here, at around 35–40 m below the surface, the temperature drops from 28 °C to about 16 °C. The organisms of the epipelagic zone differ fundamentally between the tropical and temperate zones, although in the deep ocean there is high similarity of biota between zones. Thus, moving across this boundary is as great as moving several thousand km equatorward or poleward.

Low latitude ocean masses have been defined by Longhurst (2007) as the "Trades Biome," where the mixed-layer depth is forced by geostrophic adjustment on an ocean-basin scale to local or distant wind forcing. Tropical ocean ecosystems are thus divided into: (i) a small-amplitude response to trade wind seasonality and (ii) large-amplitude response to monsoon reversal of trade winds. The first case is appropriate to that part of the ocean at low latitudes forming 22% of the global ocean that lies under the influence of trade winds, where seasonality is weakest and where changes in depth of discontinuity layers are imposed principally by geostrophic adjustment of the zonal equatorial current systems. Mixed-layer depth is therefore relatively constant, so that primary production and phytoplankton biomass are low. The nutricline is shallower than the photic depth, so vertically integrated production is not light-limited, but limited by nutrients and trace elements. This situation occurs year-round and the pattern breaks down with deepening of the mixed layer only briefly and only in some regions. The second case describes the dynamics of those parts of the trade-wind zone where strong seasonality is a feature of local wind forcing. These are the monsoonal regions where, in the extreme case of the NW Indian

Ocean, seasonal wind reversal results in seasonal reversal of the monsoon currents and seasonal alternation between eutrophic and oligotrophic ecosystems. The small zonal dimension of the tropical Atlantic permits a strong seasonal response to the surge of southerly trades across the equator that is analogous to the effect of monsoon reversal in the Arabian Sea.

The "trade winds" regime has several physical characteristics that makes it unique and different to regimes at higher latitudes (Longhurst 2007): (i) the radiation balance is such that there is a mean annual positive downward heat flux across the surface, (ii) the seasonal radiation flux is such that surface mixed layer is maintained continuously, (iii) the level of solar radiation is such that autotrophs are less commonly light-limited than they are at higher latitudes, (iv) the Rossby radius of internal deformation increases into the tropical zone due to diminishing Coriolis force, (v) baroclinic timescales are weeks in the tropics rather than years at higher latitudes, (vi) the equatorward-diminishing Coriolis parameter means that the slope of the sea surface required to force horizontal motion diminishes, (vii) the pycnocline is coincident with permanent tropical thermocline and has very high stability equatorward of latitude at which winter convective mixing becomes trivial, and (viii) the tropical Ekman layer lies above a very sharp density discontinuity where turbulence is weak.

An increase in the rate of primary production is normally accompanied by an accumulation of biomass as growth and losses are linked. It is in the trade winds biome that the smallest cell-size fractions of the phytoplankton dominate biomass and production; even eukaryotic cells are of small size while the submicron picoplankton dominates. Less than 90% of plant biomass and <80% of phytoplankton production may be attributable to these small cell sizes. Perennially low nutrient concentrations have induced the evolution of special strategies, such as symbiotic relationships of heliozoans and their autotrophic plastids, and some large cells have evolved special forms that enhance the rate of uptake of scarce nutrients across their cell walls, minimising their sinking rate to the deep ocean.

It is in tropical open ocean waters that diel vertical migration extends over greater depth intervals and is most ubiquitous. In all seasons and in all areas, a substantial fraction of the zooplankton and nekton, especially copepods, euphausiids, myctophid fish, and squids, will arrive at or close under the surface ocean soon after dusk and descend again to 200–300 m at dawn. As pointed out by Longhurst (2007) "in this biome, the pelagic ecosystem is at its most taxonomically diverse and represents the climax community of the pelagos. All other pelagic ecosystems can be viewed as derivatives from it, constrained or stressed in different ways that suppress some of the complexities of the tropical pelagos." In the tropical open ocean, the overall dominance of copepods is weakest as copepods are reduced to only about 33% of total plankton biomass; several other groups make up the total plankton biomass. Pelagic fish reach their peak abundance and biomass in the tropical open ocean; a great variety of shoaling clupeids, schooling tuna, and other scombroids and solitary sharks are characteristic of tropical seas, such that there are multiple food webs, each more complex and longer than those of higher latitudes. Diel migration is now being affected by the increase in OMZs due to global warming. Some pelagic organisms, such as the euphausiid, *Euphausia distinguenda*, have varied the diel migration patterns in the eastern tropical Pacific off Mexico in response to this phenomenon (Herrera et al. 2019). This euphausiid was found in the mixed layer at night and near the centre of the oxygen minimum zones (OMZ) during the daytime between 200 and 350 m depth. It withstands the low oxygen levels (0.2 ml O_2 l^{-1}) by slowing its metabolic rate.

Discussion of tropical open ocean ecosystems below will follow the global province system (Figure 7.9) of Longhurst (2007):

NATR, the North Atlantic Gyral Province, comprises that part of the North Atlantic gyre lying south of the subtropical convergence running zonally across the ocean at about 30 °N. The NATR has a consistently low and rather featureless surface chlorophyll field with a seasonal cycle of small magnitude and much irregularity despite the significant seasonal change in irradiance. Chlorophyll analyses of this province clearly show the distinct characteristics of tropical phytoplankton in which there is a relatively large cyanobacterial component and a relatively small diatom component. Microzooplankton biomass and abundance are as small as anywhere in the oceans and 5–10 times smaller than in neighbouring subtropical provinces. There is a major chlorophyll enhancement in the extreme SW corner of NATR, and it is in this region that the "Typical Tropical Structure" is found. The key characteristic of this feature is the fact that the depth of the deep chlorophyll maximum is determined by pycnocline, nutricline, and irradiance, and the profiles of heterotrophic bacteria and herbivores match that of autotrophs in a manner suggestive of consumption and recycling in situ of organic material. The NATR is the southern part of a region of very sparse zooplankton that occupies the central part of the North Atlantic gyre. Macrozooplankton and nekton numbers in NATR are 5–10 times lower than in the mesotrophic and eutrophic areas of the Canary Current. These assemblages are comprised of siphonophores (*Chelophyes*), pteropods (*Clio*), chaeognaths, euphausiids (*Stylocheiron, Euphausia*), and fish (*Cyclothone*). The vertical distribution of individual organisms is rather dispersed. Daytime residence depths of diel migrants widely vary compared with the depths entered at night, which are consistently within the upper 50–75 m. The bathypelagic fish, *Cyclothone,* remains at depth day and night. Euphausiids are active diel migrants, except for the genus *Stylocheiron* which resides day and night at species-specific depth ranges within the upper 200 m. The summer migrations of the tropical yellowfin (*Thunnus albacares*) and skipjack (*Katsuwonus pelamis*) tuna occur within this province. These large predators were abundant until the 1950s and 1960s and now consist of small populations.

CARB is the Caribbean Province which comprises the Gulf of Mexico and the northern Caribbean Sea. The seasonal signal of primary productivity is small in magnitude, but there appears to be chlorophyll maxima in both winter and summer. Primary productivity ranges from 10 to 17 mg $C\,m^{-2}\,d^{-1}$. In summer, there is a clear difference between oligotrophic regions having crystal clear surface water and areas of chlorophyll enhancement, whereas in winter this distinction is more obscure because the entire province then experiences slight but uniform chlorophyll enhancement. Deep oligotrophic chlorophyll profiles are characteristic of the Gulf of Mexico where there is significant nonuniformity of the depth of the pycnocline due to complex flow structure at the mesoscale. Maxima of chlorophyll, phytoplankton carbon, and microzooplankton carbon coincide within the pycnocline and above a nitrocline. The vertical distribution of mesozooplankton follows the pattern of the oligotrophic profiles; the 150 μm fraction is dominated by copepods that are concentrated in the mixed layer. These are a typical suite of warm water genera (*Clausocalanus, Euchaeta, Scolecithrix, Nannocalanus*). Diel migrants move between daytime depths of 300–400 m and the photic zone at night. The mesozooplankton respond positively to upwelling along the coast of northern Yucatan where the dominant copepod, *Temora stylifera,* may attain abundances of >20 000 ind. m^{-3}. There are strong effects of cyclonic and anticyclonic eddies in this province. The distribution of zooplankton and micronekton biomass reflect nutrient enhancement so that relatively strong

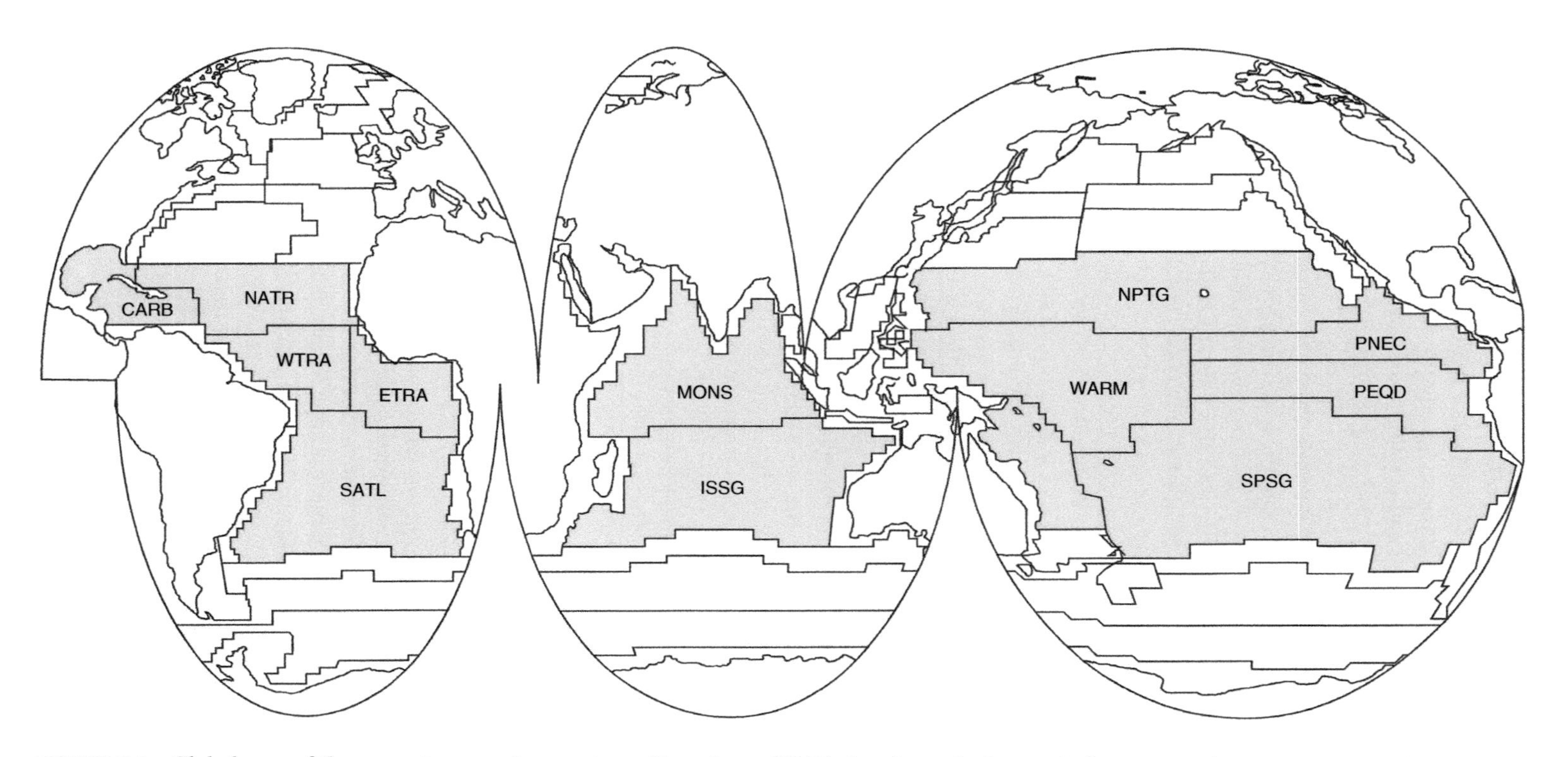

FIGURE 7.9 Global map of the ecosystem province system of Longhurst (2007) showing only the tropical ocean provinces. *Source:* Longhurst (2007), figure 1, p. 543. © Elsevier.

phytoplankton growth occurs in cyclonic eddies and in the confluences of flow between eddy pairs. This phenomenon translates to the highest trophic levels so that whales and other cetaceans are not only concentrated along the shelf break, but also associated with cyclones, and flow confluence between eddy pairs. Such a preferential distribution has also been related to an observed relationship between the distribution of zooplankton biomass and cephalopod larval numbers. Cephalopods are a mainstay of the diets of the sperm whales, pygmy sperm whales, pilot whales, and other cetaceans.

WTRA, the Western Tropical Atlantic Province, comprises the tropical Atlantic west of 15 °W, lying between NATR to the north and SATL to the south (Figure 7.9). The dominant physical processes in this complex province are (i) divergence at the equator that penetrates seasonally into the western half of the ocean, (ii) a zonal field of tropical instability waves in the western ocean that propagate into the North Brazil Current, and (iii) the tropical cyclonic gyre or Guinea Dome in the eastern part of the province. The chlorophyll field responds most strongly to the equatorial divergence, tropical instability waves, and the Guinea Dome. The consequences of equatorial divergence are evident in the chlorophyll field as a band of chlorophyll enhancement symmetrically aligned with the equator for most of the year. Also, water and chlorophyll concentrations having an Amazon signature in salinity and dissolved organic matter is entrained offshore into the open ocean. Mesoscale features move progressively westward, that is, to the flux of Amazon discharge water towards the east. Eddy upwelling results in arcs of high chlorophyll concentration across the province. Tropical instability waves result in a vortex in which there are enhanced concentrations of chlorophyll and primary production as well as unique vertical and horizontal distributions of SST and velocity, nitrate, together with zooplankton and micronekton biomass that have been described as "whirling ecosystems." Upwelling at depth within the vortex may not imply cross-isopycnal flux that would supply new nutrients that may rather be supplied in the poleward surface flow from the equatorial upwelling. The entire region occupied by both the Guinea Dome and the North Equatorial Counter Current (NECC) exhibits enhanced chlorophyll and is an area of rich pelagic fisheries. Depending on the depth of the mixed layer, from almost 100 m in the west to 20–30 m in the Guinea Dome, the chlorophyll and productivity maxima and the subjacent nutricline lie either at the top of the pycnocline or deep within it.

The Guinea Dome exhibits a remarkable stratification of abiotic and biological attributes. The situation is sufficiently stable here that the nitrate-deficient surface layer is significantly deeper than the wind-mixed surface layer. The nutricline is therefore located at some depth within the pycnocline and is associated with very sharp maxima of phytoplankton growth and biomass and of numbers of herbivorous zooplankton. Depths of maximal mesozooplankton are usually close to the depth of maximal algal growth rate, implying a trophic link involving nutrient regeneration by excretion of zooplankton. Progressive shoaling brings oxygen-deficient water to within 75 m of the surface in the Guinea Dome, and this physical feature has important effects on the distribution of mesopelagic fish species. Myctophids dominate the biomass of this component and are represented by species having northern, subtropical, and tropical distributions, whereas other families of mesopelagic fish and some of the deeper-dwelling myctophids are absent. These anomalies may be accounted for by the unusual vertical distribution of temperature in the doming region and by larval retention within the semi-closed circulation south of the Cape Verde Frontal Zone. The mesopelagic fish are likely to be attracted to the relatively high zooplankton biomass in the NECC/Guinea Dome region. It is not surprising that this region is a favoured location for the tuna long-line fishery.

ETRA, the Eastern Tropical Atlantic Province, represents the oceanic Gulf of Guinea to the east of the boundary with WTRA. There is a rather diffuse seasonal enhancement of surface chlorophyll in the ETRA caused by a variety of responses to: (i) wind-driven linear equatorial divergence, with which is associated a chlorophyll enhancement centred upon an episodic, equatorial upwelling, (ii) Ekman upwelling above and around the Angola dome, with which is associated a second, broader zone of chlorophyll enhancement that narrows westward, (iii) the highly pigmented, shallow plume of the Congo River that lies diagonally across the tropical gyre at about 5 °S, and (iv) offshore effects of upwelling processes within the North Equatorial Counter Current-Guinea Current that occurs south of the east-west coast of tropical West Africa and thus along the zonal flow of the Guinea Current. Chlorophyll enhancement near the equator has been associated with the mixed layer. Cell size fractions are uniform with picophytoplankton contributing 80% of biomass in the oligotrophic zones and 60% at the equatorial divergence, suggesting strong top-down control of primary production by herbivores. There are few studies of higher trophic levels in the ETRA. Standing stocks of mesozooplankton are positively correlated with chlorophyll. At upper trophic levels, there is some evidence that ETRA produces more of the two shallower-swimming species of tuna, namely yellowfin and skipjack, than does the deeper and more oligotrophic mixed layer of the WTRA. Conversely, the deeper-swimming bigeye (*Thunnus obesus*) shows less east-west difference in abundance in the tropical Atlantic.

SATL, the South Atlantic Gyral Province, comprises the anticyclonic circulation of the South Atlantic south of the WTRA and ETRA. There is little modern information on the ecology of the open South Atlantic. As occurs in the northern hemisphere, prominent tropical instability waves associated with the equatorial divergence may be observed in the surface chlorophyll along the northern border of the SATL. The phytoplankton turnover rate in the oligotrophic SATL gyre is low representing < 20% of maximal expected growth. An unexpectedly high variability in phytoplankton dynamics has been noted in the oligotrophic gyre so that productivity and growth rates can vary by a factor of 8 and microbial biomass usually remains relatively constant. N supply in oligotrophic gyres is predominantly from the sub-thermocline, probably mostly due to fine scale upwelling in mesoscale eddies and frontal systems. Very striking is the differential distribution of *Prochlorococcus* in the subtropical gyre and of *Synechococcus* and picoeukaryotes, together with heterotrophic bacteria, in the more eutrophic regions north and south. Phytoplankton and mesozooplankton diversity is higher on the western, more dynamic side of the gyre. The major changes in distribution occur across the subtropical convergence zone and that change is weaker between the gyral populations and those within the equatorial zones to the north,

MONS, the Indian Monsoon Gyres province, comprises the northern half of the Indian Ocean (at 10 °S) including much of the Arabian Sea but excluding several coastal provinces (Figure 7.9). The phytoplankton responds strongly to the forcing of the reversing monsoon wind systems, the SW and NE Monsoons. Seasonality of chlorophyll biomass is dominated by the period of clear, blue seawater that succeeds the winter monsoon during April–June. This is the period of maximal solar irradiance, the minimum occurring at the height of the SW Monsoon under cloudy skies.

The Arabian Sea has higher surface chlorophyll biomass than the Bay of Bengal and the strongest response of the phytoplankton is aligned approximately below the Findlater Jet, rather in the NW quadrant of the Arabian Sea than towards its centre.

During the SW Monsoon, the region of higher chlorophyll moves progressively towards the southeast across the Arabian Sea to reach its maximal extent in the inter-monsoon period. Then, the period of minimal biomass occurs during the NE Monsoon. At the onset of the spring inter-monsoon, chlorophyll biomass rapidly decreases so that from April to June the entire offshore Arabian Sea is extremely oligotrophic. The responses of phytoplankton, regardless of season, are highly dynamic and complex across the province. During the SW and NE Monsoons, photosynthetic prokaryotes are responsible for about 25 and 50% of productivity, respectively, and whereas the growth of *Prochlorococcus* is in balance with its consumption by protists, that of *Synechococcus* largely escapes consumption. The nano- and microzooplankton form a relatively constant biomass while mesozooplankton biomass fluctuates more strongly with the seasons and responds to the onset of upwelling. There is an increase of zooplankton and mesopelagic fish during the SW Monsoon compared with the spring inter-monsoon. Heterotrophic nanoflagellate biomass exceeds that of the microheterotroph biomass, dominated by aloricate ciliates and dinoflagellates. These organisms are an order of magnitude more abundant than sarcodines and crustacean nauplii.

During and after the SW Monsoon, in the open gyre, consumption rates of microzooplankton grazing represent up to 20–30% of daily primary production. During the inter-monsoons, daily consumption represents 30–50% of a rather smaller productivity. Mesozooplankton herbivory represents about 25–30% of that of microzooplankton. Copepod mesozooplankton biomass ranges from 10 to 50 mg C m^{-2} and these copepods ingest 10–40 mg C m^{-2} of food daily, representing 10–60% of daily primary production; lowest rates of ingestion/primary production are found during the NE Monsoon. The seasonal response of copepod biomass to autotrophic productivity is rapid, so that in the central gyre during the spring inter-monsoon, this is of the order 20–50 mg C m^{-2}, whereas during the SW Monsoon it increases to 30–200 mg C m^{-2}.

The OMZ at about 200–500 m of the Arabian Sea extends over much of the northern Indian Ocean, suppressing the diel vertical migration of all size classes of zooplankton. Abundant crustacea occur in this zone; pelagic decapod shrimps (*Gennades, Segia, Eupasiphae*) and portunid crabs (*Charybdis smithii*) live largely within the anoxic layer and must have major respiratory modifications. Feeding on mesozooplankton occurs principally within the anoxic layer and is continuous day and night. Populations of pelagic crabs are at times very dense and the species appear to have an annual life cycle in which the pelagic phase is returned passively to the shelf regions for reproduction. Tunas are capable of long-distance migrations and respond to the relative productivity of different parts of the province. Yellowfin, skipjack, and bigeye tuna all avoid the Bay of Bengal, except the area of the coastal current along eastern India and Sri Lanka. The central Arabian Sea is largely avoided, probably because of the OMZ; only yellowfin penetrate the Arabian Sea in large numbers, concentrating mostly in the Oman upwelling zone. On average, catch per unit effort for tuna has been negatively correlated with the IOD with a periodicity centred around four years. During positive IOD events, SSTs are relatively higher, net primary productivity is lower, and catch per unit effort decreases and catch distributions are restricted to the northern and western margins of the western Indian Ocean. During negative IOD events, SSTs are lower and net primary productivity is higher as is catch per unit effort, particularly in the Arabian Sea and seas surrounding Madagascar, and catches expand into central regions of the western Indian Ocean.

ISSG, the Indian South Subtropical Gyre Province, extends from about 10 °S (where there is a hydrochemical front) to the Subtropical Convergence at about 40 °S. Water clarity is persistently high over much of this province for much of the year, and low chlorophyll is common. There are four regions of higher chlorophyll: (i) a zonal region south of the equator across the entire ocean corresponding to the South Equatorial Current (SEC), (ii) in the eddy field to the west of Australia, (iii) to the east of southern Madagascar, and (iv) at mid-ocean locations of shoal water and banks. In the SEC, production/loss balance suggests that coupling between herbivores and phytoplankton is extremely close as accumulation tracks the rise in rate of primary production only very briefly and then declines well before productivity slackens. The seasonal reproductive migrations of southern bluefin tuna carry them into the region of 20 °S to the west of Australia, where they encounter sufficiently high abundance of squid and other large nekton and sufficiently abundant mesozooplankton and nekton to support the growth of their larvae and young-of-the-year. Fishing effort for yellowfin and bigeye tuna continues higher than background in a linear zone across the southern Indian Ocean that corresponds to the SEC. Concentration of fish is highest at about 90–100 °E or at about the major reflux into the interior of the gyre.

NPTG, the North Pacific Tropical Gyre Province, is the equatorward part of the North Pacific central or subtropical gyre lying between the Subtropical Convergence at about 30–32 °N in mid-ocean and 20–25 °N in the west and the northern Doldrum Front at about 10–11 °N. It spans the width of the deep Pacific Ocean. Vertical ecological structure is modified by the development of a seasonal shallow thermocline. The deep, NO_3^--depleted euphotic zone of this province in summer, reaching even to 100 m, lies above a permanent nutricline. The summer deep chlorophyll maximum contains maximum values of around 0.2–0.4 mg chl m^{-3}, whereas the winter deep chlorophyll maximum is a weaker feature, usually about 0.1 mg chl m^{-3}. Primary production is maximal rather shallower in the mixed layer at 30–60 m. Primary production rates increase in summer but the interannual differences are at least as great as seasonal differences.

Profiles of microplankton biomass (mostly protists, monads, flagellates, and naked dinoflagellates) do not have a subsurface maximum corresponding to the deep chlorophyll maximum but rather a broad depth range of relatively high abundance above a deeper zone of lower abundance. Separation of vertical habitat occurs also among mesozooplankton, some species of which specialise in the deep chlorophyll maximum, whereas others occupy the upper mixed layer where primary production is maximal. Grazing by microplankton in the deep chlorophyll maximum (70–120 m) in the western part of the province contributes 60–100% of total consumption and balanced by production within a time scale of a few days. Mesozooplankton contribute only <5% of all consumption; at shallower locations than the deep chlorophyll maximum, these organisms are largely secondary predators, consuming microplankton, though at the deep chlorophyll maximum they contribute to the consumption of algae. The major consumers of the small, dominant phytoplankton remain protists with which the mesozooplankton cannot compete. Small copepods remain abundant and represent 80% of all zooplankton biomass by day and 77% by night. Chaetognaths and larger crustacea (euphausiids, decapods, amphipods) dominate the large size fractions.

PNEC, the North Pacific Equatorial Counter Current Province, lies between the northern and southern Doldrum salinity fronts, around 5–10 °N in the central Pacific, widening along the American continent (Figure 7.9). The flow across the province of the North Equatorial Counter Current (NECC) originates in the Mindanao Dome at the termination of the western boundary current and terminates in the cyclonic flow around the Costa Rica Dome off Central America. A seasonally and annually variable region of chlorophyll enhancement characterises the NECC region and may occasionally be traced clear across the Pacific Ocean to the origin of the NECC. Here, enhancement has been attributed to upwelling associated with current meandering. The surface chlorophyll enhancement in the eastern part of the NECC does not normally extend west of about 110 °W, consistent with the progressive deepening of the mixed layer. Surface chlorophyll is significantly lower in El Niño years. The NECC bloom exists in summer when it is absent farther to the west. In boreal summer, the NECC bloom is truncated at 120 °W but extends westward to about 145 °W.

Divergent upwelling in the Costa Rica Dome may be so strong as to remove the entire mixed layer laterally and so expose the strong tropical thermocline and its associated nutricline at the surface. The Costa Rica Dome is delimited by values higher than background for NO_3^-, productivity and zooplankton biomass as it is for blue whale sightings. The main gradient of the thermocline mixed layer lay at 20–50 m and the Costa Rica Dome at 15–20 m with a maximum concentration of 3.0 mg m^{-3} and primary production is maximal at 5–15 m with a maximal rate of 8.0 mg C m^{-3} h^{-1}.

Copepods are the principal mesozooplankton grazers with their density maxima at the layer of maximal primary production, from the surface to about 50 m, below which layers of diel migrants by day occur at 200–250 m, well above the top of the OMZ. West of the Costa Rica Dome, waters are less productive and mesozooplankton are partitioned among small herbivorous copepods whose highest abundance is just shallower than the deep chlorophyll maximum. Large *Eucalanus* spp. lay mostly below the deep chlorophyll maximum yet within the thermocline and omnivorous copepods and ostracods (*Conchoecia* spp.) lay below the thermocline and well into the upper oxygen-depleted zone. The cladocerans, *Evadne tergestina* and *E. spinifera*, are widely distributed in this province, occurring close to the surface (0–20 m), and are thought to consume bacterioplankton in preference to phytoplankton. Higher trophic-level organisms respond to the distribution of mesozooplankton and hence physical forcing. The distribution of jumbo flying squid (*Dosidocus gigas*) is linked to the area of active upwelling along the ridge of the pycnocline west of the Costa Rica Dome. Accordingly, the squid occur in high abundance and are spatially associated with a well-developed deep chlorophyll maximum, except during La Niña conditions. The strong salinity front that characterises the edge of the NECC retains the squid along the ridge, as well as yellowfin tuna (*Thunnus albacares*) that, like the squid, feed largely on micronekton.

PEQD, the Pacific Equatorial Divergence Province, lies between the Equatorial Convergent Front (ECF) to the north and the irregular flow of the South Equatorial Counter Current to the south. The northern boundary is the southern boundary of the eastward flowing NECC and lies above a thermocline trough at distances north of the equator that increase eastward. The PEQD has a spectacular phytoplankton bloom at the equatorial divergence. Such blooms are thought to occur at the termination of strong ENSO events. This region of the equatorial Pacific has attracted a lot of attention because it plays a critical role in the global carbon cycle. The PEQD is the largest single natural source of atmospheric CO_2 and is also the site of a major fraction of global oceanic primary production and hence of the uptake of CO_2 from the atmosphere (Section 10.9). Integrated primary production rates of 650–950 mg C m^{-2} d^{-1} have been measured from 110 to 170 °W and the biomass of autotroph fractions is relatively constant (μg l^{-1}): *Synechoccus*: 0.9–2.9, eukaryotic picoplankton: 0.5–5.0, haptophytes: 2.1–3.0, dinoflagellates: 3.7–10.7, pennate diatoms: 0.5–11.7, centric diatoms: 0.2–2.7 and ciliates with sequestered chloroplasts: 0.001–0.01. Biomass and production are usually greatest close to the equator, irrespective of El Niño. Productivity is significantly higher during the cool period later in the year than during the El Niño at the beginning of the year. Heterotrophic bacterial biomass (6–8 × 10^8 cells l^{-1}) averages 70% of autotrophic biomass at the equator at 140 °W and although relatively high heterotroph: autotroph ratios are characteristic of oligotrophic water, bacteria respond relatively slowly to upwelling, causing persistent uncoupling autotrophs. The major fluxes in the pelagic ecosystem are dominated by small cells of the microbial loop. With the small autotrophic cells occur protistan grazers. There is a tightly coupled production/consumption system in which the population size of the grazers can respond as fast as that of the autotrophs. Grazing by microzooplankton can remove the entire daily production of picophytoplankton. Mesozooplankton also heavily graze smaller cells, estimated to be <5% daily of the standing stock of autotrophs, implying that mesozooplankton largely consume microzooplankton and detritus to support their growth. More than 80% of the carbon ingested by mesozooplankton is not phytoplankton yet may represent removal of >27% of the biomass of large diatoms daily.

WARM, the Western Pacific Warm Pool Province, lies under the low-pressure region of the atmospheric Walker circulation, that is, from the date line to the western boundary current that flows NE-SW along the shelf edge on the eastern side of the Indo-Pacific archipelago (Figure 7.9). Heavy rainfall leads to an excess of precipitation over evaporation so that a lens of warm, brackish surface water is formed, to be further diluted by the eastward advection of low-salinity surface water from the Indonesian archipelago. It is in this province that the westerly wind anomalies that mark the onset of the El Niño event first develop and, during the strong negative anomalies in the Southern Oscillation, conditions characteristic of the WARM province may come to lie so far east that it is convenient to evoke the concept of a trans-Pacific WARM-ENSO province.

There are two regional blooms within the WARM province. The most important of these responds to divergence at the equator and narrows progressively westward towards the northern coast of New Guinea. The other bloom is farther to the west and much smaller, a meandering narrow chlorophyll enhancement associated with the northern edge of the NECC where this begins its flow eastward across the ocean. This has been described as a "ribbon of dark water" in the western ocean and is especially well developed during El Niño periods, when it includes higher chlorophyll concentrations than at other times. The chlorophyll enhancement has been attributed to: (i) upwelling associated with meandering, (ii) seasonal Ekman pumping, and (iii) the interannual differences in nutricline depth associated with ENSO events. Most primary production (≈75%) occurs within the deep chlorophyll maximum under low illumination. During El Niño events, the principal change is the progressive shoaling of both the thermocline and nutricline with a consequent shift in the depth of greatest production.

A deep euphotic zone generally comprises two ecologically distinct depth strata: an upper nitrate-limited, cyanobacteria-dominated zone and a deeper light-limited zone dominated by eukaryotic microalgae. The picophytoplankton are dominated by *Prochlorococcus* and *Synechococcus* and, when transient nutrient flux induces increased productivity, it is the latter that shows a small increase in abundance along with diatoms and chlorophytes. There is a strong zonal discontinuity in mesoplankton biomass at around 170 °W, at the eastern boundary of the province. Biomass to the west of this discontinuity is about one-third of that in PEQD to the east. Zooplankton biomass lies deeper, and there is a greater distinction in distribution between microzooplankton and the deeper mesozooplankton. Several genera of euphausiids migrate between great depths of around 400–500 m by day and the deep chlorophyll maximum by night.

The boundaries of the WARM province enclose the region of greatest Pacific abundance of skipjack (*Katsuwonus pelamis*) and yellowfin (*Thunnus albacares*) tuna which are characteristic of the trade wind zone. For both these species as well as bigeye tuna (*Thunnus obesus*), the greatest concentration of larvae occurs in WARM, though they are also widely distributed across the other trade wind biome provinces. It is unclear why such an oligotrophic province should house such high concentrations of tuna. Longhurst (2007) suggested that the answer lies in the unique character of this province and its extraordinarily strongly stratified water column with a boundary layer. This feature may serve to aggregate layers of food organisms or perhaps the very stable water column provides "invariant and predictable" conditions for feeding by tuna larvae. Modelling exercises predict that climate change in this province will result in bigeye tuna populations declining and shifting poleward (Section 13.7), whereas skipjack and yellowfin tuna will become more abundant (Erauskin-Extramiana et al. 2019).

ARCH, the Archipelagic Deep Basins Province, comprises the deep basins of the Indonesian archipelago lying between the Indian and Pacific Oceans, together with the South China Sea and the Coral Sea. This province also comprises the Sulu, Celebes, Molucca, Flores, Bismarck, and Solomon Seas. The circulation of this province may be divided into three components: (i) the South China Sea, (ii) the archipelagic seas and (iii) the Coral Sea. The East Asian monsoon system is the dominant feature; the intertropical convergence zone (ITCZ) between the easterly winds of the northern and southern hemisphere lies seasonally on each side of the equator. The pelagic ecosystem here responds to the strong seasonality of the reversing monsoon winds with winter phytoplankton blooms in each hemisphere. Surface chlorophyll is maximum in the Coral Sea and in adjacent basins in the smaller archipelagic seas the seasonal signal is not so clear. The Coral Sea seasonal bloom extends westward through the Banda and Flores Seas, while the Sulu and Celebes Seas follow the same pattern as the South China Sea. This pattern is a simple response to the seasonal deepening of the mixed layer in the two hemispheres and the consequent flux of nutrients into the euphotic zone. Island wakes are commonly observed in this province and uplift of the nutrocline in island wakes induces enhanced production of phytoplankton and microzooplankton. Most of the primary production occurs shallower than the midday 20% isolume and little is lost to sinking; there is a balance between production and consumption.

The Coral Sea and the other marginal seas share other ecological characteristics with the oligotrophic South Pacific, including the persistent occurrence of blooms of N_2-fixing cyanobacteria. Phytoplankton of the Coral Sea are dominated by the pico-size fraction (<2 μm), although quite high productivity (1–3 g C m^{-2} d^{-1}) has been measured adjacent to the continental margins. In Indonesian seas, chlorophyll biomass at the surface is enhanced during the southeast monsoon from 0.25 to 2.5 mg m^{-3} above levels occurring during the NE monsoon. During ENSO and positive IOD, upwelling strength increases. This seasonal buildup of biomass, associated with an increase in the rate of primary production, is greatest in the eastern seas (Banda, Flores and Seram) and decreases westward through the archipelago, becoming a very slight effect in the Sulawesi Sea and the Makassar Strait.

Downwelling–upwelling events have been observed in the Banda Sea with predictable consequences for surface nitrate and seasonal primary production. The strength of upwelling decreases during La Niña and negative IOD. The degree of seasonal chlorophyll enhancement throughout the archipelago seems to be correlated with SSTs and hence the relative strength of the seasonal upwelling. In the Banda Sea, chlorophyll is five times higher in austral winter during the SE monsoon. In August during the SE monsoon (upwelling and low irradiance), the surface layer is homogeneous down to 60 m and primary production is highest at the surface, whereas in February, under the influence of the northerly monsoon (downwelling and high irradiance), stratification is established and a deep chlorophyll maximum forms at 40–80 m. At these depths, *Synechococcus* occur at high concentrations. Primary productivity is higher during the southerly than the northerly monsoon (1.85 and 0.9 g C m^{-2} d^{-1}, respectively).

Upwelling occurs along the south coasts of most the islands of the archipelago, including along the south coast of the island of New Guinea. The mesoplankton biomass of the Banda Sea almost doubles at the onset of the upwelling, when chlorophyll accumulation reaches 3–4 mg m^{-3} due to the rise from depth of large populations of *Calanoides philippinensis* and *Rhincalanus nasutus*, which reach an abundance of 4000 ind. m^{-2}, grazing on phytoplankton and microzooplankton. Tropical mesozooplankton herbivores account for only a small part of primary production (5–25% daily) or of standing stock (2–6% daily). Micronekton biomass responds similarly but with smaller magnitude; there is a difference of only about 50% between the two monsoon seasons.

The South China Sea is less oligotrophic than the adjacent waters of the western Pacific with surface chlorophyll almost twice that of the western Pacific. Physical processes forced by the reversing monsoon winds control primary production in the South China Sea, where there are three zones of upwelling: off the east coast of Vietnam, to the east of the Philippine island of Luzon, and along the northern edge of the Sunda Shelf. All upwelling zones have significant tuna fisheries and significant cetacean sightings.

SPSG, the South Pacific Subtropical Gyre Province, comprises the central and southern part of the subtropical gyre of the South Pacific Ocean (Figure 7.9). A well-defined winter maximum occurs in August–September and the rate of seasonal change in chlorophyll values follows very closely the seasonal change in environmental forcing; here, equilibrium between irradiance, mixing, nutrient flux, and growth appears to be constantly maintained. The peripheral regions of the province are everywhere adjacent to areas of higher productivity. This province is in a state of permanent oligotrophy, and there is a weak winter maximum accompanied by unusually high numbers of picoeukaryotes. There is a general dominance of cyanobacteria and an extensive occurrence of *Trichodesmium* blooms, especially in the SW of the province and especially in summer. A subsurface maximum chlorophyll layer lies just shallower than the nitracline and on the upper part of the density gradient acting to trap any NO_3^- mixed up across the upper pycnocline. Maximum primary production likely occurs much shallower and N_2-fixing organisms are prominent; the microbial loop is highly active and mediated by bacterioplankton, pico-fraction cyanobacteria, prochlorophytes, and microflagellates. Most consumption is by a complex protist community. Mesozooplankton are small, except for diel migrant metridiids, which rise at night, some to the deep chlorophyll maximum and some to graze on protists in the mixed layer. The distribution of tuna is somewhat contrary, with highest abundance being concentrated marginally along the northern and southern reaches of the province, adjacent to more productive regions.

References

Aerts, K., Vanagt, T., Degraer, S. et al. (2004). Macrofaunal community structure and zonation of an Ecuadorian sandy beach (Bay of Valdivia). *Belgian Journal of Zoology* 134: 17–24.

Aké-Castillo, J.A. and Vázquez, G. (2008). Phytoplankton variation and its relation to nutrients and allochthonous organic matter in a coastal lagoon on the Gulf of Mexico. *Estuarine, Coastal and Shelf Science* 78: 705–714.

Albaret, J.J. and Laé, R. (2003). Impact of fishing on fish assemblages in tropical lagoons: the example of the Ebrié lagoon, west Africa. *Aquatic Living Resources* 16: 1–9.
Aller, J.Y. (1995). Molluscan death assemblages on the Amazon shelf: implication for physical and biological controls on benthic populations. *Palaeogeography, Palaeoclimatology and Palaeoecology* 118: 181–200.
Aller, J.Y. and Aller, R.C. (1986). General characteristics of benthic faunas on the Amazon inner continental shelf with comparison to the shelf off the Changjiang River, East China Sea. *Continental Shelf Research* 6: 291–310.
Aller, J.Y. and Aller, R.C. (2004). Physical disturbance creates bacterial dominance of benthic biological communities in tropical deltaic environments of the Gulf of Papua. *Continental Shelf Research* 24: 2395–2416.
Aller, J.Y. and Stupakoff, I. (1996). The distribution and seasonal characteristics of benthic communities on the Amazon shelf as indicators of physical processes. *Continental Shelf Research* 16: 717–737.
Aller, J.Y., Alongi, D.M., and Aller, R.C. (2008a). Biological indicators of sedimentary dynamics in the central Gulf of Papua: seasonal and decadal perspectives. *Journal of Geophysical Research: Earth Science* 113: F01S08. https://doi.org/10.1029/2007JF000823.
Aller, R.C., Blair, N.E., and Brunskill, G.J. (2008b). Early diagenetic cycling, incineration, and burial of sedimentary organic carbon in the central Gulf of Papua (Papua New Guinea). *Journal of Geophysical Research, Earth Science* 113: F10S09. https://doi.org/10.1029/2006JF000689.
Alongi, D.M. (1990). The ecology of tropical soft-bottom benthic ecosystems. *Oceanography and Marine Biology: An Annual Review* 28: 381–496.
Alongi, D.M. (1998). *Coastal Ecosystem Processes*. Boca Raton, FL: CRC Press.
Alongi, D.M. (2005). Ecosystem types and processes. In: *The Sea, Volume 13, The Global Coastal Ocean, Multiscale Interdisciplinary Processes* (eds. A.R. Robinson and K. Brink), 317–352. Cambridge, MA: Harvard.
Alongi, D.M. (2008). Mangrove forests, resilience, protection from tsunamis, and responses to global climate change. *Estuarine, Coastal and Shelf Science* 76: 1–13.
Alongi, D.M. (2009). *The Energetics of Mangrove Forests. Dordrecht*, The Netherlands: Springer.
Alongi, D.M. and Robertson, A.I. (1995). Factors regulating benthic food chains in tropical river deltas and adjacent shelf areas. *Geo-Marine Letters* 15: 145–154.
Alongi, D.M., Christoffersen, P., Tirendi, F. et al. (1992). The influence of freshwater and material export on sedimentary facies and benthic processes within the Fly delta and adjacent Gulf of Papua. *Continental Shelf Research* 12: 287–303.
Alongi, D.M., Trott, L.A., and Møhl, M. (2011). Strong tidal currents and labile organic matter stimulate benthic decomposition and carbonate fluxes on the southern Great Barrier Reef. *Continental Shelf Research* 31: 1384–1395.
Alongi, D.M., Patten, N.L., McKinnon, D. et al. (2015). Phytoplankton, bacterioplankton and virioplankton structure and function across the southern Great Barrier Reef shelf. *Journal of Marine Systems* 142: 25–39.
Alvarez-Cadena, J.N. and Segura-Puertas, L. (1997). Zooplankton variability and copepod species assemblages from a tropical coastal lagoon. *Gulf Research Reports* 9: 345–355.
Ambriz-Arreola, I., Gómez-Gutiérrez, J., del Carmen Franco-Gordo, M. et al. (2018). Seasonal succession of tropical community structure, abundance, and biomass of five zooplankton taxa in the central Mexican Pacific. *Continental Shelf Research* 168: 54–67.
Andrades, R., Gomes, M.P., Pereira-Filho, G.H. et al. (2014). The influence of allochthonous macroalgae on the fish communities of tropical sandy beaches. *Estuarine, Coastal and Shelf Science* 144: 75–81.
Ansari, Z.A., Furtado, R., Badesab, P. et al. (2012). Benthic macroinvertebrate community structure and distribution in the Ayeyarwady continental shelf, Andaman Sea. *Indian Journal of Geo-Marine Sciences* 41: 272–278.
Arceo-Carranza, D. and Vega-Cendejas, M.E. (2009). Spatial and temporal characterization of fish assemblages in a tropical coastal system influenced by freshwater inputs, NW Yucatan peninsula. *Revista de Biología Tropical* 57: 89–103.

Aschenbroich, A., Michaud, E., Steiglitz, T. et al. (2016). Brachyuran crab community structure and associated sediment reworking activities in pioneer and young mangroves of French Guiana, South America. *Estuarine, Coastal and Shelf Science* 182: 60–71.
Aschenbroich, A., Michaud, E., Gilbert, F. et al. (2017). Bioturbation functional roles associated with mangrove development in French Guiana, South America. *Hydrobiologia* 794: 179–202.
Azam, F. and Malfatti, F. (2007). Microbial structuring of marine ecosystems. *Nature Reviews Microbiology* 5: 782–791.
Bancroft, G.T., Strong, A.M., and Carrington, M. (1995). Deforestation and its effects on forest-nesting birds in the Florida Keys. *Conservation Biology* 9: 835–844.
Barboza, F.R. and Defeo, O. (2015). Global diversity patterns in sandy beach macrofauna: a biogeographic analysis. *Scientific Reports* 5: 14515. https://doi.org/10.1038/srep14515.
Bar-Zeev, E., Berman-Frank, I., Stambler, N. et al. (2009). Transparent exopolymer particles (TEP) link phytoplankton and bacterial production in the Gulf of Aqaba. *Aquatic Microbial Ecology* 56: 217–225.
Berger, U. and Hildenbrandt, H. (2000). A new approach to spatially explicit modelling of forest dynamics, spacing, aging and neighbourhood competition of mangrove trees. *Ecological Modelling* 132: 287–302.
Berger, U. and Hildenbrandt, H. (2003). The strength of competition among individual trees and the biomass-density trajectories of the cohort. *Plant Ecology* 167: 89–96.
Birtles, R.A. and Arnold, P.W. (1988). Distribution of trophic groups of epifaunal echinoderms and molluscs in the soft sediment areas of the central Great Barrier Reef shelf. *Proceedings of the Sixth International Coral Reef Symposium*, Townsville, Australia (8–12 August 1988), vol. 6, pp. 325–332.
Blaber, S.J.M. (2002). 'Fish in hot water': the challenges facing fish and fisheries in tropical estuaries. *Journal of Fish Biology* 61: 1–20.
de Boer, W.F. (2007). Seagrass-sediment interactions, positive feedbacks and critical thresholds for occurrence: a review. *Hydrobiologia* 591: 5–24.
Bomer, E.J., Wilson, C.A., Hale, R.P. et al. (2020). Surface elevation and sedimentation dynamics in the Ganges-Brahmaputra tidal delta plain, Bangladesh: evidence for mangrove adaptation to human-induced tidal amplification. *Catena* 187: 104312. https://doi.org/10.1016/j.catena.2019.104312.
Breitbart, M., Middelboe, M., and Rohwer, F. (2008). Marine viruses, community dynamics, diversity, and impact on microbial processes. In: *Microbial Ecology of the Oceans* (ed. D.L. Kirchman), 443–480. New York: Wiley.
Buchanan, J.B. (1958). The bottom fauna across the continental shelf off Accra, Gold Coast. *Proceedings of the Zoological Society of London* 130: 1–56.
Calbet, A. and Landry, M.R. (2004). Phytoplankton growth, microzooplankton grazing, and carbon cycling in marine systems. *Limnology and Oceanography* 49: 51–59.
Campanyà-Llovet, N., Snelgrove, P.V., and Parrish, C.C. (2017). Rethinking the importance of food quality in marine benthic food webs. *Progress in Oceanography* 156: 240–251.
Capone, D.G., Dunham, S.G., Horrigan, S.G. et al. (1992). Microbial nitrogen transformations in shallow, unconsolidated carbonate sediments. *Marine Ecology Progress Series* 80: 75–88.
Cardoso, R.S. and Defeo, O. (2003). Geographical patterns in reproductive biology of the Pan-American sandy beach isopod *Excirolana braziliensis*. *Marine Biology* 143: 573–581.
Cardoso, R.S. and Defeo, O. (2004). Biogeographic patterns in life history traits of the Pan-American sandy beach isopod *Excirolana braziliensis*. *Estuarine, Coastal and Shelf Science* 61: 559–568.
Caswell, B.A., Dissanayake, N.G., and Frid, C.L.J. (2020). Influence of climate-induced biogeographic range shifts on mud flat ecological functioning in the subtropics. *Estuarine. Coastal and Shelf Science* 237: 106692. https://doi.org/10.1016/j.ecss.2020.106692.
Chagas, G.G. and Suzuki, M.S. (2005). Seasonal hydrochemical variation in a tropical coastal lagoon (Açu Lagoon, Brazil). *Brazilian Journal of Biology* 65: 597–607.

Chan, B.K.K. (2006). Ecology and biodiversity of rocky intertidal barnacles along a latitudinal gradient, Japan, Taiwan and Hong Kong. *Publications of the Seto Marine Biological Laboratory, Special Publication Series* 8: 1–10.
Cordero-Oviedo, C., Correa-Metrio, A., Urrego, L.E. et al. (2019). Holocene establishment of mangrove forests in the western coast of the Gulf of Mexico. *Catena* 180: 212–223.
Crase, B., Vesk, P.A., Liedloff, A. et al. (2015). Modelling both dominance and species distribution provides a more complete picture of changes to mangrove ecosystems under climate change. *Global Change Biology* 21: 3005–3021.
Crosbie, N.D. and Furnas, M.J. (2001). Net growth rates of picocyanobacteria and nano-/microphytoplankton inhabiting shelf waters of the central (17° S) and southern (20° S) Great Barrier Reef. *Aquatic Microbial Ecology* 24: 209–224.
Danin, A.P.F., Pombo, M., Martinelli-Lemos, J.M. et al. (2020). Population ecology of the hermit crab *Clibanarius symmetricus* (Anomura: Diogenidae) on an exposed beach of the Brazilian Amazon. *Regional Studies in Marine Science* 33: 100944. https://doi.org/10.1016/j.rsma.2019.100944.
Defeo, O. and Cardoso, R. (2002). Macroecology of population dynamics and life history traits of the mole crab *Emerita brasiliensis* in Atlantic sandy beaches of South America. *Marine Ecology Progress Series* 239: 169–179.
Defeo, O. and Cardoso, R. (2004). Latitudinal patterns in abundance and life-history traits of the mole crab *Emerita brasiliensis* on South American sandy beaches. *Diversity and Distribution* 10: 89–98.
Defeo, O. and McLachlan, A. (2005). Patterns, processes and regulatory mechanisms in sandy beach macrofauna: a multi-scale analysis. *Marine Ecology Progress Series* 295: 1–20.
Defeo, O. and McLachlan, A. (2013). Global patterns in sandy beach macrofauna, species richness, abundance, biomass and body size. *Geomorphology* 199: 106–114.
Dinh, Q.M., Tran, L.T., Tran, T.T.M. et al. (2020). Variation in diet composition of the mudskipper, *Periophthalmodon septemradiatus*, from Hau River, Vietnam. *Bulletin of Marine Science* 96: 487–500.
Dittmann, S. (2002). Benthic fauna in tropical tidal flats – a comparative perspective. *Wetlands Ecology and Management* 10: 189–195.
Duarte, C.M. (1989). Temporal biomass variability and production/biomass relationships of seagrass communities. *Marine Ecology Progress Series* 51: 269–279.
Duarte, C.M., Merino, M., and Gallegos, M. (1995). Evidence of iron deficiency in seagrasses growing above carbonate sediments. *Limnology and Oceanography* 40: 1153–1158.
Ducklow, H.W. (2000). Bacterial production and biomass in the oceans. In: *Microbial Ecology of the Oceans* (ed. D.L. Kirchman), 85–120. New York: Wiley.
Ecoutin, J.M., Richard, E., Simier, M. et al. (2003). Spatial versus temporal patterns in fish assemblages of a tropical estuarine coastal lake, The Ebrié Lagoon (Ivory Coast). *Estuarine, Coastal and Shelf Science* 64: 623–635.
Erauskin-Extramiana, M., Arrizabalaga, H., Hobday, A.J. et al. (2019). Large-scale distribution of tuna species in a warming ocean. *Global Change Biology* 25: 2043–2060.
Erftemeijer, P.L.A., Stapel, J., Smekens, M.J.E. et al. (1994). The limited effect of *in situ* phosphorus and nitrogen additions to seagrass beds on carbonate and terrigenous sediments in south Sulawesi, Indonesia. *Journal of Experimental Marine Biology and Ecology* 182: 123–135.
Farjalla, V.F., Faria, B.M., and Esteves, F.A. (2002). The relationship between DOC and planktonic bacteria in tropical coastal lagoons. *Archiv fur Hydrobiologie* 156: 97–119.
Faunce, C.H. and Serafy, J.E. (2006). Mangroves as fish habitat: 50 years of field studies. *Marine Ecology Progress Series* 318: 1–18.
Feller, I.C., Lovelock, C.E., Berger, U. et al. (2010). Biocomplexity in mangrove ecosystems. *Annual Review of Marine Science* 2: 395–417.
Fernandes, V., Ramaiah, N., Sardessai, S. et al. (2008). Strong variability in bacterioplankton abundance and production in central and western Bay of Bengal. *Marine Biology* 153.
Filho, J.S.R., Gomes, T.P., Almeida, M.F. et al. (2011). Benthic fauna of macrotidal sandy beaches along a small-scale morphodynamic gradient on the Amazon coast (Algodoal Island, Brazil). *Journal of Coastal Research* 64: 435–439.

Fitzpatrick, B.M., Harvey, E.S., and Heyward, A.J. (2012). Habitat specialization in tropical continental shelf demersal fish assemblages. *PloS One* 7: e39634. https://doi.org/10.1371/journal.pone.0039634.
Friess, D.A. (2017). J.G. Watson, inundation classes, and their influence on paradigms in mangrove forest ecology. *Wetlands* 37: 603–613.
Friess, D.A., Rogers, K., Lovelock, C.E. et al. (2019). The state of the world's mangrove forests: past, present, and future. *Annual Review of Environment and Resources* 44: 89–115.
Fromard, F., Puig, H., Mougin, E. et al. (1998). Structure, above-ground biomass and dynamics of mangrove ecosystems: new data from French Guiana. *Oecologia* 115: 39–53.
Fromard, F., Vega, C., and Proisy, C. (2004). Half a century of dynamic coastal change affecting mangrove shorelines of French Guiana. A case study based on remote sensing data analyses and field surveys. *Marine Geology* 208: 265–280.
Gattuso, J.-P., Pichon, M., Jaubert, J. et al. (1996). Primary production, calcification and air-sea CO_2 fluxes in coral reefs: organism, ecosystem and global scales. *Bulletin de l'Institut Oceanographique Monaco* 14: 39–46.
Gauns, M., Mochemadkar, S., Pratihary, A. et al. (2020). Phytoplankton associated with seasonal oxygen depletion in waters of the western continental shelf of India. *Journal of Marine Systems* 204: 103308. https://doi.org/10.1016/j.marsys.2020.103308.
Glud, R.N., Eyre, B.D., and Patten, N. (2008). Biogeochemical responses to mass coral spawning at the Great Barrier Reef: effects on respiration and primary production. *Limnology and Oceanography* 53: 1014–1024.
Gocke, K., Hernandez, C., Giesenhagen, H. et al. (2004). Seasonal variations of bacterial abundance and biomass and their relation to phytoplankton in the hypertrophic tropical lagoon Ciénaga Grande de Santa Marta, Colombia. *Journal of Plankton Research* 26: 1429–1439.
Gray, J.S. and Elliot, M. (2009). *Ecology of Marine Sediments, 2e*. Oxford, UK: Oxford University Press.
Hacking, N. (2007). Effects of physical state and latitude on sandy beach macrofauna of eastern and southern Australia. *Journal of Coastal Research* 23: 899–910.
Hall, L.M., Hanisak, M.D., and Virnstein, R.W. (2006). Fragments of the seagrasses *Halodule wrightii* and *Halophila johnsonii* as potential recruits in Indian River Lagoon, Florida. *Marine Ecology Progress Series* 310: 109–117.
Hanamura, Y., Stow, R., Chee, P.E. et al. (2009). Seasonality and biological characteristics of the shallow-water mysid *Mesopodopsis orientalis* (Crustacea, Mysida) on a tropical sandy beach, Malaysia. *Plankton and Benthos Research* 4: 53–61.
Harada, Y., Fry, B., Lee, S.Y. et al. (2020). Stable isotopes indicate ecosystem restructuring following climate-driven mangrove dieback. *Limnology and Oceanography* 65: 1251–1263.
Harkantra, S.N. and Parulekar, A.H. (1987). Benthos off Cochin, SW coast of India. *Indian Journal of Marine Sciences* 16: 57–59.
Hatcher, B.G. (1988). Coral reef primary productivity: a beggar's banquet. *Trends in Ecology and Evolution* 3: 106–111.
Hatcher, B.G. (1990). Coral reef primary productivity: a hierarchy of pattern and process. *Trends in Ecology and Evolution* 5: 149–152.
Hedgpeth, J.W. (1955). Bottom communities of the Gulf of Mexico. *United States Fish and Wildlife Service Fisheries Bulletin* 55: 203–214.
Hemminga, M.A. and Duarte, C.M. (2000). *Seagrass Ecology*. Cambridge University Press: Cambridge, UK.
Heng, P.L., Lim, J.H., and Lee, C.W. (2017). *Synechococcus* production and grazing loss rates in nearshore tropical waters. *Environmental Monitoring and Assessment* 189: 117. https://doi.org/10.1007/s10661-017-5838-1.
Hernández-Guevara, N.A., Pech, D., and Ardisson, P.L. (2008). Temporal trends in benthic macrofauna composition in response to seasonal variation in a tropical coastal lagoon, Celestun, Gulf of Mexico. *Marine and Freshwater Research* 59: 772–779. https://doi.org/10.1071/MF07189.
Herrera, A. and Bone, D. (2011). Influence of riverine outputs on sandy beaches of Higuerote, central coast of Venezuela. *Latin American Journal of Aquatic Research* 39: 56–70.

Herrera, I., Yebra, L., Antezana, T. et al. (2019). Vertical variability of *Euphausia distinguenda* metabolic rates during diel migration into the oxygen minimum zone of the eastern tropical Pacific off Mexico. *Journal of Plankton Research* 41: 165–176.

Hogarth, P.J. (2015). *The Biology of Mangroves and Seagrasses*. Oxford, UK: Oxford University Press.

Hutchinson, N. and Williams, G.A. (2003). Disturbance and subsequent recovery of mid-shore assemblages on seasonal, tropical, rocky shores. *Marine Ecology Progress Series* 249: 25–38.

Jourde, J., Christine, D., Nguyen, H. et al. (2017). Low benthic macrofauna diversity in dynamic, tropical tidal mud flats: migrating banks on Guiana's coast, South America. *Estuaries and Coasts* 40: 1159–1170.

Jyothibabu, R., Balachandran, K.K., Jagadeesan, L. et al. (2018). Mud banks along the SW coast of India are not too muddy for plankton. *Scientific Reports* 8: 2544. https://doi.org/10.1038/s41598-018-20667-9.

Kaehler, S. and Williams, G.A. (1996). Distribution of algae on tropical rocky shores: spatial and temporal patterns of non-coralline encrusting algae in Hong Kong. *Marine Biology* 125: 177–187.

Kaniewska, P., Alon, S., Karako-Lampert, S. et al. (2015). Signaling cascades and the importance of moonlight in coral broadcast mass spawning. *eLife* 4: e09991. https://doi.org/10.7554/eLife.09991.

Kathiresan, K. and Bingham, B.L. (2001). Biology of mangroves and mangrove ecosystems. *Advances in Marine Biology* 40: 81–251.

Kimirei, I.A., Nagelkerken, I., Mgaya, Y.D. et al. (2013). The mangrove nursery paradigm revisited: otolith stable isotopes support nursery-to-reef movements by Indo-Pacific fish. *PLoS One* 8: e66320. https://doi.org/10.1371/journal.pone.0066320.

Kinsey, D.W. (1985). Open-flow systems. In: *Handbook of Phycological Methods*, 4. Ecological Field Studies, Macroalgae (eds. M.M. Littler and D.S. Littler), 427–460. Cambridge, UK: Cambridge University Press.

Kouadio, K.N., Diomandé, D., Koné, Y.J.M. et al. (2011). Distribution of benthic macroinvertebrate communities in relation to environmental factors in the Ebrié Lagoon (Ivory Coast, West Africa). *Vie et Milieu* 61: 59–69.

Krauss, K.W. and Osland, M.J. (2020). Tropical cyclones and the organization of mangrove forests: a review. *Annals of Botany* 125: 213–234.

Krauss, K.W., Lovelock, C.E., McKee, K.L. et al. (2008). Environmental drivers in mangrove establishment and early development: a review. *Aquatic Botany* 89: 105–127.

Lamptey, E. and Armah, A.K. (2008). Factors affecting microbenthic fauna in a tropical hypersaline coastal lagoon in Ghana, west Africa. *Estuaries and Coasts* 31: 1006–1019.

Lastra, M., Dugan, J.E., and Hubbard, D.M. (2002). Burrowing and swash behaviour of the Pacific mole crab *Hippa pacifica* (Anomura, Hippidae) in tropical sandy beaches. *Journal of Crustacean Biology* 22: 53–58.

Laudien, J., Flint, N.S., van der Bank, F.H. et al. (2003). Genetic and morphological variation in four populations of the surf clam *Donax serra* (Roding) from southern African sandy beaches. *Biochemical Systematics and Ecology* 31: 751–772.

Le, D.Q., Fui, S.Y., Tanaka, K. et al. (2020). Feeding habitats of juvenile reef fish in a tropical mangrove-seagrass continuum along a Malaysian shallow-water coastal lagoon. *Bulletin of Marine Science* 96: 469–486.

León-Morales, R. and Vargas, J.A. (1998). Macroinfauna of a tropical fjord-like embayment: Golfo Dulce, Costa Rica. *Revista de Biología Tropical* 46: 81–90.

Lewis, J.B. (1965). A preliminary description of some marine benthic communities from Barbados, West Indies. *Canadian Journal of Zoology* 43: 1049–1074.

Lirman, D. and Cropper, W.P. (2003). The influence of salinity on seagrass growth, survivorship, and distribution within Biscayne Bay, Florida: field, experimental, and modelling studies. *Estuaries* 26: 131–141.

Longhurst, A. (1959). Benthos densities off tropical west Africa. *Journal Conseil International pour l' Exploration de la Mer* 25: 21–28.

Longhurst, A.L. (2007). *Ecological Geography of the Sea, 2e*. London: Academic Press.
Longhurst, A. and Pauly, D. (1987). *Ecology of Tropical Oceans*. San Diego, USA: Academic Press.
Lovelock, C.E. and Ball, M.C. (2002). Influence of salinity on photosynthesis of halophytes. In: *Salinity, Environment-Plants-Molecules* (eds. A. Lauchli and U. Lüttge), 315–339. Utrecht, The Netherlands: Kluwer.
Lubchenco, J., Menge, B.A., Garrity, S.D. et al. (1984). Structure, persistence, and role of consumers in a tropical rocky intertidal community (Taboguilla Island, Bay of Panama). *Journal of Experimental Marine Biology and Ecology* 78: 23–73.
Lugo, A.E. (1980). Mangrove ecosystems: successional or steady state? *Biotropica* 12: 65–72.
Lugo, A.E. (1997). Old-growth mangrove forests in the United States. *Conservation Biology* 11: 11–20.
MacKenzie, K.M., Robertson, D.R., Adams, J.N. et al. (2019). Structure and nutrient transfer in a tropical pelagic upwelling food web: from isoscapes to the whole ecosystem. *Progress in Oceanography* 178: 102145. https://doi.org/10.1016/j.pocean.2019.102145.
Manokaran, S., Khan, S.A., and Lyla, P.S. (2014). Macrobenthic composition of the southeast continental shelf of India. *Marine Ecology* 36: 1–15.
Manson, R.A., Loneragan, N.R., Skilleter, G.A. et al. (2005). An evaluation of the evidence for linkages between mangroves and fisheries: a synthesis of the literature and identification of research directions. *Oceanography and Marine Biology: An Annual Review* 43: 483–513.
Marley, G.A., Deacon, A.E., Phillip, D.A.T. et al. (2020). Mangrove or mud flat: prioritising fish habitat for conservation in a turbid tropical estuary. *Estuarine, Coastal and Shelf Science* 240: 106788. https://doi.org/10.1016/j.ecss.2020.106788.
Mathew, K.J. and Gopinathan, C.P. (2000). The study of mud banks of the Kerala coast- a retrospect. In: *Marine Fisheries, Research and Management* (eds. V.N. Pillai and N.G. Menon), 177–188. Kerala, India: Central Marine Fisheries Research Institute.
Mayakun, J., Prathep, A., and Kim, J.H. (2020). Timing of disturbance, top-down, and bottom-up driving on early algal succession patterns in a tropical intertidal community. *Phycological Research* 68: 135–143.
McKinnon, A.D., Duggan, S., Holliday, D. et al. (2015). Plankton community structure and connectivity in the Kimberley-Browse region of NW Australia. *Estuarine, Coastal and Shelf Science* 153: 156–167.
McLachlan, A. and Defeo, O. (2018). *The Ecology of Sandy Shores, 3e*. London: Academic Press.
McManus, G.B., Costas, B.A., Dam, H.G. et al. (2007). Microzooplankton grazing of phytoplankton in a tropical upwelling region. *Hydrobiologia* 575: 69–81.
Melo, S., Bozelli, R.L., and Esteves, F.A. (2007). Temporal and spatial fluctuations of phytoplankton in a tropical coastal lagoon, southeast Brazil. *Brazilian Journal of Biology* 67: 475–483.
Menge, B.A. (1991). Relative importance of recruitment and other causes of variation in rocky intertidal community structure. *Journal of Experimental Marine Biology and Ecology* 146: 69–100.
Menge, B.A. and Lubchenco, J. (1981). Community organization in temperate and tropical rocky intertidal habitats: prey refuges in relation to consumer pressure gradients. *Ecological Monographs* 51: 429–450.
Mohapatra, M., Behera, P., Kim, J.Y. et al. (2020). Seasonal and spatial dynamics of bacterioplankton communities in the brackish water coastal lagoon. *Science of the Total Environment* 705: 134729. https://doi.org/10.1016/j.scitotenv.2019.134729.
Monismith, S.G. (2007). Hydrodynamics of coral reefs. *Annual Review of Fluid Mechanics* 39: 37–55.
Mutti, M. and Hallock, P. (2003). Carbonate systems along nutrient and temperature gradients: some sedimentological and geochemical constraints. *International Journal of Earth Science* 92: 465–475.
Mwandya, A.W. (2019). Influence of mangrove deforestation and land use change on trophic organization of fish assemblages in creek systems. *African Journal of Biological Sciences* 1: 42–57.

Naqvi, S.W.A., Jayakumar, D.A., Narvekar, P.V. et al. (2000). Increased marine production of N_2O due to intensifying anoxia on the Indian continental shelf. *Nature* 408: 346–349.

Naranjo, L.G. (1997). A note on the birds of the Colombian Pacific mangroves. In: *Mangrove Ecosystem Studies in Latin America and Africa* (eds. B. Kjerfve, L.D. Lacerda and S. Diop), 64–70. Paris: UNESCO.

Nóbrega, M.F., Garcia, J. Jr., Rufener, M.C. et al. (2019). Demersal fish of the NE Brazilian continental shelf: spatial patterns and their temporal variation. *Regional Studies in Marine Science* 27: 100534. https://doi.org/10.1016/j.rsma.2019.100534.

O'Neil, J.M. and Capone, D.G. (2008). Nitrogen cycling in coral reef environments. In: *Nitrogen in the Marine Environment* (ed. D.G. Capone), 937–977. New York: Elsevier.

Pabis, K., Sobczyk, R., Siciński, J. et al. (2020). Natural and anthropogenic factors influencing abundance of the benthic macrofauna along the shelf and slope of the Gulf of Guinea, a large marine ecosystem off West Africa. *Oceanologia* 62: 83–100.

Parulekar, A.H. and Wagh, A.B. (1975). Quantitative studies on benthic macrofauna of NE Arabian Sea shelf. *Indian Journal of Marine Sciences* 4: 174–176.

Pauly, D. (1979). Theory and management of tropical multispecies stocks: a review, with emphasis on the southeast Asian demersal fisheries. *ICLARM Studies Review* 1: 1–35.

Payet, J.P., McMinds, R., Burkepile, D. et al. (2014). Unprecedented evidence for high viral abundance and lytic evidence in coral reef waters of the south Pacific. *Frontiers in Microbiology* 5: 493–504.

Picard, J. (1967). Essai de classement des grands types de peuplements marine benthiques tropicaux d'apres les observations effectivees dans les parages de Tulear (S.W. Madagascar). *Recueil des Travaux Station Marine Endoume Fasicle Hors Series Supplement* 6: 3–24.

Piersma, T., De Goeij, P., and Tulp, I. (1993). An evaluation of intertidal feeding habitats from a shorebird perspective: towards relevant comparisons between temperate and tropical mud flats. *Netherlands Journal of Sea Research* 31: 503–512.

Raja, S., Khan, S.A., Lyla, P.S. et al. (2014). Diversity of macrofauna from continental shelf off Singarayakonda (southeast coast of India). *Pakistan Journal of Biological Sciences* 17: 641–649.

Rhoads, D.C., Boesch, D.F., Tang, Z.C. et al. (1985). Macrobenthos and sedimentary facies on the Changjiang delta platform and adjacent continental shelf, East China Sea. *Continental Shelf Research* 4: 189–203.

Rougerie, F. and Wauthy, B. (1988). The endo-upwelling concept: a new paradigm for solving an old paradox. *Proceedings of the Sixth International Coral Reef Symposium*, Townsville, Australia (8–12 August 1988), vol. 3, pp. 21–25.

Rueda, M. (2001). Spatial distribution of fish species in a tropical estuarine lagoon: a geostatistical appraisal. *Marine Ecology Progress Series* 222: 217–226.

Rueda, M. and Defeo, O. (2003). Spatial structure of fish assemblages in a tropical estuarine lagoon: combining multivariate and geostatistical techniques. *Journal of Experimental Marine Biology and Ecology* 296: 93–112.

Saenger, P. (2002). *Mangrove Ecology, Silviculture and Conservation*. Dordrecht, The Netherlands: Kluwer.

Salem, M.V.A., van der Geest, M., Piersma, T. et al. (2014). Seasonal changes in mollusc abundance in a tropical intertidal ecosystem, Banc d'Arguin (Mauritania): testing the 'depletion by shorebirds' hypothesis. *Estuarine, Coastal and Shelf Science* 136: 26–34.

Sánchez-Botero, J., Garcez, D.S., Caramaschi, E.P. et al. (2009). Indicators of influence of salinity in the resistance and resilience of fish community in a tropical coastal lagoon (southeastern Brazil). *Bolentin de Investigaciones Marinas y Costeras* 38: 171–195.

Santangelo, J.M., Rocha, A.D.M., Bozelli, R.L. et al. (2007). Zooplankton responses to sandbar opening in a tropical eutrophic coastal lagoon. *Estuarine, Coastal and Shelf Science* 71: 657–668.

Santos, T.M.T. and Aviz, D. (2018). Macrobenthic fauna associated with *Dipatra cuprea* (Onuphidae: Polychaeta) tubes on a macrotidal sandy beach of the Brazilian Amazon coast. *Journal of the Marine Biological Association of the United Kingdom* 99: 751–759.

Sharp, G.D. (1988). Fish populations and fisheries. In: *Ecosystems of the World, Volume 27* (eds. H. Postma and J.J. Zijlstra), 55–186. Amsterdam: Elsevier.

Sheppard, C., Davy, S., and Pilling, G. (2018). *The Biology of Coral Reefs, 2e*. Oxford, UK: Oxford University Press.

Short, F.T., Short, C.A., and Novak, A.B. (2018). Seagrasses. In: *The Wetland Book, II. Distribution, Description, and Conservation* (eds. C.M. Finlayson, R. Milton, C. Prentice, et al.), 73–91. New York: Springer.

Sibaja-Cordero, J. and Cortés, J. (2008). Vertical zonation of rocky intertidal organisms in a seasonal upwelling area (eastern tropical Pacific), Costa Rica. *Revisita de Biologia Tropical* 56: 91–104.

Sibaja-Cordero, J.A., Camacho-García, Y.E., Azofeifa-Solano, J.C. et al. (2019). Ecological patterns of macrofauna in sandy beaches of Costa Rica: a Pacific-Caribbean comparison. *Estuarine, Coastal and Shelf Science* 223: 94–104.

Siswanto, E., Horii, T., Iskandar, I. et al. (2020). Impacts of climate changes on the phytoplankton biomass of the Indonesian Maritime Continent. *Journal of Marine Systems* 212: 103451. https://doi.org/10.1016/j.jmarsys.2020.103451.

Sivadas, S.K., Ingole, B., Ganesan, P. et al. (2012). Role of environmental heterogeneity in structuring the microbenthic community in a tropical sandy beach, west coast of India. *Journal of Oceanography* 68: 295–305.

Smith, T.J. III (1992). Forest structure. In: *Tropical Mangrove Ecosystems* (eds. A.I. Robertson and D.M. Alongi), 101–136. Washington, DC: American Geophysical Union.

Smith, K.L., Rowe, G.T., and Clifford, H. (1974). Sediment oxygen demand in an outwelling and upwelling area. *Tethys* 6: 223–229.

Soares, A.G., Scapini, F., Brown, A.C. et al. (1999). Phenotypic plasticity, genetic similarity and evolutionary inertia in changing environments. *Journal of Molluscan Studies* 65: 136–139.

Sosa-López, A., Mouillot, D., Chi, T.D. et al. (2005). Ecological indicators based on fish biomass distribution along trophic levels: an application to the Terminos coastal lagoon, Mexico. *ICES Journal of Marine Science* 62: 453–458.

Sosa-López, A., Mouillot, D., Ramos-Miranda, J. et al. (2007). Fish species richness decreases with salinity in tropical coastal lagoons. *Journal of Biogeography* 34: 52–61.

Taplin, K.A., Irlandi, E.A., and Raves, R. (2005). Interference between the macroalga *Caulerpa prolifera* and the seagrass *Halodule wrightii*. *Aquatic Botany* 83: 175–186.

Thelaus, J., Haecky, P., Forsman, M. et al. (2008). Predation pressure on bacteria increases along aquatic productivity gradients. *Aquatic Microbial Ecology* 52: 45–55.

Thiel, H. (1982). Zoobenthos of the CINECA area and other upwelling regions. *Rapports et Procés-Verbaux de Reunions International L'Exploration de la Mer* 180: 323–334.

Thorson, G. (1957). Bottom communities (sublittoral or shallow shelf). In: *Treatise on Marine Ecology and Paleoecology, Volume 1. Ecology* (ed. J.W. Hedgpeth), 461–535. Washington, DC: Memoirs of the Geological Society of America.

Touchette, B.W. (2007). Seagrass-salinity interactions: physiological mechanisms used by submersed angiosperms for a life at sea. *Journal of Experimental Marine Biology and Ecology* 350: 194–215.

Valero-Pacheco, E., Alvarez, F., Abarca-Arenas, L.G. et al. (2007). Population density and activity pattern of the ghost crab, *Ocypode quadrata,* in Veracruz, Mexico. *Crustaceana* 80: 313–325.

Vanagt, T., Beekman, E., Vincx, M. et al. (2006). ENSO and sandy beach macrobenthos of the tropical east Pacific: some speculations. *Advances in Geosciences* 6: 57–61.

Velayudham, N., Desai, D.V., and Anil, A.C. (2020). Macrobenthic diversity and community structure at Cochin Port, an estuarine habitat along the SW coast of India. *Regional Studies in Marine Science* 34: 101075. https://doi.org/10.1016/j.rsma.2020.101075.

Vermaat, J.E., Rollon, R.N., Lacap, C.D.A. et al. (2004). Meadow fragmentation and reproductive output of the S.E. Asian seagrass *Enhalus acoroides*. *Journal of Sea Research* 52: 321–328.

Vital, H., Leite, T.S., Viana, M.G. et al. (2020). Seabed character and associated habitats of an equatorial tropical shelf: the Rio Grande do Norte Shelf, NE Brazil. In: *Seafloor Geomorphology as Benthic Habitat, 2e* (eds. P. Harris and E. Baker), 587–603. New York: Elsevier.

Walcker, R., Anthony, E.J., Cassou, C. et al. (2015). Fluctuations in the extent of mangroves driven by multi-decadal changes in North Atlantic waves. *Journal of Biogeography* 42: 2209–2219.

Woodroffe, C.D., Rogers, K., McKee, K.L. et al. (2016). Mangrove sedimentation and response to relative sea-level rise. *Annual Review of Marine Science* 8: 243–266.

Yamano, H., Inoue, T., Adachi, H. et al. (2019). Holocene sea-level change and evolution of a mixed coral reef and mangrove system at Iriomote Island, Japan. *Estuarine, Coastal and Shelf Science* 220: 166–175.

Yang, K.Y., Lee, S.Y., and Williams, G.A. (2003). Selective feeding by the mudskipper (*Boleophthalmus pectinirostris*) on the microalgal assemblage of a tropical mud flat. *Marine Biology* 143: 245–256.

Zabbey, N. and Hart, A. (2014). Spatial variability of macrozoobenthic diversity on tidal flats of the Niger delta, Nigeria: the role of substratum. *African Journal of Aquatic Science* 39: 67–76.

Zettler, M.L., Boichert, R., and Pollehne, F. (2009). Macrozoobenthos diversity in an oxygen minimum zone off northern Namibia. *Marine Biology* 156: 1949–1961.

PART 3

FUNCTION

CHAPTER 8

Primary Production

8.1 Introduction

All ecosystems are ultimately fuelled by carbon fixed by autotrophs. The fixation of carbon is derived from the sun's energy via photosynthesis, and it is this energy that is responsible for the functioning of tropical marine ecosystems. Primary producers are consumed by secondary consumers and passed up the food chain and incorporated into complex food webs.

Photosynthesis is a complex process. The light-dependent reactions comprise absorption of light energy. Transfer of electrons through the photosynthesis machinery is coupled to the reduction of water to oxygen and production of adenosine triphosphate (ATP) and the reduced form of nicotinamide adenine dinucleotide phosphate (NADPH) which is then used as reducing power for the biochemical reactions in the Calvin cycle of photosynthesis. The light-dependent reactions of photosynthesis fix CO_2 into carbohydrates (sugars and their polymers). For unicellular algae, such as phytoplankton, more than half of their reduced carbon is directed into protein synthesis and the principal store of high-energy molecules is often lipids. The overall photosynthesis reaction producing carbohydrates is

$$CO_2 + 2\,H_2O + \text{light} \rightarrow CH_2O + H_2O + O_2$$

with the oxygen produced being from the reactant H_2O, not from the CO_2. The factors that control photosynthesis rates in the sea are (i) those that control the rates of reactions of the photosystems I and II (the molecular machinery) and (ii) those that control the rates of the light-dependent reactions. The former are light intensity and availability of H_2O and CO_2. These are abundant in seawater although reaction rates can be forced somewhat by CO_2 loading. The latter include temperature and nutrients including several species of N, PO_4^{3-}, various metal ions, $SiOH_4^+$ (for diatoms and chrysophytes), and sometimes vitamins.

Gross primary production (GPP or P_g) is total photosynthesis generated, and net primary production (NPP or P_n) is gross primary production minus respiration (R). It is net primary production that is available to herbivores.

Tropical Marine Ecology, First Edition. Daniel M. Alongi.

8.2 Sandy Beaches and Tidal Flats

Data on primary production on tropical sandy beaches and tidal flats are few. This is unfortunate as studies have shown that microalgae and other autotrophs are abundant on tropical intertidal sand and mud. Supratidal flats are a very poorly understood habitat and are under pressure from human development and climate change. In the wet-dry tropics of Australia, it was found that primary production and nutrient release were enhanced upon freshwater inundation of supratidal flats (Burford et al. 2016). The main driver for the whole-system primary production rates was the areal extent of inundation rather than the duration of inundation, provided inundation lasted longer than the minimum period of primary production to occur. These data show that supratidal flats may be important contributors to coastal productivity (Burford et al. 2016).

Productivity is highly site-specific. For example, there was no significant seasonal variation of benthic primary production in the polluted Santos estuary in Brazil (de Sousa et al. 1998). Mean rates of primary production varied seasonally from 40.7 to 140 mg C m^{-2} hr^{-1} and correlated with chlorophyll a, NO^{3-}, N, PO_4^{3-}, and organic matter in the sediment. In contrast, rates of respiration can vary greatly as measured on intertidal mud flats in the Fly River delta (45–306 mg C m^{-2} d^{-1}), Papua New Guinea; GPP ranged from 159 to 240 mg C m^{-2} d^{-1} with low (45–93 mg C m^{-2} d^{-1}) rates of NPP (Alongi 1991). These rates were low compared with those measured on temperate intertidal flats and were attributed to continual surface sediment disturbance and silt-laden waters in the delta. The data on nutrient flux showed that these muds were a net sink for carbon rather than a net exporter. Unpublished data from this author of net primary production from tidal sand and mud flats in northern Australia confirm that net primary productivity is low (<100 mg C m^{-2} d^{-1}) in tropical intertidal deposits, except in the case of polluted sediments, where rates can be high as seen from the Brazil data. The small amount of data suggest that bacterial activity is typically greater in tropical sediments than in temperate sediment of identical grain size (Alongi 1998). Such conclusions are problematical, but the available data from the Papua New Guinea and Australian locations indicate levels of net algal production that are not greatly different from those from temperate habitats (Alongi 1998, p. 27, table 2.4).

Primary productivity and abundance of benthic fauna and flora in tropical intertidal sediments are greatly affected by monsoonal rains (Duggan et al. 2014). Sediment chlorophyll *a* in the Norman River, northern Australia, were significantly reduced during consecutive wet season floods; primary productivity on the mud flats was below detection limits (Duggan et al. 2014). The lower salinity and burial of sediment on the mud flats were the main causes of the reduced benthic primary productivity. Similarly, benthic GPP and R were low in sand flat sediments near seagrasses and mangroves in a Bahamas lagoon (Koch and Madden 2001) during the wet season. Benthic primary productivity often remains low in the dry season due to high irradiance and low nutrient availability (Underwood 2002).

8.3 Mangrove Forests

Mangroves are successful in intertidal environments partly because of their physiological plasticity. Mangroves minimise water loss and maximise carbon gain with high water-use efficiency and low transpiration rates to maximise growth, being flexible depending on environmental conditions. This is an advantageous strategy. Some species, for example, minimise energy expenditure by opportunistically maximising growth during the wet season or during short periods of freshwater input and synchronising reproductive output during these wet periods.

The light response curves of mangroves are like other terrestrial and aquatic plants with a steep linear increase up to about 300–400 μmol photons m^{-2} s^{-1} after which saturation is reached. Maximum CO_2 assimilation rates may exceed 25 μmol CO_2 m^{-2} s^{-1} under favourable conditions of low vapour pressure deficit (<22 mbar) and low salinity (<15), but most rates lie between 5 and 20 μmol CO_2 m^{-2} s^{-1}. Due to their low stomatal conductance and intercellular CO_2 concentrations, mangrove photosynthesis reaches saturation at comparatively low light levels (Alongi 2009).

Rates of net photosynthesis vary widely among species with the major regulatory factors being soil salinity, vapour pressure deficit between leaf and surrounding air, and light intensity. A comparison of CO_2 assimilation rates between mangroves and tropical terrestrial trees (Figure 8.1) suggests higher median rates of photosynthesis in mangroves. There is great overlap in rates between and within groups owing to species-specific differences in assimilation rates, position in the canopy, tree age, environmental conditions, and nutrient availability. Shade-intolerant species among terrestrial trees have a median photosynthetic rate of 13 μmol CO_2 m^{-2} s^{-1}, which is virtually identical to mangroves (12 μmol CO_2 m^{-2} s^{-1}). Grouping all shade-tolerant and shade-intolerant terrestrial species, the median rate is 7 μmol CO_2 m^{-2} s^{-1}. Lower values for both mangrove and terrestrial trees are found in the dry tropics signifying the importance of climate.

There are very few measurements of respiration of leaves, roots, branches, and stems. Rates of dark respiration in mangrove leaves range from 0.2 to 1.4 μmol CO_2 m^{-2} s^{-1} (Table 8.1) with photosynthesis to respiration (P:R) ratios ranging from 2.1 to 11.2, which are at the upper end of the range for tropical terrestrial trees.

Golley et al. (1962) offered the first crude estimates of dark respiration from Puerto Rican forests of *Rhizophora mangle* for mangrove roots. They measured rates that averaged 169 mmol C m^{-2} prop root d^{-1}. This respiratory loss was the second largest in these forests after leaf respiration. On *Avicennia* pneumatophores, Burchett

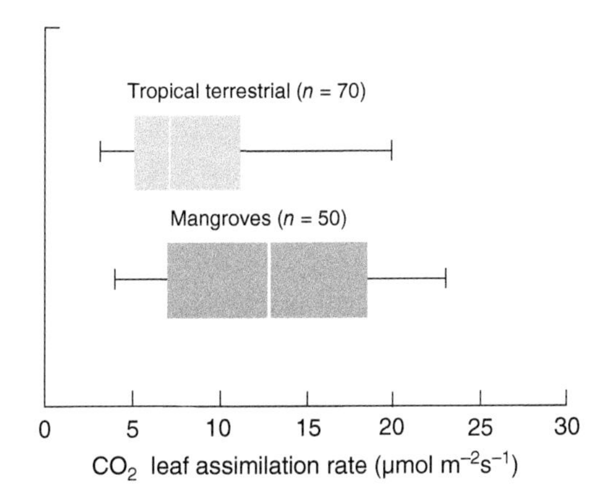

FIGURE 8.1 Comparison in CO_2 assimilation rates between various species of mature tropical mangrove and terrestrial trees. *Source*: Alongi (2009), Schmitz et al. (2012), Nguyen et al. (2015), Reef et al. (2015, 2016), and dos Santos Garcia et al. (2017). © John Wiley & Sons.

TABLE 8.1 **Mean rates of dark respiration (R, μmol CO_2 m^{-2} s^{-1}) and photosynthesis to respiration ratio (P:R) in leaves of some mangrove species.**

Species	R	P:R	Location
Rhizophora mangle	1.1	3.5	Puerto Rico
Ceriops australis	0.6	10.1	Australia
Rhizophora apiculata	1.4	7.4	Australia
Avicennia germinans	0.5	11.2	Venezuela
Conocarpus erectus (dry season)	0.2	9.4	Venezuela
C. erectus (wet season)	0.7	5.8	Venezuela
R. mangle	0.7	6.0	USA
Lumnitzera racemosa	1.0	3.4	USA
A. germinans	0.4	2.1	USA

Source: Golley et al. (1962), Lugo et al. (1975), Smith et al. (1989), Gong et al. (1992), Clough (1998) and Alongi (2009).

et al. (1984) found maximal rates of root respiration (3.2 μmol CO_2 g^{-1} FW root h^{-1}) at 25% of full-strength seawater with slower rates in freshwater (2.8 μmol CO_2 g^{-1} FW root h^{-1}) and in 50% (3.1 μmol CO_2 g^{-1} FW root h^{-1}) and 100% (2.1 μmol CO_2 g^{-1} FW root h^{-1}) seawater. Kitaya et al. (2002) measured net photosynthesis in the raised pneumatophores of *Sonneratia alba* and *Avicenna marina* and in the prop roots of *Rhizophora stylosa* of 0.6, 0.2 and 0.1 μmol CO_2 m^{-2} s^{-1}, respectively. No photosynthetic activity was detected in the tree roots of *Bruguiera gymnorrhiza*. Rates of root respiration averaged 1.3, 0.8, and 2.5 μmol CO_2 m^{-2} s^{-1} in *S. alba, A.marina,* and *R. stylosa*, respectively. These and other mangrove root respiration rates (Lovelock et al. 2006) are low compared with estimates for other angiosperm trees (Reich et al. 1998; Cheng et al. 2005).

Photosynthesis and growth vary with many factors, especially light intensity, temperature, nutrient and water availability including proximity to groundwater, wave energy and weather. Annual growth rings (Verheyden et al. 2004) as well as seasonal patterns (Krauss et al. 2008) of tree growth underscore the importance of short- and long-term consequences of climate on mangrove growth.

NPP is the balance between GPP and R and represents the amount of carbon available for growth and tissue maintenance. Rates of root production vary widely. Growth of fine roots over the upper 30 cm of soil averaged 0.3, 0.3, and 2.8 mg DW cm^{-3} a^{-1} in *Rhizophora apiculata, B. gymnorrhiza,* and *S. alba* forests, respectively, in Micronesia (Gleason and Ewel 2002). Rates of root production in other mangrove forests have ranged from 0.5 to 28.4 t DW m^{-2} a^{-1} (Cahoon et al. 2003; Sanchéz 2005; Adame et al. 2014; Cormier et al. 2015; Poungparn et al. 2015; Xiong et al. 2017; Liu et al. 2017; Ono et al. 2017; see Table 4 in Muhammad-Nor et al. 2019). Fine root turnover rate in Chinese forests ranged from 1.46 to 5.96 a^{-1}, and fine root production was 1–3.5 × higher than aboveground litter production (Xiong et al. 2017). These estimates are at the lower end of the range of measurements in tropical terrestrial forests (Clark et al. 2001; Perry et al. 2008), but most of these measurements were made in fringe stands and used short (<1 m) sediment cores. The methods used measure only fine root production, not the production of larger roots.

Wood production in contrast is well quantified with a large database available from harvesting studies. Saenger (2002) provides a complete analysis of the mean annual growth of mangrove wood. Regressing mean growth rates of mostly *R. apiculata* trees in variously aged plantation stands throughout Southeast Asia, Saenger (2002) found a complex relationship with age, with peak growth at about 15 years and declining thereafter. Across all species and ages, mean annual increment of mangrove trees range from 0.1 to 1.8 cm a^{-1} at diameter-at-breast height (DBH). Globally, mangrove wood production in natural forests correlates with stand mean DBH growth, and tree DBH growth is affected mostly by climate (Xiong et al. 2019). In mangrove plantations, the global pattern of tree DBH growth is affected mostly by taxon and stand wood production is affected mostly by taxon and density (Xiong et al. 2019).

There is no thorough perception as to how carbon is allocated within a mangrove tree. For 22 year-old *R. apiculata* trees in Malaysia, Clough et al. (1997) constructed a preliminary carbon allocation model. Of a total annual net daytime fixed carbon production of 56 t ha^{-1} a^{-1}, 22% was respired by the foliage overnight, 11% was accumulated as above-ground biomass, 8% was lost as litter, 1% accumulated as above-ground biomass, and by difference, the remaining 58% was presumably used in root turnover and in respiration of branches, stem, roots, and other woody parts. It is not yet possible to construct a robust model of carbon balance for mangrove trees owing to a lack of empirical data and the difficulty in measuring root processes and respiration of woody parts. However, the preliminary carbon model for *R. apiculata* suggests that roughly one-half of fixed carbon is eventually respired, in agreement with similar estimates for tropical terrestrial trees (Barnes et al. 1998; Clark et al. 2001; Perry et al. 2008).

Some estimates exist of the balance between net assimilation and respiration in mangrove canopies. Early data from Florida mangroves (Odum et al. 1982) showed that canopy R accounted for an average of 58% of GPP with a range of 14–86%. Alongi et al. (2004) estimated that tree R equated to 41% of GPP. Canopy photosynthesis and respiration have rarely been measured simultaneously. Extensive measurements of GPP and R throughout the mangrove canopy of a *Kandelia candel* forest on Okinawa found maximum rates of GPP and R at the top of the canopy with a two- to sevenfold decline to the bottom of the canopy (Suwa et al. 2006). Annual canopy GPP averaged 102.9 t CO_2 ha^{-1} a^{-1} and canopy respiration averaged 44 t CO_2 ha^{-1} a^{-1} for an average annual NPP of 58.9 t CO_2 ha^{-1} a^{-1} or 57% of GPP. The similarity in the percentage of fixed carbon respired among studies is likely a reflection of the physiological limits of carbon assimilation and allocation as well as thermodynamic constraints imposed on physiological processes.

Rates of forest NPP are rapid compared with other tropical estuarine and marine primary producers (Table 8.2), although caution must be applied if one considers that several methods have been used to measure different components of fixed carbon, not to mention that these forests in disparate settings are of different age and environmental conditions.

If we accept the data obtained using the light attenuation method as a reliable estimate of net primary productivity of mangroves in the daytime, the rate of NPP averages 64 t DW ha^{-1} a^{-1}. In comparison, the estimates based on incremental growth plus litterfall average 11 t DW ha^{-1} a^{-1}. The light attenuation method seems likely to overestimate NPP while the latter method seems to underestimate mangrove NPP.

TABLE 8.2 Estimates of aboveground net primary production (ANPP = t DW ha^{-1} a^{-1}) of mangrove forests in various parts of the world based on different methods.

Species	Location	ANPP	Method
R. mangle, A. germinans, L. racemosa	USA	46.0	Gas exchange
A. germinans	USA	20.5	Gas exchange
R. mangle	USA	16.9	Gas exchange
R. mangle, A. germinans, L. racemosa	USA	22.5	Gas exchange
R. mangle, A. germinans, L. racemosa	Puerto Rico	58.4	Gas exchange
R. mangle, A. germinans, L. racemosa	USA	26.1 (fringe) 8.1 (dwarf)	Demographic/allometric
Kandelia candel	Vietnam	5.3	Demographic/allometric
K. candel	Vietnam	13.4	Demographic/allometric
R. mangle, A. germinans, L. racemosa	Dominican Republic	19.7	Demographic/allometric
R. apiculata	Thailand	63.7	Light attenuation
Ceriops decandra	Thailand	48.7	Light attenuation
R. apiculata	Malaysia	112.1	Light attenuation
R. apiculata (70 years old)	Malaysia	102.2	Light attenuation
R. apiculata (18 years old)	Malaysia	65.7	Light attenuation
R. apiculata (5 years old)	Malaysia	36.5	Light attenuation
R. apiculata	Thailand	13.5	Light attenuation
R. stylosa	Australia	40.5	Light attenuation
A. marina	Australia	30.6	Light attenuation
Mixed *Rhizophora* species	Australia	29.2	Light attenuation
R. apiculata, Bruguiera parviflora	Papua New Guinea	30.5	Light attenuation
Nypa fruticans	Papua New Guinea	30.1	Light attenuation
A. marina, Sonneratia lanceolata	Papua New Guinea	24.4	Light attenuation
R. apiculata, A. marina	Indonesia	104.6	Light attenuation
R. apiculata, A. marina	Indonesia	96.9	Light attenuation
A. marina, A. officinalis	Indonesia	103.2	Light attenuation

TABLE 8.2 (*Continued*)

Species	Location	ANPP	Method
C.tagal, R. apiculata	Indonesia	106.1	Light attenuation
C.tagal, R. apiculata	Indonesia	109.4	Light attenuation
R. stylosa, S. alba	Indonesia	63.7	Light attenuation
R. apiculata, K. candel	Indonesia	74.3	Light attenuation
A. marina, R. mucronata	India	32.0	Light attentuation
B. parviflora	Malaysia	27.4	Harvest/incremental growth
R. mangle (5 years old)	Cuba	1.6	Harvest/incremental growth
A. germinans	Cuba	5.9	Harvest/incremental growth
L. racemosa	Cuba	5.4	Harvest/incremental growth
Sonneratia apetala	Bangladesh	12.5	Harvest/incremental growth
Sonneratia caseolaris	Bangladesh	26.4	Harvest/incremental growth
A. officinalis	Bangladesh	7.6	Harvest/incremental growth
A. marina	Bangladesh	4.4	Harvest/incremental growth
Avicennia alba	Bangladesh	2.1	Harvest/incremental growth
B. gymnorrhiza	Bangladesh	0.6	Harvest/incremental growth
Bruguiera sexangula	Bangladesh	0.1	Harvest/ incremental growth
Excoecaria agallocha	Bangladesh	4.7	Harvest/incremental growth
Xylocarpus moluccensis	Bangladesh	0.5	Harvest/incremental growth
Mixed species	Micronesia	4.2	Harvest/incremental growth
R. apiculata, B. gymnorrhiza	Malaysia	8.7	Harvest/incremental growth
R. apiculata	Vietnam	4.9	Harvest/incremental growth
R. apiculata	Vietnam	19.0	Incremental growth

(*continued*)

TABLE 8.2 *(Continued)*

Species	Location	ANPP	Method
R. apiculata	Thailand	15.7	Incremental growth
R. apiculata	Thailand	10.6	Incremental growth
R. apiculata	Vietnam	9.4	Litterfall
R. apiculata	Vietnam	18.7	Litterfall
R. racemosa	Gambia	18.8	Litterfall
Avicennia africana	Gambia	11.6	Litterfall
R.racemosa	Gambia	10.4	Litterfall
R. mucronata	India	14.6	Litterfall
R. apiculata	India	13.6	Litterfall
A.marina	India	6.2	Litterfall
B. sexangula	China	11.0	Litterfall
K. candel	China	13.3	Litterfall
Aegiceras corniculatum	China	11.3	Litterfall
K. candel	China	24.4	Litterfall/allometric
R. mucronata	Indonesia	23.4	Litterfall/incremental growth
R. mangle, A. germinans, L.racemosa	Guadeloupe	21.2 (fringe) 6.2 (dwarf)	Litterfall/incremental growth
R.mangle	Hawaii	29.1	Litterfall/incremental growth
R. mucronata, A.marina	Sri Lanka	11.0	Litterfall/incremental growth
R. stylosa, B. gymnorrhiza	Ishigaki Island, Japan	11.0	Litterfall/incremental growth
A. marina, Kandelia obovata, S. caseolaris, S. apetala	China	9.5	Litterfall/incremental growth
S. caseolaris, A. alba, R. apiculata, R. mucronata, Xylocarpus granatum	Thailand	11.4	Litterfall/incremental growth
R. stylosa, B. gymnorrhiza, K. obovata	Okinawa, Japan	27.9	Litterfall/incremental growth
Heritiera fomes	Bangladesh	21.0	Litterfall/incremental growth
R. apiculata	Vietnam	22.7	Growth modelling

TABLE 8.2 (*Continued*)

Species	Location	ANPP	Method
N. fruticans	Australia	60.9 (Herbert River) 49.3 (McIvor River) 54.0 (Pascoe River)	Growth modelling
N. fruticans	Papua New Guinea	23.8	Growth modelling

Source: Alongi (2009), Poungparn et al. (2012), Peng et al. (2016), Pandi et al. (2018), Kamruzzaman et al. (2017 a, b, 2019), Ohtsuka et al. (2019), Phan et al. (2019) and Robertson et al. (2020). © John Wiley & Sons.

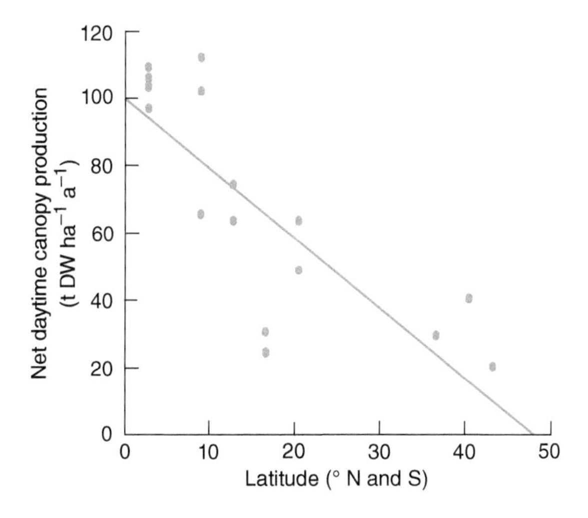

FIGURE 8.2 Latitudinal changes in net daytime canopy production of mangroves measured using the modified light interception method. *Source*: Alongi (2009), Suwa et al. (2006) and Robertson et al. (2020). © John Wiley & Sons.

Plotting the data from the light attenuation method versus latitude (Figure 8.2) gives a significant negative relationship indicating that mangrove productivity, even in a relative sense, declines with distance from the equator, mirroring the latitudinal decline in mangrove biomass and litterfall (Saenger and Snedaker 1993). However, using the arguably more reliable data of litterfall, wood, and root production, Twilley and Day (2013) found consistently high productivity grouped by latitude (Figure 8.3a), except within the 30–40° latitudinal band, although rates of fine root production increase with increasing latitude. The production: biomass ratio (Figure 8.3b) increases from a low of 0.08 at low latitudes to 0.20 at the higher tropical latitudes, meaning that the percent of total net primary production that occurs belowground increases from about 9% in equatorial latitudes to about 25% at latitudes 20–30° N or S.

How do these productivity data compare with similar data from tropical terrestrial forests? First, we must compare data obtained using identical methods.

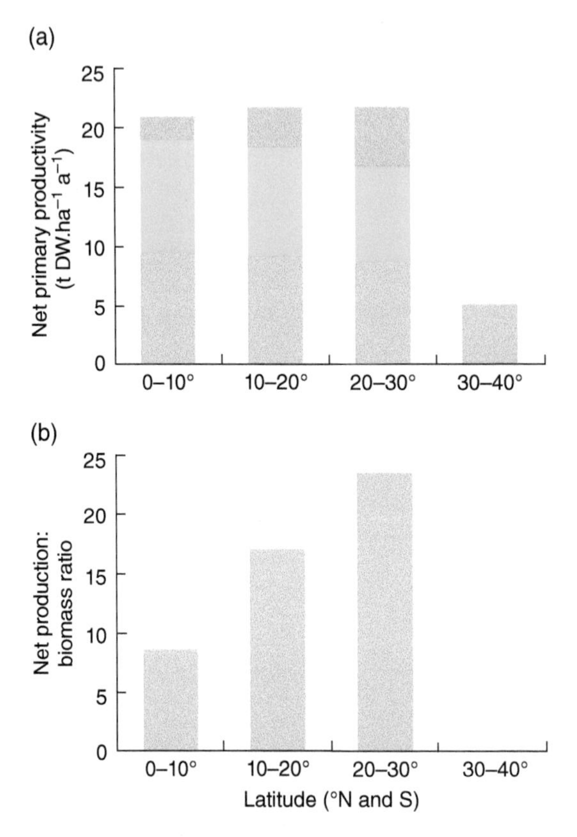

FIGURE 8.3 Mangrove net primary production (a) as litterfall, wood, and fine root production and (b) net production:biomass ratio grouped by latitude. Yellow is litterfall, blue is wood production, and green is fine root production. *Source*: Twilley and Day (2013) , figure 7.8, p. 180. © John Wiley & Sons.

The most comprehensive database for both mangroves and tropical terrestrial forests involves measurements of litterfall and incremental growth. Comparing the data in Table 8.2 and the data analysed by Clark et al. (2001) and Scurlock and Olson (2002), we find equivalent estimates of NPP (Figure 8.4). For mangroves (n = 32), the mean rate of aboveground NPP is 11.13 t DW ha^{-1} a^{-1} (=44.52 mol C m^{-2} a^{-1} assuming 48% C content of dry wood, Alongi 2014) with a median value of 8.1 and 25th and 75th percentiles of 4.6 and 19.175, respectively. For terrestrial forests, the mean rate of aboveground NPP is 11.9 t DW ha^{-1} a^{-1}; the median rate is 11.4, the 25th percentile is 8.8, and the 75th percentile is 14.4. These values are remarkably close given the differences within and between forest groups in size, age, and species differences, suggesting that rates of primary production are equivalent between mangrove and tropical terrestrial forests. It also suggests that there are similarities between forest groups in physiological limitations and ecological factors limiting production in all trees, although it must be remembered that below-ground production estimates are sparse for both forest groups. To adjust the photosynthetic rates for a true estimate of net carbon fixation, respiration of roots and woody parts is sorely needed.

Mangrove stands vary in size and age over time and therefore vary in rates of production and in the balance between production and respiration. Some studies have measured the growth dynamics of mangrove forests over time or of stands of known

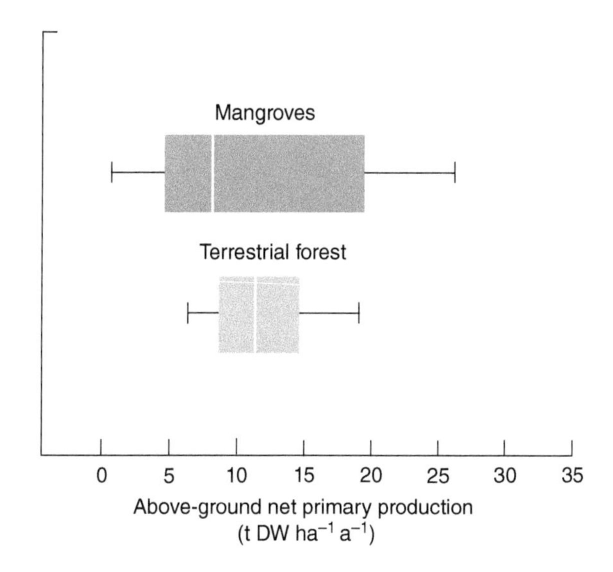

FIGURE 8.4 Comparison of above-ground net primary production in mangrove and tropical terrestrial forests based on measurements of biomass increments and litterfall. Vertical line in boxes denotes median and the boxes encompass the 25th and 75th percentiles and the outer bars denote the 5 and 95% percentiles. *Source*: Updated and adapted from Alongi (2009) using references in Table 8.3. © John Wiley & Sons.

age (Day et al. 1996; Fromard et al. 1998; Clough et al. 2000; Phan et al. 2019). Day et al. (1996) measured litterfall plus annual growth increments in zones of *Avicennia germinans* with *R. mangle* as a canopy sub-dominant, a scrub forest of *A. germinans* and a stand of mature *A. germinans*. Litterfall patterns correlated best with patterns of soil salinity, precipitation, and air temperature with these three factors explaining 74% of the variability in interannual patterns. Similarly, precipitation, temperature, and cyclone frequency explained 74% of the global trends in maximum canopy height, with other geophysical factors influencing variability at local and regional scales (Simard et al. 2019). The tallest mangrove forests were found in Gabon in equatorial Africa where trees are an average of 62.8 m in height. Mangrove ANPP was positively correlated with DBH, canopy height, basal area, and stem density, but negatively correlated with salinity; mangrove belowground NPP (BNPP) was positively linked to stem density and salinity, but negatively correlated to canopy height. The factors driving these relationships, especially for BNPP, are unclear (Ouyang and Guo 2021).

Mangroves, like other forests, change over time through successive stages of development with sequential changes in species and therefore in rates of net primary production (Fromard et al. 1998; Berger et al. 2006; Li et al. 2012). An excellent example of this change comes from the study of mangroves along the coast of French Guiana, which is greatly influenced by the Amazon outflow (Fromard et al. 1998; Proisy et al. 2021). There was a rapid increase in the growth and density of pioneering species during the first five years after colonisation, followed by maturation over about five decades with a clear decline after about 70 years (Fromard et al. 1998). However, a stable-state maturity phase is more the exception than the rule in this region as mangrove existence on this rapidly changing muddy coast depends in accretion and erosion cycles (Proisy et al. 2021). Nevertheless, some mangrove forests reach maturity, and this phase appears to be prolonged compared with what is known for most tropical terrestrial forests. The long maturity phase may represent an alternate succession state in which the clock for the climax stage was reset by each major disturbance. Similar findings were observed in neotropical mangrove forests (Berger et al. 2006) and in the Mekong delta, Cambodia (Li et al. 2012). The relationship between forest age and photosynthetic production in mangroves suggests that this prolongation, or arrested progression, happens when forests are disturbed.

Mangrove primary productivity changes as forests age. A plot of net daytime canopy production versus age of various forests of *R. apiculata* (Figure 8.5) in Southeast Asia shows log-phase production until about 20 years after which NPP levels off but does not noticeably diminish for nearly a century. Most plantations of mangroves in Southeast Asia are disturbed, and this may be the cause of a prolonged or arrested succession stage. Few forests are pristine in the tropics, but mangroves still might constitute a carbon sink for up to a century, especially if left relatively undisturbed.

Mangrove productivity can be impacted by disturbances, especially cyclones and other extreme weather events (Kauffman and Cole 2010; Barr et al. 2012; Aung et al. 2013; Long et al. 2016; Duke et al. 2017; Imbert 2018). The impact of such disturbances on mangroves depends not only on the composition and structure of the forests but also the duration, frequency, and intensity of the disturbance. Barr et al. (2012) found that in a mangrove forest recovering from a hurricane in the Florida Everglades, the effects of the storm were apparent on annual time scales, and CO_2 uptake remained lower four years after the hurricane, with dry season CO_2 uptake being relatively more affected by the hurricane than in the wet season. Night-time respiration was consistently and significantly greater after the hurricane, possibly due to the enhanced decomposition of litter and coarse woody debris generated by the hurricane.

In other regions subjected to intense storms, mangroves can be resilient or vulnerable, recovering over a period of a few or many years, or significantly or not significantly affected by the disturbance (Imbert 2018; Sippo et al. 2018). For example, Cyclone Nargis struck the Ayeyarwady delta of Myanmar causing extensive damage to some species of mangroves (mostly of the Rhizophoraceae) but not to others (Aung et al. 2013). Thus, recovery of forests varied, ranging from recovery within a

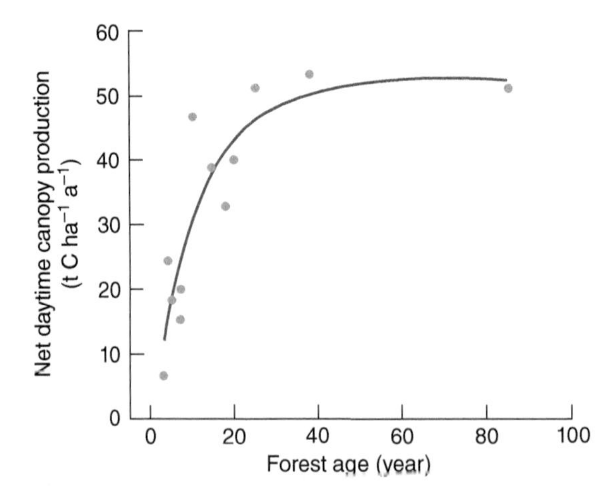

FIGURE 8.5 The relationship between forest age and net daytime canopy production of the mangrove, *Rhizophora apiculata* in Southeast Asia. *Source*: Alongi (2009), figure 2.16, p. 38. © Springer Nature Switzerland AG.

few months to little recovery after four years. The vulnerability of *Rhizophora* species was likely the result of high mortality caused by limited sprouting ability after wind-induced disturbance, erosion that occurred on the riverbank with frequent inundation and delayed phenology after the cyclone. The recovery rate was 61% during the four years after Cyclone Nargis. Recovery of mangroves from cyclones also varied by species in response to a severe typhoon in the Federated States of Micronesia (Kauffman and Cole 2010). Tree mortality ranged from 6 to 32% among stands; adaptations and responses to the typhoon varied by species and geographic location. Drought and hurricanes did not always result in mangrove dieback as found in the case of Abaco Island, The Bahamas (Rossi et al. 2020). Instead, intense grazing by leaf-crewing organisms and foliar disease were both present with herbivory facilitating the spread of disease.

Mangroves can also be severely affected by other extreme weather events (Sippo et al. 2018). For instance, a severe and sudden dieback of mangroves took place between late 2015 and early 2016 in the Gulf of Carpentaria in northern Australia, coincident with extreme temperatures, low precipitation, and a temporary drop in sea-level (Harada et al. 2020). There is normally a sustained wet season during summer months in the gulf indicating that an unusually long period of severe drought conditions due to a failed wet season can negatively impact coastal mangroves. Satellite observations between 1987 and 2015 indicate that while short-term ENSO-related sea level changes may result in dieback in the gulf, the long-term projection of an increase in sea-level is likely to lead to extension of mangroves in the landward direction (Asbridge et al. 2019).

ENSO-related events do not always lead to negative consequences for mangroves. During a major El Niño to La Niña transition in which rainfall was at its peak, litterfall production of a *R. mangle* forest thriving in a small bay in the Atrato River delta in the southern Caribbean, Colombia, was pervasively high (55 g m^{-2} d^{-1}) year-round with no temporal change (Riascos and Blanco-Libreros 2019). This phenomenon is contrary to nearly all other mangrove forests which show significant seasonality in litterfall. High freshwater input from the Atrato River due to the ENSO transition likely subsidises mangroves inhabiting the delta front sustaining the high production rates.

8.4 Seagrasses

Productivity of tropical seagrasses is regulated mainly by light, nutrients, and temperature. The dynamics of seagrass beds may be directly impacted by various disturbances, such as storms and cyclones, that result in significant loss of biomass. Seagrass photosynthesis, growth, survival, and depth distribution are linked to underwater light levels. Reduced irradiance may be caused by epiphytic and planktonic algal accumulation from excess nutrients, increased sediment runoff, and resuspension of bottom sediments.

Temperature is a major factor controlling seagrass growth with photosynthetic rates increasing in spring and summer and decreasing in autumn and winter. However, light levels and temperature often correlate and exhibit similar seasonal trends making it difficult to separate these two factors for seagrass photosynthesis and production. Elevated temperature and seagrass response to elevated DIC and CO_2 and lower carbonate (CO_3^{2-}) mimicking ocean acidification, resulted in most seagrass species utilising bicarbonate (HCO_3^-) although most were not saturated at current levels of oceanic DIC (Koch et al. 2013). Photosynthesis and growth rates of seagrasses are likely to increase under elevated concentrations of CO_2, whereas ocean acidification likely disrupts the diffusion and transport systems of H^+ and DIC controlling microenvironments that promote calcification over dissolution (Koch et al. 2013). Further, seagrasses modify seawater carbonate chemistry, and these impacts can be severe in ameliorating the impacts of ocean acidification on coral reefs (Unsworth et al. 2012).

Thermal stress due to global warming can have noticeable impacts on seagrass photosynthesis. One effect was a decrease in photosynthetic yield representing heat-induced photoinhibition related to closure of Photosystem II reaction centres. *Cymodocea rotundata, Cymodocea serrulata, Halodule uninervis*, and *Thalassia hemprichii* were more tolerant to thermal stress than *Halophila ovalis, Zostera capricorni*, and *Syringodium isoetifolium*; these latter three species are restricted to subtropical and tropical waters and their tolerance to seawater temperatures up to 40 °C is likely to be an adaptive response to high temperatures commonly occurring at low tides and peak solar irradiance. All seagrass species are likely to suffer irreparable effects from short-term or episodic changes in seawater temperatures as high as 40–45 °C (Campbell et al. 2006).

Seagrass photosynthesis controls rates of calcification and photosynthesis of calcareous macroalgae in tropical seagrass meadows (Semesi et al. 2009). Experiments show that seagrass photosynthesis increased seawater pH from 8.3–8.4 to 8.6–8.9 after 2.5 hours which in turn enhanced the rates of calcification 5.8-fold for the calcareous red alga, *Hydrolithon* sp. and 1.6-fold in two other macroalgal species. Algal calcification within seagrass meadows is considerably enhanced by the photosynthesis of the seagrasses which in turn increases seawater pH. Where light is plentiful, nutrient availability plays a significant role in regulating seagrass production. Additions of inorganic nutrients can stimulate seagrass growth suggesting nutrient limitation. Although seagrasses can access nutrients from both the sediment and the water column, effects of nutrient availability on seagrass production may vary depending on nutrient source.

Underwater light intensity is attenuated exponentially with water depth, wherein light attenuation in increasing water depth is associated with absorption and scattering processes due to phytoplankton, non-algal particulate matter and dissolved substances, and the water itself. Minimum light requirements for seagrasses have often been determined as percent light at maximal depth limit using the light attenuation coefficient, K_d. These minimum light requirements vary among species as each seagrass species has unique physiological and morphological adaptations to light availability. For *Cymococea nodosa*, the minimum light requirement is 7.3%, whereas more light was required for *Halodule wrightii* (20.7%), *Halophila engelmanni* (23.7%), *Syringodium filiforme* (23.1%) and *Thalassia testudinum* (18.4%); low light requirements have also been reported for *Halophila decipiens* (4.4–8.8%) (Lee et al. 2007).

Photosynthesis-irradiance (*P-I*) curves provide valuable information for light requirements necessary to sustain optimal growth. *P-I* curves are divided normally into three sections: (i) a light-limited region in which photosynthetic rates increase linearly with increasing light, (ii) a light-saturated region in which photosynthetic rates are independent of irradiance, and (iii) a photo-inhibited region in which photosynthesis decreases with further increases in irradiance. The curve can provide estimates for the initial slope (α, photosynthetic quantum efficiency), the light-saturated photosynthetic rate (P_{MAX}), and saturating irradiance for photosynthesis (*Ik*) where

α intersects P_{MAX}, compensation irradiance (*Ic*) where net photosynthesis is zero, and photoinhibition irradiance (*Ii*) where photosynthesis decreases with further increases in irradiance. *Ik* and *Ic* can be used to determine light requirements based on daily light saturation period (H_{sat}) and the daily light compensation period (H_{comp}), respectively. These photosynthetic parameters can vary seasonally and under different environmental conditions.

On average, P_{MAX} is greater for tropical than temperate species with rates ranging from 174 to 421 μmol O_2 g^{-1} DW h^{-1} for tropical species (Lee et al. 2007) and from 37.5 to 426.3 μmol O_2 g^{-1} DW h^{-1} for temperate species. There is wide overlap in *Ic* and *Ik* between tropical and temperate species indicating that there are similar physiological limits for both tropical and temperate species.

Seagrasses exhibit various physiological and morphological responses to limiting light conditions. These responses depend on species, light intensity and duration, and interactions between environmental conditions, such as temperature and nutrient availability. Photoadaptive responses in seagrasses to reduced light conditions include decreases in plant size, shoot density, biomass, leaf production rates, and chlorophyll content. Young plant responses to lower light conditions include lower leaf biomass and size. The tropical species, *T. testudinum,* for example, undergoes a decline in leaf biomass more rapidly than below-ground biomass due to reduced light conditions (Lee et al. 2007).

Tropical seagrasses have optimal temperatures for growth from 23 to 32 °C (Lee et al. 2007) with seasonality being the net result of increased summer temperatures. Optimal growth temperatures of tropical/subtropical species are higher (24.5–29.0 °C) than those of temperate (12.6–26.0 °C) species. Nevertheless, growth rates of tropical species can be limited by seasonally high temperatures during summer or by heavy rainfall which causes lower salinity and increased turbidity. Average optimum temperatures for photosynthesis, however, are not different between tropical (27–32.5 °C) and temperate (20.5–30.0 °C) species underscoring that the physiological limits to photosynthesis are not noticeably different between tropical and temperate species. Both respiration and photosynthesis increase with increasing water temperatures, but respiration usually increases more than photosynthesis at progressively higher temperatures thus leading to a reduction in net photosynthesis for both tropical and temperate species.

Intertidal and shallow subtidal seagrasses in the tropics are exposed to high, often extreme, temperatures, wide variations in salinity and a variety of disturbances, and changes in these factors can have a drastic impact on seagrass growth and photosynthesis. Photosynthetic efficiencies can decrease at extreme temperatures and at high or low salinity extremes. This dramatic decline at extreme conditions has been linked to chronic inhibition of photosynthesis. Tropical seagrasses are usually unable to recover from such excesses, suggesting irreversible damage to photosynthetic machinery. Chronic inhibition of photosynthesis by heat stress has been reported in seven tropical seagrass species (Campbell et al. 2006). Further, their photosynthetic responses to high temperatures suggest that photosynthesis likely undergoes irreversible damage from short-term or episodic changes in seawater temperatures as high as 40–45 °C.

The impact of storms and cyclones has significant impacts on seagrass beds, particularly the loss of biomass and thus indirectly on productivity (Cruz-Palacios and Van Tussenbroek 2005; Côté-Laurin et al. 2017; Grech et al. 2018). Sediment removal played less of a role in biomass loss than sediment burial in Caribbean seagrass beds after hurricane passage, as seagrass losses approached 70% (Cruz-Palacios and Van

Tussenbroek 2005). The extent of impact varies depending on the degree of seagrass exposure. Côté-Laurin et al. (2017) found greatest losses in the most exposed seagrasses of SW Madagascar. Seagrasses can in many instances be resilient to disturbance depending on the composition and structure of the seagrass community as well as the extent, frequency, and duration of the disturbance (Grech et al. 2018).

Seagrasses require inorganic carbon and nutrients for growth and photosynthesis. Seagrasses can utilise both CO_2 and HCO_3^- for photosynthesis, and it is inorganic N and P that are most commonly limiting for seagrass growth (Lee et al. 2007). Carbon-limited growth in seagrasses is rare but it does occur. Major sources of inorganic nitrogen include NO_3^- and NH_4^+ in the water column and NH_4^+ from the interstitial water. Assimilation of NO_3^- by seagrasses is influenced by the availability of photosynthate and/or stored carbon and so is energetically more costly than using NH_4^+. The main source of P is PO_4^{3-}. Interstitial water is probably the main source of nutrients given that the sediments hold greater concentrations than in the overlying water column.

Nutrient uptake affinities can be greater in leaves compared to roots. Seagrass leaves have developed the ability to assimilate nutrients under lower concentrations and so nutrient uptake may saturate at lower concentrations (Lee et al. 2007). Conversely, root uptake may saturate at higher concentrations than observed for leaves. Therefore, the kinetics of nutrient uptake reflects plant adaptations to life in the water column with low nutrient concentrations and relatively high nutrients in sediments. In *T. testudinum,* root uptake of NH_4^+ from sediments accounts for 50% of total plant uptake with leaf NO_3^- and NH_4^+ assimilation accounting for the remainder (Lee and Dunton 1999). Thus, although there are significant differences in nutrient concentrations between the water column and sediment, an equal contribution from both sources is highly likely. However, the respective contributions of leaf and root nutrient uptake may vary as a function of nutrient concentrations in the water column and sediment. Seagrasses allocate more biomass to below-ground roots under low sediment nutrient availability, whereas more biomass is allocated to above-ground tissue when nutrients are readily available, thus allowing for greater carbon fixation (Lee and Dunton 1999; Romero et al. 2006).

N and P may be limiting for tropical seagrasses and there is good evidence that P and Fe are limiting in carbonate-rich sediments (Burkholder et al. 2013). Typical responses when N and/or P are supplied to seagrasses are increases in biomass, productivity, and shoot size. Seagrass growth, however, is not always limited by ambient nutrient concentrations, as increased nutrient availability sometimes has a limited effect on seagrass growth. Maximum rate of photosynthesis, photosynthetic efficiency, and/or increased chlorophyll concentrations may be the net result of increased sediment nutrients. Elevated water column nutrient concentrations can adversely affect seagrasses through the stimulation of phytoplankton, epiphytic, and macroalgal productivities (Burkholder et al. 2007). Such blooms can lead to reduction in light available to seagrasses.

The role of P and N in the productivity of the tropical seagrass, *S. filiforme,* has been determined using nutrient enrichment experiments (Perez et al. 1991; Short et al. 1993; Armitage et al. 2005). After P enrichment, *S. filiforme* living in carbonate sediments showed a substantial increase in growth, biomass, and tissue P levels, likely reflecting limitation (Short et al. 1993). P limitation was also found in *C. nodosa* in carbonate sediments where increased P levels resulted in increased seagrass tissue P content, shoot growth, biomass and density, and reduced tissue N:P ratios (Perez et al. 1991). The turtlegrass, *T. testudinium*, revealed a spatial pattern in P limitation in Florida Bay, from severe limitation in the eastern bay to a balanced condition with N availability in the western bay (Armitage et al. 2005).

Seagrass production does not correlate with sediment and water column nutrients because of high nutrient uptake affinities in leaves and possible detrimental effects of high water column nutrient concentrations on seagrass growth. The lack of a plant response to changes in sediment NH_4^+ suggests that interstitial concentrations of 100 μmol l^{-1} may provide an adequate pool of inorganic N for *T. testudinum* growth (Romero et al. 2006). Although sediment nutrient concentrations in some tropical seagrass beds are indicative of nutrient growth limitation, *in situ* nutrient levels should not be confused with nutrient availability. The lack of a strong correlation between leaf productivities and sediment nutrient concentrations implies that *in situ* nutrient concentrations are not a good indicator of nutrients available for growth. Further, nutrient concentrations are not a good indicator of nutrient remineralisation; *in situ* nutrient concentrations represent a balance between inorganic nutrient production and consumption/loss to the water column. Regeneration of inorganic nutrients in sediments and turnover rates of nutrient pools are important factors that determine the degree of nutrient availability.

Nutrients essential for seagrass growth are partly derived from decomposition of organic matter in the water column and sediments. Bacterial N_2-fixation also provides substantial inputs of N to seagrass. Nutrient pools in seagrass beds often have rapid turnover rates ranging from 0.3 to 6-d and in *T. testudinum* meadows DIN turnover times in both the water column and in sediments can be <2-d (Lee and Dunton 1999; Romero et al. 2006) indicating the importance of nutrient regeneration in fuelling seagrass production.

There is no evidence of DIC limitation in tropical seagrasses, but seagrasses needing to assimilate large amounts of DIC may be restricted due to low solubility and diffusion rates in seawater. Many seagrasses have evolved the ability to utilise HCO_3^-, the most abundant form of DIC in seawater (Gavin and Durako 2019). Uncharged CO_2 may permeate cells by passive diffusion, whereas HCO_3^- acquisition usually involves extracellular carbonic anhydrase (CA) for HCO_3^- dehydration to CO_2 prior to uptake. Therefore, the affinity for CO_2 may be higher than that of HCO_3^-. However, utilisation of HCO_3^- is important in seagrasses if they are to sustain relatively high photosynthetic rates. Seagrasses are probably less efficient in the use of HCO_3^- than marine macroalgae (Lee et al. 2007). Consequently, seagrass photosynthesis may not be saturated with DIC at natural seawater concentrations. C limitation has been suggested for both temperate and tropical species (Lee et al. 2007). A common physiological feature for seagrasses may be carbon limitation of photosynthesis, and this limitation may be intensified in stagnant waters due to boundary layer resistance of CO_2 (Lee et al. 2007).

NPP of tropical seagrasses varies among species and with environmental conditions (Table 8.3). Aboveground NPP has been measured frequently compared with belowground productivity, with aboveground rates ranging from 0.01 to 75 g DW m^{-2} d^{-1} with most rates <5 g DW m^{-2} d^{-1}. Most measurements are of *T. testudinum* in the USA and of *Enhalus acoroides* in the Philippines, Indonesia, and Papua New Guinea.

In an analysis of seagrass biomass and production, Duarte and Chiscano (1999) observed that aboveground biomass tends to peak in boreal latitudes (50–60°) and belowground biomass tends to be lowest in the tropical latitudes (10°), although there is no significant trend with latitude (Figure 8.6). Aboveground NPP peaks in the boreal latitudes (50–60°) and belowground NPP peaks in the boreal and equatorial (0°) latitudes. The ratio of above-to belowground biomass tends to increase with latitude,

TABLE 8.3 Aboveground primary productivity (ANPP = g DW m^{-2} d^{-1}) of subtropical and tropical seagrasses worldwide.

Species	Location	ANPP	References
Cymodocea nodosa	Tunisia	1.0	Sghaier et al. (2011)
Cymodocea rotundata	The Philippines	0.5	Agawin et al. (2001)
C. rotundata	Kenya	2.0	Udy and Bjork (2005)
Cymodocea serrulata	Kenya	2.2	Udy and Bjork (2005)
C. serrulata	India	19.2	Kaladharan and Raj (1989)
C. serrulata	India	2-12	Govindasamy et al. (2013)
C. serrulata	Australia	1.2	Udy and Dennison (1997)
C. serrulata	Australia	1.3	Udy and Dennison (1997)
C. serrulata	Mozambique	0.6	de Boer (2000)
Enhalus acoroides	Papua New Guinea	1.8	Brouns and Heijs (1986)
E. acoroides	Papua New Guinea	2.6	Brouns and Heijs (1986)
E. acoroides	Indonesia	0.1–0.6	Erftemeijer and Herman (1994)
E. acoroides	Indonesia	0.2–0.4	Erftemeijer and Herman (1994)
E. acoroides	Indonesia	0.3	Erftemeijer et al. (1993)
E. acoroides	Indonesia	0.7	Erftemeijer et al. (1993)
E. acoroides	The Philippines	75	Terrados et al. (1999)
E. acoroides	The Philippines	60	Terrados et al. (1999)
E. acoroides	The Philippines	45	Terrados et al. (1999)
E. acoroides	The Philippines	3.1	Estacion and Fortes (1988)
E. acoroides	The Philippines	52	Terrados et al. (1999)
E. acoroides	The Philippines	0.4	Agawin et al. (2001)
Halodule univeris	Australia	2.0	Udy et al. (1999)
H. univeris	Australia	1.3	Udy et al. (1999)
H. univeris	Australia	1.2	Udy and Dennison (1997)
H. univeris	Australia	1.7	Udy and Dennison (1997)
Halodule wrightii	USA	0.01	Powell et al. (1989)
H. wrightii	Mozambique	0.2	de Boer (2000)
Halophila stipulacea	Jordan	2.2	Wahbeh (1984)
H. stipulacea	Jordan	0.2	Cardini et al. (2017)
Syringodium filiforme	Australia	0.3	Udy et al. (1999)
S. filiforme	USA	1.4	Short et al. (1993)

TABLE 8.3 *(Continued)*

Species	Location	ANPP	References
S. filiforme	USA	1.5	Short et al. (1993)
S. filiforme	USA	1.8	Short et al. (1993)
Syringodium isoetifolium	India	3-14	Govindasamy et al. (2013)
Thalassia hemprichii	Kenya	2.9	Udy and Bjork (2005)
T. hemprichii	Kenya	5.4	Udy and Bjork (2005)
T. hemprichii	Indonesia	1.0-5.1	Erftemeijer and Herman (1994)
T. hemprichii	Indonesia	4.7	Erftemeijer et al. (1994)
T. hemprichii	The Philippines	4.1	Agawin et al. (2001)
T. hemprichii	Indonesia	3.5	Erftemeijer et al. (1993)
T. hemprichii	Indonesia	3.6	Erftemeijer et al. (1993)
T. hemprichii	Tanzania	2.0	Githaiga et al. (2016)
T. hemprichii	Mozambique	1.1	Larsson (2009)
Thalassia testudinum	Belize	0.7	Tomasko and Lapointe (1991)
T. testudinum	Belize	0.5	Tomasko and Lapointe (1991)
T. testudinum	Belize	1.3	Tomasko and Lapointe (1991)
T. testudinum	Belize	0.9	Tomasko and Lapointe (1991)
T. testudinum	Costa Rica	3.3	Paynter et al. (2001)
T. testudinum	USA	0.3	Tomasko and Lapointe (1991)
T. testudinum	USA	1.1	Tomasko and Lapointe (1991)
T. testudinum	USA	0.8	Tomasko and Lapointe (1991)
T. testudinum	USA	1.4	Tomasko and Lapointe (1991)
T. testudinum	USA	3.9	Lapointe et al. (1994)
T. testudinum	USA	3.5	Lapointe et al. (1994)
T. testudinum	USA	4.2	Lapointe et al. (1994)
T. testudinum	USA	5.6	Lapointe et al. (1994)
T. testudinum	USA	8.2	Lapointe et al. (1994)

(continued)

TABLE 8.3 *(Continued)*

Species	Location	ANPP	References
T. testudinum	USA	1.5	Lapointe et al. (1994)
T. testudinum	USA	3.5	Lapointe et al. (1994)
T. testudinum	USA	3.7	Lapointe et al. (1994)
T. testudinum	USA	4.3	Lapointe et al. (1994)
T. testudinum	USA	5.1	Lapointe et al. (1994)
T. testudinum	USA	1.8	Powell et al. (1989)
T. testudinum	USA	0.5–2.0	Kaldy and Dunton (2000)
T. testudinum	USA	0.4–3.5	Kaldy and Dunton (2000)
T. testudinum	USA	0.1–0.8	Kaldy and Dunton (2000)
T. testudinum	USA	0.1–1.7	Kaldy and Dunton (2000)
T. testudinum	USA	1.4	Lee and Dunton (1999)
T. testudinum	USA	2.5	Lee and Dunton (2000)
T. testudinum	USA	0.07–5.9	Lee and Dunton (1996)
T. testudinum	USA	4.3	Lee and Dunton (1999)
T. testudinum	USA	4.9–9.4	Lee and Dunton (2000)
T. testudinum	USA	1.8–5.9	Heck et al. (2000)
T. testudinum	USA	3.6–8.5	Long et al. (2015)
Thalassodendron ciliatum	Kenya	3.7	Udy and Bjork (2005)
T. ciliatum	Kenya	4.9	Hemminga et al. (1995)
T. ciliatum	Kenya	9.5	Hemminga et al. (1995)
T. ciliatum	Kenya	7.5	Hemminga et al. (1995)
T. ciliatum	Kenya	3.4	Udy and Bjork (2005)
Zostera capensis	Mozambique	0.2	de Boer (2000)

whereas the ratio of above-to belowground NPP peaks at 10° S or N (Figure 8.6). Turnover rates of seagrass biomass average 2.6% d^{-1} and 0.77% d^{-1} for above- and belowground, respectively, indicating that leaf biomass turns over almost 3× faster than belowground biomass. There is a tendency for some tropical species to sustain faster aboveground turnover rates than temperate species, but there is no clear latitudinal pattern in above- or belowground turnover rates. The data for all species indicate that mean above- and belowground biomass is similar, 223.9 and 237.4 g DW m^{-2}, respectively, with a general trend of equal distribution of biomass between leaves and roots plus rhizomes. Thus, aboveground biomass and productivity underestimate total seagrass biomass and production.

Primary production of belowground roots and rhizomes of tropical species can be large, despite the trend of an increasing ratio of aboveground to belowground

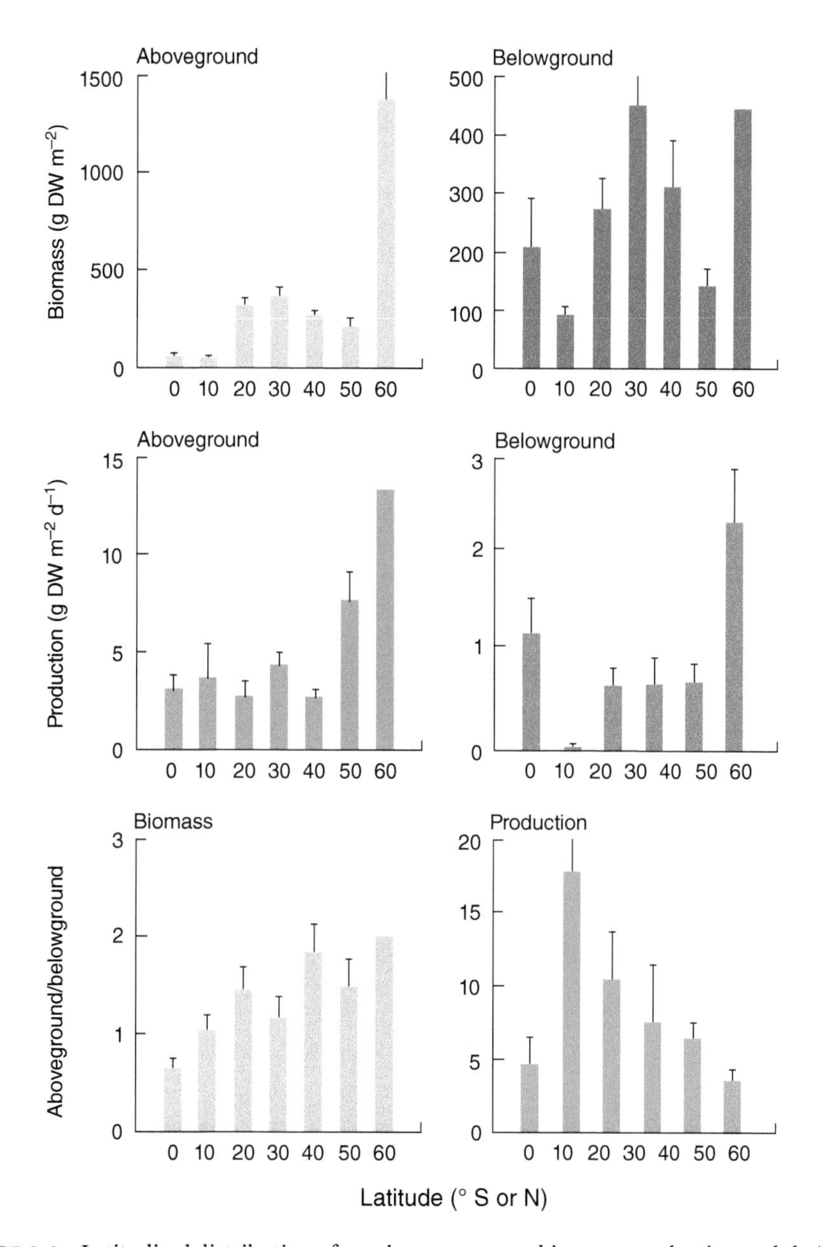

FIGURE 8.6 Latitudinal distribution of maximum seagrass biomass, production and their above- to below-ground allocation. Bars depict mean + 1 standard error. *Source*: Duarte and Chiscano (1999), figure 4, p. 168. © Elsevier.

production with decreasing latitude. Net rhizome production in *T. hemprichii* beds averaged 0.1–0.3 g C m^{-2} d^{-1} with aboveground NPP <500 mg C m^{-2} d^{-1} (Erftemeijer et al. 1993). Aboveground leaf production of *S. filiforme* ranged from 502 to 664 g DW m^{-2} a^{-1}, and root and rhizome production ranged from 81–108 to 145–152 g DW m^{-2} a^{-1}, respectively, in the subtropical Indian River Lagoon in Florida (Short et al. 1993). Biomass was even more skewed towards aboveground allocation with aboveground biomass varying from 1.93 to 9.42 g dry plant^{-1} and roots and rhizome biomass varying from 0.19 to 1.14 and from 0.46 to 2.74 g dry plant^{-1}, respectively, suggesting a faster turnover of belowground biomass than leaves. Above- and belowground production of *T. testudinum* in the Lower Laguna Madre in Texas averaged a leaf production value of 614 g DW m^{-2} a^{-1} and a belowground production of 339 g DW m^{-2} a^{-1} (Kaldy and Dunton (2000). On an annual basis, rhizome and root production accounted for 35% of total plant production. Rhizome growth was seasonal, correlating with underwater irradiance, daylength, and temperature. Total biomass ranged from 750 to 1500 g DW m^{-2}, with belowground biomass accounting for 80–90% of the total suggesting again that there is greater turnover belowground than aboveground.

Epiphytes and other autotrophs contribute substantially to total seagrass productivity, but leaf epiphytes can considerably reduce photosynthesis by their coverage of leaf area. Benthic microalgae accounted for 20% of total benthic NPP with seagrasses contributing the remaining 80% on Grand Bahama Bank (Dierssen et al. 2010). Epiphytes may contribute up to 36% to total seagrass productivity in meadows in eastern Indonesia; epiphyte productivity averaged 230 mg C m^{-2} leaf surface d^{-1} assuming an epiphyte coverage of 40% of total leaf area (Lindeboom and Sandee 1989). The maximum rate of net epibenthic production on *S. isoetifolium* averaged 2 g C g chl*a*$^{-1}$ h^{-1} and was attained only at light intensities above the leaf canopy because of shading (Pollard and Kogure 1993). Epibenthic and epiphytic production averaged 4.2 g C m^{-2} d^{-1} and 11.5 g C m^{-2} d^{-1} over the study period; the net epiphytic and seagrass productivities were equal and positively correlated. The net seagrass leaf production was only 25% of the total seagrass net production. Epiphytic algae accounted for more than four times aboveground NPP. Epiphytes contributed 2–9% of the total annual mean aboveground NPP in seagrass beds of Papua New Guinea (Brouns and Heijs 1986) in which annual mean production of epiphytes ranged from 39 to 65 mg DW m^{-2} d^{-1}. Annual mean biomass of epiphytes and seagrass leaves in other Papua New Guinea meadows ranged from 54 g DW m^{-2} in a community of *C. rotundata* to 169 g DW m^{-2} in a community of *S. isoetifolium* (Heijs 1985). The contribution of the epiphytes to total aboveground biomass ranged from 22 to 24%. Epiphyte production contributed 35–44% of total aboveground production of four distinct seagrass communities, being highest (2.1 g DW m^{-2} sediment surface d^{-1}) on leaves of *Halodule univervis*.

8.5 Coral Reefs

The photosynthetic machinery of corals is exceedingly complex and regulated by a wide variety of factors and is intimately linked to production of $CaCO_3$ (Gattuso et al. 1999a,b). Coral reefs display the greatest diversity and abundance of $CaCO_3$-depositing organisms that carry out photosynthesis (calcareous algae) or harbour photosynthetic symbionts (hermatypic corals, foraminifera, molluscs). DIC is used by the animal host to deposit skeletal $CaCO_3$ and by the endosymbiont for its photosynthesis (Furla et al. 2000; Sheppard et al. 2018). Calcification, respiration, and photosynthesis take place simultaneously:

$$CO_2 + H_2O \rightarrow CH_2O + O_2 \text{ (photosynthesis)}$$

$$CH_2O + O_2 \rightarrow CO_2 + H_2O \text{ (respiration)}$$

$$Ca^{2+} + 2HCO_3^- \rightarrow CaCO_3 + CO_2 + H_2O \text{ (calcification)}$$

Photosynthesis and calcification both consume inorganic carbon, but both processes can be thought of as mutually supporting because the CO_2 generated by calcification can be used in photosynthesis During most of the daylight period, photosynthesis is higher than respiration with the resulting process being:

$$Ca^{2+} + 2HCO_3^- \rightarrow CaCO_3 + CH_2O + O_2$$

At night, the processes combine to:

$$Ca^{2+} + 2\,HCO_3^- + CH_2O + O_2 \rightarrow CaCO_3 + 2CO_2 + 2\,H_2O$$

The ratio of CO_2 released to $CaCO_3$ precipitated is approximately 0.6 in seawater due to its buffering capacity. Thus, the amount of CO_2 generated by calcification that can potentially be used in photosynthesis is lower than suggested by the equations. Calcification is carried out by the ectodermal cells of the aboral layers called the calicoblastic epithelium. Photosynthesis is conducted by zooxanthellae, which are located mostly in the endothermal cells of the oral layers (Baums et al. 2014).

The coelenteric fluid is composed of constituents of photosynthesis, calcification, and advective exchange of seawater, which occurs through the mouth and/or by transepithelial transport mechanisms (Allemand et al. 2011). Whether or not passage is active or passive, transepithelial ion fluxes occur through transcellular and/or paracellular pathways and are required to provide material fluxes to and from the sites of calcification and photosynthesis. The transcellular pathway involves carriers (membrane proteins) and at least one energy-dependent step, either uptake by or efflux from the cell. The paracellular pathway is driven by molecular diffusion through lateral cell junctions that enable attachment of cells.

As for all invertebrates, the mechanisms of Ca^{2+} transport for calcification in corals are poorly understood. The incorporation of Ca^{2+} into the coral skeleton is an active process, following a saturable kinetics with respect to external Ca^{2+} concentration that implies an enzyme-mediated process (Yuan et al. 2018). At normal salinity, the rate of calcification is saturated in some coral species, but not all. Thus, Ca^{2+} limitation can occur. There is only one transcellular energy-dependent step in Ca^{2+} transport, located in the calicoblastic epithelium; the other steps result from rapid paracellular pathways (Sevilgen et al. 2019). Ca^{2+} does not diffuse freely into the cell, and specialised transport proteins are required to mediate Ca^{2+} entry across the lipid bilayer of the aboral ectoderm. Ca^{2+} ions, once inside the cell, must be sequestered in vesicles or organelles and/or bound to Ca^{2+}-binding proteins to maintain a low free intracellular concentration. Ca^{2+} may exit the calicoblastic cells via two mechanisms: Na^+/Ca^{2+} exchanges and Ca^{2+} ATPase.

External HCO_3^- is the preferred DIC substrate for coral photosynthesis (Furla et al. 2000). Seawater is the DIC reservoir used by free-living algae while the immediate source used by zooxanthellae is in the coral host cell which is ultimately derived from external seawater. At ambient DIC concentrations, carbon supply is saturated. A carbon concentrating mechanism, which actively absorbs HCO_3^- to sustain photosynthesis, has been suggested to exist because of the relatively high affinity of photosynthetic O_2 release for inorganic carbon. Rubisco (ribulose biphosphate carboxylase-oxygenase) is an enzyme that catalyses net photosynthetic carbon fixation that also exhibits a high affinity for oxygen which decreases its photosynthetic efficiency. The uptake of HCO_3^- by the host cell involves two anion carriers that are sensitive to DIDS, an inhibitor of anion transport. One of them is either a

Na^+-dependent Cl^-/HCO_3^- exchanger or a Na^+/HCO_3^- cotransporter; the nature of the second carrier is unclear. Within the endodermal cell, HCO_3^- is dehydrated into CO_2.

The source and transport of inorganic carbon for coral calcification are poorly understood, although DIC from seawater can be incorporated into the coral skeleton (Furla et al. 2000). There is some evidence that DIC from internal sources is used. There is direct evidence that some CO_2 generated by host respiration is deposited in the skeleton. Approximately 40% of the carbon supply is from seawater DIC and 60% from recycled metabolic CO_2. More than one HCO_3^- transport mechanism is probably involved in calcification.

Interactions between photosynthesis and calcification are complex. Calcification in the light is significantly greater than calcification in the dark, suggesting a clear link between the two processes. The ratio of light: dark calcification ranges from negative to 127, but the median is 3.0 (Allemand et al. 2011). These data were collected over a wide range of environmental conditions, both *in situ* and in the field, which probably accounts of the wide range of values. Nevertheless, calcification in the light is linked to photosynthesis. Several mechanisms have been invoked to explain higher calcification in the light. These include uptake by zooxanthellae of animal metabolic wastes or substances interfering with $CaCO_3$ precipitation, the translocation of photosynthate to fuel active transport mechanisms or to synthesise the organic matrix, or the increase of $CaCO_3$ saturation by photosynthetic CO_2 uptake and the maintenance of an oxidised environment (Gattuso et al. 1999a,b). It is unclear which of these mechanisms is the dominant process.

There are two hypotheses regarding the interactions between calcification and photosynthesis (Gattuso et al. 1999a,b). In the first hypothesis, photosynthetic CO_2 uptake lowers the extracellular CO_2 partial pressure in the coral tissue, increasing carbonate saturation and favouring precipitation of $CaCO_3$. The second hypothesis suggests that Ca^{2+}-ATPase supplies Ca^{2+} to and removes H^+ from the site of calcification. This mechanism favours carbonate precipitation by maintaining an elevated pH of the extracytoplasmic calcifying fluid and photosynthesis by increasing the CO_2 reservoir. Both hypotheses imply a high concentration of HCO_3^- in the coelenteron to buffer calcification-induced H^+. Very few data, however, are available on the relationship between calcification and photosynthesis in zooxanthellate corals.

The relationship between calcification and photosynthesis is slightly >1 although the data are scant, and it is not always clear whether gross photosynthesis (G) or net photosynthesis (P_n) is being reported (Gattuso et al. 1999b). $G:P_n$ relates to the amount of CO_2 that can potentially be supplied by calcification to the CO_2 required by net photosynthesis. The median ratio of $G:P_n$ is 1.3 indicating that CO_2 generated by $CaCO_3$ deposition could potentially supply 78% of the inorganic carbon required for photosynthesis (Gattuso et al. 1999b; Falter et al. 2013). However, the ratio varies widely from −8 to 17, implying that this contribution may not occur in all coral species.

At the community level, coral/algal reef flats exhibit wide ranges of community gross primary production (GPP: 79–584 mol C m^{-2} a^{-1}), respiration (R: 76–538 mol C m^{-2} a^{-1}), and net calcification (G: 5–126 mol $CaCO_3$ m^{-2} a^{-1}). Differences in community composition play a major role in the variability of the data. Nevertheless, G and P_g are significantly correlated and the slope of the regression is 0.2 which is the ratio of $G:P_g$ for all communities including algal-dominated zones, whole reefs, lagoons, sediments, reef flats, and algal pavements (Hatcher 1990). Except for algal-dominated areas, no significant correlations were found when these communities were examined separately. In fact, an opposite trend (decrease in G with increasing P_g) was observed in lagoons, algal pavements, and whole reefs. Nevertheless, most daily $G{:}\,P_g$ ratios are within the range of 0.1–0.4 and rarely exceed 0.5; the idea that $G{:}\,P_g = 1$ in photosynthetic and calcifying communities is not supported by the literature.

Daily calcification and net community production at a variety of sites are not significantly correlated. When the water residence time is short, such as on reef flats, the CO_2 generated by community calcification is not the major source of DIC sustaining net community production (Albright et al. 2013, 2015). The large decrease in the partial pressure of CO_2 (pCO_2) measured in such systems during the day demonstrates that DIC is drawn from the seawater reservoir and to a much lesser extent, from atmospheric CO_2. Organic production controls calcification when the residence time is longer. pH and $CaCO_3$ saturation rise when the net community production exceeds the influx of atmospheric CO_2. Calcification is therefore stimulated, increases pCO_2, and reduces the CO_2 influx by the net community carbon metabolism. The net air-sea CO_2 flux is close to zero in such systems due to the tight interaction between the inorganic carbon and organic carbon metabolism.

Most reef flats are sources of CO_2 to the atmosphere due to their low net fixation of CO_2 by photosynthesis and rather large release of CO_2 by precipitation of $CaCO_3$ (Albright et al. 2013, 2015). The one notable exception is algal-dominated reef communities, which exhibit a larger community excess production and/or a lower community calcification and are sinks for atmospheric CO_2.

The complex dynamics of photosynthesis, respiration, and calcification result in widely varying rates of GPP and NPP among different coral reefs, different zones of individual coral reefs, and different organisms. Measurements of organic production and calcification have been carried out in several coral reef ecosystems (reviewed by Kinsey 1985; see Table 2 in Shamberger et al. 2011). Community gross primary production (GPP or P_g) is frequently high with an average value of 7 g C m^{-2} d^{-1} (Kinsey 1985). Community respiration (R) is of the same order of magnitude as P_g. Consequently, the community net primary productivity ('excess production') is close to zero (0 ± 0.7 g C m^{-2} d^{-1}, Crossland et al. 1991). Net carbon fluxes in coral reefs are thus of minor significance except on some epilithic turf mats and the reef excess production is $\approx$ 0.05% of the net CO_2 fixation rate of the global ocean (Crossland et al. 1991). The average net calcification of coral reef flats is 5.0 ± 0.7 kg $CaCO_3$ m^{-2} a^{-1} (Falter et al. 2013). The global carbon fixation by coral reefs via calcification was calculated at about 0.1 Gt a^{-1} (Kinsey and Hopley 1991). However, the fact that calcification is also a source of CO_2 to the water column and atmosphere was not considered in these calculations.

Hatcher (1990) summarized the metabolic measurements of various reef zones and entire reefs (Figure 8.7). Primary productivity can vary dramatically among reefs. The gross carbon fixation rate of 7 g C m^{-2} d^{-1} as established by Kinsey (1985), and a P:R ratio of 1 applies primarily to windward reef flats near sea level in the Pacific Ocean; the gross primary production of reef flat communities at different locations and at different stages of development often deviates from these averages. Differences in the mean productivities derived from different reef zones are often significant, leading Kinsey (1985) to identify three modes of production: coral/algal, algal pavement, and sediment with gross production rates of 20, 5 and 1 g C m^{-2} d^{-1}, respectively. At the ecosystem scale, reef margins in the eastern Caribbean appear to be more productive than those in the Pacific. Many Caribbean reefs are still active, with upward growth as they catch up or keep up with sea level. This contrasts with Pacific reefs that have reached sea level a few thousand years ago and that grow mainly on their peripheries.

As for patterns within reefs, zones of high and low primary productivity alternate based on wave forcing, sea level, community composition, and geomorphology (Falter et al. 2013). Lagoonal sands exhibit productivities typically <1 g C m^{-2} d^{-1} and are located downstream of productive reef flats (Figure 8.7). This spatial separation has led to the view that different reef zones contribute differently to the overall function of an entire coral reef. Reef structure is a major determinant in the net production of coral reef ecosystems. The back reef, for example (Figure 8.7) exhibits a wide range of values for P_n (NPP) from $^-$1.7 to 27 g C m^{-2} d^{-1} compared with the reef crest (P_n = 0.3–1.5 g C m^{-2} d^{-1}). However, despite these different ranges, entire reefs have a P_g range of 2.1–6.4 g C m^{-2} d^{-1}, a P_n range of 0.01–0.20 g C m^{-2} d^{-1}, and a P_g: R ratio of approximately 1.

Reef zones often show no clear differences in productivity due to strong seasonality (Cardini et al. 2016). In a seasonally variable fringing coral reef in the Red Sea, P_g, R, and P_n varied widely across different reef zones (Table 8.4). Seasonal changes were large due to dramatic seasonal changes in water mass movements and temperature. Only a relatively depauperate sand belt had low production and respiration rates. As pointed out by Hatcher (1988), coral reefs are metabolically speaking

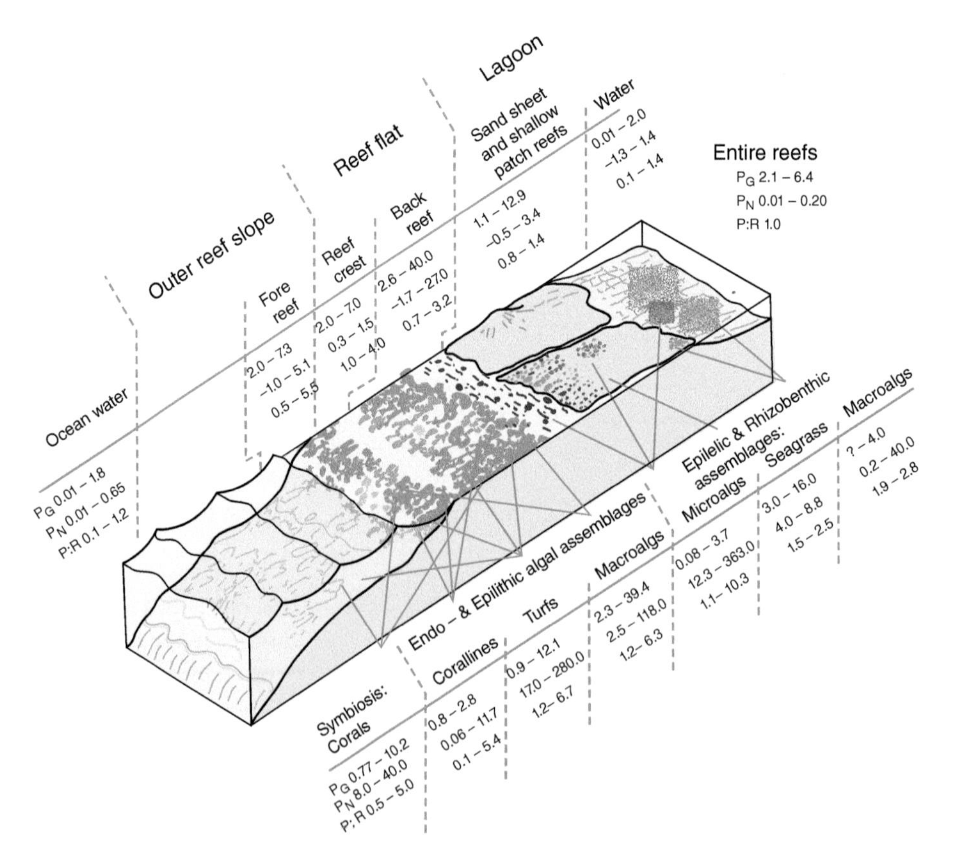

FIGURE 8.7 Primary productivity on different coral reef zones. The minimum and maximum diel gross (P_g) and net (P_n) community primary production (g C m^{-2} d^{-1}) and the diel community gross production to respiration (P:R) ratio. Along the bottom of the diagram is the minimum and maximum diel gross primary productivity (g C m^{-2} d^{-1}), diel weight-specific net primary productivity (mg C g dry weight^{-1} d^{-1}) and diel gross P:R ratio for each major component of reef productivity. *Source*: Updated and adapted from Hatcher (1990) using references in this section. © John Wiley & Sons.

TABLE 8.4 **Rates (mmol C m^{-2} d^{-1}) of gross primary production, respiration, and net primary production in different zones of a highly seasonal fringing coral reef, Red Sea. Values are means and ranges in parentheses.**

Zone	P_g	R	P_n
Fore reef	327 (284–359)	190 (167–223)	137 (116–187)
Reef crest	246 (222–266)	137 (121–165)	109 (88–139)
Reef flat	205 (178–240)	109 (86–145)	96 (88–112)
Sand belt	149 (117–173)	85 (75–97)	65 (42–95)
Transition zone	268 (234–303)	154 (134–186)	114 (95–156)
Water column	5 (3–7)	46 (30–55)	—

Source: Adapted from Cardini et al. (2016). © John Wiley & Sons.

a "beggar's banquet" in which the "food is plain and hard to find, and at the end of the day nobody leaves with much." Maximum areal rates rival the best agricultural systems, but most organic production is conserved and recycled.

Upwelling increases NPP in coral reefs. Under upwelling conditions on a coral reef on the Pacific coast of Costa Rica, P_n of all algal taxa remained comparatively uniform despite high nutrient levels, but P_n of corals increased by 70% and reef P_g increased by >40% (Stuhldreier et al. 2015). Due to high benthic coverage, corals contributed 60% to total reef P_n and P_g and turf algae contributed 25% to primary production. A stimulatory response was also measured on Caribbean coral reefs on the Colombian coast (Eidens et al. 2014). Hermatypic corals showed highest P_n and P_g rates during the non-upwelling season, but corals and algal turfs dominated the primary production during upwelling. At the ecosystem level, corals contributed most to primary production, but during upwelling, corals were significant contributors only at an exposed site. On both reef systems, primary productivity was uniformly high and stable year-round, indicating that these reefs are acclimated to seasonal variations in nutrient concentrations and water conditions such as temperature, O_2, and light.

Coral bleaching has an impact on coral reef calcification and production. Net organic carbon production and net inorganic carbon production declined in the massive hermatypic corals, *Favites* sp. and *Porites* sp., when subjected to bleaching in mesocosms (Fujimura et al. 2001). Bleaching resulted in the decrease in calcification in daytime and an increase in carbonate dissolution at night. The *Pg:R* ratio remained nearly constant (1.2) suggesting that the biological activity of zooxanthellae was not significantly altered by bleaching. Net ecosystem calcification after bleaching of a coral reef at Lizard Island, Australia, was 40–46% lower than prior to bleaching (McMahon et al. 2019).

As for individual contributors, endo- and epilithic algal assemblages are the most productive components, even more productive than symbiotic corals (Hatcher 1990; Falter et al. 2013). On the Great Barrier Reef, crustose coralline algae were highly productive with net productivities in the range of 15–132 mmol O_2 m^{-2} d^{-1} (Chisholm 2003). Multiplying these estimates by the actual surface area, the productivity yields estimated contributions to reef organic production of 0.9–5 g C (net) planar m^{-2} d^{-1} over the depth interval of 0–18 m, suggesting that crustose coralline algae make a larger contribution to coral reef net primary productivity that previously thought. Endolithic algae were major primary producers in dead carbonate substrates on Hawaiian coral reefs (Tribollet et al. 2006) with a P_n of 2.3 g C m^{-2} d^{-1}. The contribution of endolithic communities to community net production ranged from 32 to 46% of the entire reef NPP. Similarly, epilithic algal communities (EAC) on reefs in the Gulf of Aqaba and on the Great Barrier Reef were highly productive, with high maximal rates of net photosynthesis of 8–25 mmol O_2 m^{-2} h^{-1} and dark respiration of 1.3–3.3 mmol O_2 m^{-2} h^{-1} (Larkum et al. 2003). The EAC were a shade-adapted community; pH increased from 8.2 in the overlying seawater to 8.9 at the EAC surface suggesting that little carbon in the form of CO_2 occurs at the EAC surface. The major source of DIC must be in the form of HCO_3^-. Due to the diffusive boundary layer, DIC is likely to be a major limiting factor for EAC photosynthesis.

Soft sediments are a small source of fixed carbon on coral reefs. Algae on soft lagoonal sediments in a barrier reef flat in French Polynesia accounted for only 3–4 % of total reef P_n (Boucher et al. 1998) but accounted for 21–32% of total respiration on the reef. Phytoplankton similarly contributed little to total reef productivity. Lagoonal phytoplankton production in the Great Barrier Reef (Furnas et al. 1990) ranged from 0.2 to 1.6 g C m^{-2} d^{-1}, but biomass-specific production within individual reef lagoons was high (range: 41–90 mg chl^{-1} d^{-1}); production was inversely correlated with lagoon flushing rates and standing crop. Primary production in the Great Astrolabe Lagoon in Fiji averaged 1.3 g C m^{-2} d^{-1} with 53% due to phytoplankton (Charpy and Blanchot 1999). *Synechococcus* was the most abundant group at all sites, followed by *Prochlorococcus* and picoeukaryotes. In the upper 40 m of surrounding oceanic water, phytoplankton structure was different, with diatoms being present, although *Prochlorococcus* was the most abundant phytoplankton group. In Takapoto Atoll on the Tuamotu Archipelago in French Polynesia (Delesalle et al. 2001), P_g averaged 8 g C m^{-2} d^{-1} whereas P_n averaged 0.7 g C m^{-2} d^{-1}. No significant long-term changes in biomass or primary productivity were observed over a 24 year period, but there was a shift in composition from diatoms, dinoflagellates and coccolithophorids to pico- and nanophytoplankton, mainly chlorophytes, prymnesiophytes, and dinoflagellates; the microplankton were absent 22 years later. Picoplankton was dominant, representing >60% of total biomass and >50% of primary productivity. It was unclear why the phytoplankton shifted to a smaller size range as there were no obvious changes in environmental conditions.

Plankton productivity and P_n of reef organisms and substrates in coral reefs vary depending on stratified and mixed conditions. Although P_g remained similar between stratified and mixed conditions on a coral reef in the northern Red Sea (Tilstra et al. 2018), P_n decreased from mixed to stratified conditions. Heterotrophy increased significantly during the stratified period. In coral reefs in the Gulf of Aqaba, northern Red Sea, different species and substrates differed in response to stratified or mixed conditions (Table 8.5). Most organisms and substrates tended to be more productive under mixed conditions, although only P_n rates of the hermatypic coral, *Stylophora* sp., and the sponge, *Mycale* sp. were significantly less under stratified conditions. Coral reefs and their associated organisms tend to flourish when mass water movements through the reef system are vigorous and oxygenated.

Macrophytes contribute significantly to primary productivity of reef lagoons. Seagrass of a Caribbean reef lagoon in Mexico contributed 59% to total lagoonal production with a nearly equal percentage contributed by macroalgae, including *Halimeda* (Naumann et al. 2013). Microflora contributed only a tiny percentage to lagoonal primary production. In a reef lagoon in the Bahamas, all seagrasses had

TABLE 8.5 **Rates of NPP (μmol O_2 cm^{-2} d^{-1}) of different organisms and substrates under mixed and stratified conditions on coral reefs in the Gulf of Aqaba.**

Organism/substrate	Mixed	Stratified
Acropora sp. (hermatypic coral)	14.44±0.78	13.14±1.01
Stylophora sp. (hermatypic coral)	14.09±1.13	10.77±0.94
Pocillophora sp. (hermatypic coral)	11.11±1.32	11.93±1.27
Goniastrea sp. (hermatypic coral)	14.91±1.11	14.58±1.04
Turf algae	10.86±0.52	14.27±1.36
Mycale sp. (sponge)	−3.76±0.50	−7.78±1.05
Xeniidae (octocorals)	10.16±0.84	8.58±0.49
Sarcophyton sp. (octocoral)	4.88±0.63	5.31±0.79
Carbonate sand	6.49±1.63	4.23±0.50
Microbial mat	16.94±1.51	15.51±1.30
Caulerpa sp. (green alga)	23.75±2.38	19.29±1.29
Lobophora sp. (brown alga)	7.73±0.54	6.69±0.45

Values are mean ± 1 SE.*Source:* Bednarz et al. (2015a,b), Cardini et al. (2016), Rix et al. (2015) and Tilstra et al. (2018).

high NPP rates (1.65–2.29 g C m^{-2} d^{-1}) in meadows adjacent to mangroves (Koch and Madden 2001). NPP rates of epiphytes (5.22 g C kg^{-1} d^{-1}) and prop-root algae (8.54 g C kg^{-1} d^{-1}) approached those of seagrasses (10.49–13.18 g C kg^{-1} d^{-1}). In a large coral reef lagoon in New Caledonia, mean GPP was 12.06 mol C m^{-2} a^{-1} and mean R was 13.68 mol C m^{-2} a^{-1} for a P_g:R ratio of 0.88, indicating net heterotrophy (Clavier and Garrigue 1999). There were, however, large differences along a land-ocean gradient. Muddy bottoms were net heterotrophic, whereas the P_g:R ratio in sandy bottoms was either near 1 or slightly positive, indicating net autotrophy in some parts of the lagoon.

8.6 Coastal Lagoons, Estuaries, and Tidal Waterways

Tropical tidal waterways, estuaries, and coastal lagoons are sites of high primary productivity (Robertson and Blaber 1992; McKinnon et al. 2017). Production is linked to flushing times, nutrient availability, degree of pollution, groundwater discharge, and the extent of terrestrial input. For example, in India where inshore waters are highly polluted, waters are eutrophic with high rates of NPP (> 1 g C m^{-2} d^{-1}). In the poorly flushed coastal lagoons of the Ivory Coast, NPP may be up to 5 g C m^{-3} d^{-1}. Such high production rates occur in lagoons which are not only poorly flushed but that also receive significant quantities of N and P from adjacent human populations, usually in the form of sewage and agricultural runoff. However, even in systems that are not eutrophic, phytoplankton production can still be high. For example, daily production can be up to 2.4 g C m^{-3} d^{-1} in the coastal lagoons of Mexico. Phytoplankton production appears to be significantly lower in estuarine mangrove waters than in lagoons or open bays fringed by mangroves. For instance, daily production rates ranged from 22 to 693 mg C m^{-2} in the Fly River delta, Papua New Guinea (Robertson et al. 1998), similar to rates measured in Malaysian estuarine mangrove waters (Table 8.6). Although NPP varies widely among lagoons and tidal waterways (Table 8.6), they are consistently high compared with rates measured in similar temperate habitats (Cloern et al. 2014).

Chlorophyll concentrations are also highly variable. For instance, in pristine mangrove creeks in Missionary Bay, Australia and in the Fly River, Papua New Guinea, there is a wide range of chlorophyll concentrations (0.15–5.07 μg l^{-1}). In contrast, chlorophyll concentrations may reach 60 μg l^{-1} in polluted areas close to large human populations or where large monsoonal rainfall delivers high concentrations of nutrients to enclosed mangrove lagoons.

Phytoplankton production can contribute significantly to total ecosystem net primary productivity. For instance, phytoplankton production contributed 50% to total ecosystem production in Terminos Lagoon, Mexico, whereas mangroves and seagrasses contributed 43 and 7% to total production, respectively (Yáñez-Arancibia and Day 1982). There are large areas of relatively clear, shallow water in this lagoon, conducive to phytoplankton production. A similar situation has been observed in lagoonal ecosystems along the Ivory Coast and in the Cochin backwaters in India (Table 8.6). In estuarine mangrove ecosystems typical of most regions of Southeast Asia, Central America, and tropical South America, high turbidity, large fluctuations in salinity, and a relatively small ratio of open waterway to mangrove forest area all ensure that the contribution of phytoplankton to total ecosystem productivity is likely to be relatively small. Phytoplankton and mangroves contributed 20 and 80%, respectively, to total ecosystem production in the Fly River estuary in Papua New Guinea (Robertson and Alongi 1995). However, considering the poor nutritional quality of mangrove detritus, phytoplankton production may play a more important role in supporting higher trophic levels in estuarine mangroves than has been previously acknowledged.

Benthic primary production can be a significant fraction of NPP in coastal lagoons, estuaries, and tidal waterways. de Sousa et al. (1998) measured NPP of benthic microflora of 40.7–140 mg C m^{-2} h^{-1} in a polluted estuary in Brazil. There was a significant positive correlation between primary production and chlorophyll *a* concentration, with lower correlations with NO_3^-, N, P, and organic matter in the sediment. In contrast, seagrasses were the largest contributor to total ecosystem production in three Mexican coastal lagoons: in Celestun lagoon, 40% seagrass, 26% phytoplankton, and 34% macroalgae; in Chelem Lagoon, 43% seagrass, 28% phytoplankton, and 29% macroalgae; and in Dzilam Lagoon, 53% seagrass, 40% phytoplankton, and 7% macroalgae (Herrera-Silveira et al. 2002). The Celestun Lagoon exhibits a tendency towards eutrophication, whereas Chelem Lagoon is oligotrophic. Dzilem Lagoon is like Celestun Lagoon in terms of trophic status, but with longer water residence times. These lagoons are greatly influenced by groundwater discharge. Benthic NPP in a subtropical, hypersaline coastal lagoon in SE Brazil (Knoppers et al. 1996) was much higher (185–406 mg C m^{-2} d^{-1}) than pelagic NPP (20–53 mg C m^{-2} d^{-1}). As in most hypersaline lagoons, primary production is largely driven by benthic communities which thrive at the sediment–water interface as patchy algal films or consolidated (up to

TABLE 8.6 **Net primary production (NPP) and chlorophyll *a* concentration ($\mu g\ l^{-1}$) in tropical tidal waterways, estuaries, and lagoons. NPP is expressed as either mg C $m^{-3}\ d^{-1}$ or mg C $m^{-2}\ d^{-1}$ (italics).**

Location	Habitat	Chlorophyll *a*	NPP
Gambia	Estuarine mangroves	0.3–8.2	1–445
Ghana	Coastal lagoon, surface		626–1992
Ghana	Coastal lagoon, bottom		222–600
Ghana	Coastal lagoon, surface		385–1420
Ghana	Coastal lagoon, bottom		120–320
Ivory Coast	Coastal lagoons	10.3–16.4	*200–5000*
Mauritania	Mangrove embayment	0.46–3.60	580
Mauritania	Mangrove creek	0.20–1.07	215
Guadeloupe	Mangrove channel	10–60	8–1700
Colombia	Coastal lagoon	6.0–182.0	
Costa Rica	Mangrove creek		*120–2760*
Costa Rica	Estuarine waters	2.7–20.0	*3014–6043*
Mexico	Coastal lagoons	0.3–8.2	*420–1440*
Brazil	Estuarine mangroves	1.08–19.26	110–500 (*100–800*)
Mexico	Coastal lagoon, Pacific		*2450*
Bangladesh	Mangrove estuary		*15–86*
India	Estuarine waters	2.5–14.0	*232–1211*
India	Mangrove waters		*266–833*
India	Coastal lagoons	4.36–39.8	60–662
India	Estuarine mangroves	2.1	*190–1540*
Thailand	Embayment		*560–2410*
Pakistan	Mangrove creeks	1.0–40.0	*200–1000*
Malaysia	Estuarine mangroves		*274–959*
Taiwan	Coastal lagoon	5.8–300.2	*26.6–5036*
Taiwan	Estuarine waters	0.3–4.4	596–618
Malaysia	Estuarine mangroves	0.53–21.2	10–1068
Papua New Guinea	Estuarine mangroves	0.25–5.07	22–693
Australia	Mangrove creek	1.3	45–820
Australia	Polluted mangrove creeks	0.4–49.8	*1572–2988*
Australia	Estuarine waters	0.2–37.4	147–594
Australia	Mangrove estuary, creek		78

(*continued*)

TABLE 8.6 *(Continued)*

Location	Habitat	Chlorophyll *a*	NPP
USA	Subtropical lagoon	0.5–23.4	*50–2800*
Vietnam	Estuarine waters	0.80–22.88	*32–3514*

Source: Robertson and Blaber (1992), Harrison et al. (1997), Herrera-Silveira (1998), Pennock et al. (1999), Gocke et al. (2001a,b, 2004), McKinnon et al. (2002), Sridhar et al. (2006), Senthilkumar et al. (2008), Biswas et al. (2010), Rochelle-Newall et al. (2011), Hsieh et al. (2012), Kumar and Perumal (2012), Smith et al. (2012), Dix et al. (2013), Kanuri et al. (2013), Pamplona et al. (2013), Nandan et al. (2014), Rahman et al. (2014), Seguro et al. (2015), Pan et al. (2016), Saifullah et al. (2016), McKinnon et al. (2017) and Vargas-Zamora et al. (2018).

many cm thick) algal-bacterial mats. Pelagic and benthic primary production in most of these environments is limited by P due to P binding in $CaCO_3$-rich sediments.

Salinity and water residence times are important in determining the relative contributions of benthic and pelagic primary producers. Benthic primary production was low compared with pelagic production in Terminos Lagoon (Yáñez-Arancibia and Day 1982), but phytoplankton productivity, respiration, and chlorophyll levels were higher in turbid, low-salinity, river-influenced areas than in clear waters of the lagoon. Phytoplankton productivity ranged from 0.87 to 5.5 g O_2 m^{-3} d^{-1}. Seagrass productivity was highest during the dry season and in clear, high-salinity waters (0.68–4.6 g cm^{-2} d^{-1}). Benthic microalgal productivity was low (0.3 g cm^{-2} d^{-1}) and limited to a narrow zone nearshore; GPP was 2.6 g O_2 m^{-2} d^{-1}. There was a seasonal pattern in chlorophyll levels with the lowest mean values in the dry season and the highest values during and following the rainy season. Freshwater input also affects mangrove productivity with higher rates of production in the riverine areas.

Comparatively few measurements have been made of benthic and pelagic metabolism in tropical estuaries. One of the most comprehensive studies was conducted in two Indian estuaries, the Mandovi and Zuari Rivers, and adjoining coastal waters during the pre-monsoon, monsoon, and post-monsoonal seasons. At the estuarine stations, net ecosystem production showed monthly variation and a transition from net autotrophy during the pre-monsoon and post-monsoon seasons to net heterotrophy in the monsoon season (Table 8.7). Seasonal monsoon-driven changes, such as increased allochthonous inputs, resulted in enhanced heterotrophic respiration and reduced primary production. In adjacent coastal waters, the monthly variation in net ecosystem production was not significant and net heterotrophy was prevalent, thereby serving as a net source of CO_2 to the atmosphere. Excess organic matter from these tropical rivers supports heterotrophy in adjacent coastal waters. There is thus seasonal switching in net ecosystem production in these two estuaries.

The monsoon period is the time for rapid runoff of riverine materials and a period of high turbidity fostering net heterotrophy. Similar results were obtained in another study of the Mandovi-Zuari estuarine complex (Kumari et al. 2002) in which monsoonal NPP was lower (9.53 mmol C m^{-2} d^{-1}) than during the pre-monsoon (19.48 mmol C m^{-2} d^{-1}) and post-monsoon (20.61 mmol C m^{-2} d^{-1}) periods. This pattern suggests that maximum riverine runoff during the monsoon season imparts high turbidity to estuarine waters thus limiting light for phytoplankton. In an Australian estuary with similarly distinct wet and dry seasons, freshwater inputs did not increase nutrient and chlorophyll *a* concentrations and phytoplankton

TABLE 8.7 **Median and range of the integrated primary production (NPP) and community respiration (R) and production to respiration ratio (NPP:R) in the Mandovi and Zuari estuaries and adjacent coastal waters.**

Stations	NPP (mmol C m^{-2} d^{-1})	R (mmol C m^{-2} d^{-1})	NPP:R
Mandovi			
Pre-monsoon	112.4 (72.2–188.2)	75.3 (36.2–83.0)	1.49
Monsoon	23.0 (10.1–40.1)	83.3 (40.2–159.4)	0.28
Post-monsoon	100.4 (33.3–146.7)	42.3 (37.5–69.5)	2.37
Zuari			
Pre-monsoon	122.2 (86.2–148.2)	48.3 (46.9–89.2)	2.53
Monsoon	61.7 (23.6–153.2)	91.6 (46.9–148.2)	0.67
Post-monsoon	72.9 (26.5–116.1)	44.2 (21.4–53.7)	1.65
Coastal			
Pre-monsoon	90.3 (79.6–127.3)	468.3 (337.7–692.5)	0.19
Post-monsoon	154.7 (49.7–188.9)	339.5 (161.0–386.7)	0.46

Source: Ram et al. (2003). © American Society of Limnology and Oceanography, Inc.

productivity rates decreased, probably due to low water residence times (Burford et al. 2012). There was no evidence of post-flood stimulation of chlorophyll *a* concentration, with net export of nutrients in the wet and dry seasons. N and light appeared to be the key limiting factors for phytoplankton growth and productivity.

Dry and wet monsoonal differences in plankton production have been observed in most tropical estuaries. Most detailed evidence comes from India showing that phytoplankton production is limited during the wet season due to changes in salinity and low light levels (Madhu et al. 2010a,b; Pednekar et al. 2011; Acharyya et al. 2012; Thaw et al. 2017). Large increases in primary production in the Cochin estuary indicated mesotrophic conditions during the pre-monsoon season and was attributed to a large increase in nanoplankton numbers (Madhu 2010a,b); these conditions extended to the coastal waters off the estuary. In some Indian estuaries, the damming of rivers upstream and subsequent reduced discharge has given rise to intense phytoplankton blooms (Pednekar et al. 2011; Acharyya et al. 2012). The end of the dry season is the most productive off the estuaries of the Myanmar coast with low productivity during the rainy season because of highly turbid conditions. The turbidity is partly due to soil erosion from deforestation and mangrove deterioration within the adjacent estuaries (Thaw et al. 2017).

Metabolic processes in tropical estuaries are greatly affected by seasonal differences in rainfall and runoff (Sarma et al. 2009, 2011; Chaudhuri et al. 2012). High bacterial respiration resulted in high pCO_2 concentrations during peak river discharge in the Gautami Godavari estuarine system in India, whereas respiration was much lower during the dry period (Sarma et al. 2009). In 27 Indian estuaries, the emission of CO_2 to the atmosphere was 4–5× higher during the wet than in the dry season indicating greater wet season pelagic respiration (Sarma et al. 2011). The mean pCO_2 and POC concentrations showed positive correlations with rates of river discharge suggesting availability of higher quantities of organic matter that led to enhanced microbial decomposition.

Allochthonous input into tropical estuaries can produce autotrophic-heterotrophic switching in the water column. For example, bacterial productivity (BP) and respiration (BR) were examined in relation to primary productivity in the shallow Cochin estuary in India (Thottathil et al. 2008). The degree of dependence of bacterial productivity (BP = 6.3–199.7 μg C l^{-1} d^{-1}) and bacterial respiration (BR= 6.6–430.4 μg C l^{-1} d^{-1}) on primary productivity (NPP = 2.1–608.0 μg C l^{-1} d^{-1}) was weak. The BP:NPP (0.05–8.5) and NPP:BR (0.03–7.9) ratios varied widely in the estuary depending on season and location. There was a seasonal shift from autotrophy to heterotrophy due to terrestrial inputs which enhanced bacterial heterotrophic activity and produced high pCO_2 levels. The heterotrophic zones were characterised by low NPP, but high BP and BR led to oxygen under-saturation and exceptionally high pCO_2 concentrations. Thottathil et al. (2008) proposed that the CO_2 supersaturation caused by increased BR was a result of bacterial degradation of allochthonous organic matter, indicating that sources other than planktonic components are needed to explain carbon cycling in this estuary.

Further studies in the Godavari estuary have examined the influence of river discharge during the monsoon season on planktonic metabolic rates. During the peak discharge period, significant amounts of DIN and DIP along with suspended materials were found with net heterotrophy and low GPP. Conversely, high productivity was sustained for about 1.5 months during October–November when net community production turned from net heterotrophy to autotrophy. The mean NPP:R ratio was 2.38, but the GPP:R ratio was only 0.14 suggesting that primary production was not enough to support water-column heterotrophic activity. Carbon demand was likely met by inputs and utilization of allochthonous organic matter. The additional terrestrial organic carbon needed to support total bacterial activity was 97–99% during peak discharge period and 40–75% during the dry period, suggesting that large amounts of terrestrial organic carbon were being decomposed in the estuary.

Allochthonous material is necessary to support bacterial and primary production in other tropical estuaries. In the upper Gulf of Nicoya, an extremely productive, phytoplankton-dominated estuary with an annual GPP of 1037, NPP of 610, and R of 427 g C m^{-2} a^{-1} (Gocke et al. 2001b), highest NPP occurred during the dry season and at the beginning of the rainy season. High water turbidity reduced the euphotic zone to 4–5 m depth making the system light limited. On an annual basis, 41% of the organic carbon produced in the estuary was consumed in the euphotic layer and 79% was consumed in the water column. The ecosystem receives a considerable amount of organic material from the mangrove forests bordering the gulf, so a surplus of organic carbon is exported from the upper Gulf of Nicoya, enhancing the overall productivity of the lower gulf and the adjacent coastal zone.

Phytoplankton and bacterioplankton production (BP) varied between the two seasons and along the transects in the highly turbid, tropical Bach Dang estuary in Vietnam (Rochelle-Newall et al. 2011). This estuary is polluted with mercury (Hg), and the Hg cycle is tightly coupled to phytoplankton dynamics. BP greatly exceeded NPP in the estuary indicating that DOC fuelling BP originated from sources other than plankton. In contrast, at the high salinity site offshore, the BP:NPP ratio was much lower, meaning that more DOC was being produced during primary production than was required for bacterial biomass production, potentially leading to an accumulation of DOC in the water column. Phytoplankton community structure in the estuary was likely influenced by concentrations of inorganic mercury, methyl mercury, and tri, di- and mono-butyl tin, potentially affecting NPP. Differences in rates of NPP between the estuary and offshore may have been due to shifting phytoplankton diversity due to the metal pollution in the estuary. Such conditions need to be considered as to how pollution affects productivity in tropical rivers and estuaries given that many tropical estuaries are no longer pristine.

8.7 Shelf Seas

Rates of primary production are often higher on tropical continental shelves than on temperate shelves, but primary production is greater in some areas of the boreal shelves and in the great upwelling regions (Walsh 1988; Alongi 1998; Jahnke 2010). Peak primary productivity on tropical shelves is most often associated with 'estuarisation' of the inner and middle shelf and with coastal upwelling along the shelf edge. 'Estuarisation' occurs when monsoonal rains are delivered onto the shelf proper by rivers; the best data exist, not surprisingly, for the large river plumes. The Amazon plume during high discharge periods extends to the outer shelf and can have profound effects on primary productivity and benthic processes (Gouveia et al. 2019). Chlorophyll concentrations were greatest (up to 25.5 $\mu g\ l^{-1}$) in a zone located outside the turbid, high nutrient, low salinity, riverine waters (Smith and DeMaster 1996), but shoreward of the clear, high salinity, low nutrient waters. Maximum chlorophyll levels occurred in the upper 5 m of water characterised by reduced salinities and elevated nutrients at the surface and low nutrient, high salinity water below. Primary productivity on the Amazon shelf was greatest in the transition zone and can occasionally exceed 8 g C $m^{-2}\ d^{-1}$, whereas productivity in the turbid, nutrient-rich waters and the clear, offshore regions often average 2.1 and 0.8 g C $m^{-2}\ d^{-1}$, respectively (Smith and DeMaster 1996). Photosynthesis in Amazon waters on the shelf appears to be limited by low light levels inshore, whereas offshore primary production is nutrient-limited. The narrow zone of high primary production is supported by the riverine input of nutrients and the dynamics of sediment flocculation.

A schematic representation of primary productivity in relation to salinity for the large tidal rivers (Alongi 1998; Dagg et al. 2004) reveals two patterns (Figure 8.8): algal blooms occur either within the river plumes (the Amazon and Fly Rivers) or offshore (Changjiang, Mekong, and Zaire Rivers). The difference in shelf location is related to shelf topography. At or near the mouths of the Amazon and Fly Rivers, a shoal area results in rapid sedimentation and lower turbidity, enough to permit rapid phytoplankton growth inshore. In contrast, sediments remain in suspension at much higher salinities off the Changjiang and Zaire Rivers owing to a deeper water column off the mouth of both estuaries. Despite clear differences in spatial and seasonal patterns of primary production and end-member concentrations of nutrients, the range of rates of primary production are similar among the rivers.

Primary productivity near large tidal rivers is regulated by availability of light and nutrients, but it has been difficult to reconcile which factor plays the dominant role and why. Most studies have shown that in highly turbid riverine waters, low light limits primary productivity (Robertson et al. 1998; Furnas and Carpenter 2016; McKinnon et al. 2017). The situation becomes more complex, however, in less turbid

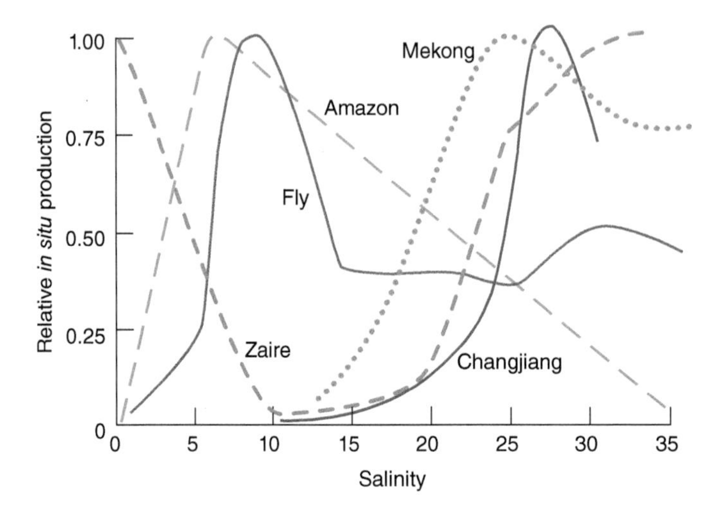

FIGURE 8.8 Patterns of primary production in relation to salinity in the plumes of the Amazon, Changjiang, Zaire, Mekong, and Fly Rivers. *Source*: Dagg et al. (2004) Robertson (1993, 1998) and Grosse et al. (2010). © Elsevier.

waters where nutrient concentrations often deviate from conservative mixing either due to biological uptake and/or uptake or release from particles. On the Amazon shelf, light limits production within the inshore plume, but on the outer shelf, NO_3^- and water residence time apparently regulate production (Brunskill 2010). In river/ocean mixing zones off the Amazon, buoyancy-driven, shelf-edge upwelling drives plankton production and N cycling. The strength of upwelling varies with intensity of river discharge. During rising river discharge, algal blooms develop coinciding with peak NO_3^- flux; long residence time allows phytoplankton to maximise growth and production within the region of relatively low turbidity and high nutrients. Inshore, sediment resuspension is a primary mechanism for nutrient flux to the water column, but on the shelf proper, the chief source of N enhancing productivity is likely to be the shelf-edge upwelling rather than the river per se.

Rates of heterotrophic respiration and bacterial activity on shelves influenced by large rivers are often high, owing to high nutrient concentrations and rates of primary productivity (Robertson et al. 1998; Dagg et al. 2004). Within the Fly River delta and adjacent Gulf of Papua, bacterial respiration accounted for most of the total pelagic respiration (McKinnon et al. 2007). Inshore waters were net heterotrophic and microbially dominated; rates of respiration far exceeded those of primary productivity. Pelagic food chains in the Fly delta and Gulf of Papua are fuelled primary by phytoplankton fixed carbon and by detritus advected from the extensive freshwater marshes and estuarine mangroves within the delta. It remains to be seen whether pelagic food chains of other large tropical and subtropical rivers are like those on the Brazilian and Papuan shelves, being fuelled by carbon fixed by phytoplankton and by tidal export of allochthonous detrital carbon.

On tropical shelves without significant upwelling, pelagic dynamics is different. On the tropical Atlantic continental shelf off Sierra Leone, phytoplankton production was balanced by grazers, but only in the wet season (Longhurst 1983). On the shelf and in the dry season estuary, phytoplankton production exceeded demand by 70–90%. Regional production was dominated by the dry season estuary where diatom blooms supported large populations of the clupeid fish, *Ethmalalosa fimbriata*. Large amounts of inshore organic carbon production may have been available for export, or burial, or both, either in inshore mud banks or at the shelf edge by slumping. In Peninsular Malaysia waters, higher riverine nutrient input and a well-mixed water column in the wet season resulted in primary production up to 83 μg C l^{-1} h^{-1} (Hee et al. 2020). During the dry season, primary production was lower (7 μg C l^{-1} h^{-1}) and driven by nutrient inputs of offshore southern waters and from the Gulf of Thailand in offshore northern waters.

Intrusions of cold, nutrient-rich waters due to coastal upwelling, especially along eastern boundary upwelling systems, stimulate phytoplankton production (Messié and Chavez 2015). Upwelling along the coasts of the eastern tropical Pacific foster enhanced rates of primary productivity (Pennington et al. 2006). Coastal boundaries are characterized by high standing stocks of phytoplankton and rates of phytoplankton production, with the Peruvian coastal upwelling region being by far the richest in terms of plankton biomass and production (Pennington et al. 2006). The Gulfs of Tehuantepac, Papagayo, and Panama are areas of enhanced primary production due to nutrients supplied by wind-driven upwelling and near surface mixing. Wind jets, mixing, eddies, and the Costa Rica Dome enhance shelf and open ocean biological productivity in what otherwise would be an oligotrophic eastern Pacific warm pool. Coastal upwelling off NE New Guinea results in high rates of primary production, although rates are low close to the mouth of the Sepik River where light levels are low from high suspended sediment loads (Stanley et al. 2010). Episodic upwelling occurs on the Great Barrier Reef shelf and along the northern coast of Australia in the Timor and Arafura Seas, driving periodic blooms of phytoplankton (Furnas and Carpenter 2016).

Monsoons trigger upwelling along other tropical continental margins. The best-studied example is the northern Arabian Sea where equatorial, open sea, and coastal upwelling all conjoin to foster high rates of primary productivity (Moffett and Landry 2020). Upwelling occurs in the Arabian Sea off the coast of Oman around to the southern tip of India and the Somali upwelling regime farther south in the equatorial western Indian Ocean is also generated by the onset of monsoons. Such monsoon-driven events have a significant impact on pelagic food webs and nutrient cycles in shelf waters. Measurements suggest high rates of primary production in the western Arabian Sea during the SW monsoon; rates of production and standing stocks decline eastwards partly due to iron limitation until coastal influences, such as from the Indus delta and smaller rivers along the west coast of India, and coastal upwelling facilitate an increase in plankton production in planktonic and benthic production. Off the Indian coast, red tides result from phytoplankton blooms that occur during both monsoon seasons. The intrusions of upwelled water during the SW monsoon may penetrate river mouths, stimulating high abundance of phytoplankton and zooplankton.

In the eastern Arabian Sea, air–sea interactions, coastal circulation, and primary production exhibit an annual cycle (Gupta et al. 2016). During June–September, strong SW winds promote sea surface cooling through surface heat loss and vertical mixing in the central Arabian Sea and force the West India Coastal Current equatorward. Ekman pumping in the northern Arabian Sea is promoted by positive wind stress curl induced by the jet stream, and equatorward directed along shore wind stress induces upwelling which lowers sea surface temperature along the SW shelf of India and enhances phytoplankton biomass by >70% as compared to that in the central Arabian Sea. During the winter monsoon, dry and weak NE winds enhance convective cooling of the upper ocean and deepens the mixed layer, thereby increasing the vertical flux of nutrients which promote phytoplankton blooms in the northern Arabian Sea. The primary production rate averaged for the entire western Indian shelf increases from winter to summer monsoon from 0.8 to 2.3 g C m^{-2} d^{-1} and from 0.3 to 24 mg m^{-2} d^{-1}, respectively. Coastal Kelvin waves from the Bay of Bengal propagate into the coastal Arabian Sea, which modulate circulation patterns along the western Indian shelf. Seasonal monsoonal forcing and remotely forced waves modulate the circulation and primary production in the eastern Arabian Sea. The primary productivity of the Bay of Bengal is forced by a different set of events. The large freshwater influx during summer inhibits the upward transport of nutrients into the eutrophic zone by wind mixing thereby curtailing phytoplankton production. Mesoscale eddies which are ubiquitous in the Bay of Bengal control the bulk of the observed biological productivity. Tropical cyclones occurring during the spring and autumn inter-monsoons also trigger high primary productivity over short time scales (Kumar et al. 2007).

What do these complex processes mean for the net community production in the northern Indian Ocean? The monthly/seasonal variability in net community production is controlled by the monsoon either through increasing nutrients to the euphotic zone or an increase in river runoff. Net autotrophy is found during the SW monsoon in the Arabian Sea, and heterotrophy is found during the spring and autumn inter-monsoon periods. There is large spatial heterogeneity in net production with autotrophy in the west and heterotrophy in the eastern Arabian Sea. In contrast to the Arabian Sea, net heterotrophy dominates in the Bay of Bengal during the SW monsoon. Bacterial biomass and productivity are higher during the SW monsoon in the Bay of Bengal. Increased river discharge leads to net heterotrophy because of enhanced community respiration, resulting in higher pCO_2 and thus flux of CO_2 to the atmosphere in the Bay of Bengal during the SW monsoon season. The annual mean net community production to the northern Indian Ocean amounts to 300–318 Tg C (Sarma et al. 2009, 2019). Fixed carbon that sinks to the aphotic zone is enough to support bacterial carbon demand in the deep ocean.

The average primary production in the inshore and the offshore waters of the Bay of Bengal ranged from 252 to 350 mg C m^{-2} d^{-1} (Sarma et al. 2019). Subsurface chlorophyll maxima were observed at 20–50 m in the spring inter-monsoon, whereas in summer and winter, the chlorophyll maxima occurred in the upper layers of the euphotic zone. The vertical distribution of phytoplankton was restricted to the pycnocline and the bulk of phytoplankton were within the surface mixed layer. Diatoms were the major group (>85%) in total phytoplankton abundance. During the spring inter-monsoon, large patches of *Trichodesmium* occurred in the coastal zone.

Subtropical and tropical continental shelves can be categorised based on primary productivity, location, and major river systems (Table 8.8). The data show that low-latitude upwelling areas on the eastern boundaries of the ocean are often the most productive shelf ecosystems and that subtropical and tropical river dominated shelves on the ocean's western boundaries are nearly as productive as shelves in temperate seas. Unfortunately, few of these margins are well understood, despite their importance to humans.

Three generalisations have been made about continental margins in low latitudes:

- Export of materials from the continent is usually restricted to the coastal zone by complex physical, chemical, and geological processes, but this material greatly influences inshore food chains and nutrient cycles.
- Impingement of nutrient-rich, open ocean water enhances primary and secondary productivity on the outer shelf and shelf edge, including important fisheries, particularly along boundary currents.
- Little organic matter is exported from the continental margin.

TABLE 8.8 Categorisation of the world's subtropical and tropical continental shelves based on location, major river, and primary productivity (NPP = $g\ C\ m^{-2}\ a^{-1}$).

Region	Major river	NPP
Eastern Boundary Current		
Ecuador-Chile	—	1000–2000
SW Africa	—	1000–2000
NW Africa	—	200–500
Eastern Tropical Atlantic	—	157
Caribbean	—	190
Baja California	—	600
Somali coast	Juba	175
Arabian Sea	Indus	200
Indian Ocean Monsoon Gyre	—	105
Western Boundary Current		
Brazil	Amazon	90
Gulf of Guinea	Congo	130
Oman/Persian Gulfs	Tigris	80
Bay of Bengal	Ganges	110
Andaman Sea	Ayeyarwady	50
Guinea Current Coastal	—	495
Guiana Current Coastal		699
Western India Continental Shelf	—	369
East Africa Coastal	—	190
Australia-Indonesia Coastal	—	199
Sunda-Arafura Seas Coastal	—	328
South Pacific Subtropical Gyre	—	87
Java/Banda Seas	Brantas	110
Western Tropical Atlantic	—	130
Timor Sea	Fitzroy	100
Coral Sea	Fly	20–175
Arafura Sea	Mitchell	150
Red Sea	Awash	35
Mozambique Channel	Zambesi	100–150
South China Sea	Mekong	215–317
Caribbean Sea	Orinoco	66–139

TABLE 8.8 *(Continued)*

Region	Major river	NPP
Central America	Magdalena	180
West Florida shelf	Appalachicola	30
South Atlantic Bight	Altamaha	130–350

Source: Longhurst et al. (1995), Alongi (1998), Pennington et al. (2006) and Stanley et al. (2010).

The most important factors determining to what extent a shelf ecosystem validates or violates these generalisations include the presence or absence of rivers, presence or absence of upwelling, the location on ocean boundaries, and shelf width. All four factors are influenced by climate.

Tropical continental shelves exhibit physical, chemical, and geological characteristics different from those of temperate and boreal latitude (Chapters 3 and 4), including shelves composed of gradients of terrigenous to carbonate facies. The Great Barrier Reef (GBR) is the exemplar of such a shelf margin and where primary productivity is highly variable, ranging from 129 to 1972 mg C m^{-2} d^{-1} (Furnas and Carpenter 2016). Phytoplankton primary production rates per unit area generally increase with latitude on the GBR shelf. Shelf areas are frequently disturbed by cyclones, after which primary production rates between 230 and 4300 mg C m^{-2} d^{-1} have been measured. Total shelf production was estimated to be between 217 and 260 t C d^{-1}, equivalent to an average annual shelf primary production of 354–424 g C m^{-2} with phytoplankton contributing 58–65% of total production. Seasonality of pelagic community respiration and net community production on the shelf was the greatest determinant of respiration rate with median respiration rates of 1.85 mmol O_2 m^{-3} d^{-1} in the dry season and 2.87 mmol O_2 m^{-3} d^{-1} in the wet season (McKinnon et al. 2013). Net community production normally ranged up to 9.16 mmol O_2 m^{-3} d^{-1} depending on depth and phytoplankton biomass. Both respiration and net community production were higher inshore. Shelf waters were net autotrophic with a median GPP:R ratio of 1.5. Peaks in net community production were associated with floods, upwelling of nutrient-rich, sub-thermocline Coral Sea water and localised phytoplankton blooms, but otherwise showed few seasonal or spatial trends. Respiration was comparable to rates in other oligotrophic oceanic waters, but net community production was more typical of other shelf ecosystems.

Size fractionation of phytoplankton in Great Barrier Reef shelf waters indicated that growth rates of diatoms exceeded those of pico-cyanobacteria when dissolved inorganic nitrogen (DIN) concentrations were $\geq$0.05 µmol l^{-1} (Crosbie and Furnas 2001). The pico-cyanobacteria achieved higher growth rates than diatoms when DIN concentrations were $\leq$0.05 µmol l^{-1}. Growth rates of pico-cyanobacteria were not nutrient limited under ambient DIN concentrations and were of similar order of magnitude to those measured in the equatorial Pacific Ocean where NO_3^- concentrations are far above growth-saturating levels. In contrast, growth rates for nano- and micro-phytoplankton appeared to be N-limited at DIN concentrations $<$0.1 µmol l^{-1}, supporting the notion that pico-cyanobacteria dominate oligotrophic waters because of their ability to grow at low nutrient concentrations.

Not all tropical shelves are productive. Many coastal and shelf areas in the dry tropics are unproductive. For example, NPP was <25 mg C m^{-3} d^{-1} and chlorophyll *a* was 0.2–0.3 mg m^{-3} in Exmouth Gulf in arid Western Australia (Ayukai and Miller 1998). The low levels of phytoplankton biomass and production were ascribed to the small terrestrial inputs to the gulf. In dilution experiments, the proportion of phytoplankton grazed ranged from 79 to 155%. On the eastern side of Exmouth Gulf, grazer biomass ranged between 4.6 and 8.8 mg C m^{-3}, not much less than the estimated phytoplankton biomass of 6–15 mg C m^{-3}; this grazer community appeared to consume more organic matter than the phytoplankton biomass could produce. This relatively high grazing pressure and disproportionately large grazer biomass may have been due to inputs of additional organic matter from benthic macroalgal communities and/or from mangroves and salt flats inhabiting the eastern side of the gulf.

Phytoplankton biomass and productivity in tropical arid Western Australia have been linked to ENSO events (Furnas 2007). There were episodic intrusions of upper thermocline waters and subsurface chlorophyll maxima on the outer shelf during the 1997/1998 ENSO event. Despite similarities in irradiance, temperature, and wind stress, large differences were observed in phytoplankton biomass, community structure, and productivity between the summers of 1997/1998 and 1998/1999. Phytoplankton and bacterial production were two to fourfold higher during the summer of 1997/1998 than in 1998/1999. Larger phytoplankton during the ENSO event were composed mostly of diatoms. There were no signs of upwelling, but NPP reached levels (3–8 g C m^{-2} d^{-1}) characteristic of eastern boundary upwelling zones. Bacterial production was 0.6–145% of primary production indicating intense bacterioplankton–phytoplankton coupling.

Phytoplankton production along northern Australia is also stimulated by monsoonal and cyclone activity. Primary production and chlorophyll levels in the monsoonal Gulf of Carpentaria were similar within the coastal boundary layer between wet and dry seasons where waters are well-mixed year-round (Burford and Rothlisberg 1999; Rothlisberg and Burford 2016). NPP was low (557 mg C m^{-2} d^{-1}) in well-mixed deeper (>20 m) waters of the gulf compared with stratified waters during the summer (955 mg C m^{-2} d^{-1}). The low production was due to high turbidity and light attenuation. Nutrient additions and ^{15}N experiments showed no N limitation. Light, rather than nutrients, limited phytoplankton production during winter in offshore waters (Rothlisberg and Burford 2016). Rothlisberg et al. (1994) earlier measured high rates of phytoplankton activity during the summer wet season. Mean NPP was 914 mg C m^{-2} d^{-1} and was 1430 mg C m^{-2} d^{-1} in the shallow coastal waters and 660 mg C m^{-2} d^{-1} in the central gulf. The highest rates of growth occurred within a very narrow light regime.

Intense storms and cyclones can trigger high rates of primary productivity as has been found off the coast of Zanzibar Island off the east coast of Africa (Lugomela et al. 2001). During the rainy season, NPP ranged from 204 to 4142 mg C m^{-2} d^{-1}. About 77% of primary production was funnelled through heterotrophic flagellates, ciliates, and heterotrophic dinoflagellates indicating intense grazing. In the South China Sea, satellite data were used to estimate the effect of a tropical cyclone on primary production (Lin et al. 2003). In July 2000, Cyclone Kai-Tak passed over the shelf edge causing an average 30-fold increase in surface chlorophyll *a* concentration. The contribution of tropical cyclones to annual new production in the South China Sea may be as much as 20–30%. Nearly 80% of this tropical cyclone-induced annual primary production was the result of the biogeochemical response to the 30% strongest tropical cyclones (Menkes et al. 2016). The passage of a cyclone off Western Australia similarly increased phytoplankton abundance, biomass, and primary productivity (McKinnon et al. 2003). Net primary production off Peninsular Malaysia ranged from 8.49 to 55.95 μg C l^{-1} h^{-1}, whereas bacterial production ranged from 0.17 to 70.66 μg C l^{-1} h^{-1} (Lee and Bong 2008) with higher rates over very intense storms during the wet season.

Even short-lived wind events can trigger intense blooms of phytoplankton activity. In the Gulf of Tehuantepec off Mexico, productivity rates after a wind event were comparable to those measured in upwelling-driven ecosystems; NPP ranged from 0.07 to 1.43 g C m^{-2} d^{-1}. The Gulf of Tehuantepac acts as a nutrient and phytoplankton-carbon pump enriching the offshore waters of the eastern tropical Pacific (Pennington et al. 2006).

8.8 Open Ocean

Rates of primary production in the tropical open ocean waters vary greatly, depending on ocean physics and chemistry. Upwelling, downwelling, fronts, eddies, and gyres are but some of the physical processes influencing the structure of primary producers and their productivity (Longhurst 2007). A satellite image of global primary productivity (Figure 8.9) shows hot spots of productivity along the eastern boundary currents, particularly along the West African coast where the Guinea Dome lies and off Peru, as well as equatorial upwelling along a fine line across the tropical Pacific. The Costa Rica Dome is also a hot spot of primary productivity as well as deep water seas within the Indonesian archipelago and offshore areas of the Arabian Sea and off southern India and Sri Lanka.

Tropical instability waves play an important role in influencing primary productivity (Longhurst 2007; Vichi et al. 2008). Such waves have the characteristics of Rossby waves and are most evident within a band approximately 5° latitude N and

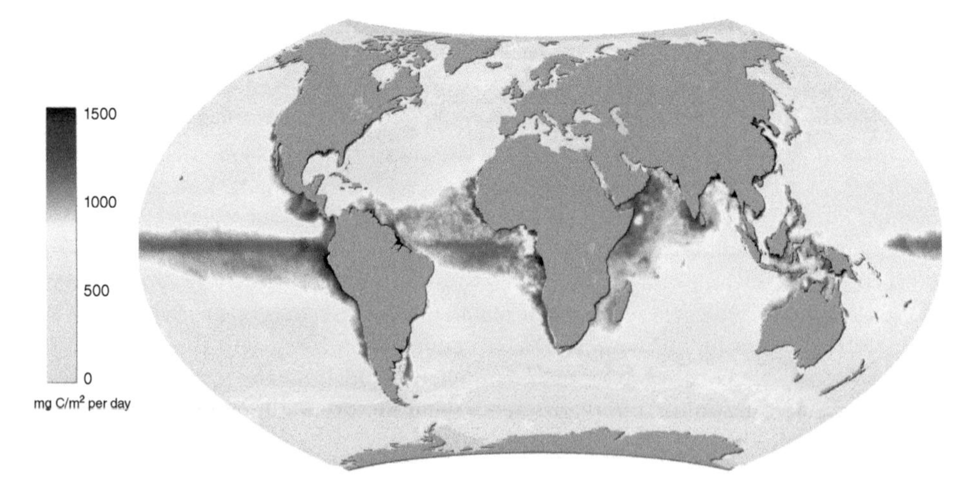

FIGURE 8.9 Satellite image of the global distribution of annual marine net primary productivity (mg C m^{-2} d^{-1}), 2017, based on the updated cbpm algorithm from VIIRS (GSM) data, prepared by Oregon State University. Monthly files have been averaged. *Source*: Image produced by Hugo Ahlenius and used with permission from Nordpil (acquired 18 June 2018). © Nordpil.

S of the equator. They propagate westwards with velocities of 50 km d^{-1}, a wavelength about a 1000 km, and a period about 20 d. These waves and their modulation have been described as a function of shear between the North Equatorial Counter Current, Equatorial Undercurrent, and the South Equatorial Current (SEC) and variability in the SEC and the equatorial SST front. Their magnitude is stronger north of the equator, and they are not observed during strong El Niño events. Conversely, they may be particularly visible during the onset of La Niña when the trade winds and equatorial currents regain strength. Tropical instability waves have also been described as westward-propagating trains of off-equator, anticyclonic vortices that exhibit large horizontal and vertical velocities which are likely to transport significant amounts of nutrients and biomass. These waves transport both the surface expression of the equatorial upwelling tongue and the cool nutrient-rich source waters that are upwelled to generate the phytoplankton bloom (Vichi et al. 2008). Thus, there should also be a component of enhanced production due to enhanced nutrient flux to the euphotic zone. That is, the very intense upwelling at the trailing edge of the cool SST anomaly entrains subsurface nutrients into the euphotic zone and stimulates phytoplankton growth, greater than in the existing bloom conditions of the equatorial upwelling tongue.

Primary productivity thus varies greatly temporally and spatially depending upon complex physical and chemical gradients. Tropical intra-seasonal variability is an important process, and a notable form of such variability is the MJO (Section 2.6), which is characterized by large-scale eastward-propagating disturbances in the tropics. The MJO produces systematic and significant variations in ocean surface chlorophyll (and likely primary productivity) in several regions across the tropical Pacific and Indian Oceans, including the northern Indian Ocean, a broad expanse of the NW tropical Pacific Ocean, and several near-coastal areas in the far eastern Pacific Ocean (Chang et al. 2019). Wind-induced vertical entrainment at the base of the ocean mixed layer appears to play an important role in modulating ocean chlorophyll.

In the tropical Atlantic, primary productivity may be P and N limited; nitrogen fixation may in turn be limited by atmospheric deposition of Fe (Chien et al. 2016). Nutrient fluxes are highest in the eastern tropical Atlantic owing to the proximity of West Africa and Europe and lowest in the western tropical Atlantic (Baker et al. 2007). Phytoplankton biomass and production across the eastern tropical Atlantic appear to be driven by major changes in nutrient supply across the thermocline, which average 0.99 mmol N m^{-2} d^{-1} and 0.13 mmol P m^{-2} d^{-1} and are particularly high in the upwelling area off NW Africa. From 2° N to 5° S, the equatorial divergence results in a shallowing of the pycnocline and the presence of relatively high NO_3^- concentrations in surface waters. Primary production can vary from about 200 mg C m^{-2} d^{-1} in less productive, subtropical gyres to about 1300 mg C m^{-2} d^{-1} in the equatorial divergence area (Pérez et al. 2005). Because of the relatively high primary production rates in the equatorial upwelling region, only a moderate rise in phytoplankton biomass occurs as compared to nearby nutrient-depleted areas. Picophytoplankton are the main contributors (>60%) to both biomass and production in this region. Equatorial upwelling does not alter the phytoplankton size structure typically found in the tropical open ocean, which suggests a strong top-down control of primary producers by zooplankton. The entire region shows net autotrophic community metabolism since respiration accounts for only about half of GPP in the euphotic zone. Below the euphotic zone lies a moderate OMZ; mesoscale eddies with close to anoxic oxygen concentrations are located just below the mixed layer. In these eddies, primary productivity is enhanced, and carbon uptake rates are up to 3× higher than in surrounding waters. Carbon uptake rates below the euphotic zone correlates with the presence of *Prochlorococcus*. The high primary productivity fuels export of production and supports enhanced respiration below the mixed layer (Löscher et al. 2015). There is significant seasonal variation in primary production in the eastern tropical Atlantic with the thermocline shoaling dramatically in response to seasonal intensification of the trade winds in the western tropical Atlantic. The dynamic uplift of the thermocline combined with mixing by local winds may be the physical mechanism controlling vertical nutrient fluxes and therefore phytoplankton blooms.

In the southern Caribbean Sea, annual primary production rates can reach levels of 200–400 g C m^{-2}, especially in waters influenced by discharge from the Orinoco River (López et al. 2013). The frequency is greater, and the area of blooms larger, during periods of strong wind. High mean pigment concentrations around Margarita than off central Venezuela reflect more intense upwelling near the broad continental shelf, where the horizontal, cross-wind scale of coastal upwelling may be smaller than over the narrow shelf. Production and pigment concentrations are often high near the capes and headlands. The Orinoco plume is located between Trinidad and Tobago during the first half of the year and during July–November the plume engulfs Tobago, Grenada, and St. Vincent. It is during the latter period that phytoplankton blooms occur reflecting high rates of primary productivity.

Across the tropical Pacific, there are several factors that control primary productivity. Fe plays a key role in regulating phytoplankton growth in both high NO_3^--low chlorophyll and oligotrophic waters near the equator and further south, whereas N and zooplankton grazing are the primary factors that regulate production in the north (Behrenfeld et al. 2006). Productivity in the tropical Pacific basin may be 1.2–2.5 Pg C a^{-1} lower than previous estimates; this difference is comparable to global productivity changes during El Niño to La Niña transitions.

In the eastern tropical Pacific, large-scale spatial patterns of primary production are determined by large-scale physical processes, and local-scale patterns depend on NO_3^- and PO_4^{3-} supply from below the thermocline (Stramma et al. 2013). Where the thermocline is shallow and intersects the euphotic zone, primary production is enhanced. Enhanced rates of macronutrient supply maintain levels of primary production above those of the oligotrophic subtropical gyres. Lack of Fe limits phytoplankton growth and nitrogen fixation over large areas of the eastern tropical Pacific, depressing rates of primary production and resulting in high NO_3^--low chlorophyll conditions (Li et al. 2015), with weak seasonal cycles over much of the open ocean. The ENSO fluctuation, however, is an exceedingly important source of interannual variability in this region. ENSO events result in a depressed thermocline and thus reduced rates of macronutrient supply and primary production (Turk et al. 2011).

The multidecadal PDO is also a source of variability (Section 2.6), with the 'El Viejo' phase of the PDO resulting in warmer and lower nutrient and productivity conditions like El Niño. The PDO influences both pelagic and benthic food webs. The PDO-related interannual SST anomaly in spring significantly affects periodic life cycle events of phytoplankton assemblages (Chiba et al. 2012). Phytoplankton blooms occur either later or earlier in years of positive (cold conditions) or negative (warm) PDO.

On average, the eastern tropical Pacific is moderately productive, occupying about 18% of Pacific Ocean area and accounting for 22–23% of its productivity. It occupies about 9% of the global ocean and accounts for 10% of global productivity (Pennington et al. 2006). Depth-integrated, regional primary productivity over decades has averaged 0.7 Gt C a^{-1} in the eastern boundary current, 0.9 Gt C a^{-1} in the equatorial upwelling, 1.3 Gt C a^{-1} in the South Equatorial Current, 1.1 in the South Pacific Subtropical Gyre, 0.4 Gt C a^{-1} in the North Equatorial Counter Current, 0.1 Gt C a^{-1} in the 10° N Thermocline Ridge, 0.2 Gt C a^{-1} in the North Equatorial Current, and 0.9 Gt C a^{-1} in coastal boundaries for a total of 6 Gt C a^{-1} for the entire eastern tropical Pacific (Pennington et al. 2006).

In the Indian Ocean, aside from the IOD, MJO, and ENSO phenomena, there is also strong monsoonal wind forcing that affects primary productivity (Pandey et al. 2019). The wind forcing is strongest in the western Indian Ocean leading to strong coastal and open ocean upwelling resulting in coastal divergence of Ekman transport and Ekman pumping, supplying nutrients to the surface, and supporting elevated rates of primary production. The western Indian Ocean is also the region with the largest warming trend in tropical SSTs during the past century. Roxy et al. (2016) point to an alarming decrease of up to 20% in phytoplankton production and biomass in the western Indian Ocean over the past six decades, driven by enhanced ocean stratification due to rapid warming in the Indian Ocean, which suppresses nutrient mixing from subsurface layers. In the central Indian Ocean, primary production in areas of the equatorial divergence and in waters of the Monsoon and Trade Wind Current has ranged from 0.3 to 1.5 g C m^{-2} d^{-1} (Madhupratap et al. 2003). Primary production in some surface waters was up to 15–20 mg C m^{-3} d^{-1} and in areas of convergence it decreased from 0.1 to 0.2 g C m^{-2} d^{-1}. Deep maxima of accumulation of active phytoplankton has been recorded at the upper boundary of the thermocline.

The area of highest sustained primary productivity in the Indian Ocean lies within the Arabian Sea (Figure 8.9). Phytoplankton productivity measurements and growth rates suggest that phytoplankton are not strongly limited by either irradiance or nutrient supply (Marra and Barber 2005). Grazing rates match phytoplankton net photosynthetic rates and growth. Variations in phytoplankton biomass have been explained by variations in mixing at diurnal and seasonal scales, and spatial variations caused by upwelling near the Omani coast and the presence of mesoscale eddies. Vertical mixing regulates the supply of irradiance and nutrients, but vertical mixing is never deep enough to limit phytoplankton productivity, and N does not appear to limit phytoplankton growth. Vertical mixing also affects grazing by dilution of micro-grazers along with phytoplankton. Marra and Barber (2005) argue that mixed layer deepening allows phytoplankton to escape grazing losses and grow, and thereby creating the observed variability in phytoplankton biomass. The most productive season in the Arabian Sea occurs during the SW monsoon when productivity peaks at 1476 mg C m^{-2} d^{-1}. Productivity is high year-round and during the NE monsoon it reaches $\approx$ 1344 mg C m^{-2} d^{-1} (Vase et al. 2018). During the 1995 SW monsoon, strong, narrow, and meandering current filaments extended from the region of coastal upwelling to about 700 km offshore. These filaments had levels of biomass, primary productivity, and diatom abundance that were elevated relative to other locations during the SW monsoon. The SW monsoon is usually the most productive period, but SW monsoon primary productivity has been lower than predicted because efficient grazing by mesozooplankton has kept diatoms from accumulating the biomass necessary for achieving high rates of primary productivity (Moffett and Landry 2020). The high rates of chlorophyll-specific productivity observed during the 1995 SW monsoon, together with the observed dust flux and Fe concentrations, indicate that the Arabian Sea is more Fe replete than the equatorial Pacific.

In the less productive Bay of Bengal, there are periods of enhanced primary productivity (Sarma et al. 2019). Kumar et al. (2004, 2007) proposed eddy pumping as a possible mechanism of vertical transfer of nutrients across the halocline to the oligotrophic euphotic zone during the summer monsoon when the upper ocean is highly stratified. This would induce rapid biological uptake and primary production. In the northern Bay of Bengal, riverine input acts as an additional source of nutrients and augments the subsurface nutrient injection to the euphotic zone by eddy pumping. However, low rates of primary production in the northern Bay are likely due to light limitation from huge riverine discharge.

In Southeast Asia, the Indonesian throughflow (Figure 3.2) is responsible for most coastal and offshore upwellings throughout Indonesian seas. Upwelling and subsequent enhanced primary productivity occurs in the deep Banda Sea, located west of the island of New Guinea, and along the south coasts of many of the islands of Indonesia, including Timor, Sumatra, Bali, and Java. In the nearby South China Sea and in the Sulu Sea, seasonal distributions of phytoplankton relate closely to the East Asian Monsoon (Ning et al. 2004; Chen 2005; Chen et al. 2006; Ferrera et al. 2017). The summer SW monsoon induces upwelling along the China and Vietnam coasts. Several mesoscale cyclonic cold eddies and anticyclonic warm pools occur in both seasons. In the summer, the upwelling and cold eddies, both associated with rich nutrients, low dissolved oxygen, high chlorophyll *a*, and primary production, are found in the areas off the central Vietnamese coast, southeast of Hainan Island and north of the Sunda shelf, whereas in the winter they form a cold trough over the deep basin aligning from SW to NE. The warm pools with poor nutrients, high dissolved O_2, low chlorophyll, and low primary productivity are found during the summer southeast of Vietnam, east of Hainan Island and west of the Philippine island of Luzon. Phytoplankton stock and primary production are lower in summer due to PO_4^{3-} depletion. In the upwelling zones, peak primary production was 684 mg C m^{-2} d^{-1} (Chen et al. 2006). For the whole basin, annual mean productivity averaged 354 mg C m^{-2} d^{-1} with rates varying with season: 550 mg C m^{-2} d^{-1} in winter, 260 mg C m^{-2} d^{-1} in spring, 190 mg C m^{-2} d^{-1} in summer, and 280 mg C m^{-2} d^{-1} in autumn. In the Sulu Sea off the southern Philippines, there are high and low productivity areas associated with upwelling, vertical mixing between smaller islands, and bottom-intensified flow SW of Mindoro and Panay Straits (central Philippines). Net primary productivity in the low production areas ranged from 262 to 325 mg C m^{-2} d^{-1} and in the high production areas from 353 to 755 mg C m^{-2} d^{-1} (Ferrera et al. 2017).

Over the past century, there has been a decline in global phytoplankton abundance and productivity (Boyce et al. 2010). There have been interannual and decadal variations in phytoplankton abundance at local, regional, and global scales since 1899. Boyce et al. (2010) and Gregg and Rousseaux (2019) have estimated a global decline of about 1% of the global median per year with interannual and decadal phytoplankton fluctuations superimposed on long-term trends. These fluctuations are strongly correlated with basin-scale climate indices, whereas long-term declining trends are related to increasing SSTs. Such fluctuations were found in the equatorial Atlantic, the tropical Pacific but not in the Indian Ocean where there has been a slight increase in phytoplankton abundance.

An analysis of global marine primary production that constrains fisheries catches (Chassot et al. 2010) suggests that upwelling at eastern boundary areas shows high levels of phytoplankton production required (PPR) for fisheries catches, especially in the Canary, Humboldt, and Benguela ecosystems. The high PPR can be attributed to the high catches of small low trophic level pelagic species, such as anchovies, sardines, and sardinellas and the low transfer efficiency of this type of ecosystem. There are no differences in PPR between temperate and tropical ecosystems.

References

Acharyya, T., Sarma, V.V.S.S., Sridevi, B. et al. (2012). Reduced river discharge intensifies phytoplankton blooms in Godavari estuary, India. *Marine Chemistry* 132-133: 15–22.

Adame, M.F., Teutli, C., Santini, N.S. et al. (2014). Root biomass and production of mangroves surrounding a karstic oligotrophic coastal lagoon. *Wetlands* 34: 479–488.

Agawin, N.S.R., Duarte, C.M., Fortes, M.D. et al. (2001). Temporal changes in the abundance, leaf growth and photosynthesis of three co-occurring Philippine seagrasses. *Journal of Experimental Marine Biology and Ecology* 260: 217–239.

Albright, R., Langdon, C., and Anthony, K.R.N. (2013). Dynamics of seawater carbonate chemistry, production, and calcification of a coral reef flat, central Great Barrier Reef. *Biogeosciences* 10: 6747–6758.

Albright, R., Benthuysen, J., Cantin, N. et al. (2015). Coral reef metabolism and carbon chemistry dynamics of a coral reef flat. *Geophysical Research Letters* 42: 3980–3988.

Allemand, D., Tambutté, É., Zoccola, D. et al. (2011). Coral calcification, cells to reefs. In: *Coral Reefs: An Ecosystem in Transition* (eds. Z. Dubinsky and N. Stambler), 119–150. Dordrecht, The Netherlands: Springer.

Alongi, D.M. (1991). The role of intertidal mudbanks in the diagenesis and export of dissolved and particulate materials from the Fly delta, Papua New Guinea. *Journal of Experimental Marine Biology and Ecology* 149: 81–107.

Alongi, D.M. (1998). *Coastal Ecosystem Processes*. Boca Raton, FL: CRC Press.

Alongi, D.M. (2009). *The Energetics of Mangrove Forests*. Dordrecht, The Netherlands: Springer.

Alongi, D.M., Sasekumar, A., Chong, V.C. et al. (2004). Sediment accumulation and organic material flux in a managed mangrove ecosystem: estimates of land-ocean-atmosphere exchange in peninsular Malaysia. *Marine Geology* 208: 383–402.

Armitage, A.R., Frankovich, T.A., Heck, K.L. et al. (2005). Experimental nutrient enrichment causes complex changes in seagrass, microalgae, and macroalgae community structure in Florida Bay. *Estuaries* 28: 422–434.

Asbridge, E.F., Bartolo, R., Findlayson, C.M. et al. (2019). Assessing the distribution and drivers of mangrove dieback in Kakadu National Park, northern Australia. *Estuarine, Coastal and Shelf Science* 228: 106353. https://doi.org/10.1016/j.ecss.2019.106353.

Aung, T.T., Mochida, Y., and Than, M.M. (2013). Prediction of recovery pathways of cyclone-disturbed mangroves in the mega delta of Myanmar. *Forest Ecology and Management* 293: 103–113.

Ayukai, T. and Miller, D. (1998). Phytoplankton biomass, production and grazing mortality in Exmouth Gulf, a shallow embayment on the arid, tropical coast of Western Australia. *Journal of Experimental Marine Biology and Ecology* 225: 239–251.

Baker, A.R., Weston, K., Kelly, S.D. et al. (2007). Dry and wet deposition of nutrients from the tropical Atlantic atmosphere: links to primary productivity and nitrogen fixation. *Deep-Sea Research I* 54: 1704–1720.

Barnes, B.V., Zak, D.R., Denton, S.R. et al. (1998). *Forest Ecology*, 5e. New York: Wiley.

Barr, J.G., Engel, V., Smith, T.J. III et al. (2012). Hurricane disturbance and recovery of energy balance, CO_2 fluxes and canopy structure in a mangrove forest of the Florida Everglades. *Agricultural and Forest Meteorology* 153: 54–66.

Baums, I.B., Devlin-Durante, M.K., and Lajeunesse, T.C. (2014). New insights into the dynamics between reef corals and their associated dinoflagellate endosymbionts from population genetic studies. *Molecular Ecology* 23: 4203–4215.

Bednarz, V.N., Cardini, U., van Hoytema, N. et al. (2015a). Seasonal variations in dinitrogen fixation and oxygen fluxes associated with two dominant zooxanthellae soft corals from the northern Red Sea. *Marine Ecology Progress Series* 519: 141–152.

Bednarz, V.N., van Hoytema, N., Cardini, U. et al. (2015b). Dinitrogen fixation and primary productivity by carbonate and silicate reef sand communities of the northern Red Sea. *Marine Ecology Progress Series* 527: 47–57.

Behrenfeld, M.J., Worthington, K., Sherrell, R.M. et al. (2006). Controls on tropical Pacific Ocean productivity revealed through nutrient stress diagnostics. *Nature* 442: 1025–1028.

Berger, U., Adams, M., Grimm, V. et al. (2006). Modelling secondary succession of neotropical mangroves: causes and consequences of growth reduction in pioneer species. *Perspectives in Plant Ecology, Evolution and Systematics* 7: 243–252.

Biswas, H., Dey, M., Ganguly, D. et al. (2010). Comparative analysis of phytoplankton composition and abundance over a two-decade period at the land-ocean boundary of a tropical mangrove ecosystem. *Estuaries and Coasts* 33: 384–394.

de Boer, W.F. (2000). Biomass dynamics of seagrasses and the role of mangrove and seagrass vegetation as different nutrient sources for an intertidal ecosystem. *Aquatic Botany* 66: 225–239.

Boucher, G., Clavier, J., Hily, C. et al. (1998). Contribution of soft-bottoms to the community metabolism (primary production and calcification) of a barrier reef flat (Moorea, French Polynesia). *Journal of Experimental Marine Biology and Ecology* 225: 269–283.

Boyce, D.G., Lewis, M.R., and Worm, B. (2010). Global phytoplankton decline over the past century. *Nature* 466: 591–596.

Brouns, J.J.W.M. and Heijs, F.M.L. (1986). Production and biomass of the seagrass *Enhalus acoroides* (L.f.) Royle and its epiphytes. *Aquatic Botany* 25: 21–45.

Brunskill, G.J. (2010). Tropical margins. In: *Carbon and Nutrient Fluxes in Continental Margins: A Global Synthesis* (eds. K.-K. Liu, L. Atkinson, R. Quinones, et al.), 423–493. Dordrecht, The Netherlands: Springer.

Burchett, M.D., Field, C.D., and Pulkownik, A. (1984). Salinity, growth and root respiration in the grey mangrove, *Avicennia marina*. *Plant Physiology* 60: 113–118.

Burford, M.A. and Rothlisberg, P.C. (1999). Factors limiting phytoplankton production in a tropical continental shelf ecosystem. *Estuarine, Coastal and Shelf Science* 48: 541–549.

Burford, M.A., Webster, I.T., Revill, A.T. et al. (2012). Controls on phytoplankton productivity in a wet-dry tropical estuary. *Estuarine, Coastal and Shelf Science* 113: 141–151.

Burford, M.A., Valdez, D., Curwen, G. et al. (2016). Inundation of saline supratidal mud flats provides an important source of carbon and nutrients in an aquatic system. *Marine Ecology Progress Series* 545: 21–33.

Burkholder, J.M., Tomasko, D.A., and Touchette, B.W. (2007). Seagrasses and eutrophication. *Journal of Experimental Marine Biology and Ecology* 350: 46–72.

Burkholder, D.A., Fourqurean, J.W., and Heithaus, M.R. (2013). Spatial pattern in seagrass stoichiometry indicates both N-limited and P-limited regions of an iconic P-limited subtropical bay. *Marine Ecology Progress Series* 472: 101–115.

Cahoon, D.R., Hensel, P., Rybczyz, J. et al. (2003). Mass tree mortality leads to mangrove peat collapse at Bay Islands, Honduras after Hurricane Mitch. *Journal of Ecology* 91: 1093–1105.

Campbell, S.J., McKenzie, L.J., and Kerville, S.P. (2006). Photosynthetic responses of seven tropical seagrasses to elevated seawater temperatures. *Journal of Experimental Marine Biology and Ecology* 330: 455–468.

Cardini, U., Bednarz, V.N., van Hoytema, N. et al. (2016). Budget of primary production and dinitrogen fixation in a highly seasonal Red Sea coral reef. *Ecosystems* 19: 771–785.

Cardini, U., van Hoytema, N., Bednarz, V.N. et al. (2017). N_2 fixation and primary productivity in a Red Sea *Halophila stipulacea* meadow exposed to seasonality. *Limnology and Oceanography* 63: 786–798.

Chang, C.W.J., Hsu, H.H., Cheah, W. et al. (2019). Madden–Julian Oscillation enhances phytoplankton biomass in the Maritime Continent. *Scientific Reports* 9: 1–9.

Charpy, L. and Blanchot, J. (1999). Picophytoplankton biomass, community structure and productivity in the Great Astrolabe Lagoon, Fiji. *Coral Reefs* 18: 255–262.

Chassot, E., Bonhommeau, S., Dulvy, N. et al. (2010). Global marine primary production constrains fisheries catches. *Ecology Letters* 13: 495–505.

Chaudhuri, K., Manna, S., Sarma, K.S. et al. (2012). Physiochemical and biological factors controlling water column metabolism in Sundarbans estuary, India. *Aquatic Biosystems* 8: 26. https://doi.org/10.1186/2046-9063-8-26.

Chen, Y.-L.L. (2005). Spatial and seasonal variations of nitrate-based new production and primary production in the South China Sea. *Deep-Sea Research I* 52: 319–340.
Chen, C.-C., Shiah, F.-W., Chung, S.-W. et al. (2006). Winter phytoplankton blooms in the shallow mixed layer of the South China Sea enhanced by upwelling. *Journal of Marine Systems* 59: 97–110.
Cheng, W., Fu, S., Susfalk, R.B. et al. (2005). Measuring tree root respiration using ^{13}C natural abundance: rooting medium matters. *New Phytology* 167: 297–307.
Chiba, S., Batten, S., Sasaoka, K. et al. (2012). Influence of the Pacific Decadal Oscillation on phytoplankton phenology and community structure in the western North Pacific. *Geophysical Research Letters* 39: GL052912. https://doi.org/10.1029/2012GL052912.
Chien, C.T., Mackey, K.R., Dutkiewicz, S. et al. (2016). Effects of African dust deposition on phytoplankton in the western tropical Atlantic Ocean off Barbados. *Global Biogeochemical Cycles* 30: 716–734.
Chisholm, J.R.M. (2003). Primary productivity of reef-building crustose coralline algae. *Limnology and Oceanography* 48: 1376–1387.
Clark, D.A., Brown, S., Kicklighter, D.W. et al. (2001). Net primary production in tropical forests: an evaluation and synthesis of existing field data. *Ecological Applications* 11: 371–384.
Clavier, J. and Garrigue, C. (1999). Annual sediment primary production and respiration in a large coral reef lagoon. *Marine Ecology Progress Series* 191: 79–89.
Cloern, J.E., Foster, S.Q., and Kleckner, A.E. (2014). Phytoplankton primary production in the world's estuarine-coastal ecosystems. *Biogeosciences* 11: 2477–2501.
Clough, B.F. (1998). Mangrove forest productivity and biomass accumulation in Hinchinbrook Channel, Australia. *Mangroves and Salt Marshes* 2: 191–198.
Clough, B.F., Ong, J.E., and Gong, W.K. (1997). Estimating leaf area index and photosynthetic production in canopies of the mangrove *Rhizophora apiculata*. *Marine Ecology Progress Series* 159: 285–292.
Clough, B.F., Tan, D.T., Phuong, D.X. et al. (2000). Canopy leaf area index and litterfall in stands of the mangrove *Rhizophora apiculata* of different age in the Mekong Delta, Vietnam. *Aquatic Botany* 66: 311–320.
Cormier, N., Twilley, R.R., Ewel, K.C. et al. (2015). Fine root productivity varies along nitrogen and phosphorus gradients in high-rainfall mangrove forests of Micronesia. *Hydrobiologia* 750: 69–87.
Côté-Laurin, M.-C., Benbow, S., and Erzini, K. (2017). The short-term impacts of a cyclone on seagrass communities in SW Madagascar. *Continental Shelf Research* 138: 132–141.
Crosbie, N.D. and Furnas, M.J. (2001). Net growth rates of picocyanobacteria and nano-/microphytoplankton inhabiting shelf waters of the central (17° S) and southern (20° S) Great Barrier Reef. *Aquatic Microbial Ecology* 24: 209–224.
Crossland, C.J., Hatcher, B.G., and Smith, S.V. (1991). Role of coral reefs in global ocean production. *Coral Reefs* 10: 55–64.
Cruz-Palacios, V. and Van Tussenbroek, B. (2005). Simulation of hurricane-like disturbances on a Caribbean seagrass bed. *Journal of Experimental Marine Biology and Ecology* 324: 44–60.
Dagg, M., Benner, R., Lohrenz, S. et al. (2004). Transformation of dissolved and particulate materials on continental shelves influenced by large rivers: plume processes. *Continental Shelf Research* 24: 833–858.
Day, J.W. Jr., Coronada-Molina, C., Vera-Herrera, F.R. et al. (1996). A 7 years record of above-ground net primary production in a southeastern Mexican mangrove forest. *Aquatic Botany* 55: 39–60.
Delesalle, B., Sakka, A., Legendre, L. et al. (2001). The phytoplankton of Takapoto Atoll (Tuamotu Archipelago, French Polynesia): time and space variability of biomass, primary production and composition over 24 years. *Aquatic Living Resources* 14: 175–182.
Dierssen, H.M., Zimmerman, R.C., Drake, L.A. et al. (2010). Benthic ecology from space: optics and net primary production in seagrass and benthic algae across the Grand Bahama Bank. *Marine Ecology Progress Series* 411: 1–15.

Dix, N., Philips, E., and Suscy, P. (2013). Factors controlling phytoplankton biomass in a subtropical coastal lagoon: relative scales of influence. *Estuaries and Coasts* 36: 981–996.
dos Santos Garcia, J., Dalmolin, Â.C., França, M.G.C. et al. (2017). Different salt concentrations induce alterations both in photosynthetic parameters and salt gland activity in leaves of the mangrove *Avicennia schaueriana*. *Ecotoxicology and Environmental Safety* 141: 70–74.
Duarte, C.M. and Chiscano, C.L. (1999). Seagrass biomass and production: a reassessment. *Aquatic Botany* 65: 159–174.
Duggan, M., Connolly, R.M., Whittle, M. et al. (2014). Effects of freshwater flow extremes on intertidal biota of a wet-dry tropical estuary. *Marine Ecology Progress Series* 502: 11–23.
Duke, N.C., Kovacs, J.M., Griffiths, A.D. et al. (2017). Large-scale dieback of mangroves in Australia's Gulf of Carpentaria: a severe ecosystem response, coincidental with an unusually extreme weather event. *Marine and Freshwater Research* 68: 1816–1829.
Eidens, C., Bayraktarov, E., Hauffe, T. et al. (2014). Benthic primary production in an upwelling-influenced coral reef, Colombian Caribbean. *PeerJ* 2: e554. https://doi.org/10.7717/peerj.554.
Erftemeijer, P.L.A. and Herman, P.M.J. (1994). Seasonal changes in environmental variables, biomass, production and nutrient contents in two contrasting tropical intertidal seagrass beds in south Sulawesi, Indonesia. *Oecologia* 99: 45–59.
Erftemeijer, P.L.A., Osinga, R., and Mars, A.E. (1993). Primary production of seagrass beds in south Sulawesi (Indonesia): a comparison of habitats, methods and species. *Aquatic Botany* 46: 67–90.
Erftemeijer, P.L.A., Stapel, J., Smekens, M.J.E. et al. (1994). The limited effect of *in situ* phosphorus and nitrogen additions to seagrass beds on carbonate and terrigenous sediments in south Sulawesi, Indonesia. *Journal of Experimental Marine Biology and Ecology* 182: 123–135.
Estacion, J.S. and Fortes, M.D. (1988). Growth rates and primary production of *Enhalus acoroides* (L.f.) Royle from Lag-it, North Bais Bay, the Philippines. *Aquatic Botany* 29: 347–356.
Falter, J.L., Lowe, R.J., Zhang, Z. et al. (2013). Physical and biological controls on the carbonate chemistry of coral reef waters: effects of metabolism, wave forcing, sea level, and geomorphology. *PLoS One* 8: e53303. https://doi.org/10.1371/journal.pone.0053303.
Ferrera, C.M., Jacinto, G.S., Chen, C.-T.A. et al. (2017). Carbonate parameters in high and low productivity areas of the Sulu Sea, Philippines. *Marine Chemistry* 195: 2–14.
Fromard, F., Puig, H., Mougin, E. et al. (1998). Structure, above-ground biomass and dynamics of mangrove ecosystems: new data from French Guiana. *Oecologia* 115: 39–53.
Fujimura, H., Oomori, T., Maehira, T. et al. (2001). Change of coral carbon metabolism influenced by coral bleaching. *Galaxea* 3: 41–50.
Furla, P., Galgani, I., Durand, I. et al. (2000). Sources and mechanisms of inorganic carbon transport for coral calcification and photosynthesis. *Journal of Experimental Biology* 203: 3445–3457.
Furnas, M.J. (2007). Intra-seasonal and interannual variations in phytoplankton biomass, primary production and bacterial production at North West Cape, Western Australia: links to the 1997-1998 El Niño event. *Continental Shelf Research* 27: 958–980.
Furnas, M.J. and Carpenter, E.J. (2016). Primary production in the tropical continental shelf seas bordering northern Australia. *Continental Shelf Research* 129: 33–48.
Furnas, M.J., Mitchell, A.W., Gilmartin, M. et al. (1990). Phytoplankton biomass and primary production in semi-enclosed reef lagoons of the central Great Barrier Reef, Australia. *Coral Reefs* 9: 1–10.
Gattuso, J.-P., Allemand, D., and Frankignoulle, M. (1999a). Photosynthesis and calcification at cellular, organismal and community levels in coral reefs, a review on interactions and control by carbonate chemistry. *American Zoologist* 39: 160–183.
Gattuso, J.-P., Frankignoulle, M., and Smith, S.V. (1999b). Measurement of community metabolism and significance in the coral reef CO_2 source-sink debate. *Proceedings of the National Academy of Sciences* 96: 13012–13022.
Gavin, N.M. and Durako, M.J. (2019). Carbon acquisition mechanisms in *Halophila johnsonii* and *Thalassia testudinum*. *Aquatic Botany 152*: 64–69.

Githaiga, M.N., Gilpin, L., Kairo, J.G. et al. (2016). Biomass and productivity of seagrasses in Africa. *Botanica Marina* 59: 173–186.

Gleason, S.M. and Ewel, K.C. (2002). Organic matter dynamics on the forest floor of a Micronesian mangrove forest: an investigation of species composition shifts. *Biotropica* 34: 190–198.

Gocke, K., Cortés, J., and Murillo, M.M. (2001a). Planktonic primary production in a tidally influenced mangrove forest on the Pacific coast of Costa Rica. *Revisita Biologia Tropical (Supplement)* 2: 279–288.

Gocke, K., Cortés, J., and Murillo, M.M. (2001b). The annual cycle of primary productivity in a tropical estuary, the inner regions of the Golfo de Nicoya, Costa Rica. *Revisita Biologia Tropical (Supplement)* 2: 289–306.

Gocke, K., Hernandez, C., Giesenhagen, H. et al. (2004). Seasonal variations of bacterial abundance and biomass and their relation to phytoplankton in the hypertrophic tropical lagoon Ciénaga Grande de Santa Marta, Colombia. *Journal of Plankton Research* 26: 1429–1439.

Golley, F.B., Odum, H.T., and Wilson, R.F. (1962). The structure and metabolism of a Puerto Rican red mangrove forest in May. *Ecology* 43: 9–19.

Gong, W.K., Ong, J.E., and Clough, B.F. (1992). Photosynthesis in different aged stands of a Malaysian mangrove ecosystem. In: *Third ASEAN Science and Technology Week Conference Proceedings, Volume 6, Marine Science, Living Coastal Resources* (eds. L.M. Chou and C.R. Wilkinson), 345–351. Singapore: National University of Singapore and National Science Technology.

Gouveia, N.A., Gherardi, D.F.M., Wagner, F.H. et al. (2019). The salinity structure of the Amazon River plume drives spatiotemporal variation of oceanic primary productivity. *Journal of Geophysical Research: Biogeosciences* 124: 147–165.

Govindasamy, C., Arulpriya, M., Ruban, P. et al. (2013). Seasonal variations in seagrass biomass and productivity in Palk Bay, Bay of Bengal, India. *International Journal of Biodiversity and Conservation* 5: 408–417.

Grech, A., Hanert, E., McKenzie, L. et al. (2018). Predicting the cumulative effect of multiple disturbances on seagrass connectivity. *Global Change Biology* 24: 3093–3104.

Gregg, W.W. and Rousseaux, C.S. (2019). Global ocean primary production trends in the modern ocean color satellite record (1998-2015). *Environmental Research Letters* 14: 124011. https://doi.org/10.1088/1748-9326/ab4667.

Grosse, J., Bombar, D., Doan, H.N. et al. (2010). The Mekong River plume fuels nitrogen fixation and determines phytoplankton species distribution in the South China Sea during low and high discharge season. *Limnology and Oceanography* 55: 1668–1680.

Gupta, G.V.M., Sudheesh, V., Sudharma, K.V. et al. (2016). Evolution to decay of upwelling and associated biogeochemistry over the southeastern Arabian Sea shelf. *Journal of Geophysical Research: Biogeosciences* 121: 159–175.

Harada, Y., Fry, B., Lee, S.Y. et al. (2020). Stable isotopes indicate ecosystem restructuring following climate-driven mangrove dieback. *Limnology and Oceanography* 65: 1251–1263.

Harrison, P.J., Khan, N., Yin, K. et al. (1997). Nutrient and phytoplankton dynamics in two mangrove tidal creeks of the Indus River delta, Pakistan. *Marine Ecology Progress Series* 157: 13–19.

Hatcher, B.G. (1988). Coral reef primary productivity: a beggar's banquet. *Trends in Ecology and Evolution* 3: 106–111.

Hatcher, B.G. (1990). Coral reef primary productivity: a hierarchy of pattern and process. *Trends in Ecology and Evolution* 5: 149–152.

Heck, K.L. Jr., Pennock, J.R., Valentine, J.F. et al. (2000). Effects of nutrient enrichment and small predator density on seagrass ecosystems: an experimental assessment. *Limnology and Oceanography* 45: 1041–1057.

Hee, Y.Y., Suratman, S., and Weston, K. (2020). Nutrient cycling and primary production in Peninsular Malaysia waters; regional variation and its causes in the South China Sea. *Estuarine, Coastal and Shelf Science* 245: 106947. https://doi.org/10.1016/j.ecss.2020.106947.

Heijs, F.M.L. (1985). Some structural and functional aspects of the epiphytic component of four seagrass species (Cymodoceoideae) from Papua New Guinea. *Aquatic Botany* 23: 225–247.

Hemminga, M.A., Gwada, P., Slim, F.J. et al. (1995). Leaf production and nutrient contents of the seagrass *Thallassodendron ciliatum* in the proximity of a mangrove forest (Gazi Bay, Kenya). *Aquatic Botany* 50: 159–170.
Herrera-Silveira, J.A. (1998). Nutrient-phytoplankton production relationships in a groundwater-influenced tropical coastal lagoon. *Aquatic Ecosystem Health and Management* 1: 373–385.
Herrera-Silveira, J.A., Medina-Gomez, I., and Colli, R. (2002). Trophic status based on nutrient concentration scales and primary producer community of tropical coastal lagoons influenced by groundwater discharges. *Hydrobiologia* 475 (476): 91–98.
Hsieh, W.-C., Chen, C.-C., Shiah, F.-K. et al. (2012). Community metabolism in a tropical lagoon: carbon cycling and autotrophic ecosystem induced by a natural nutrient pulse. *Environmental Engineering Science* 29: 776–782.
Imbert, D. (2018). Hurricane disturbance and forest dynamics in east Caribbean mangroves. *Ecosphere* 9: e02231. https://doi.org/10.1002/ecs2.2231.
Jahnke, R.A. (2010). Global synthesis. In: *Carbon and Nutrient Fluxes in Continental Margins: A Global Synthesis* (eds. K.-K. Liu, L. Atkinson, R. Quinones, et al.), 597–615. Dordrecht, The Netherlands: Springer.
Kaladharan, P. and Raj, D.I. (1989). Primary production of seagrass *Cymodocea semlata* and its contribution to productivity of Amini atoll. *Indian Journal of Marine Sciences* 18: 215–216.
Kaldy, J.E. and Dunton, K.H. (2000). Above- and below-ground production, biomass and reproductive ecology of *Thalassia testudinum* (turtle grass) in a subtropical coastal lagoon. *Marine Ecology Progress Series* 193: 271–283.
Kamruzzaman, M., Ahmed, S., and Osawa, A. (2017a). Biomass and net primary productivity of mangrove communities along the oligohaline zone of Sundarbans, Bangladesh. *Forest Ecosystems* 4: 16. https://doi.org/10.1186/s40663-017-0104-0.
Kamruzzaman, M., Osawa, A., Deshar, R. et al. (2017b). Species composition, biomass, and net primary productivity of mangrove forest in Okukubi River, Okinawa Island, Japan. *Regional Studies in Marine Science* 12: 19–27.
Kamruzzaman, M., Mouctar, K., Sharma, S. et al. (2019). Comparison of biomass and net primary productivity among three species in a subtropical mangrove forest at Manko Wetland, Okinawa, Japan. *Regional Studies in Marine Science* 25: 100475. https://doi.org/10.1016/j.rsma.2018.100475.
Kanuri, V.V., Muduli, P.R., Robin, R.S. et al. (2013). Plankton metabolic processes and its significance on dissolved organic carbon pool in a tropical brackish water lagoon. *Continental Shelf Research* 61-62: 52–61.
Kauffman, J.B. and Cole, T.G. (2010). Micronesian mangrove forest structure and tree responses to a severe typhoon. *Wetlands* 30: 1077–1084.
Kinsey, D.W. (1985). Open-flow systems. In: *Handbook of Phycological Methods, 4. Ecological Field Studies, Macroalgae* (eds. M.M. Littler and D.S. Littler), 427–460. Cambridge, UK: Cambridge University Press.
Kinsey, D.W. and Hopley, D. (1991). The significance of coral reefs as global carbon sinks-response to greenhouse. *Palaeogeography, Palaeoclimatology, Palaeoecology* 89: 363–377.
Kitaya, Y., Yabuki, K., and Kiyota, M. (2002). Gas exchange and oxygen concentration in pneumatophores and prop roots of four mangrove species. *Trees* 16: 155–158.
Knoppers, B., de Souza, F.L., de Souza, M.F.L. et al. (1996). In situ measurements of benthic primary production, respiration and nutrient fluxes in a hypersaline coastal lagoon of SE Brazil. *Revista Brasilian Oceanographica* 44: 155–165.
Koch, M.S. and Madden, C.J. (2001). Patterns of primary production and nutrient availability in a Bahamas lagoon with fringing mangroves. *Marine Ecology Progress Series* 219: 109–119.
Koch, M., Bowes, G., Ross, C. et al. (2013). Climate change and ocean acidification effects on seagrasses and marine macroalgae. *Global Change Biology* 19: 103–132.
Krauss, K.W., Lovelock, C.E., McKee, K.L. et al. (2008). Environmental drivers in mangrove establishment and early development: a review. *Aquatic Botany* 89: 105–127.
Kumar, C.S. and Perumal, P. (2012). Studies on phytoplankton characteristics in Ayyampattinam coast, India. *Journal of Environmental Biology* 33: 585–589.

Kumar, S.P., Nuncio, M., Narvekar, J. et al. (2004). Are eddies nature's trigger to enhance biological productivity in the Bay of Bengal? *Geophysical Research Letters* 31: L007309. https://doi.org/10.1029/2003GL019274.
Kumar, S.P., Nunico, M., Ramaiah, N. et al. (2007). Eddy-mediated biological productivity in the Bay of Bengal during fall and spring inter-monsoons. *Deep-Sea Research I* 54: 1619–1640.
Kumari, L.K., Bhattathiri, P.M.A., Matondkar, S.G.P. et al. (2002). Primary productivity in Mandovi-Zuari estuaries in Goa. *Journal of the Marine Biological Association of India* 44: 1–13.
Lapointe, B.E., Tomasko, D.A., and Matzie, W.R. (1994). Eutrophication and trophic state classification of seagrass communities in the Florida Keys. *Bulletin of Marine Science* 54: 696–717.
Larkum, A.W.D., Koch, E.-M., and Kühl, M. (2003). Diffusive boundary layers and photosynthesis of the epilithic algal community of coral reefs. *Marine Biology* 142: 1073–1082.
Larsson, S. (2009). The production of the seagrass *Thalassia hemprichii* in relation to epiphytic biomass. MSc thesis. University of Gothenberg.
Lee, C.-W. and Bong, C.-W. (2008). Bacterial abundance and production, and their relation to primary production in tropical coastal waters of Peninsular Malaysia. *Marine and Freshwater Research* 59: 10–21.
Lee, K.-S. and Dunton, K.H. (1996). Production and carbon reserve dynamics of the seagrass *Thalassia testudinum* in Corpus Christi Bay, Texas, USA. *Marine Ecology Progress Series* 143: 201–210.
Lee, K.-S. and Dunton, K.H. (1999). Influence of sediment nitrogen availability on carbon and nitrogen dynamics in the seagrass *Thalassia testudinum*. *Marine Biology* 134: 217–226.
Lee, K.-S. and Dunton, K.H. (2000). Effects of nitrogen enrichment on biomass allocation, growth and leaf morphology of the seagrass *Thalassia testudinum*. *Marine Ecology Progress Series* 196: 39–48.
Lee, C.-W., Lee, C.-W., and Bong, C.-W. (2007). Bacterial respiration, growth efficiency and protist grazing rates in mangrove waters in Cape Rachado, Malaysia. *Asian Journal of Water, Environment and Pollution* 1: 11–16.
Li, Z., Saito, Y., Tamura, T. et al. (2012). Mid-Holocene mangrove succession and its response to sea-level change in the upper Mekong River delta, Cambodia. *Quaternary Research* 78: 386–399.
Li, Q., Legendre, L., and Jiao, N. (2015). Phytoplankton responses to nitrogen and iron limitation in the tropical and subtropical Pacific Ocean. *Journal of Plankton Research* 37: 306–319.
Lin, I., Liu, W.T., Wu, C.-C. et al. (2003). New evidence for enhanced ocean primary production triggered by tropical cyclone. *Geophysical Research Letters* 30: 1718. https://doi.org/10.1029/2003GL017141.
Lindeboom, H.J. and Sandee, A.J.J. (1989). Production and consumption of tropical seagrass fields in eastern Indonesia measured with bell jars and microelectrodes. *Netherlands Journal of Sea Research* 23: 181–190.
Liu, X., Xiong, Y., and Liao, B. (2017). Relative contributions of leaf litter and fine roots to soil organic matter accumulation in mangrove forests. *Plant and Soil* 421: 493–503.
Long, M.H., Berg, P., and Falter, J.L. (2015). Seagrass metabolism across a productivity gradient using the eddy covariance, Eulerian control volume, and biomass addition techniques. *Journal of Geophysical Research: Oceans* 120: 3624–3639.
Long, J., Giri, C., Primavera, J. et al. (2016). Damage and recovery assessment of the Philippines' mangroves following Super Typhoon Haiyan. *Marine Pollution Bulletin* 109: 734–743.
Longhurst, A. (1983). Benthic-pelagic coupling and export of organic carbon from a tropical Atlantic continental shelf-Sierra Leone. *Estuarine, Coastal and Shelf Science* 17: 261–285.
Longhurst, A.L. (2007). *Ecological Geography of the Sea*, 2e. London: Academic Press.
Longhurst, A., Sathyendranath, S., Platt, T. et al. (1995). An estimate of global primary production in the ocean from satellite radiometer data. *Journal of Plankton Research* 117: 1245–1271.
Lopez, R., Lopez, J.M., Morell, J. et al. (2013). Influence of the Orinoco River on the primary production of eastern Caribbean surface waters. *Journal of Geophysical Research: Oceans* 118: 4617–4632.

Löscher, C.R., Fischer, M.A., Neulinger, S.C. et al. (2015). Hidden biosphere in an oxygen-deficient Atlantic open-ocean eddy: future implications of ocean deoxygenation on primary production in the eastern tropical north Atlantic. *Biogeosciences* 12: 7467–7482.

Lovelock, C.E., Ruess, R.W., and Feller, I.C. (2006). Fine root respiration in the mangrove *Rhizophora mangle* over variation in forest structure and nutrient availability. *Tree Physiology* 26: 1601–1606.

Lugo, A.E., Evink, G., Brinson, M.M. et al. (1975). Diurnal rates of photosynthesis, respiration, and transpiration in mangrove forests of south Florida. In: *Tropical Ecological Systems: Trends in Terrestrial and Aquatic Research* (eds. F.B. Golley and E. Medina), 335–350. Berlin: Springer.

Lugomela, C., Wallberg, P., and Nielsen, T.G. (2001). Plankton composition and cycling of carbon during the rainy season in a tropical coastal ecosystem, Zanzibar, Tanzania. *Journal of Plankton Research* 23: 1121–1136.

Madhu, N.V., Jyothibabu, R., and Balachandran, K.K. (2010a). Monsoon-induced changes in the size-fractionated phytoplankton biomass and production rate in the estuarine and coastal waters of SW coast of India. *Environmental Monitoring and Assessment* 166: 521–528.

Madhu, N.V., Balachandran, K.K., Martin, G.D. et al. (2010b). Short-term variability of water quality and its implications on phytoplankton production in a tropical estuary (Cochin backwaters-India). *Environmental Monitoring and Assessment* 170: 287–300.

Madhupratap, M., Gauns, M., Ramaiah, N. et al. (2003). Biogeochemistry of the Bay of Bengal: physical, chemical and primary productivity characteristics of the central and western Bay of Bengal during summer monsoon 2001. *Deep-Sea Research II* 50: 881–896.

Marra, J. and Barber, R.T. (2005). Primary productivity of the Arabian Sea: a synthesis of JGOFS data. *Progress in Oceanography* 65: 159–175.

McKinnon, A.D., Trott, L.A., Alongi, D.M. et al. (2002). Water column production and nutrient characteristics in mangrove creeks receiving shrimp farm effluent. *Aquaculture Research* 33: 55–73.

McKinnon, A.D., Meekan, M.G., Carleton, J.H. et al. (2003). Rapid changes in shelf waters and pelagic communities on the southern NW Shelf, Australia, following a tropical cyclone. *Continental Shelf Research* 23: 93–111.

McKinnon, A.D., Carleton, J.H., and Duggan, S. (2007). Pelagic production and respiration in the Gulf of Papua during May 2004. *Continental Shelf Research* 27: 1643–1655.

McKinnon, A.D., Logan, M., Castine, S.A. et al. (2013). Pelagic metabolism in the waters of the Great Barrier Reef. *Limnology and Oceanography* 58: 1227–1242.

McKinnon, A.D., Duggan, S., Logan, M. et al. (2017). Plankton respiration, production, and trophic state in tropical coastal and shelf waters adjacent to Northern Australia. *Frontiers in Marine Science* 4: 346. https://doi.org/10.3389/fmars.2017.00346.

McMahon, A., Santos, I.R., Schulz, K.G. et al. (2019). Coral reef calcification and production after the 2016 bleaching event at Lizard Island, Great Barrier Reef. *Journal of Geophysical Research: Oceans* 124: 4003–4016.

Menkes, C.E., Lengaigne, M., Levy, M. et al. (2016). Global impact of tropical cyclones on primary production. *Global Biogeochemical Cycles* 30: 767–786.

Messié, M. and Chavez, F.P. (2015). Seasonal regulation of primary production in eastern boundary upwelling systems. *Progress in Oceanography* 134: 1–18.

Moffett, J.W. and Landry, M.R. (2020). Grazing control and iron limitation of primary production in the Arabian Sea: implications for anticipated shifts in SW Monsoon intensity. *Deep-Sea Research II* 176: 104687. https://doi.org/10.1016/j.dsr2.2019.104687.

Muhammad-Nor, S.M., Huxham, M., Salmon, Y. et al. (2019). Exceptionally high mangrove root production rates in the Kelantan Delta, Malaysia; an experimental and comparative study. *Forest Ecology and Management* 444: 214–224.

Nandan, B.S., Jayachandran, P.R., and Sreedevi, O.K. (2014). Spatio-temporal pattern of primary production in a tropical coastal wetland (Kodungallur-Azhikode estuary), south west coast of India. *Journal of Coastal Development* 17: 2. https://doi.org/10.4172/1410-5217.1000392.

Naumann, M.S., Jantzen, C., Haas, A.F. et al. (2013). Benthic primary production budget of a Caribbean reef lagoon (Puerto Morelos, Mexico). *PLoS One* 8: e82923. https://doi.org/10.1371/journal.pone.0082923.

Nguyen, H.T., Stanton, D.E., Schmitz, N. et al. (2015). Growth responses of the mangrove *Avicennia marina* to salinity: development and function of shoot hydraulic systems require saline conditions. *Annals of Botany* 115: 397–407.

Ning, X., Chai, F., Xue, H. et al. (2004). Physical-biological coupling influencing phytoplankton and primary production in the South China Sea. *Journal of Geophysical Research* 109: C10005. https://doi.org/10.1029/2004JC002365.

Odum, W.E., McIvor, C.C., and Smith, T.J. III (1982). *The Ecology of Mangroves of South Florida: A Community Profile*. FWS/OBS-81/24. Washington, DC: U.S. Fish and Wildlife Service.

Ohtsuka, T., Tomotsune, M., Suchewaboripont, V. et al. (2019). Stand dynamics and aboveground net primary productivity of a mature subtropical mangrove forest on Ishigaki Island, south-western Japan. *Regional Studies in Marine Science* 27: 100516. https://doi.org/10.1016/j.rsma.2019.100516.

Ono, K., Fujimoto, K., Tabuchi, R. et al. (2017). Estimation of fine root production and decomposition rates in tropical and subtropical mangrove forests. *JGU-AGU Joint Meeting*, https://pdfs.semanticscholar.org/a56b/c4edc30941f7cb3ad092b1e7823f3443c708.pdf ()

Ouyang, X. and Guo, F. (2021). Patterns of mangrove productivity and support for marine fauna. In: *Handbook of Halophytes* (ed. M.-N. George), 1–20. Gland, Switzerland: Springer Nature.

Pamplona, F.C., Paes, E.T., and Nepomuceno, A. (2013). Nutrient fluctuations in the Quatipuru River: a macrotidal estuarine mangrove system in the Brazilian Amazonian basin. *Estuarine, Coastal and Shelf Science* 133: 273–284.

Pan, C.-W., Chuang, Y.-L., Chou, L.-S. et al. (2016). Factors governing phytoplankton biomass and production in tropical estuaries of western Taiwan. *Continental Shelf Research* 118: 88–99.

Pandey, S., Bhagawati, C., Dandapat, S. et al. (2019). Surface chlorophyll anomalies associated with Indian Ocean Dipole and El Niño Southern Oscillation in North Indian Ocean: a case study of 2006–2007 event. *Environmental Monitoring and Assessment* 191: 807. https://doi.org/10.1007/s10661-019-7754-z.

Pandi, S.P., Anna, G.G., Purvaja, R. et al. (2018). Spatial assessment of net canopy photosynthetic rate and species diversity in Pichavaram mangrove forest, Tamil Nadu. *Indian Journal of Ecology* 45: 717–723.

Paynter, C.K., Cortés, J., and Engels, M. (2001). Biomass, productivity and density of the seagrass *Thalassia testudinum* at three sites in Cahuita National Park, Costa Rica. *Revista de Biología Tropical* 49: 265–272.

Pednekar, S.M., Prabhu Matondkar, S.G., Gomes, H.D.R. et al. (2011). Fine-scale responses of phytoplankton to freshwater influx in a tropical monsoonal estuary following the onset of SW monsoon. *Journal of Earth Systems Science* 120: 545–556.

Peng, C.J., Qian, J.W., Guo, X.D. et al. (2016). Vegetation carbon stocks and net primary productivity of the mangrove forests in Shenzhen, China. *The Journal of Applied Ecology* 27: 2059–2065. (in Chinese).

Pennington, J.T., Mahoney, K.L., Kuwahara, V.S. et al. (2006). Primary production in the eastern tropical Pacific: a review. *Progress in Oceanography* 69: 285–317.

Pennock, J.R., Boyer, J.N., Herrera-Silveira, J.A. et al. (1999). Nutrient behaviour and phytoplankton production in Gulf of Mexico estuaries. In: *Biogeochemistry of Gulf of Mexico Estuaries* (eds. T.S. Bianchi, J.R. Pennock and R.R. Twilley), 109–132. New York: Wiley.

Perez, M., Romero, J., and Duarte, C.M. (1991). Phosphorus limitation of *Cymodocea nodosa* growth. *Marine Biology* 109: 129–133.

Pérez, V., Fernández, E., Marañón, E. et al. (2005). Latitudinal distribution of microbial plankton abundance, production, and respiration in the equatorial Atlantic in autumn 2000. *Deep-Sea Research I* 52: 861–880.

Perry, D.A., Oren, R., and Hart, S.C. (2008). *Forest Ecology*, 2e. Baltimore, USA: Johns Hopkins.

Phan, S.M., Nguyen, H.T.T., Nguyen, T.K. et al. (2019). Modelling above ground biomass accumulation of mangrove plantations in Vietnam. *Forest Ecology and Management* 432: 376–386.

Pollard, P.C. and Kogure, K. (1993). The role of epiphytic and epibenthic algal productivity in a tropical seagrass, *Syringodium isoetifolium* (Aschwers.) Dandy, community. *Australian Journal of Marine and Freshwater Research* 44: 141–154.
Poungparn, S., Komiyama, A., Sangteian, T. et al. (2012). High primary productivity under submerged soil raises net ecosystem productivity of a secondary mangrove forest in eastern Thailand. *Journal of Tropical Ecology* 28: 303–306.
Poungparn, S., Charoenphonphakdi, T., Sangtiean, T. et al. (2015). Fine root production in three zones of secondary mangrove forest in eastern Thailand. *Trees* 30: 467–474.
Powell, G.V.N., Kenworthy, W.J., and Fourqurean, J.W. (1989). Experimental evidence for nutrient limitation of seagrass growth in a tropical estuary with restricted circulation. *Bulletin of Marine Science* 44: 324–340.
Proisy, C., Walcker, R., Blanchard, E. et al. (2021). Mangroves: a natural early-warning system of erosion on open muddy coasts in French Guiana. In: *Dynamic Sedimentary Environments of Mangrove Coasts* (eds. F. Sidik and D.A. Friess), 47–66. Amsterdam: Elsevier.
Rahman, F., Rahman, M.T., Rahman, M.S. et al. (2014). Organic production of Korojol, Passur River system of the Sundarbans. *Bangladesh. Asian Journal of Water, Environment and Pollution* 11: 95–103.
Ram, A.S.P., Nair, S., and Chandramohan, D. (2003). Seasonal shift in net ecosystem production in a tropical estuary. *Limnology and Oceanography* 48: 1601–1607.
Reef, R., Winter, K., Morales, J. et al. (2015). The effect of atmospheric carbon dioxide concentrations on the performance of the mangrove *Avicennia germinans* over a range of salinities. *Physiologia Plantarum* 154: 58–368.
Reef, R., Slot, M., Motro, U. et al. (2016). The effects of CO_2 and nutrient fertilisation on the growth and temperature response of the mangrove *Avicennia germinans. Photosynthesis Research* 129: 159–170.
Reich, P.B., Walters, M.B., Tjoelker, M.G. et al. (1998). Photosynthesis and respiration rates depend on leaf and root morphology in nine boreal tree species differing in relative growth rate. *Functional Biology* 12: 395–405.
Riascos, J.M. and Blanco-Libreros, J.F. (2019). Pervasively high mangrove productivity in a major tropical delta throughout an ENSO cycle (Southern Caribbean, Colombia). *Estuarine, Coastal and Shelf Science* 227: 106301. https://doi.org/10.1016/j.ecss.2019.106301.
Rix, L., Bednarz, V.N., Cardini, U. et al. (2015). Seasonality in dinitrogen fixation and primary productivity by coral reef framework substrates from the northern Red Sea. *Marine Ecology Progress Series* 533: 79–92.
Robertson, A.I. and Alongi, D.M. (1995). Role of riverine mangrove forests in organic carbon export to the tropical coastal ocean: a preliminary mass balance for the Fly delta (Papua New Guinea). *Geo-Marine Letters* 15: 134–139.
Robertson, A.I. and Blaber, S.J.M. (1992). Plankton, epibenthos and fish communities. In: *Tropical Mangrove Ecosystems* (eds. A.I. Robertson and D.M. Alongi), 173–224. Washington DC: American Geophysical Union.
Robertson, A.I., Daniel, P.A., Dixon, P. et al. (1993). Pelagic biological processes along a salinity gradient in the Fly delta and adjacent river plume (Papua New Guinea). *Continental Shelf Research* 13: 205–224.
Robertson, A.I., Dixon, P., and Alongi, D.M. (1998). The influence of fluvial discharge on pelagic production in the Gulf of Papua, northern Coral Sea. *Estuarine, Coastal and Shelf Science* 46: 319–331.
Robertson, A.I., Dixon, P., Daniel, P.A. et al. (2020). Primary production in forests of the mangrove palm *Nypa fruticans. Aquatic Botany* 167: 103288. https://doi.org/10.1016/j.aquabot.2020.103288.
Rochelle-Newall, E.J., Chu, V.T., Pringault, O. et al. (2011). Phytoplankton distribution and productivity in a highly turbid, tropical coastal system (Bach Dang Estuary, Vietnam). *Marine Pollution Bulletin* 62: 2317–2329.
Romero, J., Lee, K.S., Pérez, M. et al. (2006). Nutrient dynamics in seagrass ecosystems. In: *Seagrasses: Biology, Ecology and Conservation* (eds. A.W.D. Larkum, R.J. Orth and C.M. Duarte), 227–254. Dordrecht, The Netherlands: Springer.

Rossi, R.E., Archer, S.K., Giri, C. et al. (2020). The role of multiple stressors in a dwarf red mangrove (*Rhizophora mangle*) dieback. *Estuarine, Coastal and Shelf Science* 237: 106660. https://doi.org/10.1016/j.ecss.2020.106660.

Rothlisberg, P.C. and Burford, M.A. (2016). Biological oceanography of the Gulf of Carpentaria, Australia: a review. In: *Aquatic Microbial Ecology and Biogeochemistry: A Dual Perspective* (eds. P.M. Glibert and T.M. Kana), 51–260. Switzerland: Springer.

Rothlisberg, P.C., Pollard, P.C., Nichols, P.D. et al. (1994). Phytoplankton community structure and productivity in relation to the hydrological regime of the Gulf of Carpentaria, Australia, in summer. *Australian Journal of Marine and Freshwater Research* 45: 265–282.

Roxy, M.K., Modi, A., Murtugudde, R. et al. (2016). A reduction in marine primary productivity driven by rapid warming over the tropical Indian Ocean. *Geophysical Research Letters* 43: 826–833.

Saenger, P. (2002). *Mangrove Ecology, Silviculture and Conservation*. Dordrecht, The Netherlands: Kluwer.

Saenger, P. and Snedaker, S.C. (1993). Pantropical trends in mangrove above-ground biomass and annual litterfall. *Oecologia* 96: 293–299.

Saifullah, A.S.M., Kamal, A.H.M., Idris, M.H. et al. (2016). Phytoplankton in tropical mangrove estuaries: role and interdependency. *Forest Science and Technology* 12: 104–113.

Sanchéz, B.G. (2005). Belowground productivity of mangrove forests in southwest Florida. PhD Dissertation. Louisiana State University.

Sarma, V.V.S.S., Gupta, S.N.M., Babu, P.V.R. et al. (2009). Influence of river discharge on plankton metabolic rates in the tropical monsoon driven Godavari estuary, India. *Estuarine, Coastal and Shelf Science* 85: 515–524.

Sarma, V.V.S.S., Kumar, N.A., Prasad, V.R. et al. (2011). High CO_2 emissions from the tropical Godavari estuary (India) associated with monsoon river discharges. *Geophysical Research Letters* 38: L08601. https://doi.org/10.1029/2011GL046928.

Sarma, V.V.S.S., Rao, D.N., Rajula, G.R. et al. (2019). Organic nutrients support high primary production in the Bay of Bengal. *Geophysical Research Letters* 46: 6706–6715.

Schmitz, N., Egerton, J.J.G., Lovelock, C.E. et al. (2012). Light-dependent maintenance of hydraulic function in mangrove branches: do xylary chloroplasts play a role in embolism repair? *New Phytologist* 195: 40–46.

Scurlock, J.M.O. and Olson, R.J. (2002). Terrestrial net primary productivity – a brief history and a new worldwide database. *Environmental Research* 10: 91–109.

Seguro, I., García, C.M., Papaspyrou, S. et al. (2015). Seasonal changes of the microplankton community along a tropical estuary. *Regional Studies in Marine Science* 2: 189–202.

Semesi, I.S., Beer, S., and Bjork, M. (2009). Seagrass photosynthesis controls rates of calcification and photosynthesis of calcareous macroalgae in a tropical seagrass meadow. *Marine Ecology Progress Series* 382: 41–47.

Senthilkumar, B., Purvaja, R., and Ramesh, R. (2008). Seasonal and tidal dynamics of nutrients and chlorophyll *a* in a tropical mangrove estuary, southeast coast of India. *Indian Journal of Marine Sciences* 37: 132–140.

Sevilgen, D.S., Venn, A.A., Hu, M.Y. et al. (2019). Full in vivo characterization of carbonate chemistry at the site of calcification in corals. *Science Advances* 5: eaau7444. https://doi.org/10.1126/sciadv.aau7447.

Sghaier, Y.R., Zakhama-Sraieb, R., and Charfi-Cheikrouha, F. (2011). Primary production and biomass in a *Cymodocea nodosa* meadow in the Ghar El Melh lagoon, Tunisia. *Botanica Marina* 54: 411–418.

Shamberger, K.E.F., Feely, R.A., Sabine, C.L. et al. (2011). Calcification and organic production on a Hawaiian coral reef. *Marine Chemistry* 127: 64–75.

Sheppard, C., Davy, S., and Pilling, G. (2018). *The Biology of Coral Reefs*, 2e. Oxford, UK: Oxford University Press.

Short, F.T., Montgomery, J., Zimmermann, C.F. et al. (1993). Production and nutrient dynamics of a *Syringodium filiforme* Kütz seagrass bed in Indian River lagoon, Florida. *Estuaries* 16: 323–334.

Simard, M., Fatoyinbo, L., Smetanka, C. et al. (2019). Mangrove canopy height globally related to precipitation, temperature and cyclone frequency. *Nature Geoscience* 12: 40–45.

Sippo, J.Z., Lovelock, C.E., Santos, I.R. et al. (2018). Mangrove mortality in a changing climate: an overview. *Estuarine, Coastal and Shelf Science* 215: 241–249.

Smith, W.O. Jr. and DeMaster, D.J. (1996). Phytoplankton biomass and productivity in the Amazon River plume: correlation with seasonal river discharge. *Continental Shelf Research* 16: 291–319.

Smith, J.A.C., Popp, M., Luttge, U. et al. (1989). Ecophysiology of xerophytic and halophytic vegetation of a coastal alluvial plain in northern Venezuela. VI. Water relations and gas exchange of mangroves. *New Phytologist* 111: 293–307.

Smith, J., Burford, M.A., Revill, A.T. et al. (2012). Effect of nutrient loading on biogeochemical processes in tropical tidal creeks. *Biogeochemistry* 108: 359–380.

de Sousa, E.C.P., Tommasi, L.R., and David, C.J. (1998). Microphytobenthic primary production, biomass, and nutrients and pollutants of Santos estuary (24°S, 46°20′W), São Paulo, Brazil. *Brazilian Archives in Biology and Technology* 41: 25–34.

Sridhar, R., Thangaradjou, T., Kumar Senthil, S. et al. (2006). Water quality and phytoplankton characteristics in the Palk Bay, southeast coast of India. *Journal of Environmental Biology* 27: 561–566.

Stanley, R.H.R., Kirkpatrick, J.B., Cassar, N. et al. (2010). Net community production and gross primary production rates in the western equatorial Pacific. *Global Biogeochemical Cycles* 24: GB4001. https://doi.org/10.1029/2009GB003651.

Stramma, L., Bange, H.W., Czeschel, R. et al. (2013). On the role of mesoscale eddies for the biological productivity and biogeochemistry in the eastern tropical Pacific Ocean off Peru. *Biogeosciences* 10: 7293–7306.

Stuhldreier, I., Sánchez-Noguera, C., Roth, F. et al. (2015). Upwelling increases net primary production of corals and reef-wide gross primary production along the Pacific coast of Costa Rica. *Frontiers in Marine Science* 2: 113. https://doi.org/10.3389/fmars.2015.00113.

Suwa, R., Khan, M.N.I., and Hagihara, A. (2006). Canopy photosynthesis, canopy respiration and surplus production in a subtropical mangrove *Kandelia candel* forest, Okinawa Island, Japan. *Marine Ecology Progress Series* 320: 131–139.

Terrados, J., Agawin, N.S.R., Duarte, C.M. et al. (1999). Nutrient limitation of the tropical seagrass *Enhalus acoroides* (L.) Royle in Cape Bolinao, NW Philippines. *Aquatic Botany* 65: 123–139.

Thaw, M.-S.-H., Obara, S., Matsuoka, K. et al. (2017). Seasonal dynamics influencing coastal primary production and phytoplankton communities along the southern Myanmar coast. *Journal of Oceanography* 73: 345–364.

Thottathil, S.D., Balachandran, K.K., and Gupta, G.V.M. (2008). Influence of allochthonous input on autotrophic–heterotrophic switch-over in shallow waters of a tropical estuary (Cochin Estuary), India. *Estuarine, Coastal and Shelf Science* 78: 551–562.

Tilstra, A., van Hoytema, N., Cardini, U. et al. (2018). Effects of water column mixing and stratification on planktonic primary production and dinitrogen fixation on a northern Red Sea coral reef. *Frontiers in Microbiology* 9: 2351. https://doi.org/10.3389/fmicb.2018.02351.

Tomasko, D.A. and Lapointe, B.E. (1991). Productivity and biomass of the seagrass *Thalassia testudinum* as related to water column nutrient availability and epiphyte levels: field observations and experimental studies. *Marine Ecology Progress Series* 75: 9–17.

Tribollet, A., Langdon, C., Golubic, S. et al. (2006). Endolithic microflora are major primary producers in dead carbonate substrates of Hawaiian coral reefs. *Journal of Phycology* 42: 292–303.

Turk, D., Meinen, C.S., Antoine, D. et al. (2011). Implications of changing El Niño patterns for biological dynamics in the equatorial Pacific Ocean. *Geophysical Research Letters* 38: L23603. https://doi.org/10.1029/2011GL049674.

Twilley, R.R. and Day, J.W. Jr. (2013). Mangrove wetlands. In: *Estuarine Ecology*, 2e (eds. J.W. Day Jr., B.C. Crump, W.M. Kemp, et al.). Hoboken, USA: Wiley-Blackwell.

Udy, J.W. and Bjork, M. (2005). Productivity aspects of three tropical seagrass species in areas of different nutrient levels in Kenya. *Estuarine, Coastal and Shelf Science* 63: 407–420.

Udy, J.W. and Dennison, W.C. (1997). Growth and physiological responses of three seagrass species to elevated sediment nutrients in Moreton Bay, Australia. *Journal of Experimental Marine Biology and Ecology* 217: 253–277.

Udy, J.W., Dennison, W.C., Lee Long, W.J. et al. (1999). Responses of seagrass to nutrients in the Great Barrier Reef, Australia. *Marine Ecology Progress Series* 185: 257–271.

Underwood, G.J.C. (2002). Adaptations of tropical marine microphytobenthic assemblages along a gradient of light and nutrient availability in Suva Lagoon, Fiji. *European Journal of Phycology* 37: 449–462.

Unsworth, R.K.F., Collier, C.J., Henderson, G.M. et al. (2012). Tropical seagrass meadows modify seawater carbon chemistry: implications for coral reefs impacted by ocean acidification. *Environmental Research Letters* 7: 024026.

Vargas-Zamora, J.A., Acuña-González, J., Sibaja-Cordero, J.A. et al. (2018). Water parameters and primary productivity at four marine embayments of Costa Rica (2000-2002). *Revista de Biologia Tropical* 66 (Suppl. 1): S211–S230.

Vase, V.K., Dash, G., Sreenath, K.R. et al. (2018). Spatio-temporal variability of physico-chemical variables, chlorophyll a, and primary productivity in the northern Arabian Sea along India coast. *Environmental Monitoring and Assessment* 190: 1–14.

Verheyden, A., Kairo, J.G., Beeckman, H. et al. (2004). Growth rings, growth ring formation and age determination in the mangrove *Rhizophora mucronata*. *Annals of Botany* 94: 59–66.

Vichi, M., Masina, S., and Nencioli, F. (2008). A process-oriented model study of equatorial Pacific phytoplankton: the role of iron supply and tropical instability waves. *Progress in Oceanography* 78: 147–162.

Wahbeh, M.I. (1984). The growth and production of the leaves of the seagrass *Halophila stipulacea* (Forsk.) Aschers. from Aqaba, Jordan. *Aquatic Botany* 20: 33–41.

Walsh, J.J. (1988). *On the Nature of Continental Shelves*. San Diego, USA: Academic Press.

Xiong, Y., Liu, X., Guan, W. et al. (2017). Fine root functional group-based estimates of fine root production and turnover rate in natural mangrove forests. *Plant and Soil* 413: 83–95.

Xiong, Y., Cakir, R., Phan, S.M. et al. (2019). Global patterns of tree stem growth and stand aboveground wood production in mangrove forests. *Forest Ecology and Management* 444: 382–392.

Yáñez-Arancibia, A. and Day, J.W. Jr. (1982). Ecological characterization of Terminos Lagoon, a tropical lagoon-estuarine system in the southern Gulf of Mexico. *Oceanologica Acta* 8: 431–440.

Yuan, X., Cai, W.J., Meile, C. et al. (2018). Quantitative interpretation of vertical profiles of calcium and pH in the coral coelenteron. *Marine Chemistry 204*: 62–69.

CHAPTER 9

Secondary Production

9.1 Introduction

A variable proportion of carbon fixed by plants is consumed by various secondary consumers such as pelagic and benthic invertebrates and fish. These consumers are ordinarily heterotrophic, requiring food in various forms such as microbes, detritus and fresh animal and plant material. Heterotrophic bacteria are an example of secondary consumers, but one must remember that not all bacteria are heterotrophic; many types are chemoautotrophic or photoautotrophic. Higher life forms, such as invertebrates, are secondary consumers and they form complex food webs based either directly or indirectly on autotrophic production. The productivity of these organisms and heterotrophic bacteria is the focus of this chapter.

Other aspects of these secondary consumers in addition to their growth and production will be examined. The data on secondary production are scant compared with primary productivity because methods to measure secondary production are time-consuming and complicated by the fact that many tropical invertebrates reproduce continuously, that is, they have life histories in which it is difficult to distinguish year classes or cohorts. A good example of this limitation is the smaller invertebrates, such as nematodes, in which it is difficult if not impossible to distinguish juveniles from adults. Nevertheless, data do exist for some life forms, such as heterotrophic bacterioplankton, zooplankton, benthic macrofauna, and fish.

9.2 Heterotrophic Bacterioplankton

Most data on the production of heterotrophic bacterioplankton production originate from studies in temperate and boreal latitudes, and it is only in the past decade or so that measurements have been made in tropical estuarine, coastal, and shelf waters. The data (Table 9.1) suggest that tropical heterotrophic bacterioplankton are highly productive compared with those communities of higher latitude (Ducklow 2000), fuelled by equally high rates of primary productivity (Chapter 8). Specific growth rates are variable, but usually rapid, especially in eutrophic conditions such as in estuaries and coastal waters of India and Brazil.

Bacterial production is a variable percentage of primary production, but average percentages are high, indicating a tight coupling between bacterioplankton and phytoplankton. For example, the average percentage was 80% in Ebrié Lagoon on the Ivory Coast (Torréton et al. 1989). What this means is that, assuming a 50% growth efficiency

Tropical Marine Ecology, First Edition. Daniel M. Alongi.

TABLE 9.1 **Rates of heterotrophic bacterioplankton production in subtropical and tropical estuarine, coastal, and shelf waters. Note that production numbers in normal text are in volume-specific units and those in italicised text are in area-specific units.**

Location	Production (mg C m^{-3} d^{-1}) *mg C m^{-2} d^{-1}*	Specific growth rate (d^{-1})	References
Bay of Bengal	1.58–2.10	0.24–0.68	Fernandes et al. (2008)
Coastal Red Sea		0.32–2.02	Silva et al. (2019)
Southern Great Barrier Reef shelf, Coral Sea	*77–1643*	0.2–5.7	Alongi et al. (2015)
Northern South China Sea shelf	0.5–39.4 (summer) 0.7–3.6 (winter)	0.04–1.15 (summer) 0.03–0.28 (winter)	Austria et al. (2018), Li et al. (2019a)
Pearl River estuary, China	3.93–144.0		Li et al. (2019b)
Cross River estuary, Nigeria	7.15–98.47		Antai et al. (2013)
Dona Paula Bay, India	6.48–86.4	0.049–0.58	Bhaskar and Bhosie (2008)
Godavari estuary, India	0.4–33.3		Gawade et al. (2017)
Cochin estuary, India	19.70–149.76		Thottathil et al. (2008), Parvathi et al. (2013), Jasna et al. (2017)
Chika Lagoon, India	11.5–186.3		Robin et al. (2016)
Gulf of Aqaba, Red Sea	0.09–0.568	0.01–0.15	Grossart and Simon (2002)
Guanabara Bay, SE Brazil	4.8–316.32		Paranhos et al. (2001), Guenther et al. (2008), Signori et al. (2018)
Conceição Lagoon, SE Brazil		0.63–0.79	Fontes et al. (2018)
Recife Harbor, NE Brazil	8.6–115.0		Guenther et al. (2017)
Zuari estuary, India	35–70		de Souza et al. (2003)
Tuamotu Archipelago, French Polynesia	0.33–43.6	0.03–1.54	Torréton et al. (2002)
South-West Lagoon, New Caledonia	1.1–34.0		Torréton et al. (2010)
Vale Grande Channel, SE Brazil	36–528	1.87–6.67	Barrera-Alba et al. (2008)
Great Astrolabe Lagoon, Fiji	2.57–4.43	0.15–0.31	Torréton (1996)
Ebrié Lagoon, Ivory Coast	11.0–91.0	0.17–0.33	Torréton et al. (1989)

for bacteria, heterotrophic requirements represented 160% of autotrophic production. Thus, heterotrophic bacterial activity relies largely on allochthonous organic inputs, such as organic matter derived from pollution and terrigenous inputs from land. Many tropical coastal estuaries, bays, and inshore habitats are polluted and eutrophic in the densely populated tropics with an abundant supply of allochthonous organic matter, as in the Ebrié Lagoon.

Tropical estuarine and coastal waters do not have to be polluted to have an allochthonous supply of organic material. On the southern Great Barrier Reef shelf where waters are unpolluted, bacterial growth rates were rapid with a mean generation time of 10.5 hours or 2.3 doublings d^{-1} (Alongi et al. 2015). These short generation times are generally greater on average than those measured on most temperate shelves (Ducklow 2000) but are at the high end of the range of bacterial production and growth rates measured in subtropical and tropical waters (Table 9.1). The bacterioplankton and phytoplankton data imply that phytoplankton production is insufficient to meet bacterial demand on the southern Great Barrier Reef shelf. Across-shelf differences in plankton function and structure are driven by changes in mixing intensity, sediment resuspension, and the relative contributions of terrestrial, reef, and oceanic nutrients. Discarded larvacean houses and coral mucus are important in the detrital food web, as are mangrove, seagrass, and macroalgal detritus. Production of aggregates by flocculation and aggregation of picoplankton is a key driver for high bacterial production. Viral numbers and biomass correlated with bacterioplankton numbers and production on the southern Great Barrier Reef shelf, and bacterial growth and respiration correlated with net primary production, suggesting close virus–bacteria–phytoplankton interactions. Strong vertical mixing facilitated tight coupling of pelagic and benthic shelf processes in these highly turbid waters. Virus-mediated lysis of bacterioplankton plays a similarly important role in the loss of bacteria in the Cochin estuary, India (Parvathi et al. 2013).

Bacterial production is usually highest inshore probably due to sediment resuspension and terrestrial inputs. Such is the case for the southern Great Barrier Reef shelf (Alongi et al. 2015), the Bay of Bengal (Fernandes et al. 2008), Kenyan coastal waters (Goosen et al. 1997), the Great Astrolabe Lagoon, Fiji (Torréton 1996), and in Guanabara Bay, Brazil (Andrade et al. 2003). In the Bay of Bengal, the shallower western edge of the bay is more productive than the central deep section. The mean bacterial production to primary production ratio is about 30% which infers that there must be extensive allochthonous organic inputs into the bay fuelling bacterial demand. The allochthonous matter that is brought in by rivers during periods of high continental runoff might be altering the relative importance of phytoplankton as a source of nutrition for heterotrophic bacteria in the bay. Heterotrophic bacteria efficiently take up low-molecular weight dissolved organic matter (DOM) and colonise particulate aggregates throughout the water column. Turnover times of bacterioplankton averaged three days in the central bay and nine days in the western part of the bay.

The need for allochthonous matter to help fuel bacterioplankton growth and production has been observed in different environments. For instance, there is strong influence of allochthonous organic material flux for the bacterioplankton communities in the Cochin estuary, India (Thottathil et al. 2008), where bacterial productivity and respiration were examined in relation to primary productivity. The degree of dependence of bacterial production and respiration on primary production was found to be extremely weak. The ratios of BP:NPP (0.05–8.5) and NPP:R

(0.02–7.9) varied widely in the estuary depending on the season and location. There was a seasonal shift in net primary production from net autotrophy to heterotrophy due to terrestrial organic matter input from rivers enhancing bacterial heterotrophic activity. In the estuarine lagoon system of Cananéia-Iguape, Brazil, bacterial productivity rates were among the highest recorded in tropical coastal waters, indicating that this system is highly heterotrophic due to high loads of allochthonous carbon derived from bordering mangrove forests (Barrera-Alba et al. 2008), and implying that bacterial production rates exceed rates of primary production.

Even in coral reef atolls, allochthonous carbon may be required to fuel rates of bacterial productivity. Bacterial carbon demand exceeding phytoplankton carbon production was measured in some of the most oligotrophic atoll lagoons (Torrèton 1996; Torrèton et al. 2002). Only in coral reef waters are bacterial production and turnover times slower than in other tropical coastal waters (Pakulski et al. 1998; Torrèton et al. 2002, 2010). In fact, photoinhibition of bacterial production was found in a subtropical coral reef near Key Largo, Florida (Pakulski et al. 1998), at all depths during daylight. Bacterial production and abundance exhibited recovery at night. The recovery of bacterial production and respiration during a second day of exposure suggested photoinduced selection for light tolerant cells and/or physiological adaptation to ambient light regimes. Clearly, clear waters on coral reefs induce photoinhibition of bacterial activity in the water column, although the mechanism(s) is unclear.

In other coastal waters that are turbid and laden with particles, bacterioplankton production is often linked to particle dynamics in the water column. In Guanabara Bay, southeast Brazil, a close association between bacterioplankton production and the abundance of pelagic particles was observed and linked to tidal cycle oscillations (Guenther et al. 2008). The input of allochthonous dissolved organic carbon (DOC) and temperature were the main factors controlling bacterial carbon metabolism. In surface waters of the bay, particulate carbon production was higher than bacterial respiration, and bacterioplankton acted equally as particulate organic carbon (POC) producers and DOM remineralisers. In bottom waters, R was equivalent to total POC production and higher than bacterioplankton production. Thus, bacteria acted mostly as a DOM sink. Bacterial production in Dona Paula Bay in India was similarly linked to particle-associated carbohydrates; the bulk of total bacterial production and glucosidase activity was associated with particles, suggesting that carbohydrates derived from sediments may serve as an important alternative carbon source sustaining bacterial carbon demand. The dynamics of particle-associated bacteria in the Zuari estuary, India, shows that particle-associated bacteria varied seasonally and accounted for 20–80% of total bacterial abundance (de Souza et al. 2003). Bacterial carbon demand was higher than the primary production and was met by allochthonous input.

9.3 Zooplankton

Several studies have measured secondary production of zooplankton in subtropical and tropical waters (Table 9.2). While these studies are not as plentiful as those from temperate and boreal latitudes, rates of secondary production in tropical

TABLE 9.2 Secondary production and production to biomass ratio (P:B) of subtropical and tropical zooplankton communities. Note that production rates in normal text are volume-based units and rates in italicised text are area-specific units. Single values are means and ranges are in parentheses.

Location	Species/Group	Production (mg C m^{-3} d^{-1}) *(mg C m^{-2} d^{-1})* (mg DW m^{-3} d^{-1})	P:B	References
Continental shelf, Ivory Coast, West Africa	Entire copepod community	*(17.9–103)*		Binet (1979)
Gulf of Guinea, West Africa	Entire zooplankton community		0.34–2.3	Le Borgne (1982)
Coastal waters off Goa, India	Entire zooplankton community	*(39.9–62.6)*		Goswami and Padmavati (1996)
Dharamtar, Bombay Harbour, India	Entire zooplankton community	(0.083–52.60)		Tiwari and Nair (1991, 1993)
Mandovi-Zuari estuaries, India	Entire zooplankton community	**(3.3–29.7)**		Selvakumar et al. (1980)
Kingston, Jamaica	Entire copepod community	*21.5*	0.56	Chisholm and Roff (1990), Hopcroft and Roff (1998a, b), Hopcraft et al. (1998), Persad et al. (2003), Rose et al. (2004)
	Centropages velificatus	*(1.9–2.3)*		
	Paracalanus aculeatus	*(3.5–4.0)*		
	Temora turbinata	*(1.2–1.5)*		
	Penilia avirostris	*8.76*		
	Entire larvacean community	155		
	Entire ctenophore and medusae communities	(116–200)		
		(6.8–19.6)		
	Bolinopsis vitra	137		
	B.ovata	(25.9–53.8)		
	Gracilis	9.9		
	L. tetraphylla			
	Clytia spp.			
West and east coasts, India	Entire zooplankton community	*(1.4–57.3)*		Mathew et al. (1990)

Tuamotu Archipelago, French Polynesia	Entire zooplankton community		1.02	Le Borgne et al. (1989)
	Undinula vulgaris		0.34	
	Thalia democratica		8.16	
Kavaratti Atoll, Lakshadweep, India	Entire zooplankton community	*(6.6–44.8)*		Goswami (1983)
North West Cape, Australia	Entire zooplankton community	*9.3*		McKinnon and Duggan (2003)
Parangipettai estuary, southeast India	Entire tintinnid community	(0.02–2.5)		Godhantaraman (2002)
Hooghly River estuary, India	Entire tintinnid community	(0.04–3.13)		Rakshit et al. (2014)
Coral reef, Malaysia	Entire zooplankton community	(1.8–2.5)		Nakajima et al. (2014)
Andaman Sea, Thailand	Entire copepod community	*(20.1–26.9)*	(0.06–0.07)	Satapoomin et al. (2004)
	Entire calanoid copepod community	*(10.5–19.4)*	0.08	
		(2.0–2.7)	(0.02–0.05)	
	Entire cyclopoid copepod community	*(0.2–0.6)*	(0.03–0.06)	
	Entire harpacticoid copepod community			
Cananéia Lagoon, Brazil	Entire copepod community	5.249	(0.20–0.38)	Ara (2001, 2002, 2004)
	Temora turbinata			
	Acartia lilljeborgi	(0.002–1.115)	(0.17–0.45)	
			(0.18–1.05)	
		(0.176–7.401)		
Uvea Atoll, New Caledonia	Entire zooplankton community	10.4	1.14	Le Borgne et al. (1997)
Great Barrier Reef, Australia	Entire copepod community	*(1.4–15.5)*		McKinnon et al. (2005), McKinnon et al. (2015a, b)
	Entire zooplankton community			
		(0.18–3.39)	(0.07–0.78)	
Kimberley coast, Australia	Entire zooplankton community	(0.68–30.25)	(0.12–1.36)	McKinnon et al. (2015a, b)

(continued)

TABLE 9.2 (*Continued*)

Location	Species/Group	Production ($mg\,C\,m^{-3}\,d^{-1}$) (*$mg\,C\,m^{-2}\,d^{-1}$*) ($mg\,DW\,m^{-3}\,d^{-1}$)	P:B	References
Andaman-Nicobar Islands, India	Entire zooplankton community	(*0.09–24.89*)		Antony et al. (1997)
Ebrié Lagoon, Ivory Coast	*Acartia clausi*	(3.2–26.8)	(0.12–0.48)	Pagano and Saint-Jean (1989)
Grand-Lahou Lagoon, Ivory Coast	*Oithona brevicornis*	(0.0–0.34)		Etilé et al. (2012)
Zanzibar, coastal	Entire copepod community	*88*	0.88	Lugomela et al. (2001)
Jamaica, oceanic	Entire copepod community	*11*	0.2	Webber and Roff (1995a, b)
Hawaii	Entire copepod community	*25*	0.09	Roman et al. (2002)
Bermuda	Entire copepod community	*10*	0.07	Roman et al. (2002)
Arabian Sea	Entire copepod community	*156*	0.12	Roman et al. (2000)
Coastal and oceanic tropical SW Atlantic Ocean, Brazil	Entire copepod community	(0.17–163.20)		Dias et al. (2015)
Taperaçu estuary, Brazil	*Acartia lilljeborgii* *Acartia tonsa*	(6.91–486.29) (0.48–33.49)		Magalhães et al. (2013)
Patos Lagoon, Brazil	Entire copepod community *Acartia tonsa*	(7.9–12.5) 1.17 (0.4–3.65)		Avila et al. (2012), Muxagata et al. (2012), Teixeira-Amaral et al. (2017)
Macaé, São João, Bracuí, Perequê-Açu estuaries, SE Brazil	Entire copepod community	(3.0–13.5)	(0.4–0.55)	Araujo et al. (2017)
Guanabara Bay, Rio de Janeiro, Brazil	Entire copepod community	22.0	(0.15–1.20)	Suchy et al. (2016)
Hong Kong	Entire copepod community	2.73 (0.19–16.64)		Lie et al. (2013)
Off Ubatuba, Brazil	Entire larvacean community	(*11.4–12.7*)		Miyashita and Lopes (2011)

zooplankton communities are lower, or at best equivalent, to those of temperate systems, whereas P:B ratios (Steinberg and Landry 2017) are higher. A tendency towards lower zooplankton productivity in tropical waters may be ascribed to several factors: (i) tropical zooplankton expend more energy on higher metabolism than on growth and reproduction, (ii) food is generally less abundant in the tropics, and (iii) the phytoplankton are dominated by pico- and nano-planktonic size groups which are too small to be directly consumed by zooplankton (except for those that get trapped in aggregates).

Specific respiration rates (d^{-1}) of tropical mesozooplankton communities are many times higher than those of mesozooplankton communities of higher latitude (Hernández-León and Ikeda 2005). Area-weighted respiration (mg C m^{-2} d^{-1}) is higher between 10 °S and 20 °N latitude than in other latitudinal bands, except for the Southern Ocean between latitudes 40 °S and 60 °S (Hernández-León and Ikeda 2005), although mesozooplankton biomass is lowest between 10°–40 °S and 20–40 °N latitude (Hernández-León and Ikeda 2005). P:R ratios for tropical zooplankton are at the high end of the range for estuarine and marine zooplankton.

Huntley and Lopez (1992) examined the published estimates of generation times for 33 species of marine copepods at temperatures ranging from −1.7 °C to 30.7 °C and found that temperature alone explains >90% of the variance in growth rates supporting the idea that subtropical and tropical species grow faster than their boreal and temperate counterparts. Even if tropical production is low, growth rates may still be high. However, other studies (Kleppel et al. 1996; McKinnon 1996) suggest that food limitation counteracts the exponential increase in growth rates predicted by Huntley and Lopez (1992). Growth rates of copepods are determined by several factors besides temperature, including food limitation and trophic factors (Kleppel et al. 1996; McKinnon 1996; Steinberg and Landry 2017). Not all copepod species show clear responses of growth rates and production to changes in temperature and food availability (Dam 2013; Doan et al. 2019). Sea surface temperatures (SSTs) above 30 °C, for example, impair growth and productivity of the tropical marine copepod, *Pseudodiaptomus annandalei*, which inhabits shallow coastal regions of Southeast Asia where mean SST hovers around 30 °C (Doan et al. 2019), implying that this copepod appears to be living at or near its thermal optimum.

Clearly, several factors play a role in determining zooplankton growth rates. At high temperatures off Kingston, Jamaica (Hopcroft and Roff 1998a, b; Hopcroft et al. 1998), total naupliar development times were short (three to four days) inshore and (four to five days) offshore. Mean instantaneous growth rates (g) ranged from 0.90 d^{-1} for *Paravocalanus crassirostris* to 0.41 d^{-1} for *Corycaeus* spp. Nauplii of cyclopoid copepods appeared to grow more slowly than those of similar-sized calanoids. In a follow-up experiment, growth rates were determined for copepodites of the genera *Acartia, Centropages, Corpcaeus, Oithona, Paracalanus, Parvocalanis*, and *Tempora* (Hopcroft et al. 1998). At high temperatures, total copepodite development times were short (four to five days) and mean instantaneous growth rates (g) ranged from 0.1 to 1.2 d^{-1}. Some zooplankton groups in Kingston Harbour, such as the cladoceran, *Penilia avirostris,* were highly productive with correspondingly high growth rates (Rose et al. 2004). Development time averaged 20.5 hours for juveniles and 41.4 hours for adult females. Growth rate appeared to be exponential, with somatic growth rates averaging 0.26 and 0.34 d^{-1} in adult females. The zooplankton community, especially when small species were considered, had rates of secondary production higher than previously estimated for species in other tropical waters, but comparable to other eutrophic tropical bays and many productive temperate ecosystems.

The effects of temperature on zooplankton growth can be ameliorated by diet (Mathews et al. 2018). Reduced survival of nauplii and copepodites of the copepod, *Parvocalanus crassirostris*, at a high temperature (32 °C) was modified when fed P-replete food. Reproduction and rate of development of the harpacticoid copepod, *Nitocra affinis* f. *californica,* was maximal at a salinity of 30–35 and a temperature of 30 °C; overall reproduction was highest and development rate was shortest under lowest light intensity (Matias-Peralta et al. 2005). In laboratory studies, naupliar survival of the cyclopoid copepod, *Oithonia davisae*, was reduced to about 60% at a low food concentration (Almeda et al. 2010). The calanoid copepod, *Temora turbinata,* had similarly rapid growth rates in the food rich, Cananéia Lagoon, Brazil, equivalent to other members of the family Temoridae in temperate areas (Chisholm and Roff 1990).

There are seasonal changes in amplitude in zooplankton production between tropical and temperate waters. Total copepod production estimates in Jamaica were comparable to estimates in temperate coastal regions (Middlebrook and Roff 1988). Zooplankton production off Jamaica (Table 9.2) is continuous, while in temperate zones, zooplankton show much greater modulation of their annual abundance and production cycles (Middlebrook and Roff 1988) with a unimodal summer peak and reductions in winter. A neritic copepod community on a Malaysian coral reef exhibited seasonally sustained secondary production which on an annual basis was comparable to their temperate counterparts (Nakajima et al. 2014, 2017).

Of course, not all tropical zooplankton communities exhibit high rates of production and not all exhibit seasonal constancy. Zooplankton production was low (Table 9.2) in coral reef waters at Kavaratti Atoll (Lakshadweep), but specific growth rates were high, indicative of low biomass but rapid turnover. A similar situation was observed in an atoll lagoon in New Caledonia (Le Borgne et al. 1997). However, production can be higher on some reefs to the extent that allochthonous detritus is necessary to fuel secondary production (Nakajima et al. 2014).

Secondary production of tropical zooplankton in some coastal regions is dependent on the onset of the monsoon season. For example, secondary production was high in September–October in coastal waters off Goa, India, and there was a gradual decrease until February when the values increased again during March–April. Higher secondary production in September–October was due to swarms of the cladoceran, *Evadne tergestina*, and the pteropod, *Cresis acicula*, common in the ecosystem (Goswami and Padmavati 1996). The secondary peak in March–April was due mainly to the abundance of carnivores, such as the siphonophores, *Lensia* sp. and *Diphyes* sp., and the chaetognath, *Sagitta* sp. There was no obvious correlation between zooplankton production and temperature and salinity, as higher production was recorded during low salinities and temperature. In the Dharamtar Creek adjoining Bombay Harbour, rates of secondary production were highest from August to November, with low rates of production in January–March. Low rates of secondary productivity were inevitably observed for the pre-monsoon season, whereas monsoon and post-monsoon periods showed higher rates of production. Low secondary productivity of zooplankton was linked to the lowering of salinity at the onset of the monsoon (Selvakumar et al. 1980; Pagano and Saint-Jean 1989; Antony et al. 1997; Ara 2001; Dias et al. 2015). The seasonal average of zooplankton production in the Picharavam mangrove waterways on the Vellar estuary was highest in the post-monsoon summer and lowest during the monsoon; lower production was linked to lower salinity in the wet monsoon season.

There are exceptions to the generalisation that declines in secondary production occur during the rainy season. In the Taperaçu estuary, Brazil, high productivity of the copepods, *Acartia lilljeborgii* and *Acartia tonsa,* peaked during the rainy season when hypersaline conditions were alleviated (Magalhães et al. 2013). However, in most estuaries, peak secondary production occurs during the pre-monsoon or post-monsoon periods when the water column is most stable.

While copepods are ordinarily the major taxonomic group for tropical zooplankton, such is not always the case and some other groups can be even more productive than copepods. Off Kingston Harbour, Jamaica, the larvacean community are important secondary producers. While copepod biomass was 10 X higher than that of the larvaceans, copepod growth rates were only one-third those of larvaceans (Hopcroft and Roff 1998a, b). Larvacean secondary production was at least 30% of that of the copepods due to their rapid growth rates and at least 50% that of the copepods when house production is considered. The productivity of the larvacean, *Oikopleura dioica,* may exceed the production of copepods (Hopcroft and Roff 1990). Larvacean biomass was equivalent to <10% of copepod biomass off SE Brazil, but larvacean production comprised about 77% of that of copepods, whereas the production of discarded houses and faecal pellets comprised up to 2800% of larvacean secondary production, confirming the potential trophic significance of larvaceans in subtropical and tropical waters (Miyashita and Lopes 2011).

There are large-scale regional differences in secondary production of zooplankton with highest rates measured in upwelling regions. Zooplankton production is high on the west coast of India and fish landings are also great along this coast. This regional phenomenon is attributed to coastal upwelling and the greater primary productivity in the Arabian Sea than in the Bay of Bengal (Table 9.2). Such regional variations may occur in other tropical areas, but data are few. Zooplankton production and respiration are many times greater in the macro-tidal Kimberley coast of NW Australia than in the Great Barrier Reef along the east coast of Australia (McKinnon et al. 2015a, b). Area-specific respiration by > 73 μm zooplankton was sevenfold higher in the Kimberley than in the Great Barrier Reef; production by > 150 μm zooplankton was 278 mg C m^{-2} d^{-1} in the Kimberley and 42 mg C m^{-2} d^{-1} on the Great Barrier Reef. These regional differences were attributed to greater physical forcing and higher tides on the Kimberley coast.

Egg production rates of tropical planktonic copepods are very variable among species and locations (Table 9.3). There are no clear latitudinal differences, but egg production is highest in polar seas during seasonal phytoplankton blooms (Madsen et al. 2008). Annual egg production of calanoid copepod species relates best to temperature, salinity, mixed layer depth, dissolved O_2, and chlorophyll *a* concentration, linked to seasonal and episodic upwelling–downwelling processes. In coastal waters off Mexico, few species maintained continuous reproduction given the relatively high temperatures and strong fluctuations in food availability (Figure 9.1). Some species such as *Temora discaudata* showed significantly higher egg production rates during stratified conditions. Temperature, chlorophyll *a,* and salinity were the stronger drivers of zooplankton production, linked to seasonal and episodic upwelling–downwelling processes on the Mexican continental shelf of the eastern tropical Pacific (Kozak et al. 2017).

TABLE 9.3 **Egg production rates (eggs female^{-1} d^{-1}) of subtropical and tropical zooplankton in estuaries, lagoons, and coastal waters. Values in parentheses are ranges and single values are means.**

Location	Species/Group	Egg production	References
NW Gulf of Mexico	*Centropages furcatus*	13.3	Checkley et al. (1992)
Fukyama Harbor, Japan	*Oithona davisae*	2.6–11.6	Uye and Sano (1995)
Exmouth Gulf, Western Australia	*Oithona attenuata* *Oithona simplex*	3.2 2.4	McKinnon and Ayukai (1996)
Shelf waters off Jamaica, West Indies	Entire copepod community *Oithona plumifera* *Oithona nana* *Corycaeus amazonicus* *Euterpina acutifrons*	(3.2–88) 12.9 (17–20) 43 (18–20)	Hopcroft and Roff (1996, 1998a, b)
Six mangrove estuaries, NE Australia	*Acartia sinjiensis* *Oithona* sp.1 *Oithona* sp. 2 *Oithona aruensis* *Parvocalanus crassirostris* *Bestiolina similis*	(1.3–14.9) (2.3–15.3) (0.08–11.85) (0.8–11.3) (1.3–36.2) (3.6–51.4)	McKinnon and Klumpp (1998)
Bahía Magdalena, Mexico	*Acartia clausi* *Acartia lilljeborgii* *Paracalanus parvus*	2.7 6.2 (4–23)	Gómez-Gutiérrez et al. (1999)
Bahíade La Paz, Mexico	*Acartia clausi* *A. lilljeborgii* *Centropages furcatus* *Labidocera acuta*	12 13.5 23 116	Palomares-García et al. (2003)
Andaman Sea, Indian Ocean	*Oithona plumifera*	2.4 (0.2–11.2)	Satapoomin et al. (2004)
Straits of Malacca, Malaysia	*Acartia pacifica*	6.2	Yoshida et al. (2012)
Grand-Lahou Lagoon, Ivory Coast	*Oithona brevicornis*	(0.11–3.33)	Etilé et al. (2012)
Great Barrier Reef waters, NE Australia	Entire copepod community *O. attenuata* *Oithona dissimilis*	(2.6–8.5) (0.22–1.44) (1.69-3.34)	McKinnon et al. (2005), Zamora-Terol et al. (2014)

TABLE 9.3 *(Continued)*

Location	Species/Group	Egg production	References
Eastern Pacific shelf off Mexico	*Acartia tonsa*	3.3	Kozak et al. (2017)
	Candacia catula	4.9	
	Centropages furcatus	16.2	
	Centropages gracilis	10.1	
	Labidocera acuta	71.8	
	Nannocalanus minor	11.6	
	Pontellopsis lubbocki	22.8	
	Pontellina sobrina	13.4	
	Subeucalanus subcrassus	2.2	
	Temora discaudata	17.5	

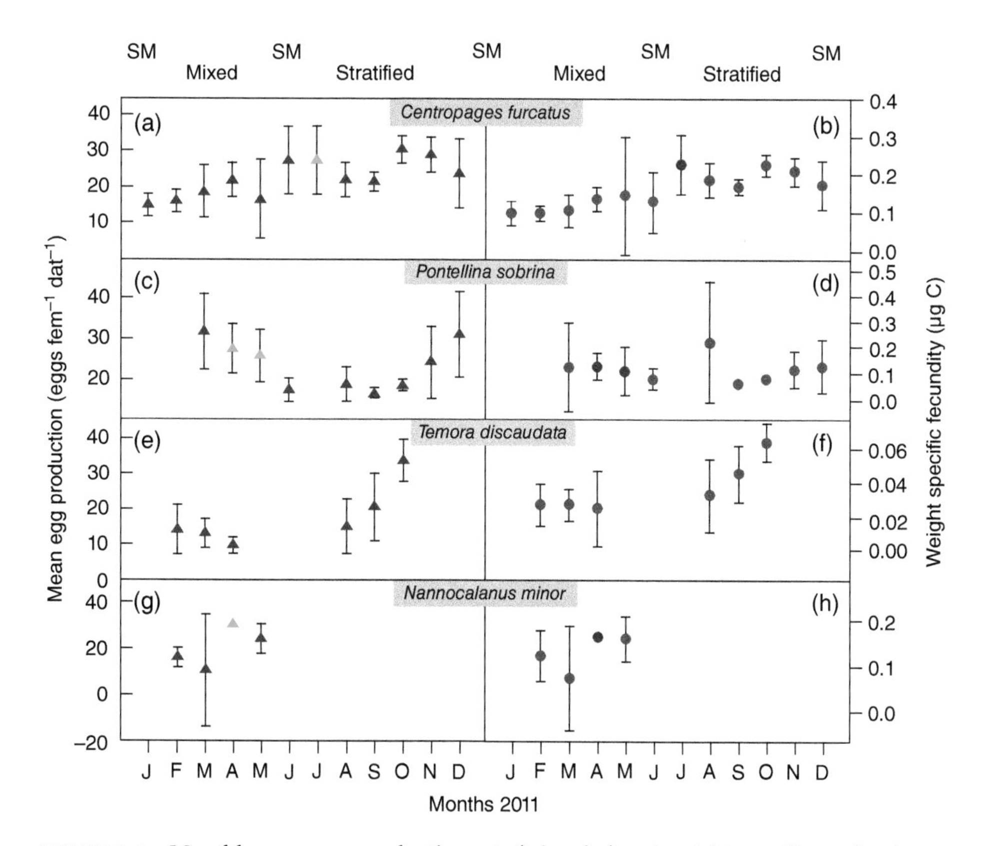

FIGURE 9.1 Monthly mean egg production rate (triangles) and weight-specific production rates (circles) showing the 95% confidence intervals of the means of the copepods, *Centropages furcatus* (a, b), *Pontellina sobrina* (c, d), *Temora discaudata* (e, f), and *Nannocalanus minor* (g, h) from the eastern tropical Pacific off Mexico. Unfilled symbols indicate where incubations fell below target of 10 per month. Seasonal periods are expressed as mixed, semi-mixed (SM) and stratified. *Source*: Kozak et al. (2017), figure 4, p. 146. © Elsevier.

9.4 Benthos

Our knowledge of benthic microbial activity in the tropics is based on measurements of respiration (oxygen consumption) and bacterial productivity, but no information is available on secondary productivity of protozoans and fungi. Oxygen consumption in tropical sediments spans the entire range of rates measured with latitude (Figure 9.2). Peak O_2 uptake rates occur at about the equator and at 40 and 55 °N and 30 °S (Boynton et al. 2018) and are probably linked to high rates of primary productivity. Why rates tend to be so variable in tropical latitudes is problematic considering that few measurements have been made compared with those of higher latitude.

In contrast, bacterial production (Alongi 1990), mostly from Australian habitats, indicates high rates of productivity in sediments. However, these data are based on rates of ^3H-thymidine uptake (deoxyribonucleic acid [DNA] synthesis), a method no longer favoured for sediment environments as some doubt exists as to whether DNA synthesis corresponds directly to growth, and calculations depend greatly on poorly understood conversion factors. Even if incorrect by some considerable margin,

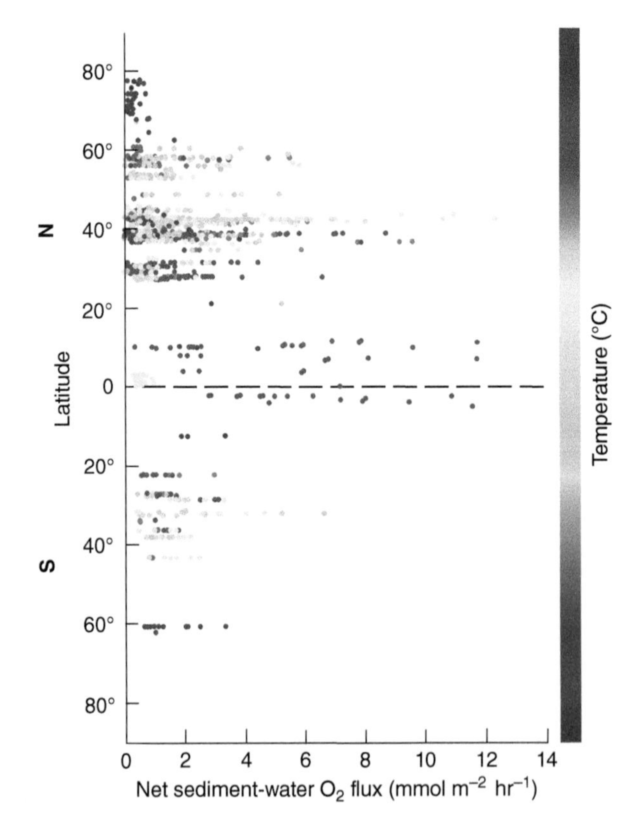

FIGURE 9.2 Net rates of sediment O_2 consumption versus latitude. Dots are colour coded for bottom water or sediment temperature at the time of measurement. The original figure in Boynton et al. (2018) has been updated with subtropical and tropical data from Aller et al. (2004, 2008), Alongi (1989, 1990, 1991, 1996), Alongi et al. (1992, 1999, 2000a, b, 2004, 2005, 2006, 2007, 2008a, b, 2011, 2012, 2013), Clavier and Garrigue (1999), Kristensen et al. (2000), Eyre and Ferguson (2005), Clavier et al. (2014), Grenz et al. (2019), Johnson et al. (2019). *Source*: Boynton et al. (2018), figure 7, p. 317. © Springer Nature Switzerland AG.

they nonetheless point to high rates of bacterial activity in tropical and subtropical marine deposits. Turnover times are rapid, on the order of hours to few days, and are inversely related to temperature, especially in intertidal sediments, although the range of temperatures is narrow. These turnover times are, on an average, greater than those measured in temperate habitats (Kemp 1987).

Production of meiofauna, like fungi and protozoans, is unknown but there is a comparatively large database for macrobenthos from a range of tropical and subtropical habitats (Table 9.4). Most data were obtained for individual species on exposed sandy beaches in southeast Brazil. Except in upwelling, mangrove, and eutrophic habitats (Table 9.4), secondary production of macrofauna in the tropics is low to moderate compared with that in temperate regions (Cusson and Bourget 2005). Biomass is ordinarily low leading to moderate to high P:B ratios for tropical macrobenthos (Ricciardi and Bourget 1999; Bissoli and Bernardino 2018). Globally, crustaceans exhibit P:B ratios from 0.08 to 54.1 with an average P:B ratio of 8.3 indicating rapid turnover and short lifespans (Ricciardi and Bourget 1999). Annelids have the second highest P:B ratios averaging 4.3 (range: 1.0–9.7). Molluscs have the third highest P:B ratios, ranging from 0.2 to 36.9 and averaging 3.4. In terms of total secondary production and again excluding upwelling, eutrophic and mangrove habitats, taxa exhibiting the highest average rates of secondary production are crustaceans (16.4 g DW m^{-2} a^{-1}) followed by molluscs (at 7.2 g DW m^{-2} a^{-1}), annelids (2.8 g DW m^{-2} a^{-1}), and echinoderms (1.35 g DW m^{-2} a^{-1}), with total macroinfauna production ranging from 0.5 to 151.5 g DW m^{-2} a^{-1}. Of course, these data are skewed in favour of exposed sandy beaches which by their very nature are oligotrophic.

The first extensive measurements of macrofaunal secondary production were made by Edwards (1973a,b) in Venezuela and by Ansell et al. (1978) in India. Edwards (1973a, b) used biomass and respiration estimates to estimate production using the relationship between total annual population respiration and production. Using production–elimination methods and the P:R relationship, Ansell et al. (1972b) estimated production of two species of the wedge clam, *Donax incarnates* and *Donax spiculum*, from exposed sandy beaches at Shertallai and Cochi, India. They estimated production rates at the lower end of the range estimated for other tropical molluscs (Table 9.4). Ansell et al. (1978) proposed that a large proportion of the production was preyed upon by demersal fish and crabs. Production of some species was low due to high mortality and disruption caused by monsoons. These early studies were single sampling occasions, so the extent to which production may vary among populations and yearly cohorts is unknown. Much higher production of *Donax serra* (167–637 g DW m^{-2} a^{-1}) was measured on exposed sandy beaches influenced by upwelling (Laudien et al. 2003).

Production is significantly greater in eutrophic and upwelling areas. Along the west coast of Africa, coastal and open ocean upwelling enhances primary production, some of which impinges on sandy beaches up the coast. The same is true for other organically enriched habitats, such as mangrove forests, where enough food exists to foster high rates of secondary production. Berry and bin Othman (1983) studied the life cycle of the gastropod, *Umbonium vestiarium*, on eutrophic sandy shores in north Penang, Malaysia, and found that heavy recruitment was observed in June–July with the cohort growing to full size by January–March of the following year. This weakly seasonal annual cycle was probably keyed to changes in wave action and salinity caused by the NE monsoon.

Macroinfaunal production of tropical and temperate beaches was originally compared by Ansell et al. (1978) who concluded that an equivalent biomass in the tropics produces a rate of biomass turnover 10 times faster than in temperate habitats. This is expressed in higher growth rates and mortality and in the P:B ratio.

TABLE 9.4 **Annual production estimates (g DW m^{-2} a^{-1}) and production to biomass (P:B) ratios for subtropical and tropical macrobenthos. Single numbers are means and ranges are in parentheses.**

Species	Habitat	Production	P:B	References
		Mollusca		
Anadara tuberculosa	Mangrove mud and muddy sand	54.1		Felix-Pico et al. (2015)
Anodontia alba	Subtidal muddy sand		1.8	Moore (1972)
Anomalocardia brasiliana	Mid-intertidal sand flat	2.89	0.61	Corte et al. (2017)
Bivalvia San Luis	Subtidal muddy sand	0.7	4.0	Edwards (1973b)
Bivalvia Las Maritas	Subtidal muddy sand	10.7	2.5	Edwards (1973b)
Bivalvia San Luis	Intertidal sand	5.3	0.5	Edwards (1973b)
Bivalvia Las Maritas	Intertidal sand	2.1	0.3	Edwards (1973b)
Bullia melanoides	Exposed sandy beach	1.5	13.5	Ansell et al. (1978)
Cardita affinis	Sheltered sandy beach	1.9 (La Niña) 1.6 (post-La Niña)	0.3 (La Niña) 0.2 (post-La Niña)	Riascos et al. (2008)
Cerithium atratum	Intertidal muddy sand and sand		3.9	Denadai et al. (2004)
C. atratum	Sheltered sandy beach	4.6	1.0	Cardoso and Cabrini (2016)
Chione cancellata	Subtidal muddy sand		0.6	Moore (1972)
Donax denticulatus	Exposed sandy beach	31.4	5.8	Velez et al. (1985)
Donax spiculum	Exposed sandy beach	1.5	10.3	Ansell et al. (1978)
Donax faba	Exposed sandy beach		(0.3–0.4)	Singh et al. (2011)
Donax hanleyanus	Exposed sandy beach	(0.8–3.7)	(1.5–1.6)	Cardoso and Veloso (2003)
Donax incarnatus	Exposed sandy beach	1.8	5.9	Ansell et al. (1978)
Donax serra	Exposed sandy beach, upwelling	(167–637)	(1.2–1.6)	Laudien et al. (2003)
Donax striatus	Exposed sandy beach	(2.9–6.1)	(3.5–5.1)	Ocaña (2015)

TABLE 9.4 *(Continued)*

Species	Habitat	Production	P:B	References
		Mollusca		
Gafrarium pectinatum	Exposed sandy beach	3.2	4.9	Ansari et al. (1986)
Heleobia australia	Polluted subtidal mud and muddy sand	(28.3–49.4)		Figueiredo-Barros et al. (2006)
Keletistes rhizoecus	Mangrove mud	11.2	2.9	Zabbey et al. (2010)
Mactra olorina	Exposed sandy beach	0.3	7.3	Ansell et al. (1978)
Mesodesma donacium	Exposed upwelled sandy beach	453.1	2.6	Arntz et al. (1987)
Mytilidae and Solecurtidae community	Mangrove and tidal flat	(0.028–0.035)		Bissoli and Bernardino (2018)
Nassarius vibex	Sheltered sandy beach	0.7	1.2	Cardoso and Cabrini (2016)
Olivancillaria vesica vesica	Exposed sandy beach	(0.1–0.2)	(1.0–1.1)	Caetano et al. (2003)
Saccostrea palmula	Mangrove mud and muddy sand	117.5		Felix-Pico et al. (2015)
Strombus canarium	Estuarine sand and muddy sand	0.6	1.5	Cob et al. (2009)
Tagelus divisus	Subtidal muddy sand		1.6	Moore (1972)
Tellina cuspis	Subtidal estuarine mud	3.1	0.7	Edward and Ayyakkannu (1991)
Tellina martinicensis	Subtidal muddy sand		2.4	Moore (1972)
Tellina nobilis	Subtidal estuarine mud	5.8	1.1	Edward and Ayyakkannu (1991)
Timoclea imbricata	Exposed sandy beach	0.2	36.9	Ansell et al. (1978)
Tivela mactroides	Intertidal and subtidal sand and muddy sand	(110.9–148.6)	(1.0–1.2)	Turra et al. (2015)
Umbonium vestiarium	Moderately exposed mud flat	82.1	1.4	Berry and Bin Othman (1983)

(continued)

TABLE 9.4 (*Continued*)

Species	Habitat	Production	P:B	References
Annelida				
Euzonus furciferus	Exposed sandy beach	0.5	2.1	de Souza and Borzone (2007)
Glycera alba	Exposed sandy beach	0.05	5.8	Ansell et al. (1978)
Lumbriconereis laireilli	Exposed sandy beach	0.02	6.8	Ansell et al. (1978)
Neanthes glandicincta	Intertidal mud	8.0	6.2	Shin (2003)
Nereis oligohalina	Subtropical marsh mud	17.9	4.5	Pagliosa and Lana (2000)
Onuphis eremita	Exposed sandy beach	0.3	9.7	Ansell et al. (1978)
Polychaetes San Luis	Subtidal mud	6.2	2.9	Edwards (1973a, b)
Polychaetes Las Maritas	Subtidal mud	13.9	2.5	Edwards (1973a, b)
Polychaetes San Luis	Intertidal sand	0.3	1.0	Edwards (1973a, b)
Polychaetes Las Maritas	Intertidal sand	0.3	1.0	Edwards (1973a, b)
Scolelepis squamata	Exposed sandy beach	0.6	2.7	de Souza and Borzone (2000)
Scoloplos marsupialis	Exposed sandy beach	0.03	6.5	Ansell et al. (1978)
Other polychaetes	Exposed sandy beach	0.01	7.6	Ansell et al. (1978)
Thoracophelia furcifera	Exposed sandy beach	6.6	1.2	Otegui et al. (2012)
Crustacea				
Amphioplus coniatodes	Subtidal mud		2.3	Moore (1972)
Callichirus major	Exposed sandy beach	11.2	1.3	de Souza et al. (1998)
Callinectes sp.	Subtidal mud	0.5	4.8	Edwards (1973a, b)
Clibanarius symmetricus	Exposed sandy beach	0.7	2.44	Danin et al. (2020)

TABLE 9.4 *(Continued)*

Species	Habitat	Production	P:B	References
		Crustacea		
Crustacea San Luis	Subtidal mud	1.5	15.2	Edwards (1973a, b)
Crustacea Las Maritas	Subtidal mud	3.4	13.6	Edwards (1973a, b)
Crustacea San Luis	Intertidal sand	10.7	2.8	Edwards (1973a, b)
Crustacea Las Maritas	Intertidal sand	0.6	2.7	Edwards (1973a, b)
Elamenopsis kempi	Subtidal muddy sand	13.2	5.9	Ali and Salman (1998)
Emerita brasiliensis	Semi-exposed sandy beach	(39.9–156.1)	(6.8–9.6)	Petracco et al. (2003)
Emerita holthuisi	Exposed sandy beach	7.0	19.3	Ansell et al. (1978)
Entire crustacean community	Mangrove and tidal flat	(0.04–0.1)		Bissoli and Bernardino (2018)
Eurydice sp.	Exposed sandy beach	0.3	41.4	Ansell et al. (1978)
Other crustaceans	Exposed sandy beach	0.05	54.1	Ansell et al. (1978)
Excirolana armata	Exposed sandy beach	(15.6–17.3)	(3.1–3.6)	Petracco et al. (2012)
Excirolana braziliensis	Exposed sandy beach	(0.1–0.6)	(1.6–2.2)	Caetano et al. (2006)
Moira atropus	Subtidal mud		0.8	Moore (1972)
Monokalliapseudes schubarti	Coastal lagoon	10.6	5.44	Pennafirme and Soares-Gomes (2017)
Ophionephthys limicola	Subtidal mud		2.3	Moore (1972)
Pseudorchestoidea brasiliensis	Exposed sandy beach	0.3	2.2–2.3	Cardoso and Veloso (1996)
Uca cumulanta	Mangrove muddy sand	31.7	4.5	Koch and Wolff (2002)
Uca lactea annulipes	Mangrove mud and muddy sand	4.8	0.08	Mokhtari et al. (2008)
Uca rapax	Mangrove muddy sand	120.8	4.5	Koch and Wolff (2002)

(continued)

TABLE 9.4 (*Continued*)

Species	Habitat	Production	P:B	References
		Crustacea		
Uca rapax	Mangrove sand	13.4	15.1	Costa and Soares-Gomes (2011)
Uca rapax	Mangrove mud and muddy sand	(120.4–166.8)	(1.4–2.9)	Costa and Soares-Gomes (2015)
Uca vocator	Mangrove muddy sand	210.7	6.5	Koch and Wolff (2002)
		Echinodermata		
Ophiuroidea, Las Maritas	Subtidal mud	1.5	0.8	Edwards (1973a)
Starfish, Las Maritas	Subtidal mud	1.2	2.4	Edwards (1973a)
		Macroinfauna		
Entire macroinfauna community	Subtidal estuarine mud	9.2		Ansari and Parulekar (1994)
Entire macroinfauna community	Subtidal mud and muddy sand	(0.5–7.2)		Parulekar et al. (1982)
Entire macroinfauna community	Subtidal mud and muddy sand	7.1	0.82	Melake (1993)
Entire macroinfauna community	Eutrophic sandy beach	(40.2–151.5)		Veloso et al. (2003)
Entire macroinfauna community	Subtidal mud and muddy sand	37.8		Santhanam et al. (1993)

There is clear overlap, but their analysis of increasing ratios in relation to the major zoogeographic regions indicates a trend of increasing P:B ratios towards subtropical and tropical latitudes (Ansell et al. 1978; Cusson and Bourget 2005). The latitudinal trend is clearest for the bivalves probably because more production estimates are available for this group (Table 9.4). Tropical species exhibit greater mobility, faster growth rates, higher rates of mortality, shorter life spans, and greater P:B ratios than their temperate counterparts (Ansell et al. 1978; Cusson and Bourget 2005).

Tropical species exhibit faster growth rates, but does this derive from continuous reproduction compared with seasonal reproduction of temperate macrobenthos? The evidence is equivocal with some evidence of both discontinuous and continuous recruitment and reproduction in the tropics. Several tropical species exhibit continuous recruitment and reproduction (Edward and Ayyakkannu 1991; Cardoso and Veloso 1996, 2003; Ali and Salman 1998; de Souza and Borzone 2000; Riascos et al. 2008; Costa and Soares-Gomes 2011; Otegui et al. 2012; Petracco et al. 2012; Turra et al. 2015) while more exhibit discontinuous or seasonal peaks (de Souza et al. 1998; Pagliosa and Lana 2000; Petracco et al. 2003; Caetano et al. 2003, 2006;

Veloso et al. 2003; Denadai et al. 2004; de Souza and Borzone 2007; Mokhtari et al. 2008; Cob et al. 2009; Singh et al. 2011; Felix-Pico et al. 2015; Ocaña 2015; Costa and Soares-Gomes 2015; Cardoso and Cabrini 2016). Reproduction tends to be continuous in equatorial regions, but seasonality increases with distance from the equator. Discontinuous reproduction is linked to seasonality in temperature or to the onset of the monsoon whereas closer to the equator, temperature is nearly constant and monsoonal rainfall is more frequent.

9.5 Fisheries

The production of phytoplankton, zooplankton, and benthic macrofauna forms the basis for estimation of fisheries yields, as there is a direct correlation between primary production and fisheries yields in the global coastal ocean (Mann 1993; Chassot et al. 2010). Other factors play a role in sustaining fish production, including structural diversity, predation, river discharge, and upwelling. Coastal regions comprising large rivers, wetlands, and estuaries tend to support high yields of economically important fish, crustaceans, and molluscs. Many viable hypotheses have been offered to explain this phenomenon, including (i) high rates of primary productivity, especially macrophytes, which suggest high availability of pelagic and/or benthic prey, (ii) high level of habitat diversity, structural complexity, and turbidity, offering increased partitioning of resources and shelter from predation, and (iii) physiological and behavioural attraction to river discharge and precipitation, that is, lower salinity preferences for part or the whole of the life cycle. The relative significance of each hypothesis probably varies with the uniqueness of each coastal habitat and associated harvestable communities. Considering the large temporal and spatial variations characteristic of tidal estuaries, bays, and lagoons, it would be ironic indeed if the long-term predictability of such changes is one of the major factors driving the high fisheries yields in the tropical coastal zone.

More intensive yield of harvestable organisms in the marine biosphere compared with freshwater ecosystems may be indirectly caused by greater spatial and temporal variations of physical energy inputs, such as upwelling–downwelling, vertical mixing, tidal exchanges, tidal fronts, and stratification–destratification of water masses (Nixon 1988; Mann 1993). These physical events may enhance transfer of materials and energy up the food chain. A simple example would be when vertical mixing, caused by wind and tidal currents, erodes the benthic boundary layer, permitting more rapid release of nutrients to the water column to enhance primary production, eventually making more food available to fish. Variations in physical forces, such as solar heating, freshwater input, temperature, and wind strength, may lead to variations in fish production. The extent to which physical factors are a dominant force in controlling fish stocks in the tropical ocean is less understood, although several studies have indicated enhanced fish production in proximity to river discharge (de Silva 1986; Mann 1993), monsoonal upwelling (Lin et al. 1995), or a combination of both (Dwivedi 1993).

A comparison of marine capture fish yields to humans for boreal, temperate, upwelling, and tropical shelf ecosystems reveals clear latitudinal trends. Uncorrected Fisheries and Agricultural Organisation (FAO) data (Table 9.5) show continued growth of wild marine fish catch in tropical countries, especially in South and Southeast Asia, whereas global wild marine capture landings have been relatively stable since the late 1990s, although there was an increase in 2018 to 84.4 million

TABLE 9.5 **Annual yield (Mt) of wild capture fisheries by tropical region, 1950–2018.**

1950	1960	1970	1980	1990	2000	2010	2018
Africa (including Middle East)							
0.45	0.82	1.51	1.66	2.89	4.14	4.56	4.71
South America (including the Caribbean)							
0.38	0.69	1.46	3.48	3.55	3.71	4.11	4.32
Oceania							
0.007	0.018	0.037	0.089	0.158	0.230	0.344	0.350
South and Southeast Asia							
1.81	3.95	7.49	9.73	13.738	18.919	24.807	30.289
Total Yields							
2.65	**5.49**	**10.49**	**14.97**	**20.341**	**26.995**	**33.829**	**39.668**

Mt = million tonnes. *Source*: Modified from FAO (2020). © John Wiley & Sons.

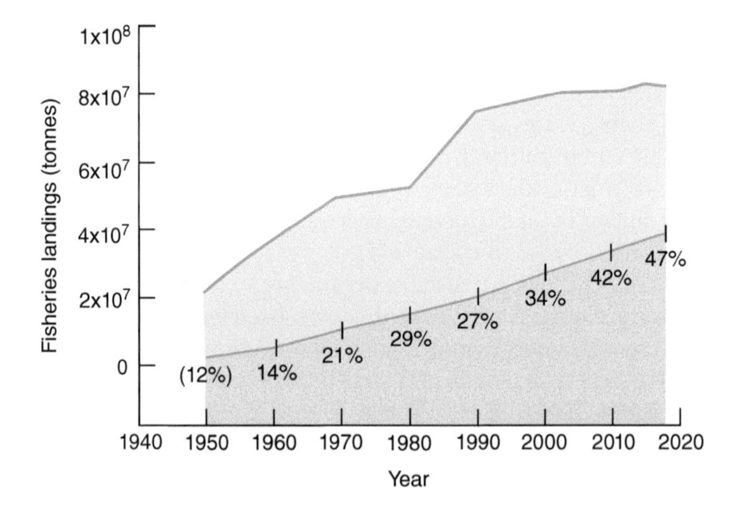

FIGURE 9.3 Landings (tonnes) of global (blue) and tropical (green) wild capture fisheries from 1950 to 2018. Percentages are the proportion of global catch that is composed of tropical catch. *Source*: Based on FAO website (http://www.fao.org/statistics/databases/en/) (accessed 9 October 2020) and FAO (2020). © John Wiley & Sons.

tonnes (Figure 9.3). Considering that tropical continental shelves constitute 30% of the total area of the world's continental shelves and 36% of global open ocean area and correcting the FAO data using the correction factors in Pauly and Zeller (2016), tropical fisheries (including tropical upwelling areas) made up 47% of the global wild capture fisheries catch in 2018, continuing a proportional rise since 1950 (Figure 9.3).

Tropical fisheries may be proportionately more productive than those in boreal and temperate latitudes, but the number of fishermen and women has increased as well (FAO 2020), suggesting an increase in fishing effort in the tropics. While fisheries landings have declined or levelled off in some tropical countries, many countries are still showing large increases in landings, such as Indonesia (6.71 Mt in 2018), India (3.62 Mt), Vietnam (3.19 Mt), Philippines (1.89 Mt), Thailand (1.51 Mt), and Mexico (1.47 Mt). The trend of increasing catches continued in 2017 and 2018, with catches in the Indian Ocean and in the Pacific Ocean reaching the highest levels recorded at 12.3 and 13.5 Mt, respectively (FAO 2020).

As a percentage of major catch, large pelagic fish, namely, tuna, mackerels, and bonitos, occur in 48% of total tropical catch, with snappers second in relative abundance with 18%, shrimps and prawns with 14%, and sardinella and croakers with 10% of the major catch.

India and Indonesia are by far the most populous nations with the greatest demand for fish protein (Table 9.6). The total catch of India is expected to stagnate and ultimately decline in the future due to overfishing and environmental degradation (Bhathal and Pauly 2008), while some South and Southeast Asian countries have overfished their resources, including Myanmar, Thailand, and the Philippines (Stobutzki et al. 2006).

Tunas and their congeners dominate the world's largest ecosystems and sustain some of the most valuable fisheries, especially in the tropics. The impacts of fishing on these species have given rise to several conjectures as to the sustainability of tuna fisheries in the global ocean especially as to the scale and extent of the impacts of fisheries in pelagic ecosystems. Juan-Jordá et al. (2011) have assessed age-structured stocks to evaluate the adult biomass trajectories and exploitation status of 26 populations of tunas and their relatives from 1954 to 2006. The analysis suggests that populations have declined by an average of 60% over the past 50 years, but the decline is lower in total adult biomass, driven by fewer abundant populations. The steepest declines were exhibited by two distinct groups: the largest, longest-lived, highest value temperate tunas and the smaller short-lived mackerels, both with most of their populations being overexploited. Tropical tunas have been fished down to approximately maximum sustainable yield levels, preventing further expansion of catches of these species.

TABLE 9.6 **Top subtropical and tropical fishing nations in the top 25 major producers in 2018. The Peruvian and Chilian upwelling systems are included in italics**

Country	Catch (Mt a^{-1})
Peru	*7.15*
Indonesia	6.71
India	3.62
Vietnam	3.19
Chile	*2.49*
Philippines	1.89
Thailand	1.51
Mexico	1.47
Myanmar	1.14
Mauritania	0.95

Mt a^{-1} = million tonnes per year (2018).*Source*: Modified from FAO (2020). © John Wiley & Sons.

Overcapacity of tuna fisheries is jeopardising their long-term sustainability. Tropical tunas (yellowfin, skipjack, and bigeye tunas) are shorter-lived, reproduce earlier, and have a longer breeding time than their temperate counterparts, making them more accessible and therefore more productive. The largest decline in biomass has occurred in two groups of species with distinct life histories: the large and less productive temperate tunas and the small and more productive mackerel species. The less productive temperate tunas have been affected the most by fishing, exhibiting steeper and larger declines than the more productive tropical species, suggesting that low productivity and slower life histories are an important factor in determining their vulnerability to fishing. Currently, skipjack and yellowfin tuna fisheries are considered healthy, unlike most of the temperate tuna stocks which are overfished (Juan-Jordá et al. 2011). As expected, tropical tuna and mackerel species exhibit a higher tolerance for warmer waters and a lower tolerance for colder waters. The increase in temperature tolerance range with increasing body size among tunas is consistent with the theory that tunas were originally inshore tropical fish which, through endothermic adaptations, became less vulnerable to environmental variability and reduced their competition with other species. The theory that tunas were originally tropical species is further supported by the need of their larvae for extremely warm waters (Boyce et al. 2008).

Catches have been increasing steadily in the eastern Indian Ocean since the 1980s, with catches of small pelagics, large pelagics (tunas and billfish), and shrimps accounting for most of the increase. It is unclear whether the increase was caused by increased fishing effort, productivity, or an artefact in reporting methods. Stocks of toli shad, croaker, and rums, hairtails, catfish, sardinellas, and Indian oil sardine are probably overfished, but anchovies, hilsa shad, Indian mackerel, scads, banana prawn, giant tiger prawn, squids, and cuttlefish are being fished sustainably. In the western Indian Ocean, total landings also continue to increase, but the main shrimp stocks in the SW show clear signs of overfishing. In the western central Pacific Ocean, tuna and tuna-like species accounted for most of the increase in catches, with skipjack tuna in particular increasing from 1.0 to > 1.8 Mt since 2000, with the other main species groups remaining stable, or in the case of small pelagics, decreasing in recent years (FAO 2020). In the eastern central Pacific, catches have remained static, ranging between 1.6 Mt and 2.0 Mt a^{-1} with a large proportion of landings being small- to medium-sized pelagic fish (mainly California pilchard, anchovy, and Pacific jack mackerel), squids, and shrimps. These stocks of short-lived species are naturally more susceptible to changes in oceanographic conditions. In contrast, the eastern central Atlantic Ocean has seen almost continuous increases in catches, reaching 5.5 Mt in 2018, the highest yield recorded. In the western central Atlantic, many stocks have become static, with important stocks such as the Gulf menhaden (*Brevoortia patronus*), round sardinella (*Sardinella aurita*), and skipjack tuna showing decreasing, but sustainable, catches. In the Gulf of Mexico, valuable invertebrates such as the Caribbean spiny lobster and queen conch appear to be fully fished; shrimp stocks have also declined throughout the Caribbean and the Guianas shelf. The southeast Atlantic has seen decreased landings since the early 1970s; horse mackerel and hake are the largest fisheries in the region and the sardinella stocks off Namibia and Angola remain at sustainable levels.

In 2017, 34.2% of the world's fish stocks were classified as overfished. The southeast Pacific region had the highest percentage (54.5%) of stocks fished at unsustainable levels, followed closely by the SW Atlantic (53.3%). The two most important species in the southeast Pacific are Peruvian anchoveta (*Engraulis ringens*) and jumbo flying squid. In contrast, the eastern central Pacific, SW Pacific, and western central Pacific regions had the lowest proportion (13–22%) of stocks fished at unsustainable levels (FAO 2020). Among the seven principal tuna species (albacore, bigeye, skipjack, yellowfin, and three species of bluefin), 33.3% of the stocks were fished at unsustainable levels as of 2017; three stocks have seen their status improve from unsustainable to sustainable, including eastern and western Pacific bigeye tuna and eastern Pacific yellowfin tuna.

Global marine primary production ultimately drives fish production (Ursin 1984). Chassot et al. (2010) have shown that global fisheries catch since 1950 has been increasingly constrained by the amount of primary production and assert that global primary production appears to be declining with concomitant declines in fisheries catches. However, an analysis by Worm and Branch (2012) pointed to the fact that only developed countries have adequate management policies in place and that regions, especially in the developing tropics where there are biodiversity hotspots, still show increasing fisheries catch despite poor or no management policies. This supports the notion that these ecosystems are still productive, but that exploitation rates are probably increasing and will lead to lower biomass over time, especially in regions where the capacity to manage fisheries is low.

One consequence of this situation is that improved management in the developing world may not reduce global fishing pressure but may instead lead to a redistribution of excess fishing effort to countries with less controlled management policies. Illegal, underreported, and unsustainable (IUU) fishing effort has been increasing in the tropics due to poor management and lack of enforcement, leading to eventual declines in many tropical fish stocks. A pressing challenge lies in obtaining stock status in developing countries and in contributing effective management solutions in areas known as "fisheries-conservation hotspots." Hotspots are defined as areas such as the large marine ecosystems (LMEs) with the poorest management policies combined with high biodiversity, such as the Indonesian LME or with large increases in catch levels (the Red Sea LME). These ecosystems typically harbour rapidly increasing human population growth with high dependence on fishing for food security and livelihoods; such conditions can promote "Malthusian overfishing" where immediate needs for food and income overrides long-term sustainability. A common factor in these LMEs is the fact that foreign fleets may be driving the increased catches and exploitation rates. On top of these concerns of overfishing by foreign fleets and Malthusian overfishing is the problem of ocean warming in global fisheries catch (Section 13.7). While marine fish and invertebrates can respond to ocean warming by expanding their latitudinal distributions to higher latitudes and deeper waters, fisheries may become affected by "tropicalisation" of catch, that is, increasing dominance of warm water species in higher latitudes (Cheung et al. 2012, 2013).

References

Ali, M.H. and Salman, S.D. (1998). Production of the crab, *Elamenopsis kempi* Hymenosomatidae, in the Garmat-Ali Region, Basrah, Iraq. *Marine Ecology* 19: 67–75.

Aller, R.C., Hannides, A., Heilbrun, C. et al. (2004). Coupling of early diagenetic processes and sedimentary dynamics in tropical shelf environments, the Gulf of Papua deltaic complex. *Continental Shelf Research* 24: 2455–2486.

Aller, J.Y., Alongi, D.M., and Aller, R.C. (2008). Biological indicators of sedimentary dynamics in the central Gulf of Papua: seasonal and decadal perspectives. *Journal of Geophysical Research: Earth Science* 113: F01S08. https://doi.org/10.1029/2007JF000823.

Almeda, R., Calbet, A., Alcaraz, M. et al. (2010). Effects of temperature and food concentration on the survival, development, and growth rates of naupliar stages of *Oithona davisae* (Copepoda, Cyclopoida). *Marine Ecology Progress Series* 410: 97–109.

Alongi, D.M. (1989). Benthic processes across mixed terrigenous-carbonate sedimentary facies on the central Great Barrier Reef continental shelf. *Continental Shelf Research* 9: 629–663.

Alongi, D.M. (1990). The ecology of tropical soft-bottom benthic ecosystems. *Oceanography and Marine Biology: An Annual Review* 28: 381–496.

Alongi, D.M. (1991). The role of intertidal mudbanks in the diagenesis and export of dissolved and particulate materials from the Fly delta, Papua New Guinea. *Journal of Experimental Marine Biology and Ecology* 149: 81–107.

Alongi, D.M. (1996). The dynamics of benthic nutrient pools and fluxes in tropical mangrove forests. *Journal of Marine Research* 54: 123–148.

Alongi, D.M., Boto, K.G., and Robertson, A.I. (1992). Nitrogen and phosphorus cycles. In: *Tropical Mangrove Ecosystems* (eds. A.I. Robertson and D.M. Alongi), 251–292. Washington DC: American Geophysical Union.

Alongi, D.M., Tirendi, F., Dixon, P. et al. (1999). Mineralization of organic matter in intertidal sediments of a tropical semi-enclosed delta. *Estuarine, Coastal and Shelf Science* 48: 451–467.

Alongi, D.M., Tirendi, F., Trott, L.A. et al. (2000a). Benthic decomposition rates and pathways in plantations of the mangrove *Rhizophora apiculata* in the Mekong Delta, Vietnam. *Marine Ecology Progress Series* 194: 87–101.

Alongi, D.M., Wattayakorn, G., Ayukai, T. et al. (2000b). An organic carbon budget for mangrove fringed Sawi Bay, southern Thailand. *Phuket Marine Biological Center Special Publication* 22: 79–85.

Alongi, D.M., Sasekumar, A., Chong, V.C. et al. (2004). Sediment accumulation and organic material flux in a managed mangrove ecosystem: estimates of land-ocean-atmosphere exchange in peninsular Malaysia. *Marine Geology* 208: 383–402.

Alongi, D.M., Pfitzner, J., Trott, L.A. et al. (2005). Rapid sediment accumulation and microbial mineralization in forests of the mangrove *Kandelia candel* in the Jiulongjiang estuary, China. *Estuarine, Coastal and Shelf Science* 63: 605–618.

Alongi, D.M., Pfitzner, J., and Trott, L.A. (2006). Deposition and cycling of carbon and nitrogen in carbonate mud of the lagoons of Arlington and Sudbury Reefs, Great Barrier Reef. *Coral Reefs* 25: 123–143.

Alongi, D.M., Trott, L.A., and Pfitzner, J. (2007). Deposition, mineralization, and storage of carbon and nitrogen in sediments of the far northern and northern Great Barrier Reef shelf. *Continental Shelf Research* 27: 2595–2622.

Alongi, D.M., Trott, L.A., Undu, M.C. et al. (2008a). Benthic microbial metabolism in seagrass meadows along a carbonate gradient in Sulawesi, Indonesia. *Aquatic Microbial Ecology* 51: 141–152.

Alongi, D.M., Trott, L.A., and Pfitzner, J. (2008b). Biogeochemistry of inter-reef sediments on the northern and central Great Barrier Reef. *Coral Reefs* 27: 407–420.

Alongi, D.M., Trott, L.A., and Møhl, M. (2011). Strong tidal currents and labile organic matter stimulate benthic decomposition and carbonate fluxes on the southern Great Barrier Reef. *Continental Shelf Research* 31: 1384–1395.

Alongi, D.M., Wirasantosa, S., Wagey, T. et al. (2012). Early diagenetic processes in relation to river discharge and coastal upwelling in the Aru Sea, Indonesia. *Marine Chemistry* 140: 10–23.

Alongi, D.M., Brinkman, R., Trott, L.A. et al. (2013). Enhanced benthic response to upwelling of the Indonesian Throughflow onto the southern shelf of Timor-Leste, Timor Sea. *Journal of Geophysical Research: Biogeosciences* 118: 158–170.

Alongi, D.M., Patten, N.L., McKinnon, D. et al. (2015). Phytoplankton, bacterioplankton and virioplankton structure and function across the southern Great Barrier Reef shelf. *Journal of Marine Systems* 142: 25–39.

Andrade, L., Gonzelez, A.M., Araujo, F.V. et al. (2003). Flow cytometry assessment of bacterioplankton in tropical marine environments. *Journal of Microbiological Methods* 55: 841–850.

Ansari, Z.A. and Parulekar, A.H. (1994). Ecological energetics of benthic communities of an estuarine system of the west coast of India. *Proceedings of the Indian National Science Academy* B60: 99–106.

Ansari, Z.A., Chatterji, A., and Parulekar, A.H. (1986). Growth and production of the benthic bivalve *Gafrarium pectinatum* (Linn.) from west coast of India. *Indian Journal of Marine Sciences* 15: 262–264.

Ansell, A.D., McLusky, D.S., Stirling, A. et al. (1978). Production and energy flow in the macrobenthos of two sandy beaches in south west India. *Proceedings of the Royal Society of Edinburgh, Section B* 76: 269–296.

Antai, E.E., Asitok, A.D., and Akaninyene, P. (2013). Bacterial growth efficiency, respiration and secondary production in the Cross River estuary, Nigeria and the near coast. *International Journal of Science and Research* 4: 1791–1798.

Antony, G., Kurup, K.N., Naomi, T.S. et al. (1997). Zooplankton abundance and secondary production in the seas around Andaman-Nicobar Islands. *Indian Journal of Fisheries* 44: 141–154.

Ara, K. (2001). Temporal variability and production of the planktonic copepods in the Cananéia Lagoon estuarine system, São Paulo, Brazil. II. *Acartia illjeborgi. Plankton Biology and Ecology* 48: 35–45.

Ara, K. (2002). Temporal variability and production of *Temora turbinata* (Copepoda, Calanoida) in the Cananéia Lagoon estuarine system, São Paulo, Brazil. *Scientia Marina* 66: 399–406.

Ara, K. (2004). Temporal variability and production of the planktonic copepod community in the Canaenéia Lagoon estuarine system, São Paulo, Brazil. *Zoological Studies* 43: 179–186.

Araujo, A.V., Dias, C.O., and Bonecker, S.L.C. (2017). Effects of environmental and water quality parameters on the functioning of copepod assemblages in tropical estuaries. *Estuarine, Coastal and Shelf Science* 194: 150–161.

Arntz, W.E., Brey, T., Tarazona, J. et al. (1987). Changes in the structure of a shallow sandy beach community in Peru during an El Nino event. *South African Journal of Marine Science* 5: 645–658.

Austria, E.S., Lai, C.-C., Ko, C.-Y. et al. (2018). Growth-controlling mechanisms on heterotrophic bacteria in the South China Sea shelf: summer and winter patterns. *Terrestrial, Atmospheric and Oceanic Sciences* 29: 441–453.

Avila, T.R., de Souza Machado, A.A., and Bianchini, A. (2012). Estimation of zooplankton secondary production in estuarine waters: comparison between the enzymatic (chitobiase) method and mathematical models. *Journal of Experimental Marine Biology and Ecology* 416-417: 144–152.

Barrera-Alba, J.J., Gianesella, S.M.F., Moser, G.A.O. et al. (2008). Bacterial and phytoplankton dynamics in a sub-tropical estuary. *Hydrobiologia* 598: 229–246.

Berry, A.J. and Bin Othman, Z. (1983). An annual cycle of recruitment, growth and production in a Malaysian population of the trochacean gastropod *Umbonium testiarium* (L.). *Estuarine. Coastal and Shelf Science* 17: 375–363.

Bhaskar, P.V. and Bhosle, N.B. (2008). Bacterial production, glucosidase activity and particle-associated carbohydrates in Dona Paula Bay, west coast of India. *Estuarine, Coastal and Shelf Science* 80: 413–424.

Bhathal, B. and Pauly, D. (2008). 'Fishing down marine food webs' and spatial expansion of coastal fisheries in India, 1950-2000. *Fisheries Research* 91: 26–34.

Binet, D. (1979). Estimation de la production zooplanctonique sue le plateau continental ivoirien. *Côtière d'Abidjan* 10: 81–97.

Bissoli, L.B. and Bernardino, A.F. (2018). Benthic macrofaunal structure and secondary production in tropical estuaries on the Eastern Marine Ecoregion of Brazil. *PeerJ* 6: e4441. https://doi.org/10.7717/peerj.4441.

Boyce, D.G., Tittensor, D.P., and Worm, B. (2008). Effects of temperature on global patterns of tuna and billfish richness. *Marine Ecology Progress Series* 355: 267–276.

Boynton, W.R., Ceballos, M.A.C., Bailey, E.M. et al. (2018). Oxygen and nutrient exchanges at the sediment-water interface: a global synthesis and critique of estuarine and coastal data. *Estuaries and Coasts* 41: 301–333.

Caetano, C.H.S., Veloso, V.G., and Cardoso, R.S. (2003). Population biology and secondary production of *Olivancillaria vesica vesica* (Gmelin, 1791) (Gastropoda; Olividae) on a sandy beach in southeastern Brazil. *Journal of Molluscan Studies* 69: 67–73.

Caetano, C.H.S., Cardoso, R.S., Veloso, V.G. et al. (2006). Population biology and secondary production of *Excirolana braziliensis* (Isopoda, Cirolanidae) in two sandy beaches of southeastern Brazil. *Journal of Coastal Research* 22: 825–835.

Cardoso, R.S. and Cabrini, T.M.B. (2016). Population dynamics and secondary production of gastropods on a sheltered beach in south-eastern Brazil: a comparison between an herbivore and a scavenger. *Marine and Freshwater Research* 68: 87–94.

Cardoso, R.S. and Veloso, V.G. (1996). Population biology and secondary production of the sandhopper *Pseudorchestoidesa brasiliensis* (Amphipoda, Talitridae) at Prainha Beach, Brazil. *Marine Ecology Progress Series* 142: 111–119.

Cardoso, R.S. and Veloso, V.G. (2003). Population dynamics and secondary production of the wedge clam *Donax hanleyanus* (Bivalvia, Donacidae) on a high-energy, subtropical beach of Brazil. *Marine Biology* 142: 153–162.

Chassot, E., Bonhommeau, S., Dulvy, N. et al. (2010). Global marine primary production constrains fisheries catches. *Ecology Letters* 13: 495–505.

Checkley, D.M. Jr., Dagg, M.J. Jr., and Uye, S. (1992). Feeding, excretion and egg production by individuals and population of the marine, planktonic copepods, *Acartia* spp. and *Centropages furcatus*. *Journal of Plankton Research* 14: 71–97.

Cheung, W.W., Meeuwig, J.J., Feng, M. et al. (2012). Climate-change induced tropicalisation of marine communities in Western Australia. *Marine and Freshwater Research* 63: 415–427.

Cheung, W.W., Sarmiento, J.L., Dunne, J. et al. (2013). Shrinking of fish exacerbates impacts of global ocean changes on marine ecosystems. *Nature Climate Change* 3: 254–258.

Chisholm, L.A. and Roff, J.C. (1990). Abundances, growth rates and production of tropical neritic copepods off Kingston, Jamaica. *Marine Biology* 106: 79–89.

Clavier, J. and Garrigue, C. (1999). Annual sediment primary production and respiration in a large coral reef lagoon. *Marine Ecology Progress Series* 191: 79–89.

Clavier, J., Chauvaud, L., Amice, E. et al. (2014). Benthic metabolism in shallow coastal ecosystems of the Banc d'Arguin, Mauritania. *Marine Ecology Progress Series* 50: 11–23.

Cob, Z.C., Ghaffar, M.A., Arshad, A. et al. (2009). Exploring the use of empirical methods to measure the secondary production of *Strombus canarium* (Gastropoda, Strombidae) population in Johor Straits, Malaysia. *Sains Malaysiana* 38: 817–825.

Corte, G.N., Coleman, R.A., and Amaral, A.C.Z. (2017). Environmental influence on population dynamics of the bivalve *Anomalocardia brasiliana*. *Estuarine, Coastal and Shelf Science* 187: 241–248.

Costa, T. and Soares-Gomes, A. (2011). Population dynamics and secondary production of *Uca rapax* (Brachyura, Ocypodidae) in a tropical coastal lagoon, southeast Brazil. *Journal of Crustacean Biology* 31: 66–74.

Costa Tde, M.M. and Soares-Gomes, A. (2015). Secondary production of the fiddler crab *Uca rapax* from mangrove areas under anthropogenic eutrophication in the western Atlantic, Brazil. *Marine Pollution Bulletin* 101: 533–538.

Cusson, M. and Bourget, E. (2005). Global patterns of macroinvertebrate production in marine benthic habitats. *Marine Ecology Progress Series* 297: 1–14.

Dam, H.G. (2013). Evolutionary adaptation of marine zooplankton to global change. *Annual Reviews in Marine Science* 5: 349–370.

Danin, A.P.F., Pombo, M., Martinelli-Lemos, J.M. et al. (2020). Population ecology of the hermit crab *Clibanarius symmetricus* (Anomura: Diogenidae) on an exposed beach of the Brazilian Amazon. *Regional Studies in Marine Science* 33: 100944. https://doi.org/10.1016/j.rsma.2019.100944.

Denadai, M.R., Amaral, A.C.Z., and Turra, A. (2004). Biology of a tropical intertidal population of *Cerithium atratum* (Born, 1778) (Mollusca, Gastropoda). *Journal of Natural History* 38: 1695–1710.
Dias, C.O., Araujo, A.V., Vianna, S.C. et al. (2015). Spatial and temporal changes in biomass, production and assemblage structure of mesozooplanktonic copepods in the tropical southwest Atlantic Ocean. *Journal of the Marine Biological Association of the United Kingdom* 95: 483–496.
Doan, N.X., Vu, M.T.T., Pham, H.Q. et al. (2019). Extreme temperature impairs growth and productivity in a common tropical marine copepod. *Scientific Reports* 9: 4550.
Ducklow, H.W. (2000). Bacterial production and biomass in the oceans. In: *Microbial Ecology of the Oceans* (ed. D.L. Kirchman), 85–120. New York: Wiley.
Dwivedi, S.N. (1993). Long-term variability in the food chains, biomass yield and oceanography of the Bay of Bengal ecosystem. In: *Large Marine Ecosystems: Stress, Mitigation and Sustainability* (eds. K. Sherman, L.M. Alexander and B.D. Gold), 475–495. Washington, DC: American Association for the Advancement of Science.
Edward, J.K.P. and Ayyakkannu, K. (1991). Temporal variation in annual production of *Tellina nobilis* and *Tellina cuspis* in a tropical estuarine environment. *Mahasagar* 24: 21–29.
Edwards, R.R.C. (1973a). Production ecology of two Caribbean marine ecosystems. I. Physical environment and fauna. *Estuarine. Coastal and Marine Science* 1: 303–318.
Edwards, R.R.C. (1973b). Production ecology of two Caribbean marine ecosystems. II. Metabolism and energy flow. *Estuarine. Coastal and Marine Science* 1: 319–333.
Etilé, R.N., Aka, M.N., Kouassi, A.M. et al. (2012). Spatiotemporal variations in the abundance, biomass, fecundity, and production of *Othona brevicornis* (Copepoda: Cyclopoida) in a West African tropical coastal lagoon (Grand-Lahou, Côte d'Ivoire). *Zoological Studies* 51: 627–643.
Eyre, B.D. and Ferguson, A.J.P. (2005). Benthic metabolism and nitrogen cycling in a subtropical east Australian estuary (Brunswick): temporal variability and controlling factors. *Limnology and Oceanography* 50: 81–96.
FAO (2020). The state of world fisheries and aquaculture 2020. *Sustainability in Action*. Rome: FAO. https://doi.org/10.4060/ca9229en (accessed 19 November 2020).
Felix-Pico, E.F., Ramirez-Rodriguez, M., and Lopez-Rocha, J.A. (2015). Secondary production of bivalve populations in the mangrove swamps. In: *The Arid Mangrove Forest of Baja California* (eds. R.R. Rodriguez and A.F. Gonzalez-Acosta), 151–173. New York: Nova Science.
Fernandes, V., Ramaiah, N., Sardessai, S. et al. (2008). Strong variability in bacterioplankton abundance and production in central and western Bay of Bengal. *Marine Biology* 153.
Figueiredo-Barros, M.P., Lear, J.J.F., Esteves, F.A. et al. (2006). Life cycle, secondary production and nutrient stock in *Heleobia australis* (d'Orbigny 1835) (Gastropoda, Hydrobiidae) in a tropical coastal lagoon. *Estuarine, Coastal and Shelf Science* 69: 87–95.
Fontes, M.L.S., Fernandes, H., Brandáo, M. et al. (2018). Bacterioplankton activity in a mesoeutrophic subtropical coastal lagoon. *International Journal of Microbiology* 2018: 3209605. https://doi.org/10.1155/2018/3209605.
Gawade, L., Sarma, V.V.S.S., Rao, Y.V. et al. (2017). Variation of bacterial metabolic rates and organic matter in a monsoon-affected tropical estuary. *Geomicrobiology Journal* 34: 628–640.
Godhantaraman, N. (2002). Seasonal variations in species composition, abundance, biomass and estimated production rates of tintinnids at tropical estuarine and mangrove waters, Parangipettai, southeast coast of India. *Journal of Marine Systems* 36: 161–171.
Goosen, N.K., Van Rijswijk, P., De Bie, M. et al. (1997). Bacterioplankton abundance and production and nanozooplankton abundance in Kenyan coastal waters (western Indian Ocean). *Deep-Sea Research II* 44: 1235–1250.
Goswami, S.C. (1983). Production and zooplankton community structure in the lagoon and surrounding sea at Kavaratti Atoll (Lakshapweep). *Indian Journal of Marine Sciences* 12: 31–35.
Goswami, S.C. and Padmavati, G. (1996). Zooplankton production, composition and diversity in the coastal waters of Goa. *Indian Journal of Marine Sciences* 25: 91–97.

Grenz, C., Origel Moreno, M., Soetaert, K. et al. (2019). Spatio-temporal variability in benthic exchanges at the sediment-water interface of a shallow tropical coastal lagoon (south coast of Gulf of Mexico). *Estuarine, Coastal and Shelf Science* 218: 368–380.

Grossart, H.-P. and Simon, M. (2002). Bacterioplankton dynamics in the Gulf of Aqaba and the northern Red Sea in early spring. *Marine Ecology Progress Series* 239: 263–276.

Guenther, M., Paranhos, R., Rezande, C.E. et al. (2008). Dynamics of bacterial carbon metabolism at the entrance of a tropical eutrophic bay influenced by tidal oscillation. *Aquatic Microbial Ecology* 50: 123–133.

Guenther, M., Gonzalez-Rodriguez, E., Flores-Montes, M. et al. (2017). High bacterial carbon demand and low growth efficiency at a tropical hypereutrophic estuary: importance of dissolved organic matter remineralization. *Brazilian Journal of Oceanography* 65: 382–391.

Gómez-Gutiérrez, J., Palomares-García, R., De Silva-Dávila, R. et al. (1999). Copepod daily egg production and growth rates in Bahía Magdalena, Mexico. *Journal of Plankton Research* 21: 2227–2244.

Hernández-León, S. and Ikeda, T. (2005). A global assessment of mesozooplankton respiration in the ocean. *Journal of Plankton Research* 27: 153–158.

Hopcroft, R.R. and Roff, J.C. (1990). Phytoplankton size fractions in a tropical neritic ecosystem near Kingston, Jamaica. *Journal of Plankton Research* 12: 1069–1088.

Hopcroft, R.R. and Roff, J.C. (1996). Zooplankton growth rates: diel egg production in the copepods *Oithona, Euterpina* and *Corycaeus* from tropical waters. *Journal of Plankton Research* 18: 789–803.

Hopcroft, R.R. and Roff, J.C. (1998a). Production of tropical larvaceans in Kingston Harbour, Jamaica: are we ignoring an important secondary producer? *Journal of Plankton Research* 20: 557–569.

Hopcroft, R.R. and Roff, J.C. (1998b). Zooplankton growth rates: the influence of female size and resources on egg production of tropical marine copepods. *Marine Biology* 132: 79–86.

Hopcroft, R.R., Roff, J.C., and Lombard, D. (1998). Production of tropical copepods in Kingston Harbour, Jamaica: the importance of small species. *Marine Biology* 130: 593–604.

Huntley, M.E. and Lopez, M.D.G. (1992). Temperature-dependent production of marine copepods: a global synthesis. *The American Naturalist* 140: 201–242.

Jasna, V., Parvathi, A., Ram, A.S.P. et al. (2017). Viral-induced mortality of prokaryotes in a tropical monsoonal estuary. *Frontiers in Microbiology* 8: 895. https://doi.org/10.3389/fmicb.2017.00895.

Juan-Jordá, M.J., Mosqueira, I., Cooper, A.B. et al. (2011). Global population trajectories of tunas and their relatives. *Proceedings of the National Academy of Sciences* 108: 20650–20655.

Kemp, P.F. (1987). Potential impact on bacteria of grazing by a macrofaunal deposit-feeder, and the fate of bacterial production. *Marine Ecology Progress Series* 36: 151–161.

Kleppel, G.S., Davis, C.S., and Carter, K. (1996). Temperature and copepod growth in the sea: a comment on the temperature-dependent model of Huntley and Lopez. *The American Naturalist* 148: 397–406.

Koch, V. and Wolff, M. (2002). Energy budget and ecological role of mangrove epibenthos in the Caeté estuary, North Brazil. *Marine Ecology Progress Series 228*: 119–130.

Kozak, E.R., Franco-Gordo, C., Palomares-Garcia, R. et al. (2017). Annual egg production rates of calanoid copepod species on the continental shelf of the Eastern Tropical Pacific off Mexico. *Estuarine, Coastal and Shelf Science* 184: 138–150.

Kristensen, E., Andersen, F.Ø., Holmboe, N. et al. (2000). Carbon and nitrogen mineralization in sediments of the Bangrong mangrove area, Phuket, Thailand. *Aquatic Microbial Ecology* 22: 199–213.

Laudien, J., Brey, T., and Arntz, W.E. (2003). Population structure, growth and production of the surf clam *Donax serra* (Bivalvia, Donacidae) on two Namibian sandy beaches. *Estuarine, Coastal and Shelf Science* 58S: 105–115.

Le Borgne, R. (1982). Zooplankton production in the eastern tropical Atlantic Ocean: net growth efficiency and P:B in terms of carbon, nitrogen, and phosphorus. *Limnology and Oceanography* 27: 681–698.

Le Borgne, R., Blanchot, J., and Charpy, L. (1989). Zooplankton of Tikehau atoll (Tuamotu archipelago) and its relationship to particulate matter. *Marine Biology* 102: 341–353.

Le Borgne, R., Rodier, M., Le Bouteiller, A. et al. (1997). Plankton biomass and production in an open atoll lagoon, Uvea, New Caledonia. *Journal of Experimental Marine Biology and Ecology* 212: 187–210.

Li, X., Xu, J., Shi, Z. et al. (2019a). Regulation of protist grazing on bacterioplankton by hydrological conditions in coastal waters. *Estuarine, Coastal and Shelf Science* 218: 1–8.

Li, X., Xu, J., Shi, Z. et al. (2019b). Response of bacterial metabolic activity to the river discharge in the Pearl River estuary: implications for CO_2 degassing fluxes. *Frontiers in Microbiology* 10: 1026. https://doi.org/10.3389/fmicb.2019.01026.

Lie, A.Y.A., Wong, L.C., and Wong, C.K. (2013). Phytoplankton community size structure, primary production and copepod production in a subtropical coastal inlet in Hong Kong. *Journal of the Marine Biological Association of the United Kingdom* 93: 2155–2166.

Lin, C., Xu, B., and Huang, S. (1995). Long-term variations in the oceanic environment of the East China Sea and their influence of fisheries resources. In: *Climate Change and Northern Fish Populations* (ed. R.J. Beamish), 307–327. Ottawa, Canada: National Research Council of Canada.

Lugomela, C., Wallberg, P., and Nielsen, T.G. (2001). Plankton composition and cycling of carbon during the rainy season in a tropical coastal ecosystem, Zanzibar, Tanzania. *Journal of Plankton Research* 23: 1121–1136.

Madsen, S.J., Nielsen, T.G., Tervo, O.M. et al. (2008). Importance of feeding for egg production in *Calanus finmarchicus* and *C. glacialis* during the Arctic spring. *Marine Ecology Progress Series* 353: 177–190.

Magalhães, A., Nobre, D.S.B., Besa, R.S.C. et al. (2013). Diel variation in the productivity of *Acartia lilljeborgii* and *Acartia tonsa* (Copepoda, Calanoida) in a tropical estuary (Taperaçu, northern Brazil). *Journal of Coastal Research* 65: 1164–1169.

Mann, K.H. (1993). Physical oceanography, food chains and fish stocks: a review. *ICES Journal of Marine Science* 50: 105–132.

Mathew, K.J., Naomi, T.S., Antony, G. et al. (1990). Studies on zooplankton biomass and secondary production and tertiary production of the EEZ of India. *Proceedings of the First Workshop on Scientific Results of FORV Sagar Sampada*, New Delhi, India. (5–7 June 1989). New Delhi: Department of Ocean Development.

Mathews, L., Faithful, C.L., Lenz, P.H. et al. (2018). The effects of food stoichiometry and temperature on copepods are mediated by ontogeny. *Oecologia* 188: 75–84.

Matias-Peralta, H., Yusoff, F.M., Shariff, M. et al. (2005). Effects of some environmental parameters on the reproduction and development of a tropical marine harpacticoid copepod *Nitocra affinis* f. *californica* Lang. *Marine Pollution Bulletin* 51: 722–728.

McKinnon, A.D. (1996). Growth and development in the subtropical copepod *Acrocalanus gibber*. *Limnology and Oceanography* 41: 1438–1447.

McKinnon, A.D. and Ayukai, T. (1996). Copepod egg production and food resources in Exmouth Gulf, Western Australia. *Marine, Freshwater Research* 47: 595–603.

McKinnon, A.D. and Duggan, S. (2003). Summer copepod production in subtropical waters adjacent to Australia's North West Cape. *Marine Biology* 143: 897–907.

McKinnon, A.D. and Klumpp, D.W. (1998). Mangrove zooplankton of North Queensland, Australia. *Hydrobiologia* 362: 145–160.

McKinnon, A.D., Duggan, S., and De'ath, G. (2005). Mesozooplankton dynamics in nearshore waters of the Great Barrier Reef. *Estuarine, Coastal and Shelf Science* 63: 497–511.

McKinnon, A.D., Duggan, S., Holliday, D. et al. (2015a). Plankton community structure and connectivity in the Kimberley-Browse region of NW Australia. *Estuarine, Coastal and Shelf Science* 153: 156–167.

McKinnon, A.D., Doyle, J., Duggan, S. et al. (2015b). Zooplankton growth, respiration and grazing on the Australian margins of the tropical Indian and Pacific Oceans. *PLoS One* 10: e0140012. https://doi.org/10.1371/journal.pone.0140012.

Melake, K. (1993). Ecology of macrobenthos in the shallow coastal areas of Tewalit (Massawa), Ethiopia. *Journal of Marine Systems* 4: 31–44.

Middlebrook, K. and Roff, J.C. (1988). Comparison of methods for estimating annual productivity of the copepods *Acartia hudsonica* and *Eurytemora herdmani* in Passamaquoddy Bay, New Brunswick. *Canadian Journal of Fisheries and Aquatic Sciences* 43: 656–664.

Miyashita, L.K. and Lopes, R.M. (2011). Larvacean (Chordata, Tunicata) abundance and inferred secondary production off southeastern Brazil. *Estuarine, Coastal and Shelf Science* 92: 367–375.

Mokhtari, M., Savari, A., Rezai, H. et al. (2008). Population ecology of fiddler crab, *Uca lactea annulipes* (Decapoda, Ocypodidae) in Sirik mangrove estuary, Iran. *Estuarine, Coastal and Shelf Science* 76: 273–281.

Moore, H.B. (1972). An estimate of carbonate production by macrobenthos in some tropical soft-bottom communities. *Marine Biology* 17: 145–148.

Muxagata, E., Amaral, W.J.A., and Barbosa, C.N. (2012). *Acartia tonsa* production in the Patos Lagoon estuary, Brazil. *ICES Journal of Marine Science* 69: 475–482.

Nakajima, R., Yoshida, T., Othman, B.H.R. et al. (2014). Biomass and estimated production rates of metazoan zooplankton community in a tropical coral reef of Malaysia. *Marine Ecology* 35: 112–131.

Nakajima, R., Yamazaki, H., Lewis, L.S. et al. (2017). Planktonic trophic transfer in a coral reef ecosystem – grazing versus microbial food web and the production of mesozooplankton. *Progress in Oceanography* 156: 104–120.

Nixon, S.W. (1988). Physical energy inputs and the comparative ecology of lake and marine ecosystems. *Limnology and Oceanography* 33: 103–1023.

Ocaña, F.A. (2015). Growth and production of *Donax striatus* (Bivalvia, Donacidae) from Las Balsas beach, Gibara, Cuba. *Revista de Biologia Tropical* 63: 639–646.

Otegui, M.B.P., Blankensteyn, A., and Pagliosa, P.R. (2012). Population structure, growth and production of *Thoracophelia furcifera* (Polychaeta, Opheliidae) on a sandy beach in southern Brazil. *Helgoland Marine Research* 66: 479–488.

Pagano, M. and Saint-Jean, L. (1989). Biomass and production of the calanoid copepod *Acartia clausi* in a tropical coastal lagoon, Lagune Ebrié, Ivory Coast. *Scientia Marina* 53: 617–624.

Pagliosa, P.R. and Lana, P.C. (2000). Population dynamics and secondary production of *Nereis oligohalina* (Nereididae, Polychaeta) from a subtropical marsh in southeast Brazil. *Bulletin of Marine Science* 67: 259–268.

Pakulski, J.D., Aas, P., Jeffrey, W. et al. (1998). Influence of light on bacterioplankton production and respiration in a subtropical coral reef. *Aquatic Microbial Ecology* 14: 137–148.

Palomares-García, R., Martínez-Lopez, A., and De Silva-Dávila, R. (2003). Winter egg production rates of four calanoid copepod species in Bahía de la Paz, Mexico. *Contributions to the Study of East Pacific Crustaceans* 2: 139–152.

Paranhos, R., Andrade, L., Mendonça-Hagler, L.C. et al. (2001). Coupling bacterial abundance and production in a polluted tropical coastal bay. In: *Aquatic Microbial Ecology in Brazil* (eds. B.M. Faria, V.F. Farjalla and F.A. Esteves), 117–132. Vol. IX, PPGE-UFRJ. Rio de Janiero: Series Oecologia Brasiliensis.

Parulekar, A.H., Harkantra, S.N., and Ansai, Z.A. (1982). Benthic production and assessment of demersal fishery resources of the Indian Seas. *Indian Journal of Marine Sciences* 11: 107–114.

Parvathi, A., Jasna, V., Haridevi, K.C. et al. (2013). Diurnal variations in bacterial and viral production in Cochin estuary, India. *Environmental Monitoring Assessment* 185: 8077–8088.

Pauly, D. and Zeller, D. (eds.) (2016). *Global Atlas of Marine Fisheries. A Critical Appraisal of Catches and Ecosystem Impacts.* New York: Island Press.

Pennafirme, S. and Soares-Gomes, A. (2017). Population dynamics and secondary production of a key benthic tanaidacean, *Monokalliapseudes schubarti* (Mañé-Garzón, 1949) (Tanaidacea, Kalliapseudidae), from a tropical coastal lagoon in southeastern Brazil. *Crustaceana* 90: 1483–1499.

Persad, G., Hopcraft, R.R., Webber, M.K. et al. (2003). Abundance, biomass and production of ctenophores and medusae off Kingston, Jamaica. *Bulletin of Marine Science* 73: 379–396.

Petracco, M., Veloso, V.G., and Cardoso, R.S. (2003). Population dynamics and secondary production of *Emerita brasiliensis* (Crustacea, Hippidae) at Prainha Beach, Brazil. *P.S.Z.N. Marine Ecology* 24: 231–245.

Petracco, M., Cardoso, R.S., Turra, A. et al. (2012). Production of *Excirolana armata* (Dana, 1853) (Isopoda, Cirolanidae) on an exposed sandy beach in southeastern Brazil. *Helgoland Marine Research* 66: 265–274.

Rakshit, D., Biswas, S.N., Sarkar, S.K. et al. (2014). Seasonal variations in species composition, abundance, biomass and production rate of tintinnids (Ciliata: Protozoa) along the Hooghly (Ganges) River estuary, India: a multivariate approach. *Environmental Monitoring and Assessment* 186: 3063–3078.

Riascos, J.M., Heilmayer, O., and Laudien, J. (2008). Population dynamics of the tropical bivalve *Cardita affinis* from Málaga Bay, Colombian Pacific related to La Niña 1999-2000. *Helgoland Marine Research* 62: S62–S71.

Ricciardi, A. and Bourget, E. (1999). Global patterns of macroinvertebrate biomass in marine intertidal communities. *Marine Ecology Progress Series* 185: 21–35.

Robin, R.S., Kanuri, V.V., Muduli, P.R. et al. (2016). CO_2 saturation and trophic shift induced by microbial metabolic processes in a river dominated ocean margin (tropical shallow lagoon, Chilika, India). *Geomicrobiology Journal* 33: 513–529.

Roman, M.A., Smith, S., Wishner, K. et al. (2000). Mesozooplankton production and grazing in the Arabian Sea. *Deep-Sea Research II* 47: 1423–1450.

Roman, M.R., Adolf, H.A., Landry, M.R. et al. (2002). Estimates of oceanic mesozooplankton production: a comparison using the Bermuda and Hawaii time-series data. *Deep-Sea Research II* 49: 175–192.

Rose, K., Roff, J.C., and Hopcraft, R.R. (2004). Production of *Penilia avirostris* in Kingston Harbour, Jamaica. *Journal of Plankton Research* 26: 605–615.

Santhanam, R., Srinivasan, A., and Devaray, M. (1993). Trophic models of an estuarine ecosystem at the SW coast of India. In: *Trophic Models of Aquatic Ecosystems* (eds. V. Christensen and D. Pauly), 230–233. ICLARM Contributions 28. Penang, Malaysia: ICLARM.

Satapoomin, S., Nielsen, T.G., and Hansen, P.J. (2004). Andaman Sea copepods: spatio-temporal variations in biomass and production, and role in the pelagic food web. *Marine Ecology Progress Series* 274: 99–122.

Selvakumar, R.A., Nair, V.R., and Madhupratap, M. (1980). Seasonal variations in secondary production of the Mandovi-Zuari estuarine system of Goa. *Indian Journal of Marine Sciences* 9: 7–9.

Shin, P.K.S. (2003). Population dynamics and secondary production of *Neanthes glandicinecta* (Polychaeta, Nereididae) from a subtropical mud flat. *Asian Marine Biology* 18: 117–127.

Signori, C.N., Valentin, J.L., Pollery, R.C.G. et al. (2018). Temporal variability of dark carbon fixation and bacterial production and their relations with environmental factors in a tropical estuarine system. *Estuaries and Coasts* 41: 1089–1101.

de Silva, A.J. (1986). River runoff and shrimp abundance in a tropical coastal ecosystem- the example of the Sofala Bank (central Mozambique). In: *The Role of Freshwater Outflow in Coastal Marine Ecosystems* (ed. S. Skreslet), 329–338. Berlin: Springer.

Silva, L., Calleja, M.L., Huete-Stauffer, T.M. et al. (2019). Low abundances but high growth rates of coastal heterotrophic bacteria in the Red Sea. *Frontiers in Microbiology* 9: 3244. https://doi.org/10.3389/fmicb.2018.03244.

Singh, Y.T., Krishnamoorthy, M., and Thippeswamy, S. (2011). Population ecology of the wedge clam *Donax faba* (Gmelin) from the Panambur beach, near Mangalore, south west coast of India. *Journal of Theoretical and Experimental Biology* 7: 171–182.

de Souza, J.R.B. and Borzone, C.A. (2000). Population dynamics and secondary production of *Scolelepis squamata* (Polychaeta, Spionidae) in an exposed sandy beach of southern Brazil. *Bulletin of Marine Science* 67: 221–233.

de Souza, J.R.B. and Borzone, C.A. (2007). Population dynamics and secondary production of *Euzonus furciferus* Ehlers (Polychaeta, Opheliidae) in an exposed sandy beach of southern Brazil. *Revista Brasileira de Zoologia* 24: 1139–1144.

de Souza, J.R.B., Borzone, C.A., and Brey, T. (1998). Population dynamics and secondary production of *Callichirus major* (Crustacea, Thalassinidea) on a southern Brazilian sandy beach. *Archiv Fisheries Marine Research* 46: 151–164.

de Souza, M.-J.B.D., Nair, S., Loka Bharathi, P.A. et al. (2003). Particle-associated bacterial dynamics in a tropical tidal plain (Zuari estuary, India). *Aquatic Microbial Ecology* 33: 29–40.

Steinberg, D.K. and Landry, M.R. (2017). Zooplankton and the ocean carbon cycle. *Annual Review in Marine Science* 9: 413–444.

Stobutzki, I.C., Silvestre, G.T., Abu Talib, A. et al. (2006). Decline in demersal coastal fisheries in three developing Asian countries. *Fisheries Research* 78: 130–142.

Suchy, K.D., Avila, T.R., Dower, J.F. et al. (2016). Short-term variability in chitobiase-based planktonic crustacean production rates in a highly eutrophic tropical estuary. *Marine Ecology Progress Series* 545: 77–89.

Teixeira-Amaral, P., Amaral, W.J.A., de Ortiz, D.O. et al. (2017). The mesozooplankton of the Patos Lagoon estuary, Brazil: trends in community structure and secondary production. *Marine Biology Research* 13: 48–61.

Thottathil, S.D., Balachandran, K.K., and Gupta, G.V.M. (2008). Influence of allochthonous input on autotrophic–heterotrophic switch-over in shallow waters of a tropical estuary (Cochin Estuary), India. *Estuarine, Coastal and Shelf Science* 78: 551–562.

Tiwari, L.R. and Nair, V.R. (1991). Contribution of zooplankton to the fishery of Dharamtar Creek, adjoining Bombay Harbour. *Journal of the Indian Fisheries Association* 21: 15–19.

Tiwari, L.R. and Nair, V.R. (1993). Zooplankton composition in Dharamtar Creek adjoining Bombay Harbour. *Indian Journal of Marine Sciences* 22: 63–69.

Torréton, J.-P. (1996). Biomass, production and heterotrophic activity of bacterioplankton in the Great Astrolabe Reef Lagoon (Fiji). *Coral Reefs* 18: 43–53.

Torréton, J.-P., Guiral, D., and Arfi, R. (1989). Bacterioplankton biomass and production during destratification in a monomictic eutrophic bay of a tropical lagoon. *Marine Ecology Progress Series* 57: 53–67.

Torréton, J.-P., Torréton, J.-P., Pagès, J. et al. (2002). Relationships between bacterioplankton and phytoplankton biomass, production and turnover rate in Tuamotu atoll lagoons. *Aquatic Microbial Ecology* 28: 267–277.

Torréton, J.-P., Rochelle-Newall, E., Pringault, O. et al. (2010). Variability of primary and bacterial production in a coral reef lagoon (New Caledonia). *Marine Pollution Bulletin* 61: 335–348.

Turra, A., Petracco, M., Amaral, A.C.Z. et al. (2015). Population biology and secondary production of the harvested clam *Tivela mactroides* (Born, 1778) (Bivalvia, Veneridae) in southeastern Brazil. *Marine Ecology* 36: 221–234.

Ursin, E. (1984). The tropical, the temperate and the Arctic seas as media for fish production. *Dana* 3: 43–56.

Uye, S. and Sano, K. (1995). Seasonal reproductive biology of the small cyclopoid copepod *Oithona davisae* in a temperate eutrophic inlet. *Marine Ecology Progress Series* 118: 121–128.

Velez, A., Venables, B.J., and Fitzpatrick, L. (1985). Growth production of the tropical beach clam *Donax denticulatus* (Tellinidae) in eastern Venezuela. *Caribbean Journal of Science* 21: 63–73.

Veloso, V.G., Cardoso, R.S., and Petracco, M. (2003). Secondary production of the intertidal macrofauna of Prainha Beach, Brazil. *Journal of Coastal Research* 35: 385–391.

Webber, M.K. and Roff, J.C. (1995a). Annual structure of the copepod community and its associated pelagic environment off Discovery Bay, Jamaica. *Marine Biology* 123: 467–479.

Webber, M.K. and Roff, J.C. (1995b). Annual biomass and production of the oceanic copepod community off Discovery Bay, Jamaica. *Marine Biology* 123: 481–495.

Worm, B. and Branch, T.A. (2012). The future of fish. *Trends in Ecology and Evolution* 27: 594–599.

Yoshida, T., Eio, E.J.E., Toda, T. et al. (2012). Food size dependent feeding and egg production of *Acartia pacifica* from a tropical strait. *Bulletin of Marine Science* 88: 251–266.

Zabbey, N., Hart, A.I., and Wolff, W.J. (2010). Population structure, biomass and production of the West African Lucinid *Keletistes rhizoecus* (Bivalvia, Mollusca) in Sivibilagbara swamp at Bodo Creek, Niger delta, Nigeria. *Hydrobiologia* 654: 193–203.

Zamora-Terol, S., McKinnon, A.D., and Saiz, E. (2014). Feeding and egg production of *Oithona* spp. in tropical waters of North Queensland, Australia. *Journal of Plankton Research* 36: 1047–1059.

CHAPTER 10

Food Webs and Carbon Fluxes

10.1 Introduction

Photoautotrophs are powered by the sun, the ultimate energy source for the biosphere. Energy flows out of the ecosystem in the form of heat (respiration) and in other transformed or processed forms, such as dissolved inorganic carbon (DIC) and detritus. Being necessary for all life, water, gases, and nutrients constantly enter and leave the ecosystem as do some biota (e.g. fish, birds, plankton, seeds). These ecosystem components can be illustrated by a simple ecosystem model, emphasising the flow of energy and materials, such as carbon (Figure 10.1). Unlike nutrients and water, energy cannot be recycled. Being a unidirectional flow that can be transformed by the biological community (e.g. converted to organic matter), most energy passes out of the ecosystem as heat.

Marine ecosystems are composed of a series of benthic and pelagic food webs, composed of photoautotrophs (phytoplankton, algae, mangroves, seagrasses, and kelps) that fix the sun's energy and use simple inorganic compounds to make complex organic structures, and heterotrophs that use, transform, and decompose the organic matter fixed by autotrophs and imported into the system. This 'producer-consumer' concept is overly simplistic, but usually represents the bulk of energy and material flow through most ecosystems.

Benthic and pelagic food webs are dominated numerically and functionally by microbes. Benthic food webs are a mixture of small and large invertebrates known collectively as meiofauna and macrofauna with smaller microbial groups such as protozoans, bacteria, archaea, viruses, and fungi making up the microbenthos. A virtual menagerie of organisms is encountered walking, running, crawling, sliding, swimming, and hiding among the interstices of silt, clay, and sand particles within the seabed. Pelagic processes such as plankton productivity and the transport of dissolved and particulate detritus are functionally important, but it is on and within the seabed that many energetic processes of the greatest magnitude within marine food webs take place. Many epibenthic and infauna organisms harvest a wide range of foods, from dissolved and particulate detritus to bacteria, fungi, microalgae, and meiofauna. It is this catholicity that makes it so difficult to categorise benthic biota in terms of their trophic position; the problem of separating biota from fine particles is what makes it so difficult to categorise them structurally and functionally.

Tropical Marine Ecology, First Edition. Daniel M. Alongi.

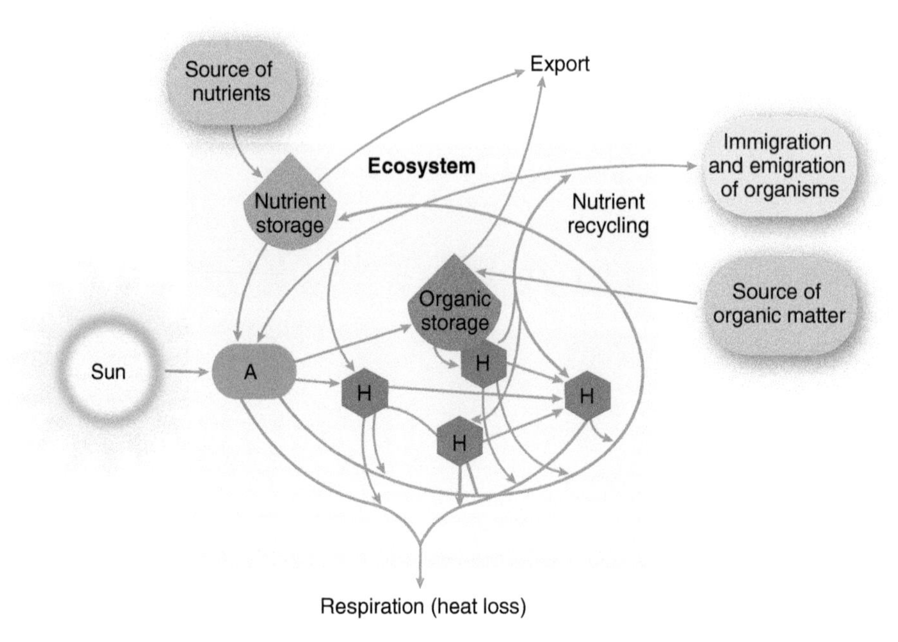

FIGURE 10.1 Conceptual model of the flow of energy and materials through an idealised ecosystem.

A large proportion of dissolved organic matter (DOM) is funnelled through highly diverse and actively growing assemblages of archaea, bacteria, ciliates, flagellates, and amoebae, of which many have mixed trophic states, and subsequently transferred to higher consumers in a chain of grazers in what essentially is a microbial loop, the concept of which has evolved to recognise the functional complexity of the delicate microbial machinery as more of a microbial hub than a loop. Production and consumption within the microbial hub are largely passed to higher consumers via multiple trophic transfers or lost via remineralisation. Whether or not such a microbial hub operates in the seabed is arguable as there is little direct evidence; again, it is difficult to separate benthic microorganisms from the sediment particles, so experimental evidence, especially for tropical benthic food webs, is lacking. As discussed in Chapter 9, some evidence indicate that bacterial productivity is recycled mainly within the microbial hub rather than being transferred up to higher consumers. Bacterial productivity in the benthos is high, much higher than macrofaunal productivity, and there are no data on meiofauna productivity. Bacteria in the seabed mainly drive the biogeochemical cycles, processing organic matter using many different trophic types, including sulphate-reducing, methanogenic, and metal-reducing bacteria. Aerobic and anaerobic bacteria are fed upon by larger consumers, including protozoans and meiofauna, and by macroinfauna, such as polychaetes within their tubes and burrow walls (Figure 10.2). These higher consumers then form a more classical food chain to demersal fish and shrimps up to humans. In deep sediments beneath the upper biologically active zone where bioturbation dominates, there are unlikely to be trophic links between anaerobic microbes (mostly bacteria and archaea) and larger consumers (meiofauna and macroinfauna). The anaerobic microbes in deep sediment layers thus function as a trophic "dead-end." In a sense, this reflects the duality of the nature of life in sediments.

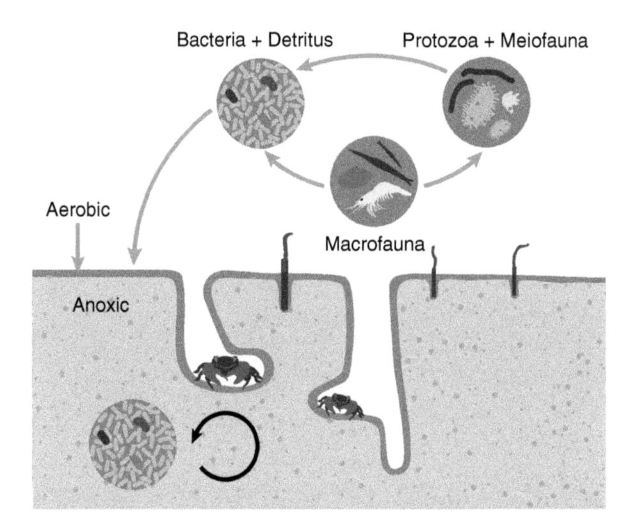

FIGURE 10.2 Conceptual model of benthic food web interactions in aerobic and anaerobic sediments. Most interactions take place under aerobic and partially aerobic conditions in surface deposits and in biogenic structures such as tubes and burrows. In anaerobic sediments, most interactions and organic matter recycling are among microbial consortia (depicted as a circular black arrow).

In this chapter, the food web dynamics in different tropical environments are examined with a view to understanding just how little we know of the topic in the tropics compared to how much we do know in the higher latitudes.

10.2 Sandy Beaches and Tidal Flats

Life in and on sandy beaches is dominated by microbes, meiofauna, macroinfauna, and seabirds. The actual trophic linkages among these groups have not been demonstrated; indeed, some evidence suggests that they are functionally divorced from each other due to overwhelming physical forces. Bacterial productivity in tropical tidal flat sediments is high, but the temporal dynamics of sediment bacterial production has not been linked to the temporal dynamics of consumers (Alongi 1988). Meiofauna were able to consume only a small percentage of the bacterial flora on these Australian tidal flats. Benthic diatoms, however, can accumulate as large patches on some mesotidal beaches and are an important food for suspension feeders. However, some mesotidal dissipative sandy beaches may sustain only transient, but highly productive, food webs (de Oliveira da Rocha Franco et al. 2018).

Wave-exposed sandy beaches are open systems connected to adjacent coastal waters only in terms of what matter and energy flows from the sea to the shore; in truth, there are a variety of ecological connections among land, the shore, and the sea. Not all intertidal habitats are open ecosystems. Most tidal flats oscillate between being open and closed with respect to energy and material exchange. Adjacent surf zones must be considered an integral part of sandy beach ecosystems. These beach surf-zone habitats are essentially semi-closed or closed ecosystems, being fuelled by surf-zone diatom blooms (Oliveira-Santos et al. 2016; Gonzalez and Jùnior 2017).

Reflective sandy beaches are open ecosystems in which food webs receive most of their energy input from either the land or sea, or both. These open systems can fall into two categories: (i) sandy beaches receiving significant quantities of plant detritus from adjacent seagrass meadows, kelp beds, mangroves, or reefs, and (ii) sandy beaches receiving little or no detritus, dependent for energy subsidies from carrion, phytoplankton, or other small filterable particles or DOM delivered from offshore (McLachlan and Defeo 2018). On beaches receiving large inputs of macrophyte detritus, this material is fragmented within the surf by a variety of scavengers at the drift line, driving macroscopic and microbial food webs. The rate of fragmentation and incorporation of this material into food webs depends on several factors, such as the nutritional quality of the detritus, the location and rate of input, and the physical environment, that is, variations in wave and tidal energy (McLachlan and Defeo 2018).

Macrophyte detritus, detached and exported from mangrove forests, reefs and salt marshes, strand on many beaches, but only the input of kelp and seagrass detritus to subtropical and tropical beaches on the west coasts of Australia and southern Africa have been well studied. A carbon flow model of a beach/surf-zone wrack system in Western Australia shows that the main utilisation pathway is via colonising microbes and large populations of the surf-zone amphipod, *Allorchestes compressa* (Figure 10.3). These dense populations of amphipods feed on most of the small particles produced by fragmentation and decay, processing roughly one-third of the detritus per year. Most of the macrophyte detritus is processed by bacteria and other microbes and respired. An impoverished infauna consisting mainly of donacid bivalves and hippid and ocypodid crabs occupies the drift line, supported by faeces from the amphipods and carrion (dead ascidians and sponges) from nearby reefs. Phytoplankton and zooplankton have a minor role in this system. The dense swarms of amphipods attract wading birds, crabs, and fish, to the extent that these wrack deposits are important nursing grounds for several species of fish.

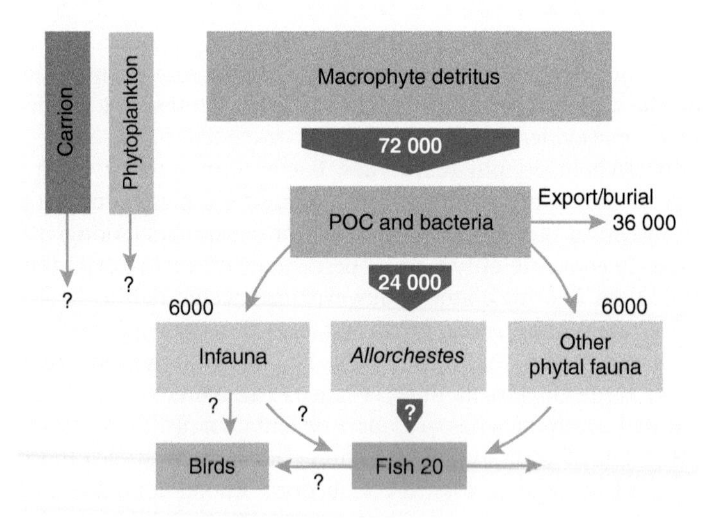

FIGURE 10.3 C flow model of the reflective sandy beach/ surf-zone ecosystem on the southwest coast of Western Australia which receives large amounts of macrophyte detritus. Flows are g C m^{-2} a^{-1} and standing stocks are g C m^{-2}. *Source*: Alongi (1998), figure 2.3, p. 18. © Taylor & Francis.

Amphipods and bacteria have a large role to play on other Australian sandy intertidal areas receiving plant detritus (Poovachiranon et al. 1986). On the tropical Queensland coast, macerated mangrove litter settles as wrack on many sheltered sand flats. These piles of litter shelter large populations of bacteria, protozoa, meiofauna, and the amphipod, *Parhyale hawaiensis*. Densities of these organisms and bacterial growth rates were significantly greater in the litter than in clean sand nearby. In laboratory experiments, Poovachiranon et al. (1986) found that the amphipod preferred mangrove litter when offered a choice of foods, ingesting litter at higher rates than other foods. Many sheltered sand flats and mud flats adjacent to mangrove forests, seagrass beds, and reefs receive significant inputs of detritus derived from these habitats and may be important sites for nutrient regeneration and energy flow.

The fate of wrack detritus is dependent on its position along the surf-zone–beach–shore continuum (McLachlan and Defeo 2018). In contrast to the sandy beach ecosystem of Western Australia, kelp detritus transported to the beaches on the western coast of South Africa remains on the beach rather than in the surf zone because of a greater tidal range. The rate of detrital input is greater due to proximity to the kelp beds. The kelp is cast high upon the beach, so most macroconsumers are largely semi-terrestrial or terrestrial species. A carbon model of these beaches shows that talitrid amphipods were major consumers of the deposited kelp, with kelp-fly larvae and herbivorous beetles consuming lesser amounts of the detritus (Figure 10.4). These scavengers fall prey to birds, carnivorous beetles, and isopods. Carrion and phytoplankton debris contribute minor amounts of detritus to the beach and are consumed largely by carnivorous isopods and bivalve molluscs. Most of the kelp detritus is processed initially by scavengers with the remaining detritus decomposed by bacteria, percolated through the sand interstices as DOM or particulate organic matter (POM), or washed back out to sea. Ultimately, the bulk of the detritus is mineralised by sediment bacteria, protozoa, and meiofauna only 0.5% of the kelp detritus goes to terrestrial organisms, with the remaining 99.5% either respired or exported back to sea.

The fate of organic material filtering into the sand is therefore an important pathway of energy flow. Adding 100 g C of kelp detritus to a sandy microcosm resulted in 70% being respired, 20% converted back into bacterial biomass, and 2% lost to invertebrate grazers; only 0.2% was lost as leachate (Koop and Lucas 1983). Thus, 90% of the leachate was consumed by bacteria. The bulk of kelp detritus deposited onto these sandy beaches was lost via respiration.

Temporal and spatial use of beach wrack by macrofauna is complex, with some species using the wrack as food and others as a refuge. A study in tropical Somalia found that there was a successional change in beach wrack colonisation, with molluscs invading wrack during the first days of deposition and amphipods thereafter (Colombini et al. 2000). Most species exploited the wrack in different ways both in space and time. Thus, even over a short period, there is a rapid turnover of species highly specialised in the exploitation of the wrack when the microclimatic conditions become optimal for that species. Analysis of the tidal component of species activity in the wrack showed that some species moved at ebbing tide while others moved at rising tide. Also, differences were found in the mean hours of tidal activity with differences during day and night. The zonation of each species showed that in some cases, wrack deposits were closely followed by the fauna as their position changed during the semi-lunar phase. Early colonisation of wrack deposits by the amphipod, *Talorchestia martensii*, shows their tendency to forage on freshly stranded material. Mean zonation showed that the amphipods were the group closest to the sea, with gradual changes from landward to seaward zones and vice versa.

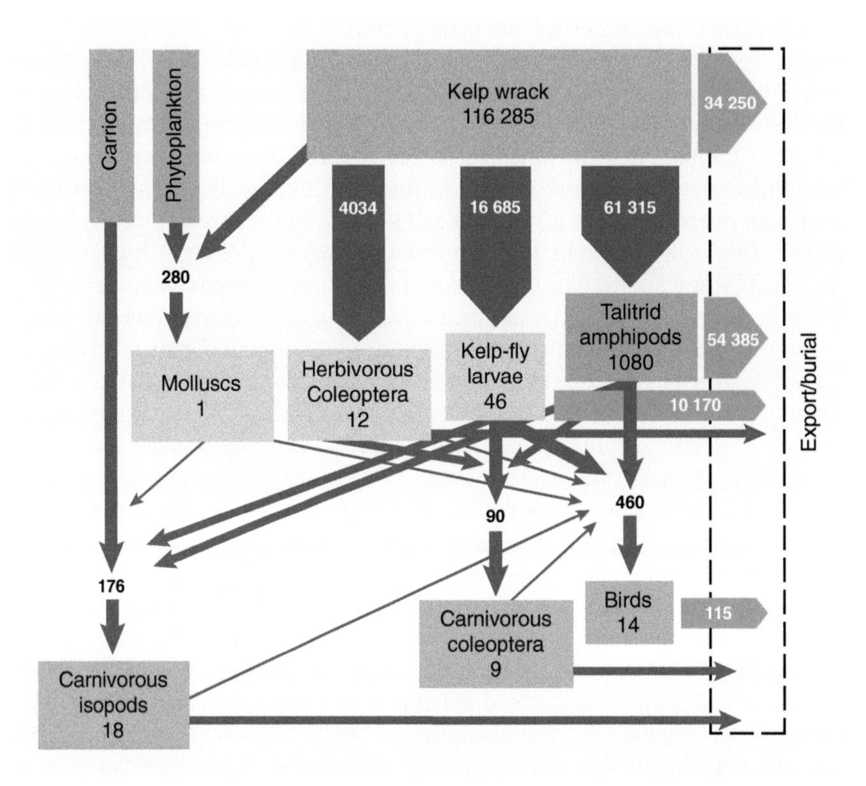

FIGURE 10.4 C flow model of an exposed sandy beach on the west coast of Cape Peninsula, South Africa, which receives large quantities of kelp detritus. Flows are g C m^{-2} a^{-1}. Standing stocks are g C m^{-2}. *Source*: Alongi (1998), figure 2.4, p. 20. © Taylor & Francis.

Ghost crabs may fill the niche of the amphipods on other subtropical sandy beaches. Morrow et al. (2014) examined the trophic links on sandy beaches with wrack material and found that ghost crabs feed primarily on wrack while those on another beach consume wrack-associated amphipods, suggesting that the wrack material serves as an important food subsidy and underlies a dietary shift observed for ghost crab consumers on some beaches. Beach morphology may also contribute to variability in trophic structure on subtropical sandy beaches.

Drift algae may represent a more nutritious source of food for fish and may influence the structure of fish communities in the surf zones of tropical sandy beaches (Andrades et al. 2014). Fish biomass, density, and species richness increased from pre-drift to drift periods of the algae on a sandy shore in Brazil. The density of young-of-the-year fish also increased significantly from pre-drift to drift periods, suggesting the importance of drift algae in the diets of young shore-associated fish. Gut contents of the two most abundant fish species demonstrated the importance of macrophyte-associated amphipods in their diets suggesting that drift algae may be used as feeding habitats for these fish. In contrast, measurements in Johor Strait, Singapore, found that calanoid copepods were the main food of fish in turbid beach zones. Only three main guilds were identified, with seasonal switching of fish guilds where one fish is in one guild (calanoid copopod feeders) for part of the year, and in another guild (polychaete feeders) during a different season.

On sandy beaches not receiving large quantities of macrophyte detritus or without significant in situ primary production, sources of organic matter include small particles of POM transported from offshore, phytoplankton advected from adjacent coastal waters, and mass stranding (Pinotti et al. 2014). The latter two sources are not thought to be major sources of energy for sandy beaches. However, surf-zone diatoms can be an important food resource for many higher consumers (Odebrecht et al. 2014). The accumulation of high biomass of surf-zone diatoms is a characteristic feature of sandy beaches where wave energy is sufficiently high and where surf-zone diatoms can thrive in such an environment. Their occurrence is controlled by physical and chemical factors, such as wave energy, tides, beach slope and length, water circulation patterns in the surf zone, and nutrient availability. The main forces driving the temporal variability in diatom accumulations is meteorological; over hours, the action of wind stress and wave energy controls diatom biomass accumulation, whereas over weeks and months, seasonal onshore winds and storm events are important. Anthropogenic disturbances that influence the beach ecosystem as well as large-scale events such as El Niño-Southern Oscillation (ENSO) may lead to significant changes in the long term. Surf-zone diatoms obviously form a short and very productive food web on tropical sandy beaches, although exactly how they link with microbial groups and higher consumers deserves further study.

Microalgae play an important part in trophic dynamics on tidal mud and sand flats. Selective feeding by the mudskipper, *Boleophthalmus pectinirostris*, on the microalgal assemblage of a tropical mud flat in Hong Kong indicated that diatoms >50 μm in length were selectively ingested, whereas smaller species were not eaten (Yang et al. 2003). Due to selective feeding and seasonal variations in microalgal abundance, this mudskipper exerted a significant impact on the abundance and structure on microphytobenthos on the mud flat. A similar study of the mudskipper, *Periophthalmus waltoni,* showed that foraging activity is largely independent of environmental conditions only being inhibited by high wind speeds and surface temperatures below 15 °C (Clayton and Snowden 2000). To increase their efficiency, mudskippers foraged successively in different areas of the mud flat because their prey remains in their burrows longer than normal following a hunt in their immediate vicinity (Dinh et al. 2020).

Dense concentrations of the surf-zone diatom, *Asterionellopsis glacialis*, are at the base of the food web on subtropical sandy shores in southern Brazil, sustaining a wide variety of consumers, including those at the tertiary level (de Oliveira da Rocha Franco et al. 2018). Trophic relationships, however, varied seasonally and with the morphology of the beach (Pinotti et al. 2014). The bivalves, *Amarilladesma mactroides*, *Donax hanleyanus*, and the hippid crab, *Emerita brasiliensis,* accounted for >95% of all the secondary production in the surf zone. In the surf zone of a tropical sandy beach in the equatorial SW Atlantic, micro- and mesozooplankton communities were abundant but of small size or in their developmental stages. High and constant temperatures and oligotrophic conditions explained their dominance, but also their high diversity (Oliveira-Santos et al. 2016).

Interactions between meiofauna and larger infauna as well as trophic interactions with the infauna of tropical tidal flats (Vargas 1988; Dittmann 1993, 1996, 2002; Carlén and 'Olafsson 2002) have indicated intense interrelationships. To investigate the impact of feeding activities of the soldier crab, *Mictyris longicarpus,* on small infauna, a series on caging experiments were conducted on a tropical mud flat in northern Australia (Dittmann 1993). Comparison of meiofaunal abundances in undisturbed surface sediment and sediment handled by soldier crabs indicated consumption of nematodes, copepods, and platyhelminthes with significantly reduced number of these groups in the pseudofaecal pellets discarded by crabs foraging on the surface and in the sediments deposited as hummocks by subsurface feeding crabs. Significant increases in the abundance of all meiofaunal taxa were found when soldier crabs were excluded. The reduced numbers of meiofauna were unlikely due to disturbance, as mechanical disturbance of the sediment to simulate crab scaping had no impact on the meiofauna. Foraging soldier crabs are therefore important predators on meiofauna.

More meiofauna occur within burrows than in adjacent tidal flat sediments (Dittmann 1996). Nematodes, copepods, and platyhelminthes were significantly higher in burrows of fiddler crabs (*Uca* spp.), the mud shrimp, *Callianassa australiensis*, and the brachiopod, *Lingula anatina* on this same Australian tidal flat. Significantly lower meiofauna abundance was observed when *C. australiensis* was excluded by cages. The effect of the shrimp exclusion on macrofauna was less pronounced but after a year, total numbers of macrofauna were significantly lower in the exclusions, indicating that positive interactions can play an important role in structuring tropical tidal flat communities. Caging experiments conducted in the Gulf of Nicoya, Costa Rica, resulted in insignificant changes in abundance and species numbers of benthos inside cages (Vargas 1988). These results indicate that the role of macropredators (birds, fish, and crabs) in community structure may be relatively unimportant.

Contrary results were observed in East African mud flats. Carlén and 'Olafsson (2002) tested if the gastropod, *Terebralia palustris,* through food competition and/or surface sediment disturbance effect infauna densities and assemblages. After nine weeks with gastropods either enclosed or excluded from experimental cages, high densities of gastropods clearly reduced meiofaunal densities compared to control cages without the gastropod. The gastropod had no significant effect on the macroinfauna. On a tropical mud flat in Amazonia (Pascal et al. 2019), feeding rates of large predators (crabs, shrimps, and fish) increased with the availability of meiofauna and macrofauna. The predators preferred macrofauna, while meiofauna were consumed only during early life stages or in the absence of large food items. Thus, feeding relationships among various trophic groups are often complex and may even be group- or species-specific.

Trophic relationships on tropical mud flats may or may not be linked to adjacent habitats. Mangrove-derived carbon was predominantly assimilated by invertebrates and fish within the mangrove forest of the Orinoco River estuary, whereas these organisms derived <21% of their carbon from mangrove sources while feeding on the adjacent mud flat, where microphytobenthos and phytoplankton were their main foods (Marley et al. 2019). However, highly mobile predators such as osprey, snowy egret, and spectacled caiman, ate the same foods in both habitats showing connectivity at higher trophic levels. Similarly, the utilisation of various carbon sources varied among macrobenthos in mangrove and mud flat habitats in Daya Bay, south China (Arbi et al. 2018). Microphytobenthos dominated the mud flat food web, whereas a mixture of foods was eaten within the mangrove forests; mangrove leaf litter transported to the mud flats were not a major food source.

Feeding relationships have been deduced for macrobenthos on sandy beaches. The Pacific mole crab, *Hippa pacifica,* is a conspicuous scavenger in the swash zone of exposed sandy beaches on Hawaii (Lastra et al. 2016). Rapid attraction and aggregation of large crabs to bait stations contrasted with the population samples of small crabs collected in the absence of bait, suggesting that small crabs were excluded by larger crabs when competing for carrion. Large crabs were more efficient than small crabs in scavenging in the swash zone. Abundance of crabs was correlated with the amount of edible organic matter on the beach, suggesting that the mole crab is a highly opportunistic species able to feed on a wide range of foods. On a Brazilian sandy beach, Arruda et al. (2003) distinguished six feeding guilds for the molluscs, with suspension feeder and deposit feeder guilds more abundant than herbivores and carnivore/scavengers. The distribution of the feeding guilds correlated best with salinity, particle size distribution, percent silt-clay, and POM content in the beach sand.

Fish communities often play an important role in the trophodynamics of tropical sandy beaches and intertidal flats. In a stable isotope study of fish diets, Claudino et al. (2015) investigated the relative importance of basal food resources to fish and decapod crustaceans. There were marked shifts in C and N isotope values for primary producers, fish, and crustaceans, with the mixing model indicating that in the inner section of the estuary where the sandy beaches and tidal flat are located, consumers tended to assimilate nutrients derived mainly from mangrove and macroalgae, whereas near the mouth of the estuary and in the adjacent marine area, they assimilated nutrients derived mainly from macroalgae, seagrass, and organic matter in the sediment. These findings support the idea that the relative importance of basal food resources to fish reflects the dominant autochthonous primary producer. Tropical mud flats support high fish diversity and several feeding guilds. Shrimp-feeding fish were the dominant trophic group on a large intertidal mud flat in Klang, Malaysia, and the other seven dietary guilds fed secondarily on penaeid and sergestid shrimps (Lee et al. 2019). Stable isotopes revealed that four to five trophic levels in a relatively complex food web were fuelled by benthic microalgae and phytoplankton. Mud flat primary production and the relatively high number of trophic levels were likely sustained by shallow waters and strong tidal mixing.

The importance of mangrove forests, mud and sand flats, and seagrass beds as feeding areas for juvenile fish was examined in Chwaka Bay, Zanzibar (Lugendo et al. 2006), where gut contents and stable isotopes were used to parse the diets of several juvenile fish species. Gut content analysis revealed that almost the same foods were consumed by fish regardless of habitat. Crustaceans were the preferred prey items for most zoobenthivores and omnivores, while fish and algae were the preferred food for piscivorous and herbivorous species, respectively. The importance of mangrove and seagrass as a primary source of food to higher trophic levels was limited. Some fish species appeared to show a connectivity with respect to feeding between different habitats. In the Indian Sundarbans, the feeding guild structure is as complex as it is in other sandy beach and intertidal flat habitats. Chaudhuri et al. (2014) found five feeding guilds of fish in the Sundarbans complex: planktonic-benthivores, herbivores, detritivores, omnivores, and carnivores. Teleost fish and decapods were the main prey item of the carnivorous fish. Food did not appear to be limiting on the mud flats.

Scavenging of carrion by shorebirds is an underappreciated trophic pathway on sandy beaches and intertidal sand and mud flats (Piersma et al. 1993; Kober and Bairlein 2006; Schlacher et al. 2013). Piersma et al. (1993) found that trophic use by shorebirds was greater and more varied in temperate than tropical mud flats. Shorebirds of the Bragantinian Peninsula, northern Brazil, have broad diets with overlapping species, with no clear clustering evident according to dietary preference. A generalist foraging strategy was most favourable. Larger species were slightly more restricted to large biomass prey. The food stock was scarce and temporally variable, so a generalist and opportunistic strategy was most favourable. An opportunistic foraging strategy might be an appropriate strategy at other tidal flats in the tropics under similar conditions. In any event, shorebirds respond opportunistically and rapidly to pulsed carrion at the land–ocean interface. Stranded marine carrion supplies high-quality food to scavengers, but the role of animal carcasses is generally underreported in sandy beach food webs. Detection of carrion effects at several trophic levels suggests that feeding links arising from carcasses shape the architecture and dynamics of food webs at the land–beach–ocean interface.

Sediment metabolism has been measured in a few tropical intertidal mud flats (Alongi et al. 1999; Kristensen et al. 2000; Clavier et al. 2014; Bento et al. 2017) with rates indicating rapid microbial respiration. Rates of benthic metabolism across the sediment–water interface on Australian mud flats ranged from 45 to 61 mmol O_2 m^{-2} d^{-1} and 14–15 mmol CO_2 m^{-2} d^{-1}, respectively in winter and sediment–air fluxes ranged from 3 to 4 mmol O_2 m^{-2} d^{-1} and 1–2 mmol CO_2 m^{-2} d^{-1}, respectively, in summer. These rates were lower than in adjacent mangrove forests (Alongi et al. 1999). On two Thai mud flats similarly adjacent to mangroves, rates of benthic metabolism across the sediment–water interface ranged from 24 to 41 mmol O_2 m^{-2} d^{-1} and 32–62 mmol CO_2 m^{-2} d^{-1}, respectively, but with no clear difference with metabolic rates in mangrove sediments (Kristensen et al. 2000). On an intertidal sand flat of the Banc d'Arguin, gross primary production (GPP) ranged from 33 to 42 mmol C m^{-2} d^{-1}, respiration (R) ranged from 35 to 55 mmol C m^{-2} d^{-1} with net primary production (NPP) ranging from 2 to −13 mmol C m^{-2} d^{-1} and were less than those measured in adjacent intertidal seagrass beds (Clavier et al. 2014). On a Brazilian salt flat, highest GPP and R (−7.6 and 4.7 mmol C m^{-2} d^{-1}, respectively) were measured when it rained as salt flats lack water for much of the year. The salt flat was undergoing heterotrophic metabolism on dry days, but at low rates (<2 mmol C m^{-2} d^{-1}), reflecting the importance of water content on CO_2 fluxes in such environments.

10.3 Rocky Intertidal Shores

The dominant space-occupying primary producers on most tropical rocky shores are low-lying, encrusting algae and turf algae. Studies from tropical Africa (John and Lawson 1991), Central America (Lubchenco et al. 1984), and from the tropical western Pacific (Kaehler and Williams 1996) all highlight the importance of encrusting algae as a dominant space occupant in tropical rocky intertidal food webs. In the food web of the tropical rocky shore on Taboguilla Island, Panama (Figure 10.5) encrusting algae were central to the feeding habits of sea urchins, herbivorous gastropods, crabs, and omnivorous fish. There were intense trophic interactions within the food web (Macusi and Ashoka Deepananda 2013), with omnivorous fish eating predatory and herbivorous gastropods which, respectively, eat and compete for space with herbivorous fish and crabs (Figure 10.5). Menge et al. (1986) found that diets were broad and overlapping and that 30% of the consumers were omnivorous. Each consumer group had strong effects on prey occurring at lower trophic levels: (i) fish and crabs reduced the abundance of herbivorous molluscs, predaceous snails, foliose algae, and sessile invertebrates, (ii) predaceous gastropods reduced the abundance of herbivorous molluscs and sessile invertebrates, and (iii) herbivorous molluscs reduced the abundance of foliose algae and young sessile invertebrates. The encrusting algae, normally the dominant space occupier, proved to be an inferior competitor for space with other sessile organisms, but escaped some consumers due to their anti-herbivore defences, and competed for space despite intense grazing.

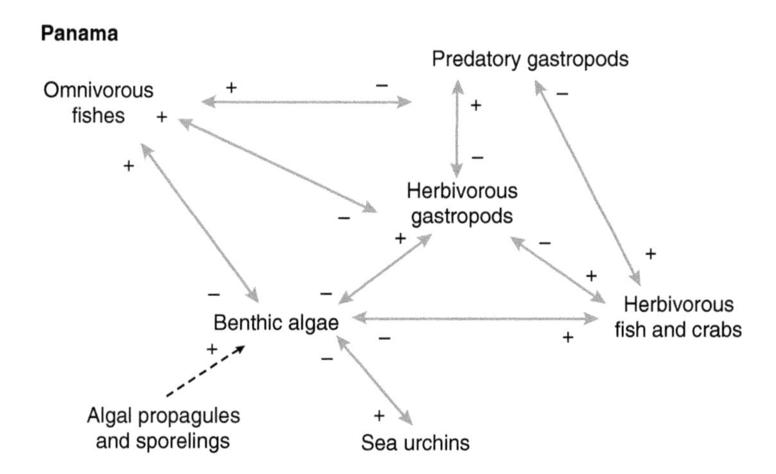

FIGURE 10.5 Trophic model of interactions among food web members in the rocky intertidal of Panama. Positive interactions are denoted with a plus sign, negative interactions with a minus sign. *Source*: Macusi and Ashoka Deepananda (2013), figure 2a, p. 498. Reproduced with permission from IJSPR. © IJSRP Inc.

Similarly, fish were the dominant top consumers along with gastropods and crustaceans in the tropical rocky shores of southeast Brazil (Vinagre et al. 2018). Primary consumers were mostly crab, gastropods, and bivalves; among the top consumers were two invasive crustaceans, the Japanese peppermint shrimp, *Lysmata lipkei*, and the Indo-Pacific swimming crab, *Charybdis helleri*. Unlike many tropical rocky intertidal shores, this Brazilian food web has a low dependence on the benthic pathway, with the top consumers dependent mostly on the macroalgal and pelagic energy pathways. In Hong Kong, rocky intertidal shores were similarly dominated by encrusting algae forming the crux of the food web (Figure 10.6). Encrusting algae and a variety of benthic plants were fed upon by herbivorous fish, gastropods, crabs, and sea urchins; herbivorous crabs and gastropods were fed upon by predatory crabs and herbivorous fish were outcompeted by herbivorous gastropods (Macusi and Ashoka Deepananda 2013).

The Panamanian food web tends to support the Menge-Sutherland model of community organisation in physically benign habitats: (i) community structure will be more strongly affected by predation, (ii) the effect of predation will increase with a decrease in trophic position in the food webs, (iii) trophically intermediate species will be influenced by both competition and predation, and (iv) competition will occur among prey species which successfully escape consumers. It appears that the concept of keystone species may not apply to tropical rocky intertidal shores due to indirect effects, overlapping food requirements, and a diffuse predation due to a diverse set of assemblages that may mask the effect of one species (Macusi and Ashoka Deepananda 2013).

The Hong Kong assemblages are subject to more intense physical stress than those in Panama. There are seasonal changes in physical factors in Hong Kong as the shores are subjected to cool, dry winters, and hot, wet summers. At the low shore, assemblages dominated by encrusting algae typical of rocky shores in Hong Kong developed only in the presence of herbivores and only during the hot season. During the cool season, erect *Corallina* spp. became dominant, while exclusion of herbivores during either season resulted in the development of ephemeral, erect algal

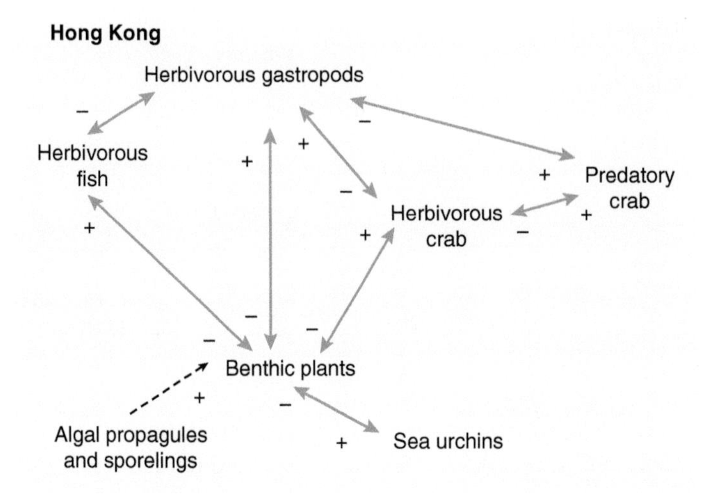

FIGURE 10.6 Trophic model of interactions among food web members in the rocky intertidal of Hong Kong. Positive interactions are denoted with a plus sign, negative interactions with a minus sign. *Source*: Macusi and Ashoka Deepananda (2013), figure 2B, p. 498. © IJSRP Inc.

communities (Kaehler and Williams 1998). During summer, free space availability was greatest due to the seasonal die-off of species and negligible recruitment, as new species were unable to survive the extreme physical conditions (Hutchinson and Williams 2001). Compared to non-local processes such as recruit supply and seasonal variation in physical stress, the role of grazing appears to be of secondary importance in structuring assemblages on Hong Kong shores (Williams et al. 2000). In the mid-intertidal rocky shore of eutrophic Guanabara Bay, Brazil, yearly changes in the benthic community related best to minimum water and air temperature, minimum concentration of *Enterococcus* spp., length of air exposure, and level of pollution (Puga et al. 2019). The dominant species were the green macroalga, *Ulva* spp. which varied seasonally, the mangrove oyster, *Crassostrea rhizophorae,* the hooded oyster, *Saccostrea cucculata*, the ribbed barnacle, *Tetraclita stalactifera*, and the acorn barnacle, *Amphibalanus amphitrita*. Niche overlap was observed among the oysters and with the acorn barnacle. The presence of all these species were strongly related to the presence of *Ulva* spp. as the mid-intertidal zone is characterised by the absence of algae due to long periods of air exposure caused by tides. The two indigenous species, *Amphibalanus amphitrite* and *S. cucculata,* exhibited distinct long-term patterns with *A. amphitrite* (a species introduced a long time ago) slowly disappearing and *S. cucculata* (a recently introduced species) increasing its percent cover as did the indigenous oyster, *C. rhizophorae.*

Other forms of physical stress play an important role in structuring tropical rocky shores. For instance, El Niño plays a strong role on Galápagos rocky shores (Vinueza et al. 2006). The shore is under natural grazing conditions dominated either by encrusting algae or by red algal turf and articulated corallines. Algae fluctuate as follows in response to El Niño. During an early phase, crustose *Gymnogongrus* and/or red algal turf dominate. During El Niño, grazers have limited effects on algal cover but greatly influence algal sizes. Most algae, especially edible forms, are scarce or decline, although warm water ephemeral species (*Giffordia mitchelliae*) flourish, increasing diversity and overgrowing encrusting algae. Mortality of iguanas is high and crab densities are low. When normal conditions return, warm water ephemerals decline, crab densities rise, and grazers have significant, but site-specific effects on algae. Differences between study sites and large-scale temporal changes associated with El Niño indicate that these rocky shores are not as constant in time and space as those in Panama. Mobile grazers reduced abundance of algal communities, but far greater effects were attributable to inter-site differences and temporal shifts in oceanographic conditions. El Niño events reduced nutrients, intensified wave action, and raised sea levels, affecting food availability for herbivores and their influence on benthic algae (Vinueza et al. 2006).

Acidification plays a role in structuring rocky intertidal communities as shown by Hossain et al. (2019) in a tropical estuary on the island of Borneo. Acidification of the Sungai Brunei River and the Brunei Bay, NW Borneo, is generated by acidic water discharge from catchment soils and groundwater, biologically elevated CO_2 levels upstream and nearby urban areas, with pH and carbonate saturation state steeply increasing seawards along the estuary. The most abundant species are the tanaid, *Tanais* sp., the amphipod, *Corophium* sp., chironomid insects, and the bivalve, *Xenostrobus* sp. The epifaunal community pattern confirmed the primacy of the estuarine salinity gradient with three distinct communities in the upper, middle, and lower estuary. The community shifted from tanaid–polychaete dominance in the upper estuary to bivalve–dipteran dominance in the middle estuary to bivalve–dipteran–amphipod dominance in the lower estuary. Most of the organisms that were abundant in acidified, low carbonate waters were weakly calcifying, small polychaetes, amphipods, and tanaids. Strongly calcifying species were less abundant in the upper estuary. These results are similar those found at CO_2 seep systems, in which a few opportunistic species benefit from acidified conditions, but most species do not.

10.4 Seagrass Meadows

Seagrass meadows also offer a rich home to a variety of invertebrates and fish, both permanent and temporary residents (Short et al. 2018). Fish are encountered at a variety of trophic levels, being herbivores, carnivores, omnivores, piscivores, and detritivores. Fish are attracted to seagrasses for protection from predators as well as for the variety of foods on offer. Most herbivorous fish graze on the epiphytes on seagrass blades and rarely, if ever, feed on the seagrasses themselves. Many omnivores pursue a mixed diet, and there are a many fish species that mix epiphytes with seagrass detritus in their diet. Parrotfish are one of the few fish that feed on seagrass tissue, and it has been shown that as much as 80% of above-ground biomass can be grazed down by these fish. Parrotfish often travel from nearby coral reefs to graze on seagrass; this has been found to occur quite commonly in the Caribbean. Seagrasses can deter herbivores and have some defences against overconsumption. Many seagrasses produce noxious chemicals to deter consumption (Sieg and Kubanek 2013). Seagrasses also translocate nutrients to non-digestible parts to conserve nutrients as a form of compensatory growth. On the other hand, seagrasses may benefit from the presence of herbivores by the wastes they excrete (Allgeier et al. 2013). Herbivores also often graze on more nutritious younger leaves avoiding older leaves richer in tannins or other noxious compounds.

Predatory fish are common in seagrass meadows, feeding on other fish, shrimps, amphipods, worms, and other small invertebrates. Predation can be a difficult activity within seagrasses owing to the complexity of the habitat, but there are ambush predators such as pipefish and sea horses that use the vegetation for concealment. Large predators such as rays are common and can impact the seagrasses directly by burrowing and disturbing the seabed. By these actions, they can affect seagrass colonisation and succession. Predation on sessile invertebrates reduces invertebrate abundance, richness, and diversity in tropical seagrass meadows but does not in temperate communities, supporting the notion that predation is stronger in the tropics (Freestone et al. 2020).

The fish fauna of seagrass meadows varies greatly over time and space and is often diverse. They are so variable that no generalisations can be made about their structure and composition other than to note their diversity and differences from one meadow to another and even within a meadow over time (Hogarth 2015). Nevertheless, fish are abundant in seagrass meadows and people fish within meadows to the extent that some areas are in danger of being overfished. Coupled to other problems such as pollution, fisheries in seagrass meadows are on the wane in many tropical locations. Another problem is that climate change is allowing 'tropicalisation' and transformation of temperate seagrass meadows. It is expected that tropical herbivorous fish will establish in temperate seagrass meadows, followed later by megafauna (Hyndes et al. 2016). Tropical seagrasses will likely establish later as they are limited by more restricted dispersal abilities. Seagrass food webs are likely to shift from detritus-based to more direct consumption-based systems, thereby affecting various ecosystem services, including their nursery habitat role for fishery species.

Tropical seagrass meadows are species-rich with many organisms feeding on bacteria and seagrass detritus and others feeding more directly on seagrass tissue. Benthic and pelagic food webs are complex and productive, resulting in clear trophic links with commercially valuable fish species (Unsworth et al. 2019). For instance, seagrass detritus is an important food for fauna in tropical Indonesian seagrass beds as seagrass is consumed by many species and important to a large portion of the food web (Vonk et al. 2008b; Du et al. 2016). Dugongs and turtles are major consumers of seagrass, often grazing down significant portions of a meadow.

The structure of seagrass food webs is influenced by habitat complexity because of changes in the abundance of small and/or larger animals (Jinks et al. 2019). Small invertebrates to large vertebrates utilise seagrass communities which provide food, shelter, and protection against what can often be a hostile environment. Most species are not endemic to seagrasses but are simply utilising a more favourable habitat compared to the surrounding environment. Many species which use seagrass habitat migrate between habitats such as mangroves or coral reefs for some portion of their life cycle or simply utilise seagrass for feeding or for shelter.

In most food webs of seagrass meadows, many consumers feeding on a variety of foods show mixed trophic importance. For example, there are many organisms (sea urchins, gastropods, and fish) in Indonesian seagrass meadows that feed and assimilate both epiphytes and *Sargassum* sp. (Vonk et al. 2008a,b). So, no fauna is dependent solely on phytoplankton and POM for sustenance. Although seagrass is an important food source, it is avoided by many consumers. Mixed inputs of seagrass, epiphytes, and macroalgae fuel secondary production in a pristine seagrass meadow in Shark Bay, Western Australia (Belicka et al. 2012), for example, with seagrass detritus supporting many fish species, but through benthic intermediates. Epiphytes were consumed mostly by snails, whereas seagrass detritus, macroalgae, gelatinous zooplankton, and phytoplankton all contributed to higher trophic levels, including sea turtles and sharks.

The extent of herbivory varies greatly both within and among species, ranging from negligible values to as much as 50% of leaf production removed by some species (Cebrián and Duarte 1998). Differences among species in their consumption of leaf production are associated with differences in leaf-specific growth rates, pointing to herbivore selective feeding upon faster-growing species with higher nutritional quality. This selection seems to be independent of leaf nutrient concentrations, suggesting that nutrient levels as such are a poor descriptor of seagrass palatability as nutrients can be bound to indigestible fibre. Cebrián and Duarte (1998) found no relationship between herbivory intensity and leaf-specific growth rates among populations of a single species, but differences among species in the areal flux of production transferred to herbivores seemed to be related to differences in the level of production attained.

Globally, rates of production and consumption of seagrass are greatest in the tropics, decreasing with increasing latitude (Vergés et al. 2018). Nutritional quality of seagrass was lowest in the tropics, where fish remove an average of about 30% of primary production.

Many seagrass food webs are detritus-based, with little seagrass production grazed directly (Valentine and Duffy 2006), but many other meadows are heavily grazer-based. Dugongs, sea turtles, waterfowl, and fish have locally intense impacts on seagrass biomass, but the most abundant primary consumers are smaller invertebrate grazers that feed on algal epiphytes. The major components of a seagrass food web include seagrasses and their attached algae, invertebrate grazers, detritivores, vertebrate grazers, small invertebrate-feeding predators, piscivorous predators, and humans (Figure 10.7). The relative dominance of seagrasses, macroalgae, and microalgae are directly and indirectly controlled by interactions among these functional groups and by environmental factors such as eutrophication.

Direct grazing of seagrasses by herbivores can represent a significant transfer of carbon and energy to food webs (Valentine and Duffy 2006; Vonk et al. 2008a,b; Belicka et al. 2012). The variegated sea urchin, *Lytechinus variegatus*, consumed 50–100% of above-ground seagrass biomass in some areas of the eastern Gulf of Mexico and Caribbean Sea. This sea urchin can significantly reduce the above-ground biomass of turtle grass habitats to permanently barren seabed, if grazing persists throughout the winter and spring. Turtle grass persists in summer under severe grazing pressure and regrows to levels that either equal or exceed the standing crop of ungrazed seagrasses nearby. Sea urchins consume significant quantities of seagrasses in the tropical Pacific and Indian Oceans (Klumpp et al. 1993; Kasim 2009; Lyimo et al. 2011) as do thalassinidean shrimps (Kneer et al. 2008), molluscs (Holzer et al. 2011) and parrotfish (Gullström et al. 2011). A carbon flow model (Figure 10.8) of a Philippine seagrass meadow found that the sea urchins, *Tripneustes gratilla* and *Salmacis sphaeroides,* consumed mostly seagrass, although *S. sphaeroides* also consumed detached seagrass debris, the red alga, *Amphiroa fragilissima,* algal-coated sediment and rubble in proportions that varied with the availability of preferred food types (Klumpp et al. 1993). Estimated annual grazing rate was equivalent to 24% of annual seagrass production, but owing to large variations in population density, grazing impact varied from 5 to 100% at different times of the year. Epiphytes were abundant, accounting for as much biomass as seagrass, with most being grazed down by snails and crustaceans. Thus, the seagrass–sea urchin and the periphyton–epifauna grazing pathways are both important in this tropical seagrass food web.

Grazing by green turtles (*Chelona mydas*) on *Thalassia testudinium* meadows in Little Cayman, Cayman Islands, had spatially variable effects on macroalgal diversity. Lower macroalgal diversity was the result of lower densities of all seagrass

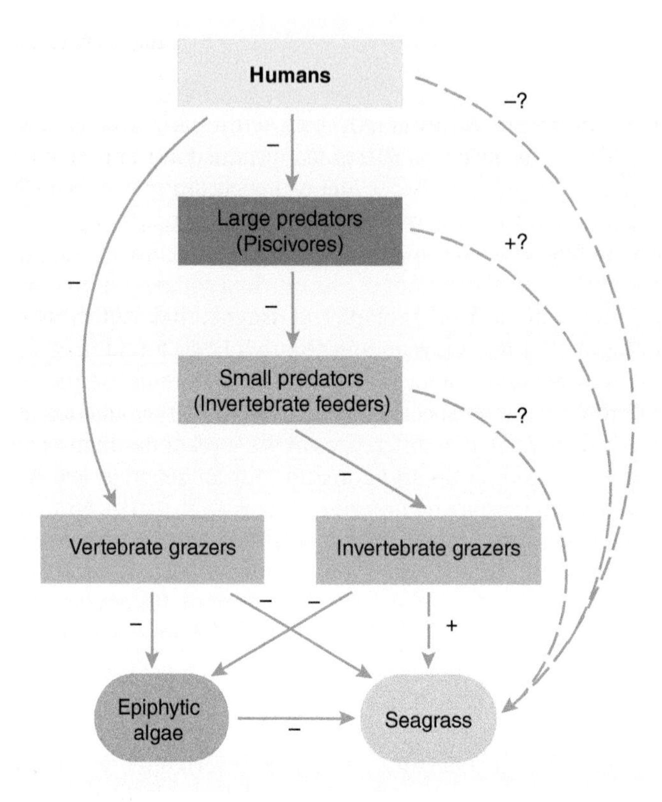

FIGURE 10.7 Model of a simplified seagrass bed food web. Solid and broken arrows indicate direct and indirect effects, respectively. + and – signs indicate positive and negative interaction, respectively, that one group of organisms has on another in the direction of the arrow. Indirect, cascading effects of humans and other predators on seagrasses are indicated by question marks. *Source*: Valentine and Duffy (2006), figure 3, p. 467. © Springer Nature.

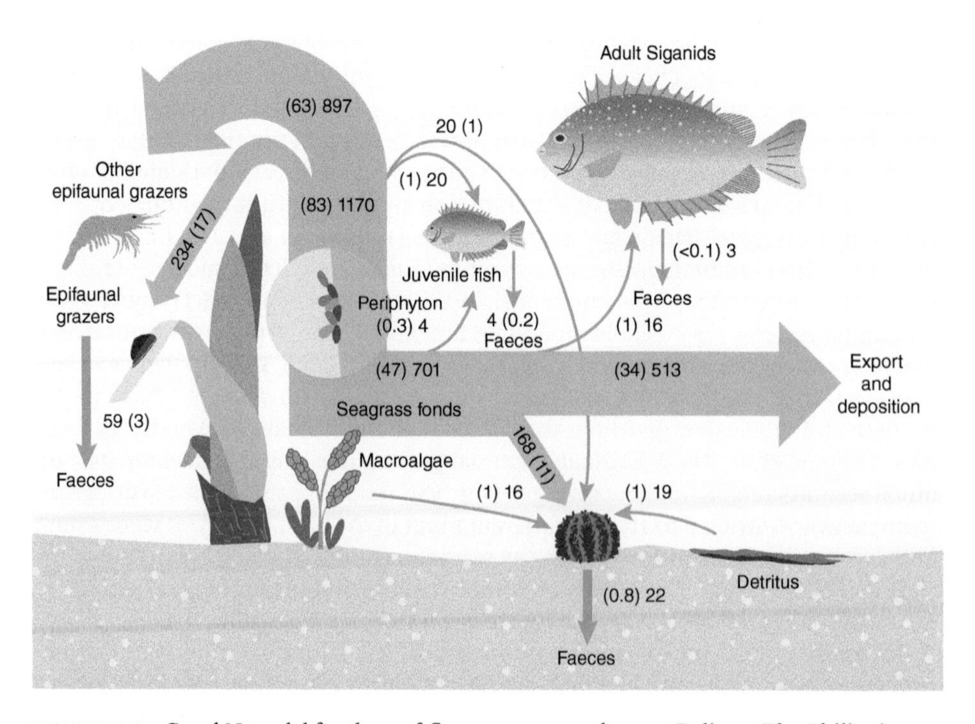

FIGURE 10.8 C and N model for the reef flat seagrass meadows at Bolinao, The Philippines. Units are mg C m^{-2} d^{-1}. Units in parentheses are mg N m^{-2} d^{-1}. *Source*: Klumpp et al. (1993), figure 4, p. 214. © Elsevier.

and macroalgae among grazed areas in one bay, but not in an adjacent bay (Hearne et al. 2019). Diversity was positively correlated with seagrass and macroalgal densities across all meadows. Grazing by green turtles can affect seagrass ecosystem metabolism (Johnson et al. 2020); rates of net ecosystem production ranged from 56 to 96% lower in grazed than in ungrazed *T. testudinium* meadows across the Greater Caribbean and Gulf of Mexico.

Grazers exert an important bottom-up control over the nutritional status, abundance, and composition of higher consumers. This is especially true for the small invertebrate grazers, the key link between primary producers and higher trophic levels. Grazing directly on seagrass biomass can be intense and controlled by both 'top-down' and 'bottom-up' factors, such as nutritional quality of the material and human overharvesting of large vertebrate grazers (Clores and Cuesta 2019). These controls are not mutually exclusive, as poor nutritional quality could deter many herbivores, whereas human hunting reduce the densities of manatees, dugongs, and sea turtles capable of grazing seagrasses.

In the tropics, however, herbivorous fish, not sea urchins, are the dominant herbivores. More than 30 species of Caribbean fish, predominantly parrotfish and surgeonfish, are significant consumers of living seagrass tissue, and numerous herbivorous fish of the Indo-West Pacific have also been found to be dominant grazers of seagrass. The luderick, *Girella tricuspidata* can completely graze beds of *Zostera muelleri* along the eastern coast of Australia (Wendländer et al. 2020). Large numbers of small and large juvenile fish can rapidly populate and feed in intertidal seagrass beds, arriving at the onset of flood tides, as observed in the Philippines (Espadero et al. 2020), where juvenile labrids, lethrinids, lutjanids, and siganids were the most abundant fish, whereas large juveniles of the latter three families were first to invade intertidal seagrass beds with the incoming tide. Species of the Labridae, Lethrinidae, Siganidae, and Lutjanidae families foraged during the rising tide, underscoring the importance of intertidal seagrasses as a feeding ground (Espadero et al. 2020).

Detritus food webs play a crucial role in the trophic structure of tropical seagrass meadows. It is estimated that a large, but variable, percentage of seagrass production enters the detritus food web. There are essentially three pathways leading to detritus. First, the leaf may remain on the plant until it falls off and is broken down over a period of months by bacteria and fungi which have the capability of digesting the cellulose structure of the leaf. Second, the leaf may be consumed by a large herbivore which digests its epiphytes and cell contents but is unable through a lack of cellulose-digesting enzymes to break down the leaf structure. The fragmentation of the leaf in feeding and subsequent defaecation, however, undoubtedly speeds the breakdown process by bacteria in the sediments. Dense populations of benthic, pelagic, and epiphytic bacterial assemblages are associated with seagrass leaves and their root and rhizome systems (Williams et al. 2009). These bacteria form a major source of food for detritus feeders living within the sediments, including a rich variety of meiofauna, polychaetes, crustaceans, bivalves, and gastropods. These organisms strip the microbes and other epiphytes from leaf detritus, and faecal pellets are subsequently re-ingested following recolonisation by bacteria and fungi. Some fish and several large invertebrates consume detritus from the surface of seagrasses and sediments. Third, leaf detritus can be buried in anoxic sediment where decomposition is hampered. This may lead to substantial carbon sequestration (Miyajima and Hamagushi 2019).

Some carbon flow models illustrate the importance of detrital pathways. A carbon budget of leaves of the tropical intertidal seagrass, *Thalassia hemprichii*, shows that 80% of leaf production in Taiwanese beds was shunted through the detrital pathway with 20% consumed by herbivores and further on to fish and sea urchins (Figure 10.9). Once in the detrital pool, 44% was decomposed, 32% was exported to adjacent food webs, and 4% was stored in sediments (Chiu et al. 2013). A carbon flow model of the subtropical seagrass beds at Dongsha Island in the South China Sea (Figure 10.10) shows that 96% of leaf production was shunted into the detrital pathway with 83% of the detritus decomposed; below-ground production was also shunted to the decomposer pathway with only 0.6% shunted to herbivores (Huang et al. 2015). There was little storage (4.3%) and only 10.7% was exported to adjacent habitats. Grazing and detritus pathways thus co-occur in tropical seagrass meadows. In these seagrass beds, the most diverse group of detritivores was the crustaceans, followed by abundant nemerteans, sipunculids, and ophiuroids. The most dominant group of suspension feeders were the bivalves. Omnivorous species included nereidid polychaetes and a variety of decapod crustaceans. Drilling gastropods were the most common carnivores. Only one species of fish (*Sparisoma radians*) was herbivorous, feeding exclusively on fresh *Thalassia testudinum* blades. Fourteen fish species were classified as

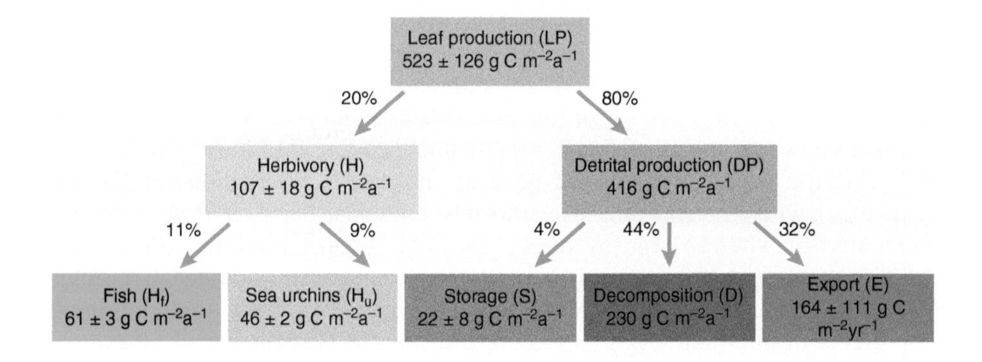

FIGURE 10.9 The annual budget and fate of leaf-derived carbon in an intertidal seagrass bed of *Thalassia hemprichii* in southern Taiwan (mean ± 1 standard error). Values outside the boxes represent the percent of leaf production. *Source*: Chiu et al. (2013), figure 7, p. 33. © Elsevier.

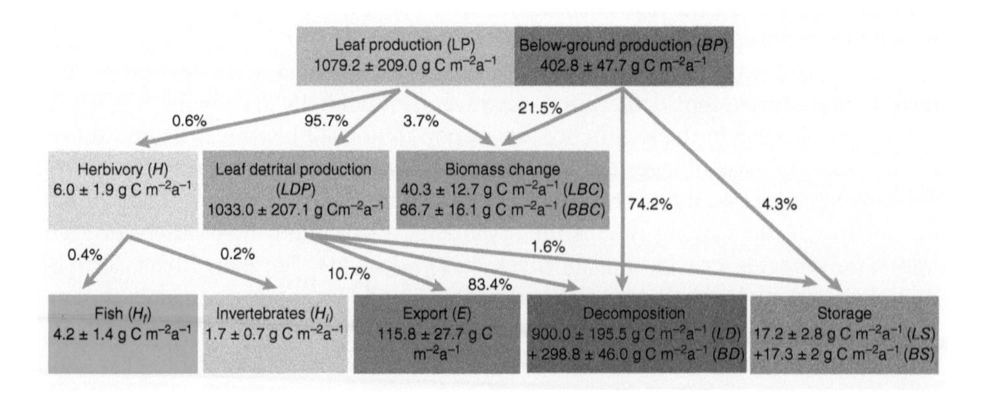

FIGURE 10.10 Annual budget of seagrass carbon in a lagoon on the northern coast of Dongsha Island, South China Sea. Values are mean ± 1 standard error. Values outside the boxes represent the percentage of leaf or below-ground production. *Source*: Huang et al. (2015), figure 8a, p. 99. © Elsevier.

omnivorous, consuming fresh and/or detrital blades or algae. Absorption efficiencies for organic matter ranged from 64 to 71% for urchins consuming a mixture of fresh and detrital blades and egesting a mixture of faecal chunks and pellets. Thus, the sea urchin plays a dual role in this seagrass meadow, being major grazers on seagrass leaves and their faeces being a potential food source for detritivores.

Thalassinidean shrimp play a key role in detritus and grazing food webs of tropical seagrasses (Kneer et al. 2008). Seagrass blades offered to the thalassinidean shrimps, *Neaxius acanthus* and *Corallianassa coutierei*, in seagrass meadows of the Spermonde Archipelago, Indonesia, were pulled underground, cut into pieces, and incorporated into their burrow walls. The diet of *N. acanthus,* its commensal *Austrolethaps wardi,* and *C. coutierei* was mainly derived from detrital seagrass leaves with a small contribution from sediment organic matter and seagrass epiphytes.

Most gastropods inhabiting seagrass meadows are believed to feed on epiphytes rather than directly on living seagrasses. However, Holzer et al. (2011) demonstrated that the emerald neritid, *Smaragdia viridis*, feeds preferentially on seagrasses common to the Caribbean and Bermuda: *T. testudinum, Halodule wrightii* and *Syringodium filiforme*. Similarly, gastropods and several other invertebrates feed preferentially on seagrass tissue and its epiphytes in a tropical seagrass meadow in Indonesia (Du et al. 2016). For 14 of the 17 fish species, seagrass and their epiphytes were the main contributors of nutrition. Seagrass and their epiphytes were consumed by most of the fish and invertebrates present and were important for the food web. These results are supported by a study in seagrasses of the Pulai River estuary, Johor, Peninsular Malaysia (Mukhtar et al. 2016), where three primary producers were found to be important for the seagrass food web: phytoplankton, *Ulva* sp. and the seagrasses, *Enhalus acoroides, Halophila ovalis,* and *Halophila spinulosa*. Cnidarians, molluscs, echinoderms, arthropods, and fish all showed stable isotope signatures indicating that various mixtures of all three primary producers were ingested.

Epiphytes indeed play an important trophic role in tropical seagrass meadows. In a northern Gulf of Mexico seagrass system, epiphytic algae were responsible for 46 to 60% of total system and benthic primary production, respectively, with the seagrass *H. wrightii* accounting for only 13% to total ecosystem NPP (Moncreiff and Sullivan 2001). Virtually all consumers had $\delta^{13}C$ values that fell within the range for epiphytic algae and sand-associated benthic microalgae, and virtually no signature for seagrasses, macroalgae, and phytoplankton. The stable isotope data indicated that the epiphytic algae were the primary food source for higher trophic levels. A similar study conducted in Fiji in a *Syringodium isoetifolium*-dominated seagrass bed showed that epiphytic cyanobacteria, not seagrass leaves and its detritus, were the primary source of organic matter for heterotrophic consumers (Yamamuro 1999). These results suggest also that inorganic N released through breakdown of cyanobacteria by heterotrophs may support the continued production of seagrass. The extent to which epiphytes play a role in herbivory depends on the amounts of nutrients a seagrass bed may receive. In some subtropical seagrass beds, nutrient loading causes a decline in seagrasses and a proliferation of epiphytic algae (Campbell et al. 2018).

Fish play an important higher-level trophic role in tropical seagrass meadows (Manickchand-Heileman et al. 1998; Nagelkerken et al. 2006; Unsworth et al. 2007a). Nagelkerken et al. (2006) found that the seagrass fish community of a Caribbean meadow consisted of three trophic levels: herbivores; omnivores, zoobenthivores and zooplanktivores; and piscivores. Herbivores partitioned food by specialising on seagrass epiphytes, seagrass leaves, or macroalgae. Fish at the second trophic level exhibited temporal segregation in feeding habits between fish families, while species within families showed segregation by food type and source. Most fish species exhibited a very narrow diet breadth with significant segregation into resource use.

Fish can show diel trophic structuring in tropical seagrass beds (Unsworth et al. 2007a; Du et al. 2018). In Indonesia, Unsworth et al. (2007a) studied diel migrations of fish into and out of seagrass beds and found that fish were more abundant and diverse at night suggesting migration into the seagrass beds from adjacent reefs, mangrove forests, and deep water. Fish assemblages were dominated during both day and night by invertebrate and fish feeders. However, a major diel change in trophic structure occurred in the abundance of omnivores. During the day omnivores were abundant, but they were replaced at night by invertebrate feeders. Unsworth et al. (2007b) proposed that diel changes in seagrass fish assemblages are predominantly structured by food availability.

Seagrasses, including their detritus, are often linked to adjacent mangrove forests and coral reefs. Heck et al. (2008) found that seagrass ecosystems provide a large subsidy to both near and distant locations through the export of POM and living plant and animal biomass. Seagrasses are thus part of a greater seascape containing both marine and terrestrial components, each linked to the other via the foraging of consumers, and the drift of seagrass and seagrass-associated algal detritus. Trophic subsidy from seagrasses can assist in maintaining the abundance of coral reef fish and shelf communities. Commercially valuable, penaeid shrimp populations annually transfer large amounts of biomass from seagrass meadows to the adjacent continental shelf in tropical and subtropical latitudes. Mangroves can be a source of carbon for adjacent seagrasses (Chen et al. 2017) and linked to seagrasses in terms of faunal connectivity (Skilleter et al. 2017).

A remarkable three-stage mutualism (Section 6.6) exists between seagrass, lucinid bivalves, and their sulphide-oxidising gill bacteria, forming the basis of seagrass ecosystems (van der Heide et al. 2012). The gill bacteria in the lucinid bivalves reduce sulphide stress for seagrasses. The bivalve–sulphide-oxidiser symbiosis reduces sulphide levels and enhances seagrass production (Figure 6.6). In turn, the bivalves and their endosymbionts profit from organic matter accumulation and radial oxygen release from the seagrass roots. Paleo records suggest that lucinid bivalves and their endosymbionts date back to the Silurian, but that they increasingly diversified since the evolutionary emergence of seagrasses in the late Cretaceous. Seagrasses later became widespread in the Eocene, and lucinid remains frequently occur in association with their deposits. Indirect support for the hypothesis that this symbiosis is crucial for the survival of seagrass meadows was provided by a meta-analysis that revealed a relationship covering 11 out of 12 seagrass genera and at least 18 genera of Lucinidae (van der Heide et al. 2012). The association spans six out of seven continents. The bivalves are present in 97% of tropical seagrass sites, 90% of subtropical meadows, and 56% of temperate seagrass beds surveyed, indicating that the association may be dependent on temperature-related sulphide production. These findings elucidate the long-term success of seagrasses in subtropical and tropical waters.

Primary production in seagrass beds is partitioned among seagrasses, benthic microalgae, phytoplankton, epiphytes, and macroalgae (Mateo et al. 2006). A carbon budget for subtropical seagrass beds in a coastal lagoon (Kaldy et al. 2002) indicates that macroalgae accounted for 33–42% of total ecosystem net primary production (TENPP) and seagrasses accounted for about 33–38%. Phytoplankton contribution was about 15–20% and the contribution from the benthic microalgae varied between 8 and 30% of TENPP. The water column over the seagrass bed was net heterotrophic, and consequently, a carbon sink consuming between 5 and 22% of TENPP.

Macroalgae and microalgae contributed 50–60% of TENPP, and seagrass may be more important structurally than as a source of organic matter to the food web. Thus, some seagrass beds are dominated by seagrasses and some are dominated by algae.

Comparing carbon fluxes in temperate and tropical seagrass meadows, Duarte et al. (2010) found that tropical communities tend to support greater rates of GPP (252.3 mmol O_2 m^{-2} d^{-1}) than temperate meadows (165.52 mmol O_2 m^{-2} d^{-1}). R and P:R ratios also tend to be greater in tropical (R = 217.48 mmol O_2 m^{-2} d^{-1}; P:R = 1.61) than in temperate communities (R = 129.76 mmol O_2 m^{-2} d^{-1}; P:R = 1.44), although the opposite is true for net community production (NCP) (temperate = 33.47 mmol O_2 m^{-2} d^{-1} versus tropical = 23.73 mmol O_2 m^{-2} d^{-1}). Sediment respiration rates in tropical seagrass beds are often high (Alongi et al. 2008; Clavier et al. 2014). Benthic metabolism has more recently been measured in intertidal *Zostera noltii* and shallow subtidal *Cymodocea nodosa* meadows in the Banc d'Arguin, Mauritania (Clavier et al. 2014), a region influenced by upwelling. The *Zostera* bed was autotrophic during November–January with mean daily NPP of 71.3 mmol C m^{-2} d^{-1}; the *C. nodosa* bed was heterotrophic in November (−96.2 mmol C m^{-2} d^{-1}) but autotrophic in January (33.4 mmol C m^{-2} d^{-1}). Benthic respiration rates were higher in the *C. nodosa* bed (201.5–273.4 mmol C m^{-2} d^{-1}) than in the *Z. noltii* meadow (77.9–107.0 mmol C m^{-2} d^{-1}). In some seagrass meadows, however, community metabolism is not high due to constraining influences such as strong nutrient and temperature gradients as found in the Red Sea (Anton et al. 2020). The seagrasses, *Halophila uninervis, H. ovalis, Halophila stiplacea, T. hemprichii*, and *Thalassodendron ciliatum* inhabiting the Saudi Arabia coast were net autotrophic (GPP:R range = 1.6–2.7), but rates of GPP were low, averaging 136, 85, 121, 108, and 141 mmol O_2 m^{-2} d^{-1}, respectively. Rates of R and NPP averaged 49, 35, 72, 61, and 33 mmol O_2 m^{-2} d^{-1} and 58, 50, 52, 47 and 51 mmol O_2 m^{-2} d^{-1}, respectively. These seagrass meadows may likely be P- and Fe-limited in this oligotrophic sea.

Nonetheless, the data suggest faster metabolism in tropical than in temperate seagrass meadows. Tropical seagrass habitats are probably significant sites of carbon sequestration and carbon burial, although such estimates are constrained by the lack of data on global area of tropical seagrasses (Fourqurean et al. 2012). Rates of CO_2 and CH_4 fluxes from seagrass sediments may be impacted by global warming, specifically increased SSTs (Burkholz et al. 2020). Benthic CO_2 and CH_4 fluxes from *Halophila stipulacea* meadows in the Red Sea were 6× and 10–100× higher, respectively, than in adjacent unvegetated sediments. With gradual warming from 25 to 37 °C, respective CO_2 and CH_4 fluxes averaged 10.42 mmol m^{-2} d^{-1} and 88.11 µmol m^{-2} d^{-1} with prolonged darkness leading to an increase in CO_2 fluxes but a decrease in CH_4 fluxes in seagrass sediments. In the Red Sea, Garcias-Bonet and Duarte (2017) measured highly variable rates of CH_4 production in meadows dominated by *H. stipulacea/Halodule univervis* (24.5–111.1 µmol m^{-2} d^{-1}), *T. ciliatum* (0.1–6.9 µmol m^{-2} d^{-1}), *T. hemprichii* (0.3–16 µmol m^{-2} d^{-1}), *Halophila decipiens* (0.7–1.9 µmol m^{-2} d^{-1}), *E. acoroides* (−11.9–270.1 µmol m^{-2} d^{-1}), *Caulerpa serrulata/H. univervis* (135.5–565.3 µmol m^{-2} d^{-1}), and *H. univervis* (25.4–59.9 µmol m^{-2} d^{-1}) with no daily or seasonal patterns. Globally, Garcias-Bonet and Duarte (2017) estimated that CH_4 production and emissions by seagrass ecosystems could range from 0.09 to 2.7 Tg a^{-1}.

Seagrasses are biogeochemically linked to adjacent mangroves and coral reefs (Macklin et al. 2019). CO_2 fluxes across a mangrove–seagrass–coral reef continuum in a pristine embayment on the island of Bali, Indonesia, showed that the CO_2 source in the mangrove-dominated upper bay (mean CO_2 flux = 18.1 mmol m^{-2} d^{-1}) was associated with delayed groundwater inputs and a shifting CO_2 source-sink in the lower bay was driven by the uptake of CO_2 (mean flux = 2.5 mmol m^{-2} d^{-1}) by *Cymodocea rotundata*, *H. ovalis*, and *Enhalus acroides* and mixing with oceanic waters. The mouth of the bay was a minor source of CO_2 (mean flux = 0.3 mmol m^{-2} d^{-1}) due to calcification of adjacent coral reefs (Macklin et al. 2019).

10.5 Mangrove Forests

A paradigm was established in the early 1970s that stressed the dominance of mangrove production in supporting nearshore secondary production via detritus-based food webs (Odum et al. 1982). The paradigm stressed that the preservation of mangrove forests rests on the fact that carbon and energy fixed by mangrove vegetation is the most important nutritive source for mangrove-associated organisms. We now know that this tells only part of the story (Odum et al. 1982), as algae are an important and often dominant part of animal nutrition in mangrove food webs (Alongi 2009). These latter studies used carbon isotopes to show that the number of organisms in coastal waters adjacent to mangroves dependent on mangrove carbon is smaller than was suggested by the early paradigm.

Mangrove food webs are composed of both algal grazing and detritus food chains (Bouillon et al. 2002). The grazing food chain relies on phytoplankton, macroalgae, and benthic microalgae as well as fallen timber and mangrove litter. The pelagic autotrophs are fed upon by herbivorous zooplankton that are fed upon in turn by omnivorous and carnivorous zooplankton with the microbial hub operating as an essential intermediate. On the forest floor, microalgae and macroalgae are fed upon by herbivorous meiofauna, protozoans, and various higher consumers, especially by ocypodid and sesarmid crabs (Figure 10.2). This benthic food chain may be "short-circuited" by carnivorous fish that feed directly on crabs (Sheaves and Molony 2000), whereby a substantial fraction of mangrove productivity sequestered by sesarmid crabs is exported offshore by migration of these predatory fish.

Mangrove forests have a complex arboreal food web in which various invertebrates, insects, algae, spiders, bats, birds, monkeys, and other mammals play an important role in trophic dynamics (Masagca 2011; Sousa and Dangremond 2011). Mangrove leaves, flowers, stems, and propagules are fed upon by a variety of such organisms. Folivory, especially by insects, is more common than often supposed and has been underestimated as has stem boring by wood-eating isopods and insects. A wide array of organisms, such as scale insects and planthoppers, feed on leaf or stem sap (Cannicci et al. 2008). Herbivores damage flowers in addition to leaves and other tree parts, including propagules. The extent of damage to propagules varies with the species of mangrove and predator (Section 6.9). The consequence of this damage for propagule survival, establishment as a seedling, and seedling growth also varies. For some species of mangroves, the damage is severe enough to kill the embryo. For most species, the effects are limited; grapsid crabs are the most damaging herbivores on propagules (Cannicci et al. 2008; van Nedervelde et al. 2015).

Arguably, the most dominant trophic pathway in mangrove forests is the herbivory by sesarmid crabs and insects. The leaf-eating sesarmid crabs are ecosystem engineers, constructing extensive burrows and feeding extensively on mangrove leaves and propagules. The role of crabs in leaf litter removal is well documented. An average of 57% (Sousa and Dangremond 2011) of the annual biomass of leaves, propagules, and other fine litter that falls to the forest floor is consumed or buried by crabs. These rates vary widely by species among sites, tidal heights, and locations. Crabs of the family Sesarmidae are the primary consumers of leaf litter in the Indo-West Pacific; genera include *Sesarma, Neosarmatium, Perisesarma, Neoepisesarma*, and *Chiromantes*. The crabs *Ucides cordatus cordatus, Ucides cordatus occidentalis* (Ocypodoidea: Ucididae), and *Goniopsis cruentata* (Grapsoidea: Grapsidae) are the important litter consumers in New World mangrove forests (Cannicci et al. 2008; Wellens et al. 2015).

The crab–litter pathway plays an important role in organic matter decomposition. Leaves that have been dragged down crab burrows instead of exported by tides give microbes a chance for decomposition and thus enlarging the leaf-driven microbial pathway of nutrient recycling and retention within mangrove forests. Furthermore, crabs egest a large part of the consumed litter in the form of smaller particles that, in turn, can be used by other detritivores and microbes (Cannicci et al. 2008). The availability of this otherwise unavailable food is thus increased for many detritivores and is a major energy source in an ecosystem with a large supply of leaf litter whose nutritional availability largely depends on microbial decomposition. The transformation by crabs therefore sustains populations of smaller detritivores (Scharler 2011; Sousa and Dangremond 2011).

Most mangrove crabs forage on the forest floor, feeding on detritus, macroalgae, living invertebrates, carrion as well as microbes and diatoms on fine soil. Some of these species will also feed on freshly fallen propagules and green leaves and on stems of mangrove seedlings. Other invertebrates such as the intertidal gastropod, *Melampus coffeus*, also forage on leaf litter at low tide, climbing tree trunks to avoid inundation during high tide. Unlike many grazers, these gastropods can assimilate mangrove leaf material. A field experiment, (Proffitt and Devlin 2005) in Floridian mangroves showed that snail grazing greatly increased the rate of *Rhizophora mangle* and *Avicennia germinans* mangrove leaf litter breakdown. Grazing by the snail resulted in 90% weight loss in <1month for *A. germinans* and 7 weeks for *R. mangle* compared to slower breakdown in litter bags. Another experiment also showed that there was greater leaf litter accumulation in plots there the snail was excluded. *M. coffeus* consumed an estimated 40% of mangrove leaf fall, while 20% of leaf litter was exported as particulate or dissolved material (Proffitt and Devlin 2005). Sesarmid crabs are similarly dominant in the initial processing of litter in the Pichavaram mangroves of India (Ravichandran et al. 2007), where sesarmid crabs strongly preferred 40-day-old decomposing *Avicennia marina* leaves over *Rhizophora mucronata* and *Acanthus ilicifolius* leaves of the same age. This agrees with other studies that have found that sesarmid crabs prefer leaves with less tannins and comparatively more N. In northern Brazil, the large mangrove crab, *Ucides cordatus cordatus,* fed on mangrove leaves, unidentified plant material and detritus, roots, soil, bark, and some animal material. Unlike sesarmid crabs, this crab preferred leaves of *R. mangle* over *A. germinans* despite a lower N content and higher tannin content in the former leaves. The idea that leaf ageing may help with digestion of the leaves was rejected for this crab as C:N ratio and the microbial abundance did not differ significantly between senescent leaves and leaves taken from burrows (Nordhaus and Wolff 2007). The sesarmid crabs, *Episesarma singaporense* and *Episesarma versicolor,* meet their N requirements by assimilating N from *Rhizophora apiculata* leaves and using internal reserves (Herbon and Nordhaus 2013).

An apparent paradox is the fact that mangrove crabs seem to derive most of their nutrition from leaves containing little N. Stable isotope analysis of C and N does not seem to support the argument for assimilation of low-quality leaves. Experiments by

Bui and Lee (2014) found that the mangrove grapsid crab, *Parasesarma erthodactyla,* feeding on mangrove leaf litter or mangrove litter plus microphytobenthos developed a significantly higher hepatosomatic index than those with access to only soil. Lipid biomarker and feeding experiments using ^{13}C and ^{15}N-enriched leaf litter measured rapid assimilation of mangrove C and N, confirming the ability of this grapsid crab to process low-quality mangrove organic matter. In other feeding experiments, the leaf-eating crab, *Neoepisesarma versicolor,* preferred eating high-quality fish meat, indicating that this crab can meet its N demands by ingesting animal tissue (Kristensen et al. 2010). The diet of this crab consisted of ≈60% leaves with the balance from animal tissue and benthic microbes. Animal foods may contribute up to half of the N in the diet of *N. versicolor.* In further experiments with sesarmid crabs, leaf litter in the form of brown leaves was always the most important C source, while animal tissue in the form of live and dead prey or benthic diatoms at the soil surface were the dominant N sources (Kristensen et al. 2017). Like most heterotrophs, crabs may utilise a variety of foods to maintain a balanced diet.

Crabs and other higher consumers ingest algae to supplement their diets (Meziane and Tsuchiya 2000; Bouillon et al. 2002; Hsieh et al. 2002; Kieckbusch et al. 2004; Pape et al. 2008). A variety of mangrove fiddler crabs and gastropods on mangrove tidal flats on Okinawa, Japan, fed on bacteria, diatoms, and macroalgae (Meziane and Tsuchiya 2000), with bacteria and green macroalgae being the primary foods. Using stable isotopes, Bouillon et al. (2002) found that mangrove-derived organic matter was detectable in only a limited number of species in mangroves along the Indian coast suggesting that local and imported algal sources are the major source of C for benthic invertebrates. Similar results were found for intertidal polychaetes and crabs in a mangrove estuary in northern Taiwan (Hsieh et al. 2002). Mangrove litter did not appear to be a major food either for various invertebrates in fringe mangroves of the subtropical Atlantic/Caribbean (Kieckbusch et al. 2004). Algal-based diets thus appear to be the norm for most mangrove-associated invertebrates.

The feeding habits of the mangrove gastropod, *T. palustris,* are, like most benthic invertebrates, complex and vary with age (Pape et al. 2008). This gastropod often dominates soil surfaces on intertidal flats and mangrove forests, where they physically disturb surface particles. The diet and behaviour of juvenile and adult gastropods differ, and a spatial segregation was observed between juveniles and adults. On an intertidal mud flat, juveniles avoided sediment patches characterised by high saline water in intertidal pools and high mud content, while adults tended to dwell on sediments covered with high amounts of mangrove litter (Pape et al. 2008). Stable isotope analysis revealed selective feeding by juveniles on surface diatoms.

Sex differences have also been observed in dietary preferences. Males of the sesarmid crab, *Perisesesarma bidens,* prefer algae to leaves and propagules, whereas the females prefer algae and leaves equally, but not propagules (Mchenga and Tsuchiya 2010). Fatty acid analyses and comparisons of tissues and faeces indicate that these crabs efficiently assimilate essential fatty acids from algae to a greater extent than from leaves and propagules. Sesarmid crabs eat mangrove leaves, but macroalgae can be an important N source for *P. bidens.*

Higher consumers have more direct access to mangrove tissues as a food source, but several recent studies indicate that mangrove invertebrates exhibit much more diverse patterns of resource utilisation than previously expected (Bouillon et al. 2008). The degree to which they rely on mangrove-derived C varies across mangrove ecosystems, with a higher reliance on forests where more mangrove detritus is retained and where there is less exchange of materials with adjacent systems. Despite the potentially large-scale movement of organic matter, evidence is emerging that many invertebrates have a small home-range and derive most of their diet from locally available foods. A stable isotope study of invertebrates within forests, creek banks, and in forest gaps found different diets depending on location (Bouillon et al. 2008). Invertebrates from tidal creeks and inside forests showed isotopic signatures like that of mangrove leaf litter, whereas invertebrate tissues from forest gaps were close to microphytobenthos values, highlighting the fact that invertebrates are restricted to using foods available in their microhabitats. Another study similarly showed that the primary C sources that support food webs depends on location (Thimdee et al. 2004).

Mangrove invertebrates show plasticity of diets depending on life cycle stage, food availability, and location within the forest (Thimdee et al. 2001; Oakes et al. 2010; Poon et al. 2010; Tue et al. 2014). The diet of the mangrove crab, *Scylla serrata,* changes with age with small-sized crabs being omnivorous, feeding on small crabs and plant materials, and medium- and large-sized crabs being carnivores, feeding on worms, crabs, and molluscs (Thimdee et al. 2001). Spatial and temporal variations in diets of the crabs, *Metopograpsus frontalis* and *P. bidens,* were found in mangrove forests of Hong Kong (Poon et al. 2010). *M. frontalis* was omnivorous, eating animal and plant materials supplemented with sediment. *P. bidens* was a detritivore with plant materials and soil dominating the gut contents. The diet of *M. frontalis* shows more algal material in winter and *P. bidens* shows more soil in summer, but diets were similar between sexes in both species. The diets of these crabs thus appear to be a function of the interplay between seasons, climate, and biological factors, especially food availability. The pattern of utilisation of algae and mangrove detritus by mangrove invertebrates is different as feeding experiments showed that addition of isotope-labelled mangrove detritus resulted in enrichment of the crabs, *Parasesarma erythrodactyla* and *Australoplax tridentata,* and the foram, *Ammonia beccarii* (Oakes et al. 2010). However, microphytobenthos contributed 93% of the nutrition for *A. tridentata* and 33% of the nutrition of *P. erythrodactyla* and provided more nutrition to *A. beccarii.*

Mangrove plankton and nekton are, like their benthic and canopy-living counterparts, key players in the flow of materials and energy in mangrove ecosystems (Saifullah et al. 2019). Trophic relationships within and between microbial assemblages are virtually unknown, but presumably intense, as protists such as ciliates and flagellates are voracious consumers of bacteria and are known to graze heavily on bacterioplankton. Mangrove phytoplankton communities, in contrast, are thought to be species-poor due to the inhibitory effects of high concentrations of soluble tannins and other polyphenolics derived from mangrove litter and from the chemical defences of the trees. Phytoplankton abundance, diversity, and productivity range widely in mangroves, usually in relation to light availability, temperature, salinity, dissolved oxygen concentration, pH, nutrients, and flushing rates of waterways (Saifullah et al. 2019).

Zooplankton are a crucial link between microbes, penaeid shrimps, and zooplanktivorous fish. Zooplankton can be extremely abundant (10^5 ind. m^{-3}) and higher in abundance in mangrove tidal waters than in adjacent nearshore waters. The main factor controlling zooplankton abundance and species composition is the seasonal change in salinity, with the onset of the monsoon season the prime stimulus for changes in abundance and composition. Zooplankton consist of four trophic groups in mangroves: (i) a stenohaline marine group that penetrates as far as the estuarine mouth or further upstream if there is full salinity, (ii) a euryhaline group that is found throughout mangrove waterways that are estuarine, (iii) a true estuarine group, and (iv) a freshwater group if freshwater is extant in the upper reaches of the mangrove waterway (Kathiresen and Bingham 2001). The most dominant zooplankton in mangroves are members of the cyclopoid copepod family Oithonidae. This family is conspicuous because it may have a selective advantage of small size to avoid predators as well as several strategies to maximise reproduction and growth. Larger zooplankton populations constitute an important trophic link with higher consumers (fish) but play an equally crucial role in being able to help structure microzooplankton.

Pelagic food webs in tropical mangrove creeks and waterways are as complex as their benthic counterparts. In mangrove waters in Cape Rachado, Malaysia, protists consumed 22% of bacterioplankton production, and adding nutrients did not significantly increase growth rates of bacterioplankton (Lee and Bong 2007). Carbon consumed by the bacteria was estimated at $585\,g\,C\,m^{-3}\,a^{-1}$ and only $8\,g\,C\,m^{-3}\,a^{-1}$ was channelled into protists. Thus, the role of bacteria is essentially that of a remineraliser and a sink for carbon.

Plankton–zooplankton interactions are clouded in controversy in the sense that some cases show weak interactions and at other times, strong interactions in mangrove waters (Chew et al. 2012). For instance, phytoplankton fuels energy flow from zooplankton to small nekton in turbid mangrove waters in the Matang estuary, Peninsular Malaysia. The stable isotope composition and C:N ratios of fine seston in the estuary indicated the importance of phytoplankton over detritus to zooplankton nutrition with a trophic contribution of 70–84%, whereas mangrove detritus contributed <11%. In adjacent coastal waters, zooplankton grazed both phytoplankton and benthic diatoms. Diatoms are abundant in the seston but do not appear to be consumed or assimilated by zooplankton. Gut content analysis of young and small nekton in the estuary showed significant consumption of zooplankton, especially copepods, sergestids, and mysid shrimps, while stable isotope values indicated an increasing importance of mangrove carbon to juvenile fish nutrition. The range of $\delta^{15}N$ values from primary producers to small predatory fish indicates four trophic levels in the Matang estuary, with zooplankton at the second and third level. In the Gautami Godavari estuary, Andhra Pradesh, India, despite large amounts of detritus available to pelagic food webs, the locally produced phytoplankton appear to be a more important carbon source for the zooplankton (Bouillon et al. 2000).

Prawns function in regulating the abundance of smaller plankton and nekton as mid-level and top omnivores and spend their post-larval and juvenile stages in mangrove estuaries until emigrating offshore when they spawn in the wet season. The primary factor controlling juvenile prawn abundance in mangroves is larval supply and post-larval settlement. Mangroves may be a sink for settlement and early growth of shrimp and prawns but may also be a source for larvae that are transported to other habitats (Kathiresan and Bingham 2001).

Crab larvae show a complex suite of feeding choices that cause an overlap in planktonic and benthic trophic connections. For instance, the zoeae of *Aratus pisonii*, a grapsid crab, ingested large centric diatoms, mangrove detritus, and zooplankton in the Itamaracá estuary in NE Brazil (Schwamborn et al. 2006). And although detritus seemed to be readily ingested, *A. pisonii* zoeae preferred large diatoms and occasionally ingested zooplankton. The tidal channels of mangrove islands such as at Twin Cays, Belize, support a productive and diverse microzooplankton community. This community in turn supports large copepod populations that form dense aggregations in the prop-root environment along the margins of these channels (Buskey et al. 2004). Grazing by the swarm-forming copepod, *Dioithona oculata*, indicated that ingestion rates were highest on ciliates and autotrophic dinoflagellates, and that copepod populations were capable of grazing about 10% of the protozoan population each day, indicating a weak trophic link between zooplankton and protists. Ciliates are the second most abundant form of heterotrophic microzooplankton which graze between 60 and 90% of phytoplankton production.

Microzooplankton can, however, play a key role in planktonic food webs in mangrove waters. The gut contents of fish and copepods are often full of microzooplankton, such as tintinnids, cirriped nauplii, veliger larvae, foraminifera, and naked ciliates, and these trophic links often vary temporally with seasonal peaks in microzooplankton abundance. Zooplankton can also feed on a component of suspended organic matter (SOM). However, there is often a strong seasonal signal for SOM and so locally produced phytoplankton appears to be a more important carbon source for the mesozooplankton (Bouillon et al. 2000).

Fish are usually organised into feeding guilds. Mangrove fish of the Red River estuary, Vietnam, were clearly separated into five feeding guilds: detritivores, omnivores, piscivores, zoobenthivores, and zooplanktivores (Tue et al. 2014). Mangrove carbon contributed a small proportion to the diets of the fish with dominant food sources being ocypodid and grapsid crabs, penaeid shrimps, bivalves, gastropods, and polychaetes. Dietary plasticity and resource partitioning can also play a role in the trophic positioning of mangrove fish. Two estuarine sparid fish, *Acanthopagrus australis* and *Acanthopagrus pacificus,* both consumed a wide variety of prey including bivalves, gastropods, crustaceans, and polychaetes, but there were clear differences in diets among mangrove locations (Sheaves et al. 2014). Diets at a sandy bay site were composed largely of benthic infauna, whereas fish inhabiting a mangrove-lined site contained large amounts of mangrove-associated prey. Species such as *A. pacificus* have the potential both to link different food webs and to add to the complexity to tropical mangrove food webs.

Some fish species show tidal differences in feeding. For instance, the sea catfish, *Sciades herzbergii,* varied its diet depending on the tidal cycle in an equatorial west Atlantic mangrove creek (Krumme et al. 2008a). At spring tides, *Uca* spp. and the grapsid crab, *Pachygrapsus gracilis,* dominated the catfish's diet with other important food items being insects and the semi-terrestrial crab, *Ucides cordatus*. At neap tides, capitellid polychaetes dominated the diet. Thus, the spring–neap tide pulse is likely to be a major driver of temporal variations in feeding of intertidal fish in mangrove creeks in macrotidal and mesotidal estuaries.

Fish diets also vary depending on the range of habitats they inhabit. For example, reef fish inhabiting seagrass beds and adjacent mangrove forests feed on invertebrates and larger prey items (de la Morinière et al. 2003; Nagelkerken and van der Velde 2004; Lugendo et al. 2006). However, in a pristine fringing mangrove ecosystem, Heithaus et al. (2011) found no evidence that mangrove productivity directly supports local fish populations that also inhabit adjacent seagrass beds. The food webs of a mangrove forest and adjacent seagrasses in Gazi Bay, Kenya, indicated several structural differences: (i) species occurring solely within seagrass meadows in close vicinity of the mangrove swamps, (ii) species migrating between mangrove forests and the seagrass beds together with species occurring throughout the entire seagrass area, and (iii) species that use the seagrass meadows as a habitat (Marguillier et al. 1997). The results showed that seagrass meadows were the main feeding grounds for most fish. Stable nitrogen isotope signatures distinguished three trophic levels: (i) fish species feeding on seagrasses and macroalgae, (ii) fish feeding on zoobenthos and plankton, and (iii) fish feeding on other fish and/or large crustaceans.

Trophic position of a given population or species clearly depends on location. The primary source of nutrition for several species of penaeid shrimps often depends on their location within a mangrove estuary. Stable isotopic carbon signatures of juvenile prawns in seagrass beds are usually close to those of seagrass carbon and their epiphytes, whereas for juvenile prawns in an upstream mangrove creek the signature is frequently intermediate between mangrove and seagrass. While a considerable amount of mangrove detritus is exported to adjacent coastal waters, it is unlikely to contribute to offshore food webs supporting adult shrimps (Thimdee et al. 2004). For example, mangrove fish along the Tanzanian coast had stable carbon signatures close to mangroves, while those further offshore showed a mixed diet (Lugendo et al. 2007). Shrimps and fish feed on mangrove-associated food items for at least part of their life cycle while in mangrove estuaries. The diets of decapod crustaceans and fish change along mangrove-ocean gradients from the Mamanguape River estuary in NE Brazil (Claudino et al. 2015). In the inner estuary, consumers had stable isotope signatures like that of mangrove detritus and macroalgae, whereas nearer the mouth of the estuary and in adjacent coastal waters, consumers assimilated nutrients derived from macroalgae, seagrass, and sediment organic matter. The trophic signal of mangrove detritus faded quickly from the source.

There may or may not be several parallel food webs operating in mangrove forests and their waterways. Stable isotope ratios in five penaeid shrimp species inhabiting waters of Saco da Inhaca, southern Mozambique, indicated that these shrimps derived their carbon either from detritus, plankton remains, or benthic organisms (Macia 2004). The carbon signature shifted as the shrimps got larger, suggesting a change in diet with growth. No significant differences were found in the stable isotope signal indicating that the shrimps belong in the same trophic position. There was evidence, however, of a high level of diet overlap between fish species in a near-pristine mangrove estuary in tropical north Queensland, Australia (Abrantes et al. 2015). Nonetheless, five main trophic pathways were identified: one based on both mangrove and microphytobenthos, one on plankton, two on both microphytobenthos and seagrass, and one mainly on seagrass. Some fish and shrimps in wet–dry tropical mangrove estuaries have stable isotope signals compatible with mangrove material (Abrantes et al. 2015). Thus, mangrove detritus is the most important food item in estuaries with the greatest mangrove cover.

There is some evidence that mangrove material can still be important in estuaries with less mangrove cover, underscoring the importance of mangroves in supporting estuarine nurseries and fisheries productivity. In contrast, Al-Maslamani et al. (2012) found that in an arid zone coastal embayment, food sources other than mangroves were important for consumers. Stable isotope signatures of shrimp tissue identified seagrass as a major source of carbon and nitrogen for two penaeid shrimps, *Penaeus semisulcatus* and *Metapenaeus ensis*. Microbial mats were also identified as a food source for early-stage *M. ensis* post-larvae. Food web segregation structures a mangrove creek of the Curuçá estuary in northern Brazil into three relatively distinct food webs: (i) a mangrove food web where vascular plants contribute directly or indirectly via POM to the most $\delta^{13}C$-depleted consumers, (ii) a mixed food web where the consumers use carbon from different sources, and (iii) an algal food web where benthic algae are eaten directly by consumers (Giarrizzo et al. 2011). Some consumers feed on both pelagic and benthic food items. For instance, the Atlantic anchoveta, *Cetengraulis edentulus,* is a phytoplanktivorous filter feeder whose principal foods are pelagic and benthic diatoms (Krumme et al. 2008b). Foraging occurs close to the substratum and throughout the turbid water column.

Several ecosystem models of trophic flows exist for tropical mangrove ecosystems (Vega-Cendejas and Arreguín-Sánchez 2001; Wolff 2006; Simon and Raffaelli 2016). A trophic flow model of the Caeté mangrove estuary of northern Brazil shows that 99% of total system biomass consists of mangrove forests which contribute 60% to ecosystem primary production (Wolff 2006). The remaining biomass was distributed between pelagic and benthic biomass in proportions of 10 and 90%, respectively. Mangroves served as the main food source via litterfall, which was consumed directly by herbivores or detritivores. The dominance of mangrove epibenthos was attributed to the fact that a large part of the ecosystem production remained within the mangrove forest as material export was restricted to spring tides. Total system throughput and mean transfer efficiency (10%) calculated by the model fit well into the range reported for other tropical coastal ecosystems.

An ECOPATH model for a mangrove coastal lagoon in Yucatán Peninsula, Mexico (Vega-Cendejas and Arreguín-Sánchez 2001) showed that a great proportion of the primary production was exported to adjacent waters, while 45% was grazed and only 7% shunted to detritus. Nevertheless, detritus played an important role within the mangrove forests with 64% of the flows being utilised and transferred to juvenile fish via small crustaceans. Mixed trophic impacts showed that detritus and low trophic levels had a positive influence on most groups, while a negative impact occurred with increasing biomass at high trophic compartments. A network model of trophic interactions in the Cameroon estuary (Simon and Raffaelli 2016) revealed that 98% of total throughput was confined to the first and second trophic levels with the overall system strongly dependent on the consumption of mangrove detritus. The distribution of energy and the mixed trophic impacts showed that detritus and low trophic levels had a positive influence on most groups.

Generalisations among mangrove ecosystems need to be tempered with caution as differences in trophic relationships among ecosystems must be considered. Wolff (2006) compared ECOPATH models of the Caeté estuary, Brazil and the Gulf of Nicoya, Costa Rica and found that both ecosystems shared little in common regarding the main trophic pathways due to differences in topography, tidal regime, and mangrove cover. While the mangrove forests in Gulf of Nicoya are exposed to semidiurnal tides and an efficient daily water exchange with a strong export of mangrove matter to the adjacent gulf water, the mangrove forest of the Caeté estuary is flushed each fortnight only, leaving the largest part of mangrove production within the forest. In the Gulf of Nicoya, detritus exported from the mangrove forests to the estuary feeds an aquatic food web with shrimps and other aquatic detritivores in the centre, while in the Caeté estuary, most energy remains in the benthic domain of the mangrove forest where it is transferred to a great biomass of leaf-consuming mangrove crabs.

A carbon mass balance of the world's mangrove forests (Figure 10.11) shows that ≈64% of GPP is respired by the canopy with NPP vested nearly equally in litter, wood and below-ground root production (Alongi 2020). About 41% of litter production is exported to adjacent tidal waters, and as estimated by difference, about 40% is buried and 44% incorporated and eventually decomposed in the massive soil C_{ORG} pool. C_{ORG} burial equates to about 12% of NPP. Of the combined total carbon decomposition (subsurface DIC production + CO_2 respiration at the soil surface), 25% is released at the forest floor surface in a separate process, and about all subsurface DIC production is exported to adjacent tidal waters in the form of DOC (≈30%), dissolved CH_4 (<0.2%) and DIC (≈70%). A considerable, but unquantified, amount of exported DIC, DOC, and CH_4 is derived from groundwater derived from adjacent upland (Maher et al. 2018), so it is unclear exactly how much dissolved carbon exported from mangroves is derived from soil decomposition.

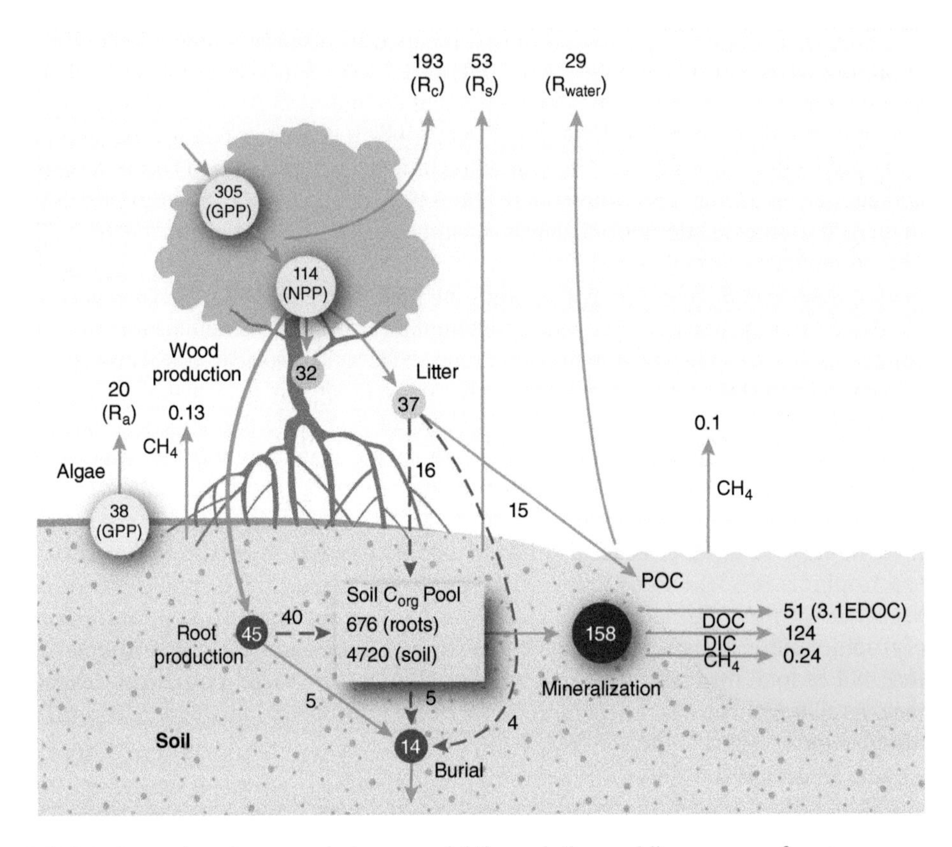

FIGURE 10.11 A carbon mass balance model through the world's mangrove forests assuming a global area of 83 495 km² (Hamilton and Casey 2016). All values are in Tg C a^{-1}. Solid blue arrows represent mean values based on empirical data (see text for explanation). Dashed red arrows represent mean values estimated indirectly (by difference). The C_{org} pool (both roots and soil) in soils to a depth of 1 m is presented as a box in the forest floor with units of Tg C. Unquantified inputs of dissolved carbon from land-derived groundwater and organic matter inputs from adjacent marine and catchments are not depicted. Abbreviations: DIC = dissolved inorganic carbon, DOC = dissolved organic carbon, GPP = gross primary production, NPP = net primary production, POC = particulate organic carbon, R_a = algal respiration, R_c = canopy respiration, R_s = soil respiration, R_{H2O} = waterway respiration. *Source*: Reproduced from Alongi (2020), figure 5, p. 9. Licensed under CC BY 4.0. © MDPI.

Supersaturation of mangrove waters leads to significant CO_2 and CH_4 release to the atmosphere (Alongi 2020). The rates of soil decomposition imply that the turnover time of the entire soil C_{ORG} pool is ≈25 years. This time frame is supported by empirical findings that mangrove roots decompose slowly, and that mangrove soil organic matter is composed mostly of allochthonous, highly refractory, plant-derived material that is high in lignocellulose and hemicellulose derived mostly from leaves that decompose slowly. About 58% of soil carbon is mangrove-derived, a value that comes from stable isotope signatures of mangrove soils (Kristensen et al. 2008a) and about one-third of the total soil carbon pool is composed of dead roots that also decompose slowly (Alongi 2014). Assuming 4 Tg C a^{-1} of litter is buried and that all POC export is derived from litter, then the remaining 16 Tg C a^{-1} of litter produced must fall to the forest floor where it is presumably incorporated into detritus food webs and eventually decomposed in situ. Wood that falls to the forest floor may be eventually incorporated into the soil pool, but decomposition is slow and likely a minor flux, so it is not included in the mass balance. Also not included are benthos and zooplankton production and chemical defences. Total soil decomposition equates to approximately 140% of mangrove NPP. This anomaly suggests that: (i) inputs from allochthonous marine and terrestrial sources are necessary to balance the decomposition outputs (158 Tg C a^{-1}), (ii) a large proportion of the soil pool and its subsequent decomposition is derived from the intertidal mud flat prior to mangrove colonisation, (iii) wood, algae and fauna contribute to the soil pool, and/or (iv) the subsurface soil decomposition rates and subsequent export data are overestimates. It is also conceivable that root production is underestimated given that the empirical dataset is small and that there are some methodological shortcomings in deriving production estimates.

About 42% of soil organic matter may be derived from external sources. Measurements of radiogenic and stable isotopes in a subtropical Australian mangrove indicate that century-old, sequestered carbon is still susceptible to remineralisation and tidal export (Maher et al. 2017), supporting the idea that organic carbon deposited prior to mangrove colonisation continues to be decomposed, as all mangroves colonise intertidal mud flats that have considerable amounts of soil C_{ORG}. Mangrove DIC export contributes nearly 60% of DIC and 27% of DOC discharged from the world's tropical rivers to the coastal ocean, based on comparison with tropical riverine export values in Huang et al. (2012). Mangroves inhabit only 0.3% of global coastal ocean area but contribute 55% of air–sea exchange compared with the global average (Chen et al. 2013), 28% of DIC export, 14% of C burial, and 13% of DOC + POC export, compared to global averages in (Bauer et al. 2013) for the world's coastal ocean. Mangrove ecosystems thus contribute a disproportionate share to carbon cycling in tropical seas and in the global coastal ocean.

The flux of CH_4 from mangrove soils is important in terms of climate change as CH_4 is a powerful greenhouse gas with significant global warming potential. CH_4 emissions vary greatly among mangrove locations ranging from undetectable or net uptake in mangrove plantations in the Mekong delta (Alongi et al. 2000a) to rapid rates in organically polluted soils in China, India, and Brazil (Zheng et al. 2018). Rates of CH_4 and CO_2 fluxes from mangrove soils are regulated by several drivers, including the degree of wetting, microbial composition, soil texture, redox, temperature, organic content, bioturbation, below-ground root activities, organic pollution, and climate. CH_4 fluxes worldwide show that mangroves affected by anthropogenic activities (median = 1267.2 μmol m^{-2} d^{-1}; range = 4.6–124,046.9 μmol m^{-2} d^{-1}) had emission rates 10–14× greater than pristine mangroves (median = 85.7 μmol m^{-2} d^{-1}; range = −145.2–1896 μmol m^{-2} d^{-1}).

Emission rates tend to be lower in low-nutrient and disturbed environments. In oligotrophic environments such as in the Red Sea, CH_4 fluxes from soils of stunted mangroves ranged from 0.9 to 13.3 μmol m^{-2} d^{-1}; rates of CO_2 flux were also low (−3.4–7.3 mmol m^{-2} d^{-1}) indicating low rates of organic carbon decomposition (Sea et al. 2018). Deforested and semi-arid mangrove areas are sites of low gas emissions. In semi-arid mangroves in NE Brazil, gas emissions ranged from 8.9 to 24.2 mmol CO_2 m^{-2} d^{-1} and 1.05 to 13.2 μmol CH_4 m^{-2} d^{-1} with higher rates associated with mangroves receiving shrimp farm and urban wastes (Nóbrega et al. 2016). In the Philippines, Castillo et al. (2017) measured emissions of CO_2 and CH_4 in degraded mangrove soils that were three to seven times lower than intact forests. No or low CH_4 emissions were measured in hydrologically disturbed mangroves in SW Florida (Cabezas et al. 2018). In three oceanic mangrove forests on north Sulawesi, Indonesia, CO_2 and CH_4 fluxes (−32.2–93.1 mmol CO_2 m^{-2} d^{-1} and −8.4–14.6 mmol CH_4 m^{-2} d^{-1}) were not high given the rich organic carbon and nitrogen in these deposits, leading to accumulation of organic matter in these forests (Chen et al. 2014). CO_2 and CH_4 soil fluxes ranged from 31.3 to 187.5 mmol m^{-2} d^{-1} and 39.4 to 428.1 μmol m^{-2} d^{-1}, respectively, in *Rhizophora* spp. forests in New Caledonia (Jacotot et al. 2019), showing strong seasonality with higher fluxes in the summer. CO_2 fluxes were greater at night reflecting the uptake of CO_2 during daylight by benthic autotrophs. There was some evidence that surface soil biofilms slowed gas transfer across the soil–air interface. In subtropical *Kandelia obovata* forests in Taiwan (Lin et al. 2020), no seasonal differences were found in CH_4 emissions (1.3–1326.6 μmol m^{-2} d^{-1}) or with emissions measured in an adjacent mud flat (45.2–334.5 μmol m^{-2} d^{-1}), but there were significant seasonal differences in CH_4 emissions (10.1–7606.6 μmol m^{-2} d^{-1}) in tropical *A. marina* forests where CH_4 fluxes were significantly higher than in adjoining mud flats (6.6–192.7 μmol m^{-2} d^{-1}). The pneumatophores of *A. marina* played a more important role than soil properties in affecting soil CH_4 fluxes, whereas soil water and organic matter content were the main factors regulating fluxes in the *K. obovata* forests.

There were tidal differences in rates of gas efflux from soils in Tanzanian mangrove forests (Kristensen et al. 2008b). Lowest emissions were during high tide (1–6 mmol CO_2 m^{-2} d^{-1}; 10–80 μmol CH_4 m^{-2} d^{-1}) and highest during low tide (30–80 mmol CO_2 m^{-2} d^{-1}; 100–350 μmol CH_4 m^{-2} d^{-1}). In waters overlying the mangrove forest floor, gas emissions were highly variable, driven by tidal changes as found in the same New Caledonia forests measured for soil–air exchanges. Jacotot et al. (2018) measured mean CO_2 and CH_4 emissions of 3.4 ± 3.6 mmol C m^{-2} h^{-1} and 18.3 ± 27.7 μmol C m^{-2} h^{-1}, respectively, with an inverse relationship between the magnitude of the emissions and the thickness of the water column above the mangrove soil. $\delta^{13}CO_2$ values suggested a mixing between enriched porewater CO_2 and incoming tidal waters. Emissions were significantly higher during ebb tides due mainly to the progressive enrichment of the overlying water by diffusive fluxes as its residence time over the forest floor increased. Emissions were also higher during spring tides suggesting a higher contribution of gas from the soil. In mangrove tidal creeks, the main driver of CH_4 and CO_2 fluxes is tides as are waters within the forests.

In mangrove creeks of the Andaman Islands, Bay of Bengal, mean tidal creek emissions were about 23–173 mmol CO_2 m^{-2} d^{-1} and 0.11–0.47 mmol CH_4 m^{-2} d^{-1} (Linto et al. 2014). CH_4 and pCO_2 concentrations correlated negatively with tidal height and were always highly supersaturated, consistent with a tidal pumping response to hydrostatic pressure change.

In regions of the South Asian monsoon, gas fluxes from mangrove soils can be greatly impacted by the ebb and onset of the monsoon and extreme precipitation (Cameron et al. 2021). In the Ayeyarwady delta, Myanmar, rates of CO_2 release were among the highest recorded in an eight-year-old rehabilitated site (540.5 mmol CO_2 m^{-2} d^{-1}) during the February dry season. These rapid fluxes were attributed to several factors such as a rapid tree growth, especially below-ground roots, and rapid microbial turnover of fine roots (Figure 10.12, left panel). During an extreme 2019 wet season, rates of CO_2 efflux at all sites ranged from 198.7 mmol CO_2 m^{-2} d^{-1} in a dry disused pond to 740.4 mmol CO_2 m^{-2} in the eight-year-old site. Elevated effluxes at all sites in October were attributed to lowered porewater salinity and deposited new alluvium, stimulating high rates of autotrophic and heterotrophic productivity (Figure 10.12, right panel).

Lateral carbon fluxes between mangroves and adjacent seagrasses and coral reefs are not well understood, although these habitats are frequently adjacent to one another. Along a mangrove–seagrass–coral continuum on the coast of Iriomote Island, Japan, measurements of air–water CO_2 fluxes found that mangrove, seagrass, and coral reef waters acted as strong, moderate, and weak sources of atmospheric

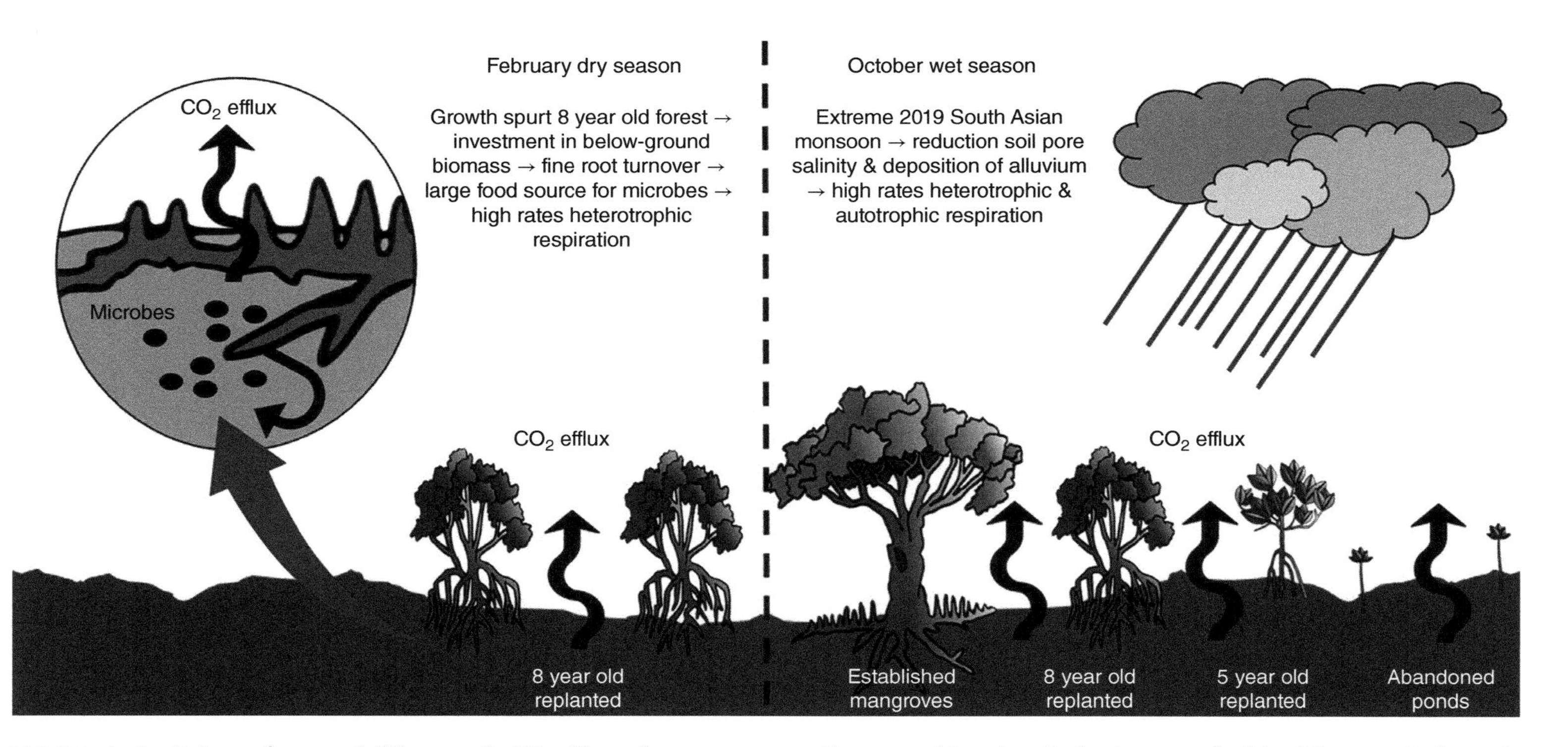

FIGURE 10.12 Drivers of seasonal differences in CO_2 effluxes from mangrove soils at several locations in the Ayeyarwady delta, Myanmar. Left panel denotes drivers of soil gas efflux in the February dry season and the right panel depicts factors affecting gas effluxes during the extreme rains of the 2019 South Asian monsoon. *Source*: Cameron et al. (2021a), figure 1, p.1. © Elsevier.

CO_2, respectively (Akhand et al. 2021). The mangroves acted as a net source for DIC, DOC, and total alkalinity, but as a net sink for POC (Figure 10.13). Mangrove organic matter originated from rivers and mangroves illustrating efficient sediment trapping within the forest. Sediments further seaward were composed of proportionally greater coastal material. The seagrasses acted as a net source of all carbon species and total alkalinity, whereas the coral reefs acted as a net sink of total alkalinity, DIC and DOC. However, the lateral transport of carbon from mangrove and rivers offset atmospheric CO_2 uptake in the seagrass zone. Due to DIC uptake by autotrophs (mainly in the coral zone), the entire continuum was a net DIC sink and atmospheric CO_2 evasion was lowered. Thus, lateral transport of riverine and mangrove derived carbon affect CO_2 dynamics and air–water fluxes in adjacent seagrass and coral reef ecosystems (Akhand et al. 2021).

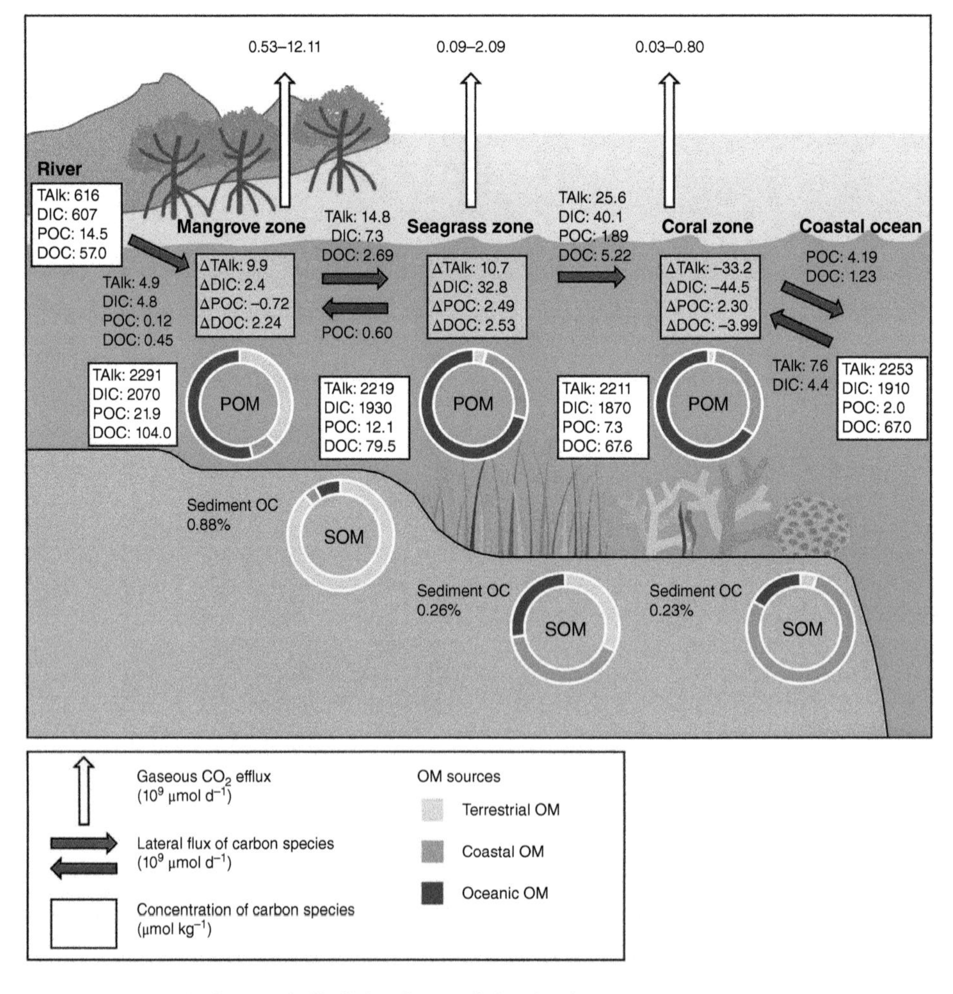

FIGURE 10.13 Carbon and alkalinity fluxes of riverine inputs across a mangrove–seagrass–coral reef continuum out to the coastal ocean on Iriomote Island, Japan. CO_2 effluxes were area integrated. Fluxes of total alkalinity (TAlk), dissolved organic carbon (DOC), particulate organic carbon (POC), and dissolved inorganic carbon (DIC) across the continuum were estimated by using a biogeochemical mass balance model. *Source*: Akhand et al. (2021), figure 5, p.9. Licensed under CC BY 4.0. © Elsevier.

10.6 Coral Reefs

The food webs of coral reefs are extraordinarily complex, with different reef zones dominated by a menagerie of food webs. The distribution and abundance of pelagic and benthic communities on coral reefs ultimately mirror the distribution and availability of essential nutrients. Light and temperature play a role, but zonation patterns of organisms has been attributed primarily to gradations in sediment granulometry, aggregations of essential nutrients, rates of water turbulence and current flow, and detritus deposition (Mumby and Steneck 2018). Attempts to measure rates of detritus deposition and transport on coral reefs have been complicated by variations in water motion. The available data indicate high spatial and temporal heterogeneity in rates of detritus flux on a coral reef, with rates in coral reef lagoons ranging from 9 to 1500 mg C m^{-2} d^{-1} (Sheppard et al. 2018). The more enclosed the lagoon, the more likely there is efficient retention and longer turnover time of detritus. Rates of detritus deposition appear to be lower in more open lagoons.

Turf algae fragmented, eroded, grazed, and exported as faecal material constitutes much of the material deposited into the reef lagoon. Most zooplankton advected onto the reef is consumed at the reef front, and coral mucus is likely to be an intermediate contributor given the moderate rates of mucus production by corals. Although not quantified, some carbon may also be imported from waters impinging on the reef. The proportion of net primary production on the reef front and reef flat that is transported to the lagoon varies seasonally and spatially.

Bacteria do most of the organic matter decomposition but there is comparatively little data on the role of bacteria in coral reefs compared with other ecosystems (Knowlton and Rohwer 2003). So, how much detritus and other material gets consumed by bacteria on coral reefs? The answer is that most detritus ultimately gets decomposed by bacteria, and many other planktonic and benthic organisms are involved in the process (Sheppard et al. 2018). The structure of coral reef food webs is such that many organisms are involved in the transfer of material and energy through reef ecosystems. Microbes may dominate trophic flow through lagoonal sediments, but this may not be true for all lagoons where some detritivores and filter-feeders account for much more detrital carbon than do bacteria. Most studies that have examined this question have not considered the large temporal and spatial variations in rates of detritus flux, bacterial activity and distribution and abundance of higher consumers within the reef. Bacteria may even be unimportant trophically for brief periods of time, although even if metazoans consume a large share of the detritus pool, ultimately, waste and most unassimilated material will either be respired or recycled by microbes.

The trophic fate of bacteria and other microbes cannot be readily deduced owing to the dichotomy between oxidised and anoxic environments and the lack of selectivity of most, especially benthic, detritivores. Benthic deposit-feeders, such as holothurians, acquire food by swallowing large volumes of sediment relative to their body size and weight, processing both microbial and non-living carbon for food. Detritivores are functionally omnivores, feeding on living and dead heterotrophic and autotrophic biomass. The relative contributions of each to their diet depend on relative availability, caloric value, and nutritional quality. The rate of detritus utilisation also depends on the rate of detritus supply and source. Because most reef detritus is algal-derived, a large proportion of this material is likely decomposed without the need for microbial enrichment as algal detritus is more readily assimilated than vascular plant detritus.

Bacterial, algal, and detrital foods are utilised rapidly by both planktonic and benthic organisms on coral reefs (Sheppard et al. 2018). This idea is supported by the fact that such organisms are highly abundant in both the water column and on the seabed. There are few data on consumption rates of such foods by reef organisms, as nearly all studies have focussed on the most obvious organisms, such as holothurians and gastropods. Grazing by these organisms can reduce bacterial and detrital mass by roughly 10–40% indicating their reliance on these foods for sustenance (Hansen and Skilleter 1994).

Field experiments have been equivocal and the mechanisms responsible for a positive or negative effect of grazing have not been readily discerned. For example, the gastropod, *Rhinoclavis aspera,* feeds on microalgae and detritus in One Tree Reef on the southern Great Barrier Reef, but caging experiments found no consistent effects of the gastropod on either microalgal biomass or rates of bacterial growth, although bacterial numbers were lower in cages with high densities of gastropods (Hansen and Skilleter 1994). The complexity of the problem is that physical disturbance alone can produce similar effects, so it is difficult to discern if the effect is due to ingestion alone or to the combined effects of ingestion plus the physical effect of disrupting the sediment particles. Clear demonstration of consumption is extremely difficult under natural conditions as finely balanced gradients within sediments are nearly always disrupted. Bacteria may or may not be tightly coupled to higher consumers, yet the fact remains that they are highly abundant and productive, especially in unconsolidated, carbonate reef sand, and mud deposits.

Benthic and pelagic food webs are tightly coupled on coral reefs which is unsurprising given the extensive mixing between the seabed and overlying waters on the reef crest and back reef. Benthic algal fragments, algal and coral exudates, amorphous aggregates, and associated microflora are continually being produced and settle to the benthos from the water column and resuspended. Faecal pellets, dead and living plankton, marine flocs, and other suspended particles either settle passively to the bottom or are actively filtered by a variety of suspension-feeding benthic organisms, including corals; much of this material that is not readily consumed is re-suspended in the water. A seemingly endless variety of pelagic organisms feed on living cells and detrital material advected onto reefs from surrounding waters or produced in situ.

The activity of pico- and nano-planktonic communities is enhanced by the release of DOM from corals and other benthos (Silveira et al. 2017). Coral reef waters are no different than other parcels of water in the tropics in that cycling of energy, materials, and nutrients takes place through microbial networks. The proportion of energy and nutrients that flows through coral reef waters is unknown, yet it is likely to be considerable, but variable, from reef to reef and from reef zone to reef zone. The composition of reef plankton is dominated by coccoid cyanobacteria and by viruses much smaller in size. On Davies Reef, Australia, the plankton carbon biomass (Ayukai 1995) was composed of autotrophic picoplankton (46%), nano-phytoplankton (32%), net-sized phytoplankton (13%), bacteria and bacteriovores (7%), and ciliates and zooplankton (3%), all varying in abundance across reef zones and between reefs and surrounding tropical waters. There was often a consistent pattern of lower densities of phytoplankton, bacteria, nanoflagellates, and ciliates over leeward reef flats in open ocean water or on reef fronts (Ayukai 1995). This decline in numbers is equivalent to the retention of nearly all NCP. This decline probably reflects not only consumption in the water column, but also consumption by benthic filter feeders as the retention of pelagic organisms may be one mechanism by which coral reefs conserve nutrients.

The consumption of bacteria by protozooplankton and zooplankton has rarely been quantified in reef waters, but presumably their trophic importance depends on their growth rates and whether the cells are free-living or attached to particles. A greater proportion of bacteria, on average, are attached to particles in reef waters compared with bacteria in surrounding waters. Ducklow (1990) concluded that consumption of bacterioplankton appears to be a minor sink for reef bacterial production as far more bacteria are consumed by benthic filter-feeders. Zooplankton seem to be differentially removed at the reef front by the 'wall of mouths' resulting in the dominance of phytoplankton and bacteria in the water column inside a coral reef. Indeed, it has been suggested that phytoplankton and bacteria are not effectively removed but are exported from coral reefs. However, the most consistent observation on coral reefs is that bacteria and other microbial cells are retained. Bacteria are linked to corals via the production of mucus. Hermatypic corals may release from 40 to 60% of their photosynthetically fixed carbon as mucus. This material is rich in wax esters, triglycerides, fatty acids, and other nutrient-rich compounds, serving as a rich food for bacteria and other animals such as benthic infauna, commensals, and some specialised fish. Rates of mucus production vary greatly, but the value of mucus in food webs on coral reefs appears certain.

Reef organisms display a variety of feeding modes, including uptake and assimilation of DOM, yet most benthic and pelagic organisms eat a variety of foods to maintain a balanced diet. Reef organisms can at most times be herbivorous but may supplement their diet by turning carnivorous or using DOM to obtain vital essential elements. Cnidarians and poriferans are particularly adept at defying classification of their feeding status. Some soft corals exhibit mixotrophy depending on location within a coral reef and may depend on both phototrophy and heterotrophy to supply their nutritional requirements. The energetic costs of growth, reproduction, and mucus production for soft corals are virtually unknown, yet under normal light conditions, phototrophy is insufficient to meet their nutritional needs.

Like herbivory and mixotrophy, the effect of carnivory on energy and materials flow within coral reef food webs is unclear. Large and small predators, especially fish, exist in great abundance on most reefs and are undoubtedly important to the trophic organisation of reef food webs. Conversely, the rich diversity and abundance of life on coral reefs attracts and sustains many large and solitary oceanic predators that may not survive without them. The trophic importance of top predators to the structure and functioning of coral reef food webs is likely to be disproportionate to their energetic significance, considering the attenuation of energy from lower to higher trophic levels; such assumptions, however, remain untested. While top predators feed at higher trophic levels, they also feed at many lower trophic levels such that the 'average' trophic level of feeding can be low. Planktivorous fish on the windward face of a reef remove most of the zooplankton from the water prior to its impingement onto the reef. Carnivory has been underestimated on coral reefs and may be responsible for a greater proportion of energy and material flow than previously believed.

Coral reefs have many trophic levels and transfers in their food webs due mainly to their high primary productivity. A comparatively simple food web model for the coral reefs of French Frigate Shoals, Hawaii, suggested that the main primary producers were benthic algae with the main pathway to heterotrophic benthos, which in turn were fed upon by reef fish (Polovina 1984). The main flux of biomass of reef fish was recycled as well as being consumed by monk seals, sharks, jacks, and scombrids (Figure 10.14). The pelagic grazing pathway was dominated by zooplankton grazing on

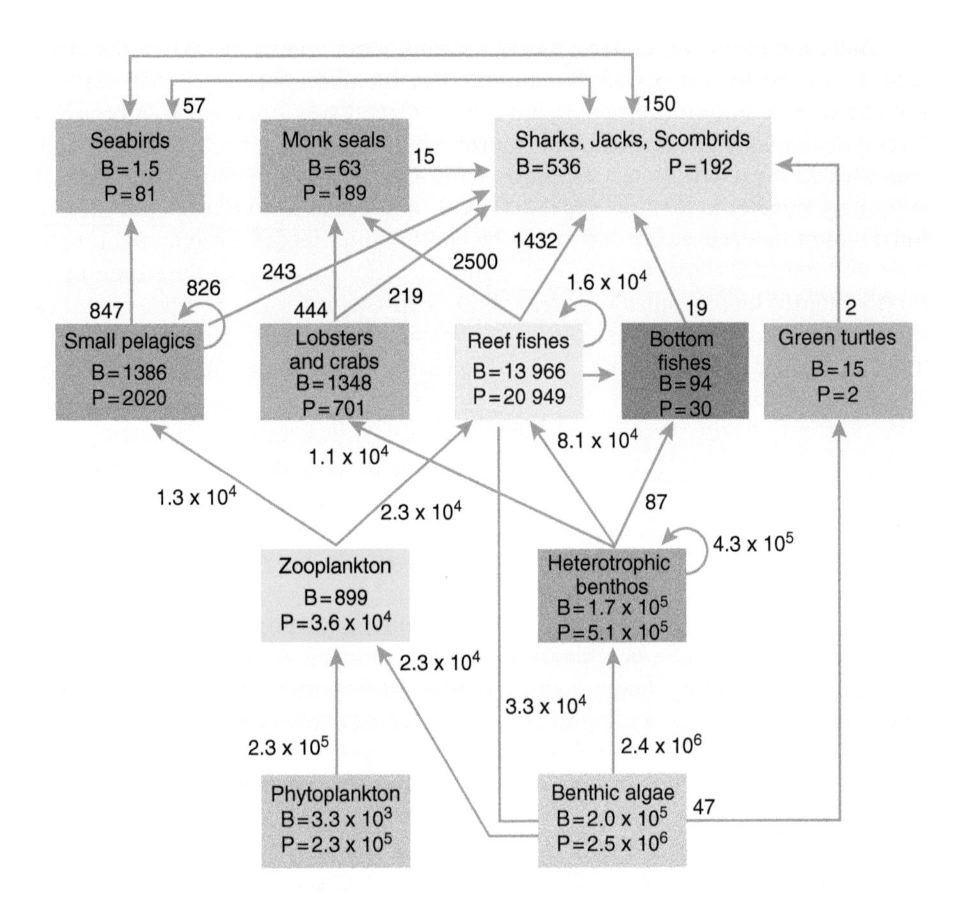

FIGURE 10.14 Food web dynamics in coral reefs of French Frigate Shoals. Annual production denoted as P and mean annual biomass as B with values in units of kg km^{-2} on a habitat area of 1200 km. *Source*: Polovina (1984), figure 1, p.8. © Springer Nature.

phytoplankton which in turn were fed upon by reef fish and small predators. The high internal predation among reef fish and the benthic heterotrophs suggested that each of these groups probably represents two trophic levels. The schematic food web indicates an ecosystem composed of six trophic levels. In contrast, Carreón-Palau et al. (2013) found that phytoplankton were the major food for all fish on a Gulf of Mexico coral reef, and that dietary energy from mangrove detritus was transferred only to juvenile fish, while seagrass detritus was transferred to the entire food web.

Grazing by reef-associated herbivores along the coral reef–seagrass interface can be intense. In the Florida Keys, Valentine et al. (2007) documented the impacts of piscivore density on herbivory along the coral reef–seagrass interface, finding that grazing rates were greater than seagrass production near reefs in the upper Keys, but <48% of production in the lower Keys. These large differences were not related to regional differences in either herbivore density or species composition. Seagrass biomass was also lower near reefs in the upper Keys, where estimates of grazing were highest. Piscivores were dominated by transient predators whose densities varied with region and distance from the reef. Predation risk rather than direct reductions in density may limit grazers to intense feeding on seagrasses adjacent to the upper Keys. Thus, seagrasses are often important to the trophic dynamics in coral reefs.

There are often complex pathways on coral reefs involving herbivore, strong corallivore, and carnivore interactions regardless of the dominant primary producer. In a community-wide analysis of a coral reef off the Pacific coast of Panama, Glynn (2004) modelled the trophic interactions in the complex food webs, identifying 4 trophic levels with 31 inter-guild links and feeding interrelationships with more than 287 species linked in a coral–corallivore subweb. The importance of various trophic groups changed greatly over annual to decadal-scale periods. Such changes may be coupled to strong ENSO events, such that intermittently intense herbivory by echinoids and corallivory by gastropod molluscs, COTS, hermit crabs, and various fish species result in high levels of coral mortality and reef bioerosion. A notable feature of this Panama coral reef food web was the large number of coral–corallivore pathways and the interannual variability of corallivory. At least 8 invertebrate and 11 fish species consumed corals on the reef. This can be attributed to the numerous consumer species that are adapted to feed on pocilloporid corals, which are notably abundant and among the main reef builders in this reef. Coral mortality resulting from ENSO warming also alters the relative abundances of coral prey and their spatial relationships relative to corallivore feeding activities. Surviving coral populations may be subject to increased corallivory during and following ENSO events.

The complexity of the lower trophic levels in coral reef food webs was demonstrated by the ECOPATH model of the fringing and barrier reefs of Tiahura, Moorea Island, French Polynesia, in which a total of 43 and 46 trophic groups were identified in the two reef types, respectively (Arias-González et al. 1997). The model output indicated that most primary productivity was processed and recycled (59–69% of NPP) within the food web through detritus-based, microbially mediated processes, with a substantial but secondary flux through grazer-based food webs, producing long pathways with low efficiencies at the higher trophic levels. The trophic structure of both reef habitats efficiently conserved energy and materials within the reef ecosystem through two forms of internal recycling: (i) a relatively large cycle produced through a detritus-based microbial food web and (ii) a relatively short cycle driven by predation. The model suggested that bottom-up and top-down controls are both ecologically important in the reef habitats.

A trophic model of the coral reef lagoon of Uvea Atoll, New Caledonia, showed the importance of the detritus pathway (Bozec et al. 2004). Phytoplankton production approximately equalled benthic primary production and benthic biomass accounted for >80% of the total living biomass. The benthos required input of food from the planktonic food web (mainly zooplankton) and from adjacent areas to sustain the biomass of predatory fish. Predation pressure was found to be a major force structuring the food webs, but it was also shown that water circulation within the lagoon influenced plankton and benthic microalgal biomass and detritus standing stocks. A high level of detritus recycling was inherent in this model as has been found in models of other coral reef ecosystems.

The removal of top predatory fish by humans has a key top-down role for the remaining reef food web. Extraction of fish by humans fishing on East African coral reefs resulted in the elimination of piscivores and invertivores (McClanahan 2008b). Fishing of piscivores kept sea urchins at low levels resulting in an increase in herbivorous fish, a reduction in algae, and an increase in coral and reef growth. The ecological effects of fishing were less benign as algal weight doubled and net productivity was reduced; an eventual increase in reef growth due to calcification by algae alone with little contribution from corals was predicted. Both piscivores and invertivores in these

Kenyan reefs were fished at a level above where the switch to sea urchin dominance was predicted. These reefs still produced fish, but the dominant fishable species were the seagrass-feeding rabbitfish and parrotfish, and other sand and seagrass-associated generalists and pelagic species that were not highly dependent on the coral bottom.

Food web structure of two reef slopes in semi-protected areas and a third reef slope subjected to more intense exploitation in the Mexican Caribbean found that: (i) production was always lowest for the unprotected reef slope, (ii) net primary production was 3X higher for the semi-protected slopes than for the unprotected slope, (iii) total catch in the unprotected slope was 3–8× higher than for the two semi-protected slopes, (iv) food chain length increased as total catch increased, (v) the calculated trophic level of catch was relatively lower on the unprotected slope, and (vi) catch per unit net primary production (i.e. gross efficiency) was higher on the unprotected reef slope than on both semi-protected slopes (Arias-González et al. 1997).

Fishing on coral reefs relaxes trophic cascades. A rapid shift in fish community structure in Belize accompanied a marked decline in grouper and snapper abundance and a switch towards smaller, less desirable, herbivorous parrotfish (Mumby et al. 2012). Over a six to seven-year period, large-bodied grouper declined significantly, and the biomass of carnivorous snappers underwent a sevenfold decline and the inclusion of parrotfish in fish catches at a nearby atoll increased. Parrotfish biomass subsequently declined with a major decline in the large and dominant herbivore, *Sparisoma viride*. Several indirect effects of parrotfish fishing were observed. The biomass of predators increased dramatically by 880%, a response attributed to a release from predation and constraints to foraging behaviour imposed by large snappers. The density of adult damselfish also decreased, attributed to elevated predation by the increased densities of predators which prey on damselfish. No change in damselfish densities was found at two control locations where fishing was prohibited. In coral reefs of the Marshall Islands, the biomass of sharks, large-bodied piscivores, and secondary invertebrate consumers was expectedly larger in the absence of major human populations (Houk and Musburger 2013). There was a doubling in the density and biomass of small-bodied fish in the presence of humans, whereas densities of large-bodied parrotfish were halved. These trends provide evidence for prey release of small fish but also indicate a reduction in grazing. Apex predator biomass may enhance the abundance of large-bodied herbivores with ensuing benefits to calcifying benthic substrates and coral diversity.

Serious consequences can result from fishing exploitation on the structure of coral reef food webs (Dulvy et al. 2004). Predatory reef fish densities, coral-eating starfish densities, and coral reef trophic structure were examined along a 13-island gradient of fisheries exploitation in Fiji. Along the fishing intensity gradient, predator densities declined by 61% and starfish densities increased by three orders of magnitude. Due to starfish, predation corals and coralline algae declined by 35% and were replaced by filamentous algae. Starfish population growth was negative under light fishing intensity, with high predator densities on islands with higher fishing intensity and low predator densities suggesting the depletion of functionally important consumers by exploitation can indirectly influence food web structure and function.

Whether human exploitation can affect carbon fluxes on coral reefs is unknown, but carbon flow on coral reefs can be sensitive to changes in trophic structure and location. Analyses of reef fish with putative diets composed of zooplankton, benthic macroalgae, reef-associated detritus, and coral tissue confirmed that all four groups contributed to fish production; a single end member often dominated dietary carbon assimilation of a given species, even for highly mobile, generalist top predators

(McMahon et al. 2016). An important secondary carbon source for most species was microbially reworked detritus. Location played a key role in structuring resource utilisation patterns as, for instance, *Lutjanus ehrenbergii* showed a significant shift from a benthic macroalgal food web on shelf reefs to a phytoplankton-based food web on oceanic reefs. Thus, diverse carbon sources play an important role in the structure and function of reef food webs.

Sponges play an important role in carbon cycling on coral reefs. The 'sponge loop' plays a crucial role in converting DOC into detritus (Rix et al. 2016, 2018; Mumby and Steneck 2018) as mucus derived from corals is transformed into particulate detritus with sponges playing a key role in transferring energy and DOM to higher trophic levels, demonstrating a direct trophic link between corals and sponges (Figure 10.15). About 21–40% of mucus C and 32–39% of the N assimilated by sponges was released as detritus (Rix et al. 2016, 2018). This provides a mechanism by which a large proportion of carbon fixed and released by seaweeds and corals as DOC is returned as POC to the benthic fauna, retaining biomass in the system (Pawlik and McMurray 2020). This POC is then eaten by benthic detritivores and suspension feeders and eventually passed to higher trophic levels. DOC processing by sponges is probably microbially mediated. The return of DOC to POC by sponges is on par with the daily GPP of the entire coral reef ecosystem. The sponge loop is likely to be a dominant component in carbon cycling for coral reef ecosystems. Much of the DOM and POM on coral reefs is of high quality which explains why coral reef food webs recycle these materials so efficiently (Pawlik and McMurray 2020).

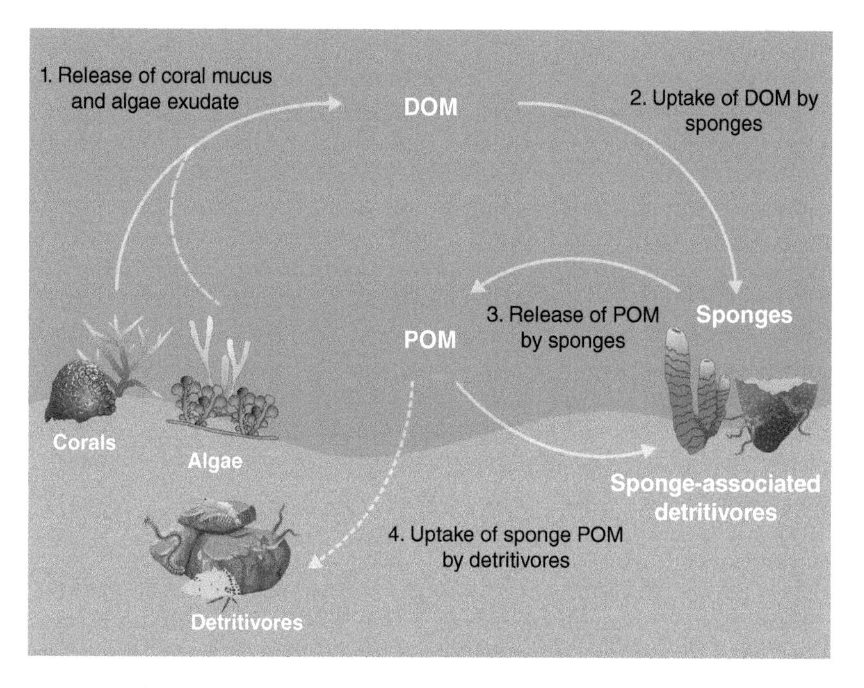

FIGURE 10.15 The sponge loop in relation to carbon and detritus cycling on a coral reef. Steps of the pathway: (1) corals and algae release exudates as dissolved organic matter (DOM), (2) sponges take up DOM, (3) sponges release detrital particulate organic matter (POM), and (4) sponge detritus (POM) is taken up by sponge-associated and free-living detritivores. *Source*: Rix et al. (2018), figure 3, p. 92. Licensed under CC BY 4.0. © Inter-Research Science Publisher.

Organic matter produced on a coral reef can be exported to adjacent coastal ecosystems. POM equivalent to about 6–7% of GPP on Shiraho coral reef, Ishigaki Island, Japan, was exported offshore with 14–20% of the exported POC and 0.2% of GPP transported to ≈ 1 km off the reef (Hata et al. 2002). Casareto et al. (2000) found that POC export from Bora Bay reefs was greater than the import of material from the ocean. The bulk of the transported POC (>90%) was micro-seston, including microbes and a large proportion of non-living particles.

Not all coral reefs export organic matter. Some may depend on allochthonous supply of POM. High uptake rates of allochthonous POM were measured in Ningaloo Reef, Western Australia (Wyatt et al. 2013), but net release of POM was observed over the reef flat with gross rates indicating that the release of autochthonous POM may be of similar magnitude to allochthonous uptake. These POM fluxes highlight the dependence of coral reefs on allochthonous POM supply as well as the potential for autochthonous POM production to supply nutrients to benthic and pelagic communities downstream.

Coral reef ecosystems, mostly reef flats, are sources of CO_2 to the atmosphere because their net fixation of CO_2 via photosynthetic processes are exceeded by large releases of CO_2 from precipitation of calcium carbonate (Gattuso et al. 1999). Some reef flats may be sinks for atmospheric CO_2, but such conclusions are hampered by the techniques used and the limited datasets available and are not consistent with reef sediment geochemistry. Most studies suggesting that reefs may be CO_2 sinks were conducted on fringing reefs, which are more likely to be subject to anthropogenic stress (Gattuso et al. 1999). There are an increasing number of reefs shifting from coral dominance to algal dominance because of human-induced activities, so the effect of these changes may lead to an increase in excess primary production and a decrease in community calcification, possibly shifting the ecosystem from a CO_2 source to a CO_2 sink. However, CO_2 flux on coral reefs is subject to seasonal change and the annual balance of air–sea CO_2 flux. Coral reefs in Bermuda, for example, act as sources of CO_2 to overlying water (Bates 2002). The magnitude of this source varies seasonally in response to changes in the reef community between coral- and macroalgae-dominated states. Whether the Bermuda coral reefs act as an oceanic sink of CO_2 not only depends on seasonal variability but also on the pre-existing air–sea CO_2 disequilibrium of open waters surrounding the reef.

An analysis of carbon fluxes in Indo-Pacific coral reefs showed that net organic carbon production averaged 608 mg C m^{-2} d^{-1} with values ranging from −5 to 1968 mg C m^{-2} d^{-1} (Suzuki and Kawahata 2004). Net $CaCO_3$ production in these reefs averaged 1172 mg C m^{-2} d^{-1} with rates ranging from 30 to 3036 mg C m^{-2} d^{-1}. Higher rates of carbon production occurred on reefs with water residence times of <6 hours. The P_G:R ratio averaged 1.2 and ranged from 1.0 to 1.5. Most (77%) of the coral reefs were net sources of CO_2. Suzuki and Kawahata (2004) noted that the system-level net organic to inorganic carbon production ratio (R_{OI}) is a major parameter for controlling the carbon cycle in coral reefs, including the sink–source behaviour for atmospheric CO_2. They found that a reef system with a R_{OI} < 0.6 has the potential for releasing CO_2; surface pCO_2 concentrations in the lagoons of atolls and barrier reefs were consistently higher than those in offshore waters, with depletion in total alkalinity, indicating predominant carbonate production relative to net organic carbon production. Reef topography has a secondary effect on the magnitude of the offshore lagoon pCO_2 difference. Terrestrial influence in coastal reefs, including Great Barrier Reef lagoons and a fringing reef of the Ryuku Islands (Suzuki and Kawahata 2004), results in high carbon input enhancing CO_2 efflux to the atmosphere because of high C:P ratios. Coral reefs act as an alkalinity sink and a CO_2 source due to carbonate precipitation and the input of land-derived carbon (Suzuki and Kawahata 2004; Cyronak et al. 2018). These carbon fluxes can be altered by ocean acidification (Section 13.5).

The direction and magnitude of reef CO_2 flux at the air–sea interface depends on the difference in pCO_2 between both phases and not on the difference in total inorganic CO_2 concentrations (Gattuso et al. 1999). $CaCO_3$ deposition increases pCO_2 and drives CO_2 to the atmosphere. Gattuso et al. (1996) investigated the respective effects of CO_2 uptake by photosynthesis and CO_2 release by respiration and calcification on the carbon budget of a coral reef. Measurements of GPP and R showed that the reef displayed a small excess production. Net calcification was positive both during the day and night. The measured air–sea CO_2 fluxes were close to zero in the open ocean but displayed a strong daily pattern at the reef front and the back reef indicating that this reef was source of CO_2 to the atmosphere. Albright et al. (2013, 2015) measured diel and seasonal trends in carbonate chemistry, production, and calcification across the reef flat at Davies Reef, Great Barrier Reef, and found that seawater carbonate chemistry was highly variable over both diel and seasonal cycles. Net ecosystem calcification was sensitive to changes in aragonite saturation state indicating that relatively small shifts in aragonite saturation state may drive measurable shifts in calcification rates, and hence carbon budgets.

Coral reefs are an important source of $CaCO_3$ accumulating in the global ocean. Total $CaCO_3$ accumulation in coral reefs ($154\,Tg\,a^{-1}$), tropical carbonate banks ($14\,Tg\,a^{-1}$), and other tropical shelf environments ($48\,Tg\,a^{-1}$) account for 48% of $CaCO_3$ accumulation in the global ocean (Smith and MacKenzie 2016). About 80% of reef carbonate is preserved and total $CaCO_3$ dissolution on coral reefs, tropical carbonate banks, and shelf areas account for only 8% of total $CaCO_2$ dissolution in the global ocean. $CaCO_3$-mediated CO_2 flux from coral reefs ($-94\,Tg\,a^{-1}$) accounts for 96% of $CaCO_3$-mediated CO_2 fluxes from tropical sedimentary environments.

10.7 Estuaries and Coastal Lagoons

10.7.1 Food Webs

Food web structure and carbon flow in tropical estuaries and lagoons are greatly dependent on the mixing of seawater and freshwater, largely driven by tides, waves, and buoyancy effects from river runoff. The balance between precipitation and evaporation fosters dramatic changes in salinity and exerts an equally dramatic change in pelagic and benthic food webs. The onset of the wet season presages seasonality and causes the most dramatic change in a tropical estuary or coastal lagoon. Thus, seasonality is not affected by changes in temperature, but rather by the monsoon cycle. This seasonality primarily affects carbon flow and food web structure within the microbial hub. Microbes respond to temperature and salinity changes like higher trophic levels, but strong wet season runoff can scour the seabed, especially in intertidal areas, resulting in resuspension of sediments and attached benthic microbes. Higher nutrient inputs from land through river flow may increase pelagic primary production and respiration resulting in greater rates of CO_2 release from estuarine and lagoonal waters during the monsoon (Sarma et al. 2009, 2011)

Spatial and temporal variations in the quantity and quality of dissolved and particulate organic and inorganic matter in estuaries or lagoons is controlled by the onset of the monsoon season (Alongi et al. 2014). Seasonality is thus imparted on a given estuary or lagoon based on precipitation and subsequent runoff. For instance, the microtidal estuary of the Cochin backwaters, India, underwent a characteristic transformation from being river dominated during the summer monsoon to tidal dominated during the pre-monsoon season (Martin et al. 2008). The increased tidal activity during the pre-monsoon changed the estuary into a vertically mixed and flow-restricted system supporting enhanced organic production which peaked during the monsoon. After the monsoon, river flow weakened and flushing of the estuary diminished; anthropogenic N and P loads and sediment re-suspension during and immediately after the monsoon altered N:P stoichiometry substantially. This implies that monsoon-induced hydrology plays an important role in regulating nutrient composition, and productivity in the estuary. In recent times, reduced river discharge caused by a decrease in precipitation has altered flushing times and nutrient patterns, intensifying phytoplankton blooms (Acharyya et al. 2012).

In most tropical estuaries and lagoons, the nutrient supply is autochthonous in the dry season. In the Brantas River estuary on Java, Indonesia, a major part of the fluvial nutrient input was trapped in the inner estuary (Jennerjahn et al. 2004) with seasonal differences in the quantity and origin of suspended sediments and organic matter transported by the river, which exports large inputs derived from agricultural soils to adjacent coastal waters during the monsoon season. A similar situation occurs in the Dumai estuary, Sumatra, Indonesia, where the river is black with humic substances during the monsoon; pH and concentrations of dissolved inorganic nutrients and total suspended matter were low in the river and increased towards the sea (Jennerjahn et al. 2004). DOC concentrations in the river originate from extensive peat soils which account for a large fraction of the catchment soils. Leaching of DOC may be a significant source of carbon to adjacent coastal seas. C_4 plant inputs from the catchment of the Tana estuary, northern Kenya, contributed about 50% to the POC pool in the estuary, but mangroves and other C_3 plants provided a substantially higher contribution to the DOC pool (Bouillon et al. 2007a). Consequently, the water column in the freshwater part of the river and tidal creeks in the estuarine delta was strongly heterotrophic as the net result of strong interactions with the large intertidal areas where respiration in the sediments and in the overlying water during tidal inundation left a marked signature on the water column. Export of DOC from blackwater rivers may be quantitatively more significant for the global DOC input into the ocean than previously believed.

Tropical estuaries are often affected by aperiodic events, such as short-term floods. The Wanquan River on Hainan Island, China, is a small tropical river and estuary that was impacted by Cyclone Kammuri in 2008 and a strong-rain event in 2009 (Wu et al. 2013). Both DOC and POC concentrations decreased with increasing salinity in the estuary with minimal seasonal or annual variations, except when impacted by the cyclone and the strong-rain event. Stable isotope values combined with C:N ratios reflected a mixture of terrestrial organic matter and in situ production. The amplitudes of the POC and DOC variations due to the rain event were larger than those observed during the cyclone, demonstrating the importance of one-off rain episodes on small rivers and estuaries along tropical coasts.

Sources of organic matter may be spatially partitioned within an estuary. For instance, terrestrial organic matter was detected only in the rivers or at the upper river–lagoon boundaries in the tropical estuarine–lagoon system of Mundaú-Manguaba,

NE Brazil, (Costa et al. 2011). Throughout the lagoon, fresh and recently produced autochthonous POM was detected, most likely originating from diatoms. Cyanobacteria were detected in the freshwater Manguaba lagoon indicating spatially segregated sources of organic matter. These differences showed up in the sediments using lipid biomarkers (Carreira et al. 2011).

Spatial and temporal physicochemical changes in tropical estuaries and lagoons are also the norm for planktonic and benthic food webs. Planktonic food webs are greatly impacted by grazing by microzooplankton. In an analysis of phytoplankton growth, microzooplankton grazing and carbon cycling in marine systems, Calbet and Landry (2004) found that microzooplankton consumption is the main source of phytoplankton mortality, accounting for 75% in daily growth in tropical and subtropical waters compared with 67% for the full dataset and 60% in coastal and estuarine environments, suggesting that microzooplankton are the major grazers on phytoplankton in tropical estuaries and lagoons.

Metazooplankton are also significant grazers in planktonic food webs. Bouvy et al. (2006) examined the trophic impact of metazooplankton predators (two calanoid copepods and barnacle larvae) on microbial plankton in the Senegal River estuary, West Africa, using microcosms and observed that removal of bacterial predators increased the net growth rate of heterotrophic nanoflagellates. The same results were observed in the presence of the copepod, *Temora stylifera,* suggesting trophic control on protozoans by this copepod, permitting an increase in heterotrophic nanoflagellate growth rates through indirect effects. The presence of bacteria and a high abundance of heterotrophic nanoflagellates suggest an efficient heterotrophic pathway within the microbial hub of the Senegal estuary. In contrast, Bong and Lee (2011) observed that heterotrophic nanoflagellates were not significant grazers of bacteria in Malaysian waters. Heterotrophic nanoflagellate grazing rates increased with bacterial production, but it only accounted for about one-third of bacterial production, suggesting a bacterial production-grazing imbalance with other loss factors such as viral lysis, sedimentation or the presence of benthic filter feeders possibly accounting for the balance.

Bacterial growth efficiency is a key factor in understanding bacterial dynamics on carbon flow in tropical estuaries and lagoons. Bacterial growth efficiency in the tropical Mandovi and Zuari estuaries in SW India and in adjoining coastal waters of the Arabian Sea ranged from 3 to 61% and showed clear temporal variability with significantly higher values in the estuaries than coastal waters (Ram et al. 2009). Greater variability within the estuaries suggests some systematic response to nutrient composition and the variability of DOM pools, as bacterial growth efficiency is governed by bacterial production. Monsoonal rains brought significant variability in bacterial growth efficiency and to the bacterial production to primary productivity (BP:PP) ratio when compared to non-monsoon seasons in the estuaries and coastal waters. High BP:PP ratios (>1) together with high carbon flux through bacteria (>100% of PP) suggest that bacterioplankton consumed DOC in amounts greater than that produced in situ by phytoplankton.

The supply of allochthonous organic matter can have profound effects on bacterioplankton in tropical estuaries and coastal lagoons. Most such waters rely on allochthonous inputs for fuelling the microbial hub. For instance, gradients of bacterial biomass and activity as well as seston exist along the salinity gradient in a subtropical estuary in Brazil (Barrera-Alba et al. 2009) with strong decreasing gradients of seston and dissolved inorganic nutrients occurring from the river–estuary boundary to the estuary–coastal interface. Bacterial production was inversely correlated with salinity and positively correlated with temperature, organic matter, exopolymer particles, and particulate-attached bacteria. Bacterial production was much higher than primary production during the rainy season and values of BP:PP >20 were recorded during high freshwater input, suggesting that bacterial activity was predominantly supported by allochthonous inputs of organic carbon. Similarly, plankton metabolic processes in response to the DOC pool were measured in Chilika Lagoon, India, to estimate the percentage use by bacterioplankton of photosynthetically derived DOC and particulate organic carbon pools (Kanuri et al. 2013). Bacterioplankton assimilated the non-humic DOC pool at higher rates than the humic DOC. Annually, DOC directly derived from photosynthesis satisfied 30% of bacterial carbon requirements in the lagoon. Despite a relatively high percentage of extracellular DOC release observed during the monsoon season, the measured uptake was much lower than the estimated bacterial carbon requirements, suggesting a weak link between phytoplankton exudation and bacterial metabolism. In addition to phytoplankton exudates, the heterotrophic bacteria must use other sources of DOC to achieve their carbon requirements.

Freshwater inputs and the subsequent mixing of freshwater and seawater can have observable effects on bacterioplankton activity. Cissoko et al. (2008) collected freshwater and seawater samples from the Senegal River, West Africa, and experimentally subjected freshwater to abrupt mixing with seawater. They found that production rates of freshwater bacteria and viruses sharply declined immediately after seawater addition, followed by a sharp recovery of the surviving bacterial and viral populations. Conversely, neither bacteria nor viruses were significantly affected by mixing seawater with freshwater, suggesting that the turbulent front between ascending tidal seawater and outwelling of freshwater was a more favourable environment for marine bacterioplankton.

Freshwater inputs also have a profound impact on phytoplankton and microzooplankton. In the river dominated estuary of the Paraiba do Sul River, Brazil, phytoplankton were represented by freshwater species reflecting the strong influence of the river (Costa et al. 2009). Remarkable shifts in composition and biomass occurred from the low to the high flushing seasons, due much more to river discharge than to nutrient availability. The higher river flow supported a lower phytoplankton biomass composed mainly of fast-growing nanoplankton, which maintained biomass even in high flush conditions. The lower river flow led to the coexistence of large organisms, including slow-growing populations typically found in mesotrophic lakes. In the Cochin backwaters, microzooplankton numbers declined with the onset and peak of the monsoon (Jyothibabu et al. 2009). Species diversity and richness of microzooplankton showed similar trends as that of abundance and biomass. Grazing experiments showed that microzooplankton consumed half of the daily phytoplankton standing stock during the high saline pre-monsoon conditions, with less grazing occurring during the low-salinity monsoon periods. This resulted in a weak transfer of primary and bacterial carbon to higher trophic levels during the monsoon, leaving behind much unconsumed food in the estuary. A major portion of the primary carbon was either deposited or transported to the adjacent coastal zone during the rainy season. High flushing of Cochin backwaters also facilitated faster removal of primary producers to the coastal zone during the monsoon season.

Trophic dynamics of planktonic food webs are greatly influenced by large climatic events such as El Niño. The trophic dynamics of plankton in the Gulf of Nicoya, Pacific coast of Costa Rica, were significantly affected by the El Niño 1997/1998 event (Brugnoli-Olivera and Morales-Ramirez 2008). The study detected

El Niño impacts in the inner gulf ('the Punta Morales zone') as changes in the physicochemical water characteristics. Some nutrients were correlated with salinity, and significant differences were observed between the transition and rainy season in the phytoplankton biomass. Changes in predation and grazing pressure induced by El Niño affected the trophic structure of phytoplankton and zooplankton populations in the inner zone of the gulf.

Freshwater flow has profound impacts on larger biota, such as macrobenthos and fish. Fish collected near forested areas in intermittently connected estuaries of tropical Australia had low stable carbon signatures, suggesting a high incorporation of C_3 terrestrial material (Abrantes and Sheaves 2010). A seasonal variation in $\delta^{13}C$ was detected indicating a greater incorporation of terrestrial carbon after the wet season. Negative seasonal shifts in fish stable carbon represented a greater dependence on carbon from riparian vegetation (C_3 *Juncus* sp.) in the post-wet season. In Hilo Bay, Hawaii, the contribution of different POM sources to the diets of zooplankton and juvenile fish differed between low and high river flow conditions (Atwood et al. 2012). Diets of zooplankton and juvenile fish were affected by river flow, but the magnitude and the change in basal resources depended on the location of the sampling site in the estuary relative to the ocean and the river mouths. Consumers from the station mostly isolated from the ocean and with groundwater and overland inputs utilised a combination of estuarine and terrestrial POM during both low and high river flow and exhibited less variability in their basal resources than sites with direct ocean exchange. Consumers from sites in Hilo Bay affected by both ocean exchange and river inputs utilised a combination of estuarine, terrestrial, and marine POM during low flow conditions but shifted to terrestrial sources during high flow conditions. Factors suspected to affect POM source(s) to consumers in Hilo Bay are GPP, availability of exported terrestrial organic matter, and estuarine bacteria biomass, all of which are affected by river flow.

Salinity fluctuations have an important role in structuring benthic macrofaunal assemblages. The intertidal fauna at the mouth of a tropical Brazilian estuary were greatly different between low and high freshwater flow conditions, suggesting an importance of habitats and an increase in connectivity between adjacent estuarine and coastal zone habitats (Lacerda et al. 2014). Moreover, sandy beaches in the estuary provided an alternative nursery and a protected shallow-water area for the initial development phase of many marine and estuarine species. In the hypersaline shallow Keta Lagoon, Ghana, the macrofauna was low in density and species diversity, and numerically dominated by bivalves and capitellid polychaetes, but they appeared able to withstand physical disturbance when water levels were extremely low and osmotic stress when salinities were extremely high and tended to redistribute along environmental gradients (Lamptey and Armah 2008). Salinity, percent clay, pH, and turbidity are the major environmental variables structuring the benthic assemblages in Keta Lagoon.

Long-term temporal and spatial differences in benthic community structure occur in subtropical and tropical estuaries. Long-term variability of benthic macrofauna in two subtropical estuaries in southern Brazil was strongly influenced by both ENSO and PDO events (Silva Francisco and Netto 2020). The benthic fauna of the river dominated estuary was influenced by ENSO, but the main modulating force of decadal variability of the benthos was the PDO (19%). Thus, climatic phenomena have a clear influence on benthic processes in such estuaries. In the semi-enclosed subtropical estuary of Port Curtis, Australia, benthic species richness and abundance progressively declined 72% over a six-year period and subsequently recovered by 68%

(Currie and Small 2005). Similar temporal trends were displayed by filter feeders, deposit feeders, scavengers, and predators, and it appears that drivers underpinning the observed changes had a consistent influence on most trophic levels. Both species richness and abundance were highly correlated with turbidity, suggesting that high levels of turbidity may promote recruitment and growth of benthic organisms in Port Curtis. Strong correlations between regional rainfall, freshwater inflow, nutrient, and chlorophyll *a* concentration further support the idea that recent changes in benthic productivity within the estuary were principally the result of long-term climatic cycles, including El Niño events.

Strong seasonal salinity changes may have trophic consequences for higher organisms of commercial importance, such as fish and shorebirds. A weak seasonal signal was, however, found for fish diversity and community composition in the Pueblo Viejo Lagoon, Mexico (Castillo-Rivera et al. 2002). Monthly changes in salinity, turbidity, and precipitation, as well as the presence or absence of submerged vegetation, were the most important environmental variables determining the observed variability in fish community composition. Salinity plays a significant role in fostering fish species richness as salinity was significantly and consistently related to fish species richness in Terminos Lagoon, Mexico (Sosa-Lopez et al. 2007). Significant negative correlations between fish species richness and salinity were found, reflecting the penetration of freshwater fish into estuarine areas following freshwater discharge and the dominance of estuarine taxa more able to tolerate low than high salinity.

Food sources for estuarine fish are diverse and change dramatically over time and space. Organic detritus contributed >50% of the stomach content of the large-scale mullet, *Liza macrolepis*, in mangrove creeks and a lagoon in Taiwan (Lin et al. 2007), but consumed items were distinct between both habitats and corresponded to the habitats in which they reside. The consumed items in the lagoon were more diverse than those observed in the mangrove creeks. The diet composition of the mullet in the creeks was determined by season, not by body size. There were no clear seasonal or size-dependent grouping patterns for dietary composition in the lagoon. Instead, there were significant seasonal and spatial variations in stable isotopic C and N of potential food sources and the mullet, indicating that benthic microalgae were the most important food in both seasons for this fish in the creeks, whereas a greater reliance on microalgal and macroalgal periphyton was observed in the lagoon. A similar study of food items of many commercially exploited fish species in an estuarine tropical marine protected area (MPA) in Senegal (Faye et al. 2011) partitioned primary consumers into pelagic copepods, oysters, and mussels which feed mainly on suspended POM, and intermediate consumers feeding on freshly deposited POM and benthic microalgae. Secondary consumers were divided into three groups. The first group included mullets that graze on benthic microalgae. The second group was most heterogeneous and fed mainly on pelagic components. The third group included piscivores and invertebrate predators, which dominated the top of the food web. Food webs were dominated by secondary consumers. The fish food web varied with season in faunal composition and food chain length.

Diets of many estuarine fish species change as they age. Ontogenetic changes in food preference were distinguished in nine species of estuarine fish in the mangrove estuary of the Urauchi River, Iriomote Island, Japan (Nanjo et al. 2008). Juveniles of these species fed mostly on small crustaceans or detritus. With they grew, larger prey items became a dominant part of their diets. The fish assemblages comprised eight trophic groups: plant, detritus, zooplankton, small benthic crustacean, large benthic crustacean, polychaete, insect, and fish. Of these, large and small benthic crustacean feeders, which consumed mainly crabs and amphipods, respectively, were the most abundant in terms of species, whereas polychaete and insect feeders were each represented by only two species.

Stable isotopes are useful to identify patterns of trophic dynamics and trophic levels in tropical estuaries. Abrantes et al. (2014) used such metrics to investigate patterns in trophic structure in five East African estuaries that differ in size, sediment yield, and catchment vegetation cover (C_3/C_4 plants): the Zambesi estuary in Mozambique, the Tana estuary in Kenya, and the Rianila, Betsiboka and Pangalanes estuaries in Madagascar. Trophic length of food chains varied between three–six (Madagascar) and four–seven levels (Zambesi), but did not vary seasonally for any estuary. Among the four open estuaries, the Betsiboka (C_4 dominated) had lower trophic diversity than the Zambesi and Rianila (C_3 dominated), probably due to limited availability of aquatic sources by high suspended sediment loads. Trophic redundancy decreased from the pre-wet to the post-wet monsoon seasons probably due to higher variability and availability of sources after the wet season, which allowed diets to diversify. Thimdee et al. (2004) similarly investigated stable isotope signatures for biota in the mangrove-fringed estuary of Khung Krabaen Bay, Thailand, where the $\delta^{13}C$ values of animals collected from the fringing mangroves were more negative than those of animals collected from the bay and offshore. The primary carbon sources that supported the food webs were clearly dependent on location. The contribution of mangroves was limited to only animals taken from mangrove forests, but a mixture of macroalgae and plankton was a major carbon source for organisms in the inner bay area. Offshore organisms clearly derived their carbon from the planktonic food web. The $\delta^{15}N$ values indicated that some animals have capacity to change their feeding habits with location and availability of foods. As a result, individuals of the same species were assigned to different trophic levels at different locations within the bay.

A variety of trophic network models have been applied to investigate trophic dynamics of food webs in several tropical estuaries and coastal lagoons. A trophic model of the Gulf of Paria between Venezuela and Trinidad suggested ecosystem impacts of fishing (Manickchand-Heileman et al. 2004). The model indicated that the food web was dominated by detritus which was also exported out of the gulf. Mixed trophic impacts showed that detritus and lower trophic levels have a significant bottom-up positive impact of other trophic groups. The ecosystem therefore appears to be relatively mature but may experience some degree of instability of fisheries and large seasonal variation in salinity. Model simulations indicated that increased fishing causes a decline in fish biomass and an increase in biomass of invertebrates, such as shrimps and crabs. If fishing pressure is relaxed, fish biomass recovers but crab biomass declines. The model indicates that fisheries cause a possible shift towards an ecosystem dominated by lower trophic levels. An ECOPATH model of commercial fishing in the Sundarbans estuary, Bay of Bengal, indicated that penaeid shrimps, small pelagics, and medium pelagics are the most exploited groups in the estuary. Consequently, the coastal northern Bay of Bengal food web is dominated by detritivores and planktivores, while top predators are absent from the ecosystem (Dutta et al. 2017a).

Other network models show different effects of exploitation on food web structure. For instance, a network analysis of Laguna Alvarado, western Gulf of Mexico, distinguished 18 fish groups, 7 invertebrate groups, and 1 group each of sharks and rays, marine mammals, phytoplankton, seagrasses, and detritus (Cruz-Escalona et al. 2007). Total system production was higher than consumption with NPP higher than R. However, the model suggested that primary production and consumption are in overall balance, due mainly to the presence of highly productive seagrass meadows which are more important than detritus pathways. Ching (2015) conducted a comparative study of the Sine-Saloum (Senegal) and Gambia estuaries and found that fish were highly resilient to exploitation, despite exposure to vast hydrodynamic variations and stress. Taxonomically related and morphologically similar species did not necessarily play similar ecological roles in these two systems. High production and consumption rates of some groups in both ecosystems indicated high system productivity. Elevated productivity may have been due to higher abundance of juvenile fish in those groups that utilise the estuaries as refuge and/or nursery zones. Both ecosystems are phytoplankton-driven, and differences in group trophic and ecological roles are mainly due to adaptive responses of the species to seasonal and long-term climate and anthropogenic stressors.

Estuaries and lagoons differ greatly in terms of their main food sources. Ebrié Lagoon, Ivory Coast, for instance, is dominated by phytoplankton while the Lake Nokoué ecosystem, Benin, is detritus-driven (Villanueva et al. 2006). Both ecosystems are highly productive, with high abundance of juveniles in most groups utilising these systems as refuge zones and nurseries. An ECOPATH analysis of Terminos Lagoon, Mexico, measured a high detritivory to herbivory ratio indicating that most primary production is recycled through the detritus-based food web (Manickchand-Heileman et al. 1998) with benthic invertebrates having a significant role in transferring energy from detritus to higher trophic levels. Analyses of mixed trophic levels indicate that fish have little impact on the other compartments due to their relatively low biomass and consumption. Detritus and lower trophic levels had significant positive impacts on other groups in the lagoon, suggesting 'bottom-up' control of the food web. A trophic model for the geologically young lagoon Chantuto-Panzacola, Mexico, showed that detritus is the most important ecosystem component with diverse groups of macrobenthos linking detritus to upper trophic groups such as fish, birds, and crocodilians (Lopez-Vila et al. 2019). In contrast, macroinvertebrates and fish represented over 90% of the total biomass in a subtropical coastal lagoon in the Gulf of California (Muro-Torres et al. 2019) where four food webs were identified. The first food web was based on phytoplankton which is preyed upon by zooplankton, bivalves, anchovies, and sardines. The second food web was detritus- and seston-based, associated mainly with mangrove forests but also receiving other autochthonous and allochthonous organic matter. A third food web was supported by benthic macroalgae, which is fed upon by snails and mullets. The fourth food web was associated with the mangrove forests.

Different estuaries and lagoons are clearly different in terms of the degree to which their food webs are altered by humans and this shift depends on whether their food webs are phytoplankton or detritus-based. Chiku Lagoon in Taiwan, for instance, is a highly productive tropical lagoon with high fishery yields. Network analysis showed that the lagoon is more dependent on phytoplankton than detritus and periphyton to generate food for consumers (Lin et al. 2006, 2007). Transfer efficiency is high at lower trophic levels but declines at higher levels due to high fishery pressure. Thus, only a small fraction of organic matter is recycled. The model indicated that the high fishery yield in this lagoon can be attributed to high planktonic activity induced by the high rate of nutrient loading and the straight-through pathways of the food web.

A comparative analysis of trophic flow structure in the world's estuaries (Lira et al. 2018) indicates differences in ecosystem structure from tropical to temperate estuaries. Trophic structure in temperate and subtropical estuaries is based on higher flows of detritus and export, whereas tropical estuaries have greater biomass, respiration, and consumption rates. Omnivory is more common in tropical and subtropical estuaries, implying more complex food webs. While none of the estuaries examined were classified as fully mature ecosystems, the tropical systems were considered more mature than the subtropical and temperate estuaries. Thus, the trophic flow structure of estuaries of different latitudes is fundamentally different.

10.7.2 Carbon Dynamics

Carbon budgets demonstrate the highly productive nature of tropical estuaries and lagoons. An organic carbon budget for a semi-enclosed estuary, Hinchinbrook Channel, Australia, indicates that mangroves, seagrasses, benthic microalgae, and phytoplankton contribute 55, 7, 1, and 10%, respectively, to total organic carbon inputs into the channel with riverine input contributing the remainder (Alongi et al. 1998). Much of the organic carbon in adjacent coastal sediment was of mangrove origin, indicating that mangroves were the largest contributor of POM to the coastal zone. Much of this material is trapped within a coastal boundary layer which only breaks down during significant flood events in the summer wet season. An organic carbon mass balance for the estuarine Gulf of Nicoya, Costa Rica (Gómez-Ramírez et al. 2019), indicated that the highly productive planktonic community incorporated large amounts of organic carbon into the estuary, as GPP ranged from 1.5 to 7.2 g C m^{-2} d^{-1} in the inner gulf during the dry season. Thus, the contribution of pelagic primate production to the carbon mass balance was ≈ 1030 and 2059 t C d^{-1} during the respective rainy and dry seasons. The input of allochthonous organic carbon to the gulf was small, but mangroves were an important source of organic carbon. POC flux to the seabed ranged from 21 to 180% of pelagic GPP during the dry season. The fraction of organic carbon flux to the sediment was highly variable within the estuary, but the fraction accumulated for the whole inner estuary was about 60% in both seasons, which is high compared to other estuaries. Total C burial in the inner gulf was equivalent to 2 and 4% of the daily CO_2 emissions from the entire gulf (Gómez-Ramírez et al. 2019). At the mouth of the Cross River estuary off the Gulf of Guinea, Nigeria, primary production was high (25.9 g C m^{-2} a^{-1}), but during dredging activities in the estuary, NPP declined to 20.7 g C m^{-2} a^{-1} due to high turbidity (Ewa-Oboho et al. 2008). Pelagic copepod grazing rates were higher before dredging than after but grazing consumed a high percentage of primary production. Phytoplankton, detritus, and faecal pellets accounted for 40, 33, and 25%, respectively, of the total annual POM flux within the estuary (Ewa-Oboho et al. 2008).

Small- to medium-sized tropical rivers and their estuaries and lagoons can export significant quantities of carbon derived from mangroves, seagrasses, algae, and estuarine phytoplankton to the adjacent coastal zone, depending on drivers such as climate, tidal amplitude, ratio of wetland to total catchment area and wetland productivity. The bulk of wetland material exported to the coastal zone appears to be refractory due to high rates of mineralisation of labile fractions (with subsequent efflux of CO_2 to the atmosphere) within these estuaries and coastal lagoons.

Large tropical estuaries and lagoons with significant coastal mangrove forests and seagrass beds, such as the Mekong, Ayeyarwady, and Fly deltas, export large quantities of terrestrially derived carbon to the coastal ocean. The contribution of coastal wetlands has been quantified only for the Fly River delta. Like many large tropical estuaries, the vegetation within the Fly delta, Papua New Guinea, is dominated by mangroves, with community composition governed by salinity regime. The Fly River contributed 1.7 Tg C a^{-1} to the Fly delta and mangroves contributed 0.6 Tg C a^{-1} (Robertson and Alongi 1995). Respiratory losses (1.0 Tg C a^{-1}) and sedimentation on accreting mud banks (0.6 Tg C a^{-1}) left 0.7 Tg C a^{-1} available for export to the adjacent Gulf of Papua. Export of mangrove POC and DOC (0.3 Tg C a^{-1}) equated to 43% of total organic carbon export. Organic carbon being mineralised in offshore sediments 2–5 km from the delta mouth was about 40% terrestrial and about 60% marine, with a clear decrease in terrestrial influence further offshore (Aller and Blair 2008). Robertson and Alongi (1995) noted large quantities of microdetritus floating offshore and sinking to the adjacent deep Coral Sea. However, isotopic signatures in gulf sediments indicate that these amounts are minor compared with marine-derived organic carbon and terrestrial fine POM.

Tropical estuaries and lagoons are sources of CO_2 to the atmosphere, but there is little understanding of the mechanisms and origin of DIC export and of CO_2 outgassing for most river–estuary–ocean margins. This is unfortunate as two global models (Bouillon et al. 2007b; Alongi 2014) of mangrove carbon indicated that a large portion of net mangrove ecosystem production was unaccounted for, most probably lost as DIC from the forest floor via lateral or groundwater transport to adjacent coastal waters. A similar mechanism has been proposed for temperate coastal ecosystems (Cai 2011). This pathway of carbon loss would explain the supersaturated state of coastal waters and rapid rates of CO_2 efflux to the atmosphere, especially in tropical estuarine and lagoonal environments where carbon preservation is low. Maher et al. (2013) confirmed that the 'missing carbon' from mangrove forests is exported as DIC from subsurface forest respiration and groundwater (or porewater) exchange driven by tidal pumping. DIC and DOC were exported from and POC was imported to the mangrove forests of an Australian tidal creek during all tidal cycles; 93–99% of the DIC and 89–92% of the DOC exports were driven by groundwater advection in the creek. DIC export averaged 3 g C m^{-2} d^{-1} and was an order of magnitude higher than DOC export, like the global estimates of the missing mangrove carbon. This finding was confirmed by Taillardat et al. (2018) and Call et al. (2019a,b) who assessed the contribution of porewater discharge in carbon export and CO_2 release from mangrove tidal creeks. In the Can Gio mangrove forest, Vietnam, a clear peak of DIC, DOC, and pCO_2 was observed at low tide, especially during tidal cycles of large amplitude, indicating discharge of mangrove porewater to the adjacent creek (Taillardat et al. 2018, 2019). A mass balance model revealed that the tidal creek was a net exporter of dissolved carbon to adjacent coastal waters, with a significant contribution (38%) coming from DIC derived from porewater discharge. In a macrotidal mangrove creek, Amazonia, tidally driven porewater exchange drove surface water pCO_2 and CH_4 concentrations leading to high concentrations at low tide; peak concentrations occurred after tidal flushing of infrequently inundated high intertidal flats, coinciding with the first flush of stored porewaters (Call et al. 2019a). Emissions of CO_2 and CH_4 from creek water to the atmosphere were among the highest reported from mangrove waters, suggesting that mangroves in other macrotidal estuaries may be a greater source of gaseous carbon emissions than in mangroves in micro- and mesotidal settings.

Tropical estuaries are often a source of CH_4 to the atmosphere, depending on a variety of factors, such as the extent of excess organic pollution and the degree of eutrophication. Oligotrophic estuaries usually emit less CH_4 than eutrophic ones. In the estuaries of the Mekong delta, CO_2 and CH_4 emissions to the atmosphere were high reflecting intense organic matter decomposition due to intense phytoplankton blooms and nutrient enrichment from shrimp farming ponds (Borges et al. 2018). In the organic-rich Indian Sundarbans, total annual CH_4 emissions from the intertidal environments (mostly mangroves) to the atmosphere was estimated to be 10.8 Gg a^{-1} (Dutta et al. 2017b). From the waters of the Sundarbans, 0.9 Gg a^{-1} was released to the atmosphere and 0.5 Gg a^{-1} of CH_4 was exported to the Bay of Bengal. Of the total CH_4 flux to the atmosphere (10.9 Gg a^{-1}), about 85% was photo-oxidised within the atmospheric boundary layer of the Sundarbans.

CH_4 emissions are similarly high in other Indian estuaries, nearly all of which are disturbed to some degree by industrialisation, habitat loss, and eutrophication (Araujo et al. 2018a,b). In the Mandovi–Zuari estuarine system of Goa, waters were supersaturated with CH_4 exhibiting significant temporal and spatial variability. Estuarine sediments were the main source of CH_4 in the Mandovi estuary (Araujo et al. 2018b), with benthic fluxes exhibiting an increasing trend from 4.71 μmol m^{-2} d^{-1} at the marine end to 16.01 μmol m^{-2} d^{-1} in the brackish water zone and 93.90 μmol m^{-2} d^{-1} at the freshwater end, mirroring the patterns in the water column. In both waters and sediments, CH_4 oxidation rates were high with an increasing trend up estuary indicating salinity control on methanogenesis. Similar oversaturation (2221–3819% with respect to atmospheric equilibrium) occurred in several rivers and lagoons along the Ivory Coast, West Africa (Koné et al. 2010), where diffusive air–water CH_4 fluxes were rapid, with largest oversaturation and fluxes from lagoons that were permanently stratified where anoxic bottom waters were favourable for CH_4 production. In coastal lagoons surrounded by mangroves in Yucatán, Mexico, the polluted Chelem Lagoon had higher CH_4 concentrations and fluxes than the relatively pristine Celestún Lagoon (Chuang et al. 2017). In the polluted Guanabara Bay, Brazil, calculated air–sea CH_4 fluxes were well above those of most tropical estuaries and embayments (Cotovicz et al. 2016). In the Lupar and Saribas estuaries in NW Borneo where the catchments are dominated by tropical peatlands, oil palm plantations, and other crops, CH_4 fluxes were very low with CH_4 distribution driven by tides rather than freshwater input (Müller et al. 2016). In the neighbouring Rajang, Maludam, Sebuyau and Simunjan Rivers and estuaries of NW Borneo, CH_4 fluxes across the water–air interface ranged widely over time and space but were rapid, equivalent to 0.1–1% of global riverine and estuarine CH_4 emissions (Bange et al. 2019). In pristine mangrove-dominated estuaries in tropical Australia, seasonal CO_2 and CH_4 emissions varied, driven by groundwater and riverine carbon inputs in the wet season (Rosentreter et al. 2018a). Rosentreter et al. (2018b) advocated that high CH_4 evasion rates as measured in Australian estuaries have the potential to partially offset carbon burial rates in mangrove sediments by an average of 20% using the 20 years global warming potential (Table 10.1).

Monsoons appear to explain the organic carbon dynamics in two contrasting mangrove creeks in Palau, Micronesia, while tidal pumping appears to drive DIC dynamics in both creeks (Call et al. 2019b). The creek located within a semi-enclosed bay exported more particulate and dissolved carbon than the other creek located adjacent to fringing coral reefs. There is thus considerable heterogeneity in mangrove creeks near reefs and subject to weather conditions.

The export and outgassing of DIC, CO_2, and CH_4 have been measured in several different estuaries and all point to tropical estuaries and lagoons as being a source of carbon to the adjacent coastal ocean and the atmosphere (Mukhopadhyay et al. 2002; Biswas et al. 2004; Sarma et al. 2011; Muduli et al. 2012, 2013; Noriega and Araujo 2014; Müller et al. 2015; Samanta et al. 2015; Rao and Sarma 2017; Borges et al. 2018; Dutta et al. 2019; Volta et al. 2020; Akhand et al. 2021). CO_2 fluxes are not always from the estuary to the atmosphere as seasonality of the monsoon affects carbon fluxes (Pattanaik

TABLE 10.1 Mean global CH_4 emissions, C burial rates, and offsets in mangroves, assuming a global mangrove forest area of 83 495 km^2 (Hamilton and Casey 2016), a mean sequestration rate of 227.2 ± 30.67 g C m^{-2} a^{-1}, and a mean global C burial rate (Tg a^{-1}) of 31.3 ± 4.22 (references in Rosentreter et al. 2018b). GWP_{20} = global warming potential at 20 years time frame; GWP_{100} = global warming potential at century timescale. Values are means ± 1 SE.

	CH_4 flux rate (μmol m^{-2} d^{-1})	Global CH_4 emission (Tg CH_4 a^{-1})	Global CH_4 emission (Tg C a^{-1})	CH_4 emission (CO_2 eq.) GWP_{20} (Tg C a^{-1})	CH_4 emission (CO_2 eq.) GWP_{100} (Tg C a^{-1})	Mean offset (%)
Air–water flux	288.0 ± 73.2	0.141 ± 0.036	0.105 ± 0.027	3.29 ± 0.83	1.30 ± 0.33	17.3
Air–sediment flux	391.2 ± 153.4	0.191 ± 0.075	0.143 ± 0.056	4.47 ± 1.76	1.77 ± 0.42	23.6
Global average[a]	339.6 ± 106.2	0.1632 ± 0.05	0.124 ± 0.024	3.89 ± 0.75	1.53 ± 0.29	20.5

[a] = ± scaled error.

Source: Data from Rosentreter et al. (2018b). © John Wiley & Sons.

et al. 2020). For instance, the Hooghly estuary, India, acts as a CO_2 source in the pre-monsoon but acts as a CO_2 sink during the monsoon (Mukhopadhyay et al. 2002). The Hooghly is a heterotrophic turbid estuary and acts on an annual basis as a CO_2 source to the atmosphere. Intra-annual variation of air–water CO_2 fluxes differ between the Dharma and Mahanadi estuaries of the Bay of Bengal, with the Mahanadi acting as a CO_2 source during the monsoon and the Dharma acting as a CO_2 sink during pre-monsoon months and monsoon months (Pattanaik et al. 2020). These differences were attributed to different seasonal salinity, turbidity, and carbon mineralisation patterns.

CO_2 fluxes in the major estuaries flowing into the tropical Atlantic (Table 10.2) show significant differences between periods of low and high river discharge and averaged annual fluxes. The Orinoco, Amazon, São Francisco, Paraiba do Sul, Volta, Niger, and Congo Rivers deliver about 0.1 Pg C a^{-1} of DOC and DIC into the tropical Atlantic Ocean, representing 27.3% of the global DOC and 13.2% of the global DIC delivered by rivers to the global ocean (Araujo et al. 2014). Air–sea CO_2 fluxes (Table 10.2) indicate a slightly higher atmospheric liberation from the African estuaries (10.6 mmol m^{-2} d^{-1}) than from the South American estuaries (5.4 mmol m^{-2} d^{-1}). During high discharge periods, all systems acted as a CO_2 source to the atmosphere (12 mmol m^{-2} d^{-1}), except the Orinoco River estuary. During low discharge periods, the mean CO_2 efflux decreased to 5.2 mmol m^{-2} d^{-1}.

The sink–source debate is complex, and a given parcel of estuarine water can at any given time be either a source or sink for CO_2. The Sundarbans in the Bay of Bengal is a source of CO_2 during the pre-monsoon and early monsoon periods but a sink during the post-monsoon period (Biswas et al. 2004). Although 59% of the emitted CO_2 is removed from the atmosphere by biological processes, on an annual basis, the mangrove forest supplies CO_2 to the atmosphere. Respiration by marine plankton is one of the possible sources for DIC, as is dissolved carbon discharged from mangrove porewater (Dutta et al. 2019).

TABLE 10.2 **Mean CO_2 flux (mmol C m^{-2} d^{-1}) in the estuaries of the major rivers of the tropical Atlantic during low and high river discharge periods and averaged annual fluxes. Positive fluxes indicate that the system acts as a source of CO_2 for the atmosphere and negative fluxes indicate that the estuaries behave as a sink of atmospheric CO_2.**

River	Low discharge	High discharge	Annual average
South American Rivers			
Orinoco	−8.3	−0.4	−0.3
Amazon	−4.2	6.5	0.6
São Francisco	7.3	10.3	6.3
Paraiba do Sul	17.5	16.4	14.8
African Rivers			
Volta	8.8	18.9	11.5
Niger	14.5	20.3	15.3
Congo	0.8	12.2	5.1

Source: Data from Araujo et al. (2014). © John Wiley & Sons.

The sink–source issue can also have a spatial component as discovered by Muduli et al. (2012, 2013) in Chilka Lagoon, India. Partial pressure of CO_2 differs depending on location within the lake with discernible gradients during the monsoon season, while the entire lagoon was a source of carbon to the atmosphere. About 15% of CO_2 efflux from the lake during the monsoon originated from rivers debouching into the lake and the rest was contributed by in situ heterotrophic activity. Analysis of partial pressure of CO_2 and CO_2 flux showed that the northern part of the lake maintained a high CO_2 efflux to the atmosphere associated with the peak monsoon season. Higher bacterial abundance and bacterial respiration during the high flow period suggest rapid organic carbon decomposition. In contrast, the southern part of the lagoon was least affected by river discharge, with low CO_2 flux in the dry period. On average, the air–water CO_2 flux from the entire lagoon is relatively high compared to mean CO_2 emissions from subtropical and tropical estuaries.

Tropical rivers and estuaries are important for outgassing of CO_2 to the atmosphere, but this is not true in every case. For example, the Nyong River in Cameroon sequesters significant amounts of carbon annually. The combined DOC, DIC, and CO_2 fluxes from the Nyong River accounts for only 3% of the input on an annual basis (Brunet et al. 2009), including low CO_2 outgassing comparable to 115% of the annual flux of DOC and 4× greater than the flux of DIC.

Over large spatial scales, tropical estuaries are sources of CO_2 to the atmosphere. The outgassing of CO_2 from an estuary or lagoon is often derived from water column activity in addition to riverine and mangrove inputs (Borges et al. 2018). High rates of bacterial respiration in an estuary result in CO_2 supersaturation (Sarma et al. 2011). Noriega and Arajo (2014) found that in 28 estuarine environments in the north and NE regions of Brazil, all exhibited a net release of CO_2. In addition, a negative correlation between O_2 saturation and partial pressure of CO_2 was observed, suggesting control by biological processes, especially by organic matter decomposition. The range of partial pressures of CO_2 were like those measured in inner estuaries in other parts of the world, except for a few semi-arid estuaries in which record low levels of pCO_2 have been detected (Noriega and Araujo 2014; Leopold et al. 2017). An analysis of biogeochemical fluxes in large tropical rivers and their associated estuaries in South America and Africa (Araujo et al. 2014) showed that air–sea CO_2 fluxes have a slightly higher atmospheric efflux from the African systems than from the South American estuaries. The fluxes remained positive in all systems during the high river discharge periods, except at the mouth of the Orinoco, which continued to act as a CO_2 sink. During the periods of low river discharges, mean CO_2 effluxes decreased.

Carbonate dissolution and export from rivers and estuaries may also contribute to the efflux of CO_2. For example, a significant amount of DIC and dissolved calcium is produced within the Hooghly River estuary at salinity >10, particularly during the monsoon season (Samanta et al. 2015). Based on mass balance calculations and a strong positive correlation between the 'excess' DIC and 'excess' calcium, the probable source of DIC generated within the estuary is carbonate dissolution in conjunction with decomposition of organic carbon. In the Shark River estuary, Florida, $CaCO_3$ dissolution similarly produces DIC (Volta et al. 2020). Riverine inputs of nutrients and sediment to estuaries in Princess Charlotte Bay, Australia, dominate as flood pulses in the wet season, whereas carbonate input is influenced mostly by groundwater discharge (Crosswell et al. 2020). Carbonate input counteracts the minimum buffering zone that would otherwise occur at low salinities, thereby decreasing air–water CO_2 fluxes.

Peatlands in Southeast Asia may contribute to DIC efflux to the tropical ocean. Carbon dynamics in two estuaries surrounded by peatlands in Sarawak, Malaysia, indicated that peat-draining rivers transport terrestrial organic carbon to the estuaries where a large fraction of this dissolved carbon was respired (Müller et al. 2015). Unlike smaller peat-draining estuaries which tend to transport most carbon downstream, estuaries in Sarawak function as an efficient filter for organic carbon, releasing large amounts of CO_2 to the atmosphere.

10.8 Coastal Bays and Continental Shelves

10.8.1 Trophic Dynamics

Trophic dynamics and food webs on tropical continental shelves, including inner shelf areas, such coastal marine bays, and inshore areas, are not well understood although several models have been used to describe shelf food webs. A preliminary trophic model of the continental shelf off SW Gulf of Mexico suggested an imbalance between primary production and consumption, with only about 10% of primary production being consumed in the water column (Manickchand-Heileman et al. 1998). Most of the primary production was exported to the detritus pool, forming the basis of the food web. Benthic invertebrates played a significant role in transferring energy from detritus to higher trophic levels with eight trophic groups distinguished. Analysis of mixed trophic impacts showed that detritus and lower trophic levels had a significant positive impact on other food web components. Seven per cent of NPP was required to sustain current fish catch levels. An ECOPATH model of the coastal ecosystem of Sri Lanka found 39 functional groups representing all trophic levels in the food web (Haputhantri et al. 2008). Simulations were conducted to determine the impact of the 1998 El Niño event on key functional groups. The results showed that the time needed for any impacted functional group to recover to its initial abundance increased with trophic level, but the present level of exploitation of small pelagic fishery resources was not sustainable. Fishing effort simulations show that high fishing intensity by small gillnets contributes to a decline of small pelagic catch. A mass balance trophic model of the Bay of Bengal found the system fully exploited by humans (Ullah et al. 2012). Phytoplankton and detritus were the main food resources; energy transfer from lower trophic groups was found to be high.

Trophic modelling of continental shelves provides realistic scenarios of food web complexity. For instance, an ECOPATH model of the continental shelf off the southern Gulf of Mexico revealed 33 functional groups and a detritus pool (Cruz-Escalona et al. 2007). Barracudas and sharks were the main predators, and detritus provided half of the system's primary energy supply, having significant positive trophic impacts on several functional groups. A constant supply of organic matter via continental discharge reinforced the important role of detritus in the ecosystem. A high degree of omnivory in the food web also formed intricate trophic relationships. According to ecosystem indices, the ecosystem is well organised with flows within compartments greater than flows between them. These compartments include heterogeneous functional groups that are central to the resilience of the ecosystem. The complexity of the food web is responsible for its resilience and stability. An ECOPATH model of the marine food web in the Bitung area, north Sulawesi, Indonesia, as well as stable isotope analyses found that the food web consisted of about 50 functional groups spread over four trophic levels; the base of the food web was composed of phytoplankton, macro/calcareous algae, seagrass, mangroves, and detritus which is the largest pool (Du et al. 2020). The main groups in the food web were groupers, coral trout, rays, butterflyfish, medium-sized pelagics, reef fish, demersal fish, small and large planktivores, macroalgal grazing fish, scraping grazers, detritivorous fish, shrimps, squids, sea cucumbers, crabs, and other invertebrates (Du et al. 2020).

Trophic models have also been used to assess the impact of MPAs on ecosystem health of continental shelves. An ECOPATH model showed that the Banc d'Arguin MPA on the Mauritanian shelf contributed about 9–13% to total consumption, supporting about 23% of the total production, 18% of the total catch, and up to 50% of coastal fish on the Mauritanian shelf (Guénette et al. 2014). Of 29 exploited groups, 15 depend on the MPA for >30% of their direct and indirect consumptions. Fishing pressure increased between 1991 and 2006 leading to a decrease in biomass and the catch of higher trophic levels, confirming their overexploitation. Model simulations showed that adding a new fishing fleet in Banc d'Arguin would have dramatic impacts on the species with a high reliance on the Banc d'Arguin for food, resulting in a 23% decrease in catches outside the MPA.

Tropical continental shelves support diverse fish communities that are highly evolved and complex in their trophic relationships. Simulated changes in the biomass of small pelagic fish in a tropical upwelling ecosystem on the Caribbean coast of Colombia resulted in reallocation of the biomass of higher trophic level organisms, but not of lower trophic level plankton (Duarte and Garcia 2004). This was attributed to bottom-up control exerted by small pelagic fish on pelagic predatory fish. Plankton biomass remained almost unchanged, although plankton was the main food of small pelagic fish. The results indicate that small pelagic fish play an important role in this ecosystem because perturbations in their biomass brought about by fishing propagate through the upper levels of the food web. Guedes and Araújo (2008) studied the diets of five flatfish species in a tropical bay in southeast Brazil to test the hypothesis that resource partitioning along spatial and size dimensions enables coexistence. Two species (*Achirus lineatus* and *Trinectes paulistanus*) showed narrow niche width indicating a specialised feeding strategy. Three other species (*Citharichthy spilopterus, Etropus crossotus* and *Symphurus tessellatus*) showed broader niche width and a generalised feeding strategy. These dietary differences along with spatial and size changes in the use of available resources contribute to allowing these species to coexist.

Niche partitioning has been found on other tropical shelves. In the South China Sea (SCS) off Thailand, fish show either extraordinarily wide diets or narrow diets (Hajisamae 2009). Most species have high food intake with shrimp being the main food (32%) followed by calanoid copepods (17%), fish (13%) and gammarid amphipods (8%). Shelf fish communities were categorised into six different trophic guilds and further subdivided into four categories: piscivore, zooplanktivore, zoobenthivore, and miscellaneous/opportunist, with fish more likely to group according to species rather than size.

Stable isotopes have distinguished food preferences among fish species. There was considerable overlap in the stable carbon signatures between fish, seagrasses, seagrass epiphytes, and macroalgae in Gazi Bay, Kenya (Nyunja et al. 2009). Nevertheless, the signatures for most primary producers were sufficiently distinct to indicate that the dominant carbon sources for these fish were mainly derived from seagrass and their associated epiphytes and possibly macroalgae. Mangrove-derived organic matter contributed only marginally to the fish diets. Carbon supporting these fish communities was derived directly through grazing by herbivorous and some omnivorous fish, or indirectly through the benthic food web. Fish from the mangrove creeks had stable carbon signatures different from those collected from adjacent seagrass beds indicating that these habitats are used as distinct sheltering and feeding zones with minimal degree of exchange within the fish communities.

Pelagic food webs on tropical shelves have been poorly studied in terms of energy flow and trophic interrelationships, but some studies suggest intense trophic interactions, especially in relation to the microbial hub (Wallberg et al. 1999; McManus et al. 2007; Chen et al. 2009; Conroy et al. 2016). Off Zanzibar Island, Tanzania, in situ growth rates of ciliates and heterotrophic nanoflagellates were measured during rainy and dry seasons, with bacterial and phytoplankton activities as well as growth rates of heterotrophic nanoflagellates being significantly greater during the rainy season (Wallberg et al. 1999). There was a shift towards a larger sized community during the wet season. Due to the relatively lower production of larger diatoms, however, heterotrophic microplankton may be relatively more important as a food source for higher trophic levels, such as copepods, during the dry season. Microzooplankton grazing in a tropical upwelling region (e.g. the South Brazil Bight) could account for most phytoplankton mortality (McManus et al. 2007). Both phytoplankton growth and microzooplankton grazing in the bight were higher during the summer upwelling season than in winter. Grazing did not correlate with microzooplankton biomass suggesting that most of the grazing was done by nano-sized zooplankton. Nanoflagellate–bacteria interactions and potential trophic levels were identified within the microbial food web in the oligotrophic SCS (Chen et al. 2009). Removing grazers by filtration relieved grazing pressure on lower trophic levels influencing net growth rates of picoplankton. Size fractionation had limited influence on net growth rates of low DNA heterotrophic bacteria but strongly affected high DNA heterotrophic bacteria. It thus appears that there were five trophic levels corresponding to different size fractions, confirming the existence of multiple trophic levels within the microbial hub.

Trophic linkages between micrograzers and their prey in coastal waters do not appear to differ substantially from those observed in the open ocean. Meso- and microzooplankton grazing in the western tropical North Atlantic Ocean near the Amazon River plume were measured, with mesozooplankton grazing rates highest in plume waters and dominated by smaller size classes (Conroy et al. 2016). Microzooplankton consumed 68% of bulk phytoplankton growth. Comparison of meso- and microzooplankton grazing suggests a transition in food web dynamics from a mesozooplankton-dominated structure in the plume transitioning to a microzooplankton-dominated structure in mesohaline and oceanic waters.

Tropical coastal and shelf phytoplankton communities are usually dominated by nanoplankton and picoplankton except in eutrophic waters near some African and Asian coasts and in upwelling ecosystems where net plankton dominate. Above the massive mud banks off the west and east coasts of India, exceptionally high standing stocks of phytoplankton exist, fuelled by coastal upwelling and resuspension of nutrients from the seabed (Jyothibabu et al. 2018). Seasonal upwelling on tropical shelves can sustain large phytoplankton and zooplankton communities, such as off the west coasts of Mexico and Panama in the tropical eastern Pacific (MacKenzie et al. 2019), where the pelagic ecosystem may have a relatively long trophic chain with inefficient nutrient transfer between low and high trophic levels, although nutrient turnover times can be rapid.

ENSO events cause significant changes in tropical plankton communities. An analysis of phytoplankton communities over a 22 years period in Ilha Grande Bay, SW Atlantic Ocean (Barrera-Alba et al. 2019), found that the strong 1997/98 El Niño led to a shift in dominance from nanoplankton to micro-phytoplankton, especially diatoms, whereas higher nanoplankton abundances were associated with La Niña events. Over the short term, the seasonal rainfall regime influenced the phytoplankton via greater dissolved inorganic nutrient availability, especially NO_3^- and $SiOH_4^+$, but over the long term, ENSO events are responsible for large-scale phytoplankton variability.

10.8.2 Carbon Cycling

The balance between pelagic production and respiration in tropical coastal and shelf waters varies greatly with location, although the tropical ocean is under sampled compared with temperate waters. McKinnon et al. (2017) measured plankton community respiration (CR), NCP, GPP and environmental variables in 14 regions and three ecosystem types (coastal, coral reef, and shelf-open ocean) from Australia, Papua New Guinea, and Indonesia. They found that these regions were strongly autotrophic, although some coastal and oceanic locations were intermittently heterotrophic. The average GPP:CR ratio was 2.14 with Scott Reef, an open ocean system off Western Australia, having the lowest GPP:CR ratio (0.84) to waters of the macrotidal Kimberley coast having the highest GPP:CR ratio (5.21). Temperature was the most important driver of CR in oceanic and coral reef ecosystems, but less so in coastal waters; chlorophyll concentration and sampling depth were more important in regulating GPP than temperature.

Muddy deposits on tropical shelves act as 'incinerators' of sedimentary organic matter; these muds are sites where nearly all organic carbon deposited to the seabed is mineralised. These deposits are usually suboxic (no free oxygen) and non-sulphidic (no free sulphides) as discovered off the Amazon (Aller and Blair 2006) and the Fly delta (Aller et al. 2004, 2008). A combination of tides, wind-driven waves, and coastal currents off the Amazon form massive fluid mud banks and mobile surface sediment layers about 0.5–2 m thick, which are dynamically refluxed and frequently re-oxidised. The seabed functions as a periodically mixed, batch reactor, efficiently re-mineralising organic matter in a gigantic sedimentary incinerator of global importance. Within the Amazon delta, both terrestrial and marine organic matter contribute substantially to early diagenetic remineralisation, although reactive marine substrate dominates. As sedimentary organic carbon is depleted during transit, marine sources become virtually the exclusive substrate for remineralisation, except close to the mangrove shoreline. On the Amazon shelf and in adjacent inshore mobile mud banks, net benthic CO_2 production over the upper 1–2 m of deposits averaged >50 mmol m^{-2} d^{-1} with highest rates during periods of low or falling river flow. Most organic matter was predominantly of planktonic origin (Aller and Blair 2006). Anaerobic metabolism, especially Fe reduction, accounted for most benthic remineralisation. There was a substantial loss of remineralised C to either authigenic sedimentary carbonate formation or non-steady-state diagenesis. In the Amazon-Guianas mobile mud belt, >90% of sedimentary organic carbon was similarly remineralised, with both terrestrial and marine organic matter contributing to early diagenesis (Aller and

Blair 2006). Marine plankton was the primary source of remineralised organic matter within 1 km of shore, as fresh marine material was constantly entrained into mobile deposits and increasingly drives early diagenesis. Comparatively refractory terrestrial organic carbon was lost more slowly but steadily during sedimentary remobilisation and suboxic diagenesis.

Tropical mobile mud belts off the Fly delta in the Gulf of Papua similarly represent a major class of biogeochemical and diagenetic systems characterised by extensive and frequent physical reworking of fine-grained, organic-rich muds underlying oxygenated waters. These muds are typically suboxic and non-sulphidic with remineralisation rates independent of the relative dominance of terrestrial and marine carbon. These conditions reflect coupling between delivery of oxide-rich terrestrial debris, remobilisation, and re-oxidation of deposits, and repetitive entrainment/remineralisation of both labile and refractory organic carbon. Terrestrial C inshore gradually changed to marine C offshore, demonstrating cross-shelf exchange. The sediments of the topset bed off the Fly River processed >50% of sedimentary organic C, functioning like Amazon shelf sediments as a suboxic incinerator. In the water column of the Gulf of Papua, community respiration (CR) in the Fly River plume averaged 712 mg C m^{-2} d^{-1} and NCP averaged 626 mg C m^{-2} d^{-1} with an average P:R ratio of 1.97 indicating net autotrophy. Further offshore, waters were finely metabolically balanced (P:R ratio = 1.27), alternating between autotrophy and heterotrophy (McKinnon et al. 2007). Both pelagic and benthic microbial communities appear to mineralise most of the labile marine and terrestrial organic carbon in the Gulf of Papua.

Detritus from extensive mangrove forests in proximity to both small and large river deltas also contribute greatly to remineralisation in sedimentary deposits on the adjacent continental shelf. Mangrove carbon introduced into the coastal ocean globally accounts for about 11% of the total input of terrestrial carbon into the ocean and 15% of the total carbon accumulating in modern marine sediments (Jennerjahn and Ittekkot 2002). A separate set of calculations underscore the importance of mangrove forests to the tropical coastal ocean (Alongi and Mukhopadhyay 2015). Mangrove forests occupy <2% of the world's coastal ocean area yet they account for about 5% of the net primary production, 12% of ecosystem respiration, and about 30% of carbon burial on all continental margins in subtropical and tropical seas. Globally, mangrove waters release to the atmosphere more than 2.5 times the amount of CO_2 emitted from all other subtropical and tropical coastal waters.

Mangrove outwelling to adjacent shelf areas is a significant source of oceanic exchangeable organic carbon, as DOC. Exchangeable DOC exports to the coastal ocean were equivalent to 11% of total DOC exports (Sippo et al. 2017). Based on previous global DOC export estimates, and exchangeable DOC:DOC ratios, carbon export from mangrove ecosystems is equivalent to 60% of the global exchangeable DOC flux to the coastal ocean (Sippo et al. 2017). Export of particulate mangrove litter contributes disproportionately to carbon burial on tropical shelves. Mangrove litter exported to the adjacent shelf is highly refractory and of poor nutritional quality which appears to be a contributory factor in preventing the establishment of large-sized benthic infauna and in perpetuating the dominance of small-sized, pioneering benthic assemblages. Litter is not widely dispersed but there is often close coupling between mangrove forests and adjacent seagrass beds (but not reefs) in terms of particulate organics.

The dispersal of mangrove litter and the extent of export of mangrove-derived CO_2 depend greatly on geomorphological settings. For example, in Gazi Bay, Kenya, one creek acted as a mangrove litter retention site where export of mangrove material was limited to the contiguous intertidal area, while another creek acted as a 'flow-through' system from which mangrove litter spreads into the bay, especially during the rainy season (Signa et al. 2017). Suspended POM showed the important contribution of mangroves across the entire bay, up to a coral reef, as an effect of the strong ebb tide. Overall, mixing models pointed to a widespread contribution of both allochthonous and autochthonous organic matter sources across Gazi Bay.

Organic carbon budgets of tropical coastal systems point to the importance of a variety of organic carbon sources to coastal bays and adjacent nearshore areas. A carbon mass balance for subtropical Moreton Bay in Queensland, Australia, indicates that outputs exceeded inputs by about 15–20%, with the largest input of carbon being primary production, while respiration and ocean exchange were the largest outputs (Eyre and McKee 2002). There was significant recycling of phytoplankton organic carbon in the bay. An organic carbon budget for mangrove fringed, Sawi Bay in Thailand (Alongi et al. 2000b), identified that mangrove canopy production was the largest source of organic carbon followed by phytoplankton production and import of organic carbon from small creeks lining the bay. The largest losses were pelagic respiration, mangrove respiration, and export from the bay, but carbon burial was limited as noted for Moreton Bay.

On the Great Barrier Reef shelf, organic carbon inputs originate from many sources, including rivers, plankton, benthic microalgae, mangroves, seagrasses, coral reefs, and intermittent upwelling from the Coral Sea. Median rates of pelagic GPP and NPP on the shelf are 1.62 g C m^{-2} d^{-1} and 0.60 g C m^{-2} d^{-1}, respectively, with a median GPP:CR ratio of 1.5 (McKinnon et al. 2013). Rates were more rapid in the wet than in the dry season, with pelagic respiration rates averaging 1.2 g C m^{-2} d^{-1} in the dry season and 1.6 g C m^{-2} d^{-1} in the wet season. Peaks in NPP were associated with floods, intrusions of nutrient-rich, sub-thermocline Coral Sea water, and localised phytoplankton blooms. Respiration rates were comparable to those measured in oligotrophic oceanic waters, but NPP was more typical of shelf ecosystems (McKinnon et al. 2013). River-derived material accounted for 40–80% of the carbon mineralised in the water column and in the seabed (Alongi and McKinnon 2005).

A budget of benthic carbon flow at three inner shelf locations (Figure 10.16) illustrates the relative contributions of mangroves, seagrasses, and plankton detritus, benthic microalgal production and fluvial discharge fuelling various microbial metabolic pathways in these deposits. In Missionary Bay which is lined by extensive mangrove forests (Figure 10.16a), mangrove litter was the largest source of organic carbon, followed by seagrass and phytoplankton detritus and a smaller amount of advected material from offshore or discharge from nearby rivers. Only 8% of total C was buried, with the remainder mineralised by sulphate reducers, iron and manganese reducers, aerobic heterotrophs and denitrifiers. In Rockingham Bay, the contribution from all sources was equivalent, but with proportionally less carbon buried in the sediment (Figure 10.16b); most mineralisation was by sulphate, iron, and manganese reducers. In shelf sediments in the far northern and northern regions of the Great Barrier Reef margin, about 81–94% of organic carbon depositing to the seabed was mineralised and sulphate reduction accounted for about 30% and denitrification for 5% of total C mineralisation; about 25% of ΣCO_2 flux was involved in carbonate mineral formation (Alongi et al. 2007). The relative proportion of phytoplankton production remineralised on the seafloor was in the range of 30–50% which is at the high end of the range found on other shelves. The highly reactive nature of these sediments was attributed to the deposition of high-quality detritus, the shallowness

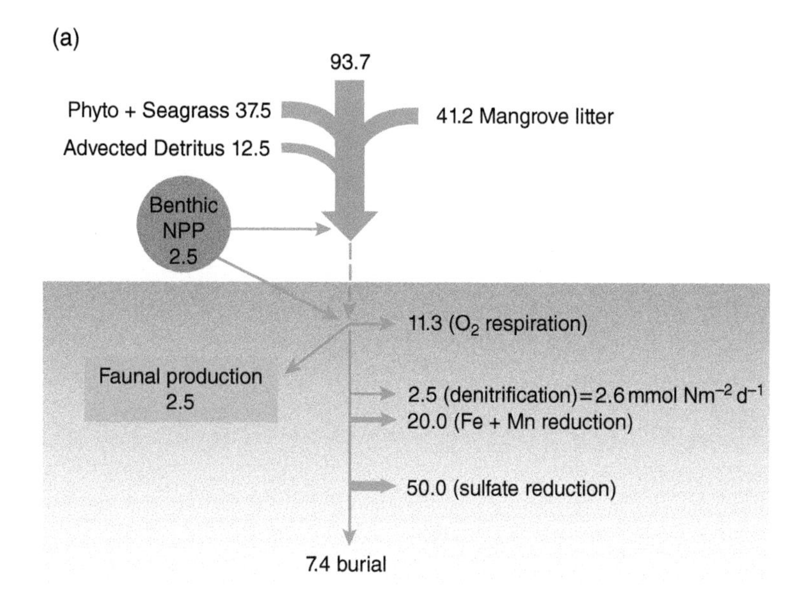

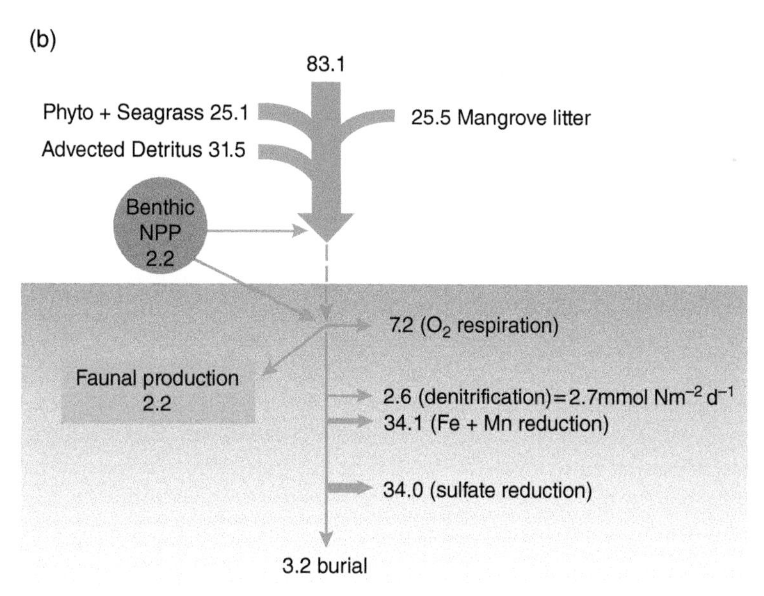

FIGURE 10.16 C budget (mmol C m^{-2} d^{-1}) for inner shelf sediments in the central Great Barrier Reef lagoon at (a) Missionary Bay and (b) Rockingham Bay. *Source*: Alongi and McKinnon (2005), figure 3, p. 243. © Elsevier.

of the shelf, warm temperatures year-round, and a variety of physical disturbances (cyclones, trawling) fostering physicochemical conditions favourable for maintaining rapid rates of benthic metabolism.

On the southern Great Barrier Reef shelf, pelagic and benthic productivity and metabolism are greater than on the central, northern, and far northern sections of the shelf (Alongi et al. 2011, 2015). Bacterioplankton production ranged from 77 to 1643 mg C m^{-2} d^{-1} with specific growth rates (μ) ranging from 0.2 to 5.7 d^{-1}. Net phytoplankton production ranged from 131 to 1785 mg C m^{-2} d^{-1} and pelagic respiration rates ranged from 58 to 1874 mg C m^{-2} d^{-1}. Primary production, respiration, and bacterioplankton production peaked inshore and at the outer shelf margin. These differences were the result of several interrelated drivers that are expressed as strong gradients across the shelf: (i) organic matter quality, which changes from mostly terrestrial plant detritus inshore to increasingly labile reef/oceanic organic material offshore and (ii) strong turbulent mixing driven by tides which change from a large prism inshore to a reduced tidal range offshore. The ultimate driver of pelagic processes is the complex oceanography of the southern shelf. Flux rates of DIC, O_2, and Mn across the sediment–water interface were rapid compared with shelf sediments further north (Alongi et al. 2011). About 20% of remineralised DIC was involved in carbonate dissolution, whereas 10% was involved in authigenic mineral formation on the outer shelf and shelf edge. High tidal energy on the outer shelf leads to increased exposure of high-quality organic matter to microbes and oxygen, fostering rapid rates of aerobic respiration, manganese reduction and carbonate dissolution, supporting the notion that sediments of tropical shelves act as efficient incinerators of organic matter.

The western Indian continental shelf is one of the most productive systems in the global ocean (Pratihary et al. 2014) but differs from many other tropical shelves in that there is significant burial of organic carbon in sediments. This shelf undergoes extreme changes in oxygen regime, being oxic from November to May and suboxic/anoxic from June to October, owing to the biogeochemical response to cyclical monsoonal influences. Benthic oxidation of organic matter under anoxia, lower temperature, and reduced benthic faunal activity appeared to decrease in inner shelf sediments from April to October, suggesting higher C preservation during the late SW monsoon. There was some CH_4 accumulation in shelf waters during seasonal anoxia off Goa (Shirodkar et al. 2018). Low organic loading arising from lower productivity and consequently weak sedimentary methanogenesis may largely account for the comparatively low CH_4 concentrations in anoxic bottom water over the shelf.

Continental shelves of western and eastern boundary upwelling systems are areas where cold and nutrient-rich waters are entrained to the sea surface where they fuel high rates of primary production and rapid rates of organic matter recycling in both waters and sediments. And while they cover only 0.2% of the ocean, these upwelling areas account for at least 50% of global fish catch (Chavez and Messie 2009). Several of these ecosystems are in subtropical and tropical latitudes, such as in the Arabian Sea (including Pakistan, Somalia, Yemen, and the west coast of India), northern Benguela (from the Gulf of Guinea to Namibian coast), off Mauritania in NW Africa and the upwelling ecosystem off Peru. Intermittent, seasonal episodes of upwelling occur in other regions, such as off some islands of Southeast Asia, especially in Indonesia, off Tanzania and off the Great Barrier Reef margin.

The Arabian Sea is one of the most productive regions in the global ocean, driven by upwelling systems along the SW coast of India and off the coasts of Somalia and Arabia during the SW (summer) monsoon season (Sarma et al. 2003). In the eastern Arabian Sea, sulphate reduction was the major pathway of organic matter mineralisation in inner shelf sediments (Naik et al. 2017). These rates were much lower than those reported from other upwelling areas, especially off Namibia and Peru. With an average organic carbon accumulation rate of 5.6 mol C m^{-2} a^{-1}, the bulk of organic matter was likely buried in shelf deposits. Most carbon flow occurs in relation to an extensive oxygen minimum layer. Oxygen profiles and flux measurements suggested that most organic carbon was remineralised within the oxygen minimum zone (OMZ) rather than in slope and deep-sea sediments (Breuer et al. 2009).

The northern Benguela upwelling ecosystem (NBUS) off the west coast of Africa is a distinct region of the larger Benguela upwelling ecosystem in relation to seasonality of wind forcing and properties of upwelling water, productivity, and oxygen status (Emeis et al. 2018). The northern boundary of the NBUS is the Angola Benguela frontal zone at about 15–17 °S, while the southern boundary is the Lüderitz Upwelling Cell at 27 °S which separates the northern and southern sectors of the Benguela upwelling system. Winds in the NBUS are strongest during July–September (winter) and the interaction between the coastal circulation and the wind field causes Ekman transport and curl-driven upwelling along the entire Namibian coast. Upwelling subsides in austral summer when the trade winds are weak. Vertical POC fluxes (1.4–3.7 mmol C m^{-2} d^{-1}) often vary more temporally than spatially (Osma et al. 2014).

A carbon flow model (Heymans and Baird 2000) indicated that phytoplankton were the main primary producer with intense relations with the microbial hub. Primary consumers include mesozooplankton, macrozooplankton, and jellyfish, while the fish community consisted of anchovy, pilchard, lanternfish, gobies, horse mackerel, hake, snoek, benthic feeding fish, and other carnivorous fish. The top consumers were seabirds and seals. Jellyfish were important due to their high biomass and consumption of about 10% of primary production. Organic carbon produced by phytoplankton transferred through the food web with an efficiency of 10–20% per trophic level, with the remaining 80–90% released as CO_2, DOC, or POC. Zooplankton and fish contributed strongly to the carbon flux from the euphotic zone to deeper waters. Calanoid copepods such as *Calanoides carinatus* and *Nannocalanus minor* contributed 83 and 5%, respectively, to total consumption by calanoid copepods with the former species capable of locally removing up to 90% of diatom biomass (Schukat et al. 2013). Dead and moribund organisms, faecal pellets, moults of crustaceans, and blooms of jellyfish and thaliaceans occasionally cause major carbon export (Ekau et al. 2018).

Over the past four decades, the northern Benguela ecosystem has changed dramatically and not for the better as the ecosystem has been overfished (Cury and Shannon 2004; Heymans et al. 2004). In the 1970s, the system sustained high catches and had large populations of a few planktivorous fish. In the 1980s, planktivorous fish species expanded with high catches, although anchovy and sardine biomass was reduced. By the 1990s, the system was severely stressed, catches were much lower and omnivory was reduced. Some of these changes were caused by the Benguela Niño that occurred in 1995. In the 2000s, the ecosystem shifted to a new regime in which jellyfish became extremely abundant; off Namibia, the jellyfish, *Chrysaora hysoscella* and *Aequorea aequorea*, reached abundances such that fishing practices have been disrupted (Cury and Shannon 2004). The low biomass of sardine and other small pelagic fish may resulted in underused phytoplankton production leading to low oxygen conditions which in turn were unfavourable for sardine spawning. This may explain why the northern Benguela ecosystem is now in a low biomass state. The NBUS is nevertheless a significant net CO_2 source (14.8 Tg C a^{-1}) to the atmosphere (Emeis et al. 2018).

The Peruvian upwelling margin forms part of the boundary current system of the eastern tropical South Pacific and is one of the most productive regions in the world (Pennington et al. 2006). In subsurface waters, rapid pelagic respiration of organic matter led to the development of an extensive and perennial OMZ, with dissolved O_2 concentrations being below detection limits from the shelf down to 400 m (Bohlen et al. 2011). Primary production usually ranges from 73 to 122 mmol m^{-2} d^{-1} and POC flux to the benthos varies from 3 to 80 mmol m^{-2} d^{-1} with the highest rates on the inner shelf and the lowest rates in and below the OMZ (Dale et al. 2015). Sediment organic carbon content can be >15% indicating that much of the organic carbon deposited to the seabed was preserved. Mass sediment accumulation rates ranged from 768 to 1800 g m^{-2} a^{-1} on the inner shelf to 128–600 g m^{-2} a^{-1} on the outer shelf to 44–370 g m^{-2} a^{-1} in the OMZ to 259–540 g m^{-2} a^{-1} below the OMZ. POC accumulation rates varied from 29 to 60 g C m^{-2} a^{-1} on the inner shelf to 11–41 g C m^{-2} a^{-1} on the outer shelf to 2–46 g C m^{-2} a^{-1} in the OMZ to 10–32 g C m^{-2} a^{-1} below the OMZ. Rates of sediment and organic carbon accumulation therefore decrease from the inner shelf to offshore. Carbon burial efficiency in sediments was ordinarily high, ranging from 17 to 47% of POC flux on the inner shelf, to 24–74% on the outer shelf to 19–71% in the OMZ to 46–81% below the OMZ. DIC fluxes across the sediment–water interface follow the same across shelf pattern, with rates ranging from 8 to 66 mmol m^{-2} d^{-1} on the inner shelf to 3–20 mmol m^{-2} d^{-1} on the outer shelf to 2–8 mmol m^{-2} d^{-1} in the OMZ to 1–3 mmol m^{-2} d^{-1} below the OMZ (Dale et al. 2015). These carbon flux rates are rapid compared to non-upwelling margins and equivalent to those measured in other upwelling systems, reflecting the high rates of primary productivity.

The Peruvian upwelling ecosystem was well known for its high fish yields until a decline in the mid-twentieth century. Landings of anchovy peaked in the late 1960s ($>10 \times 10^6$ Mt a^{-1}) declining to lows of about 2.8×10^6 Mt a^{-1} between 1983 and 1985 due to overfishing and El Niño events (Nixon and Thomas 2001). The Peruvian anchovy, *Engraulis ringens*, is still the most ecologically and economically important pelagic fish species in the ecosystem and the main prey for marine mammals, seabirds, fish, and fishers. A reanalysis of the trophodynamics of the position of the anchovy in the Peruvian food web showed that whereas phytoplankton largely dominated anchovy diets comprising >99% of ingested prey items, the carbon content of prey items indicated that zooplankton was by far the most important dietary component, with euphausiids contributing 68% of dietary carbon followed by copepods (26%); anchovies feed mainly during daylight (Espinoza and Bertrand 2008).

Off NW Africa, the Mauritanian upwelling ecosystem is one of the four major eastern boundary upwelling systems of the world ocean. The pelagic and benthic food webs of the Mauritanian shelf are correspondingly rich, with small pelagic fish yields exceeding 500 000 t a^{-1}, composed of the European anchovy (*Engraulis encrasicolus*), European sardine (*Sardina pilchardus*), round sardinella (*Sardinella aurita*), flat sardinella (*Sardinella maderensis*), West-African horse mackerel (*Trachurus trecae*) and jack mackerel (*Caranx rhonchus*), European horse mackerel (*Trachurus trachurus*) and Atlantic mackerel (*Scomber japonicus*). Micronekton communities form a strong link in mesopelagic food webs (Czudaj et al. 2020), but there are significant regional differences in the food web structure and vertical trophic interactions due to geographic differences in oxygen availability. Within the low-oxygen upwelling zone, isotopic tracers suggested enhanced competition with an increased importance of food from lower trophic levels (non-crustacean and/or gelatinous prey) for fish compared with equatorial waters further north. Connectivity between offshore upwelling and inshore food webs was weak with the inshore food webs supported mainly by benthic primary production in the Banc d'Arguin (Carlier et al. 2015).

Like the other upwelling centres, upwelled waters off NW Africa are autotrophic with rapid rates of NCP and pelagic respiration ranging from 20.3–1.5 to 1.8–5.7 g C m^{-2} d^{-1}, respectively (Kitidis et al. 2014). Benthic aerobic respiration was highest on the inner shelf (0.4–1.0 g C m^{-2} d^{-1}) with lower rates on the mid-shelf (0.1–0.4 g C m^{-2} d^{-1}) and at the shelf break (0.1–0.3 g C m^{-2} d^{-1}) reflecting high abundances (mean = 8743 ind. m^{-2}; range = 3010–16 380 ind. m^{-2}) and standing biomass (mean

= 16.8 g AFDW m^{-2}; range = 6.2–42.1 g AFDW m^{-2}) of benthic infauna with highest abundances and biomass on the inner shelf (Duineveld et al. 1993). This region is a regional hotspot of CH_4 emissions, with seasonally adjusted rates in the range of 1.6–2.9 Gg CH_4 a^{-1} (Kock et al. 2008).

Upwelling and river discharge throughout the islands of Indonesia and Timor Leste result in enhanced responses by benthic microbial communities (Alongi et al. 2013). The southern shelf of the island of Timor Leste receives upwelled water from the Indonesian Throughflow during the winter SE monsoon season. Subsurface upwelling occurs west of 126°25′E and surface upwelling occurs at the far eastern end of the shelf. As reflected by high concentrations of sediment chlorophyll *a* and phaeopigment, deposition of high-quality POC was sufficient to fuel rapid rates of benthic O_2 consumption (90–142 mmol m^{-2} d^{-1}) and DIC release (108–149 mmol m^{-2} d^{-1}) across the sediment–water interface; DIC and NH_4^+ production from incubated sediments ranged from 95–142 to 13–20 mmol m^{-2} d^{-1}, respectively (Alongi et al. 2013). Molar ratios of DIC:NH_4^+ (range = 6.6–7.7) were lower in fine-grained sediments under the subsurface upwelling regime than in sandy, possibly scoured, sediments under surface upwelling (range = 11.9–21.2) where there was no evidence of benthic enrichment. These zones of high biological activity attract and support large populations of tuna and cetaceans that have been fished for centuries along this portion of the shelf.

East of Timor Leste lies the island of New Guinea, the SW coast of which delivers massive sediment discharge (427×10^6 t a^{-1}) onto a wide shelf in the Aru Sea (Alongi et al. 2012). This region is a rich fishing ground with total catch exceeding 220 000 t a^{-1}, mostly in areas enriched by coastal upwelling from the adjacent Banda Sea. Early diagenesis in shelf deposits was dominated by Fe and Mn reduction, with rates of O_2 uptake (9–26 mmol m^{-2} d^{-1}) and DIC release (6–29 mmol m^{-2} d^{-1}) across the sediment–water interface and depth-integrated production rates of DIC and NH_4^+ highest where upwelling enriches phytoplankton and fisheries production. Sedimentary organic matter was reactive as evidenced by rapid decomposition rates. Massive discharge of highly weathered debris, physical disturbance, rapid deposition of vascular plant detritus inshore, and extensive trawling foster oxic and suboxic diagenesis under non-steady-state conditions. Rates of C (14–48 mmol m^{-2} d^{-1}) and N (1.4–4.2 mmol m^{-2} d^{-1}) mineralisation were equivalent to rates measured on other tropical shelves, further supporting the notion that tropical margins are efficient incinerators of organic matter, even in the Aru Sea where organic matter recycling due to upwelling is decoupled from that resulting from massive river inputs.

In the OMZ on the continental shelf off Guatemala (eastern tropical North Pacific), large CH_4 pools exist in a 300 m-thick layer. Shelf sediments are a source of CH_4 to the overlying OMZ, in which rates of methanogenesis averaged 262 nmol m^{-2} d^{-1}, peaking at 1007 nmol m^{-2} d^{-1} at 500 m and slowing to 162 nmol m^{-2} d^{-1} at 300 m (Chronopoulou et al. 2017). CH_4 efflux was greater in sediments from locations where the OMZ intersected the shelf compared with those with oxygenated bottom waters. There is potential for both aerobic and anaerobic CH_4 oxidation in the waters within and above the OMZ potentially controlling the release of CH_4 emissions from the OMZ. On tropical shelves with no discernible OMZ, such as in the SCS and the Western Philippine Sea (WPS), the water column was also a source of CH_4 to the atmosphere (Tseng et al. 2017). The SCS and WPS emit CH_4 to the atmosphere at mean rates of 8.6 and 4.9 μmol m^{-2} d^{-1}, respectively, with higher emissions over the continental shelf (11.0 μmol m^{-2} d^{-1}) than over the deep ocean (6.1 μmol m^{-2} d^{-1}) due to greater biological productivity and closer benthic–pelagic coupling on the shelf.

The SCS emits on average 30.1×10^6 mol CH_4 d^{-1} to the atmosphere and exports an average of 1.82×10^6 mol CH_4 d^{-1} to the WPS. Along the eastern continental shelf of India in the Bay of Bengal, CH_4 emissions across the water–air interface ranged widely from 13.6 to 248.5 μmol m^{-2} d^{-1} with a grand mean flux of 38.9 μmol m^{-2} d^{-1} in the wet season, over 35× greater than fluxes (1.1 μmol m^{-2} d^{-1}) in the dry season (Rao and Sarma 2017). These rapid rates were attributed to the immense discharge of freshwater from the Ganges, Godavari, Krishna, and Mahanadi Rivers (22 839 m^3 s^{-1}); these systems all release CH_4 to the atmosphere at rates ranging from 12 to 1312 μmol m^{-2} d^{-1}. In contrast, shelf waters off western India in the Arabian Sea release CH_4 to the atmosphere at much lower rates (1.37 μmol m^{-2} d^{-1} in the dry season and 38.9 μmol m^{-2} d^{-1} in the wet season). From a whole-shelf perspective, the eastern shelf in the Bay of Bengal releases much more CH_4 (0.3×10^9 g a^{-1} in the dry season and 15.7×10^9 g a^{-1} in the wet season) than the western shelf in the Arabian Sea (0.2×10^9 g a^{-1} in the dry season and 1.6×10^9 g a^{-1} in the wet season).

Like tropical estuaries and coastal lagoons, continental shelves at low latitudes are sources of CO_2 to the atmosphere, unlike shelves of high latitudes which are sinks for atmospheric CO_2 (Cai et al. 2006; Roobaert et al. 2019). The sources of CO_2 on tropical shelves are high rates of microbial respiration driven by terrestrial organic carbon input, export of respired CO_2 from estuaries and CO_2 release during carbonate formation. Several studies have measured CO_2 dynamics on tropical margins (Chen and Borges 2009; Cai 2011; Lefèvre et al. 2017; Mayer et al. 2018; Robbins et al. 2018; Hung et al. 2019; Crosswell et al. 2020). Cai et al. (2006) estimated that CO_2 emissions from low-latitude shelves to the atmosphere is ≈ 0.11 Pg C a^{-1}, while shelves of mid to high latitude are a CO_2 sink for ≈ 0.33 Pg C a^{-1}. The latitudinal contrast in the direction of shelf air–sea fluxes may be associated with differences in temperatures, the extent of freshwater input to shelves, and the quality and quantity of terrestrial inputs of organic carbon. Organic carbon inputs from 0 to 30° low-latitude non-upwelling shelves accounts for 60% of the global inputs to shelf zones. Cai et al. (2006) state that "it has been believed that most of the terrestrial organic carbon is decomposed in continental margins. If only ½ of the terrestrial organic carbon supplied to the low latitudes is decomposed in the shelf, it is sufficient to support the CO_2 release there."

Not all tropical continental margins are sources of CO_2 to the atmosphere. Guanabara Bay off Rio de Janeiro is a CO_2 sink due to strong eutrophication (Cotovicz et al. 2015). Partial pressure of CO_2 in the water mirrors the trends in dissolved oxygen and chlorophyll *a* indicating that eutrophic conditions drive CO_2 fluxes. In contrast to estuaries and coastal embayments worldwide, the sequestration of carbon in Guanabara Bay was enhanced by strong radiation intensity, thermal stratification, and high availability of nutrients, which often promotes phytoplankton development, net autotrophy, and carbon deposition in sediments. A similar scenario with a strong undersaturation of surface water CO_2 has been repeatedly observed in the Amazon River plume (Körtzinger 2003; Cooley and Yager 2006). The dramatic change of the CO_2 saturation state from highly supersaturated river waters to undersaturated plume waters was probably caused by a combination of CO_2 outgassing from river water, mixing between river and ocean water, and strong biological carbon drawdown in the plume. There was biological drawdown due to significant production of the N_2-fixing cyanobacteria, *Trichodesmium* spp. and *Richelia intracellularis* (Cooley and Yager 2006). Physical dilution by river water also dominated DIC inventories, driving CO_2 partial pressure well below atmospheric levels. If other large river plumes are similarly strong sinks of CO_2, the amount of CO_2 that is released from tropical shelves may be lower than previously suggested.

Some tropical shelves oscillate spatially and temporally as CO_2 sinks or sources. On the Sunda shelf (Gulf of Thailand, Malacca Straits, and Java Sea), there is significant temporal variability in air–sea CO_2 fluxes (Mayer et al. 2018). The Gulf of Thailand changes from an atmospheric CO_2 sink in winter to a source in summer due to higher water temperatures, whereas other subregions as well as the entire Sunda shelf act as a continuous source of CO_2 to the atmosphere. However, with rising atmospheric CO_2 concentrations and seawater temperatures, climate simulations indicated that the entire Sunda shelf may turn into a permanent sink for atmospheric CO_2 within the next 30–35 years (Mayer et al. 2018). The Kaoping Submarine Canyon off SW Taiwan is a CO_2 sink and CO_2 source further offshore, especially in summer (Hung et al. 2019). This difference was attributed to different controlling factors, such as small-scale variations in temperature, nutrients, and chlorophyll *a*. Such variability has been observed on the west Florida shelf where shelf waters fluctuate between being a weak source to a weak sink of carbon (Robbins et al. 2018). Spatially, nearshore waters are a year-round source of CO_2,while waters associated with the 40–200 m isobath in the northern and southern regions of the shelf are weak sinks, except for summer when they act as sources Along a mangrove–seagrass transect in a tropical bay on Bali, Indonesia, embayment waters surrounded by mangroves released CO_2 to the atmosphere at an average rate of 18 mmol m^{-2} d^{-1} (Macklin et al. 2019). The adjacent seagrass beds released CO_2 at lower rates (3 mmol m^{-2} d^{-1}) although in some instances, the seagrasses were a CO_2 sink. A CO_2 source in the mangrove-dominated upper bay was delayed groundwater inputs, and a shifting CO_2 source-sink in the lower bay was driven by CO_2 uptake by seagrass and mixing with oceanic waters.

The entire intertidal zone of the southern SCS plays an important role in the dynamics of CO_2 in the region. Yusup et al. (2018) measured CO_2 flux in intertidal zones using the eddy covariance method and found that these intertidal areas take up CO_2 at an annual average of 2.1 mol C m^{-2} a^{-1}. The role of these zones as a CO_2 sink decreased by 60% during the SW monsoon and autumn transitional monsoon phases due to increased precipitation.

10.9 Open Ocean

Food webs inhabiting the tropical open ocean are intricate, composed mostly by nano- and pico-size classes of autotrophic and heterotrophic plankton, including small sizes of zooplankton, and larger size classes of zooplankton, fish, and mammals (Longhurst 2007). While tropical open ocean organisms have been well studied, actual linkages between and among various trophic groups has not received much attention. For example, it is well known that phytoplankton are consumed by a complex web of microbes and zooplankton, but how these organisms such as those of the microbial hub relate to larger zooplankton and fish is not well known.

Picophytoplankton (cyanobacteria and pico-eukaryotes), nanophytoplankton (diatoms, dinoflagellates, haptophytes, and pelagophytes), and microphytoplankton (diatoms, dinoflagellates, and filamentous cyanobacteria) compose the primary level of the oceanic food web, whereas nanozooplankton (heterotrophic flagellates and small ciliates), microzooplankton (radiolarians, foraminiferans, tintinnids, and larval copepods), and mesozooplankton (copepods, chaetognaths, larvaceans, ostracods, doliolids, and fish larvae) comprise the secondary level. The higher trophic levels are comprised of the macrozooplankton (pteropods, heteropods, siphonophores, jellyfish, and salps), micronekton (small fish, amphipods, cephalopods, and shrimp), and nekton (larger fish and mammals). Grazing is ordinarily intense in tropical open ocean food webs. Members of the microbial hub are fed upon by the microzooplankton and mesozooplankton, which in turn are fed upon by macrozooplankton, micronekton, and nekton. Richardson et al. (2004) and Landry et al. (2011) studied the trophic dynamics of plankton in the eastern equatorial Pacific and found that (i) primary production of the larger phytoplankton size classes was most often dominated by the prymnesiophytes and pelagophytes and not by diatoms, (ii) 70% of primary production was utilised by protistan herbivores with 30% consumed by mesozooplankton, and (iii) some loss of production was expected as export of sinking phytoplankton cells and aggregates, but was likely only a small percentage of daily production due to the apparent balance between production and grazing.

Primary productivity and chlorophyll biomass in the equatorial Pacific are greatly affected by grazing and nutrient dynamics. This oceanic environment is oligotrophic due to permanent stratification of the water column that limits input of nutrients to the photic zone from deeper water; here, mesoplankton grazing removed only 1–9% d^{-1} of the biomass and 1–12% of the primary production within the euphotic zone (Dam et al. 1995), but budget calculations suggested that production and grazing were in balance. Most of the phytoplankton were <2 μm in size and presumably too small to be consumed by mesozooplankton. However, of the >2 μm phytoplankton, the mesozooplankton removed about 27% d^{-1} of the >2 μm standing stock. Mesoplankton grazing can balance production of the >2 μm phytoplankton during periods of El Niño. Dam et al. (1995) inferred that (i) >80% of the carbon ingested by mesozooplankton was not phytoplankton, (ii) mesozooplankton excretion supported <7% of the N demands of phytoplankton, (iii) the flux of carbon passing through the mesozooplankton was equivalent to 23% of the primary production, (iv) mesozooplankton faecal carbon accounted for the sinking POC flux, and (v) a significant fraction of the microzooplankton probably passed through the mesozooplankton.

There are, however, latitudinal variations in mesozooplankton grazing and metabolism in the Central Pacific. Mesozooplankton grazing is higher in equatorial regions (5 °S–5 °N) than at higher latitudes (5–12 °S, 5–12°N), although there are significant differences between normal and El Niño conditions (Dam et al. 1995). During El Niño, mesozooplankton grazing is higher but biomass is lower. Mesozooplankton grazing was equivalent to an average daily removal of 2–3% of the total chlorophyll standing stock in the euphotic zone. However, mesozooplankton grazing was equivalent to daily removal of 36–47% of the >5 μm chlorophyll standing stock. Mesozooplankton biomass and rates of ingestion, egestion, and production exhibited an increase downstream of the upwelling centre as the zooplankton populations had a delayed response to enhanced phytoplankton production (Roman et al. 2002). Zooplankton grazing on phytoplankton is low in the equatorial Pacific Ocean, with the zooplankton consisting of omnivores that consume microzooplankton and detritus. Zooplankton growth rates in the equatorial upwelling zone are high, thus the response of equatorial zooplankton to increases in phytoplankton production is more rapid than normally occurs in subtropical and temperate waters.

Meso- and microzooplankton grazing in the Amazon River plume and the western tropical Atlantic varies with the strength of the Amazon outflow and the fate of the plume (Conroy et al. 2016). Mesozooplankton grazing is highest in plume-influenced surface waters and usually dominated by smaller size classes.

Microzooplankton grazing equates to about 68% of phytoplankton growth. A comparison of micro- and mesozooplankton grazing suggests a transition in food web dynamics from a mesozooplankton-dominated structure in the plume transitioning to a microzooplankton-dominated structure at mesohaline and oceanic waters. Seasonality is induced by changes in river discharge with low mesozooplankton grazing during spring peak discharge followed by higher grazing rates during low river discharge in autumn.

Bacterioplankton abundance and production are also affected by the complex physics and chemistry of water motion. In the NE equatorial Pacific, surface water convergence and divergence have significant impacts of bacterioplankton dynamics (Hyun and Yang 2005). Bacterial abundance, production, and turnover times as well as grazing by heterotrophic nanoflagellates in the well-mixed South Equatorial Current (SEC)/North Equatorial Counter Current (NECC) were consistently higher than in the highly stratified, North Equatorial Current (NEC) area. Greater bacterial production and nanoflagellate grazing in the SEC/NECC region implies that heterotrophic bacteria are a significant trophic link between DOC and higher trophic levels in the microbial hub. Changes in water column structure in relation to convergence and divergence play a key role in enhancing the amount of available nutrients and chlorophyll, thereby stimulating bacterial abundance and production, and intensifying the trophic link between heterotrophic bacteria and heterotrophic protozoa. Trophic links can thus be altered by oceanographic processes.

Microbial dynamics in the equatorial Pacific are greatly affected in regions of divergence and convergence (Brown et al. 2003; Landry et al. 2003; Yang et al. 2004). Phytoplankton biomass is enhanced in the divergence zone, but heterotrophic microbial biomass is patchy across the region. Prokaryotic species (*Prochlorococcus* spp., *Synechococcus* spp., and heterotrophic bacteria) show similar patterns of abundance, with the main feature being their distributional asymmetry south of the equator. Both autotrophic and heterotrophic biomasses are enriched in the convergent zone at 4–5 °N, between the SEC and the NECC. Heterotrophic biomass exceeds phytoplankton biomass in the more nutrient-impoverished waters north and in the branch of a tropical instability wave eddy. The dynamics of *Prochlorococcus* indicates balanced growth and grazing and therefore close control by microzooplankton. About 70% of primary production was utilised by protist herbivores within the microbial community and 30% was consumed by mesoplankton (Landry et al. 2011). Microzooplankton nearly consumed all production of picophytoplankton. Among groups of larger eukaryotic taxa, micro-grazers consumed 51–62% of production, with the remainder consumed by mesozooplankton. Although microbial and zooplankton biomass are generally small, there is rapid recycling and turnover in these plankton communities in the equatorial Pacific. Protistivory, rather than herbivory, is the most important feeding mode for mesozooplankton in waters of the Costa Rica Dome, an open ocean upwelling ecosystem (Stukel et al. 2018). Vertical N export is facilitated by mesozooplankton, primarily through active transport of N consumed in the surface layer and excreted at depth, accounting for 36–46% of total export. Consequently, the ratio of passively sinking particle export to phytoplankton production is low in this upwelling ecosystem.

Like the tropical Pacific and Indian Oceans, the tropical Atlantic is an area of contrasts, between the productive eastern margin, which includes the NW African upwelling that is affected by the Guinea Dome, and the oligotrophic central and western areas, which include unproductive gyres (Armengol et al. 2019). The higher chlorophyll concentrations near the NW African coast result from upwelling caused by the Guinea Dome that supplies abundant nutrients to support high rates of primary productivity. In contrast, low phytoplankton biomass and production is associated with low nutrient supply in the central tropical Atlantic where microzooplankton consume about 80% of primary production.

These regional differences are associated with differences in food web structure. A shorter food web and more efficient energy transfer to upper trophic levels exists in productive areas, whereas a microbially dominated food web exists in oligotrophic areas of the tropical Atlantic (Armengol et al. 2019). Plankton biomass is dominated by heterotrophs, and community respiration and bacterial carbon requirements exceed NPP (Agusti et al. 2001; Armengol et al. 2019) in the most unproductive areas of the tropical Atlantic suggesting a significant allochthonous supply of organic carbon. Potential external sources of N, which appears to be limiting, include advected surplus N and atmospheric deposition. Excess N from the NW African upwelling area and the Guinea Dome can be exported to the deficient areas through the clockwise current system of the Canary and North Equatorial Currents, transporting water south and then west along 10–15 °N, which border the subtropical North Atlantic gyre. Nutrients may be also be exported by Ekman surface drift. Bacterioplankton in the oligotrophic subtropical gyre are fed upon by plastidic (various taxonomic groups including the Prymnesiophyceae and Chrysophyceae) and aplastidic protists at comparable rates (Hartmann et al. 2012). Because of their higher abundance, it is the plastidic protists rather than the aplastidic forms that control bacterivory in these waters. The tightness of biotic carbon cycling can be explained by the presence of mixotrophs since CO_2 fixation as well as predation and respiration are concomitantly performed by the same cells. Such tight intercellular coupling of production and respiration contributes to the stability of oligotrophic ecosystems in the absence of seasonal or episodic disturbances. These findings change our basic understanding of food web function in the open ocean, because plastidic protists should now be considered as the main bacteriovores as well as the main CO_2 fixers in the oligotrophic gyres (Hartmann et al. 2012). Pyrosomes, the largest pelagic colonial tunicates, are important grazers in the tropical eastern Pacific, exhibiting substantial omnivory and carnivory as well as significant grazing on POM in the surface mixed layer, representing an additional, distinct pathway for carbon transfer up the pelagic food web (Décima et al. 2019). This evidence from the tropical Atlantic and Pacific Oceans suggests that mixotrophy is crucial to the functioning of planktonic food webs.

Vertically migrating animals swim to depth in the daytime, presumably to avoid predators, but diel migration in the tropical open ocean is often affected by the depletion of oxygen in subsurface waters. The depth of diel migration follows coherent large-scale patterns (Bianchi et al. 2013), but oxygen concentration is the best single predictor of migration depth at the global scale. In OMZs, migratory animals generally descend as far as the upper margins of the low-oxygen waters. Diel vertical migration and respiration by migrants intensifies oxygen depletion in the upper margin of OMZs (Bianchi et al. 2013). This has severe consequences for food web dynamics and fisheries as the regions of OMZs are currently expanding in the open ocean.

The depth distribution of suitable habitats is important to the ecology of pelagic oceanic fish. Little data exist to characterise how most pelagic species utilise water depths or how oceanographic features affect their distribution. In the eastern tropical Pacific and Atlantic Oceans, large areas of cold hypoxic water occur as distinct strata due to high productivity initiated by intense upwelling and recent increases in global ocean temperatures. This stratum restricts the depth distribution of tropical pelagic marlins, sailfish, and tunas by compressing the available physical habitat into a narrower surface layer (Section 13.7). The surface layer extends downward to a variable boundary defined by a shallow thermocline above a barrier of cold hypoxic water. The cold hypoxic lower boundary is restricted to the eastern tropical Pacific, whereas dissolved oxygen is not limiting in the western tropical North Atlantic. Nevertheless, eastern Atlantic sailfish are larger than those in the western Atlantic. The larger sizes may reflect enhanced foraging opportunities afforded by the closer proximity of predators and prey in a compressed habitat space, as well as by higher productivity driven by upwelling (Prince and Goodyear 2006). However, the narrower band of oxygenated habitat makes the top predator more vulnerable to fishing. The long-term landings of tropical pelagic tunas from areas of habitat compression have been far greater than in areas with oxygenated habitat. Many tropical pelagic species in the Atlantic Ocean are currently fully exploited or overfished and these pressures may increase with the expansion of OMZs.

The spatial and temporal distribution of the two most economically important species of tunas in the tropical Pacific Ocean, yellowfin tuna, *Thunnus albacares*, and bigeye tuna, *Thunnus obesus*, varies with ocean temperature fluctuations that occur during ENSO events (Lu et al. 2001). These ENSO episodes cause large alterations in tuna fishing in the Pacific Ocean. In the western tropical Pacific, warm water is relocated during ENSO events, which leads to the skipjack tuna, *Katsuwonus pelamis*, fishing ground being displaced by over 6000 km (Lima and Naya 2011). The poor fishing success in the eastern Pacific during 1982/1983 was also due an extended El Niño. Relatively high catches of both yellowfin and bigeye tunas have been mostly associated with regions where SSTs increased during El Niño or La Niña years. Conversely, low catches were associated with regions where SSTs fell during La Niña years. The catches varied concurrently or within three months after the onset of an ENSO event, as determined by the Southern Oscillation Index. The parcel of water preferred by bigeye tuna during El Niño periods is extended in western regions of the eastern tropical Pacific and compressed in the eastern regions, displacing bigeye from east to west. Yellowfin tuna appear during La Niña events to undergo a "meridian displacement" extending their preferred range northwards (Lu et al. 2001).

Yellowfin and bigeye tuna, swordfish, striped marlin, dolphinfish, lancetfish and pelagic thresher have been examined for their diets in the eastern tropical Pacific (Moteki et al. 2001; Olson et al. 2014), and it was found that the choice of prey were similar among species. Similarity of diets was, however, low between striped marlin and lancetfish, dolphinfish and pelagic thresher, and lancetfish and pelagic thresher. These seven predator species can be divided into three trophic groups: (i) generalist feeders from surface to midwater (yellowfin tuna and striped marlin), (ii) epipelagic feeders (dolphinfish), and (iii) mesopelagic feeders (bigeye tuna, swordfish, lancetfish and pelagic thresher). The most important prey fish belong to the Sternoptychidae, Phosichthyidae, Paralepididae, Omosudidae, Myctophidae, Exocoetidae, Hemiramphidae, Bramidae, Gempylidae, and Scombridae. Although these predators extensively use these prey fish, different dominant families and feeding depths of the predators are thought to reduce competition for food in the eastern tropical Pacific.

Trophic partitioning of food resources among large pelagic predators also occurs in the western tropical Indian Ocean. The trophic relations of yellowfin and bigeye tuna differed by depth at which the prey was caught (Potier et al. 2004). Crustaceans were the almost exclusive food source of surface-swimming bigeye tuna, with the stomatopod, *Natosquilla investigatoris*, being the sole prey item. The diet of deep-swimming yellowfin tuna was balanced between epipelagic fish, crustaceans, and cephalopods. Bigeye tuna fed predominantly on cephalopods and mesopelagic fish.

Tunas are heavily exploited between November and January in the eastern tropical Atlantic. The fishery in this region consists of small size skipjack (71%), bigeye (15%) and yellowfin tunas (14%) due to the fish aggregating devices used. The main prey of all small tunas was the mesopelagic *Vinciguerria nimbaria* (Photichthyidae) which concentrates in dense schools in upper layers during daylight. The major catch area is the South Sherbro region (0–5 °N, 10–20 °W), located north of the equatorial divergence, where strong upwellings occur during boreal summer (June–September); the area is characterised by tropical instability waves. These oscillations involve the SEC, the NECC, and the Equatorial Under Current and have an important role in seasonal fluctuations in tuna concentrations (Ménard et al. 2000).

Congregations of whales, dolphins, and seabirds are similarly associated with physical oceanographic features. For instance, the blue whale (*Balaenoptera musculus*) is attracted to both Baja California and the Costa Rica Dome with additional sightings made near the Galapagos Islands, the coasts of Ecuador and northern Peru, and off southern Sri Lanka (de Vos et al. 2014) where cool upwelled waters favour high abundances of euphausiids. The sightings off western Baja California occur mostly during the spring peak of upwelling and biological production. The Costa Rica Dome is occupied year-round suggesting either a resident population or that whales from the Northern and Southern hemisphere overlap in their visits (Fiedler et al. 2017).

Mesoscale features of the eastern tropical Pacific (including the Costa Rica Dome) that have significant influences on cetaceans and seabirds include the Equatorial Front and the counter current thermocline ridge (Balance et al. 2006; Sutton et al. 2019). The Costa Rica Dome is significant not only for blue whales, but also short-beaked common dolphins. The Equatorial Front is a feature of significance for planktivorous seabirds and the counter current thermocline ridge is a significant feature for flocking seabirds that associate with mixed-species schools of spotted and spinner dolphins and yellowfin tuna. Some seabirds and cetaceans have species-specific preferences for surface currents, but more common are associations with distinct water masses. Some of the complexity in species–habitat relationship patterns may be accounted for by the interactions between seasonal and interannual variation in oceanographic conditions with seasonal patterns in the biology of seabirds and cetaceans (Balance et al. 2006; Sutton et al. 2019). Seabirds in the eastern tropical Pacific feeding on fish and squids are associated with habitats characterised by deep and strong thermoclines, while planktivorous species prefer habitat characterised by shallower thermoclines with cooler surface temperatures and a less stratified water column. Thus, seabirds in different feeding guilds within the eastern tropical Pacific have clear and distinct habitat preferences. Thermocline topography is a key variable in predicting the distribution and abundance of seabirds in this region, probably due to its influence on the availability of prey (Vilchis et al. 2006).

The feeding habits of seabirds of the eastern tropical Pacific have been deduced, with their diet consisting of 83% fish, 17% cephalopods and an insignificant amount of non-cephalopod invertebrates (Spear et al. 2007). Based on behaviour, the seabirds sorted into two groups: (i) 15 species that fed solitarily and (ii) 15 species that generally fed in multispecies flocks. Diets of the two seabird groups differ: the solitary group feeds mostly at night, mainly on mesopelagic fish and the flocking group feeds primarily on flying fish and squid (*Sthenoteuthis oualaniensis*) caught when feeding diurnally in association with tuna. The avifauna utilises four feeding strategies: (i) in association with surface-feeding fish, namely tuna, (ii) nocturnal feeding on diel, vertically migrating mesopelagic prey, (iii) scavenging dead cephalopods, and (iv) feeding diurnally on non-cephalopod invertebrates (e.g. scyphozoans, molluscs, crustaceans) and fish eggs. Prey partitioning among the avifauna is dramatic, being a function of sex, body size, feeding behaviour, habitat, and species.

Microbial turnover is rapid in tropical open ocean ecosystems and this rapid recycling translates into high rates of carbon flow. The turnover rate of members of the microbial hub is on the order of hours to days, especially in cool upwelled waters of the eastern boundary zones. The high rates of microbial productivity thus match the high rates of primary productivity. However, even in non-upwelled waters, microbial turnover can be rapid compared with rates in open ocean waters of higher latitude, suggesting that factors such as stratification due to continually warm sea surface temperatures and complex water movements due to fronts, eddies, and gyres play a role in controlling turnover rates (Longhurst 2007). Rapid microbial turnover translates into rapid rates of microbial respiration leading in turn to rapid rates of CO_2 flux. In the northern SCS, for example, integrated GPP, NCP, and bacterial R were all higher in the monsoonal summer than in winter, ranging from 851, 435 and 243 mg C m^{-2} d^{-1} to 5032, 10 707 and 7547 mg C m^{-2} d^{-1}, respectively (Hung et al. 2020).

In the upwelling regime of the eastern and central Arabian Sea, a net supply of 25 Tg C a^{-1} was estimated to the upper 1000 m water column by physical pumping of water. Photosynthetic activity (234 Tg C a^{-1}) did not support total C demands (1203 Tg C a^{-1}) by bacteria and microzooplankton and mesozooplankton in surface waters. Most of the carbon demand might be met from internal cycling involving zooplankton grazing/excretion activities, although the demand requires organic carbon nearly double that of the total living biomass production rate (644 Tg C a^{-1}). Sinking of organic carbon from surface waters (69 Tg C a^{-1}) accounted for 30% of the total photosynthetic production rate, indicating intense remineralisation in the surface layers of the Arabian Sea, which emits CO_2 (32 Tg C a^{-1}) to the atmosphere.

A global map of air–sea CO_2 fluxes (Figure 10.17) shows that the tropical open ocean, like tropical continental shelf waters, is either a source of CO_2 to the atmosphere or in equilibrium. As noted earlier, highest effluxes are associated with the eastern boundary currents and with some areas such as the Arabian Sea, mirroring the high rates of biological productivity and the upwelling of waters rich in DIC. Except for some areas of the Southern Ocean and the Bering Sea, waters of temperate and polar latitudes are net sinks for CO_2, especially off the east coasts of Japan and Canada, and some areas of the Arctic Ocean (Figure 10.17a). A global biogeochemical modelling exercise (Lacroix et al. 2020) suggested that large CO_2 outgassing occurs mostly in proximity to major tropical rivers with substantial outgassing further offshore, especially in the tropical Atlantic. The addition of riverine loads of nutrients and carbon in the model resulted in a strong increase in NPP in the tropical west Atlantic, Bay of Bengal, and the East China Sea.

There is nevertheless large spatial and seasonal variability in air–water CO_2 fluxes within any given open ocean region (Figure 10.17b). The SCS is a good example of such complex region, encompassing a large variety of oceanographic and biogeochemical processes involving deep basins, large riverine inputs, and extensive shelf margins. In a detailed study of diffusive air–water CO_2 fluxes, Zhai et al. (2013) sampled four contrasting physical-biogeochemical domains over the entire expanse of the SCS (Figure 10.18): Domain A was adjacent to the Pearl River estuary (near

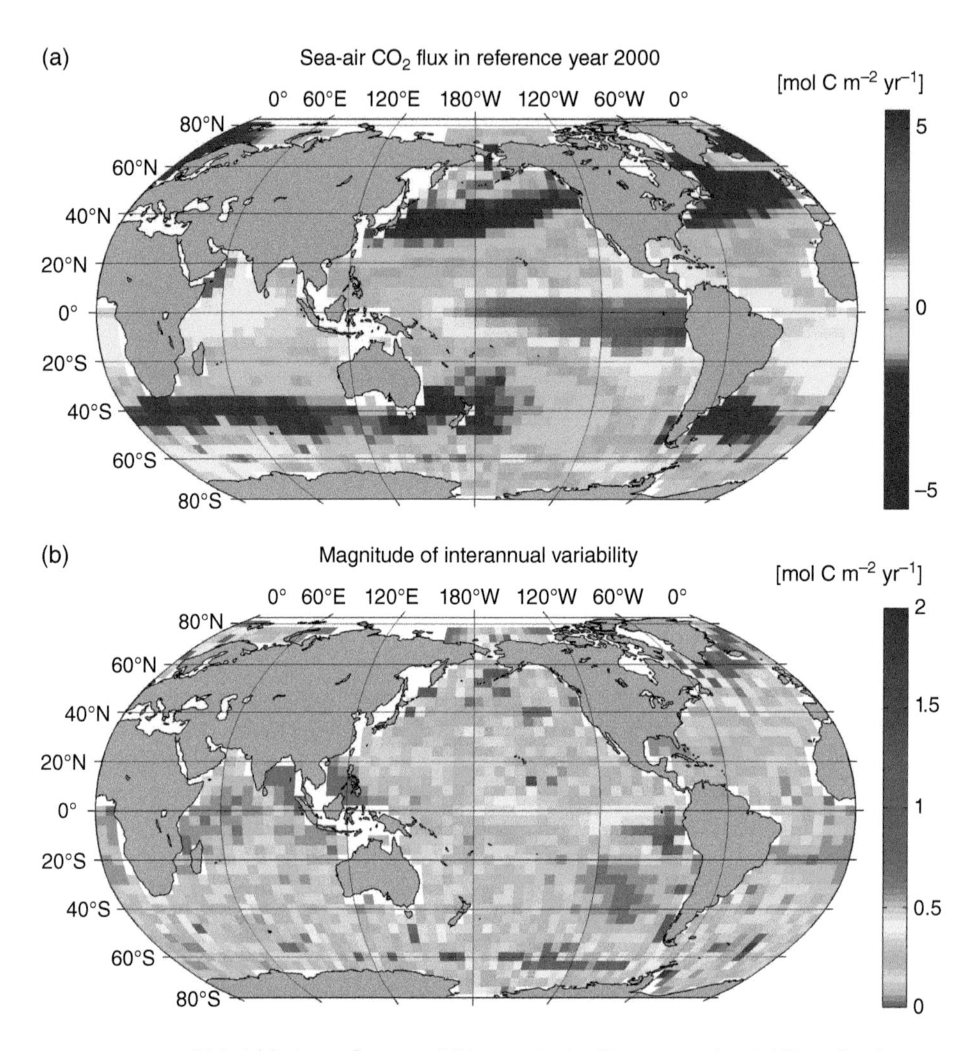

FIGURE 10.17 Global (a) air-sea fluxes and (b) magnitude of interannual variability of carbon dioxide in 2000. Negative values denote fluxes from the atmosphere into the ocean and positive values denote fluxes from the ocean to the atmosphere. Image courtesy of the PMEL Carbon Group, NOAA. Source: Park et al. (2010), figure 5, p. 359. Licensed under CC BY 3.0. Copyright NOAA.

Hong Kong) and was influenced by the summer estuarine plume; Domain B covered the slope and deep basin areas in the northern SCS which is oligotrophic and low productivity; Domain C covered a large portion of the SCS basin and is a tropical oligotrophic environment; and Domain D was located west of the Luzon Strait and was impacted by the Kuroshio Current intrusions which generate various mesoscale eddies and internal waves. In Domain A, the ocean moderately absorbed CO_2 from the atmosphere in winter, while the sea–air CO_2 exchanges were nearly in equilibrium in spring, summer, and autumn; annual exchanges were nearly in equilibrium (−0.44 to −0.51 mol m^{-2} a^{-1}). In Domain B, sea–air exchanges were nearly in equilibrium in winter and autumn, but there was net release to the atmosphere in spring and summer; annual sea–air CO_2 exchange in the domain was estimated at 0.46–0.53 mol m^{-2} a^{-1}. Domain C released CO_2 to the atmosphere throughout the year with annual average sea–air fluxes estimated at 1.37–1.58 mol m^{-2} a^{-1} (Figure 10.18). In Domain

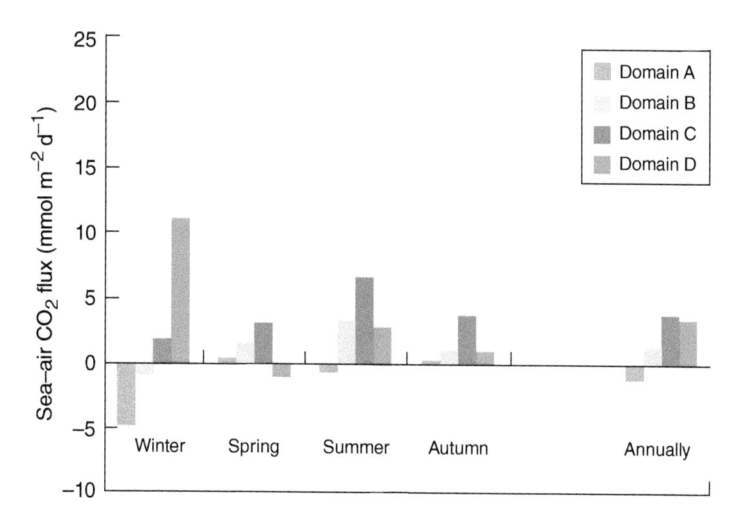

FIGURE 10.18 Mean seasonal and annual variations in estimated air–sea CO_2 fluxes (mmol m^{-2} d^{-1}) in four physical-biogeochemical domains (a–d) in the South China Sea. *Source*: Zhai et al. (2013), figure 8, p. 7786. Licensed under CC BY 3.0. © Copernicus Publications.

D, sea–air fluxes varied from weak CO_2 uptake in spring to weak or moderate CO_2 release in summer and were nearly in equilibrium in autumn; in winter, however, seas within this domain varied very much from a typically weak CO_2 sink in early winter to a significant CO_2 source associated with a monsoon-driven upwelling event (Zhai et al. 2013). Thus, the entire SCS serves as a weak to moderate source of atmospheric CO_2 despite dramatic variations in pCO_2 concentrations over time and space due to factors such as phytoplankton blooms, upwelling and river plume events, eddies, and variations in SSTs and SSSs.

Equatorial upwelling in the Pacific brings up subsurface water with relatively high concentrations of DIC leading to the outgassing of >1 Pg C a^{-1}, making the equatorial Pacific the largest natural source of CO_2 to the atmosphere. The diagnostic model of Park et al. (2010) yielded an interannual variability of ± 0.14 Pg C a^{-1} with a 26 years mean of about 1.48 Pg C a^{-1}. Most of the CO_2 evasion takes place in the upwelling-influenced eastern Pacific, whereas atmospheric and sea surface partial pressures of CO_2 are near equilibrium in the warm pool of the western Pacific. Large changes in the surface area of the upwelling region, which occur due to the ENSO phenomenon, accounts for up to 70% of the interannual variability in the net air–sea flux CO_2 flux. On average, the export of biological production is equal to CO_2 evasion (0.8–1.5 Pg C a^{-1}), but there is less temporal variability due to very slow uptake of new macronutrients in the equatorial ecosystem (Le Borgne et al. 2002).

As in other tropical ecosystems, both the warm pool and upwelling areas of the tropical Pacific are characterised by the dominance of picophytoplankton and a steady-state equilibrium, achieved by the balance between predation and growth. Larger phytoplankton are more abundant in the nutrient-replete waters of the upwelling region with the result that biomass, mean organism size, and export fluxes are greater than in the nutrient-depleted regions of the warm pool. Because of nutrient, particularly Fe, limitation of primary production in the upwelling zone, the difference in export flux of carbon between zones is modest. The upwelling and warm pool zones show an essentially constant flux of carbon to the deep ocean, known as the 'biological pump.' These fluxes vary due to the passage of equatorial Kelvin waves and tropical instability waves in the upwelling region through horizontal advection and possible inputs of nutrients from the deeper layers.

ENSO events play an important role in the seasonal and interannual variations in air–sea CO_2 fluxes. The lowest fluxes were measured in the equatorial Pacific during the 1991/1994 and 1997/1998 El Niño events (Feeley et al. 2002). The low fluxes were the result of the combined effects of interconnected large-scale and locally forced physical processes, such as the development of a low-salinity surface cap as part of the formation of the warm pool in the western and central equatorial Pacific, a deepening of the thermocline by propagating Kelvin waves in the eastern Pacific, and the weakening of the winds in the eastern half of the tropical Pacific. These processes reduced pCO_2 values in the central and eastern equatorial Pacific towards near equilibrium at the peak of the warm ENSO phase. In the western equatorial Pacific, there is usually a small but significant increase in seawater pCO_2 during strong El Niño events, and little or no change during weak El Niño events. The net effect is a lower-than-normal flux of CO_2 (0.2–0.4 Pg C a^{-1}) to the atmosphere during El Niño.

Eastern tropical Atlantic waters are also a source of CO_2 to the atmosphere. However, an unexpected area of low CO_2 concentrations in the South Equatorial Counter Current with pCO_2 values close to equilibrium or even slightly undersaturated have been measured (Lefevre et al. 2008). Conversely, pCO_2 values increase at the beginning of the upwelling season and continue through July to September. The monthly CO_2 fluxes ranged from 1.2 mmol m^{-2} d^{-1} in June to a maximum of 8.4 mmol m^{-2} d^{-1} in October when high pCO_2 values are maintained by the warming of surface water. Most of the pCO_2 distribution can be explained by physical processes and a strong relationship between pCO_2 and sea surface temperature (SST) during the upwelling season. The productive area off the NW African upwelling acts as a source of CO_2 to the atmosphere (0.5 mmol C m^{-2} d^{-1}), driven by high DIC inputs from the thermocline's interior, whereas the less productive areas south and north of the upwelling zone act as net CO_2 sinks (2.5 mmol C m^{-2} d^{-1}). The carbon sink in these unproductive areas may be supported largely by a downward DOC flux, which is partially supported by DOC excretion from phytoplankton cell lysis, which is twice as high as the average upward DIC flux (Agusti et al. 2001). In contrast, off northern Brazil, CO_2 is taken up by the ocean at the Amazon River plume which is advected into the surface circulation of the tropical Atlantic (Körtzinger 2003). A conservative estimate of the plume-related CO_2 sink yielded a net air–sea flux of 0.014 Pg C a^{-1} with a corresponding average CO_2 flux of 1.35 mmol m^{-2} d^{-1} in waters outside the plume but in the opposite direction. The dramatic change of the CO_2 saturation state from highly supersaturated river waters to markedly undersaturated surface waters in the plume can be explained by a combination of CO_2 outgassing from river water, mixing between river and ocean water, and strong biological carbon drawdown in the plume, especially the sinking of C fixed by N_2-fixers in the plume (Subramaniam et al. 2008).

In the tropical Indian Ocean, the largest area of CO_2 efflux is associated with the upwelling and complex oceanography of the Arabian Sea, although most of the southern Indian Ocean is a sink for CO_2 (Figure 10.17). The net annual CO_2 flux for the entire Indian Ocean was calculated as a net uptake of −0.15 Pg C. The relatively small net flux resulted from different surface water pCO_2 distributions and seasonal variations between the northern and southern Indian Ocean. The equatorial and northern regions have pCO_2 values that are generally above atmospheric values. During the SW monsoon, pCO_2 values in the Arabian Sea coastal upwelling region are among the highest recorded in the global ocean. However, the upwelling is seasonal in strength and only affects a relatively small area (Figure 10.17). The equatorial region of the Indian Ocean generally has pCO_2 values slightly above atmospheric values. In the southern Indian Ocean, the strongest undersaturation was observed in austral winter, with summer values reaching near or slightly above atmospheric values. The Indian Ocean is a net overall source of CO_2 (Bates et al. 2006), with an estimated net sea to air CO_2 flux of 237 Tg C a^{-1}. Regionally, the Arabian Sea, Bay of Bengal, and the 10 °N–10 °S zones are perennial sources of CO_2. In the 10–35 °S zone, the CO_2 sink or source status of the surface ocean shifts seasonally, although the region is a net oceanic sink of atmospheric CO_2.

High rates of microbial respiration play a role in determining whether a given parcel of water is a CO_2 sink or source. Protists and other members of the microbial hub are important contributors to total community R (Calbet and Landry 2004). Regionally averaged estimates of protist respiration range from 35 to 43% of daily primary production for the first level of consumer or 49–59% of primary production for three trophic transfers (Calbet and Landry 2004). The estimated contributions of microbial grazers to total community respiration are the same order of magnitude as bacterial respiration. Further, the proportion of primary production consumed by micrograzers varies among marine habitats and regions, ranging from 75% for tropical–subtropical regions to 59% for temperate–polar ecosystems (Calbet and Landry 2004). Microbe-sized grazers thus play a disproportionate role in carbon flow in tropical open ocean ecosystems. This is also reflected in the ratio of open ocean GPP: NCP which is small. Thus, most of the GPP in the tropical open ocean is lost by microbial and possibly zooplankton respiration (del Giorgio and Duarte 2002; Steinberg and Landry 2017).

References

Abrantes, K.G., Barnett, A., and Bouillon, S. (2014). Stable isotope-based community metrics as a tool to identify patterns in food web structure in east African estuaries. *Functional Ecology* 28: 270–282.

Abrantes, K.G., Johnston, R., Connolly, R.M. et al. (2015). Importance of mangrove carbon for aquatic food webs in wet-dry tropical estuaries. *Estuaries and Coasts* 38: 383–399.

Abrantes, K.G. and Sheaves, M. (2010). Importance of freshwater flow in terrestrial-aquatic energetic connectivity in intermittently connected estuaries of tropical Australia. *Marine Biology* 157: 2071–2086.

Acharyya, T., Sarma, V.V.S.S., Sridevi, B. et al. (2012). Reduced river discharge intensifies phytoplankton blooms in Godavari estuary, India. *Marine Chemistry* 132-133: 15–22.

Agusti, S., Duarte, C.M., Vaqué, D. et al. (2001). Food-web structure and elemental (C, N and P) fluxes in the eastern tropical north Atlantic. *Deep-Sea Research II* 48: 2295–2321.

Akhand, A., Watanabe, K., Chanda, A. et al. (2021). Lateral carbon fluxes and CO_2 evasion from a subtropical mangrove-seagrass-coral continuum. *Science of the Total Environment* 752: 142190. https://doi.org/10.1016/j.scitotenv.2020.142190.

Albright, R., Benthuysen, J., Cantin, N. et al. (2015). Coral reef metabolism and carbon chemistry dynamics of a coral reef flat. *Geophysical Research Letters* 42: 3980–3988.

Albright, R., Langdon, C., and Anthony, K.R.N. (2013). Dynamics of seawater carbonate chemistry, production, and calcification of a coral reef flat, central Great Barrier Reef. *Biogeosciences* 10: 6747–6758.

Aller, J.Y., Alongi, D.M., and Aller, R.C. (2008). Biological indicators of sedimentary dynamics in the central Gulf of Papua: seasonal and decadal perspectives. *Journal of Geophysical Research: Earth Science* 113: F01S08. https://doi.org/10.1029/2007JF000823.

Aller, R.C. and Blair, N.E. (2006). Carbon remineralization in the Amazon-Guianas tropical mobile mudbelt: a sedimentary incinerator. *Continental Shelf Research* 26: 2241–2259.
Aller, R.C. and Blair, N.E. (2008). Early diagenetic remineralization of sedimentary organic C in the Gulf of Papua deltaic complex (Papua New Guinea): net loss of terrestrial C and diagenetic fractionation of C isotopes. *Geochimica et Cosmochimica Acta* 68: 1815–1825.
Aller, R.C., Hannides, A., Heilbrun, C. et al. (2004). Coupling of early diagenetic processes and sedimentary dynamics in tropical shelf environments, the Gulf of Papua deltaic complex. *Continental Shelf Research* 24: 2455–2486.
Allgeier, J.E., Yeager, L.A., and Layman, C.A. (2013). Consumers regulate nutrient limitation regimes and primary production in seagrass ecosystems. *Ecology* 94: 521–529.
Al-Maslamani, I., Walton, M.E.M., Kennedy, H. et al. (2012). Sources of primary production supporting food webs in an arid coastal embayment. *Marine Biology* 159: 1753–1762.
Alongi, D.M. (1988). Bacterial productivity and microbial biomass in tropical mangrove sediments. *Microbial Ecology* 15: 59–79.
Alongi, D.M. (1998). *Coastal Ecosystem Processes*. Boca Raton, FL: CRC Press.
Alongi, D.M. (2009). *The Energetics of Mangrove Forests*. Dordrecht, The Netherlands: Springer.
Alongi, D.M. (2014). Carbon cycling and storage in mangrove forests. *Annual Review of Marine Science* 6: 195–220.
Alongi, D.M. (2020). Carbon cycling in the world's mangrove ecosystems revisited: significance of non-steady state diagenesis and subsurface linkages between the forest floor and the coastal ocean. *Forests* 11: 977. https://doi.org/10.3390/f11090977.
Alongi, D.M., Ayukai, T., Brunskill, G.J. et al. (1998). Sources, sinks, and export of organic carbon through a tropical, semi-enclosed delta (Hinchinbrook Channel. Australia). *Mangroves and Salt Marshes* 2: 237–242.
Alongi, D.M., Bouillon, S., Duarte, C. et al. (2014). Carbon and nutrient fluxes across tropical river-coastal boundaries. In: *Biogeochemical Dynamics at Major River-Coastal Interfaces: Linkages with Global Change* (eds. T.S. Bianchi, M.A. Allison and W.-J. Cai), 373–394. Cambridge, UK: Cambridge University Press.
Alongi, D.M., Brinkman, R., Trott, L.A. et al. (2013). Enhanced benthic response to upwelling of the Indonesian Throughflow onto the southern shelf of Timor-Leste, Timor Sea. *Journal of Geophysical Research: Biogeosciences* 118: 158–170.
Alongi, D.M. and McKinnon, A.D. (2005). The cycling and fate of terrestrially derived sediments and nutrients in the coastal zone of the Great Barrier Reef shelf. *Marine Pollution Bulletin* 51: 239–252.
Alongi, D.M. and Mukhopadhyay, S.K. (2015). Contribution of mangroves to coastal carbon cycling in low latitude seas. *Agricultural and Forest Meteorology* 213: 266–272.
Alongi, D.M., Patten, N.L., McKinnon, D. et al. (2015). Phytoplankton, bacterioplankton and virioplankton structure and function across the southern Great Barrier Reef shelf. *Journal of Marine Systems* 142: 25–39.
Alongi, D.M., Tirendi, F., Dixon, P. et al. (1999). Mineralization of organic matter in intertidal sediments of a tropical semi-enclosed delta. *Estuarine, Coastal and Shelf Science* 48: 451–467.
Alongi, D.M., Tirendi, F., Trott, L.A. et al. (2000a). Benthic decomposition rates and pathways in plantations of the mangrove *Rhizophora apiculata* in the Mekong Delta, Vietnam. *Marine Ecology Progress Series* 194: 87–101.
Alongi, D.M., Trott, L.A., and Møhl, M. (2011). Strong tidal currents and labile organic matter stimulate benthic decomposition and carbonate fluxes on the southern Great Barrier Reef. *Continental Shelf Research* 31: 1384–1395.
Alongi, D.M., Trott, L.A., and Pfitzner, J. (2007). Deposition, mineralization, and storage of carbon and nitrogen in sediments of the far northern and northern Great Barrier Reef shelf. *Continental Shelf Research* 27: 2595–2622.
Alongi, D.M., Trott, L.A., Undu, M.C. et al. (2008). Benthic microbial metabolism in seagrass meadows along a carbonate gradient in Sulawesi, Indonesia. *Aquatic Microbial Ecology* 51: 141–152.

Alongi, D.M., Wattayakorn, G., Ayukai, T. et al. (2000b). An organic carbon budget for mangrove fringed Sawi Bay, southern Thailand. *Phuket Marine Biological Center Special Publication* 22: 79–85.

Alongi, D.M., Wirasantosa, S., Wagey, T. et al. (2012). Early diagenetic processes in relation to river discharge and coastal upwelling in the Aru Sea, Indonesia. *Marine Chemistry* 140: 10–23.

Andrades, R., Gomes, M.P., Pereira-Filho, G.H. et al. (2014). The influence of allochthonous macroalgae on the fish communities of tropical sandy beaches. *Estuarine, Coastal and Shelf Science* 144: 75–81.

Anton, A., Baldry, K., Coker, D.J. et al. (2020). Drivers of the low metabolic rates of seagrass meadows in the Red Sea. *Frontiers in Marine Science* 7: 69. https://doi.org/10.3389/fmars.2020.00069.

Araujo, J., Naqvi, S.W.A., Naik, H. et al. (2018a). Biogeochemistry of methane in a tropical monsoonal estuarine system along the west coast of India. *Estuarine, Coastal and Shelf Science* 207: 435–443.

Araujo, J., Pratihary, A., Naik, R. et al. (2018b). Benthic fluxes of methane along the salinity gradient of a tropical monsoonal estuary: implications for CH_4 supersaturation and emission. *Marine Chemistry* 202: 73–85.

Araujo, M., Noriega, C., and Lefèvre, N. (2014). Nutrients and carbon fluxes in the estuaries of major rivers flowing into the tropical Atlantic. *Frontiers in Marine Science* 1: 10. https://doi.org/https://doi.org/10.3389/fmars.2014.00010.

Arbi, I., Liu, S., Zhang, J. et al. (2018). Detection of terrigenous and marine organic matter flow into a eutrophic semi-enclosed bay by $\delta^{13}C$ and $\delta^{14}N$ of intertidal macrobenthos and basal food sources. *Science of the Total Environment* 613-614: 847–860.

Arias-González, J.E., Delesalle, B., Salvat, B. et al. (1997). Trophic functioning of the Tiahura reef sector, Moorea island, French Polynesia. *Coral Reefs* 16: 231–246.

Armengol, L., Calbet, A., Franchy, G. et al. (2019). Planktonic food web structure and trophic transfer efficiency along a productivity gradient in the tropical and subtropical Atlantic Ocean. *Scientific Reports* 9: 2044. https://doi.org/10.1038/s41598-019-38507-9.

Arruda, E.P., Domaneschi, O., and Amaral, A.C.Z. (2003). Mollusc feeding guilds on sandy beaches in São Paulo State, Brazil. *Marine Biology* 143: 691–701.

Atwood, T.B., Wiegner, T.N., and Mackenzie, R.A. (2012). Effects of hydrological forcing on the structure of a tropical estuarine food web. *Oikos* 121: 277–289.

Ayukai, T. (1995). Retention of phytoplankton and planktonic microbes on coral reefs within the Great Barrier Reef, Australia. *Coral Reefs* 14: 141–147.

Balance, L.T., Pitman, R.L., and Fiedler, P.C. (2006). Oceanographic influences on seabirds and cetaceans of the eastern tropical Pacific: a review. *Progress in Oceanography* 69: 360–390.

Bange, H.W., Sim, C.H., Bastian, D. et al. (2019). Nitrous oxide (N_2O) and methane (CH_4) in rivers and estuaries of NW Borneo. *Biogeosciences* 16: 4321–4335.

Barrera-Alba, J.J., Abreu, P.C., and Tenenbaum, D.R. (2019). Seasonal and interannual variability in phytoplankton over a 22 years period in a tropical coastal region in the SW Atlantic Ocean. *Continental Shelf Research* 176: 51–63.

Barrera-Alba, J.J., Gianesella, S.M.F., Moser, G.A.O. et al. (2009). Influence of allochthonous organic matter on bacterioplankton biomass and activity in a eutrophic, sub-tropical estuary. *Estuarine Coastal and Shelf Science* 82: 84–94.

Bates, N.R. (2002). Seasonal variability of the effect of coral reefs on seawater CO_2 and air-sea CO_2 exchange. *Limnology and Oceanography* 47: 43–52.

Bates, N.R., Pequignet, A.C., and Sabine, C.L. (2006). Ocean carbon cycling in the Indian Ocean. 1. Spatiotemporal variability of inorganic carbon and air-sea CO2 gas exchange. *Global Biogeochemical Cycles* 20: GB3020. https://doi.org/10.1029/2005GB002491.

Bauer, J.E., Cai, W.-J., Raymond, P.A. et al. (2013). The changing carbon cycle of the coastal ocean. *Nature* 504: 61–70.

Belicka, L.L., Burkholder, D., Fourqurean, J.W. et al. (2012). Stable isotope and fatty acid biomarkers of seagrass, epiphytic, and algal organic matter to consumers in a pristine seagrass ecosystem. *Marine and Freshwater Research* 63: 1085–1097.

Bento, L., Masuda, L.S.M., and Peixoto, R.B. (2017). Regulation of metabolism and community structure of a tropical salt flat after rainfall. *Journal of Coastal Research* 33: 304–308.
Bianchi, D., Galbraith, E.D., Carozza, D.A. et al. (2013). Intensification of open-ocean oxygen depletion by vertically migrating animals. *Nature Geoscience* 6: 545–548.
Biswas, H., Mukhopadhyay, S.K., De, T.K. et al. (2004). Biogenic controls on the air-water carbon dioxide exchange in the Sundarbans mangrove environment, NE coast of Bay of Bengal, India. *Limnology and Oceanography* 49: 95–101.
Bohlen, L., Dale, A.W., Sommer, S. et al. (2011). Benthic nitrogen cycling traversing the Peruvian oxygen minimum zone. *Geochimica et Cosmochimica Acta* 75: 6094–6111.
Bong, C.W. and Lee, C.W. (2011). The contribution of heterotrophic nanoflagellate grazing towards bacterial mortality in tropical waters: comparing estuaries and coastal ecosystems. *Marine and Freshwater Research* 62: 414–420.
Borges, A.V., Abril, G., and Bouillion, S. (2018). Carbon dynamics and CO_2 and CH_4 outgassing in the Mekong delta. *Biogeosciences* 15: 1093–1114.
Bouillon, S., Borges, A.V., Castañeda, E. et al. (2007b). Mangrove production and carbon sinks: a revision of global budget estimates. *Global Biogeochemical Cycles* 22: GB2013. https://doi.org/10.1029/2007GB003052.
Bouillon, S., Chandra Mohan, P., Sreenivas, N. et al. (2000). Sources of suspended organic matter and selective feeding by zooplankton in an estuarine mangrove ecosystem as traced by stable isotopes. *Marine Ecology Progress Series* 208: 79–92.
Bouillon, S., Connolly, R.M., and Lee, S.Y. (2008). Organic matter exchange and cycling in mangrove ecosystems: recent insights from stable isotope studies. *Journal of Sea Research* 59: 44–58.
Bouillon, S., Dehairs, F., Schiettecatte, L.-S. et al. (2007a). Biogeochemistry of the Tana estuary and delta (northern Kenya). *Limnology and Oceanography* 52: 46–59.
Bouillon, S., Koedam, N., Raman, A.V. et al. (2002). Primary producers sustaining macro-invertebrate communities in intertidal mangrove forests. *Oecologia* 130: 441–448.
Bouvy, M., Pagano, M., M'Boup, M. et al. (2006). Functional structure of microbial food web in the Senegal River estuary (west Africa): impact of metazooplankton. *Journal of Plankton Research* 28: 195–207.
Bozec, Y.-M., Gascuel, D., and Kulbicki, M. (2004). Trophic model of lagoonal communities in a large open atoll (Uvea, Loyalty islands, New Caledonia). *Aquatic Living Resources* 17: 151–162.
Breuer, E.R., Law, G.T.W., Woulds, C. et al. (2009). Sedimentary oxygen consumption and microdistribution at sites across the Arabian Sea oxygen minimum zone (Pakistan margin). *Deep-Sea Research II* 56: 296–304.
Brown, S.L., Landry, M.R., Neveux, J. et al. (2003). Microbial community abundance and biomass along a 180° transect in the equatorial Pacific during an El Niño-Southern Oscillation cold phase. *Journal of Geophysical Research* 108: 8139. https://doi.org/10.1029/2001JC000817.
Brugnoli-Olivera, E. and Morales-Ramirez, A. (2008). Trophic planktonic dynamics in a tropical estuary, Gulf of Nicoya, Pacific coast of Costa Rica during El Niño 1997 event. *Revista de Biología Marina y Oceanografía* 43: 75–89.
Brunet, F., Dubois, K., Veizer, J. et al. (2009). Terrestrial and fluvial carbon fluxes in a tropical watershed: Nyong basin, Cameroon. *Chemical Geology* 265: 563–572.
Bui, T.H.H. and Lee, S.Y. (2014). Does 'you are what you eat' apply to mangrove crabs? *PLoS One* 9: e89074. https://doi.org/10.1371/journal.pone.0089074.
Burkholz, C., Garcias-Bonet, N., and Duarte, C.M. (2020). Warming enhances carbon dioxide and methane fluxes from Red Sea seagrass (*Halophila stipulacea*) sediments. *Biogeosciences* 17: 1717–1730.
Buskey, E.J., Hyatt, C.J., and Speekmann, C.L. (2004). Trophic interactions within the planktonic food web in mangrove channels of Twin Cays, Belize, Central America. *Atoll Research Bulletin* 529: 1–22.
Cabezas, A., Mitsch, W.J., MacDonnell, C. et al. (2018). Methane emissions from mangrove soils in hydrologically disturbed and reference mangrove tidal creeks in SW Florida. *Ecological Engineering* 114: 57–65.

Cai, W.-J. (2011). Estuarine and coastal ocean carbon paradox: CO_2 sinks or sites of terrestrial carbon incineration? *Annual Review of Marine Science* 3: 123–146.
Cai, W.-J., Dai, M., and Wang, Y. (2006). Air-sea exchange of carbon dioxide in ocean margins: a province-based synthesis. *Geophysical Research Letters* 33: L12603. https://doi.org/10.1029/2006GL026219.
Calbet, A. and Landry, M.R. (2004). Phytoplankton growth, microzooplankton grazing, and carbon cycling in marine systems. *Limnology and Oceanography* 49: 51–59.
Call, M., Sanders, C.J., Macklin, P.A. et al. (2019b). Carbon outwelling and emissions from two contrasting mangrove creeks during the monsoon storm season in Palau, Micronesia. *Estuarine, Coastal and Shelf Science* 218: 340–348.
Call, M., Santos, I.R., Dittmar, T. et al. (2019a). High pore-water derived CO_2 and CH_4 emissions from a macro-tidal mangrove creek in the Amazon region. *Geochimica et Cosmochimica Acta* 247: 106–120.
Cameron, C., Hutley, L.B., Munksgaard, N.C. et al. (2021). Impact of an extreme monsoon on CO_2 and CH_4 fluxes from mangrove soils of the Ayeyarwady Delta, Myanmar. *Science of the Total Environment* 760: 143422. https://doi.org/10.1016/j.scitotenv.2020.143422.
Campbell, J.E., Altieri, A.H., Johnston, L.N. et al. (2018). Herbivore community determines the magnitude and mechanism of nutrient effects on subtropical and tropical seagrasses. *Journal of Ecology* 106: 401–412.
Cannicci, S., Burrows, D., Fratini, S. et al. (2008). Faunal impact on vegetation structure and ecosystem function in mangrove forests: a review. *Aquatic Botany* 89: 186–200.
Carlén, A. and 'Olafsson, E. (2002). The effects of the gastropod *Terebralia palustris* on infaunal communities in a tropical tidal mud-flat in east Africa. *Wetlands Ecology and Management* 10: 303–311.
Carlier, A., Chauvaud, L., van der Geest, M. et al. (2015). Trophic connectivity between offshore upwelling and the inshore food web of Banc d'Arguin: new insights from isotopic analysis. *Estuarine, Coastal and Shelf Science* 165: 149–158.
Carreira, R.S., Araújo, M.P., Costa, T.L.F. et al. (2011). Lipids in the sedimentary record as markers of the sources and deposition of organic matter in a tropical Brazilian estuarine-lagoon system. *Marine Chemistry* 127: 1–11.
Carreón-Palau, L., Parrish, C.C., del Angel-Rodriguez, J.A. et al. (2013). Revealing organic carbon sources fueling a coral reef food web in the Gulf of Mexico using stable isotopes and fatty acids. *Limnology and Oceanography* 58: 593–612.
Casareto, B.E., Suzuki, Y., Fukami, K. et al. (2000). Particulate organic carbon budget and flux in a fringing coral reef at Miyako Island, Okinawa, Japan in July 1996. *Proceedings Ninth International Coral Reef Symposium*, Bali, Indonesia (23–27 October 2000), vol. 1, pp. 1–6.
Castillo, J.A.A., Apan, A.A., Maraseni, T.N., and Salmo, S.G. III (2017). Soil greenhouse gas fluxes in tropical mangrove forests and in land uses on deforested mangrove lands. *Catena* 159: 60–69.
Castillo-Rivera, M., Zavala-Hurtado, J.A., and Zárate, R. (2002). Exploration of spatial and temporal patterns of fish diversity and composition in a tropical estuarine system of Mexico. *Reviews in Fish Biology and Fisheries* 12: 167–177.
Cebrián, J. and Duarte, C.M. (1998). Patterns in leaf herbivory on seagrasses. *Aquatic Botany* 60: 67–82.
Chaudhuri, A., Mukherjee, S., and Homechaudhuri, S. (2014). Food partitioning among carnivores within feeding guild of fish inhabiting a mud flat ecosystem of Indian Sundarbans. *Aquatic Ecology* 48: 35–51.
Chavez, F.P. and Messie, M. (2009). A comparison of eastern boundary upwelling ecosystems. *Progress in Oceanography* 83: 80–96.
Chen, B., Liu, H., and Wang, Z. (2009). Trophic interactions within the microbial food web in the South China Sea revealed by size-fractionation method. *Journal of Experimental Marine Biology and Ecology* 368: 59–66.
Chen, C.-T.A. and Borges, A.V. (2009). Reconciling opposing views on carbon cycling in the coastal ocean: continental shelves as sinks and near-shore ecosystems as sources of atmospheric CO_2. *Deep-Sea Research II* 56: 578–590.

Chen, C.-T.A., Huang, T.-H., Chen, Y.-C. et al. (2013). Air-sea exchanges of CO_2 in the world's coastal seas. *Biogeosciences* 10: 6509–6544.

Chen, G.C., Azkab, M.H., Chmura, G.L. et al. (2017). Mangroves as a major source of soil carbon storage in adjacent seagrass meadows. *Scientific Reports* 7: 42406. https://doi.org/10.1038/srep42406.

Chen, G.C., Ulumuddin, Y., Pramudji, S. et al. (2014). Rich soil carbon and nitrogen but low atmospheric greenhouse gas fluxes from north Sulawesi mangrove swamps in Indonesia. *Science of the Total Environment* 487C: 91–96.

Chew, L.L., Chong, V.C., Tanaka, K. et al. (2012). Phytoplankton fuel the energy flow from zooplankton to small nekton in turbid mangrove waters. *Marine Ecology Progress Series* 469: 7–24.

Ching, V.M. (2015). Contrasting tropical estuarine ecosystem functioning and stability: a comparative study. *Estuarine, Coastal and Shelf Science* 155: 89–103.

Chiu, S.-H., Huang, Y.-H., and Lin, H.-J. (2013). Carbon budget of leaves of the tropical intertidal seagrass *Thalassia hemprichii*. *Estuarine, Coastal and Shelf Science* 125: 27–35.

Chronopoulou, P.-M., Shelley, F., Pritchard, W.J. et al. (2017). Origin and fate of methane in the Eastern Tropical North Pacific oxygen minimum zone. *The ISME Journal* 11: 1386–1399.

Chuang, P.-C., Young, M.B., Dale, A.W. et al. (2017). Methane fluxes from tropical coastal lagoons surrounded by mangroves, Yucatán, Mexico. *Journal of Geophysical Research: Biogeosciences* 122: 1156–1174.

Cissoko, M., Desnues, A., Bouvy, M. et al. (2008). Effects of freshwater and seawater mixing on virio- and bacterioplankton in a tropical estuary. *Freshwater Biology* 53: 1154–1162.

Claudino, M.C., Pessanha, A.L.M., and Araújo, F.G. (2015). Trophic connectivity and basal food resources sustaining tropical aquatic consumers along a mangrove to ocean gradient. *Estuarine, Coastal and Shelf Science* 167: 45–55.

Clavier, J., Chauvaud, L., Amice, E. et al. (2014). Benthic metabolism in shallow coastal ecosystems of the Banc d'Arguin, Mauritania. *Marine Ecology Progress Series* 50: 11–23.

Clayton, D.A. and Snowden, R. (2000). Surface activity in the mudskipper, *Periophthalmus waltoni* Koumans 1941 in relation to prey activity and environmental factors. *Tropical Zoology* 13: 239–249.

Clores, M.A. and Cuesta, M.A. (2019). Trophic models of seagrass ecosystems in Maquesda Channel, Caramoan Peninsula, Philippines. *Applied Environmental Research* 41: 14–31.

Colombini, I., Aloia, A., Fallaci, M. et al. (2000). Temporal and spatial use of stranded wrack by the macrofauna of a tropical sandy beach. *Marine Biology* 136: 531–541.

Conroy, B.J., Steinberg, D.K., Stukel, M.R. et al. (2016). Meso- and microzooplankton grazing in the Amazon River plume and western tropical north Atlantic. *Limnology and Oceanography* 61: 825–840.

Cooley, S.R. and Yager, P.L. (2006). Physical and biological contributions to the western tropical north Atlantic Ocean carbon sink formed by the Amazon River plume. *Journal of Geophysical Research, Oceans* 111: JC002954. https://doi.org/10.1029/2005JC002954.

Costa, L.S., Huszar, V.L.M., and Ovalle, A.R. (2009). Phytoplankton functional groups in a tropical estuary: hydrological control and nutrient limitation. *Estuaries and Coasts* 32: 508–521.

Costa, M.F., Silva-Cavalcanti, J.S., Barbosa, C.C. et al. (2011). Plastics buried in the inter-tidal plain of a tropical estuarine ecosystem. *Journal of Coastal Research SI* 64: 339–343.

Cotovicz, L.C. Jr., Knoppers, B.A., Brandini, N. et al. (2015). A strong CO_2 sink enhanced by eutrophication in a tropical coastal embayment (Guanabara Bay, Rio de Janeiro, Brazil). *Biogeosciences* 12: 6125–6146.

Cotovicz, L.C. Jr., Knoppers, B.A., Brandini, N. et al. (2016). Spatio-temporal variability of methane (CH_4) concentrations and diffusive fluxes from a tropical coastal embayment surrounded by a large urban area (Guanabara Bay, Rio de Janeiro, Brazil). *Limnology and Oceanography* 61: S238–S252.

Crosswell, J.R., Carlin, G., and Steven, A. (2020). Controls on carbon, nutrient, and sediment cycling in a large, semiarid estuarine system; Princess Charlotte Bay, Australia. *Journal of Geophysical Research: Biogeosciences* 125: e2019JG005049. https://doi.org/10.1029/2019JG005049.

Cruz-Escalona, V.H., Arreguin-Sánchez, F., and Zetina-Rejón, M. (2007). Analysis of the ecosystem structure of Laguna Alvarado, western Gulf of Mexico, by means of a mass balance model. *Estuarine, Coastal and Shelf Science* 102: 1–13.

Currie, D.R. and Small, K.J. (2005). Macrobenthic community responses to long-term environmental change in an east Australian sub-tropical estuary. *Estuarine, Coastal and Shelf Science* 63: 315–331.

Cury, P. and Shannon, L. (2004). Regime shifts in upwelling ecosystems: observed changes and possible mechanisms in the northern and southern Benguela. *Progress in Oceanography* 60: 223–243.

Cyronak, T., Andersson, A.J., Langdon, C. et al. (2018). Taking the metabolic pulse of the world's coral reefs. *PLoS One* 13: e0190872. https://doi.org/10.1371/journal.pone.0190872.

Czudaj, S., Giesemann, A., Hoving, H.-J. et al. (2020). Spatial variation in the trophic structure of micronekton assemblages from the eastern tropical North Atlantic in two regions of differing productivity and oxygen environments. *Deep-Sea Research I* 163: 103275. https://doi.org/10.1016/j.dsr.2020.103275.

Dale, A.W., Sommer, S., Lomnitz, U. et al. (2015). Organic carbon production, mineralisation and preservation on the Peruvian margin. *Biogeosciences* 12: 1537–1559.

Dam, H.G., Zheng, X., Butler, M. et al. (1995). Mesozooplankton grazing and metabolism at the equator in the central Pacific: implications for carbon and nitrogen fluxes. *Deep-Sea Research II* 42: 735–756.

de la Moriniére, E.C., Pollux, B.J.A., Nagelkerken, I. et al. (2003). Ontogenetic dietary changes of coral reef fish in the mangrove-seagrass-reef continuum, stable isotopes and gut-content analysis. *Marine Ecology Progress Series* 246: 279–289.

de Oliveira da Rocha Franco, A., de Oliveira Soares, M., and Moreira, M.O.P. (2018). Diatom accumulations on a tropical meso-tidal beach: environmental drivers on phytoplankton biomass. *Estuarine Coastal Shelf Science* 207: 414–421.

de Vos, A., Pattiaratchi, C.B., and Harcourt, R.G. (2014). Interannual variability in blue whale distribution off southern Sri Lanka between 2011 and 2012. *Journal of Marine Science and Engineering* 2: 534–550.

Décima, M., Stukel, M.R., López-Lopez, L. et al. (2019). The unique ecological role of pyrosomes in the eastern tropical Pacific. *Limnology and Oceanography* 64: 728–743.

del Giorgio, P.A. and Duarte, C.M. (2002). Respiration in the open ocean. *Nature* 420: 379–384.

Dinh, Q.M., Tran, L.T., Tran, T.T.M. et al. (2020). Variation in diet composition of the mudskipper, *Periophthalmodon septemradiatus*, from Hau River, Vietnam. *Bulletin of Marine Science* 96: 487–500.

Dittmann, S. (1993). Impact of foraging soldier crabs (Decapoda, Mictyridae) on meiofauna in a tropical tidal flat. *Revista de Biologia Tropical* 41: 627–637.

Dittmann, S. (1996). Effects of microbenthic burrows on infaunal communities in tropical tidal flats. *Marine Ecology Progress Series* 134: 119–130.

Dittmann, S. (2002). Benthic fauna in tropical tidal flats – a comparative perspective. *Wetlands Ecology and Management* 10: 189–195.

Du, J., Christianus Makatipu, P., Tao, L.S.R. et al. (2020). Comparing trophic levels estimated from a tropical marine food web using an ecosystem model and stable isotopes. *Estuarine, Coastal and Shelf Science* 233: 106518. https://doi.org/10.1016/j.ecss.2019.106518.

Du, J., Wang, Y., Peristiwady, T. et al. (2018). Temporal and spatial variation of fish community and their nursery in a tropical seagrass meadow. *Acta Oceanologica Sinica* 37: 63–72.

Du, J., Zheng, X., Peristiwady, T. et al. (2016). Food sources and trophic structure of fish and benthic macroinvertebrates in a tropical seagrass meadow revealed by stable isotope analysis. *Marine Biology Research* 12: 748–757.

Duarte, C.M., Marbá, N., Gacia, E. et al. (2010). Seagrass community metabolism: assessing the carbon sink capacity of seagrass meadows. *Global Biogeochemical Cycles* 24: GB4032. https://doi.org/10.1029/2010GB003793.

Duarte, L.O. and Garcia, C.B. (2004). Trophic role of small pelagic fish in a tropical upwelling ecosystem. *Ecological Modelling* 172: 323–338.

Ducklow, H.W. (1990). The biomass, production and fate of bacteria in coral reefs. In: *Ecosystems of the World 25* (ed. Z. Dubinsky), 265–289. Amsterdam: Elsevier.
Duineveld, G.C.A., de Wilde, P.A.W.J., Berghuis, E.M. et al. (1993). The benthic infauna and benthic respiration off the Banc d'Arguin (Mauritania, NW Africa). *Hydrobiologia* 258: 107–117.
Dulvy, N.K., Freckleton, R.P., and Polunin, N.V. (2004). Coral reef cascades and the indirect effects of predator removal by exploitation. *Ecology Letters* 7: 410–416.
Dutta, M.K., Bianchi, T.S., and Mukhopadhyay, S.K. (2017b). Mangrove methane biogeochemistry in the Indian Sundarbans: a proposed budget. *Frontiers in Marine Science* 4: 187. https://doi.org/10.3389/fmars.2017.00187.
Dutta, M.K., Kumar, S., Mukherjee, R. et al. (2019). Diurnal carbon dynamics in a mangrove-dominated tropical estuary (Sundarbans, India). *Estuarine, Coastal and Shelf Science* 229: 106426. https://doi.org/10.1016/j.ecss.2019.106426.
Dutta, S., Chakraborty, K., and Hazra, S. (2017a). Ecosystem structure and trophic dynamics of an exploited ecosystem, of Bay of Bengal, Sundarbans estuary, India. *Fisheries Science* 83: 145–159.
Ekau, W., Auel, H., Hagen, W. et al. (2018). Pelagic key species and mechanisms driving energy flows in the northern Benguela upwelling ecosystem and their feedback into biogeochemical cycles. *Journal of Marine Systems* 188: 49–62.
Emeis, K., Anja, E., Anita, F. et al. (2018). Biogeochemical processes and turnover rates in the northern Benguela upwelling system. *Journal of Marine Systems* 188: 63–80.
Espadero, A.D.A., Nakamura, Y., Uy, W.H. et al. (2020). Tropical intertidal seagrass beds: an overlooked foraging habitat for fish revealed by underwater videos. *Journal of Experimental Marine Biology and Ecology* 526: 151353. https://doi.org/10.1016/j.jembe.2020.151353.
Espinoza, P. and Bertrand, A. (2008). Revisiting Peruvian anchovy (*Engraulis ringens*) trophodynamics provides a new vision of the Humboldt Current system. *Progress in Oceanography* 79: 215–227.
Ewa-Oboho, I., Oladimeiji, O., and Asuquo, F.E. (2008). Effect of dredging on benthic-pelagic production in the mouth of the Cross River estuary (off the Gulf of Guinea), S.E. Nigeria. *Indian Journal of Marine Sciences* 37: 291–297.
Eyre, B.D. and McKee, L.J. (2002). Carbon, nitrogen and phosphorus budgets for a shallow subtropical coastal embayment (Moreton Bay, Australia). *Limnology and Oceanography* 47: 1043–1055.
Faye, D., Tito de Morais, L., Raffray, J. et al. (2011). Structure and seasonal variability of fish food webs in an estuarine tropical marine protected area (Senegal): evidence from stable isotope analysis. *Estuarine, Coastal and Shelf Science* 92: 607–617.
Feeley, R.A., Boutin, J., Cosca, C.E. et al. (2002). Seasonal and interannual variability of CO_2 in the equatorial Pacific. *Deep-Sea Research II* 49: 2443–2469.
Fiedler, P.C., Redfern, J.V., and Balance, L.T. (2017). *Oceanography and Cetaceans of the Costa Rica Dome Region*, NOAA Technical Memorandum NMFS. Washington D.C., USA: NOAA.
Fourqurean, J.W., Duarte, C.M., Kennedy, H. et al. (2012). Seagrass ecosystems as a globally significant carbon stock. *Nature Geoscience* 5: 505–509.
Freestone, A.L., Carroll, E.W., Papacostos, K.J. et al. (2020). Predation shapes invertebrate diversity in tropical but not temperate seagrass communities. *Journal of Animal Ecology* 89: 323–333.
Garcias-Bonet, N. and Duarte, C.M. (2017). Methane production by seagrass ecosystems in the Red Sea. *Frontiers in Marine Science* 4: 340. https://doi.org/10.3389/fmars.2017.00340.
Gattuso, J.-P., Allemand, D., and Frankignoulle, M. (1999). Photosynthesis and calcification at cellular, organismal and community levels in coral reefs, a review on interactions and control by carbonate chemistry. *American Zoologist* 39: 160–183.
Gattuso, J.-P., Pichon, M., Jaubert, J. et al. (1996). Primary production, calcification and air-sea CO_2 fluxes in coral reefs: organism, ecosystem and global scales. *Bulletin de l'Institut Oceanographique Monaco* 14: 39–46.
Giarrizzo, T., Schwamborn, R., and Saint-Paul, U. (2011). Utilization of carbon sources in a northern Brazilian mangrove ecosystem. *Estuarine, Coastal and Shelf Science* 95: 447–457.

Glynn, P.W. (2004). High complexity food webs in low-diversity eastern Pacific reef-coral communities. *Ecosystems* 7: 358–367.

Gonzalez, J.G. and Jùnior, T.V. (2017). Feeding ecology of the beach silverside, *Atherinella blackburni* (Atherinopsidae), in a tropical sandy beach, southeastern Brazil. *Brazilian Journal of Oceanography* 65: 346–355.

Guedes, A.P.P. and Araújo, F.G. (2008). Trophic resource partitioning among five flatfish species (Actinopterygii, Pleuronectiformes) in a tropical bay in south-eastern Brazil. *Journal of Fish Biology* 72: 1035–1054.

Guénette, S., Meissa, B., and Gascuel, D. (2014). Assessing the contribution of marine protected areas to the trophic functioning of ecosystems: a model for the Banc d'Arguin and the Mauritanian shelf. *PLoS One* 9: e94742. https://doi.org/10.1371/journal.pone.0094742.

Gullström, M., Berkström, C., Öhman, M.C. et al. (2011). Scale-dependent patterns of variability of a grazing parrotfish (*Leptoscarus vaigiensis*) in a tropical seagrass-dominated seascape. *Marine Biology* 158: 1483–1495.

Gómez-Ramírez, E.H., Corzo, A., Garcia-Robledo, E. et al. (2019). Benthic-pelagic coupling of carbon and nitrogen along a tropical estuarine gradient (Gulf of Nicoya, Costa Rica). *Estuarine, Coastal and Shelf Science* 228: 106362. https://doi.org/10.1016/j.ecss.2019.106362.

Hajisamae, S. (2009). Trophic ecology of bottom fish assemblage along coastal areas of Thailand. *Estuarine, Coastal and Shelf Science* 82: 503–514.

Hamilton, S.E. and Casey, D. (2016). Creation of a high spatio-temporal resolution global database of continuous mangrove forest cover for the 21st century (CGMFC-21). *Global Ecology and Biogeography* 25: 729–738.

Hansen, J.A. and Skilleter, G.A. (1994). Effects of the gastropod *Rhinoclavis aspera* (Linnaeus, 1758) on microbial biomass and productivity in coral reef flat sediments. *Marine and Freshwater Research* 45: 569–584.

Haputhantri, S.S.K., Villanueva, M.C.S., and Moreau, J. (2008). Trophic interactions in the coastal ecosystem of Sri Lanka: an ECOPATH preliminary approach. *Estuarine, Coastal and Shelf Science* 76: 304–318.

Hartmann, M., Grob, C., Tarran, G.A. et al. (2012). Mixotrophic basis for Atlantic oligotrophic ecosystems. *Proceedings of the National Academy of Sciences* 109: 5756–5760.

Hata, H., Kudo, S., Yamano, H. et al. (2002). Organic carbon flux in Shiraho coral reef (Ishigaki Island, Japan). *Marine Ecology Progress Series* 232: 129–140.

Hearne, E.L., Johnson, R.A., Gulick, A.G. et al. (2019). Effects of green turtle grazing on seagrass and macroalgae diversity vary spatially among seagrass meadows. *Aquatic Botany* 152: 10–15.

Heck, K.L. Jr., Carruthers, T.J.B., Duarte, C.M. et al. (2008). Trophic transfers from seagrass meadows subsidize diverse marine and terrestrial consumers. *Ecosystems* 11: 1198–1210.

Heithaus, E.R., Heithaus, P.A., Heihaus, M.R. et al. (2011). Trophic dynamics in a relatively pristine subtropical fringing mangrove community. *Marine Ecology Progress Series* 428: 49–61.

Herbon, C. and Nordhaus, I. (2013). Experimental determination of stable carbon and nitrogen isotope fractionation between mangrove leaves and crabs. *Marine Ecology Progress Series* 490: 91–105.

Heymans, J.J. and Baird, D. (2000). A carbon flow model and network analysis of the northern Benguela upwelling system, Namibia. *Ecological Modelling* 126: 9–32.

Heymans, J.J., Shannon, L.J., and Jarre, A. (2004). Changes in the northern Benguela ecosystem over three decades: 1970s, 1980s, and 1990s. *Ecological Modelling* 172: 175–195.

Hogarth, P.J. (2015). *The Biology of Mangroves and Seagrasses*. Oxford, UK: Oxford University Press.

Holzer, K.K., Rueda, J.L., and McGlathery, K.J. (2011). Caribbean seagrasses as a food source for the emerald neritid *Smaragdia viridis*. *American Malacological Bulletin* 29: 63–67.

Hossain, M.B., Marshall, D.J., and Hall-Spencer, J.M. (2019). Epibenthic community variation along an acidified tropical estuarine system. *Regional Studies in Marine Science* 32: 100888. https://doi.org/10.1016/j.rsma.2019.100888.

Houk, P. and Musburger, C. (2013). Trophic interactions and ecological stability across coral reefs in the Marshall Islands. *Marine Ecology Progress Series* 488: 23–34.
Hsieh, H.-L., Chen, C.-P., Chen, Y.-G. et al. (2002). Diversity of benthic organic matter flows through polychaetes and crabs in a mangrove estuary: $\delta^{13}C$ and $\delta^{34}S$ signals. *Marine Ecology Progress Series* 227: 145–155.
Huang, T.H., Fu, Y.H., Pan, P.Y. et al. (2012). Fluvial carbon fluxes in tropical rivers. *Current Opinion in Environmental Sustainability* 4: 162–169.
Huang, Y.-H., Lee, C.-L., Chung, C.-Y. et al. (2015). Carbon budgets of multispecies seagrass beds at Dongsha Island in the South China Sea. *Marine Environmental Research* 106: 92–102.
Hung, J.-J., Wang, Y.-J., and Tseng, C.-M. (2019). Regulation and linkages of metabolic states and atmospheric CO_2 fluxes in a tropical coastal sea off SW Taiwan. *Journal of Sea Research* 150-151: 24–32.
Hung, J.-J., Wang, Y.-J., Tseng, C.-M. et al. (2020). Controlling mechanisms and cross linkages of ecosystem metabolism and atmospheric CO_2 flux in the northern South China Sea. *Deep-Sea Research I* 157: 103205. https://doi.org/10.1016/j.dsr.2019.103205.
Hutchinson, N. and Williams, G.A. (2001). Spatio-temporal variation in recruitment on a seasonal, tropical rocky shore: the importance of local versus non-local processes. *Marine Ecology Progress Series* 215: 57–68.
Hyndes, G.A., Heck, K.L. Jr., Vergés, A. et al. (2016). Accelerating tropicalization and the transformation of temperate seagrass meadows. *Bioscience* 66: 938–948.
Hyun, J.-H. and Yang, E.J. (2005). Meso-scale spatial variation in bacterial abundance and production associated with surface convergence and divergence in the NE equatorial Pacific. *Aquatic Microbial Ecology* 41: 1–13.
Jacotot, A., Marchand, C., and Allenbach, M. (2018). Tidal variability of CO_2 and CH_4 emissions from the water column within a *Rhizophora* mangrove forest (New Caledonia). *Science of the Total Environment* 631: 334–340.
Jacotot, A., Marchand, C., and Allenbach, M. (2019). Biofilm and temperature controls on greenhouse gas (CO_2 and CH_4) emissions from a *Rhizophora* mangrove soil (New Caledonia). *Science of the Total Environment* 650: 1019–1028.
Jennerjahn, T.C. and Ittekkot, V. (2002). Relevance of mangroves for the production and deposition of organic matter along tropical continental margins. *Naturwissenschaften* 89: 23–30.
Jennerjahn, T.C., Ittekkot, V., Klopper, S. et al. (2004). Biogeochemistry of a tropical river affected by human activities in its catchment, Brantas River estuary and coastal waters of Madura Strait, Java, Indonesia. *Estuarine, Coastal and Shelf Science* 60: 503–514.
Jinks, K.I., Brown, C.J., Rasheed, M.A. et al. (2019). Habitat complexity influences the structure of food webs in Great Barrier Reef seagrass meadows. *Ecosphere* 10: e02928. https://doi.org/10.1002/ecs2.2928.
John, D.M. and Lawson, G.W. (1991). Littoral ecosystems of tropical western Africa. In: *Intertidal and Littoral Ecosystems* (eds. A.C. Mathieson and P.H. Nienhuis), 297–323. Amsterdam: Elsevier.
Johnson, R.A., Gulick, A.G., Constant, N. et al. (2020). Seagrass ecosystem metabolism carbon capture in response to green turtle grazing across Caribbean meadows. *Journal of Ecology* 108: 1101–1114.
Jyothibabu, R., Balachandran, K.K., Jagadeesan, L. et al. (2018). Mud banks along the SW coast of India are not too muddy for plankton. *Scientific Reports* 8: 2544. https://doi.org/10.1038/s41598-018-20667-9.
Jyothibabu, R., Madhu, N.V., Jayalakshmi, K.V. et al. (2009). Impact of freshwater influx on microzooplankton-mediated food web in a tropical estuary (Cochin backwaters – India). *Estuarine, Coastal and Shelf Science* 69: 505–518.
Kaehler, S. and Williams, G.A. (1996). Distribution of algae on tropical rocky shores: spatial and temporal patterns of non-coralline encrusting algae in Hong Kong. *Marine Biology* 125: 177–187.
Kaehler, S. and Williams, G.A. (1998). Early development of algal assemblages under different regimes of physical and biotic factors on a seasonal tropical rocky shore. *Marine Ecology Progress Series* 172: 61–71.

Kaldy, J.E., Onuf, C.P., Eldridge, P.M. et al. (2002). Carbon budget for a subtropical seagrass dominated coastal lagoon: how important are seagrasses to total ecosystem net primary production? *Estuaries* 25: 528–539.

Kanuri, V.V., Muduli, P.R., Robin, R.S. et al. (2013). Plankton metabolic processes and its significance on dissolved organic carbon pool in a tropical brackish water lagoon. *Continental Shelf Research* 61-62: 52–61.

Kasim, M. (2009). Grazing activity of the sea urchin *Tripneustes gratilla* in tropical seagrass beds of Buton Island, southeast Sulawesi, Indonesia. *Journal of Coastal Development* 13: 19–27.

Kathiresan, K. and Bingham, B.L. (2001). Biology of mangroves and mangrove ecosystems. *Advances in Marine Biology* 40: 81–251.

Kieckbusch, D.K., Koch, M.S., Serafy, J.E. et al. (2004). Trophic linkages among primary producers and consumers in fringing mangroves of subtropical lagoons. *Bulletin of Marine Science* 74: 271–285.

Kitidis, V., Tilstone, G.H., Serret, P. et al. (2014). Oxygen photolysis in the Mauritanian upwelling: implications for net community production. *Limnology and Oceanography* 59: 299–310.

Klumpp, D.W., Salita-Espinosa, J.S., and Fortes, M.D. (1993). Feeding ecology and the trophic role of sea urchins in a tropical seagrass community. *Aquatic Botany* 45: 205–216.

Kneer, D., Asmus, H., and Vonk, J.A. (2008). Seagrass as the main food source of *Neaxius acanthus* (Thalassinidae, Strahlaxiidae), its burrow associates, and of *Corallianassa coutierei* (Thalassinidea, Callianassidae). *Estuarine, Coastal and Shelf Science* 79: 620–630.

Knowlton, N. and Rohwer, F. (2003). Multispecies microbial mutualisms on coral reefs: the host as a habitat. *The American Naturalist* 162: S51–S62.

Kober, K. and Bairlein, F. (2006). Shorebirds of the Bragantinian Peninsula. II. Diet and foraging strategies of shorebirds at a tropical site in northern Brazil. *Ornitologia Neotropical* 17: 549–562.

Kock, A., Gebhardt, S., and Bange, H.W. (2008). Methane emissions from the upwelling area off Mauritania (NW Africa). *Biogeosciences* 5: 1119–1125.

Koné, Y.J.M., Abril, G., Delille, B. et al. (2010). Seasonal variability of methane in the rivers and lagoons of Ivory Coast (West Africa). *Biogeochemistry* 100: 21–37.

Koop, K. and Lucas, M.I. (1983). Carbon flow and nutrient regeneration from the decomposition of macrophyte debris in a sandy beach microcosm. In: *Sandy Beaches as Ecosystems* (eds. A. McLachlan and T. Erasmus), 249–261. The Hague, The Netherlands: Junk.

Körtzinger, A. (2003). A significant CO_2 sink in the tropical Atlantic Ocean associated with the Amazon River plume. *Geophysical Research Letters* 30: 2287. https://doi.org/10.1029/2003GL018841.

Kristensen, D.K., Kristensen, E., and Mangion, P. (2010). Food partitioning of leaf-eating mangrove crabs (Sesarminae): experimental and stable (13C and 15N) evidence. *Estuarine, Coastal and Shelf Science* 87: 583–590.

Kristensen, E., Andersen, F.Ø., Holmboe, N. et al. (2000). Carbon and nitrogen mineralization in sediments of the Bangrong mangrove area, Phuket, Thailand. *Aquatic Microbial Ecology* 22: 199–213.

Kristensen, E., Bouillon, S., Dittmar, T. et al. (2008a). Organic carbon dynamics in mangrove ecosystems: a review. *Aquatic Botany* 89: 201–219.

Kristensen, E., Flindt, M.R., Ulomi, S. et al. (2008b). Emission of CO2 and CH4 to the atmosphere by sediments and open waters in two Tanzanian mangrove forests. *Marine Ecology Progress Series* 370: 53–67.

Krumme, U., Brenner, M., and Saint-Paul, U. (2008a). Spring-neap cycle as a major driver of temporal variations in feeding of intertidal fish: evidence from the sea catfish *Sciandes herzbergii* (Ariidae) of equatorial west Atlantic mangrove creeks. *Journal of Experimental Marine Biology and Ecology* 367: 91–99.

Krumme, U., Keuthen, H., Barletta, M. et al. (2008b). Resuspended intertidal microphytobenthos as major diet component of planktivorous Atlantic Anchoveta *Cetengraulis edentulus* (Engraulidae) from equatorial mangrove creeks. *Ecotropica* 14: 121–128.

Lacerda, C.H.F., Barletta, M., and Dantas, D.V. (2014). Temporal patterns in the intertidal faunal community at the mouth of a tropical estuary. *Journal of Fish Biology* 85: 1571–1602.

Lacroix, F., Ilyina, T., and Hartmann, J. (2020). Oceanic CO_2 outgassing and biological production hotspots induced by pre-industrial river loads of nutrients and carbon in a global modelling approach. *Biogeosciences* 17: 55–88.

Lamptey, E. and Armah, A.K. (2008). Factors affecting microbenthic fauna in a tropical hypersaline coastal lagoon in Ghana, west Africa. *Estuaries and Coasts* 31: 1006–1019.

Landry, M.R., Brown, S.L., Neveux, J. et al. (2003). Phytoplankton growth and microzooplankton grazing in high-nutrient, low-chlorophyll waters of the equatorial Pacific: community and taxon-specific rate assessments from pigment and flow cytometric analyses. *Journal of Geophysical Research* 108: 8142. https://doi.org/10.1029/2000JC000744.

Landry, M.R., Selph, K.E., Taylor, A.G. et al. (2011). Phytoplankton growth, grazing and production balances in the HNLC equatorial Pacific. *Deep-Sea Research II* 58: 524–535.

Lastra, M., Lopez, J., Troncoso, J. et al. (2016). Scavenger and burrowing features of *Hippa pacifica* (Dana 1852) on a range of tropical sandy beaches. *Marine Biology* 163: 212–221.

Le Borgne, R., Feely, R.A., and Mackey, D.J. (2002). Carbon fluxes in the equatorial Pacific: a synthesis of the JGOFS programme. *Deep-Sea Research II* 49: 2425–2442.

Lee, S.L., Chong, V.C., and Then, A.Y.-H. (2019). Fish trophodynamics in tropical mud flats: a dietary and isotopic perspective. *Estuaries and Coasts* 42: 868–889.

Lefèvre, N., da Silva Dias, F.J., de Torres, A.R. Jr. et al. (2017). A source of CO_2 to the atmosphere throughout the year in the Maranhense continental shelf (2°30′S, Brazil). *Continental Shelf Research* 141: 38–50.

Leopold, A., Marchand, C., Deborde, J. et al. (2017). Water biogeochemistry of a mangrove-dominated estuary under a semi-arid climate (New Caledonia). *Estuaries and Coasts* 40: 773–791.

Lima, M. and Naya, D.E. (2011). Large-scale climatic variability affects the dynamics of tropical skipjack tuna in the western Pacific Ocean. *Ecogeography* 34: 597–605.

Lin, C.-W., Kao, Y.-C., Chou, M.-C. et al. (2020). Methane emissions from subtropical and tropical mangrove ecosystems in Taiwan. *Forests* 11: 470. https://doi.org/10.3390/f11040470.

Lin, H.-J., Dai, X.-X., Shao, K.-T. et al. (2006). Trophic structure and functioning in a eutrophic and poorly flushed lagoon in SW Taiwan. *Marine Environmental Research* 62: 61–82.

Lin, H.-J., Kao, W.-Y., and Wang, Y.-T. (2007). Analyses of stomach contents and stable isotopes reveal food sources of estuarine detritivorous fish in tropical/subtropical Taiwan. *Estuarine, Coastal and Shelf Science* 73: 527–537.

Linto, N., Barnes, J., Ramachandran, R. et al. (2014). Carbon dioxide and methane emissions from mangrove-associated waters of the Andaman Islands, Bay of Bengal. *Estuaries and Coasts* 37: 381–398.

Lira, A., Angelini, R., Le Loc'h, F. et al. (2018). Trophic flow structure of a neotropical estuary in NE Brazil and the comparison of ecosystem model indicators of estuaries. *Journal of Marine Systems* 182: 31–45.

Longhurst, A.L. (2007). *Ecological Geography of the Sea*, 2e. London: Academic Press.

Lopez-Vila, J.M., Schmitter-Soto, J.J., Velázquez-Velázquez, E. et al. (2019). Young does not mean unstable: a trophic model for an estuarine lagoon system in the southern Mexican Pacific. *Hydrobiologia* 827: 225–246.

Lu, H.-J., Lee, K.-T., Lin, H.-L. et al. (2001). Spatio-temporal distribution of yellowfin tuna *Thunnus albacares* and bigeye tuna *Thunnus obesus* in the tropical Pacific Ocean in relation to large-scale temperature fluctuation during ENSO episodes. *Fisheries Science* 67: 1046–1052.

Lubchenco, J., Menge, B.A., Garrity, S.D. et al. (1984). Structure, persistence, and role of consumers in a tropical rocky intertidal community (Taboguilla Island, Bay of Panama). *Journal of Experimental Marine Biology and Ecology* 78: 23–73.

Lugendo, B.R., Nagelkerken, I., Kruitwagen, G. et al. (2007). Relative importance of mangroves as feeding habitats for fish: a comparison between mangrove habitats with different settings. *Bulletin of Marine Science* 80: 497–512.

Lugendo, B.R., Nagelkerken, I., Van Der Velde, G. et al. (2006). The importance of mangroves, mud and sand flats, and seagrass beds as feeding areas for juvenile fish in Chwaka Bay, Zanzibar: gut content and stable isotope analyses. *Journal of Fish Biology* 69: 1639–1661.

Lyimo, T.J., Mamboya, F., Hamisi, M. et al. (2011). Food preference of the sea urchin *Tripneustes gratilla* (Linnaeus, 1758) in tropical seagrass habitats at Dar es Salaam, Tanzania. *Journal of Ecology and the Natural Environment* 3: 415–423.

Macia, A. (2004). Primary carbon sources for juvenile penaeid shrimps in a mangrove-fringed bay of Inhaca Island, Mozambique: a dual carbon and nitrogen isotope analysis. *Western Indian Ocean Journal of Marine Science* 3: 151–161.

MacKenzie, K.M., Robertson, D.R., Adams, J.N. et al. (2019). Structure and nutrient transfer in a tropical pelagic upwelling food web: from isoscapes to the whole ecosystem. *Progress in Oceanography* 178: 102145. https://doi.org/10.1016/j.pocean.2019.102145.

Macklin, P.A., Suryaputra, I.G.N.A., Maher, D.T. et al. (2019). Drivers of CO_2 along a mangrove-seagrass transect in a tropical bay: delayed groundwater seepage and seagrass uptake. *Continental Shelf Research* 172: 57–67.

Macusi, E.D. and Ashoka Deepananda, K.H.M. (2013). Factors that structure algal communities in tropical rocky shores: what have we learned? *International Journal of Scientific and Research Publications* 3: 490–503.

Maher, D.T., Call, M., Santos, I.R. et al. (2018). Beyond burial: lateral exchange is a significant carbon sink in mangrove forests. *Biology Letters* 14: 20180220.

Maher, D.T., Santos, I.R., Golsby-Smith, L. et al. (2013). Groundwater-derived dissolved inorganic and organic carbon exports from a mangrove tidal creek: the missing mangrove carbon sink? *Limnology and Oceanography* 58: 475–488.

Maher, D.T., Santos, I.R., Schulz, K.G. et al. (2017). Blue carbon oxidation revealed by radiogenic and stable isotopes in a mangrove system. *Geophysical Research Letters* 44: 4889–4896.

Manickchand-Heileman, S., Arreguin-Sánchez, F., Lara-Dominguez, A. et al. (1998). Energy flow and network analysis of Terminos Lagoon, SW Gulf of Mexico. *Journal of Fish Biology* 53: 179–197.

Manickchand-Heileman, S., Mendoza-Hill, J., Kong, A.L. et al. (2004). A trophic model for exploring possible ecosystem impacts of fishing in the Gulf of Paria, between Venezuela and Trinidad. *Ecological Modelling* 172: 307–322.

Marguillier, S., van der Velde, G., Dehairs, F. et al. (1997). Trophic relationships in an interlinked mangrove-seagrass ecosystem as traced by $\delta^{13}C$ and $\delta^{15}N$. *Marine Ecology Progress Series* 151: 115–121.

Marley, G.A., Lawrence, A.J., Phillip, D.A.T. et al. (2019). Mangrove and mud flat food webs are segregated across four trophic levels yet connected by highly mobile top predators. *Marine Ecology Progress Series* 632: 13–25.

Martin, G.D., Vijay, J.G., Lalurai, C.M. et al. (2008). Fresh water influence on nutrient stoichiometry in a tropical estuary, SW coast of India. *Applied Ecology and Environmental Research* 6: 57–64.

Masagca, J.T. (2011). Occurrence of arboreal climbing grapsids and other brachyurans in two mangrove areas of southern Luzon, Philippines. *Biotropia* 18: 61–73.

Mateo, M.A., Cebrián, J., Dunton, K. et al. (2006). Carbon fluxes in seagrass ecosystems. In: *Seagrasses: Biology, Ecology and Conservation* (ed. A.W.D. Larkum), 159–192. Dordrecht, The Netherlands: Springer.

Mayer, B., Rixen, T., and Pohlmann, T. (2018). The spatial and temporal variability of air-sea CO_2 fluxes and the effect of net coral reef calcification in the Indonesian seas: a numerical sensitivity study. *Frontiers in Marine Science* 5: 116. https://doi.org/10.3389/fmars.2018.00116.

McClanahan, T.R. (2008). Food-web structure and dynamics of east African coral reefs. In: *Food Webs and the Dynamics of Marine Reefs* (eds. T.R. McClanahan and G.M. Branch), 162–184. Oxford, UK: Oxford University Press.

Mchenga, I.S.S. and Tsuchiya, M. (2010). Feeding choice and the fate of organic materials consumed by sesarmid crabs *Perisesarma bidens* (De Haan) when offered different diets. *Journal of Marine Biology* 2010: 201932. https://doi.org/10.1155/2010/201932.

McKinnon, A.D., Carleton, J.H., and Duggan, S. (2007). Pelagic production and respiration in the Gulf of Papua during May 2004. *Continental Shelf Research* 27: 1643–1655.

McKinnon, A.D., Duggan, S., Logan, M. et al. (2017). Plankton respiration, production, and trophic state in tropical coastal and shelf waters adjacent to Northern Australia. *Frontiers in Marine Science* 4: 346. https://doi.org/10.3389/fmars.2017.00346.

McKinnon, A.D., Logan, M., Castine, S.A. et al. (2013). Pelagic metabolism in the waters of the Great Barrier Reef. *Limnology and Oceanography* 58: 1227–1242.
McLachlan, A. and Defeo, O. (2018). *The Ecology of Sandy Shores*, 3e. London: Academic Press.
McMahon, K.W., Thorrold, S.R., Houghton, L.A. et al. (2016). Tracing carbon flow through coral reef food webs using a compound-specific stable isotope approach. *Oecologia* 180: 809–821.
McManus, G.B., Costas, B.A., Dam, H.G. et al. (2007). Microzooplankton grazing of phytoplankton in a tropical upwelling region. *Hydrobiologia* 575: 69–81.
Ménard, F., Fonteneau, A., Gaertner, D. et al. (2000). Exploitation of small tuna by a purse-seine fishery with fish aggregating devices and their feeding ecology in an eastern tropical Atlantic ecosystem. *ICES Journal of Marine Science* 57: 525–530.
Menge, B.A., Lubchenco, J., Gaines, S.D. et al. (1986). A test of the Menge-Sutherland model of community organization in a tropical rocky intertidal food web. *Oecologia* 71: 75–89.
Meziane, T. and Tsuchiya, M. (2000). Fatty acids as tracers of organic matter in sediment and food web of a mangrove/intertidal flat ecosystem, Okinawa, Japan. *Marine Ecology Progress Series* 200: 49–57.
Miyajima, T. and Hamaguchi, M. (2019). Carbon sequestration in sediment as an ecosystem function of seagrass meadows. In: *Blue Carbon in Shallow Coastal Ecosystems* (eds. T. Kuwae and M. Hori), 33–71. Singapore: Springer.
Moncreiff, C.A. and Sullivan, M.J. (2001). Trophic importance of epiphytic algae in subtropical seagrass beds: evidence from multiple stable isotope analysis. *Marine Ecology Progress Series* 215: 93–106.
Morrow, K., Bell, S.S., and Tewfik, A. (2014). Variation in ghost crab trophic links on sandy beaches. *Marine Ecology Progress Series* 502: 197–206.
Moteki, M., Aral, M., Tsuchiya, K. et al. (2001). Composition of piscine prey in the diet of large pelagic fish in the eastern tropical Pacific Ocean. *Fisheries Science* 67: 1063–1074.
Muduli, P.R., Kanuri, V.V., Robin, R.S. et al. (2012). Spatio-temporal variation of CO_2 emission from Chilika Lake, a tropical coastal lagoon, on the east coast of India. *Estuarine, Coastal and Shelf Science* 113: 305–313.
Muduli, P.R., Kanuri, V.V., Robin, R.S. et al. (2013). Distribution of DIC and net ecosystem production in a tropical brackish water lagoon, India. *Continental Shelf Research* 64: 75–87.
Mukhopadhyay, S.K., Biswas, H., De, T.K. et al. (2002). Seasonal effects on the air-water carbon dioxide exchange in the Hooghly estuary, NE coast of Bay of Bengal, India. *Journal of Environmental Monitoring* 4: 549–552.
Mukhtar, A., Zulkifli, S.Z., Mohamat-Yusuf, F. et al. (2016). Stable isotope ($\delta^{13}C$ and $\delta^{15}N$) analysis as a tool to quantify the food web structure in seagrass area of Pulai River estuary, Johor, peninsular Malaysia. *Malayan Nature Journal* 68: 91–102.
Müller, D., Bange, H.W., Warneke, T. et al. (2016). Nitrous oxide and methane in two tropical estuaries in a peat-dominated region of NW Borneo. *Biogeosciences* 13: 2415–2428.
Müller, D., Warneke, T., Rixen, T. et al. (2015). Fate of peat-derived carbon and associated CO_2 and CO emissions from two southeast Asian estuaries. *Biogeochemical Discussions* 12: 8299–8340.
Mumby, P.J. and Steneck, R.S. (2018). Paradigm lost: dynamic nutrients and missing detritus on coral reefs. *Bioscience* 68: 487–495.
Mumby, P.J., Steneck, R.S., Edwards, A.J. et al. (2012). Fishing down a Caribbean food web relaxes trophic cascades. *Marine Ecology Progress Series* 445: 13–24.
Muro-Torres, V.M., Soto-Jiménez, M.F., Green, L. et al. (2019). Food web structure of a subtropical coastal lagoon. *Aquatic Ecology* 53: 407–430.
Nagelkerken, I. and van der Velde, G. (2004). Are Caribbean mangroves important feeding grounds for juvenile reef fish from adjacent seagrass beds? *Marine Ecology Progress Series* 274: 143–151.
Nagelkerken, I., van der Velde, G., Verberk, W.C.E.P. et al. (2006). Segregation along multiple resources axes in a tropical seagrass fish community. *Marine Ecology Progress Series* 308: 79–89.
Naik, R., Naqvi, S.W.A., and Araujo, J. (2017). Anaerobic carbon mineralization through sulphate reduction in the inner shelf sediments of eastern Arabian Sea. *Estuarine, Coastal and Shelf Science* 40: 134–144.

Nanjo, K., Kohno, H., and Sano, M. (2008). Food habits of fish in the mangrove estuary of Urauchi River, Iriomote Island, southern Japan. *Fisheries Science* 74: 1024–1033.

Nixon, S.W. and Thomas, A. (2001). On the size of the Peru upwelling ecosystem. *Deep-Sea Research I* 48: 2521–2528.

Nordhaus, I. and Wolff, M. (2007). Feeding ecology of the mangrove crab *Ucides cordatus* (Ocypodidae): food choice, food quality and assimilation efficiency. *Marine Biology* 151: 1665–1681.

Noriega, C. and Araujo, M. (2014). Carbon dioxide emissions from estuaries of northern and northeastern Brazil. *Scientific Reports* 4: 1–9.

Nyunja, J., Ntiba, M., Onyari, J. et al. (2009). Carbon sources supporting a diverse fish community in a tropical coastal ecosystem (Gazi Bay, Kenya). *Estuarine, Coastal and Shelf Science* 83: 333–341.

Nóbrega, G.N., Ferreira, T.O., Siqueira Neto, M. et al. (2016). Edaphic factors controlling summer (rainy season) greenhouse gas emissions (CO_2 and CH_4) from semiarid mangrove soils (NE-Brazil). *Science of the Total Environment* 542: 685–693.

Oakes, J.M., Connolly, R.M., and Revill, A.T. (2010). Isotope enrichment in mangrove forests separates microphytobenthos and detritus as carbon sources for animals. *Limnology and Oceanography* 55: 393–402.

Odebrecht, C., Du Preez, D.R., Abreu, P.C. et al. (2014). Surf zone diatoms: a review of the drivers, patterns and role in sandy beach food chains. *Estuarine, Coastal and Shelf Science* 150: 24–35.

Odum, W.E., McIvor, C.C., and Smith, T.J. III (1982). *The Ecology of Mangroves of South Florida: A Community Profile*. FWS/OBS-81/24. Washington, DC: U.S. Fish and Wildlife Service.

Oliveira-Santos, N.M., Martins-Garcia, T., and de Oliveira-Soares, M. (2016). Micro- and mesozooplankton communities in the surf zone of a tropical sandy beach (Equatorial SW Atlantic). *Latin American Journal of Aquatic Research* 44: 247–255.

Olson, R.J., Duffy, L.M., Kuhnert, P.M. et al. (2014). Decadal diet shift in yellowfin tuna *Thunnus albacares* suggests broad-scale food web changes in the eastern tropical Pacific Ocean. *Marine Ecology Progress Series* 497: 157–178.

Osma, N., Fernández-Urruzola, I., Packard, T.T. et al. (2014). Short-term patterns of vertical particle flux in northern Benguela: a comparison between sinking POC and respiratory carbon consumption. *Journal of Marine Systems* 140: 150–162.

Pape, E., Muthumbi, A., Kamanu, C.P. et al. (2008). Size-dependent distribution and feeding habits of *Terebralia palustris* in mangrove habitats of Gazi Bay, Kenya. *Estuarine, Coastal and Shelf Science* 76: 797–808.

Park, G.-H., Wanninkhof, R., Doney, S.C. et al. (2010). Variability of global net sea-air CO_2 fluxes over the last three decades using empirical relationships. *Tellus* 62B: 352–368.

Pascal, P.-V., Bocher, P., Lefrançois, C. et al. (2019). Meiofauna versus macrofauna as a food resource in a tropical intertidal mud flat. *Marine Biology* 166: 144. doi:0.1107/s00227-019-3588-z.

Pattanaik, S., Chanda, A., Kumar Sahoo, R. et al. (2020). Contrasting intra-annual inorganic carbon dynamics and air-water CO_2 exchange in Dhamra and Mahanadi estuaries of northern Bay of Bengal, India. *Limnology* 21: 129–138.

Pawlik, J.R. and McMurray, S.E. (2020). The emerging ecological and biogeochemical importance of sponges on coral reefs. *Annual Review of Marine Science* 12: 315–337.

Pennington, J.T., Mahoney, K.L., Kuwahara, V.S. et al. (2006). Primary production in the eastern tropical Pacific: a review. *Progress in Oceanography* 69: 285–317.

Piersma, T., De Goeij, P., and Tulp, I. (1993). An evaluation of intertidal feeding habitats from a shorebird perspective: towards relevant comparisons between temperate and tropical mud flats. *Netherlands Journal of Sea Research* 31: 503–512.

Pinotti, R.M., Minasi, D.M., Colling, L.A. et al. (2014). A review of microbenthic trophic relationships along subtropical sandy shores in southernmost Brazil. *Biota Neotropica* 14: e20140069. https://doi.org/10.1590/1676-0603214006914.

Polovina, J.J. (1984). Model of a coral reef ecosystem. I. The ECOPATH model and its application to French Frigate Shoals. *Coral Reefs* 3: 1–11.

Poon, D.Y.N., Chan, B.K.K., and Williams, G.A. (2010). Spatial and temporal variation in diets of the crabs *Metopograpsus frontalis* (Grapsidae) and *Perisesarma bidens* (Sesarmidae): implications for mangrove food webs. *Hydrobiologia* 638: 29–40.

Poovachiranon, S., Boto, K.G., and Duke, N.C. (1986). Food preference studies and ingestion rate measurements of the mangrove amphipod *Parhyale hawaiensis* (Dana). *Journal of Experimental Marine Biology and Ecology* 98: 19–28.

Potier, M., Marsac, F., Lucas, V. et al. (2004). Feeding partitioning among tuna taken in surface and mid-water layers: the case of yellowfin (*Thunnus albacares*) and bigeye (*T. obesus*) in the western tropical Indian Ocean. *Western Indian Ocean Journal of Marine Science* 3: 51–62.

Pratihary, A.K., Naqvi, S.W.A., Narvenkar, G. et al. (2014). Benthic mineralization and nutrient exchange over the inner continental shelf of western India. *Biogeosciences* 11: 2771–2791.

Prince, E.D. and Goodyear, C.P. (2006). Hypoxia-based habitat compression of tropical pelagic fish. *Fisheries Oceanography* 15: 451–464.

Proffitt, C.E. and Devlin, D.J. (2005). Grazing by the intertidal gastropod *Melampus coffeus* greatly increases mangrove litter degradation rates. *Marine Ecology Progress Series* 296: 209–218.

Puga, C.A., Torres, A.S.S., Paiva, P.C. et al. (2019). Multi-year changes of a benthic community in the mid-intertidal rocky shore of a eutrophic tropical bay (Gunanabara Bay, RJ- Brazil). *Estuarine, Coastal and Shelf Science* 226: 106265. https://doi.org/10.1016/j.ecss.2019.106265.

Ram, A.S.P., Nair, S., and Chandramohan, D. (2009). Bacterial growth efficiency in a tropical estuary: seasonal variability subsidized by allochthonous carbon. *Microbial Ecology* 53: 591–599.

Rao, G.D. and Sarma, V.V.S.S. (2017). Influence of river discharge on the distribution and flux of methane in the coastal Bay of Bengal. *Marine Chemistry* 197: 1–10.

Ravichandran, S., Anthonisamy, A., Kannupandi, T. et al. (2007). Leaf choice of herbivorous crabs. *Research Journal of Environmental Sciences* 1: 26–30.

Richardson, T.L., Jackson, G.A., Ducklow, H.W. et al. (2004). Carbon fluxes through food webs of the eastern equatorial Pacific: an inverse approach. *Deep-Sea Research I* 51: 1245–1274.

Rix, L., de Goeij, J.M., Mueller, C.E. et al. (2016). Coral mucus fuels the sponge loop in warm- and cold-water coral reef ecosystems. *Scientific Reports* 6: 18715. https://doi.org/10.1038/srep18715.

Rix, L., de Goeij, J.M., van Oevelen, D. et al. (2018). Reef sponges facilitate the transfer of coral-derived organic matter to their associated fauna via the sponge loop. *Marine Ecology Progress Series* 589: 85–96.

Robbins, L.L., Daly, K.L., Barbero, L. et al. (2018). Spatial and temporal variability of pCO_2, carbon fluxes, and saturation state on the West Florida shelf. *Journal of Geophysical Research: Oceans* 123: 6174–6188.

Robertson, A.I. and Alongi, D.M. (1995). Role of riverine mangrove forests in organic carbon export to the tropical coastal ocean: a preliminary mass balance for the Fly delta (Papua New Guinea). *Geo-Marine Letters* 15: 134–139.

Roman, M.R., Adolf, H.A., Landry, M.R. et al. (2002). Estimates of oceanic mesozooplankton production: a comparison using the Bermuda and Hawaii time-series data. *Deep-Sea Research II* 49: 175–192.

Roobaert, A., Laruelle, G.G., Landschützer, P. et al. (2019). The spatiotemporal dynamics of the sources and sinks of CO_2 in the global coastal ocean. *Global Biogeochemical Cycles* 33: 1693–1714.

Rosentreter, J.A., Maher, D.T., Erler, D.V. et al. (2018a). Factors controlling seasonal CO_2 and CH_4 emissions in three tropical mangrove-dominated estuaries in Australia. *Estuarine, Coastal and Shelf Science* 215: 69–82.

Rosentreter, J.A., Maher, D.T., Erler, D.V. et al. (2018b). Methane emissions partially offset 'blue carbon' burial in mangroves. *Science Advances* 8: eaao4985. https://doi.org/10.1126/sciadv.aao4985.

Saifullah, A.S.M., Kamal, A.H.M., Idris, M.H. et al. (2019). Community composition and diversity of phytoplankton in relation to environmental variables and seasonality in a tropical mangrove estuary. *Regional Studies in Marine Science* 32: 100826. https://doi.org/10.1016/j.rsma.2019.100826.

Samanta, S., Dalai, T.K., Pattanaik, J.K. et al. (2015). DIC and its $\delta^{13}C$ in the Ganga (Hooghly) River estuary, India: evidence of DIC generation via organic carbon degradation and carbonate dissolution. *Geochimica et Cosmochimica Acta* 165: 226–248.

Sarma, V.V.S.S., Gupta, S.N.M., Babu, P.V.R. et al. (2009). Influence of river discharge on plankton metabolic rates in the tropical monsoon driven Godavari estuary, India. *Estuarine, Coastal and Shelf Science* 85: 515–524.

Sarma, V.V.S.S., Kumar, N.A., Prasad, V.R. et al. (2011). High CO_2 emissions from the tropical Godavari estuary (India) associated with monsoon river discharges. *Geophysical Research Letters* 38: L08601. https://doi.org/10.1029/2011GL046928.

Sarma, V.V.S.S., Swathi, P.S., Dileep Kumar, M. et al. (2003). Carbon budget in the eastern and central Arabian Sea: an Indian JGOFS synthesis. *Global Biogeochemical Cycles* 17: 1102. https://doi.org/10.1029/2002GB001978.

Scharler, U.M. (2011). Whole food-web studies: mangroves. In: *Treatise on Estuarine and Coastal Science* (eds. E. Wolanski and D. McClusky), 275–286. Amsterdam: Elsevier.

Schlacher, T.A., Strydom, S., Connolly, R.M. et al. (2013). Donor-control of scavenging food webs at the land-ocean interface. *PLoS One* 8: e68221. https://doi.org/10.1037/journal.pone.0068221.

Schukat, A., Teuber, L., Hagen, W. et al. (2013). Energetics and carbon budgets of dominant calanoid copepods in the northern Benguela upwelling system. *Journal of Experimental Marine Biology and Ecology* 442: 1–9.

Schwamborn, R., Ekau, W., Silva, A.P. et al. (2006). Ingestion of large centric diatoms, mangrove detritus, and zooplankton by zoeae of *Aratus pisonii* (Crustacea, Brachyura, Grapsidae). *Hydrobiologia* 560: 1–13.

Sea, M.A., Garcias-Bonet, N., Saderne, V. et al. (2018). Carbon dioxide and methane fluxes at the air-sea interface of Red Sea mangroves. *Biogeosciences* 15: 5365–5375.

Sheaves, M. and Molony, B. (2000). Short-circuit in the mangrove food chain. *Marine Ecology Progress Series* 199: 97–109.

Sheaves, M., Sheaves, J., Stegemann, K. et al. (2014). Resource partitioning and habitat-specific dietary plasticity of two estuarine sparid fish increase food-web complexity. *Marine and Freshwater Research* 65: 114–124.

Sheppard, C., Davy, S., and Pilling, G. (2018). *The Biology of Coral Reefs*, 2e. Oxford, UK: Oxford University Press.

Shirodkar, G., Wajih, S., Naqvi, A. et al. (2018). Methane dynamics in the shelf waters of the west coast of India during seasonal anoxia. *Marine Chemistry* 203: 55–63.

Short, F.T., Short, C.A., and Novak, A.B. (2018). Seagrasses. In: *The Wetland Book, II. Distribution, Description, and Conservation* (eds. C.M. Finlayson, R. Milton, C. Prentice, et al.), 73–91. New York: Springer.

Sieg, R.D. and Kubanek, J. (2013). Chemical ecology of marine angiosperms: opportunities at the interface of marine and terrestrial systems. *Journal of Chemical Ecology* 39: 687–711.

Signa, G., Mazzola, A., Kairo, J. et al. (2017). Small-scale variability in geomorphological settings influences mangrove-derived organic matter in a tropical bay. *Biogeosciences* 14: 617–629.

Silva Francisco, A. and Netto, S.A. (2020). El Niño-Southern Oscillations and Pacific Decadal Oscillation as drivers of the decadal dynamics of benthic macrofauna in two subtropical estuaries (southern Brazil). *Ecosystems* 23: 1380–1394.

Silveira, C.B., Cavalcanti, G.S., Walter, J.M. et al. (2017). Microbial processes driving coral reef organic carbon flow. *FEMS Microbiology Reviews* 41: 575–595.

Simon, L.N. and Raffaelli, D. (2016). A trophic model of the Cameroon estuary mangrove with simulations of mangrove impacts. *International Journal of Scientific and Technology Research* 5: 137–155.

Sippo, J.Z., Maher, D.T., Tait, D.R. et al. (2017). Mangrove outwelling is a significant source of exchangeable organic carbon. *Limnology and Oceanography Letters* 2: 1–8.
Skilleter, G.A., Loneragan, N.R., Olds, A. et al. (2017). Connectivity between seagrass and mangroves influences nekton assemblages using nearshore habitats. *Marine Ecology Progress Series* 573: 25–43.
Smith, S.V. and Mackenzie, F.T. (2016). The role of $CaCO_3$ reactions in the contemporary oceanic CO_2 cycle. *Aquatic Geochemistry* 22: 153–175.
Sosa-López, A., Mouillot, D., Ramos-Miranda, J. et al. (2007). Fish species richness decreases with salinity in tropical coastal lagoons. *Journal of Biogeography* 34: 52–61.
Sousa, W.P. and Dangremond, E.M. (2011). Trophic interactions in coastal and estuarine mangrove forest ecosystems. In: *Treatise on Estuarine and Coastal Science* (eds. E. Wolanski and D. McClusky), 43–93. Amsterdam: Elsevier.
Spear, L.B., Ainley, D.G., and Walker, W.A. (2007). Foraging dynamics of seabirds in the eastern tropical Pacific Ocean. *Studies in Avian Biology* 35: 1–99.
Steinberg, D.K. and Landry, M.R. (2017). Zooplankton and the ocean carbon cycle. *Annual Review in Marine Science* 9: 413–444.
Stukel, M.R., Décima, M., Landry, M.R. et al. (2018). Nitrogen and isotope flow through the Costa Rica Dome upwelling ecosystem: the crucial mesozooplankton role in export flux. *Global Biogeochemical Cycles* 32: 1815–1832.
Subramaniam, A., Yager, P.L., Carpenter, E.J. et al. (2008). Amazon River enhances diazotrophy and carbon sequestration in the tropical north Atlantic Ocean. *Proceedings of the National Academy of Sciences* 105: 10460–10465.
Sutton, A.L., Jenner, K.C.S., and Jenner, M.-N.M. (2019). Habitat associations of cetaceans and seabirds in the tropical eastern Indian Ocean. *Deep-Sea Research II* 166: 171–186.
Suzuki, A. and Kawahata, H. (2004). Reef water CO_2 system and carbon production of coral reefs: topographic control of system-level performance. In: *Global Environmental Change in the Ocean and on Land* (eds. M. Shiyomi, H. Kawahata, H. Koizumi, et al.), 229–248. Tokyo: Terrapub.
Taillardat, P., Willemsen, P., Marchand, C. et al. (2018). Assessing the contribution of porewater discharge in carbon export and CO_2 evasion in a mangrove tidal creek (Can Gio, Vietnam). *Journal of Hydrology* 563: 303–316.
Taillardat, P., Ziegler, A.D., Friess, D.A. et al. (2019). Assessing nutrient dynamics in mangrove porewater and adjacent tidal creek using nitrate dual-stable isotopes: a new approach to challenge the outwelling hypothesis? *Marine Chemistry* 214: 103662. https://doi.org/10.1016/j.marchem.2019.103662.
Thimdee, W., Deein, G., Sangrungruang, C. et al. (2004). Analysis of primary food sources and trophic relationships of aquatic animals in a mangrove-fringed estuary, Khung Krabaen Bay (Thailand) using duel stable isotope techniques. *Wetlands Ecology and Management* 12: 135–144.
Tseng, H.-C., Chen, C.-T.A., Borges, A.V. et al. (2017). Methane in the South China Sea and the western Philippine Sea. *Continental Shelf Research* 135: 23–34.
Tue, N.T., Hamaoka, H., Quy, T.D. et al. (2014). Dual isotope study of food sources of a fish assemblage in the Red River mangrove ecosystem, Vietnam. *Hydrobiologia* 733: 71–83.
Ullah, M.H., Rashed-Un-Nabi, M., and Al-Mamun, M.A. (2012). Trophic model of the coastal ecosystem of the Bay of Bengal using mass balance Ecopath model. *Ecological Modelling* 225: 82–94.
Unsworth, R.K.F., Nordlund, L.M., and Cullen-Unsworth, L.C. (2019). Seagrass meadows support global fisheries production. *Conservation Letters* 12: e12566. https://doi.org/10.1111/conl.12566.
Unsworth, R.K.F., Taylor, J.D., Powell, A. et al. (2007b). The contribution of scarid herbivory to seagrass ecosystem dynamics in the Indo-Pacific. *Estuarine, Coastal and Shelf Science* 74: 53–62.
Unsworth, R.K.F., Wylie, E., Smith, D.J. et al. (2007a). Diel trophic structuring of seagrass bed fish assemblages in the Wakatobi Marine Natural Park, Indonesia. *Estuarine, Coastal and Shelf Science* 72: 81–88.

Valentine, J.F. and Duffy, J.E. (2006). The central role of grazing in seagrass ecology. In: *Seagrasses: Biology, Ecology and Conservation* (ed. A.W.D. Larkum), 463–501. Dordrecht, The Netherlands: Springer.

Valentine, J.F., Heck, K.L. Jr., Blackmon, D. et al. (2007). Food web interactions along seagrass-coral reef boundaries: effects of piscivore reductions on cross-habitat energy exchange. *Marine Ecology Progress Series* 333: 37–50.

van der Heide, T., Govers, L.L., de Fouw, J. et al. (2012). A three-stage symbiosis forms the foundation of seagrass ecosystems. *Science* 336: 1432–1434.

van Nedervelde, F., Cannicci, S., Koedam, N. et al. (2015). What regulates crab predation on mangrove propagules? *Acta Oecologia* 63: 63–70.

Vargas, J.A. (1988). Community structure of macrobenthos and the results of macropredator exclusion on a tropical intertidal mud flat. *Revista de Biologica Tropical* 36: 287–308.

Vega-Cendejas, M.E. and Arreguin-Sánchez, F. (2001). Energy fluxes in a mangrove ecosystem from a coastal lagoon in Yucatan Peninsula, Mexico. *Ecological Modelling* 137: 119–133.

Vergés, A., Doropoulos, C., Czarnik, R. et al. (2018). Latitudinal variation in seagrass herbivory: global patterns and explanatory mechanisms. *Global Ecology and Biogeography* 27: 1968–1079.

Vilchis, L.I., Balance, L.T., and Fiedler, P.C. (2006). Pelagic habitat of seabirds in the eastern tropical Pacific: effects of foraging ecology on habitat selection. *Marine Ecology Progress Series* 315: 279–292.

Villanueva, M.C., Laléyé, P., Albaret, J.-J. et al. (2006). Comparative analysis of trophic structure and interactions of two tropical lagoons. *Ecological Modelling* 197: 461–477.

Vinagre, C., Mendonça, V., Flores, A.A.V. et al. (2018). Complex food webs of tropical intertidal rocky shores (SE Brazil) – an isotopic perspective. *Ecological Indicators* 95: 485–491.

Vinueza, L.R., Branch, G.M., Branch, M.L. et al. (2006). Top-down herbivory and bottom-up El Niño effects on Galápagos rocky-shore communities. *Ecological Monographs* 76: 111–131.

Volta, C., Ho, D.T., Maher, D.T. et al. (2020). Seasonal variations in dissolved carbon inventory and fluxes in a mangrove-dominated estuary. *Global Biogeochemical Cycles* 34: GB006515. https://doi.org/10.1029/2019GB006515.

Vonk, J.A., Christianen, M.J.A., and Stapel, J. (2008b). Redefining the trophic importance of seagrasses for fauna in tropical Indo-Pacific meadows. *Estuarine, Coastal and Shelf Science* 79: 653–660.

Vonk, J.A., Pijnappels, M.H.J., and Stapel, J. (2008a). In situ quantification of *Tripneustes gratilla* grazing and its effects on three co-occurring tropical seagrass species. *Marine Ecology Progress Series* 360: 107–114.

Wallberg, P., Jonsson, P.R., and Johnstone, R. (1999). Abundance, biomass and growth rates of pelagic microorganisms in a tropical coastal ecosystem. *Aquatic Microbial Ecology* 18: 175–185.

Wellens, S., Sandrini-Neto, L., Gonzalez-Wangűemert, M. et al. (2015). Do the crabs *Goniopsis cruentata* and *Ucides cordatus* compete for mangrove propagules? A field-based experimental approach. *Hydrobiologia* 757: 117–128.

Wendländer, N.S., Lange, T., Connolly, R.M. et al. (2020). Assessing methods for restoring seagrass (*Zostera muelleri*) in Australia's subtropical waters. *Marine and Freshwater Research* 71: 996–1005.

Williams, C.J., Jaffé, R., Anderson, W.T. et al. (2009). Importance of seagrass as a carbon source for heterotrophic bacteria in a subtropical estuary (Florida Bay). *Estuarine, Coastal and Shelf Science* 85: 507–514.

Williams, G.A., Davies, M.S., and Nagarkar, S. (2000). Primary succession on a seasonal tropical rocky shore: the relative roles of spatial heterogeneity and herbivory. *Marine Ecology Progress Series* 203: 81–94.

Wolff, M. (2006). Biomass flow structure and resource potential of two mangrove estuaries: insights from comparative modelling in Costa Rica and Brazil. *Revista de Biología Tropical* 54: 69–86.

Wu, Y., Bao, H.-Y., Unger, D. et al. (2013). Biogeochemical behaviour of organic carbon in a small tropical river and estuary, Hainan, China. *Continental Shelf Research* 57: 32–43.
Wyatt, A.S.J., Lowe, R.J., Humphries, S. et al. (2013). Particulate nutrient fluxes over a fringing coral reef: source-sink dynamics inferred from carbon to nitrogen ratios and stable isotopes. *Limnology and Oceanography* 58: 409–427.
Yamamuro, M. (1999). Importance of epiphytic cyanobacteria as food sources for heterotrophs in a tropical seagrass bed. *Coral Reefs* 18: 263–271.
Yang, E.J., Choi, J.L., and Hyun, J.-H. (2004). Distribution and structure of heterotrophic protist communities in the NE equatorial Pacific Ocean. *Marine Biology* 146: 1–15.
Yang, K.Y., Lee, S.Y., and Williams, G.A. (2003). Selective feeding by the mudskipper (*Boleophthalmus pectinirostris*) on the microalgal assemblage of a tropical mud flat. *Marine Biology* 143: 245–256.
Yusup, Y., Alkarkhi, A.F.M., Kayode, J.S. et al. (2018). Statistical modelling the effects of microclimate variables on carbon dioxide flux at the tropical coastal ocean in the southern South China Sea. *Dynamics of Atmospheres and Oceans* 84: 10–21.
Zhai, W.-D., Dai, M.-H., Chen, B.-S. et al. (2013). Seasonal variations of sea-air CO_2 fluxes in the largest tropical marginal sea (South China Sea) based on multiple-year underway measurements. *Biogeosciences* 10: 7775–7791.
Zheng, X., Guo, J., Song, W. et al. (2018). Methane emission from mangrove wetland soils is marginal but can be stimulated significantly by anthropogenic activities. *Forests* 9: 738. https://doi.org/10.3390/f9120738.

CHAPTER 11

Nutrient Biogeochemistry

11.1 Introduction

Macronutrients such as N and P and micronutrients such as Fe and copper (Cu) are essential elements necessary to sustain life. The elemental composition of marine plankton was first suggested by Redfield et al. (1963) as uniformly close to the ratios of dissolved nutrients in the deep ocean and defined the Redfield molar ratio of $C_{106}N_{16}P_1$. A subsequent study extended the Redfield ratio to include trace elements, with the average composition of phytoplankton being $C_{124}N_{16}P_1S_{1.3}K_{1.7}Mg_{0.56}Ca_{0.5}Fe_{0.0075}Zn_{0.0008}Cu_{0.00038}Cd_{0.00021}Co_{0.00019}$ (Quigg et al. 2003). Recent studies have demonstrated systematic variation in the Redfield ratio across ocean regions and seasons, refuting the idea of a constant plankton Redfield ratio (Moreno and Martiny 2018). The C:N, C:P, and N:P ratios of POM are below or near Redfield proportions in high-latitude and equatorial upwelling regions, but commonly above Redfield proportions in the oligotrophic gyres; seasonality has been observed with the three ratios higher in the summer and fall and lower in the winter and spring (Moreno and Martiny 2018).

The movement of N between the biosphere, atmosphere, and geosphere in different forms is described by the N cycle, one of the major biogeochemical cycles. Like the C cycle, the N cycle (Figure 11.1) consists of various storage pools and processes in both anaerobic and aerobic conditions by which N is cycled. Typically, seven main processes transform N: the assimilatory pathways of N_2 fixation and NO_3^- and NH_4^+ uptake and the dissimilatory pathways of mineralisation (ammonification), nitrification, denitrification, dissimilatory nitrite reduction to ammonium (DNRA), and anaerobic ammonium oxidation (anammox). Bacteria and archaea perform nearly all principal N transformations. As microbially mediated processes, these N transformations tend to occur faster than geological processes, involve multiple oxidation states and phases, and are affected by environmental factors that influence microbial activity, such as temperature, moisture, and resource availability. In contrast, P cycling is somewhat simpler (Figure 11.2) in that it largely lacks changes in oxidation state and is dominated by mineralisation into soluble reactive phosphorus (SRP) and uptake thereof. Most P is tied up in sedimentary reservoirs, but there are atmospheric and fluvial inputs of P into marine systems. In the water column, SRP is rapidly

Tropical Marine Ecology, First Edition. Daniel M. Alongi.

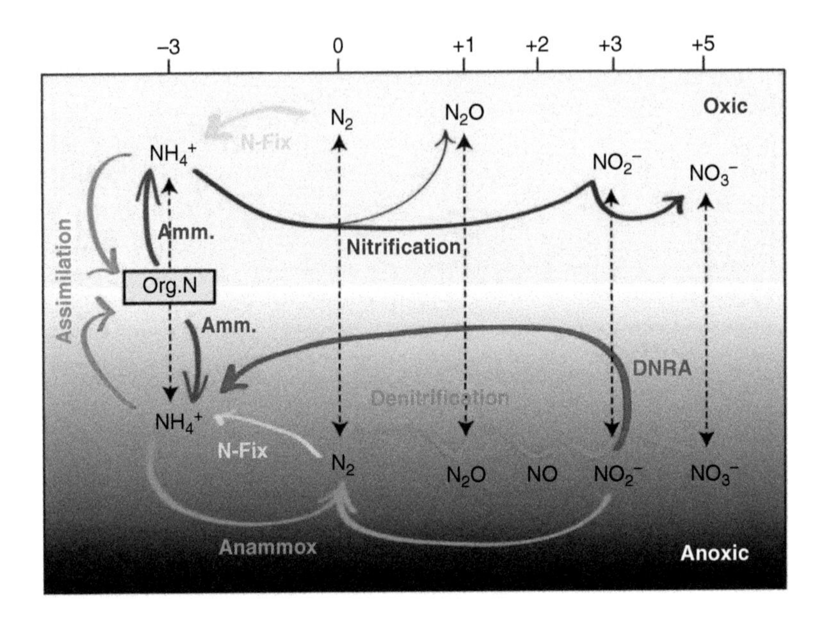

FIGURE 11.1 The N cycle under aerobic and anaerobic conditions, showing the main seven transformations involved in the cycle. Coloured arrows represent N transformations: N_2 fixation (yellow), nitrification (purple), DNRA (red), denitrification (orange), assimilation (green), ammonification (brown), and anammox (blue). Positive and negative oxidation states are at the top of the diagram.

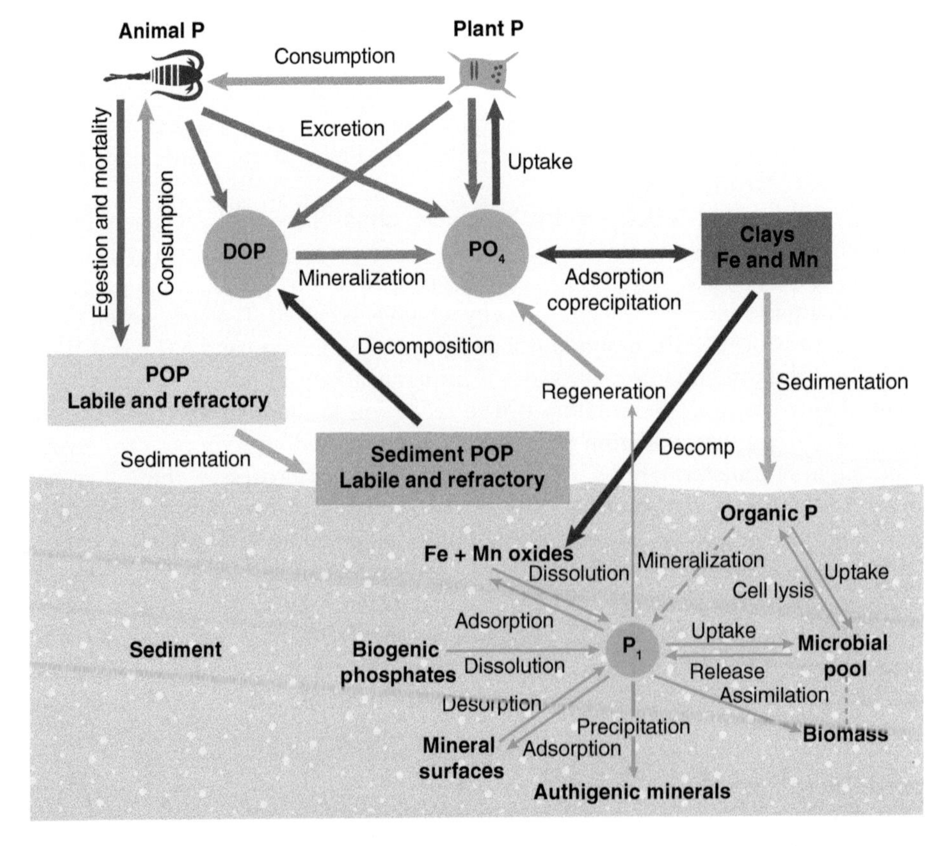

FIGURE 11.2 The P cycle in the coastal and marine environment. Major transformation processes between dissolved and solid phases and included sorption processes are shown.

recycled through the microbial hub while in sediments, P is involved in several reactions, including dissolution/absorption with Fe and Mn oxides, mineral surfaces and biogenic phosphates, precipitation as authigenic minerals such as vivianite, and involvement in biological processes such as mineralisation, uptake and release, cell lysis and assimilation into living biomass. SRP, typically as orthophosphate (PO_4^{3-}) and dissolved organic phosphorus (DOP), are efficiently taken up by marine plants from the interstitial water after being desorbed and unbound from the sedimentary particles. During the decomposition of organic matter, organic P is partly assimilated and partly released as dissolved inorganic phosphorus (DIP) by microheterotrophs, with the degree of partitioning primarily dependent on the C:P ratio of the source material. DIP in sediments is available as either $H_2PO_4^-$ or HPO_4^{2-} depending on pH, but readily forms insoluble precipitates with several divalent cations, particularly Ca^{2+} and Fe^{2+}. Under aerobic (oxic) conditions, PO_4^{3-} also adsorbs onto positively charged clay surfaces and oxides of Fe^{3+} and Al^{3+}. Under anaerobic (anoxic) conditions, Fe^{3+} is reduced to Fe^{2+} and other ferrous minerals (e.g. siderite = Fe carbonate) that are less adsorptive; P is subsequently released into the interstitial water.

N and P are essential nutrients, often limiting to marine plants and other organisms. Micronutrients such as Fe and Cu are also often limiting, especially in carbonate-rich environments. The role of macro- and micronutrients in controlling productivity is well understood. Major sources of nutrients to coastal ecosystems include atmospheric deposition and terrestrial export, upwelling, fertilisers from agriculture and aquaculture, and sewage. Tidal exchange can introduce nutrients such as marine N, but this source can sometimes be relatively minor in bays and estuaries receiving high nutrient loads from land via rivers. Both DON and DOP can be the major forms of N and P in marine systems, but only a small fraction of these nutrient species has been chemically characterised.

Excess inorganic nutrient loading from coastal watersheds results in increases in primary production, changes in habitat structure and function, acidification, and trophic dynamics of the receiving waters. By comparison, little is known about the cycling and bioavailability of organic nutrients in nearshore ecosystems, even though organic nutrient inputs to these systems can sometimes be higher than inorganic loadings. The retention and transport of nutrients through tropical marine and estuarine ecosystems have received far less attention than in temperate systems. A major difference in nutrient cycling processes within tropical ecosystems is the large influence that benthic primary producers have on nutrient transformations in shallow coastal areas. Exchanges across the sediment–air interface play a key role in nutrient cycling in tropical coastal ecosystems because the water column is well mixed and small in volume relative to the larger and more abundant sediment pool. It is this coastal margin interface that makes the biogeochemical cycles of nutrients complex.

11.2 Sandy Beaches, Tidal Flats, and Rocky Intertidal Shores

Nutrient cycling in tropical intertidal sand, mud, and rocky shores is poorly understood. There are some data for sandy beaches, but none for tidal flats and rocky intertidal shores.

A N budget exists for a subtropical sandy beach in South Africa (Figure 11.3) which indicates rapid uptake and utilisation of N by biota (Cockcroft and McLachlan 1993). Phytoplankton were the major users of N as macrophytes were absent. Inputs of N occurred via groundwater inflow and from an estuary, with rainfall and airborne detritus being of minor importance. Loss of N to the sediment due to beach accretion or via denitrification was negligible because of sediment reworking and the highly oxidised state of the surf zone. Sorting, shifting, and disturbance of sediments by benthic organisms resulted in changes to the redox state of the deposits and to the availability of particulate and dissolved nutrients. The amounts of N recycled by the macrofauna, meiofauna, and the microbial hub were large. The budget indicated that (i) inorganic N inputs (groundwater, rain, and estuary) into the surf zone supply 13% of the phytoplankton requirements for primary production, (ii) the three trophic groups together recycle 99% of the N required for NPP, and (iii) DON and particulate organic nitrogen (PON) recycled by the macrofauna can supply 24% of the N requirements for the microbial hub. The concept that the surf zone is a self-sustaining, semi-closed ecosystem is supported by the fact that the fauna regenerate almost all the N required for phytoplankton production (Cockcroft and McLachlan 1993).

There is other evidence that the macrofauna can support the microbial hub on sandy beaches. N regeneration and subsequent excretion by some macrofauna can be rapid (Cockcroft et al. 1988; Cockcroft 1990). N regeneration by two surf-zone mysid shrimps, *Mesopodopsis slabberi* and *Gastrosaccus psammodytes,* was rapid, constituting 10% of total phytoplankton N requirements in the surf zone. N excretion by the surf zone bivalves, *Donax serra* and *D. sordidus,* accounted for 2% of total phytoplankton N requirements. Adding together all the macrofauna excretion in the surf zone, nearly all N requirements of the phytoplankton were accounted for.

N cycling in tropical tidal flats, mud banks, and rocky intertidal shores is poorly understood but presumably rapid given the rapid turnover of carbon on these shores. In intertidal mud banks in the Fly delta, Alongi (1991) found that bacterial densities were extraordinarily high, and that bacterial production was very rapid. Fluxes of dissolved N and P across the sediment–water interface were mostly

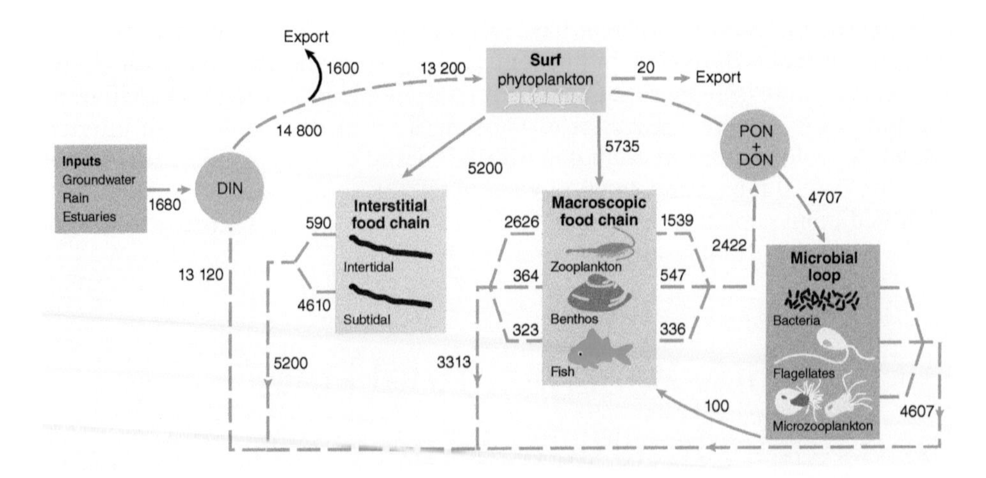

FIGURE 11.3 N budget and recycling in the Sundays beach/surf zone in South Africa. Solid lines indicate grazing, broken lines denote recycling or nutrient flux. All values in g N m^{-1} d^{-1}. *Source:* Cockcroft and McLachlan (1993), figure 3, p. 294. © Inter-Research Science Publisher.

undetectable, indicating possible storage of nutrients in the mud banks and/or uptake at the sediment–water interface by benthic microalgae. Rapidly growing bacterial communities in these sediments sequestered and mineralised most of the labile organic matter on these mud banks which acted as nutrient sinks rather than as net exporters of nutrients. That N cycling is presumably rapid in tropical intertidal shores can be gleaned from the rapid rates of denitrification measured on two intertidal mud flats in northern Australia (Alongi et al. 1999). Unlike in most other sedimentary environments, denitrification in these mud flats accounted for 56–89% of total carbon mineralisation.

Other evidence supports the notion that intertidal habitats can act as either nutrient sinks or sources of N and P for adjacent coastal habitats. Whether a particular habitat is a sink or source depends greatly on the depositional setting: quiescent tidal flats where sediments are depositing rapidly are nutrient sinks, whereas energetic tidal flats are more likely to be net sources. Gontharet et al. (2014) found that most of the organic matter on energetic intertidal mud banks in French Guiana was mostly controlled by suspended particulate matter associated with variable amounts of POM derived from mangroves and microphytobenthos. The rapid accumulation of these materials forms highly unstable shore-attached mud banks which temporally emerge and then rapidly colonised and stabilised by microphytobenthos and mangroves, mainly by *Avicennia germinans*. These mud banks, being unstable, are preferential sites for accumulation and significant remineralisation of organic matter due to intense erosion/deposition cycles and potential biological colonisation.

Bacterial communities in tidal flat sediments are highly active leading to significant mineralisation of N and P. In tidal flats near Jamaican mangroves, Nedwell et al. (1994) found that these deposits were active sites of subsurface organic matter mineralisation with turnover times of NH_4^+ on the order of a few days. In Thai tidal flats close to seagrasses and mangroves, Kristensen et al. (2000) similarly found that the sediments were primarily aerobic, with aerobic respiration being responsible for 45–65% of total mineralisation with 10% decomposed via sulphate reduction and the remainder by iron reduction. Both sand flat and mud flat mineralisation rates were quite different than in the seagrasses and mangroves due to the presence of plant roots and infauna, although the shift in dominance of electron acceptors was thought to be also a function of grain size, organic content, and Fe content. Nitrogenase activity was measured by Lyimo and Lugomela (2006) in intertidal sediments along the Tanzanian coast where N_2-fixing ability was highly active, contributing a substantial amount of N to the Tanzanian coastline. The available data, although small, point to rapid N and P cycling in tropical intertidal sediments.

11.3 Seagrass Meadows

N and P cycling is several times faster in seagrass sediments than in adjacent unvegetated sediments due to high densities of bacteria, especially N_2-fixing and denitrifying bacteria, that reside in plant roots and rhizomes. Seagrasses have a major impact on sediment nutrient cycling by leakage of O_2 and DOM from the roots. Both processes are photosynthetically driven. Oxygen produced by leaf photosynthesis is transported to the roots to support aerobic respiration via a well-developed lacunal system, a series of air channels comprising up to 60% of the total plant volume. The

O_2 that is not respired is released into the rhizosphere, creating oxidised zones in an otherwise reducing environment. Seagrasses simultaneously release DOM from the roots, typically as simple organic compounds that are recent products of photosynthesis. Diurnal cycles of rhizophore metabolism in seagrasses are tightly coupled to light/dark cycles.

Bacterial productivity in the rhizosphere has been linked to these organic exudates. Moriarty et al. (1986) showed that 11% of recent photosynthate was released within six hours into the rhizosphere of the tropical seagrass, *H. wrightii*. Seasonal and light-induced variation in bacterial activity and a subsurface peak in bacterial productivity in the sediments where root biomass was highest, indicated that root exudates stimulated bacterial productivity.

For some seagrass species, O_2 released from roots and the creation of oxidised zones around the rhizosphere stimulates nitrification, which produces NO_3^- to support enhanced denitrification. This stimulation may be linked to seagrass growth and to variations in O_2 release. Enhanced denitrification may occur in sediments vegetated by several species, especially in tropical species such as *T. testudinum*, which allocate more biomass to roots and rhizomes than temperate species, and which tend to be strongly P-limited in carbonate sediments. Organic matter concentrations and bacterial oxygen demand tend to be lower in tropical carbonate sediments, which may make oxygenation of the rhizosphere more efficient. There are relatively few measurements of denitrification in tropical seagrass-vegetated carbonate sediments (Table 11.1), but rates tend to be high (Shieh and Yang 1997; Alongi et al. 2008a). The available data suggest that both denitrification and N_2 fixation rates are higher in tropical seagrass meadows than in temperate meadows (McGlathery et al. 2004; Garcias-Bonet et al. 2018). These differences may partly be due to increased DOM release from the greater root/rhizome biomass of many tropical seagrasses.

N_2 fixation activity is enhanced in tropical seagrass sediments (Iizumi and Yamamuro 2000; Alongi et al. 2008a; Lyimo and Hamisi 2008; Hamisi et al. 2009; Raja et al. 2012; Garcias-Bonet et al. 2018) relative to unvegetated substrate. The availability of organic material is a major factor controlling N_2 fixation rates in seagrasses. Many studies have shown increases in N_2 fixation by addition of organic compounds, and seasonal and diel variations that are consistent with the role of photosynthetic exudates stimulating N_2-fixing bacteria (McGlathery et al. 2004; Garcias-Bonet et al. 2018). Much of this activity is associated with sulphate reducers which are also believed to be stimulated by root exudates. Autotrophic, heterocystous cyanobacterial epiphytes on seagrass leaves also contribute to N_2 fixation rates, especially in warm tropical environments, although on an area basis the activity is lower than that of heterotrophic bacteria in the sediments.

Few studies have measured N_2 fixation and denitrification simultaneously in seagrass beds to determine if N gains by fixation can compensate for N losses via denitrification. In Jamaican seagrasses, Blackburn et al. (1994) observed that N_2 fixation rates account for <50% of N loss by denitrification. Similarly, Alongi et al. (2008a) measured rapid rates of denitrification compared with rates of N_2 fixation in Indonesian seagrasses. Rates of ammonification were greater than denitrification, suggesting substantial intermediate transformation via nitrification supporting denitrification.

The uptake of DON, NH_4^+, and PO_4^{3-} by leaves and roots of tropical seagrasses can be equally rapid (Stapel et al. 1996; Gras et al. 2003; Vonk et al. 2008c; Alexandre et al. 2014). Stapel et al. (1996) and Alexandre et al. (2014) found that the uptake of nutrients by *T. hemprichii* and *H. stipulacea* can be described by Michaelis-Menten

TABLE 11.1 **Rates of nitrogen fixation and denitrification in some tropical/subtropical seagrass meadows.**

Species	Location	N_2 fixation	Denitrification	Reference
Enhalus acoroides	Gulf of Carpentaria, Australia	1.8		Moriarty and O'Donohue (1993)
E. acoroides	Awerange Bay, Indonesia	0.04–1.3	1.0–1.9	Alongi et al. (2008a)
E. acoroides	Central Red Sea, Saudi Arabia	0.4	2.5	Garcias-Bonet et al. (2018)
Halophila ovalis	Moreton Bay, Australia	0.6		Carlson-Perret et al. (2019)
Syringodium isoetifolium	Gulf of Carpentaria, Australia	1.1–3.4		Moriarty and O'Donohue (1993)
Thalassia testudinum	Oyster Bay, Jamaica	1.0	2–4	Blackburn et al. (1994)
T. testudinum	Florida Bay, Florida, United States	0.7–4	1.6–3.0	Kemp and Cornwell (2001)
T. testudinum	St. Joseph's Bay, Florida, United States		2.1–5.1	Hoffman et al. (2019)
T. hemprichii	Gulf of Carpentaria, Australia	1.1		Moriarty and O'Donohue (1993)
Zostera capricorni	Moreton Bay, Australia	0.7–2.9		O'Donohue et al. (1991)
Zostera muelleri	Moreton Bay, Australia	0.9–1.8		Carlson-Perret et al. (2018, 2019)

Rates are in mmol N m^{-2} d^{-1}. Single values are means and multiple values are ranges.*Sources:* Based on Shieh and Yang (1997)and Alongi et al. (2008a).

kinetics. Calculations suggested that nutrient uptake by roots was probably diffusion-limited rather than by uptake capacity, while leaves clearly have a significant ability for nutrient uptake, and that in some situations, nutrient uptake by leaves may even be essential in meeting plant nutrient requirements. Indeed, Alexandre et al. (2014) and Carlson-Perret et al. (2018) found that leaf uptake was greater than uptake by roots. Seagrasses take up DON in the form of urea (Vonk et al. 2008c) and seagrass leaves prefer urea, NH_4^+ and NO_3^- over amino acids, with differences between species. Seagrass roots, however, take up amino acids at rates comparable to NH_4^+, whereas uptake rates of urea and NO_3^- are much lower. The ability of tropical seagrasses to take up DON enables seagrasses to shortcut N cycling and gives them access to additional N resources. N uptake, assimilation, translocation, and storage under a range of organic N sources through above- and below-ground tissues of *C. serrulata* and *T. hemprichii* when enriched with high N concentrations showed that uptake rates of *T. hemprichii* were higher than *C. serrulata* in leaves and rhizomes, whereas root uptake was higher in *C. serrulata* (Viana et al. 2020). Acropetal (upwards from the base to the apex) and basipetal (downwards to the roots) translocation occurred in both species. *C. serrulata* tended to immediately use the available N, whereas *T. hemprichii* allocated more N to assimilation and storage. These results suggest that tropical seagrasses can adapt to N enrichment.

Leaf die-off is rapid and probably results in some retention of N in seagrass meadows. In a field experiment, Stapel et al. (2001) enriched shoots of the tropical seagrass, *T. hemprichii,* with ^{15}N by brief exposure of the leaves. Thereafter, the ^{15}N absorbed in the seagrass declined rapidly with loss of ^{15}N in detached leaf fragments. Of the lost fragments, 19% were recovered within the experimental plots and 25% deposited outside the plots but within the meadow. The limited N retention in the plots was ascribed to the combined effects of major allocation of N to leaf production, restricted N resorption from senescent leaves and a dynamic environment facilitating detachment and export of leaf fragments from the experimental plots.

At the scale of an entire meadow, N conservation via detritus could be of considerable significance. The input of N into litter of the seagrasses, *T. hemprichii, H. uninervis,* and *C. rotundata,* during decomposition and the uptake of ^{15}N-enriched seagrass leaf material is rapid, with the seagrass meadows efficiently taking up the released N (Vonk and Stapel 2008). Canopy thickness of the seagrass meadow played a role in the uptake efficiency of N released from litter, increasing N retention in these Indo-Pacific meadows. The high litter decomposition rates, substantial input of exogenous N onto litter and efficient uptake of N released from litter suggest that N cycling, outside the living plant but within the meadow via the detrital pathway is important to retain N in these nutrient-poor meadows.

There is considerable evidence for N and P limitation in tropical and subtropical seagrass meadows, especially those inhabiting carbonate deposits (Touchette and Burkholder 2000). Johnson et al. (2006) found evidence to suggest both N and P limitation with greater P limitation in two bays in the northern Gulf of Mexico, while N limitation was observed in other Gulf of Mexico seagrass beds (Lee and Dunton 2000); further east in south Florida, both N and P limitations in seagrass meadows were indicated (Ferdie and Fourqurean 2004). In *H. pinifolia* and *C. serrulata* meadows in Palk Bay, India, N:P stoichiometry suggested P deficiency (Thangaradjou et al. 2015). The available data thus strongly indicate that P is limiting for seagrasses in carbonate environments (Jensen et al. 1998, 2009; McGlathery et al. 2001; Herbert and Fourqurean 2009; Thangaradjou et al. 2015).

The forms and availability of P are influenced by both O_2 and DOM release by seagrass roots, and by sediment Fe content (McGlathery et al. 2004). The formation of Fe oxides that effectively bind P in the solid phase are the result of oxidation of Fe in the rhizosphere. The redox conditions of the sediment and the formation of autogenic P minerals influence the release of P to overlying waters. In carbonate sediments that commonly dominate tropical seagrass systems, P turnover rates are generally high in the porewaters relative to N, indicating that P is preferentially removed (Alongi et al. 2008a). The mechanisms responsible for this phenomenon include surface adsorption onto carbonate grains, direct precipitation of Ca-P mineral phases, and uptake by P-limited primary producers. In vegetated carbonate sediments in Bermuda, Jensen et al. (1998) found that only 2% of total sediment P was in the loosely adsorbed pool that could be readily released to the porewater; 5–20% was more strongly adsorbed to the sediment surface, and the remainder (80%) was bound in the mineral matrix of the sediments, probably in the form of carbonate fluorapatite and other carbonates.

An early supposition was that carbonate sediments are a permanent sink for P in these tropical environments and that mineral P is not available for seagrasses.

However, modern studies suggest that seagrass metabolism facilitates the dissolution of carbonate minerals in the rhizosphere releasing bound P (Burdige et al. 2010). Mechanisms for this may be the decrease in pH resulting from root respiration or the stimulation of bacterial respiration by root DOM release. The oxidation of sulphides in the rhizosphere may also decrease pH. Another mechanism is the release of organic acids from the roots that also acidify the rhizosphere, causing the dissolution of carbonate minerals.

Fe associated with shallow-water carbonate minerals (typically 0.06–4.86% by sediment DW; Best et al. 2007) can serve as sorption sites for inorganic P and can react with sedimentary sulphides to precipitate iron sulphide minerals that reduce sulphide toxicity to primary producers. Chambers et al. (2001) found that Fe additions increased the total sediment P pool and reduced exposure of the seagrass *T. testudinum* to free sulphide but found no growth response probably because most of the sediment P was bound in unavailable forms. One might expect a more significant effect of Fe additions on seagrasses in carbonate sediments where P is not as strongly limiting, that is, where P is in more readily exchangeable forms.

On an ecosystem scale, tropical seagrass meadows can be important temporary sinks for nutrients. The sink–source role varies seasonally, depending on the growth requirements of the seagrasses. Seagrasses enhance nutrient concentrations in the sediments both in particulate and dissolved forms, and by an increased input of organic matter from plant biomass and imported material. Roots and rhizomes typically decay in situ, and some above-ground material may remain in the system rather than be exported to adjacent coastal waters. However, it appears that much, if not most, of the organic matter decomposed in seagrass sediments is imported. Fine, organic-rich particles settle out as the water flow is slowed by the seagrass canopy, and the roots and rhizomes also help to stabilise the sediments and prevent resuspension. As a result, organic matter and nutrient concentrations are typically higher in vegetated than in unvegetated sediments. Nutrient accumulation may occur in seagrass meadows over the timescale of years, but this source–sink role may in part depend on the trophic status of the system.

11.4 Mangrove Forests

11.4.1 N Cycling

Like seagrasses, N cycling in mangrove forests is rapid and efficient, to the extent that the only dissolved or particulate N exported from the system is highly refractory. N is conserved in mangroves as it is often a limiting element. N is taken up via fine tree roots in the form of NO_3^- and/or NH_4^+. Organic N derived from roots and inputs from upstream and marine sources (Figure 11.4) support high rates of ammonification in mangrove soils (Table 11.2). Ammonification is a key process in the N cycle, whereby N principally in the form of proteins and nucleotides is hydrolysed and catabolised by ammonifying bacteria to be liberated as NH_4^+. Ammonification is not easily or accurately measured in saline soils, particularly those containing substantial amounts of dead and live fine roots.

An analysis of the available soil N cycling data indicated that gross rates of soil ammonification are significantly greater than net ammonification rates

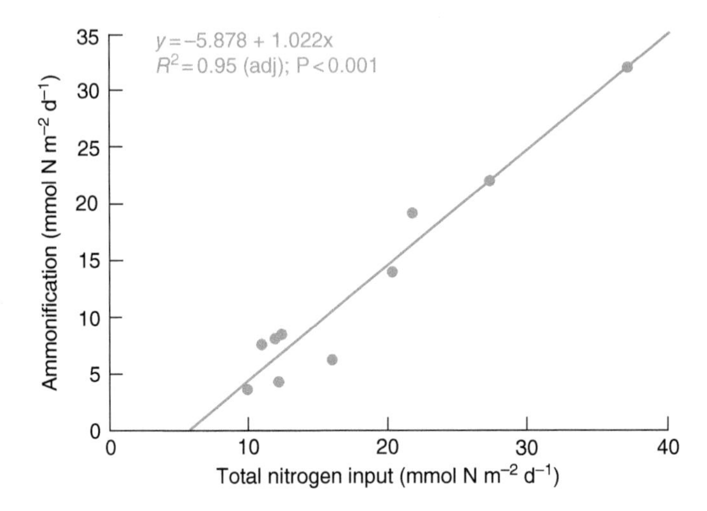

FIGURE 11.4 The relationship between ammonification rate and rate of total input of N to soils in mangrove forests of Thailand, China, and Malaysia. *Source:* Alongi (2009), figure 5.11, p. 120. © Springer Nature.

TABLE 11.2 **Rates of N transformation processes (mg N m^{-2} d^{-1}) in mangrove soils.**

Transformation process	Number of observations	Mean ± 1 SE	Median	Range
Denitrification	165	26.25 ± 3.34	3.90	0–443.52
Gross ammonification	17	301.63 ± 50.90	267.40	77.40–898.80
Net ammonification	52	31.40 ± 6.19	15.40	0.17–200.00
Gross nitrification	25	15.08 ± 5.81	4.74	0–141.00
Net nitrification	37	5.80 ± 1.20	1.93	0–30.80
Anammox	35	22.11 ± 5.49	4.35	0–99.4
DNRA	21	18.19 ± 6.77	4.54	0.01–108.64

All rates are to soil depths of 5–20 cm. Abbreviations: DNRA = dissimilatory nitrate reduction to ammonium; anammox = anaerobic ammonium oxidation; N_2O = nitrous oxide; NO = nitric oxide. Original references in Alongi (2020b). Reproduced from Alongi (2020b) under the Creative Commons Attribution License (http://creativecommons.org/licenses/by/4.0/).*Source:* Alongi (2020b). Licensed under CC BY 4.0. © MDPI.

(Alongi 2020b). Neither were there significant differences between gross and net rates of soil nitrification, although mean rates of gross nitrification were nearly three times greater than net rates (Table 11.2), indicating significant N immobilisation. Although it was originally thought that anammox would be a minor transformation process in mangrove soils compared to denitrification, the available data showed no significant difference between rates of both processes. Rates of microbial N metabolism depend on many drivers, including temperature, soil water content and fertility, microbial community structure, plant metabolism and root activities, bioturbation, intertidal position, and salinity. Other N-transforming bacterial and archaeal groups are present in mangrove soils, but no rate data are available for methane denitrification, nitrite oxidation, or phototrophic nitrate oxidation.

N_2 fixation rates were highest on algal/bacterial crusts growing on tree stems (Table 11.3), although the few measurements preclude meaningful statistical analysis. Differences between rates measured on the soil surface and the other forest vegetation were significantly different (Alongi 2020b). Rates of N_2 fixation on pneumatophores and prop roots were significantly greater than on the soil surface and vegetation components, except for belowground roots and rhizomes. Although few measurements have been made on vegetation and cyanobacterial mats, N_2-fixing microbial communities associated with roots, stem surfaces, and cyanobacterial mats may provide significant inputs of N to mangrove forests.

High rates of N_2 fixation may also be occurring deep within the extensive root systems of mangroves. These N_2 fixers are in a mutualistic relationship with non-N_2-fixing bacteria and the trees providing the bulk of N for immediate plant use (Holguin and Bashan 1996; Bashan et al. 1998; Naidoo et al. 2008). N_2 fixation by the bacterium, *Azospirillum brasilense*, was enhanced when cultured in the presence of a mangrove rhizobacterium, *Staphlococcus* sp., suggesting metabolic byproducts produced by rhizobacteria benefit the growth of N_2 fixers (Holguin and Bashan 1996). Transfer of fixed N to mangrove roots from filamentous cyanobacteria similarly enhanced the growth of mangrove seedlings (Bashan et al. 1998). Benthic fauna also influences rates of soil N_2 fixation. In arid-zone stands of *A. marina* in the central Red Sed where crab densities were high, N_2 fixation was significantly less than in stands with lower crab densities (Qashqari et al. 2020) likely because of aeration of the soil and/or grazing of cyanobacteria by crabs; N_2 fixation was higher in mature than in juvenile forests, but lower at 35 °C than at 28 °C.

TABLE 11.3 **Rates of nitrogen fixation (mg N m^{-2} d^{-1}) at the mangrove soil surface, cyanobacteria mats, aboveground roots (pneumatophores and prop roots), belowground roots and rhizomes, litter and senescent leaves lying on the forest floor, and microbial crusts on the bark of tree stems.**

Component	Number of observations	Mean ± 1 SE	Median	Range
Soil surface	57	8.22 ± 1.69	3.22	0–58.92
Cyanobacteria mats	28	9.69 ± 2.62	3.43	0–60.42
Aboveground roots	18	31.78 ± 5.63	26.25	2.77–73.4
Belowground roots + rhizomes	15	6.21 ± 1.65	4.50	1.04–26.20
Litter	9	1.16 ± 0.40	0.45	0.21–3.30
Senescent leaves	7	0.75 ± 0.13	0.83	0.39–1.20
Stem bark	5	100.95 ± 36.42	100.95	17.40–201.20

Original references in Alongi (2020b).

Denitrification has been measured often in mangrove soils, mostly within the upper 5–20 cm (Table 11.2). Denitrification is regulated mainly by NO_3^- availability (oxygen), temperature, salinity, and soil organic matter content. NO_3^- availability is the prime regulatory factor, as increasing the supply of NO_3^- increases the rate of denitrification in mangrove soils (Rivera-Monroy and Twilley 1996; Joye and Lee 2004; Lee and Joye 2006). Denitrification and nitrification are often linked but may also be co-linked to the Fe cycle (Fernandes et al. 2013). A plot of the relationship of soil denitrification versus age of *R. apiculata* forests in Southeast Asia (Figure 11.5) showed a significant positive relationship, but the regression was skewed towards the measurements from the oldest forest. The data do not suggest a relationship within the forest age span of 3–35 years suggesting the importance of co-factors; denitrification did not relate either to rates of ammonification or rates of N input.

Factors other than NO_3^- availability, therefore, come into play in regulating the rate of denitrification. The presence of benthic microalgal mats, for instance, stimulates denitrification (Joye and Lee 2004; Lee and Joye 2006). On Twin Cays off Belize, benthic mats composed of filamentous, heterocystous, and coccoidal cyanobacteria, purple sulphur bacteria and heterotrophic bacteria, are dense within dwarf mangrove stands, playing an important role in N cycling. In these mats, NO_3^- availability was an important limiting factor for denitrification, and N_2 fixation was regulated mainly by the sensitivity of the nitrogenase enzyme to oxygen inhibition. The size and thus the overall contribution of the mats to ecosystem N cycling was controlled by the seasonal and tidal frequency of wetting as well as elevation.

Only two measurements at one site have been made of NO fluxes across the soil–air interface (Table 11.4) and both indicated negligible release. Rates of N_2O flux show a net mean flux to the atmosphere, although some measurements indicated net uptake by mangrove soils (Table 11.4).

Rates of DON, $NO_2^- + NO_3^-$, and NH_4^+ (Table 11.4) flux varied widely among forests, due mostly to differences between fluxes measured in light (clear) versus dark (opaque) chambers. Most measurements from light bottles showed net uptake of solutes indicating utilisation of dissolved nutrients in overlying tidal waters by microalgae, cyanobacteria, and other autotrophs on the mangrove soil surface. This

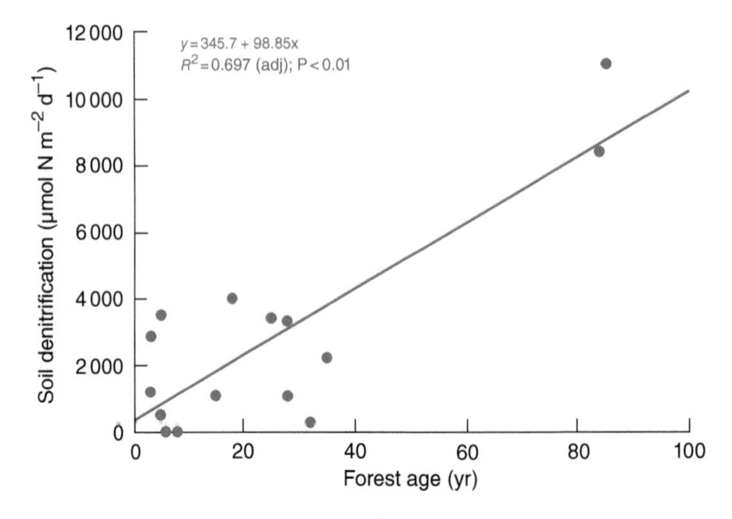

FIGURE 11.5 The relationship between rates of soil denitrification and age of *Rhizophora apiculata* forests in Southeast Asia. *Source:* Alongi (2009), figure 5.12, p. 121. © Springer Nature.

TABLE 11.4 Estimates of net NO and N_2O gas and dissolved nitrogen fluxes across the mangrove soil surface.

Gas and solute soil fluxes	Number of observations	Mean ± 1 SE	Median	Range
N_2O soil–air flux	69	0.60 ± 0.17	0.22	−2.03–9.01
NO soil–air flux	2	0.05 ± 0.009	0.05	0.04–0.06
DON soil–water flux	41	−18.29 ± 18.14	0.00	−743–19.6
$NO_2^- + NO_3^-$ soil–water flux	66	−4.46 ± 1.05	−0.24	−29.04–3.36
NH_4^+ soil–water flux	78	−1.48 ± 2.20	−0.16	−88.7–55.19

Units = mg N m^{-2} d^{-1}. Negative values indicate fluxes into the soil and positive values indicate fluxes from the soil to the overlying tidal water/atmosphere. Original references in Alongi (2020b). Reproduced from Alongi (2020b) under the Creative Commons Attribution License (http://creativecommons.org/licenses/by/4.0/).*Source:* Alongi (2020b). Licensed under CC BY 4.0. © MDPI.

fact is reflected in the mean rates of exchange being negative for all three dissolved N forms. Due to high variation, all three solutes appear to be utilised equally by benthic autotrophs.

N_2O fluxes in mangrove tidal waters have been measured in more than 30 mangrove-fringed estuaries, tidal creeks, and waterways (Alongi 2020b) with a mean net flux to the atmosphere, averaging (±SE) 0.11 ± 0.03 mg N m^{-2} d^{-1} with a median of 0.02 mg N m^{-2} d^{-1} and ranging from net uptake (−0.06 mg N m^{-2} d^{-1}) to net release (1.32 mg N m^{-2} d^{-1}).

Rates of dissolved N exchange between forests and adjacent tidal waterways vary greatly within and among sites, as reflected in the wide range of estimates for NH_4^+, $NO_2^- + NO_3^-$, and DON flux (Table 11.5). Despite the variability, net exchange was net import for all DN species based on mean rates; median rates indicated little net exchange (Table 11.5). The direction of DN transport between tidal waters and mangroves depends on a wide range of factors, including tidal prism, geomorphology, climate, seasonal weather patterns, the ratio of forest to waterway area, temperature, salinity, pH, dissolved oxygen levels, and plankton metabolism. The net direction of DN exchange depends on tidal range, extent of groundwater discharge, ratio of evaporation to precipitation, rates of primary productivity, salinity, turbidity, pH, dissolved oxygen concentrations, and rates of microbial assimilation. Another driver often overlooked is the extent to which porewater N concentrations exceed the demands of primary producers. Simply, an ecosystem will tend to export nutrients if there are more nutrients released than needed for utilisation. Conversely, nutrients will be imported into the ecosystem if not enough is available. Anthropogenic changes sustained by estuaries may also lead to shifts in patterns of nutrient and material exchange.

A N budget for the world's mangrove forests (Alongi 2020b) has been constructed (Figure 11.6). The mean values used are not absolute, and the budget was constructed based on several assumptions: (i) global mangrove area is 86 495 km^2, (ii) although mangrove tidal creeks and waterways are only a small fraction of total mangrove area, the total global mangrove area was used to calculate all tidal exchanges because the data include many measurements taken from mangrove-fringed estuaries that

TABLE 11.5 **Rates of NH_4^+, DON, and $NO_2^- + NO_3^-$ exchange between mangrove forests and adjacent tidal waterways, including mangrove-fringed estuaries.**

Component	Number of observations	Mean ± 1 SE	Median	Range
NH_4^+ import	26	−9.81 ± 4.35	−1.57	−100.76 – −0.006
NH_4^+ export	23	8.00 ± 3.63	0.39	0.03–62.40
NH_4^+ net exchange	49	−1.45 ± 3.12	−0.01	−100.76–62.4
DON import	9	−4.69 ± 2.17	−2.87	−21.84 – −0.91
DON export	10	1.45 ± 0.33	1.37	0.08–3.27
DON net exchange	19	−1.46 ± 1.24	0.08	−21.84–3.27
$NO_2^- + NO_3^-$ import	24	−3.62 ± 1.52	−0.73	−28.80 – −0.003
$NO_2^- + NO_3^-$ export	27	1.84 ± 0.55	0.29	0.05–11.90
$NO_2^- + NO_3^-$ net exchange	51	−0.73 ± 0.85	0.08	−28.8–11.90

Import into the mangroves is shown as a negative value and export to adjacent tidal waters is a positive value. Original references in Alongi (2020b). Reproduced from Alongi (2020b) under the Creative Commons Attribution License (http://creativecommons.org/licenses/by/4.0/). Units = mg N m^{-2} water d^{-1}.*Source:* Alongi (2020b). Licensed under CC BY 4.0. © MDPI.

are likely in toto to be of equivalent area, (iii) litter export (PON) was estimated by converting litter C export to N assuming a litter C/N of 79.4 (Alongi 2020a), (iv) wood, litter, root, and benthic microalgal NPP were estimated by converting the C values in Alongi (2020a) using the C/N ratios (g/g) in Table 1 in Alongi (2020a) and using a microalgal C/N ratio (g/g) of 12 (Alongi 2020a, b), (v) standing stocks of soil and below-ground roots and above-ground forest were derived from Table 2 in Alongi (2020a), (vi) the input of the total below-ground root + soil pool (441 150 Gg N) to N burial (1127 Gg a^{-1}) was estimated by the difference between the mean N burial value (1239 Gg a^{-1}) minus the inputs from root production (62 Gg a^{-1}) and litter production (50 Gg a^{-1}), (vii) wet N deposition was estimated from the only two studies available (Alongi et al. 1992b; Cerón et al. 2015), and (viii) the differences between gross and net ammonification and nitrification are assumed to represent N immobilisation.

The budget does not include: (i) direct inputs from groundwater and upstream, (ii) marine and terrigenous particle flux and deposition at the soil surface, (iii) pelagic and benthic production, (iv) dry deposition, (v) consumption and assimilation by biota, (vi) N_2 fixation on tree stems, cyanobacterial mats, above-ground roots, senescent leaves and litter; the latter are not included because of the inability to extrapolate these rates to more than a small area given the lack of knowledge of their areal coverage in a 'typical' mangrove forest, and (vii) rates of soil N transformations such as ammonification and denitrification likely account for only a part of total N flux in soils as most studies measured these rates only to soil depths of 5–20 cm. C mineralisation is active to a soil depth of 1 m (Alongi 2020a), so it is likely that significant N mineralisation occurs in deeper soils.

Further, mangroves differ in their location and climate, and N cycle pathways are significantly affected by physicochemical and biological factors, such as temperature,

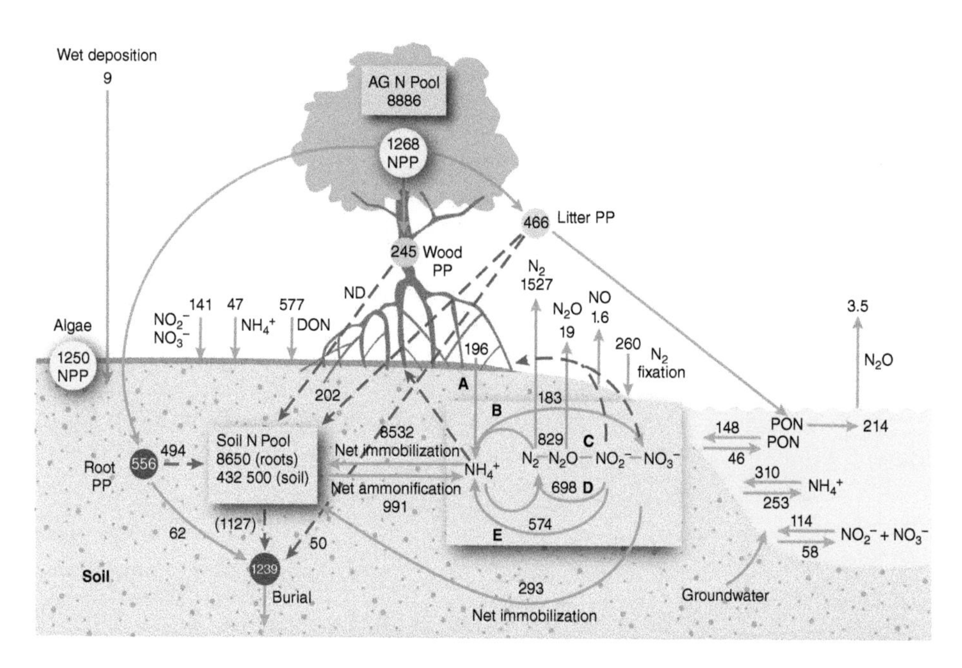

FIGURE 11.6 N cycling in the world's mangrove ecosystems. Mean fluxes = Gg N a^{-1}; mean standing stocks = Gg N. The model assumes a global mangrove area of 86495 km^2 (Hamilton and Casey 2016). Soil N transformations are lettered as: (A) root + rhizome N_2 fixation; (B) net nitrification; (C) denitrification; (D) anammox; (E) dissimilatory nitrate reduction to ammonium. Dashed red arrows represent mean values estimated indirectly (by difference); solid blue arrows represent mean values based on empirical measurements. The N pool (both roots + soil) in soils to a depth of 1 m is presented as a box on left in the forest floor. Unquantified inputs and outputs of dissolved N from land-derived groundwater and organic matter inputs from adjacent marine waters and catchments are not represented. Abbreviations: ND = no data; AG N Pool = aboveground forest N biomass pool; PP = primary production; NPP = net primary production; $NO_2^- + NO_3^-$ = nitrite plus nitrate; NH_4^+ = ammonium; N_2 = gaseous nitrogen; N_2O = nitrous oxide; NO = nitric oxide; PON = particulate organic nitrogen. *Source:* Alongi (2020b). Licensed under CC BY 4.0. © MDPI.

grazing, presence of biogenic structures, tides, pH, soil nutrients, soil salinity and water content, soil type and rate of N input. Such variations have not, of course, been included in the mass balance, but must be kept in mind when considering the mangrove N cycle. Individual mangrove forests and ecosystems vary significantly from the global averages used here.

Despite these significant shortcomings, the budget does suggest rapid rates of N cycling and transformations in mangrove forests. Burial in soil equates to about 29% of total N inputs. Net tidal exchange of dissolved N is import or near zero, indicating N conservation. Anammox is as important a transformation process as denitrification. Denitrification equates to 35% of total N input, within the range found in other coastal ecosystems (Seitzinger et al. 2006). Net N immobilisation is the single largest transformation process, signifying its importance in N conservation. After trees, benthic microalgae appear to be the most important primary producers, although cyanobacterial mats are not represented. Net uptake of dissolved N by soils from overlying tidal waters is a significant N conservation mechanism. Alongi (1996) measured high rates of dissolved N uptake by stems, logs, prop roots, and twigs, supporting the idea that microbial communities on these surfaces also function as an N conservation mechanism.

The budget indicates a N burial efficiency (burial/total input × 100) of 29% and a N mineralisation efficiency (gross ammonification-net ammonification/total input × 100) of 141% which would be more realistically lower when other inputs (N_2 fixation from stems, etc.) are eventually included in the budget. Soil N is very efficiently mineralised, as measured in Asian mangroves. In Chinese mangroves (*K. candel*), burial efficiencies ranged from 8 to 31% and mineralisation efficiencies from 69 to 92% (Alongi et al. 2005). In Thai mangroves, N burial efficiencies ranged from 4 to 12% and N mineralisation efficiencies from 68 to 88% (Alongi et al. 2002) and in Malaysian forests (Alongi et al. 2004) burial and mineralisation efficiencies ranged from 10 to 29% and 67 to 81%, respectively.

A mass balance of all inputs and outputs (Table 11.6) indicates a net positive gain of 957.9 Gg N a^{-1} which equals 12% of all inputs and outputs, well within the

TABLE 11.6 **A nitrogen mass balance of the world's mangrove ecosystems.**

Inputs		**Outputs**	
Component	**Gg N a^{-1}**	**Component**	**Gg N a^{-1}**
Net primary production		**Burial**	1239
Wood	245	**Soil NO release**	1.6
Litter	466	**Soil N_2O release**	19
Roots	556	**Denitrification/Anammox**	1527
Microalgae	1250	**Water–air exchange**	
N_2 fixation		N_2O	3.5
Roots	196	**Tidal exchange**	
Soil	260	DON	58
Precipitation	9	$NO_2^- + NO_3^-$	46
Tidal exchange		NH_4^+	253
DON	148	PON	214
$NO_2^- + NO_3^-$	114		
NH_4^+	310		
Soil–Water exchange			
DON	577		
$NO_2^- + NO_3^-$	141		
NH_4^+	47		
Total	4319	Total	3361.1
Inputs–Outputs = 957.9 (Net Gain)			

Units = Gg N a^{-1}. Reproduced from Alongi (2020b) under the Creative Commons Attribution License (http://creativecommons.org/licenses/by/4.0/).*Source:* Alongi (2020b). Licensed under CC BY 4.0. © MDPI.

sum of systematic errors of the many measurements made. Considering that several important processes were not included, it is likely that N flow in mangrove ecosystems are in approximate balance. However, considering the high rates of N_2 fixation on tree stems, cyanobacterial mats, above-ground roots, senescent leaves and litter, ecosystem N inputs will need to be revised upwards when further studies with a proper sampling design incorporating the great spatial variability of this process will warrant their inclusion in the budget. The budget would be unbalanced with a larger net positive gain, but consumption and assimilation of these unquantified N_2 fixers by mangrove-associated fauna would perhaps redress the imbalance. Many organisms such as gastropods and other benthic invertebrates commonly dwell on tree stems, cyanobacterial mats, prop roots, pneumatophores, leaves, and litter and readily consume organic particles, micro- and macroalgae, bacteria, and detritus on these surfaces. Further, as pointed out by Reis et al. (2017), mangrove N cycling becomes unbalanced when forests receive anthropogenic inputs, such as effluents from aquaculture ponds and human and animal sewage, so only relatively pristine mangroves would be expected to exhibit balanced N flow.

At the global scale, mangroves contribute variable percentages of N to the coastal ocean. Mangrove PON export (2474 kg N km^{-2} a^{-1}) equates to 95% of PON export (2612 kg N km^{-2} a^{1}) from the world's major tropical rivers. However, most rivers in low latitudes have not been measured for PON export (Alvarez-Cobelas et al. 2008), so the true contribution is probably considerably smaller. Mangrove PON export accounts for only 1.5% of the entire world's river discharge (Beusen et al. 2016). The contributions of mangrove N_2O emissions, denitrification, and burial to the global coastal ocean (Fowler et al. 2013) are modest (0.4, 0.5–2.0 and 6%, respectively) but are disproportionate relative to their small area (0.31%) (Alongi 2020b).

Several mechanisms operate to conserve N. First, soil N cycling is rapid, to the extent that the bulk of dissolved N is taken up by the trees. Comparatively little (about 5% of total N input to the soil) is lost via denitrification and by efflux during flood tides. The microbe–soil–root complex thus rapidly recycles N via mortality, decomposition, uptake, and growth of organisms, thus retaining N. Second, feeding and bioturbation by crabs assist in minimising the loss of litter and maximising N gain. Third, tree stems, roots, logs, and other mangrove wood on the forest floor provide space to maximise colonisation by N_2 fixers and thus the rate of atmospheric input. Finally, the C:N ratio of dissolved and particulate material leaving the forest is high and high in concentrations of humic and fluvic acids and polyphenolics, indicative of an advanced state of decomposition.

Nutrient limitation occurs in some mangrove forests for often complex reasons. In the field, mangroves have demonstrated a variety of responses to added N and P, differing in relation to soil type and texture, salinity, frequency of tidal inundation and species composition. In one of the early direct tests to determine whether N and/or P are limiting to mangroves, Boto and Wellington (1983) added N or P to mixed *Rhizophora* forests in northern Australia and found that while N limitation occurred across the intertidal, P limitation was problematical, being apparent only in the high-intertidal forest where soils contain naturally low levels of P. The level of complexity of the nutrient limitation issue can be discerned from the extensive field studies conducted by Feller and colleagues (Feller 1995; Feller et al. 2003, 2007; Lovelock et al. 2006; Almahasheer et al. 2016). In Belizean mangroves in oceanic settings with minimal terrestrial input, forests were limited by comparatively low P concentrations, and to a lesser extent, by N, although there was a switch from N to P limitation on small islands from the shore to the interior of each island. The most consistent pattern for Caribbean mangroves was that *Rhizophora mangle* was N-limited seawards, dwarf trees in the interior were P-limited, and trees across the transition from low to high intertidal were limited by N and P. Trees under P limitation are often water-deficient, showing more pronounced changes in structure and function when P deficiency is eased than those trees under a regime of N limitation (Lovelock et al. 2006). In Florida, however, mangroves were limited only by N across the intertidal seascape (Feller et al. 2003). The data for these fertilisation studies demonstrated either N or P limitation, or both, depending on species composition, intertidal position, extent of terrigenous input, soil fertility and composition, soil redox status, and salinity, to name but few of the factors that likely regulate mangrove–nutrient relations. Mangrove growth can be limited by micronutrients (Alongi 2010, 2017; Almahasheer et al. 2016). Early growth of several species was found to be limited by dissolved Fe and Cu in microcosms (Alongi 2010, 2017), whereas field and experimental studies with stunted stands of the mangrove *A. marina* in the central Red Sea demonstrated that Fe was the primary limiting nutrient; the latter results were attributed to the dominance of carbonate sediments and the lack of riverine sources of Fe. Several studies have also emphasised the importance of flooding and draining of tidal waters on mangrove growth and nutrient use as tidal changes relate to rates of sediment and nutrient input (Krauss et al. 2006, 2007a, b). Thus, what nutrient (s) is (are) limiting to a particular forest is ultimately a function of environmental setting.

The high photosynthetic rates of mangroves are supported by a high requirement for nutrients, implying high nutrient-use efficiencies and high rates of leaf resorption (Reef et al. 2010). In comparing mangroves to fast-growing tropical plantation trees and evergreen and deciduous trees, Alongi (2009) found that while the range of nutrient-use efficiency for N and P for all forest types was wide, rates of N-use efficiency for mangroves (median: 210 $g\,g^{-1}$ N; range: 150–280 $g\,g^{-1}$N) were at the upper end of the range for tropical terrestrial forests (median: 120 $g\,g^{-1}$ N; range: 75–180 $g\,g^{-1}$N). Rates of P-use efficiency in mangroves (median: 2500 $g\,g^{-1}$ P; range: 1250–4500 $g\,g^{-1}$ P) however were well within the range for all tropical forests (median: 2000 $g\,g^{-1}$ P; range: 1100–3500 $g\,g^{-1}$ P; Alongi 2009) indicating similar physiological needs among tropical forests.

Mechanisms to conserve limiting nutrients are clearly advantageous. The apparent strategy of using nutrients efficiently relates well to the low nutrient concentrations in mangrove leaves and other tree parts, and the generally high efficiency with which N and P is resorbed by leaves (Reef et al. 2010). In comparing mangroves with other forested ecosystems, Alongi (2009) found that N was resorbed by mangrove trees (median: 63%; range: 58–71%) at a level of efficiency at the higher end of the range for tropical forests (median: 48%; range: 38–62%). Efficiency of P resorption by mangroves (median: 55%; range: 35–72%) was within the mid-range of other tropical forests (median: 53%; range: 29–68%). As in other forested ecosystems, the high rates of nutrient-use efficiency in mangroves are reflected in high rates of nutrient productivity and in shorter N and P residence times.

11.4.2 P Cycling

Mangrove soils, due to their normally high organic content, may contain a high proportion of organic-bound P, up to 75–80% of total extractable P in some locations (Alongi et al. 1992). There are quite variable proportions between organic and inorganic fractions in relation to grain size and origin (terrestrial vs. marine) and stage of forest development, with many soils exceeding 50% of P within the inorganic fractions. In Micronesian forests, soluble reactive P concentrations related to redox conditions and forest species composition, the latter relationship engendered either by the trees or to micro-scale differences among stands composed of different species (Gleason et al. 2003).

The availability of P to mangroves is not regulated by biological processes, but by geochemical reactions. Readily available P is rapidly incorporated onto clay particles and metal oxyhydroxides and immobilised by precipitation as Ca, Fe, Al salts, limiting the available P pool. Organic phosphates, mainly P esters originating from living cell tissue, are often resistant to hydrolysis and therefore limiting to microbes and plants. Early (Boto and Wellington 1983; Alongi et al. 1992) and more recent (Tam and Wong 1996; Holmboe et al. 2001) experiments showed that when PO_4^{3-} is added to mangrove soils, it is quickly immobilised onto Fe and Al oxyhydroxides and into easily exchangeable fractions. Mangrove soils with such a high adsorption capacity function as P sinks.

An overview of P cycling in the world's largest mangrove ecosystem, the Sundarbans of India (Figure 11.7), showed that mangrove biomass was the largest reservoir of P with comparatively little stored in soils (Ray et al. 2018). The largest input was from atmospheric dry deposition followed by estuarine, land, and ocean inputs. The largest exchanges were between mangroves and soils. The largest loss from the system was export to adjacent coastal waters, followed by aerosol emission. The total flux to the soil was 16.06 Gg P a^{-1} and total removal of P was 14.7 Gg P a^{-1} with 'missing' P estimated at 1.36 Gg P a^{-1} which corresponds to 8.5% of the total input of P to the ecosystem. Given that these are estimates and extrapolations, the P budget for the Sundarbans appears to be in overall balance.

The uptake of soluble P by mangroves involves mutualistic interrelationships among bacteria, fungi, and tree roots. Arbuscular mycorrhizal fungi in the mangrove rhizosphere benefit from O_2 translocated by the trees to their roots. The presence of vesicles (nutrient storage organs) in the root cells of some mangrove species (Kothamasi et al. 2006) suggests that fungal symbionts play a role in nutrient uptake.

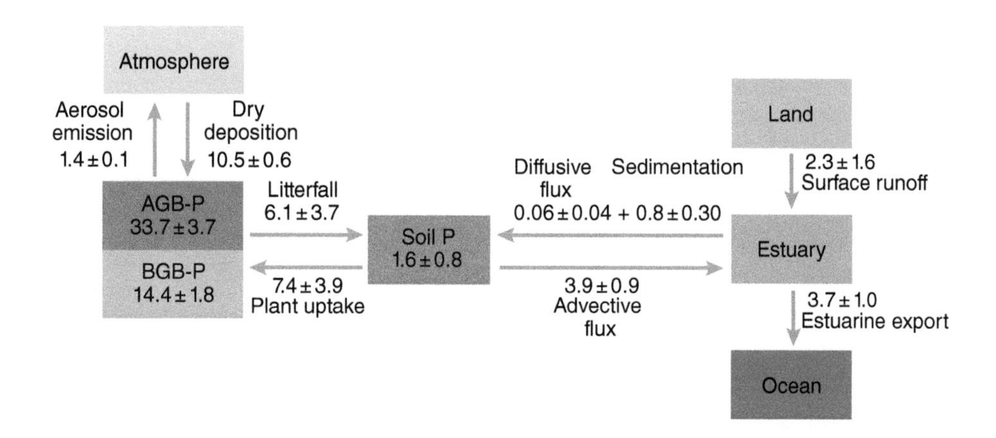

FIGURE 11.7 Annual P budget for the Indian Sundarbans mangrove ecosystem. Standing stocks = Gg P, fluxes = Gg P a^{-1}. Values depict mean ± 1 standard deviation. Abbreviations: ABG-P = P in above-ground biomass; BGB-P = P in below-ground biomass. *Source:* Ray et al. (2018), figure 6, p. 10. © Springer Nature.

PO_4^{3-}-solubilising bacteria associated with the roots and fungi may release PO_4^{3-} that could be taken up by the fungal hyphae and transferred to the host or taken up directly by the roots. PO_4^{3-}-solubilising bacteria have been discovered in the roots of several mangrove species (Vazquez et al. 2000; Rojas et al. 2001; Bashan and Holguin 2002; Kothamasi et al. 2006), and their presence has been shown to increase rates of other bacterial processes, such as N_2 fixation (Rojas et al. 2001). How these bacteria solubilise PO_4^{3-} is unclear, but culture experiments suggest that organic acids produced by the bacteria may dissolve $CaPO_4^{3-}$ (Vazquez et al. 2000). As suggested by Bashan and Holguin (2002), such plant growth-promoting bacteria and fungi can be used as a tool for reforestation. In any event, further research is needed to obtain a holistic picture of the role of P in mangrove ecosystems.

Disturbances such as hurricanes can have a significant impact on P dynamics in mangrove forests. For example, the passing of Hurricane Irma across the Florida Everglades in September 2017 resulted in storm deposits that were 7–14x greater than natural rates of soil accretion (Castañeda-Moya et al. 2020). Total P inputs doubled concentrations in underlying soils, contributing 49–98% to the soil nutrient pool. Three months after Irma's passage, all mangroves showed a significant increase in litter foliar P and soil porewater DIP concentrations. Mean TP loading rates were 5x greater in mangrove-dominated estuaries underscoring the positive role of hurricanes as a natural fertilisation mechanism influencing mangrove productivity.

11.5 Coral Reefs

Coral reefs, despite low nutrient concentrations, are highly productive ecosystems due to the rapid turnover of their N and P pools primarily via the microbial hub. The supply of essential nutrients supporting high productivity of coral reefs comes from a variety of sources: advection, upwelling and endo-upwelling, groundwater, rain, guano, terrestrial runoff, and immigrant organisms (Atkinson 2011). High rates of N_2 fixation also fuel N cycling (O'Neil and Capone 2008). The external nature of these sources attests to the importance of the connections between coral reefs and surrounding waters, and their proximity to other coastal and oceanic habitats. Connections between the various zones of a coral reef are also vital to the sustainability of an entire reef. Seawater impinging upon and moving over reef structures ensures that food webs and nutrient cycles within the different zones of a coral reef are interlinked, helping to maintain a balance between production and consumption. Reef geomorphology undoubtedly plays a role in determining the residence time of water, nutrients, and organisms and thus the flux of energy and material transfer within a coral reef.

Coral reefs are thought to be limited by the availability of nutrients, yet little consensus exists as to whether N or P or Fe are limiting or co-limiting and under what conditions such limitation occurs (Alongi 1998; Atkinson 2011). Nutrient limitation has rarely been directly tested on coral reefs but has been inferred from the fact that reefs are bathed in oligotrophic waters and demonstrate high rates of N_2 fixation (Table 11.7). Such inferences may be incorrect. High rates of N_2 fixation do not preclude further increases in productivity, and coral reefs have nutrient requirements like those communities in the surrounding seas. Part of the problem is that studies of nutrient limitation have utilised physiological, bioassay, mathematical, and stoichiometric approaches that are all different in emphasis and scale.

TABLE 11.7 Range of rates of various N transformation processes in coral reef sediments from various locations. Units are $\mu mol\ m^{-2}\ h^{-1}$.

Transformation rate	Method	Location	Reference
NH_4^+ release			
0.7–1.0	Diffusive modelling	Fringing reef, Puerto Rico	Corredor and Morell (1985)
0–0.8	Diffusive modelling	Tikehau lagoon, French Polynesia	Capone et al. (1992)
50–704	Chambers	Anoxic sediments, Tikehau Lagoon	Capone et al. (1992)
0.4	Diffusive modelling	Hydrolab, St. Croix	Fisher et al. (1990)
35	Dark chamber	Hydrolab, St. Croix	Fisher et al. (1990)
3	Chambers	Back-reef, Tague Bay, St. Croix	Williams et al. (1985)
0.4–4	Diffusive modelling	Various reefs, Great Barrier Reef	Capone et al. (1992)
−0.7–1.2	Chambers	Lagoon, Ishigaki Island	Miyajima et al. (2001)
0.8–23	Dark chamber	Arlington Reef, Great Barrier Reef	Alongi et al. (2006)
0–17	Dark chamber	Sudbury Reef, Great Barrier Reef	Alongi et al. (2006)
0.6–7	Dark chamber	Various inter-reef areas, Great Barrier Reef	Alongi et al. (2008b)
−5.0–70.0	Core incubation (light)	Carbonate sands, SW New Caledonia lagoon	Grenz et al. (2010)
774 ± 486 (dark) 3528 ± 1302 (light)	Chambers	Enhanced flux in presence of upside-down jellyfish, *Cassiopea* sp., Red Sea coral reef	Jantzen et al. (2010)
6.7–13.7	Light chambers	Back-reef, Reunion Island	L'Helguen et al. (2014)
−267.5–210.0	Diffusive modelling	Florida Key reefs, United States	Lisle et al. (2014)
12.05	Light chambers	Beneath pearl oyster culture, deep atoll lagoon, French Polynesia	Lacoste and Gaertner-Mazouni (2016)
−2.5 −25	Mass transfer modelling	Tallon Island reefs, Western Australia	Gruber et al. (2019)
Ammonification			
4.5	Direct, Tube pack	Tague Bay, St. Croix	Williams et al. (1985)

(*continued*)

TABLE 11.7 *(Continued)*

Transformation rate	Method	Location	Reference
4.4–50	Core incubation	Various reefs, Great Barrier Reef	Capone et al. (1992)
39–219	Core incubation	Arlington Reef, Great Barrier Reef	Alongi et al. (2006)
70–129	Core incubation	Sudbury Reef, Great Barrier Reef	Alongi et al. (2006)
8.8–68	Core incubation	Various inter-reef areas, Great Barrier Reef	Alongi et al. (2008b)
Denitrification			
19	Acetylene blockage	Lagoonal sediment, Bermuda	Seitzinger and D'Elia (1985)
50–100	Acetylene blockage	Fringing reefs, Puerto Rico	Corredor and Capone (1985)
0.1–13	Acetylene blockage	Various reefs, Great Barrier Reef	Capone et al. (1992)
1.7–6.5	Acetylene blockage	Lagoon, Ishigaki Island	Miyajima et al. (2001)
8–18	Chamber incubation, Direct N_2 release	Arlington Reef, Great Barrier Reef	Alongi et al. (2006)
3.5–41	Chamber incubation, Direct N_2 release	Sudbury Reef, Great Barrier Reef	Alongi et al. (2006)
2.5–186	Dark chamber incubation, Direct N_2 release	Various inter-reef areas, Great Barrier Reef	Alongi et al. (2008b)
34–92 (pre-spawning) 96–480 (post-spawning)	Dark chamber, N_2:Ar	Carbonate sand, Heron Island, Great Barrier Reef	Eyre et al. (2008)
20–105	Chamber incubation, Direct N_2 release	Various reefs, Southern Great Barrier Reef lagoon	Alongi et al. (2011)
70–336	Core incubation	Carbonate sand, Heron Island, Great Barrier Reef	Santos et al. (2012)
2.1–3.5	Sediment slurries, $\delta^{15}N\text{-}NH_4^+$ and $\delta^{15}N\text{-}N_2$	Various reefs, Southern Great Barrier Reef lagoon	Erler et al. (2013)

TABLE 11.7 (*Continued*)

Transformation rate	Method	Location	Reference
54(dark), 66 (light), winter 80–214 (dark), 2–14 (light), summer 82 (dark), 344 (light), spring	Chambers, N_2:Ar	Carbonate sand, Heron Island, Great Barrier Reef	Eyre et al. (2013)
N_2 fixation			
1.7–12.7	Acetylene reduction	Central Great Barrier Reef	Wilkinson et al. (1984)
6.5–12.8	Acetylene reduction	Carbonate muds, Puerto Rico	Corredor and Capone (1985)
0.4–3.6	Acetylene reduction	Carbonate sands, Puerto Rico	Corredor and Capone (1985)
3–12	Acetylene reduction	Carbonate muds, Bermuda	O'Neil and Capone (1989)
0.4–11.5	Acetylene reduction	Carbonate sands, Bermuda	O'Neil and Capone (1989)
6	Acetylene reduction	Hydrolab, St, Croix	King et al. (1990)
4–10	Acetylene reduction, $^{15}N_2$	Southern Great Barrier Reef	O'Donohue et al. (1991)
4.1–6.8	Acetylene reduction	Lagoon, Ishigaki Island	Miyajima et al. (2001)
0–10	Acetylene reduction	Various inter-reef areas, Great Barrier Reef	Alongi et al. (2008b)
0.18–0.78	Acetylene reduction	Carbonate sand, Heron Island, Great Barrier Reef	Werner et al. (2008)
11.7 (coral rock) 14.7 (turf algae)	Acetylene reduction	Gulf of Aqaba reef, Northern Red Sea	Rix et al. (2015)
0.03–6.3	Acetylene reduction	Mayotte, Tulear, La Réunion Island reefs, Indian Ocean	Charpy et al. (2012)
6.7–38.3	Acetylene reduction	Gulf of Aqaba reef, Northern Red Sea	Cardini et al. (2016)
38.6–503.3	Acetylene reduction	Benthic cyanobacterial mats, Coral reefs, Curaçao	Brocke et al. (2018)
Nitrification			
24–42	Chamber incubation	Arlington Reef, Great Barrier Reef	Alongi et al. (2006)
45–58	Chamber incubation	Sudbury Reef, Great Barrier Reef	Alongi et al. (2006)

Source: Based on O'Neil and Capone (2008). © John Wiley & Sons.

Some aspects of the N cycle in reef systems are better understood than others. Rarely has the entire cycle been deduced for entire benthic and pelagic components of a coral reef. Some N transformation processes in surface carbonate sediments on reefs have been measured in individual reefs of the Great Barrier Reef (GBR) (Capone et al. 1992; Alongi et al. 2006). Capone et al. (1992) measured N fluxes on three reefs and all systems showed rapid rates of ammonification in relation to pool size, suggesting highly dynamic N pools and turnover times of <1-day. Alongi et al. (2006) also found that ammonification was rapid enough to fuel rapid cycling of the NH_4^+ and NO_3^- pools in carbonate muds of Arlington (Figure 11.8a) and Sudbury Reefs (Figure 11.8b), GBR. Denitrification rates were great enough to ensure a very rapid turnover (h) of the NO_3^- pool. Of total N inputs, roughly 10% of the N was buried in the lagoonal muds with most NH_4^+ being shunted to the bound fraction. A moderate amount of NH_4^+ and NO_3^- were lost via efflux to the overlying water.

Most measured rates of N cycling processes in reef sediments (Table 11.7) show great variability because of different methods used, differences in sediment grain size and organic content, variable rates of particulate organic N input and seasonality. Regardless of these differences and difficulties, the data indicate that these small N pools turnover rapidly, as found in other tropical marine deposits.

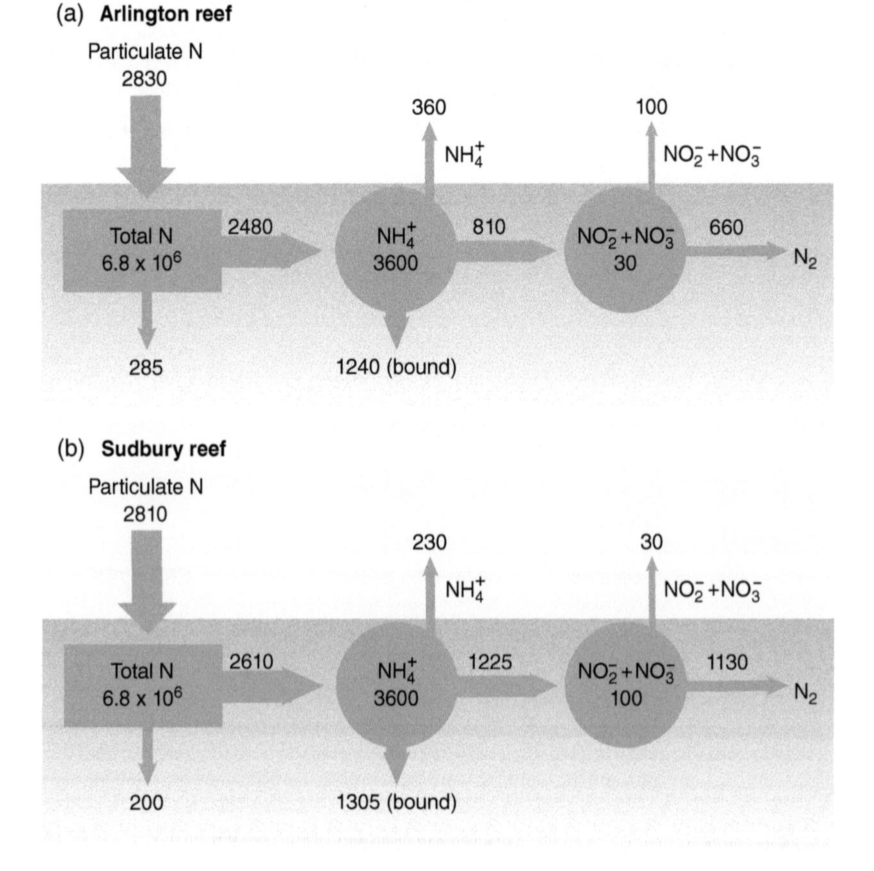

FIGURE 11.8 Benthic N cycling in surface (0–20 cm) sediments of (a) Arlington Reef and (b) Sudbury Reef, northern Great Barrier Reef. Standing stocks are μmol N m^{-2}. Fluxes are μmol N m^{-2} d^{-1}. Values are means of seasonally averaged figures. *Source:* Alongi et al. (2006). © Springer Nature.

Nitrification usually occurs in aerobic zones and can occur at high rates in coral reefs sediments (O'Neil and Capone 2008). Nitrification in reef lagoons can oxidise NH_4^+ recently released from N_2 fixation or regenerative processes. Nitrification was first measured on Eniwetok Atoll when net NO_3^- production was first noted. This was taken as strong evidence for nitrification as a significant process on coral reefs. NH_4^+ oxidation and NO_2^- oxidation or reduction were presumed to be coupled as there was no build-up of NO_2^-. Nitrifying bacteria have been identified as sponge symbionts with highest rates of NO_3^- released associated with sponges that had cyanobacterial endosymbionts, suggesting that sponge-mediated nitrification may be common in tropical reef systems.

Denitrification measured in coral reef habitats (Table 11.7) indicates a high level of variability within and among reefs, coral species, reef zones, water depths, and with differences in sediment texture (sand vs. mud). Some of the highest rates have been measured in carbonate muds on the GBR and in deposits receiving organic matter from coral spawning. The importance of anammox has yet to be considered on coral reefs and may be the source of N_2 release rather than denitrification. In the Red Sea corals, *Acropora hemprichii, Millapora dichotoma* and *Pleuractis granulosa*, the *nirS* marker gene, a proxy for denitrification potential, as well as denitrification, was found in all three coral species (Tilstra et al. 2019). *A. hemprichii* exhibited the highest rates of denitrification (0.38 nmol N cm^{-2} d^{-1}), followed by *M. dichotoma* (0.17 nmol N cm^{-2} d^{-1}) and *P. granulosa* (0.05 nmol N cm^{-2} d^{-1}). N_2 fixation rates were highest in *M. dichotoma* (0.23 nmol N cm^{-2} d^{-1}), followed by *A. hemprichii* (0.21 nmol N cm^{-2} d^{-1}) and lowest in *P. granulosa* (0.04 nmol N cm^{-2} d^{-1}). There was a positive correlation between denitrification and N_2 fixation among the corals, suggesting that denitrification may counterbalance N input from N_2 fixation in the coral holobiont and perhaps may be limited by photosynthates released by endosymbiotic dinoflagellates and prokaryotes (Tilstra et al. 2019).

Corals themselves thus constitute a "coral ecosphere" (Weber et al. 2019) in which a rich variety of microbial types are associated with and influenced by hermatypic corals. Seawater samples taken in proximity to the corals, *Porites astreoides, Orbicella faveolata, Montastraea cavenosa, Pseudodiploria strigose*, and *Acropora cervicornis* on a Cuban coral reef generally harboured copiotrophic-type bacteria, and its bacterial and archaeal composition was influenced by coral species and the local environment. Picoplankton abundances either decreased or increased away from the coral colonies based on coral species and picoplankton functional group. Genes characteristic of surface-attached and potentially virulent microbes were enriched in water near corals compared to water distal to them. Weber et al. (2019) also found a prominent association between the coral, *P. astreoides* and the coral symbiont, *Endozoicomonas*, suggesting recruitment and/or shedding of these cells into the surrounding seawater. These findings support the fact the corals are linked functionally with a wide diversity of microbial assemblages, including those associated with nutrient transformation processes and cycling.

Recycling of nutrients between coral hosts and symbiotic dinoflagellates is a key process on coral reefs (Benavides et al. 2017). Zooxanthellae take up the NH_4^+ excreted by the coral host whereby this waste N is not released to the wider system. Recycling of N produced by the coral host could satisfy all the N needs of the symbiotic algae. Uptake of N for growth has been relatively well studied for the coral/algal symbiosis, but other organisms in the reef environment including macroalgae, benthic microalgae, and phytoplankton as well as many heterotrophic prokaryotes have the capacity to take up N. Oxidised forms of N such as NO_3^- need to first be reduced to NH_4^+ before their incorporation into biomass.

N_2 fixation is one of the principal mechanisms by which coral reefs obtain and utilise N for metabolic processes. N_2 fixation rates in coral reef sediments are highly variable like all other N transformation processes (Table 11.7) depending on the measurement method used and a variety of biological and environmental factors, but the process is essential for N cycling on coral reefs (O'Neil and Capone 2008). Sediments are normally prime sites for nutrient transformations, but the extreme spatial complexity and diversity of coral reefs make it likely that other reef components, previously thought to play a minor role in nutrient cycling, may play a larger role. Comparatively recent discoveries show that photosynthetic symbionts in some hermatypic corals and sponges can fix N and can nitrify (Benavides et al. 2017). These energetically demanding processes in diazotrophs are fuelled by high light flux and recycled P. Heterotrophic diazotrophic activity is supported by a combination of recycled P and organic matter.

An analysis of the contribution of the main benthic coral reef components to the input of new N in a Red Sea reef via N_2 fixation (Figure 11.9) shows that cyanobacterial mats have the highest N_2 fixation rates followed by seagrasses (if present) with much lower rates in hermatypic corals, other cnidarians, sponges, macroalgae, turf algae, limestone/dead coral, and carbonate sediments (Cardini et al. 2016). Nitrogenase activity on limestone reef surfaces, including exposed atoll rim areas, has been attributable to cyanobacteria. In Tikehau Lagoon, French Polynesia, N_2 fixation on these surfaces accounted for 25–28% of the total N demand for benthic primary productivity (Charpy-Roubaud et al. 2001; Charpy-Roubaud and Larkum 2005). Rates of N_2 fixation in sediments were lower than rates in localised areas or in cyanobacterial mats, but given the huge surface area of unconsolidated sediments, make a significant contribution to the overall reef ecosystem. Low ^{15}N values in reef macrophytes and coral tissue provide further evidence that much of the N in coral reefs is derived from N_2 fixation (Benavides et al. 2017).

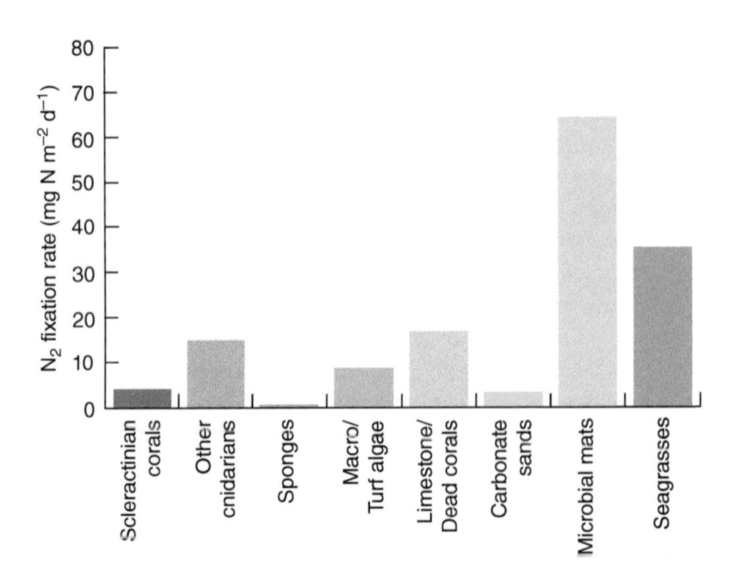

FIGURE 11.9 Contribution of the main benthic coral reef components to the input of new N in the reef via N_2 fixation. Mean rates are mg N m^{-2} d^{-1}. *Source:* Based on Cardini et al. (2016), figure 2, p. 1709. © John Wiley & Sons.

Rates of N_2 fixation can be affected by the presence of grazers (O'Neil and Capone 2008). Grazing by COTS on corals results in high rates of N_2 fixation on the coral skeletons. Similarly, the sea urchin, *Diadema antillanum,* significantly increases N_2 fixation when grazing on algal turf as compared to ungrazed areas. Algal turf assemblages contain a significant proportion of cyanobacteria in addition to diatoms and dinoflagellates. Grazers also aid in excreting significant quantities of nitrogenous waste products in the form of urea and NH_4^+ that stimulate the growth of turf algae. Fish grazing can also be important in maintaining rates of N_2 fixation on coral reefs, by keeping benthic algae in check. Cyanobacteria may have less competition for growing space and may expand into areas where more palatable species have been removed by grazers, as cyanobacteria tend to be less nutritionally complete, lacking some essential fatty acids.

N_2 fixation remains underestimated and under-sampled in coral reef environments (Benavides et al. 2017). Invertebrate diazotrophic symbioses have been reported from reef systems, including reef sponges that acquire fixed N via associated cyanobacteria. Symbiotic cyanobacteria and bacteria are found in almost all marine sponges and may form mutualistic associations with hosts, especially if the symbiont provides fixed C or N. The cyanobacterium, *Oscillatoria spongieliae,* has been found within several species of sponges including the reef sponge, *Dysidea herbacea* (Ridley et al. 2005). *Synechococcus* has also been described from several species of sponge, which are genetically distinct from free-living planktonic species of *Synechococcus.* In some instances, metabolic byproducts are translocated from symbiont to host (Ridley et al. 2005).

N_2 fixation in coral reefs may be affected by climate change. A laboratory study investigating the short-term effects of ocean acidification on N_2 fixation associated with the hermatypic coral, *Seriatopora hystrix,* found that under high pCO_2 conditions, N_2 fixation declined rapidly and significantly (53%) compared to corals under present-day pCO_2 conditions (Rädecker et al. 2014). Moreover, N_2 fixation associated with the coral holobiont showed a positive exponential relationship with its calcification rates, suggesting that that even small declines in calcification rates under high CO_2 conditions may result in decreased N availability and may trigger a negative feedback loop on coral productivity.

Given the rapid recycling of N on coral reefs, most of this recycling must come about via ammonification from the decomposition of organic N (Table 11.7). Ammonification can also occur in coral skeletons, and other hard-bottom structures may be important sources of regeneration of nutrients. Up to 70% of reef volume can be attributed to the myriad of nooks and crannies that trap sediment, including faecal pellets and decaying organic matter (Szmant 2002). Regenerated inorganic nutrients can then percolate through these porous structures. Elevated nutrient concentrations have been measured in reef cavities and interstices which can be introduced to the reef through percolation of nutrient-enriched water through holes and fissures that provide reef organisms with pulses of nutrients. Cavities within the reef structure are areas of high bacterial activity and nutrient regeneration. In a study in Curacao, open water bacterial productivity was limited by both N and P, whereas reef crevice bacteria were limited only by P (Scheffers et al. 2005). N limitation was probably circumvented in these interstices through remineralisation processes which can, in turn, increase inorganic N concentrations in the overlying reef waters through exchange with the ambient water. Rasheed et al. (2002) determined that nutrient concentrations in these regenerative spaces within the reef framework are 1–3x higher than surrounding water. Nutrients within crevice sediments were 15–80x higher than in the water column, underscoring the importance of these crevices in trapping organic matter and regenerating nutrients.

N transformations in reef waters are less understood than in sediments, complicated by high spatial and temporal heterogeneity of nutrient concentrations owing to complex patterns of water mixing and variable residence times. Availability of nutrients may be limited by exchanges with benthic surfaces, although this is not always true. Hopkinson et al. (1987) found that rates of NH_4^+ regeneration in the water column on Davies Reef in the central GBR were highly variable in winter. They did find, however, that regeneration rates are more important in terms of supplying regenerated N to plankton from the reef front to the back lagoon.

Benthic–pelagic coupling is evidently facilitated, and perhaps partly mediated, by demersal organisms that feed at the sediment–water interface. In several reef lagoons within the GBR, Bishop and Greenwood (1994) measured rates of N excretion by some demersal macrozooplankton equating to 29% of N regeneration reported for microheterotrophs and 9% of N utilisation by phytoplankton. Compared to the benthos, these excretion rates were equivalent to 10–13% of benthic NH_4^+ release and from 5 to 32 and 2 to 13% of benthic ammonification and utilisation rates, respectively. N excretion by some demersal macrozooplankton may even exceed excretion by some benthic infauna. NH_4^+ is released through grazing processes in which fish and other biota, including detritivores and bacteria, recycle organic N and excrete it. Coral reef fish excrete products rich in N with NH_4^+ as the most important source of DIN and with excretory products that are enriched in P. Coral growth is higher in the presence of reef fish, particularly resting fish such as grunts, probably due to the stimulatory effects of fish excretion.

The multitude of algal–invertebrate symbioses that have evolved in reef environments to efficiently recycle nutrients is testament to the strategy of sustained high productivity in oligotrophic seas. N transformations are the crux of these associations, with the symbiotic dinoflagellates providing C for translocation from photosynthesis to coral host, and the transfer of N from animal waste to the symbiotic algae. Less well-characterised associations are known, such as those between sponges and their symbionts as well as giant clams, anemones, jellyfish, and ascidians, as well as N transfer from endosymbiotic, diazotrophic cyanobacteria to symbiotic dinoflagellates within a coral (Lesser et al. 2007).

Corals can take up both DIN and DON as well as ingest PON (Mills et al. 2004). The amount of N available to corals is a function of both particulate feeding as well as inorganic N uptake which can be affected by nutritional history (Piniak and Lipschultz 2004). Corals and other sessile organisms acquire essential nutrients by harvesting zooplankton, microbes, and detritus directly from the water column using mucus-netting strategies, or indirectly by capture of protozoa that graze on bacteria. These trapping processes are so efficient that they permit coral reefs to support high biomass in oligotrophic waters. Corals can gain up to 70% of their N from particulate sources and microbes which account for 30–45% of the incorporated particulate matter. The second pathway is the uptake of DIN by the symbiont. This N may itself be derived from recycled animal wastes which are then retained by the symbionts in exchange for photosynthate released back to the host (Piniak and Lipschultz 2004). Alternatively, the symbiont may obtain inorganic N either in the form of NH_4^+ or NO_3^- directly from seawater. The latter is often a light-dependent process (Grover et al. 2003). It has been further demonstrated that both intact corals as well as isolated zooxanthellae can take up both NO_3^- and NH_4^+. This net uptake can account for approximately 30% of the daily N requirement for tissue growth, gamete production, and DON production.

Yet other pathways of N acquisition may exist (Rädecker et al. 2015). Symbionts can take up dissolved free amino acids (DFAAs), with amino acid uptake accounting for as much as 10% of the N demands of the coral. This may be an important episodic source of N for corals during times of local fish migrations and the activity of other grazers that result in algal cell breakage or exudation which increase DON concentration on the reef. Concentration-dependent uptake of ^{15}N-urea was demonstrated in the coral, *Stylophora pistillata* (Grover et al. 2006), which suggested adaptation of the coral to low urea concentrations; when compared to uptake rates of ambient NH_4^+ and NO_3^- concentrations, urea was preferred to NO_3^-. Urea may therefore be an episodic source of N.

Coral mucus plays an important role in nutrient recycling on reefs. Up to 50% of the C assimilated by zooxanthellae may be exuded as mucus, which protects the coral colony against fouling, desiccation, and sedimentation. Mucus efficiently traps suspended particulate matter, including bacteria known to fix both N and C, thereby increasing the nutritional quality in coral communities. Wild et al. (2004) found that coral mucus could increase both its C and N content up to three orders of magnitude in two hours. Bacteria are much better than corals at assimilating nutrients at low concentrations (Knowlton and Rohwer 2003) and coral-associated bacteria may scavenge limiting nutrients, including Fe and vitamins, which are then harvested by the corals themselves by eating their own mucus. These enriched mucus-particle aggregates subsequently provide 10–20% of the total organic C supply to the sediments, maintaining nutrients for heterotrophs and remineralisation processes (Wild et al. 2004). Coral mucus and organic flux from phenomena such as mass spawning events may also affect sediment biogeochemistry by enhancing sediment organic content and microbial processes and nutrient processing (Wild et al. 2008).

P regeneration within reef waters and between reef waters and the benthos is similarly rapid (Sorokin 1990, 1992). On reefs within the Capricornia section of the GBR, Sorokin (1990, 1992) used ^{32}P to trace the flow of P and found that although dissolved P concentrations were low, turnover was rapid in the water column, ranging from two to eight days. Bacterioplankton was the major consumer of inorganic P, accounting for 50–90% of total P uptake. Bacterioplankton transfer of P when consumed by benthos and the release of P by benthos roughly equalled the amount consumed, resulting in no or little net uptake. There was a small net release of dissolved P from sediments.

The modes and conditions under which coral reefs cycle P are not well understood. Atkinson and his colleagues (Atkinson 1988, 1992; Atkinson and Bilger 1992; Bilger and Atkinson 1992) and the ORSTOM program in French Polynesia (Charpy and Charpy-Roubaud 1988) initiated experiments to detail elements of P cycling on coral reefs. In the lagoonal atoll of Tikehau, Charpy and Charpy-Roubaud (1988) found that P concentrations in lagoon waters were slightly greater than in the surrounding water with excretion by pelagic heterotrophs supplying 75% of phytoplankton P requirements; P required for benthic primary production was met by release from the sediments. Their budget suggested that P was exported from the lagoon in particulate form but equated to only 3% of phytoplankton production.

The mass transfer limitation hypothesis, proposed by Atkinson and his colleagues (Atkinson 1988, 1992; Bilger and Atkinson 1992), predicts that the extent to which P is regulated within a coral reef is not dependent on concentration but is in fact dependent on the rate at which P is transported onto the reef. Nutrient regulation and limitation on coral reefs are therefore dependent on the rate of water motion into and across a coral reef. To test their hypothesis, Bilger and Atkinson (1992) used an engineering approach to calculate the ratio of the uptake rate to advection rate (known to engineers as the Stanton number, a dimensionless figure) on a simplified reef flat. Their calculations yielded an anomalously high mass transfer of P due to the difference between a smooth surface and the real three-dimensional benthic surfaces of a coral reef. Atkinson and Bilger (1992) performed a series of flume experiments on reef communities which confirmed higher than predicted rates of mass transfer. Increasing water flow velocity increased the rate of P uptake, indicating that uptake is regulated by diffusive boundary layers near the surfaces of the organisms and that the natural geometry of reef surfaces coupled with physical flow conditions facilitates transfer of nutrients. In most other respects, the data were consistent with the engineering model of mass transfer, being somewhat influenced by water temperature, solar insolation, and nutrient loading. Applying the mass transfer limitation idea to other P data, Atkinson (1988, 1992) concluded that there is a maximal rate of P uptake under highest water flow conditions, but the residence time of the water over the reef is shortest, and the amount of P removed from the water is only about 5% of the amount that passes over the reef; this may explain the inability in earlier studies to observe P uptake over a short distance.

The idea of coral reefs being limited by mass transfer of nutrients is intriguing as it agrees with the strong water flow dependence observed for growth of many reef organisms. But it is unlikely that any ecosystem conforms closely to engineering equations, and the hypothesis requires further testing to determine the scope of its applicability. Yet the view that P is mass transfer limited is not necessarily at odds with the view that recycling is an important regulator on coral reefs. Under low-flow conditions, recycling mechanisms may be the pre-eminent means of conserving and modulating the P cycle. Under high-flow conditions, supply rate may outpace the rate of P recycling on coral reefs. The mass transfer idea is an elegant refinement of Smith's (1984) earlier supposition that the rate of hydrodynamic flux determines whether N or P is limiting to coral reefs. The conditions and the extent to which this is true remain to be further explored. It is highly unlikely that any one factor limits energy and nutrient flow under all climatic conditions. More recent findings (Cuet et al. 2011; Wyatt et al. 2012) support the notion of mass transfer limitation. For instance, Wyatt et al. (2012) found that NO_2^-, NO_3^- and PO_4^{3-} dynamics over Ningaloo Reef, Western Australia, depends on oceanographic forcing. Gross uptake rates increased significantly with increasing oceanic concentrations and wave energy dissipation and were 35–80% higher over the reef crest and were significantly correlated with independent estimates of POM-mediated gross uptake, supporting both mass transfer-limited uptake and the strong role of oceanic POM supply. The relative supply of NO_2^- and NO_3^- and POM was linked to the seasonal dynamics of a regional current regime.

Uncertainty about the relative importance of recycling and the mass transfer limitation of nutrients does not mean that coral reefs require a large supply of 'exotic' nutrients to explain the high rates of nutrient flux, as suggested by the endo-upwelling concept of Rougerie and Wauthy (1988). This hypothesis states that geothermal heat deep within the limestone and basaltic framework underlying Pacific atolls facilitates upwelling and emergence of nutrient-rich, Antarctic intermediate water to the shallow surface waters to fuel high rates of C fixation. Endo-upwelling may be a significant geochemical process, particularly on Pacific atolls, but it is unlikely to be a significant mechanism for nutrient supply to living continental shelf reefs.

P limitation of coral growth has been examined experimentally (Suzumura et al. 2002; Godinot et al. 2009, 2011; Dunn et al. 2012). Godinot et al. (2009) assessed the uptake rates of PO_4^{3-} by the coral, *S. pistillata,* and its symbiotic zooxanthellae. They found that the velocity of PO_4^{3-} absorption increased in the light, suggesting a relationship with the photosynthetic activity of the zooxanthellae; a correlation was found between PO_4^{3-} uptake rates and the organic or inorganic feeding history of the corals. The same coral was subjected to uptake experiments by Godinot et al. (2011) who showed that P is limiting for this coral. Dunn et al. (2012) conducted a series of experiments with the coral, *Acropora muricata,* with increasing doses of PO_4^{3-}. They found that tissue growth increased with increasing amounts of PO_4^{3-}. They suggested that the PO_4^{3-} enhanced growth was due to increased zooxanthellae populations and photosynthetic production within the coral. Skeletal density reduction was found indicating a more brittle skeleton. Under thermal stress, experiments with the hard corals, *Turbinaria reniformis, Pocillopora damicornis, Pavona cactus, Galaxea fascicularis* and the soft coral, *Heteroxenia fuscescens* showed that N acquisition rates decreased but P uptake rates increased in most species (Ezzat et al. 2016). P was required to maintain symbiont density and photosynthetic rates as well as to enhance the translocation and retention of carbon within the host tissue, highlighting the importance of P for symbiont health.

Coral reef sponges play a crucial role in P cycling on coral reefs. The giant barrel sponge, *Xestospongia muta*, found at Conch Reef in Key Largo, Florida, produced polyphosphate which was stored by sponge endosymbionts in the form of polyphosphate granules (Zhang et al. 2015a, b). While few studies have determined the change in concentration of dissolved and particulate P between ambient waters and sponge effluent, reef sponges do show a net release of DIP in the form of PO_4^{3-} (Pawlik and McMurray 2020). The accumulation of polyphosphate granules might be used for energy storage or used by endosymbiotic bacteria during times of P shortage. PO_4^{3-} release would lead to a temporal offset between P uptake and release by sponges. The main overall impact of P cycling on coral reefs would still relate to the conversion of less labile organophosphorus forms to easily used PO_4^{3-}. However, if a large amount of sponge polyphosphate is released in particulate form and deposits in sediments, then this could be a major sink for P. Zhang et al. (2015b) noted that polyphosphate could serve as a precursor phase, which converts directly to the mineral apatite or decomposes to promote apatite supersaturation and precipitation within the sponge tissue or within sediments. These results may cast sponges in a dual role regarding reef P cycling. The release of polyphosphate through the metabolism of particulate and dissolved organic compounds restores P to a more bioavailable form and supports reef primary production.

Fe may be limiting for some coral reefs. Entsch et al. (1983) found that some algae in symbiotic associations on corals may suffer Fe deficiency and concluded that the degree and extent of Fe stress in primary producers may influence growth rates, biomass, and distribution of species. Ferrier-Pagés et al. (2001) similarly found that growth of corals may be Fe-limited, examining whether the addition of Fe alone or in combination with NO_3^- affects growth and photosynthesis of the coral, *S. pistillata.* Fe addition induced a significant increase in the areal density of zooxanthellae and gross photosynthetic rates increased following Fe enrichment.

Excess nutrient loading has severe consequences for coral reefs. In a meta-analysis, Shantz and Burkepile (2014) found that over a wide range of concentrations, excess N reduced coral calcification by 11%, on average, but enhanced metrics of coral photobiology, such as photosynthetic rate. In contrast, P enrichment increased average calcification rates by 9%, probably through direct impacts on calcification. There were few synergistic impacts of combined N and P on corals, as the nutrients impacted corals via different pathways. Polluted groundwater was found to play in important role in N cycling on a nearshore reef in Bermuda, where sewage entered the groundwater through untreated cesspits (Sims et al. 2020). Tissue $\delta^{15}N$ of the dominant reef coral, *Porites astreoides,* was elevated by about 3‰ on the nearshore reef compared to a control site indicating that corals were assimilating groundwater-derived N. Coral skeletal density and calcification rates were inversely correlated with NO_3^- concentration and $\delta^{15}N$ indicating an overall negative response to groundwater-borne inputs. The response of corals to increasing nutrient availability is context-dependent, varying with coral taxa and morphology, enrichment source and nutrient.

11.6 Estuaries and Coastal Lagoons

Highly variable salinities, high-salinity plugs and tidal trapping and time lags in estuarine circulation, and high rates of precipitation and evaporation are just some of the environmental characteristics peculiar to or dominant in tropical estuarine and inshore waters (Section 3.4). These physicochemical features are unique in toto and have important consequences for biogeochemical cycling.

Most tropical estuaries and coastal lagoons are polluted and eutrophic due to heavy human use and abuse. A large database exists for nutrient concentrations in tropical estuaries and coastal lagoons, but much less data are available on nutrient fluxes and biogeochemical cycling. For instance, Ebrié Lagoon, Ivory Coast, is the largest lagoonal ecosystem in West Africa and one of the most polluted. Two-thirds of its biological oxygen demand and 95% of its total N and P loads are sewage. Total annual N loads to the lagoon were 45% from urban sources, 42% from land runoff, and 13% from atmospheric deposition. Estimates for P inputs were 30% from urban sources, 48% from land runoff, and 13% from the atmosphere (Scheren et al. 2004). Scenario analysis has shown that population growth without pollution reduction measures should result in an estimated fivefold increase in nutrient inputs to the lagoon over the period 1980–2050. An extreme outlier of human-induced impacts is the Brantas River estuary and the coastal waters of Madura Strait on Java (Jennerjahn et al. 2004) where there is intensive agricultural activity through the catchment. Nutrient concentrations were high in the river and then decreased rapidly seaward. Runoff from agricultural lands was a major source for nutrients. A high abundance of phytoplankton, despite NO_3^- depletion and a N:P ratio < 4, indicated additional N sources for phytoplankton growth. Stable isotope studies indicated an autochthonous origin of suspended organic matter. There were seasonal differences in the quantity and origin of suspended sediments and organic matter transported by the river with high inputs to coastal waters during the wet season.

Nutrient concentrations and processes are often linked to anthropogenic activities. In the Ohuira Lagoon, NW Mexico, declines in freshwater supply due to the impoundment of a local river led to a dramatic change in the lagoon's nutrient status (Ruiz-Fernández et al. 2007). Sedimentary C, N, and P concentrations have increased over the past century due to land clearing, water impoundment, and agricultural practices in the catchment. C:N:P ratios and $\delta^{13}C$ suggested an estuarine system responsive to increased C loading from a N-limited phytoplankton community, whereas

$\delta^{15}N$ values indicated a transition from an estuarine to a more marine environment. In the polluted Muthupet Lagoon, SE India, terrestrial inputs through runoff mixing and residual fluxes were dominant forcing mechanisms in maintaining nutrient concentrations. Nutrient exchange was rapid with non-conservative fluxes of dissolved inorganic and organic P and N indicating that these nutrients were removed from the system, probably by microbial processes (Gupta et al. 2006).

Denitrification can help alleviate nutrient over-enrichment by removing N from a given estuary or lagoon, although not all systems show rapid rates of denitrification. In Little Lagoon, a groundwater-influenced coastal lagoon, Gulf of Mexico, rates of N_2 fixation exceeded rates of denitrification in summer, when NO_3^- concentrations were low and sulphide concentrations were high (Bernard et al. 2014). Sulphidic sediments resulted in denitrification being supported mainly by water column NO_3^- rather than nitrification. Submarine groundwater discharge rates of N were 42x larger than benthic rates; both the seabed and groundwater provided dissolved organic N to the water column. In the tropical lagoon in Chiku, SW Taiwan, the residence time of nutrients varied with water exchange time (Hung and Kuo 2002). Non-conservative flux of dissolved inorganic P from the lagoon was around -0.1 mol m^{-2} a^{-1}, equivalent to an internal organic carbon sink of 11 mol C m^{-2} a^{-1}. The lagoon is net autotrophic. Despite net N_2 fixation being observed during some periods, denitrification exceeded N_2 fixation indicating a net loss of N from the lagoon via microbial pathways.

Anthropogenic pollution often coincides with oceanographic features, resulting in some unique biogeochemical cycles. For instance, the coastal waters of the Cochin backwaters of India are unique in that anoxia occurs in the estuary's bottom waters due to the presence of organic pollution coupled with upwelling from the Arabian Sea which brings cool, high saline, oxygen-deficient, and nutrient-rich waters into the estuary mouth (Martin et al. 2010). Increased stratification coupled with low mixing and the presence of high organic matter have resulted in anoxia which has led to H_2S production. The reduction of NO_3^- and formation of NO_2^- indicate strong denitrification in the estuary. The persistence of anoxia and formation of H_2S has led to an increase in greenhouse gas emissions from the estuary. Nutrient stoichiometry plays an important role in N transformations in the Cochin estuary, as heterotrophic diazotrophs are controlled mainly by the N:P ratio. N_2 fixation rates in Cochin waters were higher during the post-monsoon and lower during the monsoon (Jabir et al. 2020) and appeared to be primarily controlled by the DIN:DIP ratio.

N cycling has been measured in some tropical coastal lagoons and estuaries. In the Ria Lagartos Lagoon, Mexico, nitrification rates were higher than denitrification (Valdes and Real 2004). High concentrations of particulate N and P in the sediments suggested that this lagoon may be a sink for nutrients. In the heavily impacted, San José Lagoon in Puerto Rico, direct measurements of net N_2 production and inorganic nutrient fluxes showed that fluxes were high compared with those measured in most other tropical lagoons and estuaries (Pérez-Villalona et al. 2015). Net N_2 production was an important sink in the N balance of this lagoon.

The factors controlling N dynamics in tropical coastal lagoons are not well understood, but likely include climate, the size of the opening of the lagoon relative to the open ocean, water residence time, circulation, and the composition of the pelagic and benthic fauna and flora. Assessing the regulatory factors controlling N transformations in several tropical Brazilian lagoons, Enrich-Prast et al. (2016a, b) found low rates of denitrification, with denitrification and various N oxidation and reduction processes significantly correlated with oxygen consumption. The contribution of nitrification to total O_2 consumption decreased from 9% at sites with the lowest sediment mineralisation rates to <0.1% at sites with the highest rates. NH_4^+ oxidation and NO_2^- reduction by anammox contributed up to 3% of the total N loss in sediments with the lowest sediment mineralisation rates. Low denitrification rates were ascribed to low NO_3^- concentrations and little NO_3^- supply from nitrification. These patterns could also be explained by variations in the microbial environments from largely stable, oxic conditions at low mineralisation rates to more variable, anaerobic conditions at high sediment mineralisation rates. Algal and microbial mats affected all N processes, but the activity of NO_3^--reducing microorganisms was not dependent on NO_3^- availability.

NO_3^- reduction to NH_4^+ is an important process in sediments of some tropical estuaries. In three tropical estuaries in Thailand, Indonesia, and Fiji during wet, dry, and intermediate seasons, Dong et al. (2011) found that dissimilatory reduction of NO_3^- (DNRA) was the dominant N process with no measurable anammox. Denitrification was either zero or extremely low, driven by water column NO_3^-. N_2O flux was not formed during denitrification, so these estuaries are unlikely to be significant sources of N_2O. Benthic NO_3^- reduction was NO_3^- limited. The predominance of DNRA over denitrification may have been due to an energetic advantage of NO_3^- ammonifiers over denitrifiers when competing for limited NO_3^- and to higher affinity for NO_3^- by the NO_3^- ammonifiers.

Nitrification may be an important process in coastal lagoons and estuaries, although it is probably uncoupled to denitrification. Significant nitrification was measured in the Pearl River estuary, China, (Dai et al. 2008) with a dramatic decrease in NH_4^+, an increase in NO_3^-, but no significant change in NO_2^- in the upper estuary. However, species and concentrations of inorganic N changed with season: in winter with low runoff, the upper reaches of the estuary showed relatively low rates of NH_4^+ oxidation and NO_2^- oxidation accompanied by extremely high concentrations of NH_4^+ and NO_3^-; in summer, the upper estuary had higher nitrification rates with lower amounts of NH_4^+ and NO_3^-. Temperature appeared to control nitrification which contributed significantly to the consumption of dissolved oxygen and CO_2 production in the upper estuary. In the Brunswick estuary, Australia, nutrient cycling exhibited four different states (Ferguson et al. 2004; Eyre and Ferguson 2005). During high flow, freshwater residence times were <1 day, internal cycling processes were bypassed and virtually all dissolved and most particulate nutrients were delivered to the continental shelf. During the growth phase of a phytoplankton bloom, enhanced recycling occurred as this phase supplied up to 50% of phytoplankton DIN demands. In post-bloom conditions, DIN uptake by phytoplankton decreased in the autumn wet season when biomass doubling times began to exceed residence times. In the dry season when flows decreased further, there was tight recycling and phytoplankton blooms and uptake by the sediments accounted for the entire DIN load to the estuary resulting in complete removal of DIN from the water column, with the ocean a potentially important source of DIN to the estuary. Net N_2 effluxes were controlled by a complex interaction between the supply of NO_3^- from the water column and nitrification, the supply and decomposition of labile C, benthic productivity, and macrofauna abundance. About 22% of the remineralised N was permanently removed via denitrification. This active competition for limited N resources between heterotrophs, autotrophs, and chemoautotrophs appears to be a mechanism by which N-limited oligotrophic, subtropical estuaries tightly recycle and conserve N.

Tropical estuaries often show nutrient behaviours much different than those in temperate estuaries. Eyre and Balls (1999) examined nutrient behaviour along the salinity gradient of several tropical and temperate estuaries and found that the most distinct difference was the timing and variability of the major physical forces on the estuary (e.g. river flow, insolation). The tropical estuaries were much more episodic than their temperate counterparts, with a much more dynamic salinity structure and more variable river water concentrations, so that delivery of material to the estuary was dominated by short-lived flood events. However, seawater concentrations were more stable in the tropical estuaries due to higher solar insolation, resulting in high rates of biological activity. There was biological removal of DIP in the low-salinity region of some tropical estuaries and a low-salinity input of NO_3^- in one tropical estuary most likely due to nitrification. Similar effects were not observed in the temperate estuaries. One similarity between the temperate and tropical estuaries was the phenomenon of mid-estuary inputs of NH_4^+, although they occurred under vastly contrasting conditions of low river discharge and periods of flood, respectively. In a preliminary nutrient budget for the tropical Moresby estuary and catchment in north Queensland, Australia, Eyre (1995) found that nutrient behaviour was greatly impacted by the summer wet season. Fertiliser application on cane land in the catchment was the dominant source of P and N contributing 88 times more P and N to the catchment than is supplied by other sources. However, only about 4% of the P and about 11% of the N added to the catchment was transferred to the estuary. About 82% of the P and about 90% of the N flux from the catchment occurred in the wet season. Once in the estuary, only small amounts of P and N were retained with most of the P being transferred to the ocean and it was suggested that most of the N may be used biologically.

Some tropical estuaries and lagoons are conduits for materials and water flow to the sea, although some are more closely linked to the coastal ocean than others. For Darwin Harbour, a tropical macrotidal estuary in northern Australia, the largest new source of N is from the ocean (Burford et al. 2008). Atmospheric inputs via N_2 fixation on the intertidal flats and subtidal sediments were lower than oceanic inputs. The rivers feeding into the harbour and sewage were minor N inputs. N demand by primary producers was high relative to available N inputs, suggesting that N recycling within the water column and mangrove forests must be important processes.

Of course, some tropical estuaries, especially those associated with medium and large rivers, export dissolved and particulate nutrients to the coastal ocean. For example, the Hooghly estuary which is bounded by the Sundarbans and discharges water originating from the Ganges (Mukhopadhyay et al. 2006) exports DIC, DIN, and DIP to the Bay of Bengal. Out of total transport, monsoonal runoff accounted for 64% of the DIC, 71% of the DIN, and 76% of the DIP. During estuarine transport, only 8% of DIC was removed. Light-limited conditions and autotrophic–heterotrophic coupling within the estuary resulted in regeneration of nutrients from organic matter originating from external sources. Other Indian estuaries export nutrients, but they also show active nutrient recycling. In the Mandovi estuary, the sediment acts as a net source of DIN in the dry season but changes to a net sink in the monsoon period (Pratihary et al. 2009). The sediment was a huge reservoir of NH_4^+ and PO_4^{3-} and acted as a net sink of combined N because of high rates of denitrification that removed 22% of riverine DIN influx. Benthic supply of DIN and PO_4^{3-} met 49 and 25% of algal N and P demand, respectively, in the estuary. It was hypothesised that benthic NH_4^+ efflux sustained high estuarine productivity in the NO_3^--depleted, dry season. Substantial contribution of bacterial N to PON indicated N enrichment on terrestrial POM during the monsoon season (Fernandes 2011). Transport of terrestrial POM enriched with bacterial organic matter to Indian coastal waters influences coastal productivity and ecosystem functioning during the monsoon season. In the Cochin estuary in western India, the sediments sustained higher uptake rates of NH_4^+ and NO_3^- compared with adjacent coastal waters. Despite high DIN concentrations, the estuary also had higher N_2 fixation rates than coastal waters. NH_4^+ was the preferred substrate for phytoplankton growth indicating the significance of regenerative processes in primary production. A significant negative correlation between the total N:P ratio and NH_4^+ uptake as well as N_2 fixation suggests that nutrient stoichiometry played a major role in modulating N transformation rates (Bhavya et al. 2016). In the Mandovi and Zuari estuaries, pelagic N_2 fixation was rapid, greatly influenced by the intrusion of upwelled water from the adjacent coastal zone (Ahmed et al. 2019).

N_2O concentrations and emissions can be significant in some tropical estuaries and lagoons. In six estuaries in NW Borneo, N_2O concentrations were very variable, but correlated positively with rainfall. N_2O fluxes ranged from −28.51 to 190.1 μmol m^{-2} d^{-1} across the estuaries with total emissions of 1.09 Gg N_2O a^{-1} (Bange et al. 2019). In the nearby Lupar and Saribas estuaries, flux rates ranged from 3.6 to 32.9 μmol m^{-2} d^{-1} with the wet season driving variability of N_2O concentrations and fluxes (Müller et al. 2016).

Si, trace metals, and vitamins are important limiting factors of diatom growth in tropical estuaries and the adjacent coastal ocean. In tropical estuaries of western Taiwan, variations in SiO_2 concentrations had a direct effect on phytoplankton biomass and production and an indirect effect by changing community composition (Pan et al. 2016). In some estuaries, such as in Guanabara Bay, Brazil, community composition shifts from cyanobacteria to diatoms and flagellates when Si concentrations increase (Odebrecht et al. 2018); high Si concentrations also favoured the predominance of diatoms in the subtropical Patos Lagoon estuary. Similarly, high Si:P (22.6) and Si:N (4.9) ratios in the Cochin estuary, India, suggest that this Si excess can be one of the factors for the massive proliferation of diatoms throughout most of the year (Madhu et al. 2017). In Indian estuaries, Si variability is controlled by lithogenic supply, diatom uptake, mixing of seawater, and the extent of land use (Mangalaa et al. 2017) with the influence of lithogenic control significantly higher during the wet season due to particle supply through monsoonal discharge. Significant control by diatoms on the Si cycle only occurs during the dry season.

11.7 Coastal Bays and Continental Shelves

Tropical nearshore and shelf environments receive and process large amounts of nutrients and other elements, sediments, and water discharged from land (Chapter 4). Most studies have described spatial and temporal variations in nutrient concentrations rather than measuring nutrient transformations and fluxes. The most detailed studies of N and P fluxes have been done on the GBR shelf, on the Amazon shelf, the subtropical–tropical western and eastern boundary upwelling ecosystems (the northern Benguela, the Peruvian, the Mauritanian, and the Arabian Sea) and on the continental margins of India. The data from each of these margins point to rapid rates of pelagic and benthic nutrient cycling.

On the GBR shelf, both benthic and pelagic microbial communities process N very rapidly (Alongi and McKinnon 2005; Furnas et al. 2011). Sediment N budgets revealed several patterns of N flow. Denitrification was driven by equally rapid rates of coupled NH_4^+ production and nitrification. Some NH_4^+ was lost to the water column, but there was little, if any, N_2 fixation. Turnover times of the small porewater N pools were on the order of hours to one to two days. N budgets indicate sediment burial rates of 11% in Missionary Bay in the inner central GBR shelf and 14% in Princess Charlotte Bay in the northern inshore GBR lagoon. The rapid rates of N mineralisation appear reasonable if considered in relation to the rates of carbon mineralisation. In Princess Charlotte Bay, where the sediment C:N ratio averages 7, C:N mineralisation rates averaged 6. In Missionary Bay, C:N mineralisation rates averaged 29 in sediments with an average C:N ratio of 16 (Alongi and McKinnon 2005). N fluxes in mid-shelf sediments of the GBR lagoon can be compared to the rates and pathways of coastal N flow. In these mixed terrigenous-carbonate sediments, N input, ammonification and nitrification rates were similar, but more dissolved N was lost via efflux than via denitrification. However, turnover rates of the particulate and dissolved N pools were equally rapid.

Measurements of P fluxes are few for the GBR shelf. Most data consist of rates of PO_4^{3-} and DOP flux from the seabed. Extrapolating from mass sediment accumulation rates for Princess Charlotte Bay and the central GBR shelf, burial rates of P were 18% of total P input to Princess Charlotte Bay sediments and 3% in Missionary Bay muds. PO_4^{3-} release equated to 24 and 1–2% of the estimated total P input for Princess Charlotte Bay and Missionary Bay, respectively. These are crude estimates but do point to equally rapid processing of P in these coastal deposits.

N and P budgets for the northern and central sections of the GBR shelf were constructed by Furnas et al. (2011). The budgets did not balance within seasonal and annual time frames. System-level N and P sources (rainfall, runoff, upwelling, and N_2 fixation) were less than estimated losses (denitrification, cross-shelf break mixing, and burial). Nutrient dynamics were dominated by tightly coupled uptake and mineralisation of soluble N and P in the water column and during sedimentation-resuspension of particulate detritus. On an area-averaged basis, internal cycling fluxes were an order of magnitude greater than input–output fluxes. Denitrification in shelf sediments was a significant sink for N, while lateral mixing was both a P source and sink.

Rapid rates of N and P cycling have been measured in other Australian bays and shelf regions. In the Gulf of Carpentaria, high rates of N_2 fixation and depleted $\delta^{15}N$ ratios in the particulate matter in the water column were measured during a summer bloom. A N budget indicated that river N was unlikely to be a major contributor to primary productivity (Burford et al. 2009). During summer, bottom N concentrations increased and $\delta^{15}N$ ratios became more depleted, suggesting extensive benthic mineralisation in the gulf. During winter, wind-driven mixing resulted in N from bottom waters reaching the euphotic zone to fuel primary productivity. On the NW shelf of Australia, uptake rates of $^{15}NH_4^+$ measured in the water column indicated that ambient DIN was cycled in <3 hours (Furnas and Mitchell 1999). Water column stocks of particulate matter turned over 1–2 times daily by sedimentation and resuspension. N and P budgets constructed for Moreton Bay, Queensland, showed that much of the N and P were either exported or recycled in the sediments and water column (Eyre and McKee 2002). N_2 fixation was the dominant source of N and point sources were the dominant source of P to the bay. About 41% of the N and 70% of the P entering the bay was exported to the ocean, and about 56% of the N was lost via denitrification. The high percentage loss of P to the ocean was related to the short residence time of water in the bay. In contrast, the loss of N to the ocean was low compared to other coastal systems due to the high percentage loss through denitrification associated with autotrophic sediments. The distinct difference between subtropical Moreton Bay and temperate systems is the dominance of biological over physical inputs and losses of N. The budgets suggest that the bay may to P-limited.

Off western India, the continental shelf has been increasingly affected by eutrophication leading to intensifying anoxia. This intensification has resulted in one of the largest low-oxygen zones in the ocean. The zone develops during late summer and autumn, resulting in high accumulations of H_2S and N_2O in open coastal waters (Naqvi et al. 2000; Bardhan and Naqvi 2020). Increased N_2O production was caused by the addition of anthropogenic NO_3^- and its subsequent denitrification, which is favoured by hypoxic conditions. The highest N_2O concentrations occur in suboxic waters with extremely high NO_2^- levels, in contrast to the trend seen in the open ocean suboxic zones where NO_2^- build up is almost invariably associated with depletion of N_2O. The occurrence of NO_2^- oxidation was indicated by a decoupling of stable N and O isotopes of NO_3^-. Low N_2O concentrations in NO_2^--bearing waters are also found especially close to the bottom in the mid-shelf region, which often results in a mid-depth N_2O maximum. However, high N_2O levels were not confined to the mid-shelf, extending to the inner shelf during periods when sulphate reduction does not occur. Near-bottom depletion of N_2O was noticed less frequently in this zone and concentrations varied greatly with time. Denitrification was likely the major process responsible for the unusual N_2O accumulation. Shelf sediments were a perennial source of DIN and PO_4^{3-} to the overlying water column with about 75% of NH_4^+ flux being nitrified, of which about 77% remained coupled to denitrification. Fifty-eight per cent of NH_4^+ flux was lost in active coupled nitrification–denitrification resulting in substantial N loss. Shelf sediments switch from being a NO_3^- source during the oxic regime to a NO_3^- sink during the anoxic regime (Pratihary et al. 2014). N loss continues even under sulphidic conditions possibly through chemolithoautotrophic denitrification. PO_4^{3-} flux increases 4x under sulphidic conditions due to dissolution of metal oxides.

Nutrient budgets for Bandon Bay, Thailand, similarly show rapid cycling and export of N and P. Wattayakorn et al. (2001) found that while most of the dissolved N and P loading was taken up in the estuarine section of the bay, a significant portion was exported to the Gulf of Thailand. Nutrient export was greater during the wet than the dry season. The export flux of DIN was dominated by NH_4^+, greater than the export of DON, while DIP was exported at about the same rate of DOP. Bandon Bay appears to be in balance metabolically, suggesting a high efficiency of recycling organic material.

On the Amazon continental margin, rapid recycling of nutrients is the norm. NO_3^- is the dominant form of DIN throughout most of the river/ocean mixing zone, but on the outer shelf where NO_3^- is depleted by biological production, NO_3^- occurs at concentrations equivalent to those of NH_4^+, $NO_2^-NO_2^-$ and urea (DeMaster and Pope 1996). Nearshore, high turbidity limits phytoplankton production, whereas on the outer shelf, NO_3^- appears to be limiting growth. Nutrient uptake was observed, but most was probably regenerated in shelf bottom water because little biogenic material is buried in the seabed. The Amazon shelf is not an efficient nutrient trap. Diffusive nutrient fluxes from Amazon shelf deposits are low relative to riverine supply rates. What is normally described as estuarine circulation is pushed out onto the open shelf by the large river source. The shoreward movement of upwelled water reflects this estuarine circulation regime and buoyancy flux. Nutrient flux from shoreward movement may be as much as 5–10x greater than the annual flow of the Amazon. This shoreward flux accounts for about 80% of the externally supplied NH_4^+, 52% of the PO_4^{3-}, and 38% of the NO_3^- reaching the outer shelf, with the remainder coming from the river. Thus, outer shelf algal blooms are supported by inputs of offshore, subsurface waters.

N and P budgets have been constructed for the Amazon shelf (Table 11.8) and show that ≈ 15 and 30% of total N and P inputs were buried in shelf sediments. Export to the ocean (and in the case of N also to the atmosphere) was roughly equivalent to river inputs. The largest input was from river and the next largest input was from advection offshore. Production was the amount of N and P utilised in NPP and was the largest pool of nutrients. In the Amazon plume, substantial SiO_2 drawdown was detected where coastal diatoms (*Skeletonema marinoi*, *Pseudonitzchia* spp., *Thalasiossira alienii*, and *Chaetoceros* spp.) dominate the biomass and where diatom-diazotroph associations (DDA) such as *Hemiaulus hauckii-Richelia intracellularis* are common (Weber et al. 2017). DDA bloom formation is stimulated in the N-poor and Si-rich water of the Amazon plume by decreased grazing pressure when mesozooplankton concentrations decrease (Stukel et al. 2014). In addition to diatom productivity, reverse weathering in sediments was significant in the Amazon River estuary, during which the formation of authigenic aluminosilicate from the dissolving biogenic Si alters the composition of Si in the sediment. The non-conservative behaviour of $SiOH_4^+$ most likely reflects the preferential utilisation of Si during diatom growth in the outer parts of the Amazon River plume (Zhang et al. 2020). In tropical deltas such as the Amazon and Fly, there is rapid and almost complete alteration of initial opal to new forms, most likely authigenic clay (Rahman et al. 2017). Globally, 4.5–4.9 Tmol a^{-1} Si may be trapped in marine nearshore depocentres as rapidly formed clay which helps to close the existing imbalance in the marine Si cycle (Rahman et al. 2017).

Within the Mekong plume off Vietnam, N_2 fixation rates were about 10x higher during the monsoon season, but such is not the case in a 40–50-km-wide upwelling strip along the coast. The stability of the water column, micronutrients, and/or trace metals in the Mekong plume may be responsible for enhanced cyanobacterial growth, especially diatom-associated symbionts like those associated with other river plumes such as the Amazon (Voss et al. 2006). DDAs that occur in proximity to large river plumes require Si and can be Si-limited.

TABLE 11.8 **N and P budgets for the Amazon shelf.**

Source of N and P	mol N m^{-2} a^{-1}	mol P m^{-2} a^{-1}
River input	2.3	0.22
Advection from offshore	0.53	0.08
Atmospheric input	0.01	–
Nearshore export	0.08	0.02
Accumulation of total N/total P	0.42	0.09
Accumulation of total organic N/total organic P	0.28	0.02
Ocean or atmospheric export	2.3	0.19
Net ocean export	1.77	0.11
Production	8.8	0.55

Source: Brunskill (2010). © Springer Nature.

On the Yucatán shelf, the largest in the Gulf of Mexico, N cycling is problematic due to the complexity of the shelf's dynamic physical processes, including coastal upwelling, coastal-trapped waves, and the influence of the Yucatán Current via bottom Ekman transport and dynamic uplift (Estrada-Allis et al. 2020). A coupled physical-biogeochemical model indicated that the main input of DIN was through the southern and eastern margins; total N was then advected to the deep oligotrophic Bay of Campeche and central Gulf of Mexico. The inner shelf efficiently processes N as all DIN imported into the shelf was consumed by phytoplankton. Twenty per cent of the total N that enters the inner shelf comes from groundwater discharge, while 53% of total N was removed by denitrification. Variability of total N fluxes in the southern margin was modulated by fluxes from the Yucatán Current leaning against the shelf break on the eastern margin, helping to maintain upwelling at the tip of the Yucatán peninsula, bringing nutrient-rich water into the euphotic layer. Export of total N was modulated by coastal-trapped waves. Overall, N fluxes from the inner to the outer shelf are weak, but the outer shelf is the main N supplier of the inner shelf.

Measurements during project GENUS (Geochemistry and Ecology of the Namibian Upwelling System) in the northern Benguela upwelling ecosystem (NBUS) off NW Africa found that pelagic and benthic denitrification and benthic PO_4^{3-} release were minor influences on NO_3^-:PO_4^{3-} ratios, and excess PO_4^{3-} in aged upwelling water was derived from upwelling source water. The low N:P ratios do not trigger N_2 fixation in the edges of the upwelling, probably due to the lack of *Trichodesmium*. However, nutrient regeneration was intense in the mixed layer. As this system is highly productive, immense amounts of organic matter accumulate in a mud belt in relatively shallow depths of between 100 and 200 m. In these mud deposits, sediment–water NH_4^+ and PO_4^{3-} fluxes were rapid and attributed to the activity of large sulphur bacteria (Neumann et al. 2016). Benthic nutrient fluxes of the mud belt contributed $< 5\%$ to the nutrient budget of the shelf. There was a NO_3^- deficit generated in the OMZ over the shelf. Nagel et al. (2013) calculated that denitrification accounted for only 10% of N loss. Sediment contribution to dissolved P in upwelling water was roughly 60% (Flohr et al. 2014).

The Peruvian upwelling ecosystem forms part of the boundary current system of the eastern tropical South Pacific and is one of the most biologically productive regions in the world (Pennington et al. 2006). The highest rates of primary productivity (1.8–3.6 g C m^{-2} d^{1}) are six months out of phase with upwelling intensity due to deepening of the mixed layer during the upwelling period. Microbial respiration in subsurface waters leads to the development of an extensive and perennial OMZ which lies from the shelf down to 400 m. Sediments within the OMZ have organic carbon contents $> 15\%$, much higher than the average continental margin of $<2\%$ and almost entirely of marine origin (Burdige 2006). Shelf sediments are mostly diatomaceous, rapidly accumulating muds. Equatorward winds engender upwelling of nutrient-rich, equatorial subsurface water along the Peruvian coast, with most intense upwelling where the shelf narrows between 5 and 15 °S. Austral winter and spring is the main upwelling period, with interannual variability imposed by ENSO. The mean depth of the upper boundary of the OMZ on the shelf at 11–12 °S °S is about 50 m but deepens to more than 200 m or more during ENSO years.

At a time series station on the outer Peruvian shelf (150 m water depth), conversion of fixed N to N_2 by denitrification varied seasonally, with a long-term loss for the shelf (Dale et al. 2017). Rates of N_2 fixation were correspondingly high and controlled positively by organic matter content and negatively by free sulphide (Gier et al. 2016). Dissimilatory NO_3^- reduction to NH_4^+ (DNRA) was a major transformation process in the benthic N cycle, producing NH_4^+ at twice the rate of ammonification. DNRA was attributed to filamentous NO_3^- storing bacteria (*Marithioploca* and *Beggiatoa*) and to denitrification by foraminifera (Dale et al. 2019). The OMZ sediments were a source of NO_2^- and NH_4^+ to the bottom water. Anammox enhanced N_2 fluxes by a factor of 2–3, accounting for 70% of fixed N loss to N_2. The relative contribution of anammox to N_2 production across the shelf ranged from 3 to 47% with an average of 34% (Rich et al. 2020). Moreover, because most of the porewater NH_4^+ was generated by DNRA, up to two-thirds of biologically transported NO_3^- may have been lost to N_2. Porewater DON accumulated with increasing sediment depth suggesting an imbalance between DOM production and remineralisation. Diffusion-driven fluxes of DON were rapid but highly variable, reflecting active microbial DOM mineralisation at the sediment–water interface, potentially stimulated by NO_3^- and NO_2^- (Loginova et al. 2020).

Benthic P cycling in the OMZ off Peru is correspondingly high, with rapid rates of dissolved P flux across the sediment–water interface; a lack of oxygen likely promotes intensified release of dissolved P while preserving the POC:TPP burial ratio (Lomnitz et al. 2016). Benthic dissolved P fluxes were always higher than the TPP rain rate to the seabed, which is supposedly caused by transient P release by bacterial mats that stored P during previous periods when bottom waters were less reducing. At the outer rim of the OMZ, dissolved P was taken up by sediments indicating phosphorite formation (Lomnitz et al. 2016).

In continental margin sediments off Pakistan where the OMZ of the Arabian Sea impinges, total benthic N loss declines over water depths from 360 to 1430 m. While denitrification rates mirrored this pattern, anammox rates increased (Sokoll et al. 2012). Benthic N loss may be responsible for about half of the total N loss in the Arabian Sea. Benthic P concentrations are high in this region, with the high fractions of organic and biogenic P and low molar C_{org}:P_{org} ratios suggesting accumulation under high surface production and low residence time of labile forms of P due to high sedimentation rates as well as significant P transformation (Babu and Nath 2005).

The Mauritanian upwelling system off NW Africa is characterised by complex topography and circulation and a wide shelf area over which upwelled water produces filaments and eddies. The system is separated into two regimes: the shelf between 15 and 20 °N which undergoes periodic upwelling during winter and spring, and the region north of 20 °N which is characterised by continual coastal upwelling with maximal intensity during summer and autumn. Pelagic N cycling is rapid, mirroring high rates of primary productivity, with high rates of NH_4^+ and NO_3^- assimilation reflecting the importance of regenerated N in supporting phytoplankton N requirements (Clark et al. 2016). Highest rates of NH_4^+ regeneration coincided with the highest ambient NH_4^+ concentrations and abundance of phytoplankton, mostly flagellates, implying a role for microflagellates in NH_4^+ regeneration. Twenty per cent of regenerated NH_4^+ was nitrified. Rates of NH_4^+ and NO_2^- oxidation were 0.30–8.75 $nmol\,l^{-1}\,h^{-1}$ and 25.55–81.11 $nmol\,l^{-1}\,h^{-1}$, respectively, resulting in an average turnover of 0.3 and 0.2 d^{-1} for NO_2^- and NO_3^-, respectively. The complex circulation in the region assists N cycling, as vertical circulation pumps additional nutrients supporting high new production rates (Hosegood et al. 2017). Sub-mesoscale circulations such as cold filaments are likely responsible for anomalously high levels of primary productivity (8.2 $g\,C\,m^{-2}\,d^{-1}$) by resupplying nutrients to the euphotic zone. P cycling is similarly rapid in these waters, with rates of DIP uptake by particulate matter ranging from 5.4 to 19.9 $nmol\,P\,l^{-1}\,d^{-1}$; this transformation is biologically mediated. Hence, a substantial fraction of the P pool must be recycled through a particle-associated microbial pool (Sokoll et al. 2017). Such rapid rates of pelagic activity eventually translate into similarly rapid rates in the seabed. Extrapolating over the entire area, denitrification amounted to 995 kt a^{-1} (mean = 3.6 mmol m^{-2} d^{-1}). Benthic N_2 fixation under the OMZ was highest on the shelf and lowest below 412 m, with part of the N_2 fixation linked to sulphate- and iron-reducing bacteria (Gier et al. 2017). High N_2 fixation coincided with bioturbation caused by burrowing infauna. Bioturbation activity enhanced PO_4^{3-} and Fe^{2+} release as did deoxygenation in the Mauritanian OMZ (Schroller-Lomnitz et al. 2019).

11.8 Open Ocean

Rapid turnover of microbial assemblages results in equally rapid rates of nutrient cycling in tropical open ocean water. N, P, S, and Fe are often, but not always, limiting to oceanic phytoplankton (Moore et al. 2013). There are two broad regimes of phytoplankton nutrient limitation in the euphotic zone: (i) N availability tends to limit productivity throughout much of the surface tropical ocean, where the supply of nutrients from the subsurface is relatively slow, and (ii) Fe often limits productivity where subsurface nutrient supply is enhanced, including within the main oceanic upwelling regions of the Southern Ocean and the eastern equatorial Pacific. Phytoplankton productivity may be co-limited by P, vitamins, and micronutrients other than Fe. A key determinant for nutrient limitation is the variability in the stoichiometries of nutrient supply and biological demand.

In the tropical Pacific, Fe, N, P, and Si can be limiting and show active and intense recycling. Using isotopic NO_3^- as a tracer, Rafter et al. (2012) found that constraints of NO_3^- concentrations and stable isotope signatures required (i) lateral exchange between the high-latitude source waters and the zones of denitrification in the eastern tropical Pacific and (ii) the accumulation of remineralised nutrients at depth. There was rapid transport of NO_3^- sources within the equatorial zone. Rafter et al. (2012) proposed that the remineralised products of N_2 fixation, at the source of the Southern Subsurface Counter Current in the western South Pacific, were the origin of low $\delta^{15}N$ of NO_3^- in the tropical Pacific. Nitrification was also a source of NO_3^- with a pattern of increasing nitrification rates from the oligotrophic North Pacific Subtropical Gyre to the more productive eastern Tropical North Pacific (Sutka et al. 2004).

Co-limitation of nutrients can be complex. In the equatorial Pacific upwelling zone, production and sinking of diatoms dominated particulate N export at $SiOH_4^+$ concentrations > 4 μM, whereas below this level, $SiOH_4^+$ was preferentially retained (Dunne et al. 1999). While inorganic N was completely utilised, $SiOH_4^+$ remained constant between concentrations of 1–2 μM and was completely exhausted only under non-steady state conditions. Retention of Si relative to N may be due to a combination of new production by non-diatoms, dissolution of silica frustules after grazing, Fe limitation, and steady state upwelling. Thus, Si and N were tightly coupled only at periods of high nutrient concentration and non-steady state conditions. Complex circulation, a conveyor belt generated by upwelling at the equator and downwelling some degrees south and associated with biological in situ remineralisation of NH_4^+ and NO_3^-, appeared to be a very efficient means for recycling inorganic N in the euphotic layer and thus for supporting high regenerated production levels (Raimbault et al. 1999). On the other hand, high NO_3^-:$SiOH_4^+$ ratios in the upwelling waters seem to indicate that $SiOH_4^+$ was not efficiently recycled because of its low regeneration rate and rapid sinking of diatom cell walls (Leynaert et al. 2001).

In the tropical North Atlantic, N is the proximal limiting nutrient for primary production, followed by P and Fe; net chlorophyll synthesis was stimulated by addition of N and further stimulated by addition of P (Davey et al. 2008). A more complex response was found when examining picophytoplankton responses: cellular red fluorescence, an index of cell chlorophyll content, increased in *Prochlorococcus*, *Synechococcus,* and picoeukaryotes in response to addition of NH_4NO_3 but was not affected by single or combined additions of P and Fe. In contrast, cell abundances increased only after combined N and P or N and Fe additions. Thus, chlorophyll synthesis and primary production were limited by the availability of N, while a net increase in cell abundance was co-limited by N and P or N and Fe in most picophytoplankton communities (Davey et al. 2008). The role of atmospheric N inputs to the epipelagic food web in the eastern tropical Atlantic was investigated by measuring N isotopes of different functional groups of zooplankton along 23 °W (17 °N–4 °S) and 18 °N (20–24 °W). Atmospheric N input was highest (61% of total N) in the strongly stratified and oligotrophic region between 3 and 7 °N, which featured abundant *Tricodesmium* colonies, strong thermocline stratification, and low zooplankton $\delta^{15}N$ (Sandel et al. 2015). The equatorial region received considerable atmospheric N input (35%) with vertical diffusive N fluxes on the order of 8 mmol m^{-2} d^{-1}, which was an order of magnitude higher than in any other area. N_2 fixation was the major source of atmospheric N input as indicated by a close correlation between cyanobacterial abundance and zooplankton $\delta^{15}N$.

Oceanic N_2 fixation has been attributed to the filamentous cyanobacterium *Trichodesmium*, but molecular techniques now indicate that unicellular diazotrophic cyanobacteria are abundant, globally widespread, and contribute significantly to N_2 fixation in many ocean basins. For instance, in the mesopelagic zone of the Bismarck and Solomon Seas in the SW Pacific, N_2 fixation rates correlated positively with transparent exopolymer particles suggesting a dependence of diazotroph activity on organic matter. Moreover, an analysis of *nifH* genes suggested the presence of alpha-, beta-, gamma-, and delta-proteobacteria and a potentially anaerobic cluster indicating that N_2 fixation was partially supported by presumably particle-attached diazotrophs sustained by organic matter (Benavides et al. 2015).

It has been assumed that primary production in tropical oceanic waters is fuelled by the upward flux of NO_3^- and by NH_4^+ generated by mineralisation of organic matter in the euphotic zone. However, it is likely that nitrification in the water column plays a significant role in fostering oceanic new production (Yool et al. 2007). For much of the world ocean a substantial fraction of the NO_3^- taken up is generated through nitrification near the surface. At the global scale, and including tropical oceanic waters, nitrification accounts for about half of the NO_3^- consumed by growing phytoplankton.

In the Arabian Sea and NW Indian Ocean, N assimilation rates by phytoplankton in the euphotic zone varied from 1.1 to 23.6 mmol N m^{-2} d^{-1} with generally higher rates near the Arabian coast and at about 8 °N. NH_4^+ was the preferred N substrate for phytoplankton, with the average integrated assimilation rate of NH_4^+ being 3.7 mmol N m^{-2} d^{-1} compared with 1.6 and 1.8 mmol N m^{-2} d^{-1} for urea and NO_3^-, respectively (Watts and Owens 1999). Not surprisingly, the Arabian Sea shows intense N cycling, driven in part by high rates of atmospheric Fe inputs through dust deposition. (Guieu et al. 2019). This region is a significant source of N_2O and a major sink for fixed N mainly due to enhanced rates of denitrification that occur in the OMZ. Additions of fixed N via N_2 fixation are small compared with rates of pelagic denitrification. Consequently, the fixed N budget of the Arabian Sea was dominated by an advective supply from the south, and by the sink arising from pelagic denitrification. Inputs of fixed N from runoff and from the atmosphere have significant impacts on surface waters and on the coastal waters of western India (Bange et al. 2005). The eastern Arabian Sea witnesses the highest rates of N_2 fixation (1300–2500 μmol $N m^{-2} d^{-1}$) among the world's oceans, sustained by excess PO_4^{3-} (Singh et al. 2019). The eastern Arabian Sea gains about 92% of its new N through N_2 fixation (Kumar et al. 2017). There is no doubt that denitrification is the dominant mode of N loss in the Arabian Sea (Ward et al. 2009). Oceanic OMZs are responsible for about 35% of oceanic N_2 production and up to half of that occurs in the Arabian Sea. In contrast, anammox is the dominant mode of N loss in the eastern tropical South Pacific. However, Dalsgaard et al. (2012) reported that denitrification in the OMZ of the eastern South Pacific accounted for most N losses, being responsible for up to 72% of total N_2 production and 77% of the total removal of fixed N including N_2O production. Anammox activity was highest just below the oxic–anoxic interface and declines exponentially with depth, whereas denitrification was highly patchy.

In the eastern tropical South Pacific, the N cycle is greatly influenced by the presence of the OMZ (Lam et al. 2009). Anammox was the dominant process of N loss in this zone, obtaining 67% or more of NO_2^- from NO_3^- reduction, and 33% or less from aerobic ammonia oxidation. Dissimilatory NO_3^- reduction to NH_4^+ occurred throughout the OMZ and could satisfy a substantial part of the NH_4^+ requirement for anammox. The remaining NH_4^+ came from remineralisation via NO_3^- reduction and probably from microbial respiration. Altogether, deep-sea NO_3^- accounted for about 50% of the N loss from the eastern tropical South Pacific.

The OMZ in the Costa Rica Dome is also an area of intense N cycling. NO_3^- reduction occurs at a maximum rate at the top of the OMZ, at the same depth as the maximum rate of NO_2^- reduction. NO_2^- oxidation occurred at maximum rates above the secondary NO_2^- maximum, but also in the secondary NO_2^- maximum, within the OMZ. Anammox contributed to NO_2^- oxidation but did not account for all N loss. NO_3^- isotope profiles suggest a NO_3^- sink by denitrification (Buchwald et al. 2015). NO_2^- oxidation was an important process in the OMZ with implications for the distribution and physiology of NO_2^--oxidising bacteria and for total N loss in the largest marine OMZ.

N_2 fixation is the major process by which new N is provided in the tropical open ocean. N_2 fixation, especially in subtropical and tropical open ocean waters, has a major role in the global marine N budget (Karl et al. 2002). However, N_2 fixation is not always a dominant process. For instance, in the eastern tropical South Pacific, Knapp et al. (2016) found that N_2 fixation supported only about 20% of export production. Although euphotic zone-integrated short-term N_2 fixation rates were higher than in suboxic waters at depth, N_2 fixation was a minor source of new N to surface waters of the eastern tropical South Pacific.

In other tropical open ocean waters, however, rates of N_2 fixation are rapid and can account for most of the new N to the euphotic zone. In the eastern tropical North Pacific, N_2 fixation rates in the summer varied from 15 to 70 μmol $m^{-2} d^{-1}$, with specific blooms contributing as much as 795 μmol $m^{-2} d^{-1}$ (White et al. 2013). The contribution of N_2 fixation to particle export was highly variable, but the blooms of diatom-*Richelia* symbioses accounted for as much as 44% of the summer carbon flux at 100 m. Temporal and spatial variability in N_2 fixation rates was high and greatly affected by eddies, gyres, and El Niño and La Niña events. N_2 fixation rates in the eastern tropical South Pacific range from 0.01 to 0.88 nmol $N l^{-1} d^{-1}$ with higher rates during El Niño conditions (Dekaezemacker et al. 2013). High N_2 fixation rates were found in the

OMZ indicating that inputs of new N can occur in parallel with N loss processes. N_2 fixation rates respond positively to Fe and N additions indicating nutrient limitation. Dekaezemacker et al. (2013) hypothesised that N_2 fixation is directly limited by inorganic nutrient availability, or indirectly through the stimulation of primary production and the subsequent excretion of DOM and/or the formation of microenvironments favourable for heterotrophic N_2 fixation. In the western tropical Pacific, N_2 fixation rates were low (0.06 to 2.8 nmol l^{-1} d^{-1}) in high nutrient-low chlorophyll waters, higher in the warm pool (0.11 to 18.2 nmol l^{-1} d^{-1}), and extremely high (38 to 610 nmol l^{-1} d^{-1}) close to the island of New Guinea (Bonnet et al. 2009). Seventy-four per cent of total activity was accounted for by the <10 µm size fraction. In the western South Pacific between New Caledonia and Tahiti, rates of N_2 fixation were more variable and associated with Fe inputs (Knapp et al. 2018). In the eastern tropical North Pacific, N_2 fixation rates were associated with oxygen, light, and nutrient gradients within and adjacent to the OMZ. In oxygen-deplete waters, N_2 fixation was largely undetectable, but in suboxic waters near volcanic islands N_2 fixation was measurable to 3000 m (Selden et al. 2019). N_2 fixation was highest in overlying euphotic waters near the continent, exceeding 500 µmol N m^{-2} d^{-1}, but overall, regional N_2 fixation appears insufficient to compensate for N loss in the OMZ.

In the tropical Atlantic, N_2 fixation rates vary by an order of magnitude (Voss et al. 2004). The highest rates occur off NW Africa and in the Amazon plume, contributing 1–12% of the N demand of primary production. N_2 fixation rates correlated with dissolved Fe concentrations suggesting Fe limitation. High atmospheric Fe inputs combined with a shallow nutricline make the tropical Atlantic a favourable environment for N_2 fixers. This has been substantiated by Capone et al. (2005) who measured rates of N_2 fixation that clearly indicate that fixation by *Trichodesmium* is a major source of new N in the tropical North Atlantic rather than the upward flux of NO_3^- from subsurface waters.

In the Indian Ocean, N_2 fixation in the Arabian Sea (24.6–47.1 µmol N m^{-2} d^{-1}) was significantly higher than in the equatorial and southern Indian Ocean (6.3–16.6 µmol N m^{-2} d^{-1}) indicating that Fe may control diazotrophy in the Indian Ocean (Shiozaki et al. 2014). Most diazotrophs belong to the proteobacteria and cyanobacterial diazotrophs were absent in all areas except the Arabian Sea. The low cyanobacterial diazotrophy was attributed to the shallow nitracline, which is rarely observed in the Pacific and Atlantic oligotrophic regions. The shallower nitracline favoured enhanced upward NO_3^- flux, but the competitive advantage of cyanobacterial diazotrophs over non-diazotrophic phytoplankton was not as significant as it is in other oligotrophic regions (Shiozaki et al. 2014). N_2 fixation in the Indian Ocean is primarily controlled by Fe fluxes from atmospheric dust. Heterotrophic bacteria likely play a major role in N_2 fixation in this ocean. In contrast, N_2 fixation was not detected in waters of the Bay of Bengal from 10 to 560 m water depth which covers the water column between the deep chlorophyll maximum and the OMZ (Löscher et al. 2020). The absence of N_2 fixation may be a consequence of micronutrient limitation or of an oxygen sensitivity of the OMZ diazotrophs in the bay; modelling suggested that OMZ-based N_2 fixation may be sensitive to small changes in water column stratification. Further south, in the oligotrophic central Indian Ocean, N cycling is rapid, with uptake rates of NO_3^-, NH_4^+, and urea all < 1.5 nmol N l^{-1} h^{-1} south of 15 °S but with increased rates for NO_3^-, NH_4^+, and urea of 3.4, 9.0, and 21 nmol N l^{-1} h^{-1} between 1.5 and 6.5 °N (Baer et al. 2019); over that same area, primary production mirrored the same pattern of N uptake with a local maximum of 7.4–10.8 nmol N l^{-1} h^{-1}.

P cycling in tropical open ocean waters is equally rapid. In the northern Atlantic Ocean subtropical gyre, which receives Fe-rich dust from the Sahara, waters are PO_4^{3}-deleted, although NPP is comparable to NPP in the South Atlantic subtropical gyre which is not PO_4^{3-}-deleted (Mather et al. 2008). However, during boreal spring up to 30% of primary production in the North Atlantic gyre is supported by DOP, with shorter residence times than in the South Atlantic gyre. The asymmetry of DOP cycling in the two gyres may be a consequence of enhanced N_2 fixation in the North Atlantic gyre, which forces the ecosystem towards P limitation. Across the tropical Atlantic Ocean, most of the total P pool is present as DOP. In areas influenced by the Amazon, soluble reactive P and particulate organic P are elevated, but DOP is not. The turnover time of the PO_4^{3-} pool was more rapid in the western (<10-h) than the eastern (>100-h) tropical Atlantic (Sohm and Capone 2010). Fast turnover times may be indicative of P deficiency and there was a trend of an east to west increase in P deficiency. *Trichodesmium* spp. contribute substantially to total alkaline phosphatase activity but are out-completed for PO_4^{3-} by other phytoplankton; colonies can satisfy their P needs by supplementing DIP uptake with P cleaved from DOP via alkaline phosphatase (Sohm and Capone 2006). However, N_2 fixation stimulated by any excess atmospheric Fe supply and phytoplankton utilisation of atmospheric nutrient inputs will therefore tend to drive the ecosystem towards P limitation.

References

Ahmed, A., Naik, H., Pratirupa Bardhan, S.A.S. et al. (2019). Nitrogen fixation and carbon uptake in a tropical estuarine system of Goa, western India. *Journal of Sea Research* 144: 16–21.

Alexandre, A., Georgiou, D., and Santos, R. (2014). Inorganic nitrogen acquisition by the tropical seagrass *Halophila stipulacea*. *Marine Ecology* 35: 387–394.

Almahasheer, H., Duarte, C.M., and Irigoien, X. (2016). Nutrient limitation in central Red Sea mangroves. *Frontiers in Marine Science 3*: 271. https://doi.org/10.3389/fmars.2016.00271.

Alongi, D.M. (1991). The role of intertidal mudbanks in the diagenesis and export of dissolved and particulate materials from the Fly delta, Papua New Guinea. *Journal of Experimental Marine Biology and Ecology* 149: 81–107.

Alongi, D.M. (1996). The dynamics of benthic nutrient pools and fluxes in tropical mangrove forests. *Journal of Marine Research* 54: 123–148.

Alongi, D.M. (1998). *Coastal Ecosystem Processes*. Boca Raton, FL: CRC Press.

Alongi, D.M. (2009). *The Energetics of Mangrove Forests*. Dordrecht, The Netherlands: Springer.

Alongi, D.M. (2010). Dissolved iron supply limits early growth of estuarine mangroves. *Ecology* 91: 3229–3241.

Alongi, D.M. (2017). Micronutrients and mangroves: experimental evidence for copper limitation. *Limnology and Oceanography* 62: 2759–2772.

Alongi, D.M. (2020a). Carbon cycling in the world's mangrove ecosystems revisited: significance of non-steady state diagenesis and subsurface linkages between the forest floor and the coastal ocean. *Forests* 11: 977. https://doi.org/10.3390/f11090977.

Alongi, D.M. (2020b). Nitrogen cycling and mass balance in the world's mangrove forests. *Nitrogen* 1: 167–189.

Alongi, D.M. and McKinnon, A.D. (2005). The cycling and fate of terrestrially derived sediments and nutrients in the coastal zone of the Great Barrier Reef shelf. *Marine Pollution Bulletin* 51: 239–252.

Alongi, D.M., Boto, K.G., and Robertson, A.I. (1992). Nitrogen and phosphorus cycles. In: *Tropical Mangrove Ecosystems* (eds. A.I. Robertson and D.M. Alongi), 251–292. Washington DC: American Geophysical Union.

Alongi, D.M., Tirendi, F., Dixon, P. et al. (1999). Mineralization of organic matter in intertidal sediments of a tropical semi-enclosed delta. *Estuarine, Coastal and Shelf Science* 48: 451–467.

Alongi, D.M., Trott, L.A., Wattayakorn, G. et al. (2002). Below-ground nitrogen cycling in relation to net canopy production in mangrove forests of southern Thailand. *Marine Biology* 140: 855–864.

Alongi, D.M., Sasekumar, A., Chong, V.C. et al. (2004). Sediment accumulation and organic material flux in a managed mangrove ecosystem: estimates of land-ocean-atmosphere exchange in peninsular Malaysia. *Marine Geology* 208: 383–402.

Alongi, D.M., Pfitzner, J., Trott, L.A. et al. (2005). Rapid sediment accumulation and microbial mineralization in forests of the mangrove *Kandelia candel* in the Jiulongjiang estuary, China. *Estuarine, Coastal and Shelf Science* 63: 605–618.

Alongi, D.M., Pfitzner, J., and Trott, L.A. (2006). Deposition and cycling of carbon and nitrogen in carbonate mud of the lagoons of Arlington and Sudbury Reefs, Great Barrier Reef. *Coral Reefs* 25: 123–143.

Alongi, D.M., Trott, L.A., Undu, M.C. et al. (2008a). Benthic microbial metabolism in seagrass meadows along a carbonate gradient in Sulawesi, Indonesia. *Aquatic Microbial Ecology* 51: 141–152.

Alongi, D.M., Trott, L.A., and Pfitzner, J. (2008b). Biogeochemistry of inter-reef sediments on the northern and central Great Barrier Reef. *Coral Reefs* 27: 407–420.

Alongi, D.M., Trott, L.A., and Møhl, M. (2011). Strong tidal currents and labile organic matter stimulate benthic decomposition and carbonate fluxes on the southern Great Barrier Reef. *Continental Shelf Research* 31: 1384–1395.

Alvarez-Cobelas, M., Angeler, D.G., and Sánchez-Carrillo, S. (2008). Export of nitrogen from catchments: a worldwide analysis. *Environmental Pollution* 156: 261–269.

Atkinson, M.J. (1988). Are coral reefs nutrient-limited? *Proceedings of the Sixth International Coral Reef Symposium*, Townsville, Australia (8–12 August 1988), vol. 1, pp. 157–162.

Atkinson, M.J. (1992). Productivity of Enewetak Atoll reef flats predicted from mass transfer relationships. *Continental Shelf Research* 12: 799–810.

Atkinson, M.J. (2011). Biogeochemistry of nutrients. In: *Coral Reefs: An Ecosystem in Transition* (eds. Z. Dubinsky and N. Stambler), 199–206. Dordrecht, The Netherlands: Springer.

Atkinson, M.J. and Bilger, R.W. (1992). Effects of water velocity on phosphate uptake in coral reef-flat communities. *Limnology and Oceanography* 37: 261–271.

Babu, C.P. and Nath, B.N. (2005). Processes controlling forms of phosphorus in surficial sediments from the eastern Arabian Sea impinged by varying bottom water oxygenation conditions. *Deep-Sea Research II* 52: 1965–1980.

Baer, S.E., Rauschenberg, S., Garcia, C.A. et al. (2019). Carbon and nitrogen productivity during spring in the oligotrophic Indian Ocean along the GO-SHIP IO9N transect. *Deep-Sea Research II* 161: 81–91.

Bange, H.W., Wajih, S., Naqvi, A. et al. (2005). The nitrogen cycle in the Arabian Sea. *Progress in Oceanography* 65: 145–158.

Bange, H.W., Sim, C.H., Bastian, D. et al. (2019). Nitrous oxide (N_2O) and methane (CH_4) in rivers and estuaries of NW Borneo. *Biogeosciences* 16: 4321–4335.

Bardhan, P. and Naqvi, S.W.A. (2020). Nitrogen and phosphorus cycling over the western continental shelf of India during seasonal anoxia: a stable isotope approach. *Journal of Marine Systems* 207: 1031144. https://doi.org/10.1016/j.marsys.2019.01.003.

Bashan, Y. and Holguin, G. (2002). Plant growth-promoting bacteria: a potential tool for arid mangrove reforestation. *Trees* 16: 159–166.

Bashan, Y., Puente, M.E., Myrold, D.D. et al. (1998). in vitro transfer of fixed nitrogen from diazotrophic filamentous cyanobacteria to black mangrove seedlings. *FEMS Microbiology Ecology* 26: 165–170.

Benavides, M., Moisander, P.H., Berthelot, H. et al. (2015). Mesopelagic N_2 fixation related to organic matter composition in the Solomon and Bismarck Seas (SW Pacific). *PLoS One* 10: e0143775. https://doi.org/10.1371/journal.pone.0143775.

Benavides, M., Bednarz, V.N., and Ferrier-Pagés, C. (2017). Diazotrophs: overlooked key players within coral symbiosis and tropical reef ecosystems? *Frontiers in Marine Science* 4: 10. https://doi.org/10.3389/fmars.2017.00010.

Bernard, R.J., Mortazavi, B., Wang, L. et al. (2014). Benthic nutrient fluxes and limited denitrification in a sub-tropical groundwater-influenced coastal lagoon. *Marine Ecology Progress Series* 504: 13–26.

Best, M.M., Ku, T.C., Kidwell, S.M. et al. (2007). Carbonate preservation in shallow marine environments: unexpected role of tropical siliciclastics. *The Journal of Geology* 115: 437–456.

Beusen, A.H.W., Bouwman, A.F., Van Beck, L.P.H. et al. (2016). Global riverine N and P transport to ocean increased during the 20th century despite increased retention along the aquatic continuum. *Biogeosciences* 13: 2441–2451.

Bhavya, P.S., Kumar, S., Gupta, G.V.M. et al. (2016). Nitrogen uptake dynamics in a tropical eutrophic estuary (Cochin, India) and adjacent coastal waters. *Estuaries and Coasts* 39: 54–67.

Bilger, R.W. and Atkinson, M.J. (1992). Anomalous mass transfer of phosphate on coral reef flats. *Limnology and Oceanography* 37: 261–270.

Bishop, J.W. and Greenwood, J.G. (1994). Nitrogen excretion by some demersal macrozooplankton in Heron and One Tree Reefs, Great Barrier Reef, Australia. *Marine Biology* 120: 447–457.

Blackburn, T.H., Nedwell, D.B., and Wiebe, W.J. (1994). Active mineral cycling in a Jamaican seagrass sediment. *Marine Ecology Progress Series* 110: 233–239.

Bonnet, S., Biegala, I.C., Dutrieux, P. et al. (2009). Nitrogen fixation in the western equatorial Pacific: rates, diazotrophic cyanobacterial size class distribution, and biogeochemical significance. *Global Biogeochemical Cycles* 23: GB3012. https://doi.org/10.1029/2008GB003439.

Boto, K.G. and Wellington, J.T. (1983). Phosphorus and nitrogen nutritional status of a northern Australian mangrove forest. *Marine Ecology Progress Series* 11: 63–69.

Brocke, H.J., Piltz, B., Herz, N. et al. (2018). Nitrogen fixation and diversity of benthic cyanobacterial mats on coral reefs in Curaçao. *Coral Reefs* 37: 861–874.

Brunskill, G.J. (2010). Tropical margins. In: *Carbon and Nutrient Fluxes in Continental Margins: A Global Synthesis* (eds. K.-K. Liu, L. Atkinson, R. Quinones, et al.), 423–493. Dordrecht, The Netherlands: Springer.

Buchwald, C., Santoro, A.E., Stanley, R.H.R. et al. (2015). Nitrogen cycling in the secondary nitrite maximum of the eastern tropical north Pacific off Costa Rica. *Global Biogeochemical Cycles* 29: 2061–2081.

Burdige, D.J. (2006). *Geochemistry of Marine Sediments*. Princeton, NJ: Princeton University Press.

Burdige, D.J., Hu, X., and Zimmerman, R.C. (2010). The widespread occurrence of coupled carbonate dissolution/reprecipitation in surface sediments on the Bahamas Bank. *American Journal of Science* 310: 492–521.

Burford, M.A., Alongi, D.M., McKinnon, A.D. et al. (2008). Primary production and nutrients in a tropical macrotidal estuary, Darwin Harbour, Australia. *Estuarine, Coastal and Shelf Science* 79: 440–448.

Burford, M.A., Rothlisberg, P.C., and Revill, A.T. (2009). Sources of nutrients driving production in the Gulf of Carpentaria, Australia: a shallow tropical shelf system. *Marine and Freshwater Research* 60: 1044–1053.

Capone, D.G., Dunham, S.G., Horrigan, S.G. et al. (1992). Microbial nitrogen transformations in shallow, unconsolidated carbonate sediments. *Marine Ecology Progress Series* 80: 75–88.

Capone, D.G., Burns, J.A., Montoya, J.P. et al. (2005). Nitrogen fixation by *Trichodesmium* spp.: an important source of new nitrogen to the tropical and subtropical north Atlantic Ocean. *Global Biogeochemical Cycles* 19: GB2024. https://doi.org/10.1029/2004GB002331.

Cardini, U., Bednarz, V.N., van Hoytema, N. et al. (2016). Budget of primary production and dinitrogen fixation in a highly seasonal Red Sea coral reef. *Ecosystems* 19: 771–785.

Carlson-Perret, N.L., Erler, D.V., and Eyre, B.D. (2018). Dinitrogen (N_2) fixation rates in a subtropical seagrass meadow measured with a direct ^{15}N-N_2 tracer method. *Marine Ecology Progress Series* 605: 87–101.

Carlson-Perret, N.L., Erler, D.V., and Eyre, B.D. (2019). Comparison of dinitrogen fixation rates in two subtropical seagrass communities. *Marine Chemistry* 209: 62–69.

Castañeda-Moya, E., Rivera-Monroy, V.H., Chambers, R.M. et al. (2020). Hurricanes fertilize mangrove forests in the Gulf of Mexico (Florida Everglades, USA). *Proceedings of the National Academy of Sciences* 117: 4831–4841.

Cerón, R.M., Cerón, J.G., Muriel, M. et al. (2015). Spatial and temporal distribution of throughfall deposition of nitrogen and sulfur in the mangrove forests associated with Terminos Lagoon. In: *Current Air Quality Issues* (ed. F. Nejadkoorki), 147–164. London: Intech Open Ltd.

Chambers, R.M., Fourqurean, J.W., Macko, S.A. et al. (2001). Biogeochemical effects of iron availability on primary producers in a shallow marine carbonate environment. *Limnology and Oceanography* 46: 1278–1286.

Charpy, L. and Charpy-Roubaud, C. (1988). Phosphorus budget in an atoll lagoon. *Proceedings of the Sixth International Coral Reef Symposium*, Townsville, Australia (8–12 August 1988), vol. 2, pp. 547–551.

Charpy, L., Palinska, K.A., Abed, R.M.M. et al. (2012). Factors influencing microbial mat composition, distribution and dinitrogen fixation in three western Indian Ocean coral reefs. *European Journal of Phycology* 47: 51–66.

Charpy-Roubaud, C. and Larkum, A.W.D. (2005). Dinitrogen fixation by exposed communities on the rim of Tikehau atoll (Tuomotu Archipelago, French Polynesia). *Coral Reefs* 24: 622–628.

Charpy-Roubaud, C., Charpy, L., and Larkum, A.W.D. (2001). Atmospheric dinitrogen fixation by benthic communities of Tikehau Lagoon (Tuamotu Archipelago, French Polynesia) and its contribution to benthic primary production. *Marine Biology* 139: 991–997.

Clark, D.R., Widdicombe, C.E., Rees, A.P. et al. (2016). The significance of nitrogen regeneration for new production within a filament of the Mauritanian upwelling system. *Biogeosciences* 13: 2873–2888.

Cockcroft, A.C. (1990). Nitrogen excretion by the surf zone bivalves *Donax serra* and *D. sordidus*. *Marine Ecology Progress Series* 60: 57–65.

Cockcroft, A.C. and McLachlan, A. (1993). Nitrogen budget for a high-energy ecosystem. *Marine Ecology Progress Series* 100: 287–299.

Cockcroft, A.C., Webb, P., and Wooldridge, T. (1988). Nitrogen regeneration by two surf-zone mycids, *Mesopodopsis slabberi* and *Gastrosaccus psammodytes*. *Marine Biology* 99: 75–82.

Corredor, J.E. and Capone, D.G. (1985). Studies on nitrogen diagenesis in coral reef sands. *Proceedings Fifth Coral Reef Symposium*, Tahiti, French Polynesia (27 May–1 June 1985), vol. 3, pp. 395–399.

Corredor, J.E. and Morell, J.M. (1985). Inorganic nitrogen in coral reef sediments. *Marine Chemistry* 16: 379–384.

Cuet, P., Atkinson, M.J., Blanchot, J. et al. (2011). CNP budgets of a coral-dominated fringing reef at La Réunion, France: coupling of oceanic phosphate and groundwater nitrate. *Coral Reefs* 30: 45–54.

Dai, M., Wang, L., Guo, X. et al. (2008). Nitrification and inorganic nitrogen distribution in a large perturbed river/estuarine system, the Pearl River estuary, China. *Biogeosciences Discussions* 5: 1545–1585.

Dale, A.W., Graco, M., and Wallmann, K. (2017). Strong and dynamic benthic-pelagic coupling and feedbacks in a coastal upwelling system (Peruvian shelf). *Frontiers in Marine Science* 4: 29. https://doi.org/10.3389/fmars.2017.00029.

Dale, A.W., Bourbonnais, A., Altabet, M. et al. (2019). Isotopic fingerprints of benthic nitrogen cycling in the Peruvian oxygen minimum zone. *Geochimica et Cosmochimica Acta* 245: 406–425.

Dalsgaard, T., Thamdrup, B., Farias, L. et al. (2012). Anammox and denitrification in the oxygen minimum zone of the eastern south Pacific. *Limnology and Oceanography* 57: 1331–1346.

Davey, M., Tarran, G.A., Mills, M.M. et al. (2008). Nutrient limitation of picophytoplankton photosynthesis and growth in the tropical north Atlantic. *Limnology and Oceanography* 53: 1722–1733.

Dekaezemacker, J., Bonnet, S., Grosso, O. et al. (2013). Evidence of active dinitrogen fixation in surface waters of the eastern tropical south Pacific during El Niño and La Niña events and evaluation of its potential nutrient controls. *Global Biogeochemical Cycles* 27: 768–779.

DeMaster, D.J. and Pope, R.H. (1996). Nutrient dynamics in Amazon shelf waters: results from AMASSEDS. *Continental Shelf Research* 16: 263–277.

Dong, L.F., Naqasima Sobey, M., Smith, C.J. et al. (2011). Dissimilatory reduction of nitrate to ammonium, not denitrification or anammox, dominates benthic nitrate reduction in tropical estuaries. *Limnology and Oceanography* 56: 279–291.

Dunn, J.G., Sammarco, P.W., and LaFleur, G. Jr. (2012). Effects of phosphate on growth and skeletal density in the hermatypic coral *Acropora muricata:* a controlled experimental approach. *Journal of Experimental Marine Biology and Ecology* 411: 34–44.

Dunne, J.P., Murray, J.W., Aufdenkampe, A.K. et al. (1999). Silicon-nitrogen coupling in the equatorial Pacific upwelling zone. *Global Biogeochemical Cycles* 13: 715–726.

Enrich-Prast, A., Figueiredo, V., de Assis Esteves, F. et al. (2016a). Controls of sediment nitrogen dynamics in tropical coastal lagoons. *PLos One* 11: e0155586. https://doi.org/10.1371/journal.pone.0155586.

Enrich-Prast, A., Santoro, A.L., Coutinho, R.S. et al. (2016b). Sediment denitrification in two contrasting tropical shallow lagoons. *Estuaries and Coasts* 39: 657–663.

Entsch, B., Sim, R.G., and Hatcher, B.G. (1983). Indications from photosynthetic components that iron is a limiting nutrient in primary producers on coral reefs. *Marine Biology* 73: 17–30.

Erler, D.V., Trott, L.A., Alongi, D.M. et al. (2013). Denitrification, anammox and nitrate reduction in sediments of the southern Great Barrer Reef lagoon. *Marine Ecology Progress Series* 478: 57–70.

Estrada-Allis, S.N., Pardo, J.S., de Souza, J.M.A.C. et al. (2020). Dissolved inorganic nitrogen and particulate organic nitrogen budget in the Yucatán shelf: driving mechanisms through a physical-biogeochemical coupled model. *Biogeosciences* 17: 1087–1111.

Eyre, B.D. (1995). A first-order nutrient budget for the tropical Moresby estuary and catchment, north Queensland, Australia. *Journal of Coastal Research* 11: 717–732.

Eyre, B.D. and Balls, P. (1999). A comparative study of nutrient behaviour along the salinity gradient of tropical and temperate estuaries. *Estuaries* 22: 313–326.

Eyre, B.D. and Ferguson, A.J.P. (2005). Benthic metabolism and nitrogen cycling in a subtropical east Australian estuary (Brunswick): temporal variability and controlling factors. *Limnology and Oceanography* 50: 81–96.

Eyre, B.D. and McKee, L.J. (2002). Carbon, nitrogen and phosphorus budgets for a shallow subtropical coastal embayment (Moreton Bay, Australia). *Limnology and Oceanography* 47: 1043–1055.

Eyre, B.D., Glud, R.N., and Patten, N. (2008). Mass coral spawning: a natural large-scale nutrient addition experiment. *Limnology and Oceanography* 53: 997–1013.

Eyre, B.D., Santos, I.R., and Maher, D.T. (2013). Seasonal, daily and diel N_2 effluxes in permeable carbonate sediments. *Biogeosciences* 10: 1–15. https://doi.org/10.5194/bg-10-1-2013.

Ezzat, L., Maguer, J.-F., Grover, R. et al. (2016). Limited phosphorus availability is the Achilles heel of tropical reef corals in a warming ocean. *Scientific Reports* 6: 31768. https://doi.org/10.1038/srep31768.

Feller, I.C. (1995). Effects of nutrient enrichment on growth and herbivory of dwarf red mangrove (*Rhizophora mangle* L). *Ecological Monographs* 65: 477–505.

Feller, I.C., Whigham, D.F., McKee, K.L. et al. (2003). Nitrogen limitation of growth and nutrient dynamics in a mangrove forest, Indian River Lagoon, Florida. *Oecologia* 134: 405–414.

Feller, I.C., Lovelock, C.E., and McKee, K.L. (2007). Nutrient addition differentially affects ecological processes of *Avicennia germinans* in nitrogen versus phosphorus limited mangrove ecosystems. *Ecosystems* 10: 347–359.

Ferdie, M. and Fourqurean, J.W. (2004). Responses of seagrass communities to fertilization along a gradient of relative availability of nitrogen and phosphorus in a carbonate environment. *Limnology and Oceanography* 49: 2082–2094.

Ferguson, A., Eyre, B., and Gay, J. (2004). Nutrient cycling in the sub-tropical Brunswick estuary, Australia. *Estuaries* 27: 1–17.
Fernandes, L. (2011). Origin and biochemical cycling of particulate nitrogen in the Mandovi estuary. *Estuarine, Coastal and Shelf Science* 94: 291–298.
Fernandes, S.O., Gonsalves, M.J., Michotey, V.D. et al. (2013). Denitrification activity is closely linked to the total ambient Fe concentration in mangrove sediments of Goa, India. *Estuarine, Coastal and Shelf Science* 131: 64–74.
Ferrier-Pagés, C., Schoelzke, V., Jaubert, J. et al. (2001). Response of a hermatypic coral, *Stylophora pistillata*, to iron and nitrate enrichment. *Journal of Experimental Marine Biology and Ecology* 259: 249–261.
Fisher, T.R., Morrissey, K.M., Smith, L.K. et al. (1990). Final science report of NURP Hydrolab mission 85-1. Washington, DC: NOAA.
Flohr, A., van der Plas, A.K., Emeis, K.-C. et al. (2014). Spatiotemporal patterns of C:N:P ratios in the northern Benguela upwelling system. *Biogeosciences* 11: 885–897.
Fowler, D., Coyle, M., Skiba, U. et al. (2013). The global nitrogen cycle in the twenty-first century. *Philosophical Transactions of the Royal Society B* 368: 20130164. https://doi.org/10.1098/rstb.2013.0164.
Furnas, M.J. and Mitchell, A.W. (1999). Wintertime carbon and nitrogen fluxes on Australia's NW shelf. *Estuarine, Coastal and Shelf Science* 49: 165–175.
Furnas, M.J., Alongi, D.M., McKinnon, A.D. et al. (2011). Regional-scale nitrogen and phosphorus budgets for the northern (14°S) and central (17°S) Great Barrier Reef shelf ecosystem. *Continental Shelf Research* 31: 1967–1990.
Garcias-Bonet, N., Fusi, M., Ali, M. et al. (2018). High denitrification and anaerobic ammonium oxidation contribute to net nitrogen loss in a seagrass ecosystem in the central Red Sea. *Biogeosciences* 15: 7333–7346.
Gier, J., Sommer, S., Löscher, C.R. et al. (2016). Nitrogen fixation in sediments along a depth transect through the Peruvian oxygen minimum zone. *Biogeosciences* 13: 4065–4080.
Gier, J., Löscher, C.R., Dale, A.W. et al. (2017). Benthic dinitrogen fixation traversing the oxygen minimum zone off Mauritania (NW Africa). *Frontiers in Marine Science* 4: 390. https://doi.org/10.3389/fmars.2017.00390.
Gleason, S.M., Ewel, K.C., and Hue, N. (2003). Soil redox conditions and plant-soil relationships in a Micronesian mangrove forest. *Estuarine, Coastal and Shelf Science* 56: 1065–1074.
Godinot, C., Ferrier-Pagés, C., and Grover, R. (2009). Control of phosphate uptake by zooxanthellae and host cells in the hermatypic coral *Stylophora pistillata*. *Limnology and Oceanography* 54: 1627–1633.
Godinot, C., Grover, R., Allemand, D. et al. (2011). High phosphate uptake requirements of the hermatypic coral *Stylophora pistillata*. *Journal of Experimental Biology* 214: 2749–2754.
Gontharet, S., Mathieu, O., Lévêque, J. et al. (2014). Distribution and sources of bulk organic matter (OM) on a tropical intertidal mud bank in French Guiana from elemental and isotopic proxies. *Chemical Geology* 376: 1–10.
Gras, A.F., Koch, M.S., and Madden, C.J. (2003). Phosphorus uptake kinetics of a dominant tropical seagrass *Thalassia testudinum*. *Aquatic Botany* 76: 299–315.
Grenz, C., Denis, L., Pringault, O. et al. (2010). Spatial and seasonal variability of sediment oxygen consumption and nutrient fluxes at the sediment water interface in a sub-tropical lagoon (New Caledonia). *Marine Pollution Bulletin* 61: 399–412.
Grover, R., Maguer, J.F., Reynaud-Vaganay, C. et al. (2003). Uptake of ammonium by the hermatypic coral *Stylophora pistillata:* effects of feeding, light, and ammonium concentration. *Limnology and Oceanography* 47: 782–790.
Grover, R., Maguer, J.F., Allemand, D. et al. (2006). Urea uptake by the hermatypic coral *Stylophora pistillata*. *Journal of Experimental Marine Biology and Ecology* 332: 216–225.
Gruber, R.K., Lowe, R.J., and Falter, J.L. (2019). Tidal and seasonal forcing of dissolved nutrient fluxes in reef communities. *Biogeosciences* 16: 1921–1935.
Guieu, C., Al Azhar, M., Aumont, O. et al. (2019). Major impact of dust deposition on the productivity of the Arabian Sea. *Geophysical Research Letters* 46: 6736–6744.

Gupta, G.V.M., Natesan, U., Ramana Murthy, M.V. et al. (2006). Nutrient budgets for Muthupet lagoon, southeastern India. *Current Science* 90: 967–972.

Hamilton, S.E. and Casey, D. (2016). Creation of a high spatio-temporal resolution global database of continuous mangrove forest cover for the 21st century (CGMFC-21). *Global Ecology and Biogeography* 25: 729–738.

Hamisi, M.I., Lyimo, T.J., Muruke, M.H.S. et al. (2009). Nitrogen fixation by epiphytic and epibenthic diazotrophs associated with seagrass meadows along the Tanzanian coast, western Indian Ocean. *Aquatic Microbial Ecology* 57: 33–42.

Herbert, D.A. and Fourqurean, J.W. (2009). Phosphorus availability and salinity control productivity and demography of the seagrass *Thalassia testudinum* in Florida Bay. *Estuaries and Coasts* 32: 188–201.

Hoffman, D.K., McCarthy, M.J., Newell, S.E. et al. (2019). Relative contributions of DNRA and denitrification to nitrate reduction in *Thalassia testudinum* seagrass beds in coastal Florida (USA). *Estuaries and Coasts* 42: 1001–1014.

Holguin, G. and Bashan, Y. (1996). Nitrogen-fixation by *Azospirillum brasilense* is promoted when co-cultured with a mangrove rhizosphere bacterium (*Staphylococcus* sp.). *Soil Biology and Biochemistry* 28: 1651–1660.

Holmboe, N., Kristensen, E., and Andersen, F.Ø. (2001). Anoxic decomposition in sediments from a tropical mangrove forest and the temperate Wadden Sea: implications of N and P addition experiments. *Estuarine, Coastal and Shelf Science* 53: 125–140.

Hopkinson, C.S. Jr., Sherr, B.F., and Ducklow, H.W. (1987). Microbial regeneration of ammonium in the water-column of Davies Reef, Australia. *Marine Ecology Progress Series* 41: 147–155.

Hosegood, P.J., Nightingale, P.D., Rees, A.P. et al. (2017). Nutrient pumping by submesoscale circulations in the Mauritanian upwelling system. *Progress in Oceanography* 159: 223–236.

Hung, J.-J. and Kuo, F. (2002). Temporal variability of carbon and nutrient budgets from a tropical lagoon in Chiku, SW Taiwan. *Estuarine, Coastal and Shelf Science* 54: 887–900.

Iizumi, H. and Yamamuro, M. (2000). Nitrogen fixation activity by periphytic blue-green algae in a seagrass bed on the Great Barrier Reef. *Japan Agricultural Research Quarterly* 24: 69–73.

Jabir, T., Veettil Vipindas, P., Jesmi, Y. et al. (2020). Nutrient stoichiometry (N:P) controls nitrogen fixation and distribution of diazotrophs in a tropical eutrophic estuary. *Marine Pollution Bulletin* 151: 110799. https://doi.org/10.1016/j.marpolbul.2019.110799.

Jantzen, C., Wild, C., Rasheed, M. et al. (2010). Enhanced pore-water nutrient fluxes by the upside-down jellyfish *Cassiopea* sp. in a Red Sea coral reef. *Marine Ecology Progress Series* 411: 117–125.

Jennerjahn, T.C., Ittekkot, V., Klopper, S. et al. (2004). Biogeochemistry of a tropical river affected by human activities in its catchment, Brantas River estuary and coastal waters of Madura Strait, Java, Indonesia. *Estuarine, Coastal and Shelf Science* 60: 503–514.

Jensen, H.S., McGlathery, K.J., Marino, R. et al. (1998). Forms and availability of sediment phosphorus in carbonate sand of Bermuda seagrass beds. *Limnology and Oceanography* 43: 799–810.

Jensen, H.S., Nielsen, O.I., Koch, M.S. et al. (2009). Phosphorus release with carbonate dissolution coupled to sulphide oxidation in Florida Bay seagrass sediments. *Limnology and Oceanography* 54: 1753–1764.

Johnson, M.W., Heck, K.L. Jr., and Fourqurean, J.W. (2006). Nutrient content of seagrasses and epiphytes in the northern Gulf of Mexico: evidence of phosphorus and nitrogen limitation. *Aquatic Botany* 85: 103–111.

Joye, S.B. and Lee, R.Y. (2004). Benthic microbial mats: important sources of fixed nitrogen and carbon to the Twin Cays, Belize ecosystem. *Atoll Research Bulletin* 53: 1–24.

Karl, D., Michaels, A., Bergman, B. et al. (2002). Dinitrogen fixation in the world's oceans. *Biogeochemistry* 57 (58): 47–98.

Kemp, W.M. and Cornwell, J.C. (2001). Role of benthic communities in the cycling and balance of nitrogen in Florida Bay. Final Report to the U.S. Environmental Protection Agency, Region 4, Atlanta, GA.

King, G.M., Carlton, R.G., and Sawyer, T.E. (1990). Anaerobic metabolism and oxygen distribution in the carbonate sediments of a submarine canyon. *Marine Ecology Progress Series* 58: 275–285.

Knapp, A.N., Casciotti, K.L., Berelson, W.M. et al. (2016). Low rates of nitrogen fixation in eastern tropical south Pacific surface waters. *Proceedings of the National Academy of Sciences* 113: 4398–4403.

Knapp, A.N., McCabe, K.M., Grosso, O. et al. (2018). Distribution and rates of nitrogen fixation in the western tropical South Pacific Ocean constrained by nitrogen isotope budgets. *Biogeosciences* 15: 2619–2628.

Knowlton, N. and Rohwer, F. (2003). Multispecies microbial mutualisms on coral reefs: the host as a habitat. *The American Naturalist* 162: S51–S62.

Kothamasi, D., Kothamasi, S., Bhattacharayya, A. et al. (2006). Arbuscular mycorrhizae and phosphate-solubilizing bacteria of the rhizosphere of the mangrove ecosystem of Great Nicobar Island, India. *Biology and Fertility of Soils* 42: 358–361.

Krauss, K.W., Doyle, T.W., Twilley, R.R. et al. (2006). Evaluating the relative hydroperiod and soil fertility on growth of south Florida mangroves. *Hydrobiologia* 569: 311–324.

Krauss, K.W., Keeland, B., Allen, J. et al. (2007a). Effects of season, rainfall, and hydrogeomorphic setting on mangrove tree growth in Micronesia. *Biotropica* 39: 161–170.

Krauss, K.W., Young, P.J., Chambers, J.L. et al. (2007b). Sap flow characteristics of neotropical mangroves in flooded and drained soils. *Tree Physiology* 27: 775–783.

Kristensen, E., Andersen, F.Ø., Holmboe, N. et al. (2000). Carbon and nitrogen mineralization in sediments of the Bangrong mangrove area, Phuket, Thailand. *Aquatic Microbial Ecology* 22: 199–213.

Kumar, P.K., Singh, A., Ramesh, R. et al. (2017). N_2 fixation in the Eastern Arabian Sea: probable role of heterotrophic diazotrophs. *Frontiers in Marine Science* 4: 80. https://doi.org/10.3389/fmars.2017.00080.

L'Helguen, S., Chauvaud, L., Cuet, P. et al. (2014). A novel approach using the ^{15}N tracer technique and benthic chambers to determine ammonium fluxes at the sediment-water interface and its application in a back-reef zone on Reunion Island (Indian Ocean). *Journal of Experimental Marine Biology and Ecology* 452: 143–151.

Lacoste, E. and Gaertner-Mazouni, N. (2016). Nutrient regeneration in the water column and at the sediment-water interface in pearl oyster culture (*Pinctada margaritifera*) in a deep atoll lagoon (Ahe, French Polynesia). *Estuarine, Coastal and Shelf Science* 182: 304–309.

Lam, P., Lavik, G., Jensen, M.M. et al. (2009). Revising the nitrogen cycle in the Peruvian oxygen minimum zone. *Proceedings of the National Academy of Sciences* 106: 4752–4757.

Lee, K.-S. and Dunton, K.H. (2000). Effects of nitrogen enrichment on biomass allocation, growth and leaf morphology of the seagrass *Thalassia testudinum*. *Marine Ecology Progress Series* 196: 39–48.

Lee, R.Y. and Joye, S.B. (2006). Seasonal patterns of nitrogen fixation and denitrification in oceanic mangrove habitats. *Marine Ecology Progress Series* 307: 127–141.

Lesser, M.P., Falcón, L.I., Rodríguez-Román, A. et al. (2007). Nitrogen fixation by symbiotic cyanobacteria provides a source of nitrogen for the scleractinian coral *Montastraea cavernosa*. *Marine Ecology Progress Series* 346: 143–152.

Leynaert, A., Tréguer, P., Lancelot, C. et al. (2001). Silicon limitation of biogenic silica production in the equatorial Pacific. *Deep-Sea Research I* 48: 639–660.

Lisle, J., Reich, C., and Halley, R. (2014). Aragonite saturation states and nutrient fluxes in coral reef sediments in Biscayne National Park, FL, USA. *Marine Ecology Progress Series* 509: 71–85.

Loginova, A.N., Dale, A.W., Le Moigne, F.A.C. et al. (2020). Sediment release of dissolved organic matter to the oxygen minimum zone off Peru. *Biogeosciences* 17: 4663–4679.

Lomnitz, U., Sommer, S., Dale, A.W. et al. (2016). Benthic phosphorus cycling in the Peruvian oxygen minimum zone. *Biogeosciences* 13: 1367–1386.

Löscher, C.R., Mohr, W., Bange, H.W. et al. (2020). No nitrogen fixation in the Bay of Bengal? *Biogeosciences* 17: 851–864.

Lovelock, C.E., Feller, I.C., Ball, M.C. et al. (2006). Differences in plant function in phosphorus and nitrogen-limited mangrove ecosystems. *New Phytologist* 172: 514–522.

Lyimo, T.J. and Hamisi, M.I. (2008). Cyanobacteria occurrence and nitrogen fixation rates in the seagrass meadows of the east coast of Zanzibar: comparisons of sites with and without seaweed farms. *Western Indian Ocean Journal of Marine Science* 7: 45–55.

Lyimo, T.J. and Lugomela, C. (2006). Nitrogenase activity in intertidal sediment along the Tanzanian coast, western Indian Ocean. *Western Indian Ocean Journal of Marine Science* 5: 133–140.

Madhu, N.V., Martin, G.D., Haridevi, C.K. et al. (2017). Differential environmental responses of tropical phytoplankton community in the SW coast of India. *Regional Studies in Marine Science* 16: 21–35.

Mangalaa, K.R., Cardinal, D., Brajard, J. et al. (2017). Silicon cycle in Indian estuaries and its control by biogeochemical and anthropogenic processes. *Continental Shelf Research* 148: 64–88.

Martin, G.D., Muraleedharan, K.R., Vijay, J.G. et al. (2010). Formation of anoxia and denitrification in the bottom waters of a tropical estuary, SW coast of India. *Biogeosciences Discussions* 7: 1751–1782.

Mather, R.L., Reynolds, S.E., Wolff, G.A. et al. (2008). Phosphorus cycling in the north and south Atlantic Ocean subtropical gyre. *Nature Geoscience* 1: 439–443.

McGlathery, K.J., Berg, P., and Marino, R. (2001). Using porewater profiles to assess nutrient availability in seagrass-vegetated carbonate sediments. *Biogeochemistry* 56: 239–263.

McGlathery, K.J., Sundbäck, K., and Anderson, I.C. (2004). The importance of primary producers for benthic nitrogen and phosphorus cycling. In: *Estuarine Nutrient Cycling: The Influence of Primary Producers* (eds. S. Nielsen, G. Banta and M. Pedersen), 231–261. Dordrecht, The Netherlands: Kluwer.

Mills, M.M., Lipschultz, F., and Sebens, K.P. (2004). Particulate matter ingestion and associated nitrogen uptake by four species of scleractinian corals. *Coral Reefs* 23: 311–323.

Miyajima, T., Suzumura, M., Umezama, Y. et al. (2001). Microbiological nitrogen transformation in carbonate sediments of a coral-reef lagoon and associated seagrass beds. *Marine Ecology Progress Series* 217: 273–286.

Moore, C.M., Mills, M.M., Arrigo, K.R. et al. (2013). Processes and patterns of nutrient limitation. *Nature Geoscience* 6: 701–710.

Moreno, A.R. and Martiny, A.C. (2018). Ecological stoichiometry of ocean plankton. *Annual Reviews in Marine Science* 10: 43–69.

Moriarty, D.W. and O'Donohue, M.J. (1993). Nitrogen fixation in seagrass communities during summer in the Gulf of Carpentaria, Australia. *Australian Journal of Marine and Freshwater Research* 44: 117–125.

Moriarty, D.W., Iverson, J.R., and Pollard, P.C. (1986). Exudation of organic carbon by the seagrass *Halodule wrightii* and its effect on bacterial growth in the sediment. *Journal of Experimental Marine Biology and Ecology* 96: 115–126.

Mukhopadhyay, S.K., Biswas, H., De, T.K. et al. (2006). Fluxes of nutrients from the tropical River Hooghly at the land-ocean boundary of Sundarbans, NE coast of Bay of Bengal, India. *Journal of Marine Systems* 62: 9–21.

Müller, D., Bange, H.W., Warneke, T. et al. (2016). Nitrous oxide and methane in two tropical estuaries in a peat-dominated region of NW Borneo. *Biogeosciences* 13: 2415–2428.

Nagel, B., Emeis, K.-C., Flohr, A. et al. (2013). N-cycling and balancing of the N-deficit generated in the oxygen minimum zone over the Namibian shelf – an isotope-based approach. *Biogeosciences* 118: 361–371.

Naidoo, Y., Steinke, T.D., Mann, F.D. et al. (2008). Epiphytic organisms on the pneumatophores of the mangrove *Avicennia marina*: occurrence and possible function. *African Journal of Plant Science* 1: 12–15.

Naqvi, S.W.A., Jayakumar, D.A., Narvekar, P.V. et al. (2000). Increased marine production of N_2O due to intensifying anoxia on the Indian continental shelf. *Nature* 408: 346–349.

Nedwell, D.B., Blackburn, T.H., and Wiebe, W.J. (1994). Dynamic nature of the turnover of organic carbon, nitrogen and sulphur in the sediments of a Jamaican mangrove forest. *Marine Ecology Progress Series* 110: 223–231.

Newman, M., Alexander, M.A., Ault, T.R. et al. (2016). The Pacific Decadal Oscillation, revisited. *Journal of Climate* 29: 4399–4427.

O'Donohue, M.J., Moriarty, D.W., and MacRae, I.C. (1991). Nitrogen fixation in sediments and the rhizosphere of the seagrass *Zostera capricorni*. *Microbial Ecology* 22: 53–64.

O'Neil, J.M. and Capone, D.G. (1989). Nitrogenase activity in tropical carbonate marine sediments. *Marine Ecology Progress Series* 56: 145–156.

O'Neil, J.M. and Capone, D.G. (2008). Nitrogen cycling in coral reef environments. In: *Nitrogen in the Marine Environment* (ed. D.G. Capone), 937–977. New York: Elsevier.

Odebrecht, C., Villac, M.C., Abreu, P.C. et al. (2018). Flagellates versus diatoms: phytoplankton trends in tropical and subtropical estuarine-coastal ecosystems. In: *Plankton Ecology of the SW Atlantic* (eds. M.S. Hoffmeyer, M.E. Sabatini, F.P. Brandini, et al.), 249–269. Cham, Switzerland: Springer.

Pan, C.-W., Chuang, Y.-L., Chou, L.-S. et al. (2016). Factors governing phytoplankton biomass and production in tropical estuaries of western Taiwan. *Continental Shelf Research* 118: 88–99.

Pawlik, J.R. and McMurray, S.E. (2020). The emerging ecological and biogeochemical importance of sponges on coral reefs. *Annual Review of Marine Science* 12: 315–337.

Pennington, J.T., Mahoney, K.L., Kuwahara, V.S. et al. (2006). Primary production in the eastern tropical Pacific: a review. *Progress in Oceanography* 69: 285–317.

Pérez-Villalona, H., Cornwell, J.C., Ortiz-Zayas, J.R. et al. (2015). Sediment denitrification and nutrient fluxes in the San José Lagoon, a tropical lagoon in the highly urbanized San Juan Bay estuary, Puerto Rico. *Estuaries and Coasts* 38: 2259–2278.

Piniak, G.A. and Lipschultz, F. (2004). Effects of nutritional history on nitrogen assimilation in congeneric temperate and tropical hermatypic corals. *Marine Biology* 145: 1085–1096.

Pratihary, A.K., Naqvi, S.W.A., Naik, H. et al. (2009). Benthic flux in a tropical estuary and its role in the ecosystem. *Estuarine, Coastal and Shelf Science* 85: 387–398.

Pratihary, A.K., Naqvi, S.W.A., Narvenkar, G. et al. (2014). Benthic mineralization and nutrient exchange over the inner continental shelf of western India. *Biogeosciences* 11: 2771–2791.

Qashqari, M.S., Garcias-Bonet, N., Fusi, M. et al. (2020). High temperature and crab density reduce atmospheric nitrogen fixation in Red Sea mangrove sediments. *Estuarine, coastal and Shelf Science* 232: 106487. https://doi.org/10.1016/j.ecss.2019.106487.

Quigg, A., Finkel, Z.V., Iwin, A.J. et al. (2003). The evolutionary inheritance of elemental stoichiometry in marine phytoplankton. *Nature* 425: 291–294.

Rädecker, N., Meyer, F.W., Bednarz, V.N. et al. (2014). Ocean acidification rapidly reduces dinitrogen fixation associated with the hermatypic coral *Seriatopora hystrix*. *Marine Ecology Progress Series* 511: 297–302.

Rädecker, N., Pogoreutz, C., Voolstra, C.R. et al. (2015). Nitrogen cycling in corals: the key to understanding holobiont functioning? *Trends in Microbiology* 23: 490–497.

Rafter, P.A., Sigman, D.M., Charles, C.D. et al. (2012). Subsurface tropical Pacific nitrogen isotopic composition of nitrate: biogeochemical signals and their transport. *Global Biogeochemical Cycles* 26: GB1003. https://doi.org/10.1029/2010GB003979.

Rahman, S., Aller, R.C., and Cochran, J.K. (2017). The missing silica sink: revisiting the marine sedimentary Si cycle using cosmogenic ^{32}Si. *Global Biogeochemical Cycles* 31: 1559–1578.

Raimbault, P., Slawyk, G., Boudjellal, B. et al. (1999). Carbon and nitrogen uptake and export in the equatorial Pacific at 150 °W: evidence of an efficient regenerated production cycle. *Journal of Geophysical Research* 104: 3341–3356.

Raja, S., Thangaradjou, T., Sivakumar, K. et al. (2012). Rhizobacterial population density and nitrogen fixation in seagrass community of Gulf of Mannar, India. *Journal of Environmental Biology* 33: 1033–1037.

Rasheed, M., Badran, M.I., Richter, C. et al. (2002). Effect of reef framework and bottom sediment on nutrient enrichment in a coral reef of the Gulf of Aqaba, Red Sea. *Marine Ecology Progress Series* 239: 277–285.

Ray, R., Majumder, N., Chowdhury, C. et al. (2018). Phosphorus budget of the Sundarbans mangrove ecosystem: box model approach. *Estuaries and Coasts* 41: 1036–1049.

Redfield, A.C., Ketchum, B.H., and Richards, F.A. (1963). The influence of organisms on the composition of seawater. In: *The Sea, Volume 2. The Composition of Seawater Comparative and Descriptive Oceanography* (ed. M.N. Hill), 26–77. Cambridge, USA: Harvard University Press.

Reef, R., Feller, I.C., and Lovelock, C.E. (2010). Nutrition of mangroves. *Tree Physiology* 30: 1148–1160.

Reis, C.R.G., Nardoto, G.B., and Oliveira, R.S. (2017). Global overview on nitrogen dynamics in mangroves and consequences of increasing nitrogen availability for these systems. *Plant and Soil* 410: 1–19.

Rich, J.J., Arevalo, P., Chang, B.X. et al. (2020). Anaerobic ammonium oxidation (anammox) and denitrification in Peru margin sediments. *Journal of Marine Systems* 207: 103122. https://doi.org/10.1016/j.marsys.2018.09.007.

Ridley, C.P., Faulkner, D.J., and Haygood, M.G. (2005). Investigation of *Oscillatoria spongeliae*-dominated bacterial communities in four dictyoceratid sponges. *Applied and Environmental Microbiology* 71: 7366–7375.

Rivera-Monroy, V.H. and Twilley, R.R. (1996). The relative role of denitrification and immobilization in the fate of inorganic nitrogen in mangrove sediments (Terminos Lagoon, Mexico). *Limnology and Oceanography* 41: 284–296.

Rix, L., Bednarz, V.N., Cardini, U. et al. (2015). Seasonality in dinitrogen fixation and primary productivity by coral reef framework substrates from the northern Red Sea. *Marine Ecology Progress Series* 533: 79–92.

Rojas, A., Holguin, G., Glick, B.R. et al. (2001). Synergism between *Phyllobacterium* sp. (N_2-fixer) and *Bacillus licheniformis* (P-solubilizer), both from a semi-arid mangrove rhizosphere. *FEMS Microbiology and Ecology* 35: 181–191.

Rougerie, F. and Wauthy, B. (1988). The endo-upwelling concept: a new paradigm for solving an old paradox. *Proceedings of the Sixth International Coral Reef Symposium*, Townsville, Australia (8–12 August 1988), vol. 3, pp. 21–25.

Ruiz-Fernández, A.C., Frignani, M., Tesi, T. et al. (2007). Recent sedimentary history of organic matter and nutrient accumulation in the Ohuira Lagoon, NW Mexico. *Archives of Environmental Contamination and Toxicology* 53: 159–167.

Sandel, V., Kiko, R., Brandt, P. et al. (2015). Nitrogen fuelling of the pelagic food web of the tropical Atlantic. *PLoS One* 10: e1031258. https://doi.org/10.1371/journal.pone.0131258.

Santos, I.R., Eyre, B.D., and Glud, R.N. (2012). Influence of porewater advection on denitrification in carbonate sands: evidence from repacked sediment column experiments. *Geochimica et Cosmochimica Acta* 96: 247–258.

Scheffers, S.R., Bak, R.P.M., and van Duyl, F.C. (2005). Why is bacterioplankton growth in coral reef framework cavities enhanced? *Marine Ecology Progress Series* 299: 89–99.

Scheren, P.A.G.M., Kroeze, C., Janssen, F.J.J.G. et al. (2004). Integrated water pollution assessment of the Ebrié Lagoon, Ivory Coast, west Africa. *Journal of Marine Systems* 44: 1–17.

Schroller-Lomnitz, U., Hensen, C., Dale, A.W. et al. (2019). Dissolved benthic phosphate, iron and carbon fluxes in the Mauritanian upwelling system and implications for ongoing deoxygenation. *Deep-Sea Research I* 143: 70–84.

Seitzinger, S.P. and D'Elia, C. (1985). Preliminary studies of denitrification on a coral reef. In: *The Ecology of Coral Reefs* (ed. M.L. Reaka), 199–208. Washington, DC: NOAA Symposium Series Undersea Research.

Seitzinger, S., Harrison, J.A., Böhlke, J.K. et al. (2006). Denitrification across landscapes and waterscapes: a synthesis. *Ecological Applications* 16: 2064–2090.

Selden, C.R., Mulholland, M.R., Bernhardt, P.W. et al. (2019). Dinitrogen fixation across physico-chemical gradients of the eastern tropical North Pacific oxygen deficient zone. *Global Biogeochemical Cycles* 33: 1187–1202.

Shantz, A.A. and Burkepile, D.E. (2014). Context-dependent effects of nutrient loading on the coral-algal mutualism. *Ecology* 95: 1995–2005.

Shieh, W.Y. and Yang, J.T. (1997). Denitrification in the rhizosphere of the two seagrasses *Thalassia hemprichii* (Ehrenb.) Aschers and *Halodule uninervis* (Forsk.) Aschers. *Journal of Experimental Marine Biology and Ecology* 218: 229–241.

Shiozaki, T., Ijichi, M., Kodama, T. et al. (2014). Heterotrophic bacteria as major nitrogen fixers in the euphotic zone of the Indian Ocean. *Global Biogeochemical Cycles* 28: 1096–1110.

Sims, Z.C., Cohen, A.L., Luu, V.H. et al. (2020). Uptake of groundwater nitrogen by a nearshore coral reef community in Bermuda. *Coral Reefs* 39: 215–228.

Singh, A., Gandhi, N., and Ramesh, R. (2019). Surplus supply of bioavailable nitrogen through N_2 fixation to primary producers in the eastern Arabian Sea during autumn. *Continental Shelf Research* 181: 103–110.

Smith, S.V. (1984). Phosphorus and nitrogen limitation in the marine environment. *Limnology and Oceanography* 29: 1149–1165.

Sohm, J.A. and Capone, D.G. (2006). Phosphorus dynamics of the tropical and subtropical north Atlantic, *Trichodesmium* spp. versus bulk plankton. *Marine Ecology Progress Series* 317: 21–28.

Sohm, J.A. and Capone, D.G. (2010). Zonal differences in phosphorus pools, turnover and deficiency across the tropical north Atlantic Ocean. *Global Biogeochemical Cycles* 24: GB2008. https://doi.org/10.1029/2008GB003414.

Sokoll, S., Holtappels, M., Lam, P. et al. (2012). Benthic nitrogen loss in the Arabian Sea off Pakistan. *Frontiers in Microbiology* 3: 395. https://doi.org/10.3389/fmicb.2012.00395.

Sokoll, S., Ferdelman, T.G., Holtappels, M. et al. (2017). Intense biological phosphate uptake onto particles in sub-eutrophic continental margin waters. *Geophysical Research Letters* 44: 2825–2834.

Sorokin, Y.I. (1990). Phosphorus metabolism in coral reef communities: dynamics in the water column. *Australian Journal of Marine and Freshwater Research* 41: 775–783.

Sorokin, S.Y. (1992). Phosphorus metabolism in coral reef communities: exchange between the water column and bottom biotopes. *Hydrobiologia* 242: 105–113.

Stapel, J., Aarts, T.L., van Duynhoven, B.H.M. et al. (1996). Nutrient uptake by leaves and roots of the seagrass *Thalassia hemprichii* in the Spermonde Archipelago, Indonesia. *Marine Ecology Progress Series* 134: 195–206.

Stapel, J., Hemminga, M.A., Bogert, C.G. et al. (2001). Nitrogen (^{15}N) retention in small *Thalassia hemprichii* seagrass plots in an offshore meadow in south Sulawesi, Indonesia. *Limnology and Oceanography* 46: 24–37.

Stukel, M.R., Coles, V.J., Brooks, M.T. et al. (2014). Top-down, bottom-up and physical controls on diatom-diazotroph assemblage growth in the Amazon River plume. *Biogeosciences* 11: 3259–3278.

Sutka, R.L., Ostrom, N.E., Ostrom, P.H. et al. (2004). Stable nitrogen isotope dynamics of dissolved nitrate in a transect from the north Pacific subtropical gyre to the eastern tropical north Pacific. *Geochimica et Cosmochimica Acta* 68: 517–527.

Suzumura, M., Miyajima, T., Hata, H. et al. (2002). Cycling of phosphorus maintains the production of microphytobenthic communities in carbonate sediments of a coral reef. *Limnology and Oceanography* 47: 771–781.

Szmant, A.M. (2002). Nutrient enrichment on coral reefs: is it a major cause of coral reef decline? *Estuaries* 25: 743–766.

Tam, N.F.Y. and Wong, Y.S. (1996). Retention of wastewater-borne nitrogen and phosphorus in mangrove soils. *Environmental Technology* 17: 851–859.

Thangaradjou, T., Prasad, M.B.K., Subhashini, P. et al. (2015). Biogeochemical processes in tropical seagrass beds and their role in determining the productivity of the meadows. *Geochemistry International* 53: 473–486.

Tilstra, A., El-Khaled, Y.C., Roth, F. et al. (2019). Denitrification aligns with N_2 fixation in Red Sea corals. *Scientific Reports* 9: 19460. https://doi.org/10.1038/s41598-019-55408-z.

Touchette, B.W. and Burkholder, J.M. (2000). Review of nitrogen and phosphorus metabolism in seagrasses. *Journal of Experimental Marine Biology and Ecology* 250: 133–167.

Valdes, D.S. and Real, E. (2004). Nitrogen and phosphorus in water and sediments at Ria Lagartos coastal lagoon, Yucatan, Gulf of Mexico. *Indian Journal of Marine Sciences* 53: 338–345.

Vazquez, P., Holguin, G., Puente, M.E. et al. (2000). Phosphate-solubilizing microorganisms associated with the rhizosphere of mangroves in a semiarid coastal lagoon. *Biology and Fertility of Soils* 30: 460–468.

Viana, I.G., Moreira-Saporiti, A., and Teichberg, M. (2020). Species-specific trait responses of three tropical seagrasses to multiple stressors: the case of increasing temperature and nutrient enrichment. *Frontiers in Plant Science* 11: 571363. https://doi.org/10.3389/fpls.2020.5711363.

Vonk, J.A. and Stapel, J. (2008). Regeneration of nitrogen (^{15}N) from seagrass litter in tropical Indo-Pacific meadows. *Marine Ecology Progress Series* 368: 165–175.

Vonk, J.A., Middelburg, J.J., Stapel, J. et al. (2008). Dissolved organic nitrogen uptake by seagrasses. *Limnology and Oceanography* 53: 542–548.

Voss, M., Croot, P., Lochte, K. et al. (2004). Patterns of nitrogen fixation along 10 °N in the tropical Atlantic. *Geophysical Research Letters* 31: L23S3. https://doi.org/10.1029/2004GL020127.

Voss, M., Bombar, D., Loick, N. et al. (2006). Riverine influence on nitrogen fixation in the upwelling region off Vietnam, South China Sea. *Geophysical Research Letters* 33: L07604. https://doi.org/10.1029/2005GL025569.

Ward, B.B., Devol, A.H., Rich, J.J. et al. (2009). Denitrification as the dominant nitrogen loss process in the Arabian Sea. *Nature* 461: 78–82.

Wattayakorn, G., Prapong, P., and Noichareon, D. (2001). Biogeochemical budgets and processes in Bandon Bay, Suratthani, Thailand. *Journal of Sea Research* 46: 133–142.

Watts, L.J. and Owens, N.J.P. (1999). Nitrogen assimilation and the *f*-ratio in the NW Indian Ocean during an intermonsoon period. *Deep-Sea Research II* 46: 725–743.

Weber, S.C., Carpenter, E.J., Coles, V.J. et al. (2017). Amazon River influence on nitrogen fixation and export production in the western tropical North Atlantic. *Limnology and Oceanography* 62: 618–631.

Weber, L., Gonzalez-Díaz, P., Armenteros, M. et al. (2019). The coral ecosphere: a unique coral reef habitat that fosters coral-microbial interactions. *Limnology and Oceanography* 64: 2373–2388.

Werner, U., Blazejak, A., Bird, P. et al. (2008). Microbial photosynthesis in coral reef sediments (Heron Island, Australia). *Estuarine, Coastal and Shelf Science* 76: 876–888.

White, A.E., Foster, R.A., Benitez-Nelson, C.R. et al. (2013). Nitrogen fixation in the Gulf of California and the eastern tropical north Pacific. *Progress in Oceanography* 109: 1–17.

Wild, C., Huettel, M., Klueter, A. et al. (2004). Coral mucus functions as an energy carrier and particle trap in the reef ecosystem. *Nature* 428: 66–70.

Wild, C., Jantzen, C., Struck, U. et al. (2008). Biogeochemical responses following coral mass spawning on the Great Barrier Reef: pelagic-benthic coupling. *Coral Reefs* 27: 123–132.

Wilkinson, C.R., Williams, D.M., Sammarco, P.W. et al. (1984). Rates of nitrogen fixation rates on coral reefs across the continental shelf on the central Great Barrier Reef. *Marine Biology* 80: 255–262.

Williams, S.L., Yarish, S.M., and Gill, I.P. (1985). Ammonium distributions, production, and efflux from backreef sediments, St. Croix, US Virgin Islands. *Marine Ecology Progress Series* 24: 57–64.

Wyatt, A.S.J., Falter, J.L., Lowe, R.J. et al. (2012). Oceanographic forcing of nutrient uptake and release over a fringing coral reef. *Limnology and Oceanography* 57: 401–419.

Yool, A., Martin, A.P., Fernández, C. et al. (2007). The significance of nitrification for oceanic new production. *Nature* 447: 999–1002.

Zhang, W., Leung, Y., and Fraedrich, K. (2015a). Different El Niño types and intense typhoons in the western North Pacific. *Climate Dynamics* 44: 2965–2977.

Zhang, F., Blasiak, L.C., Karolin, J.O. et al. (2015b). Phosphorus sequestration in the form of polyphosphate by microbial symbionts in marine sponges. *Proceedings of the National Academy of Sciences* 112: 4381–4386.

Zhang, Z., Cao, Z., Grasse, P. et al. (2020). Dissolved silicon isotope dynamics in large river estuaries. *Geochimica et Cosmochimica Acta* 273: 367–382.

PART 4

HUMAN IMPACTS

CHAPTER 12

Pollution

12.1 Introduction

Most tropical estuarine and coastal environments are impacted by humankind; only some remote habitats are pristine (Halpern et al. 2008). Populations and communities of fish and other organisms are declining; warming is acidifying the open ocean (Section 13.5), and large expanses of the tropical marine environment are increasingly polluted by an array of chemicals and trash, including plastic bags, bottles, and lost fishing nets. There have been many acute and chronic oil spills. And, many abnormalities in marine animals are due to sublethal effects of chemicals manufactured and disposed of by humans. Sediments, water, and a wide variety of organisms are contaminated by wastes released into the coastal zone (Islam and Tanaka 2004).

These problems are increasing in the tropics as human populations continue to grow in developing countries, leading to more urban development such as manufacturing and industrial estates. The coastal ocean is long been exploited for food, recreation, transport, waste disposal, and other needs, such as sand and gravel dredging, oil, gas and mineral exploration, timber, tourism, harbours, and marinas. It is partially for these reasons that more than half of humankind live within 100 km of a coast. Increasing urbanisation, industrialisation, and affluence foreshadow greater pressure on the living and non-living resources of the tropical coastal ocean.

Major problems in the world's coastal ocean include eutrophication, coastal development including aquaculture, habitat destruction, and alteration (Chapter 14), disruption of coastal hydrological cycles, including river discharge, point and non-point source release of toxins and pathogens, introduction of exotic species, fouling by plastic litter, build-up of chlorinated hydrocarbons, shoreline erosion or siltation, unsustainable or uncontrolled exploitation of resources, noise pollution, and global climate change (Chapter 13). Arguably, the greatest concerns are coastal development and accompanying modification and destruction of habitats, eutrophication, microbial pathogens, fouling by plastics, and persistent increases in chlorinated hydrocarbons. Due to national and international environmental regulations and improvements in technology, most trace elements (particularly lead), radionuclides, and crude and refined oil now present less of a problem. However, other urgent problems have not been addressed, such as global climate change, the decline of marine mammal populations, overfishing and the collapse of fisheries, fish and shellfish diseases, and crashes in seabird populations.

Tropical Marine Ecology, First Edition. Daniel M. Alongi.

Most pollutants enter the sea from land and the atmosphere, with minor pathways of delivery via maritime transportation, dumping, and offshore gas and oil production. Both land and atmospheric sources account for roughly two-thirds of total pollutant inputs. The large contributions from land and atmosphere reinforce the importance of their connection to the coastal zone and the urgent need to understand better the nature of these relationships. There is increasing recognition that coastal pollution problems are linked to contamination of adjacent watersheds and catchments, and aerosols in the lower atmosphere.

12.2 Hydrocarbons

Oil pollution is a worldwide problem, and although declining in frequency since the advent of new technologies and enforcement of national and international agreements, hydrocarbons of various types persist for long periods of time, even in the tropics where oil degradation is faster than in higher latitudes. There have been several major accidents in the past that still provide lessons for today, and there are still minor spills that occur from time to time.

Oil spills have caused major damage to sandy beaches, tidal flats, coral reefs, seagrass beds, mangrove forests, and to coastal waters and the seabed. The extent of damage depends on the type of oil and the size of the spill. As with all disturbances, the speed of recovery depends on the frequency, intensity, and areal extent of the spill. Some habitats and their associated organisms have not been greatly affected by small oil spills, whereas others have been greatly impacted by spillage of oil and use of dispersants. The mixture of oil and dispersants can be more toxic than the oil alone.

Data for oil spills on tropical sandy beaches, sand and mud flats, and rocky intertidal shores indicate that damage was minimal and recovery was rapid; however, these were all small spills. On a subtropical sand flat of Taiwan, an extensive oil spill decreased sediment chlorophyll *a*, affecting ecological functioning by suppressing or even stopping microalgal production, increasing bacterial respiration, and causing a trophic shift from net autotrophy to net heterotrophy (Lee and Lin 2013). Effects of the spill on the macrofauna were more severe than on benthic microalgae, affecting sedentary infauna more than motile epifauna. Despite these impacts, the affected area appeared to return to normal in about 23 days. In subtropical Paranaguá Bay, Brazil, the benthic macrofauna was acutely affected by an experimental diesel oil spill, but the recovery to pre-disturbance population levels was extremely fast (Egres et al. 2012). An increase in benthic faunal densities after the spill was the result of an increase in densities of oligochaetes and ostracods. The fauna was thus highly resilient to the small-scale disturbance. On a tropical rocky shore, the impact of the *Chitra* oil spill was not so devastating (Sukumaran et al. 2014). Diversity indices indicate that the ecological status of the site was affected briefly during the spill in August 2010, but a healthy environment prevailed thereafter. The oil spill did not have long-term deleterious effects on the macrobenthos. Rapid recovery was attributed to the small size of the spill, rapid breakdown of the oil in a tropical climate, and physical breakdown by waves and tides.

In a much larger oil spill in Bodo Creek in the Niger delta, Nigeria, infauna recovery in intertidal sand and sandy mud flats was slow, with a post-spill shift in dominance from crustacea to more pollution-tolerant polychaetes, such as *Capitella capitata*, although there was a gradual increase in species richness but not to pre-spill values (Nwipie et al. 2019). The massive 1991 gulf war oil spill had an initially deadly impact on the shoreline biota of the Saudi Arabian Gulf (Jones et al. 1998). By December 1991, between 50 and 100% mortality of biota was observed on the upper shore. By 1995, however, species diversity on the lower shore was like that found on unpolluted shores; in the upper intertidal and at the top of the shore, diversity ranged from normal to 83% of that found on control shores. Abundances of individual species also increased during the survey period. Recovery rates for Saudi Arabian shores were within the time scale for shores worldwide. Twelve years on, a large-scale risk assessment found that 67% of samples still had significant amounts of crude oil (Bejarano and Michel 2010).

Long-term effects of oil pollution have been found on Curaço (Nagelkerken and Debrot 1995). Mollusc communities in shores massively inundated by oil seven years earlier still showed evidence of a negative impact of the oil spill as these shores retained high concentrations of tar. Average mollusc densities were only 21% of those encountered in unpolluted sites. Mollusc densities and species diversity correlated negatively with percent tar cover.

Oiling of sandy beach invertebrates results in highly variable recovery in time and space, depending on several factors, including site-specific physical properties and processes, for example, sand grain size and beach exposure, the degree of oiling, depth of oil burial, and biological factors such as species-specific life history traits of fauna (Figure 12.1). Recovery of affected communities ranges from several weeks to several years, with longer recoveries generally associated with physical factors that facilitate oil persistence, or when clean-up activities are absent on heavily oiled beaches (Bejarano and Michel 2016).

Seagrasses have been oiled in various tropical locales, and the effects have been equivocal. In 1994, the barge *Morris J. Berman* spilled 713 269 gallons of fuel oil on the seagrasses and beaches of Puerto Rico, but there were no significant impacts on the plants (NOAA 1996). Similarly, there was little effect of oil pollution on the seagrasses in Sharm El-Moyia Bay in the Egyptian Red Sea. Oil pollution led to localised destruction of seagrass-associated macroinvertebrates, but it did not cause any degradation of the meadow (Gab-Alla 2000). A bunker fuel spill off Panay Gulf, The Philippines, resulted in oil covering the shores and seagrasses of a marine protected area, Taklong Island National Marine Reserve; most shorelines were heavily covered for weeks before manual clean-up was initiated (Nievales 2009). There was a decrease in seagrass cover and shoot density, and lower seagrass cover, shoot and blade densities, and above-ground biomass within a year after the spill. Obviously, the degree of impact on seagrass meadows by oil pollution depends on the size of the spill, the extent to which the seagrasses are covered, and the type of oil.

Over time, saltmarshes, mangroves, and seagrass meadows in the Saudi Arabian Gulf have accumulated massive amounts of petroleum (Ashok et al. 2019). However, despite being enormous, the spill had little apparent effect on the seagrass meadows along the shoreline (Durako et al. 1993). Photosynthetic and respiratory responses of leaf tissues of the seagrasses, *H. ovalis, H. stipulacea,* and *H. uninervis* exposed to unweathered Kuwait crude oil indicated no effect of exposure to the water-soluble fraction of the oil (Durako et al. 1993). The gulf war oil spill primarily impacted intertidal communities rather than the submerged plant communities of the northern gulf region. Although most mangroves survived the massive spill, some mangroves, such as *A. marina,* showed sublethal effects. Two years after the spill, many trees developed a high number of branched pneumatophores and adventitious roots.

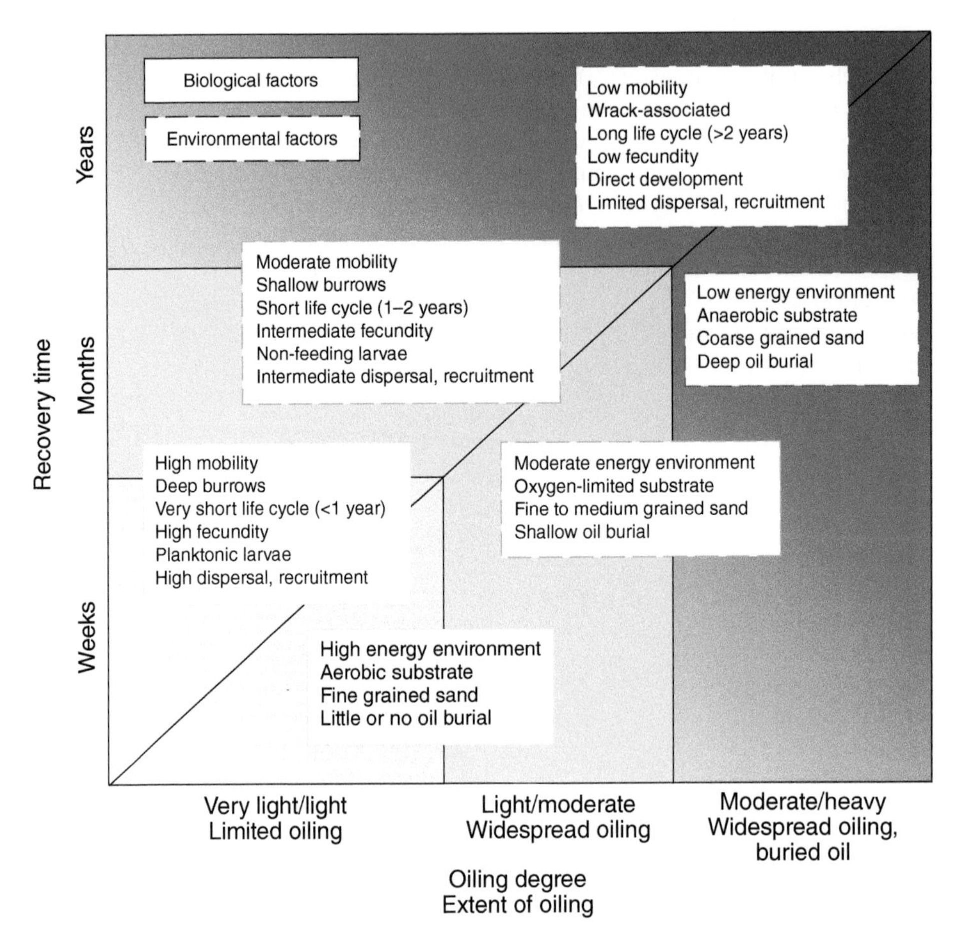

FIGURE 12.1 Factors that may contribute to the recovery of invertebrate communities on sandy beaches following an oil spill. The anticipated recovery times are a continuum rather than exact times and are a function of site-specific physical and biological factors and their continuum. *Source:* Bejarano and Michel (2016), figure 3, p. 717. © Elsevier.

In 1986, >9.3 × 10^6 l of crude oil spilled into a 40 km² region of mangroves, seagrass beds, and coral reefs just east of the Caribbean entrance to the Panama Canal (Jackson et al. 1989). This was the largest recorded spill into coastal habitats in the tropical Americas. Intertidal fauna and flora were covered by oil and died soon after. Extensive mortality of shallow subtidal reef corals and infauna of seagrass beds was observed. Dead mangroves occurred along 27 km of the coast. 1.5 years after the spill, recovery of prop root fauna and flora facing the open sea was patchy. In channels deep within the mangroves, there was little recovery of biota. The mangrove-fringe root habitat was largely destroyed. Entire beds of intertidal *Thalassia* were killed on some heavily oiled reef flats, but subtidal *Thalassia* survived after the spill, although leaves became brown and were heavily fouled by algae for several months. The effects on coral reefs were complex, although most lethality was expressed in the intertidal reef flats where cover of microalgae, fleshy red algae, crustose corallines, and *Halimeda opuntia* was greatly reduced. There were sublethal impacts on subtidal coral reefs. Most of the hard corals still alive in <3 m showed signs of stress including bleaching or swelling of tissues, conspicuous production of mucus, recently dead areas devoid of coral tissue, and globules of oil. Assessing the long-term effects of the spill on the subtidal coral reefs, Guzman et al. (2020) found that short-term effects on corals could be unequivocally linked to the spill, but assessment of mid- and long-term impacts was complicated by cumulative different stressors, including diseases, bleaching, warming, and additional accidental oil discharges.

Biological effects of oil spills are usually most severe in low-energy environments where oil accumulates and is retained in sediments, where in high-energy environments, it is soon washed away. This pattern matches the greater and more persistent disturbance to riverine and channel mangrove root communities than to those along open coasts and the rapid recovery of many organisms at the seaward edge of the reef flat at the *Galeta* oil spill (Jackson et al. 1989). Some results of the *Galeta* oil spill were unexpected. For instance, before the spill, it was believed that oil pollution does not significantly affect coral reefs, as extensive mortality of subtidal corals and infauna of seagrasses had not been demonstrated before. Subtidal effects may have been partly due to the use of dispersants and were more extensive and more important in the long-term than the initial mortality. Injury of corals allowed colonisation of algae and other sessile organisms that overgrew colonies that survived the initial spill. Corals stressed by oil are also probably more susceptible to epidemic disease and likely to grow and reproduce more slowly than unaffected colonies. Any combination of these effects may further reduce overall abundance of corals. Thirty years after the spill, parts of the ecosystem have still not recovered.

The *Galeta* oil spill has had long-term effects on the mangrove communities. Changes in the physical structure of the mangrove fringe after oiling included defoliation, limb loss, and eventual collapse of dead trees (Garrity and Levings 1993; Garrity et al. 1994; Levings et al. 1994). Five years after the spill, the length of shore fringed by mangroves was reduced at oiled relative to unoiled sites. Surviving trees at oiled sites had fewer and shorter submerged prop roots and a higher proportion of dead roots than trees at control sites. These changes reduced the surface area of submerged prop roots by 33% on oiled open coast, 38% in channels, and 74% upstream. Oil-induced reductions to the biota on the submerged prop roots were still negatively affected five years after the spill. Areas most affected were low- and mid-intertidal zones of the mangrove, *R. mangle,* and the extent and location of deforestation were site- and spill-specific, notably influenced by prevailing wind and tidal conditions (Duke et al. 1997, 1999). Some areas were deforested twice. Canopy leaf biomass decreased in open canopy sites where oil concentrations in surrounding sediments were high. Natural seedlings in oiled habitats showed higher leaf node production, indicating that remaining oil did not depress growth and was possibly greater because of higher light levels in more open, oil-deforested light gaps (Duke et al. 1999).

Experimental studies have shown that the effects of oil are complex and variable depending on how the mangroves are oiled, what species are present and the type of oil (Grant et al. 1993; Duke et al. 1998, 2000; Youssef 2002; Zhang et al. 2007a). In glasshouse and field studies, Grant et al. (1993) measured the survival of *A. marina* seedlings when exposed to weathered crude oil. They found that 96% of seedlings died within two weeks, while all untreated seedlings survived. A further field experiment found that establishment and survival were not enhanced by light gaps and canopy effects, but crude oil in the sediment inhibited establishment and decreased the number of seedlings surviving for several years. In contrast, the physiological responses of *A. marina* seedlings to the phytotoxic effects of the water-soluble fraction of light Arabian crude oil were short-lived (Youssef 2002). Irregular stomatal behaviour and weak control over transpiration were observed during the first day of exposure. After six weeks, however, all treated plants showed no significant difference in their relative growth rate or in the net assimilation rate compared to control plants. Disturbed stomatal behaviour may have been due to the venting of the volatile low molecular weight aromatic hydrocarbons through the stomata; uptake of water-soluble hydrocarbons was equally possible through both the root system and the foliage.

Mangroves retain petroleum hydrocarbons in tissues and soils, and this contamination may be passed on to biota. In the mangrove forests on Hainan Island, China, the standing accumulation, annual absorption, annual net retention, annual return, and the turnover period of PAHs in all mangrove tissues were 2228 $\mu g\,m^{-2}$, 869 $\mu g\,m^{-2}\,a^{-1}$, 206 $\mu g\,m^{-2}\,a^{-1}$, 663 $\mu g\,m^{-2}\,a^{-1}$, and 3.4 a^{-1}, respectively (Qiu et al. 2018). Bioaccumulation was also discovered in *A. marina* stands along the Arabian Gulf with the trend of total PAH concentrations in the following order: roots > leaves > soils (Orif and El-Maradny 2018). In experimental tests, Naidoo and Naidoo (2018) investigated the uptake and accumulation of PAHs in seedlings of *A. marina* and *R. mucronata* and found that PAH concentrations were higher in *A. marina* than in *R. mucronata*. Most PAHs (mostly phenanthrene) were in roots (96% in *A. marina* and 98% in *R. mucronata*). The greater susceptibility of *A. marina* appeared to be due to its greater root length which permits more exposure to oil than *R. mucronata*. Other contributory factors include larger air spaces in roots, lower suberisation of root epidermal cells, lower concentrations of tannins, lignin, and a less efficient anti-oxidative system. Mangrove animals can also bioaccumulate hydrocarbons as found by Numbere (2019) in the West African red mangrove crab, *Goniopsis pelii,* in highly polluted mangrove forests in the Niger delta, Nigeria. There were no significant differences in THCs in the body parts of crabs, although concentrations tended to be higher in intestine and gut and in males than in females, probably due to their greater size. Internal parts of crab had higher concentrations than external parts indicating high bioaccumulation.

In addition to the effects of the *Galeta* oil spill, other data suggest a detrimental influence of oil pollution on coral reefs. Long-term field studies in the Red Sea and laboratory experiments have witnessed detrimental effects of oil spills on reef corals, such as a complete lack of colonisation by hermatypic corals in reef areas chronically polluted, decrease in colony viability damage to the reproductive system of corals, lower life expectancy of planulae, and abnormal behavioural responses of planulae and corals (May et al. 2020). Other detrimental effects on corals and their associated biota, mainly reported from the Caribbean, include lower growth rates, direct damage to tissues, thinning of cell layers and disruption of cell structure, damage to tactile stimuli and normal feeding mechanisms, excessive mucus production leading to enhanced bacterial growth, and eventual coral destruction. For example, reef fish communities have undergone large declines in species richness (38%) and species diversity (26%) six years after the 2010 *Deepwater Horizon* oil spill in the northern Gulf of Mexico (Lewis et al. 2020). Initial changes were driven by widespread declines across trophic guilds with uneven recovery among guilds and taxa. Densities of small demersal invertivores, small demersal browsers, generalist carnivores, and piscivores remained persistently low with little recovery seven-years after the spill. Lack of recovery was hampered by predation from the recently arrived, invasive lionfish (*Pterois* spp.).

Dispersants enhance the damage done to corals. Harmful effects of five third-generation oil dispersants were observed on planula larvae of the Red Sea stony coral, *S. pistillata,* and the soft coral, *Heteroxenia fuscescens* (Epstein et al. 2000). Larvae were exposed to Egyptian oil water-soluble fractions, dispersed oil water-accommodated fractions, and dispersants dissolved in seawater. While oil water-soluble fractions resulted in reductions in planulae settlement only, treatments by all dispersants revealed a further decrease in settlement rates and additional high toxicity. Dispersed oil exposures resulted in a dramatic increase in toxicity to both coral larvae species. Toxicity was observed for the larvae of the hermatypic reef corals, *Acropora tenuis, Goniastrea aspera,* and *Platygyra sinensis* (Lane and Harrison 2000) and for the coral, *A. millepora* (Negri and Hayward 2000). The water-soluble fraction of crude oil has detrimental effects on the stony coral, *Pocillopora damicornis* (Rougée et al. 2006). Significant changes were seen in cytochrome P450 I-class, cytochrome P450-2 class, glutathione-S-transferase-pi, and cnidarian multi-xenobiotic resistance protein-I biomarkers which are involved in the cellular response to, and manipulation and excretion of, toxic compounds. Similar physiological effects have been observed for the hard coral, *Porites lobata,* after the MV *Kyowa Violet* fuel oil spill (Downs et al. 2006).

Bioremediation using probiotic microbiota can reduce the impact of oil spills on coral reefs. In an experimental study, a probiotic bacterial consortium was produced from the coral, *Mussismilia harttii,* and was used to degrade water-soluble fractions of crude oil (Santos et al. 2015). The use of the bacterial consortium as a bioremediation agent was evaluated with the bacteria responsible for highly efficient degradation of petroleum hydrocarbons; they also minimised the effects of water-soluble fractions on coral health, as indicated by raised photosynthetic efficiencies. Moreover, the impact of the water-soluble fractions on the coral microbiome was diminished by the introduced bacterial consortium. Thus, the bacteria have a dual function, promoting oil degradation and improving coral health with its probiotic features.

The impacts of oil pollution on coral reef environments can be exacerbated by environmental conditions such as ultraviolet radiation and climate (Nordborg et al. 2020). Exposure of shallow reefs to high levels of UV radiation can substantially increase the toxicity of some oil components through phototoxicity by 7.2-fold in all tested conducted. Co-exposure to elevated temperature and low pH, both within the range of current daily and seasonal fluctuations and/or projected climate change, can increase oil toxicity by 3- and 1.3-fold, respectively.

Tropical estuaries, bays, and the shelf margins have not been investigated as thoroughly as mangrove, coral reef, and seagrass habitats. Oil-related activities on the Campeche shelf in the southern Gulf of Mexico negatively impacted benthic communities (Hernández Arana et al. 2005). A pattern of contamination existed, with increased levels of contaminants at stations close to oil rigs and in areas of high oil platform densities. At these sites, macroinfauna densities and biomass were reduced. Increased variability in community composition was linked to oil-related disturbance. Near a Brazilian oil refinery, there were serious effects of oil contamination with a decrease in macrofauna density, number of species, and diversity related to the occurrence of high aliphatic hydrocarbon and polycyclic aromatic hydrocarbon concentrations associated with fine sediments (Venturini et al. 2008). Their results showed that chronic oil pollution in Todos os Santos Bay in Brazil was responsible for a depauperate benthic infauna consisting mostly of opportunistic species that are well-adapted to polluted environments. Along the shelf and slope of the Gulf of Guinea of West Africa, the structure of benthic communities was most affected by petroleum hydrocarbons and total hydrocarbons (Pabis et al. 2020) with low to moderate densities of polychaetes, amphipods, and bivalves.

The effects of oil pollution can be detected even out in the open sea. Plankton and nekton in the Bay of Bengal have detectable amounts of petroleum hydrocarbons

(Ansari et al. 2012). Twenty-seven commercially important fish, crustacean, and cephalopod species from the Orissa coast were examined; several species, including the cephalopod, *Loligo,* recorded the maximum total petroleum hydrocarbons, while the penaeid shrimp, *Metapenaeus dobsoni,* recorded the maximum total hydrocarbons among the crustaceans. These concentrations were higher than values reported in other studies in the Bay of Bengal but are comparable with those from Indian coastal waters.

Coastal lagoons are among the most heavily polluted environments in the tropical coastal zone. Lagoons receive various types and concentrations of pollutants from petroleum hydrocarbons and heavy metals to excess nutrients leading to eutrophication, all of which have an impact on biota (Uwadiae 2016). In the Ebrié Lagoon, Ivory Coast, both sediments and overlying waters are polluted with heavy metals and PAHs (Affian et al. 2009) that have severe impacts on planktonic and benthic organisms. Benthic communities are also affected by spatial and temporal changes in salinity, dissolved oxygen, and other environmental variables making it difficult to ascribe benthic community changes to just one variable (Kouadio et al. 2011).

Multiple contaminants are also a problem in tropical estuaries, mangroves, coral reefs, seagrass meadows, and continental shelf environments. This is the result of numerous uses of these water bodies and the unregulated and illegal dumping of contaminants from a wide variety of sources. In most estuaries and lagoons, there is a long history of contamination that is decades old and is still happening. Heavy metals, nutrients, petroleum hydrocarbons, plastics and other debris, and organic wastes such as pesticides, herbicides, and organochlorine chemicals (mostly DDTS-related pesticides, PCBs, and dioxins) are all types of industrial and agricultural waste that wash into streams and rivers and ultimately down into estuaries and the adjacent coastal zone.

12.3 Metals

Metals come from the earth's crust, found in rocks, soils and sediments, but they can become contaminants when they are concentrated by industrial activity. Being chemical elements, they cannot break down but instead get taken up by biota. Like other contaminants, metals bioaccumulate within living organisms and can biomagnify up the food chain.

Metals released from mining and industrial processes are among the major contaminants of the tropical coastal zone. Mercury (Hg), cadmium (Cd), copper (Cu), zinc (Zn), chromium (Cr), and silver (Ag) are major contaminants from industrial processes, including effluents from power plants. Hg deposited from the atmosphere is a significant fraction of the Hg entering coastal waters and accounts for ≈90% of the Hg fraction in the open ocean (Strode et al. 2007). Some metals such as Cu and Zn are micronutrients, but some like Hg play no normal biological role. Most contamination originates from land-based industrial sources, but some metals are used in antifouling paints for ships. One of the most toxic ingredients in antifouling paints has been tributyltin (TBT) which was first developed in the 1940s.

Lead (Pb) can also be a metal of concern as it comes in runoff from road surfaces during rain, from its previous use as a petrol additive. Pb remains in the environment and like other metals does not break down, so some coastal environments have elevated amounts in their sediments due to decades of deposition. Other metals, such as Cd, Cr,

Zn, and Cu, can similarly deposit and become an environmental problem. Metals bind to sediment particles from which they are available to varying degrees to marine organisms, particularly benthos, from which the metals move up the food chain.

Bioavailability of sediment-bound metals is a critical issue for their toxicity. The toxicity of metals depends on their form in sediments. Acid-volatile sulphide (AVS) in sediments binds metals and has been used to predict toxicity of some metals, including Cu, Cd, nickel (Ni), Pb, arsenic (As), Hg, and Zn. AVS in sediments reacts with metal to form an insoluble metal sulphide that is relatively unavailable for biological uptake. Sediments tend to have high levels of sulphide and thus relatively low bioavailability of sediment-bound metals. Elevating the oxygen in overlying water or mixing of sediments decreases AVS thus increasing the bioavailability of metals.

In tropical intertidal environments, metal concentrations are higher than in the overlying water and some are more toxic than others to marine life. For instance, the toxic effects of heavy metals on the intertidal crab, *Scylla serrata,* were evaluated in India (Krishnaja et al. 1987). Phenyl mercuric acetate was found to be the most toxic and lead nitrate the least toxic. Marked histopathological changes were identified in the hepatopancreas and gill exposed to these metals as well as mercuric chloride, cadmium chloride, selenium dioxide, and arsenic trioxide. The free amino acid pool from muscle and hepatopancreas did not show any reduction in number or quantity, but sublethal concentrations of Hg and Cd brought about degenerative changes in the hepatopancreas and gill of exposed animals. Similarly, the tissues of the mussel, *Perna viridis,* were negatively affected by exposure to Cd, Cu, Fe (iron), Pb, Ni, and Zn in the Johore Straits, Malaysia (Yap et al. 2006).

Sandy beaches can be contaminated by metals derived from industry and brought down by rivers, with seasonal variations in metal concentrations tied to the monsoon. Beaches along the Tamil Nadu coast of southeast India, for example, are contaminated by Fe, Mn (manganese), Cr, Cd, Zn, Cu, and Ni due to direct input from various coastal industries (Lakshumanan et al. 2010). On La Escobilla beach, Mexico, both metals and metalloids have been found in whole blood and tissues of Olive Ridley turtles (Cortés-Gómez et al. 2014). They were contaminated with Cd, while Pb levels showed a decline due to the gradual disuse of leaded petrol. The beach wedge clam, *Donax faba,* was contaminated with heavy metals on Panambur beach near industrial areas in India (Singh et al. 2012). *D. faba* showed seasonal variations of metal accumulation with maximum concentrations in the post-monsoon season, with metal concentrations in the clam in the order of Fe > Cu > Cr > Ni > Pb. High concentrations of metals in the whole tissue of the wedge clams were likely derived from the industrial areas nearby. Metals can be found in tar balls and in crude oil washed up on sandy beaches. The behaviour of metals on sandy beaches or on intertidal sand and mud flats is highly variable owing to the monsoon season and to spatial and temporal changes in metals discharged from industrial activities.

Many species of tropical intertidal organisms are contaminated by heavy metals that affect their physiological state (Capparelli et al. 2016). The effects of metal contamination on osmoregulation and oxygen consumption in the mud flat fiddler crab, *Uca rapax,* from Brazil were examined (Capparelli et al. 2016). The contamination led to bioaccumulation and induced biochemical and physiological changes, with crabs from contaminated sites exhibiting stronger hyper- and hypo-osmotic regulatory abilities and greater gill enzyme activities than crabs from a pristine site. Oxygen consumption was higher in high bioaccumulation crabs but increased in crabs with low to moderate bioaccumulation levels. Thus, fiddler crabs chronically contaminated *in situ* exhibit compensatory biochemical and physiological adjustments.

Tropical seagrasses and mangroves are sites of metal contamination in some regions. In a mangrove creek and adjacent seagrass meadow on Mactan Island, the Philippines, six of eight heavy metals measured had significantly higher concentrations in creek bank deposits than in seagrass sediments (Dy et al. 2005). In contaminated sites in Brazil, the seagrass, *H. wrightii,* had high concentrations of Cd and Zn compared with control sites, indicating that metal mobilisation from contaminated sediments through dredging activities was transferred via accumulation by the seagrass (Filho et al. 2004). The contaminated seagrass had less biomass than those in non-impacted sites. In the Lakshadweep Islands of India, heavy metal enrichment was found in seven species of seagrasses (Thangaradjou et al. 2013). Mg and Al were present in high concentrations that varied by species. *H. decipiens* was a strong accumulator, followed by *H. uninervis* and *H. pinifolia*. Small-leaved seagrasses were found to be more efficient accumulators of heavy metals than large-leaved species.

Mangroves are highly tolerant to metal contamination and act as an efficient trap for metals. Mangrove trees store metals mainly in their roots (MacFarlane et al. 2003). In *A. marina,* Cu and Pb accumulated in root tissue to levels higher than surrounding soil levels and Zn accumulated to levels equivalent to levels in soils. Pb showed little mobility to leaf tissue, whereas Zn accumulated in *A. marina* leaves. Increasing soil concentrations of Pb and Zn resulted in a greater accumulation of Pb in both root and leaf tissue. In tissues of *R. mucronata*, elevated metal concentrations occurred in the following order: Zn > Pb > Cu > Cd (Ganeshkumar et al. 2019). Bioaccumulation and translocation of Pb from the soil to the plant tissues were relatively higher than for the other metals.

High concentrations of metals in mangrove tissue do not necessarily lead to increased mortality, but there are often sublethal effects as accumulated metals can induce subcellular biochemical changes (Gonzalez-Mendoza et al. 2007; Zhang et al. 2007b; Liu et al. 2009; Huang and Wang 2010a, b; Huang et al. 2011; Xie et al. 2013). For instance, subcellular changes in *A. marina* exposed to a range of Cu, Pb, and Zn concentrations were discovered under laboratory conditions (MacFarlane and Burchett 2001). Limited Cu uptake by leaves was observed at low soil Cu levels, with visible toxicity at soil levels > 400 $\mu g\, g^{-1}$. Leaf Pb concentrations remained low over a range of Pb soil concentrations up to 400 $\mu g\, g^{-1}$, above which it appeared that unrestricted transport of Pb occurred, although no visible signs of Pb toxicity were observed. Significant increases in enzyme activity and decreases in photopigments were found with Cu and Zn at concentrations lower than those inducing visible toxicity. Significant increases in enzyme activity were found when plants were exposed to Pb.

Metals at sublethal doses can induce biochemical changes in mangroves. The leaves and roots of the mangroves, *K. candel* and *B. gymnorrhiza,* exposed to different concentrations of Pb, Cu and Hg showed different responses to heavy metal stress (Zhang et al. 2007b). An increase in enzyme activities demonstrated that *K. candel* was more tolerant to heavy metals than *B. gymnorrhiza*. Lipid peroxidation was enhanced only in leaves of heavy metal-stressed *B. gymnorrhiza,* indicating that antioxidative enzyme activities may play an important role in both species and that cell membranes in leaves and roots of *K. candel* have greater stability than those of *B. gymnorrhiza*.

Mangroves, like other plants, have evolved a suite of mechanisms that control and respond to the uptake and accumulation of heavy metals. One of the major defence mechanisms involves the chelation and sequestration of heavy metals by heavy metal-binding ligands, such as phytochelatins (PCs) and metallothioneins (MTs). MTs are characterised as low molecular weight, cysteine-rich, metal-binding proteins. The role of MTs in organisms has been proposed as detoxification of toxic heavy metals and being involved in homoeostasis of trace elements. Despite the confirmation of the presence of MT genes in various plants, the role and function of MTs in plants remain unclear. Huang and Wang (2010a, b) and Huang et al. (2011) assessed the role of MTs in heavy metal tolerance by analysing the expression level of type 2 metallothionein gene in leaves of *B. gymnorrhiza* exposed to heavy metals. They demonstrated that gene expression was regulated by Zn, Cu, Cd, and Pb, but that the regulation pattern was different among metals. Similar gene expression was found in *A. germinans* exposed to Cd and Cu (Gonzalez-Mendoza et al. 2007) and in *A. marina* (Huang and Wang 2010a). The transfer of metals to mangroves also depends on the metal speciation. In *K. obovata* forests in southern China, speciation analysis of Hg in non-rhizosphere soils showed that residual, oxidizable, and volatile Hg were dominant fractions, with volatile Hg being most important for Hg bioaccumulation in the tree (Chai et al. 2020). Iron plaques on roots acted as a natural barrier to limit transfer of Hg from soil to the tree.

Mangroves may also respond to the presence of metals by translocating oxygen to their roots and diffusing oxygen to their rhizosphere (known as radial oxygen loss, ROL). This allows the plant to oxidise reduced substances such as phytotoxins and to significantly alter the immediate environment around the roots, significantly altering both microbial and chemical processes in rhizosphere sediments, such as nitrification, nutrient availability, aerobic degradation, and transformation of environmental pollutants. The effects of Pb, Zn, and Cu on growth, ROL, and the spatial pattern of ROL were investigated in seedlings of *A. corniculatum*, *A. marina*, and *B. gymnorrhiza* (Liu et al. 2009). Heavy metals inhibited seedling growth and led to decreased ROL and changes in the tight barrier spatial pattern of ROL. There was a significant positive correlation between the amount of ROL from the roots of seedlings and metal tolerance. *B. gymnorrhiza,* the species with the highest ROL amount, was the most tolerant to the presence of heavy metals. Cheng et al. (2010) investigated root anatomy, ROL, and Zn uptake and tolerance in seedlings of *A. corniculatum, B. gymnorrhiza,* and *R. stylosa* and found that *B. gymnorrhiza* took up the least amount of Zn and showed the highest Zn tolerance. Furthermore, Zn significantly decreased the ROL of all three plants by inhibition of root permeability, which included an obvious thickening of the outer cortex and significant increase of lignification in cell walls. The results of scanning electron microscopy further confirmed that such an inducible, low permeability of roots was a likely adaptive strategy to metal stress by direct prevention of excessive Zn entering the root.

The physiological and biochemical responses in the leaves of *K. candel* and *B. gymnorrhiza* exposed to multiple heavy metals were investigated (Huang and Wang 2010b) and an increase in Cd, Pb, and Hg content was measured in the leaves of both species, whereas higher metal levels led to a breakdown of leaf chlorophyll. The content of proline (a compound that normally accumulates in response to environmental stress), glutathione (GSH), and phytochelatins (PCs-SH) in the leaves of both species exhibited a significant increase in response to exposure to these heavy metals. Increased contents of proline, GSH, and PCs-SH in metal-treated plants suggest that metal tolerance in both species might be associated with the efficiency of these antioxidants. The exposure of these seedlings to heavy metals also resulted in an increase in exudation from the roots of low-molecular-weight (LMW) organic acids and amino acids. In laboratory growth experiments with *K. obovata,* exposure to Cd resulted in a release in oxalic, acetic, l-malic, tartaric acid, tyrosine, methionine, cysteine, isoleucine, and arginine from the roots (Xie et al. 2013). Both organic acids and amino acids excreted from the plant played a key role in resistance to Cd toxicity.

Heavy metals can bioaccumulate in mangrove-associated biota. In the Sundarbans in Bangladesh, significant variations in contamination were recorded in mud crabs, mudskippers, and gastropods (Figure 12.2). Bioconcentration factors indicated that metals were highly bioaccumulated and biomagnified in benthic fauna (Ahmed et al. 2011). Heavy metal contamination was similarly recorded in mangrove biota in Dar Es Salaam, Tanzania (Mremi and Machiwa 2003) where crabs contained higher concentrations of heavy metals (Pb, Zn, and Cu) compared with soil and mangrove plant parts. Cu enrichment in crabs was more than 6x the concentrations in mangrove soil. Cd contamination was found to be a concern for edible shellfish in the mangroves of Senegal (Bodin et al. 2013). Trace metals exhibited higher concentrations in shellfish compared with soil. However, strong differences of metal bioavailability and bioaccumulation in biota were demonstrated. Accumulation of metals in the crab, *Ucides cordatus* and *R. mangle,* was found in a heavily polluted mangrove site in Brazil (Pinheiro et al. 2012). The accumulation of Cd, Cr, Cu, Hg, Mn, and Pb was examined in different organs of *U. cordatus* and in different maturation stages of *R. mangle* leaves. Cd, Cu, Cr, and Mn in the crab were concentrated in the hepatopancreas, followed by the gills. The highest accumulation of metals in leaves occurred in pre-abscission, senescent, and green mature leaves. Patterns of bioaccumulation between the crab and the leaves differed for each metal, probably due to the specific requirements of each organism. However, there was a close and direct relationship between metal accumulation in the mangrove trees and in the crabs feeding on the leaves.

Bioaccumulation of metals, particularly Hg, commonly occurs in mangrove-associated coastal fish, as found by Le et al. (2018) along the east coast of Peninsular Malaysia. Out of 20 species observed, the highest Hg concentrations were found among the carnivore/invertebrate feeders, followed by omnivores with the lowest concentration in herbivores. Not only diet but also foraging habitat is a clear discriminator of the extent of Hg contamination in fish. Examination of Hg and stable isotope ratios of $\delta^{13}C$ and $\delta^{15}N$ in fish muscle belonging to 23 different species from the Senegalese coast showed that vertical (water-column distribution) and horizontal habitat (distance from the coast) led to differential Hg accumulation among species (Le Croizer et al. 2019). Coastal and demersal fish were more contaminated than offshore and pelagic species. The main carbon source was significantly correlated with Hg concentration, revealing a higher accumulation for fish feeding in nearshore and benthic habitats. Mass mortality of estuarine fish has been recorded, especially in Asia. In the Adyar estuary, south India, mass mortality was attributed to high levels of Cr, with histology of gills and liver of the flathead grey mullet, *Mugil cephalus*, indicating cellular necrosis, epithelial lifting, hyperplasia, oedema, mucous cell proliferation in the gills, cytoplasmic vacuolation of hepatocytes, and liver degeneration (Raja et al. 2019).

Despite mangroves being a taxonomically diverse group, there are similar patterns of bioaccumulation among species (Kulkarni et al. 2018). Patterns of metal accumulation are similar for all metals regardless of whether species are salt secreters or non-secreters. Metals accumulate in roots to concentrations like those in the soil with root bioconcentration factors (BCF: ratio of root metal to sediment metal concentration) of <1. Roots BCFs are constant across all metals, while metal concentrations in leaves are normally half that of roots or lower. Essential metals such as Cu and Zn show greater mobility than non-essential metals, such as Cd and Pb.

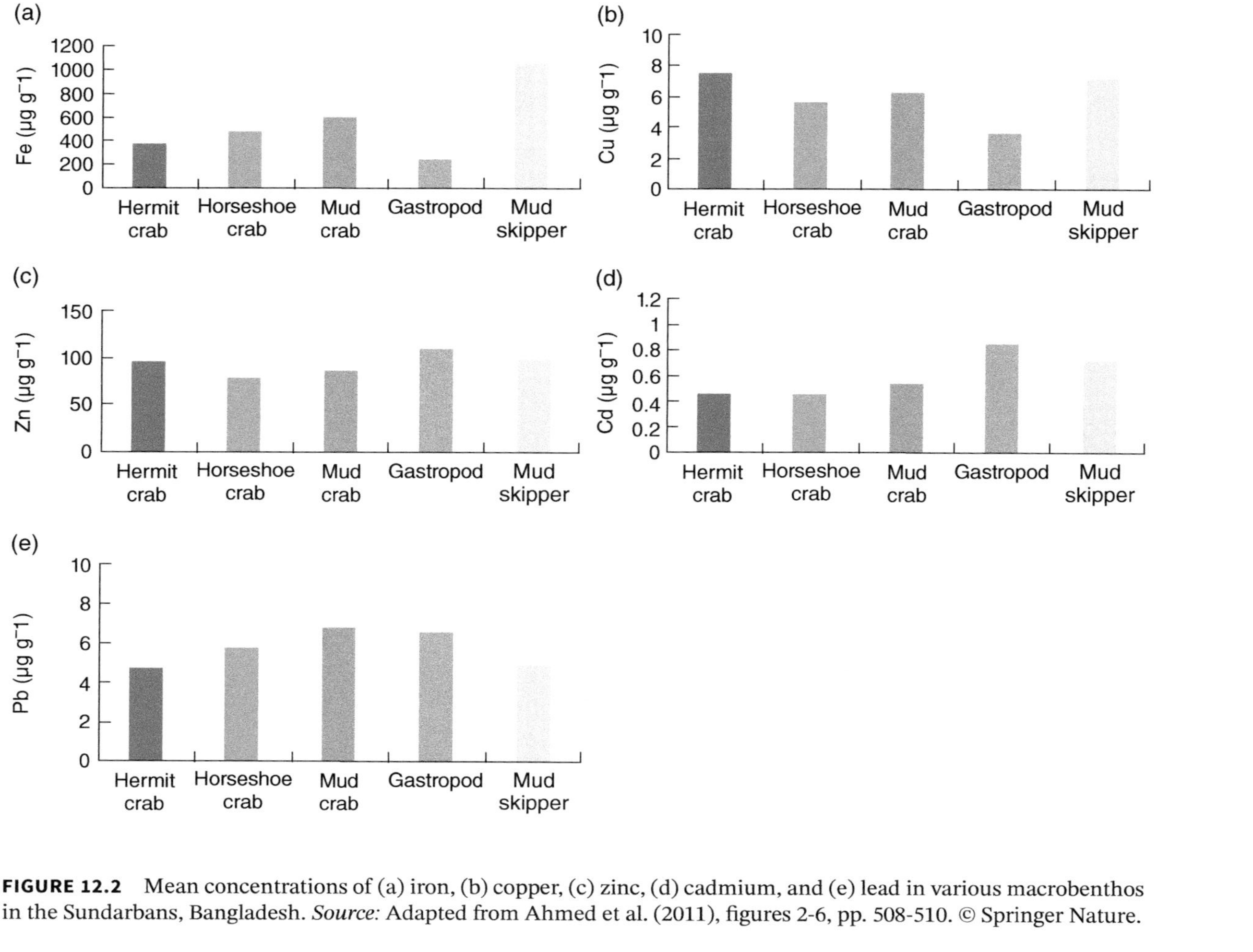

FIGURE 12.2 Mean concentrations of (a) iron, (b) copper, (c) zinc, (d) cadmium, and (e) lead in various macrobenthos in the Sundarbans, Bangladesh. *Source:* Adapted from Ahmed et al. (2011), figures 2-6, pp. 508-510. © Springer Nature.

Leaf BCFs decrease for essential metals as environmental concentrations increase; non-essential metals such as Pb are normally excluded from leaf tissue regardless of environmental concentrations. Thus, mangroves tend to operate as excluders for non-essential metals and regulators of essential metals. Mangroves are a sink for metals as shown by metal budgets in mangrove forests showing little export of metals in litter or decomposition processes (Ramos e Silva et al. 2006). Further, the roots of many mangrove species have Fe plaques on their external surface that help to regulate or exclude various essential and non-essential metals (Machado et al. 2005).

Corals and reef-associated biota are less tolerant than mangrove and seagrass communities to metal contamination, exhibiting both lethal and sublethal effects. Metal concentrations in coral tissues can be especially elevated on coastal reefs close to ports and other industrial developments. For example, heavy metals in hermatypic coral tissues from the industrialised Bocas del Toro Archipelago, Panama, were elevated compared to those in adjacent sediments and metal concentrations in the coral tissue varied by location and species. In the Caribbean, the highest metal concentrations have been measured in *Porites furcata* tissues, with Cu and Hg concentrations significantly higher in *P. furcata* than in a neighbouring species, *Agaricia tenuifolia* (Berry et al. 2013); *P. furcata* has a higher affinity for metal storage and accumulation. Hard coral cover was lowest at reefs in closest proximity to port facilities, suggesting that metal pollution from port-related activities was influencing hard coral abundance. Hg levels in coral reefs along the Caribbean coast of Central America (Figure 12.3a) were high (Guzmán and Garcıa 2002), originating from erosion, runoff, flooding, mining, overuse of agrochemicals, industrial waste, ports, and refineries. Widespread distribution suggests that Hg has been carried long distances within the region as it has been measured in high concentrations even in pristine reefs a considerable distance from the coast. A survey of heavy metals in the skeleton of the coral, *Siderastrea siderea,* and in reef sediments along the Caribbean coast of Costa Rica and Panama indicates high levels of metal pollution (Figure 12.3b). Central American coastal areas are exposed to a larger range of metal pollution than ever before due to increasing contamination from sewage discharges, oil spills from refineries and tankers, the misuse of agrochemicals and fertilizers and topsoil erosion (Guzmán and Garcıa 2002).

The uptake and partitioning of metals in corals, including coral gametes, larvae, and symbionts, involve the interactive effects of variables such as temperature and the machinery of the coral microbiome (Kuzminov et al. 2013; Metian et al. 2015; da Silva Fonseca et al. 2019; Gissi et al. 2019). Symbiotic zooxanthellae may play an important role in metal accumulation and regulation as they accumulate metals in greater concentration than coral tissue (Metian et al. 2015). Using various biochemical techniques, Kuzminov et al. (2013) found that the photosynthetic machinery of symbiotic dinoflagellates was impaired by heavy metals. When the coral, *Acropora muricata,* was exposed to Ni and Cu separately, bleaching was observed when the coral was exposed to high concentrations, but after 96 hours, significant discolouration was only observed with high Ni concentrations with no changes in the composition of their microbiome. In contrast, exposure to Cu not only resulted in bleaching but also altered the composition of both the eukaryote and bacterial communities of the coral's microbiomes (Gissi et al. 2019). Under the combined effects of increasing temperature and Cu exposure, the coral, *Mussismilia harttii,* showed temporary inhibitory effects on energy metabolism enzymes, while prolonged exposure led to strong deleterious effects on coral metabolism due to the combination of stressors (da Silva Fonseca et al. 2019).

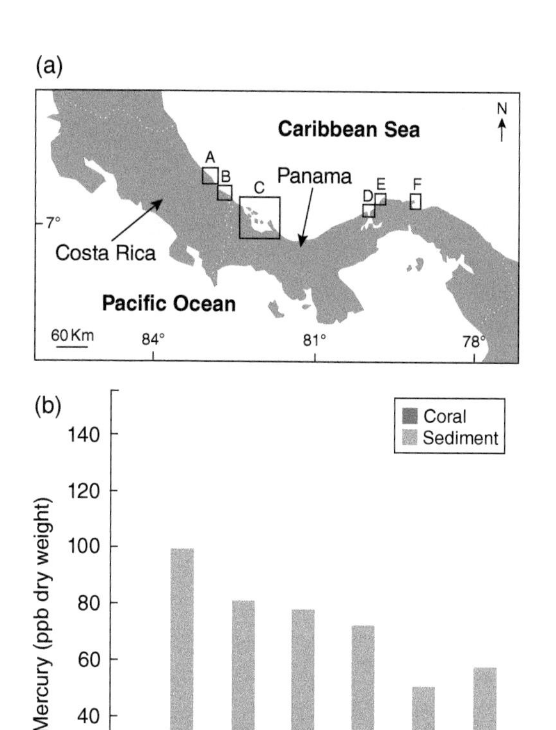

FIGURE 12.3 (a) Map of Costa Rica and Panama showing study sites and (b) mean mercury concentrations in reef sediments and in the skeleton of the coral *Siderastrea sidereal* grouped by study sites along the Caribbean coast of Costa Rica and Panama. *Source:* Adapted from Guzmán and Garcıa (2002). © John Wiley & Sons.

Coral larvae are impaired by heavy metal concentrations (Reichelt-Brushett and Harrison 2005), implying that fertilization success of corals can be substantially reduced by metals. Reichelt-Brushett and Harrison (2005) studied the effects of various concentrations of Cu, Pb, Zn, Cd, and Ni on fertilization success of gametes of the corals, *Goniastrea aspera, G. retiformis, Acropora tenuis, and A. longicyathus.* Pb was much less toxic than Cu for most species, and fertilization responses to Zn and Ni were variable. A significant reduction in fertilization success was observed only at high Cd concentrations. Some trace metals impaired the fertilization success of gametes from faviid and acroporiid corals. The use of antifouling paints containing Cu and TBT has a similar effect on fertilization and larval metamorphosis in corals (Inoue et al. 2004).

Several studies examining *Porites* coral cores have established the history and nature of metal contamination of corals (Ramos et al. 2004; Al-Rousan et al. 2007; Edinger et al. 2008; Tanaka et al. 2013; Chen et al. 2015). Ramos et al. (2004) detected high Pb/Ca ratios in coral cores from a coral reef from an urbanised area in Okinawa, Japan, whereas a similar contaminant history was revealed for cores from reefs in the Jordanian Gulf of Aqaba (Al-Rousan et al. 2007). In the latter study, there was a 35 year record of contamination from a suite of trace metals; most metals showed a dramatic increase in growth band sections younger than 1965 suggesting extensive contamination of the coastal area since the mid-1960s (Figure 12.4). These data represent the beginning of a period that witnessed increasing coastal activities, construction and urbanization and a concomitant decline in coral growth. Mining activities also have been recorded in *Porites* cores (Edinger et al. 2008). A two-century record of concentrations of Al, Zn, and Pb was reconstructed from a massive *Porites* coral skeleton from Hong Kong to evaluate the impacts of anthropogenic activity on corals and the marine environment. Zn:Ca and Pb:Ca ratios fluctuated synchronously from the early nineteenth century to the present indicating the influence of industrialization, whereas the Al:Ca ratio was elevated from 1900 to 1950, reflecting the effects of land reclamation, mining, and ship-building activities. Pb:Ca levels in Hong Kong, Guandong and Hainan corals imply a continuous supply of Pb-based contamination in southern China. Similarly, *Porites* cores have been used to hindcast contamination history in the Gulf of Thailand (Tanaka et al. 2013). High V:Cd ratios found in the Thai coral records anthropogenic V inputs due to fuel oil pollution in the gulf since the late 1990s. The historical variation in Pb: Ca ratios in the coral skeletons suggests the exposure of anthropogenic Pb was a result of discharge from 1984 to 1998 which has been gradually reduced since Thailand prohibited the use of leaded petrol in the late 1990s.

Coral cores and metal ratios can also reflect oceanographic conditions. Decadal variations in metal concentrations in a *Porites* coral from the South China Sea reflect oceanic and climatic control of metal variations (Chen et al. 2015). The input of Mn

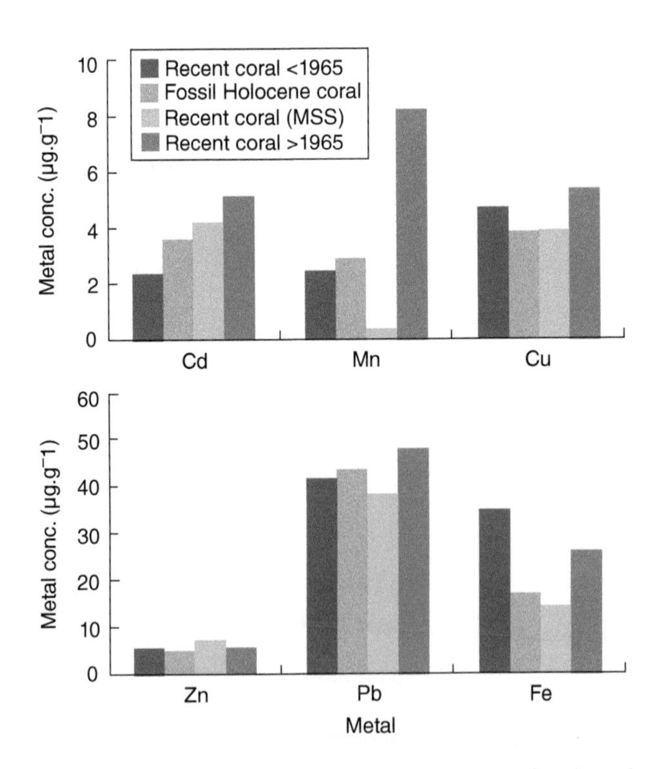

FIGURE 12.4 Comparison of mean heavy metal concentrations in fossil coral sections, recent coral sections (<1965 and >1965), and recent coral used from unimpacted reef (MSS). *Source:* Al-Rousan et al. (2007), figure 6, p. 1919. © Elsevier.

to reef water was partly controlled by the PDO which controls precipitation and river runoff. Surface water concentrations of Cu were controlled by summer upwelling. These results imply that in the South China Sea, ocean-atmosphere climate fluctuations are important factors influencing long-term variability of metals in seawater, by controlling precipitation-related river runoff and the strength of upwelling systems.

Metals, of course, also affect coral reef biota. Many members of the microbial community associated with coral reef sponges are sensitive to free Cu (Webster et al. 2001). Sponges exposed to Cu^{2+} became highly necrosed after two days. The echinoid, *Diadema setosum,* was contaminated by metals on reefs near the port of Singapore, no doubt reflecting port, ships, and other urban activities. A decreasing north–south gradient of metal concentrations was observed in *D. setosum* populations correlating with distance from Singapore (Flammang et al. 1997). Coral reef food webs are negatively impacted by metal pollution, as observed in coral reefs in the SW lagoon of New Caledonia (Briand et al. 2018). The benthic food webs in both the lagoon and on the reefs were contaminated by metals originating from mining activities with Ni, cobalt (Co), Cr, Fe, and Mn preferentially integrated into these food webs. Metals linked to agro-industrial (As, Hg, Zn) and urban (Ag, Cd, Cu, Pb, selenium [Se], V) activities were preferentially accumulated into the reef benthic food web and to a lesser extent into the detrital benthic food web of adjacent seagrasses. Most of the metals bioaccumulated with increasing trophic levels along trophic networks.

Coastal lagoons are some of the most polluted environments in the tropics, and numerous studies have documented a wide range of heavy metal concentrations. In contrast, few studies have examined the effect of metals in lagoonal biota. Bioaccumulation and depuration of Zn and Cd in mangrove oysters have been examined in a contaminated tropical coastal lagoon in Brazil (do Amaral et al. 2005). Transplanting oysters to and from a non-polluted lagoon and a polluted lagoon, do Amaral et al. (2005) found that after three-month exposure to the polluted site, the oyster, *Crassostrea rhizophorae,* transplanted from the non-polluted site accumulated Zn fourfold without reaching the concentration level of resident oysters; Cd concentrations had a slight but significant decrease during the same period. Oysters transplanted to the non-polluted site showed threefold Zn depuration, while Cd had no significant variation.

One of the most polluted environments is Lagos Lagoon in Nigeria. Pb content has been consistently high in edible fish in the lagoon (Ajagbe et al. 2012). Fish species with high level of Pb include Senegal jack, Bobo croaker, Cassava croaker, crayfish, edible mussels, and crabs. The mean level of Pb in biota from Lagos Lagoon were about 8–24 × higher than the WHO daily allowable standards for Pb in sea foods and are unsafe for human consumption.

Several tropical estuaries are polluted with heavy metals derived from industrialisation and other anthropogenic activities (Balachandran et al. 2005; Vineetha et al. 2020). In some monsoonal estuaries, such as the Cochin estuary in India, wet season flooding can flush out significant quantities of heavy metals to the extent that zooplankton and benthos respond favourably to the reduced pollution (Vineetha et al. 2020). A much more extensive literature has catalogued the effects of heavy metals on tropical estuarine fish. Bioaccumulation of toxic metals was examined in 17 species of fish from the Cochin backwaters of India (Nair et al. 2006). Regardless of the tissue type, accumulation was maximal for Zn and minimal for Cd. There was a positive relationship between Zn and Pb in fish tissues which is indicative of the anthropogenic impact in the estuary. Biomarkers of metal toxicity and histology of the mollusc, *P. viridis,* were measured in samples collected from the Ennore estuary and the less polluted Kovalam coast of India (Vasanthi et al. 2012). The results showed marked differences between the two sites as well as significant differences among the tissues. Pb and Cd measured in the soft tissues were in low concentration, but higher concentrations were recorded in adductor muscle and gills for other metals. The metal accumulation in the gills and digestive gland was found to be quite high compared with the adductor muscle.

Among the most toxic metal compounds is methylmercury (CH_3Hg) which has been found in several tropical estuaries. Within Guanabara estuary, Brazil, CH_3Hg, Hg, Se, Cd, Pb, and Cu were detected in the microbial hub composed in this estuary of three size classes of autotrophic and heterotrophic microbes (Kehrig et al. 2009). The trophic transfer of CH_3Hg and Se was observed between trophic levels from prey to predator. The successive amplification of the ratios of CH_3Hg: Hg with increasing trophic levels from seston (43%) to microplankton (59%) and mesoplankton (77%) indicates that biomagnification was occurring within the microbial food web. Se efficiently accumulated by organisms through trophic transference was biomagnified along the food web, while Hg, Cd, Pb, and Cu did not show the same behaviour, suggesting the importance of the role of trophic levels and microorganism size in regulating metal transfers. CH_3Hg and Hg contamination has also been examined in larger organisms in the same estuary as well as in other Brazilian estuaries (Baêta et al. 2006; Seixas et al. 2014a, b). Baêta et al. (2006) examined CH_3Hg and total Hg content in the muscle, liver, and gonad of carnivorous, omnivorous, and iliophagous fish in Guanabara Bay. In all cases, the liver was the preferential organ for Hg bioaccumulation. Total Hg in muscle was higher and more variable in carnivorous than in omnivorous and iliophagous fish. Carnivorous and omnivorous fish presented a similar percentage (97–99%) of CH_3Hg in the muscle. The different feeding habits of the studied species were important for the accumulation of Hg and CH_3Hg, both of which were biomagnified up the food chain. Seixas et al. (2014a) showed that feeding habit influenced the accumulation of Hg, CH_3Hg and Se and that the pelagic food web has higher CH_3Hg biomagnification potential than benthic food webs. Seixas et al. (2014b) found highly significant differences between Hg and CH_3HgCH_3Hg concentrations among microplankton, shrimp, fish, and dolphins. CH_3Hg increased with increasing trophic position, while Hg did not present the same pattern. CH_3Hg was more efficiently biomagnified over the entire food web unlike Hg.

Hg contamination was examined in other Central and South American estuaries. Variations of Hg in sediments and biota from Coatzacoalcos estuary in the Gulf of Mexico indicated highest levels in sediments closest to the industrialised area of the estuary (Ruelas-Inzunza et al. 2009). In most fish tissue, the sequence of Hg concentration was liver > muscle > gills with significant differences in concentrations with season. In Cartagena Bay, a tropical estuary on the north coast of Colombia, Olivero-Verbel et al. (2009) examined Hg concentrations in fish belonging to several trophic levels in the estuary. They found that carnivorous fish presented the highest mean Hg concentrations, followed by omnivorous and detritivorous species.

Metal contamination can be found even in offshore biota. Cd and Hg accumulation was found in the blue swimmer crab, *Portunus pelagicus*, in the northern Bay of Bengal (Karar et al. 2019). The metals concentrated in the hepatopancreas, whereas Cu concentrated in the gill. All metal concentrations were higher in the monsoon season than in the non-monsoon, implying that metals were exported via river runoff to offshore waters.

12.4 Eutrophication

Eutrophication is the oversupply of inorganic and organic nutrients to coastal waters, especially coastal inlets and bays, estuaries, and lagoons, resulting in enhanced primary production and often resulting in oxygen depletion, causing severe impacts on the biota. Overloading of PO_4^{3-}, NO_3^-, NH_4^+, and organic wastes is the main cause of eutrophication. These nutrients are derived from sewage, fertilizers, livestock waste, and fossil fuel emissions. Perhaps more than any other type of coastal and marine pollution problem, eutrophication has an immediate and demonstrative impact on ecosystems. The most heavily affected areas are estuaries, bays, and lagoons across tropical latitudes as nearly all such water bodies have been affected at one time or another by eutrophication.

Enhanced growth of primary producers responding to the increased nutrient supply initially leads to an alteration of species composition and, secondarily, to toxic or nuisance blooms and behavioural effects on organisms such as plankton and benthic vertebrates and invertebrates, including fish. Rates of autotrophic production then exceed the rate of consumption, resulting in the settling of the excess organic material, the decomposition of which leads to extreme effects such as oxygen depletion (hypoxia and anoxia), build-up of toxins (e.g. sulphides), smothering, and mortality of benthic and pelagic species. Ultimately, anaerobic conditions develop in normally oxidised surface sediments and overlying waters, resulting in mass death of many communities and the onset of tolerant, low diversity, opportunistic species assemblages.

Many such events are seasonal, accentuated by peak temperatures and a warm climate. Such eutrophic episodes are increasing in both frequency, severity, and areal extent in many coastal areas, which has led to the development of many dead zones. For example, the coasts of India, Southeast Asia, and the Caribbean are becoming increasingly eutrophic. Now and in the coming decades, coastal waters of Asia, Latin America, and Africa will experience such events with increasing frequency and severity. In all these areas, most attention has focused on the most obvious symptoms of coastal eutrophication: toxic algal blooms, enhanced hypoxia and anoxia, and fish kills.

Much less obvious is the encroachment into rivers and their watersheds of subtle increases in inputs of nutrients, sediments, and organic matter caused by deforestation and land-clearing for crops and grazing, accelerated use and loss of fertilizers and pesticides, sewage release and increased domestic and industrial waste discharge. A detailed analysis (Peierls et al. 1991) of nutrient concentrations and fluxes from 42 of the world's major rivers established that the mean annual concentrations of NO_3^- and NO_3^- exports to the coastal ocean are positively related to human population size within the river catchments. This implies that rapid population growth in the tropics will only worsen eutrophication. These results support other evidence that atmospheric deposition, deforestation, and agricultural inputs have significant impacts on the amounts of NO_3^- exported to the coastal zone; from 10 to over 50% of coastal N is derived from wet and dry atmospheric deposition. This atmospheric material is an important source of 'new' N to many coastal ecosystems and is especially critical considering that estuarine and coastal primary producers are generally N-limited and thus sensitive to N enrichment.

While extensive databases exist for eutrophic temperate ecosystems, no such similarly comprehensive data exist for polluted rivers and estuaries in the tropics.

The best studied tropical estuaries are arguably those in India, where very productive systems in proximity to large cities, such as Mumbai, have undergone considerable deterioration since industrialisation and urbanisation (and increases in sewage discharge) began increasing early last century.

Most efforts to investigate the impacts of pollution in tropical coastal waters have naturally focused on structural changes in planktonic and benthic communities rather than on ecosystem processes. An exception is the study by Ramaiah et al. (1995) of the estuary draining Mumbai Harbour where pelagic microbial activity was not obviously stimulated by river inputs or organic wastes draining into the estuary but were lower than rates reported for other tropical estuaries. They suggested that the industrial wastes dumped into this system were toxic to life, inhibiting the growth of most microbial assemblages.

The river systems of Asia are among the most polluted on earth, ascribed directly to human population growth and the resultant growth of domestic and industrial waste discharge and agricultural runoff. Untreated sewage is the biggest problem as nearly all medium and large Asian rivers are organically polluted. In addition to organic pollutants, two other major threats to Asian rivers are degradation of the watershed and drainage basins, due to deforestation and overgrazing by livestock, and uncoordinated river and drainage basin control and regulation. The first problem leads directly to sediment loss, siltation, and excessive flooding during the monsoon season. As for the second problem, dams and other massive projects have been widely constructed for decades, but the scale and number of such projects have been increasing exponentially in pace with human population growth, greater affluence, industrialisation, urbanisation, and domestication. The clearest example of such problems is the massive projects underway and planned for the Mekong River. These projects are and will further alter tidal regimes and monsoonal flooding, to which many tropical organisms are adapted, including mating and feeding migrations. These problems have led to severe declines in the number of reptiles, river dolphins, macrophytes, fish and invertebrates, and terrestrial animals associated with the watersheds and drainage basins.

Tropical sandy beaches and sand and mud flats, especially in Brazil and India, are subject to eutrophication to the extent that beaches must often be closed due to high faecal coliform counts. Even on intertidal flats, eutrophication can be a serious problem. For instance, Thane Creek on the west coast of India is so polluted that intertidal polychaete worms show poor diversity with only two species being abundant and obviously tolerant to the high degree of pollution (Quadros et al. 2009). Enhancement of total N and organic loading and hypoxic levels of dissolved oxygen is present. Fish kills are frequent in shallow intertidal and subtidal zones in many places where organic pollution occurs and have the additional problems of bacterial disease and harmful blooms of algae, including macroalgae (Glibert et al. 2002). In some polluted bays, a strong CO_2 sink may be enhanced by eutrophication (Cotovicz et al. 2015). In other intertidal flats, the benthic infauna is dominated by surface-dwelling, opportunistic polychaete species that are symptomatic of organically enriched conditions, namely *Capitella capitata*, *Polydora* spp., and *Streblospio gynobranchiata* (Botter-Carvalho et al. 2014). Nearly every system responds differently to eutrophication. Physical boundary conditions play an important role in the manifestation of the ultimate effect of local eutrophication. Also, the transformation and retention of nutrients in estuarine and coastal systems contribute to ecosystem-specific responses. Depending on all these different conditions, site-specific responses to site-specific problems may occur. How eutrophication will affect one estuary or tidal flat may not occur in the same way in another system.

Seagrass meadows and coral reefs are especially vulnerable to eutrophication to the extent that it is one of the major causes of their decline (Burkholder et al. 2007). For seagrasses, the most common cause of decline is the reduction of light through stimulation of overgrowth by rapidly growing epiphytes and macroalgae (Figure 12.5). Also contributing are direct physiological responses, such as NH_4^+ toxicity and water column NO_3^- inhibition through internal carbon limitation (Burkholder et al. 2007). Indirect and feedback mechanisms are involved, manifested as sudden shifts in seagrass abundance rather than continuous, gradual changes in parallel with rates of increased nutrient over-enrichment. High salinity, high temperature, and low light have been found to exacerbate the adverse effects of eutrophication. Re-suspension of sediment from seagrass loss, increased system respiration and resulting oxygen stress, and depressed advective water exchange from thick macroalgal overgrowth can accelerate seagrass decline.

Also adding to the problems of eutrophication are sediment alterations, such as anoxia with increased sulphide concentrations and internal nutrient loading via enhanced nutrient fluxes from sediments to the overlying water. Loss of herbivores may be critically important through increased hypoxia/anoxia as well as indirect effects of trophic structure. The loss of herbivores is critical as they are important in controlling the overgrowth of epiphytes and macroalgae, as is the shift favouring exotic grazers that out-complete seagrasses for space. Natural seagrass populations shifts are disrupted, slowed, or indefinitely blocked by eutrophication. It is not yet possible to develop a conceptual model of the effects of eutrophication on seagrasses owing to the lack of long-term data that is consistent in quality.

Decline of seagrass by eutrophication nonetheless is a definitive process as there is sufficient data to quantify the effects of nutrient over-enrichment. Recovery is not as clear a process, and there are few examples of seagrass meadow recovery following nutrient reduction. One of the best documented recoveries was in Tampa Bay, Florida, where estimates of localized seagrass decline in parts of Tampa Bay exceeded 90% between 1948 and 1982. Improved water clarity and lower phytoplankton biomass as well as a 50% reduction in N loading in the bay resulted in recovery of more

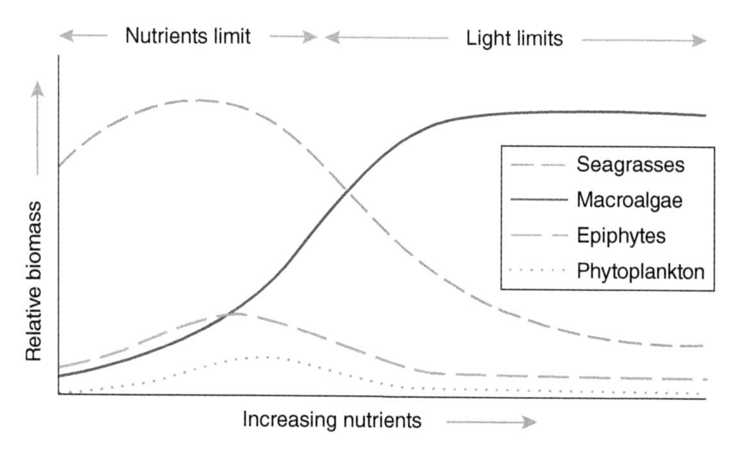

FIGURE 12.5 Changes in the biomass of various primary producers in shallow coastal ecosystems with increasing nutrient concentrations. Under oligotrophic conditions, nutrients limit primary production. As eutrophication progresses, light becomes the primary limiting factor. While seagrasses decline with increasing eutrophication, macroalgae increase. *Source:* Burkholder et al. (2007), figure 1, p. 48. © Elsevier.

extensive seagrass coverage (Burkholder et al. 2007). While such local successes do occur, globally, seagrass meadows are declining throughout both the temperate and tropical biomes. In east and Southeast Asia, seagrasses have been declining due to eutrophication with aquaculture effluents being a main factor (Thomsen et al. 2020). An examination of some seagrass meadows in eastern Asia visited in 2008/2009 and revisited after nine-years of effluent exposure found that during that period seagrass aboveground biomass declined by 87% and species diversity decreased with increasing effluent exposure. $\Delta^{15}N$ values of seagrass leaves identified pond effluents as the driver of the observed eutrophication (Thomsen et al. 2020).

A meta-analysis of the relative effects of grazers and nutrients on seagrasses has implicated eutrophication or reductions of epiphyte grazers in large-scale declines of seagrass. Hughes et al. (2004) examined seagrass studies to compare the relative strength of 'top-down' grazer effects and 'bottom-up' nutrient effects on epiphyte biomass and seagrass aboveground growth rate, aboveground and belowground biomass and shoot density. The meta-analysis found that seagrass growth and biomass are limited *in situ* by sediment nutrients, whereas light limitation has been emphasised in the literature to date. Water-column enrichments have strong negative effects on seagrass biomass. Overall, grazing has a positive effect on shoot density, but negligible effects on seagrass biomass and growth rate. The positive effects of epiphyte grazers are comparable in magnitude to the negative impacts of water column nutrient enrichment, suggesting that the two factors should not be considered separately (Hughes et al. 2004).

For coral reefs, four effects of eutrophication have been recognised: (i) weakening of coral skeletons, (ii) stimulation of overgrowth by benthic micro-and macroalgae smothering living corals, (iii) reduction in water clarity by algal (including phytoplankton) blooms, and (iv) settlement and mass death of primary producers, leading to oxygen depletion (Fabricius 2011). Much has been learned of the effects of eutrophication on coral reefs from the ENCORE experiment (Koop et al. 2001) which studied the responses of coral reef organisms and processes to controlled additions of DIN and DIP on an offshore reef in the Great Barrier Reef. Nutrients were added in two pulses. Most processes were unaffected by the initial low-loading nutrient pulse, except for coral reproduction which was affected in all nutrient treatments. In *Acropora longicyathus* and *A. aspera*, fewer successfully developed embryos were formed and in *A. longicyathus* fertilisation rates and lipid levels decreased. In the second, high-loading nutrient pulse, a variety of significant biotic responses occurred. Encrusting algae incorporated virtually none of the added nutrients. Corals and giant clams, organisms containing endosymbiotic zooxanthellae, assimilated dissolved nutrients rapidly and were responsive to added nutrients. Coral mortality became evident with increased nutrient dosage, particularly in *Pocillopora damicornis*. The addition of N stunted coral growth but added P had variable effects. Rates of coral calcification and linear extension increased with added P, but skeletal density was reduced, making coral more susceptible to damage. N treatments resulted in a reduction of all coral larvae, but conversely, P treatments resulted in enhanced larval settlement from brooded species. In N and N + P treatments, recruitment of stomatopods and benthic crustaceans living in coral rubble was reduced, while grazing rates of fish were not affected by nutrient treatments. Microbial communities showed enhanced transformations of N with N_2-fixation significantly increased in P treatments and denitrification increased in all N treatments. Bioerosion and grazing rates were unaffected by nutrient addition. On average, effects were affected by dose level and whether N or P was added; effects were often species-specific. Impacts were generally sublethal and subtle, and the treated reefs at the end of the experiment were visually like the control reefs.

Other small-scale experiments showed profound sublethal and lethal effects of eutrophication on living corals and other coral-associated organisms. Chronic exposure to additional nutrients led to an increase in prevalence and severity of coral disease and bleaching (Thurber et al. 2014). However, the effects were usually species-specific. At the termination of enrichment experiments, *Siderastrea siderea* corals in enrichment plots had a twofold increase in both the prevalence and severity of disease compared with corals in control plots, whereas *Agaricia* spp. exposed to nutrients suffered a 3.5-fold increase in bleaching frequency relative to control corals, providing experimental support for a link between coral bleaching and eutrophication (Thurber et al. 2014). However, one year after the enrichment experiments were terminated, there were no differences in incidence of coral disease or coral bleaching between previously enriched and control treatments.

Enrichment experiments have also found that agents and rates of bioerosion can be altered. In reef zones on Reunion Island in the Indian Ocean receiving high nutrient inputs, the development of encrusting calcareous algae and macroalgae was associated with the lowest grazing and macroboring rates recorded among sites. In contrast, high microboring rates were found in enriched areas in association with high macroalgal cover (Chazottes et al. 2002).

Eutrophication can cause sublethal effects on coral reefs including a change in benthic communities. Coral reefs fringing the coast of the Gulf of Aqaba have experienced a drastic change in coral cover due to changes in nutrient concentrations. From 2010 to 2013, live hard coral cover declined by 12% at a current-sheltered location. Hard coral cover decline was significantly and highly correlated with a substantial increase in turf algae cover, replacing hard corals as the dominant space occupiers. These changes correlated significantly with ambient PO_4^{3-} and NH_4^+ increases since 2010 (Naumann et al. 2015) suggesting a direct link between eutrophication and the replacement of hard corals with turf algae. On a coral reef in the Red Sea subjected to DIN additions over eight-weeks, hard corals, soft corals, and turf algae incorporated fertilizer N by significant increases in $\delta^{15}N$ by 8, 27, and 28%, respectively (Karcher et al. 2020). Organic carbon content increased in sediments (24%) and turf algae (33%) suggesting that only turf algae were limited by N availability and thus benefited most from N addition. Turf algae thus potentially gain a competitive advantage over hard corals.

Nutrients that suppress coral health can come from any number of sources. For instance, nutrient supply from fish facilitated the growth of macroalgae and suppressed corals on a coral reef in the Florida Keys (Burkepile et al. 2013). Fish excretion supplied 25× more N to forereefs than all other biotic and abiotic sources combined. With fish excretion, macroalgal cover increased considerably with herbivore biomass also showing a negative correlation with macroalgal cover suggesting strong interactions of 'top-down' and 'bottom-up' controls (Figure 12.6). The supply of nutrients by fish also showed a negative relationship with the density of juvenile coral density, likely mediated by competition between coral and macroalgae, suggesting that fish excretion may dampen coral recovery following large-scale coral loss (Burkepile et al. 2013).

The effects of eutrophication can be exacerbated by the abundance and activity of herbivores as found on Caribbean coral reefs (McClanahan et al. 2003; Burkepile and Hay 2009; Vermeij et al. 2010). On Glover's Reef off Belize, herbivory negatively affected algal biomass, whereas turf algal cover and the cover of total algae were

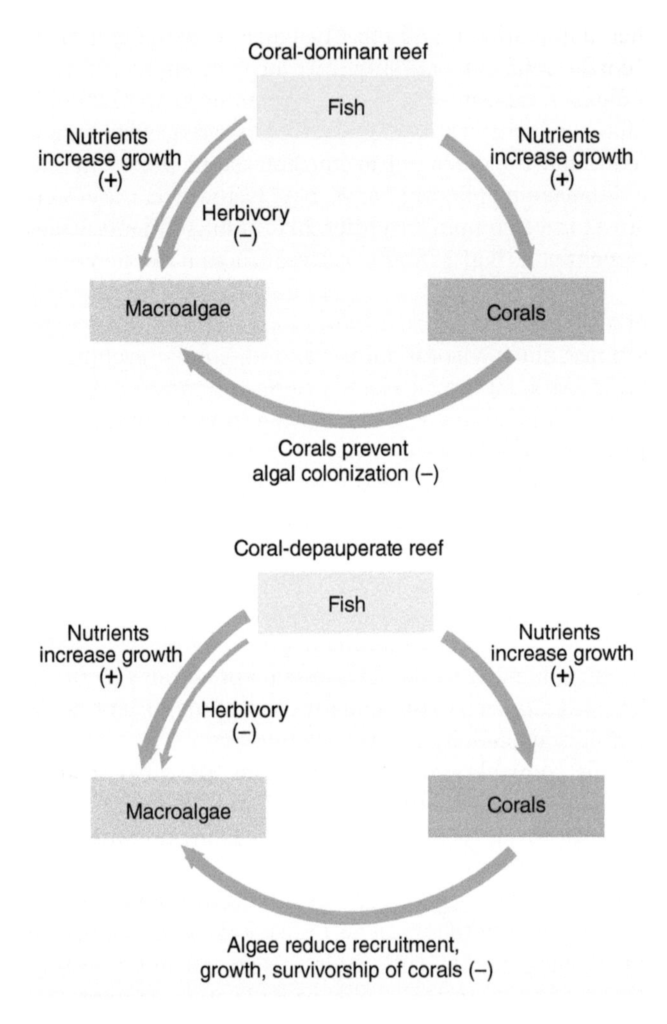

FIGURE 12.6 Conceptual model of the effects of fish excretion on (top) coral-dominant reef and (bottom) coral-depauperate reef. Line thickness represents the strength of the interaction. *Source:* Burkepile et al. (2013), figure 1, p. 2. © Springer Nature.

positively influenced by nutrients and negatively by herbivory. There were more algal taxa and higher dominance in low compared to high herbivory treatments. In contrast, turf algae did best under conditions of low herbivory and high nutrients (McClanahan et al. 2003). On another Caribbean reef, herbivore exclusion increased algal cover and algal biomass but suppressed cover of crustose coralline algae by 80–100% (Burkepile and Hay 2009). Nutrient enrichment had no effect on the cover and biomass of upright algae, but species composition was altered by the suppression of cyanobacteria and the facilitation of red macroalgae in the absence of herbivores. On nutrient-enriched plots, feeding by herbivorous fish increased compared with control plots. Herbivores selectively targeted more nutritional prey, and the data suggest that herbivores suppressed macrophyte accumulation at sites with increased nutrient availability. Vermeij et al. (2010) investigated the effects of increased nutrients on the interaction between the Caribbean coral, *Montastraea annularis,* and turf algae at their boundaries, finding that turf algae caused overgrowth and reduced fitness of the coral. Coral can overgrow turf algae at ambient nutrient concentrations, but turf algae overgrew coral when nutrient levels were increased. Exclusion of herbivores had no effect on the rate that turf algae overgrew the coral, but turf algae always reduced the photochemical efficiency of neighbouring corals, regardless of nutrient or herbivore conditions. Herbivores were not capable of controlling the abundance of turf algae and nutrient enrichment resulted in turf algae having a competitive advantage over corals, with serious implications for the health of Caribbean coral reefs.

The 'Relative Dominance Model' provides a framework to assess the relative dominance of phase shifts among the benthic functional groups on tropical coral reefs, that is, the microalgal turf algae and frondose macroalgae versus reef-building corals and calcareous coralline algae. Experiments and analyses of existing communities tested hypotheses that their relative dominance is mediated by two main controlling factors: nutrients ('bottom-up' control) and herbivory ('top-down' control). Results from several studies showed that reduced nutrients alone do not preclude fleshy algal growth when herbivory is low, and high herbivory alone does not prevent fleshy algal growth when nutrients are elevated. 'Bottom-up' nutrient controls and their interactions stimulate harmful fleshy algal blooms; conversely, elevated nutrients inhibit the growth of beneficial reef-building corals. The results show even further complexity in that nutrients also act directly as either limiting factors or as stimulatory mechanisms as well as functioning indirectly by influencing competitive outcomes (Littler et al. 2006).

Aside from the complexity of herbivory as an additional factor of concern in assessing the effects of eutrophication, overfishing can similarly induce complex effects on coral reefs. Overfishing and land-derived eutrophication are major threats to Red Sea coral reefs and may affect the outcome of benthic community dominance (Jessen et al. 2013). Investigating the combined effects of inorganic nutrient enrichment (simulating eutrophication) and herbivore exclusion (simulating overfishing) on benthic algal development at an offshore patch reef of Saudi Arabia, Jessen et al. (2013) found that nutrient enrichment alone did not affect benthic algal dynamics. In contrast, herbivore exclusion significantly increased algal dry mass up to 300-fold and in conjunction with nutrient enrichment, this total increased to 500-fold. Exclusion of herbivores significantly increased the relative abundance of filamentous algae on light-exposed tiles and reduced crustose coralline algae and non-coralline red crusts on shaded tiles. These results suggest that herbivore reduction, particularly with nutrient enrichment, favours non-calcifying filamentous algae growth with high biomass production, thoroughly out completing the encrusting algae that dominates in undisturbed conditions. And, the healthy reef may experience rapid shifts in benthic community composition if overfishing is not controlled. On a coral reef in the Gulf of Thailand, simulated overfishing and eutrophication using herbivore exclusion cages and slow-release fertiliser found that simulated eutrophication did not significantly alter response parameters, but simulated overfishing positively affected dry mass, turf algal height and fleshy macroalgae (Stuhldreier et al. 2015). Settlement of crustose coralline algae decreased, while abundances of ascidians increased compared with controls, suggesting that herbivory is a controlling factor on the benthic community.

The competitive responses of corals to eutrophication in the face of herbivory and overfishing are species-specific, with some coral species being more affected than others. In the Spermonde Archipelago in Indonesia, Sawall et al. (2011) assessed the nutritional status and metabolism of the hermatypic coral, *Stylophora subseriata*, along an eutrophication gradient. Coral fragments were incubated in light and dark chambers to measure photosynthesis, respiration, and calcification along the gradient. There was a weak correlation between calcification and respiration and photosynthesis and a lack of metabolic stress. The latter suggests that part of the energetic gains through auto-and heterotrophy were spent on metabolic expenditures, such as mucus production. Coastal eutrophication is always deleterious to coral reefs, but the results from these experiments showed that the effect on corals may not always to negative, at least for this common species. Similarly, Ferrier-Pagés et al. (2000) found that under nutrient-enriched conditions, photosynthesis rates of *S. pistillata* were often higher during enriched conditions than during nutrient-poor periods.

In contrast to seagrasses and coral reefs, mangroves are tolerant to eutrophication to the extent that they are often used as biofilters for domestic sewage and aquaculture wastes (Gautier et al. 2001; McKinnon et al. 2002; Costanzo et al. 2004; Lambs et al. 2011; Shiau et al. 2016). In a mixed shrimp pond-mangrove system on the Caribbean coast of Colombia, shrimp farm effluents were partially recirculated through a mangrove forest; the mangroves efficiently eliminated suspended solids from the effluent, but the dissolved nutrients were higher in the mangroves partly due to the presence of a large marine bird community (Gautier et al. 2001). In north Queensland, Australia, small-scale discharges in mangrove creeks receiving shrimp pond effluent did not elevate dissolved nutrient concentrations but did elevate concentrations of particulate nutrients, chlorophyll, and suspended solids proximal to the site of discharge (McKinnon et al. 2002). Rates of primary and bacterial production downstream from the discharge site exceeded rates in the shrimp ponds because of the synergistic effects of turbulent mixing and eutrophication. In the lower reaches of the creeks and immediately offshore, standing stocks of particulate material and rates of primary and bacterial production were within the range of values found in non-discharge areas, suggesting that pelagic biological processes strip nutrients from shrimp farm effluent. Discharge of C and N during shrimp harvest periods did not cause eutrophication further downstream probably due to a combination of intensive tidal flushing and biological nutrient transformations by pelagic microbes and their subsequent grazing by micro-zooplankton and fish. However, elevated $\delta^{15}N$ signatures of mangroves and macroalgae indicated a broader influence of shrimp pond effluent, extending to the lower reaches of the farm discharge creek (Costanzo et al. 2004). Using stable isotopes as a tracer, Lambs et al. (2011) measured an increase in mangrove growth in forests receiving sewage on the island of Mayotte, a small French colony in the Indian Ocean.

In a mangrove estuary on Hainan Island, south China, higher $\delta^{15}N$ values of NH_4^+, NO_3^-, and suspended matter were measured in the pond-covered inner estuary than further upstream indicating a strong influence of untreated pond effluents (Herbeck et al. 2021). Fish and benthic invertebrates of the inner estuary showed an identical pattern, reflecting direct or indirect uptake of ^{15}N-enriched effluents by phytoplankton and benthic algae. Small-size fish were a major part of the artisanal catches from the estuary which were used as feed in the aquaculture operations. As a result, estuarine fish incorporating aquaculture-based food web components were harvested and recycled as feed in the ponds whose effluents sustain the local food web. The $\delta^{15}N$ was at the high end of the global range on all trophic levels indicating an "anthropogenic nitrogen loop" (Figure 12.7) in which some portion of the reactive N initially introduced into the ponds is continuously recycled and affecting the estuarine food web. This loop constitutes a shortcut in the normally inefficient N sink

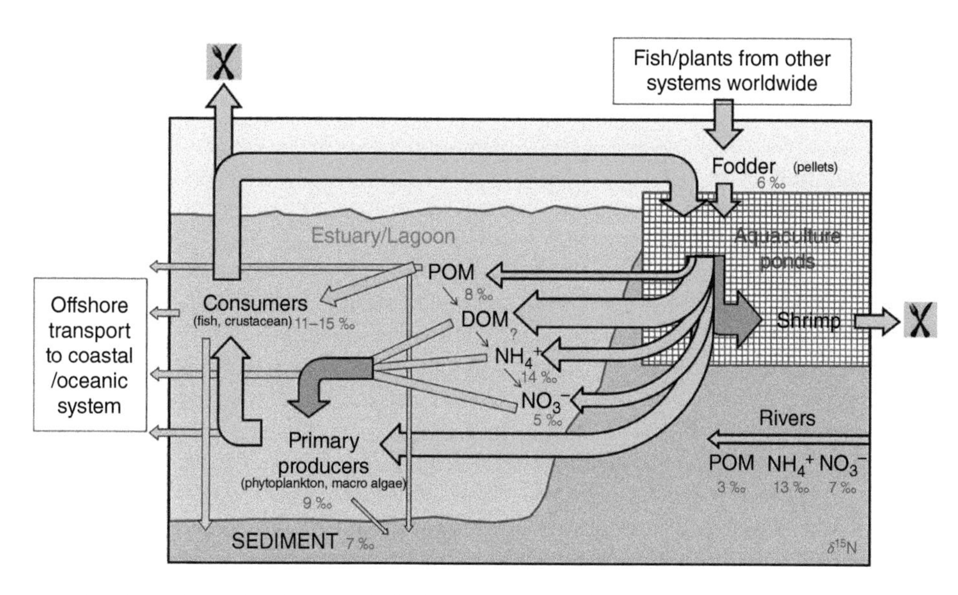

FIGURE 12.7 Flow of shrimp pond-derived nitrogen in the Wenchang/Wenjiao estuary on Hainan Island, China showing a nitrogen loop between estuary and aquaculture. Red numbers below yellow arrows are the percentages of $\delta^{15}N$ accounted for by each process. Width of the yellow arrows indicates the relative quantitative importance of the N flux. Pond effluents have a high $\delta^{15}N$ and enhance the $\delta^{15}N$ of the different components of the estuary. Further fractionation during uptake by primary producers generates a high $\delta^{15}N$ within the estuarine food web. *Source:* Herbeck et al. (2021), figure 10, p. 13. © Elsevier.

function in tropical estuaries. The effect of reactive N from aquaculture facilities may thus be larger than previously assumed (Herbeck et al. 2021).

A lack of large-scale eutrophication was observed in mangrove waterways subjected to discharge from fish cage aquaculture in peninsular Malaysia (Alongi et al. 2003). Concentrations of dissolved inorganic and particulate nutrients were usually greater in cage versus adjacent, non-cage waters, although most variability in water-column chemistry related to water depth and tides. There were few consistent differences in plankton abundance, production or respiration between cage and non-cage sites. Rates of primary production were low compared to rates of pelagic mineralisation reflecting high suspended loads coupled with large inputs of organic matter from mangrove forests, fishing villages, fish cages, pig farms, and other industries within the catchment. There was no large-scale eutrophication due to the cages. An estimate of the contribution of fish cage inputs to the mangrove estuaries showed that fish cages contributed only about 2% C but greater percentages of N (32–36%) and P (83–99%) to these waters relative to phytoplankton and mangrove inputs. Isolating and detecting impacts of cage culture in such heavily used waterways typical of most mangrove estuaries in Southeast Asia are constrained by a background of large, highly variable fluxes of organic material derived from extensive mangrove forests and other human activities.

Mangroves remove N not only by taking up DIN but also by denitrification. Shiau et al. (2016) investigated the effects of temperature, salinity, NO_3^-, and organic carbon on denitrification enzyme activity (DEA). Potential DEA was lowest in soil collected from the site with lowest amount of soil organic C and total N. DEA was enhanced by increased soil temperature and negatively correlated with salinity. Potential DEA was stimulated by labile organic carbon, such as glucose and sucrose, but was not affected by lactose, acetate, or mannitol (Figure 12.8). This study showed that mangrove forest soils can reach maximum rate of treating NO_3^- from sewage effluent if the inflow concentrations are <3 mM. The treatment efficiency of NO_3^- appears to vary depending on availability of labile organic C, soil redox, and temperature.

Benthic infauna, including meiofauna, can be affected by eutrophication. Behavioural responses of the mangrove fiddler crabs, *Uca annulipes* and *U. inversa,* were modified by urban sewage loading (Bartolini et al. 2009). During their activity period, crabs inside contaminated mesocosms seemed to satisfy their feeding demand faster than those in control mesocosms, spending a significantly longer time on courtship and territorial defence. Such reduced foraging depresses their sediment bioturbation activities.

Other crabs, however, contribute in other ways to dealing with the effects of eutrophication. The tidal flat mangrove crab, *Helice formosensis,* extensively reworks sediments in tidal flats associated with mangroves, having a major impact on sediment properties, biogeochemical cycling, and pollutant redistribution. Mchenga and Tsuchiya (2008) compared sediments that were and were not subjected to the activities of the crab. Agricultural and domestic waste discharge induced the growth of green macroalgae and bacteria on the tidal flat. Sediments without crabs exhibited more than twice the NH_4^+ release than sediments with crabs. A significant correlation existed between NO_2^- and NO_3^- concentrations in crab burrow sediments suggesting that the burrow wall provides ideal conditions for denitrification. At a Brazilian site, the fiddler crab, *Uca rapax,* was negatively affected by eutrophication with lowest production at the most polluted location (Costa and Soares-Gomes 2015). Biomass was higher, but a lower P:B ratio found at the more eutrophic site was most likely the result of a high mortality rate and an aged population. In experimental treatments, Penha-Lopes et al. (2009a) tested if fiddler crabs are potentially useful ecosystem engineers in mangroves used as biofilters. They tested the effect of different sewage concentrations and absence or presence of mangrove trees on the survival, bioturbation activities, and burrow morphology of fiddler crabs. They found that crabs inhabiting pristine control conditions achieved higher survival than those living in

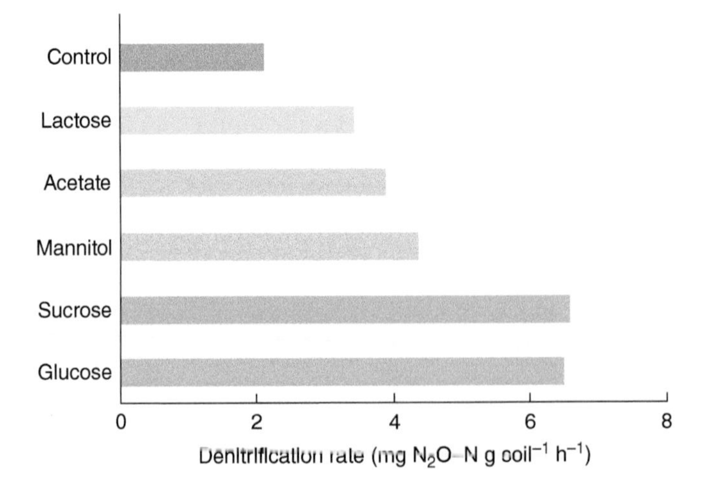

FIGURE 12.8 Mean rates of mangrove soil denitrification simulated by labile organic carbon substrates (lactose, acetate, mannitol, sucrose, and glucose). *Source:* Shiau et al. (2016), figure 6, p. 627. © Springer Nature.

sewage-exposed conditions. At the highest sewage loading, fiddler crabs processed less sediment during feeding and excavated slightly more sediment than in controls. Fiddler crabs had moderate mortality levels in these artificial mangrove wetlands, but mainly in sewage impacted cells. However, they still functioned as ecosystem engineers through bioturbation and burrow construction.

Mangroves can be negatively affected by eutrophication. In the Mahakam delta on Borneo, mangrove foliage is negatively impacted by eutrophication (Fauzi et al. 2014). Anthropogenic disturbances at the Mahakam site resulted in eutrophication, lowering levels of foliar and soil nutrients. Path analysis suggested that increasing soil N reduced soil pH which in turn reduced the levels of foliar and soil base cations in mangroves. In various mangrove forests in the Caribbean and the Indo-Pacific, Lovelock et al. (2009) observed enhanced mangrove mortality due to experimental nutrient enrichment at sites where high soil salinity was coincident with low rainfall and low humidity. Mortality occurred because nutrient enrichment favoured growth of shoots relative to roots, thus enhancing growth rates but increasing vulnerability to environmental stresses that adversely affect plant-water relations. High salinity and low humidity required greater investment in roots to meet the demands for water by the shoots. Thus, the benefits of increased mangrove growth in response to eutrophication can be offset by the costs of decreased resilience due to mortality during drought, with mortality increasing with soil water salinity along gradients from the dry to wet tropics.

Tropical lagoons and estuaries in proximity to large population centres are nearly always eutrophic and a large database exists documenting this eutrophication, particularly resultant harmful algal blooms (Prabhudessai et al. 2019; Viana et al. 2020). Harmful algal blooms and high densities and productivity of phytoplankton are the norm in these environments. The extent of eutrophication is the product not only of the amounts of waste delivered into the system but also the degree of tidal flushing. Poor flushing results in the worst-case scenarios. The Ebrié Lagoon on the Ivory Coast and poorly flushed lagoons in Taiwan offer the best documented cases of coastal eutrophication in the tropics. Tapong Bay, a tropical lagoon in SW Taiwan with only one inlet and poor flushing is an important site for oyster culture. The bay is also surrounded by aquaculture ponds producing fish and shrimps and receives waste discharges from two mangrove-lined creeks which drain the surrounding ponds, providing nutrient-rich water to the bay (Lin et al. 2006). ECOPATH and network analysis indicated that phytoplankton and periphyton were the most-influential living compartments in the ecosystem. Comparative analyses with the eutrophic Chiku Lagoon showed that high nutrient loadings stimulated the growth and accumulation of phytoplankton and periphyton and therefore supported high fishery yields. However, net primary production, total biomass, fishery yields per unit area, and mean transfer efficiency of Tapong Bay were lower than those of Chiku Lagoon. The lower transfer efficiency likely resulted from the low mortality of cultured oysters and invasive bivalves from predation or the lower density of benthic feeders constrained by hypoxic bottom waters due to poor tidal flushing. The hypoxic bottom water reduced the recycling of the detritus entering the food web. In contrast to many estuaries and tropical lagoons, poor flushing of this eutrophic tropical lagoon induced a shift from detritivory to herbivory.

Eutrophication in tropical estuaries, bays, and lagoons affects pelagic and benthic food webs. Bacterioplankton growth rates and productivity are high in such environments. Some environments are so polluted that phytoplankton and bacterioplankton activity are dampened and bottom waters become anoxic or hypoxic leading to N loss via denitrification. For example, in Mumbai Harbour, India, microbial heterotrophic activity is extremely low; low diversity and low rates of ^{14}C uptake by the natural microbial assemblages suggest a general deterioration of biological processes in the inner harbour (Ramaiah et al. 1995). The health of Indian estuaries is linked to monsoonal rainfall and dam-regulated discharge. In the Godavari estuary, with a decrease in rainfall over three-years, reduced water discharge slowed the flushing of the estuary from 1.2 to 6.3-days, respectively. The subsequent stability of the water-column and reduced suspended material load gave rise to intense phytoplankton blooms (Acharyya et al. 2012). In Cochin backwaters, the formation of bottom-water anoxia also resulted in N being lost via denitrification (Martin et al. 2010). The prevalent upwelling in the Arabian Sea brings cool, high saline, oxygen-deficient, and nutrient-rich waters into the bottom waters of the Cochin estuary. High freshwater discharge in the surface layers brings high amounts of nutrients and makes the estuary highly productive. Increased stratification coupled with low ventilation and the presence of high organic matter have resulted in an anoxic condition 2–6 km away from the mouth of the estuary, leading to the production of H_2S. The reduction of NO_3^- and formation of NO_2^- within the oxygen deficient waters indicate strong denitrification intensity in the estuary. The expansion of the oxygen deficient zone, denitrification, and H_2S formation led to a decline in biodiversity and an increase of greenhouse gas emissions from the estuary. The microbial communities of the estuary also showed high uptake rates of NH_4^+ and NO_3^- compared to adjacent coastal waters (Bhavya et al. 2016). NH_4^+ was the preferred substrate for phytoplankton growth illustrating the significance of regenerative processes in primary production.

Eutrophication also influences benthic communities in tropical coastal lagoons and estuaries. In the Cochin backwaters, industrial activities are high in the northern and central part of the estuary, with the central zone characterised by organic enrichment, low species diversity, and increased pollution-tolerant species (Martin et al. 2011). The deterioration of the estuary was caused by a sixfold increase in nutrient and chlorophyll levels during the last few decades. Flow restrictions in the lower estuary led to a fourfold increase in sediment organic carbon over three decades. These changes have caused a reduction in benthic diversity followed by an invasion of opportunistic polychaetes, such as *Capitella capitata,* indicating eutrophic stress in the estuary. In the equally eutrophic Ebrié Lagoon, Ivory Coast, benthic diversity was low in the more polluted areas of the lagoon, but benthic community composition was different not only between lagoonal areas but also between seasons (Kouadio et al. 2011). Four distinct benthic assemblages were identified between which diversity and abundances were significantly different. Dissolved oxygen, mud, and coarse sand were the primary environmental variables most influential in structuring the macroinfauna.

12.5 Pesticides and Industrial Organic Chemicals

Pesticides and other organic chemicals used in agriculture, industrial estates, lawns, and gardens wash into lakes, streams, and rivers and ultimately down into estuaries and out to the coastal zone. These chemicals are designed to kill agricultural pests and weeds on land where they are sprayed and washed into water bodies and soil when it rains; these chemicals kill or produce sublethal effects on estuarine and marine organisms. Chemicals of greatest concern are those that are persistent, that bioaccumulate in organisms, and that are toxic at low concentrations (Beiras 2018). Some of these, such as DDTs, are banned in many countries, but they nevertheless persist in estuarine and marine sediments. They are still used in some countries and continue to run off into aquatic environments. Organochlorine chemicals, mostly DDT-related pesticides, PCBs, and dioxins, have been studied intensively for decades in tropical marine habitats. Dichlorodiphenyltrichloroethane (DDT) is the most powerful pesticide known and can kill hundreds of different kinds of estuarine and marine organisms (Beiras 2018). DDTs and related chemicals are fat-soluble and highly persistent, with low water solubility. Because of these properties, chlorinated hydrocarbons tend to accumulate in sediments and in animal tissues. Related pesticides including aldrin, dieldrin, chlordane, heptachlor, and toxaphene can cause fish kills when applied near water. Chlorinated hydrocarbons remain in the environment for decades, so they continue to be found long after they have been banned. They can be moved by winds and currents far from their site of origin or application. At sea, commercial bottom trawling churns up contaminated coastal and shelf sediments, releasing pollutants so that they can be taken up by animals that bioaccumulate them.

Organisms can take up contaminants from water, sediments, and from their food and may accumulate in tissues at levels much greater than those in the environment. They also biomagnify through food webs, increasing from one step to the next, concentrating in fatty tissues. Biomagnification under field conditions is delineated by trophic magnification factors (TMFs), which are measured as per trophic level change in log-concentration of a chemical. Walters et al. (2016) synthesised > 1500 TMFs to identify organic chemicals predisposed to biomagnify and found that the highest TMFs were for organic compounds that are slowly metabolised by organisms and are moderately hydrophobic. TMFs are more variable in marine than freshwaters, unrelated to latitude and highest in food webs containing endotherms. Each trophic level will have greater concentrations than the level below it, so that the highest concentrations are in the top carnivores, big fish, predatory birds, marine mammals, and humans. Large carnivorous fish may have hazardous levels of contaminants because of biomagnification.

Newer chemicals are less persistent than chlorinated hydrocarbons and generally do not kill fish, but can cause sublethal effects such as endocrine disruption, altered development and behaviour, and reduced growth. These 'second generation' pesticides are short-lived organophosphates and carbamates. However, if spraying coincides with the time of reproduction and early life stages of susceptible organisms, they can cause considerable harm. In aquatic environments, organophosphates break down in a matter of weeks. They were developed from chemical compounds similar to nerve gas, and thus they affect chemicals in the body that are important for the transmission of nerve impulses. They degrade rapidly in the environment, especially in moist soil, and have relatively low toxicity to estuarine organisms. They are usually broken down within a few weeks by water and sunlight and by bacteria in soil and water, but they can affect non-target estuarine organisms before they are completely broken down. Many different pesticides (fungicides, herbicides, and insecticides) are found in water, sediments, and aquatic organisms of estuaries adjacent to agricultural areas, even though they are less persistent that earlier pesticides.

'Third generation' pesticides are more specialised in their toxicity and much less toxic to birds, mammals, and fish. Some of these newer pesticides target the molt cycle of insect larvae by mimicking their specific biology or hormones. Larvicides target the insect's larval stages and are less harmful to non-target organisms and are generally more effective and specific than chemicals that focus on adults. While few impacts have been observed in non-target aquatic organisms, these chemicals might have harmful effects on crustaceans which are closely related to insects.

Toxic effects can be either lethal or sublethal. Effects can be documented in laboratory exposures or observed in organisms in the field. Effects can be studied at various levels of biological organisation, from the molecular to the organism to the population to the community level. Effects on the immune system, endocrine system, nervous system, and reproductive system are critical. Of great concern are chemicals with the ability to cause embryonic malformations, genetic alterations, endocrine disruption, or cancer. Chlorinated hydrocarbons such as DDTs, which biomagnify and take a long time to break down, can prevent marine larvae from developing normally, reduce respiration and metabolism, and impair growth and salt and water balance. Organophosphate and carbamate pesticides, the second-generation pesticides, still have harmful effects.

Polychlorinated biphenyls (PCBs) are chlorinated hydrocarbons that are chemically related to organochlorine pesticides and are more toxic with more chlorine atoms than PCBs with less chlorine. Like chlorinated hydrocarbons, they persist in the environment, have low water solubility, and accumulate in fat. PCBs are suspected of causing cancers and have been linked to male sterility and birth defects. In birds and fish, they decrease egg hatchability, alter behaviour, and decrease immune response. Unlike pesticides, PCBs were never intentionally applied to the environment, but got into the environment through carelessness during their manufacture and use. They can still be released into the environment through poorly maintained hazardous waste sites, illegal or improper dumping of PCB wastes, leaks from electrical transformers containing PCBs, and disposal of PCB-containing products into landfills not designed for hazardous waste. PCBs may also be released into the environment by the burning of some wastes. They can be carried long distances and are found all over the world in areas far away from where they were released. The lighter the PCB in terms of chlorine atoms attached, the further it can be transported. PCBs accumulate in estuarine and marine biota, including plankton and fish, and like chlorinated pesticides and CH_3Hg, they can biomagnify. Thus, organisms higher in the food web have higher concentrations than those lower in the food web.

Dioxins and furans are among the most toxic chemicals to have been released into the marine environment (Beiras 2018). These chemicals are not manufactured but are formed as unintentional by-products of industrial processes that use chlorine, such as chemical and pesticide manufacturing, pulp and paper mills that use chlorine bleach, the production of polyvinyl chloride (PVC) plastics, the production of chlorinated chemicals, and incineration of wastes containing plastics. Dioxin is a contaminant in the herbicide Agent Orange which was used as a defoliant during the Vietnam War in the Mekong delta. Dioxins, like PCBs, are organic molecules with varying numbers and arrangements of chlorine atoms and are toxic to the immune system and to developing embryos. They are known to alter hormones and to cause reproductive problems, liver damage, wasting syndrome, and cancer. Dioxins, like chlorinated hydrocarbons, are persistent in estuarine and marine sediments, where their effects continue long after they were banned. They also biomagnify through food webs, and long-term effects are not well known. Fish embryos are highly susceptible and develop a syndrome that prevents normal development.

Many tropical estuarine and marine environments are contaminated by pesticides, although there is less data on the effects of these contaminants on biota. There is some evidence that pesticides and other organic contaminants dissipate more quickly under the wet and warm conditions in the tropics, with most of the dissipation occurring through hydrolysis in water and biological degradation in water, soil, and sediments (Sanchez-Bayo and Hyne 2011). Pesticides are even found in sea-surface microlayers in addition to sediments and overlying waters (Wurl and Obbard 2005). Biomarkers have been used to monitor this contamination. For example, sequential measurements of hemolymph cholinesterase activities have been used as a non-invasive biomarker of seasonal organophosphate/carbamate exposure in the tropical scallop, *Euvola ziczac* (Owen et al. 2002a). Sampling of hemolymph from these scallops showed acetylcholinesterase and butyrylcholinesterase inhibition. This inhibition did not relate to biochemical or physiological changes associated with gonad maturation and spawning, but rather reflects diffuse contamination of the marine environment by cholinesterase inhibitors or increased bioavailability of such inhibitors.

Sediments and biota of tidal flats have rarely been examined for contamination by pesticides. In benthic fauna from three estuarine mud flats along the Malay Peninsula (Ao Ban Don and Pattani Bay, Thailand and Jeram, Malaysia), organisms from Pattani Bay contained the highest PCB concentrations, particularly in bivalves and shrimp (Everaarts et al. 1991). The bivalves showed significant interspecific differences in their PCB concentrations. In all species studied, the concentrations of the organochlorine pesticides dieldrin, DDTs and its metabolites DDE and DDD, were approximately one order of magnitude higher than those of γ-HCH (Lindane), penta- and hexachlorobenzene. Interspecific differences were found in concentration levels, with crabs showing the highest accumulation of dieldrin and DDE. Contamination is greater in mud flats closer to cities and industrial and agricultural estates.

Seagrass meadows have been examined more frequently for the effects of pesticides, especially herbicides, than tidal flats (Haynes et al. 2000a, b; McMahon et al. 2005; Wahedally et al. 2012; Flores et al. 2013; Negri et al. 2015; Wilkinson et al. 2015, 2017). The impact of the herbicide diuron (3-[3',4'-dichlorophenyl]-1,1-dimethylurea) on the tropical seagrasses, *C. serrulata, H. ovalis,* and *Z. capricorni,* was assessed by Haynes et al. (2000a) in outdoor aquaria. Exposure to diuron resulted in a decline in photosynthetic performance within two hours of exposure. Photosynthetic performance in *H. ovalis* and *Z. capricorni* was significantly depressed at all diuron concentrations after five days of exposure, whereas it was significantly lower in *C. serrulata* exposed to highest diuron concentrations. Depression of photosynthetic performance was present five days after plants were returned to fresh seawater. In the field, diuron concentrations along the Great Barrier Reef coast range from 0.2 to 10.1 $\mu g\,kg^{-1}$, indicating that seagrasses along the coast may be affected by this herbicide (Haynes et al. 2000b). Diuron was detected in sediments of seagrass meadows of Hervey Bay, Australia, where no known photosynthetic stress was detected during low river flow. However, with moderate river flow, nearshore seagrasses are at risk of being exposed to herbicide concentrations that are known to inhibit photosynthesis (McMahon et al. 2005). That diuron is an effective herbicide was confirmed by the work of Wahedally et al. (2012) who found that the tropical seagrass, *T. ciliatum,* was stressed by diuron but not by the herbicides Fusillade (Forte) and 2,4-D amine. Diuron is toxic to *T. ciliatum* at a concentration which can be found in polluted environments.

Toxicity of four Photosystem II (PS II) herbicides to tropical seagrasses was examined by Flores et al. (2013) on the Great Barrier Reef. They found that the seagrasses, *Z. muelleri* and *H. univervis,* were more sensitive to the PS II herbicides diuron, atrazine, hexazinone, and tebuthiuron than corals and tropical microalgae. The PS II herbicides caused rapid decline in photosynthetic performance for both *Z. muelleri* (Figure 12.9a) and *H. univervis* (Figure 12.9b). There is a strong likelihood that seagrass productivity would be greatly affected in the GBR lagoon when PS II herbicides are combined with the negative effects of light limitation from flood plumes. Diuron has lethal and sublethal effects on seagrasses (Negri et al. 2015). *Z. muelleri* and *H. univervis* were exposed to elevated diuron concentrations over 79-days followed by a two-week recovery period in fresh seawater. There was a rapid effect on photosystem II in both seagrass species at the lowest concentration, with significant inhibition of photosynthetic efficiency and inactivation of photosystem II. Significant mortality and reductions in growth were only observed at the highest diuron concentration. During the two-week recovery period, the photosynthetic capacity of the seagrasses improved with only plants from the highest diuron treatment still exhibiting chronic damage. Although tropical seagrasses may survive prolonged exposure to diuron, concentrations > 0.6 $\mu g\,l^{-1}$ cause measurable impacts that may leave them vulnerable to other environmental stressors. Wilkinson et al. (2015) exposed the seagrass, *H. ovalis,* to mixtures of up to ten different PS II herbicides and found that the effects

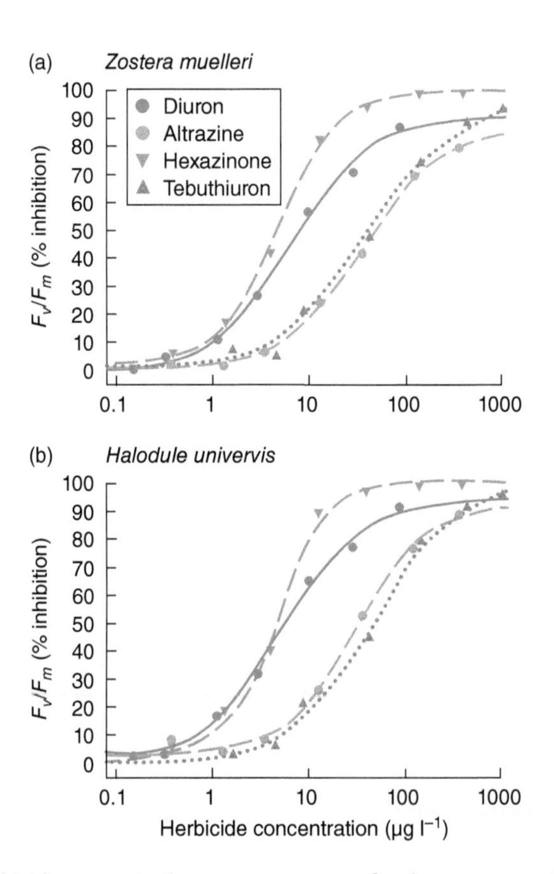

FIGURE 12.9 Herbicide concentration-response curves for the seagrasses, (a) *Zostera muelleri* and (b) *Halodule univervis.* Mean percent inhibition relative to control for maximum potential yields (F_V/F_M). *Source:* Negri et al. (2015), figure 2, p. 6. © Elsevier.

were additive, while individual herbicides exhibited a broad range of toxicities, with inhibition of photosynthetic performance by 50%. In follow-up experiments, Wilkinson et al. (2017) found that the effects of diuron on *H. ovalis* were compounded by temperature. Diuron can increase the sensitivity of seagrasses to thermal stress. The thermal optimum for photosynthetic efficiency in *H. ovalis* was 31 °C, while lower and higher temperatures reduced photosynthetic performance as do elevated concentrations of diuron. Both stressors caused negative responses, and the sum of the responses was greater than that caused by temperature or diuron alone.

Organochlorine pesticides and a variety of other organic contaminants have been detected in mangrove soil, water, and biota, especially in mangrove forests close to urban and industrialised areas (Qiu et al. 2019; Miranda et al. 2021). Information on the effects of pesticides on mangroves is scant, as few studies have demonstrated the impact of pesticides on mangrove growth and survivability. Mangroves are sensitive to root application of Photosystem II-inhibiting herbicides (Bell and Duke 2005). Seedlings of four mangrove species, two salt-excreting (*A. marina* and *A. corniculatum*) and two salt-excluding species (*R. stylosa* and *C. australis*) were treated with a range of concentrations of the herbicides diuron, ametryn, and atrazine. Salt-excreting species were more susceptible to all herbicides than salt-excluding species, indicating that root physiology was a major factor in the uptake of herbicides. Submergence of leaves facilitated herbicide uptake, having serious implications for seedling recruitment under natural conditions. The herbicides were most damaging to least effective in the following order: duiron > ametryn > atrazine. *A. marina* was the most sensitive species to the herbicides. Shete et al. (2009) also found that *A. marina* was sensitive to herbicides, with the overall pattern of accumulation of organochlorine pesticides observed as HCHs > DDTs > endosulfan > aldrin. Leaf and root samples showed better uptake of organochlorine pesticides, whereas lower concentration levels of pesticides were observed in the case of soil samples, with the highest levels of pesticides found in root samples. *A. marina* thus bioaccumulates pesticides.

The best example of the effect of herbicides on mangroves comes from the widespread application of Agent Orange in the Mekong delta during the Vietnam War. The use of Agent Orange, a mixture of phenoxyl herbicides, over the delta led to the decimation of mangrove forests. Arnaud-Haond et al. (2009) studied the genetic recovery of mangroves from the application of Agent Orange on *A. alba* stands. The results showed that genetic diversity of the *A. alba* population is still increasing in the Mekong delta decades since the end of the Vietnam War, but is reaching an asymptotic level. This might be a sign of genetic recovery, but may also reveal a limitation, either of genetic enrichment due to current predominance of auto-recruitment or of demographic increase due to intraspecific competition. Taking a different approach, Phan et al. (2015) used aerial photographs and satellite images to determine land cover changes during the period 1953–2011 in the Mekong delta. The mangrove area declined drastically from approximately 71 345 ha in 1953 to 33 083 ha in 1992 and then rose to 46 712 ha in 2011. Losses due to herbicide attacks during the Vietnam War, overexploitation, and conversion to agriculture and aquaculture encouraged by land management policies are being partially counteracted by natural regeneration and replanting, especially a gradual increase in plantations of mixed mangrove-shrimp pond farming systems.

Persistent organic pollutants have been found in a variety of mangrove biota and to transfer through food webs. For instance, organochlorine pesticides have been detected in soils, subsurface water, sea-surface microlayers, and 24 species of biota at two mangrove habitats in Singapore (Bayen et al. 2005). Data confirmed the ubiquity of persistent organic pollutants including polybrominated diphenyl ethers (PBDEs). Biomagnification was observed among the species collected, with thunder crabs and fish displaying the highest levels of pollutants. Profiles of PBDEs varied among mangrove biota, suggesting that different metabolic pathways exist for flame retardants. Similarly, crabs showed an ability to metabolise chlordane insecticide.

Bioaccumulation and biomagnification occur in other mangrove environments such as in south China (Sun et al. 2015), India (Shete et al. 2009), Brazil (Miranda et al. 2021), and in West and East Africa (Bodin et al. 2011; Oyo-Ita et al. 2014; Sturve et al. 2016). In Mozambique, the use of pesticides is increasing along with the development of agriculture, being delivered to the coastal zone via rivers (Sturve et al. 2016); all species sampled at polluted sites showed inhibited cholinesterase activities indicating that a variety of organisms were affected by pesticides along the coastal zone, including mangrove forests. Organochlorine pesticides have been found in several mollusc species in Senegal, West Africa (Bodin et al. 2011). Higher levels of PCBs and DDTs were observed in the wet season indicating delivery to the coastal zone via river runoff. Exploited molluscs were exposed to the same contaminants as those measured in soils; significantly higher levels of PCBs were measured in soft tissues of bivalves revealing a higher bioaccumulation potential mainly due to the lipophilicity of these compounds. In contrast, the use of pesticides in southeast Nigeria resulted in a serious health risk to humans who eat mangrove biota (Oyo-Ita et al. 2014). Biota-soil accumulation factors for DDTs and HCHs varied among the organisms tested. Food web magnification was found for PCBs, DDTs, PBDEs, and DP (dechlorane plus) for mangrove biota from the Pearl River estuary, south China (Sun et al. 2015).

Different transfer pathways of a persistent organochlorine pesticide, chlordecone, were measured in adjacent mangroves, seagrass beds, and coral reefs on the eastern coast of Basse-Terre in Guadeloupe (Dromard et al. 2018). In mangrove food webs close to the pollution source, lower TMFs indicated that bioconcentration prevailed over bioamplification, whereas the opposite was true in food webs of one of two seagrass habitats and at all coral reefs in which the concentration of the contaminant increased along the food chain, from lower to higher trophic groups. Per- and polyfluoroalkyl substances (PFASs) were found to bioaccumulate in bivalves, crustaceans, and fish in the mangrove food web of the Subaé River estuary, eastern Brazil (Miranda et al. 2021). PFASs have not previously been measured in tropical estuaries, but the Brazilian study suggests that these fluorinated chemicals can bioaccumulate in estuarine organisms via multiple pathways, underscoring their danger to both animals and humans.

Organochlorine pesticides are very toxic to corals (Owen et al. 2002b; Jones and Kerswell 2003; Jones et al. 2003; Jones 2005; Negri et al. 2005, 2011; Cantin et al. 2007; Markey et al. 2007; Ross et al. 2015). The antifouling herbicide irgarol 1051 is prevalent in tropical marine ecosystems and a potent inhibitor of coral photosynthesis; reduction in net photosynthesis of corals was found at concentrations of $100\,\mathrm{ng\,l^{-1}}$ with little or no photosynthesis at concentrations exceeding $1000\,\mathrm{ng\,l^{-1}}$ after two to eight-hour exposure (Owen et al. 2002b). Similar effects were found on the Great Barrier Reef for the herbicides diuron and atrazine (Jones et al. 2003). In detailed experiments, diuron was found to be more toxic than atrazine in short-term toxicity tests. Time-course experiments found the effects of diuron to be rapid and reversible, depending on the dosage (Jones et al. 2003). Doses of both diuron (Figure 12.10a) and atrazine (Figure 12.10b) were toxic for the symbiotic dinoflagellates within the tissues of the corals, *Acropora formosa*, *Montipora digitata*, *Porites*

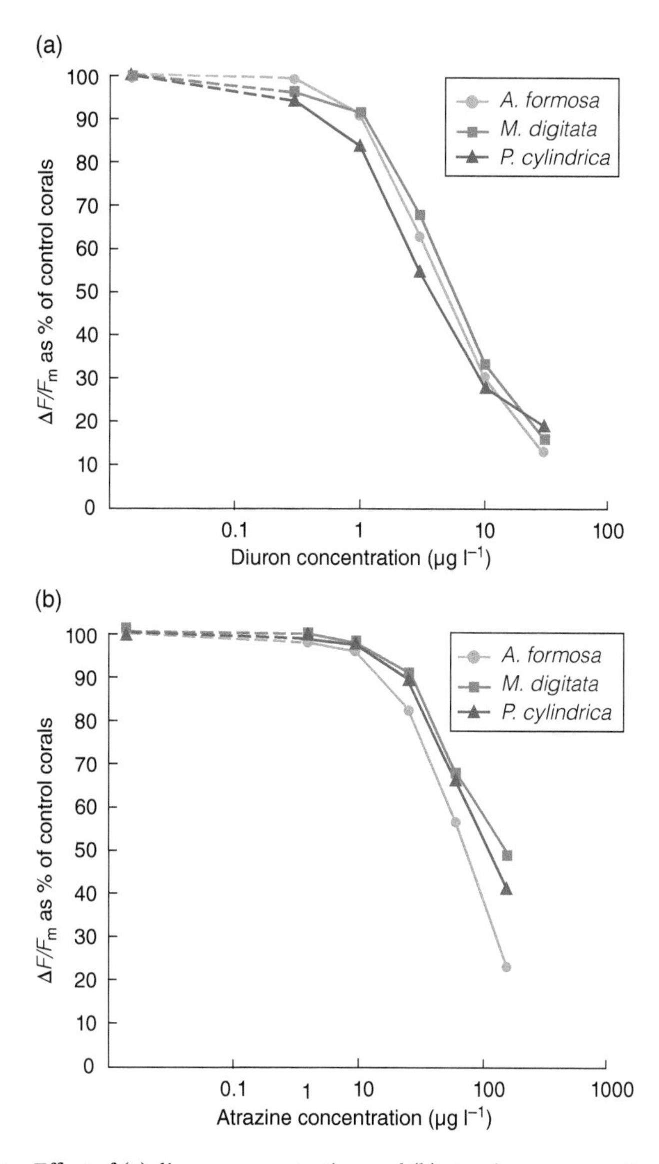

FIGURE 12.10 Effect of (a) diuron concentration and (b) atrazine concentration on symbiotic zooflagellates in the corals *Acropora formosa, Montipora digitata* and *Porites cylindrica*. Effective quantum yield ($\Delta F/F_M$) of symbiotic dinoflagellates as percentage of control corals. All measurements made at the end of a 10 hour light period. *Source:* Adapted from Jones and Kerswell (2003). © John Wiley & Sons.

cylindrica, and *S. pistillata*. At high doses, there was pronounced tissue retraction, causing the corals to pale or bleach.

The effects of herbicides on corals are a function not only of dosage but also the type of herbicide, as absolute and relative toxicities vary depending on the herbicide and its length of exposure. The effects of herbicides varied over two orders of magnitude for the coral, *Seriatopora hystrix* with decreasing toxicity in the following order: irgarol 1051 > ametryn > diuron > hexazinone > atrazine > simazine > tebuthiuron > ionynil (Jones and Kerswell 2003). Recovery was rapid in the diuron-exposed treatments when returned to fresh seawater, but much slower for the irgarol 1051 treatments.

PSII herbicides appear able to penetrate coral tissues and rapidly (within minutes) reduce the photochemical efficiency of the intracellular algal symbionts (Jones 2005). The symbiotic dinoflagellates appear to be as sensitive to PSII herbicides as other phototrophs with photosynthesis being affected at exceptionally low concentrations ($ng\,l^{-1}$ range). At these levels and over short exposure periods, the effects can be fully reversible and vary according to type of herbicide. However, when exposed to higher concentrations in the light or over longer exposure periods, a long-term sustained reduction of the photochemical efficiency of the algae occurs, symptomatic of chronic photoinhibition. This can result in bleaching which is a common but nevertheless significant sublethal stress response. Reliance of corals on an endosymbiotic photoautotrophic energy source, together with a predilection for the symbiosis to dissociate when photosynthesis of the algae is affected, renders coral particularly susceptible to changes in environmental conditions and especially phytotoxins such as PSII herbicides (Jones 2005).

Herbicides can affect the earliest life history stages of coral. The effects of the herbicide diuron on the early life history stages of broadcast spawning and brooding corals were examined by Negri et al. (2005) who found that fertilisation of *A. millepora* and *M. aequituberculata* oocytes was not inhibited at diuron concentrations of up to $1000\,\mu g\,l^{-1}$. Metamorphosis of symbiont-free *A. millepora* larvae was only significantly inhibited at $300\,\mu g\,l^{-1}$. *P. damicornis* larvae which contained symbiotic dinoflagellates were able to undergo metamorphosis after 24 hour exposure to diuron at $1000\,\mu g\,l^{-1}$. Two-week old *P. damicornis* recruits on the other hand were as susceptible to diuron as adult colonies with bleaching evident at $10\,\mu g\,l^{-1}$ diuron after 96 hour exposure. Reversible metamorphosis was observed at high diuron concentrations, with fully bleached polyps escaping from their skeletons. The dark-adapted corals also declined in quantum efficiency indicating chronic photoinhibition and damage to photosystem II in the corals. Cantin et al. (2007) further examined the importance of energy derived from photosynthesis to the gametogenesis of corals following long-term exposures to diuron. Two broadcast spawning corals, *A. tenuis* and *A. valida,* and the brooding coral, *P. damicornis,* were exposed to control, low and moderate diuron treatments prior to spawning or planulation. Diuron caused photoinhibition in all species. *A. valida* and *P. damicornis* were sensitive to chronic diuron-induced photoinhibition, becoming severely bleached. At a moderate concentration, *A. valida* sustained both partial and full colony mortality. *A. tenuis* was more resistant to diuron and neither bleached nor sustained partial mortality. A two and one half to fivefold reduction in total lipid content was measured in all species in the presence of diuron indicating significant use of storage lipid to meet nutritional demands under conditions of chronic photoinhibition. Polyp fecundity was reduced sixfold in *A. valida,* and both *A. valida* and *P. damicornis* were unable to spawn or planulate following long-term exposure to low concentrations of diuron.

Fungicides are a potent inhibitor of the early life stages of corals. In laboratory experiments, the fungicide MEMC (2-methoxyethylmercuric chloride) affected all life-history stages of *A. millepora*; both fertilisation and metamorphosis were inhibited, polyps became withdrawn, and photosynthetic efficiency was slightly reduced at $1\,\mu g\,l^{-1}$ (Markey et al. 2007). At $10\,\mu g\,l^{-1}$ of MEMC, branches bleached and some host tissue died. Runoff from agricultural fields containing fungicides may be having a deleterious effect on inshore corals.

Thermal stress can also play a role in how corals respond to the presence of pesticides. Ross et al. (2015) exposed *P. astreoides* larvae to various concentrations of the pesticides naled and permethrin at seawater temperatures elevated 3.5 °C. The effects of pesticide exposure and elevated temperature were interactive, but overall, there were few sublethal or lethal effects at elevated temperature. Contrary results were found by Negri et al. (2011) who exposed corals and crustose coralline algae to three PSII herbicides at four temperatures. Herbicides increased the negative effects of thermal stress on coral at 31 and 32 °C. The effect of either diuron or atrazine in combination with higher SSTs on chronic photoinhibition was distinctly greater than additive.

Tropical coastal lagoons are persistently polluted by pesticides and other organic contaminants. For example, the sediments, water, and biota of the Altata-Ensenada del Pabellon Lagoon, Mexico, indicated the presence of organochlorine and organophosphorus pesticides (Carvalho et al. 2002a). Concentrations of chlorpyrifos approached acute toxic levels for shrimp in lagoon water, suggesting that drainage from agricultural fields during high runoff may on occasion cause mass mortality of shrimp and fish. Experiments with ^{14}C-labelled chlorpyrifos and parathion showed that they may be stabilised for relatively long time periods through sediment-water partitioning, increasing their potential for impact on the lagoon biota. Similarly, pesticide and PCB residues have been found in the Terminos Lagoon, Mexico (Carvalho et al. 2009), where residues of chlorinated compounds were present in sediments and in biota. The more widespread contaminants were residues of DDTs, PCBs, endosulfan, and lindane.

Several studies point to the pollution of other coastal lagoons in Mexico (Labrada-Martagón et al. 2011; Granados-Galvin et al. 2015). Organochlorine pesticide residues have been discovered in fillets of the snappers, *Lutjanus colorado, L. argentiventris*, and *L. novemfasciatus,* from the Navachiste Lagoon (Granados-Galvin et al. 2015). In two lagoons in the Baja California peninsula, East Pacific green turtles have been contaminated by organochlorine pesticides (Labrada-Martagón et al. 2011). In one lagoon, concentrations of pesticides correlated with the weight of individuals, but clearly in both lagoons, the turtles have been subject to oxidative stress from contaminants.

Other tropical lagoons are polluted by pesticides. In coastal lagoons of Nicaragua, river discharge into the lagoons is the main transport pathway of pesticide residues, whereas atmospheric deposition is likely to be the main pathway for the introduction of PCBs (Carvalho et al. 2002b). There was widespread contamination with soluble organophosphorus compounds and organochlorine pesticides. Concentrations of organochlorine compounds in soft tissues of clams correlated with the concentrations of the same compounds in sediments, indicating that the sediments were the source of contamination to biota.

Pesticides can persist in tropical marine environments for extraordinarily long times. In experiments conducted in large open tanks under different light scenarios and with and without sediments, Mercurio et al. (2016) found that all PSII herbicides persisted under control conditions with half-lives of 300-days for atrazine, 499-days for diuron, 1994-days for hexazinone, and 1766-days for tebuthiuron, while non-PSII herbicides were less persistent at 147-days for metolachlor and 59-days for 2,4-D. Herbicide degradation was 2–10 × more rapid in the presence of a diurnal light cycle and coastal sediments, except for 2,4-D which degraded more slowly in the presence of light. Half-lives were >100 days for the PSII herbicides despite more rapid decomposition observed for most herbicides in the presence of light and sediments, the effects of which were likely due to their influence on microbial community composition and their ability to utilise the herbicides as a carbon source.

The geographical distribution and accumulation features of organochlorine residues in fish in tropical Asia and Oceania indicate that DDTs are the predominantly identified compounds in most locations (Kannin et al. 1995). In general, concentrations of organochlorines in tropical fish are lower than those in fish in temperate regions, with residual levels in tropical fish showing little spatial variability. This is different from the patterns observed for air and water in which higher concentrations occur in tropical latitudes compared to mid-latitudes. The limited data on organochlorine residue levels in fish in tropical Asian waters indicate little temporal variability due to low accumulation levels. A short residence time of semi-volatile organochlorines results in lower levels of residues in tropical fish.

Pesticides have even been found in tropical bays and inner shelf environments. In Paranaguá Bay, southern Brazil, DDTs have been measured in bivalves, fish, and sponges (Liebezeit et al. 2011). The usage of DDTs was widespread around the bay, with DDTs degrading to DDE and DDD within approximately five months. In the Pearl River delta, south China, although a production ban of HCHs and DDTs was imposed in 1983, sedimentary DDT fluxes displayed increasing trends or strong rebounds in the 1990s. Enhanced soil runoff in the process of large-scale land transformation, as well as higher river flows in the early 1990s, mobilised these pesticides from soil to delta sediments (Zhang et al. 2002). In Singapore Strait, 90 hydrophobic organic contaminants were analysed in fish, sediments, and water along with bioaccumulation factors, biota-sediment accumulation factors, organism-environment media fugacity ratios, and TMFs, all indicating that legacy persistent organic pollutants and PBDE 47 showed bioaccumulation behaviour, while triclosan, tonalide, and other pydrophobic chemicals did not (Zhang and Kelly 2018). Legacy use of DDTs and HCH and more recent use of lindane have been observed on the west coast of Unguja Island, Tanzania, in sediments and the pollution-indicating polychaete, *Capitella capitata* (Mwevura et al. 2020); organochlorine pesticides and PAHs were also found to be bioavailable and taken up by the benthic food web via *C. capitata*.

Patterns of distribution indicate how pesticides are distributed across long distances. For example, across an east-to-west transect in the tropical Atlantic Ocean, Lohmann et al. (2012) found that air-water exchange gradients of selected PCBs suggest net volatilisation to the atmosphere. The use of passive samplers also enabled the detection of DDTs and its transformation products across the tropical Atlantic indicating net deposition. There were clear differences between the southern and northern hemispheres in atmospheric concentrations, with much higher concentrations of HCBs in the northern hemisphere. For large swaths of the tropical Atlantic, neither PCBs nor organochlorine pesticide concentrations varied much longitudinally, probably due to efficient mixing by ocean currents. Dissolved concentrations reflect the influence of river plumes and major ocean currents in transport away from the continents.

HCH, DDT, and HCB concentrations differ markedly along the east and west coasts of India, reflecting differing agricultural and urban usage (Sarkar et al. 2008). Measured concentrations of HCHs were lower than DDT residues that might be due to higher water solubility, vapour pressure, and biodegradability. HCH and DDT residues in fish in India are lower than those in temperate countries indicating a lower accumulation in tropical fish, which might be related to rapid volatilisation of these insecticides in the tropical environment. On average, concentrations of pesticides are lower in the tropics than in temperate latitudes and this is consistent with more rapid degradation due to higher temperatures, microbial activity, and rainfall in the tropics.

12.6 Plastics and Other Marine Debris

Plastics are synthetic organic polymers that are manufactured in enormous quantities (up to 250 Mt annually) and are lightweight, strong, durable, and cheap (Peng et al. 2020). It is for these same reasons that plastics are such a serious hazard to marine life and human health when disposed of in the marine environment (Figure 12.11). In sedimentary environments worldwide, median microplastic particle concentrations (particles kg^{-1} DW sediment) are highest in fjords (7000), followed by estuarine environments (300), beaches (200), shallow coastal habitats (200), and continental shelves (50) (Harris 2020).

Microplastics and nanoplastics are especially easy to transfer across different trophic levels of food webs and even across different tissues inside contaminated organisms, causing symptoms such as malnutrition, inflammation, chemical poisoning, growth thwarting, decrease of fecundity, and death due to damages at individual, organ, tissue, cell, and molecular levels (Auta et al. 2017; Peng et al. 2020). Nanoplastics especially have the potential to penetrate different biological barriers, including the gastrointestinal tract and the brain blood barrier and have been detected in organs such as the brain, the circulatory system, and the liver.

The most widely used synthetic plastics are low- and high-density polyethylene (PE), polypropylene (PP), polyvinyl chloride (PVC), polystyrene (PS), and polyethylene terephthalate (PET). These plastics in total represent 90% of the total world production. Most synthetic polymers are buoyant in water. Consequently, substantial quantities of plastic debris that are buoyant enough to float in seawater are transported and eventually washed ashore (Law 2017). The polymers that are denser than seawater (e.g. PVC) tend to settle near the point where they entered the environment. However, they can still be transported considerable distances due to tides and currents. Also, microbial films rapidly develop on submerged plastics, changing their physicochemical properties. If these plastics sink, they settle on the seabed where

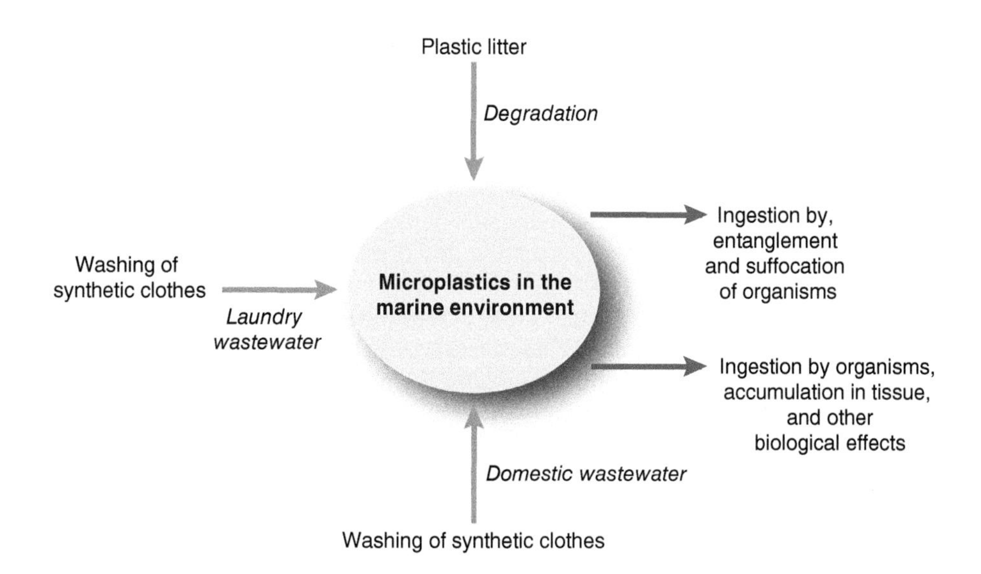

FIGURE 12.11 Sources and pathways of microplastics in the ocean. Source: Auta et al. (2017), figure 1, p. 168. Copyright Elsevier. © Elsevier.

they may persist for decades to possibly centuries. Additionally, hydrophobic pollutants available in seawater may adsorb onto plastic debris; most of these pollutants are persistent, bioaccumulating, and toxic (Law 2017). Thus, they are of concern for human and environmental health. Plastics can thus serve as vehicles for other pollutants to marine biota and humans.

Small plastics enter the environment directly, whereas larger pieces are continuously fragmenting (Law 2017). Microplastics (particles ≤ 5 mm) are directly released to the environment in the form of small (μm) pellets that are used as abrasives in industrial and domestic applications. They can also be released by spilling virgin plastic pellets (mm). Facial cleansers that are used by millions of people contain PS particles (μm) that directly enter sewage and thus adjacent coastal waters. Larger plastics eventually undergo some degradation and subsequent fragmentation, leading to the formation of small pieces. In any event, plastics enter the marine environment in various sizes from microbial-sized particles to large shipping containers.

Degradation is a chemical change that reduces the average molecular weight of polymers, and the most-used polymers (PE and PP) have high molecular weights and are non-biodegradable. However, once in the water, they start to undergo photo-oxidative degradation by UV solar radiation, followed by thermal and/or chemical degradation, rendering the plastic susceptible to further microbial degradation which inevitably weakens the plastic. The material becomes brittle enough to fall apart into smaller fragments when subjected to currents and waves.

Plastics have spread across the entire global ocean (Auta et al. 2017; Law 2017). Microplastics may accumulate in the centres of subtropical gyres (Law et al. 2010), but their means of movement and transport in the ocean are largely unknown. Plastics are thus available to every level of the food web, from primary producers to higher carnivores. Individuals who ingest plastic, especially microplastics, may suffer physical harm, such as internal abrasion and blockage. Impacts are possible at the population and community level but are largely unknown. From rivers alone, 1.2–12.7 Mt of plastic waste enters the ocean annually, with over 74% of this export occurring between May and October. The top 20 polluted rivers, located mostly in Asia, account for 67% of the global total (Lebreton et al. 2017).

Most plastic enters the global ocean from land and secondarily from ships and to a much lesser degree, the atmosphere (Figure 12.11). At least 100 million macroplastic particles and 51 trillion particles of microplastic are currently floating on the surface of the global ocean (Agamuthu et al. 2019). An earlier estimate concluded that a minimum of 5.25 trillion particles weighing 268 940 t is floating in the global ocean (Eriksen et al. 2014). This estimate was derived from 24 expeditions across all five subtropical gyres, coastal Australia, the Bay of Bengal, and the Mediterranean Sea. A tremendous loss of microplastics is observed from the sea surface compared to expected rates of fragmentation, suggesting that there are mechanisms at play that remove the <4.75 mm-sized plastic particles from the ocean surface. Aside from land, the world's fishing fleet alone dumped about 135 400 t of plastic fishing gear and 23 600 t of synthetic packing material into the sea in 1975. Merchant ships dump about 639 000 plastic containers daily; thus, ships are a major source of plastic debris to the global ocean (Derraik 2002). Recreational fishing and boats are also responsible for dumping considerable amounts of marine debris. According to the US Coast Guard, they dispose approximately 52% of all rubbish dumped into American coastal waters (Derraik 2002).

Most debris entering the ocean is plastic, with the remainder being metal, wood, glass, rubber, and paper products. Major inputs of plastic litter from land-based sources in densely populated and industrialised areas are mostly in the form of packaging. Less conspicuous, small plastic pellets and granules are found in large quantities on beaches and are the raw material for the manufacture of plastic products that end up in the marine environment through accidental spillage during transport and handling, not as litter or waste as for other plastics. Plastic pellets can be found in high quantities on remote and non-industrialised islands such as Tonga, Rarotonga, and Fiji. A global inventory of small floating plastic debris found that the accumulated number of microplastic particles ranges from 15 to 51 trillion particles weighing between 93 and 236 thousand, which is only about 1% of global plastic waste estimated to enter the world ocean (Agamuthu et al. 2019).

Plastics fragment in the marine environment, forming micro- and nano-pieces, but no long-term studies have been undertaken to estimate the actual residence time of these fragments (do Sul and Costa 2014). Microplastics have a larger surface area to volume ratio than macro-plastics and are more susceptible to contamination by several airborne pollutants. Because plastics are made of highly hydrophobic materials, the chemical pollutants are concentrated in and onto their surfaces and microplastics act as reservoirs of toxic chemicals in the environment. Weathering significantly changes the superficial characteristics of virgin plastic pellets. Additionally, coloured plastics and different types of polymers may adsorb persistent organic pollutants (POP) from the environment differently. Microplastics transport pollutants over large areas and contaminate the marine biota when ingested. Organisms at every level of the marine food web ingest microplastics, but those inhabiting industrialised areas are exposed to higher amounts and may be more contaminated (Wright et al. 2013). However, the quantities of contaminants vary significantly among fragments; consequently, the toxicity of pollutants and incorporation into bodily tissues vary for each species. Some groups such as holothurians apparently ingest microplastics with specific colours and shapes; if those polymers adsorb higher quantities of pollutants, the consequences are most likely greater. There is no doubt that lower organisms ingest microplastics and attenuate these particles to higher up the food web (Wright et al. 2013).

The geographical spread of microplastics is ubiquitous, but the major areas of congregation are poorly understood (Law 2017). Some large-scale convergence zones of plastic debris have been identified, but there is still the urgency to standardise common methodologies to measure and quantify plastics in seawater and sediments. Temporal trends are poorly understood, geographical distribution and the global cycle of plastic is still illusive, and these shortcomings can have important management implications when defining the origin, possible drifting tracks, and ecological consequences of plastic pollution. Microplastics contaminate all shorelines worldwide, with more material in densely populated areas, but there is no clear relationship between the abundance of microplastics and the mean size-distribution of natural particulates.

An important source of microplastic appears to be sewage contaminated by fibres from washing clothes (Browne et al. 2011). Forensic evaluation of microplastics in sediments shows that the proportions of polyester and acrylic fibres in clothing resemble those found in habitats that receive sewage discharges and sewage effluent. Experiments sampling wastewater from domestic washing machines demonstrate that a single garment can produce > 1900 fibres per wash, suggesting that the washing of clothes can be a major process by which microplastics enter the marine environment (Browne et al. 2011).

Several tropical sandy beaches have been surveyed for debris, and nearly all have reported a preponderance of plastics, including microplastic pellets. Along the NE

Brazilian coast, floating debris leaving polluted coastal bays has accumulated on nearby pristine beaches (Santos et al. 2009). Plastics accounted for 76% of the sampled items, followed by styrofoam (14%). Across 150 km of relatively undeveloped, tropical beaches in Costa do Dendê, Brazil, small plastic fragments resulting from the breakdown of larger pieces were ubiquitous. Because the dominant littoral drift is southward, average beach debris densities along Costa do Dendê are threefold higher than densities observed on beaches north near Salvador City. River dominated and stable beaches have higher debris quantities than unstable, erosional beaches. Areas immediately south of the major regional bays are preferential accumulation sites, indicating that rivers draining populous areas are the major source of debris. In a snapshot of another Brazilian beach, Costa et al. (2010) registered a surface density of 0.3 virgin plastic pellets and plastic fragments cm^{-2} of the beach strandline. The main source of fragments (98%) was attributed to the breaking down of larger plastic items deposited on the beach. In the case of virgin plastic pellets (3%), the main sources are the marine environment and possibly nearby port facilities. An analysis of the riverine contribution of solid waste contamination on an isolated beach in NE Brazil found an exceptionally high level of contamination by plastics of urban origin (Araújo and Costa 2007). The main items found were household and hospital wastes. Surveys have found similarly high densities of plastic debris on sandy beaches and sandy tidal flats in Malaysia, Hawaii, Colombia, Indonesia, Hong Kong, and in the Gulf of Guinea.

Comparison of plastic waste abundance between a recreational beach and fish-landing beaches was conducted to determine if the plastic waste differed by beach usage. In Malaysia, Fauziah et al. (2015) observed that the type of plastic was related to beach function. Recreational beaches accumulated abundant quantities of plastic film, foamed plastic, including polystyrene and plastic fragments, whereas fish-landing beaches accumulated line and foamed plastic. A total of 2542 pieces of small plastic debris were collected from all beaches. In east Malaysia, during the NE monsoon, plastics were the most numerous debris (91%) followed by wood, rubber, glass, metal, and cloth. Twenty-four percent of items collected were objects that were directly associated with marine resources. Out of the 21 major objects identified as marine debris, 87% were composed of ropes, oil bottles, packaging, and cigarette lighters. Wrappers, shopping bags, cardboard cartons, aluminium cans, and clothes contributed a total of 98% of domestic wastes. In Colombia, microplastic pellets were the main form of plastic waste on urban beaches (Acosta-Coley and Olivero-Verbel 2015). Most waste was polyethylene followed by polypropylene, and most particles were white, representing virgin surfaces with little oxidation.

Most plastic pieces contain fractures, horizontal notches, flakes, pits, and grooves ideal for chemical weathering (Po et al. 2020). Some particles are triangular and have highly oxidised surfaces such as those found on the beaches of Hong Kong where it is thought that the triangular shapes may be caused by fragmentation by macrofauna. On the beaches of the Atlantic coast of Colombia, a total of 7597 items were found from 26 beaches surveyed, with vegetation and plastic debris being the primary source of litter corresponding to litter transported by rivers, mainly the Magdalena River, together with human activities related to tourism (Rangel-Buitrago et al. 2017). In sand flat sediments of the Gulf of Guinea, microplastic distribution was heterogenous with a total of 3424 particles m^{-2} found within the drift and high waterlines, mostly in the form of fragments with polyethylene terephthalate as the major polymer type (Benson and Fred-Ahmadu 2020).

Microplastics and other fragments are readily consumed by a variety of estuarine planktonic and benthic organisms (Wang et al. 2016) and may negatively affect protozoan microplankton in the microbial hub by accidental ingestion of microplastic particles during the feeding process (Geng et al. 2021). Distribution patterns of microplastics are like that of the plankton in some tropical estuaries (Lima et al. 2014). In the Goiana estuary, Brazil, the total density of microplastics equates to half of the total fish larvae density and is comparable to fish egg density. Soft and hard plastics, threads, and paint chips were found in the plankton samples, and their origins were probably the river basin, the sea, and fisheries. The highest amount of microplastics was observed during the late rainy season during highest river flow. The distribution of microplastics in the estuary mirrored the distribution of plankton, suggesting that microplastics were readily available for ingestion. In the Goiana estuary, the meeting of the salt wedge and freshwater fronts resulted in the retention of microplastics in the upper and lower estuary most part of the year. During the rainy season, the high freshwater inflow flushes microplastics, together with the biota, seaward. During this season, a microplastic maximum was observed, followed by a fish larvae maximum.

Comparable densities in the water-column increase the chances of interaction between microplastics and fish larvae, including the ingestion of smaller fragments, whose shape and colour are like zooplankton prey (Lima et al. 2014). In a north Brazilian coastal zone, megafauna consumed plastics. A large fish, *Trichiurus lepturus*, the turtle, *Chelonia mydas,* and the dolphins, *Pontoporia blainvillei* and *Sotalia guianensis,* had plastic fragments in their stomachs (Di Beneditto and Awabdi 2014). *T. lepturus* and *S. guianesis* showed the lowest percent frequencies of debris ingestion, followed by the benthic predator, *P. blainvillei*, and the benthic herbivore, *C. mydas*. The debris found in turtle stomachs was opportunistically ingested while feeding on local macroalgal banks. The seabed accumulated more debris than the water-column, and so benthic/demersal feeders were more susceptible to encounters and ingestion. In other tropical Brazilian estuaries and sandy beaches (Dantas et al. 2020), gerreid and urban beach-inhabiting fish were found to ingest plastic fragments and microplastics. The main plastic ingested was blue nylon fragments originating from fishing ropes. The fish may have mistakenly identified nylon fragments as prey items, but there are other possible pathways: (i) from fragments that the fish prey have already ingested, (ii) through ingestion of fragments along with sediment that is sucked in during feeding, and (iii) through ingestion of organisms that have consumed fragments.

Seabirds also ingest plastics. On the southern Great Barrier Reef, the wedge-tailed shearwater, *Ardenna pacifica,* was found ingesting plastics to the extent that 21% of surveyed chicks were fed plastic fragments by their parents (Verlis et al. 2013). The most common colours of ingested plastic fragments were off white and green. This ingestion of plastics has a considerable impact on the nutrition of the birds. At Yongxing Island in the South China Sea, four of nine seabirds and shorebirds ingested plastic debris with microplastics accounting for 93% of the total items (Zhu et al. 2019). Most items were blue, and the primary shape was thread, with most fragments composed of polypropylene-polyethylene copolymer.

Plastic fragments, including microplastics, contain persistent organic pollutants, and these chemicals can have a deleterious effect on biota (Law 2017). The ability of chemicals to transfer from plastics to animals upon ingestion has been clearly demonstrated in laboratory animals (Law 2017). In a study of organics on plastic debris from the open ocean and remote and urban beaches, Hirai et al. (2011) detected PCBs, DDTSs, PBDEs, alkylphenols, and bisphenol A on the fragments. PCBs were most likely derived from legacy pollution, while nonylphenol, bisphenol A, and PBDEs probably came from additives. Plastic debris is a trap for persistent organic pollutants as other studies have found high concentrations of organics linked to plastic fragments. Samples taken from the North Pacific Gyre and selected sites in California, Hawaii and from Guadalupe Island, Mexico, were found to contain significant concentrations of PCBs, DDTSs, and PAHs and aliphatic hydrocarbons (Rios et al. 2007).

Like other tropical environments, tidal flats and mangroves accumulate plastic debris. Mangroves are particularly traps for marine litter (Martin et al. 2019a; Deng et al. 2021; van Bijsterveldt et al. 2021). In an intertidal flat of NE Brazil, heavily weathered plastic fragments of different sizes and shapes (soft and hard plastics, nylon) were found, distributed in space and time, and varying in size from 1 mm to 160 cm^2. The most likely sources were the river basin, the communities that inhabit the margins of the estuary, fishery activities, and the mangrove forest. Nearer to the mangrove forest, there was a larger accumulation of plastic.

Microplastics have also been found in the mangroves of Singapore (Hazimah and Obbard 2013), the Pearl River estuary of south China (Zuo et al. 2020), the north coast of Java, Indonesia (van Bijsterveldt et al. 2021), and in Central and South American forests (Deng et al. 2021). The majority of microplastics were fibrous and <20 μm in the Singaporean mangroves, containing the polymers polyethylene, polypropylene, nylon, and polyvinyl chloride. The presence of microplastics is likely due to the degradation of marine plastic debris accumulating in the mangroves. The abundance of microplastics in mangrove sediments in the Pearl River estuary ranged from 100 to 7900 items kg^{-1} DW, with highest values in mangroves closest to high population centres; microplastics (<500 μm in size) dominated, with polypropylene–polyethylene copolymer, green/black and fibre fragments being the dominant type, colour, and shape. Throughout mangrove forests of southern China, microplastic concentrations ranged from 227 to 2249 items kg^{-1} DW, with highest concentrations near rivers. The predominant shape, colour, and size of microplastics were fibrous, white-transparent, and 500–5000 μm, respectively, with most fragments composed of polypropylene, polyethylene, and polystyrene (Li et al. 2020a). Microplastics have also been observed in Malaysian mangrove waters (Barasarathi et al. 2014) where about 85% of surface water samples contained microplastic particles. Being of the same size spectrum as phytoplankton, the chance of being ingested by marine organisms is high. Plastics may be incorporated in larger species up the food web, such as marine mammals, seabirds, and even humans. Micro-size polystyrene foam and plastic fragments were the most abundant types of microplastics. Macroplastics can result in negative impacts, as field experiments in Javanese mangroves demonstrated that large sheets of plastic waste induce extreme aerial root growth when partially covered and tree death when roots are completely covered (van Bijsterveldt et al. 2021).

Fragmentation may have a seasonality component. An experiment observed the behaviour of selected tagged plastic items deliberately released in different habitats of a tropical mangrove patch in NE Brazil (do Sul et al. 2014). Significant differences were not recorded among seasons, but marine debris retention varied among habitats in relation to characteristics such as hydrodynamics, vegetation density, and tree height. The high intertidal zone retained significantly more items compared with the mid- and low-intertidal zones and creek bank. The balance between items retained and items lost was positive, demonstrating that mangroves are net sinks for plastic. A similar scenario was found in Bootless Bay, Papua New Guinea, where surveys revealed exceptionally high concentrations of plastic debris in mangrove-dominated, depositional areas. The worse affected area returned more than 37 000 items with a combined weight of 889 kg. Plastics comprised by far most debris across all sites (90%). In the São Vincente estuary, Brazil, the predominant litter type in terms of density was plastic and, by weight, wood. The greatest deposition of articles was associated with low mangrove densities, indicating that the presence of obstacles is not critical for retaining floating residues.

With the ubiquity of plastic fragments in mangrove habitats, it is inevitable that organisms will ingest plastics, especially microplastic particles. In tropical mangrove estuaries, there are seasonal and spatial patterns of ingestion. In two estuarine drums (Sciaenidae), plastic fragments were found in 8% of individuals (Dantas et al. 2012). Nylon fragments occurred in 9% of *Stellifer stellifer* and 7% of *Stellifer brasiliensis*. The highest number of nylon fragments ingested was observed in adults during the late rainy season. The ingestion of mostly blue nylon fragments probably occurred during normal feeding activities. During the rainy season, the discharge of freshwater transports nylon fragments to the main channel where they are more available to fish.

Consumption of plastic fragments may occur relative to phases of the moon, as has been found in the Goiana estuary (Lima et al. 2016). The full moon had a positive influence on the abundance of the estuarine fish, *Gobionellus oceanicus, Cynoscion acoupa,* and *Atherinella brasiliensis,* and the new moon on *Ulaema lefroyi*. The full and new moons influenced the number of hard and soft plastic debris available to these species. Hard microplastics were positively associated with different moon phases such that different fish species fed on different plastic debris depending on their synchronicity with the life stages of the fish. Juveniles of the estuarine fish, *Oreochromis mossambicus, Terapon jarbua, ambassis dussumieri,* and *Mugil* sp. inhabiting mangroves in Kwazulu-Natal, South Africa, were found to contain microplastics (52% of fish), mostly blue fibres and fragments (Naidoo et al. 2020). Fish ingested rayon (70%), polyester (10%), nylon (5%), and polyvinylchloride (3%). Similar results were found in the Ciénaga Grande de Santa Marta estuary, Colombia (Calderon et al. 2019) where 12% of fish contained microplastics and in the Talisayan Harbour, East Kalimantan, Indonesia, where individuals of the anchovy, *Stolephorus* spp., contained many particles, mostly in the shape of microfilm ranging in size of 50–500 μm (Ningrum et al. 2019). In the Amazon River estuary, 30% of fish examined contained microplastics in their digestive tracts, mostly pellets ranging in size from 0.38 to 4.16 mm (de Souza e Silva Pegado et al 2018).

Limited data are available on the ingestion of microplastics by mangrove invertebrates. For example, the deposit-feeding mangrove fiddler crab, *Uca rapax,* ingests microplastics and translocates them to their digestive organs including the gills, hepatopancreas, and stomach (Brennecke et al. 2015). North of Beibu Gulf, China, the mangrove sea snail, *Ellobium chinense*, had microplastics in its organs, but there was no relationship between the levels of microplastic in the organs and the levels in the soil or tidal water (Li et al. 2020b). Microplastics in snail organs mainly originate from the mangrove pore water.

Ingestion of microplastics can impact corals in several ways (de Oliveira Soares et al. 2020), including reduced growth, impairment of reproduction, tissue inflammation, decreased calcification, necrosis, and reduction of fitness (Figure 12.12). Hard coral colonies are severely damaged by derelict fishing gear. Diving and clean up surveys in the Florida Keys National Marine Sanctuary (Chiappone et al. 2005) confirm the threat to coral reefs. In the sanctuary, 14 tonnes of derelict fishing gear were removed. Trawl netting was the most frequent debris type encountered (88%) and the greatest debris component by weight (35%), followed by monofilament gillnet

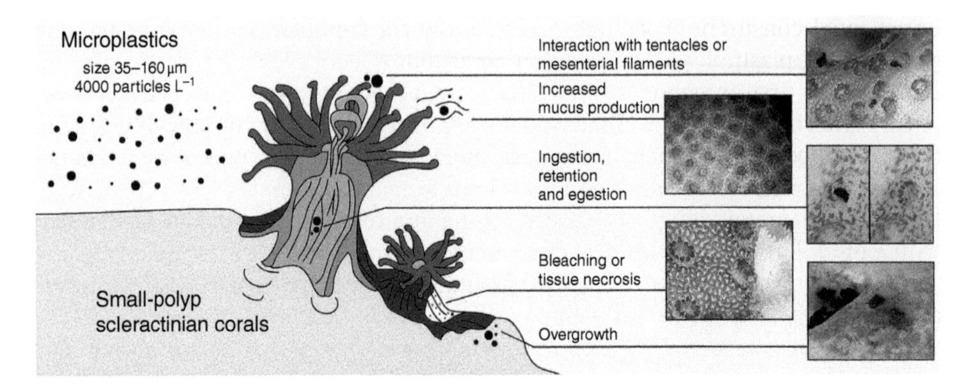

FIGURE 12.12 Schematic indicating pathways and impacts of microplastics on reef building corals. Source: Reichert et al. (2018), graphical abstract. © Elsevier.

(34%) and maritime line (23%). Most debris had light or no fouling, suggesting that the debris was recent. At French Frigate Shoals, Morishige et al. (2007) calculated deposition rates of plastics and investigated relationships among deposition and year, season, ENSO, and La Niña events; 52 442 items were collected with plastic comprising 71% of all items. Debris deposition was significantly greater during El Niño events than during La Niña for reasons not understood. At the Florida Keys National Marine Sanctuary, 63 offshore coral reef and hard-bottom sites were surveyed to quantify the impacts of lost fishing gear to coral reef sessile invertebrates (Chiappone et al. 2005). Lost hook-and-line fishing gear accounted for 87% of all debris and was responsible for partial individual or colony mortality. Branching gorgonians were the most frequently affected, followed by milleporid hydrocorals and sponges. Factors affecting the impacts of lost fishing gear included coral densities and the amount and type of gear lost.

Other types of debris, including plastics, may also have deleterious effects on coral reefs as shown in the Marshall Islands (Richards and Beger 2011) where sections of the Majuro lagoon have the second highest standing stock of macro-debris recorded to date of any benthic marine habitat in the world. Most debris originated from household sources, with peak abundance recorded in areas of medium affluence. Marine debris in the lagoon caused suffocation, shading, tissue abrasion, and mortality of corals; a significant negative correlation was observed between the level of hard coral cover and coverage of the debris. The standing stock of debris will likely persist for centuries given their slow decomposition rates.

With so much plastic available on coral reefs, it is unsurprising that ingestion by coral reef organisms and contamination by persistent organic pollutants is now relatively common. In the Red Sea, three coral species removed plastic particles through suspension-feeding at rates ranging from 0.25×10^3 to 14.8×10^3 microplastic particles polyp^{-1} h^{-1} (Martin et al. 2019b). However, this was only 2% of the total removal rate, with passive removal through adhesion to coral surfaces being 40 × higher than active removal. Passive adhesion of plastic to coral reef surfaces is therefore the major sink for microplastics suspended in the water column. When the coral reef fish, *Pseudochromis fridmani,* was exposed to food-grade plastic, it took up toxic quantities of nonylphenol (NP). Dottybacks held in food-grade polyethylene bags accumulated high body concentrations of NP. Although labelled food grade, these bags leached highly toxic levels of NP (Hamlin et al. 2015). Allen et al. (2017)

investigated consumption of weathered, unfouled, biofouled, pre-production, and microbe-free plastic by a scleractinian coral that relies on chemosensory cues for feeding. The first experiment found that corals ingested many plastic types while mostly ignoring organic-free sand, suggesting that the plastic contained phagostimulants. A second experiment found that corals ingested more plastic that was not covered in a microbial biofilm than plastics that had a biofilm (Figure 12.13). Additionally, corals retained 8% of ingested plastic and retained particles stuck to corals (Allen et al. 2017). Unfouled plastic triggered a stronger feeding response in coral. These experimental results suggest that chemoreception drives plastic consumption in hard corals.

Corals mistake microplastics for prey and consume up to about 50 μg plastic cm^{-2} h^{-1}, rates that are similar to the consumption of plankton (Hall et al. 2015). Ingested microplastics were found wrapped within the gut cavity, indicating that ingestion of high concentrations of microplastics may impair coral health. On pristine coral reefs in the Maldives, 95% of the corals, *Porites lutea, Pocillopora verrucosa*, and *Pavona varians*, were contaminated with five phthalates esters used as plasticiser, which is used to increase flexibility, transparency, and longevity on plastics (Montano et al. 2020). These esters were found at similar concentrations in individuals of the corals regardless of depth or reef exposure, indicating the ubiquity of these compounds in coral reef environments.

Seagrass meadows and waters and offshore habitats also contain substantial amounts of various sized plastics. On remote Turneffe Atoll off Belize, the tropical seagrass, *T. testudinum* was found to accumulate microplastics via its epibiont communities (Goss et al. 2018). Most *Thalassia* blades had encrusted microplastics, with microfibres occurring more than microbeads and chips. Grazers consumed seagrasses with higher densities of epibionts, suggesting that microplastics may in this way transfer up the food web. Potential mechanisms for microplastic accumulation include

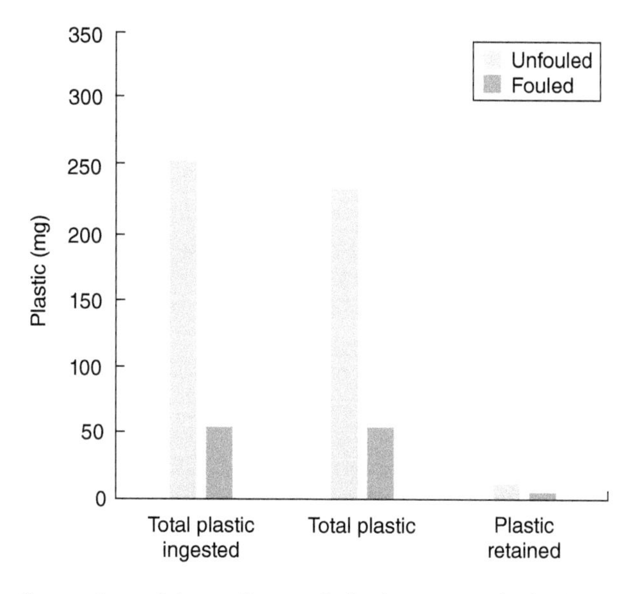

FIGURE 12.13 Comparison of the median total plastic consumed, plastic egested, and plastic retained (mg) by the coral, *Astrangia poculata* in unfouled and fouled plastic treatments per 30 minute trials. *Source:* Allen et al. (2017), figure 4, p. 5. © Elsevier.

entrapment by epibionts or attachment via biofilms. In the open ocean around the Canary Islands, the Atlantic chub mackerel, *Scomber colias*, has been found to ingest microplastics, mostly as fibres. This high incidence may be related to untreated wastewater discharged from the islands (Herrera et al. 2019).

Microplastic ingestion by pelagic and demersal fish of the eastern central Atlantic Ocean off the coast of Ghana occurs in the following order: *Sardinella maderensis* > *Dentex angolensis* > *Sardinella aurita* (Adika et al. 2020). Industrially produced pellets were the most dominant type followed by microbeads, burnt film plastics, and unidentified fragments. The order of frequency of occurrence of microplastics in these species can be linked to their distribution and movement patterns: *S. maderensis* and *D. angolensis* most commonly reside on the continental shelf near polluted coastal waters, whereas *S. aurita* prefers fully marine conditions, moving closer inshore only when the upwelling season begins and thereafter returning in winter to deeper waters. In coastal waters off Kochi, India, peak abundance of microplastics (mostly white and blue fragments) was observed in surface water during the monsoon season (James et al. 2020). Gut content analysis of pelagic and demersal fish species indicated microplastics in *Sardinella longiceps, S. gibbosa, Stolephorus indicus, Rastrelliger kanagurta,* and *Cynoglossus macrostomus*, with polyethylene and polypropylene as the main polymer types.

12.7 Biological Pollution

12.7.1 Sewage and Microbial Diseases

Biological entities are often responsible for the pollution of tropical marine ecosystems, most often in the guise of sewage microbes and invasive species. Microbial pollution comes from sewage that has not been properly treated, and faecal contamination, especially on beaches, is a concern because some microbial assemblages cause disease in marine organisms and humans. In effluent from sewage plants that do not have secondary treatment, high levels of pathogens may be present. Cities that have combined sewer systems with storm water drainage and domestic and industrial waste often discharge untreated sewage during the wet season when the total water volume exceeds the capacity of the system. As many as 20 million people every year become ill from contaminated drinking water containing pathogens from untreated waste that entered the water upstream from potable water sources. Another means of contamination is non-point runoff into coastal waters from animal faeces, livestock operations, or dense concentrations of farm animals. Hepatitus A virus and pathogenic bacteria, such as *Salmonella* spp., *Listeria monocytogenes, Vibrio cholerae,* and *Vibrio parahaemolyticus* commonly occur in tropical coastal waters.

Microbial contamination can build up in marine life (e.g. shellfish) when sewage is released to coastal waters. Bacteria and viruses from humans and animals usually attached to fine particles can affect bathing water quality. Millions of people worldwide become sick each year by swimming in waters containing pathogenic microbes. The problem is increasing, especially in developing countries. In some urban areas, some bacteria have become resistant to antibiotics. The resistant ones include potentially pathogenic strains of *Pseudomonas, Acinetobacter, Proteus*, and *Escherichia.*

What kind of diseases may result from exposure? Microbial contamination by pathogens from sewage or animal waste is not only a major concern for drinking water but is also an issue in coastal waters where swimmers may become ill after wet season runoff along the coastline. When illness is severe, it is often difficult to detect or to attribute a particular case to coastal water pollution. Gastrointestinal and respiratory infections are often associated with recreational water. As such, bathing beaches may be closed after major rain events when elevated bacterial pollution is detected.

Another concern is when shellfish and other eatable marine organisms accumulate pathogenic microbes, causing severe illness in people who eat them. Gastroenteritis and hepatitis A are commonly transmitted through shellfish, although cholera and typhoid fever are often the first to be linked to eating contaminated shellfish. These diseases create insurmountable problems in developing countries which generally do not have the detection facilities or the money to restrict runoff onto contaminated beaches. The incidence of illness is therefore greater in developing countries than in developed states.

Most studies of faecal contamination in tropical inshore communities have necessarily focused on beaches. It is on beaches that sewage and faecal contamination are most prevalent. Beach water quality monitoring efforts have focused on levels of faecal indicators in beach sand. High levels of faecal indicators may be caused by the loss or the reduced activities of common environmental stresses controlling die-off in the sand. To test this hypothesis, Feng et al. (2010) conducted microcosm experiments to compare the effects of biotic stresses from autochthonous sand bacteria, protozoa, and viruses in *Escherichia coli* and *Enterococcus faecalis* in two tropical beach sands. Inhibition of protozoan activities by cycloheximide did not significantly affect the die-off of *E. Coli,* indicating that protozoan predation was an ineffective control mechanism. The contribution from phage infection to *E. coli* die-off was also negligible. Consequently, autochthonous bacteria were identified as the predominant biotic stress to the die-off of *E. coli*. When compared with various environmental samples, the beach sand had low protozoan concentrations and low protozoan growth potential. Subsequent experiments demonstrated that bacterial antagonistic effects may be widespread and that *E. faecalis* exhibits a much longer survival in beach sand than *E. coli*.

Sewage is the major source of human pathogens in tropical marine waters. Betancourt et al. (2014) investigated the occurrence of the intestinal protozoan parasites, *Giardia* and *Cryptosporidium* in recreational marine waters receiving sewage. A Monte Carlo uncertainty analysis was performed to determine the probability distribution of risks. Higher risks for *Giardia* than for *Cryptosporidium* were found. In Senegal, West Africa, most samples failed to meet the quality levels for the faecal indicator bacteria (FIB) defined by the European Union (Bouvy et al. 2008). The high prevalence of FIB during both wet and dry months suggested chronic pollution and a potential risk to recreational swimmers and fish consumers in Hann Bay near Dakar. In Darwin Harbour, Australia, high concentrations of faecal indicators were detected in the wet season with vibrios more prevalent in biota during heavy rainfall. Virulent strains of *Vibrio parahaemolyticus* were detected in the mangrove snails, *Telescopium telescopium* and *Nerita balteata*. Moreover, there was no correlation between *E. coli* and targeted faecal pathogens in biota indicating that traditional faecal markers are not suitable surrogates for faecal pollution, at least not in the tropics (Padovan et al. 2020).

Sophisticated genetic tests have been developed to determine the risks posed by faecal indicator bacteria. Betancourt and Fujioka (2009) evaluated the presence of enterococcal surface protein genes as markers of sewage contamination in tropical recreational waters in Hawaii. They found that genes highly associated with enterococci from sewage were able to be distinguished from non-point sources of contamination. An innovative strategy using alternative faecal indicators was tested in Hawaii to determine how reliable these markers are for detecting cesspool discharge pollution in streams and receiving coastal waters (Vithanage et al. 2011). Using both traditional and advanced genetic tests, they were able to detect lower concentrations of faecal indicator bacteria using the advanced techniques, providing more reliable data on faecal contamination. Genotyping of bacteria and viruses provided additional evidence that human cesspool contamination was occurring within the watershed.

One problem in testing for faecal contamination on tropical beaches is the growth of antibiotic resistant strains. Using multiple biomarkers to test for pollution indicators along the tropical south Indian coast, Vignesh et al. (2016) found that a significant number of bacteria showed antibiotic resistance. Pollution indices and antibiotic resistance index ratios indicated that high bacterial and antibiotic loads were released into the coastal environment.

Aside from beaches, seagrass meadows, mangrove forests, and corals reefs have been greatly affected by sewage pollution to varying degrees. The research record for seagrasses is weak, as only a handful of studies have dealt with sewage contamination. In several seagrass species colonising different beaches and tidal flats near Toliara (SW Madagascar), an increase in N concentrations and δ^{15} N values demonstrated the importance of sewage coming directly onto the intertidal zone on the N cycles of the seagrasses (Lepoint et al. 2008). This influence was restricted to the upper littoral zone and was not the main cause of seagrass-die-off. On the other hand, near an adjacent mangrove forest, δ^{15} N values did not correlate with N concentrations in *Halodule* sp. or *T. hemprichii*, suggesting that natural δ^{15} N variability is driven by factors other than the δ^{15} N of N sources. Inter-individual variability of δ^{15} N values was greater than inter-specific or inter-site variability, making the δ^{15} N results difficult to interpret in the context of human influence on the N cycle of tropical seagrasses. Contrary to the author's hypothesis, variation between sites and location on the tidal flat was limited, suggesting limited impact of δ^{13} C values of sewage, emersion duration, and mechanisms on HCO_3^- incorporation.

Ambiguous results for seagrasses have also been obtained in Curaçao and Bonaire in the Caribbean, where water measurements provided only a momentary snapshot, but nutrient concentrations in *T. testudinum* tissues indicated long-term exposure to nutrient loads (Govers et al. 2014). Nutrient levels in most of the Caribbean bays did not raise any concerns, but high P values in leaves of *Thalassia* in Piscadera Bay and Spanish Water Bay showed that seagrasses may be threatened by eutrophication. Govers et al. (2014) indicated that seagrasses may be threatened, and measures should be taken, to prevent loss of these important nursery areas due to eutrophication.

Less ambiguous is the fact that many seagrasses have bioremediation qualities and produce natural biocides. The seagrasses of inhabited atolls near Sulawesi, Indonesia, can ameliorate seawater pollution from bacteria of human origin. This effect extended to potential pathogens of marine invertebrates and fish (Lamb et al. 2017). Reefs fringing the seagrass meadows showed significantly less impact from coral and fish disease. Seagrasses are known to be associated with natural biocide production but have not been evaluated for their ability to remove microbiological contamination. When seagrass meadows were present, there was a 50% reduction in the relative abundance of potential bacterial pathogens capable of causing disease in humans and marine organisms (Lamb et al. 2017). In field surveys of more than 8000 reef-building corals located adjacent to the seagrasses, there were twofold reductions in disease levels compared to corals at sites without seagrasses, highlighting the importance of seagrass ecosystems to the health of other organisms.

While mangroves show no such ameliorating qualities with respect to human and organism health, they have a high tolerance to withstand sewage pollution and associated microbial disease. Some projects have utilized mangroves to soak up sewage contamination and have found either no decline in mangrove growth rates or an increase in growth. This finding also extends to mangrove biota. Penha-Lopes et al. (2011) evaluated the potential anthropogenic disturbance on populations of the crustacean, *Palaemon concinnus,* from a peri-urban (domestic sewage impacted) waterway and two pristine mangrove creeks. The shrimps in peri-urban waters were larger, experienced longer reproductive times, presented a higher proportion of ovigerous females and better embryo quality than those shrimps in the pristine creeks. Similarly, the behavioural responses of the mangrove fiddler crabs, *Uca annulips* and *U. inversa,* to urban sewage loadings were superior to those in non-sewage controls (Bartolini et al. 2009). During their active periods, crabs inside contaminated mesocosms satisfied their feeding demands faster than those of the control crabs, spending a significantly longer time in other activities, such as courtship and territorial defence.

A comparison of fecundity, embryo loss, and fatty acid composition of mangrove crab species in sewage contaminated and pristine mangroves in Mozambique found that fecundity, egg quality, and potential fertility of the fiddler crab, *Uca annulipes,* in sewage sites were found to exceed the reproductive qualities of the crab in pristine mangroves (Penha-Lopes et al. 2009a). Similarly, the effects of vegetation and sewage load on mangrove crabs using experimental mesocosms were examined by Amaral et al. (2009) who introduced ocypodid crabs to three vegetation treatments (bare substratum, *A. marina* and *R. mucronata* seedlings) subjected to 0, 20, 60, and 100% sewage loads. Overall, crabs coped well with the administered sewage loads. Sewage enhanced crab condition in the bare substratum and *R. mucronata* treatments probably because of enhanced food availability. *Uca inversa* was more sensitive to sewage pollution than *U. annulipes*. In the *A. marina* treatment, no difference in crab condition was observed between sewage loads and this species yielded the best reduction in sewage impact.

Not all studies show a positive result of the effects of sewage pollution on mangrove biota. Bartolini et al. (2011) found that crabs exposed to sewage spent less time bioturbating the soil compared to in a pristine site. This lessened bioturbating effect may have a negative consequence for the growth of mangrove seedlings as anoxia is likely to be greater in non-bioturbated soils.

These results raise the issue of whether fiddler crabs are potentially useful ecosystem engineers in mangrove wastewater wetlands. Penha-Lopes et al. (2009b) studied the effect of different organic-rich sewage concentrations (0, 20 and 60% loading) and absence or presence of mangrove trees on the survival, bioturbation activities, and burrow morphology of two species of fiddler crabs. After six months, males of both species showed higher survival than females. Crabs inhabiting pristine conditions achieved higher survival than those living in sewage-exposed mesocosms. At 60% loading, fiddler crabs processed less sediment during feeding and excavated slightly more sediment than in pristine conditions; burrows were shallower in mesocosms loaded with sewage. For the gastropod, *Terebralia palustris,* the effect of different sewage concentrations, vegetation (bare, *A. marina, R. mucronata*), and immersion periods resulted in survival rates higher than 70% in all treatments.

Growth rate decreased significantly with increasing sewage concentrations and longer immersion periods. The treatments with sewage resulted in a three to fourfold decrease in the amount of sediment disturbed by the gastropod.

Sewage has a profound effect on growth dynamics of mangrove saplings. Growing one-year old *A. marina* and *R. mucronata* saplings in treatments with 0, 25, 75, and 100% domestic sewage, Nyomora and Njau (2012) observed that treated mangroves grew better than untreated plants and *A. marina* outperformed *R. mucronata*. P and N removal by plants was highest in 50 and 75% sewage. In another series of experiments, benthic metabolism and C oxidation pathways were evaluated in mangrove mesocosms subjected daily to seawater or 60% sewage in the presence or absence of trees and biogenic structures (Penha-Lopes et al. 2010). Total CO_2 emissions from darkened sediments devoid of biogenic structures under pristine conditions were comparable during inundation and air exposure but increased two to sevenfold in sewage-contaminated mesocosms. Biogenic structures increased low-tide CO_2 emissions under contaminated and pristine conditions. When sewage was loaded into the mesocosms under unvegetated and planted conditions, Fe reduction gave way to sulphate reduction.

The role of constructed mangrove wetlands plays an important part in understanding the role of mangroves in sewage treatment. Wu et al. (2008) studied the effect of intermittent subsurface flow in treating primary settled municipal wastewater collected from a local sewage treatment plant in Hong Kong. The study was carried out in a greenhouse and without any tidal flushing or tidal cycles, with half the tanks planted with *K. candel* and the other half without any plants. The mangroves removed significant quantities of sewage nutrients, suggesting that it is feasible to use constructed mangrove wetlands without tidal flushing as a secondary treatment process.

The effect of wastewater discharge on mangrove soils can be effective in ameliorating gas emissions. The effects of wastewater on atmospheric fluxes of N_2O, CH_4, and CO_2 from mangrove soils in shrimp pond wastewater, livestock wastewater, and municipal sewage treatments were compared (Chen et al. 2011). Gas emissions were significantly enhanced after wastewater irrigation, and the highest emissions of N_2O and CO_2 were measured in shrimp pond waste. High N_2O emissions were measured in the municipal sewage treatment. Soil analyses indicated that the high N_2O emissions from mangrove soils receiving shrimp pond and municipal wastes were attributable to denitrification and nitrification, respectively. The highest CH_4 efflux was measured in the livestock waste treatment which also had the highest CO_2 flux.

Mangrove forests can remove excess N through denitrification. Research conducted under laboratory conditions found that denitrification enzyme activity (DEA) was lowest in soil with lowest soil organic C and total N (Shiau et al. 2016). DEA correlated negatively with salinity, but was stimulated by glucose and sucrose, but was unaffected by lactose, acetate, and mannitol. CH_4 in mangrove soils can also have a significant impact on metabolising excess labile C and N from sewage. Pristine mangrove soils not receiving sewage often have low CH_4 concentrations, production rates, and no emissions, but when the soils are supplemented with aquaculture wastes, high rates of CH_4 efflux and porewater concentrations are usually produced (Strangmann et al. 2008).

Roots and rhizomes have a crucial role to play in the amelioration by mangroves of sewage. Fe plaques formed on mangrove roots increase in density with wastewater discharge (Pi et al. 2011). For *B. gymnorrhiza, E. agallocha,* and *A. ilicifolius*, Fe plaques increased rapidly with time of exposure to municipal sewage and these plaques had some bearing on the ability of these species to withstand sewage. The concentrations of heavy metals and the amounts of P immobilised were positively correlated with the amounts of Fe plaque formed. The performance of mangroves in wastewater treatments was related to Fe plaques formed and immobilising wastewater-borne pollutants. Effects of wastewater discharge on radial oxygen loss (ROL) and formation of Fe plaques on root surfaces of *B. gymnorrhiza* and *E. agallocha* were investigated (Pi et al. 2010). ROL along a lateral root increased more rapidly in control than in strong wastewater, but less plaques were formed in control plants. However, municipal sewage had an inhibitory effect on symbiosis between mangroves and arbuscular mycorrhizal fungi in the rhizosphere (Wang et al. 2014). This implies that sewage discharge reduces the potential beneficial effects of arbuscular mycorrhizal fungi in mangrove ecosystems.

The use and uptake of NH_3-N by mangroves have clear effects on the role of wastewater in supporting N cycling. In a series of tanks planted with *K. candel* under two tidal regimes, total N, inorganic N, and C derived from sewage were completely removed (Tam et al. 2009). The mass balance of N showed that the discharge of NH_4^+-rich wastewater enhanced microbial N transformations, with 15–30% of the total N inputs denitrified. Growth of *K. candel* was stimulated by NH_4^+ addition, and 3–7% of total N inputs were assimilated into tissues. Constructed mangrove wetlands with a short tidal regime had higher numbers of nitrifiers and significantly lower content of NH_4^+ than those with a long tidal regime. Higher populations of denitrifiers and lower NO_3^- were found in mangroves with a long tidal regime and with glucose addition (Tam et al. 2009). In an ^{15}N tracer study, Lambs et al. (2011) found that due to their high adaptation capacity, mangroves on the small island of Mayotte in the Indian Ocean can tolerate and bioremediate the high levels of N and pollutants found in sewage water. The initial results pointed to a boost in mangrove growth with sewage. The exact denitrification process was not understood, but the mass balance revealed loss of N.

While mangrove forests have a high affinity for use of sewage for growth, coral reefs are sensitive to sewage which can act as a vector for microbial diseases and harmful algal blooms. Lapointe et al. (2005a) collected tissue from bloom macroalgae for stable N isotopes on reefs at distinct depths ranging from shallow subtidal to the shelf break in the Caribbean. $\delta^{15}N$ values were significantly higher on inshore shallow reefs compared to mid and deep reefs. Values were also elevated in the southern portion of the study area where nearly 1 billion $l\ d^{-1}$ of secondarily treated wastewater was discharged into the ocean. This evidence supported the hypothesis that land-based sewage N is more important than upwelling as a N source for harmful algal blooms: (i) $\delta^{15}N$ values were highest on shallowest reefs and decreased with increasing depth, indicating land-based sources of enrichment, (ii) elevated $\delta^{15}N$ values occurred in these harmful algal blooms during the dry season, prior to the onset of the summer upwelling, and (iii) elevated NH_4^+ concentrations occurred on these reefs during both upwelling and non-upwelling periods and was preferred by macroalgae. These findings provide a case study of a coupling between increasing anthropogenic activities and the development of harmful algal blooms, including invasive species that threaten reefs in southeast Florida.

Pollution from untreated sewage is an environmental problem along several coral reef-dominated coastal bays. Emrich et al. (2017) used distributions of faecal sterols and foraminiferal assemblages to determine if human sewage is affecting the coral reefs off the coast of Caye Caulker, Belize. Inshore reef sediments contained elevated concentrations of faecal contamination, suggesting that human sewage is inducing nitrification which may in turn be promoting increases in macroalgal abundances throughout Belize. Linking sewage pollution and water quality to spatial patterns of growth anomalies of the coral, *Porites lobata,* was used to assess relationships with coral disease using enterococci concentrations and $\delta^{15}N$ *Ulva fasciata* bioassays (Yoshioka et al. 2016). Growth anomalies and enterococci concentrations were high, spatially variable, and positively related. Bioassay algal $\delta^{15}N$ showed low sewage pollution at the reef edge, while high values of resident algae indicated pollution nearshore off Hawaii. A similar experiment compared stable N isotope values from the common sea fan coral, *Gorgonia ventalina,* to test the hypothesis that sewage-derived N inputs are detectable and more severe in developed areas of the Mesoamerican barrier reef of Mexico (Baker et al. 2010). The Akumal coast was selected as the developed site since it is heavily populated versus the undeveloped shoreline south of Mahahual. Gorgonians sampled from Akumal were relatively enriched in $\delta^{15}N$ compared with the sea fans from Mahahual, water-column N concentrations were uniform around Akumal, and $\delta^{15}N$ in sea fans sampled parallel to shore were variable, indicating that sewage-derived N inputs were spotty along the coast. $\delta^{15}N$ values were positively correlated with faecal *Enterococcus* counts from seawater confirming that these enrichments were associated with sewage. Wastewater significantly increased inorganic nutrient and turbidity levels, and this degradation in water quality resulted in increased macroalgal density and species richness, lower cover of hard corals and significant declines in fish abundance. Therefore, the effects of nutrient pollution and turbidity can cascade across several levels of ecological organization to change key properties of the benthos and fish on coral reefs (Reopanichkul et al. 2009).

While nutrient pollution affects coral growth and health, a crucial question is whether there is a relationship between proximity of corals to sewage effluent and the prevalence of disease. The prevalence of black-band disease (BBD) and white plague type II (WP) within a St. Croix coral community was observed to vary with relative exposure to sewage outflow (Kaczmarsky et al. 2005). During sewage discharge, faecal coliform and *Enterococci* data indicated that impacts were limited to one of the sampling sites (Figure 12.14). There was significantly more disease at the impacted site with 14% of colonies of locally susceptible species infected versus the unimpacted site that had 4% infection rates, clearly suggesting a relationship between a high prevalence of BBD and WP type II and exposure to sewage (Kaczmarsky et al. 2005).

More direct studies have shown that human pathogens cause disease in the threatened Elkhorn coral, *A. palmata* (Sutherland et al. 2011). Infections by the human pathogen, *Serratia marcescens,* contributed to precipitous losses of *A. palmata,* providing the first example of the transmission of a human pathogen to a marine invertebrate. Other studies have confirmed the transmission of human pathogens derived from human sewage to corals. Hernández-Delgado et al. (2008) conducted a study of the ecological condition of Elkhorn coral assemblages located across a non-point source sewage gradient along the SW Puerto Rico shelf. Non-point pollution was a key stressor structuring local coral reef communities. Coral reef degradation was already beyond the point of recovery at most inshore habitats as long-term phase shifts favoured dominance by macroalgae and non-reef building taxa inshore, whereas crustose coralline algae were dominant at offshore remote reefs. Further, human sewage was identified as the likely source of white pox disease. The faecal enterobacterium, *Serratia marcescens,* was identified as a pathogen in both sewage and white pox lesions, linking the bacterium to the disease (Sutherland et al. 2010). White plague type II, considered to be one of the major diseases of

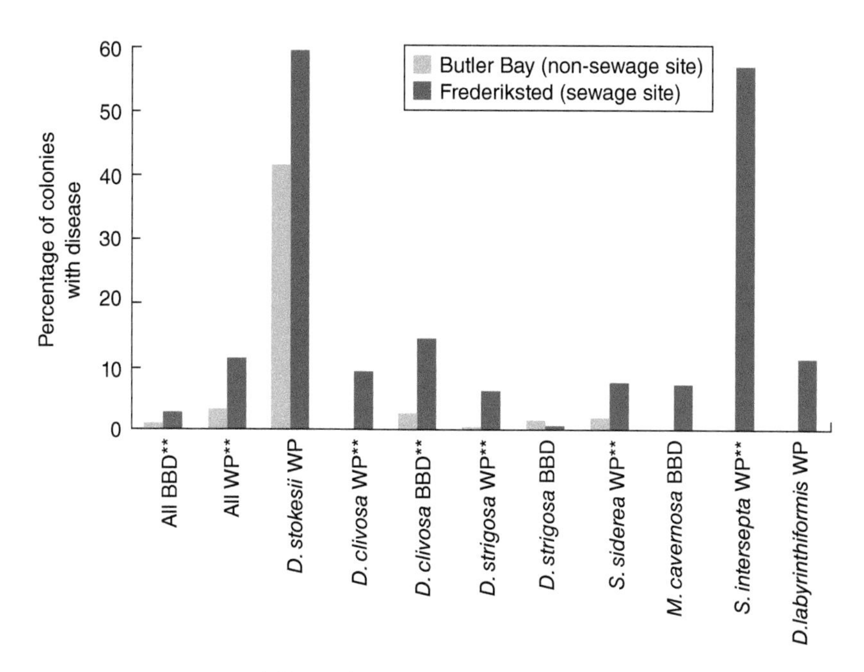

FIGURE 12.14 Percentage of diseased coral colonies of common species. Only species that were susceptible to white plague type II (WP type II) and/or black band disease (BBD) in non-sewage and sewage sites. **A significant difference between sites. *Source:* Kaczmarsky et al. (2005), figure 2, p. 129. © University of Puerto Rico at Mayagüez.

Caribbean coral species and affecting a wide range of coral hosts and causing rapid and widespread tissue loss, is caused by a single pathogen, the bacterium, *Aurantimonas coralicida* (Remily and Richardson 2006). A temperature-induced outbreak of white-band disease in the threatened staghorn coral, *Acropora cervicornis*, off Miami beach, Florida, occurred when SSTs reached 29 °C (Gignoux-Wolfsohn et al. 2020). Coral colonies showed pronounced atrophy, necrosis, and lysing of surface and basal body wall and polyp structures. The only bacteria evident microscopically was a *Rickettsiales*-like organism (RLO) occupying mucocytes and genetic sequencing also identified bacteria belonging to the order *Rickettsiales* in all fragments. Diseased fragments had more diverse bacterial communities made up of potential primary pathogens and opportunistic colonisers, as interactions among SSTs, the coral host and pathogens of the diseased microbiome all contributed to the white-band disease.

Viruses also play a role in coral disease associated with human sewage. In the thermally stressed corals, *Pavona danai, Acropora formosa, Stylophora pistillata*, and *Zoanthus* sp. and their zooxanthellae, all organisms produced numerous virus-like particles (VLPs) in the animal tissue, zooxanthellae, and the surrounding seawater (Davy et al. 2006). Infection of the corals by viruses was determined, but not the precise origin of the VLPs. Viruses are common in coral reef environments. 40% of water and coral mucus samples collected from throughout the Florida Keys and the Dry Tortugas contained genetic material from one or more human enteric viruses (Lipp et al. 2007).

Groundwater can play an important role in the spread of microbial diseases among corals. Samples of surface water, groundwater, and coral along a transect near Key Largo showed a high level of frequency such that enteric viruses were evenly distributed across the transect stations (Futch et al. 2010). Both virus types were detected twice as frequently in coral compared with surface water or groundwater. Offshore, viruses were found in groundwater, especially during the summer wet season, suggesting that polluted groundwater may be moving to the outer reefs of the Florida Keys. Conventional methods of microbial water quality cannot be used to discriminate between different faecal pollution sources. Faecal coliforms, enterococci, and human-specific *Bacteroides*, general *Bacteroides-Prevotella*, and *Clostridium coccoides* PCR assays were used to test for the presence of non-point source faecal contamination across the SW Puerto Rico shelf. Inshore waters were turbid, receiving faecal pollution from variable sources. Signals were also detected offshore; phylogenetic analysis showed that most isolates were of human faecal origin. The geographic extent of non-point source faecal pollution was large and impacted extensive coral reef systems (Bonkosky et al. 2009).

White plague (WP)-like diseases in tropical corals have been implicated in reef decline worldwide, although their aetiological cause is generally unknown. Studies have focused on bacterial or eukaryotic pathogens as the causes of these diseases, but no studies have examined the role of viruses. Molecular tests comparing viral metagenomes generated from *Montastraea annularis* corals showed signs of WP disease (Soffer et al. 2014). No bacteria were visually identified within diseased coral tissues, but viral particles were abundant in WP-diseased tissues. *Herpesviridae* gene signatures dominated healthy tissues, corroborating reports that herpes-like viruses can infect corals. Nucleocytoplasmic large DNA virus (NCLDV) was most common in bleached corals, implying that these NCLDV viruses may have a role in bleaching. Therefore, a specific group of viruses is associated with diseased Caribbean corals and highlights the potential for viral disease in regional coral reef decline.

Water mold has been implicated in causing disease in tropical estuarine fish. Six coastal fish species common in Kerala, India, were found to be infected by the water mold, *Aphanomyces invadans* under post-flood conditions (Sumithra et al. 2020), causing epizootic ulcerative syndrome. The presence of the zoonotic bacterial pathogens, *Aeomonas veronii, Shewanella putrefaciens, Vibio vulnificus,* and *V. parahaemolyticus*, in affected fish tissues indicates that the infected fish may pose a public health hazard.

The prevalence of coral decline and the rise in coral diseases has raised concern that coral reefs are now being subject to dead zones where extreme hypoxic conditions have resulted in mass mortalities (Altieri et al. 2017). Little is known about the potential threat of hypoxia in the tropics, even though the known risk factors, including elevated temperatures and eutrophication, are becoming increasingly common. Altieri et al. (2017, 2019) documented an unprecedented hypoxic event on the Caribbean coast of Panama and assessed the risk of dead zones to coral reefs worldwide. The event off Panama caused bleaching and massive mortality of corals and other reef-associated organisms, but not all coral species were equally sensitive to mass die-off. An analysis of global databases showed that coral reefs are associated with more than half of the known tropical dead zones worldwide, with >10% of all coral reefs at elevated risk for hypoxia based on local and global risk factors. Hypoxic events in the tropics and associated mortality events have likely been underreported.

12.7.2 Invasive Species

Another form of biological pollution is invasive species. These organisms move to a new environment where natural controls are unlikely to exist to keep population numbers of the invader in check. Most such species increase rapidly as there is often no control by predators, parasites, or disease, to the point where they can take over their new environment and harm native species. Invasive species are transported by various means, such as in ship ballast or by attaching to hulls, or as hitchhikers clinging to boots or scuba gear, or as consignments of live organisms traded to provide live bait or food, or as symbionts or parasites carried by other organisms.

Evidence suggests fewer successful invasions in the tropics than in temperate seas (Freestone et al. 2013; Wells et al. 2019). This pattern may be due to stronger biotic resistance in the tropics than in higher latitudes as such resistance can limit invasion and perhaps even prevent establishment (Figure 12.15). Freestone et al. (2013) provided the first experimental test of this hypothesis, conducting predator-exclusion experiments on marine epifaunal communities of a heavily invaded system focusing on non-native tunicates as a model fauna. The effect of predation on species richness of non-native tunicates was >3x greater at sites in tropical Panama than in temperate Connecticut, consistent with the hypothesis of stronger biotic resistance in the tropics. Singapore harbour, one of the world's busiest ports, has been regarded as being at high risk for invasive marine species. However, an assessment of 3650 marine invertebrates, fish, and plants collected in the harbour revealed that only 22 species were non-indigenous (Wells et al. 2019), suggesting that biodiverse marine ecosystems in the tropical Indo-Pacific may be less susceptible to invasions than previously believed.

One of the most common invasive species is the lionfish, *Pterois volitans/miles*, which is native to the Indo-Pacific and available in the tropical fish trade. This species was first spotted in the early 1990s off the coast of Florida and was believed to have been either released from home aquaria or when Hurricane Andrew flooded aquaria and pet stores near the coast. Once introduced, they voraciously took food and habitat from native fish important to the local economy and the ecology of the region. Lionfish have no natural predators and are now found in large numbers in the tropical Atlantic from the southeast United States to the South American coast (Pimiento

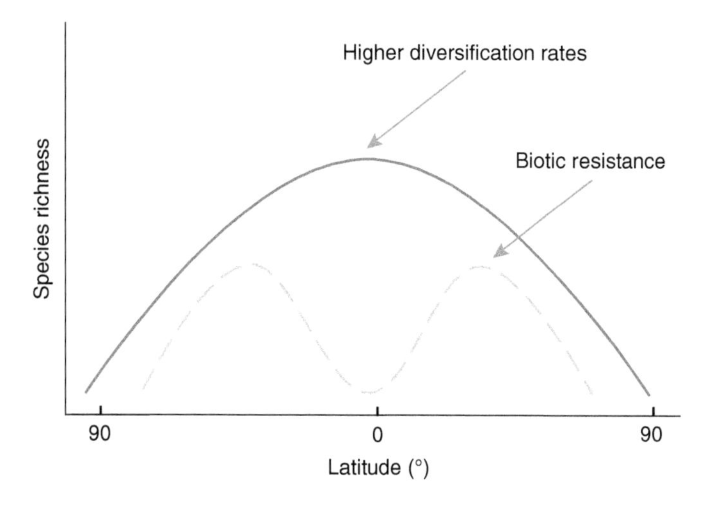

FIGURE 12.15 Model of native (solid line) and non-native (dashed line) species richness patterns and processes across latitude. Native species richness peaks in the tropics due to strong species interactions, while non-native species richness follows a bimodal pattern of peak richness at temperate latitudes, while low richness in the tropics is likely due to strong biotic resistance. *Source:* Freestone et al. (2013), figure 1, p. 1371. © John Wiley & Sons.

et al. 2013). Lionfish have spread swiftly across the western Atlantic, with unparalleled speed and magnitude (Green et al. 2012). In many areas, their numbers have surpassed native reef fish species and they grow larger and are more abundant in the invaded areas than in their native Pacific region. This invasion has had severe ecological impacts, such as disrupting the structure and function of native fish communities and coral reefs, commercial fishing, and tourism. For instance, lionfish eat important species such as parrotfish that keep algae in check on coral reefs. Lionfish have invaded mangroves, coral reefs and seagrass meadows and are opportunistic foragers moving readily between habitats (Pimiento et al. 2013). However, lionfish can selectively forage on prey and manifest strong condition-dependent foraging behaviour, as lionfish with large body size are more likely to exhibit selective foraging behaviour (Ritger et al. 2020).

Lionfish are the most dramatic example of invasive species in the tropics, but there are many other species that are considered invasive to many regions. The seagrass, *Halophila stipulacea,* for example is native to the Red Sea but has invaded the Caribbean (Willette and Ambrose 2012; Willette et al. 2014). The likely vectors for introduction were recreational and commercial ships. This species is capable of rapid expansion with rapid displacement of native species in a few weeks. Data from the eastern Caribbean show that *H. stipulacea* supports larger fish and equal or higher abundances of different trophic groups relative to the native seagrass, *Syringodium filiforme* (Willette and Ambrose 2012). The species has continued its range expansion and now spans a latitudinal distance of 6° (Willette et al. 2014), particularly in eutrophic habitats. Globally, at least 56 invasive and four non-native species have been introduced to seagrass beds (Williams 2007). Most of these species are invertebrates and seaweeds, introduced through boating and shipping and aquaculture. These species have negative consequences for the structure and function of seagrass beds.

On subtidal rocky reefs, the pest corals, *Tubastraea coccinea* and *T. tagusensis,* are common in the SW Atlantic and produce chemical defences against fish predators (Lages et al. 2010). These chemicals also reduce or enhance the settlement of other organisms. The range expansion for these two species suggest that they are becoming an increasing pest in the SW Atlantic (da Silva et al. 2014). The invasive sun corals, *Tubastraea tagusensis* and *T. coccinea,* have severely impacted the growth of the zoantharian, *Palythoa caribaeorum,* inhabiting tropical rocky shores on the Brazilian coast (Guilhem et al. 2020). After 184 days in caging experiments with four treatments (empty, skeleton, live *T. coccinea*, and live *T. tagusensis*) *P. caribaeorum* had the highest growth rate in the empty treatment and the lowest growth rate in the treatment with *T. coccinea*, unlike *T. tagusensis* which did not affect the growth rate of *P. caribaeorum*. The negative interactions between *P. caribaeorum* and the invasive *Tubastraea* spp. were species-specific and *P. caribaeorum* exhibited some resistance to *T. tagusensis* due to competition for space, probably mediated through physical and/or chemical mechanisms and rapid occupation of space. The Central and South American mussel, *Mytella strigata* has been recently reported to have expanded exponentially in the last two-years from the Philippines and Singapore to the inner Gulf of Thailand where densities have reached 40 800 ind. m^{-2} (Sanpanich and Wells 2019). This species is a potential threat to the region, especially Asian green mussel aquaculture and pearl aquaculture farms in Australia. Further, there has been a concurrent finding of the invasive false mussel *Mytilopsis sallei* in Thailand underscoring the threat faced by marine ecosystems in Southeast Asia.

On coral reefs, invasive seaweeds can often be a problem, the net result of increased eutrophic conditions (Lapointe et al. 2005b; Vermeij et al. 2009). One of the most successful and potentially threatening invasive macroalga is the green alga, *Codium isthmocladum,* in the Caribbean. Invasive blooms of the alga occur on Caribbean reefs and are considered a sign of coastal eutrophication. Nutrient enrichment promotes survival and dispersal of drifting fragments of invasive tropical macroalgae. In Hawaii where drifting fragments of the invasive red alga, *Hypnea musciformis*, smother native reef communities and forms localized blooms, increased nutrient availability increases fragment survival indicating that increased nutrient availability increases the dispersal potential of this red alga. Contributing to its success as an invader is its ability to use resources for survival and maintenance rather than growth, resulting in long periods of optimal recruitment (Vermeij et al. 2009).

Some mangrove species can be invasive to native forests. A good example of this type of plant invasion is the incursion of alien *Sonneratia apetala* and *S. caseolaris* into south China (Xin et al. 2013; Li et al. 2016). These two species have significantly higher photosynthetic abilities than native mangroves as both species have a higher photosynthetic energy-use efficiency compared to native species. Invasive mangroves can also facilitate the introduction of opportunistic exotics. For instance, *R. mangle* mangroves introduced to the Hawaiian Islands facilitated colonization by the barnacles, *Chthamalus proteus*, *Balanus reticulatus* and *B. amphitrite,* and the introduced sponges, *Suberites zeteki, Sigmadocia caerulea*, and *Gelloides fibrosa* (Demopoulos and Smith 2010). Invasive mangroves are uncommon worldwide but do nevertheless show that some species have an opportunistic life history.

References

Acharyya, T., Sarma, V.V.S.S., Sridevi, B. et al. (2012). Reduced river discharge intensifies phytoplankton blooms in Godavari estuary, India. *Marine Chemistry* 132-133: 15–22.

Acosta-Coley, I. and Olivero-Verbel, J. (2015). Microplastic resin pellets on an urban tropical beach in Colombia. *Environmental Monitoring and Assessment* 187: 435–440.

Adika, S.A., Mahu, E., Crane, R. et al. (2020). Microplastic ingestion by pelagic and demersal fish species from the eastern central Atlantic Ocean, off the coast of Ghana. *Marine Pollution Bulletin* 153: 110998. https://doi.org/10.1016/j.marpolbul.2020.110998.

Affian, K., Robin, M., Maanan, M. et al. (2009). Heavy metal and polycyclic aromatic hydrocarbons in Ebrié lagoon sediments, Côte d'Ivoire. *Environmental Monitoring Assessment* 159: 531–541.

Agamuthu, P., Mehran, S.B., Norkhairah, A. et al. (2019). Marine debris: a review of impacts and global initiatives. *Waste Management & Research* 37: 987–1002.

Ahmed, K., Mehedi, Y., Haque, R. et al. (2011). Heavy metal concentrations in some macrobenthic fauna of the Sundarbans mangrove forest, south west coast of Bangladesh. *Environmental Monitoring and Assessment* 177: 505–514.

Ajagbe, F.E., Osibona, A.O., and Otitoloju, A.A. (2012). Diversity of the edible fish of the Lagos Lagoon, Nigeria and the public health concerns based on their lead (Pb) content. *International Journal of Fisheries and Aquaculture* 2: 55–62.

Allen, A.S., Seymour, A.C., and Rittschof, D. (2017). Chemoreception drives plastic consumption in a hard coral. *Marine Pollution Bulletin* 124: 198–205.

Alongi, D.M., Chong, V.C., Dixon, P. et al. (2003). The influence of fish cage aquaculture on pelagic carbon flow and water chemistry in tidally dominated mangrove estuaries of peninsular Malaysia. *Marine Environmental Research* 55: 313–333.

Al-Rousan, S.A., Al-Shloul, R.N., Al-Horani, F.A. et al. (2007). Heavy metal contents in growth bands of *Porites* corals: record of anthropogenic and human developments from the Jordanian Gulf of Aqaba. *Marine Pollution Bulletin* 54: 1912–1922.

Altieri, A.H., Harrison, S.B., Seemann, J. et al. (2017). Tropical dead zones and mass mortalities on coral reefs. *Proceedings of the National Academy of Sciences* 114: 3660–3665.

Altieri, A.H., Nelson, H.R., and Gedan, K.G. (2019). Tropical ecosystems – corals, seagrasses, and mangroves. In: *Oxygen Deoxygenation: Everyone's Problem – Causes, Impacts, Consequences, and Solutions* (eds. D. Laffoley and J.M. Baxter), 401–429. Gland, Switzerland: IUCN.

Amaral, V., Penha-Lopes, G., and Paula, J. (2009). Effects of vegetation and sewage load on mangrove crab condition using experimental mesocosms. *Estuarine, Coastal and Shelf Science* 84: 300–304.

Ansari, Z.A., Desilva, C., and Badesab, S. (2012). Total petroleum hydrocarbon in the tissue of some commercially important fish of the Bay of Bengal. *Marine Pollution Bulletin* 64: 2564–2575.

Araújo, M.C. and Costa, M. (2007). An analysis of the riverine contribution to the solid wastes contamination of an isolated beach at the Brazilian Northeast. *Management of Environmental Quality* 18: 6–12.

Arnaud-Haond, S., Duarte, C.M., Teixeira, S. et al. (2009). Genetic recolonization of mangroves: genetic diversity still increasing in the Mekong Delta 30 years after Agent Orange. *Marine Ecology Progress Series* 390: 129–135.

Ashok, A., Cusack, M., Saderne, V. et al. (2019). Accelerated burial of petroleum hydrocarbons in Arabian Gulf blue carbon repositories. *Science of the Total Environment* 669: 205–212.

Auta, H.S., Emenike, C.U. and Fauzial, S.H. (2017). Distribution and importance of microplastics in the marine environment: a review of sources, fate, effects and potential solutions. *Environment International* 102: 165–176.

Baêta, A.P., Kehrig, H.A., Malm, O. et al. (2006). Total mercury and methylmercury in fish from a tropical estuary. *WIT Transactions on Biomedicine and Health* 10: 183–192.

Baker, D.M., Jordán-Dahlgren, E., Maldonado, M.A. et al. (2010). Sea fan corals provide a stable isotope baseline for assessing sewage pollution in the Mexican Caribbean. *Limnology and Oceanography* 55: 2139–2149.

Balachandran, K.K., Lalu Raj, C.M., Nair, M. et al. (2005). Heavy metal accumulation in a flow restricted, tropical estuary. *Estuarine, Coastal and Shelf Science* 65: 361–370.

Barasarathi, J., Agamuthu, P., Emenike, C.U. et al. (2014). Microplastic abundance in selected mangrove forest in Malaysia. *Proceedings of the ASEAN Conference on Science and Technology*, Bogor, Indonesia (18–20 August 2014). Manila: University of the Philippines.

Bartolini, F., Penha-Lopes, G., Limbu, S. et al. (2009). Behavioural responses of mangrove fiddler crabs (*Uca annulipes* and *U. inversa*) to urban sewage loadings: results of a mesocosm approach. *Marine Pollution Bulletin* 58: 1860–1867.

Bartolini, F., Cimó, F., Fusi, M. et al. (2011). The effect of sewage discharge on the ecosystem engineering activities of two east African fiddler crab species: consequences for mangrove ecosystem functioning. *Marine Environmental Research* 71: 53–61.

Bayen, S., Wurl, O., Karuppiah, S. et al. (2005). Persistent organic pollutants in mangrove food webs in Singapore. *Chemosphere* 61: 303–313.

Beiras, R. (2018). *Marine Pollution: Sources, Fate and Effects of Pollutants in Coastal Ecosystems*. Amsterdam, The Netherlands: Elsevier.

Bejarano, A.C. and Michel, J. (2010). Large-scale risk assessment of polycyclic aromatic hydrocarbons in shoreline sediments from Saudi Arabia: environmental legacy after twelve years of the Gulf war oil spill. *Environmental Pollution* 158: 1561–1569.

Bejarano, A.C. and Michel, J. (2016). Oil spills and their impacts on sand beach invertebrate communities: a literature review. *Environmental Pollution* 218: 709–722.

Bell, A.M. and Duke, N.C. (2005). Effects of Photosystem II inhibiting herbicides on mangroves – preliminary toxicology trials. *Marine Pollution Bulletin* 51: 297–307.

Benson, N.U. and Fred-Ahmadu, O.H. (2020). Occurrence and distribution of microplastics-sorbed phthalic acidesters (PAEs) in coastal psammitic sediments of tropical Atlantic Ocean, Gulf of Guinea. *Science of the Total Environment* 730: 139013. https://doi.org/10.1016/j.scitotenv.2020.139013.

Berry, K.L.E., Seemann, J., Dellwig, O. et al. (2013). Sources and spatial distribution of heavy metals in hermatypic coral tissues and sediments from the Bocas del Toro Archipelago, Panama. *Environmental Monitoring and Assessment* 185: 9089–9099.

Betancourt, W.Q. and Fujioka, R.S. (2009). Evaluation of enterococcal surface protein genes as markers of sewage contamination in tropical recreational waters. *Water Science and Technology* 60: 261–266.

Betancourt, W.Q., Duarte, D.C., Vasquez, R.C. et al. (2014). *Cryptosporidium* and *Giardia* in tropical recreational marine waters contaminated with domestic sewage: estimation of bathing-associated disease risks. *Marine Pollution Bulletin* 85: 268–273.

Bhavya, P.S., Kumar, S., Gupta, G.V.M. et al. (2016). Nitrogen uptake dynamics in a tropical eutrophic estuary (Cochin, India) and adjacent coastal waters. *Estuaries and Coasts* 39: 54–67.

van Bijsterveldt, C.E.J., van Wesenbeeck, B.K., Ramadhani, S. et al. (2021). Does plastic waste kill mangroves? A field experiment to assess the impact of macro plastics on mangrove growth, stress response and survival. *Science of the Total Environment* 756: 143826. https://doi.org/10.1016/j.scitotenv.2020.143826.

Bodin, N., N'Gom-Kâ, R., Loc'h, F.L. et al. (2011). Are exploited mangrove molluscs exposed to persistent organic pollutant contamination in Senegal, West Africa? *Chemosphere* 84: 318–327.

Bodin, N., N'Gom-Kâ, R., Kâ, S. et al. (2013). Assessment of trace metal contamination in mangrove ecosystems from Senegal, West Africa. *Chemosphere* 90: 150–157.

Bonkosky, M., Hernández-Delgado, E.A., Sandoz, B. et al. (2009). Detection of spatial fluctuations of non-point source faecal pollution in coral reef surrounding waters in Southwest Puerto Rico using PCR-based assays. *Marine Pollution Bulletin* 58: 45–54.

Botter-Carvalho, M.L., Carvalho, P.V.V.C., Valença, A.P.M.C. et al. (2014). Estuarine mud flat responses to continuous *in situ* nutrient addition on a tropical mud flat. *Marine Pollution Bulletin* 83: 214–223.

Bouvy, M., Briand, E., Boup, M.M. et al. (2008). Effects of sewage discharges on microbial components in tropical coastal waters (Senegal, west Africa). *Marine and Freshwater Research* 59: 614–626.

Brennecke, D., Ferreira, E.C., Costa, T.M.M. et al. (2015). Ingested microplastics (>100 μm) are translocated to organs of the tropical fiddler crab *Uca rapax*. *Marine Pollution Bulletin* 96: 491–495.

Briand, M., Bustamante, P., Bonnet, X. et al. (2018). Tracking trace elements into complex coral reef trophic networks. *Science of the Total Environment* 612: 1091–1104.

Browne, M.A., Crump, P., Niven, S.J. et al. (2011). Accumulation of microplastic on shorelines worldwide: sources and sinks. *Environmental Science and Technology* 45: 9175–9179.

Burkepile, D.E. and Hay, M.E. (2009). Nutrient versus herbivore control of macroalgal community development and coral growth on a Caribbean reef. *Marine Ecology Progress Series* 389: 71–84.

Burkepile, D.E., Allgeier, J.E., Shantz, A.A. et al. (2013). Nutrient supply from fish facilitates macroalgae and suppresses corals in a Caribbean coral reef ecosystem. *Scientific Reports* 3: 1493. https://doi.org/10.1038/srep01493.

Burkholder, J.M., Tomasko, D.A., and Touchette, B.W. (2007). Seagrasses and eutrophication. *Journal of Experimental Marine Biology and Ecology* 350: 46–72.

Calderon, E.A., Hansen, P., Rodríguez, A. et al. (2019). Microplastics in the digestive tracts of four fish species from the Ciénaga Grande de Santa Marta estuary in Colombia. *Water, Air and Soil Pollution* 230: 257. https://doi.org/10.1007/s11270-019-4313-8.

Cantin, N.E., Negri, A.P., and Willis, B.L. (2007). Photoinhibition from chronic herbicide exposure reduces reproductive output of reef-building corals. *Marine Ecology Progress Series* 344: 81–93.

Capparelli, M.V., Abessa, D.M., and McNamara, J.C. (2016). Effects of metal contamination *in situ* on osmoregulation and oxygen consumption in the mud flat fiddler crab *Uca Rapax* (Ocypodidae, Brachyura). *Comparative Biochemistry and Physiology, Part C* 185-186: 102–111.

Carvalho, F.P., Gonzalez-Farias, F., Villeneuve, J.-P. et al. (2002a). Distribution, fate and effects of pesticide residues in tropical coastal lagoons of NW Mexico. *Environmental Technology* 23: 1257–1270.

Carvalho, F.P., Villeneuve, J.-P., Cattini, C. et al. (2002b). Ecological risk assessment of pesticide residues in coastal lagoons of Nicaragua. *Journal of Environmental Monitoring* 4: 778–787.

Carvalho, F.P., Villeneuve, J.-P., Cattini, C. et al. (2009). Pesticide and PCB residues in the aquatic ecosystems of Laguna de Terminos, a protected area of the coast of Campeche, Mexico. *Chemosphere* 74: 988–995.

Chai, M., Li, R., Qiu, Z. et al. (2020). Mercury distribution and transfer in sediment-mangrove system in urban mangroves of fast-developing coastal region, southern China. *Estuarine, Coastal and Shelf Science* 240: 106770. https://doi.org/10.1016/j.ecss.2020.106770.

Chazottes, V., Le Campion-Alsumard, T., Peyrot-Clausade, M. et al. (2002). The effects of eutrophication-related alterations to coral reef communities on agents and rates of bioerosion (Reunion Island, Indian Ocean). *Coral Reefs* 21: 375–390.

Chen, G.C., Tam, N.F.Y., Wong, Y.S. et al. (2011). Effect of wastewater discharge on greenhouse gas fluxes from mangrove soils. *Atmospheric Environment* 45: 1110–1115.

Chen, X., Wei, G., Deng, W. et al. (2015). Decadal variations in trace metal concentrations on a coral reef: evidence from a 159 year record of Mn, Cu, and V in a *Porites* coral from the northern South China Sea. *Journal of Geophysical Research, Oceans* 120: 405–416.

Cheng, H., Liu, Y., Tam, N.F.Y. et al. (2010). The role of radial oxygen loss and root anatomy on zinc uptake and tolerance in mangrove seedlings. *Environmental Pollution* 158: 1189–1196.

Chiappone, M., Dienes, H., Swanson, D.W. et al. (2005). Impacts of lost fishing gear on coral reef sessile invertebrates in Florida Keys National Marine Sanctuary. *Biological Conservation* 121: 221–230.

Cortés-Gómez, A.A., Fuentes-Mascorro, G., and Romeo, E. (2014). Metals and metalloids in whole blood and tissues of Olive Ridley turtles (*Lepidochelys olivacea*) from La Escobilla beach (Oaxaca, Mexico). *Marine Pollution Bulletin* 89: 367–375.

Costa Tde, M.M. and Soares-Gomes, A. (2015). Secondary production of the fiddler crab *Uca rapax* from mangrove areas under anthropogenic eutrophication in the western Atlantic, Brazil. *Marine Pollution Bulletin* 101: 533–538.

Costa, M.F., do Sul, J.A.I., Silva-Cavalcanti, J.S. et al. (2010). On the importance of size of plastic fragments and pellets on the strandline: a snapshot of a Brazilian beach. *Environmental Monitoring Assessment* 168: 299–304.

Costanzo, S.D., O'Donohue, M.J., and Dennison, W.C. (2004). Assessing the influence and distribution of shrimp pond effluent in a tidal mangrove creek in north-east Australia. *Marine Pollution Bulletin* 48: 514–525.

Cotovicz, L.C. Jr., Knoppers, B.A., Brandini, N. et al. (2015). A strong CO_2 sink enhanced by eutrophication in a tropical coastal embayment (Guanabara Bay, Rio de Janeiro, Brazil). *Biogeosciences* 12: 6125–6146.

Dantas, D.V., Barletta, M., and da Costa, M.F. (2012). The seasonal and spatial patterns of ingestion of polyfilament nylon fragments by estuarine drums (Sciaenidae). *Environmental Science and Pollution Research* 19: 600–606.

Dantas, N.C.F.M., Duarte, O.S., Ferreira, W.C. et al. (2020). Plastic intake does not depend on fish eating habits: identification of microplastics in the stomach contents of fish on an urban beach in Brazil. *Marine Pollution Bulletin* 153: 110959. https://doi.org/10.1016/j.marpolbul.2020.110959.

Davy, S.K., Burchett, S.G., Dale, A.L. et al. (2006). Viruses: agents of coral disease? *Diseases of Aquatic Organisms* 69: 101–110.

Demopoulos, A.W.J. and Smith, C.R. (2010). Invasive mangroves alter macrofaunal community structure and facilitate opportunistic exotics. *Marine Ecology Progress Series* 404: 51–67.

Deng, H., He, J., Feng, D. et al. (2021). Microplastics pollution in mangrove ecosystems: a critical review of current knowledge and future directions. *Science of the Total Environment* 753: 142041. https://doi.org/10.1016/j.scitotenv.2020.142041.

Derraik, J.G.B. (2002). The pollution of the marine environment by plastic debris: a review. *Marine Pollution Bulletin* 44: 842–852.

Di Beneditto, A.P.M. and Awabdi, D.R. (2014). How marine debris ingestion differs among megafauna species in a tropical coastal area. *Marine Pollution Bulletin* 88: 86–90.
do Amaral, M.C.R., de Freitas Rebelo, M., Torres, J.P.M. et al. (2005). Bioaccumulation and depuration of Zn and Cd in mangrove oysters (*Crassostrea rhizophorae,* Guilding, 1828) transplanted to and from a contaminated tropical coastal lagoon. *Marine Environmental Research* 59: 277–285.
do Sul, J.A.I. and Costa, M.F. (2014). The present and future of microplastic pollution in the marine environment. *Environmental Pollution* 185: 352–364.
Downs, C.A., Richmond, R.H., Mendiola, W.J. et al. (2006). Cellular physiological effects of the MV *Kyowa Violet* fuel-oil spill on the hard coral, *Porites lobata. Environmental Toxicology and Chemistry* 25: 3171–3180.
Dromard, C.R., Bouchon-Navaro, Y., Cordonnier, S. et al. (2018). Different transfer pathways of an organochlorine pesticide across marine tropical food webs assessed with stable isotope analysis. *PLoS One* 13: e0191335. https://doi.org/10.1371/journal.pone.0191335.
Duke, N.C., Pinzón, Z.S., and Prada, M.C. (1997). Large-scale damage to mangrove forests following two large oil spills in Panama. *Biotropica* 29: 2–14.
Duke, N.C., Burns, K.A., and Dalhaus, O. (1998). Effects of oils and dispersed oils on mangrove seedlings in planthouse experiments: a preliminary assessment of results two months after oil treatments. *APPEA Journal* 1998: 631–636.
Duke, N.C., Pinzon, Z.S., and Prada, M.C. (1999). Recovery of tropical mangrove forest following a major oil spill: a study of recruitment and growth, and the benefits of planting. In: *Ecosistemas de Manglar en America Tropical Instituto de Ecologia A.C. Mexico UICN/ORMA* (eds. A. Yanez-Arancibia and A.L. Lara-Dominguez), 231–254. Silver Spring, USA: NOAA/NFMS.
Duke, N.C., Burns, K.A., Swannell, R.P.J. et al. (2000). Dispersant use and a bioremediation strategy as alternate means of reducing impacts of large oil spills on mangroves: the Gladstone field trials. *Marine Pollution Bulletin* 41: 403–412.
Durako, M.J., Kenworthy, W.J., Fatemy, S.M.R. et al. (1993). Assessment of the toxicity of Kuwait crude oil on the photosynthesis and respiration of seagrasses of the northern Gulf. *Marine Pollution Bulletin* 27: 223–227.
Dy, D.T., Uy, F.A., and Casiño, E. (2005). Spatial distribution of heavy metals in surface sediments of a mangrove creek and its adjacent seagrass bed (Cordova, Mactan Island, central Philippines). *The Philippine Scientist* 42: 130–143.
e Ramos, Silva, C.A., da Silva, A.P., and de Oliveira, S.R. (2006). Concentration, stock and transport rate of heavy metals in a tropical red mangrove, Natal, Brazil. *Marine Chemistry* 99: 2–11.
Edinger, E.N., Azmy, K., Diegor, W. et al. (2008). Heavy metal contamination from gold mining recorded in *Porites lobata* skeletons, Buyat-Ratototok district, north Sulawesi, Indonesia. *Marine Pollution Bulletin* 56: 1553–1569.
Egres, A.G., Martins, C.C., de Oliveira, V.M. et al. (2012). Effects of an experimental *in situ* diesel oil spill on the benthic community of unvegetated tidal flats in a subtropical estuary (Paranaguá Bay, Brazil). *Marine Pollution Bulletin* 64: 2681–2691.
Emrich, K., Martinez-Colon, M., and Alegria, H. (2017). Is untreated sewage impacting coral reefs of Caye Caulker, Belize? *Journal of Foraminiferal Research* 47: 20–33.
Epstein, N., Bak, R.P.M., and Rinkevich, B. (2000). Toxicity of third generation dispersants and dispersal Egyptian crude oil on Red Sea coral larvae. *Marine Pollution Bulletin* 40: 497–503.
Eriksen, M., Lebreton, L.C.M., Carson, H.S. et al. (2014). Plastic pollution in the world's oceans: more than 5 trillion plastic pieces weighing over 250,000 tons afloat at sea. *PLoS One* 9: e111913. 10/1371/journal.pone.0111913.
Everaarts, J.M., Bano, N., Swennen, C. et al. (1991). Cyclic chlorinated hydrocarbons in benthic invertebrates from three coastal areas in Thailand and Malaysia. *Journal of the Scientific Society of Thailand* 17: 31–49.
Fabricius, K.E. (2011). Factors determining the resilience of coral reefs to eutrophication: a review and conceptual model. In: *Coral Reefs: An Ecosystem in Transition* (eds. Z. Dubinsky and N. Stambler), 493–505. Dordrecht, The Netherlands: Springer.
Fauzi, A., Skidmore, A.K., Heitkönig, I.M.A. et al. (2014). Eutrophication of mangroves linked to depletion of foliar and soil base cations. *Environmental Monitoring and Assessment* 186: 8487–8498.

Fauziah, S.H., Liyana, I.A., and Agamuthu, P. (2015). Plastic debris in the coastal environment: the invincible threat? Abundance of buried plastic debris on Malaysian beaches. *Waste Management and Research* 33: 812–821.
Feng, F., Goto, D., and Yan, T. (2010). Effects of autochthonous microbial community on the die-off of faecal indicators in tropical beach sand. *FEMS Microbiology Ecology* 74: 214–225.
Ferrier-Pagés, C., Gattuso, J.-P., Dallot, S. et al. (2000). Effect of nutrient enrichment on growth and photosynthesis of the zooxanthellate coral *Stylophora pistillata*. *Coral Reefs* 19: 103–113.
Filho, G.M.A., Creed, J.C., Andrade, L.R. et al. (2004). Metal accumulation by *Halodule wrightii* populations. *Aquatic Botany* 80: 241–251.
Flammang, P., Warnau, M., Temara, A. et al. (1997). Heavy metals in *Diadema setosum* (Echinodermata, Echinoidea) from Singapore coral reefs. *Journal of Sea Research* 38: 35–45.
Flores, F., Collier, C.J., Mercurio, P. et al. (2013). Phytotoxicity of four Photosystem II herbicides to tropical seagrasses. *PLoS One* 8: e75798. https://doi.org/10.1371/journal.pone.0075798.
Freestone, A.L., Ruiz, G.M., and Torchin, M.E. (2013). Stronger biotic resistance in tropics relative to temperate zone: effects of predation on marine invasion dynamics. *Ecology* 94: 1370–1377.
Futch, J.C., Griffin, D.W., and Lipp, E.K. (2010). Human enteric viruses in groundwater indicate offshore transport of human sewage to coral reefs of the Upper Florida Keys. *Environmental Microbiology* 12: 964–974.
Gab-Alla, A.A.F.A. (2000). Ecological status of seagrass community in Sharm El-Moyia Bay, Egyptian Red Sea after its oil pollution in 1999. *Egyptian Journal of Biology* 2: 34–41.
Ganeshkumar, A., Arun, G., Vinothkumar, S. et al. (2019). Bioaccumulation and translocation efficacy of heavy metals by *Rhizophora mucronata* from tropical mangrove ecosystem, southeast coast of India. *Ecohydrology & Hydrobiology* 19: 66–74.
Garrity, S.D. and Levings, S.C. (1993). Effects of an oil spill on some organisms living on mangrove (*Rhizophora mangle* L.) roots in low wave-energy habitats in Caribbean Panama. *Marine Environmental Research* 35: 251–271.
Garrity, S.D., Levings, S.C., and Burns, K.A. (1994). The *Galeta* oil spill. I. Long-term effects on the physical structure of the mangrove fringe. *Estuarine. Coastal and Shelf Science* 38: 327–348.
Gautier, D., Amador, J., and Newmark, F. (2001). The use of mangrove wetland as a biofilter to treat shrimp pond effluents: preliminary results of an experiment on the Caribbean coast of Colombia. *Aquaculture Research* 32: 787–799.
Geng, X., Wang, J., Zhang, Y. et al. (2021). How do microplastics affect the marine microbial loop? Predation of microplastics by microzooplankton. *Science of the Total Environment* 758: 144030. https://doi.org/10.1016/j.scitotenv.2020.144030.
Gignoux-Wolfsohn, S.A., Precht, W.F., Peters, E.C. et al. (2020). Ecology, histopathology, and microbial ecology of a white-band disease outbreak in the threatened staghorn coral *Acropora cervicornis*. *Diseases in Aquatic Organisms* 137: 217–237.
Gissi, F., Reichelt-Brushett, A.J., Chariton, A.A. et al. (2019). The effect of dissolved nickel and copper on the adult coral *Acropora muricata* and its microbiome. *Environmental Pollution* 250: 792–806.
Glibert, P.A., Lansberg, J.H., Evans, J.J. et al. (2002). A fish kill of massive proportion in Kuwait Bay, Arabian Gulf, 2001: the roles of bacterial disease, harmful algae, and eutrophication. *Harmful Algae* 1: 215–231.
Gonzalez-Mendoza, D., Moreno, A.Q., and Zapata-Perez, O. (2007). Coordinated responses of phytochelatin synthase and metallothionein genes in black mangrove, *Avicennia germinans*, exposed to cadmium and copper. *Aquatic Toxicology* 83: 306–314.
Goss, H., Jaskiel, J., and Rotjan, R. (2018). *Thalassia testudinum* as a potential vector for incorporating microplastics into benthic marine food webs. *Marine Pollution Bulletin* 135: 1085–1089.
Govers, L.L., Lamers, L.P.M., Bouma, T.J. et al. (2014). Eutrophication threatens Caribbean seagrasses- an example from Curaçao and Bonaire. *Marine Pollution Bulletin* 89: 481–486.

Granados-Galvin, I.A., Rodriguez-Meza, D.G., Luna-González, A. et al. (2015). Human health risk assessment of pesticide residues in snappers (*Lutjanus*) fish from the Navachiste Lagoon complex, Mexico. *Marine Pollution Bulletin* 97: 178–187.

Grant, D.L., Clarke, P.J., and Allaway, W.G. (1993). The response of grey mangrove (*Avicennia marina* (Forsk.) Vierh.) seedlings to spills of crude oil. *Journal of Experimental Marine Biology and Ecology* 171: 273–295.

Green, S.J., Akins, J.L., Maljković, A. et al. (2012). Invasive lionfish drive Atlantic coral reef fish declines. *PLoS One* 7: e32596. https://doi.org/10.1371/journal.pone.0032596.

Guilhem, I.F., Masi, B.P., and Creed, J.C. (2020). Impact of invasive *Tubastraea* spp. (Cnidaria:Anthozoa) on the growth of the space dominating tropical rocky-shore zoantharian *Palythoa caribaeorum* (Duchassaing and Michelotti, 1860). *Aquatic Invasions* 15: 98–113.

Guzmán, H.M. and Garcıa, E.M. (2002). Mercury levels in coral reefs along the Caribbean coast of Central America. *Marine Pollution Bulletin* 44: 1415–1420.

Guzman, H.M., Kaiser, S., and Weil, E. (2020). Assessing the long-term effects of a catastrophic oil spill on subtidal coral reef communities off the Caribbean coast of Panama (1985-2017). *Marine Biodiversity* 50: 28. https://doi.org/10.1007/s12526-020-0157-9.

Hall, N.M., Berry, K.L.E., Rintoul, L. et al. (2015). Microplastic ingestion by hermatypic corals. *Marine Biology* 162: 725–732.

Halpern, B.S., Walbridge, S., Selkoe, K.A. et al. (2008). A global map of human impact on marine ecosystems. *Science* 319: 948–952.

Hamlin, H.J., Marciano, K., and Downs, C.A. (2015). Migration of nonylphenol from food-grade plastic is toxic to the coral reef fish species *Pseudochromis fridmani*. *Chemosphere* 139: 223–228.

Harris, P.T. (2020). The fate of microplastic in marine sedimentary environments: a review and synthesis. *Marine Pollution Bulletin* 158: 111398. https://doi.org/10.1016/j.marpolbul.2020.111398.

Haynes, D., Ralph, P., Prange, J. et al. (2000a). The impact of the herbicide diuron on photosynthesis in three species of tropical seagrass. *Marine Pollution Bulletin* 41: 288–293.

Haynes, D., Müller, J., and Carter, S. (2000b). Pesticide and herbicide residues in sediments and seagrasses from the Great Barrier Reef World Heritage Area and Queensland coast. *Marine Pollution Bulletin* 41: 279–287.

Hazimah, N. and Obbard, J. (2013). Microplastics in Singapore's coastal mangrove ecosystems. *Marine Pollution Bulletin* 79: 278–283.

Herbeck, L.S., Krumme, U., Nordhaus, I. et al. (2021). Pond aquaculture effluents feed an anthropogenic nitrogen loop in a SE Asian estuary. *Science of the Total Environment* 756: 144083. https://doi.org/10.1016/j.scitotenv.2020.144083.

Hernández Arana, H.A., Warwick, R.M., Attrill, M.J. et al. (2005). Assessing the impact of oil-related activities on benthic macroinfauna assemblages of the Campeche shelf, southern Gulf of Mexico. *Marine Ecology Progress Series* 289: 89–107.

Hernández-Delgado, E.A., Sandoz, B., Bonkosky, M. et al. (2008). Impacts of non-point source sewage pollution on Elkhorn coral, *Acropora palmata* (Lamarck), assemblages of the SWern Puerto Rico shelf. *Proceedings of the 11th International Coral Reef Symposium*, Ft. Lauderdale, FL (7–11 July 2008), vol. 2, pp. 747–753.

Herrera, I., Yebra, L., Antezana, T. et al. (2019). Vertical variability of *Euphausia distinguenda* metabolic rates during diel migration into the oxygen minimum zone of the eastern tropical Pacific off Mexico. *Journal of Plankton Research* 41: 165–176.

Hirai, H., Takada, H., Ogata, Y. et al. (2011). Organic micropollutants in marine plastics debris from the open ocean and remote and urban beaches. *Marine Pollution Bulletin* 62: 1683–1692.

Huang, G.-Y. and Wang, Y.-S. (2010a). Expression and characterization analysis of type 2 metallothionein from grey mangrove species (*Avicennia marina*) in response to metal stress. *Aquatic Toxicology* 99: 86–92.

Huang, G.-Y. and Wang, Y.-S. (2010b). Physiological and biochemical responses in the leaves of two mangrove plant seedlings (*Kandelia candel* and *Bruguiera gymnorrhiza*) exposed to multiple heavy metals. *Journal of Hazardous Materials* 182: 848–854.

Huang, G.-Y., Wang, Y.-S., and Ying, G.-G. (2011). Cadmium-inducible BgMT2, a type 2 metallothionein gene from mangrove species (*Bruguiera gymnorrhiza*): its encoding protein shows metal-binding ability. *Journal of Experimental Marine Biology and Ecology* 405: 128–132.
Hughes, A.R., Jun Bando, K., Rodriguez, L.F. et al. (2004). Relative effects of grazers and nutrients on seagrasses: a meta-analysis approach. *Marine Ecology Progress Series* 282: 87–99.
Inoue, M., Suzuki, A., Nohara, M. et al. (2004). Coral skeletal tin and copper concentrations at Pohnpei, Micronesia: possible index for marine pollution by toxic anti-fouling paints. *Environmental Pollution* 129: 399–407.
Islam, M.S. and Tanaka, M. (2004). Impacts of pollution on coastal and marine ecosystems including coastal and marine fisheries and approach for management: a review and synthesis. *Marine Pollution Bulletin* 48: 624–649.
Jackson, J.B.C., Cubit, J.D., Keller, B.D. et al. (1989). Ecological effects of a major oil spill on Panamanian coastal marine communities. *Science* 243: 37–44.
James, K., Vasant, K., Padua, S. et al. (2020). An assessment of microplastics in the ecosystem and selected commercially important fish off Kochi, southeastern Arabian Sea, India. *Marine Pollution Bulletin* 154: 111027. https://doi.org/10.1016/j.marpolbul.2020.111027.
Jessen, C., Roder, C., Lizcano, J.F.V. et al. (2013). *In-situ* effects of simulated overfishing and eutrophication on benthic coral reef algae growth, succession, and composition in the central Red Sea. *PLoS One* 8: e66992. https://doi.org/10.1371/journal.pone.0066992.
Jones, R. (2005). The ecotoxicological effects of Photosystem II herbicides on corals. *Marine Pollution Bulletin* 51: 495–506.
Jones, R.J. and Kerswell, A.P. (2003). Phytotoxicity of Photosystem II (PSII) herbicides to coral. *Marine Ecology Progress Series* 261: 149–159.
Jones, D.A., Plaza, J., Watt, I. et al. (1998). Long-term (1991-1995) monitoring of the intertidal biota of Saudi Arabia after the 1991 Gulf War oil spill. *Marine Pollution Bulletin* 36: 472–489.
Jones, R.J., Muller, J., Haynes, D. et al. (2003). Effects of herbicides diuron and atrazine on corals of the Great Barrier Reef, Australia. *Marine Ecology Progress Series* 251: 153–167.
Kaczmarsky, L.T., Draud, M., and Williams, E.H. (2005). Is there a relationship between proximity to sewage effluent and the prevalence of coral disease? *Caribbean Journal of Science* 41: 124–137.
Kannin, K., Tanabe, S., and Tatsukawa, R. (1995). Geographical distribution and accumulation features of organochlorine residues in fish in tropical Asia and Oceania. *Environmental Science and Technology* 29: 2673–2683.
Karar, S., Hazra, S., and Das, S. (2019). Assessment of the heavy metal accumulation in the blue swimmer crab (*Portunus pelagicus*), northern Bay of Bengal: role of salinity. *Marine Pollution Bulletin* 143: 101–108.
Karcher, D.B., Roth, F., Carvalho, S. et al. (2020). Nitrogen eutrophication particularly promotes turf algae in coral reefs of the central Red Sea. *PeerJ* 8: e8737. https://doi.org/10.7717/peerj.8738.
Kehrig, H.A., Palermo, E.F.A., Seixas, T.G. et al. (2009). Trophic transfer of methylmercury and trace elements by tropical estuarine seston and plankton. *Estuarine, Coastal and Shelf Science* 85: 36–44.
Koop, K., Booth, D., Broadbent, A. et al. (2001). ENCORE: the effect of nutrient enrichment on coral reefs. Synthesis of results and conclusions. *Marine Pollution Bulletin* 42: 91–120.
Kouadio, K.N., Diomandé, D., Koné, Y.J.M. et al. (2011). Distribution of benthic macroinvertebrate communities in relation to environmental factors in the Ebrié Lagoon (Ivory Coast, West Africa). *Vie et Milieu* 61: 59–69.
Krishnaja, A.P., Rege, M.S., and Joshi, A.G. (1987). Toxic effects of certain heavy metals (Hg, Cd, Pb, As and Se) on the intertidal crab *Scylla serrata*. *Marine Environmental Research* 21: 109–119.
Kulkarni, R., Deobagkar, D., and Zinjarde, S. (2018). Metals in mangrove ecosystems and associated biota: a global perspective. *Ecotoxicology and Environmental Safety* 153: 215–228.
Kuzminov, F.I., Brown, C.M., Fadeev, V.V. et al. (2013). Effects of metal toxicity on photosynthetic processes in coral symbionts, *Symbiodinium* spp. *Journal of Experimental Marine Biology and Ecology* 446: 216–227.

Labrada-Martagón, V., Rodriguez, P.A.T., Méndez-Rodriguez, L.C. et al. (2011). Oxidative stress indicators and chemical contaminants in east Pacific green turtles (*Chelonia mydas*) inhabiting two foraging coastal lagoons in the Baja California peninsula. *Comparative Biochemistry and Physiology, Part C* 154: 65–75.

Lages, B.G., Fleury, B.G., Pinto, A.C. et al. (2010). Chemical defences against generalist fish predators and fouling organisms in two invasive ahermatypic corals in the genus *Tubastraea*. *Marine Ecology* 31: 473–482.

Lakshumanan, C., Viveganandan, S., Jonathan, M.P. et al. (2010). Acid leachable trace metals in beach sediments and its adjacent areas, central Tamil Nadu coast, south India. *Second International Conference on Chemical, Biological and Environmental Engineering*, Cairo, Egypt (2–4 November 2010), vol. 1, pp. 52–56.

Lamb, J.B., van de Water, J.A.J.M., Bourne, D.G. et al. (2017). Seagrass ecosystems reduce exposure to bacterial pathogens of humans, fish, and invertebrates. *Science* 355: 731–733.

Lambs, L., Léopold, A., Zeller, B. et al. (2011). Tracing sewage water by ^{15}N in a mangrove ecosystem to test its bioremediation ability. *Rapid Communications in Mass Spectrometry* 25: 2777–2784.

Lane, A. and Harrison, P.L. (2000). Effects of oil contaminants on survivorship of larvae of the hermatypic reef corals *Acropora tenuis, Goniastrea aspera* and *Platygyra sinensis* from the Great Barrier Reef. *Proceedings of the Ninth International Coral Reef Symposium*, Bali, Indonesia (23–27 October 2000), vol. 1, pp. 342–345

Lapointe, B.E., Barile, P.J., Littler, M.M. et al. (2005a). Macroalgal blooms on southeast Florida coral reefs. II. Cross-shelf discrimination of nitrogen sources indicates widespread assimilation of sewage nitrogen. *Harmful Algae* 4: 1106–1122.

Lapointe, B.E., Barile, P.J., Littler, M.M. et al. (2005b). Macroalgal blooms on southeast Florida coral reefs. I. Nutrient stoichiometry of the invasive green alga *Codium isthmocladum* in the wider Caribbean indicates nutrient enrichment. *Harmful Algae* 4: 1092–1105.

Law, K.L. (2017). Plastics in the marine environment. *Annual Review of Marine Science* 9: 205–229.

Law, K.L., Morét-Ferguson, S., Maximenko, N.A. et al. (2010). Plastic accumulation in the north Atlantic Subtropical Gyre. *Science* 329: 1185–1188.

Le Croizer, G., Schaal, G., Point, D. et al. (2019). Stable isotope analyses revealed the influence of forging habitat on mercury accumulation in tropical coastal marine fish. *Science of the Total Environment* 650: 2129–2140.

Le, D.Q., Satyanarayana, B., Fui, S.Y. et al. (2018). Mercury bioaccumulation in tropical mangrove wetland fish: evaluating potential risk to coastal wildlife. *Biological Trace Element Research* 186: 538–545.

Lebreton, L.C.M., van der Zwet, J., Damsteeg, J.-W. et al. (2017). River plastic emissions to the world's oceans. *Nature Communications* 8: 15611. https://doi.org/10.1038/ncomms15611.

Lee, L.-H. and Lin, H.-J. (2013). Effects of an oil spill on benthic community production and respiration on subtropical intertidal sand flats. *Marine Pollution Bulletin* 73: 291–299.

Lepoint, G., Frédérich, B., Gobert, S. et al. (2008). Isotopic ratios and elemental contents as indicators of seagrass C processing and sewage influence in a tropical macrotidal ecosystem (Madagascar, Mozambique Channel). *Scientia Marina* 72: 109–117.

Levings, S.C., Garrity, S.D., and Burns, K.A. (1994). The *Galeta* oil spill. III. Chronic reoiling, long-term toxicity of hydrocarbon residues and effects on epibiota in the mangrove fringe. *Estuarine, Coastal and Shelf Science* 38: 365–395.

Lewis, J.P., Tarnecki, J.H., Garner, S.B. et al. (2020). Changes in reef fish community structure following the *Deepwater Horizon* oil spill. *Scientific Reports* 10: 5621. https://doi.org/10.1038/s41598-020-62574-y.

Li, F.-L., Zan, Q.-J., Hu, Z.-Y. et al. (2016). Are photosynthetic characteristics and energetic cost important invasive traits for alien *Sonneratia* species in south China? *PLoS One* 11: e0157169. https://doi.org/10.1371/journal.pone.0157169.

Li, R., Yu, L., Chai, M. et al. (2020a). The distribution, characteristics and ecological risks of microplastics in the mangroves of southern China. *Science of the Total Environment* 708: 135025. https://doi.org/10.1016/j.scitotenv.2019.135025.

Li, R., Zhang, S., Zhang, L. et al. (2020b). Field study of the microplastic pollution in sea snails (*Ellobium chinense*) from mangrove forest and their relationships with microplastics in water/sediment located on the north of Beibu Gulf. *Environmental Pollution* 263: 114368. https://doi.org/10.1016/j.envpol.2020.114368.

Liebezeit, G., Brepohl, D., Rizzi, J. et al. (2011). DDT in biota of Paranaguá Bay, southern Brazil: recent input and rapid degradation. *Water, Air and Soil Pollution* 220: 181–188.

Lima, A.R.A., Costa, M.F., and Barletta, M. (2014). Distribution patterns of microplastics within the plankton of a tropical estuary. *Environmental Research* 132: 146–155.

Lima, A.R.A., Barletta, M., Costa, M.F. et al. (2016). Changes in the composition of ichthyoplankton assemblage and plastic debris in mangrove creeks relative to moon phases. *Journal of Fish Biology* 89: 619–640.

Lin, H.-J., Dai, X.-X., Shao, K.-T. et al. (2006). Trophic structure and functioning in a eutrophic and poorly flushed lagoon in SW Taiwan. *Marine Environmental Research* 62: 61–82.

Lipp, E.K., Futch, J.C., and Griffin, D.W. (2007). Analysis of multiple viral targets as sewage markers in coral reefs. *Marine Pollution Bulletin* 54: 1897–1902.

Littler, M.M., Littler, D.S., and Brooks, B.L. (2006). Harmful algae on tropical coral reefs: bottom-up eutrophication and top-down herbivory. *Harmful Algae* 5: 565–585.

Liu, Y., Tam, N.F.Y., Yang, J.X. et al. (2009). Mixed heavy metals tolerance and radial oxygen loss in mangrove seedlings. *Marine Pollution Bulletin* 58: 1843–1849.

Lohmann, R., Klanova, J., Kukucka, P. et al. (2012). PCBs and OCPs on a east-to-west transect: the importance of major currents and net volatilization for PCBs in the Atlantic Ocean. *Environmental Science and Technology* 46: 10471–10479.

Lovelock, C.E., Ball, M.C., Martin, K.C. et al. (2009). Nutrient enrichment increases mortality of mangroves. *PLoS One* 4: e5600. https://doi.org/10.1371/journal.pone.0005600.

MacFarlane, G.R. and Burchett, M.D. (2001). Photosynthetic pigments and peroxidase activity as indicators of heavy metal stress in the grey mangrove, *Avicennia marina* (Forsk.) Vierh. *Marine Pollution Bulletin* 42: 233–240.

MacFarlane, G.R., Pulkownik, A., and Burchett, M.D. (2003). Accumulation and distribution of heavy metals in the grey mangrove, *Avicennia marina,* (Forsk, Vierh.): biological indication potential. *Environmental Pollution* 123: 139–151.

Machado, W., Gueiros, B.B., Lisboa-Filho, S.D. et al. (2005). Trace metals in mangrove seedlings: role of iron plaque formation. *Wetlands Ecology and Management* 13: 199–206.

Markey, K.L., Baird, A.H., Humphrey, C. et al. (2007). Insecticides and a fungicide affect multiple coral life stages. *Marine Ecology Progress Series* 330: 127–137.

Martin, G.D., Muraleedharan, K.R., Vijay, J.G. et al. (2010). Formation of anoxia and denitrification in the bottom waters of a tropical estuary, SW coast of India. *Biogeosciences Discussions* 7: 1751–1782.

Martin, G.D., Nisha, P.A., Balachandran, K.K. et al. (2011). Eutrophication induced changes in benthic community structure of a flow-restricted tropical estuary (Cochin backwaters), India. *Environmental Monitoring and Assessment* 176: 427–438.

Martin, C., Almahasheer, H., and Duarte, C.M. (2019a). Mangrove forests as traps for marine litter. *Environmental Pollution* 247: 499–508.

Martin, C., Corona, E., Mahadik, G.A. et al. (2019b). Adhesion to coral surface as a potential sink for marine microplastics. *Environmental Pollution* 255: 113281. https://doi.org/10.1016/j.envpol.2019.113281.

May, L.A., Burnett, A.R., Miller, C.V. et al. (2020). Effect of Louisiana sweet crude oil on a Pacific coral, *Pocillopora damicornis*. *Aquatic Toxicology* 222: 105454. https://doi.org/10.1016/j.aquatox.2020.105454.

McClanahan, T.R., Sala, E., Stickels, P.A. et al. (2003). Interaction between nutrients and herbivory in controlling algal communities and coral condition on Glover's Reef, Belize. *Marine Ecology Progress Series* 261: 135–147.

Mchenga, I.S.S. and Tsuchiya, M. (2008). Nutrient dynamics in mangrove crab burrow sediments subjected to anthropogenic input. *Journal of Sea Research* 59: 103–113.

McKinnon, A.D., Trott, L.A., Alongi, D.M. et al. (2002). Water column production and nutrient characteristics in mangrove creeks receiving shrimp farm effluent. *Aquaculture Research* 33: 55–73.
McMahon, K., Nash, S.B., Eaglesham, G. et al. (2005). Herbicide contamination and the potential impact to seagrass meadows in Hervey Bay, Queensland, Australia. *Marine Pollution Bulletin* 51: 325–334.
Mercurio, P., Mueller, J.F., Eaglesham, G. et al. (2016). Degradation of herbicides in the tropical marine environment: influence of light and sediment. *PLoS One* 11: e0165890. https://doi.org/10.1371/journal.pone.0165890.
Metian, M., Hédouin, L., Ferrier-Pagès, C. et al. (2015). Metal bioconcentration in the hermatypic coral *Stylophora pistillata*: investigating the role of different components of the holobiont using radiotracers. *Environmental Monitoring and Assessment* 187: 178–187.
Miranda, D.A., Benskin, J.P., Awad, R. et al. (2021). Bioaccumulation of Per- and polyfluoroalkyl substances (PFASs) in a tropical estuarine food web. *Science of the Total Environment* 754: 142146. https://doi.org/10.1016/j.scitotenv.2020.142146.
Montano, S., Seveso, D., Maggioni, D. et al. (2020). Spatial variability of phthalates contamination in the reef-building coral *Porites lutea, Pocillopora verrucosa* and *Pavona varians*. *Marine Pollution Bulletin* 155: 111117. https://doi.org/10.1016/j.marpolbul.2020.111117.
Morishige, C., Donohue, M.J., Flint, E. et al. (2007). Factors affecting marine debris deposition at French Frigate Shoals, NW Hawaiian Islands Marine National Monument, 1990-2006. *Marine Pollution Bulletin* 54: 1162–1169.
Mremi, S.D. and Machiwa, J.F. (2003). Heavy metal contamination of mangrove sediments and the associated biota in Dar Es Salaam, Tanzania. *Tanzania Journal of Science* 29: 61–76.
Mwevura, H., Bouwman, H., Kylin, H. et al. (2020). Organochlorine pesticides and polycyclic aromatic hydrocarbons in marine sediments and polychaete worms from the west coast of Unguja Island, Tanzania. *Regional Studies in Marine Science* 36: 101287. https://doi.org/10.1016/j.rsma.2020.101287.
Nagelkerken, I.A. and Debrot, A.O. (1995). Mollusc communities of tropical rubble shores of Curaçao: long-term (7+ years) impacts of oil pollution. *Marine Pollution Bulletin* 30: 592–598.
Naidoo, G. and Naidoo, K. (2018). Uptake and accumulation of polycyclic aromatic hydrocarbons in the mangroves *Avicennia marina* and *Rhizophora mucronata*. *Environmental Science and Pollution Research* 25: 28875–28883.
Naidoo, T., Sershen, N.T., Thompson, R.C. et al. (2020). Quantification and characterisation of microplastics ingested by selected juvenile fish species associated with mangroves in KwaZulu-Natal, South Africa. *Environmental Pollution* 257: 113635. https://doi.org/10.1016/j.envpol.2019.113635.
Nair, M., Jayalakshmy, K.V., Balachandran, K.K. et al. (2006). Bioaccumulation of toxic metals by fish in a semi-enclosed tropical ecosystem. *Environmental Forensics* 7: 197–206.
Naumann, M.S., Bednarz, V.N., Ferse, S.C.A. et al. (2015). Monitoring of coastal coral reefs near Dahab (Gulf of Aqaba, Red Sea) indicates local eutrophication as potential cause for change in benthic communities. *Environmental Monitoring and Assessment* 187: 44–54.
Negri, A.P. and Hayward, A.J. (2000). Inhibition of fertilization and larval metamorphosis of the coral *Acropora millepora* (Ehrenberg, 1834) by petroleum products. *Marine Pollution Bulletin* 41: 420–427.
Negri, A.P., Vollhardt, C., Humphrey, C. et al. (2005). Effects of the herbicide diuron on the early life history stages of coral. *Marine Pollution Bulletin* 51: 370–383.
Negri, A.P., Flores, F., Röthig, T. et al. (2011). Herbicides increase the vulnerability of corals to rising sea surface temperature. *Limnology and Oceanography* 56: 471–485.
Negri, A.P., Flores, F., Mercurio, P. et al. (2015). Lethal and sub-lethal chronic effects of the herbicide diuron on seagrass. *Aquatic Toxicology* 165: 73–83.
Nievales, M.F.J. (2009). Some structural changes of seagrass meadows in Taklong Island National Marine Reserve, Guimaras, Western Visayas, Philippines after an oil spill. *Publications of the Seto Marine Biological Laboratory. Special Publication Series* 9: 37–44.
Ningrum, E.W., Patria, M.P., and Sedayu, A. (2019). Ingestion of microplastics by anchovies from Talisayan harbor, East Kalimantan, Indonesia. *Journal of Physics: Conference Series* 1402: 033072. https://doi.org/10.1088/1742-6596/1402/3/033072.

NOAA (1996). Effects of No. 6 oil and selected remedial options in tropical seagrass beds. *NOAA Technical Memorandum* NOS ORCA 101. Seattle, WA: Hazardous Materials Response and Assessment Division, NOAA.

Nordborg, F.M., Jones, R.J., Oelgemöller, M. et al. (2020). The effects of ultraviolet radiation and climate on oil toxicity to coral reef organisms – a review. *Science of the Total Environment* 720: 137486. https://doi.org/10.1016/j.scitotenv.2020.137486.

Numbere, A.O. (2019). Bioaccumulation of total hydrocarbon and heavy metals in body parts of the West African Red Mangrove Crab (*Goniopsis pelii*) in the Niger Delta, Nigeria. *International Letters of Natural Sciences* 75: 1–12.

Nwipie, G.N., Hart, A.I., Zabbey, N. et al. (2019). Recovery of infauna microbenthic invertebrates in oil-polluted tropical soft-bottom tidal flats: 7 years post spill. *Environmental Science and Pollution Research* 26: 22407–22420.

Nyomora, A.M.S. and Njau, K.N. (2012). Growth response of selected mangrove species to domestic sewage and abiotic stress. *Western Indian Ocean Journal of Marine Science* 11: 167–177.

de Oliveira Soares, M., Matos, E., Lucas, C. et al. (2020). Microplastics in corals: an emergent threat. *Marine Pollution Bulletin* 161: 111810. https://doi.org/10.1016/j.marpolbul.2020.111810.

Olivero-Verbel, J., Caballero-Gallardo, K., and Torres-Fuentes, N. (2009). Assessment of mercury in muscle of fish from Cartagena Bay, a tropical estuary at the north of Colombia. *International Journal of Environmental Health Research* 19: 343–355.

Orif, M. and El-Maradny, A. (2018). Bioaccumulation of polycyclic aromatic hydrocarbons in the grey mangrove (*Avicennia marina*) along the Arabian gulf, Saudi coast. *Open Chemistry* 16: 340–348.

Owen, R., Buxton, L., Sarkis, S. et al. (2002a). An evaluation of hemolymph cholinesterase activities in the tropical scallop, *Euvola (Pecten) ziczac,* for the rapid assessment of pesticide exposure. *Marine Pollution Bulletin* 44: 1010–1017.

Owen, R., Knap, A., Toaspern, M. et al. (2002b). Inhibition of coral photosynthesis by the antifouling herbicide Irgarol 1051. *Marine Pollution Bulletin* 44: 623–632.

Oyo-Ita, O.E., Ekpo, B.O., Adie, P.A. et al. (2014). Organochlorine pesticides in sediment-dwelling animals from mangrove areas of the Calabar River, SE Nigeria. *Environment and Pollution* 3: 56–63.

Pabis, K., Sobczyk, R., Siciński, J. et al. (2020). Natural and anthropogenic factors influencing abundance of the benthic macrofauna along the shelf and slope of the Gulf of Guinea, a large marine ecosystem off West Africa. *Oceanologia* 62: 83–100.

Padovan, A., Kennedy, K., Rose, D. et al. (2020). Microbial quality of wild shellfish in a tropical estuary subject to treated effluent discharge. *Environmental Research* 181: 108921. https://doi.org/10.1016/j.envres.2019.108921.

Peierls, B.L., Caraco, N.F., Pace, M.L. et al. (1991). Human influence on river nitrogen. *Nature* 350: 386–390.

Peng, L., Fu, D., Qi, H. et al. (2020). Micro- and nano-plastics in marine environment: source, distribution and threats – a review. *Science of the Total Environment* 698: 134254. https://doi.org/10.1016/j.scitotenv.2019.134254.

Penha-Lopes, G., Torres, P., Narciso, L. et al. (2009a). Comparison of fecundity, embryo loss and fatty acid composition of mangrove crab species in sewage contaminated and pristine mangrove habitats in Mozambique. *Journal of Experimental Marine Biology and Ecology* 381: 25–32.

Penha-Lopes, G., Bartolini, F., Limbu, S. et al. (2009b). Are fiddler crabs potentially useful ecosystem engineers in mangrove wastewater wetlands? *Marine Pollution Bulletin* 58: 1694–1703.

Penha-Lopes, G., Kristensen, E., Flindt, M. et al. (2010). The role of biogenic structures on the biogeochemical functioning of mangrove constructed wetlands sediments: a mesocosm approach. *Marine Pollution Bulletin* 60: 560–572.

Penha-Lopes, G., Torres, P., Cannicci, S. et al. (2011). Monitoring anthropogenic sewage pollution on mangrove creeks in southern Mozambique: a test of *Palaemon concinnus* Dana, 1852 (Palaemonidae) as a biological indicator. *Environmental Pollution* 159: 636–645.

Phan, L.K., van Thiel de Vries, J.S., and Stive, M.J. (2015). Coastal mangrove squeeze in the Mekong Delta. *Journal of Coastal Research* 31: 233–243.

Pi, N., Tam, N.F.Y., and Wong, M.H. (2010). Effects of wastewater discharge on formation of Fe plaque on root surfaces and radial oxygen loss of mangrove roots. *Environmental Pollution* 158: 381–387.

Pi, N., Tam, N.F.Y., and Wong, M.H. (2011). Formation of iron plaque on mangrove roots receiving wastewater and its role in immobilization of wastewater-borne pollutants. *Marine Pollution Bulletin* 63: 402–411.

Pimiento, C., Nifong, J.C., Hunter, M.E. et al. (2013). Habitat-use patterns of the invasive red lionfish *Pterois volitans*: a comparison between mangrove and reef systems in San Salvador. *Bahamas. Marine Ecology* https://doi.org/10.1111/maec.12114.

Pinheiro, M.A.A., e Silva, P.P.G., de Almeida Duarte, L.F. et al. (2012). Accumulation of six metals in the mangrove crab *Ucides cordatus* (Crustacea, Ucididae) and its food source, the red mangrove *Rhizophora mangle* (Angiosperma, Rhizophoraceae). *Ecotoxicology and Environmental Safety* 81: 114–121.

Po, B.H.-K., Lo, H.-S., Cheung, S.-G. et al. (2020). Characterisation of an unexplored group of microplastics from the South China Sea: can they be caused by macrofaunal fragmentation? *Marine Pollution Bulletin* 155: 1111151. https://doi.org/10.1016/j.marpolbul.2020.1111151.

Prabhudessai, S., Vishal, C.R., and Rivonker, C.U. (2019). Biotic interaction as the triggering factor for blooms under favourable conditions in tropical estuarine systems. *Environmental Monitoring and Assessment* 191: 54–67.

Qiu, Y.-W., Qiu, H.-L., Li, J. et al. (2018). Bioaccumulation and cycling of polycyclic aromatic hydrocarbons (PAHs) in typical mangrove wetlands of Hainan Island, south China. *Archives of Environmental Contamination and Toxicology* 75: 464–475.

Qiu, Y.-W., Qiu, H.-L., Zhang, G. et al. (2019). Bioaccumulation and cycling of organochlorine pesticides (OCPs) and polychlorinated biphenyls (PCBs) in three mangrove reserves of south China. *Chemosphere* 217: 195–203.

Quadros, G., Sukumaran, S., and Athalye, R.P. (2009). Impact of the changing ecology on intertidal polychaetes in an anthropogenically stressed tropical creek, India. *Aquatic Ecology* 43: 977. https://doi.org/10.1007/s10452-009-9229-8.

Raja, U.K.S., Ebenezer, V., Kumar, A. et al. (2019). Mass mortality of fish and water quality assessment in the tropical Adyar estuary, south India. *Environmental Monitoring and Assessment* 191: 512. https://doi.org/10.1007/s10661-019-7636-4.

Ramaiah, N., Ramaiah, M., Chandramohan, D. et al. (1995). Autotrophic and heterotrophic characteristics in a polluted tropical estuarine complex. *Estuarine, Coastal and Shelf Science* 40: 45–55.

Ramos, A.A., Inoue, Y., and Ohde, S. (2004). Metal contents in *Porites* corals: anthropogenic input of river run-off into a coral reef from an urbanized area, Okinawa. *Marine Pollution Bulletin* 48: 281–294.

Rangel-Buitrago, N., Williams, A., Anfuso, G. et al. (2017). Magnitudes, sources, and management of beach litter along the Atlantico department coastline, Caribbean coast of Colombia. *Ocean and Coastal Management* 138: 142–157.

Reichelt-Brushett, A.J. and Harrison, P.L. (2005). The effect of selected trace metals on the fertilization success of several hermatypic coral species. *Coral Reefs* 24: 524–534.

Reichert, J., Schellenberg, J., Schubert, P. et al. (2018). Responses of reef building corals to microplastic exposure. *Environmental Pollution* 237: 955–960.

Remily, E.R. and Richardson, L.L. (2006). Ecological physiology of a coral pathogen and the coral reef environment. *Microbial Ecology* 51: 345–352.

Reopanichkul, P., Schlacher, T.A., Carter, R.W. et al. (2009). Sewage impacts coral reefs at multiple levels of ecological organization. *Marine Pollution Bulletin* 58: 1356–1362.

Richards, Z.T. and Beger, M. (2011). A quantification of the standing stock of macro-debris in Majuro lagoon and its effect on hard coral communities. *Marine Pollution Bulletin* 62: 1693–1701.

Rios, L.M., Moore, C., and Jones, P.R. (2007). Persistent organic pollutants carried by synthetic polymers in the ocean environment. *Marine Pollution Bulletin* 54: 1230–1237.

Ritger, A.L., Fountain, C.T., Bourne, K. et al. (2020). Diet choice in a generalist predator, the invasive lionfish (*Pterois volitans/miles*). *Journal of Experimental Marine Biology and Ecology* 524: 151311. https://doi.org/10.1013/j.jembe.2020.151311.

Ross, C., Olsen, K., Henry, M. et al. (2015). Mosquito control pesticides and sea surface temperature have differential effects on the survival and oxidative stress response of coral larvae. *Ecotoxicology* 24: 540–552.

Rougée, L., Downs, C.A., Richmond, R.H. et al. (2006). Alteration of normal cellular profiles in the hermatypic coral (*Pocillopora damicornis*) following laboratory exposure to fuel oil. *Environmental Toxicology and Chemistry* 25: 3181–3187.

Ruelas-Inzunza, J., Páez-Osuna, F., Zamora-Arellano, N. et al. (2009). Mercury in biota and surficial sediments from Coatzacoalcos Estuary, Gulf of Mexico: distribution and seasonal variation. *Water, Air, and Soil Pollution* 197: 165–174.

Sanchez-Bayo, F. and Hyne, R.V. (2011). Comparison of environmental risks of pesticides between tropical and non-tropical regions. *Health & Ecological Risk Assessment* 7: 577–586.

Sanpanich, K. and Wells, F.E. (2019). *Mytella strigata* (Hanley, 1843) emerging as an invasive marine threat in Southeast Asia. *Bioinvasions Records* 8: 343–356.

Santos, I.R., Friedrich, A.C., and do Sul, J.A.I. (2009). Marine debris contamination along the undeveloped tropical beaches from NE Brazil. *Environmental Monitoring and Assessment* 148: 455–462.

Santos, H.F., Duarte, G.A.S., da Costa Rachid, C.T.C. et al. (2015). Impact of oil spills on coral reefs can be reduced by bioremediation using probiotic microbiota. *Scientific Reports* 5: 18268. https://doi.org/10.1038/srep18268.

Sarkar, S.K., Bhattacharya, B.D., Bhattacharya, A. et al. (2008). Occurrence, distribution and possible sources of organochlorine pesticide residues in tropical coastal environment of India: an overview. *Environment International* 34: 1062–1071.

Sawall, Y., Teichberg, M.C., Seemann, J. et al. (2011). Nutritional status and metabolism of the coral *Stylophora subseriata* along a eutrophication gradient in Spermonde Archipelago (Indonesia). *Coral Reefs* 30: 841–853.

Seixas, T.G., Moreira, I., Siciliano, S. et al. (2014a). Mercury and selenium in tropical marine plankton and their trophic successors. *Chemosphere* 111: 32–39.

Seixas, T.G., Moreira, I., Siciliano, S. et al. (2014b). Differences in methylmercury and inorganic mercury biomagnification in a tropical marine food web. *Bulletin of Environmental Contamination and Toxicology* 92: 274–278.

Shete, A., Gunale, V.R., and Pandit, G.G. (2009). Organochlorine pesticides in *Avicennia marina* from the Mumbai mangroves, India. *Chemosphere* 76: 1483–1485.

Shiau, Y.-J., Dham, V., Tian, G. et al. (2016). Factors influencing removal of sewage nitrogen through denitrification in mangrove soils. *Wetlands* 36: 621–630.

da Silva Fonseca, J., de Barros Marangoni, L.F., Marques, J.A. et al. (2019). Energy metabolism enzymes inhibition by the combined effects of increasing temperature and copper exposure in the coral *Mussismilia harttii*. *Chemosphere* 236: 124420. https://doi.org/10.1016/j.chemosphere.2019.124420.

da Silva, A.G., de Paula, A.F., Fleury, B.G. et al. (2014). Eleven years of range expansion of two invasive corals (*Tubastraea coccinea* and *Tubastraea tagusensis*) through the SW Atlantic (Brazil). *Estuarine, Coastal and Shelf Science* 141: 9–16.

Singh, Y.T., Krishnamoorthy, M., and Thippeswamy, S. (2012). Status of heavy metals in tissues of wedge clam, *Donax faba* (Bivalvia, Donacidae) collected from the Panambur beach near industrial areas. *Recent Research in Science and Technology* 4: 30–35.

Soffer, N., Brandt, M.E., Correa, A.M.S. et al. (2014). Potential role of viruses in white plague coral disease. *The ISME Journal* 8: 271–283.

de, e Souza, Silva Pegado, T., Schmid, K., Winemiller, K.O. et al. (2018). First evidence of microplastic ingestion by fish from the Amazon River estuary. *Marine Pollution Bulletin* 133: 814–821.

Strangmann, A., Bashan, Y., and Giani, L. (2008). Methane in pristine and impaired mangrove soils and its possible effect on establishment of mangrove seedlings. *Biological Fertility of Soils* 44: 511–519.

Strode, S.A., Jaeglé, L., Selin, N.E. et al. (2007). Air-sea exchange in the global mercury cycle. *Global Biogeochemical Cycles* 21: GB1017. https://doi.org/10.1029/2006GB002766.

Stuhldreier, I., Bastian, P., Schönig, E. et al. (2015). Effects of simulated eutrophication and overfishing on algae and invertebrate settlement in a coral reef of Koh Phangan, Gulf of Thailand. *Marine Pollution Bulletin* 92: 35–44.

Sturve, J., Scarlet, P., Halling, M. et al. (2016). Environmental monitoring of pesticide exposure and effects on mangrove aquatic organisms of Mozambique. *Marine Environmental Research* 121: 9–19.

Sukumaran, S., Mulik, J., Rokade, M.A. et al. (2014). Impact of *Chitra* oil spill on tidal pool microbenthic communities of a tropical rocky shore (Mumbai, India). *Estuaries and Coasts* 37: 1415–1431.

Sumithra, T.G., Arun Kumar, T.V., Swaminathan, T.R. et al. (2020). Epizootics of epizootic ulcerative syndrome among estuarine fish of Kerala, India, under post-flood conditions. *Diseases in Aquatic Organisms* 139: 1–13.

Sun, Y.-X., Zhang, Z.-W., Xu, X.-R. et al. (2015). Bioaccumulation and biomagnification of halogenated organic pollutants in mangrove biota from the Pearl River estuary, south China. *Marine Pollution Bulletin* 99: 150–156.

Sutherland, K.P., Porter, J.W., Turner, J.W. et al. (2010). Human sewage identified as likely source of white pox disease of the threatened Caribbean elkhorn coral. *Acropora palmata. Environmental Microbiology* 12: 1122–1131.

Sutherland, K.P., Shaban, S., Joyner, J.L. et al. (2011). Human pathogen shown to cause disease in the threatened elkhorn coral *Acropora palmata*. *PLoS One* 6: e23468. https://doi.org/10.1371/journal.pone.0023468.

Tam, N.F.Y., Wong, A.H.Y., Wong, M.H. et al. (2009). Mass balance of nitrogen in constructed mangrove wetlands receiving ammonium-rich wastewater: effects of tidal regime and carbon supply. *Ecological Engineering* 35: 453–462.

Tanaka, K., Ohde, S., Cohen, M.D. et al. (2013). Metal contents of *Porites* corals from Khang Khao Island, Gulf of Thailand: anthropogenic input of river runoff into a coral reef from urbanized areas, Bangkok. *Applied Geochemistry* 37: 79–86.

Thangaradjou, T., Raja, S., Subhashini, P. et al. (2013). Heavy metal enrichment in the seagrasses of Lakshadweep group of islands – a multivariate statistical analysis. *Environmental Monitoring and Assessment* 185: 673–685.

Thomsen, E., Herbeck, L.S., and Jennerjahn, T.C. (2020). The end of resilience: surpassed nitrogen thresholds in coastal waters led to severe seagrass loss after decades of exposure to aquaculture effluents. *Marine Environmental Research* 160: 104986. https://doi.org/10.1016/j.marenvres.2020.104986.

Thurber, R.L.V., Burkepile, D.E., Fuchs, C. et al. (2014). Chronic nutrient enrichment increases prevalence and severity of coral disease and bleaching. *Global Change Biology* 20: 544–554.

Uwadiae, R. (2016). Benthic macroinvertebrate community and chlorophyll *a* (chl-*a*) concentration in sediment of three polluted sites in the Lagos Lagoon, Nigeria. *Journal of Applied Science and Environmental Management* 20: 1147–1155.

Vasanthi, V., Peranandam, R., Arulvasu, C. et al. (2012). Biomarkers of metal toxicity and histology of *Perna viridis* from Ennore estuary, south east coast of India. *Ecotoxicology and Environmental Safety* 84: 92–98.

Venturini, N., Muniz, P., Bicego, M.C. et al. (2008). Petroleum contamination impact on microbenthic communities under the influence of an oil refinery: integrating chemical and biological multivariate data. *Estuarine, Coastal and Shelf Science* 78: 457–467.

Verlis, K.M., Campbell, M.L., and Wilson, S.P. (2013). Ingestion of marine debris plastic by the wedge-tailed shearwater *Ardenna pacifica* in the Great Barrier Reef, Australia. *Marine Pollution Bulletin* 72: 244–249.

Vermeij, M.J.A., Dailer, M.L., and Smith, C.M. (2009). Nutrient enrichment promotes survival and dispersal of drifting fragments in an invasive tropical macroalga. *Coral Reefs* 28: 429–435.

Vermeij, M.J.A., van Moorselaar, I., Engelhard, S. et al. (2010). The effects of nutrient enrichment and herbivore abundance on the ability of turf algae to overgrow coral in the Caribbean. *PLoS One* 5: e14312. https://doi.org/10.1371/journal.pone.0014312.

Viana, I.G., Moreira-Saporiti, A., and Teichberg, M. (2020). Species-specific trait responses of three tropical seagrasses to multiple stressors: the case of increasing temperature and nutrient enrichment. *Frontiers in Plant Science* 11: 571363. https://doi.org/10.3389/fpls.2020.5711363.

Vignesh, S., Dahms, H.-U., Muthukumar, K. et al. (2016). Biomonitoring along the tropical southern Indian coast with multiple biomarkers. *PLoS One* 11: e0154105. https://doi.org/10.1371/journal.pone.0154105.

Vineetha, G., Kripa, V., Komal Karathi, K. et al. (2020). Impact of a catastrophic flood on the heavy metal pollution status and the concurrent responses of the bentho-pelagic community in a tropical monsoonal estuary. *Marine Pollution Bulletin* 155: 111191. https://doi.org/10.1016/j.marpolbul.2020.111191.

Vithanage, G., Fujioka, R.S., and Ueunten, G. (2011). Innovative strategy using alternative fecal indicators (F + RNA/Somatic Coliphages, *Clostridium perfringens*) to detect cesspool discharge pollution in streams and receiving coastal waters within a tropical environment. *Marine Technology Society Journal* 45: 101–111.

Wahedally, S.F., Mamboya, F.A., Lyimo, T.J. et al. (2012). Short-term effects of three herbicides on the maximum quantum yield and electron transport rate of tropical seagrass *Thalassodendron ciliatum*. *Tanzania Journal of Natural and Applied Sciences* 3: 458–466.

Walters, D.M., Jardine, T.D., Cade, B.S. et al. (2016). Trophic magnification of organic chemicals: a global synthesis. *Environmental Science & Technology* 50: 4650–4658.

Wang, Y., Qiu, Q., Li, S. et al. (2014). Inhibitory effect of municipal sewage on symbiosis between mangrove plants and arbuscular mycorrhizal fungi. *Aquatic Biology* 20: 119–127.

Wang, J., Tan, Z., Peng, J. et al. (2016). The behaviors of microplastics in the marine environment. *Marine Environmental Research* 113: 7–17.

Webster, N.S., Webb, R.I., Ridd, M.J. et al. (2001). The effects of copper on the microbial community of a coral reef sponge. *Environmental Microbiology* 3: 19–31.

Wells, F.E., Tan, K.S., Todd, P.A. et al. (2019). A low number of introduced marine species in the tropics: a case study from Singapore. *Management of Biological Invasions* 10: 23–45.

Wilkinson, A.D., Collier, C.J., Flores, F. et al. (2015). Acute and additive toxicity of ten photosystem-II herbicides to seagrass. *Scientific Reports* 5: 17443. https://doi.org/10.1038/srep17443.

Wilkinson, A.D., Collier, C.J., Flores, F. et al. (2017). Combined effects of temperature and the herbicide diuron on Photosystem II activity of the tropical seagrass *Halophila ovalis*. *Scientific Reports* 7: 45404. https://doi.org/10.1038/srep45404.

Willette, D.A. and Ambrose, R.F. (2012). Effects of the invasive seagrass *Halophila stipulacea* on the native seagrass, *Syringodium filiforme*, and associated fish and epibiota communities in the eastern Caribbean. *Aquatic Botany* 103: 74–82.

Willette, D.A., Chalifour, J., Debrot, A.O.D. et al. (2014). Continued expansion of the trans-Atlantic invasive marine angiosperm *Halophila stipulacea* in the eastern Caribbean. *Aquatic Botany* 112: 98–102.

Williams, S.L. (2007). Introduced species in seagrass ecosystems: status and concerns. *Journal of Experimental Marine Biology and Ecology* 350: 89–110.

Wright, S.L., Thompson, R.C., and Galloway, T.S. (2013). The physical impacts of microplastics on marine organisms: a review. *Environmental Pollution* 178: 483–492.

Wu, Y., Chung, A., Tam, N.F.Y. et al. (2008). Constructed mangrove wetland as secondary treatment system for municipal wastewater. *Ecological Engineering* 34: 137–146.

Wurl, O. and Obbard, J.P. (2005). Chlorinated pesticides and PCBs in the sea-surface microlayer and seawater samples of Singapore. *Marine Pollution Bulletin* 50: 1233–1243.

Xie, X., Weiss, D.J., Weng, B. et al. (2013). The short-term effect of cadmium on low molecular weight organic acid and amino acid exudation from mangrove (*Kandelia obovata*) (S.L.) roots. *Environmental Science and Pollution Research* 20: 997–1008.

Xin, K., Zhou, Q., Arndt, S.K. et al. (2013). Invasive capacity of the mangrove *Sonneratia apetala* in Hainan Island, China. *Journal of Tropical Forest Science* 25: 70–78.

Yap, C.K., Ismail, A., Edward, F.B. et al. (2006). Use of different soft tissues of *Perna viridis* as biomonitors of bioavailability and contamination by heavy metals (Cd, Cu, Fe, Pb, Ni, and Zn) in a semi-enclosed intertidal water, the Johore Straits. *Toxicological and Environmental Chemistry* 88: 683–695.

Yoshioka, R.M., Kim, C.J.S., Tracy, A.M. et al. (2016). Linking sewage pollution and water quality to spatial patterns of *Porites lobata* growth anomalies in 'puako, Hawaii. *Marine Pollution Bulletin* 104: 313–321.

Youssef, T. (2002). Physiological responses of *Avicennia marina* seedlings to the phytotoxic effects of the water-soluble fraction of light Arabian crude oil. *Environmentalist* 22: 149–159.

Zhang, H. and Kelly, B.C. (2018). Sorption and bioaccumulation behavior of multi-class hydrophobic organic contaminants in a tropical marine food web. *Chemosphere* 199: 44–53.

Zhang, G., Parker, A., House, A. et al. (2002). Sedimentary records of DDT and HCH in the Pearl River delta, south China. *Environmental Science and Technology* 36: 3671–3677.

Zhang, C.G., Wong, K.K., and Tam, N.F.Y. (2007a). Germination, growth and physiological responses of mangrove plant (*Bruguiera gymnorrhiza*) to lubricating oil pollution. *Environmental and Experimental Botany* 60: 127–136.

Zhang, F.-Q., Wang, Y.-S., Lou, Z.-P. et al. (2007b). Effect of heavy metal stress on antioxidative enzymes and lipid peroxidation in leaves and roots of two mangrove plant seedlings (*Kandelia candel* and *Bruguiera gymnorrhiza*). *Chemosphere* 67: 44–50.

Zhu, C., Li, D., Sun, Y. et al. (2019). Plastic debris in marine birds from an island located in the South China Sea. *Marine Pollution Bulletin* 149: 110566. https://doi.org/10.1016/j.marpolbul.2019.110566.

Zuo, L., Sun, Y., Li, H. et al. (2020). Microplastics in mangrove sediments of the Pearl River estuary, south China: correlation with halogenated flame retardants' levels. *Science of the Total Environment* 725: 138344. https://doi.org/10.1016/j.scitotenv.2020.138344.

CHAPTER 13

Climate Change

13.1 Introduction

Human-induced changes to the earth's climate have resulted in changes to marine ecosystems (Harley et al. 2006; Doney et al. 2012, 2020). Most climate-related ecological research has focused on shifts in distribution, abundance, growth, reproduction, and physiology of tropical organisms driven directly by increasing temperatures, ocean acidification and CO_2, and sea-level. These works reveal that both abiotic changes and biological responses to climate change are and will be more complex than originally thought. For example, changes on ocean chemistry may be more important than changes in temperature for the life processes and survival of many organisms. Ocean circulation, which drives larval transport, will also change, with important consequences for the dynamics of marine populations. Climatic impacts on a few keystone species may result in large-scale changes in community structure and composition.

Rising atmospheric CO_2 concentrations are linked to shifts in temperature, circulation, stratification, nutrient input, oxygen content, and ocean acidification, with potentially wide-ranging effects on the biology and ecology of marine organisms and their communities (Doney et al. 2012, 2020). Shifts in populations and communities are occurring because of physiological intolerance to rapidly changing environments, altered dispersal patterns, and changes in species interactions. These processes, together with local climate-driven invasion and extinction, result in altered community structure and diversity. Impacts are particularly striking for the poles and tropics, especially for the survival of corals in the face of increased temperatures and altered ocean chemistry. Climate change effects may modify energy and material flows, as well as biogeochemical cycles, eventually impacting ecosystem functioning and structure (Doney et al. 2012).

In previous chapters, the physical basis of climate change (Section 2.7) and various biological aspects of the effects of climate change have been discussed. These include impacts on interspecific competition (Section 6.5), COTS outbreaks (Section 6.9), mangrove dieback (Section 8.3), and the decline in phytoplankton abundance and production due to increasing sea surface (SSTs) temperatures (Section 8.8). The effects of climate phenomena such as ENSO have been discussed throughout several chapters.

This chapter focuses on the best understood and most worrisome impacts of climate change on tropical estuarine and marine biota: coral bleaching; physiological

Tropical Marine Ecology, First Edition. Daniel M. Alongi.

stress due to increasing SSTs; mangrove displacement and survival in the face of sea-level rise (SLR); ocean acidification effects on calcifying organisms, especially corals; changing ocean currents and circulation patterns; and effects on tropical fisheries. The focus is on mangrove, seagrass, and coral reef biota as well as estuarine and coastal fisheries (Section 13.7). The biota of tropical sandy beach, tidal flat, coastal, shelf, and open ocean habitats are not considered due to a lack of information, although they may be equally vulnerable to climate change.

13.2 Rising Temperatures, Increased Storms, Extreme Weather Events, and Changes in Precipitation

Higher temperatures lead to faster metabolic rates, but there are tolerance limits to extreme temperature changes for an individual species, and species differ in their tolerance limits. Tropical organisms are considered more susceptible to temperature increases than temperate and boreal organisms because they live closer to their upper tolerance limits. The most dramatic example of the negative effects of rising temperatures on tropical marine communities is the bleaching of corals (Figure 13.1). Coral bleaching is often accompanied by an increase in coral diseases as temperatures rise, and such impacts affect other reef organisms and reef functioning.

Coral bleaching is the term used to describe the loss by corals of all or some of their symbiotic algal and microbial communities and photosynthetic pigments, with the result that the white carbonate skeleton becomes visible through the translucent tissue layer (Lough 2008; Hoegh-Guldberg 2011). Coral bleaching

FIGURE 13.1 Coral bleaching on the Great Barrier Reef. *Source:* Image taken by and used with permission of Morgan Pratchett, ARC Centre of Excellence for Coral Reef Studies, James Cook University, Australia. © Morgan Pratchett.

can also occur in response to other environmental impacts, such as low salinity, pollution, and high irradiance (Hoegh-Guldberg 2011). However, in these latter cases, coral bleaching is localised whereas bleaching due to climate change has resulted in an increasing frequency of large-scale, mass coral bleaching events where entire reefs are affected. Like other tropical marine biota, reef-building corals live near their upper thermal tolerance limits, so it takes only a small but sustainable rise in SSTs to initiate bleaching. The threshold for bleaching becomes critical during the seasonal maximum of SSTs. There is no absolute temperature at which corals will bleach, but rather the temperature threshold varies with ambient water temperatures at a given locality. Over long timescales, many coral species have adapted to local environmental conditions. Nevertheless, mass coral bleaching may affect thousands of square kilometres within a few weeks, hence becoming the greatest source of mortality for corals. A predictive analysis suggests that bleaching events will increase in their severity and that to save > 10% of coral reefs worldwide will require limiting warming to below 1.5 °C relative to preindustrial levels (Frieler et al. 2013). Even under optimistic assumptions regarding the ability of corals to adapt to a new temperature regime, one-third of the world's coral reefs are projected to be subject to long-term degradation under the most optimistic IPCC emissions scenarios. The ability of coral reefs to survive bleaching will depend on: (i) how much coral cover is lost, and which species are locally extirpated, (ii) the ability of remnant and recovering coral communities to adapt to or acclimatise to higher temperatures and other climatic factors such as ocean acidification, (iii) the changing balance between reef accumulation and bioerosion, and (iv) our ability to maintain ecosystem resilience by restoring healthy levels of herbivory, macroalgal cover, and coral recruitment (Baker et al. 2008).

Coral bleaching is caused by large-scale anomalies in SSTs. Spatial and temporal patterns with bleaching are most common in localities experiencing high-intensity and high-frequency thermal stress anomalies (Sully et al. 2019). Coral bleaching is significantly less common in localities with a high variance in SST anomalies. Geographically, the highest probability of coral bleaching occurs at tropical mid-latitudes 15–20 °N and S of the equator, despite similar heat stress levels at equatorial sites. The onset of coral bleaching in the last decade has occurred at significantly higher SSTs (about 0.5 °C) than in the previous decade, indicating that heat susceptible genotypes may have declined and/or adapted such that the remaining coral populations now have a higher temperature threshold for bleaching.

The most extensive and well-documented bleaching event on record occurred during the 1997/1998 ENSO event in which bleaching was recorded from nearly every coral reef region of the world (Sully et al. 2019). Sixteen per cent of the world's coral reefs were damaged and, of these, 40% recovered; the remainder did not. The scale and magnitude of this event were linked directly to anomalously high SSTs. There is evidence that for some reefs, the thresholds for the occurrence and intensity of bleaching may have been modified by the history of past bleaching events. In the southern Persian/Arabian Gulf, the meta-population of the coral, *Acropora downingi*, declined by over 90% due to bleaching and diseases caused by rising SSTs (Riegl et al. 2018). Population dynamics of formerly dominant table corals corresponded to temperature disturbance regimes in 1996 to 1998, 2010 to 2012, and 2013 to 2017 (Cacciapaglia and van Woesik 2020). Increased disturbance frequency and severity caused progressive reduction in coral size, cover, and population fecundity. This coral meta-population disintegrated into isolated populations, suggesting that this species is heading towards extinction due mainly to increasingly frequent temperature-induced mortality events.

The occurrence and intensity of coral bleaching can be highly variable at the local scale, both within a coral colony, between coral colonies within a reef and between reefs. This is partly due to coral species differing in their resistance to warmer temperatures as well as the fact that water temperatures vary within and between reefs. Such local-scale variations that reduce thermal stress can be linked to water movements, such as upwelling, mixing, tides, and surface and internal waves. Indeed, bleaching can be greatly increased under quiescent conditions when there is little water movement.

Corals may die, partially die, or recover from a bleaching event. Recovery depends on the coral rapidly regaining its zooxanthellae population soon after bleaching. There is mounting evidence that even for corals that recover fully, there are long-term consequences due to the thermal stress associated with the bleaching event. These consequences include reduced reproduction, reduced growth rates, and increased susceptibility to other disturbances, such as coral diseases (Baker et al. 2008). The overall effect of a bleaching event is a reduction in species richness, diversity, and a change in community structure.

Average temperatures in the tropical oceans have warmed less than half the global average since the nineteenth century, about 0.5 °C (Lough 2008). Mass bleaching is highly likely on a coral reef if sea temperatures rise by 1 °C above the long-term summer maximum temperatures (Hoegh-Guldberg 2011). Maintenance of hard coral cover requires corals to increase their upper thermal tolerance limits by 0.1–1.0 °C $decade^{-1}$ (Hoegh-Guldberg 2011). Corals can to some degree adapt or acclimate to increased SSTs. Corals can adapt to bleaching by changing the dominant type of symbiotic algae to more thermally tolerant partners. Corals contain several different types of symbiotic algae and may be able to change the relative abundances of these zooxanthellae ('symbiont shuffling'). This strategy may be at the expense of growth rates and reproduction and may only occur in a few species or may not occur rapidly enough to keep up with higher sea temperatures. An analysis of coral bleaching in the Seychelles (Graham et al. 2015) found that losses of live corals totalled 90% coral cover. Recovery was favoured on the most structurally complex reefs and in reefs in deeper water, and when densities of juvenile corals and herbivorous fish were relatively high and nutrient loads low. During the 1982/1983 El Niño, most reefs in the eastern tropical Pacific, such as those in the Galapagos Islands, collapsed and many more in the region were decimated by massive coral bleaching and mortality. However, after repeated thermal stress such as caused by the 1997/1998 El Niño, coral reefs in the eastern tropical Pacific region demonstrated persistence and resiliency (Romero-Torres et al. 2020). During the period 1970–2014, coral cover exhibited temporary reductions following major ENSO events, but no overall decline. Reef recovery patterns allowed coral to persist under these El Niño-stressed conditions, often recovering in 10–15 years, suggesting that these coral reefs adapted to thermal extremes and may have the ability to adapt to near-future climate change thermal anomalies.

Other factors help to mitigate heat stress. During the peak of the 2015/2016 El Niño throughout the Pacific, temperature fluctuations associated with internal ocean waves reduced cumulative heat exposure by up to 88% (Wyatt et al. 2020). The durations of severe thermal anomalies decreased by >36% in some areas and were prevented entirely in others. Thus, internal waves across depths on coral reefs have the potential to create and support thermal refuges such that heat stress and coral bleaching risk are ameliorated.

Upwelling also buffers the impacts of climate change on coral reefs of the eastern tropical Pacific (Randall et al. 2020). Along the coast of Panama, the Gulf of Panama in the east experiences stronger seasonal upwelling than in the Gulf of Chiriquí in the west. Historically, corals in the Gulf of Chiriquí have higher growth rates than those in the Gulf of Panama, but more recently, both gulfs have warmed considerably with the annual thermal maximum in the Gulf of Chiriquí increasing faster and ocean temperatures becoming more variable than in the recent past. Coral cover, coral survival, and coral growth rates were all significantly higher in the Gulf of Panama. Following the 2015/2016 El Niño event, corals in the Gulf of Chiriquí bleached extensively, whereas upwelling in the Gulf of Panama moderated the high temperatures caused by El Niño, allowing coral to largely escape heat stress. Localised upwelling zones may thus offer a refuge from the thermal impacts of climate change, while reef growth in the non-upwelling areas of the eastern tropical Pacific continues to decline.

Refuges from increasing environmental extremes can be found in adjacent habitats. Mangrove lagoons on Woody Isles and Howick Island, Great Barrier Reef, support populations of the dominant corals, *Porites lutea* and *Acropora millipora*, despite low pH (<7.6), low oxygen ($<1\,mg\,l^{-1}$), and highly variable temperature (>7 °C) conditions (Camp et al. 2019). Physiological plasticity and flexibility of symbiotic associations appear crucial in supporting these corals to thrive in a mangrove habitat. It is thus conceivable that mangroves provide a refuge for the survival of corals in other regions.

Multiple environmental and climate stressors play an important role in determining the extent of coral bleaching on reefs. Coral bleaching at 226 sites and 26 environmental variables that represent different mechanisms of stress responses from East Africa to Fiji were assessed by McClanahan et al. (2019) to evaluate the coral response to the 2014/2015 ENSO thermal anomaly. The importance of peak hot temperatures, the duration of cool temperatures and temperature bimodality explained about 50% of the variance compared to the common degree-heating temperature index that explained only 9% of the variance. These findings suggest that the threshold concept as a mechanism to explain bleaching alone was not as powerful as the multidimensional interactions of stresses, which include the duration and temporal patterning of hot and cold temperature extremes relative to average local conditions. Examining the combined effects of rising sea temperature and sea level on shallow-water coral reefs in the eastern Indian Ocean over a 40 year period, Brown et al. (2019) found that SLR not only promoted coral cover but also limited damaging effects of heat-induced bleaching. Greatest loss of coral cover and disruption of community structure occurred on the shallowest reef flats. Damage was less severe on the deepest reef flat where corals were subjected to less air exposure, rapid flushing, and longer submergence in turbid waters; recovery of the most damaged reefs took eight years.

The impact of rising temperatures on corals needs to be considered along with other climate changes, such as ocean acidification. In flow-through mesocosms, Bahr et al. (2020) exposed the coral, *Pocillopora acuta*, to a range of lower pH levels and increasing temperatures. Neither ocean warming nor acidification separately or in combination affected the size or abundance of coral recruits, but heating did impact subsequent health and survival of the recruits. Coral recruits in heated tanks experienced higher levels of bleaching and subsequent mortality during annual maximum temperatures, suggesting that *P. acuta* can recruit under ocean warming and acidification, but is susceptible to bleaching and mortality during warmer periods.

Bleaching is not the only outcome of warming oceans on coral reef ecosystems. The incidence of disease increases in a warmer world and there is sufficient evidence to show that the frequency of disease outbreaks in corals and other marine organisms has increased with increasing SSTs including on coral reefs subjected to bleaching. Warmer waters are increasing the severity of diseases in the tropical ocean, reducing the vitality of marine ecosystems, such as coral reefs (Baker et al. 2008; Lough 2008).

Coral bleaching results in impacts other than disease outbreaks. The breakdown of the reef framework and the loss of critical habitat for associated reef fish and other biota are associated with bleaching events (Baker et al. 2008). Secondary ecological effects, such as the concentration of predators on remnant surviving coral populations, have accelerated the pace of coral decline in some regions. Rising temperatures have a physiological impact in corals as their first response occurs at the cellular level. A summary of coral heat stress responses (Figure 13.2) at the cellular level shows that increasing temperatures trigger Ca release from the endoplasmic reticulum, which leads to changes in cell function, such as cytoskeleton rearrangement and cell adhesion disruption, through disruption of Ca homoeostasis (Cziesielski et al. 2019). Meanwhile, the metabolic rate is also increased causing an increase in reactive oxygen species (ROS) and NO. As oxidative stress from ROS and NO is experienced by the coral, apotosis or necrosis ensues. The symbionts have their own temperature tolerances and responses but also produce ROS under stress which can leak into the host and exacerbate oxidative stress (Cziesielski et al. 2019).

Rising temperatures affect coral reproduction. In a laboratory experiment elevating temperature by 2 °C prior to spawning in the coral, *Acropora digitifera*, spawning was advanced one day and egg volume decreased significantly as did sperm numbers (Paxton et al. 2015), suggesting that increasing SSTs play a vital role in

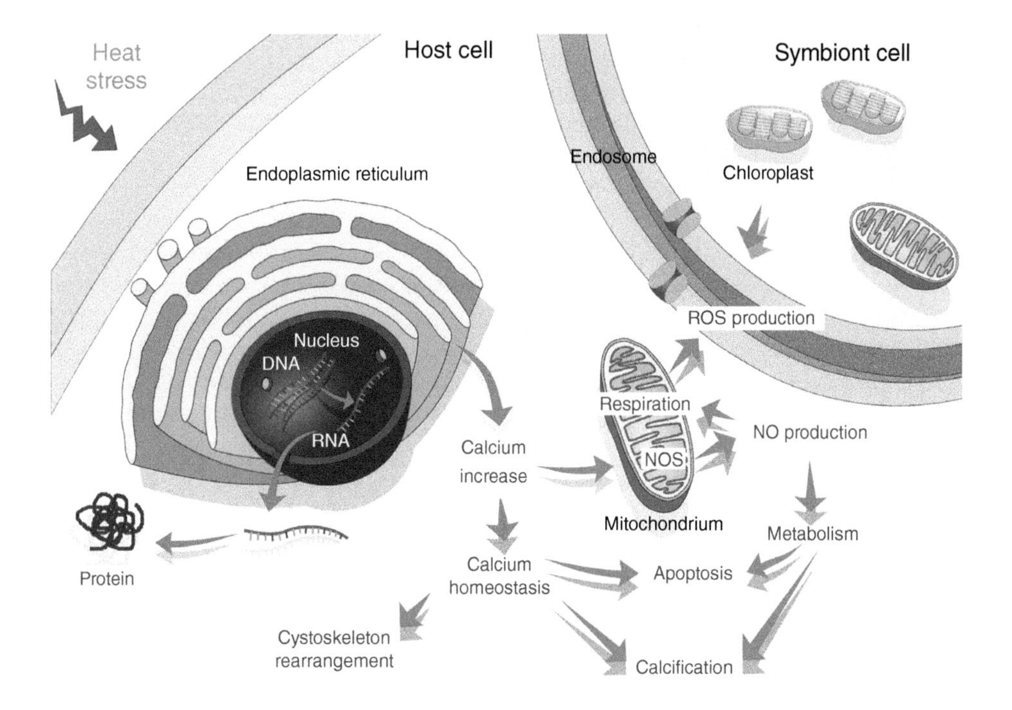

FIGURE 13.2 Representation of the impact of heat stress on coral host and symbiont cells. *Source:* Cziesielski et al. (2019), figure 2, p. 10058. Licensed under CC BY 4.0. © John Wiley & Sons.

altering reproductive timing and physiology. Dais et al. (2019) exposed several Indo-Pacific coral species to long-term increases in temperature and found that they were all sensitive to long-term exposure to 32 °C. There were interspecific differences to heat stress and each species reacted differently at each temperature treatment. The species, *Turbinaria reniformis*, *Galaxea fascicularis*, and *Psammocora contigua*, were the most resistant to heat stress, showing no damage to lipid peroxidation, catalase activity, and glutathione S-transferase at 30 °C. Unlike *G. fascicularis*, both *T. reniformis* and *P. contigua* showed no evidence of oxidative damage at 32 °C. All remaining species died at 32 °C, with *S. pistillata* and *P. damicornis* the most susceptible to heat stress.

Increasing temperatures due to global warming are seriously impacting the physiology, reproduction, and growth of coral reef fish and invertebrates. Exposing the giant clam, *Tridacna crocea*, to high temperatures led to a decrease in the density of symbiotic zooxanthellae, and the activities of superoxide dismutase was significantly lower in the group exposed to heat stress compared with the control group (Zhou et al. 2019). There were 47 significantly up-regulated and 88 significantly down-regulated genes, mostly related to unfolded protein binding and ATP binding. These results suggest that high temperature induces excessive oxidative stress through repressed antioxidant ability and furthers the collapse of the symbiosis between clam host and symbiont. Similarly, oxidative stress has been observed in populations of the coral reef damselfish, *Acanthochromis polyacanthus*, with increasing SSTs with indications that this species is highly thermally specialised. Aerobic performance indicated a limit to metabolic function at 33 °C following an increase in aerobic capacity at 31.5 °C (Rodgers et al. 2019).

Rising temperatures are leading to range expansion of some coral reef organisms, especially corals and highly mobile organisms, such as fish. Poleward expansion of *Acropora* spp. has been detected off the east coast of Australia (Baird et al. 2012) and expansion of corals has also been observed in Japan (Yamano et al. 2011). Using century-old records, Yamano et al. (2011) found that *Acropora hyacinthus, Acropora muricata, Acropora solitaryensis*, and *Pavona decussata* have expanded their range poleward since the 1930s, whereas none of the species showed range contraction or extinction. These expansions attained speeds of up to 14 km a^{-1}, possibly with the help of the Kuroshio and Tsushima warm currents. More recent analyses have suggested that range expansion for reef corals will be limited by insufficient solar radiation for symbiotic photosynthesis, especially in winter (Muir et al. 2015); Madin et al. (2016) disagree with this conclusion and argue that factors limiting coral expansion are elusive, but probably species-specific.

A larger database exists for range expansion of reef fish (Miranda et al. 2019; Kingsbury et al. 2020). Miranda et al. (2019) quantified relative growth rate, morphometric condition, and trophic niche of juvenile *Acanthurus triostegus,* a dominant surgeonfish, across 10 °C of latitude off eastern Australia. Temperate food sources were richer in N than tropical foods and stable isotope analysis of fish muscle revealed a large trophic niche breadth at the highest latitude indicating a generalist foraging strategy and more N-enriched isotopic signatures compared to tropical regions. Fish growth and body condition were similar across latitudes despite a 4 °C temperature difference, suggesting that foods of higher nutritional quality may compensate for the effects of cooler ocean temperatures. In contrast, along a 730-km latitudinal gradient, Kingsbury et al. (2020) found that tropical reef fish maintain their body condition and stomach fullness but exhibit decreased growth rates, activity levels, and feeding rates in their new temperate environment. Thus, fish may face a "growth-maintenance trade-off" under the initial phases of ocean warming, allowing them to maintain their body condition in cooler waters, but at the cost of slower growth.

For seagrasses, rises in SSTs result in increased rates of metabolism and shifts in the maintenance of a positive carbon balance (Short and Neckles 1999). These changes may result in changes in seasonal and geographic patterns of species abundance and distribution, including range expansion (Gorman et al. 2016; Beca-Carretero et al. 2020). The tropical seagrass, *Halophila decipiens*, has extended its range in the eastern Mediterranean and in the SW Atlantic coast (Brazil) in response to an increase in SSTs over the past century. Experiments attest to the phenotypic plasticity of this species, including its ability to tolerate reduced irradiance and the low temperatures of the temperate SW Atlantic (Gorman et al. 2016). Models predict that this species will keep expanding westward and northward as the Mediterranean continues to become saltier and warmer (Beca-Carretero et al. 2020).

Like corals, the direct effects of increased temperature on seagrasses will depend on the thermal tolerances of individual species and their optimum temperatures for photosynthesis, respiration, and growth. In Mediterranean *Cymodocea nodosa*, a subtropical/tropical species, oxygen production, and respiration increased with increasing temperature but with no optimum. Warm water species such as *C. nodosa* increase their photosynthesis and respiration over a wide range of temperatures, whereas temperate species show a photosynthetic optimum at temperatures below a seasonal maximum (Short and Neckles 1999). Seagrass growth is a function of factors that are subject to temperature regulation, such as nutrient uptake and enzyme-mediated processes. For species growing near their optimal temperature limits, an increase in average annual temperature will likely decrease productivity and distribution.

Higher temperatures may also alter seagrass distribution and abundance through direct effects on flowering and seed germination (Short and Neckles 1999). Temperature effects are complex, affected by changes in such co-factors as salinity and nutrient availability (Viana et al. 2020). In laboratory experiments with coexisting *T. hemprichii, C. serrulata*, and *H. stipulacea,* subjected to two temperature and two nutrient treatments, species-specific responses depending on their life history strategies were observed (Viana et al. 2020). At maximum ambient temperature, *C. serrulata* photo-acclimatised but the other two species did not. In contrast, *T. hemprichii* was resistant to nutrient over-enrichment, but *C. serrulata* suffered nutrient loss. *H. stipulacea* showed a limited response to both parameters suggesting that this species is the most tolerant.

Many seagrass species are extremely plastic in their response to temperature differences, so an increase in SSTs will probably not greatly alter recruitment from sexual reproduction. At warmer temperatures, the growth balance between seagrasses and algal epiphytes may change in favour of the algae as higher water temperatures have been found to exacerbate the impact of algal epiphytes on seagrass growth. Accelerated rates of algal production could eliminate seagrasses in tropical eutrophic estuaries where their growth and survival are marginal.

Laboratory and field data indicate that tropical seagrasses undergo thermally induced physiological stress at temperatures $> 35\,°C$ (Campbell et al. 2006; Rasheed and Unsworth 2011). For example, photosynthetic efficiency of *H. univervis* declined significantly at $35\,°C$ (Campbell et al. 2006). Temperature stress results not just from seagrass being subjected to higher seawater temperatures, but also from desiccation stress during periods of low tide during daylight hours. Shallow-water temperatures vary greatly, and it is not unusual for tropical seagrass beds to be subject to temperatures > 35 °C with resultant 'burning' of seagrass (Collier and Waycott 2014). In the Great Barrier Reef, seagrasses commonly experience water temperatures up to 43 °C at low tide. The growth dynamics of four seagrass species in response to short-duration spikes of water temperature (35, 40, and 43 °C) were examined, with increasing the temperature to 35 °C having positive effects on seagrass photosynthesis, whereas 40 °C was a critical threshold where there were strong species differences and a large impact on growth and mortality (Collier and Waycott 2014). At 43 °C, there was complete mortality after two to three days of exposure. At lower temperatures (27–33 °C), the growth of *H. univeris* was within a physiological optimum temperature range, exhibiting 2.3x higher photosynthetic rates at 33 °C than at 27 °C (Collier et al. 2011). In contrast, 33 °C exceeded the optimum temperature threshold for *Zostera muelleri*. It is likely that *H. univeris* will survive future projected temperature changes in the Great Barrier Reef, but the same is not true for *Z. muelleri*, indicating a shift in the dominance of seagrass species within the region.

A similar situation exists for intertidal seagrasses. In the macro-tidal Kimberley region of Australia, the seagrasses *T. hemprichii* and *E. acoroides* inhabit a high coralline algal terrace that is exposed at low tide where they are subjected to high pO_2, pH, and temperature (38 °C). Net photosynthesis was maximal for these species at 33 °C but declined rapidly at higher temperatures although dark respiration increased (Figure 13.3). During night-time low tides, O_2 declined sufficiently to cause shoot base anoxia. Shoots exposed to 40 °C for four hours showed recovery of net photosynthesis and dark respiration, but 45 °C caused leaf damage (Pedersen et al. 2016). These seagrasses live at the extreme margin, implying that any further increase in temperature or other environmental parameters may likely threaten their existence. Indeed, in mesocosm experiments mimicking low tide, *T. hemprichii, C. serrulata, E. acoroides*, and *Thalassodendron ciliatum* maintained full photosynthetic rates and all species but *T. ciliatum* withstood 40 °C; only at 45 °C did all species display significantly lower photosynthetic rates (George et al. 2018). During temperature stress, rates of CH_4 emission and sulphide levels increased in the seagrass sediments (George et al. 2020). Biomass, however, showed a different pattern where significant losses of above- and belowground biomass occurred in all species at temperatures above 36 °C (George et al. 2018). Although tropical seagrasses can tolerate high midday temperature stress, a few degrees increase in maximum daily temperature may cause significant losses of seagrass biomass and productivity (George et al. 2018).

Increased temperatures affect the frequency and intensity of extreme weather events, such as cyclones. Such events result in seagrass decline in many parts of the world. The direct impacts of storm disturbance on seagrass beds include erosion by wave action, shading from suspended sediments, and smothering by sediment deposition. Increased storm disturbance may result in a decline in the abundance and distribution of climax species and an increase in abundance of early colonising and mid-successional species within the seagrass community. Recolonisation for many shallow-water seagrass species occurs mainly by vegetative branching, and populations of climax species such as *T. testudinum* may take many years to recover from extreme weather events. An additional effect of increased storm activity will be increased rainfall (decreased salinity) and river discharge (sediment suspension, decline in water clarity). Long-term dynamics of seagrass meadows in north Queensland, Australia, related closely to changes in air temperature, precipitation, daytime tidal exposure, and freshwater runoff; elevated temperatures and reduced river flow were significantly correlated with periods of lower seagrass biomass

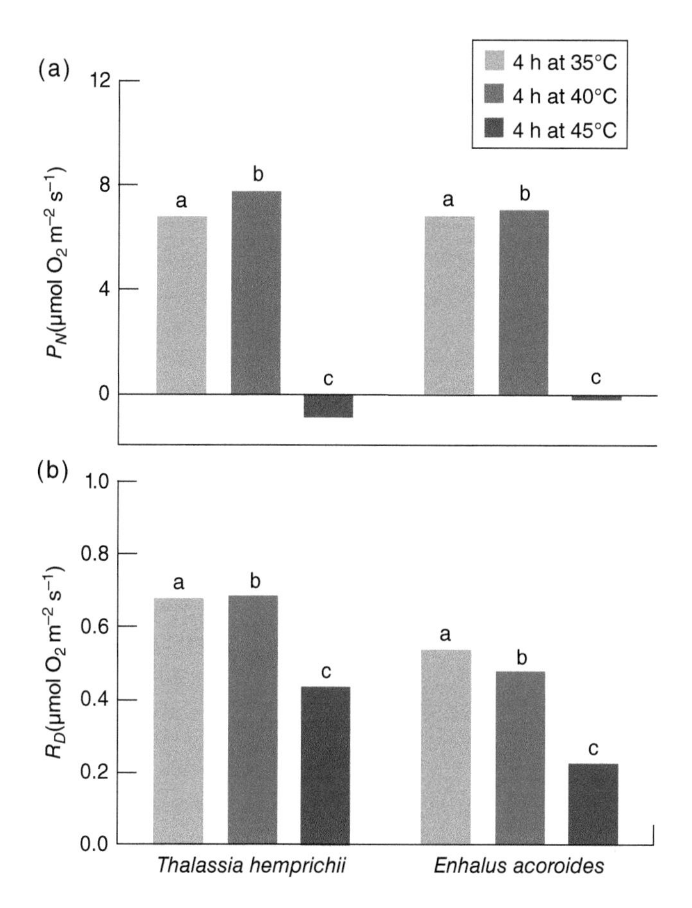

FIGURE 13.3 Recovery of (a) underwater net photosynthesis and (b) dark respiration of the intertidal seagrasses, *Thalassia hemprichii* and *Enhalus acoroides*. Whole plants were exposed to 35, 40, and 45 °C for four hours in light and then left to recover overnight at 25 °C in the dark before measurements at 35 °C were taken 12–18 hours later. Different letters indicate significant differences within each species. *Source:* Pedersen et al. (2016), figure 3, p. 1213. © John Wiley & Sons.

(Rasheed and Unsworth 2011). Not only will there be changes in precipitation rates and regional patterns of rainfall, but the frequency and intensity of extreme weather events, such as cyclonic storms and prolonged droughts, is predicted to increase. Temperature anomalies, defined as extreme temperature events more than three standard deviations from the long-term mean from 1951 to 1980, have shifted more than one standard deviation towards higher values, leading to more extreme warming events (Hansen et al. 2012).

The increased occurrence of such events is having a dramatic impact on mangroves, such as mass mortality events (Lovelock et al. 2017; Sippo et al. 2018, 2020a). The clearest example of mangrove mortality due to an extreme weather event was the massive dieback of mangroves in the Gulf of Carpentaria, Australia (Duke 2017; Sippo et al. 2018, 2020a). Mangrove forests in the gulf suffered severe dieback (6% of vegetation) during the summer of 2015/2016 along 1000 km of shoreline. The onset of the dieback was coincident with unprecedented high temperatures, low precipitation, and lack of a normal summer wet season (Duke 2017). An unusually lengthy severe drought coupled with a temporary drop in sea level contributed to mass mortality. The dieback had severe consequences for mangrove functioning as evidenced by a shift from a dominance of oceanic carbon export to increased atmospheric CO_2 emissions and decreased alkalinity export (Sippo et al. 2020b), likely driven by reduced mangrove productivity and increased oxygen soil penetration. Dieback of mangroves in the gulf resulted in a trophic shift in crabs, with a loss of litter-feeding crabs but an increase in crabs that feed on microphytobenthos (Harada et al. 2020). An examination of data since the 1960s shows about 36 000 ha of mangrove mortality worldwide, about 70% of which was attributed to high-intensity weather events (typhoons, cyclones, and hurricanes) and climate extremes (Sippo et al. 2018). Tropical cyclonic storms account for 45% of the reported mangrove mortality area since the 1960s, being the largest cause of mortality.

Recent large-scale mortality, however, associated with extreme climatic events in Australia accounts for 22% of all reported forest loss over the past six decades, suggesting the increasing importance of extreme climatic events (Sippo et al. 2018). In Mangrove Bay, Western Australia, there have been two dieback events over a 16 year period, with the most recent one coincident with the dieback in the Gulf of Carpentaria (Lovelock et al. 2017). The diebacks in Mangrove Bay were coincident with periods of low sea level due to intensification of ENSO leading to increased soil salinities and subsequent canopy loss and reduced recruitment.

The impact of any one cyclonic disturbance is often significant, with some stands taking decades to fully recover, as evidenced from mangrove forests in the Caribbean and Florida (Imbert 2018; Rivera-Monroy et al. 2019); such climatic disturbances prevent these ecosystems from ever reaching a steady state. However, not all mangroves are significantly affected by extreme climate events as found for mangroves on a Colombian Caribbean Island that appeared to be resilient to short drought events related to ENSO (Galeano et al. 2017). At the other extreme, mangroves such as *A. germinans* may be resilient to extreme freeze events due to genetically based freeze tolerances (Hayes et al. 2020). Nevertheless, increasing mortality events are likely in future considering forecasts of increasing frequency, intensity, and destructiveness of cyclonic storms and climatic extremes, such as heat waves and low and high sea-level episodes.

Mangroves show multiple responses to increasing air and water temperatures. Increased temperatures affect mangroves by: (i) altering species composition, (ii) changing phenological patterns, such as the timing of flowering and fruiting, (iii) increasing mangrove productivity where temperature does not exceed an upper threshold, and (iv) expanding mangrove ranges to higher latitudes where range is temperature-limited but not limited by other factors (Gilman et al. 2008). Rises in temperatures may also result in: (i) decreased survival in areas of increased aridity, (ii) increased water vapour pressure deficit, (iii) increased secondary production, especially microbes, and shifts in species dominance, and (iv) an increase in biodiversity (Alongi 2008). Physiologically, rates of leaf photosynthesis peak for most species at temperatures at or below 30 °C and leaf CO_2 assimilation rates of many species decline, either sharply or gradually, as temperature increases from 33 to 35 °C (Alongi 2015).

Expansion of latitudinal range at the expense of saltmarsh is underway (Cavanaugh et al. 2014, 2015; Whitt et al. 2020). Mangroves have encroached upon saltmarshes in the Gulf of Mexico, the SE United States, New Zealand, Australia, southern China, and southern Africa. Satellite imagery has shown that the area of mangrove forests has doubled at the northern end of their historic range on the east coast of Florida

(Figure 13.4a, b). Ward et al. (2016) and Osland et al. (2017, 2020) found that air temperature and rainfall best explain the distribution and abundance of the world's mangrove forests and found that the influence of climatic drivers is best characterised via regional range limits. Climatic thresholds for mangrove presence, abundance, and species richness differ among a variety of range limits in different geographical areas. Limits to the expansion of mangroves are due to the frequency and intensity of extreme cold events. An examination of 38 sites spread across the mangrove range in the Gulf of Mexico and Atlantic coasts of North America found that for *A. germinans* near their northern range limit, the temperature threshold for leaf damage was close to −4 °C with mortality thresholds closer to −7 °C (Osland et al. 2020).

Rising temperatures and expansion of mangroves into higher latitudes may impact mangrove flora and fauna. Rising temperatures also affect animal physiology in different ways. For example, the generalist fiddler crab, *Minuca rapax*, loses less water than the fiddler crab, *Leptuca thayeri* (Principe et al. 2018), due to differences in carapace permeability; survivability is higher for *M. rapax* under desiccated conditions but is more affected by temperature increase on its physiology. The fiddler crab, *L. uruguayensis*, is more sensitive to warming than *L. leptodactyla* showing physiological and behavioural differences (da Silva Vianna et al. 2020). The latter species was able to adjust its metabolic rate to temperature increase and reduce ammonia excretion, whereas the former species showed adaptation limits. Fiddler crabs inhabiting vegetated areas are more vulnerable to higher temperatures and may change their geographic range, while fiddler crabs on mud and sand flats are more tolerant to temperature increases and may have a competitive advantage as global temperatures rise. A mesocosm experiment found that an increase of 1.2 °C led to the homogenisation and flattening of mangrove root epibiont communities, leading to a 24% increase in the overall cover of algal epibionts on roots but a 33% decline in epibiont species diversity and a decrease in structural complexity (Walden et al. 2019).

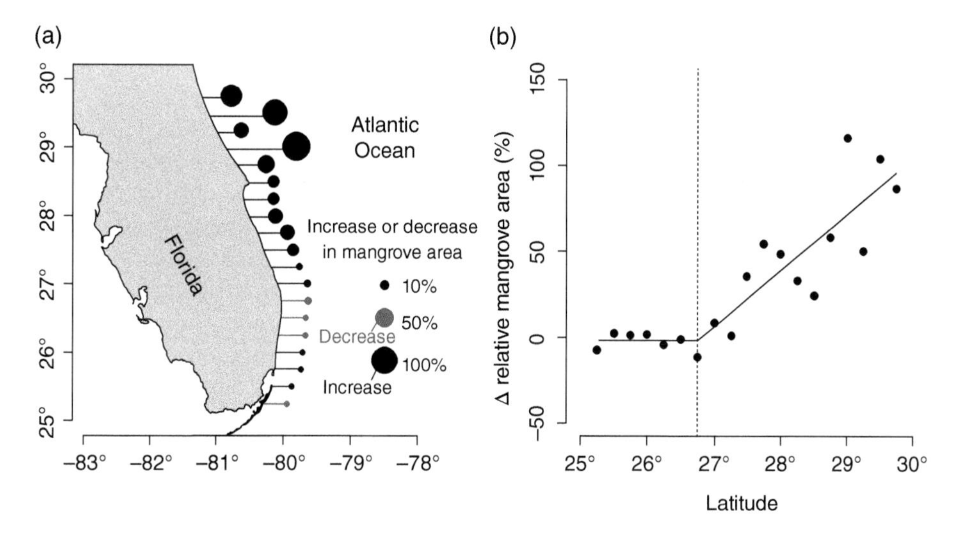

FIGURE 13.4 (a) Map of east coast of Florida showing the long-term increase (black) or decrease (red) in mangrove area for each 0.25° latitudinal band from 25.25 to 29.75 °N. (b) Relationship between latitude and relative change in mangrove area. *Source:* Cavanaugh et al. (2014), figure 1, p. 2. © National Academy of Sciences.

Along the South African coast, fiddler crabs have spread farther south, while other fauna such as the gastropod, *Terebralia palustris*, have disappeared, although there does not appear to be a decrease in diversity with an increase in latitude (Peer et al. 2018). The transition to higher latitudes and the expansion of mangroves at the expense of saltmarshes indicates that complex interactions are occurring during the transition, especially competition. Mangrove species that invade saltmarsh areas are superior competitors (Macy 2020; Manea et al. 2020). Migration may be mediated by biotic interactions and may be facilitated by increasing propagule abundance from greater reproductive rates and greater genetic variation due to outcrossing (Proffitt and Travis 2014). Surveying the Atlantic and Gulf coasts of Florida, Proffitt and Travis (2014) observed that reproductive frequencies varied significantly but increased with latitude and strongly along the gulf coast with a concomitant increase in outcrossing. Adaptation to a new environment is perpetuated and promoted by the self-enforcing nature of migration as more colonisers lead to more propagules and outcrossing leads to greater genetic variation.

While mangrove expansion and saltmarsh contraction are consistent with increasing temperatures and reduced frequency of extreme cold events, changes in precipitation patterns and nutrient availability cannot be ruled out as co-factors. Changes in patterns of precipitation will result in drastic changes in community structure and function as the wet tropics get wetter and the dry tropics get more arid (Section 2.7). Also, increases in the frequency and intense of storms, cyclones, hurricanes, and typhoons will result in changes in the disturbance regime of communities.

Seagrasses and mangroves will undergo similar changes due to changing patterns of precipitation and increased intensity of storms, but such changes are likely to co-occur with rises in sea-level, temperature, and atmospheric CO_2 concentrations. Compared to arid-zone stands, mangrove forests in the wet tropics have greater biomass and productivity, consist of less dense but taller trees, and tend to inhabit finer sediment deposits, but there are no clear species richness or diversity patterns between high and low precipitation areas; low species richness may be attributable to high variability in annual rainfall. But mangroves clearly thrive in wet environments where they can deal less stressfully with lower salinity and more available freshwater (Alongi 2015). Increasing intensity of storms and cyclones are likely to defoliate mangroves, causing increased tree mortality as well as soil erosion. Gaps and gap recruitment are likely to increase with increasing storm activity, depending on change in level of storm intensity, frequency, and stand location relative to the wind field (Alongi 2008).

In some coastal regions, marine fauna may be protected from warming due to physiological and oceanographic conditions. In the Red Sea which is naturally prone to warm temperatures and high salinity (and the warmest sea on earth), oxygen supersaturation from photosynthesis follows SST rise in the coastal zone (Giomi et al. 2019) with daily fluctuations in water temperature matching the time of maximum dissolved oxygen concentration. This phenomenon also occurs at temperate and polar latitudes (Krause-Jensen et al. 2016), suggesting that the occurrence of diel oxygen supersaturation in coastal habitats occurs globally and is not restricted to the tropics. Giomi et al. (2019) experimentally tested six common marine species living in Red Sea mangroves, seagrass beds, and coral reefs (Figure 13.5). Exposure to experimentally induced oxygen supersaturation (140% O_2 saturation) consistently enhanced lethal heat thresholds extending their survival under extreme temperatures. When exposed to supersaturated conditions, the 50% lethal temperature (LT_{50}) for all animals (representing four different phyla) increased by 1–4 °C, depending on species,

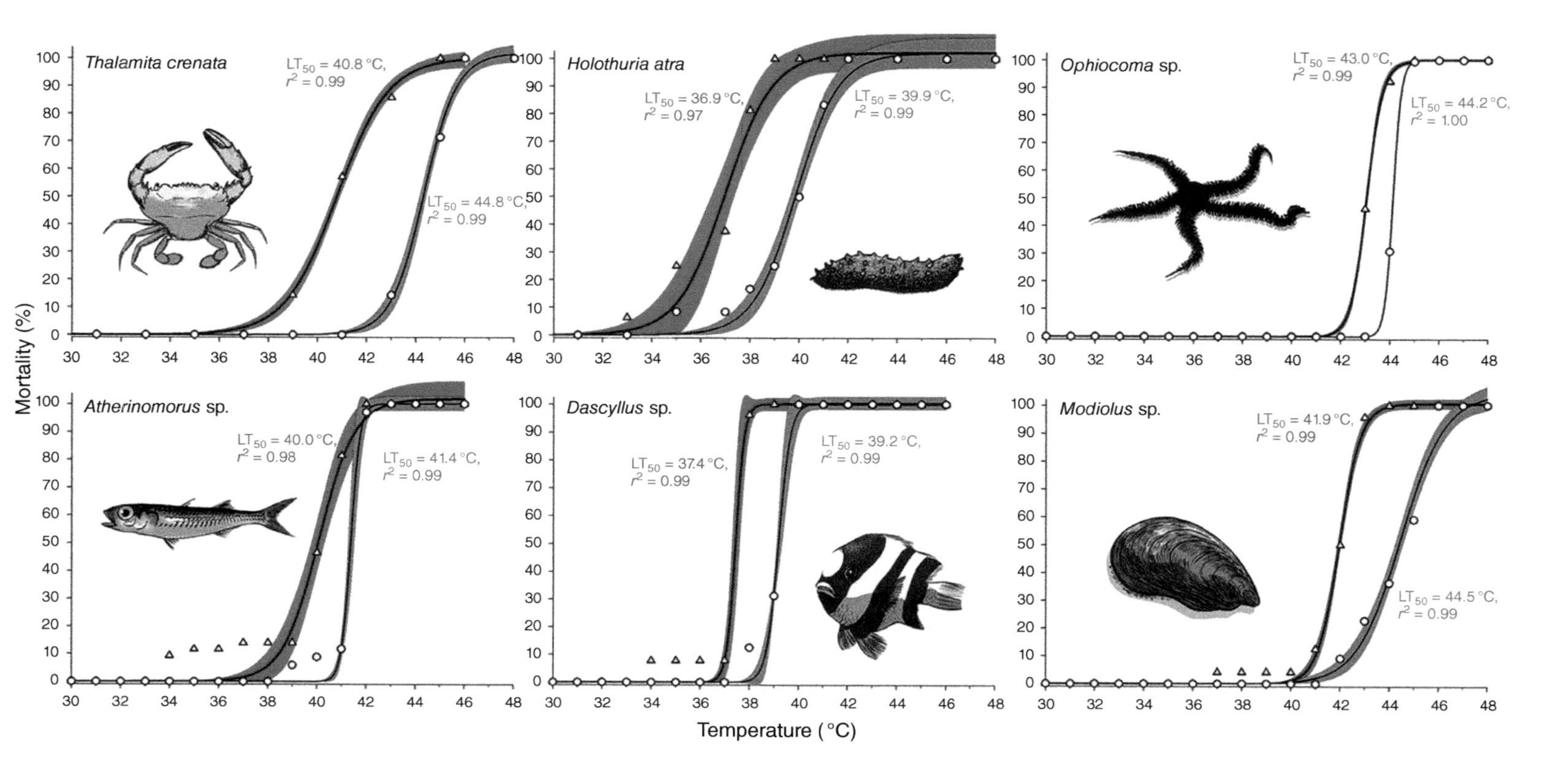

FIGURE 13.5 The relationship between water temperature and mortality for the mangrove swimming crab, sea cucumber, brittle star (top row), silverside, coral pomacentrid, and horse mussel (bottom row) under experimental conditions of normal (97%) oxygen saturation (triangles) and supersaturated (circles, 140% of oxygen saturation) treatments. The filled areas represent the 95% confidence intervals (normal conditions: red; supersaturated conditions: cyan) for each three-parameter sigmoid regression (black lines). *Source:* Giomi et al. (2019), figure 2, p. 2. © American Association for the Advancement of Science.

with respect to that under normal O_2 saturation. Thus, oxygen supersaturation at levels normally encountered in the Red Sea significantly enhanced survival of all six species, contributing to the endurance of these organisms to extreme temperatures.

13.3 Sea-level Rise (SLR)

SLR is unlikely to be a problem for coral reefs if coral growth remains vigorous (Hoegh-Guldberg 2011). However, if growth is impacted by other factors such as ocean acidification, some reef communities may struggle to maintain themselves even under the most minimal changes in sea-level. Rising sea-level may still affect coral reefs in several ways, including problems associated with coastal erosion, higher storm surges, and 'drowning' of some reefs. Positive impacts may include increased area for some reefs (Lough 2008) as some corals have grown vertically to the level where they are limited by present-day sea-level. Some rates of SLR would therefore be beneficial to such reefs, while some reefs in deeper water could eventually drown due to reduced light availability with increased water depth. SLR may affect different zones of a coral reef differently.

Observations and modelling of coral calcification in relation to SLR at Shiraho fringing reef, Ishigaki Island, Japan, indicated that calcification rates were low in nearshore and stagnant water areas due to isolation of the inner reef at low tide, but calcification rates were higher on the offshore side of the inner reef flat (Nakamura et al. 2017). Simulation modelling under different IPCC RCP scenarios indicated that SLR increases the calcification rate, particularly in the nearshore and stagnant water areas at low tide, as mass exchange of O_2 at night is enhanced between the corals and their ambient seawater due to the reduced stagnant period. When pCO_2 and SLR occur concurrently, calcification rate decreases, due to ocean acidification. However, the calcification rate in some inner reef areas will increase because of the positive effects of SLR offset the negative effects of ocean acidification, and calcification will be positive only under the best-case IPCC scenario (RCP2.6).

A more recent analysis, however, suggests that many coral reefs may not keep pace with projected increases in SLR (Perry et al. 2018; Cacciapaglia and van Woesik 2020). Calculating the vertical growth potential of >200 coral reefs in the tropical western Atlantic and Indian Ocean and comparing these against recent and projected rates of SLR under different IPCC RCP scenarios, Perry et al. (2018) concluded that while many reefs will retain accretion rates close to recent SLR trends, few will have capacity to meet SLR projections under the RCP 4.5 'lower emissions' scenario; under the RCP8.5 'business-as-usual' scenario, most reefs will experience mean water depth increases > 0.5 m by 2100. Coral cover strongly predicts the capacity of reefs to keep pace with SLR, but threshold cover levels necessary to prevent emergence are well above those observed on most reefs.

Estimating net carbonate production by considering erosional rates due to ocean acidification, increasing cyclone intensity, local pollution, fishing pressure, and projected increases in SSTs under RCP scenarios 4.5, 6.0, and 8.5, the species distribution models used by Cacciapaglia and van Woesik (2020) predict that only 4% of Indo-Pacific coral reefs, most near the equator, will keep up with SLR by 2100 under the RCP8.5 'business-as-usual' scenario. Under RCP scenarios 4.5 and 6.0, 15 and 12% of Indo-Pacific reefs have the potential to keep pace with SLR by 2100. However, fishing pressure and its cascading effects are projected to facilitate significant reef erosion, nearly halving the number of reefs able to keep up with SLR. Twenty-one to 27 per cent of coral reefs in the Indo-Pacific may survive SLR until 2100 if emissions, pollution, and fishing pressure are all drastically reduced.

As noted earlier, some reefs will survive if they can sustain high growth. Perry et al. (2015) measured the carbonate budgets of 28 reefs across the Chagos Archipelago in the Indian Ocean which experienced severe coral mortality (>90%) during the 1997/1998 warming event. Most reefs recovered rapidly with high growth rates (4.2 mm a^{-1}), rates exceeding those estimated for the mid-Holocene and close to IPCC SLR projections up to 2100. Similarly, *Porites* microatolls from reef flats in Palau (western Pacific Ocean) are keeping pace with contemporary SLR (van Woesik et al. 2015). Modelling suggests that under low-mid RCP scenarios, these microatolls will survive but will not if atmospheric CO_2 concentrations exceed 670 ppm and SSTs increase by 2.2 °C.

SLR can alter seagrass distribution, production, and community composition (Short and Neckles 1999). The greatest direct impact of SLR will be an increase in water depth and a concomitant decrease in available light. As a result, seagrasses will experience reduced distribution, productivity will decrease, community structure will be altered, and their functions reduced. A primary effect of increased water depth will be to alter the location of the maximum depth limit of seagrass growth, directly affecting seagrass distribution. Increased water depth may allow subtidal species to expand their range into areas that are currently intertidal.

The effects of elevated sea-level will have the greatest impact in areas of high turbidity and have lesser effects in areas of low turbidity. The effects of reduced light are a reduction in shoot density, leaf width, number of leaves per shoot and growth rate, and an increase in leaf length. Any rise in sea-level will reduce the amount of light reaching the seagrass community, decreasing rates of photosynthesis and productivity, and altering the bed structure resulting in a loss of function (Short and Neckles 1999). Reduced light may also alter community structure by favouring species having lower light requirements. The impact of SLR will vary regionally, depending on local topography (Keyzer et al. 2020). The migration of seagrass beds to shallower water may occur, especially in regions where there is room to shift to intertidal or shallower subtidal waters. Species with rapid recruitment capabilities (e.g. *Halophila, Halodule, Zostera*) will occupy new areas more rapidly than slower recruiting species (e.g. *Thalassia, Cymodocea*). Increased sea-level may result in uprooting of seagrass due to shoreline erosion of newly inundated but unstable sediments (Duarte 2002).

In some countries, coastal retreat and improved water quality may mitigate losses of seagrasses due to SLR. In Moreton Bay, Australia, models predict a 17% decline in seagrass bed area with a rise in sea-level of 1.1 m by the year 2100 (Saunders et al. 2013). If coastal developments are cut back, the decline is predicted to be only 5%. The predicted reduction in seagrass area could be offset by an improvement in water clarity of 30%; greater improvements in water clarity would be necessary for larger rates of SLR. On the other hand, the hardening of shorelines through coastal development poses a significant risk to seagrass habitat expansion.

A localised subsidence event in the Solomon Islands after an earthquake in 2007 in which sea-level arose by about 30–70 cm indicated that future rapid changes to sea-level can have contrasting impacts of coral reefs, seagrasses, and mangroves (Albert et al. 2017). Coral reefs in the region responded with a steady lateral growth from 2006 to 2014, resulting in a 157% increase in areal coverage. In contrast, mangroves suffered a rapid dieback of 35% after subsidence, although some forests partially recovered seven years after the earthquake, but with a different community structure. Seagrasses responded with both losses and gains in areal extent at small 10–100 m scales. Thus, the impacts of SLR will likely be highly variable in such as complex tropical landscape, with 'losers' and 'winners' varying over space and time (Albert et al. 2017).

Mangroves thrive in a tidal environment where adaptation to change in relative sea-level over long timescales is the rule rather than the exception (Alongi 2008; Ward and de Lacerda 2021). The ability of mangroves to adjust to rises in sea-level depends on accretion rate relative to the rate of sea-level change, as seen in the Solomon Islands. Are mangroves keeping pace with current rates of relative SLR? An analysis (Figure 13.6) of mangrove accretion rates versus local mean SLR suggests that mangroves (i.e. below the solid line in Figure 13.6), particularly in Australia, New Zealand, the Caribbean, Central America, and on some (but not all, Esteban et al. 2019) low coral islands and in subsiding river deltas such as the Sundarbans and in Southeast Asia are not keeping pace. In contrast, mangroves (i.e. data points located above the solid line in Figure 13.6) located in other areas of Southeast Asia and the Pacific (e.g. Papua New Guinea), South America, Africa, the Middle East, South Asia, and East Asia (China, Taiwan) are keeping pace with current SLR. Many of these latter forests occur in areas of rapid accretion due to highly impacted and populated catchments,

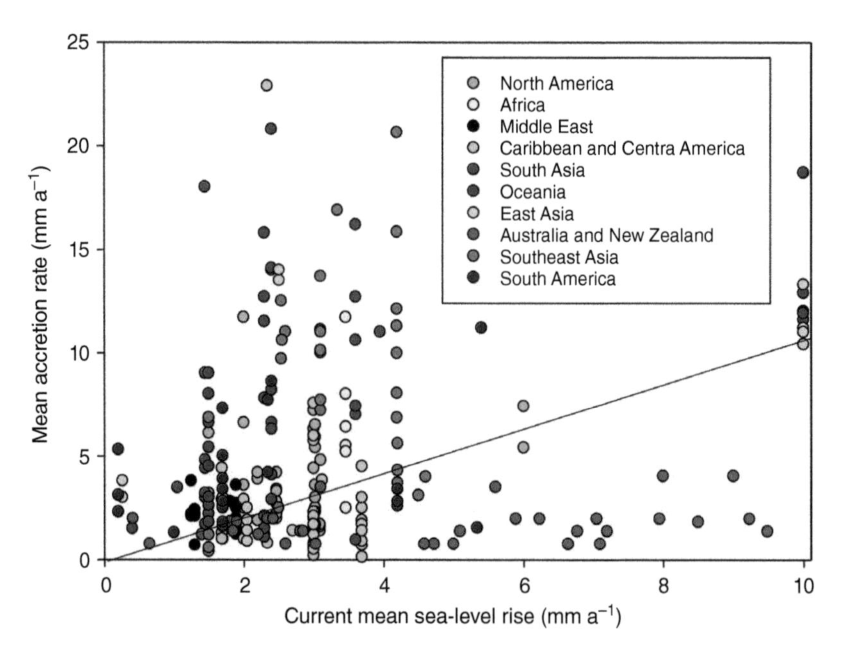

FIGURE 13.6 The relationship between measured rates of mangrove soil accretion and current rates of mean relative sea-level rise across the globe. Dotted red line delimits a 1:1 relationship and the solid blue line is the linear regression for the untransformed data. Mangrove sedimentation data are from references in Alongi (2009, 2015, 2018) and Sasmito et al. (2016) and updated from data in Hayden and Granek (2015), Hoque et al. (2015), Carnero-Bravo et al. (2016, 2018), MacKenzie et al. (2016), Marchio et al. (2016), Almahasheer et al. (2017), Phillips et al. (2017), Chappel (2018), Cusack et al. (2018), Murdiyarso et al. (2018), Pérez et al. (2018), Ruiz-Fernández et al. (2018), Saderne et al. (2018), Fu et al. (2019), Hale et al. (2019), Kusumaningtyas et al. (2019), Soper et al. (2019) Swales et al. (2019), and Bomer et al. (2020). The sea-level rise data are from satellite altimetry or tide gauge data available at http://www.nodc.noaaa.gov/General/sealevel.html (accessed 3 September 2020).

especially in China, Brazil, and India. The wide scatter of data points reflects how mangroves in disparate coastal settings in different parts of world respond so differently to the same rate of SLR. Most of the mangrove accretion rates used in Figure 13.6 were measured using radionuclides such as ^{137}Cs and ^{210}Pb and are only moderately reliable indicators of accretion compared with the use of surface elevation tables. There is also considerable uncertainty in the rates of relative SLR as there are large seasonal variations and changes in atmospheric pressure and weather (Chapter 2). The accretion rates also do not reflect the importance of changes in surface elevation gain; a mangrove forest may be rapidly accumulating soil at the surface, but the local area may be subsiding, resulting in a net decrease relative to sea-level, as is occurring in many river deltas. A detailed analysis of recent trends in mangrove surface elevation changes across the Indo-Pacific region using data from surface elevation table instruments shows that for 69% of mangroves, the current rate of SLR exceeded the soil surface elevation gain (Lovelock et al. 2015). Further model analysis suggests that mangrove forests located at low tidal range and in areas of low sediment delivery could be submerged by 2070. Mangroves have migrated consistently over geological time in synchrony with post-glacial SLR; how mangroves adjust in future depends not only on sediment availability but also local topography (Woodroffe et al. 2016).

While mangroves expanded between 9800 and 7500 years ago at a rate driven mainly by the rate of relative SLR, it was highly likely (90% probability) that they were unable to sustain accretion when relative SLR exceeded 6.1 $mm\,a^{-1}$ (Saintilan et al. 2020). This finding agrees with the data in Figure 13.6 that rates of SLR greater than 6 $mm\,a^{-1}$ represent a critical threshold for submergence. Mangrove forests are likely 'losers' with respect to SLR in regions where there is substantial subsidence (e.g. the Sundarbans, the Solomon Islands; Albert et al. 2017), a low tidal range, changes in precipitation, and/or poor ecological conditions (Cinco-Castro and Herrera-Silveira 2020). The reality is that, as they have in the past, mangroves will respond in complex ways to future SLR. Some mangroves will probably survive if the rate of SLR is slow enough, but there will likely be significant changes in community composition, forest structure, morphology, and anatomy, including changes in vascular vessel densities, fibre wall thickness, bark anatomy, formation of hypertrophied lenticels, adventitious roots, and increased aerenchyma development (Alongi 2015).

Experimental studies indicate species-specific responses to SLR and waterlogging (Cardona-Olarte et al. 2006; Chen and Wang 2017). High tolerance was exhibited by the cosmopolitan species, *A. marina*, to waterlogging, but responses were highly variable in relation to immersion depth and length of time, salinity, temperature, and other factors (Mangora et al. 2014). Another cosmopolitan species, *R. stylosa,* similarly had a high tolerance to waterlogging as experiments simulating growth of *R. stylosa* seedlings to SLR found that the species was flood-tolerant with high stem growth rate and leaf assimilation rate as well as efficient utilisation of carbohydrate reserves stored in hypocotyls of seedlings; both growth and physiology were affected by salinity and by an increase in flooding time (Chen and Wang 2017). *R. stylosa* exhibited competitive dominance at high salinity, a good adaptation of seedlings to future SLR.

The complexity of mangrove response to relative SLR was demonstrated in a modelling study developed to capture the interaction between mangrove species and hydro-sedimentary processes across a coastal profile and using numerical experiments to elucidate the response of mangroves under a range of SLR and sediment supply conditions, both in the absence and presence of anthropogenic barriers impeding inland migration (Xie et al. 2020). Mangrove coverage can increase despite SLR if sediment supply is sufficient and accommodation space landward is available (Figure 13.7), but tidal barriers are detrimental to mangrove survival and may result in species loss (Xie et al. 2020). Biomorphodynamic feedbacks may cause spatial and temporal variations in sediment delivery across the forest, leading to sediment starvation in the landward forest and reduced deposition despite longer inundation (Xie et al. 2020). Such feedbacks may decouple accretion rate from inundation time, altering habitat conditions and causing a loss of biodiversity even when forest coverage remains stable or is increasing. Further, the model reveals that vegetation-induced flow resistance linked to root density may be a major factor steering the inundation–accretion decoupling and thus species distribution (Xie et al. 2020).

Future survival of mangrove forests ultimately depends on the rapidity of future increases in relative SLR. The current prediction (Section 2.7) is that sea-level will continue to rise at a median rate of 5.3–10.5 mm a^{-1} by 2100, suggesting that most mangrove forests are at high risk of drowning by the end of the century. Mangrove

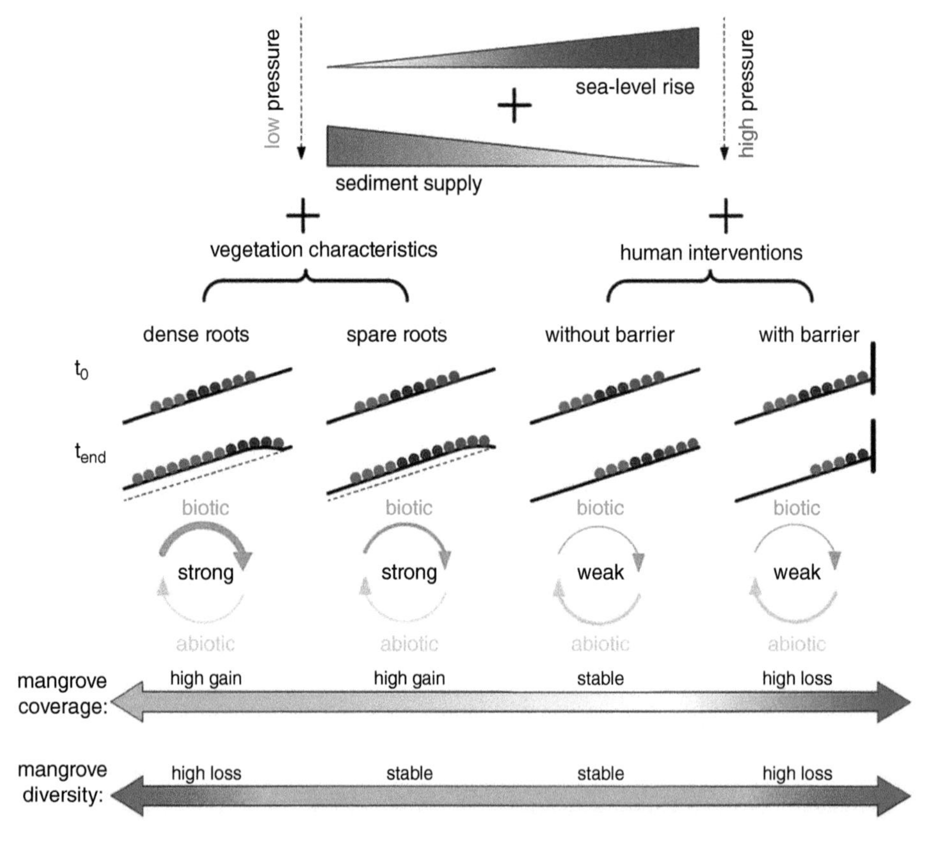

FIGURE 13.7 Representation of model results showing the complexity of mangrove response under different sea-level rise and sediment supply conditions. Under low environmental pressures, profile and vegetation distribution changes are determined by vegetation characteristics, while possible responses under high environmental pressures are mainly impacted by the presence or absence of anthropogenic barriers. The role of bio-morphodynamic feedbacks is indicated (i.e. "strong" versus "weak"), and arrow thickness represents the relative strengths of biotic and abiotic interactions. *Source:* Xie et al. (2020), figure 5, p. 8. Licensed under CC BY 4.0. © IOP Publishing Ltd.

resistance and resilience to relative SLR over timescales of years to decades are a result of four main factors: (i) the rate of change in sea level relative to the mangrove soil surface determines mangrove vulnerability, (ii) mangrove species composition affects mangrove responses, (iii) the physiographic setting, including the slope of the mangrove relative to that of the land the mangrove currently occupies, and the presence of obstacles to landward migration, affects mangrove resistance, and (iv) cumulative effects of all stressors influence mangrove resistance and resilience (Gilman et al. 2008).

Mangroves show considerable resilience to fluctuations in relative sea level due to their ability to actively modify their environment through surface elevation changes (McKee et al. 2007, 2021; Ward and de Lacerda 2021), and their ability to migrate inland over successive generations (Kumara et al. 2010; Krauss et al. 2014; Ward et al. 2016). Positive surface elevation change is facilitated by the deposition of autochthonous and allochthonous organic matter as well as by subsurface compaction and the trapping and retention of inorganic sediments (Krauss et al. 2014; McKee et al. 2021). Therefore, adaptation to global SLR is likely to be driven not only by local rates of SLR, but by available sediment, autochthonous peat production, land uplift/subsidence rates and localised sediment auto-compaction, as mangrove forests promote sediment trapping and retention (Section 3.4). Fringe mangrove forests are on average most susceptible to SLR, being at the forefront of the intertidal seascape (McKee et al. 2021; Ward and de Lacerda 2021). In contrast, high densities of mangroves can facilitate surface accretion, surface elevation change, and tree survival in areas susceptible to SLR (Kumara et al. 2010; McKee et al. 2021).

Mangroves can be driven inland if land is available; this process can be helped by oceanographic anomalies (Lopez-Medellin et al. 2011). On the Pacific coast of Mexico, a significant increase in mangrove area has occurred in backwaters of the lagoons of Magdalena Bay in Baja California during the past four decades, and especially during the El Niño anomalies of the 1980s and 1990s, while at the same time the mangrove fringe has been receding. The observed change was attributed to the combined action of the warm waters of El Niño events and SLR. These two processes were sufficient to flood large areas of tidal salt flats dispersing mangrove seedlings inland. In American Samoa, Gilman et al. (2007) found that the observed mean landward migration of mangrove seaward margins over four decades was 12–37x greater than relative SLR. It was inferred that the force of SLR relative to the mangrove surface was causing landward migration, but such migration was obstructed by coastal development. The mangroves of American Samoa could experience as high as a 50% reduction in area by 2100. A 12% reduction in mangrove area by 2100 is possible on islands of the Pacific. Indeed, the data in Figure 13.6 indicate that the mangroves of the Pacific islands are among the most susceptible to SLR. In contrast, the mangroves of Gazi Bay, Kenya, are expected to show an increase in mangrove area under a regime of SLR (Di Nitto et al. 2014). Thus, the impact of SLR on the landward migration of mangroves is highly site-specific; while seaward mangroves may be highly susceptible to drowning and loss, there may be landward migration depending upon whether there is room on the landward side for migration.

The paleoenvironmental record indicates that over millennial timescales, mangroves have been exposed to different sea-level trajectories, suggesting a broad capacity to adjust to sea-level changes (Woodroffe et al. 2016). Inorganic sediment supply and sequestration of organic matter endow mangrove forests with considerable natural resilience in response to SLR. However, there are many other anthropogenic pressures that interfere with accretion processes that decrease the capacity of mangroves to adjust (Woodroffe et al. 2016) including the fact that the current rates of SLR are greater than those in the past millennia.

13.4 Rising Atmospheric CO_2

Experimental evidence has repeatedly shown that elevated CO_2 concentrations enhance photosynthesis, growth, and leaf chlorophyll in mangroves, with responses being species-specific and variable, depending on salinity, temperature, nutrient availability, and water-use efficiency (Reef et al. 2016; Jacotot et al. 2018; Tamimia et al. 2019; Manea et al. 2020; Maurer et al. 2020). Early studies showed that the growth of *R. stylosa, R. apiculata*, and *R. mangle* was enhanced by increasing CO_2 concentrations, but only at low salinity (Alongi 2015). Snedaker and Araújo (1998) showed that *R. mangle, A. germinans, Conocarpus erectus*, and *L. racemosa* exhibited increases in transpiration efficiency but a decline in stomatal conductance and transpiration with increasing CO_2 concentrations. The response of most mangrove species to increasing CO_2 levels will be complex, with many species thriving, but some species declining or exhibiting no or little change. Species patterns within an estuary may change based on the ability of each species to respond to spatial and temporal variations in the above-mentioned drivers.

More recent experiments have supported the findings of the earlier studies. In a glasshouse study, Reef et al. (2016) examined the effects of elevated CO_2 and nutrient availability on seedlings of *A. germinans*, finding large gains in growth and photosynthesis when seedlings grown under elevated CO_2 were supplied with elevated nutrient concentrations as compared to under ambient conditions. Growth was greatly enhanced only under high nutrient conditions and elevated CO_2, and root volume doubled under low nutrient and elevated CO_2 conditions relative to ambient nutrient and CO_2 levels. Biochemical pathways play a role in mangrove responses to climate change as *A. germinans* produces the osmolyte, glycine betaine, which increases tolerance to environmental stressors. Under exposure to increasing salinity, ambient and high CO_2 concentrations and a dose of glycine betaine, *A. germinans* seedlings exhibited increased salt tolerance and higher photosynthetic rates (Maurer et al. 2020); under elevated CO_2, the temperature optimum for photosynthesis increased by 4 °C. Elevated CO_2 levels increased leaf chlorophyll *a* concentrations and root microbial biomass, with some alteration of ammonia-oxidising archaea. Further, there was a shift in carbon utilisation from the preferred carbon sources of sugars, amino acids, and carboxylic acids under ambient conditions to the use in the order of amino acids > carbohydrates > polymers > carboxylic acids > amines > phenolic acids, indicating a shift in metabolic pathways under elevated CO_2. When subjected to ambient CO_2 and a temperature of 38 °C, *R. apiculata* seedlings responded positively to the higher temperature but elevated CO_2 enhanced growth only at a lower temperature. Under conditions of high CO_2 and temperature, the seedlings nearly perished (Tamimia et al. 2019), indicating complex outcomes to elevated CO_2 concentrations when mangroves are subjected to other drivers such as increasing temperatures and flooding.

These complex responses may, nevertheless, offer a competitive advantage to mangroves as they continue to encroach upon saltmarshes. The mangroves, *A. corniculatum* and *A. marina*, grown individually and in a model saltmarsh community under increasing CO_2 and reduced salinity experienced stronger competition from saltmarsh species under elevated CO_2 (Manea et al. 2020). *A. marina* seedlings produced 48% more biomass under elevated CO_2 when grown in competition with saltmarsh species. *A. corniculatum* was not affected by elevated CO_2 but had 36% greater growth under seawater salinity compared to hypersaline conditions. Rising atmospheric CO_2 concentrations and lower salinity associated with SLR may thus enhance the establishment of mangrove seedlings in saltmarshes, facilitating mangrove encroachment.

The empirical responses of tropical seagrasses and macroalgae to elevated CO_2 concentrations are sometimes contradictory, but in some cases, are small and positive for some species (Ji and Gao 2020). Data from volcanic CO_2 seeps in Papua New Guinea (Fabricius 2011) suggest that some seagrasses may thrive under high CO_2 and low pH conditions, in agreement with some modelling and experimental exercises (Unsworth et al. 2012; Campbell and Fourqurean 2014; Ow et al. 2015; Takahashi et al. 2018). Long-term exposure to elevated pCO_2 at the Papua New Guinea vents resulted in differences in seagrass community composition, with *C. serrulata*, *C. rotundata*, and *H. univervis* abundant at high pCO_2 sites and *H. ovalis*, *T. hemprichii*, and *S. isoetifolium* occurring only at low and mid-pCO_2 sites (Takahashi et al. 2018).

Both economically important and harmful macroalgae can benefit from elevated CO_2 concentrations and temperature rise, which may be responsible for increasing events of harmful macroalgal blooms, including green blooms caused by *Ulva* spp. and golden tides caused by *Sargassum* spp. (Ji and Gao 2020). Macroalgae in the upper intertidal zone, especially those species-tolerant to dehydration during low tide, increase their photosynthetic rates under elevated CO_2 concentration, but these species might be endangered by heat waves which can expose them to temperatures above their thermal tolerance limits. The effects of UV radiation, in contrast, can be harmful as well as beneficial to many species. Moderate levels of UV-A can enhance photosynthesis of green, brown, and red algae, while UV-B is inhibitory (Ji and Gao 2020). The combined effects of elevated CO_2, temperature, and heat waves with UV radiation are little understood but are likely to be detrimental to most species. In contrast, eutrophication along with low dissolved O_2 might favour the carboxylation process by suppressing their oxygenation or photorespiration.

The effects of increased CO_2 concentrations on seagrasses depend on the degree of carbon limitation present in natural systems as well as other factors such as light and nutrient availability (Short and Neckles 1999; Ow et al. 2016). Many seagrasses use HCO_3^- in addition to CO_2 as a source of photosynthetic carbon. Seagrass photosynthesis is frequently limited by the availability of DIC under natural conditions. Carbon limitation has been attributed to the thickness of the diffusive boundary layer surrounding leaf surfaces at low current velocities or to a relatively inefficient HCO_3^- uptake system. The latter mechanism is substantiated by measurements of higher affinities for CO_2 than for HCO_3^- in several seagrasses and increases in seagrass photosynthesis with CO_2 enrichment. Generally, seagrass photosynthesis increases with increasing DIC concentrations and decreasing pH, with seagrass responses to low pH being complex and sometimes conflicting, as discussed below.

13.5 Ocean Acidification

The projected decline in ocean pH will be enough to significantly decrease the ability of corals and other marine calcifiers to form their calcium carbonate skeletons (Kleypas et al. 1999). A doubling of preindustrial atmospheric CO_2 results in decreases of up to 40% in the calcification and growth of corals and other reef calcifiers, such as red calcareous algae (Kleypas and Langdon 2006). Net carbonate accretion on coral reefs is likely to decrease below zero when carbonate ion concentrations reach 200 $\mu mol\, kg^{-1}$ of seawater, which corresponds approximately to an aragonite saturation state (Ω_{arg}) of ≤ 3.3 (Guinotte et al. 2003; Guinotte and Fabry 2008).

The effects of ocean acidification on coral reefs can be complex and often unclear, especially when combined with other climate effects such as rising temperatures and sea level. Laboratory studies (Comeau et al. 2017) show that the effects of ocean acidification on coral calcification and recruitment can be species-specific, and that coral resistance and resilience to ocean acidification partially has its basis in coral physiology. Diverse outcomes have been observed regarding responses of coral gastrovascular cavity (GVC) pH to reduced pH during light and dark incubations (Bove et al. 2020a). Measuring O_2 and pH within the GVC of the corals, *Montastraea cavernosa* and *Duncanopsammia axifuga*, in both dark and light and at three pH levels, at pH 8.2, O_2 ranged widely from 0 to >400% in both light and dark, and pH in the GVC (pH_{GVC}) was always significantly higher or lower than seawater pH with pH_{GVC} in *M. cavernosa* increasing to pH 8.4–8.7 in the light but dropping below ambient pH in the dark. pH_{GVC} for *D. axifuga* decreased linearly as ambient pH decreased in both dark and light. Calcification rates for both species were similar at pH 8.2 and 7.9, but significantly lower at pH 7.6. Calcification was thus protected from acidification by intrinsic coral physiology at pH 7.9, but not at pH 7.6. Calcification was not correlated with pH_{GVC} for *M. cavernosa* but was for *D. axifuga*, indicating diverse responses of corals to acidification.

Some corals as well as coralline algae can maintain constant rates of calcification by maintaining their carbonate chemical conditions, specifically aragonite saturation state, within the calcifying fluid (Cornwall et al. 2018). The ability to maintain $[Ca^{2+}]$ in the calcifying fluid plays a key role in the ability of corals to resist acidification (DeCarlo et al. 2018). In response to declining pH, the coral, *Pocillopora damicornis*, increased $[Ca^{2+}]_{CF}$ within the calcifying fluid to as much as 25% above that of seawater and maintained constant rates of calcification (DeCarlo et al. 2018). In contrast, the coral, *Acropora youngei*, showed less control over $[Ca^{2+}]_{CF}$ and its calcification rates declined with lower pH. Physiologically, most corals up-regulate pH at their site of calcification such that internal changes are roughly one-half those in ambient seawater (McCulloch et al. 2012). This pH-buffering capacity enables hermatypic corals to raise the saturation state of their calcifying fluid, increasing calcification rates with little additional energy. However, this ability is species-specific and not common among other calcifying organisms. In experimental mesocosms, exposure to decreasing pH resulted in a decline in calcification of the coral, *A. youngei* (Comeau et al. 2017) but there was no effect on calcification of the coral, *P. damicornis* (Figure 13.8). When exposed to current and future projections of warming and acidification, the size and abundance of recruits of the coral, *Pocillopora acuta*, were not significantly affected by rising temperature or acidification, nor their combination, but heating impacted their health and survival (Figure 13.9). Coral recruits in heated tanks experienced higher levels of bleaching and subsequent mortality (Bahr et al. 2020).

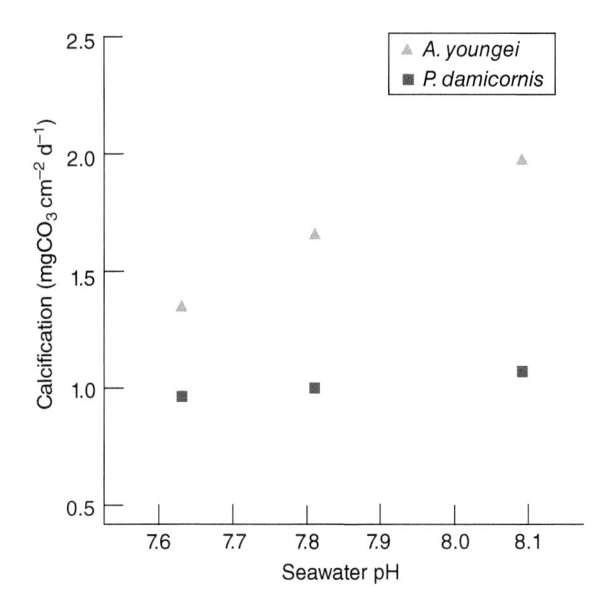

FIGURE 13.8 Experimental study showing a decline in calcification of the coral, *Acropora youngei* (grey triangles), and no effect on calcification of the coral, *Pocillopora damicornis* black dots), with decreasing seawater pH. Calcification was measured on corals that were incubated during eight weeks under pH = 8.09, 7.81, and 7.63. Values are mean ± 1SE (n =12). *Source:* Comeau et al. (2017), figure 1, p. 4. Licensed under CC BY 4.0. © Springer Nature.

Most experimental data indicate that calcification rates of hermatypic corals decline with ocean acidification, at 2x and 3x (Table 13.1) the pre-industrial level of atmospheric CO_2 although calcification rates vary widely among species. The higher concentrations of pCO_2 lead to a greater decline in calcification rates (Table 13.1).

As noted earlier, complex responses are common. The Caribbean corals, *S. siderea, P. strigosa, P. astreoides*, and *U. tenuifolia*, from the Belize Mesoamerican Barrier Reef exhibited non-linear declines in calcification rate with increasing pCO_2 (Bove et al. 2019). *S. siderea* was the most resilient to both warming and acidification owing to its ability to maintain positive calcification in all treatments. *P. strigosa* and *U. tenuifolia* were the least resilient, and *P. astreoides* was midway in the resilience spectrum among the four species. A meta-analysis of the calcification responses of Caribbean corals to experimentally induced acidification, warming, and their combination (Bove et al. 2020b) found that calcification rates declined under warming conditions alone, but acidification and the combination of both stressors did not clearly reduce calcification rates. In contrast, corals from the Florida Keys were not affected by acidification, warming, or their combination, suggesting possible regional differences in responses to warming and acidification.

Responses to ocean acidification were species-specific in two species of the calcareous tropical green algae, *Halimeda opuntia* and *H. taenicola,* of Palmyra Atoll in the central Pacific (Price et al. 2011). *H. opuntia* exhibited net dissolution and a 15% reduction in photosynthetic capacity, whereas *H. taenicola* did not calcify but did not show any change in photo-physiology. Similarly, two algal symbiont-bearing, reef-dwelling foraminifera, *Amphisorus kudakajimensis* and *Calcarina gaudichaudii*,

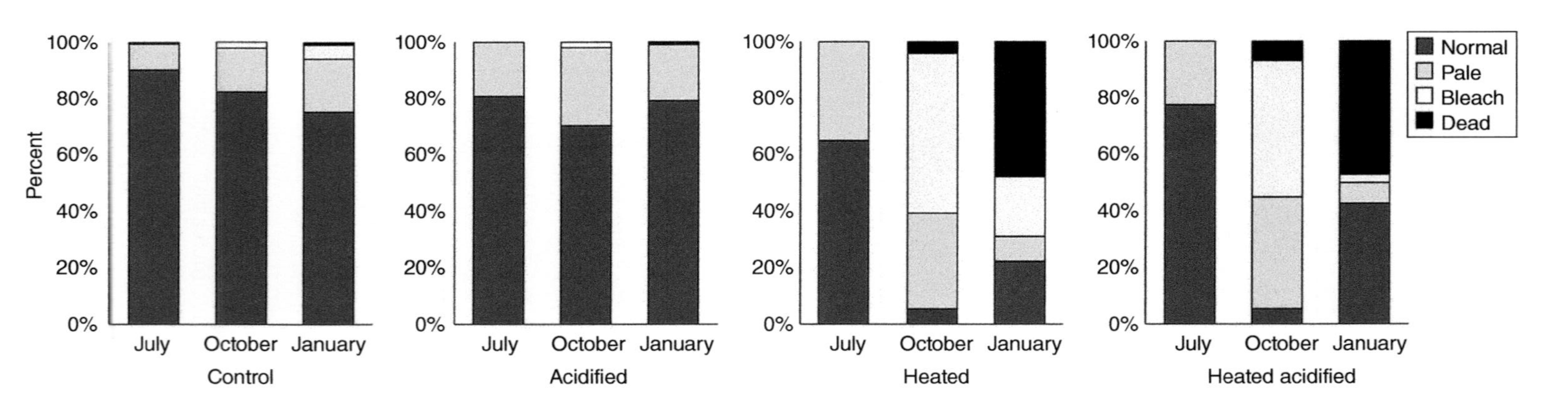

FIGURE 13.9 Mean coral recruit health assessment of each treatment (pooled across tanks) (control, acidified, heated, and heated acidified) during each survey (July 2017, October 2017, and January 2018). Coral health assessment is denoted by pigmentation of normal (red), pale (yellow), bleached (white), and dead (black). *Source:* Bahr et al. (2020), figure 5, p. 8. Licensed under CC BY 4.0. © PLOS.

TABLE 13.1 **Percentage decline in calcification of individual coral species to ocean acidification in experiments where corals were subjected to twice and three times pre-industrial atmospheric CO_2.**

Coral species	2x pre-industrial	3x pre-industrial
Acropora cervicornis	−22	nd
A. cervicornis	−40	−59
Acropora eurystoma	−55	nd
Acropora formosa	−20	nd
Acropora intermedia	−17	nd
Acropora verweyi	−12	−18
Favia fragum	−25	nd
Fungia sp.	−47	−69
Galaxea fascicularis	−19	nd
G. fascicularis	−12	−16
Madracis mirabilis	−35	nd
Montipora capitata	−101	nd
Pavona cactus	−18	−20
Porites astreoides	−78	nd
Porites compressa	−17	−25
Porites lobata	−12	nd
Porites lutea	−38	−56
P. lutea	−33	−49
Porites porites	−16	nd
Porites rus	−50	nd
Stylophora pistillata	−22	nd
S. pistillata	−14	−20
S. pistillata	−15	nd
Turbinaria reniformis	−9	−13

Calcification declines are relative to calcification at present-day pCO_2. nd = no data.*Source:* Alongi (2020a).

exhibited contrasting responses to five increasing concentrations of pCO_2 (Hikami et al. 2011). Net calcification of *A. kudakajimensis* was reduced under high pCO_2, whereas calcification of *C. gaudichaudii* increased with increasing pCO_2. The different responses were likely due to the different complexes of algal symbionts in both species.

The effects of ocean acidification on tropical fleshy and calcareous algae on Palmyra Atoll were mixed, dependent on the individual species (Table 13.2). There were

TABLE 13.2 **Ocean acidification effects on benthic reef algae from Palmyra Atoll, central Pacific.**

Species	Calcification/growth
Fleshy macroalgae	
Acanthophora spicifera	0
Avrainvillea amadelpha	–
Caulerpa serrulata (2010 experiments)	+
C. serrulata (2011 experiments)	0
Dictyota bartayresiana	+
Hypnea pannosa	+
Lobophora papenfussii	–
Upright calcareous algae	
Dichotomaria marginata	–
Galaxaura rugosa	0
Halimeda opuntia	–
Halimeda taenicola	0
Halimeda cylindracea	–
Halimeda macroloba	–
Halimeda incrassata	+
Crustose coralline algae	
Lithophyllum prototypum	–
Lithophyllum sp.	–
Hydrolithon sp.	–
Porolithon onkodes	–
Neogoniolithon sp.	+
Mixed CCA	–

+ = positive effect, – = negative effect, 0 = no effect. *Source:* Johnson et al. (2014).

several negative, positive, and no responses among species to ocean acidification (Johnson et al. 2014). Acidification reduced the ability of crustose coralline algae to calcify, but the results from these experiments suggest that conditions may favour non-calcifying algae and shift relative dominance on coral reefs under projected acidification conditions to fleshy macroalgae.

Resistance of reef calcifiers of the same species to ocean acidification may vary across distances. For instance, when the corals, *Pocillopora damicornis* and *Porites* sp., and two calcified algae, *Porolithon onkodes* and *Halimeda macroloba*, were incubated at various pCO_2 conditions in French Polynesia, Hawaii, and Okinawa, both corals and

H. macroloba were insensitive to the treatments at all locations. The effects of ocean acidification on *P. onkodes* varied among locations, with calcification of the species depressed in French Polynesia and Hawaii, but unaffected in Okinawa (Comeau et al. 2014a). Resistance of reef calcifiers is a "constitutive character" expressed across the Pacific.

Host–microbe interactions can also confer resilience to acidification. Coralline red algae (*Corallinales, Rhodophyta*) exposed to increasing pCO_2 exhibited increased photosynthetic activity and no loss of $CaCO_3$ biomass over time. The microbiome associated with these algae remained stable and healthy, but the microbial community in the water column changed with increasing pCO_2 (Cavalcanti et al. 2018). Thus, the stability of the algal microbiome is important for host resilience to acidification stress.

Several factors influence the ability of calcifying organisms to resist ocean acidification, including food supply, nutrient availability, temperature, diet, interactions with symbionts and other organisms and species. Food supply has been found to confer resistance in corals, molluscs, crustaceans, and echinoderms (Ramajo et al. 2016). N availability combined with changes in the diurnal cycle played a strong role in increasing resilience of marine diatoms to ocean acidification (Valenzuela et al. 2018). However, diet may not affect resilience to acidification in some species. For example, veligers of the slipper limpet, *Crepidula onxy*, exposed to different pH levels and fed the microalga, *Isochrysis galbana*, showed no changes in larval mortality due to pH or diet, but their interactions promoted earlier larval settlement. This slipper limpet, introduced to Hong Kong in the 1960s, was resilient to changes in pH and decreased algal nutritional value (Maboloc and Chan 2017).

Declining reef accretion has been difficult to measure in the field due to the difficulty of attributing observed changes to any one factor. De'ath et al. (2009) and Guo et al. (2020) suggest that the calcification of reef-building corals over vast areas of the Great Barrier Reef has decreased by about 13–14% since 1990. Given that bleaching events and other anthropogenic disturbances such as deteriorating water quality have been most intense on the inshore reefs of the Great Barrier Reef, the explanation that this was due to factors such as ocean acidification with/without the influence of thermal stress is plausible.

Coral reef environments will move steadily towards temperatures and carbonate ion concentrations in which reefs are unlikely to form due to excessive coral mortality rates and negative carbonate accretion (Hoegh-Guldberg 2011). Under these circumstances, reefs that are currently coral dominated will be replaced by other organisms, such as macroalgae (Figure 13.10). For coral reefs, weaker reef structures would reduce their resilience to the natural forces of erosion, and slower growth will set back recovery after bleaching and other disturbances (Lough 2008). Coral–macroalgal competition may be altered by acidification and other anthropogenic factors. In experiments with the coral, *Porites lobata* and the macroalga, *Chlorodesmis fastigiata,* a 1 °C temperature rise above ambient temperatures and a CO_2 increase of +85 ppm (RCP2.6 scenario) resulted in negative dark calcification rates; light calcification rates were negatively affected by interaction with *C. fastigiata* (Rölfer et al. 2021). Overall, the coral was negatively affected by the RCP2.6 scenario, whereas symbiont productivity was enhanced. A negative effect of the macroalga on the coral was observed for the P:R ratio under ambient conditions, but it was not enhanced under RCP2.6 conditions. However, as *P. lobata* is less affected by interactions with competing macroalgae, other coral–macroalgal species complexes may well be more significantly affected by ocean acidification and rising temperatures.

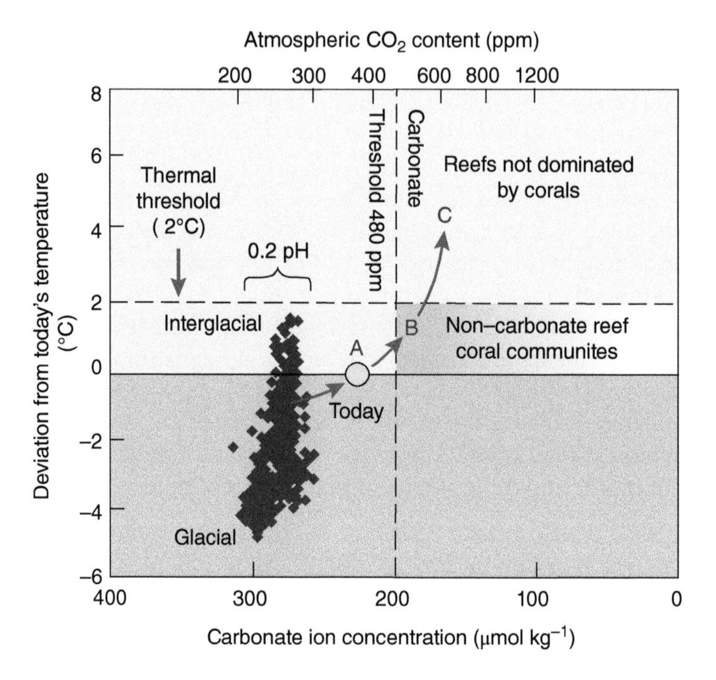

FIGURE 13.10 Interrelationships between temperature, atmospheric CO_2 concentrations, and carbonate ion concentrations reconstructed for the past 420 000 years. Red arrow points towards the right-hand top square indicating the pathway that is being followed towards an atmospheric CO_2 concentration of more than 500 ppm. The thresholds for major changes to coral communities are indicated for thermal stress (+2 °C) and carbonate ion concentrations of 200 μmol kg^{-1}. IPCC Coral Reef Scenarios CR5-A, CR5-B, and CR5-C are indicated as A, B, and C. *Source:* Hoegh-Guldberg (2011), figure 5, p. 398. © Springer Nature.

An additional threat to the ability of coral reefs to maintain their structure in an increasingly acidic environment is the sensitivity of coralline algae. These organisms contribute significantly to coral reefs by secreting skeleton that forms part of the reef structure itself and by cementing loose material together. Coralline algae calcify high-magnesium calcite at a greater metabolic cost than aragonite calcification by corals. This mineralogy makes them particularly sensitive to changes in ocean chemistry, which may not only reduce their ability to calcify but may even result in dissolution of their calcified structures, thus further destabilising coral reefs (Andersson and Glenhill 2013).

Another problem is separating the effects of natural chemical changes on coral reefs from the effects of ocean acidification (Andersson and Glenhill 2013). However, there is no doubt that the flow of carbon on reefs, especially inorganic carbon, is now being affected by acidification (Gattuso et al. 1999; Bates 2002; Suzuki and Kawahata 2004; Kleypas and Langdon 2006; Comeau et al. 2014b) despite large natural diel variations (Albright et al. 2013). For instance, diel cycles in carbonate chemistry on the reef flat of Davies Reef, Great Barrier Reef, are driven primarily by net community production (Albright et al. 2013). Warming explained 35 and 47% of the seasonal shifts in pCO_2 and pH, respectively. Daytime net community production averaged 37 mmol C m^{-2} h^{-1} in summer and 33 mmol C m^{-2} h^{-1} in winter. Night-time net community production averaged −30 and −7 mmol C m^{-2} h^{-1} in summer and winter, respectively. Daytime net ecosystem calcification averaged 11 mmol C m^{-2} h^{-1}

in summer and 8 mmol C m^{-2} h^{-1} in winter, whereas night-time net ecosystem calcification averaged 2 mmol and −1 mmol $CaCO_3$ m^{-2} h^{-1} in summer and winter, respectively. Net ecosystem calcification was sensitive to changes in Ω_{arg} for both seasons, indicating that relatively small shifts in Ω_{arg} may drive measurable shifts in calcification rates and hence carbon budgets of coral reefs throughout the year. The extent to which these processes alter the carbonate chemistry of the overlying water is a function of numerous environmental factors, including benthic community composition, physical forcing, and residence time of the water within the reef matrix.

Corals contribute to changes in the carbonate system as a function of water residence time (Anthony et al. 2011). Analyses based on flume studies found that the carbon fluxes of corals and macroalgae drove Ω_{arg} in different directions, with corals elevating pCO_2 and reducing Ω_{arg} thereby compounding the effects of ocean acidification, whereas algae drew down CO_2 and elevated Ω_{arg}, potentially offsetting ocean acidification impacts. Simulations for two CO_2 scenarios (600 and 900 ppm CO_2) suggest that a potential shift from coral to algal abundance under ocean acidification leads to improved conditions for calcification, depending on reef size, water residence time, and circulation patterns (Anthony et al. 2011). The carbon fluxes of reef communities cannot significantly counter changes in carbon chemistry at the ocean scale, but they can buffer the impacts of ocean acidification at the reef scale. Further, it is possible that some corals have the capacity to up-regulate their pH at their site of calcification such that internal changes are approximately one-half of those in ambient seawater. This species-dependent pH-buffering capacity enables corals to raise the saturation state of their calcifying medium, thereby increasing calcification rates at little additional energy cost (McCulloch et al. 2012).

Few empirical or modelling studies have addressed the long-term consequences of ocean acidification for coral reefs. At the three volcanic CO_2 seeps in Papua New Guinea, Fabricius (2011) and Takahashi et al. (2018) observed that there are organisms that gain and some that lose upon prolonged exposure to elevated CO_2. At reduced pH, reductions in coral diversity, recruitment, and abundance of reef builders, and shifts in competitive interactions between taxa were observed. Coral cover remained constant between pH 8.1 and 7.8, but reef development ceased below pH 7.7.

Adaptation, which involves selection on genetic variation to peak fitness, may serve as a mechanism to resist ocean acidification. For instance, corals from a site with naturally lower seawater pH calcified faster and maintained growth better under simulated ocean acidification than corals originating from a higher pH site (Schoepf et al. 2017). This ability was consistently linked to higher pH but lower DIC concentrations in the calcifying fluid, implying that these differences were the result of long-term acclimatisation or adaptation to naturally lower pH. Thus, high pH up-regulation with moderate DIC up-regulation may promote resistance and adaptation of coral calcification to ocean acidification (Schoepf et al. 2017).

Other calcifying organisms are negatively affected by acidification. The major calcifiers in the tropical ocean are photosynthetic calcareous algae (mostly coccolithophores), photosynthetic symbiont-bearing foraminifera and hermatypic corals which all show light-enhanced calcification (Table 13.3). Some species of molluscs, jellyfish, fish, echinoderms, and pteropods, many prominent in tropical coastal waters, showed a decline in calcification with increasing acidity (Guinotte and Fabry 2008; Kroeker et al. 2013). Early calcifying stages of benthic molluscs, such as mussels and oysters, and echinoderms, showed a strong negative response to increased seawater pCO_2 and decreased pH, $[CO_3^{2-}]$ and calcium carbonate ($CaCO_3$) saturation state. Although other benthic organisms such as crustaceans, cnidarians,

TABLE 13.3 **Percentage decline in calcification in some coccolithophore and foraminifera species to ocean acidification in experiments where species were subjected to twice and three times pre-industrial atmospheric CO_2.**

Species	2x pre-industrial	3x pre-industrial
Globigerinoides sacculifer (foraminifera)	−6	−8
Marginopora vertebralis (foraminifera)	−3	−13
Orbulina universa (foraminifera)	−8	−14
Calcidiscus leptoporus (coccolithophore)		−25
Emiliania huxleyi (coccolithophore)	−25	
E. huxleyi (coccolithophore)	−9	−18
E. huxleyi (coccolithophore)		−40
Gephyrocapsa oceanica (coccolithophore)	−29	−66

Calcification declines are relative to calcification at present-day pCO_2. *Source:* Alongi (2020a). Licensed under CC BY 4.0. © Crimson Publishers.

sponges, bryozoans, annelids, brachiopods, and tunicates possess some $CaCO_3$ in their skeletons, nothing is known of the effect of acidification on these taxa. Decreased calcification presumably compromises the fitness of these organisms and possibly shifts the competitive advantage towards non-calcifying organisms, resulting in a shift in community organisation, structure, and function. Other responses of marine fauna to ocean acidification may include shell dissolution, reduction in shell mass, growth reductions, reduced metabolism, fertility and embryo development, increased mortality, reduced thermal tolerance, reduced food intake, and increased ventilation (Guinotte and Fabry 2008; Kroeker et al. 2013).

Corresponding natural observations of tropical marine species or ecosystem changes that can be unequivocally attributed to ocean acidification are limited (Doo et al. 2020). As multiple environmental changes are co-occurring with acidification, it is difficult to detect and attribute effects in situ. There is thus a need to better evaluate and attribute acidification impacts on species and ecosystems.

Non-calcifying organisms of several phyla appear to be negatively impacted by ocean acidification under experimental conditions, including reef invertebrates and fish. The fertilisation and early larval development of the tropical sea urchin, *Echinometra lucunter,* was negatively affected by pH decrease and temperature increase (Miura Pereira et al. 2020). Fertilisation success and embryo viability declined significantly under conditions of increasing temperature (28, 30, 34, and 38 °C) and decreasing pH (8.0, 7.7, and 7.4) in combination and separately. Many species of tropical fish exhibited behavioural impairment, lower reproduction and fertility, and impairment in the ability to detect prey. At the CO_2 seeps in Papua New Guinea, fish abundances did not appear to be affected closer to the seeps, but fish exhibited abnormal behaviour such as still being able to be attracted to predator odour but being unable to distinguish between odours of different habitats (Munday et al. 2014). Similarly, elevated CO_2 and reduced pH resulted in the common coral reef meso-predator, the brown dottyback (*Pseudochromis fuscus*), to shift its behaviour from preference to avoidance of the smell of prey, suggesting a dramatic shift in predator–prey interactions when coral reefs are exposed to acidifying conditions (Cripps et al. 2011). Elevated CO_2 and ocean acidification can have severe consequences for the physiology of tropical fish species (Heuer and Grosell 2014). Notable impacts include changes in neurosensory and behavioural endpoints, otolith growth, mitochondrial function, and metabolic rates, despite some ability to compensate for or cope with altered environmental conditions.

Not all estuarine and marine organisms exhibit a negative response to ocean acidification, as a variety of organisms of different phyla show positive responses to decreasing pH and changes in pCO_2 and Ω_{arg}, including seagrasses and other macrophytes (such as brown macroalgae and kelp), sea anemones, fish, and most other non-calcifying organisms. Some calcifying organisms exhibit positive responses to increasing pCO_2. For example, when three symbiont-bearing reef foraminifers (*Baculogypsina sphaerulata*, *Calcarina gaudichaudii*, and *Amphisorus hemprichii*) were subjected to five pCO_2 levels in culture, net calcification of *B. sphaerulata* and *C. gaudichaudii* increased under intermediate levels of pCO_2 but decreased at high pCO_2 (Fujita et al. 2011). Calcification of *A. hemprichii*, however, tended to decrease at elevated pCO_2. Sensitivity of calcifying organisms to ocean acidification varies depending on species' tolerances and the degree of changes in seawater carbonate chemistry.

Seagrasses, brown macroalgae, and kelps exhibit positive responses to ocean acidification. Kelps grown under elevated pCO_2 showed enhanced growth and iodine accumulation; not only was growth of the kelp, *Saccharina japonica* enhanced, but so was growth of other edible, cultured seaweed species (Xu et al. 2018). As kelps are major iodine accumulators in the sea, these results imply that iodine levels in kelp-based coastal food webs will increase, causing changes in the biogeochemical cycles of iodine in the coastal zone.

Tropical brown macroalgae thrive under increasing acidified conditions, as shown along the natural CO_2 gradients created by the volcanic seep in Papua New Guinea (Johnson et al. 2012). Along these gradients, species of the calcifying macroalgal genus, *Padina* spp., showed reductions in $CaCO_3$ content but an increase in abundance with increasing acidified conditions closer to the seep. The success of these macroalgae may be partly explained by reduced sea urchin grazing pressure and significant increases in photosynthetic rates (Johnson et al. 2012). Generally, coralline macroalgae that deposit high-Mg calcite are most susceptible to high pCO_2, but dolomite-depositing species can acclimate to such conditions (Hofman and Bischof 2014). Although CO_2 is not likely to be limiting for photosynthesis in most macroalgae, the diffusive uptake of CO_2 is less energetically expensive than active HCO_3^- uptake, and so macroalgae using HCO_3^- likely have a selective advantage over other autotrophs under acidifying conditions (Cornwall et al. 2011). As acidified conditions become more intense such as near volcanic vents where pH declines to 6.7, macroalgal communities shift in structure and composition, with non-calcifying species thriving while calcifiers are absent (Porzio et al. 2011). At CO_2 seeps and vents, macroalgal communities are much more simplified, with less species richness. In this sense, the ecosystems associated with these seeps and vents are not climate change winners. As many macroalgal species live close to their thermal limits, they will have to up-regulate the use of HCO_3 to tolerate sublethal temperatures and to promote calcification over dissolution (Koch et al. 2013).

Tropical seagrasses respond positively to increasing CO_2 and decreasing pH, increasing their rates of photosynthesis, but growth responses to ocean acidification may depend on factors such as nutrient availability, temperature, and light. Tropical seagrasses, in turn, may also be able to modify the pH of the overlying seawater. Analyses by Unsworth et al. (2012) on the effects of seagrass productivity on seawater carbonate chemistry indicate that increases in pH of up to 0.38 units and Ω_{arg} increases of 2.9 are possible in the presence of tropical seagrass meadows (Figure 13.11), with the actual values dependent on water residence time (tidal flushing), season, and water depth.

Epiphytic communities show increased abundance under elevated CO_2 but show no effect on their community structure. Ocean acidification may take precedence over factors such as local eutrophication in altering the community structure of seagrass epiphyte assemblages (Campbell and Fourqurean 2014). *C. serrulata, H. uninervis*, and *T. hemprichii* exhibited increases in net productivity, maximum photosynthetic rates and efficiency, and an increase in the ratio of gross primary production to respiration (P_G:R) with increasing levels of pCO_2 (Ow et al. 2015). Leaf growth rates in *C. serrulata* did not increase, but those in the other two species increased significantly with increasing pCO_2.

Moreover, hermatypic coral calcification downstream of seagrasses has the potential to be about 18% greater than in habitats without seagrass, implying that coral reef resilience to ocean acidification can be enhanced by the presence of seagrass. In coral reef mesocosms with and without *T. hemprichii*, growth rates of

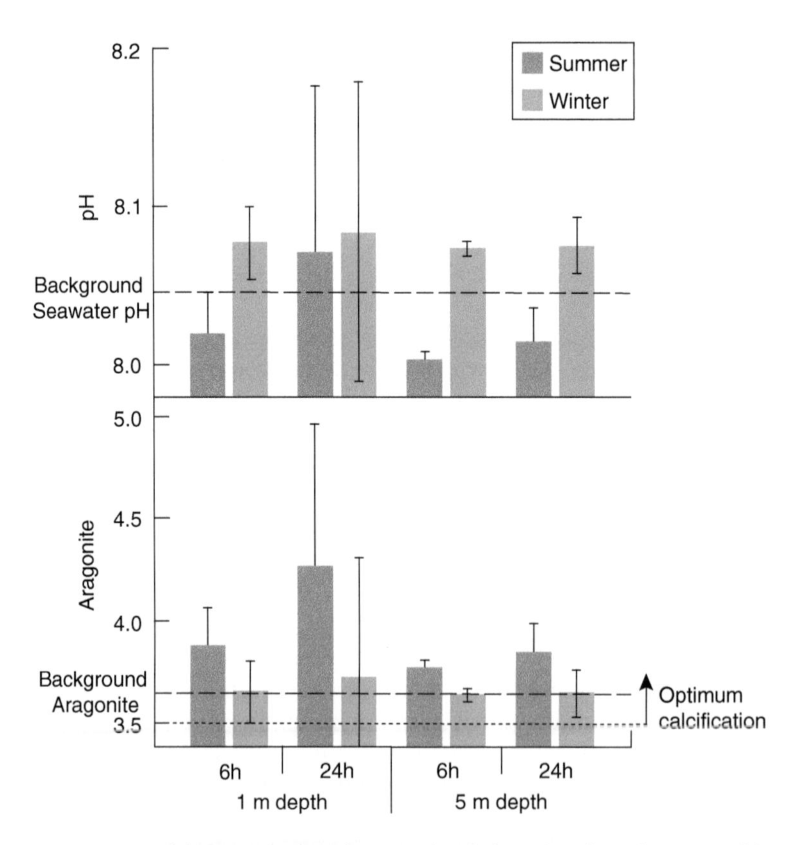

FIGURE 13.11 Seasonal (summer, winter), water depth (1 and 5 m), and water residence time (6 and 24 hours) changes of Indo-Pacific annual net seagrass productivity upon seawater inorganic carbon and the resultant pH changes. Model assumptions can be found in the source reference. *Source:* Unsworth et al. (2012), figure 3, p. 6. © IOP Publishing Ltd.

reef macroalgae were higher in seagrass-free mesocosms, but the calcification rate of the coral, *P. damicornis*, was higher in coral reef mesocosms with seagrasses under acidification conditions at 25 and 28 °C (Liu et al. 2020). Growth rates of macroalgae and coral calcification rates decreased in all acidified mesocosms when the temperature was raised to 31 °C. The presence of seagrass helped to stabilise ecosystem metabolism in response to climate change conditions as GPP, R, and NEP in the seagrass mesocosms were lower than in seagrass-free controls (Liu et al. 2020).

Seagrass invertebrate communities, however, may not always benefit from the positive responses of seagrasses to ocean acidification. At CO_2 vents off the coast of Italy, invertebrate communities associated with the Mediterranean seagrass, *Posidonia oceanica*, declined in species richness along the CO_2 gradient, but differences in community structure appeared to have been driven by indirect effects of acidification, such as changes to canopy structure and food availability (Garrard et al. 2014). However, despite the decline in species numbers, abundance of invertebrates in acidified conditions was almost double that of control sites; many heavily calcified species thrived in the high CO_2 environment.

Tropical plankton productivity, species diversity, and abundances may decline with ocean acidification depending on phyla, with shifts in community composition favouring non-calcifiers and microbes (Nagelkerken and Connell 2015). However, some tropical plankton communities exhibit complex responses to ocean acidification. In tropical Atlantic, Indian, and Pacific Ocean waters, the coccolithophore calcification to primary production ratio and cell-specific calcification were largely constant across a wide range of calcite saturation states (1.5–6.5), $[HCO_3^-]/[H^+]$ ratios, and pH (7.6–8.1), indicating that calcification by coccolithophores may be independent of carbonate chemistry. At least in tropical oceans, coccolithophore calcification may not be declining in response to ocean acidification (Marañón et al. 2016).

In the subtropical North Atlantic, colonies of the cyanobacterium, *Trichodesmium* increased N_2 fixation rates at pH 7.8 by 54% compared to ambient pH, whereas community assemblages dominated by *Prochlorococcus* and *Synechococcus* exhibited no clear response to changes in pH or/and pCO_2. Responses of these three genera may be indirect and controlled by other factors such as nutrients and temperature (Lomas et al. 2012). In the unicellular N_2-fixing cyanobacterium, *Crocospaera watsonia* from the western tropical Atlantic Ocean, the combined effects of light and CO_2 resulted in complex responses, with cyanobacteria in one treatment exhibiting no response, whereas other cultures grown under different light/CO_2 conditions showed a significant increase in both CO_2 fixation and N_2 fixation rates. Overall, cellular N retention and CO_2 fixation rates of *C. watsonia* were positively affected by elevated light and pCO_2 (Garcia et al. 2013).

In Indian waters, tropical plankton communities have exhibited mostly positive responses to ocean acidification. Along the Goa coast, an upwelling-induced, highly productive region, growth of a diatom-dominated phytoplankton community increased under increasing levels of pCO_2 (Shaik et al. 2017). In the nearby Godavari estuary, the growth of natural phytoplankton communities similarly increased in response to increased pCO_2; community dominance shifted from diatoms to cyanobacteria (Biswas et al. 2011). And in the coastal zone of the Bay of Bengal, diatom-dominated phytoplankton assemblages grew faster at increasing pCO_2 concentrations (Kumar Sahu et al. 2018), but responses were contrasting when other variables were introduced, such as light and nutrients.

The response of tropical zooplankton communities to ocean acidification is virtually unknown. At the previously examined volcanic CO_2 seep in Papua New Guinea, a threefold reduction in the biomass of demersal zooplankton was observed compared with reef sites with ambient CO_2 (Smith et al. 2016). Abundances were reduced in most taxonomic groups, but there were no dramatic shifts in community composition or in fatty acid composition, implying that ocean acidification affected food quantity but not the quality for nocturnal plankton feeders. The reduction in zooplankton abundance may be partly attributable to changes in habitat from branching to massive bouldering corals which may offer less daytime shelter.

Acclimation, which involves phenotypically plastic responses in morphology, behaviour, or physiology to maintain fitness, can also help to maintain an organism's performance in an increasingly acidified ocean. In experiments with the coccolithophore, *Emiliania huxleyi*, under conditions of elevated temperatures and declining pH, growth rates were up to 16% higher in populations acclimatised to warmer temperatures at their upper thermal tolerance limit (Schlüter et al. 2014). Particulate inorganic and organic carbon production by this species were, respectively, 101 and 55% higher under combined warming and acidification, suggesting that owing to adaptive evolution, this cosmopolitan species can resist acidifying conditions (Schlüter et al. 2014).

Adaptive evolution is a process that can involve genetic alterations that result in an increase in fitness in the face of environmental change, including ocean acidification. The coccolithophore, *E. huxleyi*, founded by single or multiple clones and exposed to increased CO_2 over time showed that around 500 generations later, this species exhibited higher growth rates under acidified conditions (Lohbeck et al. 2014). Calcification rates were lower under conditions of increased CO_2 but were up to 50% higher in adapted compared with non-adapted cultures. Further, the expression levels of 10 candidate genes putatively thought to be relevant to pH regulation, carbon transport, calcification, and photosynthesis in this species exposed to both short-term acidification and in cultures after 500 generations of high CO_2 adaptation revealed down-regulation of candidate genes and up-regulation of pH regulation and carbon transport genes (Lohbeck et al. 2014). These results indicate that adaptive evolution helps to maintain fitness in the face of ocean acidification (Lohbeck et al. 2014).

Unlike biota living on coral reefs and in the open ocean, it is questionable whether estuarine and coastal biota will be significantly affected by ocean acidification. Nearly, all tropical estuarine and nearshore waters naturally exhibit very wide variations in salinity, pH, and carbonate chemical parameters, especially pCO_2 (Table 13.4). Tropical estuarine and coastal waters are a strong source of CO_2 emissions to the atmosphere due to pCO_2 and $[CO_3^{2-}]$ oversaturation.

Oversaturation and highly variable pH are the net result of high rates of (mostly) bacterial respiration, eutrophication, and the influence of fluvial discharge, including export of alkalinity, organic matter and CO_2, deposition of anthropogenic acids and bases, intense weathering, land-use change, acid sulphate soil discharge, and acidic groundwater. Changes in water column alkalinity can also be large (Sippo et al. 2016). Duarte et al. (2013) have argued that acidification is more of an open ocean process, and that coral reefs in the coastal zone may be resilient to some degree from acidification considering that even coral reef waters can range in pH from as low as 7.63 to as high as 8.4 (see Table 2 in Duarte et al. 2013). Regulation of estuarine and coastal pH is complex compared with open ocean waters. Estuarine and nearshore environments that are metabolically intense increase Ω_{arg} due

TABLE 13.4 **Temporal and spatial variations in pH (NIST scale) and pCO_2 (μatm) concentrations, and in rates of CO_2 fluxes (mol m^{-2} a^{-1}) in tropical estuarine and nearshore waters. Salinity is expressed as PSU (practical salinity units).**

Location	pH	pCO_2	CO_2 efflux	Salinity
Piauí River estuary, NE Brazil	6.8–8.5	2200–10 000	6–21	10.2–28.5
Anai, Kuranji and Arau River estuaries, Sumatra, Indonesia	6.0–8.0	—	—	12–27
Hugli and Matla River estuaries, West Bengal, India	7.4–8.2	550–6000	2.3–32.4	0.1–20
Mangrove tidal creek, Sepetiba Bay, Brazil	7.1–7.6	—	—	13–29
Coatzacoalcos River estuary, SE Mexico	6.7–8.2	—	—	0.1–35.1
Tana River estuary and delta, N Kenya	6.51–8.58	2600–5300	—	13–34
Matang mangrove estuary, Malaysia	6.16–7.94	—	—	15.1–24.5
Agniar River estuary, SE India	7.1–8.2	—	—	5.5–34.0
Quatipuru River estuary, Amazonia, N Brazil	6.7–7.9	—	—	0–35
Nkoro River, Niger delta, Nigeria	6.1–8.5	—	—	5–17
Mulki River estuary, SW India	6.96–8.03	—	—	0.14–34.37
Ceará, Cocó, Pacoti and Pirangi River estuaries, NE Brazil	7.0–8.4	—	—	0.1–47.4
Kuala Sibuti River estuary, Sarawak, Malaysia	4.41–7.35	—	—	0.7–27.1
Kallada River and Ashtamudi estuary, SW India	5.8–8.4	—	—	18.0–24.8
Saribas and Lupar River estuaries, Sarawak, Malaysia	6.7–8.0	640–5065	14–268	0–30.6
Kodungallur-Azhikode River estuary, SW India	6.9–7.5	—	—	10.2–18.9
Taperaçu River estuary, Amazonia, N Brazil	7.2–7.8	—	—	11.0–38.7
Merbok River estuary, Malaysia	6.50–6.81	—	—	1.96–18.69
Dungun River estuary, Malaysia	6.06–8.02	—	—	0–31
Bight of Benin estuary, SW Nigeria	6.5–6.7	—	—	0.43–0.47

(*continued*)

TABLE 13.4 (*Continued*)

Location	pH	pCO_2	CO_2 efflux	Salinity
Brunei estuary system, Brunei Darussalam	5.87–8.06	—	—	0.4–28.5
Twelve estuaries, Pernambuco, Brazil	6.1–8.3	823–8907	37.5–65.0	0.6–34.5
S. Lagan, S. Mendahara River estuaries, Sumatra, Indonesia	4.0–8.1	—	—	1.5–36.4
Straits of Malacca, Malaysia	6.32–8.44	—	—	7.43–32.2
New Calabar River estuary, Niger delta, Nigeria	5.5–7.2	—	—	0–10.5
Cochin backwaters, SW India	6.61–7.51	1228–2853	23.4–100.01	0.69–19.2
Godavari River estuary, E India	6.12–8.61	221–32 763	29.2–87.6	0.09–33.52
Mandovi-Zuari River estuaries, W India	6.50–7.00	520–2700	4.0–24.5	0.07–34.59
Vellar-Coleroon River estuaries, SE India	7.2–8.4	—	—	16–34
Sundarbans, NE India	7.9–8.3	—	—	—
Sungai Brunei River estuary, Brunei Darussalam	5.78–8.3	—	—	3.58–31.2
Tapi River estuary, NW India	7.2–8.5	—	—	0.11–32
Dumai River estuary, Sumatra, Indonesia	4.0–8.7	—	—	0–27
Panguil Bay, Mindanao, The Philippines	7.3–9.2	—	—	5.6–34.7
Imo River estuary, Nigeria	5.2–8.2	—	—	0–22
Bangpakong River estuary, Thailand	6.8–7.8	—	—	0–32
15 monsoonal estuaries, Bay of Bengal, E India	6.66–8.61	263–26 521	−0.2–96.32	0.07–28.78
12 monsoonal estuaries, Arabian Sea, W India	5.98–7.51	1360–20 421	−3.24–362.45	0.05–7.32
28 estuaries, N and NE Brazil	6.60–8.20	162–8638	0.58–181.77	0.1–46.0
Pahang River estuary, Malaysia	6.6–8.4	—	—	0–32
Chilika Lake, E India	6.67–9.53	4–11 548	−19.9–271.7	0.13–35.88
Qua and Cross River estuaries, Niger delta, Nigeria	6.34–7.70	—	—	0.87–2.62
Six mangrove estuaries, N Australia	7.1–8.3	250–5000	3.2–16.8	30–40

— = not determined. *Source:* Alongi (2020a).

to high primary production, and calcification is also regulated mainly by biological processes (Hendriks et al. 2015). Further, impacts of ocean acidification must be considered with other climate change processes, such as rising SSTs, as it is likely that a combination of climate change factors will be the ultimate determinant of ecological change (Harvey et al. 2013).

Mangroves and seagrasses will likely be the most resilient tropical ecosystems to the effects of acidification. In the case of seagrasses, we have seen how individual species usually respond positively, or not at all, to lower pH. Seagrasses and other macrophytes have a capacity to modify pH within their canopy and within their habitat (Hendriks et al. 2014). Within seagrass meadows, strong diel variability in pH, DIC, Ω_{arg}, and O_2 are driven by primary productivity; changes in carbonate chemistry are related to leaf surface area available for photosynthesis (Hendriks et al. 2014). However, some organisms associated with seagrasses, such as leaf epiphytes, may not benefit from the buffering capacity of seagrasses if the meadows are declining for other reasons, such as eutrophication.

Mangroves may prove to be the most resilient in the face of coastal acidification. Mangrove soil pH is ordinarily low, within the range of 4–7, especially in the forests of south and Southeast Asia and Africa (Alongi 2009), as interstitial water is often acidic. Mangrove soils have low pH due to high rates of soil respiration, high concentrations of polyphenolic acids, and the net effects of metabolic processes associated with the trees and their root systems (Alongi 2014). Recently, it has been found that subsurface transport of groundwater derived from acidic soil waters plays a major role in carbon cycling in mangrove forests and their waterways, having important consequences for resilience to acidification (Sippo et al. 2016; Alongi 2020b).

Mangroves are apparent buffers of acidification in the tropical coastal zone (Sippo et al. 2016). An examination of DIC, dissolved CO_2, and alkalinity dynamics in six Australian mangrove tidal creeks revealed a mean export of DIC, whereas alkalinity fluxes ranged from an import of 1.2 mmol m^{-2} d^{-1} to an export of 117 mmol m^{-2} d^{-1}. A net import of free CO_2 of −11.4 mmol m^{-2} d^{-1} was measured, equivalent to one-third of the estimated air–water CO_2 flux of 33.1 mmol m^{-2} d^{-1} (Sippo et al. 2016). Upscaling these results globally, mangrove alkalinity exports (4.2 Tmol a^{-1}) are equivalent to about 14% of global river or continental shelf benthic alkalinity fluxes. The effect of DIC and alkalinity exports creates a measurable increase in pH, implying that mangroves partly counteract acidification in adjacent tropical coastal waters. Mangroves may thus be one of the largest sources of alkalinity to the tropical coastal ocean, providing buffering against acidification.

Mangrove environments can assist in the survival of other tropical organisms, including some species of corals. In the US Virgin Islands, over 30 species of corals grow in association with mangroves, including two major reef-building corals, *Colpophyllia natans* and *Diploria labyrinthiformis*. These corals thrive in low light from mangrove shading and at higher temperatures than nearby reef corals (Yates et al. 2014). A higher proportion of colonies of *C. natans* were shaded by mangroves, with no bleached colonies. And although fewer *D. labyrinthiformis* colonies shaded beneath mangroves, more unshaded colonies bleached. Mangrove habitats can therefore be a refuge for diverse coral assemblages from climate change (Yates et al. 2014).

Acidification in coastal upwelling zones can in some cases be ameliorated by biological activities. The Qiongdong upwelling (QDU), one of the primary western boundary upwelling systems, prevails during summer in the northern South China Sea. Off the east coast of Hainan Island, which is an epicentre of the QDU, profiles of carbonate chemistry and hydrographic parameters from nearshore to the offshore upwelling zone show divergent responses in pH, Ω_{arg} and changes in DIC concentrations under the coastal upwelling which were attributed to biological activities in mediating acidification in the upper 30 m across the east coast of Hainan Island (Dong et al. 2021). Offshore, enhanced photosynthesis reduced the upwelling-induced acidification by 40%, whereas aerobic respiration increased acidification in the subsurface waters nearshore by 15%, leading to a pH and Ω_{arg} minimum in upwelled waters from nearshore to offshore. Conversely, the biological contribution was negligible in surface nearshore waters due to the balance between net primary production (NPP) and calcification (Dong et al. 2021).

In summary, tropical coastal ecosystems and their associated species assemblages can be impacted by ocean acidification in complex and very variable ways. Some ecosystems such as seagrass meadows, macrophyte assemblages, and mangrove forests and their associated waterways are either positively affected or unaffected by acidification. Coral reefs are the most vulnerable ecosystems to ocean acidification, with many hermatypic corals exhibiting declining rates of calcification with decreasing pH and increasing pCO_2. Other calcifying organisms are similarly affected, including some species of foraminifera and coccolithophores. Some tropical organisms, including some corals and fleshy and calcareous algae, exhibit complex, non-linear responses to acidification. Coral communities associated with CO_2 seeps show large changes in community composition with some species responding positively, but most impacted negatively, along the CO_2 gradient. All ecosystems and their associated species assemblages exhibit various degrees of resilience and adaptation as well as adaptive evolution to declining pH and altered carbonate chemistry. The naturally wide range of pH in estuarine and nearshore environments predisposes resilience of biota, especially in estuaries and rivers inhabited by mangrove forests.

13.6 Increasing Hypoxia

As exemplified by the expansion of oxygen minimum zones and dead zones, the decline in dissolved oxygen (DO) concentrations in the global ocean is accelerating (Altieri et al. 2021). This is especially true in the tropics where warmer temperatures decrease the saturation capacity of DO in seawater, while simultaneously increasing rates of aerobic respiration that consumes and depletes oxygen. Ocean deoxygenation is being driven by increasing SSTs and eutrophication (Breitburg et al. 2018) and is thought to be widespread in tropical ecosystems, including coral reefs, mangroves, and seagrass meadows (Altieri et al. 2019), although the documentary record is poor.

There are three main drivers of oxygen depletion (Altieri et al. 2021) that make these ecosystems naturally subject to hypoxic conditions over intermittent spatial scales and time periods: (i) natural biogenic features (seagrass pools, mangrove ponds and channels, reef lagoons, and flats) reduce oxygen replenishment by increasing stratification and reducing flushing rates, all enhanced by tides, seasonality, and extreme weather or climate; (ii) their complexity and high production of organic matter results in the trapping of organic material which fuels decomposition and the depletion of oxygen; and (iii) naturally rapid rates of oxygen consumption can lead to a decline in DO which can be exacerbated by slowing of water flow and wave dampening by seagrass canopies, coral reefs, and mangrove forest structures (Altieri et al. 2021). Human-induced deoxygenation is likely to mimic these natural events.

Mangroves, seagrasses, and corals all have mechanisms to cope with and mediate oxygen levels in their environment and can to a considerable extent, counteract hypoxia. As discussed in earlier chapters, corals maintain elevated oxygen concentrations in their tissues and surrounding water by utilising oxygen produced by their endosymbiotic algae. Mangroves and seagrasses similarly possess structural and physiological mechanisms, such as aboveground roots that supply oxygen to mangrove roots and rhizosphere belowground and translocate oxygen from seagrass blades down to the rhizosphere. All three groups have the capacity to utilise oxygen stored in their tissues, create oxygenated microhabitats, absorb oxygen from the atmosphere, and/or redistribute oxygen internally to counter low-oxygen concentrations (Altieri et al. 2021). Further, trophic groups in these iconic habitats have high tolerances to hypoxia, including fish, molluscs, and a variety of benthic and pelagic invertebrates. The ability of these organisms to withstand limited oxygen conditions is facilitated by mutualistic relationships with their host corals, mangroves, and seagrasses.

These three environments naturally undergo diel changes in DO concentrations in overlying waters (Figure 13.12a, b). Diel cycling of DO levels is most evident in mangrove environments and to a lesser extent in coral reefs and seagrass beds. The wide range of values in mangrove waters mimics the naturally wide range of other physiochemical parameters such as pH, pCO_2, salinity, and rates of CO_2 flux across the water–air interface (Table 13.4). And such wide variation is likely for the same reasons, including changes in tidal range, temperature, high respiration rates, eutrophication, the influence of fluvial discharge, tidal exchange of alkalinity, organic matter and CO_2, deposition of anthropogenic acids and bases, intense weathering, land-use change, acid sulphate soil discharge, and acidic groundwater. It is possible that mangroves may be more tolerant to hypoxia than either corals or seagrasses.

As pointed out by Altieri et al. (2019), there have been no known mass mortality events in mangrove forests and seagrass meadows attributed only to natural hypoxia conditions. Some corals are similarly tolerant to low oxygen as the impact is apparently species-specific. An unprecedented hypoxic event on the Caribbean coast of Panama resulted in coral bleaching and mass mortality of some (*Agaricia lamarcki*), but not other (*Stephanocoena intercepta*), coral species (Altieri et al. 2017). Altieri et al. (2017) analysed global databases and concluded that coral reefs are associated with > 50% of the known tropical dead zones worldwide, with > 20% of all coral reefs at greater risk of hypoxia based on various risk factors.

Increasing severity of deoxygenation along with co-stressors, such as eutrophication and acidification, may likely result in a decline in survivorship at the organismal scale in mangrove, seagrass, and reef environments in future. In the coastal zone, eutrophication in particular is likely to co-occur with deoxygenation, with excess nutrients causing harmful algal blooms and the production of H_2S; the latter will result in a shift to anaerobic metabolism and mass mortality and lower tolerances in corals and seagrass, and fish kills in mangrove waterways. Further, increased hypoxia will likely result in disruption of critical mutualisms that facilitate ecosystem stability. Such changes would be a sign of destabilisation and a shift to a new equilibrium. Negative feedbacks may emerge accelerating and reinforcing the change in ecosystem state. Mortality of seagrasses would lead to the accumulation of plant detritus, smothering the sediment and further accelerating oxygen depletion,

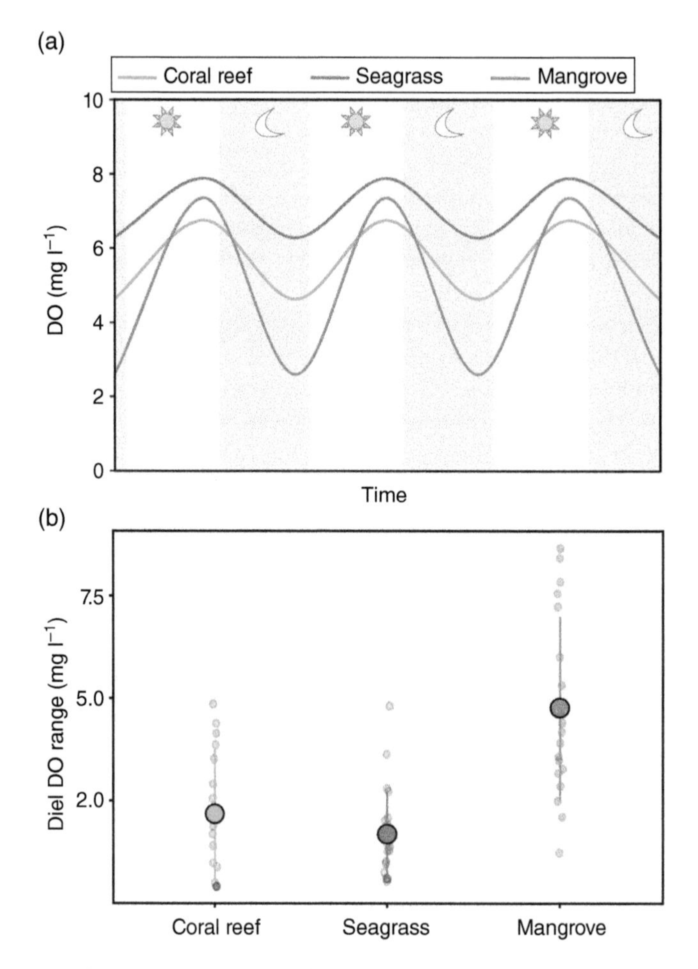

FIGURE 13.12 Diel variations in dissolved oxygen (DO) concentration in waters overlying coral reefs, seagrass meadows, and mangrove forests. (a) shows the DO variations over a day and (b) shows the diel range in DO across habitats. Large, filled circles represent the mean ± 1SD of separate diel cycles per habitat and smaller, and pale circles show the raw data points. *Source:* Altieri et al. (2021), figure 4, p. 6. © Elsevier.

H_2S, and additional plant mortality. The decay of reef organisms is similarly thought to exacerbate oxygen decline in coral reefs and in mangrove epifauna (Simpson et al. 1993; Saenger 2002). Reefs may become more susceptible to disease outbreaks with disease generating further hypoxia, accelerating its spread over coral colonies (Glas et al. 2012).

The potential to recover from hypoxia depends on a wide range of factors, including: (i) the spatial extent, duration, and intensity of the event, (ii) the number and diversity of surviving organisms, especially seeds and polyps, (iii) the extent of physical damage to the ecosystem, (iv) the degree to which community-level interactions remain intact, (v) local connectivity to source populations, and (vi) species-specific abilities to recolonise. Considering that recovery times are usually long, on the scale of years to decades, the increasing occurrence of hypoxia is of increasing concern.

13.7 Impacts on Shelf and Oceanic Ecosystems and Fisheries

Much of the physical evidence of climate change on the marine biosphere comes from continental shelf and open ocean waters, such as rising SSTs, changes in ocean circulation and carbonate chemistry, and acidification (Section 2.7). These physical and chemical changes have affected shelf and oceanic ecosystems in several ways, including 'tropicalization' of temperate environments and range expansion of many species. Many sophisticated models have forecast significant changes in phytoplankton and zooplankton biomass and productivity, leading to changes in fisheries yields.

A recent modelling exercise (called the Earth System Model) suggests twenty-first century declines in tropical primary production of between 1 and 30% (Kwiatkowski et al. 2017). Using satellite-based observations of ENSO sensitivity of tropical primary production to constrain projections of the long-term impact on primary production, tropical primary production is predicted to decline by 11% under a 'business-as-usual' scenario by 2100. Subsequent modelling by Kwiatkowski et al. (2019) projects that global phytoplankton and zooplankton biomass will decline by 6.1 and 13.6%, respectively, over the twenty-first century under a 'business-as-usual' RCP8.5 scenario. These declines will be concentrated in the low latitudes (Figure 13.13) that also have generally the highest ratio of zooplankton and phytoplankton biomass anomalies. These scenarios agree with previous modelling studies projecting that the greatest declines in primary production will be at low latitudes, apparently driven by reduced nutrient concentrations in the eutrophic zone due to increased stratification and weaker upwelling (Bopp et al. 2013; Laufkötter et al. 2015).

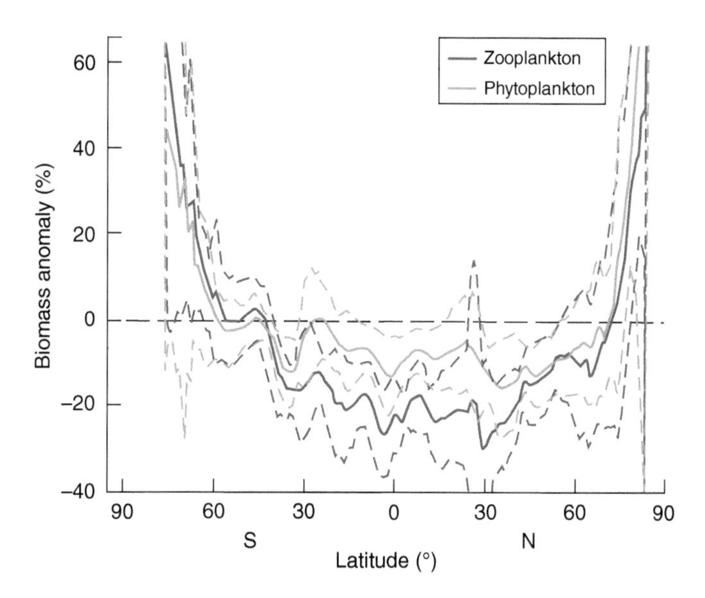

FIGURE 13.13 Model projection of mean twenty-first century changes ("anomalies") of phytoplankton and zooplankton biomass across latitudes under the IPCC RCP8.5 'business-as-usual' scenario. Note that the largest predicted changes for both trophic groups are at subtropical and tropical latitudes. Dashed lines indicate ± 1 standard deviation. *Source:* Kwiatkowski et al. (2019), figure 2, p. 232. © John Wiley & Sons.

Empirical data indicate that there have indeed been significant declines in the abundance of tropical plankton. Continuous plankton recording in the Gulf of Guinea prior to 2008 showed a significant decline in plankton biomass as SSTs increased (Kifani et al. 2018). The reproduction strategies and spawning habitat of the main fish in this region, the European sardine, round sardinella, flat sardinella, bonga shad, anchovy, Atlantic chub mackerel, horse mackerel, and false shad may have been compromised by declining pH and warming. Ocean acidification has affected fish diversity and abundance of some of these fish species off Ghana. Poleward expansion over the past decade has been recently documented for the dinoflagellate, *Gambierdiscus* spp., in the subtropical upwelling sector of the Canary Current system off Morocco, and the Canary Islands and Madeira Islands archipelagos. Further, an unprecedented massive accumulation of *Sargassum* has recently occurred along the West African coast. Fish landings within the Guinea Current system from the Congo north to the Guinea-Bissau coast have declined, most especially since 2000, partly due to SST rises and a loss of zooplankton due to environmental problems (Donkor and Abe 2012).

In the central Western Atlantic, which encompasses the Gulf of Mexico, the Caribbean Sea, and the North Brazil shelf, increasing SSTs and eutrophication have resulted in more frequent 'green tides' and toxic algal blooms and have also experienced an unprecedented influx of pelagic *Sargassum* (Oxenford and Monnereau 2018). As penaeid shrimp and weakfish, croakers, and sea catfish rely heavily on estuarine nursery habitats, their early life stages will be vulnerable to further degradation of their habitats under climate change and continuing eutrophication; adults living offshore in deeper soft bottom areas will likely be less impacted. Penaeid shrimps in the northern Gulf of Mexico where seasonal hypoxia is severe are expected to suffer significant impacts as the adults will be unable to pass through the hypoxic zones to spawn in open water (Oxenford and Monnereau 2018). Shrimp and groundfish productivity are expected to decline in the western central Atlantic, especially on the North Brazil shelf, continental countries of the Caribbean and the Gulf of Mexico. Highly migratory large pelagic species (billfish and large tunas) and smaller migrating species (dolphinfish, wahoo, mackerels, and small tunas) may be less impacted by climate change due to their high mobility, lack of biphasic life history, less vulnerable reproductive behaviour, and the fact that open ocean waters are less impacted by anthropogenic stressors. Nevertheless, some species such as bluefin tuna may abandon their spawning area in the Gulf of Mexico as SSTs warm.

Increased stratification over the Indian Ocean has led to a decline in oceanic phytoplankton abundance and productivity, as enhanced stratification has suppressed upwelling (Moustahfid et al. 2018). The models of Bopp et al. (2013) indicate that the western Indian Ocean will experience a 20–50% decrease in pelagic NPP by 2100. These changes are likely to weaken marine food webs in the region through a reduction in energy flow to higher trophic levels and a shift to a detritus-based ecosystem which may in turn result in simplified food webs and altered producer–consumer interactions (Moustahfid et al. 2018). There has been an increase in the frequency, intensity, and spatial coverage of phytoplankton blooms, especially of *Trichodesmium* sp., off the west and east coasts of India since 1900 (Kumar Sahu et al. 2018), which has been attributed to SST rises, thermal stratification, and increased eutrophication. This change in bloom-forming populations has visibly disrupted the diatom-sustained food chain which will likely put stress on fishery resources which are already being impacted by oxygen depletion, acidification and overfishing.

As variability of the distribution and abundance of tuna species are linked to ENSO and the IOD, any changes to these events will affect large pelagic fish. The

1997/1998 ENSO resulted in anomalously high SSTs in phase with a positive IOD event causing dramatic temperature and wind stress anomalies in the equatorial Indian Ocean that led to a deepening of the mixed layer in the west and a shoaling in the east. This event coincided with unusually low primary production in the western Indian Ocean and a shift in tuna stocks from the west to the east, highlighting the vulnerability of fish stocks to climate variability (Moustahfid et al. 2018). Modelling suggests that a considerable decrease in skipjack tuna abundance and associated fisheries by 2100 may be caused by habitat deterioration, driven by ocean warming.

On the opposite side of Africa, the waters off southern Angola are considered one of 24 ocean warming hotspots (Hobdey and Pecl 2014) with SSTs increasing between 0.23 and 0.8 °C per decade over the past three decades (Van der Linge and Hampton 2018). While there was some evidence of increased primary productivity, there was no clear trend in zooplankton biomass or community structure. Further south off Namibia, SSTs have similarly increased (0.2–0.5° decade^{-1}), but there has been no evidence of long-term changes in phytoplankton production, despite the apparent long-term decrease in upwelling favourable winds and temperature increases. However, there has been a major increase in copepod production and significant changes in community composition since the 1970s, with a shift from large to small copepod species (Van der Linge and Hampton 2018). Dramatic declines in pelagic fish abundance caused by overfishing have changed the food web from being dominated by sardines and anchovies to an altered trophic pathway dominated by mostly jellyfish and pelagic gobies.

While changes in fish distribution and abundance along the western and SW seaboard of Africa have not been linked to climate change, there has been a rapid shift south from Angola into Namibian waters by coastal species, particularly the sciaenid, *Argyrosomus coronus*, a species heavily fished off Angola. This shift was attributed to rapid warming of Angolan coastal waters. The most important commercial resources off Namibia (hake, sardine, anchovy, and rock lobster) have been negatively impacted by overfishing rather than by any environmental change. However, evidence suggests that the decline in rock lobster during the late 1980s to early 1990s was caused by intrusions of low-oxygen water into the fishing grounds. Presently, hake have expanded their distribution over a wider depth range, the very reduced sardine population is now confined to areas north of Walvis Bay, South Africa, while anchovy have virtually disappeared from Namibian waters.

A vulnerability assessment (Van der Linge and Hampton 2018) concluded that Angola's national economy was the most vulnerable of 132 countries to climate-induced impacts on fisheries, especially for small pelagic fish. Overfishing will be aggravated by rising SSTs and changes in ocean circulation. In a 2011 assessment (De Young et al. 2012), the demersal trawl fishery (mostly for hake) of Namibia was rated as the most vulnerable to climate change. Also rated as highly vulnerable were the small pelagic and the rock lobster fisheries, although the horse mackerel fishery was rated as one of the least vulnerable.

Along the coast of East Africa, particularly off Mozambique, artisanal fisheries have declined significantly in the past decade due to overfishing and destructive fishing practices, and to climate-related changes in the Agulhas Current (Vousden et al. 2012). The main impact of climate change on fisheries from Tanzania north to the coast of Somalia has been the destruction and degradation of fish spawning and nursery grounds and feeding areas, including the decline in coral reefs due to bleaching and a reduction in mangrove area due partly to changes in the Somali Coastal Current (Vousden et al. 2012). Inadequate data have made it difficult to assess the vulnerability of these fisheries, but increases in SSTs and sea levels are likely to negatively impact coastal resources.

The central and western Pacific Ocean supports major industrial tuna fisheries and a variety of small-scale coastal fisheries. The industrial fishery industry targets skipjack tuna and juvenile yellowfin tuna, and the longline fishery targets mature bigeye and yellowfin tuna. Coastal fisheries are the basis of fish consumption and livelihood in most Pacific Island communities and target mainly demersal fish and invertebrates associated with coral reefs, seagrass meadows and mangroves, and increasingly, tuna and other large pelagic fish nearshore (Bell et al. 2018). Reductions in nutrients in the mixed layer of the open ocean have led to a decline in phytoplankton production at various areas of the subtropical and tropical Pacific Ocean. These changes are highly variable spatially, with effects of increased SSTs and changes in ocean circulation on oceanic food webs to be different among areas of the region. Impacts are likely to be much lower in the Pacific Equatorial Divergence than in other areas due to strong upwelling. However, there could be increases in primary production in the subtropical north Pacific, but little change in the western tropical Pacific as increases in subsurface phytoplankton may offset declines in surface phytoplankton. The impact of ocean acidification on food webs supporting tuna has yet to be determined (Bell et al. 2018).

For skipjack and yellowfin tuna, an eastward and poleward shift in distribution and reductions in biomass are projected under the RCP8.5 'business-as-usual' scenario, driven mainly by changes in larval survival and spawning location. Projected decreases are particularly marked for Papua New Guinea, the Federated States of Micronesia, Nauru, and Palau for these two tuna species, but biomass is projected to increase for skipjack tuna in Vanuatu, New Caledonia, Pitcairn Islands, and French Polynesia, and for yellowfin tuna in French Polynesia. For bigeye tuna, strong decreases in biomass are forecast in all areas; for South Pacific albacore, the distributions of larvae and juveniles are expected to shift south after 2050 and densities of early life stages are projected to decrease in the Coral Sea by 2050, resulting in an adult biomass about 30% lower than in 2000. The four species of tropical tuna are expected to be of limited vulnerability across the central Pacific due to their migratory abilities. Increased stratification, however, could make the surface-dwelling skipjack and yellowfin tuna more vulnerable to capture. This forecast is based on higher catch rates for yellowfin tuna in the warm pool during El Niño when shoaling of the thermocline contracts the vertical habitat of this species. A decrease in micronekton productivity is likely to increase the risk of natural mortality and production of tuna in the region (Lehodey et al. 2011). Also, access to micronekton could become more difficult for tuna due to the increased stratification and decreased O_2 levels.

Small-scale fisheries are also expected to be affected by climate change throughout the Pacific. The abundance of coral reef fish has declined on the Great Barrier Reef, although climate change and ocean acidification are projected to have a greater range of direct and indirect effects on the distribution and abundance of demersal fish and invertebrates in the western and central Pacific Ocean (Bell et al. 2018). Higher SSTs are expected to alter metabolism, growth, reproduction, and survival of demersal fish and invertebrates, leading to changes in their abundance, size, and distribution. Changes in ocean circulation and currents are likely to affect larval dispersal and recruitment success. Climate change impacts are expected to reduce productivity of demersal fish in the region by up to 20% by 2050 and by 20–50% by 2100. Invertebrate productivity is projected to decrease by 5% by 2050 and by 10% by 2100, and their size and quality are expected to be affected by reduced levels of aragonite saturation. Coastal fisheries of the Pacific Island nations are experiencing a decline due to the loss and degradation of coastal habitats, especially coral reefs, where demersal fish (butterflyfish, wrasse, and damselfish) are heavily exploited (Pratchett et al. 2011), although the declines in fish production are likely to be minimal until 2035.

By 2100, the loss of coastal habitats is expected to reduce the number of reef-associated and other coastal habitat-dependent fish by 20–50%. Using a dynamic bioclimate envelope model (DBEM), projections were used to model impacts on marine biodiversity and potential fisheries changes in Oceania (Asch et al. 2018). The model predicts a rise in SST of ≥3 °C, a decline in $O_2 \geq 0.01\,ml\,l^{-1}$, a pH drop ≥ 0.3, and a decrease in NPP by $0.5\,g\,m^{-2}\,d^{-1}$ by 2100 under the "business-as-usual" scenario. These changes may result in rates of species extinction of >50% for fish and invertebrates declining in abundance or migrating to more suitable regions. Maximum potential catch is projected to decline by >50% over most areas, with the largest impacts in the Western Pacific Warm Pool.

South Asian fisheries in the Arabian Sea, Bay of Bengal, and the East Indian Ocean are expected to decline due to climate change-related events (Fernandes 2018). A 2 °C rise is predicted to decrease total productive potential off South and Southeast Asia by 2050, despite increased primary production of small phytoplankton. In the Arabian Sea and the Bay of Bengal, oxygen concentrations may decrease in many areas, and vertical shifts in oxygen depth distribution may potentially affect tunas and food webs. A regional study forecasted large decreases in potential catch of the two commercial species, Hilsa shad and Bombay duck, in the Bay of Bengal (Fernandes et al. 2016). Small coastal and pelagic fish such as the oil sardine (*Sardinella longiceps*) and the Indian mackerel (*Rastrelliger kanagurta*) have extended their distributional boundary to northern and eastern latitudes, while threadfin breams have shifted the timing of their spawning towards cooler months off Chennai, Southeast India. Within the Malabar upwelling zone off SW India, the abundance of oil sardine has increased over the decades, corresponding to warmer SSTs, greater wind speeds, and increases in chlorophyll concentration (Vivekanandan 2011), indicating that the current warming is beneficial to herbivorous small pelagic fish. The Indian mackerel, a plankton feeder and crucial food for large fish, has seen an increase in its contribution to bottom trawling since 1985 from about 2 to 15% by 2007. Over the last two decades, the mackerel has migrated from occupying surface and subsurface waters to deeper waters, perhaps due to warming of surface waters (Vivekanandan 2011).

Models assuming a 'business-as-usual' scenario were used to assess the impact of climate change and the vulnerability of habitats and fisheries (55 priority marine species) in the Arabian Gulf (Wabnitz et al. 2018). The modelling outputs suggest a high rate of local extinction (up to 35% of initial species richness) by 2090 relative to 2010. Projected changes (Figure 13.14) in species numbers (top), species invasion (middle), and extinction (bottom) in the Gulf by 2050 (left panels) and 2090 (right panels) relative to 2010 point to a high rate of local extinction (up to 12% of initial species richness) by the end of the century under the RCP8.5 scenario. Species invasion will remain low (up to 5% of initial species richness), but spatially these are significant differences in invasion frequency within the gulf (Figure 13.14). Local extinction may be low to moderate in 2050, with highest species losses projected along the NW coast of Bahrain and the UAE; by 2090 species loss may rise across most of the gulf, with highest numbers of species lost projected for the SW part of the gulf, off the coast of Saudi Arabia, Bahrain, Qatar, and the UAE. Species invasion, in contrast, may be similar between 2050 and 2090 and limited to the northern gulf, off the coast of Kuwait and northern Iran (Wabnitz et al. 2018).

Two of three models predicting changes in fisheries productivity in the Bay of Bengal (Figure 13.15) suggest a decline off West Bengal (Figure 13.15a) and Odisha

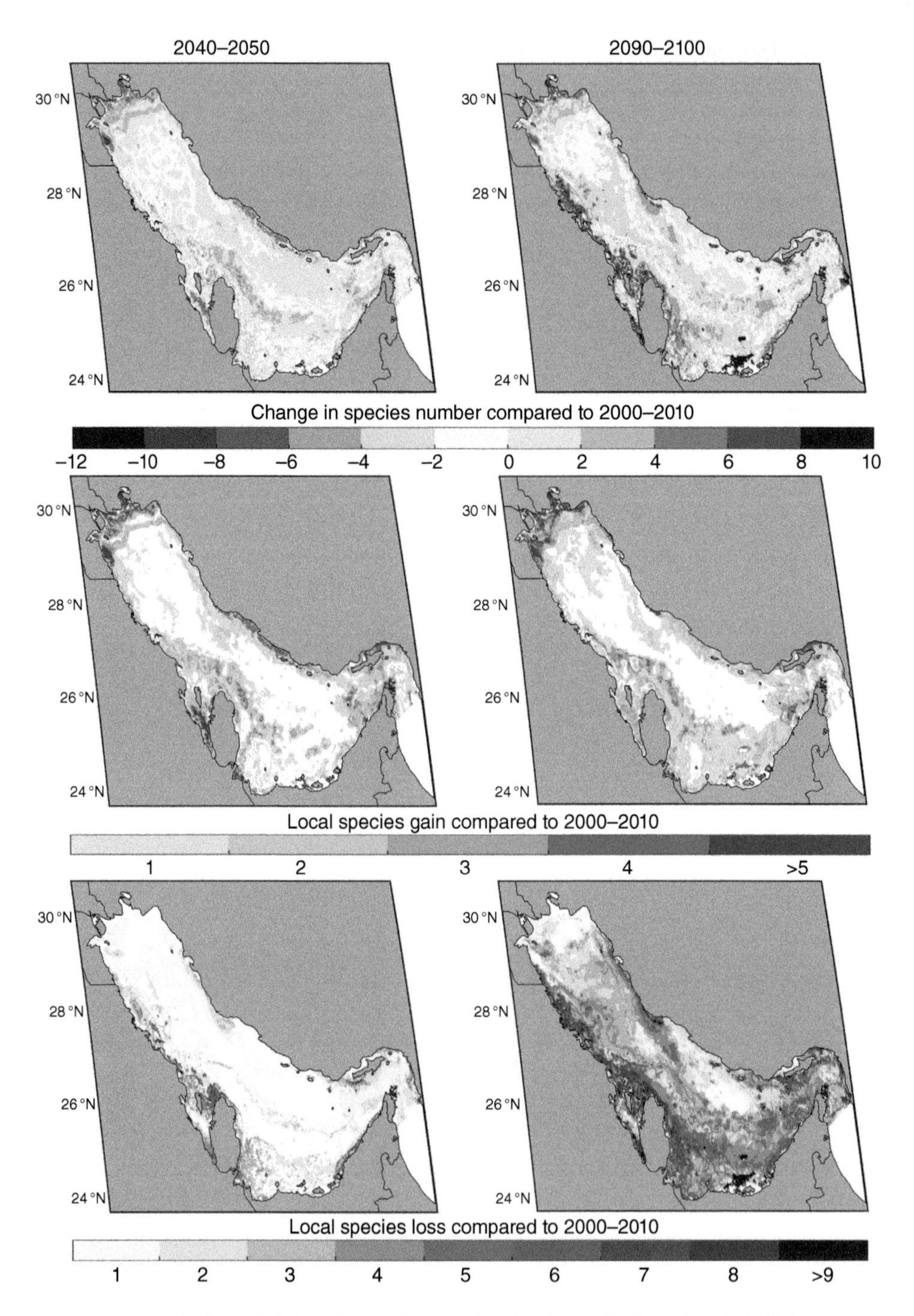

FIGURE 13.14 Projected change in species number (top), species invasion (middle), and extinction (bottom) in the Arabian Gulf by 2050 (left panel) and 2090 (right panel) relative to 2010. Results are presented for an average of three models and for the IPCC RCP8.5 'business-as-usual' scenario. *Source:* Wabnitz et al. (2018), figure 2, p. 9. Licensed under CC BY 4.0. © PLOS.

(Figure 13.15b) beginning later in this century; the HadGEM2-E5 model predicts an increase in fish production with larger increases in precipitation and temperature than the other two models (Das et al. 2020). The results are highly variable suggesting that other factors likely play a role in the Bay of Bengal fisheries. Non-climate stressors have exacerbated impacts of climate change, including pollution, dams, habitat modification and destruction, and overfishing.

Ecosystems in the southern Caribbean Sea are similarly impacted by climate change. Monthly observations from the CARIACO Ocean Time-Series Station off the coast of Venezuela between 1996 and 2010 (Taylor et al. 2012) documented significant decadal trends in SST rise (about 1 °C), intensification of stratification, reduced delivery of upwelled nutrients, and diminished intensities of phytoplankton blooms evident as overall declines in chlorophyll *a* concentration ($\Delta Chla = -2.8\,a^{-1}$) and net primary production ($\Delta NPP = -1.5\,a^{-1}$). Further, the composition of phytoplankton communities shifted from the dominance of diatoms, dinoflagellates, and coccolithophorids to smaller taxa after 2004. In contrast, mesozooplankton biomass increased and commercial landings of the planktivorous sardine, *Sardinella aurita*, declined by 87% (Taylor et al. 2012). *Sardinella aurita* had previously been very

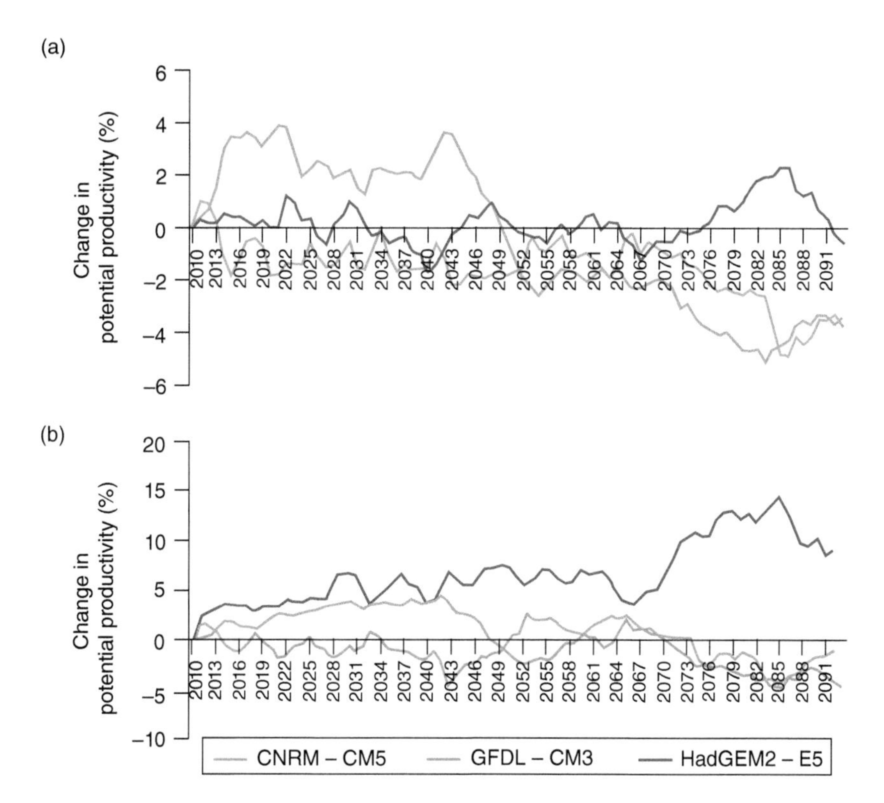

FIGURE 13.15 Projections of three models of mean twenty-first century changes in fish production in the Bay of Bengal off (a) West Bengal and (b) Odisha. The CNRM-CM5 model (blue) represented a scenario of a small increase in precipitation and a relatively small increase in temperature. The GFDL-CM3 model (green) represented a scenario of a moderate–large increase in precipitation and a moderate increase in temperature, and the HadGEM2-ES model represented a scenario of a large increase in both precipitation and temperature. All three models used the 'business-as-usual' RCP8.5 pathway to provide a strong climate signal. *Source:* Das et al. (2020), figure 8, p. 10. © Elsevier.

responsive to upwelling intensity and phytoplankton community dynamics. However, overfishing may have also contributed to the collapse of the sardine fishery. These ecosystem changes were attributed to the weakening trend in trade winds ($-1.9\%\ a^{-1}$) and trends in two climatic indices, namely the northward migration of the Azores High pressure centre, a descending branch of the Hadley cell, by 1.12 °N latitude and the NE progression of the ITCZ Atlantic centroid, an ascending branch of the Hadley cell, the March position of which shifted by about 800 km between 1996 and 2009 (Taylor et al. 2012).

Like those inhabiting coral reefs, mangroves and seagrasses, fish populations and communities in estuaries, on continental shelves and the open ocean are being impacted by climate change. In coastal and shelf waters, rising SSTs will result in several physiological changes such as faster larval growth and activity, reduced gene flow, higher pelagic larval survival, loss of habitat, and shorter and broader dispersal. Decreasing pH may lead to aberrant larval behaviour leading to increased risk of predation, while SLR is likely to only impact estuarine nursery areas. Severe weather events and reduced or more variable freshwater flow may result in changed reproductive cues and disrupted fish spawning (Booth et al. 2018). Ongoing depletion of dissolved oxygen and vertical expansion of the OMZ in regions such as the tropical NE Atlantic Ocean may restrict the usable fish habitat (Stramma et al. 2011). Electronic tagging of Atlantic blue marlin (*Makaira nigricans*) in this area confirmed compression and loss of habitat of an area of about $6 \times 10^{13}\ m^3$ or 15% for the period 1960–2010. Such habitat loss increases vulnerability to surface fishing and may partly account for the 10–50% worldwide decline of pelagic predator diversity as all tropical basins are experiencing expansion of OMZs.

Many fish species face an increased risk of extinction due to ocean warning, but temperature-sensitive species are expected to shift their ranges to higher latitudes as SSTs rise. This prognosis includes migratory fish in the open ocean, such as tuna, yellowtail, and mackerel, who have been observed or are expected to shift their spawning/feeding and overwintering grounds to cooler latitudes. Also, western boundary currents are expected to strengthen and extend further into higher latitudes, facilitating a further influx of tropical fish larvae to temperate waters. Shifts in the distribution of tropical fish have been best documented off the tropical–temperate transition zone of North America, Japan, Australia, southern and eastern Europe and southern Africa (Wu et al. 2012), changing the composition of temperate and tropical commercial and recreational fishery resources, with increasing proportions of tropical fish species in temperate fishing grounds and a decline in the proportion of subtropical fish species in tropical fishery areas (Cheung et al. 2013 a, b). The range expansion of tropical fish into temperate waters can affect local fishery resources, promoting drastic increases in the abundance of subtropical species that were rare or uncommon in a particular area with far-reaching effects on the structure, function, and services of marine ecosystems, especially fisheries yield. Some of these outcomes are undesirable, such as overgrazing of temperate seagrass meadows by tropical herbivorous fish (Heck et al. 2015).

A general decrease in the proportion of catches of subtropical species has been a phenomenon observed in many of the world's tropical fisheries, which has been associated with further reductions in abundance and body size among tropical species due to physiological responses to extreme warm waters (Cheung et al. 2012). Future impacts on tropical fisheries will be as drastic in the tropics as in the temperate latitudes (Cheung et al. 2013 a,b).

Predictions for climate change impacts on fisheries yields, animal biomass, and ecosystem structure (Booth et al. 2018; Cheng et al. 2018; Bryndum-Buchholz et al. 2019; Lam et al. 2020) all point to negative changes in the tropics; no models have yet predicted any significant global or ocean-scale positive impacts of climate change in tropical marine ecosystems, in agreement to the relatively small amount of empirical data for shelf and open ocean ecosystems. Six ecosystem models within the Fisheries and Marine Ecosystem Model Intercomparison Project (Fish-MIP) analysed the responses of animal biomass in all major ocean basins to different climate change scenarios (Bryndum-Buchholz et al. 2019). The results under the RCP8.5 scenario predicts a decline in animal biomass of 15–30% in the North and South Atlantic and Pacific and in the Indian Ocean by 2100, with the polar ocean experiencing a 20–80% increase. Biomass projections under RCP2.6 (low emissions) and RCP8.5 correlated significantly with changes in NPP and correlated negatively with projected SST increases across all ocean basins except the polar basins. Size classes are projected to decrease in the Pacific, Atlantic, and Indian Ocean basins under RCP8.5 by 2100. In the North and South Atlantic Ocean, a greater decrease in the mean biomass of medium-sized animals (17–29%) is projected compared to small (10–24.5%) and large (12–24%) animals. The reverse is predicted in the North Pacific Ocean, with projected biomass decreases of animals of all size. The overall size class trends in the South Pacific Ocean and the Indian Ocean may not change substantially. Projected declines in biomass of medium-sized animals may be explained by the decline of their smaller-sized prey. Thus, ecosystem structure is projected to change with shifts as animal biomass (Bryndum-Buchholz et al. 2019).

Current trends in fisheries yields (Section 9.5) show continued growth of wild marine fish catch in the tropics, especially in South and Southeast Asia, whereas global wild marine capture landings have been relatively stable since the late 1990s, although there was an increase in 2018 to 84.4 Mt. Fisheries landings have declined or levelled off in some tropical countries, but many nations are still showing large increases in landings. The trend of increasing catches continued in 2017 and 2018, with catches in the Indian Ocean and the Pacific Ocean reaching the highest levels recorded (FAO 2020). Projections using both the DBEM and the dynamic size-based food web model (Cheng et al. 2018) indicate a decrease in maximum catch potential by 2.8–5.3 and 2.8–4.3% under low emissions (RCP2.6) by 2050 and 2095 relative to 2000, respectively. Under the 'business-as-usual' scenario (RCP8.5), both models predict larger decreases of 7.0–12.1 and 16.2–25.2% by 2050 and 2095, respectively, relative to 2000. The largest decrease (<40%) by 2095 may be in the tropics, mostly in the South Pacific. The main reason why both models agree is that changes in potential catches are strongly driven by changes in plankton productivity, although other factors cannot be discounted. For example, consistently large decreases in catch potential in tropical regions may primarily be a result of a decline in plankton productivity, but simultaneous loss of habitat can exacerbate the forecasted decline.

Predictions of maximum catch potential (Figure 13.16) under the RCP8.5 scenario agree with forecasted biomass changes in that most of the decline of global fisheries catch will likely occur in subtropical and tropical latitudes (Booth et al. 2018; Cheng et al. 2018; Bryndum-Buchholz et al. 2019; Lam et al. 2020), with most severe declines in the eastern tropical Atlantic and in the tropical Pacific, especially in the Indo-Pacific where maximum catch potential is forecast to decrease by 47% and species turnover to decline by 36%. Throughout the tropical Pacific, more than half of coastal fish and invertebrate species important to fisheries are also projected to

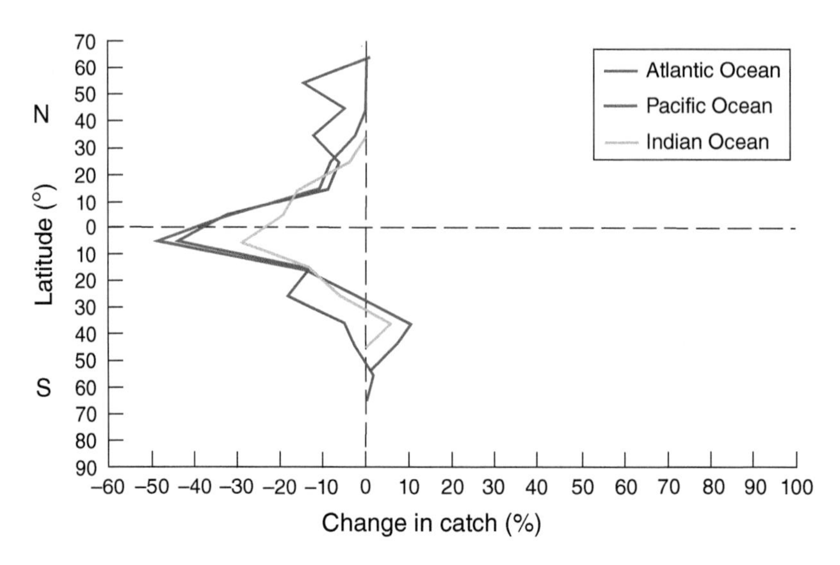

FIGURE 13.16 Mean projected changes in maximum catch potential (%) in the Atlantic, Pacific, and Indian Oceans for 2046–2060 relative to 1991–2010 under the IPCC 'business-as-usual' RCP8.5 scenario. Note that the greatest declines are predicted to occur in tropical latitudes. *Source:* Lam et al. (2020), figure 3b, p. 5. © Springer Nature.

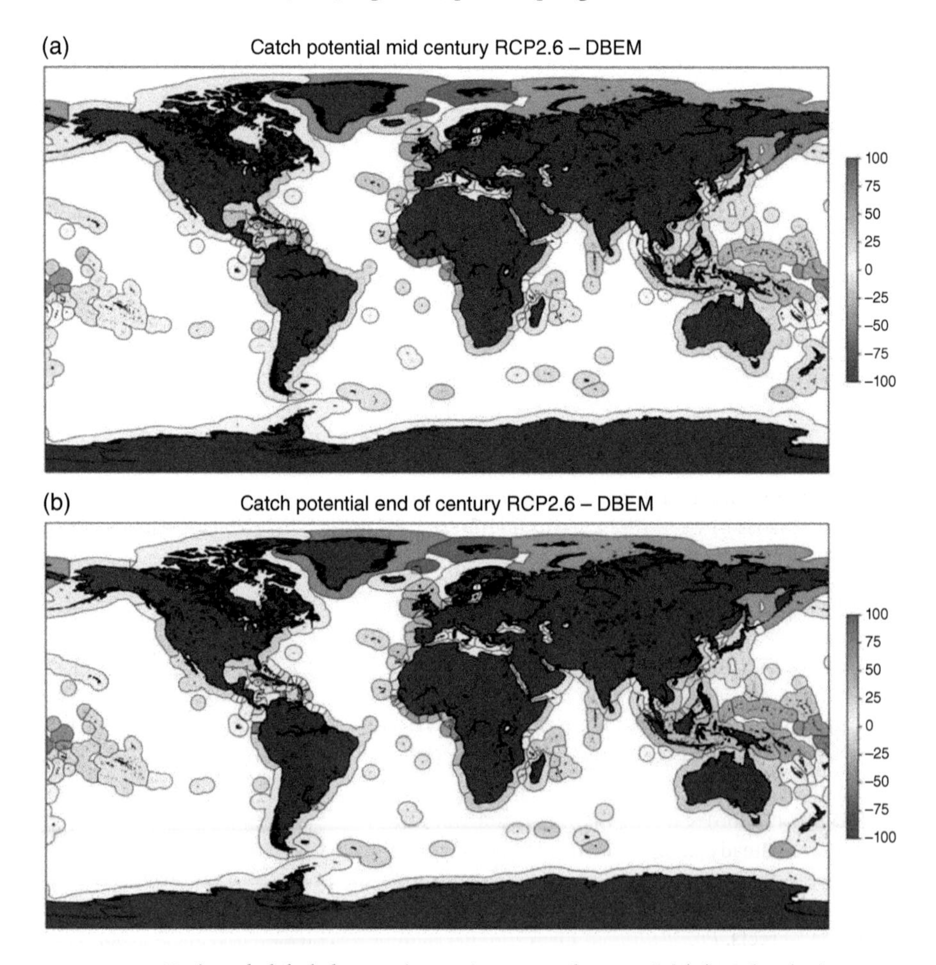

FIGURE 13.17 Projected global changes in maximum catch potential (%) under the 'strong CO_2 emission mitigation' IPCC RCP2.6 scenario at (a) 2046–2055 and (b) 2091–2100 for the dynamic bioclimate envelope model (DBEM). Gains and losses are not as great as under the RCP8.5 scenario, but the model still predicts significant losses throughout the subtropics and tropics and gains or no change in the polar oceans under a mitigation scenario. *Source:* Cheng et al. (2018), figure 4.3, p. 69. Food and Agriculture Organization of the United Nations. © FAO.

become locally extinct by 2100 under RCP8.5 (Lam et al. 2020). While many African fish species are more resilient to warming than species elsewhere, fisheries production along the east coast of Africa will likely decline due in part to the decline of coral reefs and other coastal habitats. Model simulations predict that marine fish catches in West African waters will decrease by >50% by 2050 under RCP8.5 with largest reductions near the equator (Lam et al. 2012). Forecasts for Africa are, however, model dependent with some predicting 8–26% declines in fish landings (Lam et al. 2012) and others forecasting a 24% increase in the Gulf of Guinea by 2050 (Blanchard et al. 2012; Barange et al. 2014). However, all models indicate that African fisheries are highly vulnerable to climate change.

From a global perspective and under the RCP2.6 scenario of assuming strong CO_2 emissions mitigation, projected changes in catch potential vary widely across the global ocean at mid-century (Figure 13.17a) and by the end of the century (Figure 13.17b). The largest decrease (<40%) by the end of the century are in tropical regions, mostly in the South Pacific and along the west coast of Africa. However, decreases are also forecast for the temperate Northeast Atlantic. In contrast, increases or no change in catch potential are seen near the Southern Ocean and the Arctic region (Cheung et al. 2013b; Cheng et al. 2018).

References

Albert, S., Saunders, M.I., Roelfsema, C.M. et al. (2017). Winners and losers as mangroves, coral and seagrass ecosystems respond to sea-level rise in Solomon Islands. *Environmental Research Letters* 12: 094009. https://doi.org/10.1088/1748-9326/aa7e68.

Albright, R., Langdon, C., and Anthony, K.R.N. (2013). Dynamics of seawater carbonate chemistry, production, and calcification of a coral reef flat, central Great Barrier Reef. *Biogeosciences* 10: 6747–6758.

Almahasheer, H., Serrano, O., Duarte, C.M. et al. (2017). Low carbon sink capacity of Red Sea mangroves. *Scientific Reports* 7: 9700. https://doi.org/10.1038/s41598-017-10424-9.

Alongi, D.M. (2008). Mangrove forests, resilience, protection from tsunamis, and responses to global climate change. *Estuarine, Coastal and Shelf Science* 76: 1–13.

Alongi, D.M. (2009). *The Energetics of Mangrove Forests*. Dordrecht, The Netherlands: Springer.

Alongi, D.M. (2014). Carbon cycling and storage in mangrove forests. *Annual Review of Marine Science* 6: 195–220.

Alongi, D.M. (2015). The impact of climate change on mangrove forests. *Current Climate Change Reports* 1: 30–39.

Alongi, D.M. (2018). *Blue Carbon: Coastal Sequestration for Climate Change Mitigation*, Springer Briefs in Climate Studies. Cham, Switzerland: Springer Nature.

Alongi, D.M. (2020a). Vulnerability and resilience of tropical coastal ecosystems to ocean acidification. *Examines in Marine Biology and Oceanography* 3: EIMBO000562. https://doi.org/10.31031/EIMBO.2020.03.000562.

Alongi, D.M. (2020b). Carbon cycling in the world's mangrove ecosystems revisited: significance of non-steady state diagenesis and subsurface linkages between the forest floor and the coastal ocean. *Forests* 11: 977. https://doi.org/10.3390/f11090977.

Altieri, A.H., Harrison, S.B., Seemann, J. et al. (2017). Tropical dead zones and mass mortalities on coral reefs. *Proceedings of the National Academy of Sciences* 114: 3660–3665.

Altieri, A.H., Nelson, H.R., and Gedan, K.G. (2019). Tropical ecosystems – corals, seagrasses, and mangroves. In: *Oxygen Deoxygenation: Everyone's Problem – Causes, Impacts, Consequences, and Solutions* (eds. D. Laffoley and J.M. Baxter), 401–429. Gland, Switzerland: IUCN.

Altieri, A.H., Johnson, M.D., Swaminathan, S.D. et al. (2021). Resilience of tropical ecosystems to ocean deoxygenation. *Trends in Ecology & Evolution* 36: 227–238.
Andersson, A.J. and Glenhill, D. (2013). Ocean acidification and coral reefs: effects of breakdown, dissolution, and net ecosystem calcification. *Annual Review of Marine Science* 5: 321–348.
Anthony, K.R.N., Kleypas, J.A., and Gattuso, J.-P. (2011). Coral reefs modify their seawater carbon chemistry – implications for impacts of ocean acidification. *Global Change Biology* 17: 3655–3666.
Asch, R.G., Cheung, W.W.L., and Reygondeau, G. (2018). Future marine ecosystem drivers, biodiversity, and fisheries maximum catch potential in Pacific Island countries and territories under climate change. *Marine Policy* 88: 285–294.
Bahr, K.D., Tran, T., Jury, C.P. et al. (2020). Abundance, size, and survival of recruits of the reef coral *Pocillopora acuta* under ocean warming and acidification. *PLoS One* 15: e0228168. https://doi.org/10.1371/journal.pone.0228168.
Baird, A.H., Sommer, B., and Madin, J.S. (2012). Pole-ward range expansion of *Acropora* spp. along the east coast of Australia. *Coral Reefs* 31: 1063. https://doi.org/10.1007/s00338-012-0928-6.
Baker, A.C., Glynn, P.W., and Riegl, B. (2008). Climate change and coral reef bleaching: an ecological assessment of long-term impacts, recovery trends and future outlook. *Estuarine, Coastal and Shelf Science* 40: 435–471.
Barange, M., Merino, G., Blanchard, J.L. et al. (2014). Impacts of climate change on marine ecosystems production in societies dependent on fisheries. *Nature Climate Change* 4: 211–214.
Bates, N.R. (2002). Seasonal variability of the effect of coral reefs on seawater CO_2 and air-sea CO_2 exchange. *Limnology and Oceanography* 47: 43–52.
Beca-Carretero, P., Teichberg, M., Winters, G. et al. (2020). Projected rapid habitat expansion of tropical seagrass species in the Mediterranean Sea as climate change progresses. *Frontiers in Plant Science* 11: 555376. https://doi.org/10.3389/fpls.2020.555376.
Bell, J.D., Allain, V., Sen Gupta, A. et al. (2018). Climate change impacts, vulnerabilities and adaptations: western and central Pacific Ocean marine fisheries. In: *In: Impacts of Climate Change on Fisheries and Aquaculture: Synthesis of Current Knowledge, Adaptation and Mitigation Options* (eds. M. Barange, T. Bahri, M.C.M. Beveridge, et al.), 305–324. FAO Fisheries and Aquaculture Technical Paper 627.
Biswas, H., Cros, A., Yadav, K. et al. (2011). The response of a natural phytoplankton community from the Godavari River Estuary to increasing CO_2 concentration during the pre-monsoon period. *Journal of Experimental Marine Biology and Ecology* 407: 284–293.
Blanchard, J.L., Jennings, S., Holmes, R. et al. (2012). Potential consequences of climate change for primary production and fish production in large marine ecosystems. *Philosophical Transactions of the Royal Society B* 367: 2979–2989.
Bomer, E.J., Wilson, C.A., Hale, R.P. et al. (2020). Surface elevation and sedimentation dynamics in the Ganges-Brahmaputra tidal delta plain, Bangladesh: evidence for mangrove adaptation to human-induced tidal amplification. *Catena* 187: 104312. https://doi.org/10.1016/j.catena.2019.104312.
Booth, D.J., Feary, D., Kobayashi, D. et al. (2018). Tropical marine fish and fisheries and climate change. In: *Climate Change Impacts on Fisheries and Aquaculture: A Global Analysis* (eds. B.F. Phillips and M. Pérez-Ramírez), 875–896. Cambridge, UK: Wiley.
Bopp, L., Resplandy, L., Orr, J.C. et al. (2013). Multiple stressors of ocean ecosystems in the 21st century: projections with CMIP5 models. *Biogeosciences* 10: 6225–6245.
Bove, C.B., Ries, J.B., Davies, S.W. et al. (2019). Common Caribbean corals exhibit highly variable responses to future acidification and warming. *Proceedings of the Royal Society B: Biological Sciences* 286: 20182840. https://doi.org/10.1098/rspb.2018.2840.
Bove, C.B., Whitehead, R.F., and Szmant, A.M. (2020a). Responses of coral gastrovascular cavity pH during light and dark incubations to reduced seawater pH suggest species-specific responses to the effects of ocean acidification on calcification. *Coral Reefs* 39: 1675–1691.

Bove, C.B., Umbanhowar, J., and Castillo, K.D. (2020b). Meta-analysis reveals reduced coral calcification under projected ocean warming but not under acidification across the Caribbean Sea. *Frontiers in Marine Science* 7: 127. https://doi.org/10.3389/fmars.2020.00127.

Breitburg, D., Levin, L., Olschlies, A. et al. (2018). Declining oxygen in global ocean and coastal waters. *Science* 359: eaam7240. http://doi.org/10.1126/science.aam7240.

Brown, B.E., Dunne, R.P., Somerfield, P.J. et al. (2019). Long-term impacts of rising sea temperature and sea level on shallow water coral communities over a ~ 40 year period. *Scientific Reports* 9: 8826. https://doi.org/10.1038/s41598-019-45188-x.

Bryndum-Buchholz, A., Tittensor, D.P., Blanchard, J.L. et al. (2019). 21st century climate change impacts on marine animal biomass and ecosystem structure across ocean basins. *Global Change Biology* 25: 459–472.

Cacciapaglia, C.W. and van Woesik, R. (2020). Reduced carbon emissions and fishing pressure are both necessary for equatorial coral reefs to keep up with rising seas. *Ecography* 43: 789–800.

Camp, E.F., Edmondson, J., Doheny, A. et al. (2019). Mangrove lagoons of the Great Barrier Reef support coral populations persisting under extreme environmental conditions. *Marine Ecology Progress Series* 625: 1–14.

Campbell, J.E. and Fourqurean, J.W. (2014). Ocean acidification outweighs nutrient effects in structuring seagrass epiphyte communities. *Journal of Ecology* 102: 730–737.

Campbell, S.J., McKenzie, L.J., and Kerville, S.P. (2006). Photosynthetic responses of seven tropical seagrasses to elevated seawater temperatures. *Journal of Experimental Marine Biology and Ecology* 330: 455–468.

Cardona-Olarte, P., Twilley, R.R., Krauss, K.W. et al. (2006). Responses of neotropical mangrove seedlings grown in monoculture and mixed culture under treatments of hydroperiod and salinity. *Hydrobiologia* 569: 325–341.

Carnero-Bravo, V., Sanchez-Cabeza, J.A., Ruiz-Fernández, A.C. et al. (2016). Sedimentary records of recent sea-level rise and acceleration in the Yucatan Peninsula. *Science of the Total Environment* 573: 1063–1069.

Carnero-Bravo, V., Sanchez-Cabeza, J.A., Ruiz-Fernández, A.C. et al. (2018). Sea-level rise sedimentary record and organic carbon fluxes in a low-lying tropical coastal ecosystem. *Catena* 162: 421–430.

Cavalcanti, G.S., Shukla, P., Morris, M. et al. (2018). Rhodoliths holobionts in a changing ocean: host-microbes interactions mediate coralline algae resilience under ocean acidification. *BMC Genomics* 19: 701. https://doi.org/10.1186/s12864-018-5064-4.

Cavanaugh, K.C., Kellner, J.R., Forde, A.J. et al. (2014). Poleward expansion of mangroves is a threshold response to decreased frequency of extreme cold events. *Proceedings of the National Academy of Sciences* 111: 723–727.

Cavanaugh, K.C., Parker, J.D., Cook-Patton, S.C. et al. (2015). Integrating physiological threshold experiments with climate modelling to project mangrove species' range expansion. *Global Change Biology* 21: 1928–1938.

Chappel, A.R. (2018). Soil accumulation, accretion, and organic carbon burial rates in mangrove soils of the lower Florida Keys: a temporal and spatial analysis. MS thesis. University of South Florida.

Chen, L. and Wang, W. (2017). Ecophysiological responses of viviparous mangrove *Rhizophora stylosa* seedlings to simulated sea-level rise. *Journal of Coastal Research* 33: 1333–1340.

Cheng, W.W., Bruggeman, J. and Butenschon, M. (2018). Projected changes in global and national potential marine fisheries catch under climate change scenarios in the twenty-first century. In: *Impacts of Climate Change on Fisheries and Aquaculture; Synthesis of Current Knowledge, Adaptation, and Mitigation Options* (eds. M. Barange, T. Bahri, M.C.M. Beveridge et al.), 63-86. FAO Fisheries and Aquaculture Technical Paper No. 627. Rome: FAO.

Cheung, W.W., Meeuwig, J.J., Feng, M. et al. (2012). Climate-change induced tropicalisation of marine communities in Western Australia. *Marine and Freshwater Research* 63: 415–427.

Cheung, W.W., Sarmiento, J.L., Dunne, J. et al. (2013a). Shrinking of fish exacerbates impacts of global ocean changes on marine ecosystems. *Nature Climate Change* 3: 254–258.

Cheung, W.W., Watson, R., and Pauly, D. (2013b). Signature of ocean warming in global fisheries catch. *Nature* 497: 365–369.
Cinco-Castro, S. and Herrera-Silveira, J. (2020). Vulnerability of mangrove ecosystems to climate change effects: the case of the Yucatan Peninsula. *Ocean & Coastal Management* 192: 105196. https://doi.org/10.1016/j.ocecoaman.2020.105196.
Collier, C.J. and Waycott, M. (2014). Temperature extremes reduce seagrass growth and induce mortality. *Marine Pollution Bulletin* 83: 483–490.
Collier, C.J., Uthicke, S., and Waycott, M. (2011). Thermal tolerance of two seagrass species at contrasting light levels: implications for future distribution in the Great Barrier Reef. *Limnology and Oceanography* 56: 2200–2210.
Comeau, S., Carpenter, R.C., Nojiri, Y. et al. (2014a). Pacific-wide contrast highlights resistance of reef calcifiers to ocean acidification. *Proceedings of the Royal Society of London B: Biological Sciences* 281: 20141339. https://doi.org/10.1098/respb.2014.1339.
Comeau, S., Edmunds, P.J., Spindel, N.B. et al. (2014b). Fast coral reef calcifiers are more sensitive to ocean acidification in short-term laboratory incubations. *Limnology and Oceanography* 59: 1081–1091.
Comeau, S., Cornwall, C.E., and McCulloch, M.T. (2017). Decoupling between the response of coral calcifying fluid pH and calcification to ocean acidification. *Scientific Reports* 7: 7573. https://doi.org/10.1038/s41598-017-08003-z.
Cornwall, C.E., Hepburn, C.D., Prichard, D. et al. (2011). Carbon-use strategies in macroalgae: differential responses to lowered pH and implications for ocean acidification. *Journal of Phycology* 48: 137–144.
Cornwall, C.E., Comeau, S., DeCarlo, T.M. et al. (2018). Resistance of corals and coralline algae to ocean acidification: physiological control of calcification under natural pH variability. *Proceedings of the Royal Society of London B: Biological Sciences* 285: 20181168. https://doi.org/10.1098/rspb.2018.1168.
Cripps, I.L., Munday, P.L., and McCormick, M.I. (2011). Ocean acidification affects prey detection by a predatory reef fish. *PLoS One* 6: e22736. https://doi.org/10.1371/journal.pone.0022736.
Cusack, M., Saderne, V., Arias-Ortiz, A. et al. (2018). Organic carbon sequestration and storage in vegetated coastal habitats along the western coast of the Arabian Gulf. *Environmental Research Letters* 13: 074007. https://doi.org/10.1088/1748-9326/aac899.
Cziesielski, M.J., Schmidt-Roach, S., and Aranda, M. (2019). The past, present and future of coral heat stress studies. *Ecology and Evolution* 9: 10055–10066.
Dais, M., Ferreira, A., Gouveia, R. et al. (2019). Long-term exposure to increasing temperatures on hermatypic coral fragments reveals oxidative stress. *Marine Environmental Research* 150: 104758. https://doi.org/10.1016/j.marenvres.2019.104758.
Das, I., Lauria, V., Kay, S. et al. (2020). Effects of climate change and management policies on marine fisheries productivity in the north-east coast of India. *Science of The Total Environment* 724: 138082. https://doi.org/10.1016/j.scitotenv.2020.138082.
De Young, C., Hjort, A., Sheridan, S. et al. (2012). Climate change implications for fisheries of the Benguela Current region – making the best of change. *FAO Fisheries and Aquaculture Proceedings No. 27*, Rome: FAO.
De'ath, G., Lough, J.M., and Fabricius, K.E. (2009). Declining coral calcification on the Great Barrier Reef. *Science* 323: 116–119.
DeCarlo, T.M., Comeau, S., Cornwall, C.E. et al. (2018). Coral resistance to ocean acidification linked to increased calcium at the site of calcification. *Proceedings of the Royal Society B: Biological Sciences* 285: 20180564. https://doi.org/10.1098/rspb.2018.0564.
Di Nitto, D., Neukermans, G., Koedam, N. et al. (2014). Mangroves facing climate change: landward migration potential in response to projected scenarios of sea level rise. *Biogeosciences* 11: 857–871.
Doney, S.C., Ruckelhaus, M., Duffy, J.E. et al. (2012). Climate change impacts on marine ecosystems. *Annual Review of Marine Science* 4: 11–37.

Doney, S.C., Shallin Busch, D., Cooley, S.R. et al. (2020). The impacts of ocean acidification on marine ecosystems and reliant human communities. *Annual Review of Environment and Resources* 45: 83–112.
Dong, X., Huang, H., Zheng, N. et al. (2021). Role of biological activity in mediating acidification in a coastal upwelling zone at the east coast of Hainan Island. *Estuarine, Coastal and Shelf Science* 249: 107124. https://doi.org/10.1016/j/ecss.2020.107124.
Donkor, S.M. and Abe, J. (2012). Impact of climate change in the Guinea Current large marine ecosystem. In: *Frontline Observations on Climate Change and Sustainability of Large Marine Ecosystems* (eds. K. Sherman and G. McGovern), 64–79. New York: United Nations Development Programme.
Doo, S.S., Kealoha, A., Andersson, A. et al. (2020). The challenges of detecting and attributing ocean acidification impacts on marine ecosystems. *ICES Journal of Marine Science* 77: 2411–2422.
Duarte, C.M. (2002). The future of seagrass meadows. *Environmental Conservation* 29: 192–206.
Duarte, C.M., Hendricks, I.E., Moore, T.S. et al. (2013). Is ocean acidification an open-ocean syndrome? Understanding anthropogenic impacts on seawater pH. *Estuaries and Coasts* 36: 221–236.
Duke, N.C. (2017). Mangrove floristics and biogeography revisited: further deductions from biodiversity hot spots, ancestral discontinuities, and common evolutionary processes. In: *Mangrove Ecosystems: A Global Biogeographic Perspective* (eds. V.H. Rivera-Monroy, S.Y. Lee, E. Kristensen and R.R. Twilley), 17–53. Cham, Switzerland: Springer International.
Esteban, M., Jasmero, L., Nurse, L. et al. (2019). Adaptation to sea level rise on low coral islands: lessons from recent events. *Ocean and Coastal Management* 168: 35–40.
Fabricius, K.E. (2011). Factors determining the resilience of coral reefs to eutrophication: a review and conceptual model. In: *Coral Reefs: An Ecosystem in Transition* (eds. Z. Dubinsky and N. Stambler), 493–505. Dordrecht, The Netherlands: Springer.
FAO (2020). The state of world fisheries and aquaculture 2020. *Sustainability in Action*. Rome: FAO. https://doi.org/10.4060/ca9229en (accessed 19 November 2020).
Fernandes, J.A. (2018). Climate change impacts, vulnerabilities and adaptations: southern Asian fisheries in the Arabian Sea, Bay of Bengal and east Indian Ocean. In: *Impacts of Climate Change on Fisheries and Aquaculture: Synthesis of Current Knowledge, Adaptation and Mitigation Options* (eds. M. Barange, T. Bahri, M.C.M. Beveridge, et al.), 281–304. FAO Fisheries and Aquaculture Technical Paper 627. Rome: FAO.
Fernandes, J.A., Kay, S., Hossain, M.A.R. et al. (2016). Projecting marine fish production and catch potential in Bangladesh in the 21st century under long-term environmental change and management scenarios. *ICES Journal of Marine Science* 73: 1357–1369.
Frieler, K., Meinshausen, M., Golly, A. et al. (2013). Limiting global warming to 2 °C is unlikely to save most coral reefs. *Nature Climate Change* 3: 165–170.
Fu, H., Zhang, Y., Ao, X. et al. (2019). High surface elevation gains and prediction of mangrove responses to sea-level rise based on dynamic surface elevation changes at Dongzhaigang Bay, China. *Geomorphology* 334: 194–202.
Fujita, K., Hikami, M., Suzuki, A. et al. (2011). Effects of ocean acidification on calcification of symbiont-bearing reef foraminifers. *Biogeosciences* 8: 2089–2098.
Galeano, A., Urrego, L.E., Botero, V. et al. (2017). Mangrove resilience to climate extreme events in a Colombian Caribbean Island. *Wetlands Ecology and Management* 25: 743–760.
Garcia, N.S., Fu, F.-X., Breene, C.L. et al. (2013). Combined effects of CO_2 and light on large and small isolates of the unicellular N_2-fixing cyanobacterium *Crocosphaera watsonia* from the western tropical Atlantic Ocean. *European Journal of Phycology* 48: 128–139.
Garrard, S.L., Gambi, M.C., Scipione, M.B. et al. (2014). Indirect effects may buffer negative responses of seagrass invertebrate communities to ocean acidification. *Journal of Experimental Marine Biology and Ecology* 461: 31–38.
Gattuso, J.-P., Allemand, D., and Frankignoulle, M. (1999). Photosynthesis and calcification at cellular, organismal and community levels in coral reefs, a review on interactions and control by carbonate chemistry. *American Zoologist* 39: 160–183.

George, R., Gullström, M., Mangora, M.M. et al. (2018). High midday temperature stress has stronger effects on biomass than on photosynthesis: a mesocosm experiment on four tropical seagrass species. *Ecology and Evolution* 8: 4508–4517.

George, R., Gullström, M., Mtolera, M.S.P. et al. (2020). Methane emissions and sulphide levels increase in tropical seagrass sediments during temperature stress: a mesocosm experiment. *Ecology and Evolution* 10: 1917–1928.

Gilman, E.L., Ellison, J., and Coleman, R. (2007). Assessment of mangrove response to projected relative sea-level rise and recent historical reconstruction of shoreline position. *Environmental Monitoring and Assessment* 124: 105–130.

Gilman, E.L., Ellison, J., Duke, N.C. et al. (2008). Threats to mangroves from climate change and adaptation options: a review. *Aquatic Botany* 89: 237–250.

Giomi, F., Barausse, A., Duarte, C.M. et al. (2019). Oxygen supersaturation protects coastal marine fauna from ocean warming. *Science Advances* 5: eaax1814. https://doi.org/10.1126/sciadv.aax1814.

Glas, M.S., Sato, Y., Ulstrup, K.E. et al. (2012). Biogeochemical conditions determine virulence of black band disease in corals. *ISME J* 6: 1526–1534.

Gorman, D., Turra, A., Bergstrom, E.R. et al. (2016). Population expansion of a tropical seagrass (*Halophila decipiens*) in the SW Atlantic (Brazil). *Aquatic Botany* 132: 30–36.

Graham, N.A.J., Jennings, S., MacNeil, M.A. et al. (2015). Predicting climate-driven regime shifts versus rebound potential in coral reefs. *Nature* 518: 94–97.

Guinotte, J.M. and Fabry, V.J. (2008). Ocean acidification and its potential effects on marine ecosystems. *Annals of the New York Academy of Science* 1134: 320–342.

Guinotte, J.M., Buddemeier, R.W., and Kleypas, J.A. (2003). Future coral reef habitat marginality: temporal and spatial effects of climate change in the Pacific basin. *Coral Reefs* 22: 551–558.

Guo, W., Bokade, R., Cohen, A.L. et al. (2020). Ocean acidification has impacted coral growth on the Great Barrier Reef. *Geophysical Research Letters* 47: GL086761. https://doi.org/10.1029/2019GL086761.

Hale, R.P., Wilson, C.A., and Bomer, E.J. (2019). Seasonal variability of forces controlling sedimentation in the Sundarbans National Forest, Bangladesh. *Frontiers in Earth Science* 7: 211. https://doi.org/10.3389/feart.2019.00211.

Hansen, J., Sato, M., and Ruedy, R. (2012). Perception of climate change. *Proceedings of the National Academy of Sciences* 109: E2415–E2423.

Harada, Y., Fry, B., Lee, S.Y. et al. (2020). Stable isotopes indicate ecosystem restructuring following climate-driven mangrove dieback. *Limnology and Oceanography* 65: 1251–1263.

Harley, C.D.G., Hughes, A.R., Hultgren, K.M. et al. (2006). The impacts of climate change in coastal marine systems. *Ecology Letters* 9: 228–241.

Harvey, B.P., Gwynn-Jones, D., and Moore, P.J. (2013). Meta-analysis reveals complex marine biological responses to the interactive effects of ocean acidification and warming. *Ecology and Evolution* 3: 1016–1030.

Hayden, H.L. and Granek, E.F. (2015). Coastal sediment elevation change following anthropogenic mangrove clearing. *Estuarine Coastal and Shelf Science* 165: 70–74.

Hayes, M.A., Shor, A.C., Jesse, A. et al. (2020). The role of glycine betaine in range expansions; protecting mangroves against extreme freeze events. *Journal of Ecology* 108: 61–69.

Heck, K.L. Jr., Fodrie, F.J., Madsen, S. et al. (2015). Seagrass consumption by native and a tropically associated fish species. *Marine Ecology Progress Series* 520: 165–173.

Hendriks, I.E., Olsen, Y.S., Ramajo, L. et al. (2014). Photosynthetic activity buffers ocean acidification in seagrass meadows. *Biogeosciences* 11: 333–346.

Hendriks, I.E., Duarte, C.M., Olsen, Y.S. et al. (2015). Biological mechanisms supporting adaptation to ocean acidification in coastal ecosystems. *Estuarine, Coastal and Shelf Science* 152: A1–A8.

Heuer, R.M. and Grosell, M. (2014). Physiological impacts of elevated carbon dioxide and ocean acidification on fish. *American Journal of Physiology Regulatory, Integrative and Comparative Physiology* 307: R1061–R1084.

Hikami, M., Ushie, H., Irie, T. et al. (2011). Contrasting calcification responses to ocean acidification between two reef foraminifers harbouring different algal symbionts. *Geophysical Research Letters* 38: L19601. https://doi.org/10.1029/2011GL048501.

Hobdey, A.J. and Pecl, G.T. (2014). Identification of global marine hotspots: sentinels for change and vanguards for adaptation action. *Reviews in Fish Biology and Fisheries* 24: 415–425.

Hoegh-Guldberg, O. (2011). The impact of climate change on coral reef ecosystems. In: *Coral Reefs: An Ecosystem in Transition* (eds. Z. Dubinsky and N. Stambler), 391–403. Dordrecht, The Netherlands: Springer.

Hofman, L.C. and Bischof, K. (2014). Ocean acidification effects on calcifying macroalgae. *Aquatic Biology* 22: 261–279.

Hoque, M.M., Abu Hena, M.K., Ahmed, O.H. et al. (2015). Can mangroves help combat sea-level rise through sediment accretion and accumulation? *Malaysian Journal of Science* 34: 78–86.

Imbert, D. (2018). Hurricane disturbance and forest dynamics in east Caribbean mangroves. *Ecosphere* 9: e02231. https://doi.org/10.1002/ecs2.2231.

Ji, Y. and Gao, K. (2020). Effects of climate change factors on marine macroalgae: a review. *Advances in Marine Biology* 88: 1–136.

Johnson, V.R., Russell, B.D., Fabricius, K.E. et al. (2012). Temperate and tropical brown macroalgae thrive, despite decalcification, along natural CO_2 gradients. *Global Change Biology* 18: 2792–2803.

Johnson, M.D., Price, N.N., and Smith, J.E. (2014). Contrasting effects of ocean acidification on tropical fleshy and calcareous algae. *Peer J* 2: e411. https://doi.org/10.7717/peerj.411.

Keyzer, L.M., Herman, P.M.J., Smits, B.P. et al. (2020). The potential of coastal ecosystems to mitigate the impact of sea-level rise in shallow tropical bays. *Estuarine, Coastal and Shelf Science* 246: 107050. https://doi.org/10.1016/j.ecss.2020.107050.

Kifani, S., Quansah, E., Massaki, H. et al. (2018). Climate change impacts, vulnerabilities and adaptations: eastern Central Atlantic marine fisheries. In: *Impacts of Climate Change on Fisheries and Aquaculture: Synthesis of Current Knowledge, Adaptation and Mitigation Options* (eds. M. Barange, T. Bahri, M.C.M. Beveridge, et al.), 159–184. FAO Fisheries and Aquaculture Technical Paper 627. Rome: FAO.

Kingsbury, K.M., Gillanders, B.M., Booth, D.J. et al. (2020). Range-extending coral reef fish trade-off growth for maintenance of body condition in cooler waters. *Science of the Total Environment* 703: 134598. https://doi.org/10.1016/j.scitotenv.2019.134598.

Kleypas, J.A. and Langdon, C. (2006). Coral reefs and changing seawater carbonate chemistry. In: *Coral Reefs and Climate Change: Science and Management* (eds. J.T. Phinney, O. Hoegh-Guldberg, J. Kleypas, et al.), 73–110. Washington, DC: American Geophysical Union.

Kleypas, J.A., Buddemeier, R.W., Archer, D. et al. (1999). Geochemical consequences of increased atmospheric carbon dioxide on coral reefs. *Science* 284: 118–120.

Koch, M., Bowes, G., Ross, C. et al. (2013). Climate change and ocean acidification effects on seagrasses and marine macroalgae. *Global Change Biology* 19: 103–132.

Krause-Jensen, D., Marbà, N., Sanz-Martin, M. et al. (2016). Long photoperiods sustain high pH in Arctic kelp forests. *Science Advances 2*: e1501938. https://doi.org/10.1126/sciadv.1501938.

Krauss, K.W., McKee, K.L., Lovelock, C.E. et al. (2014). How mangrove forests adjust to rising sea level. *New Phytologist* 202: 19–34.

Kroeker, K.J., Kordas, R.L., Crim, R. et al. (2013). Impacts of ocean acidification on marine organisms: quantifying sensitivities and interaction with warming. *Global Change Biology* 19: 1884–1896.

Kumar Sahu, B., Pati, P., and Panigraphy, R.C. (2018). Impact of climate change on marine plankton with special reference to Indian Seas. *Indian Journal of Geo-Marine Sciences* 47: 259–268.

Kumara, M.P., Jayatissa, L.P., Krauss, K.W. et al. (2010). High mangrove density enhances surface accretion, surface elevation change, and tree survival in coastal areas susceptible to sea-level rise. *Oecologia* 164: 545–553.

Kusumaningtyas, M.A., Hutahaean, A.A., Fischer, H.W. et al. (2019). Variability in the organic carbon stocks, sources, and accumulation rates of Indonesian mangrove ecosystems. *Estuarine Coastal and Shelf Science* 218: 310–323.

Kwiatkowski, L., Bopp, L., Aumont, O. et al. (2017). Emergent constraints on projections of declining primary production in the tropical oceans. *Nature Climate Change* 17: 355–359.

Kwiatkowski, L., Aumont, O., and Bopp, L. (2019). Consistent trophic amplification of marine biomass declines under climate change. *Global Change Biology* 25: 218–229.

Lam, V.W.Y., Cheung, W.W.L., Swartz, W. et al. (2012). Climate change impacts on fisheries in West Africa: implications for economic, food and nutritional security. *African Journal of Marine Science* 34: 103–117.

Lam, V.W.Y., Allison, E.H., Bell, J.D. et al. (2020). Climate change, tropical fisheries and prospects for sustainable development. *Nature Reviews Earth and Environment* 1: 440–454.

Laufkötter, C., Vogt, M., Gruber, N. et al. (2015). Drivers and uncertainties of future global marine primary production in marine ecosystem models. *Biogeosciences* 12: 6955–6984.

Lehodey, P., Hampton, J., Brill, R.W. et al. (2011). Vulnerability of oceanic fisheries in the tropical Pacific to climate change. In: *Vulnerability of Tropical Pacific Fisheries and Aquaculture to Climate Change* (eds. J.D. Bell, J.E. Johnson and A.J. Hobday), 433–492. Noumea: New Caledonia. Secretariat of the Pacific Community.

Liu, P.-J., Ang, S.-J., Mayfield, A.B. et al. (2020). Influence of the seagrass *Thalassia hemprichii* on coral reef mesocosms exposed to ocean acidification and experimentally elevated temperatures. *Science of the Total Environment* 700: 134464. https://doi.org/10.1016/j.j.scitotenv.2019.134464.

Lohbeck, K.T., Riebesell, U., and Reusch, T.B.H. (2014). Gene expression changes in the coccolithophore *Emiliania huxleyi* after 500 generations of selection to ocean acidification. *Proceedings of the Royal Society B: Biological Sciences* 281: 20140003. https://doi.org/10.1098/rspb.2014.0003.

Lomas, M.W., Hopkinson, B.M., Losh, J.L. et al. (2012). Effect of ocean acidification on cyanobacteria in the subtropical North Atlantic. *Aquatic Microbial Ecology* 66: 211–222.

Lopez-Medellin, X., Ezcurra, E., González-Abraham, C. et al. (2011). Oceanographic anomalies and sea-level rise drive mangroves inland in the Pacific coast of Mexico. *Journal of Vegetation Science* 22: 143–151.

Lough, J.M. (2008). 10th Anniversary Review: a changing climate for coral reefs. *Journal of Environmental Monitoring* 10: 21–29.

Lovelock, C.E., Cahoon, D.R., Friess, D.A. et al. (2015). The vulnerability of Indo-Pacific mangrove forests to sea-level rise. *Nature* 526: 559–563.

Lovelock, C.E., Feller, I.C., Reef, R. et al. (2017). Mangrove dieback during fluctuating sea levels. *Scientific Reports* 7: 1–8.

Maboloc, E.A. and Chan, K.Y.K. (2017). Resilience of the larval slipper limpet *Crepidula onxy* to direct and indirect-diet effects of ocean acidification. *Scientific Reports* 7: 12062. https://doi.org/10.1038/s41598-017-12253-2.

MacKenzie, R.A., Foulk, P.B., Klump, J.V. et al. (2016). Sedimentation and belowground carbon accumulation rates in mangrove forests that differ in diversity and land use: a tale of two mangroves. *Wetlands Ecology and Management* 24: 245–261.

Macy, A.P. (2020). Functional shifts with tropicalization in marsh-mangrove ecotones. PhD dissertation. University of South Alabama.

Madin, J.S., Allen, A.P., Baird, A.H. et al. (2016). Scope for latitudinal extension of reef corals is species specific. *Frontiers of Biogeography* 8.1 (e29328): 1–4.

Manea, A., Geedicke, I., and Leishman, M.R. (2020). Elevated carbon dioxide and reduced salinity enhance mangrove seedling establishment in an artificial saltmarsh community. *Oecologia* 192: 273–280.

Mangora, M.M., Mtolera, M.S., and Björk, M. (2014). Photosynthetic responses to submergence in mangrove seedlings. *Marine and Freshwater Research* 65: 497–504.

Marañón, E., Balch, W.M., Cermeño, P. et al. (2016). Coccolithophore calcification is independent of carbonate chemistry in the tropical ocean. *Limnology and Oceanography* 61: 1345–1357.

Marchio, D.A., Savarese, M., Bovard, B. et al. (2016). Carbon sequestration and sedimentation in mangrove swamps influenced by hydrogeomorphic conditions and urbanization in Southwest Florida. *Forests* 7: 116. doi:10.30116.390/f706.

Maurer, R., Tapia, M.E., and Shor, A.C. (2020). Exogenous root uptake of glycine betaine mitigates improved tolerance to salinity stress in *Avicenna germinans* under ambient and elevated CO_2 conditions. *The FASEB Journal* 34: 1–1.

McClanahan, T.R., Darling, E.S., Maina, J.M. et al. (2019). Temperature patterns and mechanisms influencing coral bleaching during the 2016 El Niño. *Nature Climate Change* 9: 845–851.

McCulloch, M., Falter, J., Trotter, J. et al. (2012). Coral resilience to ocean acidification and global warming through pH up-regulation. *Nature Climate Change* 2: 623–627.

McKee, K.L., Cahoon, D.R., and Feller, I.C. (2007). Caribbean mangroves adjust to rising sea level through biotic controls on change in soil elevation. *Global Ecology and Biogeography* 16: 545–556.

McKee, K.L., Krauss, K.W., and Cahoon, D.R. (2021). Does geomorphology determine vulnerability of mangrove coasts to sea-level rise? In: *Dynamic Sedimentary Environments of Mangrove Coasts* (eds. F. Sidik and D.A. Friess), 255–272. Amsterdam: Elsevier.

Miranda, T., Smit, J.A., Suthers, I.M. et al. (2019). Convictfish on the move: variation in growth and trophic niche along a latitudinal gradient. *ICES Journal of Marine Science* 76: 2404–2412.

Miura Pereira, T., Giavarini Gnocchi, K., Merçon, J. et al. (2020). The success of the fertilization and early larval development of the tropical sea urchin *Echinometra lucunter* (Echinodermata: Echinoidea) is affected by the pH decrease and temperature increase. *Marine Environmental Research* 161: 105106. https://doi.org/10.1016/j.marenvres.2020.105106.

Moustahfid, H., Marsac, F., and Gangopadhyay, A. (2018). Climate change impacts, vulnerabilities and adaptations: western Indian Ocean marine fisheries. In: *Impacts of Climate Change on Fisheries and Aquaculture: Synthesis of Current Knowledge, Adaptation and Mitigation Options* (eds. M. Barange, T. Bahri, M.C.M. Beveridge, et al.), 251–280. FAO Fisheries and Aquaculture Technical Paper 627 Rome: FAO.

Muir, P.R., Wallace, C.C., Done, T. et al. (2015). Limited scope for latitudinal extension of reef corals. *Science* 348: 1135–1138.

Munday, P.L., Cheal, A.J., Dixson, D.L. et al. (2014). Behavioural impairment in reef fish caused by ocean acidification at CO_2 seeps. *Nature Climate Change* 4: 487–492.

Murdiyarso, D., Hanggara, B.B., and Lubis, A.A. (2018). Sedimentation and soil carbon accumulation in degraded mangrove forests of North Sumatra, Indonesia. *BioRxiv*: 325191. https://doi.org/10.1101/325191.

Nagelkerken, I. and Connell, S.D. (2015). Global alteration of ocean ecosystem functioning due to increasing human CO_2 emissions. *Proceedings of the National Academy of Sciences* 112: 13272–13277.

Nakamura, T., Nadaoka, K., Watanabe, A. et al. (2017). Reef-scale modelling of coral calcification responses to ocean acidification and sea-level rise. *Coral Reefs* 37: 37–53.

Osland, M.J., Feher, L.C., Griffith, K.T. et al. (2017). Climatic controls on the global distribution, abundance, and species richness of mangrove forests. *Ecological Monographs* 87: 341–359.

Osland, M.J., Day, R.H., Hall, C.T. et al. (2020). Temperature thresholds for black mangrove (*Avicennia germinans*) freeze damage, mortality and recovery in North America: refining tipping points for range expansion in a warming climate. *Journal of Ecology* 108: 654–665.

Ow, Y.X., Collier, C.J., and Uthicke, S. (2015). Responses of three tropical seagrass species to CO_2 enrichment. *Marine Biology* 162: 1005–1017.

Ow, Y.X., Uthicke, S., and Collier, C.J. (2016). Light levels affect carbon utilisation in tropical seagrass under ocean acidification. *PLoS One* 11: e0150352. https://doi.org/10.1371/journal.pone.0150352.

Oxenford, H.A. and Monnereau, I. (2018). Climate change impacts, vulnerabilities and adaptations: western Central Atlantic marine fisheries. In: *Impacts of Climate Change on Fisheries and Aquaculture: Synthesis of Current Knowledge, Adaptation and Mitigation Options* (eds. M. Barange, T. Bahri, M.C.M. Beveridge, et al.), 185–206. FAO Fisheries and Aquaculture Technical Paper 627. Rome: FAO

Paxton, C.W., Baria, M.V.B., Weis, V.M. et al. (2015). Effect of elevated temperature on fecundity and reproductive timing in the coral *Acropora digitifera*. *Zygote* 24: 511–516.
Pedersen, O., Colmer, T.D., Borum, J. et al. (2016). Heat stress of two tropical seagrass species during low tides – impact on underwater net photosynthesis, dark respiration and diel *in situ* internal aeration. *New Phytologist* 210: 1207–1218.
Peer, N., Rajkaran, A., Miranda, N.A.F. et al. (2018). Latitudinal gradients and poleward expansion of mangrove ecosystems in South Africa: 50 years after Macnae's first assessment. *African Journal of Marine Science* 40: 101–120.
Pérez, A., Libardoni, B.G., and Sanders, C.J. (2018). Factors influencing organic carbon accumulation in mangrove ecosystems. *Biology Letters* 14: 20180237. https://doi.org/10.1098/rsbl.2018.0237.
Perry, C.T., Murphy, G.N., Graham, N.A.J. et al. (2015). Remote coral reefs can sustain high growth potential and may match future sea-level trends. *Scientific Reports* 5: 182289. https://doi.org/10.1038/srep18289.
Perry, C.T., Alvarez-Filip, L., Graham, N.A.J. et al. (2018). Loss of coral reef growth capacity to track future increases in sea-level. *Nature* 558: 396–400.
Phillips, D.H., Kumara, M.P., Jayatissa, L.P. et al. (2017). Impacts of mangrove density on surface sediment accretion, belowground biomass and biogeochemistry in Puttalam Lagoon, Sri Lanka. *Wetlands* 37: 471–483.
Porzio, L., Buia, M.C., and Hall-Spencer, J.M. (2011). Effects of ocean acidification on macroalgal communities. *Journal of Experimental Marine Biology and Ecology* 400: 278–287.
Pratchett, M.S., Munday, P.L., Graham, N.A.J. et al. (2011). Vulnerability of coastal fisheries in the tropical Pacific to climate change. In: *Vulnerability of Tropical Pacific Fisheries and Aquaculture to Climate Change* (eds. J.D. Bell, J.E. Johnson and A.J. Hobday), 493–576. Noumea, New Caledonia: Secretariat of the Pacific Community.
Price, N.N., Hamilton, S.L., Tootell, J.S. et al. (2011). Species-specific consequences of ocean acidification for the calcareous tropical green algae *Halimeda*. *Marine Ecology Progress Series* 440: 67–78.
Principe, S.C., Augusto, A., and Costa, T.M. (2018). Differential effects of water loss and temperature increase on the physiology of fiddler crabs from distinct habitats. *Journal of Thermal Biology* 73: 14–23.
Proffitt, C.E. and Travis, S. (2014). Red mangrove life history variables along latitudinal and anthropogenic stress gradients. *Ecology and Evolution* 4: 2352–2359.
Ramajo, L., Pérez-León, E., Hendriks, I.E. et al. (2016). Food supply confers calcifiers resistance to ocean acidification. *Scientific Reports* 6: 19374. https://doi.org/10.1038/srep19374.
Randall, C.J., Toth, L.T., Leichter, J.J. et al. (2020). Upwelling buffers climate change impacts on coral reefs of the eastern tropical Pacific. *Ecology* 101: e02918. https://doi.org/10.1002/ecy.2918.
Rasheed, M.A. and Unsworth, R.K.F. (2011). Long-term climate-associated dynamics of a tropical seagrass meadow: implications for the future. *Marine Ecology Progress Series* 422: 93–103.
Reef, R., Slot, M., Motro, U. et al. (2016). The effects of CO_2 and nutrient fertilisation on the growth and temperature response of the mangrove *Avicennia germinans*. *Photosynthesis Research* 129: 159–170.
Riegl, B., Johnston, M., Purkis, S. et al. (2018). Population collapse dynamics in *Acropora downingi*, an Arabian/Persian Gulf ecosystem-engineering coral, linked to rising temperature. *Global Change Biology* 24: 2447–2462.
Rivera-Monroy, V.H., Danielson, T.M., Castañeda-Moya, E. et al. (2019). Long-term demography and stem productivity of Everglades mangrove forests (Florida, USA): resistance to hurricane disturbance. *Forest Ecology and Management* 440: 79–91.
Rodgers, G.G., Rummer, J.L., Johnson, L.K. et al. (2019). Impacts of increased ocean temperatures on a low-latitude coral reef fish- processes related to oxygen uptake and delivery. *Journal of Thermal Biology* 79: 95 102.
Rölfer, L., Reuter, H., Ferse, S.C.A. et al. (2021). Coral-macroalgal competition under ocean warning and acidification. *Journal of Experimental Marine Biology and Ecology* 534: 151477. https://doi.org/10.1016/j.jembe.2020.151477.

Romero-Torres, M., Acosta, A., Palacio-Castro, A.M. et al. (2020). Coral reef resilience to thermal stress in the eastern tropical Pacific. *Global Change Biology* 26: 3880–3890.

Ruiz-Fernández, A.C., Agraz-Hernández, C.M., Sanchez-Cabeza, J.A. et al. (2018). Sediment geochemistry, accumulation rates and forest structure in a large tropical mangrove ecosystem. *Wetlands* 38: 307–325.

Saderne, V., Cusack, M., Almahasheer, H. et al. (2018). Accumulation of carbonates contributes to coastal vegetated ecosystems keeping pace with sea-level rise in an arid region (Arabian Peninsula). *Journal of Geophysical Research: Biogeosciences* 123: 1498–1510.

Saenger, P. (2002). *Mangrove Ecology, Silviculture and Conservation*. Dordrecht, The Netherlands: Kluwer.

Saintilan, N., Khan, N.S., Ashe, E. et al. (2020). Thresholds of mangrove survival under rapid sea level rise. *Science* 368: 1118–1121.

Sasmito, S.D., Murdiyarso, D., Friess, D.A. et al. (2016). Can mangroves keep pace with contemporary sea-level rise? A global data review. *Wetlands Ecology and Management* 24: 263–278.

Saunders, M.I., Leon, J., Phinn, S.R. et al. (2013). Coastal retreat and improved water quality mitigate losses of seagrass from sea-level rise. *Global Change Biology* 19: 2569–2583.

Schlüter, L., Lohbeck, K.T., Gutowska, M.A. et al. (2014). Adaptation of a globally important coccolithophore to ocean warming and acidification. *Nature Climate Change* 4: 1024. https://doi.org/https://doi.org/10.1038/NCLIMATE2379.

Schoepf, V., Jury, C.P., Toonen, R.J. et al. (2017). Coral calcification mechanisms facilitate adaptive responses to ocean acidification. *Proceedings of the Royal Society B: Biological Sciences* 284: 20172117. https://doi.org/10.1098/rspb.2017.2117.

Shaik, A.U.R., Biswas, H., and Pal, S. (2017). Increased CO_2 availability promotes growth of a tropical coastal diatom assemblage (Goa coast, Arabian Sea, India). *Diatom Research* 32: 325–339.

Short, F.T. and Neckles, H.A. (1999). The effects of global climate change on seagrasses. *Aquatic Botany* 63: 169–196.

da Silva Vianna, B., Miyai, C.A., Augusto, A. et al. (2020). Effects of temperature increase on the physiology and behavior of fiddler crabs. *Physiology & Behavior* 215: 112765. https://doi.org/10.1016/j.physbeh.2019.112765.

Simpson, C.J., Cary, J.L., and Masini, R.J. (1993). Destruction of corals and other reef animals by coral spawn slicks on Ningaloo Reef, Western Australia. *Coral Reefs* 12: 185–191.

Sippo, J.Z., Maher, D.T., Tait, D.R. et al. (2016). Are mangroves drivers or buffers of coastal acidification? Insights from alkalinity and DIC export estimates across a latitudinal transect. *Global Biogeochemical Cycles* 30: 753–766.

Sippo, J.Z., Lovelock, C.E., Santos, I.R. et al. (2018). Mangrove mortality in a changing climate: an overview. *Estuarine, Coastal and Shelf Science* 215: 241–249.

Sippo, J.Z., Santos, I.R., Sanders, C.J. et al. (2020a). Reconstructing extreme climatic and geochemical conditions during the largest natural mangrove dieback on record. *Biogeosciences* 17: 4707–4726.

Sippo, J.Z., Sanders, C.J., Santos, I.R. et al. (2020b). Coastal carbon cycle changes following mangrove loss. *Limnology and Oceanography* 65: 2642–2656.

Smith, J.N., De'ath, G., Richter, C. et al. (2016). Ocean acidification reduces demersal zooplankton that reside in tropical coral reefs. *Nature Climate Change* 6: 1124. https://doi.org/10.1038/nclimate3122.

Snedaker, S.C. and Araújo, R.J. (1998). Stomatal conductance and gas exchange in four species of Caribbean mangroves exposed to ambient and increased CO_2. *Marine and Freshwater Research* 49: 325–327.

Soper, F.M., MacKenzie, R.A., Sharma, S. et al. (2019). Non-native mangroves support carbon storage, sediment carbon burial, and accretion of coastal ecosystems. *Global Change Biology 25*: 4315–4326.

Stramma, L., Prince, E.D., Schmidtko, S. et al. (2011). Expansion of oxygen minimum zones may reduce available habitat for tropical pelagic fish. *Nature Climate Change* 2: 33–37.

Sully, S., Burkepile, D.E., Donovan, M.K. et al. (2019). A global analysis of coral bleaching over the past two decades. *Nature Communications* 10: 1264. https://doi.org/10.1038/s41467-019-09238-2.

Suzuki, A. and Kawahata, H. (2004). Reef water CO_2 system and carbon production of coral reefs: topographic control of system-level performance. In: *Global Environmental Change in the Ocean and on Land* (eds. M. Shiyomi, H. Kawahata, H. Koizumi, et al.), 229–248. Tokyo: Terrapub.

Swales, A., Reeve, G., Cahoon, D.R. et al. (2019). Landscape evolution of a fluvial sediment-rich *Avicennia marina* mangrove forest: insights from seasonal and inter-annual surface-elevation dynamics. *Ecosystems* 22: 1232–1255.

Takahashi, M., Noonan, S.H.C., Fabricius, K.E. et al. (2018). The effects of long-term *in situ* CO_2 enrichment on tropical seagrass communities at volcanic vents. *ICES Journal of Marine Science* 73: 876–886.

Tamimia, B., Wa, W.J., Nizam, M.S. et al. (2019). Elevated CO_2 concentration and air temperature impacts on mangrove plants (*Rhizophora apiculata*) under controlled environment. *Iraqi Journal of Science* 60: 1658–1666.

Taylor, G.T., Muller-Karger, F.E., Thunell, R.C. et al. (2012). Ecosystem responses in the southern Caribbean Sea to global climate change. *Proceedings of the National Academy of Science* 109: 19315–19320.

Unsworth, R.K.F., Collier, C.J., Henderson, G.M. et al. (2012). Tropical seagrass meadows modify seawater carbon chemistry: implications for coral reefs impacted by ocean acidification. *Environmental Research Letters* 7: 024026.

Valenzuela, J.J., de Lomana, A.L.G., Lee, A. et al. (2018). Ocean acidification conditions increase resilience of marine diatoms. *Nature Communications* 9: 2328. https://doi.org/10.1038/s41467-018-04742-3.

Van der Linge, C.D. and Hampton, I. (2018). Climate change impacts, vulnerabilities and adaptations: Southeast Atlantic and SW Indian Ocean marine fisheries. In: *Impacts of Climate Change on Fisheries and Aquaculture: Synthesis of Current Knowledge, Adaptation and Mitigation Options* (eds. M. Barange, T. Bahri, M.C.M. Beveridge, et al.), 219–250. FAO Fisheries and Aquaculture Technical Paper 627. Rome: FAO

Viana, I.G., Moreira-Saporiti, A., and Teichberg, M. (2020). Species-specific trait responses of three tropical seagrasses to multiple stressors: the case of increasing temperature and nutrient enrichment. *Frontiers in Plant Science* 11: 571363. https://doi.org/10.3389/fpls.2020.5711363.

Vivekanandan, E. (2011). Climate change and Indian marine fisheries. *Central Marine Fisheries Research Institute Special Publication No. 105*. Kochi: India.

Vousden, D.H., Stapley, J.R., Nggoile, M.A.K. et al. (2012). Climate change and variability of the Agulhas and Somali Current large marine ecosystems in relation to socioeconomics and governance. In: *Frontline Observations on Climate Change and Sustainability of Large Marine Ecosystems* (eds. K. Sherman and G. McGovern), 81–95. New York: UNDP.

Wabnitz, C.C.C., Lam, Y.W.Y., Reygondeau, G. et al. (2018). Climate change impacts on marine biodiversity, fisheries and society in the Arabian Gulf. *PLoS One* 13: e0194537. https://doi.org/10.1371/journal.pone.0194537.

Walden, G., Noirot, C., and Nagelkerken, I. (2019). A future 1.2° C increase in ocean temperature alters the quality of mangrove habitats for marine plants and animals. *Science of the Total Environment* 690: 596–603.

Ward, R.D. and de Lacerda, L.D. (2021). Responses of mangrove ecosystems to sea level change. In: *Dynamic Sedimentary Environments of Mangrove Coasts* (eds. F. Sidik and D.A. Friess), 235–253. Amsterdam: Elsevier.

Ward, R.D., Friess, D.A., Day, R.H. et al. (2016). Impacts of climate change on mangrove ecosystems: a region by region overview. *Ecosystem Health and Sustainability* 2: e01211. https://doi.org/10.1002/ehs2.1211.

Whitt, A.A., Coleman, R., Lovelock, C.E. et al. (2020). March of the mangroves: drivers of encroachment into southern temperate saltmarsh. *Estuarine, Coastal and Shelf Science* 240: 106776. https://doi.org/10.1016/j.ecss.2020.106776.

van Woesik, R., Golbuu, Y., and Roff, G. (2015). Keep up or drown: adjustment of western Pacific coral reefs to sea-level rise in the 21st century. *Royal Society of Open Science* 2: 150181. https://doi.org/10.1098/rsos.150181.

Woodroffe, C.D., Rogers, K., McKee, K.L. et al. (2016). Mangrove sedimentation and response to relative sea-level rise. *Annual Review of Marine Science* 8: 243–266.

Wu, L., Cai, W., Zang, L. et al. (2012). Enhanced warming over the subtropical western boundary currents. *Nature Climate Change* 2: 161–166.

Wyatt, A.S.J., Leichter, J.L., Toth, L.T. et al. (2020). Heat accumulation on coral reefs mitigated by internal waves. *Nature Geoscience* 13: 28–34.

Xie, D., Schwarz, C., Brückner, M.Z.M. et al. (2020). Mangrove diversity loss under sea-level rise triggered by bio-morphodynamic feedbacks and anthropogenic pressures. *Environmental Research Letters* 15: 114033. https://doi.org/10.1088/17548-9326/abc122.

Xu, D., Brennan, G., Xu, L. et al. (2018). Ocean acidification increases iodine accumulation in kelp-based coastal food webs. *Global Change Biology* 25: 629–639.

Yamano, H., Sugihara, K., and Nomura, K. (2011). Rapid poleward range expansion of tropical reef corals in response to rising sea surface temperatures. *Geophysical Research Letters* 38: L04601. https://doi.org/10.1029/2010GL046474.

Yates, K.K., Rogers, C.S., Herlan, J.J. et al. (2014). Diverse coral communities in mangrove habitats suggest a novel refuge from climate change. *Biogeosciences* 11: 4321–4337.

Zhou, Z., Liu, Z., Wang, L. et al. (2019). Oxidative stress, apotosis activation and symbiosis disruption in giant clam *Tridacna crocea* under high temperature. *Fish and Shellfish Immunology* 84: 451–457.

Jacotot, A., Marchand, C., and Allenbach, M. (2018). Tidal variability of CO_2 and CH_4 emissions from the water column within a *Rhizophora* mangrove forest (New Caledonia). *Science of the Total Environment* 631: 334–340.

CHAPTER 14

Habitat Destruction and Degradation

14.1 Introduction

Open ocean habitats in low latitudes are degrading due to vertically expanding oxygen minimum zones, changing circulation patterns, altering carbonate chemistry, acidification, warming, and increases in pollution, especially the increasing ubiquity of microplastics. And while the open ocean is being increasing degraded, coastal habitats such as the subtidal seabed and the water-column are also being degraded, and iconic coral reefs, seagrass meadows, and mangrove forests are being not only degraded but also destroyed. There are numerous reports of degradation of estuarine and coastal benthic and pelagic habitats, as discussed for specific locales in Chapter 12, but there is no global synthesis of these impacts in the tropics.

This final chapter therefore focuses on the degradation and destruction of coral reefs, seagrass meadows, and mangrove forests throughout the tropics as an expanding database now permits a global assessment of the extent of their degradation and destruction.

14.2 Coral Reefs

A fuller understanding of the status and trends of the world's coral reefs began in 1994 with the establishment of the International Coral Reef Initiative (ICRI), an informal consortium of 90 countries dedicated to preserving coral reefs. The ICRI established the Global Coral Reef Monitoring Network (GCRMN) in 1995 initially with the primary task of reporting on the condition of the world's coral reefs. Since then, the GCRMN has produced a range of global, regional, and thematic reports on coral reef status and trends. These reports provide a basis for a global assessment of destruction and degradation of coral reefs.

These assessments indicate both good and bad news in that some coral reefs are in a healthy state and some are in dire straits (Table 14.1). The last global assessment (Wilkinson 2008) concluded that approximately 19% of the original area of coral reefs was lost. Assuming a IPCC 'business as usual' scenario, 15% will be seriously

Tropical Marine Ecology, First Edition. Daniel M. Alongi.

TABLE 14.1 Status and trends of the world's coral reefs.

Region (countries)	Status and trends
Red Sea, Gulf of Aden	Healthy reefs with 30–50% live coral cover at most sites; >50% total on average. 30% of reefs severely damaged by 1997/1998 bleaching with rapid recovery. Severe degradation in the southern Red Sea during 2014-2017 bleaching event. Coral reefs damaged from landfilling, dredging, pore activities, pollution, and tourism.
The Persian Gulf, Gulf of Oman, Arabian Sea	Massive losses from bleaching (1996, 1998, 2002). Bahrain in danger of losing all its coral reefs due to severe natural stress and industrialization. Corals in Gulf of Oman severely damaged by Cyclone Gonu (2007); exposed corals eliminated with damage in bays, coves, and islands with signs of slow recovery. Coral reefs off Qatar virtually extinct.
East Africa (Kenya, Tanzania, Mozambique, South Africa)	Average coral cover about 40% pre-1998, then fluctuated around 30% from 1999 to 2016, a 25% decline. Many reefs had good recovery (e.g. Kenya) after massive losses in 1997/1998 bleaching event, but recovery slowed on some reefs due to fishing and COTS infestations. Coral cover has increased over the last 25 years in South Africa, but coral cover declined in Madagascar, Mauritius, and Mozambique. Severe degradation due to 2016 bleaching event.
SW Indian Ocean Islands (Comoros, Madagascar, La Réunion, Seychelles)	Before the bleaching event in 1997/1998, average coral cover was about 40%, then fluctuated at ≈30% from 1999 to 2016.
South Asia (Bangladesh, Chagos, India, Maldives, Sri Lanka)	Patchy recovery from 1997/1998 bleaching event with good recovery on Chagos; reefs in western atoll chain of Maldives and Bar Reef in Sri Lanka recovered well but little or no recovery on many Sri Lankan reefs and in the eastern atoll chain of the Maldives. The 2004 earthquake and tsunami caused significant damage on Andaman and Nicobar Islands.
Southeast Asia (Brunei, Myanmar, Cambodia, Indonesia, Thailand, Malaysia, Singapore, Philippines, Vietnam, Timor-Leste)	Between 2004 and 2008, reef conditions improved in Thailand, Philippines, Vietnam, and Singapore, but declined in Indonesia and Malaysia. 2004 tsunami caused localized damage in Indonesia, Thailand, and Malaysia.
East Asia (China, Hong Kong, Taiwan)	Coral reefs show an overall decline caused by bleaching, COTS, and typhoons plus urban stresses since 2004.
Australia and Papua New Guinea	Reefs remain in relatively good condition and the Great Barrier Reef may be in a recovery phase from previous disturbances. Some Coral Sea reefs have not recovered from storms and bleaching in 2002.
SW Pacific (Fiji, New Caledonia, Samoa, Solomon Islands, Tuvalu, Vanuatu)	Coral cover has changed positively (better management) and negatively (disturbances, coral predation, natural disasters). Average coral cover: 45% in Fiji, 27% in New Caledonia, 43% in Samoa, 30% in Solomon Islands, 65% in Tuvalu, and 26% in Vanuatu.

(*continued*)

TABLE 14.1 (*Continued*)

Region (countries)	Status and trends
Polynesia (Cook Islands, French Polynesia, Niue, Kiribati, Tonga, Tokelau, Wallis, Futuna)	90% of reefs healthy, 5% destroyed or at a critical stage, and 5% threatened. Degradation and loss only in populated areas. Severe degradation during 2014–2017 bleaching event.
Micronesia and American Samoa	Considerable reef recovery in western Micronesia from 1997/1998 bleaching event. Coral reefs in this region were amongst the most resilient in the world until severe degradation during 2014–2017 bleaching event.
Hawaii and United States Remote Islands (Baker, Howland, Palmyra, Kingman, Jarvis, Johnston, Wake)	Reefs in urban areas have suffered from pollution, fishing pressure, recreational overuse, and alien species. Most reefs were in fair to good condition until severe degradation during 2014–2017 bleaching event.
US Caribbean and Gulf of Mexico (Florida, Flower Garden Banks, Puerto Rico, Navassa, US Virgin Islands)	Most reefs in urban areas have suffered degradation, especially in southern Florida and the Florida Keys. Severe degradation during 2014-2017 bleaching event.
Northern Caribbean and the Western Atlantic Islands (Bermuda, Cayman Islands, Turks, Caicos, Jamaica, Dominican Republic, Cuba)	All northern Caribbean reefs except Bermuda were affected during the 2005 bleaching event and hurricanes. Reefs were severely damaged. 34% of corals around Jamaica bleached and half died.
Mesoamerican Barrier Reef System (Belize, Mexico Yucatan, Honduras, Guatemala)	Live coral cover has declined greatly, some reefs by >50%, in parallel with environmental changes and natural events. Regional live coral cover: 11% (11% Belize, 7.5% Mexico, Yucatan, 14% Honduras and Guatemala). Low coral cover indicates that reefs have not recovered from the 1997/1998 bleaching and Hurricane Mitch.
Lesser Antilles (French West Indies, Netherlands Antilles, Anguilla, Antigua, Grenada, Trinidad, Tobago)	Reefs were healthy until the early 1980s; coral cover has been decreasing since and algal cover increasing with most reefs having lost >50% of their corals. Very severe bleaching affected all islands in 2005 with slow recovery. Fish communities have declined through overfishing and habitat degradation and destruction.
South America (Brazil, Colombia, Costa Rica, Panama, Venezuela)	High coral cover at many locations on Caribbean (70%) and Pacific (95%) coasts with no major changes in live coral cover. Massive bleaching occurred in 2005 but severity was variable.

Source: Wilkinson (2008), Obura et al. (2017), Elliott et al. (2018), Kimura et al. (2018) and Moritz et al. (2018).

threatened during 2018–2028 and 20% will be under threat of loss during 2028–2048. Forty-six percent of coral reefs were considered as relatively healthy. Several backward steps were noted:

(i) There was considerable damage to coral reefs of the Indian Ocean due to the earthquake and tsunami on 26 December 2004.

(ii) Massive coral bleaching events are becoming more frequent as global warming continues.

(iii) Degradation of coral reefs near major population centers continues with losses of coral cover, fish, and invertebrate populations and probably biodiversity, including in the Coral Triangle.

(iv) There is an increasing evidence that climate change is having direct impacts on more and more coral reefs with clear evidence that ocean acidification will cause greater damage in future.

Coral reef declines will have alarming consequences for about 500 million people who depend on reefs for food, coastal protection, building materials, and tourism income; 30 million people are virtually totally dependent on coral reefs for their livelihoods or on the atolls they live on.

A global analysis of coral recruitment has found that recruitment of corals on settlement tiles from 1974 to 2012 declined by 82% globally and throughout the tropics by 85% at <20° latitude but increased in the subtropics by 78% at >20° latitude (Price et al. 2019). These trends indicate that a global decline in coral recruitment, the persistent reduction in equatorial latitudes, and an increase in subtropical latitudes portends current and future poleward shifts in corals and coral reefs.

Extreme hypoxic conditions ('dead zones') may lead to mass mortalities on coral reefs in the future. An unprecedented hypoxic event on the Caribbean coast of Panama in 2010 caused mass bleaching and massive mortality to corals and other reef-associated organisms; mobile organisms displayed unusual behaviors such as migration out of crevices, and dead bodies of crustaceans, gastropods, and echinoderms littered the bottom and thick mats of the bacterium, *Beggiatoa*, covered sediment surfaces (Altieri et al. 2017). Not all coral species were equally sensitive to hypoxia as demonstrated by shifts in community structure. Analyses of global databases show that coral reefs are associated that >50% of the known tropical dead zones, with >10% of all coral reefs at elevated risk of hypoxia. The growth of dead zones suggests that coral reefs in the coastal zone face a risky future.

A meta-analysis of the long-term degradation of Caribbean coral reefs (Precht et al. 2020) found that over the 1977–2001 period, four ecological shifts occurred (8.2%) because of coral loss with no change in macroalgal cover, 15% (30.6%) occurred because of macroalgal gain without coral loss, and 30% (61.2%) occurred owing to concomitant coral decline and macroalgal increase. The loss in regional coral cover (Figure 14.1) and concomitant changes to the benthic community were caused by discrete periods of white band disease and ENSO-related coral bleaching indicating the Caribbean coral reefs have been under assault from climate-related events since the 1970s (Precht et al. 2020). They have not recovered since. In the aftermath of an extensive marine heatwave on the Great Barrier in 2016, corals began to die immediately; after 8 months, reefs exposed to 4 °C–10 °C-week lost between 40 and 90% of their coral cover (Hughes et al. 2018). Further, exposure to 6 °C-week or more resulted in an unprecedented regional-scale shift in coral community composition as different taxa diverged in their response to heat stress. Roughly 29% of the Great Barrier Reef was subjected to abrupt shifts in ecological functioning and three-dimensionality; all major coral taxa collapsed in the northern third of the GBR where temperature anomalies were the most extreme (Hughes et al. 2018). On the island of Mauritius, frequent warm thermal anomalies between 2004 and 2019 resulted in declines of 40 and 83% of hard and soft corals, respectively, with a 78% increase in erect algae (McClanahan and Muthiga 2021).

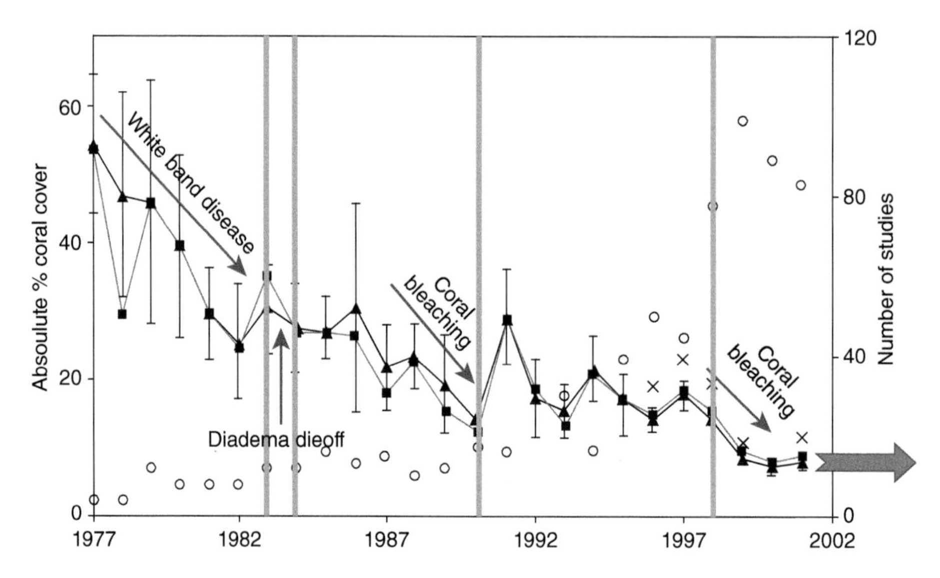

FIGURE 14.1 Long-term decline in live coral cover on degrading Caribbean reefs based on a meta-analysis of ecological studies. Solid circles are means and error bars are 95% confidence intervals. The black solid line connects the mean coral cover from 1977 to 2001. The red arrows are major punctuated disturbance events that were recorded region wide. The initial decline in 1977–1982 represents a decrease in absolute coral cover from about 50–25% and corresponds to the regional loss of acroporid coral caused by white-band disease. The declines in 1987–1990 and 1998–1999 are associated with major ENSO-induced coral bleaching events. Horizontal green bars represent the years in which there was a significantly greater number of observed shifts from coral to macroalgal dominance. Coral cover has continued to decline since 2001 associated with coral diseases, hurricanes and warm water bleaching events. *Source*: Precht et al. (2020), figure 5, p. 344. © Elsevier.

The degradation and destruction of coral reef habitats have led to concomitant changes in reef-associated biota (Wilkinson 2008; Obura et al. 2017; Elliott et al. 2018; Kimura et al. 2018; Moritz et al. 2018). A size-based ecosystem model of coral reefs was utilized to explore how predator–prey interactions and fisheries productivity respond to a gradient of reef degradation (Rogers et al. 2018). Model runs showed that during the initial stages of reef degradation, fish productivity was robust, but the composition and size structure of reef fish differed on degraded reefs, with herbivores and invertivores contributing relatively more to productivity. More significant losses associated with the erosion of structural complexity corresponded to at least a 35% decline in fisheries productivity compared to healthy reefs. Conversely, severe coral losses have been observed to shift the dynamics of coral communities. On Lizard Island, Great Barrier Reef, surveys of fish and benthos from 2003 to 2004 were compared with surveys in 2018 (Morais et al. 2020). Coral cover declined by 72–83% in forereef zones while turf cover increased by 18–100% across all zones. Reef fish communities underwent a 71% increase in standing biomass, a 41% increase in productivity, and a 37% increase in biomass consumption, mostly by herbivorous fish. However, biomass turnover rates declined 19%. These findings suggest that coral loss can drive energetic shifts leading to more productive, but "slower paced" fish communities. Although seemingly a positive outcome, the decreased turnover rates imply that the ecosystem is unable to maintain biomass replacement levels and is temporally less stable.

Coral reef degradation has resulted in changes in community structure of many organisms, especially fish, as well as the abundance of many species, leading to changes in trophic pathways and food web functioning. These changes may foster a transition of the coral reef ecosystem to an alternate state. A detailed example of this change was observed on two shallow coral reefs known as 'Limones' and 'Bonanza' off the Caribbean coast of Mexico (Morillo-Velarde et al. 2018). "Limones" has abundant colonies of *Acropora palmata* whereas Bonanza has experienced a substantial decline in live coral cover and an increase in macroalgal cover since 1985. Comparisons of the trophic structure and food chain length of associated reef communities showed that both reefs had similar food chain length and trophic structure, but different trophic pathways. On 'Limones', turf algae and epiphytes were the most important carbon sources for all consumers, whereas on 'Bonanza', particulate organic matter was the major carbon source for carnivores, suggesting that some degraded reefs may have a trophic structure robust enough to adjust up to a critical level of degradation (Morillo-Velarde et al. 2018).

14.3 Seagrass Meadows

Assessing the current global status and trends in tropical seagrass habitats is not possible given that only recently has an accurate estimate of their global area been published (McKenzie et al. 2020). The last global assessment (Waycott et al. 2009) synthesized quantitative data from 215 sites worldwide covering the period from 1879 to 2006. The analysis indicated that seagrass meadows have declined in area across the globe. There were significantly more declines in seagrasses than predicted by chance: 58% of sites declined, 25% increased, and 17% exhibited no change over the 1879–2006 period. There was a mean decline in seagrass area of 1.5% a^{-1} (median 0.9% a^{-1}). The measured area of seagrass loss was large, approximately 3370 km^2 between 1878 and 2006, equating to 29% of the maximum area measured. The difference in area lost among sites that declined was more than 10× greater than that among sites that increased. Decadal time-course analysis reveals that the rate of decline in seagrass area has accelerated over the past eight decades (Figure 14.2a) with the median rate of decline <1% a^{-1} before 1940, but 5% a^{-1} after 1980. Largest losses occurred after 1980 (Figure 14.2b) with a total loss of 35% of seagrass area. Absolute losses per number of sites increased by decade (Figure 14.2c) but accelerating rates of decline were not attributable to increased sampling effort (Waycott et al. 2009). The major causes of seagrass loss were (i) direct impacts from coastal development and dredging activities and (ii) indirect impacts from declining water quality. Other causes were marine heat waves, drought, diseases, destructive fishing practices, boat propellers, coastal engineering, cyclones, sediment runoff, aquaculture, and invasive species.

Local and regional monitoring suggest significant degradation of seagrasses in Vietnam (Chen et al. 2016), Brazil (Copertino et al. 2016), Indonesia (Unsworth et al. 2018), Réunion Island (Cuvillier et al. 2017), United States (Wilson and Dunton 2018), Solomon Islands (Moseby et al. 2020), Kosrae, Palau, Philippines, Malaysia, Australia (Short et al. 2014), and Madagascar (Côté-Laurin et al. 2017). Seagrasses have expanded at a few sites such as on Puerto Rico (León-Pérez et al. 2020). In the latter case, seagrass expansion was an actual recovery from severe losses suffered from tropical storm Baker in 1950 with recovery from remaining patchy seagrass cover.

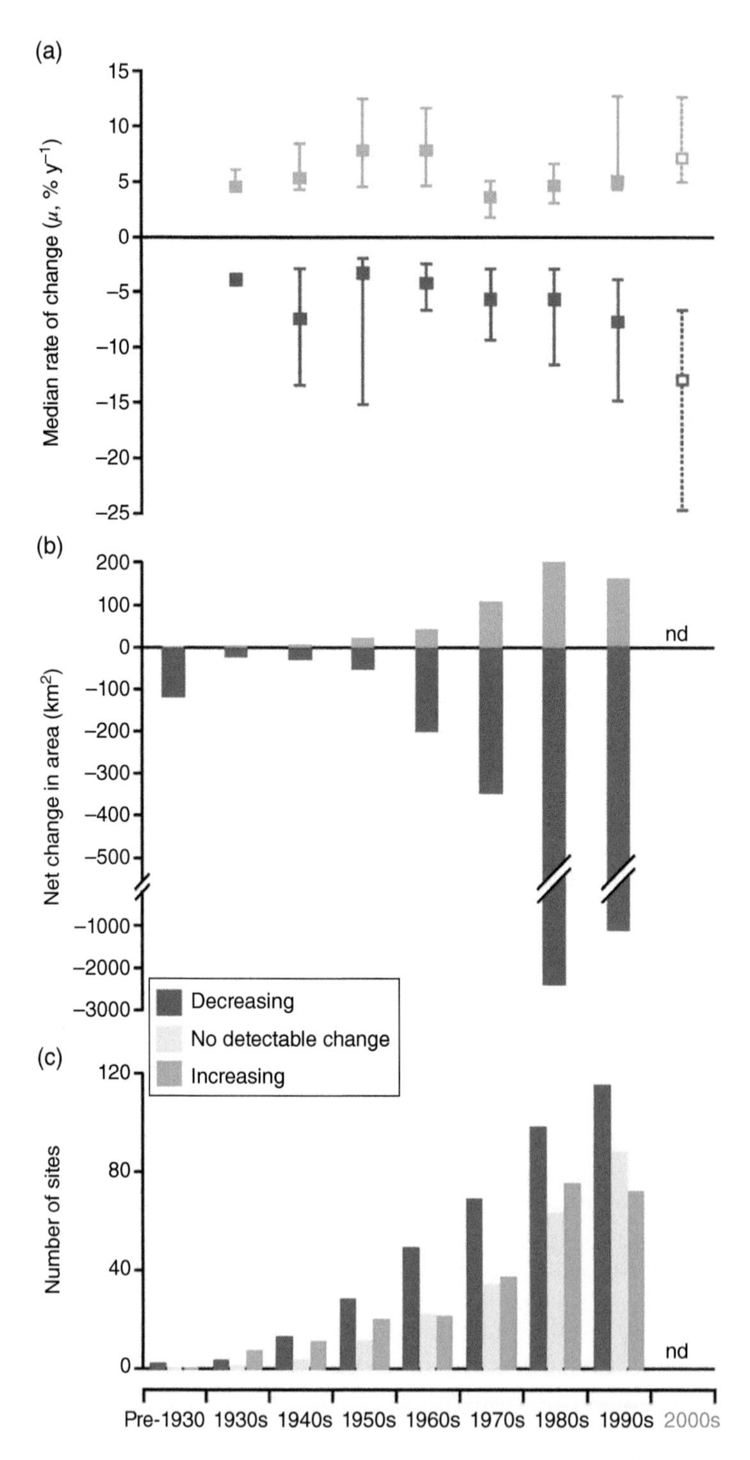

FIGURE 14.2 Decadal trends in global areal extent of seagrasses. (a) Median % rate of change (μ) by decade across sites. Error bars represent 25 and 75% quartiles. (b) Measured change in seagrass area calculated as the net change across each decade. (c) Number of monitoring sites in each category (decreasing, increasing, or no change). Values from the 2000s (dotted line in panel A) include 2000–2006 data only. nd = not determined because of incomplete data. *Source*: Waycott et al. (2009), figure 1, p. 12378. © National Academy of Sciences.

The CARICOMP monitoring network, designed to monitor seagrasses across the Caribbean from 1993 to 2007, gathered standardized data from 52 sampling stations at 22 sites, mostly *Thalassia testudinum* beds in reef systems. Community biomass and productivity varied widely, with large-scale disturbances such as hurricanes having little to no long-term effects. Only at two sites (five stations) did seagrass beds collapse due to excessive grazing by turtles or sea urchins. However, 43% of the monitoring stations showed trends indicative with changes expected from environmental degradation (Van Tussenbroek et al. 2014). Unlike in the Caribbean, cyclones negatively impacted seagrass meadows along the SW coast of Madagascar (Côté-Laurin et al. 2017). Cyclone Haruna caused significant loss in seagrass cover after passage in February 2013, especially to *Thalassia hemprichii* and *Syringodium isoetifolium* meadows. The largest negative effects were found at an exposed location previously subjected to human use and fragmentation. Cyclones have also caused extensive damage to seagrass meadows in southeast Asia, including Vietnam (Vo et al. 2020). Monitoring of the seagrasses in Van Phong Bay, Vietnam, over the past three decades has shown that 36% of the original beds were lost, with typhoons causing the destruction at open sea locations, whereas aquaculture, excavation, and sedimentation caused losses inshore.

SeagrassNet is a worldwide monitoring program that collects data on seagrass habitats, including 10 locations across the western Pacific where change at various scales were rapidly detected over the 2001–2009 period (Short et al. 2014). Seagrass cover was stable at three sites on Kosrae, the Philippines, and Australia where there were species shifts but no losses. However, the other seven sites showed declines in seagrass cover in Micronesia (0.9% a^{-1}), Palau (1.9 and 18% a^{-1}), the Philippines (1.1% a^{-1}), Malaysia (56 and 78% a^{-1}), and Indonesia (5.4% a^{-1}). Both sites in Malaysia were affected by adjacent deforestation and subsequent sediment loading that degraded the coastal zone. The Palau sites experienced wastewater impacts from a sewage plant and the Indonesia site on Komodo Island suffered nutrient loading from tourist cabins. The Kosrae and Philippines seagrasses showed less degradation but were near villages that presumably discharged sewage.

Long-term monitoring (>65 years) of a monospecific *S. isoetifolium* meadow at the Ermitage/La Saline fringing reef off Réunion Island in the southwestern Indian Ocean (Cuvillier et al. 2017) indicated seasonal and decadal variations in coverage. At an ocean-exposed station, large swell events and cyclones had a major impact on coverage, depending on their frequency, duration, and intensity. All sites were influenced by herbivory and nutrient inputs that affected the structural shape, fragmentation, and disappearance of seagrasses. Seagrass coverage strongly fluctuated with a minimum of 1680 m^2 of seagrass in 1950 to a maximum of 13 832 m^2 in 2014. There was a first area peak of 10 387 m^2 in 1966 followed by a 60% decline during the 1970s. Oscillations in coverage at the seasonal scale (up to 2016 m^2 gained or 4863 m^2 lost over a few months) can be of the same magnitude as observed at the multidecadal scale. Thus, large variability, mostly attributable to natural causes, must be considered when assessing degradation and destruction of tropical seagrass habitats. Indeed, a catastrophic earthquake and tsunami in the Solomon Islands in January 2010 resulted in the decline of seagrass beds closest to the epicenter, from 50% cover to <10% within 1 year of the event (Moseby et al. 2020). Seagrass coverage did not recover to pre-tsunami levels after eight years. Further from the epicenter, seagrass beds took longer to decline and dropped from an average coverage of 50% to <10% within 2 years with a shift in species dominance. Species richness declined from nine to four species, with some species (e.g. *S. isoetifolium*) disappearing from some monitoring sites.

14.4 Mangrove Forests

While mangroves are resilient to many types of natural disturbances, they cannot recover from extensive deforestation without manual restoration. Even then, many years must pass before a restored forest can approach the biomass, height, and biodiversity of the destroyed old-growth forest. Like seagrasses, the global area of mangrove forests is uncertain making a definite assessment of long-term losses of mangroves difficult. Crude global estimates of mangrove degradation and destruction over the past half century have varied from 20 to 50% (Alongi 2008, Polidoro et al. 2010).

Recent advances in high-resolution images from remotely sensed data (Hamilton and Casey 2016, Bunting et al. 2018) have provided a much greater understanding of mangrove area globally and their rates of change over decadal timescales. Global annual loss rates vary between 0.26 and 0.66% a^{-1} (Hamilton and Casey 2016). Between 2000 and 2016, global mangrove loss was 3363 km^2 (2.1% of total global area) for an average annual loss rate of 0.13% (Goldberg et al. 2020). Over the same period, anthropogenic causes accounted for 62% of total global mangrove loss (Figure 14.3a–g). Commodities (rice, shrimp, and oil palm cultivation) were the main drivers of mangrove loss, constituting 47% (1596 km^2) of global losses from 2000 to 2016. Nonproductive conversions caused 12% of losses with reclaimed lands for human settlements representing only 3% of deforestation (Goldberg et al. 2020). The remaining 38% of total forest loss was due to natural causes; shoreline erosion represented the second highest percentage of global losses at 27% and extreme weather events contributed 11% of losses.

Nearly 80% of direct anthropogenic loss was concentrated in only six countries: Indonesia, Myanmar, Malaysia, the Philippines, Thailand, and Vietnam (Figure 14.4a–f). Nonproductive conversion of mangroves represented more than half of national losses in 11 of the 22 African nations inhabited by mangroves (Figure 14.4a–f). The total natural loss declined from 624 km^2 in 2005 to 249 km^2 in 2016, a decreasing rate significantly less than the declining rate of anthropogenic loss (Goldberg et al. 2020). The most significant hotspot of erosional loss occurred in the Bangladesh section of the Sundarbans followed by erosion along the east coast of Brazil, most likely due to significant discharge from the Amazon and Orinoco Rivers. Extreme weather events contributed to the highest percentage losses in Oceania, with much of this loss occurring in Papua New Guinea due to Cyclone Guba in 2007, followed by significant extreme weather-related losses in Cuba, Florida, Honduras, and in the Gulf of Carpentaria, northern Australia.

Remote sensing-based studies recently have detailed losses in several regions offering a better perspective of the drivers of mangrove deforestation. These regions include Indonesia (Richards and Friess 2016, Austin et al. 2019), Myanmar (De Alban et al. 2020), the Mekong delta (Besset et al. 2019, Hong et al. 2019), Colombia (Jaramillo et al. 2018), the Niger delta (Nwobi et al. 2020), Madagascar (Scales and Friess 2019), and the northern Persian Gulf (Etemadi et al. 2021). Throughout Southeast Asia, mangrove forests were lost at an average rate of 0.18% a^{-1} between 2000 and 2012 (about 114 400 ha) with deforestation hotspots in Myanmar, Sumatra, Borneo, and Malaysia. The principal drivers of deforestation (Figure 14.5) were aquaculture, expansion of rice cultivation, urbanization, and the sustained conversion of mangroves to palm oil plantations (Richards and Friess 2016). In Indonesia, the main drivers of deforestation have been conversion to oil palm, timber plantations, and

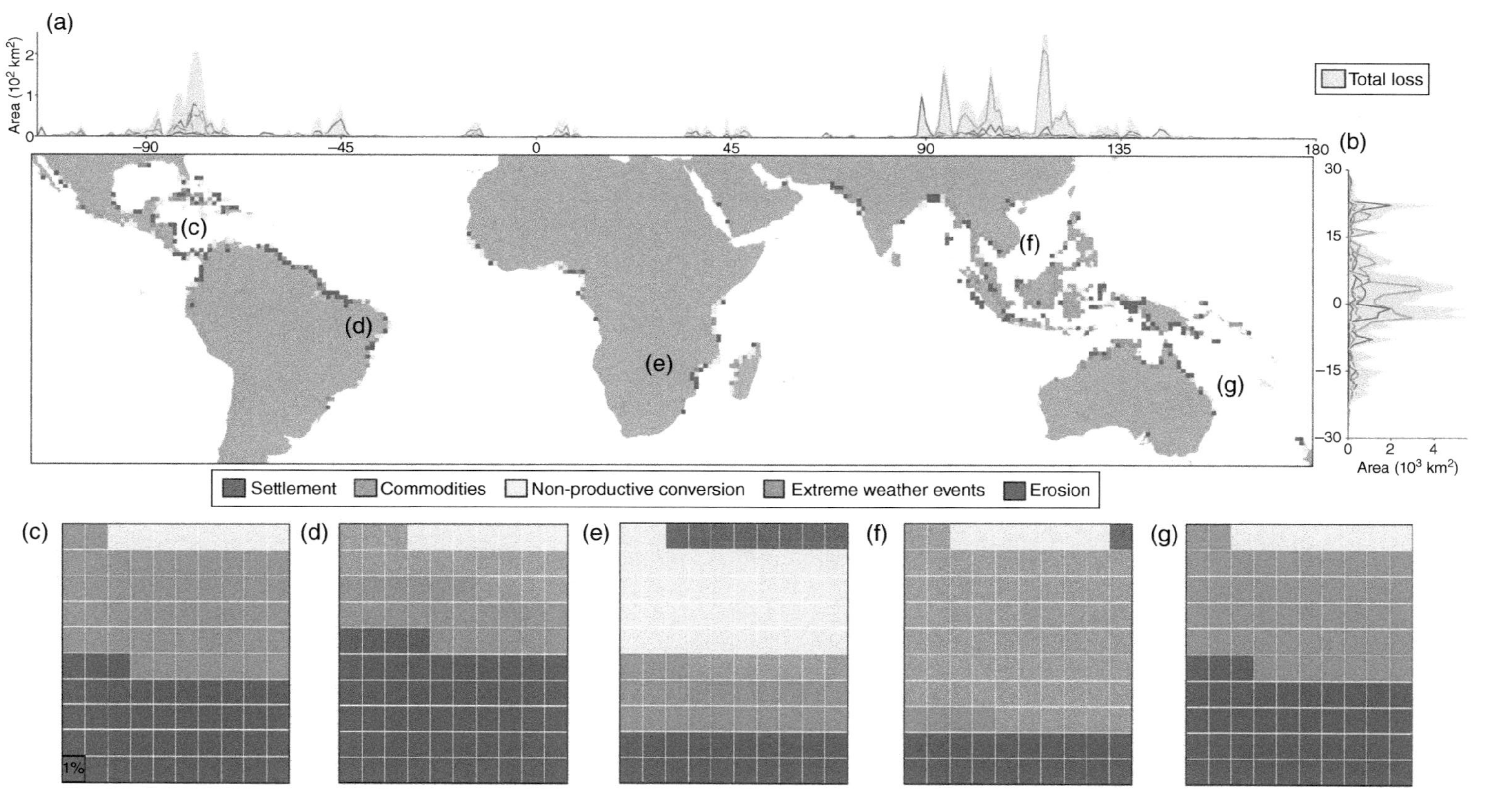

FIGURE 14.3 Global distribution of mangrove loss and its drivers from 2000 to 2016. (a) The longitudinal distribution of total mangrove loss and the relative contribution of its primary drivers. Different colors represent unique drivers of mangrove loss. (b) The latitudinal distribution of total mangrove loss and the relative contribution of its primary drivers. (c–g) Global distribution of mangrove loss and associated drivers at 1 × 1° resolution with the relative contribution (percentage) of primary drivers per continent: (c) North America, (d) South America, (e) Africa, (f) Asia, (g) Australia plus Oceania. *Source*: Goldberg et al. (2020), figure 1, p. 5847. Licensed under CC BY 4.0. © John Wiley & Sons.

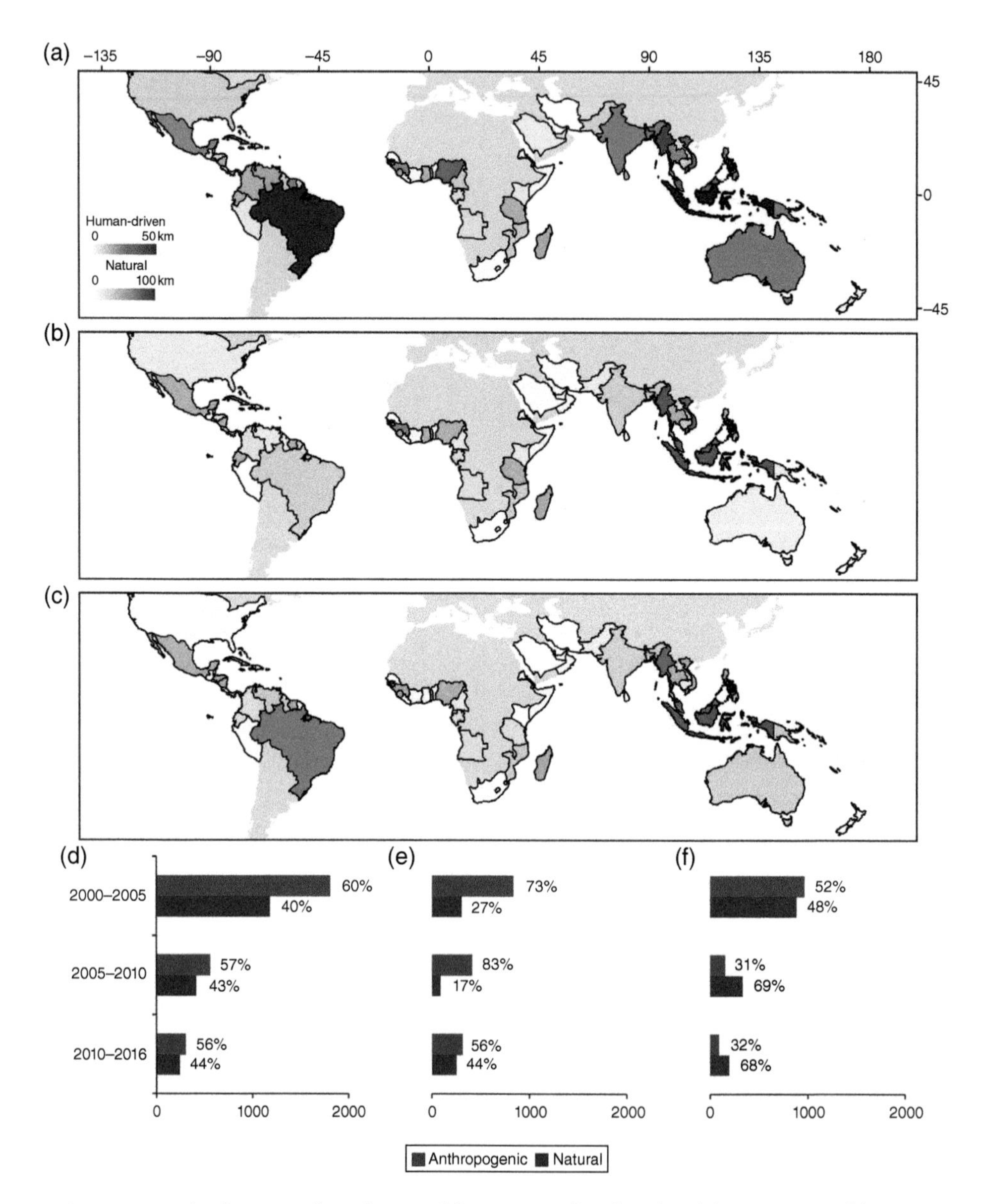

FIGURE 14.4 Anthropogenic and natural losses at regional scales: (a) 2000–2005, (b) 2005–2010, (c) 2010–2016. The primary color of the country in parts (a) to (c) corresponds to the dominant category of mangrove loss on the national level (anthropogenic or natural) per epoch, and the intensity of the colouur corresponds to the percentage of total loss driven by that category. (d) Global proportion of natural and anthropogenic loss per epoch. (e) Proportion of natural and anthropogenic loss per epoch in only Indonesia, Myanmar, Malaysia, the Philippines, Thailand, and Vietnam. (f) Proportion of natural and anthropogenic loss per epoch excluding those documented in part (e). *Source*: Goldberg et al. (2020), figure 5, p. 5851. Licensed under CC BY 4.0. © John Wiley & Sons.

grasslands/shrublands, but the importance of different drivers varies by island group (Figure 14.6) with an increase in destructive conversion to grasslands/shrublands and small-scale agriculture and other plantations during the 2001–2016 period (Austin et al. 2019). Conversion to oil palm plantations peaked in 2009, and conversion

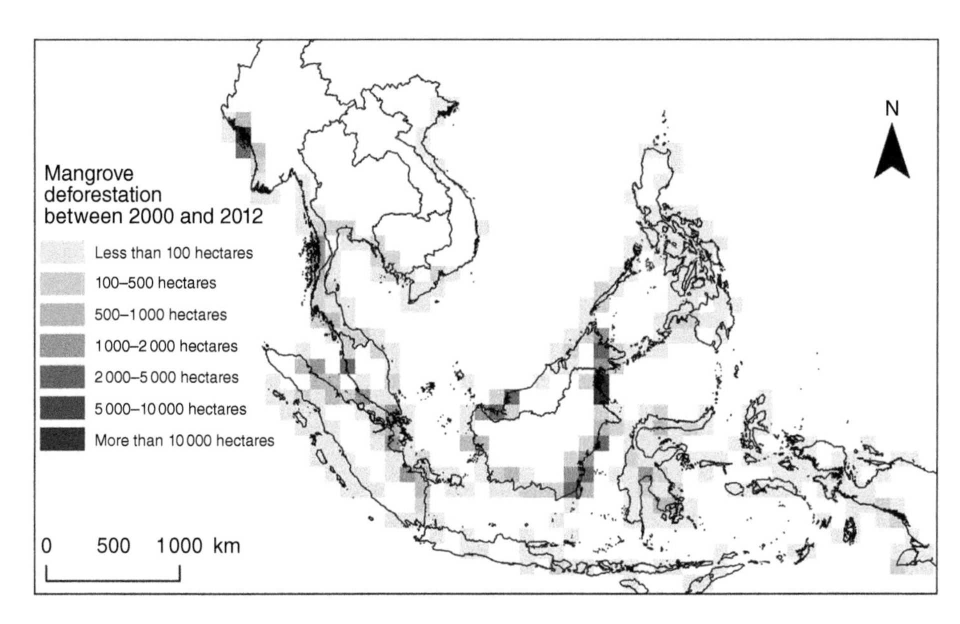

FIGURE 14.5 Percentage mangrove deforestation in Southeast Asia between 2000 and 2012, and dominant land uses of deforested areas in 2012. Land uses are summarized as the converted land use with the greatest area within each 1 decimal degree square. Circles are in the center of each gird square, and circle size represents the percentage of the mangrove area in 2000 that has been lost. *Source*: Richards and Friess (2016), figure 1, p. 345. © National Academy of Sciences.

to small-scale agriculture/plantations peaked in 2016. The major drivers of deforestation differed by region: (i) small-scale agriculture and grassland/scrubland conversion were the main drivers on Sumatra; (ii) grassland/scrubland and oil palm conversion the main causes of deforestation on Borneo; (iii) small-scale agriculture destroyed mangroves on Java, Bali, Nusa Tenggara, and Sulawesi; and (iv) logging roads and small-scale clearings were responsible for one-third of deforestation in Papua and Maluku (Austin et al. 2019).

Deforestation in Myanmar has been catastrophic, with net mangrove cover declining by 52% over the 1996–2016 period with annual net loss rates of 3.60–3.87% (De Alban et al. 2020). Most conversion was expansion for rice, oil palm, and rubber (Figure 14.7), but there were targeted transitions of mangroves to water, presumably aquaculture, and urbanization. Much of this destruction occurred in the Ayeyarwady delta. Other tropical river deltas have experienced significant losses of mangrove forest, but often because of different drivers. In the Niger delta, the *Nypa* palm (*Nypa fructicans*) has been invasive (Nwobi et al. 2020). Changes in land cover class areas (Table 14.2) in 2007 and 2017 have resulted in a massive increase in *Nypa* palm (694%) and smaller increases in build-up areas (50%) and conversion to agricultural land (11%).

Large-scale erosion has been the main cause of mangrove loss in the Mekong delta (Besset et al. 2019; Hong et al. 2019). Although no significant relationship between the change in shoreline and the width of the mangrove belt was detected between 2003 and 2012 (Besset et al. 2019), analysis of Landstat images taken from 1988 to 2018 identified seven land-use and land-cover types: dense mangroves, sparse mangroves, aquaculture farms, arable land with crop cover, arable land without crop cover, settlements, and water bodies (Hong et al. 2019). Shrimp farming has been

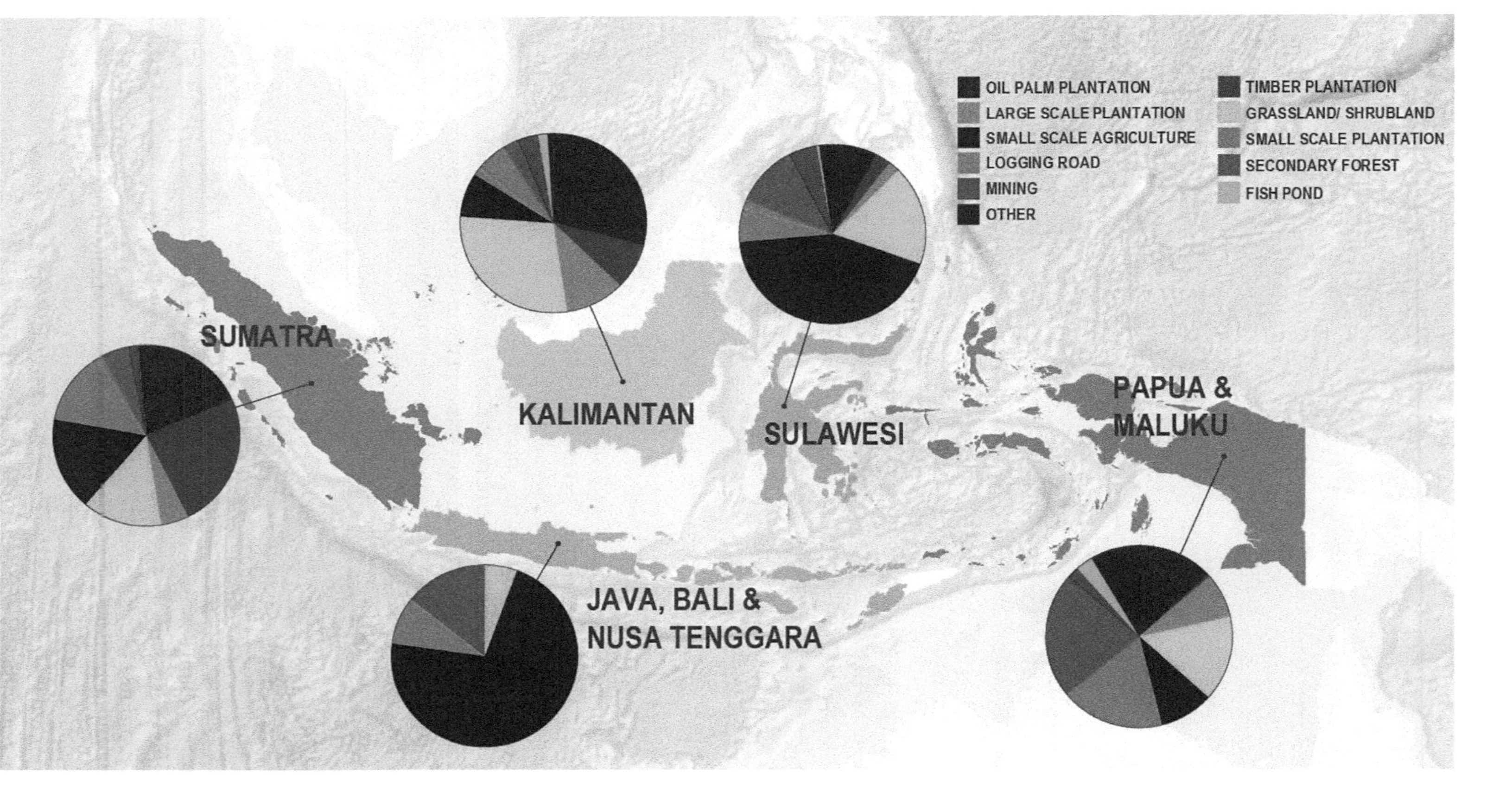

FIGURE 14.6 Proportion of mangrove deforestation, 2001–2016 by each driver category, by major region of Indonesia. *Source*: Austin et al. (2019), figure 3, p. 6. Licensed under CC BY 3.0 © IOP Publishing Ltd.

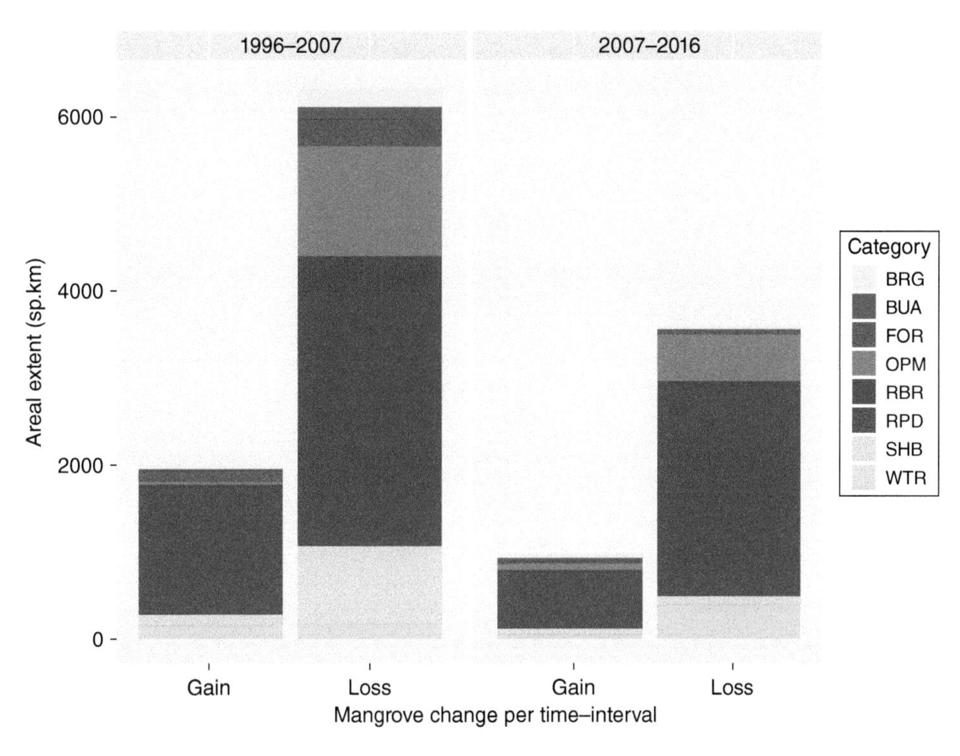

FIGURE 14.7 Mangrove area gains and losses during the 1996–2007 (left panel) and 2007–2016 periods (right panel) in Myanmar, compiled by driver category. Abbreviations: BRG = bare ground; BUA = built-up area; FOR = secondary forest; OPM = oil palm plantation; RBR = rubber plantation; RPD = rice cultivation; SHB = shrub/orchard; WTR = water body. *Source*: De Alban et al. (2020), figure 2, p. 6. Licensed under CC BY 3.0. © IOP Publishing Ltd.

one of the principal causes of rapid mangrove loss and degradation in the delta, with dense and sparse mangrove forests decreasing by 90% from 5495 to 515 ha and by 55% from 14 105 to 6289 ha, respectively, from 1988 to 2018 (Hong et al. 2019). In contrast, aquaculture farms have increased over the same period at the rate of 5024 ha a^{-1}.

Mangroves can be lost by cyclones. On the islands of Fiji, tropical cyclones accounted for 77% of loss, with highest losses along the northern coastlines of Viti Levu and Vanua Levu. Taller trees sustained more damage than fringing or scrub forests, with total losses of 1135 ha between 2001 and 2018 for an annual rate of loss of 0.11% (Cameron et al. 2021). Small-scale harvesting accounted for the remaining losses.

Some mangrove forests have experienced episodes of massive mortality due to hydrological modifications coupled to climatic changes. In the Ciénaga Grande de Santa Maria proximate to the Magdalena River on the Caribbean coast of Colombia, massive mangrove mortality starting in the 1950s has been related to a combination of drought, human alterations of hydrology, and road construction (Jaramillo et al. 2018). As a result, mangrove coverage declined from more than 500 km^2 in 1956 to 226 km^2 in 1996, with restoration efforts to increase hydrologic connectivity resulting in mangrove cover increasing to 400 km^2 as of 2015, although this level has been static since 2011. ENSO events explained most of the interannual variability in salinity in the wet season by regulating freshwater flow and in the dry season by regulating evaporation. Continuous reopening of channels resulted in large salinity

TABLE 14.2 **Mean area by land cover classes in the Niger delta between 2007 and 2017 and the changes over the decade.**

	2007		2017		2007–2017	
Land class	Area (ha)	Area (% of total area)	Area (ha)	Area (% of total area)	Change (ha)	Change (%)
Mangrove	911 548	14.31	801 774	12.6	−109 774	−12
Nypa palm	1441	0.02	11 444	0.2	10 003	694
Agricultural land	2 173 317	34.11	2 417 929	37.9	244 213	11
Tropical forest	2 889 083	45.35	2 549 919	12.6	−339 164	−12
Built-up areas	394 985	6.21	593 759	9.3	198 774	50

Source: Nwobi et al. (2020).

decreases relieving hypersaline conditions, but lack of persistent dredging slowed recovery in some areas.

In arid regions such as in the Persian Gulf, patterns of mangrove degradation have been linked to several causes, including rapid urban and industrial development, sea-level rise (SLR) and precipitation, and temperature changes (Etemadi et al. 2021). Climate change as well as anthropogenic activities associated with fluid extraction have resulted in sea-level rising by about 4 mm a^{-1} coupled by subsidence on the order of 1–2 mm a^{-1}. Some mangroves expanded between 1990 and 2002 in the Nayband region of Iran, likely in response to rising temperatures and above-average rainfall, but mangrove health and area have decreased in a nearby coastal region (Mond Protected Area) probably in response to industrial and urban development that occurred immediately after 1997 (Etemadi et al. 2021). Overall, however, SLR is considered as the main driver of mangrove degradation in the gulf.

Degradation of mangrove forests often leads to habitat fragmentation, having serious consequences for ecosystem functioning and the capacity of mangrove to provide ecosystem services, such as preventing shoreline erosion and facilitating shoreline protection (Bryan-Brown et al. 2020). The relative efficiency of such services depends in part on the size and organization of forest patches. A global analysis (Bryan-Brown et al. 2020) indicates that fragmentation of mangroves is common worldwide, but there are geographic disparities between mangrove loss and fragmentation. For example, Cambodia and the southern Caribbean have suffered relatively little loss, but their mangroves have been extensively fragmented. The conversion of mangroves to aquaculture and rice plantations is the biggest driver of fragmentation in Southeast Asia. And while conversion to oil palm plantations is responsible for >15% of all mangrove deforestation in Southeast Asia, it was only weakly correlated with fragmentation. Hotspots of fragmentation include Cambodia, Cameroon, Guatemala, Honduras, Indonesia, Malaysia, Papua New Guinea, Aruba, Grenada, and Trinidad and Tobago (Bryan-Brown et al. 2020). Mean patch sizes from largest to smallest: Malaysia (7.2 ha), Papua New Guinea (5.93 ha), Cambodia (5.42 ha), Indonesia (5.23 ha), Guatemala (5.12 ha), Cameroon (4.85 ha), Grenada (4.50 ha), Trinidad and Tobago (2.85 ha), Honduras (2.51 ha), and Venezuela

(2.29 ha). Thus, hotspots of mangrove loss are often associated with elevated levels of fragmentation.

Nevertheless, mangrove area has increased in some countries due to extensive replanting and conservation practices. In the Red River delta of Vietnam, the construction of aquaculture ponds was responsible for 49% of mangrove losses during the 2000–2006 period. Despite these losses, mangrove forest area increased overall during 1986–2019 with a change rate of 39 ha a^{-1} from 1655 ha in 1986 to 2944 ha in 2019. This increase was due to both conservation practices and natural expansion of the delta (Long et al. 2021). In Pakistan, an area of 477.22 km^2 was covered with mangrove in 1990, which increased to 1463.59 km^2 in 2020 resulting in a 3.74% annual rate of change; these habitats are fragmented but have enhanced tree canopy densities (Gilani et al. 2021).

References

Alongi, D.M. (2008). Mangrove forests: resilience, protection from tsunamis, and responses to global climate change. *Estuarine, Coastal and Shelf Science* 76: 1–13.

Altieri, A.H., Harrison, S.B., Seemann, J. et al. (2017). Tropical dead zones and mass mortalities on coral reefs. *Proceedings of the National Academy of Sciences* 114: 3660–3665.

Austin, K.G., Schwantes, A., Gu, Y. et al. (2019). What causes deforestation in Indonesia? *Environmental Research Letters* 14: 024007. https://doi.org/10.1088/1748-9326/aaf6db.

Besset, M., Gratiot, N., Anthony, E.J. et al. (2019). Mangroves and shoreline erosion in the Mekong River delta, Viet Nam. *Estuarine, Coastal and Shelf Science* 226: 106263. https://doi.org/10.1016/ecss.2019.106263.

Bryan-Brown, D.N., Connolly, R.M., Richards, D.R. et al. (2020). Global trends in mangrove forest fragmentation. *Scientific Reports* 10: 1–8.

Bunting, P., Rosenqvist, A., Lucas, R.M. et al. (2018). The Global Mangrove Watch – a new 2010 global baseline of mangrove extent. *Remote Sensing* 10: 1669. https://doi.org/10.3390/rs10101669.

Cameron, C., Maharaj, A., Kennedy, B. et al. (2021). Landcover change in mangroves of Fiji: implications for climate change mitigation and adaptation in the Pacific. *Environmental Challenges* 2: 1000018. https://doi.org/10.1016/j.envc.2020.100018.

Chen, C.-F., Lau, V.-K., Chang, N.-B. et al. (2016). Multi-temporal change detection of seagrass beds using integrated Landsat TM/ETM+/OLI imageries in Cam Ranh Bay, Vietnam. *Ecological Informatics* 35: 43–54.

Copertino, M.S., Creed, J.C., Lanari, M.O. et al. (2016). Seagrass and submerged aquatic vegetation (VAS) habitats off the coast of Brazil: state of knowledge, conservation and main threats. *Brazilian Journal of Oceanography* 64: 53–80.

Côté-Laurin, M.-C., Benbow, S., and Erzini, K. (2017). The short-term impacts of a cyclone on seagrass communities in SW Madagascar. *Continental Shelf Research* 138: 132–141.

Cuvillier, A., Villeneuve, N., Cordier, E. et al. (2017). Causes of seasonal and decadal variability in a tropical seagrass seascape (Reunion Island, south western Indian Ocean). *Estuarine, Coastal and Shelf Science* 184: 90–101.

De Alban, J.D., Jamaludin, J., de Wen, D.W. et al. (2020). Improved estimates of mangrove cover and change reveal catastrophic deforestation in Myanmar. *Environmental Research Letters* 15: 034034. https://doi.org/10.1088/1748-9326/ab666d.

Elliott, J.A., Patterson, M.R., Staub, C.G. et al. (2018). Decline in coral cover and flattening of the reefs around Mauritius (1998-2010). *PeerJ* 6: e6014. https://doi.org/10.7717/peerj.6014.

Etemadi, H., Smoak, J.M., and Abbasi, E. (2021). Spatiotemporal pattern of degradation in arid mangrove forests of the northern Persian Gulf. *Oceanologia* 63: 99–114.

Gilani, H., Naz, H.I., Arshad, M. et al. (2021). Evaluating mangrove conservation and sustainability through spatiotemporal (1990-2020) mangrove cover change analysis in Pakistan. *Estuarine, Coastal and Shelf Science* 249: 107128. https://doi.org/10.1016/j.ecss.2020.107128.
Goldberg, L., Lagomasino, D., Thomas, N. et al. (2020). Global declines in human-driven mangrove loss. *Global Change Biology* 26: 5844–5855.
Hamilton, S.E. and Casey, D. (2016). Creation of a high spatio-temporal resolution global database of continuous mangrove forest cover for the 21st century (CGMFC-21). *Global Ecology and Biogeography* 25: 729–738.
Hong, H.T.C., Avtar, R., and Fujii, M. (2019). Monitoring change in land use and distribution of mangroves in the southeastern part of the Mekong River delta, Vietnam. *Tropical Ecology* 60: 552–565.
Hughes, T.P., Kerry, J.T., Baird, A.H. et al. (2018). Global warming transforms coral reef assemblages. *Nature* 556: 492–496.
Jaramillo, F., Licero, L., Åhlen, I. et al. (2018). Effects of hydroclimatic change and rehabilitation activities on salinity and mangroves in the Ciénaga Grande de Santa Marta, Colombia. *Wetlands* 38: 755–767.
Kimura, T., Tun, K., and Chou, L.M. (2018). Status of Coral Reefs in East Asian Seas Region: 2018. Tokyo: Ministry of the Environment of Japan and Japan Wildlife Research Center.
León-Pérez, M.C., Armstrong, R.A., Hernández, W.J. et al. (2020). Seagrass cover expansion off Caja de Muertos Island, Puerto Rico, as determined by long-term analysis of historical aerial and satellite images (1950-2014). *Ecological Indicators* 117: 106561. https://doi.org/10.1016/j.ecolind.2020.106561.
Long, C., Dai, Z., Zhou, X. et al. (2021). Mapping mangrove forests in the Red River Delta, Vietnam. *Forest Ecology and Management* 483: 118910. https://doi.org/10.1016/j.foreco.2020.118910.
McClanahan, T.R. and Muthiga, N.A. (2021). Oceanic patterns of thermal stress and coral community degradation on the island of Mauritius. *Coral Reefs* 40: 53–74.
McKenzie, L.J., Nordlund, L.M., Jones, B.L. et al. (2020). The global distribution of seagrass meadows. *Environmental Research Letters* 15: 074041. https://doi.org/10.1088/1748-9326/ab7d06.
Morais, R.A., Depczynski, M., Fulton, C. et al. (2020). Severe coral loss shifts energetic dynamics on a coral reef. *Functional Ecology* 34: 1507–1518.
Morillo-Velarde, P.S., Briones-Fourzán, P., Álvarez-Filip, L. et al. (2018). Habitat degradation alters trophic pathways but not food chain length on shallow Caribbean coral reefs. *Scientific Reports* 8: 4109. https://doi.org/10.1038/s41598-018-22463-x.
Moritz, C., Vii, J., Lee, W. et al. (2018). Status and Trends of Coral Reefs of the Pacific. Global Coral Reef Monitoring Network.
Moseby, K.E., Daniels, A., Curi, V. et al. (2020). Community-based monitoring detects catastrophic earthquake and tsunami impacts on seagrass beds in the Solomon Islands. *Marine Pollution Bulletin* 150: 110444. https://doi.org/10.1016/j.marpolbul.2019.07.032.
Nwobi, C., Williams, M., and Mitchard, E.T.A. (2020). Rapid mangrove forest loss and Nipa palm (*Nypa fruticans*) expansion in the Niger Delta, 2007-2017. *Remote Sensing* 12: 2344. https://doi.org/10.3390/rs12142344.
Obura, D., Gudka, M., Rabi, F.A. et al. (2017). *Coral Reef Status Report for the Western Indian Ocean*. Global Coral reef Monitoring Network/International Coral reef Initiative.
Polidoro, B.A., Carpenter, K.E., Collins, L. et al. (2010). The loss of species: mangrove extinction risk and geographic areas of global concern. *PLoS One* 5: e10095. https://doi.org/10.1371/journal.pone.0010095.
Precht, W.F., Aronson, R.B., Gardner, T.A. et al. (2020). The timing and causality of ecological shifts on Caribbean reefs. *Advances in Marine Biology* 87: 331–360.
Price, N.N., Muko, S., Legendre, L. et al. (2019). Global biogeography of coral recruitment: tropical decline and subtropical increase. *Marine Ecology Progress Series* 621: 1–17.
Richards, D.R. and Friess, D.A. (2016). Rates and drivers of mangrove deforestation in Southeast Asia, 2000-2012. *Proceedings of the National Academy of Sciences* 113: 344–349.

Rogers. A., Blanchard, J. and Mumby, P.J. (2018). Fisheries productivity under progressive coral reef degradation. *Journal of Applied Ecology* 55: 1041–1049.
Scales, I.R. and Friess, D.A. (2019). Patterns of mangrove forest disturbance and biomass removal due to small-scale harvesting in SW Madagascar. *Wetlands Ecology and Management* 27: 609–625.
Short, F.T., Coles, R., Fortes, M.D. et al. (2014). Monitoring in the Western Pacific region shows evidence of seagrass decline in line with global trends. *Marine Pollution Bulletin* 83: 408–416.
Unsworth, R.K.F., Ambo-Rappe, R., Jones, B.L. et al. (2018). Indonesia's globally significant seagrass meadows are under widespread threat. *Science of the Total Environment* 634: 279–286.
Van Tussenbroek, B.I., Cortés, J., Collin, R. et al. (2014). Caribbean-wide, long-term study of seagrass beds reveals local variations, shifts in community structure and occasional collapse. *PLoS One* 9: e90600. https://doi.org/10.1371/journal.pone.0090600.
Vo, T.-T., Lau, K., Liao, L.M. et al. (2020). Satellite image analysis reveals changes in seagrass beds at Van Phong Bay, Vietnam during the last 30 years. *Aquatic Living Resources* 33: 4. https://doi.org/10.1051/alr/2020005.
Waycott, M., Duarte, C.M., Carruthers, T.J.B. et al. (2009). Accelerating loss of seagrasses across the globe threatens coastal ecosystems. *Proceedings of the National Academy of Sciences* 106: 12377–12381.
Wilkinson, C.R. (2008). *Status of Coral Reefs of the World: 2008*. Global Coral Reef Monitoring Network and Reef and Rainforest Research Centre, Townsville, Australia.
Wilson, S.S. and Dunton, K.H. (2018). Hypersalinity during regional drought drives mass mortality of the seagrass *Syringodium filiforme* in a subtropical lagoon. *Estuaries and Coasts* 41: 855–865.

CHAPTER 15

Epilogue

The structure and function of tropical estuarine and marine populations, communities, and ecosystems are complex, often more so than their counterparts in higher latitudes. Tropical organisms are deeply in tune with some unique chemical, geological, and physical attributes, such as high solar insolation, high temperatures, high and low rates of precipitation and evaporation, very variable salinity, and pH, 'estuarisation' of shelves by river plumes, migrating fluid mud banks, and low dissolved nutrient concentrations. It is not surprising that there is a dominance of small-size classes and individuals, a lack of deep-dwelling equilibrium benthic infauna, and intense competition and predation in the tropics. The tropics is the centre for marine biodiversity as most modern tropical biota evolved during the formation of the Tethys Sea. As such, there are large variations in species richness as well as habitat diversity.

Here is a list of the most important facts pertinent to the ecology of the marine tropics:

- There is a latitudinal diversity gradient in the marine biosphere with an increase in species richness with decreasing latitude from the poles to the tropics. However, this diversity pattern is not uniform as some tropical communities are characterised by low species richness and diversity in highly disturbed habitats.
- The greatest concentration of marine biodiversity on earth lies in the Coral Triangle, a region encompassing the nations of Malaysia, Indonesia, Papua New Guinea, the Philippines, and the Solomon Islands.
- There are true biogeographic divisions across the tropical marine biome that may not necessarily reflect evolutionary history, but certainly reflect current or recent environmental cues, and are separated by zones of rapidly changing species composition.
- The relatively high temperatures in the tropics generate and maintain high diversity of most life forms.
- Stronger biotic interactions (e.g. competition, predation) in the tropics can lead to greater species richness.
- Density-dependent behaviour in populations has been observed in tropical marine habitats and in organisms across a wide variety of size spectra.
- The best evidence for tropical marine populations having a distinct age structure comes from fish, crustacean, mollusc, and coral populations on coral reefs and in mangroves.

Tropical Marine Ecology, First Edition. Daniel M. Alongi.

- There is strong competition among and between corals and other benthic organisms with complex networks of competitive interactions rather than a simple competitive hierarchy between species.
- ENSO disturbances can greatly affect the structure and function of tropical marine ecosystems, down to the level of individual species.
- Disturbance plays a key role in structuring sessile and soft-bottom benthic populations and communities.
- The apex of mutualism in tropical seas is the symbiotic relationship between the photosynthetic dinoflagellate–microbial symbiont complex and its reef-building coral host.
- Mutualism is common in mangroves, seagrasses, and coral reefs, but evidence is lacking for inshore and coastal habitats and within pelagic food webs.
- Commensalism has been infrequently studied in the marine tropics. Most of the literature is still focused on systematics of the commensal rather than a clear delineation of the relationship.
- Parasitism appears to be strong in the tropics and is likely more common than other trophic interactions. A wide array of parasites can be found in and on a wide variety of tropical organisms, from microbes to whales.
- Predation is often intense throughout the marine tropics, with countless predator–prey interactions.
- Plant–herbivore interactions are intense throughout the marine tropics.
- Trophic cascades exist in all habitats but are best documented for coral reefs.
- Tropical rocky shores, unlike their temperate counterparts, appear superficially to be barren due to a lack of foliose macroalgae and low abundances of sessile invertebrates. The dominant space occupiers are low-lying forms, such as encrusting algae and turf algae.
- Latitudinal trends of macrofauna species richness in sandy beaches show that the number of species decreases linearly from tropical to temperate regions. Zonation patterns are more complex on tropical sandy beaches.
- Hypersaline coastal lagoons develop in many dry tropical regions, originating as wave-cut terraces when sea levels were lower during the Pleistocene glaciations. Marine terraces or 'sabkhas' surround these high-salinity lagoons.
- Tropical coastal lagoons are characterised by rich fisheries that are commercially exploited, and in most cases, overfished.
- Mangroves attain peak luxuriance in the wet tropics, with major latitudinal limits relating best to major ocean currents and the 20 °C seawater isotherm in winter.
- Mangroves possess characteristics that make them structurally and functionally unique. Such characteristics and adaptations include aerial roots, viviparous embryos, tidal dispersal of propagules, frequent absence of an understorey, wood with narrow, densely distributed vessels, highly efficient nutrient retention mechanisms, and the ability to cope with salt and to maintain water and C balance.
- A key feature of mangroves to deal with the problem of the lack of oxygen and the presence of potentially toxic metabolites in waterlogged saline soils is a series of morphological and physiological adaptations to maximise root aeration.

- Trees and bacteria constitute the bulk of forest biomass, but many other organisms originating from adjacent terrestrial and marine environments commonly inhabit mangroves.
- Like mangroves, seagrasses are composed of comparatively few plant species.
- Tropical seagrasses support a rich and diverse fauna and microflora and these organisms often show considerable connectivity with adjacent habitats, especially mangroves and coral reefs.
- Species richness of seagrasses is positively correlated with decreasing latitude with the greatest species richness occurring in Southeast Asia, especially within the Coral Triangle.
- Tropical seagrasses are often nutrient-limited, but P limitation is not a universal phenomenon; nutrient limitation and the conditions under which it happens are complex.
- Like mangroves, tropical seagrasses experience a variety of small and large disturbances. How they respond depends on the intensity, duration, areal extent, and frequency of the disturbance.
- The global distribution of coral reefs reflects the confluence of physical, chemical, and geological conditions that are most suitable for coral growth. Corals are found in the Caribbean, the Red Sea, including the Indian Ocean islands such as the Seychelles and the Indo-West Pacific.
- Corals are found in a broad band throughout the tropics, although there are extra-tropical areas where warm currents permit the existence of corals.
- Six major factors limit the development of coral reefs: temperature, light, depth, salinity, sedimentation, and exposure to air. Hermatypic corals are found in waters bounded by the 20 °C seawater isotherm with the lower temperature set at 18 °C for reef formation.
- Coral reef ecosystems consist of a high diversity of non-coral species and they are truly as species rich as any ecosystems on earth. Food webs on coral reefs are exceedingly complex and are characterised by both great abundance and diversity.
- Categorisation of the world's continental shelves reveals that: (i) low-latitude upwelling areas on the eastern boundaries of the ocean are usually the most productive shelf ecosystems, (ii) the subtropical and tropical river-dominated shelves on the ocean's western boundaries are nearly as productive as shelves at temperate latitudes, and (iii) the shelves of many cold temperate and some polar seas are comparatively unproductive.
- Densities and biomass of benthic infauna of tropical shelves are generally low compared to their temperate counterparts. Several factors mitigate against the development of high densities and large body size of benthic infauna: food limitation, warm and shallow habitats susceptible to physical and climatic disturbances (e.g. cyclones, monsoons), infringements of mass water movement, and lack of seasonal water column turnover, the latter two factors facilitating development of oxygen minimum zones.
- Tropical coastal and shelf plankton are rich in species compared to those of higher latitude and differ from temperate plankton in that microflagellates and

cyanobacteria dominate tropical phytoplankton communities more than in temperate seas where diatoms tend to flourish.

- Picoplankton assemblages play a central role in the energetics of pelagic food webs on tropical shelves.
- 'Estuarisation' of tropical shelves is common in the wet tropics, as river plumes can extend out to the outer continental margin.
- Benthic food webs on tropical river-dominated shelves are linked to sedimentary structures as controlled by factors such as rates of freshwater and sediment supply.
- Sixty-nine per cent of freshwater and 60% of sediment in the global ocean are discharged onto tropical shelves.
- Tropical shelf infauna consist mostly of pioneering assemblages dominated by bacteria and small surface deposit-feeders. In contrast, the epifauna are abundant and diverse and consist of large-sized organisms.
- Mobile mud banks or 'chakara' are common on some shelves, especially off India. These migrating mud banks and their associated fauna appear and disappear in synchrony with the onset of the monsoon or periods of high versus low river discharge.
- A unique geological characteristic of tropical shelves is that many continental margins are dominated by carbonates and various mixtures of carbonate and terrigenous sediments.
- A common feature of tropical shelves is their abundant fisheries. As in higher latitudes, tropical fisheries potential is generally related to both pelagic production and benthic standing crop.
- There are some patterns of demersal fish communities on tropical shelves: (i) an increase in diversity towards the equator, (ii) an east–west gradient in diversity with more taxa in the western Atlantic than in the eastern Atlantic, and (iii) more families in the Indo-West Pacific.
- The open ocean of the tropics comprises 37% of the area of the world's oceans ($\approx$145 million km^2).
- The trade winds regime in the open ocean has several physical characteristics that make it unique and different to regimes at higher latitudes.
- It is in the trade winds biome that the smallest cell fractions of the phytoplankton dominate biomass and production; even eukaryotic cells are of small size while the picoplankton dominates. Less than 90% of plant biomass and <80% of its production may be attributable to this latter fraction in the open ocean.
- Pelagic fish reach their peak abundance and biomass in the tropical open ocean. A great variety of shoaling clupeids, schooling tuna, and other scombroids and solitary sharks are characteristic of tropical seas, such that there are multiple food webs, each one more complex and longer than those of higher latitudes.
- Pelagic fish are more abundant and diverse than demersal fish assemblages in tropical seas.
- Rates of benthic gross primary production (GPP) and net primary production (NPP) on tropical sandy beaches and tidal flats are low compared to their temperate counterparts due to surface sediment disturbance, high temperatures, and silt-laden waters in river deltas.

- There is proportionally more C flow through microbial pathways in tropical sandy beaches than in temperate beaches.
- Rates of mangrove net photosynthesis vary widely among species with the major regulatory factors being soil salinity, temperature, vapour pressure deficit between leaf and surrounding air, and light intensity.
- A comparison of CO_2 assimilation rates between mangroves and tropical terrestrial forests suggests higher median rates of photosynthesis in mangroves.
- Rates of mangrove NPP are more rapid than other tropical estuarine and marine primary producers.
- Rates of NPP are equivalent between mangroves and tropical terrestrial forests, suggesting similar physiological and ecological constraints on photosynthesis.
- Productivity of tropical seagrasses is limited mainly by light, nutrients, and temperature.
- Seagrass photosynthesis controls rates of calcification and photosynthesis of macroalgae in tropical meadows.
- The light-saturated photosynthetic rate of seagrasses is greater for tropical than temperate species.
- Tropical seagrasses have optimal temperatures of 23–32 °C with seasonality being the net result of increased temperature in the summer wet season. Optimal growth temperatures of subtropical and tropical species are greater than those of temperate species.
- N and P may be limiting for tropical seagrasses, and there is good evidence that P and Fe are often limiting in carbonate-rich sediments.
- Aboveground production of seagrass peaks in boreal latitudes and belowground production peaks in boreal and equatorial latitudes. Aboveground biomass peaks in the boreal latitudes, whereas belowground biomass is lowest in the tropics, although there is no clear trend with latitude.
- Primary production of belowground seagrass roots and rhizomes of tropical species can be high, despite a trend of an increasing ratio of aboveground to belowground production with decreasing latitude.
- Interactions between coral reef photosynthesis and calcification are complex. Calcification in the light is linked to photosynthesis, but dark reactions are poorly understood.
- The ratio of calcification to net photosynthesis has a median value on coral reefs of 1.3 indicating that CO_2 generated by $CaCO_3$ deposition could potentially supply 78% of the inorganic carbon required for photosynthesis.
- Climate change, sea level, temperature, partial pressure of CO_2, ultraviolet radiation, hydrodynamics, sedimentation, salinity, and nutrients are the major parameters affecting coral reef structure and function.
- The complex dynamics of photosynthesis, respiration, and calcification result in widely varying rates of GPP and NPP among different coral reefs and between different zones on individual reefs.
- Community GPP on coral reefs is high and community R is of the same order of magnitude resulting in NPP being close to zero.
- Tropical tidal waterways and coastal lagoons are sites of high primary productivity. Productivity is linked to flushing times, temperature, salinity, degree of

pollution, and groundwater discharge. Phytoplankton and benthic production can contribute significantly to total ecosystem net primary productivity.

- Tropical estuaries are of great importance to the coastal zone, but few measurements have been made of their benthic and pelagic metabolism, both of which may peak during the monsoon season.
- Rates of primary production are often higher on tropical continental shelves than on temperate shelves, but primary production is greater on some boreal shelves and in the great upwelling regions. Peak productivity in the tropics is most often associated with 'estuarisation' of the inner and middle continental shelf and with coastal upwelling along the shelf edge.
- Low-latitude upwelling areas on the eastern boundaries of the ocean are usually the most productive shelf ecosystems and subtropical and tropical river-dominated shelves on the ocean's western boundaries are nearly as productive as temperate shelves.
- Three generalisations can be made about subtropical and tropical shelves: (i) export of materials from the continent is usually restricted to the coastal zone; this material greatly influences inshore food webs and nutrient cycles, (ii) impingement of nutrient-rich, open ocean water enhances primary and secondary production on the outer shelf and shelf edge, and (iii) little organic matter is exported from the continental margin to the deep ocean.
- Secondary production of heterotrophic bacterioplankton is high in tropical seas and peaks inshore compared with bacterioplankton communities of higher latitude. Turnover times are variable but are on average shorter in tropical waters.
- Hot spots of primary productivity are located within upwelling zones, such as off the western African coast where the Guinea Dome lies, and off Peru as well as equatorial upwelling along a fine line across the Pacific. The Costa Rica Dome is also a hot spot of primary productivity as are deep water areas within the Indonesian archipelago and offshore areas of the Arabian Sea.
- There is a tight coupling between bacterioplankton and phytoplankton in tropical waters.
- Subtropical and tropical zooplankton species grow faster than their counterparts at higher latitudes.
- Tropical zooplankton production is lower than in temperate regions. Lower productivity is likely the result of food levels being less abundant in the tropics and the fact that the phytoplankton is dominated by pico- and nano-planktonic size groups which are less directly available to zooplankton grazers than larger, net phytoplankton.
- Tropical zooplankton production peaks with the onset of the monsoon. Low rates of productivity are linked to the pre-monsoon season, with higher rates of production during the post-monsoon period. In some estuaries, production peaks during pre- and post-monsoon periods.
- There are large-scale, regional differences in zooplankton production with highest rates in upwelling regions.
- Benthic bacterial productivity is high in subtropical and tropical sediments in agreement with data on benthic oxygen metabolism. Turnover times are rapid, on the order of hours to days, and at the upper end of the range in temperate sediments.

- Except for upwelling, mangrove, and eutrophic habitats, secondary benthic production is low to moderate compared with temperate regions, but turnover is rapid, and lifespans are short.
- Fisheries production is positively correlated with primary production, but other factors play a role in sustaining coastal fisheries: (i) high structural diversity, (ii) abundant food, (iii) predation limited by turbidity, and (iv) enhanced production in the wake of river discharge.
- Coastal regions comprising large rivers, wetlands, and estuaries tend to support high yields of economically important fish, crustaceans, and molluscs. Hypotheses offered to explain this phenomenon include (i) high rates of primary productivity, (ii) a high level of habitat diversity, and (iii) physiological/behavioural attraction to river discharge and precipitation.
- Tropical continental shelves constitute 30% of the total area of the world's continental shelves and 36% of global open ocean area, but tropical fisheries (including tropical upwelling areas) make up 47% of the global wild capture fisheries catch, continuing a proportional rise since 1950.
- In some tropical countries, fisheries are overexploited and mismanaged. Fisheries landings have declined or levelled off in some countries, but many countries are still showing large increases in landings mainly of minor species that have less intrinsic value than high-quality fisheries, such as tuna and mackerel.
- Tropical sandy beaches and tidal flats have comparatively simple food webs where most of the C and energy flows through microbial pathways; microbes, meiofauna, and macrofauna can be functionally divorced from each other in high-energy environments.
- Encrusting algae are central to the food webs of tropical rocky shores. They are consumed by sea urchins, herbivorous gastropods, crabs, and herbivorous and omnivorous fish.
- Tropical seagrass beds are species-rich with many organisms feeding on bacteria and seagrass detritus with other organisms feeding more directly on seagrass tissue. Most seagrass tissue is grazed directly by dugongs, manatees, turtles, sea urchins, and herbivorous fish.
- Algae, rather than mangrove detritus, is the most nutritious food for most mangrove-associated organisms.
- The dominant trophic pathway in mangroves is herbivory by sesarmid crabs. Leaf-eating sesarmid crabs are ecosystem engineers by burrowing in the soil and feeding extensively on mangrove leaves and propagules.
- Mangrove C fluxes are dominated by high rates of GPP; most (64%) is lost as canopy R. Total gaseous losses from mangrove ecosystems equates to 89% of GPP.
- A large portion of mangrove net ecosystem production (NEP) is unaccounted for, but 70% of subsurface soil respiration is exported from the forest floor via lateral or groundwater discharge of dissolved inorganic carbon (DIC) to adjacent coastal waters. C burial equates to 12% of NPP. NEP is 628 g C m^{-2} a^{-1}. The ratio of P_{GPP}: R_E averaged 1.18 indicating that mangrove ecosystems are net autotrophic.
- The food webs of coral reefs are extraordinarily complex, with different reef zones dominated by different food webs.

- Coral reefs, due to their high primary productivity and structural diversity, have many trophic transfers and levels within their food webs.
- In coral reefs, there are complex pathways involving herbivore, strong corallivore, and carnivore interactions.
- The removal of top predatory fish by humans plays a key 'top-down' role in controlling coral reef food webs.
- Food web structure and function and C flow in tropical estuaries and lagoons are greatly dependent on the mixing of saline and freshwater, especially during the monsoon season, and to tidal flow.
- The onset of the monsoon season is the most dramatic change to ecosystem function and structure in tropical estuaries and lagoons.
- Large tropical estuaries and lagoons with significant mangroves and seagrasses export large quantities of C to the coastal ocean, especially during the wet season.
- Tropical rivers, estuaries, and coastal waters play a huge role in the outgassing of CO_2 to the atmosphere.
- Tropical shelf muddy deposits act as 'incinerators' of sedimentary organic matter. Little organic carbon is buried in tropical shelf deposits.
- Mangrove export to adjacent shelf areas is a significant source of oceanic exchangeable organic carbon (as exchangeable DOC). Mangroves discharge C disproportionate to their small global area, contributing 55% of air–sea exchange, 28% of DIC export, 14% of C burial, and 13% of DOC + POC export from the world's wetlands and estuaries to the coastal ocean.
- Tropical shelves are sources of CO_2 to the atmosphere, whereas shelves at higher latitudes are CO_2 sinks.
- The tropical open ocean, like tropical continental shelf waters, is either a source of CO_2 to the atmosphere or in equilibrium. The highest effluxes are associated with the eastern boundary currents and areas such as the Arabian Sea, mirroring the high rates of primary productivity and upwelling of waters high in DIC.
- Microbial respiration plays a key role in determining whether a given parcel of water is a sink or source of CO_2 to the atmosphere.
- Rapid turnover of microbial assemblages often results in equally rapid rates of nutrient cycling in tropical open ocean water.
- Tropical intertidal habitats can act as either nutrient sinks or sources of N and P.
- Both denitrification and N_2 fixation rates are higher in tropical seagrass beds than in temperate meadows.
- Tropical seagrasses can be important temporary sinks for nutrients.
- N cycling in tropical seagrass beds, coral reefs, and mangroves is rapid, to the extent that most N is conserved with relatively little, but highly refractory, N exported from these systems.
- There is close biogeochemical coupling between both mangrove trees and seagrasses with microbes and interstitial nutrients.
- A N mass balance for the world's mangrove forests shows that: (i) N burial is about 29% of total N input; (ii) N_2 fixation is at least 11% of total N input; (iii) root and litter production account for 45 and 37% of mangrove NPP (as N); (iv) tidal export equates to about 13% of total N input; and (v) denitrification and

N_2O effluxes account for 45% of total N losses. N burial equates to about 29% of total N input. Net immobilisation is the largest N flux, reflecting its significance in conserving N.

- The supply of essential nutrients supporting high productivity of coral reefs comes from advection, upwelling and endo-upwelling, groundwater, rain, guano, terrestrial runoff, and immigrant organisms.
- Connections between coral reefs and surrounding waters and their proximity to other coastal and marine habitats are essential to the sustainability of coral reefs.
- Coral reefs, like tropical seagrasses and mangroves, rapidly cycle N, P and trace elements. These nutrients are often limiting.
- Rates of biogeochemical cycling in tropical estuaries and coastal lagoons are rapid, although studies have been few.
- Tropical estuaries and coastal lagoons show nutrient behaviour much different to those in temperate estuaries and lagoons. The most distinct difference is the timing and variability of the major physical forces on the estuary (e.g. river flow, insolation) with more episodic events such as highly seasonal river inputs in the tropics.
- Nearshore and shelf environments both receive and process large amounts of nutrients and associated materials discharged from tropical estuaries and lagoons; there have been few nutrient cycling studies from tropical inshore and shelf habitats. The few studies available suggest rapid rates of N and P cycling.
- N, P, Si, and Fe are often, but not always, limiting to tropical phytoplankton in the open ocean.
- N_2 fixation is the major process by which new N is provided in the tropical open ocean and has a major role to play in the global marine N budget.
- Most tropical estuarine and marine environments are impacted by humans. Few habitats are pristine.
- The major impacts of humans on tropical estuarine and marine environments are: eutrophication, coastal development, disruption of hydrological cycles, toxins, metals, pathogens, pesticides, introduction of exotic species, plastic litter, chlorinated hydrocarbons, petroleum hydrocarbons, shoreline erosion or siltation, unsustainable exploitation, noise pollution, habitat destruction, and climate change.
- Oil pollution is declining in frequency, but oil persists for long periods of time, even in the tropics where oil degrades faster than in higher latitudes. Oil spills have caused major damage to coral reefs, subtidal benthos, seagrasses, and mangroves. Recovery from oiling depends on a many factors, including the type of oil, the amount of oil spilled, the areal extent of the spill and whether the spill environment is low, moderate, or high energy.
- Tropical intertidal environments can be contaminated by metals brought down by rivers and derived from industry, with seasonal variations due to the monsoon season.
- Compared to mangroves, corals are less tolerant to metal contamination.
- Nearly all tropical estuaries, bays, and lagoons are affected by eutrophication.
- Tropical sandy beaches and sand and mud flats are subject to eutrophication to the extent that beaches are often closed due to high faecal coliform counts.

- Seagrass meadows and coral reefs are especially vulnerable to eutrophication.
- In contrast to seagrasses and coral reefs, mangroves are usually tolerant to eutrophication to the extent that they are often used as biofilters of domestic sewage and aquaculture wastes.
- Tropical estuarine and marine environments are commonly contaminated by pesticides, but there are less data on the effects of these contaminants on their biota.
- Organochlorine pesticides are very toxic to corals.
- 1.15–2.41 Mt of plastic waste enter the ocean yearly. The top 20 polluting rivers, located mostly in Asia, account for 67% of the global total.
- Distribution patterns of microplastics mirror those of plankton in tropical estuaries.
- Plastic fragments of varying size are common in mangroves, beaches, seagrass meadows, subtidal benthic habitats, and coral reefs and are likely persist from decades to centuries.
- Sewage microbes and invasive species are often responsible for the biological pollution of tropical marine ecosystems. Evidence suggests fewer successful invasions in tropical waters than in temperate seas.
- Lionfish are the most dramatic example of an invasive species in the tropics, but there are many other species that are considered as non-native to many habitats.
- Human-induced climate change has resulted in ocean acidification, increases in temperature, greenhouse gas concentrations, and sea level, increased frequency and intensity of cyclonic storms, and changes in rainfall and ocean circulation patterns.
- Tropical organisms are thought to be more susceptible to temperature increases than temperate or polar organisms as they live closer to their upper temperature tolerance limits.
- Coral reefs are more drastically affected by climate change than mangroves and seagrasses.
- Increasing temperatures are increasing the incidence and severity of diseases, especially in corals.
- Coral bleaching is becoming more common with increasing sea surface temperatures (SSTs).
- Rises in SSTs result in increased rates of seagrass metabolism and shifts in the maintenance of a positive C balance.
- Increased temperatures may affect coral reefs, mangroves, and seagrasses by (i) changing species composition, (ii) increasing physiological stress, (iii) increasing productivity and respiration, and (iv) expanding their range to higher latitudes.
- Sea-level rise (SLR) is not considered a problem for coral reefs if their growth remains vigorous.
- A rise in sea-level can alter seagrass distribution, productivity, and community composition due to a decrease in available light.
- The ability of mangroves to adjust to SLR depends on their accretion rate relative to the rise in sea-level. Empirical measurements indicate that mangroves in Australia, New Zealand, the Caribbean, Central America, on most low coral islands, and in subsiding river deltas are not keeping pace with relative SLR.

Mangroves in some areas of SE Asia, South America, Africa, the Middle East, and South Asia are keeping pace, with the threshold rate of relative SLR being about 6.1 mm a^{-1}.

- Higher CO_2 concentrations in air can enhance the growth of mangroves, but responses are species-specific, with many responses confounded by variations in salinity, nutrient availability, and water-use efficiency.
- Higher CO_2 concentrations stimulate seagrass production, although the responses are species-specific.
- Coral reefs will be most impacted by ocean acidification (OA) as the decrease in ocean pH will be enough to significantly decrease the ability of corals to calcify. Other marine calcifying organisms are either unaffected or negatively affected by OA, depending on species. Non-calcifying biota may be negatively impacted by OA, including reef invertebrates and fish.
- Seagrass photosynthesis increases in many species with increasing DIC concentrations and decreasing pH, suggesting that some seagrasses may thrive under OA. Brown macroalgae and kelps also exhibit positive responses to OA.
- Mangroves and other estuarine and coastal communities are not expected to be significantly impacted by OA being continually exposed and likely adapted to naturally variable fluctuations in pH, salinity, pCO_2, and $[CO_3^{2-}]$.
- Several models agree that tropical phytoplankton production will decline in future, driven by reduced nutrient concentrations, increased stratification, weaker upwelling, declining pH and dissolved oxygen, expansion of oxygen minimum zones, and increased SSTs.
- Fish production and yields will decline for the same reasons in addition to continued overfishing.
- Globally, 19% of the original area of coral reefs has been lost as of 2008. A further 15% will be seriously threatened during 2018–2028, and 20% will be under threat of loss during 2028–2048; 46% of coral reefs were considered relatively healthy. Coral recruitment declined globally by 82% from 1974 to 2012.
- From 1989 to 2006, 58% of the global area of seagrasses declined, 25% increased, and 17% exhibited no change. The rate of decline in seagrass areas has accelerated with the median rate of decline $< 1\% \ a^{-1}$ before 1940, but $5\% \ a^{-1}$ after 1980. Major drivers for loss and degradation were direct impacts from coastal development and dredging, and indirect impacts from declining water quality; lesser causes were marine heat waves, drought, diseases, destructive fishing practices, boat propellers, cyclones, increased sediment runoff, aquaculture, and invasive species.
- Global loss rates of mangroves vary between 0.26 and $0.66\% \ a^{-1}$. Between 2000 and 2016, average annual loss was $0.13\% \ a^{-1}$. Anthropogenic causes accounted for 62% of forest loss, including rice, shrimp and palm oil cultivation, and coastal development. The remaining 38% of total forest loss was due to natural causes: shoreline erosion (27%) and extreme weather events (11%). Nearly, 80% of anthropogenic loss was concentrated in Southeast Asia, especially Myanmar, Indonesia, and Malaysia.

Index

Tropical Marine Ecology, First Edition. Daniel M. Alongi.

D

E

R

Y

Z